A Sampling of New Applications

Drawn from diverse fields of interest and situations that occur in the real world

BUSINESS

- Cyber Monday Sales p. 50
- Online Shoppers p. 343
- Credit Cards in China p. 465
- Online Video Advertising p. 600
- Global Defense Spending *Use the method of least squares to predict global spending on defense.* p. 601
- Waiting Time at a Bakery p. 691

CRIMINAL JUSTICE

- Corporate Fraud p. 46
- Tax Refund Fraud p. 63
- New Inmates p. 316

DEMOGRAPHICS

- Median Age of U.S. Population p. 219
- Gini Index *Use integration by parts to calculate the Gini Index.* p. 509
- Households with Someone Under 18 p. 600
- Married Males p. 704

ECONOMICS

- U.S. GDP p. 90
- Inflation p. 205
- Isoquants p. 565
- Elasticity of Demand p. 665
- Utility Functions p. 668

ENERGY & THE ENVIRONMENT

- California Emissions Caps p. 89
- Solar Panel Power Output *Use the graph of a function to describe the solar panel output over a time interval.* p. 264
- Renewable Energy p. 343
- U.S. Strategic Petroleum Reserves p. 530

ENTERTAINMENT & SPORTS

- Strike Outs p. 314
- Decline of *American Idol* p. 342
- Streaming Music p. 388
- Baseball p. 464

FINANCE

- Change in per Capita Income p. 183
- Bank Failures p. 400
- Leveraged Return p. 622
- Domar Growth Model p. 668

See the complete Index of Applications at the back of the text to find out more ...

EDITION 10

APPLIED CALCULUS
FOR THE MANAGERIAL, LIFE, AND SOCIAL SCIENCES

SOO T. TAN
STONEHILL COLLEGE

CENGAGE
Learning

Australia • Brazil • Mexico • Singapore • United Kingdom • United States

Applied Calculus for the Managerial, Life, and Social Sciences, **Tenth Edition**
Soo T. Tan

General Manager: Balraj Kalsi

Product Director: Terrence Boyle

Product Manager: Rita Lombard

Senior Content Developer: Erin Brown

Associate Content Developer: Spencer Arritt

Product Assistant: Kathryn Schrumpf

Content Developer: Andrew Coppola

Associate Marketing Manager: Ana Albinson

Content Project Manager: Cheryll Linthicum

Art Director: Vernon Boes

Manufacturing Planner: Becky Cross

Production Service: Martha Emry BookCraft

Photo and Text Researcher: Lumina Datamatics

Copy Editor: Barbara Willette

Illustrator: Graphic World, Inc.

Text Designer: Diane Beasley

Cover Designer: Irene Morris

Cover Image: Credit Cards: Shutterstock/luchunyu; Lab Sample: Veer/solarseven; Checking In: GettyImages/PhotoAlto/Eric Audras; background: Getty/Imagotres; background: Shutterstock/Ozerina Anna

Compositor: Graphic World, Inc.

For product information and technology assistance, contact us at
Cengage Learning Customer & Sales Support, 1-800-354-9706.
For permission to use material from this text or product, submit all requests online at **www.cengage.com/permissions**.
Further permissions questions can be e-mailed to
permissionrequest@cengage.com.

Library of Congress Control Number: 2015939376

Student Edition:
ISBN: 978-1-305-65786-1

Loose-leaf Edition:
ISBN: 978-1-305-95322-2

Cengage Learning
20 Channel Center Street
Boston, MA 02210
USA

Cengage Learning is a leading provider of customized learning solutions with employees residing in nearly 40 different countries and sales in more than 125 countries around the world. Find your local representative at **www.cengage.com**.

Cengage Learning products are represented in Canada by Nelson Education, Ltd.

To learn more about Cengage Learning Solutions, visit **www.cengage.com**.

Purchase any of our products at your local college store or at our preferred online store **www.cengagebrain.com**.

Printed in the United States of America
Print Number: 08   Print Year: 2021

TO PAT, BILL, AND MICHAEL

CONTENTS

CHAPTER 7

Additional Topics in Integration   501

CHAPTER 8

Calculus of Several Variables   553

CHAPTER 9

Differential Equations   643

APPENDIX A

APPENDIX B

APPENDIX C

PREFACE

Applied Calculus for the Managerial, Life, and Social Sciences is intended for use in a two-semester or three-quarter introductory calculus course for students in the managerial, life, and social sciences. In preparing the Tenth Edition, I have kept in mind two longstanding goals: (1) to write an applied text that motivates students and (2) to make the book a useful teaching tool for instructors. Underlying this is my belief that math is an integral part of everyone's daily life. One of the most important lessons I have learned from many years of teaching undergraduate mathematics courses is that most students—mathematics majors and non-mathematics majors alike—respond well when introduced to mathematical concepts and results using real-life illustrations.

I also learned from my experience teaching courses in the managerial, life, and social sciences that many students come to these courses with some degree of apprehension. This awareness led to the intuitive approach I have adopted in all of my texts. As you will see, I try to introduce each abstract mathematical concept through an example drawn from a common real-life experience. Once the idea has been conveyed, I then proceed to make it precise, thereby assuring that no mathematical rigor is lost in this intuitive treatment of the subject.

Another lesson I learned from my students is that they have a much greater appreciation of the material if the applications are drawn from their fields of interest and from situations that occur in the real world. This is one reason you will see so many exercises in my texts that are modeled on data gathered from newspapers, magazines, journals, and other media.

The Approach

Presentation

My approach is intuitive, and I state the results informally. However, I have taken special care to ensure that mathematical precision and accuracy are not compromised.

Problem-Solving Approach

A problem-solving approach is stressed throughout the text. Numerous examples and applications illustrate each new concept and result. Special emphasis is placed on helping students formulate, solve, and interpret the results of applied problems. Because students often have difficulty setting up and solving word problems, extra care has been taken to help them master these skills.

- Very early in the text, students are given guidelines for setting up word problems (see Section 2.3). This is followed by numerous examples and exercises to help students master this skill.

- Guidelines are given to help students formulate and solve related-rates problems in Section 3.6.

- In Chapter 4, optimization problems are covered in two sections. First, the techniques of calculus are used to solve problems in which the function to be optimized is given (Section 4.4); second, optimization problems that require the additional step of formulating the problem are treated (Section 4.5).

■ In Chapter 9, "Differential Equations," students are once again encouraged to set up problems involving applications (see Section 9.1) before being asked to solve these problems in Sections 9.2–9.4.

Intuitive Introduction to Concepts

Mathematical concepts are introduced with concrete, real-life examples wherever appropriate. Following are some of the topics introduced in this manner:

- **Limits:** The Motion of a Maglev
- **The algebra of functions:** The U.S. Budget Deficit
- **The Chain Rule:** The Population of Americans Aged 55 Years and Older
- **Differentials:** Calculating Mortgage Payments
- **Increasing and decreasing functions:** The Fuel Economy of a Car
- **Concavity:** U.S. and World Population Growth
- **Inflection points:** The Point of Diminishing Returns
- **Curve sketching:** The Dow Jones Industrial Average on "Black Monday"
- **Exponential functions:** Income Distribution of American Families
- **Area between two curves:** Petroleum Saved with Conservation Measures
- **Approximating definite integrals:** The Cardiac Output of a Heart
- **Sequences and series:** The Bouncing Ball

Connections

One example (the maglev) is used as a common thread throughout the development of calculus—from limits through integration. The goal here is to show students the connections between the concepts presented: limits, continuity, rates of change, the derivative, the definite integral, and so on.

Absolute Extrema on a Closed Interval

As the preceding examples show, a continuous function defined on an arbitrary interval does not always have an absolute maximum or an absolute minimum. But an important case arises often in practical applications in which both the absolute maximum and the absolute minimum of a function are guaranteed to exist. This occurs when a continuous function is defined on a *closed* interval.

Before stating this important result formally, let's look at a real-life example. The graph of the function f in Figure 58 shows the average price, $f(t)$, in dollars, of domestic airfares by days before flight. The domain of f is the closed interval $[-210, -1]$, where -210 is interpreted as 210 days before flight and -1 is interpreted as the day before flight.

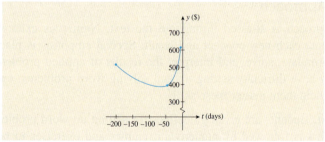

FIGURE **58**
Average price before flight
Source: Cheapair.com.

Observe that f attains the minimum value of 395 when $t = -49$ and the maximum value of 614 when $t = -1$. This result tells us that the best time to book a domestic flight is seven weeks in advance and the worst day to book a domestic flight is the day before the flight. Probably most surprising of all, booking too early can be almost as expensive as booking too late. Note that the function f is continuous on a closed interval. For such functions, we have the following theorem.

Motivation

Illustrating the practical value of mathematics in applied areas is an objective of my approach. Concepts are introduced with concrete, real-life examples wherever appropriate. These examples and other applications have been chosen from current topics and issues in the media and serve to answer a question often posed by students: "What will I ever use this for?" In this new edition, for example, the concept of finding the absolute extrema over a closed interval is introduced as shown at left.

Modeling

One important skill that every student should acquire is the ability to translate a real-life problem into a mathematical model. In Section 2.3, the modeling process is discussed, and students are asked to use models (functions) constructed from real-life data to answer questions. Additionally, students get hands-on experience constructing these models in the Using Technology sections.

New to this Edition

The focus of this revision has been the continued emphasis on illustrating the mathematical concepts in *Applied Calculus* by using more real-life applications that are relevant to the everyday life of students and to their fields of study in the managerial, life, and social sciences. Over 220 new applications have been added in the examples and exercises. A sampling of these new applications is provided on the inside front cover pages.

Many of the exercise sets have been revamped. In particular, the exercise sets were restructured to follow more closely the order of the presentation of the material in each section and to progress more evenly from easier to more difficult problems in both the rote and applied sections of each exercise set. Additional concept questions, rote exercises, and true-or-false questions were also included.

More Specific Content Changes

Chapters 1 and 2   In Section 1.4, parts (b) and (c) of Example 12 illustrate how to determine whether a point lies on a line. A new application, *Smokers in the United States,* has been added to the Self-Check Exercises in Section 1.4. The U.S. federal budget deficit graphs that are used as motivation to introduce "The Algebra of Functions" in Section 2.2 have been updated to reflect the current deficit situation. In Section 2.3, a new application of linear functions, *Erosion of the Middle Class,* has been added. New models and graphs for the *Global Warming, Social Security Trust Fund Assets,* and *Driving Costs* applications have also been provided in Section 2.3.

Chapters 3 and 4   A wealth of new application exercises has been added throughout these chapters. A new subsection on relative rates of change and a new application, *Inflation,* have been added to Section 3.4. In Section 4.1, the U.S. budget deficit (surplus) graph that is used to introduce relative extrema has been updated. The absolute extrema for the deficit function are later found in the *Federal Deficit* application in the exercise set for Section 4.4. Also in Section 4.4, a new application, *Average Fare Before a Flight,* has been added to introduce the concept of absolute extrema on a closed interval.

Chapters 5, 6, and 7   Section 5.3 has been expanded and now includes two new applications, *Investment Options* and *IRAs.* The interest rate problems in Exercise Set 5.3 have been revised to reflect the current interest rate environment. The intuitive discussion of area and the definite integral at the beginning of Section 6.3, as illustrated by the total daily petroleum consumption of a New England state, is given a firmer mathematical footing at the end of the section by demonstrating that the petroleum consumption of the state is indeed given by the area under a curve. In Section 7.1, an example and exercises have been added to show how the integration by parts formula can be applied to definite integrals.

Chapters 8, 9, and 10   In Chapter 8, the 3-D art in the text and exercises has been further enhanced. A new application, *Erosion of the Middle Class,* has been added to the Using Technology in Section 8.4. In Exercise Set 8.6, a new application, *Leveraged Return,* has been added. In Section 9.3, an applied example, *Elasticity of Demand,* has been added, and the *Domar Growth Model* and *Utility Functions* have been added to Exercise Set 9.3. In Section 10.2, the technique for finding the variance and standard deviation for grouped data is introduced, along with the applied example *Married Males.* Among the new application exercises are *Waiting Time at a Bakery* and *Expected Delivery Time.*

Chapters 11 and 12   Section 11.5, "Power Series and Taylor Series," has been reorganized for clarity. In Chapter 12, the first two sections have been rewritten. New artwork, new examples, and a discussion of periodic functions and amplitude have been added, along with corresponding exercises. In Section 12.3, a discussion of the relationship of the sine function and its derivative function, along with an additional example illustrating the technique of differentiation for the trigonometric functions, have been added.

Features

Real-World Connections

Motivating Applications

Many new applied examples and exercises have been added in the Tenth Edition. Among the topics of the new applications are Cyber Monday, family insurance coverage, leveraged return, credit card debt, tax refund fraud, salaries of married women, and online video advertising.

20. ONLINE VIDEO ADVERTISING Although still a small percentage of all online advertising, online video advertising is growing. The following table gives the projected spending on Web video advertising (in billions of dollars) through 2016:

Year	2011	2012	2013	2014	2015	2016
Spending, y	2.0	3.1	4.5	6.3	7.8	9.3

a. Letting $x = 0$ denote 2011, find an equation of the least-squares line for these data.

b. Use the result of part (a) to estimate the projected rate of growth of video advertising from 2011 through 2016.

Source: eMarketer.

Portfolios

These interviews share the varied experiences of professionals who use mathematics in the workplace. Among those included are a Senior Vice President of Supply at Earthbound Farm and an associate at JP Morgan Chase.

PORTFOLIO   Todd Kodet

TITLE   Senior Vice-President of Supply
INSTITUTION   Earthbound Farm

Earthbound Farm is America's largest grower of organic produce, offering more than 100 varieties of organic salads, vegetables, fruits, and herbs on 34,000 crop acres. As Senior Vice-President of Supply, I am responsible for getting our products into and out of Earthbound Farm. A major part of my work is scheduling plantings for upcoming seasons, matching projected supply to projected demand for any given day and season. I use applied mathematics in every step of my planning to create models for predicting supply and demand.

After the sales department provides me with information about projected demand, I take their estimates, along with historical data for expected yields, to determine how much of each organic product we need to plant. There are several factors that I have to think about when I make these determinations. For example, I not only have to consider gross yield per acre of farmland, but also have to calculate average trimming waste per acre, to arrive at net pounds needed per customer.

Some of the other variables I consider are the amount of organic land available, the location of the farms, seasonal information (because days to maturity for each of our crops varies greatly depending on the weather), and historical information relating to weeds, pests, and diseases.

I emphasize the importance of understanding the mathematics that drives our business plans when I work with my team to analyze the reports they have generated. They need to recognize when the information they have gathered does not make sense so that they can spot errors that could skew our projections. With a sound understanding of mathematics, we are able to create more accurate predictions to help us meet our company's goals.

Alli Pura, Earthbound Farm; (inset) Carterdayne/iStockphoto.com

Algebra Review

Algebra Review Gives Students a Plan of Action
A Diagnostic Test precedes the precalculus review. Each question is referenced by the section and example in the text where the relevant topic can be reviewed. Students can use this test to diagnose their weaknesses and review the material on an as-needed basis.

Diagnostic Test

1. a. Evaluate the expression:

     (i) $\left(\dfrac{16}{9}\right)^{3/2}$ **(ii)** $\sqrt[3]{\dfrac{27}{125}}$

   b. Rewrite the expression using positive exponents only: $(x^{-2}y^{-1})^3$

(Exponents and radicals, Examples 1 and 2, pages 6–7)

2. Rationalize the numerator: $\sqrt[3]{\dfrac{x^2}{yz^3}}$

(Rationalization, Example 5, page 7)

Algebra Review Where Students Need It Most
Well-placed algebra review notes, keyed to the review chapter, appear where students need them most throughout the text. These are indicated by the (x^2) icon. See this feature in action on pages 109 and 555.

EXAMPLE 6 Evaluate:

$$\lim_{h \to 0} \frac{\sqrt{1+h}-1}{h}$$

Solution   Letting h approach zero, we obtain the indeterminate form $0/0$. Next, we rationalize the numerator of the quotient by multiplying both the numerator and the denominator by the expression $(\sqrt{1+h}+1)$, obtaining

$$\frac{\sqrt{1+h}-1}{h} = \frac{(\sqrt{1+h}-1)(\sqrt{1+h}+1)}{h(\sqrt{1+h}+1)}$$

      $(\sqrt{a}-\sqrt{b})(\sqrt{a}+\sqrt{b}) = a - b$

$$= \frac{1+h-1}{h(\sqrt{1+h}+1)}$$

      (x^2) See page 19.

$$= \frac{h}{h(\sqrt{1+h}+1)}$$

$$= \frac{1}{\sqrt{1+h}+1}$$

Therefore,

$$\lim_{h \to 0} \frac{\sqrt{1+h}-1}{h} = \lim_{h \to 0} \frac{1}{\sqrt{1+h}+1} = \frac{1}{\sqrt{1}+1} = \frac{1}{2}$$

Explorations and Technology

Explore and Discuss

These optional questions can be discussed in class or assigned as homework. They generally require more thought and effort than the usual exercises. They may also be used to add a writing component to the class or as team projects.

Explore and Discuss

The average price of gasoline at the pump over a 3-month period, during which there was a temporary shortage of oil, is described by the function f defined on the interval $[0, 3]$. During the first month, the price was increasing at an increasing rate. Starting with the second month, the good news was that the rate of increase was slowing down, although the price of gas was still increasing. This pattern continued until the end of the second month. The price of gas peaked at $t = 2$ and began to fall at an increasing rate until $t = 3$.

1. Describe the signs of $f'(t)$ and $f''(t)$ over each of the intervals $(0, 1)$, $(1, 2)$, and $(2, 3)$.
2. Make a sketch showing a plausible graph of f over $[0, 3]$.

Exploring with Technology

These optional discussions appear throughout the main body of the text and serve to enhance the student's understanding of the concepts and theory presented. Often the solution of an example in the text is augmented with a graphical or numerical solution.

Exploring with **TECHNOLOGY**

Refer to Example 4. Suppose Marcus wished to know how much he would have in his IRA at any time in the future, not just at the beginning of 2014, as you were asked to compute in the example.

1. Using Formula (18) and the relevant data from Example 4, show that the required amount at any time x (x measured in years, $x > 0$) is given by

$$A = f(x) = 40,000(e^{0.05x} - 1)$$

2. Use a graphing utility to plot the graph of f, using the viewing window $[0, 30] \times [0, 200,000]$.
3. Using **ZOOM** and **TRACE**, or using the function evaluation capability of your graphing utility, use the result of part 2 to verify the result obtained in Example 4. Comment on the advantage of the mathematical model found in part 1.

Using Technology

Written in the traditional example-exercise format, these optional sections show how to use the graphing calculator as a tool to solve problems. Illustrations showing graphing calculator screens are used extensively. In keeping with the theme of motivation through real-life examples, many sourced applications are included.

A *How-To Technology Index* is included at the back of the book for easy reference to Using Technology examples.

APPLIED EXAMPLE 2 Erosion of the Middle Class The idea of a large, stable, middle class (defined as those with annual household incomes in 2010 between \$39,000 and \$118,000 for a family of three), is central to America's sense of itself. The following table gives the percentage of middle-income adults (y) in the United States from 1971 through 2011.

Year	1971	1981	1991	2001	2011
Percent, y	61	59	56	54	51

Let t be measured in decades with $t = 0$ corresponding to 1971.
a. Find an equation of the least-squares line for these data.
b. If this trend continues, what will the percentage of middle-income adults be in 2021?
Source: Pew Research Center.

Concept Building and Critical Thinking

Self-Check Exercises

Offering students immediate feedback on key concepts, these exercises begin each end-of-section exercise set and contain both rote and word problems (applications). Fully worked-out solutions can be found at the end of each exercise section. If students get stuck while solving these problems, they can get immediate help before attempting to solve the homework exercises. Applications have been included here because students often need extra practice with setting up and solving these problems.

2.6   Self-Check Exercises

1. Let $f(x) = -x^2 - 2x + 3$.
   a. Find the derivative f' of f, using the definition of the derivative.
   b. Find the slope of the tangent line to the graph of f at the point $(0, 3)$.
   c. Find the rate of change of f when $x = 0$.
   d. Find an equation of the tangent line to the graph of f at the point $(0, 3)$.
   e. Sketch the graph of f and the tangent line to the curve at the point $(0, 3)$.

2. **BANK LOSSES** The losses (in millions of dollars) due to bad loans extended chiefly in agriculture, real estate, shipping, and energy by the Franklin Bank are estimated to be

$$A = f(t) = -t^2 + 10t + 30 \qquad (0 \le t \le 10)$$

where t is the time in years ($t = 0$ corresponds to the beginning of 2007). How fast were the losses mounting at the beginning of 2010? At the beginning of 2012? At the beginning of 2014?

Solutions to Self-Check Exercises 2.6 can be found on page 153.

Concept Questions

Designed to test students' understanding of the basic concepts discussed in the section, these questions encourage students to explain learned concepts in their own words.

2.6   Concept Questions

For Questions 1 and 2, refer to the following figure.

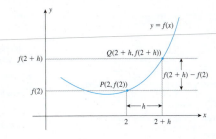

1. Let $P(2, f(2))$ and $Q(2 + h, f(2 + h))$ be points on the graph of a function f.
   a. Find an expression for the slope of the secant line passing through P and Q.
   b. Find an expression for the slope of the tangent line passing through P.

2. Refer to Question 1.
   a. Find an expression for the average rate of change of f over the interval $[2, 2 + h]$.
   b. Find an expression for the instantaneous rate of change of f at 2.
   c. Compare your answers for parts (a) and (b) with those of Question 1.

Exercises

Each section contains an ample set of exercises of a routine computational nature followed by an extensive set of modern application exercises.

2.6   Exercises

1. **AVERAGE WEIGHT OF AN INFANT** The following graph shows the weight measurements of the average infant from the time of birth ($t = 0$) through age 2 ($t = 24$). By computing the slopes of the respective tangent lines, estimate the rate of change of the average infant's weight when $t = 3$ and when $t = 18$. What is the average rate of change in the average infant's weight over the first year of life?

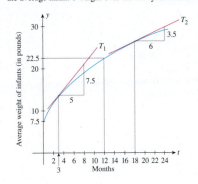

3. **TV-VIEWING PATTERNS** The following graph shows the percentage of U.S. households watching television during a 24-hr period on a weekday ($t = 0$ corresponds to 6 A.M.). By computing the slopes of the respective tangent lines, estimate the rate of change of the percent of households watching television at 4 P.M. and 11 P.M.
   Source: A. C. Nielsen Company.

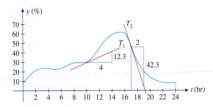

4. **CROP YIELD** Productivity and yield of cultivated crops are often reduced by insect pests. The following graph shows the relationship between the yield of a certain crop, $f(x)$, as a function of the density of aphids x. (Aphids are small insects that suck plant juices.) Here, $f(x)$ is measured in kilograms per 4000 square meters, and x is measured in

Review and Study Tools

Summary of Principal Formulas and Terms

Each review section begins with a summary, which highlights the important formulas and terms, with page numbers given for quick review.

CHAPTER 3 Summary of Principal Formulas and Terms

FORMULAS

1. Derivative of a constant	$\dfrac{d}{dx}(c) = 0$	$(c,$ a constant$)$
2. Power Rule	$\dfrac{d}{dx}(x^n) = nx^{n-1}$	
3. Constant Multiple Rule	$\dfrac{d}{dx}[cf(x)] = cf'(x)$	
4. Sum Rule	$\dfrac{d}{dx}[f(x) \pm g(x)] = f'(x) \pm g'(x)$	
5. Product Rule	$\dfrac{d}{dx}[f(x)g(x)] = f(x)g'(x) + g(x)f'(x)$	

TERMS

marginal cost (200)
marginal cost function (200)
average cost (201)
marginal average cost function (201)
marginal revenue (203)
marginal revenue function (203)

marginal profit function (204)
relative rate of change (205)
elasticity of demand (206)
elastic demand (207)
unitary demand (207)
inelastic demand (207)

second derivative of f (214)
implicit differentiation (222)
marginal rate of technical substitution (226)
related rates (227)
differential (234)

Concept Review Questions

These questions give students a chance to check their knowledge of the basic definitions and concepts given in each chapter.

CHAPTER 3 Concept Review Questions

Fill in the blanks.

1. **a.** If c is a constant, then $\dfrac{d}{dx}(c) =$ _____.
   b. The Power Rule states that if n is any real number, then $\dfrac{d}{dx}(x^n) =$ _____.
   c. The Constant Multiple Rule states that if c is a constant, then $\dfrac{d}{dx}[cf(x)] =$ _____.

   b. The demand is _____ if $E(p) > 1$; it is _____ if $E(p) = 1$; it is _____ if $E(p) < 1$.

6. Suppose a function $y = f(x)$ is defined implicitly by an equation in x and y. To find $\dfrac{dy}{dx}$, we differentiate _____ _____ of the equation with respect to x and then solve the resulting equation for $\dfrac{dy}{dx}$. The derivative of a term involving y includes _____ as a factor.

Review Exercises

Offering a solid review of the chapter material, the Review Exercises contain routine computational exercises followed by applied problems.

CHAPTER 3 Review Exercises

In Exercises 1–30, find the derivative of the function.

1. $f(x) = 3x^5 - 2x^4 + 3x^2 - 2x + 1$
2. $f(x) = 4x^6 + 2x^4 + 3x^2 - 2$
3. $g(x) = -2x^{-3} + 3x^{-1} + 2$
4. $f(t) = 2t^2 - 3t^3 - t^{-1/2}$
5. $g(t) = 2t^{-1/2} + 4t^{-3/2} + 2$
6. $h(x) = x^2 + \dfrac{2}{x}$
7. $f(t) = t + \dfrac{2}{t} + \dfrac{3}{t^2}$
8. $g(s) = 2s^2 - \dfrac{4}{s} + \dfrac{2}{\sqrt{s}}$
9. $h(x) = x^2 - \dfrac{2}{x^{3/2}}$
10. $f(x) = \dfrac{x+1}{2x-1}$
11. $g(t) = \dfrac{t^2}{2t^2+1}$

42. $3x^2y - 4xy + x - 2y = 6$
43. Find the differential of $f(x) = x^2 + \dfrac{1}{x^2}$.
44. Find the differential of $f(x) = \dfrac{1}{\sqrt{x^3+1}}$.
45. Let f be the function defined by $f(x) = \sqrt{2x^2 + 4}$.
   a. Find the differential of f.
   b. Use your result from part (a) to find the approximate change in $y = f(x)$ if x changes from 4 to 4.1.
   c. Find the actual change in y if x changes from 4 to 4.1 and compare your result with that obtained in part (b).
46. Use a differential to approximate $\sqrt[3]{26.8}$.
47. Let $f(x) = 2x^3 - 3x^2 - 16x + 3$.

Before Moving On . . .

Found at the end of each chapter review, these exercises give students a chance to determine whether they have mastered the basic computational skills developed in the chapter.

CHAPTER 3 Before Moving On . . .

1. Find the derivative of $f(x) = 2x^3 - 3x^{1/3} + 5x^{-2/3}$.

2. Differentiate $g(x) = x\sqrt{2x^2 - 1}$.

3. Find $\dfrac{dy}{dx}$ if $y = \dfrac{2x + 1}{x^2 + x + 1}$.

4. Find the first three derivatives of $f(x) = \dfrac{1}{\sqrt{x + 1}}$.

5. Find $\dfrac{dy}{dx}$ given that $xy^2 - x^2y + x^3 = 4$.

6. Let $y = x\sqrt{x^2 + 5}$.
  a. Find the differential of y.
  b. If x changes from $x = 2$ to $x = 2.01$, what is the approximate change in y?

Action-Oriented Study Tabs

Convenient color-coded study tabs make it easy for students to flag pages that they want to return to later, whether for additional review, exam preparation, online exploration, or identifying a topic to be discussed with the instructor.

Instructor Resources

TURN THE LIGHT ON WITH MINDTAP

Through personalized paths of dynamic assignments and applications, MindTap is a digital learning solution and representation of your course that turns cookie cutter into cutting edge, apathy into engagement, and memorizers into higher-level thinkers.

The Right Content: With MindTap's carefully curated material, you get the precise content and groundbreaking tools you need for every course you teach.

Personalization: Customize every element of your course—from rearranging the learning path to inserting videos and activities.

Improved Workflow: Save time when planning lessons with all of the trusted, most current content you need in one place in MindTap.

Tracking Students' Progress in Real Time: Promote positive outcomes by tracking students in real time and tailoring your course as needed based on the analytics.

Learn more at **www.cengage.com/mindtap.**

COMPLETE SOLUTIONS MANUAL by Soo T. Tan

Written by the author, the *Complete Solutions Manual* contains solutions for all exercises in the text, including Exploring with Technology and Explore and Discuss exercises. The *Complete Solutions Manual* is available on the Instructor Companion Site.

CENGAGE LEARNING TESTING POWERED BY COGNERO

Cengage Learning Testing Powered by Cognero is a flexible, online system that allows you to author, edit, and manage test bank content from multiple Cengage Learning solutions; create multiple test versions in an instant; and deliver tests from your LMS, your classroom, or wherever you want. Access to Cognero is available on the Instructor Companion Site.

INSTRUCTOR COMPANION SITE

Everything you need for your course in one place! This collection of book-specific lecture and class tools is available online at **www.cengage.com/login.** Access and download PowerPoint presentations, solutions manual, and more.

Student Resources

STUDENT SOLUTIONS MANUAL by Soo T. Tan (ISBN-13: 978-1-305-65788-5)
Giving you more in-depth explanations, this insightful resource includes fully worked-out solutions for selected exercises in the textbook, as well as problem-solving strategies, additional algebra steps, and review for selected problems.

MINDTAP FOR TAN'S APPLIED CALCULUS
MindTap is a digital representation of your course that provides you with the tools you need to better manage your limited time, stay organized, and be successful. You can complete assignments whenever and wherever you are ready to learn, with course material specially customized for you by your instructor and streamlined in one proven, easy-to-use interface. With an array of study tools, you'll get a true understanding of course concepts, achieve better grades, and set the groundwork for your future courses. Learn more at **www.cengage.com/mindtap.**

CENGAGEBRAIN.COM
Visit **www.cengagebrain.com** to access additional course materials and companion resources. At the CengageBrain.com home page, search for the ISBN of your title (from the back cover of your book) using the search box at the top of the page. This will take you to the product page where free companion resources can be found.

Acknowledgments

I wish to express my personal appreciation to each of the following reviewers, whose many suggestions have helped make a much improved book.

Paul Abraham
Kent State University—Stark

James Adair
Missouri Valley College

Faiz Al-Rubaee
University of North Florida

James V. Balch
Middle Tennessee State University

Mario Borha
Moraine Valley Community College

Jill Britton
Camosun College

Albert Bronstein
Purdue University

Debra D. Bryant
Tennessee Technological University

Kimberly Jordan Burch
Montclair State University

Michael Button
San Diego City College

Debra Carney
University of Denver

Peter Casazza
University of Missouri—Columbia

Sarah Clark
South Dakota State University

Matthew P. Coleman
Fairfield University

William Coppage
Wright State University

Lisa Cox
Texas A&M University

Mark Crawford
Waubonsee Community College

Charles Cunningham
James Madison University

Michelle Dedeo
University of North Florida

Ersin Deger
University of Cincinnati

Scott L. Dennison
University of Wisconsin—Oshkosh

Frank Deutsch
Penn State University

Christine Devena
Miles Community College

Andrew Diener
Christian Brothers University

Carl Droms
James Madison University

Bruce Edwards
University of Florida at Gainesville

Janice Epstein
Texas A&M University

Gary J. Etgen
University of Houston

Mike Everett
Santa Ana College

Stuart Farm
University of North Dakota

Kevin Ferland
Bloomsburg University

Charles S. Frady
Georgia State University

Howard Frisinger
Colorado State University

Larry Gerstein
University of California at Santa Barbara

Matthew Gould
Vanderbilt University

Harvey Greenwald
*California Polytechnic State University—
San Luis Obispo*

Tao Guo
Rock Valley College

James Hager
The Pennsylvania State University

John Haverhals
Bradley University

Ivan Haynes
University of South Carolina

Yvette Hester
Texas A&M University

George Hurlburt
Corning Community College

Mark Jacobson
Montana State University—Billings

Frank Jenkins
John Carroll University

David E. Joyce
Clark University

Herbert Kasube
Bradley University

Anton Kaul
*California Polytechnic State University—
San Luis Obispo*

Mohammed Kazemi
University of North Carolina—Charlotte

Sarah Kilby
North Country Community College

Gloria M. Kittel
University of West Georgia

Mark S. Korlie
Montclair State University

Murray Lieb
New Jersey Institute of Technology

James H. Liu
James Madison University

Lia Liu
University of Illinois at Chicago

Rebecca Lynn
Colorado State University

Norman R. Martin
Northern Arizona University

Sandra Wray McAfee
University of Michigan

Mary T. McMahon
North Central College

Daniela Mihai
University of Pittsburgh

Maurice Monahan
South Dakota State University

Dean Moore
Florida Community College at Jacksonville

Allen Muir
Pennsylvania State University

Linda E. Nash
Clayton State University

Tejinder Neelon
California State University—San Marcos

Ralph J. Neuhaus
University of Idaho

Kathy Nickell
College of DuPage

Gertrude Okhuysen
Mississippi State University

James Olsen
North Dakota State University

Lloyd Olson
North Dakota State University

Wesley Orser
Clark College

Carol Overdeep
Saint Martin's University

Mohammad Siddigne
Virginia Union University

Mari Peddycoart
Lone Star College—Kenwood

Pavel Sikorskii
Michigan State University

Katherine Pedersen
Southeastern Louisiana University

Anne Siswanto
East Los Angeles College

Shahla Peterman
University of Missouri—St. Louis

Edward E. Slaminka
Auburn University

Richard Porter
Northeastern University

Jane Smith
University of Florida

Virginia Puckett
Miami Dade College

Jennifer Strehler
Oakton Community College

Richard Quindley
Bridgewater State College

Devki Talwar
Indiana University of Pennsylvania

Mohammed Rajah
Miracosta College

Larry Taylor
North Dakota State University

Mary E. Rerick
University of North Dakota

Mary Ann Teel
University of North Texas

Dennis H. Risher
Loras College

Michael Threapleton
Centralia College

Kristi Rittby
Texas Christian University

Ray Toland
Clarkson University

Brian Rodas
Santa Monica College

Giovanni Viglino
Ramapo College of New Jersey

Thomas N. Roe
South Dakota State University

Tan Vovan
Suffolk University

Dr. Arthur Rosenthal
Salem State College

Hiroko K. Warshauer
Texas State University—San Marcos

Abdelrida Saleh
Miami Dade College

Lawrence V. Welch
Western Illinois University

Stephanie Anne Salomone
University of Portland

Jennifer Whitfield
Texas A&M University

Yvonne Sandoval
Pima Community College

Justin Wyss-Gallifent
University of Maryland at College Park

Donald R. Sherbert
University of Illinois

Lisa Yocco
Georgia Southern University

Gordon H. Shumard
Kennesaw State University

Laurie Zack
High Point University

I also wish to thank Tao Guo for the superb job he did as the accuracy checker for this text and the *Complete Solutions Manual* that accompanies the text. I also thank the editorial and production staffs of Cengage Learning—Richard Stratton, Rita Lombard, Erin Brown, Spencer Arrit, Kathryn Schrumpf, Andrew Coppola, Cheryll Linthicum, and Vernon Boes—for all of their help and support during the development and production of this edition. I also thank Martha Emry and Barbara Willette, who both did an excellent job ensuring the accuracy and readability of this edition, and Andy Bulman-Fleming for his work on the solutions manuals. Simply stated, the team I have been working with is outstanding, and I truly appreciate all of their hard work and efforts.

S. T. Tan

About the Author

SOO T. TAN received his S.B. degree from Massachusetts Institute of Technology, his M.S. degree from the University of Wisconsin–Madison, and his Ph.D. from the University of California at Los Angeles. He has published numerous papers in optimal control theory, numerical analysis, and mathematics of finance. He is also the author of a series of calculus textbooks.

1 Preliminaries

THE FIRST TWO sections of this chapter contain a brief review of algebra. We then introduce the Cartesian coordinate system, which allows us to represent points in the plane in terms of ordered pairs of real numbers. This in turn enables us to compute the distance between two points algebraically. This chapter also covers straight lines. The slope of a straight line plays an important role in the study of calculus.

How much money is needed to purchase at least 100,000 shares of the Starr Communications Company? Corbyco, a giant conglomerate, wishes to purchase a minimum of 100,000 shares of the company. In Example 11, page 21, you will see how Corbyco's management determines how much money they will need for the acquisition.

© Yuri Arcurs, 2010/ShutterStock.com

Use this test to diagnose any weaknesses that you might have in the algebra that you will need for the calculus material that follows. The review section and examples that will help you brush up on the skills necessary to work the problem are indicated after each exercise. The answers follow the test.

Diagnostic Test

1. a. Evaluate the expression:

   (i) $\left(\dfrac{16}{9}\right)^{3/2}$ **(ii)** $\sqrt[3]{\dfrac{27}{125}}$

   b. Rewrite the expression using positive exponents only: $(x^{-2}y^{-1})^3$

   (Exponents and radicals, Examples 1 and 2, pages 6–7)

2. Rationalize the numerator: $\sqrt[3]{\dfrac{x^2}{yz^3}}$

   (Rationalization, Example 5, page 7)

3. Simplify the following expressions:

   a. $(3x^4 + 10x^3 + 6x^2 + 10x + 3) + (2x^4 + 10x^3 + 6x^2 + 4x)$

   b. $(3x - 4)(3x^2 - 2x + 3)$

   (Operations with algebraic expressions, Examples 6 and 7, page 8)

4. Factor completely:

   a. $6a^4b^4c - 3a^3b^2c - 9a^2b^2$ **b.** $6x^2 - xy - y^2$

   (Factoring, Examples 8–10, pages 9–11)

5. Use the quadratic formula to solve the following equation: $9x^2 - 12x = 4$

   (The quadratic formula, Example 11, pages 12–13)

6. Simplify the following expressions:

   a. $\dfrac{2x^2 + 3x - 2}{2x^2 + 5x - 3}$ **b.** $\dfrac{(t^2 + 4)(2t - 4) - (t^2 - 4t + 4)(2t)}{(t^2 + 4)^2}$

   (Rational expressions, Example 1, page 16)

7. Perform the indicated operations and simplify:

   a. $\dfrac{2x - 6}{x + 3} \cdot \dfrac{x^2 + 6x + 9}{x^2 - 9}$ **b.** $\dfrac{3x}{x^2 + 2} + \dfrac{3x^2}{x^3 + 1}$

   (Rational expressions, Examples 2 and 3, pages 17–18)

8. Perform the indicated operations and simplify:

   a. $\dfrac{1 + \dfrac{1}{x + 2}}{x - \dfrac{9}{x}}$ **b.** $\dfrac{x(3x^2 + 1)}{x - 1} \cdot \dfrac{3x^3 - 5x^2 + x}{x(x - 1)(3x^2 + 1)^{1/2}}$

   (Rational expressions, Examples 4 and 5, pages 18–19)

9. Rationalize the denominator: $\dfrac{3}{1 + 2\sqrt{x}}$

   (Rationalizing algebraic fractions, Example 6, page 19)

10. Solve the inequalities:

   a. $x^2 + x - 12 \le 0$

   (Inequalities, Example 9, page 20)

   b. $|3x - 4| \le 2$

   (Absolute value, Examples 13 and 14, pages 22–23)

ANSWERS:

1. **a.** (i) $\dfrac{64}{27}$ (ii) $\dfrac{3}{5}$ **b.** $\dfrac{1}{x^6 y^3}$ 2. $\dfrac{x}{z\sqrt[3]{xy}}$

3. **a.** $5x^4 + 20x^3 + 12x^2 + 14x + 3$ **b.** $9x^3 - 18x^2 + 17x - 12$

4. **a.** $3a^2 b^2 (2a^2 b^2 c - ac - 3)$ **b.** $(2x - y)(3x + y)$

5. $\dfrac{2}{3}(1 - \sqrt{2}); \dfrac{2}{3}(1 + \sqrt{2})$ 6. **a.** $\dfrac{(x + 2)}{(x + 3)}$ **b.** $\dfrac{4(t^2 - 4)}{(t^2 + 4)^2}$

7. **a.** 2 **b.** $\dfrac{3x(2x^3 + 2x + 1)}{(x^2 + 2)(x^3 + 1)}$

8. **a.** $\dfrac{x}{(x + 2)(x - 3)}$ **b.** $\dfrac{x\sqrt{1 + 3x^2}(3x^2 - 5x + 1)}{(x - 1)^2}$

9. $\dfrac{3(1 - 2\sqrt{x})}{1 - 4x}$ 10. **a.** $[-4, 3]$ **b.** $\left[\dfrac{2}{3}, 2\right]$

1.1   Precalculus Review I

Sections 1.1 and 1.2 review some basic concepts and techniques of algebra that are essential in the study of calculus. The material in this review will help you work through the examples and exercises in this book. You can read through this material now and do the exercises in areas where you feel a little "rusty," or you can review the material on an as-needed basis as you study the text. The self-diagnostic test that precedes this section will help you pinpoint the areas where you might have any weaknesses.

The Real Number Line

The real number system is made up of the set of real numbers together with the usual operations of addition, subtraction, multiplication, and division.

We can represent real numbers geometrically by points on a **real number,** or **coordinate, line.** This line can be constructed as follows. Arbitrarily select a point on a straight line to represent the number 0. This point is called the **origin.** If the line is horizontal, then a point at a convenient distance to the right of the origin is chosen to represent the number 1. This determines the scale for the number line. Each positive real number lies at an appropriate distance to the right of the origin, and each negative real number lies at an appropriate distance to the left of the origin (Figure 1).

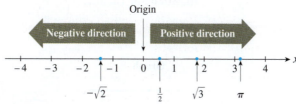

FIGURE 1
The real number line

A *one-to-one correspondence* is set up between the set of all real numbers and the set of points on the number line; that is, exactly one point on the line is associated with each real number. Conversely, exactly one real number is associated with each point on the line. The real number that is associated with a point on the real number line is called the **coordinate** of that point.

Intervals

Throughout this book, we will often restrict our attention to subsets of the set of real numbers. For example, if x denotes the number of cars rolling off a plant assembly line each day, then x must be nonnegative—that is, $x \geq 0$. Further, suppose management decides that the daily production must not exceed 200 cars. Then x must satisfy the inequality $0 \leq x \leq 200$.

More generally, we will be interested in the following subsets of real numbers: open intervals, closed intervals, and half-open intervals. The set of all real numbers that lie *strictly* between two fixed numbers a and b is called an **open interval** (a, b). It consists of all real numbers x that satisfy the inequalities $a < x < b$, and it is called "open" because neither of its endpoints is included in the interval. A **closed interval** contains *both* of its endpoints. Thus, the set of all real numbers x that satisfy the inequalities $a \leq x \leq b$ is the closed interval $[a, b]$. Notice that square brackets are used to indicate that the endpoints are included in this interval. **Half-open intervals** contain only *one* of their endpoints. Thus, the interval $[a, b)$ is the set of all real numbers x that satisfy $a \leq x < b$, whereas the interval $(a, b]$ is described by the inequalities $a < x \leq b$. Examples of these **finite intervals** are illustrated in Table 1.

TABLE 1		
Finite Intervals		
Interval	**Graph**	**Example**
Open: (a, b)		$(-2, 1)$
Closed: $[a, b]$		$[-1, 2]$
Half-open: $(a, b]$		$\left(\frac{1}{2}, 3\right]$
Half-open: $[a, b)$		$\left[-\frac{1}{2}, 3\right)$

In addition to finite intervals, we will encounter **infinite intervals.** Examples of infinite intervals are the half-lines (a, ∞), $[a, \infty)$, $(-\infty, a)$, and $(-\infty, a]$ defined by the set of all real numbers that satisfy $x > a$, $x \geq a$, $x < a$, and $x \leq a$, respectively. The symbol ∞, called *infinity*, is not a real number. It is used here only for notational purposes. The notation $(-\infty, \infty)$ is used for the set of all real numbers x, since by definition, the inequalities $-\infty < x < \infty$ hold for any real number x. Infinite intervals are illustrated in Table 2.

TABLE 2		
Infinite Intervals		
Interval	**Graph**	**Example**
(a, ∞)		$(2, \infty)$
$[a, \infty)$		$[-1, \infty)$
$(-\infty, a)$		$(-\infty, 1)$
$(-\infty, a]$		$\left(-\infty, -\frac{1}{2}\right]$

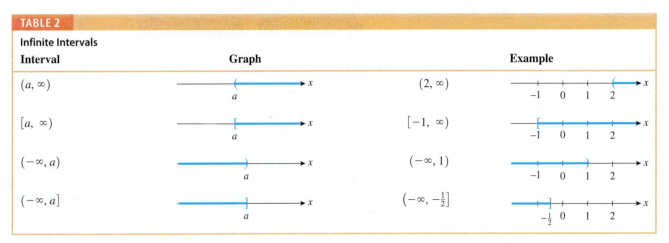

Exponents and Radicals

Recall that if b is any real number and n is a positive integer, then the expression b^n (read "b to the power n") is defined as the number

$$b^n = \underbrace{b \cdot b \cdot b \cdot \cdots \cdot b}_{n \text{ factors}}$$

The number b is called the **base,** and the superscript n is called the **power** of the exponential expression b^n. For example,

$$2^5 = 2 \cdot 2 \cdot 2 \cdot 2 \cdot 2 = 32 \quad \text{and} \quad \left(\frac{2}{3}\right)^3 = \left(\frac{2}{3}\right)\left(\frac{2}{3}\right)\left(\frac{2}{3}\right) = \frac{8}{27}$$

If $b \neq 0$, we define

$$b^0 = 1$$

For example, $2^0 = 1$ and $(-\pi)^0 = 1$, but the expression 0^0 is undefined.

Next, recall that if n is a positive integer, then the expression $b^{1/n}$ is defined to be the number that, when raised to the nth power, is equal to b. Thus,

$$(b^{1/n})^n = b$$

Such a number, if it exists, is called the ***n*th root of *b*,** also written $\sqrt[n]{b}$.

⚠️ If n is even, the nth root of a negative number is not defined. For example, the square root of -2 ($n = 2$) is not defined because there is no real number b such that $b^2 = -2$. Also, given a number b, more than one number might satisfy our definition of the nth root. For example, both 3 and -3 squared equal 9, and each is a square root of 9. So to avoid ambiguity, we define $b^{1/n}$ to be the positive nth root of b whenever it exists. Thus, $\sqrt{9} = 9^{1/2} = 3$. That's why your calculator will give the answer 3 when you use it to evaluate $\sqrt{9}$.

Next, recall that if p/q (where p and q are positive integers and $q \neq 0$) is a rational number in lowest terms, then the expression $b^{p/q}$ is defined as the number $(b^{1/q})^p$ or, equivalently, $\sqrt[q]{b^p}$, whenever it exists. For example,

$$2^{3/2} = (2^{1/2})^3 \approx (1.4142)^3 \approx 2.8283$$

Expressions involving negative rational exponents are taken care of by the definition

$$b^{-p/q} = \frac{1}{b^{p/q}}$$

Thus,

$$4^{-5/2} = \frac{1}{4^{5/2}} = \frac{1}{(4^{1/2})^5} = \frac{1}{2^5} = \frac{1}{32}$$

The rules defining the exponential expression a^n, where $a > 0$, for all rational values of n are given in Table 3.

The first three definitions in Table 3 are also valid for negative values of a. The fourth definition holds for all values of a if n is odd but only for nonnegative values of a if n is even. Thus,

$$(-8)^{1/3} = \sqrt[3]{-8} = -2 \qquad n \text{ is odd.}$$
$$(-8)^{1/2} \text{ has no real value} \qquad n \text{ is even.}$$

Finally, it can be shown that a^n has meaning for *all* real numbers n. For example, using a calculator with a $\boxed{y^x}$ key, we see that $2^{\sqrt{2}} \approx 2.665144$.

TABLE 3			
Rules for Defining a^n			
Definition of a^n $(a > 0)$	Example	**Definition of a^n $(a > 0)$**	Example
Integer exponent: If n is a positive integer, then $a^n = a \cdot a \cdot a \cdot \cdots \cdot a$ (n factors of a)	$2^5 = 2 \cdot 2 \cdot 2 \cdot 2 \cdot 2$ (5 factors) $= 32$	**Fractional exponent:** **a.** If n is a positive integer, then $$a^{1/n} \quad \text{or} \quad \sqrt[n]{a}$$ denotes the nth root of a. **b.** If m and n are positive integers, then $$a^{m/n} = \sqrt[n]{a^m} = (\sqrt[n]{a})^m$$	$16^{1/2} = \sqrt{16}$ $= 4$ $8^{2/3} = (\sqrt[3]{8})^2$ $= 4$
Zero exponent: If n is equal to zero, then $a^0 = 1$ (0^0 is not defined.)	$7^0 = 1$		
Negative exponent: If n is a positive integer, then $a^{-n} = \dfrac{1}{a^n} \quad (a \neq 0)$	$6^{-2} = \dfrac{1}{6^2}$ $= \dfrac{1}{36}$	**c.** If m and n are positive integers, then $$a^{-m/n} = \frac{1}{a^{m/n}} \quad (a \neq 0)$$	$9^{-3/2} = \dfrac{1}{9^{3/2}}$ $= \dfrac{1}{27}$

The five laws of exponents are listed in Table 4.

TABLE 4	
Laws of Exponents	
Law	**Example**
1. $a^m \cdot a^n = a^{m+n}$	$x^2 \cdot x^3 = x^{2+3} = x^5$
2. $\dfrac{a^m}{a^n} = a^{m-n} \quad (a \neq 0)$	$\dfrac{x^7}{x^4} = x^{7-4} = x^3$
3. $(a^m)^n = a^{m \cdot n}$	$(x^4)^3 = x^{4 \cdot 3} = x^{12}$
4. $(ab)^n = a^n \cdot b^n$	$(2x)^4 = 2^4 \cdot x^4 = 16x^4$
5. $\left(\dfrac{a}{b}\right)^n = \dfrac{a^n}{b^n} \quad (b \neq 0)$	$\left(\dfrac{x}{2}\right)^3 = \dfrac{x^3}{2^3} = \dfrac{x^3}{8}$

These laws are valid for any real numbers a, b, m, and n whenever the quantities are defined.

 Remember, $(x^2)^3 \neq x^5$. The correct equation is $(x^2)^3 = x^{2 \cdot 3} = x^6$.

The next several examples illustrate the use of the laws of exponents.

EXAMPLE 1 Simplify the expressions:

a. $(3x^2)(4x^3)$ **b.** $\dfrac{16^{5/4}}{16^{1/2}}$ **c.** $(6^{2/3})^3$ **d.** $(x^3 y^{-2})^{-2}$ **e.** $\left(\dfrac{y^{3/2}}{x^{1/4}}\right)^{-2}$

Solution

a. $(3x^2)(4x^3) = 12x^{2+3} = 12x^5$ Law 1

b. $\dfrac{16^{5/4}}{16^{1/2}} = 16^{5/4-1/2} = 16^{3/4} = (\sqrt[4]{16})^3 = 2^3 = 8$ Law 2

c. $(6^{2/3})^3 = 6^{(2/3)(3)} = 6^2 = 36$ Law 3

d. $(x^3y^{-2})^{-2} = (x^3)^{-2}(y^{-2})^{-2} = x^{(3)(-2)}y^{(-2)(-2)} = x^{-6}y^4 = \dfrac{y^4}{x^6}$ Law 4

e. $\left(\dfrac{y^{3/2}}{x^{1/4}}\right)^{-2} = \dfrac{y^{(3/2)(-2)}}{x^{(1/4)(-2)}} = \dfrac{y^{-3}}{x^{-1/2}} = \dfrac{x^{1/2}}{y^3}$ Law 5

We can also use the laws of exponents to simplify expressions involving radicals, as illustrated in the next example.

EXAMPLE 2 Simplify the expressions. (Assume that x, y, m, and n are positive.)

a. $\sqrt[4]{16x^4y^8}$ **b.** $\sqrt{12m^3n} \cdot \sqrt{3m^5n}$ **c.** $\dfrac{\sqrt[3]{-27x^6}}{\sqrt[3]{8y^3}}$

Solution

a. $\sqrt[4]{16x^4y^8} = (16x^4y^8)^{1/4} = 16^{1/4} \cdot x^{4/4}y^{8/4} = 2xy^2$

b. $\sqrt{12m^3n} \cdot \sqrt{3m^5n} = \sqrt{36m^8n^2} = (36m^8n^2)^{1/2} = 36^{1/2} \cdot m^4n = 6m^4n$

c. $\dfrac{\sqrt[3]{-27x^6}}{\sqrt[3]{8y^3}} = \dfrac{(-27x^6)^{1/3}}{(8y^3)^{1/3}} = \dfrac{-27^{1/3}x^2}{8^{1/3}y} = -\dfrac{3x^2}{2y}$

If a radical appears in the numerator or denominator of an algebraic expression, we often try to simplify the expression by eliminating the radical from the numerator or denominator. This process, called **rationalization,** is illustrated in the next two examples.

EXAMPLE 3 Rationalize the denominator of the expression $\dfrac{3x}{2\sqrt{x}}$.

Solution

$$\frac{3x}{2\sqrt{x}} = \frac{3x}{2\sqrt{x}} \cdot \frac{\sqrt{x}}{\sqrt{x}} = \frac{3x\sqrt{x}}{2\sqrt{x^2}} = \frac{3x\sqrt{x}}{2x} = \frac{3}{2}\sqrt{x}$$

EXAMPLE 4 Express $\dfrac{1}{2}x^{-1/2}$ as a radical, and rationalize the denominator of the expression that you obtain.

Solution

$$\frac{1}{2}x^{-1/2} = \frac{1}{2\sqrt{x}} \cdot \frac{\sqrt{x}}{\sqrt{x}} = \frac{\sqrt{x}}{2x}$$

EXAMPLE 5 Rationalize the numerator of the expression $\dfrac{3\sqrt{x}}{2x}$.

Solution

$$\frac{3\sqrt{x}}{2x} = \frac{3\sqrt{x}}{2x} \cdot \frac{\sqrt{x}}{\sqrt{x}} = \frac{3\sqrt{x^2}}{2x\sqrt{x}} = \frac{3x}{2x\sqrt{x}} = \frac{3}{2\sqrt{x}}$$

Operations with Algebraic Expressions

In calculus, we often work with algebraic expressions such as

$$2x^{4/3} - x^{1/3} + 1 \qquad 2x^2 - x - \frac{2}{\sqrt{x}} \qquad \frac{3xy + 2}{x + 1} \qquad 2x^3 + 2x + 1$$

An algebraic expression of the form $ax^m y^n$, where the coefficient a is a real number and m and n are nonnegative integers, is called a **monomial**, meaning that it consists of one term. For example, $7x^2$ is a monomial. A **polynomial** is a monomial or the sum of two or more monomials. For example,

$$x^2 + 4x + 4 \qquad x^3 + 5 \qquad x^4 + 3x^2 + 3 \qquad x^2 y + xy + y$$

are all polynomials. The degree of a polynomial is the highest power $(m + n)$ of the variables that appears in the polynomial.

Constant terms and terms containing the same variable factor are called **like,** or **similar, terms.** Like terms may be combined by adding or subtracting their numerical coefficients. For example,

$$3x + 7x = 10x \quad \text{and} \quad \frac{1}{2}xy + 3xy = \frac{7}{2}xy$$

The distributive property of the real number system,

$$ab + ac = a(b + c)$$

is used to justify this procedure.

To add or subtract two or more algebraic expressions, first remove the parentheses and then combine like terms. The resulting expression is written in order of non-increasing degree from left to right.

EXAMPLE 6

a. $(2x^4 + 3x^3 + 4x + 6) - (3x^4 + 9x^3 + 3x^2)$

$\qquad = 2x^4 + 3x^3 + 4x + 6 - 3x^4 - 9x^3 - 3x^2$ ⬥ Remove parentheses.

$\qquad = 2x^4 - 3x^4 + 3x^3 - 9x^3 - 3x^2 + 4x + 6$

$\qquad = -x^4 - 6x^3 - 3x^2 + 4x + 6$ ⬥ Combine like terms.

b. $2t^3 - \{t^2 - [t - (2t - 1)] + 4\}$

$\qquad = 2t^3 - \{t^2 - [t - 2t + 1] + 4\}$

$\qquad = 2t^3 - \{t^2 - [-t + 1] + 4\}$ ⬥ Remove parentheses and combine like terms within brackets.

$\qquad = 2t^3 - \{t^2 + t - 1 + 4\}$ ⬥ Remove brackets.

$\qquad = 2t^3 - \{t^2 + t + 3\}$ ⬥ Combine like terms within braces.

$\qquad = 2t^3 - t^2 - t - 3$ ⬥ Remove braces.

Observe that when the algebraic expression in Example 6b was simplified, the innermost grouping symbols were removed first; that is, the parentheses ( ) were removed first, the brackets [ ] second, and the braces { } third.

When algebraic expressions are multiplied, each term of one algebraic expression is multiplied by each term of the other. The resulting algebraic expression is then simplified.

EXAMPLE 7 Perform the indicated operations:

a. $(x^2 + 1)(3x^2 + 10x + 3)$ **b.** $x\left(300 - \frac{1}{4}x - \frac{1}{8}y\right) + y\left(240 - \frac{1}{8}x - \frac{3}{8}y\right)$

c. $(e^t + e^{-t})e^t - e^t(e^t - e^{-t})$

Solution

a. $(x^2 + 1)(3x^2 + 10x + 3) = x^2(3x^2 + 10x + 3) + 1(3x^2 + 10x + 3)$
$$= 3x^4 + 10x^3 + 3x^2 + 3x^2 + 10x + 3$$
$$= 3x^4 + 10x^3 + 6x^2 + 10x + 3$$

b. $x\left(300 - \dfrac{1}{4}x - \dfrac{1}{8}y\right) + y\left(240 - \dfrac{1}{8}x - \dfrac{3}{8}y\right)$

$$= 300x - \dfrac{1}{4}x^2 - \dfrac{1}{8}xy + 240y - \dfrac{1}{8}xy - \dfrac{3}{8}y^2$$

$$= -\dfrac{1}{4}x^2 - \dfrac{3}{8}y^2 - \dfrac{1}{4}xy + 300x + 240y$$

c. $(e^t + e^{-t})e^t - e^t(e^t - e^{-t}) = e^{2t} + e^0 - e^{2t} + e^0$
$$= e^{2t} - e^{2t} + e^0 + e^0$$
$$= 1 + 1 \qquad \text{Recall that } e^0 = 1.$$
$$= 2$$

Certain product formulas that are frequently used in algebraic computations are given in Table 5.

TABLE 5

Some Useful Product Formulas

Formula	Example
$(a + b)^2 = a^2 + 2ab + b^2$	$(2x + 3y)^2 = (2x)^2 + 2(2x)(3y) + (3y)^2$
	$\qquad\qquad = 4x^2 + 12xy + 9y^2$
$(a - b)^2 = a^2 - 2ab + b^2$	$(4x - 2y)^2 = (4x)^2 - 2(4x)(2y) + (2y)^2$
	$\qquad\qquad = 16x^2 - 16xy + 4y^2$
$(a + b)(a - b) = a^2 - b^2$	$(2x + y)(2x - y) = (2x)^2 - (y)^2$
	$\qquad\qquad = 4x^2 - y^2$

Factoring

Factoring is the process of expressing an algebraic expression as a product of other algebraic expressions. For example, by applying the distributive property, we may write

$$3x^2 - x = x(3x - 1)$$

To factor an algebraic expression, first check to see whether any of its terms have common factors. If they do, then factor out the greatest common factor. For example, the common factor of the algebraic expression $2a^2x + 4ax + 6a$ is $2a$ because

$$2a^2x + 4ax + 6a = 2a \cdot ax + 2a \cdot 2x + 2a \cdot 3 = 2a(ax + 2x + 3)$$

EXAMPLE 8 Factor out the greatest common factor in each expression:

a. $-3t^2 + 3t$ b. $2x^{3/2} - 3x^{1/2}$ c. $2ye^{xy^2} + 2xy^3e^{xy^2}$

d. $4x(x + 1)^{1/2} - 2x^2\left(\dfrac{1}{2}\right)(x + 1)^{-1/2}$

Solution

a. $-3t^2 + 3t = -3t(t - 1)$
b. $2x^{3/2} - 3x^{1/2} = x^{1/2}(2x - 3)$
c. $2ye^{xy^2} + 2xy^3e^{xy^2} = 2ye^{xy^2}(1 + xy^2)$

d. $4x(x + 1)^{1/2} - 2x^2\left(\dfrac{1}{2}\right)(x + 1)^{-1/2} = 4x(x + 1)^{1/2} - x^2(x + 1)^{-1/2}$

$$= x(x + 1)^{-1/2}[4(x + 1)^{1/2}(x + 1)^{1/2} - x]$$
$$= x(x + 1)^{-1/2}[4(x + 1) - x]$$
$$= x(x + 1)^{-1/2}(4x + 4 - x) = x(x + 1)^{-1/2}(3x + 4)$$

Here, we select $(x + 1)^{-1/2}$ as the greatest common factor because it is the highest power of $(x + 1)$ in each algebraic term. In particular, observe that

$$(x + 1)^{-1/2}(x + 1)^{1/2}(x + 1)^{1/2} = (x + 1)^{-1/2+1/2+1/2} = (x + 1)^{1/2}$$ ◼

Sometimes an algebraic expression may be factored by regrouping and rearranging its terms so that a common term can be factored out. This technique is illustrated in Example 9.

EXAMPLE 9 Factor:

a. $2ax + 2ay + bx + by$ **b.** $3x\sqrt{y} - 4 - 2\sqrt{y} + 6x$

Solution

a. First, factor the common term $2a$ from the first two terms and the common term b from the last two terms. Thus,

$$2ax + 2ay + bx + by = 2a(x + y) + b(x + y)$$

Since $(x + y)$ is common to both terms of the polynomial, we may factor it out. Hence

$$2a(x + y) + b(x + y) = (2a + b)(x + y)$$

b. $3x\sqrt{y} - 4 - 2\sqrt{y} + 6x = 3x\sqrt{y} - 2\sqrt{y} + 6x - 4$ Rearrange terms.

$$= \sqrt{y}(3x - 2) + 2(3x - 2)$$ Factor out common factors.

$$= (3x - 2)(\sqrt{y} + 2)$$ ◼

As we have seen, the first step in factoring a polynomial is to find the common factors. The next step is to express the polynomial as the product of a constant and/or one or more prime polynomials.

Certain product formulas that are useful in factoring binomials and trinomials are listed in Table 6.

TABLE 6	
Product Formulas Used in Factoring	
Formula	**Example**
Difference of two squares: $x^2 - y^2 = (x + y)(x - y)$	$x^2 - 36 = (x + 6)(x - 6)$ $8x^2 - 2y^2 = 2(4x^2 - y^2)$ $\qquad = 2(2x + y)(2x - y)$ $9 - a^6 = (3 + a^3)(3 - a^3)$
Perfect-square trinomial: $x^2 + 2xy + y^2 = (x + y)^2$ $x^2 - 2xy + y^2 = (x - y)^2$	$x^2 + 8x + 16 = (x + 4)^2$ $4x^2 - 4xy + y^2 = (2x - y)^2$
Sum of two cubes: $x^3 + y^3 = (x + y)(x^2 - xy + y^2)$	$z^3 + 27 = z^3 + (3)^3$ $\qquad = (z + 3)(z^2 - 3z + 9)$
Difference of two cubes: $x^3 - y^3 = (x - y)(x^2 + xy + y^2)$	$8x^3 - y^6 = (2x)^3 - (y^2)^3$ $\qquad = (2x - y^2)(4x^2 + 2xy^2 + y^4)$

The factors of the second-degree polynomial with integral coefficients

$$px^2 + qx + r$$

are $(ax + b)(cx + d)$, where $ac = p$, $ad + bc = q$, and $bd = r$. Since only a limited number of choices are possible, we can use a trial-and-error method to factor polynomials having this form.

For example, to factor $x^2 - 2x - 3$, we first observe that the only possible first-degree terms are

$$(x \quad)(x \quad) \qquad \text{\textcolor{red}{Since the coefficient of } } x^2 \text{ is 1}$$

Next, we observe that the product of the constant terms is (-3). This gives us the following possible factors:

$$(x - 1)(x + 3)$$
$$(x + 1)(x - 3)$$

Looking once again at the polynomial $x^2 - 2x - 3$, we see that the coefficient of x is -2. Checking to see which set of factors yields -2 for the coefficient of x, we find that

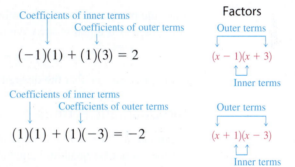

and we conclude that the correct factorization is

$$x^2 - 2x - 3 = (x + 1)(x - 3)$$

With practice, you will soon find that you can perform many of these steps mentally, and the need to write out each step will be eliminated.

EXAMPLE 10 Factor:

a. $3x^2 + 4x - 4$ **b.** $3x^2 - 6x - 24$ **c.** $-3t^2 + 192t + 195$

Solution

a. Using trial and error, we find that the correct factorization is

$$3x^2 + 4x - 4 = (3x - 2)(x + 2)$$

b. Since each term has the common factor 3, we have

$$3x^2 - 6x - 24 = 3(x^2 - 2x - 8)$$

Using the trial-and-error method of factorization, we find that

$$x^2 - 2x - 8 = (x - 4)(x + 2)$$

Thus, we have

$$3x^2 - 6x - 24 = 3(x - 4)(x + 2)$$

c. Since each term has the common factor -3, we have

$$-3t^2 + 192t + 195 = -3(t^2 - 64t - 65)$$

Using the trial-and-error method of factorization, we find that

$$(t^2 - 64t - 65) = (t - 65)(t + 1)$$

Therefore,

$$-3t^2 + 192t + 195 = -3(t - 65)(t + 1)$$

Roots of Polynomial Equations

A polynomial equation of degree n in the variable x is an equation of the form

$$a_n x^n + a_{n-1} x^{n-1} + \cdots + a_0 = 0$$

where n is a nonnegative integer and $a_0, a_1, \ldots, a_n$ are real numbers with $a_n \neq 0$. For example, the equation

$$-2x^5 + 8x^3 - 6x^2 + 3x + 1 = 0$$

is a polynomial equation of degree 5 in x.

The **roots of a polynomial equation** are precisely the values of x that satisfy the given equation.* One way to find the roots of a polynomial equation is to factor the polynomial and then solve the resulting equation. For example, the polynomial equation

$$x^3 - 3x^2 + 2x = 0$$

may be rewritten in the form

$$x(x^2 - 3x + 2) = 0 \quad \text{or} \quad x(x - 1)(x - 2) = 0$$

Since the product of two real numbers can be equal to zero if and only if one (or both) of the factors is equal to zero, we have

$$x = 0 \qquad x - 1 = 0 \qquad \text{or} \qquad x - 2 = 0$$

from which we see that the desired roots are $x = 0$, 1, and 2.

The Quadratic Formula

In general, the problem of finding the roots of a polynomial equation is a difficult one. But the roots of a quadratic equation (a polynomial equation of degree 2) are easily found either by factoring or by using the following quadratic formula.

Quadratic Formula

The solutions of the equation $ax^2 + bx + c = 0$ $(a \neq 0)$ are given by

$$x = \frac{-b \pm \sqrt{b^2 - 4ac}}{2a}$$

Note   If you use the quadratic formula to solve a quadratic equation, first make sure that the equation is in the *standard form* $ax^2 + bx + c = 0$.

EXAMPLE 11 Solve each of the following quadratic equations:

a. $2x^2 + 5x - 12 = 0$ **b.** $x^2 = -3x + 8$

Solution

a. The equation is in standard form, with $a = 2$, $b = 5$, and $c = -12$. Using the quadratic formula, we find

$$x = \frac{-b \pm \sqrt{b^2 - 4ac}}{2a} = \frac{-5 \pm \sqrt{5^2 - 4(2)(-12)}}{2(2)}$$

$$= \frac{-5 \pm \sqrt{121}}{4} = \frac{-5 \pm 11}{4}$$

$$= -4 \quad \text{or} \quad \frac{3}{2}$$

---

*In this book, we are interested only in the *real* roots of an equation.

This equation can also be solved by factoring. Thus,

$$2x^2 + 5x - 12 = (2x - 3)(x + 4) = 0$$

from which we see that the desired roots are $x = \frac{3}{2}$ or $x = -4$, as obtained earlier.

b. We first rewrite the given equation in the standard form $x^2 + 3x - 8 = 0$, from which we see that $a = 1$, $b = 3$, and $c = -8$. Using the quadratic formula, we find

$$x = \frac{-b \pm \sqrt{b^2 - 4ac}}{2a} = \frac{-3 \pm \sqrt{3^2 - 4(1)(-8)}}{2(1)}$$

$$= \frac{-3 \pm \sqrt{41}}{2}$$

That is, the solutions are

$$\frac{-3 + \sqrt{41}}{2} \approx 1.7 \quad \text{and} \quad \frac{-3 - \sqrt{41}}{2} \approx -4.7$$

In this case, the quadratic formula proves quite handy!

1.1 Exercises

In Exercises 1–6, show the interval on a number line.

1. $(3, 6)$ **2.** $(-2, 5]$ **3.** $[-1, 4)$

4. $\left[-\frac{6}{5}, -\frac{1}{2}\right]$ **5.** $(0, \infty)$ **6.** $(-\infty, 5]$

In Exercises 7–22, evaluate the expression.

7. $27^{2/3}$ **8.** $8^{-4/3}$

9. $\left(\frac{1}{\sqrt{3}}\right)^0$ **10.** $(7^{1/2})^4$

11. $\left[\left(\frac{1}{8}\right)^{1/3}\right]^{-2}$ **12.** $\left[\left(-\frac{1}{3}\right)^2\right]^{-3}$

13. $\left(\frac{7^{-5} \cdot 7^2}{7^{-2}}\right)^{-1}$ **14.** $\left(\frac{9}{16}\right)^{-1/2}$

15. $(125^{2/3})^{-1/2}$ **16.** $\sqrt[3]{2^6}$

17. $\frac{\sqrt{32}}{\sqrt{8}}$ **18.** $\sqrt[3]{\frac{-8}{27}}$

19. $\frac{16^{5/8}16^{1/2}}{16^{7/8}}$ **20.** $\left(\frac{9^{-3} \cdot 9^5}{9^{-2}}\right)^{-1/2}$

21. $16^{1/4} \cdot 8^{-1/3}$ **22.** $\frac{6^{2.5} \cdot 6^{-1.9}}{6^{-1.4}}$

In Exercises 23–32, determine whether the statement is true or false. Give a reason for your choice.

23. $x^4 + 2x^4 = 3x^4$ **24.** $3^2 \cdot 2^2 = 6^2$

25. $x^3 \cdot 2x^2 = 2x^6$ **26.** $3^3 + 3 = 3^4$

27. $\frac{2^{4x}}{1^{3x}} = 2^{4x-3x}$ **28.** $(2^2 \cdot 3^2)^2 = 6^4$

29. $\frac{1}{4^{-3}} = \frac{1}{64}$ **30.** $\frac{4^{3/2}}{2^4} = \frac{1}{2}$

31. $(1.2^{1/2})^{-1/2} = 1$ **32.** $5^{2/3} \cdot (25)^{2/3} = 25$

In Exercises 33–38, rewrite the expression using positive exponents only.

33. $(xy)^{-2}$ **34.** $3s^{1/3} \cdot s^{-7/3}$

35. $\frac{x^{-1/3}}{x^{1/2}}$ **36.** $\sqrt{x^{-1}} \cdot \sqrt{9x^{-3}}$

37. $12^0(s + t)^{-3}$ **38.** $(x - y)(x^{-1} + y^{-1})$

In Exercises 39–54, simplify the expression. (Assume that x, y, r, s, and t are positive.)

39. $\frac{x^{7/3}}{x^{-2}}$ **40.** $(49x^{-2})^{-1/2}$

41. $(x^2y^{-3})(x^{-5}y^3)$ **42.** $\frac{5x^6y^3}{2x^2y^7}$

43. $\frac{x^{3/4}}{x^{-1/4}}$ **44.** $\left(\frac{x^3y^2}{z^2}\right)^2$

45. $\left(\frac{x^3}{-27y^{-6}}\right)^{-2/3}$ **46.** $\left(\frac{e^x}{e^{x-2}}\right)^{-1/2}$

47. $\left(\frac{x^{-3}}{y^{-2}}\right)^2\left(\frac{y}{x}\right)^4$ **48.** $\frac{(r^n)^4}{r^{5-2n}}$

49. $\sqrt[3]{x^{-2}} \cdot \sqrt{4x^5}$ **50.** $\sqrt{81x^6y^{-4}}$

51. $-\sqrt[4]{16x^4y^8}$ **52.** $\sqrt[3]{x^{3a+b}}$

53. $\sqrt[6]{64x^8y^3}$ **54.** $\sqrt[3]{27r^6} \cdot \sqrt{s^2t^4}$

In Exercises 55–58, use the fact that $2^{1/2} \approx 1.414$ and $3^{1/2} \approx 1.732$ to evaluate the expression without using a calculator.

55. $2^{3/2}$ **56.** $8^{1/2}$ **57.** $9^{3/4}$ **58.** $6^{1/2}$

In Exercises 59–62, use the fact that $10^{1/2} \approx 3.162$ and $10^{1/3} \approx 2.154$ to evaluate the expression without using a calculator.

59. $10^{3/2}$ **60.** $1000^{3/2}$

61. $10^{2.5}$ **62.** $(0.0001)^{-1/3}$

In Exercises 63–68, rationalize the denominator.

63. $\dfrac{3}{2\sqrt{x}}$ **64.** $\dfrac{3}{\sqrt{xy}}$ **65.** $\dfrac{2y}{\sqrt{3y}}$

66. $\dfrac{5x^2}{\sqrt{3x}}$ **67.** $\dfrac{1}{\sqrt[3]{x}}$ **68.** $\sqrt{\dfrac{2x}{y}}$

In Exercises 69–74, rationalize the numerator.

69. $\dfrac{2\sqrt{x}}{3}$ **70.** $\dfrac{\sqrt[3]{x}}{24}$ **71.** $\sqrt{\dfrac{2y}{x}}$

72. $\sqrt[3]{\dfrac{2x}{3y}}$ **73.** $\dfrac{\sqrt[3]{x^2z}}{y}$ **74.** $\dfrac{\sqrt[3]{x^2y}}{2x}$

In Exercises 75–98, perform the indicated operations and/or simplify each expression.

75. $(7x^2 - 2x + 5) + (2x^2 + 5x - 4)$

76. $(3x^2 + 5xy + 2y) + (4 - 3xy - 2x^2)$

77. $(5y^2 - 2y + 1) - (y^2 - 3y - 7)$

78. $3(2a - b) - 4(b - 2a)$

79. $x - \{2x - [-x - (1 - x)]\}$

80. $3x^2 - \{x^2 + 1 - x[x - (2x - 1)]\} + 2$

81. $\left(\dfrac{1}{3} - 1 + e\right) - \left(-\dfrac{1}{3} - 1 + e^{-1}\right)$

82. $-\dfrac{3}{4}y - \dfrac{1}{4}x + 100 + \dfrac{1}{2}x + \dfrac{1}{4}y - 120$

83. $3\sqrt{8} + 8 - 2\sqrt{y} + \dfrac{1}{2}\sqrt{x} - \dfrac{3}{4}\sqrt{y}$

84. $\dfrac{8}{9}x^2 + \dfrac{2}{3}x + \dfrac{16}{3}x^2 - \dfrac{16}{3}x - 2x + 2$

85. $(x + 8)(x - 2)$ **86.** $(5x + 2)(3x - 4)$

87. $(a + 5)^2$ **88.** $(3a - 4b)^2$

89. $(x + 2y)^2$ **90.** $(6 - 3x)^2$

91. $(2x + y)(2x - y)$ **92.** $(3x + 2)(2 - 3x)$

93. $(2x^2 - 1)(3x^2) + (x^2 + 3)(4x)$

94. $(x^2 - 1)(2x) - x^2(2x)$

95. $6x\left(\dfrac{1}{2}\right)(2x^2 + 3)^{-1/2}(4x) + 6(2x^2 + 3)^{1/2}$

96. $(x^{1/2} + 1)\left(\dfrac{1}{2}x^{-1/2}\right) - (x^{1/2} - 1)\left(\dfrac{1}{2}x^{-1/2}\right)$

97. $100(-10te^{-0.1t} - 100e^{-0.1t})$

98. $2(t + \sqrt{t})^2 - 2t^2$

In Exercises 99–106, factor out the greatest common factor from each expression.

99. $4x^5 - 12x^4 - 6x^3$ **100.** $4x^2y^2z - 2x^5y^2 + 6x^3y^2z^2$

101. $7a^4 - 42a^2b^2 + 49a^3b$

102. $3x^{2/3} - 2x^{1/3}$ **103.** $e^{-x} - xe^{-x}$

104. $2ye^{xy^2} + 2xy^3e^{xy^2}$ **105.** $2x^{-5/2} - \dfrac{3}{2}x^{-3/2}$

106. $\dfrac{1}{2}\left(\dfrac{2}{3}u^{3/2} - 2u^{1/2}\right)$

In Exercises 107–120, factor each expression completely.

107. $6ac + 3bc - 4ad - 2bd$

108. $3x^3 - x^2 + 3x - 1$

109. $4a^2 - b^2$ **110.** $12x^2 - 3y^2$

111. $10 - 14x - 12x^2$ **112.** $x^2 - 2x - 15$

113. $3x^2 - 6x - 24$ **114.** $3x^2 - 4x - 4$

115. $12x^2 - 2x - 30$ **116.** $(x + y)^2 - 1$

117. $9x^2 - 16y^2$ **118.** $8a^2 - 2ab - 6b^2$

119. $x^6 + 125$ **120.** $x^3 - 27$

In Exercises 121–128, perform the indicated operations and simplify each expression.

121. $(x^2 + y^2)x - xy(2y)$

122. $2kr(R - r) - kr^2$

123. $2(x - 1)(2x + 2)^3[4(x - 1) + (2x + 2)]$

124. $5x^2(3x^2 + 1)^4(6x) + (3x^2 + 1)^5(2x)$

125. $4(x - 1)^2(2x + 2)^3(2) + (2x + 2)^4(2)(x - 1)$

126. $(x^2 + 1)(4x^3 - 3x^2 + 2x) - (x^4 - x^3 + x^2)(2x)$

127. $(x^2 + 2)^2[5(x^2 + 2)^2 - 3](2x)$

128. $(x^2 - 4)(x^2 + 4)(2x + 8) - (x^2 + 8x - 4)(4x^3)$

In Exercises 129–134, find the real roots of each equation by factoring.

129. $x^2 + x - 12 = 0$ **130.** $3x^2 - x - 4 = 0$

131. $4t^2 + 2t - 2 = 0$ **132.** $-6x^2 + x + 12 = 0$

133. $\dfrac{1}{4}x^2 - x + 1 = 0$ **134.** $\dfrac{1}{2}a^2 + a - 12 = 0$

In Exercises 135–140, solve the equation by using the quadratic formula.

135. $4x^2 + 5x - 6 = 0$ **136.** $3x^2 - 4x + 1 = 0$

137. $8x^2 - 8x - 3 = 0$ **138.** $x^2 - 6x + 6 = 0$

139. $2x^2 + 4x - 3 = 0$ **140.** $2x^2 + 7x - 15 = 0$

141. REVENUE OF TWO GAS STATIONS Jake owns two gas stations. The total revenue of the first gas station for the next 12 months is projected to be

$$0.2t^2 + 150t \quad (0 \le t \le 12)$$

thousand dollars t months from now. The total revenue of the second gas station for the next 12 months is projected to be

$$0.5t^2 + 200t \quad (0 \le t \le 12)$$

thousand dollars t months from now. Find an expression that gives the total revenue realized by Jake's gas stations in month t $(0 \le t \le 12)$.

142. REVENUE OF TWO GAS STATIONS Refer to Exercise 141. Find an expression that gives the amount by which the revenue of the second gas station will exceed the revenue of the first gas station in month t $(0 \le t \le 12)$.

143. DISTRIBUTION OF INCOMES The distribution of income in a certain city can be described by the mathematical model $y = (5.6 \cdot 10^{11})(x)^{-1.5}$, where y is the number of families with an income of x or more dollars.
a. How many families in this city have an income of $30,000 or more?

b. How many families have an income of $60,000 or more?

c. How many families have an income of $150,000 or more?

144. WORKER EFFICIENCY An efficiency study conducted by Elektra Electronics showed that the number of Space Commander walkie-talkies assembled by the average worker t hr after starting work at 8 A.M. is

$$-t^3 + 6t^2 + 15t \quad (0 \le t \le 4)$$

Factor the expression.

145. REVENUE OF A COMPANY Williams Commuter Air Service realizes a monthly revenue of

$$8000x - 100x^2 \quad (0 \le x \le 80)$$

dollars when the price charged per passenger is x dollars. Factor the expression.

In Exercises 146–148, determine whether the statement is true or false. If it is true, explain why it is true. If it is false, give an example to show why it is false.

146. If $b^2 - 4ac > 0$, then $ax^2 + bx + c = 0$ $(a \ne 0)$ has two real roots.

147. If $b^2 - 4ac < 0$, then $ax^2 + bx + c = 0$ $(a \ne 0)$ has no real roots.

148. $\sqrt{(a + b)(b - a)} = \sqrt{b^2 - a^2}$ for all real numbers a and b.

1.2 Precalculus Review II

Rational Expressions

Quotients of polynomials are called **rational expressions.** Examples of rational expressions are

$$\frac{6x - 1}{2x + 3} \qquad \frac{3x^2y^3 - 2xy}{4x} \qquad \frac{2}{5ab}$$

Since rational expressions are quotients in which the variables represent real numbers, the properties of real numbers apply to rational expressions as well, and operations with rational fractions are performed in the same manner as operations with arithmetic fractions. For example, using the properties of the real number system, we may write

$$\frac{ac}{bc} = \frac{a}{b} \cdot \frac{c}{c} = \frac{a}{b} \cdot 1 = \frac{a}{b}$$

where a, b, and c are any real numbers and b and c are not zero.

Similarly, using the same properties of real numbers, we may write

$$\frac{(x + 2)(x - 3)}{(x - 2)(x - 3)} = \frac{x + 2}{x - 2} \qquad (x \ne 2, 3)$$

after "canceling" the common factors.

An example of incorrect cancellation is

$$\frac{\cancel{3} + 4x}{\cancel{3}} \neq 1 + 4x$$

because 3 is not a factor of the numerator. Instead, we need to write

$$\frac{3 + 4x}{3} = \frac{3}{3} + \frac{4x}{3} = 1 + \frac{4x}{3}$$

An algebraic fraction is **simplified,** or **in lowest terms,** when the numerator and denominator have no common factors other than 1 and −1 and the fraction contains no negative exponents.

EXAMPLE 1 Simplify the following expressions:

a. $\dfrac{x^2 + 2x - 3}{x^2 + 4x + 3}$ **b.** $\dfrac{(x^2 + 1)^2(-2) + (2x)(2)(x^2 + 1)(2x)}{(x^2 + 1)^4}$

Solution

a. $\dfrac{x^2 + 2x - 3}{x^2 + 4x + 3} = \dfrac{(x + 3)(x - 1)}{(x + 3)(x + 1)} = \dfrac{x - 1}{x + 1}$

b. $\dfrac{(x^2 + 1)^2(-2) + (2x)(2)(x^2 + 1)(2x)}{(x^2 + 1)^4}$

$= \dfrac{(x^2 + 1)[(x^2 + 1)(-2) + (2x)(2)(2x)]}{(x^2 + 1)^4}$ Factor out $(x^2 + 1)$.

$= \dfrac{(x^2 + 1)(-2x^2 - 2 + 8x^2)}{(x^2 + 1)^4}$ Carry out indicated multiplication.

$= \dfrac{(x^2 + 1)(6x^2 - 2)}{(x^2 + 1)^4}$ Combine like terms.

$= \dfrac{(6x^2 - 2)}{(x^2 + 1)^3}$ Cancel the common factors.

$= \dfrac{2(3x^2 - 1)}{(x^2 + 1)^3}$ Factor out 2 from the numerator.

The operations of multiplication and division are performed with algebraic fractions in the same manner as with arithmetic fractions (Table 7).

TABLE 7

Rules of Multiplication and Division: Algebraic Fractions

Operation	Example
If P, Q, R, and S are polynomials, then	
Multiplication:	
$\dfrac{P}{Q} \cdot \dfrac{R}{S} = \dfrac{PR}{QS}$ $(Q, S \neq 0)$	$\dfrac{2x}{y} \cdot \dfrac{(x + 1)}{(y - 1)} = \dfrac{2x(x + 1)}{y(y - 1)} = \dfrac{2x^2 + 2x}{y^2 - y}$
Division:	
$\dfrac{P}{Q} \div \dfrac{R}{S} = \dfrac{P}{Q} \cdot \dfrac{S}{R} = \dfrac{PS}{QR}$ $(Q, R, S \neq 0)$	$\dfrac{x^2 + 3}{y} \div \dfrac{y^2 + 1}{x} = \dfrac{x^2 + 3}{y} \cdot \dfrac{x}{y^2 + 1} = \dfrac{x^3 + 3x}{y^3 + y}$

When rational expressions are multiplied and divided, the resulting expressions should be simplified if possible.

EXAMPLE 2 Perform the indicated operations and simplify:

$$\frac{2x - 8}{x + 2} \cdot \frac{x^2 + 4x + 4}{x^2 - 16}$$

Solution

$$\begin{aligned}
\frac{2x - 8}{x + 2} \cdot \frac{x^2 + 4x + 4}{x^2 - 16} &= \frac{2(x - 4)}{x + 2} \cdot \frac{(x + 2)^2}{(x + 4)(x - 4)} \\
&= \frac{2(x - 4)(x + 2)(x + 2)}{(x + 2)(x + 4)(x - 4)} \\
&= \frac{2(x + 2)}{x + 4} \qquad \text{Cancel the common factors} \\
&\qquad\qquad\qquad (x + 2)(x - 4).
\end{aligned}$$

For rational expressions, the operations of addition and subtraction are performed by finding a common denominator of the fractions and then adding or subtracting the numerators. Table 8 shows the rules for fractions with equal denominators.

TABLE 8

Rules of Addition and Subtraction: Fractions with Equal Denominators

Operation	Example
If P, Q, and R are polynomials, then	
Addition:	
$\dfrac{P}{R} + \dfrac{Q}{R} = \dfrac{P + Q}{R} \qquad (R \neq 0)$	$\dfrac{2x}{x + 2} + \dfrac{6x}{x + 2} = \dfrac{2x + 6x}{x + 2} = \dfrac{8x}{x + 2}$
Subtraction:	
$\dfrac{P}{R} - \dfrac{Q}{R} = \dfrac{P - Q}{R} \qquad (R \neq 0)$	$\dfrac{3y}{y - x} - \dfrac{y}{y - x} = \dfrac{3y - y}{y - x} = \dfrac{2y}{y - x}$

$$\frac{x}{2 + y} \neq \frac{x}{2} + \frac{x}{y}$$

To add or subtract fractions that have different denominators, first find a common denominator, preferably the least common denominator (LCD). Then carry out the indicated operations following the procedure described in Table 8.

To find the LCD of two or more rational expressions:

1. *Find the prime factors of each denominator.*
2. *Form the product of the different prime factors that occur in the denominators. Each prime factor in this product should be raised to the highest power of that factor appearing in the denominators.*

EXAMPLE 3 Simplify:

a. $\dfrac{2x}{x^2 + 1} + \dfrac{6(3x^2)}{x^3 + 2}$ **b.** $\dfrac{1}{x + h} - \dfrac{1}{x}$

Solution

a. $\dfrac{2x}{x^2 + 1} + \dfrac{6(3x^2)}{x^3 + 2} = \dfrac{2x(x^3 + 2) + 6(3x^2)(x^2 + 1)}{(x^2 + 1)(x^3 + 2)}$ LCD $= (x^2 + 1)(x^3 + 2)$

$= \dfrac{2x^4 + 4x + 18x^4 + 18x^2}{(x^2 + 1)(x^3 + 2)}$ Carry out the indicated multiplication.

$= \dfrac{20x^4 + 18x^2 + 4x}{(x^2 + 1)(x^3 + 2)}$ Combine like terms.

$= \dfrac{2x(10x^3 + 9x + 2)}{(x^2 + 1)(x^3 + 2)}$ Factor.

b. $\dfrac{1}{x + h} - \dfrac{1}{x} = \dfrac{x - (x + h)}{x(x + h)}$ LCD $= x(x + h)$

$= \dfrac{x - x - h}{x(x + h)}$ Remove parentheses.

$= \dfrac{-h}{x(x + h)}$ Combine like terms.

Other Algebraic Fractions

The techniques used to simplify rational expressions may also be used to simplify algebraic fractions in which the numerator and denominator are not polynomials, as illustrated in Example 4.

EXAMPLE 4 Simplify:

a. $\dfrac{1 + \dfrac{1}{x + 1}}{x - \dfrac{4}{x}}$ **b.** $\dfrac{x^{-1} + y^{-1}}{x^{-2} - y^{-2}}$

Solution

a. $\dfrac{1 + \dfrac{1}{x + 1}}{x - \dfrac{4}{x}} = \dfrac{\dfrac{x + 1 + 1}{x + 1}}{\dfrac{x^2 - 4}{x}}$ LCD for numerator is $x + 1$ and LCD for denominator is x.

$= \dfrac{x + 2}{x + 1} \cdot \dfrac{x}{x^2 - 4} = \dfrac{x + 2}{x + 1} \cdot \dfrac{x}{(x + 2)(x - 2)}$

$= \dfrac{x}{(x + 1)(x - 2)}$

b. $\dfrac{x^{-1} + y^{-1}}{x^{-2} - y^{-2}} = \dfrac{\dfrac{1}{x} + \dfrac{1}{y}}{\dfrac{1}{x^2} - \dfrac{1}{y^2}} = \dfrac{\dfrac{y + x}{xy}}{\dfrac{y^2 - x^2}{x^2y^2}}$ $x^{-n} = \dfrac{1}{x^n}$

$= \dfrac{y + x}{xy} \cdot \dfrac{x^2y^2}{y^2 - x^2} = \dfrac{y + x}{xy} \cdot \dfrac{x^2y^2}{(y + x)(y - x)}$

$= \dfrac{xy}{y - x}$

EXAMPLE 5 Perform the given operations and simplify:

a. $\dfrac{x^2(2x^2+1)^{1/2}}{x-1} \cdot \dfrac{4x^3-6x^2+x-2}{x(x-1)(2x^2+1)}$ **b.** $\dfrac{12x^2}{\sqrt{2x^2+3}} + 6\sqrt{2x^2+3}$

Solution

a. $\dfrac{x^2(2x^2+1)^{1/2}}{x-1} \cdot \dfrac{4x^3-6x^2+x-2}{x(x-1)(2x^2+1)} = \dfrac{x(4x^3-6x^2+x-2)}{(x-1)^2(2x^2+1)^{1-1/2}}$

$$= \dfrac{x(4x^3-6x^2+x-2)}{(x-1)^2(2x^2+1)^{1/2}}$$

b. $\dfrac{12x^2}{\sqrt{2x^2+3}} + 6\sqrt{2x^2+3} = \dfrac{12x^2}{(2x^2+3)^{1/2}} + 6(2x^2+3)^{1/2}$ Write radicals in exponential form.

$$= \dfrac{12x^2 + 6(2x^2+3)^{1/2}(2x^2+3)^{1/2}}{(2x^2+3)^{1/2}}$$ LCD is $(2x^2+3)^{1/2}$.

$$= \dfrac{12x^2 + 6(2x^2+3)}{(2x^2+3)^{1/2}}$$

$$= \dfrac{24x^2+18}{(2x^2+3)^{1/2}} = \dfrac{6(4x^2+3)}{\sqrt{2x^2+3}}$$

Rationalizing Algebraic Fractions

When the denominator of an algebraic fraction contains sums or differences involving radicals, we may **rationalize the denominator**—that is, transform the fraction into an equivalent one with a denominator that does not contain radicals. In doing so, we make use of the fact that

$$(\sqrt{a}+\sqrt{b})(\sqrt{a}-\sqrt{b}) = (\sqrt{a})^2 - (\sqrt{b})^2$$
$$= a - b$$

This procedure is illustrated in Example 6.

EXAMPLE 6 Rationalize the denominator: $\dfrac{1}{1+\sqrt{x}}$.

Solution   Upon multiplying the numerator and the denominator by $(1-\sqrt{x})$, we obtain

$$\dfrac{1}{1+\sqrt{x}} = \dfrac{1}{1+\sqrt{x}} \cdot \dfrac{1-\sqrt{x}}{1-\sqrt{x}}$$ $\dfrac{1-\sqrt{x}}{1-\sqrt{x}} = 1$

$$= \dfrac{1-\sqrt{x}}{1-(\sqrt{x})^2}$$

$$= \dfrac{1-\sqrt{x}}{1-x}$$

In other situations, it may be necessary to rationalize the numerator of an algebraic expression. In calculus, for example, one encounters the following problem.

EXAMPLE 7 Rationalize the numerator: $\dfrac{\sqrt{1+h}-1}{h}$.

Solution

$$\dfrac{\sqrt{1+h}-1}{h} = \dfrac{\sqrt{1+h}-1}{h} \cdot \dfrac{\sqrt{1+h}+1}{\sqrt{1+h}+1}$$

$$= \frac{(\sqrt{1+h})^2 - (1)^2}{h(\sqrt{1+h}+1)}$$

$$= \frac{1+h-1}{h(\sqrt{1+h}+1)} \qquad \begin{aligned}(\sqrt{1+h})^2 &= \sqrt{1+h} \cdot \sqrt{1+h}\\ &= 1+h\end{aligned}$$

$$= \frac{h}{h(\sqrt{1+h}+1)}$$

$$= \frac{1}{\sqrt{1+h}+1}$$

Inequalities

The following properties may be used to solve one or more inequalities involving a variable.

Properties of Inequalities

If a, b, and c are any real numbers, then

		Example
Property 1	If $a < b$ and $b < c$, then $a < c$.	$2 < 3$ and $3 < 8$, so $2 < 8$.
Property 2	If $a < b$, then $a + c < b + c$.	$-5 < -3$, so $-5 + 2 < -3 + 2$; that is, $-3 < -1$.
Property 3	If $a < b$ and $c > 0$, then $ac < bc$.	$-5 < -3$, and since $2 > 0$, we have $(-5)(2) < (-3)(2)$; that is, $-10 < -6$.
Property 4	If $a < b$ and $c < 0$, then $ac > bc$.	$-2 < 4$, and since $-3 < 0$, we have $(-2)(-3) > (4)(-3)$; that is, $6 > -12$.

Similar properties hold if each inequality sign, $<$, between a and b and between b and c is replaced by $\geq$, $>$, or $\leq$. Note that Property 4 says that an inequality sign is reversed if the inequality is multiplied by a negative number.

A real number is a *solution of an inequality* involving a variable if a true statement is obtained when the variable is replaced by that number. The set of all real numbers satisfying the inequality is called the *solution set*. We often use interval notation to describe the solution set.

EXAMPLE 8   Find the set of real numbers that satisfy $-1 \leq 2x - 5 < 7$.

Solution   Add 5 to each member of the given double inequality, obtaining

$$4 \leq 2x < 12$$

Next, multiply each member of the resulting double inequality by $\frac{1}{2}$, yielding

$$2 \leq x < 6$$

Thus, the solution is the set of all values of x lying in the interval $[2, 6)$.

EXAMPLE 9   Solve the inequality $x^2 + 2x - 8 < 0$.

Solution   Observe that $x^2 + 2x - 8 = (x + 4)(x - 2)$, so the given inequality is equivalent to the inequality $(x + 4)(x - 2) < 0$. Since the product of two real numbers is negative if and only if the two numbers have opposite signs, we solve the inequality $(x + 4)(x - 2) < 0$ by studying the signs of the two factors $x + 4$

and $x - 2$. Now, $x + 4 > 0$ when $x > -4$, and $x + 4 < 0$ when $x < -4$. Similarly, $x - 2 > 0$ when $x > 2$, and $x - 2 < 0$ when $x < 2$. These results are summarized graphically in Figure 2.

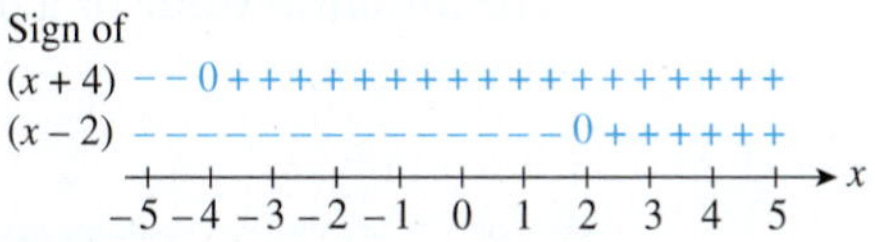

Sign of
$(x + 4)$ $-- 0 ++++++++++++++++++++$
$(x - 2)$ $-------------- 0 ++++++$

$$-5\ -4\ -3\ -2\ -1\quad 0\quad 1\quad 2\quad 3\quad 4\quad 5 \longrightarrow x$$

FIGURE 2
Sign diagram for $(x + 4)(x - 2)$

From Figure 2, we see that the two factors $x + 4$ and $x - 2$ have opposite signs when and only when x lies strictly between -4 and 2. Therefore, the required solution is the interval $(-4, 2)$.

EXAMPLE 10 Solve the inequality $\dfrac{x + 1}{x - 1} \geq 0$.

Solution The quotient $(x + 1)/(x - 1)$ is strictly positive if and only if both the numerator and the denominator have the same sign. The signs of $x + 1$ and $x - 1$ are shown in Figure 3.

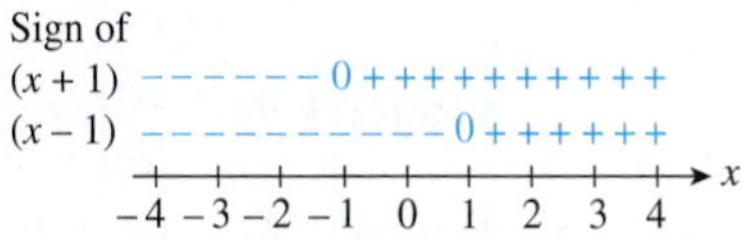

Sign of
$(x + 1)$ $------ 0 +++++++++++$
$(x - 1)$ $---------- 0 ++++++$

$$-4\ -3\ -2\ -1\quad 0\quad 1\quad 2\quad 3\quad 4 \longrightarrow x$$

FIGURE 3
Sign diagram for $\dfrac{x + 1}{x - 1}$

From Figure 3, we see that $x + 1$ and $x - 1$ have the same sign if $x < -1$ or $x > 1$. The quotient $(x + 1)/(x - 1)$ is equal to zero if $x = -1$. Therefore, the required solution is the set of all x in the intervals $(-\infty, -1]$ and $(1, \infty)$.

 APPLIED EXAMPLE 11 Stock Purchase The management of Corbyco, a giant conglomerate, has estimated that x thousand dollars is needed to purchase

$$100{,}000(-1 + \sqrt{1 + 0.001x})$$

shares of common stock of Starr Communications. Determine how much money Corbyco needs to purchase at least 100,000 shares of Starr's stock.

Solution The amount of money Corbyco needs to purchase at least 100,000 shares is found by solving the inequality

$$100{,}000(-1 + \sqrt{1 + 0.001x}) \geq 100{,}000$$

Proceeding, we find

$$-1 + \sqrt{1 + 0.001x} \geq 1$$
$$\sqrt{1 + 0.001x} \geq 2$$
$$1 + 0.001x \geq 4 \qquad \text{Square both sides.}$$
$$0.001x \geq 3$$
$$x \geq 3000$$

so Corbyco needs at least $3,000,000. (Recall that x is measured in thousands of dollars.)

Absolute Value

> **Absolute Value**
>
> The **absolute value** of a number a is denoted by $|a|$ and is defined by
>
> $$|a| = \begin{cases} a & \text{if } a \geq 0 \\ -a & \text{if } a < 0 \end{cases}$$

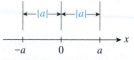

FIGURE 4
The absolute value of a number

Since $-a$ is a positive number when a is negative, it follows that the absolute value of a number is always nonnegative. For example, $|5| = 5$ and $|-5| = -(-5) = 5$. Geometrically, $|a|$ is the distance between the origin and the point on the number line that represents the number a (Figure 4).

> **Absolute Value Properties**
>
> If a and b are any real numbers, then
>
> | | | **Example** |
> |---|---|---|
> | **Property 5** | $\lvert -a \rvert = \lvert a \rvert$ | $\lvert -3 \rvert = -(-3) = 3 = \lvert 3 \rvert$ |
> | **Property 6** | $\lvert ab \rvert = \lvert a \rvert \lvert b \rvert$ | $\lvert (2)(-3) \rvert = \lvert -6 \rvert = 6 = (2)(3)$ $= \lvert 2 \rvert \lvert -3 \rvert$ |
> | **Property 7** | $\left\lvert \dfrac{a}{b} \right\rvert = \dfrac{\lvert a \rvert}{\lvert b \rvert} \quad (b \neq 0)$ | $\left\lvert \dfrac{(-3)}{(-4)} \right\rvert = \left\lvert \dfrac{3}{4} \right\rvert = \dfrac{3}{4} = \dfrac{\lvert -3 \rvert}{\lvert -4 \rvert}$ |
> | **Property 8** | $\lvert a + b \rvert \leq \lvert a \rvert + \lvert b \rvert$ | $\lvert 8 + (-5) \rvert = \lvert 3 \rvert = 3$ $\leq \lvert 8 \rvert + \lvert -5 \rvert = 13$ |

Property 8 is called the **triangle inequality**.

EXAMPLE 12 Evaluate each of the following expressions:

a. $|\pi - 5| + 3$ **b.** $|\sqrt{3} - 2| + |2 - \sqrt{3}|$

Solution

a. Since $\pi - 5 < 0$, we see that $|\pi - 5| = -(\pi - 5)$. Therefore,

$$|\pi - 5| + 3 = -(\pi - 5) + 3 = 8 - \pi$$

b. Since $\sqrt{3} - 2 < 0$, we see that $|\sqrt{3} - 2| = -(\sqrt{3} - 2)$. Next, observe that $2 - \sqrt{3} > 0$, so $|2 - \sqrt{3}| = 2 - \sqrt{3}$. Therefore,

$$|\sqrt{3} - 2| + |2 - \sqrt{3}| = -(\sqrt{3} - 2) + (2 - \sqrt{3})$$
$$= 4 - 2\sqrt{3} = 2(2 - \sqrt{3})$$

EXAMPLE 13 Solve the inequalities $|x| \leq 5$ and $|x| \geq 5$.

Solution   First, we consider the inequality $|x| \leq 5$. If $x \geq 0$, then $|x| = x$, so $|x| \leq 5$ implies $x \leq 5$ in this case. On the other hand, if $x < 0$, then $|x| = -x$, so $|x| \leq 5$ implies $-x \leq 5$ or $x \geq -5$. Thus, $|x| \leq 5$ means $-5 \leq x \leq 5$ (Figure 5a). To obtain an alternative solution, observe that $|x|$ is the distance from the point x to zero, so the inequality $|x| \leq 5$ implies immediately that $-5 \leq x \leq 5$.

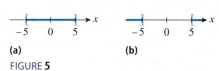

(a)   (b)

FIGURE 5

Next, the inequality $|x| \geq 5$ states that the distance from x to zero is greater than or equal to 5. This observation yields the result $x \geq 5$ or $x \leq -5$ (Figure 5b).

EXAMPLE 14 Solve the inequality $|2x - 3| \leq 1$.

Solution   The inequality $|2x - 3| \leq 1$ is equivalent to the inequalities $-1 \leq 2x - 3 \leq 1$ (see Example 13). Thus, $2 \leq 2x \leq 4$ and $1 \leq x \leq 2$. The solution is therefore given by the set of all x in the interval $[1, 2]$ (Figure 6).

FIGURE **6**
$|2x - 3| \leq 1$

1.2 Exercises

In Exercises 1–6, simplify the expression.

1. $\dfrac{x^2 + x - 2}{x^2 - 4}$

2. $\dfrac{2a^2 - 3ab - 9b^2}{2ab^2 + 3b^3}$

3. $\dfrac{12t^2 + 12t + 3}{4t^2 - 1}$

4. $\dfrac{x^3 + 2x^2 - 3x}{-2x^2 - x + 3}$

5. $\dfrac{(4x - 1)(3) - (3x + 1)(4)}{(4x - 1)^2}$

6. $\dfrac{(1 + x^2)^2(2) - 2x(2)(1 + x^2)(2x)}{(1 + x^2)^4}$

In Exercises 7–28, perform the indicated operations and simplify each expression.

7. $\dfrac{2a^2 - 2b^2}{b - a} \cdot \dfrac{4a + 4b}{a^2 + 2ab + b^2}$

8. $\dfrac{x^2 - 6x + 9}{x^2 - x - 6} \cdot \dfrac{3x + 6}{2x^2 - 7x + 3}$

9. $\dfrac{3x^2 + 2x - 1}{2x + 6} \div \dfrac{x^2 - 1}{x^2 + 2x - 3}$

10. $\dfrac{3x^2 - 4xy - 4y^2}{x^2 y} \div \dfrac{(2y - x)^2}{x^3 y}$

11. $\dfrac{58}{3(3t + 2)} + \dfrac{1}{3}$

12. $\dfrac{a + 1}{3a} + \dfrac{b - 2}{5b}$

13. $\dfrac{2x}{2x - 1} - \dfrac{3x}{2x + 5}$

14. $\dfrac{-xe^x}{x + 1} + e^x$

15. $\dfrac{4}{x^2 - 9} - \dfrac{5}{x^2 - 6x + 9}$

16. $\dfrac{x}{1 - x} + \dfrac{2x + 3}{x^2 - 1}$

17. $\dfrac{1 + \dfrac{1}{x}}{1 - \dfrac{1}{x}}$

18. $\dfrac{\dfrac{1}{x} + \dfrac{1}{y}}{1 - \dfrac{1}{xy}}$

19. $\dfrac{4x^2}{2\sqrt{2x^2 + 7}} + \sqrt{2x^2 + 7}$

20. $6(2x + 1)^2\sqrt{x^2 + x} + \dfrac{(2x + 1)^4}{2\sqrt{x^2 + x}}$

21. $5\left[\dfrac{(t^2 + 1)(1) - t(2t)}{(t^2 + 1)^2}\right]$

22. $\dfrac{2x(x + 1)^{-1/2} - (x + 1)^{1/2}}{x^2}$

23. $\dfrac{(x^2 + 1)^2(-2) + (2x)2(x^2 + 1)(2x)}{(x^2 + 1)^4}$

24. $\dfrac{(x^2 + 1)^{1/2} - 2x^2(x^2 + 1)^{-1/2}}{1 - x^2}$

25. $3\left(\dfrac{2x + 1}{3x + 2}\right)^2\left[\dfrac{(3x + 2)(2) - (2x + 1)(3)}{(3x + 2)^2}\right]$

26. $\dfrac{(2x + 1)^{1/2} - (x + 2)(2x + 1)^{-1/2}}{2x + 1}$

27.
$100\left[\dfrac{(t^2 + 20t + 100)(2t + 10) - (t^2 + 10t + 100)(2t + 20)}{(t^2 + 20t + 100)^2}\right]$

28. $\dfrac{2(2x - 3)^{1/3} - (x - 1)(2x - 3)^{-2/3}}{(2x - 3)^{2/3}}$

In Exercises 29–34, rationalize the denominator of each expression.

29. $\dfrac{1}{\sqrt{3} - 1}$

30. $\dfrac{1}{\sqrt{x} + 5}$

31. $\dfrac{1}{\sqrt{x} - \sqrt{y}}$

32. $\dfrac{a}{1 - \sqrt{a}}$

33. $\dfrac{\sqrt{a} + \sqrt{b}}{\sqrt{a} - \sqrt{b}}$

34. $\dfrac{2\sqrt{a} + \sqrt{b}}{2\sqrt{a} - \sqrt{b}}$

In Exercises 35–40, rationalize the numerator of each expression.

35. $\dfrac{\sqrt{x}}{3}$

36. $\dfrac{\sqrt[3]{y}}{x}$

37. $\dfrac{1-\sqrt{3}}{3}$

38. $\dfrac{\sqrt{x}-1}{x}$

39. $\dfrac{1+\sqrt{x+2}}{\sqrt{x+2}}$

40. $\dfrac{\sqrt{x+3}-\sqrt{x}}{3}$

In Exercises 41–44, determine whether the statement is true or false.

41. $-3 < -20$

42. $-5 \le -5$

43. $\dfrac{2}{3} > \dfrac{5}{6}$

44. $-\dfrac{5}{6} < -\dfrac{11}{12}$

In Exercises 45–62, find the values of x that satisfy the inequality (inequalities).

45. $2x + 4 < 8$

46. $-6 > 4 + 5x$

47. $-4x \ge 20$

48. $-12 \le -3x$

49. $-6 < x - 2 < 4$

50. $0 \le x + 1 \le 4$

51. $x + 1 > 4$ or $x + 2 < -1$

52. $x + 1 > 2$ or $x - 1 < -2$

53. $x + 3 > 1$ and $x - 2 < 1$

54. $x - 4 \le 1$ and $x + 3 > 2$

55. $(x + 3)(x - 5) \le 0$

56. $(2x - 4)(x + 2) \ge 0$

57. $(2x - 3)(x - 1) \ge 0$

58. $(3x - 4)(2x + 2) \le 0$

59. $\dfrac{x+3}{x-2} \ge 0$

60. $\dfrac{2x-3}{x+1} \ge 4$

61. $\dfrac{x-2}{x-1} \le 2$

62. $\dfrac{2x-1}{x+2} \le 4$

In Exercises 63–72, evaluate the expression.

63. $|-6 + 2|$

64. $4 + |-4|$

65. $\dfrac{|-12 + 4|}{|16 - 12|}$

66. $\left|\dfrac{0.2 - 1.4}{1.6 - 2.4}\right|$

67. $\sqrt{3}|-2| + 3|-\sqrt{3}|$

68. $|-1| + \sqrt{2}|-2|$

69. $|\pi - 1| + 2$

70. $|\pi - 6| - 3$

71. $|\sqrt{2} - 1| + |3 - \sqrt{2}|$

72. $|2\sqrt{3} - 3| - |\sqrt{3} - 4|$

In Exercises 73–78, suppose that a and b are real numbers other than zero and that $a > b$. State whether the inequality is true or false for all real numbers a and b.

73. $b - a > 0$

74. $\dfrac{a}{b} > 1$

75. $a^2 > b^2$

76. $\dfrac{1}{a} > \dfrac{1}{b}$

77. $a^3 > b^3$

78. $-a < -b$

In Exercises 79–84, determine whether the statement is true or false for all real numbers a and b.

79. $|-a| = a$

80. $|b^2| = b^2$

81. $|a - 4| = |4 - a|$

82. $|a + 1| = |a| + 1$

83. $|a + b| = |a| + |b|$

84. $|a - b| = |a| - |b|$

85. DRIVING RANGE OF A CAR An advertisement for a certain car states that the EPA fuel economy is 20 mpg city and 27 mpg highway and that the car's fuel-tank capacity is 18.1 gal. Assuming ideal driving conditions, determine the driving range for the car from the foregoing data.

86. Find the minimum cost C (in dollars), given that

$$5(C - 25) \ge 1.75 + 2.5C$$

87. Find the maximum profit P (in dollars) given that

$$6(P - 2500) \le 4(P + 2400)$$

88. CELSIUS AND FAHRENHEIT TEMPERATURES The relationship between Celsius (°C) and Fahrenheit (°F) temperatures is given by the formula

$$C = \frac{5}{9}(F - 32)$$

a. If the average temperature range for Montreal during the month of January is $-15° < C < -5°$, find the range in degrees Fahrenheit in Montreal for the same period.

b. If the average temperature range for New York City during the month of June is $63° < F < 80°$, find the range in degrees Celsius in New York City for the same period.

89. MEETING SALES TARGETS A salesman's monthly commission is 15% on all sales over $12,000. If his goal is to make a commission of at least $6000/month, what minimum monthly sales figures must he attain?

90. MARKUP ON A CAR The markup on a used car was at least 30% of its current wholesale price. If the car was sold for $11,200, what was the maximum wholesale price?

91. QUALITY CONTROL PAR Manufacturing manufactures steel rods. Suppose the rods ordered by a customer are manufactured to a specification of 0.5 in. and are acceptable only if they are within the *tolerance limits* of 0.49 in. and 0.51 in. Letting x denote the diameter of a rod, write an inequality using absolute value signs to express a criterion involving x that must be satisfied in order for a rod to be acceptable.

92. QUALITY CONTROL The diameter x (in inches) of a batch of ball bearings manufactured by PAR Manufacturing satisfies the inequality

$$|x - 0.1| \le 0.01$$

What is the smallest diameter a ball bearing in the batch can have? The largest diameter?

93. **MEETING PROFIT GOALS** A manufacturer of a certain commodity has estimated that her profit in thousands of dollars is given by the expression

$$-6x^2 + 30x - 10$$

where x (in thousands) is the number of units produced. What production range will enable the manufacturer to realize a profit of at least $14,000 on the commodity?

94. **CONCENTRATION OF A DRUG IN THE BLOODSTREAM** The concentration (in milligrams per cubic centimeter) of a certain drug in a patient's bloodstream t hr after injection is given by

$$\frac{0.2t}{t^2 + 1}$$

Find the interval of time when the concentration of the drug is greater than or equal to 0.08 mg/cc.

95. **COST OF REMOVING TOXIC POLLUTANTS** A city's main well was recently found to be contaminated with trichloroethylene (a cancer-causing chemical) as a result of an abandoned chemical dump that leached chemicals into the water. A proposal submitted to the city council indicated that the cost, in millions of dollars, of removing $x\%$ of the toxic pollutants is

$$\frac{0.5x}{100 - x}$$

If the city could raise between $25 million and $30 million for the purpose of removing the toxic pollutants, what is the range of pollutants that could be expected to be removed?

96. **AVERAGE SPEED OF A VEHICLE** The average speed of a vehicle in miles per hour on a stretch of Route 134 between

6 A.M. and 10 A.M. on a typical weekday is approximated by the expression

$$20t - 40\sqrt{t} + 50 \qquad (0 \le t \le 4)$$

where t is measured in hours, with $t = 0$ corresponding to 6 A.M. Over what interval of time is the average speed of a vehicle less than or equal to 35 mph?

97. **AIR POLLUTION** Nitrogen dioxide is a brown gas that impairs breathing. The amount of nitrogen dioxide present in the atmosphere on a certain May day in the city of Long Beach measured in PSI (pollutant standard index) at time t, where t is measured in hours and $t = 0$ corresponds to 7 A.M., is approximated by

$$\frac{136}{1 + 0.25(t - 4.5)^2} + 28 \qquad (0 \le t \le 11)$$

Find the time of the day when the amount of nitrogen dioxide is greater than or equal to 128 PSI.
Source: Los Angeles Times.

In Exercises 98–102, determine whether the statement is true or false. If it is true, explain why it is true. If it is false, give an example to show why it is false.

98. $\dfrac{a}{b + c} = \dfrac{a}{b} + \dfrac{a}{c}$

99. If $a < b$, then $a - c > b - c$.

100. $|a - b| = |b - a|$

101. $|a - b| \le |b| + |a|$

102. $\sqrt{a^2 - b^2} = |a| - |b|$

The Cartesian Coordinate System

The Cartesian Coordinate System

In Section 1.1, we saw how a one-to-one correspondence between the set of real numbers and the points on a straight line leads to a coordinate system on a line (a one-dimensional space).

In a similar manner, we can represent points in a plane (a two-dimensional space) by using the **Cartesian coordinate system,** which we construct as follows: Take two perpendicular lines, one of which is normally chosen to be horizontal. These lines intersect at a point O, called the **origin** (Figure 7). The horizontal line is called the **x-axis,** and the vertical line is called the **y-axis.** A number scale is set up along the x-axis, with the positive numbers lying to the right of the origin and the negative numbers lying to the left of it. Similarly, a number scale is set up along the y-axis, with the positive numbers lying above the origin and the negative numbers lying below it.

The number scales on the two axes need not be the same. Indeed, in many applications, different quantities are represented by x and y. For example, x may represent the number of smartphones sold, and y may represent the total revenue resulting from the sales. In such cases, it is often desirable to choose different number scales to represent the different quantities. Note, however, that the zeros of both number scales coincide at the origin of the two-dimensional coordinate system.

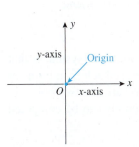

FIGURE 7
The Cartesian coordinate system

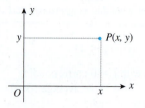

FIGURE 8
An ordered pair (x, y)

We can represent a point in the plane in this coordinate system by an **ordered pair** of numbers—that is, a pair (x, y), in which x is the first number and y is the second. To see this, let P be any point in the plane (Figure 8). Draw perpendiculars from P to the x-axis and to the y-axis. Then the number x is precisely the number that corresponds to the point on the x-axis at which the perpendicular line through P crosses the x-axis. Similarly, y is the number that corresponds to the point on the y-axis at which the perpendicular line through P crosses the y-axis.

Conversely, given an ordered pair (x, y) with x as the first number and y the second, a point P in the plane is uniquely determined as follows: Locate the point on the x-axis represented by the number x and draw a line through that point parallel to the y-axis. Next, locate the point on the y-axis represented by the number y and draw a line through that point parallel to the x-axis. The point of intersection of these two lines is the point P (see Figure 8).

In the ordered pair (x, y), x is called the **abscissa**, or **x-coordinate;** y is called the **ordinate,** or **y-coordinate;** and x and y together are referred to as the **coordinates** of the point P.

Letting $P(a, b)$ denote the point with x-coordinate a and y-coordinate b, we plot the points $A(2, 3)$, $B(-2, 3)$, $C(-2, -3)$, $D(2, -3)$, $E(3, 2)$, $F(4, 0)$, and $G(0, -5)$ in Figure 9. The fact that, in general, $P(x, y) \neq P(y, x)$ is clearly illustrated by points A and E.

The axes divide the xy-plane into four quadrants. Quadrant I consists of the points $P(x, y)$ that satisfy $x > 0$ and $y > 0$; Quadrant II, the points $P(x, y)$, where $x < 0$ and $y > 0$; Quadrant III, the points $P(x, y)$, where $x < 0$ and $y < 0$; and Quadrant IV, the points $P(x, y)$, where $x > 0$ and $y < 0$ (Figure 10).

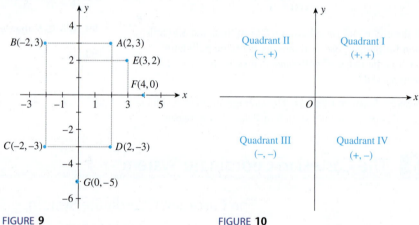

FIGURE 9
Several points in the Cartesian plane

FIGURE 10
The four quadrants in the Cartesian plane

The Distance Formula

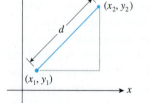

FIGURE 11
The distance d between the points (x_1, y_1) and (x_2, y_2)

One immediate benefit that arises from using the Cartesian coordinate system is that the distance between any two points in the plane may be expressed solely in terms of their coordinates. Suppose, for example, that (x_1, y_1) and (x_2, y_2) are any two points in the plane (Figure 11). Then the distance between these two points can be computed by using the following formula.

> **Distance Formula**
>
> The distance d between two points $P_1(x_1, y_1)$ and $P_2(x_2, y_2)$ in the plane is given by
>
> $$d = \sqrt{(x_2 - x_1)^2 + (y_2 - y_1)^2} \qquad (1)$$

For a proof of this result, see Exercise 51, page 32.

In what follows, we give several applications of the distance formula.

EXAMPLE 1 Find the distance between the points $(-4, 3)$ and $(2, 6)$.

Solution   Let $P_1(-4, 3)$ and $P_2(2, 6)$ be points in the plane. Then we have

$$x_1 = -4 \qquad y_1 = 3 \qquad x_2 = 2 \qquad y_2 = 6$$

Using Formula (1), we have

$$d = \sqrt{[2 - (-4)]^2 + (6 - 3)^2}$$
$$= \sqrt{6^2 + 3^2}$$
$$= \sqrt{45} = 3\sqrt{5} \approx 6.7$$

Explore and Discuss

Refer to Example 1. Suppose we label the point $(2, 6)$ as P_1 and the point $(-4, 3)$ as P_2.
(1) Show that the distance d between the two points is the same as that obtained earlier.
(2) Prove that, in general, the distance d in Formula (1) is independent of the way we label the two points.

EXAMPLE 2 Let $P(x, y)$ denote a point lying on the circle with radius r and center $C(h, k)$ (Figure 12). Find a relationship between x and y.

Solution   By the definition of a circle, the distance between $C(h, k)$ and $P(x, y)$ is r. Using Formula (1), we have

$$\sqrt{(x - h)^2 + (y - k)^2} = r$$

which, upon squaring both sides, gives the equation

$$(x - h)^2 + (y - k)^2 = r^2$$

that must be satisfied by the variables x and y.

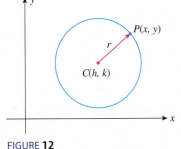

FIGURE 12
A circle with radius r and center $C(h, k)$

A summary of the result obtained in Example 2 follows.

Equation of a Circle

An equation of the circle with center $C(h, k)$ and radius r is given by

$$(x - h)^2 + (y - k)^2 = r^2 \tag{2}$$

EXAMPLE 3 Find an equation of the circle with

a. Radius 2 and center $(-1, 3)$.
b. Radius 3 and center located at the origin.

Solution

a. We use Formula (2) with $r = 2$, $h = -1$, and $k = 3$, obtaining

$$[x - (-1)]^2 + (y - 3)^2 = 2^2 \quad \text{or} \quad (x + 1)^2 + (y - 3)^2 = 4$$

(Figure 13a; see next page).

b. Using Formula (2) with $r = 3$ and $h = k = 0$, we obtain

$$x^2 + y^2 = 3^2 \quad \text{or} \quad x^2 + y^2 = 9$$

(Figure 13b).

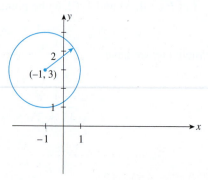

(a) The circle with radius 2 and center $(-1, 3)$

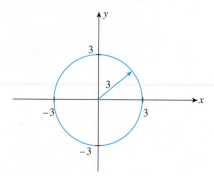

(b) The circle with radius 3 and center $(0, 0)$

FIGURE **13**

Explore and Discuss

1. Use the distance formula to help you describe the set of points in the xy-plane satisfying each of the following inequalities:
   a. $(x - h)^2 + (y - k)^2 \le r^2$
   b. $(x - h)^2 + (y - k)^2 < r^2$
   c. $(x - h)^2 + (y - k)^2 \ge r^2$
   d. $(x - h)^2 + (y - k)^2 > r^2$

2. Consider the equation $x^2 + y^2 = 4$.
   a. Show that $y = \pm\sqrt{4 - x^2}$.
   b. Describe the set of points (x, y) in the xy-plane satisfying the following equations:
      (i) $y = \sqrt{4 - x^2}$
      (ii) $y = -\sqrt{4 - x^2}$

APPLIED EXAMPLE 4 Cost of Laying Cable In Figure 14, S represents the position of a power relay station located on a straight coastal highway, and M shows the location of a marine biology experimental station on an island. A cable is to be laid connecting the relay station with the experimental station. If the cost of running the cable on land is $3.00 per running foot and the cost of running the cable under water is $5.00 per running foot, find the total cost for laying the cable.

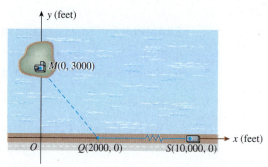

FIGURE **14**
Cable connecting relay station S to experimental station M

Solution   The length of cable required on land is given by the distance from S to Q. This distance is $(10{,}000 - 2000)$, or 8000 feet. Next, we see that the length of cable required underwater is given by the distance from M to Q. This distance is

$$\sqrt{(0 - 2000)^2 + (3000 - 0)^2} = \sqrt{2000^2 + 3000^2}$$
$$= \sqrt{13{,}000{,}000}$$
$$\approx 3606$$

or approximately 3606 feet. Therefore, the total cost for laying the cable is approximately

$$3(8000) + 5(3606) = 42{,}030$$

dollars.

Explore and Discuss

In the Cartesian coordinate system, the two axes are perpendicular to each other. Consider a coordinate system in which the x- and y-axes are not collinear and are not perpendicular to each other (see the accompanying figure).

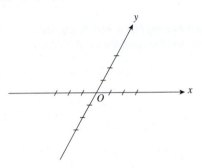

1. Describe how a point is represented in this coordinate system by an ordered pair (x, y) of real numbers. Conversely, show how an ordered pair (x, y) of real numbers uniquely determines a point in the plane.

2. Suppose you want to find a formula for the distance between two points $P_1(x_1, y_1)$ and $P_2(x_2, y_2)$ in the plane. What is the advantage that the Cartesian coordinate system has over the coordinate system under consideration? Comment on your answer.

1.3 Self-Check Exercises

1. **a.** Plot the points $A(4, -2)$, $B(2, 3)$, and $C(-3, 1)$.
   b. Find the distance between the points A and B; between B and C; between A and C.
   c. Use the Pythagorean Theorem to show that the triangle with vertices A, B, and C is a right triangle.

2. **FUEL STOP PLANNING** The figure opposite shows the location of cities A, B, and C. Suppose a pilot wishes to fly from City A to City C but must make a mandatory stopover in City B. If the single-engine light plane has a range of 650 miles, can the pilot make the trip without refueling in City B?

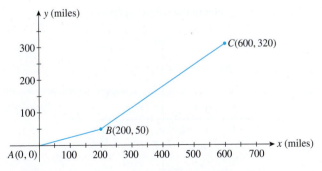

Solutions to Self-Check Exercises 1.3 can be found on page 33.

1.3 Concept Questions

1. What can you say about the signs of a and b if the point $P(a, b)$ lies in (a) the second quadrant? (b) The third quadrant? (c) The fourth quadrant?

2. Refer to the following figure.

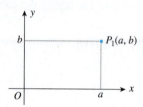

a. Given the point $P_1(a, b)$, where $a > 0$ and $b > 0$, plot the points $P_2(-a, b)$, $P_3(-a, -b)$, and $P_4(a, -b)$.

b. What can you say about the distance of the points $P_1(a, b)$, $P_2(-a, b)$, $P_3(-a, -b)$, and $P_4(a, -b)$ from the origin?

1.3 Exercises

In Exercises 1–6, refer to the following figure, and determine the coordinates of each point and the quadrant in which it is located.

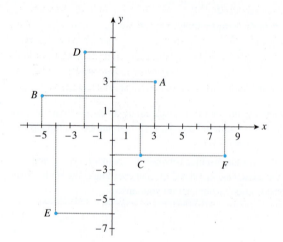

1. A
2. B
3. C
4. D
5. E
6. F

In Exercises 7–12, refer to the following figure.

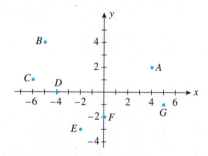

7. Which point is represented by the ordered pair $(4, 2)$?

8. What are the coordinates of point B?

9. Which points have negative y-coordinates?

10. Which point has a negative x-coordinate and a negative y-coordinate?

11. Which point has an x-coordinate that is equal to zero?

12. Which point has a y-coordinate that is equal to zero?

In Exercises 13–20, sketch a set of coordinate axes and plot each point.

13. $(-2, 5)$
14. $(1, 3)$
15. $(3, -1)$
16. $(3, -4)$
17. $\left(8, -\dfrac{7}{2}\right)$
18. $\left(-\dfrac{5}{2}, \dfrac{3}{2}\right)$
19. $(4.5, -4.5)$
20. $(1.2, -3.4)$

In Exercises 21–24, find the distance between the given points.

21. $(1, 3)$ and $(4, 7)$
22. $(1, 0)$ and $(4, 4)$
23. $(-1, 3)$ and $(4, 9)$
24. $(-2, 1)$ and $(10, 6)$

25. Find the coordinates of the points that are 10 units away from the origin and have a y-coordinate equal to -6.

26. Find the coordinates of the points that are 5 units away from the origin and have an x-coordinate equal to 3.

27. Show that the points $(3, 4)$, $(-3, 7)$, $(-6, 1)$, and $(0, -2)$ form the vertices of a square.

28. Show that the triangle with vertices $(-5, 2)$, $(-2, 5)$, and $(5, -2)$ is a right triangle.

In Exercises 29–34, find an equation of the circle that satisfies the given conditions.

29. Radius 5 and center $(2, -3)$

30. Radius 3 and center $(-2, -4)$

31. Radius 5 and center at the origin

32. Center at the origin and passes through $(2, 3)$

33. Center $(2, -3)$ and passes through $(5, 2)$

34. Center $(-a, a)$ and radius $2a$

35. TRACKING A CRIMINAL WITH **GPS**   After obtaining a warrant, the police attached a GPS tracking device to the car of a murder suspect. Suppose the car was located at the origin of a Cartesian coordinate system when the device was attached. Shortly afterward, the suspect's car was tracked going 5 mi due east, 4 mi due north, and 1 mi due west before coming to a permanent stop.
  a. What are the coordinates of the suspect's car at its final destination?
  b. What is the distance traveled by the suspect?
  c. What is the distance as the crow flies between the original position and the final position of the suspect's car?

36. PLANNING A GRAND TOUR   A grand tour of four cities begins at City A and makes successive stops at cities B, C, and D before returning to City A. If the cities are located as shown in the following figure, find the total distance covered on the tour.

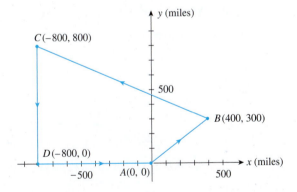

37. WILL YOU INCUR A DELIVERY CHARGE?   A furniture store offers free setup and delivery services to all points within a 25-mi radius of its warehouse distribution center. If you live 20 mi east and 14 mi south of the warehouse, will you incur a delivery charge? Justify your answer.

38. OPTIMIZING TRAVEL TIME   Towns A, B, C, and D are located as shown in the following figure. Two highways link town A to town D. Route 1 runs from Town A to Town D via Town B, and Route 2 runs from Town A to Town D via Town C. If a salesman wishes to drive from Town A to Town D and traffic conditions are such that he could expect to average the same speed on either route, which highway should he take to arrive in the shortest time?

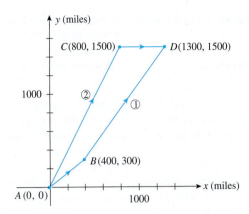

39. MINIMIZING SHIPPING COSTS FOR A FLEET OF AUTOS   Refer to the figure for Exercise 38. Suppose a fleet of 100 automobiles are to be shipped from an assembly plant in Town A to Town D. They may be shipped either by freight train along Route 1 at a cost of 66¢/mile per automobile or by truck along Route 2 at a cost of 62¢/mile per automobile. Which means of transportation minimizes the shipping cost? What is the net savings?

40. COST OF LAYING CABLE   In the following diagram, S represents the position of a power relay station located on a straight coastal highway, and M shows the location of a marine biology experimental station on an island. A cable is to be laid connecting the relay station with the experimental station. If the cost of running the cable on land is \$3/running foot and the cost of running cable under water is \$5/running foot, find an expression in terms of x that gives the total cost for laying the cable. What is the total cost when $x = 2500$? When $x = 3000$?

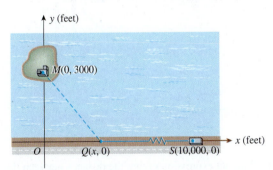

41. PURCHASING AN HDTV ANTENNA   Ivan wishes to determine which HDTV antenna he should purchase for his home. The TV store has supplied him with the following information:

Range in Miles			
VHF	**UHF**	**Model**	**Price**
30	20	A	\$50
45	35	B	60
60	40	C	70
75	55	D	80

Ivan wishes to receive Channel 17 (VHF), which is located 25 mi east and 35 mi north of his home, and

Channel 38 (UHF), which is located 20 mi south and 32 mi west of his home. Which model will allow him to receive both channels at the least cost? (Assume that the terrain between Ivan's home and both broadcasting stations is flat.)

42. DISTANCE BETWEEN TWO CRUISE SHIPS Two cruise ships leave port at the same time. Ship A sails north at a speed of 20 mph while Ship B sails east at a speed of 30 mph.
   a. Find an expression in terms of the time t (in hours) giving the distance between the two cruise ships.
   b. Using the expression obtained in part (a), find the distance between the two cruise ships 2 hr after leaving port.

43. DISTANCE BETWEEN TWO CARGO SHIPS Ship A leaves port sailing north at a speed of 25 mph. A half hour later, Ship B leaves the same port sailing east at a speed of 20 mph. Let t (in hours) denote the time Ship B has been at sea.
   a. Find an expression in terms of t giving the distance between the two cargo ships.
   b. Use the expression obtained in part (a) to find the distance between the two cargo ships 2 hr after ship A has left port.

44. WATCHING A ROCKET LAUNCH At a distance of 4000 ft from the launch site, a spectator is observing a rocket being launched. Suppose the rocket lifts off vertically and reaches an altitude of x ft (see the accompanying figure).

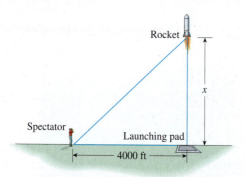

   a. Find an expression giving the distance between the spectator and the rocket.
   b. What is the distance between the spectator and the rocket when the rocket reaches an altitude of 20,000 ft?

45. a. Show that the midpoint of the line segment joining the points $P_1(x_1, y_1)$ and $P_2(x_2, y_2)$ is

$$\left(\frac{x_1 + x_2}{2}, \frac{y_1 + y_2}{2} \right)$$

   b. Use the result of part (a) to find the midpoint of the line segment joining the points $(-3, 2)$ and $(4, -5)$.

46. A SCAVENGER HUNT A tree is located 20 yd to the east and 10 yd to the north of a house. A second tree is located 10 yd to the east and 40 yd to the north of the house. The prize of a scavenger hunt is placed exactly midway between the trees.
   a. Place the house at the origin of a Cartesian coordinate system, and draw a diagram depicting the situation.
   b. What are the coordinates of the position of the prize?
   c. How far is the prize from the house?

In Exercises 47–50, determine whether the statement is true or false. If it is true, explain why it is true. If it is false, give an example to show why it is false.

47. The point $(-a, b)$ is symmetric to the point (a, b) with respect to the y-axis.

48. The point $(-a, -b)$ is symmetric to the point (a, b) with respect to the origin.

49. If the distance between the points $P_1(a, b)$ and $P_2(c, d)$ is D, then the distance between the points $P_1(a, b)$ and $P_3(kc, kd)$, $(k \neq 0)$, is given by $|k|D$.

50. The circle with equation $kx^2 + ky^2 = a^2$ lies inside the circle with equation $x^2 + y^2 = a^2$, provided that $k > 1$.

51. Let (x_1, y_1) and (x_2, y_2) be two points lying in the xy-plane. Show that the distance between the two points is given by

$$d = \sqrt{(x_2 - x_1)^2 + (y_2 - y_1)^2}$$

Hint: Refer to the accompanying figure, and use the Pythagorean Theorem.

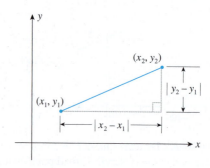

52. Show that an equation of a circle can be written in the form

$$x^2 + y^2 + Cx + Dy + E = 0$$

where C, D, and E are constants. This is called the general form of an equation of a circle.

1.3 Solutions to Self-Check Exercises

1. a. The points are plotted in the following figure:

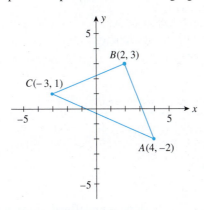

b. The distance between A and B is

$$d(A, B) = \sqrt{(2 - 4)^2 + [3 - (-2)]^2}$$
$$= \sqrt{(-2)^2 + 5^2} = \sqrt{4 + 25} = \sqrt{29}$$

The distance between B and C is

$$d(B, C) = \sqrt{(-3 - 2)^2 + (1 - 3)^2}$$
$$= \sqrt{(-5)^2 + (-2)^2} = \sqrt{25 + 4} = \sqrt{29}$$

The distance between A and C is

$$d(A, C) = \sqrt{(-3 - 4)^2 + [1 - (-2)]^2}$$
$$= \sqrt{(-7)^2 + 3^2} = \sqrt{49 + 9} = \sqrt{58}$$

c. We will show that

$$[d(A, C)]^2 = [d(A, B)]^2 + [d(B, C)]^2$$

From part (b), we see that $[d(A, B)]^2 = 29$, $[d(B, C)]^2 = 29$, and $[d(A, C)]^2 = 58$, and the desired result follows.

2. The distance between City A and City B is

$$d(A, B) = \sqrt{200^2 + 50^2} \approx 206$$

or 206 mi. The distance between City B and City C is

$$d(B, C) = \sqrt{(600 - 200)^2 + (320 - 50)^2}$$
$$= \sqrt{400^2 + 270^2} \approx 483$$

or 483 mi. Therefore, the total distance the pilot would have to cover is about 689 mi, so she must refuel in City B.

1.4 Straight Lines

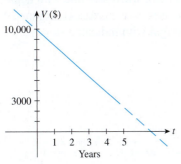

FIGURE 15
Linear depreciation of a network server

Businesses may depreciate certain assets, such as buildings, machines, furniture, vehicles, and equipment, over a period of time for income tax purposes. Linear depreciation, or the straight-line method, is often used for this purpose. The graph of the straight line shown in Figure 15 describes the book value V of a network server that has an initial value of $\$10,000$ and that is being depreciated linearly over 5 years with a scrap value of $\$3,000$. Note that only the solid portion of the straight line is of interest here.

The book value of the server at the end of year t, where t lies between 0 and 5, can be read directly from the graph. But there is one shortcoming in this approach: The result depends on how accurately you draw and read the graph. A better and more accurate method is based on finding an *algebraic* representation of the depreciation line.

Slope of a Line

To see how a straight line in the xy-plane may be described algebraically, we need to first recall certain properties of straight lines. Let L denote the unique straight line that passes through the two distinct points (x_1, y_1) and (x_2, y_2). If $x_1 \neq x_2$, we define the slope of L as follows.

> **Slope of a Nonvertical Line**
>
> If (x_1, y_1) and (x_2, y_2) are any two distinct points on a nonvertical line L (see Figure 16 on the next page), then the slope m of L is given by
>
> $$m = \frac{\Delta y}{\Delta x} = \frac{y_2 - y_1}{x_2 - x_1} \qquad (3)$$

(continued)

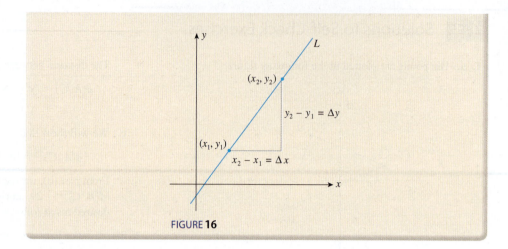

FIGURE **16**

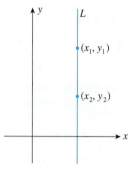

FIGURE **17**
The slope of L is undefined if $x_1 = x_2$.

If $x_1 = x_2$, then L is a vertical line (Figure 17). Its slope is undefined, since the denominator in Equation (3) will be zero and division by zero is not allowed.

Observe that the slope of a straight line is a constant whenever it is defined. The number $\Delta y = y_2 - y_1$ (Δy is read "delta y") is a measure of the vertical change in y, and $\Delta x = x_2 - x_1$ is a measure of the horizontal change in x, as shown in Figure 16. From this figure, we can see that the slope m of a straight line L is a measure of the *rate of change of y with respect to x.* Furthermore, the slope of a nonvertical straight line is constant, and this tells us that the rate of change is constant.

Figure 18a shows a straight line L_1 with slope 2. Observe that L_1 has the property that a 1-unit increase in x results in a 2-unit increase in y. To see this, let $\Delta x = 1$ in Formula (3); then $m = \Delta y$. Since $m = 2$, we conclude that $\Delta y = 2$. Similarly, Figure 18b shows a line L_2 with slope -1. Observe that a straight line with positive slope slants upward from left to right (y increases as x increases), whereas a line with negative slope slants downward from left to right (y decreases as x increases). Figure 19 shows a family of straight lines passing through the origin with indicated slopes.

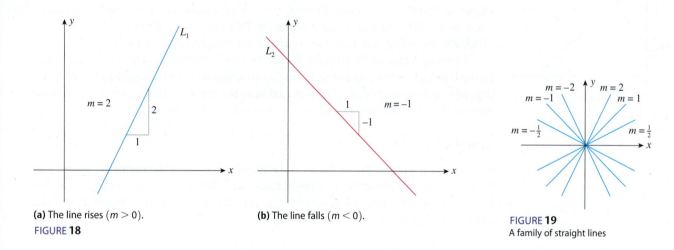

(a) The line rises ($m > 0$).

FIGURE **18**

(b) The line falls ($m < 0$).

FIGURE **19**
A family of straight lines

EXAMPLE 1 Sketch the straight line that passes through the point $(-2, 5)$ and has slope $-\frac{4}{3}$.

Solution   First, plot the point $(-2, 5)$ (Figure 20).

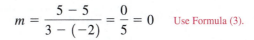

FIGURE **20**
L has slope $-\frac{4}{3}$ and passes through $(-2, 5)$.

Next, recall that a slope of $-\frac{4}{3}$ indicates that an increase of 1 unit in the x-direction produces a decrease of $\frac{4}{3}$ units in the y-direction, or equivalently, a 3-unit increase in the x-direction produces a $3\left(\frac{4}{3}\right)$, or 4, units decrease in the y-direction. Using this information, we plot the point $(1, 1)$ and draw the line through the two points. ■

> ### *Explore and Discuss*
>
> Show that the slope of a nonvertical line is independent of the two distinct points $P_1(x_1, y_1)$ and $P_2(x_2, y_2)$ used to compute it.
>
> Hint: Suppose we pick two other distinct points, $P_3(x_3, y_3)$ and $P_4(x_4, y_4)$ lying on L. Draw a picture and use similar triangles to demonstrate that using P_3 and P_4 gives the same value as that obtained using P_1 and P_2.

EXAMPLE 2 Find the slope m of the line that passes through the points $(-1, 1)$ and $(5, 3)$.

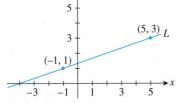

FIGURE **21**
L passes through $(5, 3)$ and $(-1, 1)$.

Solution Choose (x_1, y_1) to be the point $(-1, 1)$ and (x_2, y_2) to be the point $(5, 3)$. Then, with $x_1 = -1$, $y_1 = 1$, $x_2 = 5$, and $y_2 = 3$, we find

$$m = \frac{y_2 - y_1}{x_2 - x_1} = \frac{3 - 1}{5 - (-1)} = \frac{1}{3} \qquad \text{\textcolor{red}{Use Formula (3).}}$$

(Figure 21). Verify that the result obtained would have been the same had we chosen the point $(-1, 1)$ to be (x_2, y_2) and the point $(5, 3)$ to be (x_1, y_1). ■

EXAMPLE 3 Find the slope of the line that passes through the points $(-2, 5)$ and $(3, 5)$.

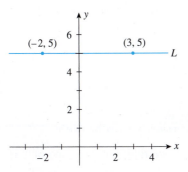

FIGURE **22**
The slope of the horizontal line L is 0.

Solution The slope of the required line is given by

$$m = \frac{5 - 5}{3 - (-2)} = \frac{0}{5} = 0 \qquad \text{\textcolor{red}{Use Formula (3).}}$$

(Figure 22). ■

Note In general, the slope of a horizontal line is zero. ■

We can use the slope of a straight line to determine whether a line is parallel to another line.

> **Parallel Lines**
> Two distinct lines are **parallel** if and only if their slopes are equal or their slopes are undefined.

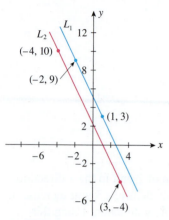

FIGURE 23
L_1 and L_2 have the same slope and hence are parallel.

EXAMPLE 4 Let L_1 be a line that passes through the points $(-2, 9)$ and $(1, 3)$, and let L_2 be the line that passes through the points $(-4, 10)$ and $(3, -4)$. Determine whether L_1 and L_2 are parallel.

Solution   The slope m_1 of L_1 is given by

$$m_1 = \frac{3 - 9}{1 - (-2)} = -2$$

The slope m_2 of L_2 is given by

$$m_2 = \frac{-4 - 10}{3 - (-4)} = -2$$

Since $m_1 = m_2$, the lines L_1 and L_2 are in fact parallel (Figure 23).

Equations of Lines

We will now show that every straight line lying in the xy-plane may be represented by an equation involving the variables x and y. One immediate benefit of this is that problems involving straight lines may be solved algebraically.

Let L be a straight line parallel to the y-axis (perpendicular to the x-axis) (Figure 24). Then L crosses the x-axis at some point $(a, 0)$ with the x-coordinate given by $x = a$, where a is some real number. Any other point on L has the form (a, y), where y is an appropriate number. Therefore, the vertical line L is described by the sole condition

$$x = a$$

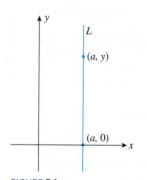

FIGURE 24
The vertical line $x = a$

and this is, accordingly, an equation of L. For example, the equation $x = -2$ represents a vertical line 2 units to the left of the y-axis, and the equation $x = 3$ represents a vertical line 3 units to the right of the y-axis (Figure 25).

Next, suppose L is a nonvertical line, so it has a well-defined slope m. Suppose (x_1, y_1) is a fixed point lying on L and (x, y) is a variable point on L distinct from (x_1, y_1) (Figure 26).

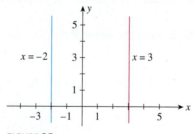

FIGURE 25
The vertical lines $x = -2$ and $x = 3$

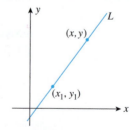

FIGURE 26
L passes through (x_1, y_1) and has slope m.

Using Formula (3) with the point $(x_2, y_2) = (x, y)$, we find that the slope of L is given by

$$m = \frac{y - y_1}{x - x_1}$$

Upon multiplying both sides of the equation by $x - x_1$, we obtain Formula (4).

> ### Point-Slope Form of an Equation of a Line
>
> An equation of the line that has slope m and passes through the point (x_1, y_1) is given by
>
> $$y - y_1 = m(x - x_1) \qquad (4)$$

Equation (4) is called the **point-slope form of an equation of a line,** since it utilizes a given point (x_1, y_1) on a line and the slope m of the line.

EXAMPLE 5 Find an equation of the line that passes through the point $(1, 3)$ and has slope 2.

Solution Using the point-slope form of the equation of a line with the point $(1, 3)$ and $m = 2$, we obtain

$$y - 3 = 2(x - 1) \qquad {\scriptstyle y - y_1 = m(x - x_1)}$$

which, when simplified, becomes

$$2x - y + 1 = 0$$

(Figure 27).

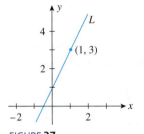

FIGURE 27
L passes through $(1, 3)$ and has slope 2.

EXAMPLE 6 Find an equation of the line that passes through the points $(-3, 2)$ and $(4, -1)$.

Solution The slope of the line is given by

$$m = \frac{-1 - 2}{4 - (-3)} = -\frac{3}{7}$$

Using the point-slope form of an equation of a line with the point $(4, -1)$ and the slope $m = -\frac{3}{7}$, we have

$$y + 1 = -\frac{3}{7}(x - 4) \qquad {\scriptstyle y - y_1 = m(x - x_1)}$$

$$7y + 7 = -3x + 12$$

$$3x + 7y - 5 = 0$$

(Figure 28).

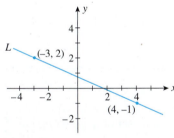

FIGURE 28
L passes through $(-3, 2)$ and $(4, -1)$.

We can use the slope of a straight line to determine whether a line is perpendicular to another line.

> ### Perpendicular Lines
>
> If L_1 and L_2 are two distinct nonvertical lines that have slopes m_1 and m_2, respectively, then L_1 is **perpendicular** to L_2 (written $L_1 \perp L_2$) if and only if
>
> $$m_1 = -\frac{1}{m_2}$$

If the line L_1 is vertical (so its slope is undefined), then L_1 is perpendicular to another line, L_2, if and only if L_2 is horizontal (so its slope is zero). For a proof of these results, see Exercise 90, page 46.

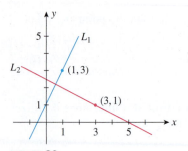

FIGURE 29
L_2 is perpendicular to L_1 and passes through $(3, 1)$.

EXAMPLE 7 Find an equation of the line that passes through the point $(3, 1)$ and is perpendicular to the line of Example 5.

Solution Since the slope of the line in Example 5 is 2, the slope of the required line is given by $m = -\frac{1}{2}$, the negative reciprocal of 2. Using the point-slope form of the equation of a line, we obtain

$$y - 1 = -\frac{1}{2}(x - 3) \qquad y - y_1 = m(x - x_1)$$
$$2y - 2 = -x + 3$$
$$x + 2y - 5 = 0$$

(Figure 29).

A straight line L that is neither horizontal nor vertical cuts the x-axis and the y-axis at, say, points $(a, 0)$ and $(0, b)$, respectively (Figure 30). The numbers a and b are called the **x-intercept** and **y-intercept,** respectively, of L.

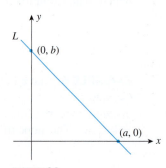

FIGURE 30
The line L has x-intercept a and y-intercept b.

Exploring with TECHNOLOGY

1. Use a graphing utility to plot the straight lines L_1 and L_2 with equations $2x + y - 5 = 0$ and $41x + 20y - 11 = 0$ on the same set of axes, using the standard viewing window.
   a. Can you tell whether the lines L_1 and L_2 are parallel to each other?
   b. Verify your observations by computing the slopes of L_1 and L_2 algebraically.
2. Use a graphing utility to plot the straight lines L_1 and L_2 with equations $x + 2y - 5 = 0$ and $5x - y + 5 = 0$ on the same set of axes, using the standard viewing window.
   a. Can you tell whether the lines L_1 and L_2 are perpendicular to each other?
   b. Verify your observation by computing the slopes of L_1 and L_2 algebraically.

Now let L be a line with slope m and y-intercept b. Using Formula (4), the point-slope form of the equation of a line, with the point $(0, b)$ and slope m, we have

$$y - b = m(x - 0)$$
$$y = mx + b$$

> **Slope-Intercept Form of an Equation of a Line**
>
> An equation of the line that has slope m and intersects the y-axis at the point $(0, b)$ is given by
>
> $$y = mx + b \qquad (5)$$

EXAMPLE 8 Find an equation of the line that has slope 3 and y-intercept -4.

Solution Using Equation (5) with $m = 3$ and $b = -4$, we obtain the required equation

$$y = 3x - 4$$

EXAMPLE 9 Determine the slope and y-intercept of the line whose equation is $3x - 4y = 8$.

Solution Rewrite the given equation in the slope-intercept form. Thus,

$$3x - 4y = 8$$
$$-4y = -3x + 8$$
$$y = \frac{3}{4}x - 2$$

Comparing this result with Equation (5), we find $m = \frac{3}{4}$ and $b = -2$, and we conclude that the slope and y-intercept of the given line are $\frac{3}{4}$ and -2, respectively.

Explore and Discuss

Consider the slope-intercept form of an equation of a straight line $y = mx + b$. Describe the family of straight lines obtained by keeping

1. The value of m fixed and allowing the value of b to vary.
2. The value of b fixed and allowing the value of m to vary.

Exploring with TECHNOLOGY

1. Use a graphing utility to plot the straight lines with equations $y = -2x + 3$, $y = -x + 3$, $y = x + 3$, and $y = 2.5x + 3$ on the same set of axes, using the standard viewing window. What effect does changing the coefficient m of x in the equation $y = mx + b$ have on its graph?
2. Use a graphing utility to plot the straight lines with equations $y = 2x - 2$, $y = 2x - 1$, $y = 2x$, $y = 2x + 1$, and $y = 2x + 4$ on the same set of axes, using the standard viewing window. What effect does changing the constant b in the equation $y = mx + b$ have on its graph?
3. Describe in words the effect of changing both m and b in the equation $y = mx + b$.

$ APPLIED EXAMPLE 10 Forecasting Sales of a Sporting Goods Store The sales manager of a local sporting goods store plotted sales (in units of ten thousand dollars) versus time for the last 5 years and found the points to lie

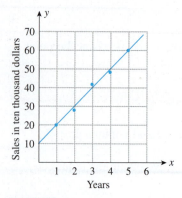

FIGURE 31
Sales of a sporting goods store

approximately along a straight line (Figure 31). By using the points corresponding to the first and fifth years, find an equation of the trend line. What sales figure can be predicted for the sixth year?

Solution   Using Formula (3) with the points $(1, 20)$ and $(5, 60)$, we find that the slope of the required line is given by

$$m = \frac{60 - 20}{5 - 1} = 10$$

Next, using the point-slope form of the equation of a line with the point $(1, 20)$ and $m = 10$, we obtain

$$y - 20 = 10(x - 1) \qquad y - y_1 = m(x - x_1)$$
$$y = 10x + 10$$

as the required equation.

The sales figure for the sixth year is obtained by letting $x = 6$ in the last equation, giving

$$y = 10(6) + 10 = 70$$

or $700,000.

APPLIED EXAMPLE 11 Appreciation in Value of a Painting Suppose a painting purchased for $50,000 is expected to appreciate in value at a constant rate of $5000 per year for the next 5 years. Use Formula (5) to write an equation predicting the value of the painting in the next several years. What will be its value 3 years from the date of purchase?

Solution   Let x denote the time (in years) that has elapsed since the date the painting was purchased, and let y denote the painting's value (in dollars). Then $y = 50,000$ when $x = 0$. Furthermore, the slope of the required equation is given by $m = 5000$, since each unit increase in x (1 year) implies an increase of 5000 units (dollars) in y. Using (5) with $m = 5000$ and $b = 50,000$, we obtain

$$y = 5000x + 50,000 \qquad y = mx + b$$

Three years from the date of purchase, the value of the painting will be given by

$$y = 5000(3) + 50,000$$

or $65,000.

Explore and Discuss

Refer to Applied Example 11. Can the equation predicting the value of the painting be used to predict long-term growth?

General Form of an Equation of a Line

We have considered several forms of an equation of a straight line in the plane. These different forms of the equation are equivalent to each other. In fact, each is a special case of the following equation.

> **General Form of a Linear Equation**
>
> The equation
>
> $$Ax + By + C = 0 \tag{6}$$
>
> where A, B, and C are constants and A and B are not both zero, is called the general form of a linear equation in the variables x and y.

We will now state (without proof) an important result concerning the algebraic representation of straight lines in the plane.

> **THEOREM 1**
>
> An equation of a straight line is a linear equation; conversely, every linear equation represents a straight line.

This result justifies the use of the adjective *linear* describing Equation (6).

EXAMPLE 12

a. Sketch the straight line represented by the equation
$$3x - 4y - 12 = 0$$
b. Does the point $\left(2, -\frac{3}{2}\right)$ lie on L?
c. Does the point $(1, -2)$ lie on L?

Solution

a. Since every straight line is uniquely determined by two distinct points, we need find only two such points through which the line passes in order to sketch it. For convenience, let's compute the *x*- and *y*-intercepts. Setting $y = 0$, we find $x = 4$; thus, the *x*-intercept is 4. Setting $x = 0$ gives $y = -3$, and the *y*-intercept is -3. A sketch of the line appears in Figure 32.

b. Substituting $x = 2$ and $y = -\frac{3}{2}$ into the left-hand side of the equation $3x - 4y - 12 = 0$ found in part (a), we obtain
$$3(2) - 4\left(-\frac{3}{2}\right) - 12 = 6 + 6 - 12 = 0$$

This shows that the equation is satisfied, and we conclude that the point $\left(2, -\frac{3}{2}\right)$ does indeed lie on L.

c. Substituting $x = 1$ and $y = -2$ into the given equation, we obtain
$$3(1) - 4(-2) - 12 = 3 + 8 - 12 = -1$$

which is not equal to zero, the number on the right-hand side of the equation. This shows that the point $(1, -2)$ does not lie on L.

Here is a summary of the common forms of the equations of straight lines discussed in this section.

> **Equations of Straight Lines**
>
> | | |
> |---|---|
> | Vertical line: | $x = a$ |
> | Horizontal line: | $y = b$ |
> | Point-slope form: | $y - y_1 = m(x - x_1)$ |
> | Slope-intercept form: | $y = mx + b$ |
> | General form: | $Ax + By + C = 0$ |

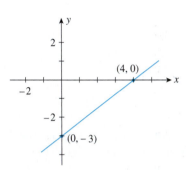

FIGURE 32
To sketch $3x - 4y - 12 = 0$, first find the *x*-intercept, 4, and the *y*-intercept, -3.

1.4 Self-Check Exercises

1. Determine the number a such that the line passing through the points $(a, 2)$ and $(3, 6)$ is parallel to a line with slope 4.

2. Find an equation of the line that passes through the point $(3, -1)$ and is perpendicular to a line with slope $-\frac{1}{2}$.

3. Does the point $(3, -3)$ lie on the line with equation $2x - 3y - 12 = 0$? Sketch the graph of the line.

4. SMOKERS IN THE UNITED STATES The following table gives the percentage of adults in the United States from 2006 through 2010 who smoked in year t. Here, $t = 0$ corresponds to the beginning of 2006.

Year, t	0	1	2	3	4
Percent, y	20.8	20.5	20.1	19.8	19.0

a. Plot the percentage of U.S. adults who smoke (y) versus the year (t) for the given years.

b. Draw the line L through the points $(0, 20.8)$ and $(4, 19.0)$.

c. Find an equation of the line L.

d. Assuming that this trend continues, estimate the percentage of U.S. adults who smoked at the beginning of 2014.

Source: Centers for Disease Control and Prevention.

Solutions to Self-Check Exercises 1.4 can be found on
page 46.

1.4 Concept Questions

1. What is the slope of a nonvertical line? What can you say about the slope of a vertical line?

2. Give (a) the point-slope form, (b) the slope-intercept form, and (c) the general form of an equation of a line.

3. Let L_1 have slope m_1 and L_2 have slope m_2. State the conditions on m_1 and m_2 so that (a) L_1 will be parallel to L_2 and (b) L_1 will be perpendicular to L_2.

4. Suppose a line L has equation $Ax + By + C = 0$.
   a. What is the slope of L if $B \neq 0$?
   b. What is the slope of L if $B = 0$ and $A \neq 0$?

1.4 Exercises

In Exercises 1–6, match the statement with one of the graphs (a)–(f).

1. The slope of the line is zero.

2. The slope of the line is undefined.

3. The slope of the line is positive, and its y-intercept is positive.

4. The slope of the line is positive, and its y-intercept is negative.

5. The slope of the line is negative, and its x-intercept is negative.

6. The slope of the line is negative, and its x-intercept is positive.

(a)

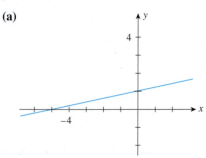

(b)

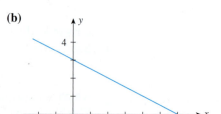

(c)

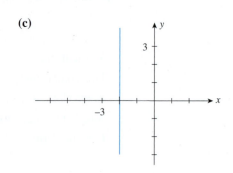

(d)

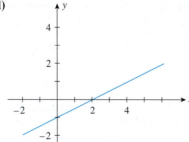

(e)

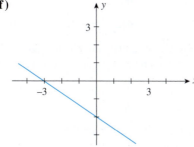

(f)

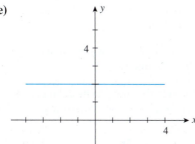

In Exercises 7–10, find the slope of the line shown in each figure.

7.

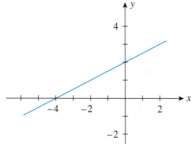

8.

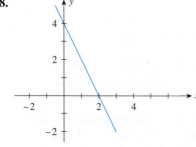

9.

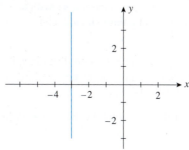

10.

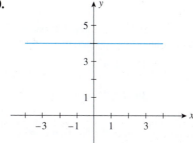

In Exercises 11–16, find the slope of the line that passes through each pair of points.

11. $(4, 3)$ and $(5, 8)$ **12.** $(4, 5)$ and $(3, 8)$

13. $(-2, 3)$ and $(4, 8)$ **14.** $(-2, -2)$ and $(4, -4)$

15. (a, b) and (c, d)

16. $(-a + 1, b - 1)$ and $(a + 1, -b)$

17. Given the equation $y = 4x - 3$, answer the following questions:
   a. If x increases by 1 unit, what is the corresponding change in y?
   b. If x decreases by 2 units, what is the corresponding change in y?

18. Given the equation $2x + 3y = 4$, answer the following questions:
   a. Is the slope of the line described by this equation positive or negative?
   b. As x increases in value, does y increase or decrease?
   c. If x decreases by 2 units, what is the corresponding change in y?

In Exercises 19 and 20, determine whether the lines AB and CD are parallel.

19. $A(1, -2)$, $B(-3, -10)$ and $C(1, 5)$, $D(-1, 1)$

20. $A(2, 3)$, $B(2, -2)$ and $C(-2, 4)$, $D(-2, 5)$

In Exercises 21 and 22, determine whether the lines AB and CD are perpendicular.

21. $A(-2, 5)$, $B(4, 2)$ and $C(-1, -2)$, $D(3, 6)$

22. $A(2, 0)$, $B(1, -2)$ and $C(4, 2)$, $D(-8, 4)$

23. If the line passing through the points $(1, a)$ and $(4, -2)$ is parallel to the line passing through the points $(2, 8)$ and $(-7, a + 4)$, what is the value of a?

24. If the line passing through the points $(a, 1)$ and $(5, 8)$ is parallel to the line passing through the points $(4, 9)$ and $(a + 2, 1)$, what is the value of a?

25. Find an equation of the horizontal line that passes through $(-4, -3)$.

26. Find an equation of the vertical line that passes through $(0, 5)$.

In Exercises 27–30, find an equation of the line that passes through the point and has the indicated slope m.

27. $(3, -4)$; $m = 2$
28. $(2, 4)$; $m = -1$

29. $(-3, 2)$; $m = 0$
30. $(1, 2)$; $m = -\dfrac{1}{2}$

In Exercises 31–34, find an equation of the line that passes through the points.

31. $(2, 4)$ and $(3, 7)$
32. $(2, 1)$ and $(2, 5)$

33. $(1, 2)$ and $(-3, -2)$
34. $(-1, -2)$ and $(3, -4)$

In Exercises 35–38, find an equation of the line that has slope m and y-intercept b.

35. $m = 3$; $b = 4$
36. $m = -2$; $b = -1$

37. $m = 0$; $b = 5$
38. $m = -\dfrac{1}{2}$; $b = \dfrac{3}{4}$

In Exercises 39–44, write the equation in the slope-intercept form and then find the slope and y-intercept of the corresponding line.

39. $x - 2y = 0$
40. $y - 2 = 0$

41. $2x - 3y - 9 = 0$
42. $3x - 4y + 8 = 0$

43. $2x + 4y = 14$
44. $5x + 8y - 24 = 0$

45. Find an equation of the line that passes through the point $(-2, 2)$ and is parallel to the line $2x - 4y - 8 = 0$.

46. Find an equation of the line that passes through the point $(2, 4)$ and is perpendicular to the line $3x + 4y - 22 = 0$.

47. Find an equation of the line that has slope -2 and passes through the midpoint of the line segment joining the points $P_1(-2, -4)$ and $P_2(3, 6)$.
Hint: See Exercise 45, page 32.

48. Find an equation of the line that passes through the midpoint of the line segment joining the points $P_1(-1, -3)$ and $P_2(3, 3)$ and the midpoint of the line segment joining the points $P_3(-2, 3)$ and $P_4(2, -3)$.
Hint: See Exercise 45, page 32.

In Exercises 49–54, find an equation of the line that satisfies the given condition.

49. The line parallel to the x-axis and 6 units below it

50. The line passing through the origin and parallel to the line joining the points $(2, 4)$ and $(4, 7)$

51. The line passing through the point (a, b) with slope equal to zero

52. The line passing through $(-3, 4)$ and parallel to the x-axis

53. The line passing through $(-5, -4)$ and parallel to the line joining $(-3, 2)$ and $(6, 8)$

54. The line passing through (a, b) with undefined slope

55. Given that the point $P(-3, 5)$ lies on the line $kx + 3y + 9 = 0$, find k.

56. Given that the point $P(2, -3)$ lies on the line $-2x + ky + 10 = 0$, find k.

In Exercises 57–62, sketch the straight line defined by the given linear equation by finding the x- and y-intercepts.
Hint: See Example 12, page 41.

57. $3x - 2y + 6 = 0$
58. $2x - 5y + 10 = 0$

59. $x + 2y - 4 = 0$
60. $2x + 3y - 15 = 0$

61. $y + 5 = 0$
62. $-2x - 8y + 24 = 0$

63. Show that an equation of a line through the points $(a, 0)$ and $(0, b)$ with $a \neq 0$ and $b \neq 0$ can be written in the form

$$\frac{x}{a} + \frac{y}{b} = 1$$

(Recall that the numbers a and b are the x- and y-intercepts, respectively, of the line. This form of an equation of a line is called the *intercept form*.)

In Exercises 64–67, use the results of Exercise 63 to find an equation of a line with the given x- and y-intercepts.

64. x-intercept 3; y-intercept 4

65. x-intercept -2; y-intercept -4

66. x-intercept $-\dfrac{1}{2}$; y-intercept $\dfrac{3}{4}$

67. x-intercept 4; y-intercept $-\dfrac{1}{2}$

In Exercises 68 and 69, determine whether the given points lie on a straight line.

68. $A(-1, 7)$, $B(2, -2)$, and $C(5, -9)$

69. $A(-2, 1)$, $B(1, 7)$, and $C(4, 13)$

70. John claims that the following points lie on a line: $(1.2, -9.04)$, $(2.3, -5.96)$, $(4.8, 1.04)$, and $(7.2, 7.76)$. Prove or disprove his claim.

71. Alison claims that the following points lie on a line: $(1.8, -6.44)$, $(2.4, -5.72)$, $(5.0, -2.72)$, and $(10.4, 3.88)$. Prove or disprove her claim.

72. TEMPERATURE CONVERSION The relationship between the temperature in degrees Fahrenheit (°F) and the temperature in degrees Celsius (°C) is

$$F = \frac{9}{5}C + 32$$

a. Sketch the line with the given equation.
b. What is the slope of the line? What does it represent?
c. What is the F-intercept of the line? What does it represent?

73. NUCLEAR PLANT UTILIZATION The United States is not building many nuclear plants, but the ones it has are running full tilt. The output (as a percent of total capacity) of nuclear plants is described by the equation

$$y = 1.9467t + 70.082$$

where t is measured in years, with $t = 0$ corresponding to the beginning of 1990.
a. Sketch the line with the given equation.
b. What are the slope and the y-intercept of the line found in part (a)?
c. Give an interpretation of the slope and the y-intercept of the line found in part (a).
d. If the utilization of nuclear power continued to grow at the same rate and the total capacity of nuclear plants in the United States remained constant, by what year were the plants generating at maximum capacity?
Source: Nuclear Energy Institute.

74. SOCIAL SECURITY CONTRIBUTIONS For wages less than the maximum taxable wage base, Social Security contributions (including those for Medicare) by employees are 7.65% of the employee's wages.
a. Find an equation that expresses the relationship between the wages earned (x) and the Social Security taxes paid (y) by an employee who earns less than the maximum taxable wage base.
b. For each additional dollar that an employee earns, by how much is his or her Social Security contribution increased? (Assume that the employee's wages remain less than the maximum taxable wage base.)
c. What Social Security contributions will an employee who earns $65,000 (which is less than the maximum taxable wage base) be required to make?
Source: Social Security Administration.

75. COLLEGE ADMISSIONS Using data compiled by the Admissions Office at Faber University, college admissions officers estimate that 55% of the students who are offered admission to the freshman class at the university will actually enroll.
a. Find an equation that expresses the relationship between the number of students who actually enroll (y) and the number of students who are offered admission to the university (x).

b. If the desired freshman class size for the upcoming academic year is 1100 students, how many students should be admitted?

76. WEIGHT OF WHALES The equation $W = 3.51L - 192$, expressing the relationship between the length L (in feet) and the expected weight W (in British tons) of adult blue whales, was adopted in the late 1960s by the International Whaling Commission.
a. What is the expected weight of an 80-ft blue whale?
b. Sketch the straight line that represents the equation.

77. THE NARROWING GENDER GAP Since the founding of the Equal Employment Opportunity Commission and the passage of equal-pay laws, the gulf between men's and women's earnings has continued to close gradually. At the beginning of 1990 ($t = 0$), women's wages were 68% of men's wages, and by the beginning of 2000 ($t = 10$), women's wages were projected to be 80% of men's wages. If this gap between women's and men's wages continued to narrow *linearly*, then women's wages were what percentage of men's wages at the beginning of 2008?
Source: Journal of Economic Perspectives.

78. DECLINING NUMBER OF PAY PHONES As cell phones proliferate, the number of pay phones continues to drop. The number of pay phones from 2004 through 2009 (in millions) are shown in the following table ($x = 0$ corresponds to 2004):

Year, x	0	1	2	3	4	5
Number of Pay Phones, y	1.30	1.15	1.00	0.84	0.69	0.56

a. Plot the number of pay phones (y) versus the year (x).
b. Draw the straight line L through the points $(0, 1.30)$ and $(5, 0.56)$.
c. Derive an equation for the line L.
d. Assuming that the trend continued, estimate the number of pay phones in 2012.
Source: FCC.

79. SPENDING ON EQUIPMENT AND SOFTWARE As the United States continues to recover slowly from the Great Recession, spending on equipment and software is projected to rise. The following table gives the percentage change in equipment and software spending, seasonally adjusted, in 2013 ($x = 0$ corresponds to the first quarter of 2013):

Quarter, x	0	1	2	3
Percent Change, y	1.3	3.8	6.0	8.2

a. Plot the percentage change (y) versus the quarter (x).
b. Draw a straight line L through the points corresponding to the first quarter and the fourth quarter.
c. Derive an equation of the line L.
d. Assuming that this trend continued, what was the percentage change in spending on equipment and software in the first quarter of 2014?
Source: U.S. Commerce Department.

80. SALES GROWTH Metro Department Store's annual sales (in millions of dollars) during the past 5 years were

Annual Sales, y	5.8	6.2	7.2	8.4	9.0
Year, x	1	2	3	4	5

   a. Plot the annual sales (y) versus the year (x).
   b. Draw a straight line L through the points corresponding to the first and fifth years.
   c. Derive an equation of the line L.
   d. Using the equation found in part (c), estimate Metro's annual sales 4 years from now ($x = 9$).

81. CORPORATE FRAUD The number of corporate fraud cases pending stood at 545 at the beginning of 2008 ($t = 0$) and was 726 cases at the beginning of 2012. The growth was approximately linear.
   a. Derive an equation of the line passing through the points $A(0, 545)$ and $B(4, 726)$.
   b. Plot the line with the equation found in part (a).
   c. Use the equation found in part (a) to estimate the number of corporate fraud cases pending at the beginning of 2016.
   Source: Federal Bureau of Investigation.

82. SOCIAL MEDIA A Nielsen survey of 3000 American moviegoers aged 12–74 years found that 27% of them used social media to chat about movies in 2010. The percentage was 29% in 2011 and 31% in 2012. Let $t = 0$, $t = 1$, and $t = 2$ correspond to the years 2010, 2011, and 2012, respectively.
   a. Explain why the three points $P_1(0, 27)$, $P_2(1, 29)$, and $P_3(2, 31)$ lie on a straight line L.
   b. Assuming that this trend continued, what was the percentage of moviegoers who used social media to chat about movies in 2014?
   c. Find an equation of L. Then use this equation to find and reconcile the result obtained in part (b).
   Source: Nielsen survey.

In Exercises 83–87, determine whether the statement is true or false. If it is true, explain why it is true. If it is false, give an example to show why it is false.

83. Suppose the slope of a line L is $-\frac{1}{2}$ and P is a given point on L. If Q is the point on L lying 4 units to the left of P, then Q is situated 2 units above P.

84. The line with equation $Ax + By + C = 0$, $(B \neq 0)$, and the line with equation $ax + by + c = 0$, $(b \neq 0)$, are parallel if $Ab - aB = 0$.

85. If the slope of the line L_1 is positive, then the slope of a line L_2 perpendicular to L_1 may be positive or negative.

86. The lines with equations $ax + by + c_1 = 0$ and $bx - ay + c_2 = 0$, where $a \neq 0$ and $b \neq 0$, are perpendicular to each other.

87. If L is the line with equation $Ax + By + C = 0$, where $A \neq 0$, then L crosses the x-axis at the point $(-C/A, 0)$.

88. Is there a difference between the statements "The slope of a straight line is zero" and "The slope of a straight line does not exist (is not defined)"? Explain your answer.

89. Show that two distinct lines with equations $a_1x + b_1y + c_1 = 0$ and $a_2x + b_2y + c_2 = 0$, respectively, are parallel if and only if $a_1b_2 - b_1a_2 = 0$.
Hint: Write each equation in the slope-intercept form and compare.

90. Prove that if a line L_1 with slope m_1 is perpendicular to a line L_2 with slope m_2, then $m_1m_2 = -1$.
Hint: Refer to the following figure. Show that $m_1 = b$ and $m_2 = c$. Next, apply the Pythagorean Theorem to triangles OAC, OCB, and OBA to show that $1 = -bc$.

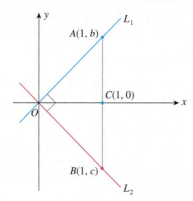

1.4   Solutions to Self-Check Exercises

1. The slope of the line that passes through the points $(a, 2)$ and $(3, 6)$ is

$$m = \frac{6 - 2}{3 - a}$$

$$= \frac{4}{3 - a}$$

Since this line is parallel to a line with slope 4, m must be equal to 4; that is,

$$\frac{4}{3 - a} = 4$$

or, upon multiplying both sides of the equation by $3 - a$,

$$4 = 4(3 - a)$$
$$4 = 12 - 4a$$
$$4a = 8$$
$$a = 2$$

2. Since the required line L is perpendicular to a line with slope $-\frac{1}{2}$, the slope of L is

$$m = \frac{-1}{-\frac{1}{2}} = 2$$

Next, using the point-slope form of the equation of a line, we have

$$y - (-1) = 2(x - 3)$$
$$y + 1 = 2x - 6$$
$$y = 2x - 7$$

3. Substituting $x = 3$ and $y = -3$ into the left-hand side of the given equation, we find

$$2(3) - 3(-3) - 12 = 3$$

which is not equal to zero (the right-hand side). Therefore, $(3, -3)$ does not lie on the line with equation $2x - 3y - 12 = 0$. (See the accompanying figure.)

Setting $x = 0$, we find $y = -4$, the y-intercept. Next, setting $y = 0$ gives $x = 6$, the x-intercept. We now draw the line passing through the points $(0, -4)$ and $(6, 0)$ as shown.

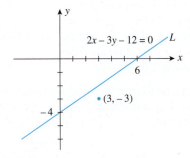

4. **a.**

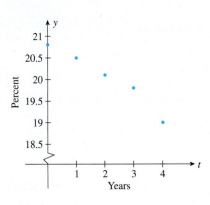

b.

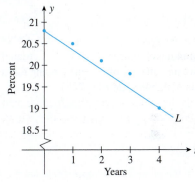

c. The slope of L is

$$m = \frac{19.0 - 20.8}{4 - 0} = -0.45$$

Using the point-slope form of the equation of a line with the point $(0, 20.8)$, we find

$$y - 20.8 = -0.45(t - 0) \quad \text{or} \quad y = -0.45t + 20.8$$

d. The year 2014 corresponds to $t = 8$, so the estimated percentage of U.S. adults who will be smoking is

$$y = -0.45(8) + 20.8 = 17.2$$

or approximately 17.2%.

CHAPTER 1 Summary of Principal Formulas and Terms

FORMULAS

1. Quadratic formula	$x = \dfrac{-b \pm \sqrt{b^2 - 4ac}}{2a}$
2. Distance between two points	$d = \sqrt{(x_2 - x_1)^2 + (y_2 - y_1)^2}$
3. Equation of a circle with center $C(h, k)$ and radius r	$(x - h)^2 + (y - k)^2 = r^2$
4. Slope of a line	$m = \dfrac{y_2 - y_1}{x_2 - x_1}$

5. Equation of a vertical line	$x = a$
6. Equation of a horizontal line	$y = b$
7. Point-slope form of the equation of a line	$y - y_1 = m(x - x_1)$
8. Slope-intercept form of the equation of a line	$y = mx + b$
9. General equation of a line	$Ax + By + C = 0$

TERMS

real number (coordinate) line (3)   infinite interval (4)   Cartesian coordinate system (25)

open interval (4)   polynomial (8)   ordered pair (26)

closed interval (4)   roots of a polynomial equation (12)   parallel lines (36)

half-open interval (4)   absolute value (22)   perpendicular lines (37)

finite interval (4)   triangle inequality (22)

CHAPTER 1   Concept Review Questions

Fill in the blanks.

1. A point in the plane can be represented uniquely by a/an _____ pair of numbers. The first number of the pair is called the _____, and the second number of the pair is called the _____.

2. **a.** The point $P(a, 0)$ lies on the _____-axis, and the point $P(0, b)$ lies on the _____-axis.
   b. If the point $P(a, b)$ lies in the fourth quadrant, then the point $P(-a, b)$ lies in the _____ quadrant.

3. The distance between two points $P_1(a, b)$ and $P_2(c, d)$ is _____.

4. An equation of the circle with center $C(a, b)$ and radius r is given by _____.

5. **a.** If $P_1(x_1, y_1)$ and $P_2(x_2, y_2)$ are any two distinct points on a nonvertical line L, then the slope of L is $m = $ _____.

   b. The slope of a vertical line is _____.
   c. The slope of a horizontal line is _____.
   d. The slope of a line that slants upward from left to right is _____.

6. If L_1 and L_2 are nonvertical lines with slopes m_1 and m_2, respectively, then L_1 is parallel to L_2 if and only if _____ and L_1 is perpendicular to L_2 if and only if _____.

7. **a.** An equation of the line passing through the point $P(x_1, y_1)$ and having slope m is _____. This form of an equation of a line is called the _____ _____.
   b. An equation of the line that has slope m and y-intercept b is _____. It is called the _____ form of an equation of a line.

8. **a.** The general form of an equation of a line is _____.
   b. If a line has equation $ax + by + c = 0$ ($b \neq 0$), then its slope is _____.

CHAPTER 1   Review Exercises

In Exercises 1–4, find the values of x that satisfy the inequality (inequalities).

1. $-x + 3 \leq 2x + 9$ 2. $-2 \leq 3x + 1 \leq 7$

3. $x - 3 > 2$ or $x + 3 < -1$

4. $2x^2 > 50$

In Exercises 5–8, evaluate the expression.

5. $|-5 + 7| + |-2|$ 6. $\left| \dfrac{5 - 12}{-4 - 3} \right|$

7. $|2\pi - 6| - \pi$

8. $|\sqrt{3} - 4| + |4 - 2\sqrt{3}|$

In Exercises 9–14, evaluate the expression.

9. $\left(\dfrac{9}{4} \right)^{3/2}$ 10. $\dfrac{5^6}{5^4}$

11. $(3 \cdot 4)^{-2}$ 12. $(-8)^{5/3}$

13. $\dfrac{(3 \cdot 2^{-3})(4 \cdot 3^5)}{2 \cdot 9^3}$ 14. $\dfrac{3\sqrt[3]{54}}{\sqrt[3]{18}}$

In Exercises 15–20, simplify the expression.

15. $\dfrac{4(x^2 + y)^3}{x^2 + y}$

16. $\dfrac{a^6 b^{-5}}{(a^3 b^{-2})^{-3}}$

17. $\dfrac{\sqrt[4]{16x^5 yz}}{\sqrt[4]{81xyz^5}}; \ x > 0, \ y > 0, \ z > 0$

18. $(2x^3)(-3x^{-2})\left(\dfrac{1}{6}x^{-1/2}\right)$

19. $\left(\dfrac{3xy^2}{4x^3 y}\right)^{-2}\left(\dfrac{3xy^3}{2x^2}\right)^3$

20. $\sqrt[3]{81x^5 y^{10}}\sqrt[3]{9xy^2}$

In Exercises 21–24, factor each expression completely.

21. $-2\pi^2 r^3 + 100\pi r^2$

22. $2v^3 w + 2vw^3 + 2u^2 vw$

23. $16 - x^2$

24. $12t^3 - 6t^2 - 18t$

In Exercises 25–28, solve the equation by factoring.

25. $8x^2 + 2x - 3 = 0$

26. $-6x^2 - 10x + 4 = 0$

27. $-x^3 - 2x^2 + 3x = 0$

28. $2x^4 + x^2 = 1$

In Exercises 29–32, find the value(s) of x that satisfy the expression.

29. $2x^2 + 3x - 2 \le 0$

30. $\dfrac{1}{x + 2} > 2$

31. $|2x - 3| < 5$

32. $\left|\dfrac{x + 1}{x - 1}\right| = 5$

In Exercises 33 and 34, use the quadratic formula to solve the quadratic equation.

33. $x^2 - 2x - 5 = 0$

34. $2x^2 + 8x + 7 = 0$

In Exercises 35–38, perform the indicated operations and simplify the expression.

35. $\dfrac{(t + 6)(60) - (60t + 180)}{(t + 6)^2}$

36. $\dfrac{6x}{2(3x^2 + 2)} + \dfrac{1}{4(x + 2)}$

37. $\dfrac{2}{3}\left(\dfrac{4x}{2x^2 - 1}\right) + 3\left(\dfrac{3}{3x - 1}\right)$

38. $\dfrac{-2x}{\sqrt{x + 1}} + 4\sqrt{x + 1}$

39. Rationalize the numerator: $\dfrac{\sqrt{x} - 1}{x - 1}$.

40. Rationalize the denominator: $\dfrac{\sqrt{x} - 1}{2\sqrt{x}}$.

In Exercises 41–43, find the distance between the two points.

41. $(-2, -3)$ and $(1, -7)$

42. $(9, 6)$ and $(6, 2)$

43. $\left(\dfrac{1}{2}, \sqrt{3}\right)$ and $\left(-\dfrac{1}{2}, 2\sqrt{3}\right)$

In Exercises 44–49, find an equation of the line L that passes through the point $(-2, 4)$ and satisfies the condition.

44. L is a vertical line.

45. L is a horizontal line.

46. L passes through the point $\left(3, \dfrac{7}{2}\right)$.

47. The x-intercept of L is 3.

48. L is parallel to the line $5x - 2y = 6$.

49. L is perpendicular to the line $4x + 3y = 6$.

50. Find an equation of the straight line that passes through the point $(2, 3)$ and is parallel to the line with equation $3x + 4y - 8 = 0$.

51. Find an equation of the straight line that passes through the point $(-1, 3)$ and is parallel to the line passing through the points $(-3, 4)$ and $(2, 1)$.

52. Find an equation of the line that passes through the point $(-3, -2)$ and is parallel to the line passing through the points $(-2, -4)$ and $(1, 5)$.

53. Find an equation of the line that passes through the point $(-2, -4)$ and is perpendicular to the line with equation $2x - 3y - 24 = 0$.

54. Does the point $P\left(-1, -\dfrac{5}{4}\right)$ lie on the line $6x - 8y - 16 = 0$? Justify your answer.

55. Given that the point $P(2, -4)$ lies on the line $2x + ky = -8$, find k.

56. Sketch the graph of the equation $3x - 4y = 24$.

57. Sketch the graph of the line that passes through the point $(3, 2)$ and has slope $-\dfrac{2}{3}$.

58. Find the minimum cost C (in dollars) given that

$$2(1.5C + 80) \le 2(2.5C - 20)$$

59. Find the maximum revenue R (in dollars) given that

$$12(2R - 320) \le 4(3R + 240)$$

60. **A Falling Stone** A stone is thrown straight up from the roof of an 80-ft building, and the height (in feet) of the stone any time t later (in seconds), measured from the ground, is given by

$$-16t^2 + 64t + 80$$

Find the interval of time when the stone is at or greater than a height of 128 ft from the ground.

61. CYBER MONDAY SALES The amount (in millions of dollars) spent on Cyber Monday for the years 2009 through 2011 is given in the following table:

Year	2009	2010	2011
Sales, y	887	1028	1251

a. Plot the Cyber Monday sales (y) versus the year (t), where $t = 0$ corresponds to 2009.

b. Draw a straight line L through the points $(0, 887)$ and $(2, 1251)$.

c. Derive an equation of the line L.

d. Assuming that the trend continues, use the equation found in part (c) to estimate the amount consumers will spend on Cyber Monday in 2016.

Source: Comscore.

62. SALES OF NAVIGATION SYSTEMS The estimated number of navigation systems (in millions) sold in North America, Europe, and Japan from 2002 through 2006 follow. Here $t = 0$ corresponds to 2002.

Systems Installed, y	3.9	4.7	5.8	6.8	7.8
Year, t	0	1	2	3	4

a. Plot the annual sales (y) versus the year (t).

b. Draw a straight line L through the points corresponding to 2002 and 2006.

c. Derive an equation of the line L.

d. Use the equation found in part (c) to estimate the number of navigation systems installed for 2005. Compare this figure with the estimated sales for that year.

Source: ABI Research.

The problem-solving skills that you learn in each chapter are building blocks for the rest of the course. Therefore, it is a good idea to make sure that you have mastered these skills before moving on to the next chapter. The Before Moving On exercises that follow are designed for that purpose. After completing these exercises, you can identify the skills that you should review before starting the next chapter.

CHAPTER 1　Before Moving On . . .

1. Evaluate:

   a. $|\pi - 2\sqrt{3}| - |\sqrt{3} - \sqrt{2}|$ b. $\left[\left(-\dfrac{1}{3}\right)^{-3}\right]^{1/3}$

2. Simplify:

   a. $\sqrt[3]{64x^6} \cdot \sqrt{9y^2x^6}$; $x > 0$ b. $\left(\dfrac{a^{-3}}{b^{-4}}\right)^2\left(\dfrac{b}{a}\right)^{-3}$

3. Rationalize the denominator:

   a. $\dfrac{2x}{3\sqrt{y}}$ b. $\dfrac{x}{\sqrt{x} - 4}$

4. Perform each operation and simplify:

   a. $\dfrac{(x^2 + 1)(\frac{1}{2}x^{-1/2}) - x^{1/2}(2x)}{(x^2 + 1)^2}$

   b. $-\dfrac{3x}{\sqrt{x + 2}} + 3\sqrt{x + 2}$

5. Rationalize the numerator: $\dfrac{\sqrt{x} + \sqrt{y}}{\sqrt{x} - \sqrt{y}}$.

6. Factor completely:

   a. $12x^3 - 10x^2 - 12x$ b. $2bx - 2by + 3cx - 3cy$

7. Solve each equation:

   a. $12x^2 - 9x - 3 = 0$ b. $3x^2 - 5x + 1 = 0$

8. Find the distance between $(-2, 4)$ and $(6, 8)$.

9. Find an equation of the line that passes through $(-1, -2)$ and $(4, 5)$.

10. Find an equation of the line that has slope $-\frac{1}{3}$ and y-intercept $\frac{4}{3}$.

2 Functions, Limits, and the Derivative

IN THIS CHAPTER, we define a *function*, a special relationship between two variables. The concept of a function enables us to describe many relationships that exist in applications. We also begin the study of differential calculus. Historically, differential calculus was developed in response to the problem of finding the tangent line to an arbitrary point on a curve. But it quickly became apparent that solving this problem provided mathematicians with a method for solving many practical problems involving the rate of change of one quantity with respect to another. The basic tool used in differential calculus is the *derivative* of a function. The concept of the derivative is based, in turn, on a more fundamental notion—that of the *limit* of a function.

Half of millenials (18- to 29-year-olds) "do not believe Social Security will exist" when they reach retirement age, according to a study released by the iOme Challenge organization. Are their concerns about the present system justified? In Example 3, page 82, we use a mathematical model constructed from data from the Social Security Administration to predict the year in which the assets of the current system will be depleted.

© EDHAR/ShutterStock.com

2.1   Functions and Their Graphs

Functions

A manufacturer would like to know how his company's profit is related to its production level; a biologist would like to know how the size of the population of a certain culture of bacteria will change over time; a psychologist would like to know the relationship between the learning time of an individual and the length of a vocabulary list; and a chemist would like to know how the initial speed of a chemical reaction is related to the amount of substrate used. In each instance, we are concerned with the same question: How does one quantity depend upon another? The relationship between two quantities is conveniently described in mathematics by using the concept of a function.

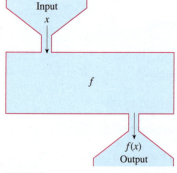

> ### Function
>
> A **function** is a rule that assigns to each element in a set A one and only one element in a set B.

The set A is called the **domain** of the function. It is customary to denote a function by a letter of the alphabet, such as the letter f. If x is an element in the domain of a function f, then the element in B that f associates with x is written $f(x)$ (read "f of x") and is called the value of f at x. The set comprising all the values assumed by $y = f(x)$ as x takes on all possible values in its domain is called the **range** of the function f.

We can think of a function f as a machine. The domain is the set of inputs (raw material) for the machine, the rule describes how the input is to be processed, and the values of the function are the outputs of the machine (Figure 1).

We can also think of a function f as a mapping in which an element x in the domain of f is mapped onto a unique element $f(x)$ in B (Figure 2).

FIGURE 1
A function machine

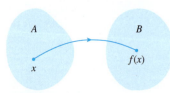

FIGURE 2
The function f viewed as a mapping

Notes

1. The output $f(x)$ associated with an input x is unique. To appreciate the importance of this uniqueness property, consider a rule that associates with each item x in a department store its selling price y. Then, each x must correspond to *one and only one y*. Notice, however, that different x's may be associated with the same y. In the context of the present example, this says that different items may have the same price.

2. Although the sets A and B that appear in the definition of a function may be quite arbitrary, in this book they will denote sets of real numbers.

An example of a function may be taken from the familiar relationship between the area of a circle and its radius. Letting x and y denote the radius and area of a circle, respectively, we have, from elementary geometry,

$$y = \pi x^2 \tag{1}$$

Equation (1) defines y as a function of x, since for each admissible value of x (that is, for each nonnegative number representing the radius of a certain circle), there corresponds precisely one number $y = \pi x^2$ that gives the area of the circle. The rule defining this "area function" may be written as

$$f(x) = \pi x^2 \tag{2}$$

To compute the area of a circle of radius 5 inches, we simply replace x in Equation (2) with the number 5. Thus, the area of the circle is

$$f(5) = \pi 5^2 = 25\pi$$

or 25π square inches.

In general, to evaluate a function at a specific value of x, we replace x with that value, as illustrated in Examples 1 and 2.

EXAMPLE 1 Let the function f be defined by the rule $f(x) = 2x^2 - x + 1$. Find:

a. $f(1)$ **b.** $f(-2)$ **c.** $f(a)$ **d.** $f(a + h)$

Solution

a. $f(1) = 2(1)^2 - (1) + 1 = 2 - 1 + 1 = 2$
b. $f(-2) = 2(-2)^2 - (-2) + 1 = 8 + 2 + 1 = 11$
c. $f(a) = 2(a)^2 - (a) + 1 = 2a^2 - a + 1$
d. $f(a + h) = 2(a + h)^2 - (a + h) + 1 = 2a^2 + 4ah + 2h^2 - a - h + 1$ ▧

 APPLIED EXAMPLE 2 Profit Functions ThermoMaster manufactures an indoor–outdoor thermometer at its Mexican subsidiary. Management estimates that the profit (in dollars) realizable by ThermoMaster in the manufacture and sale of x thermometers per week is

$$P(x) = -0.001x^2 + 8x - 5000$$

Find ThermoMaster's weekly profit if its level of production is (a) 1000 thermometers per week and (b) 2000 thermometers per week.

Solution

a. The weekly profit when the level of production is 1000 units per week is found by evaluating the profit function P at $x = 1000$. Thus,

$$P(1000) = -0.001(1000)^2 + 8(1000) - 5000 = 2000$$

or $2000.

b. When the level of production is 2000 units per week, the weekly profit is given by

$$P(2000) = -0.001(2000)^2 + 8(2000) - 5000 = 7000$$

or $7000.   ▧

Determining the Domain of a Function

Suppose we are given the function $y = f(x)$.* Then, the variable x is called the **independent variable.** The variable y, whose value depends on x, is called the **dependent variable.**

To determine the domain of a function, we need to find what restrictions, if any, are to be placed on the independent variable x. In general, if a function is defined by a rule relating x to $f(x)$ without specific mention of its domain, it is understood that the domain will consist of all values of x for which $f(x)$ is a real number. In this connection, you should keep in mind that (1) division by zero is not permitted and (2) the even root of a negative number is not a real number.

---

*It is customary to refer to a function f as $f(x)$ or by the equation $y = f(x)$ defining the function.

EXAMPLE 3 Find the domain of each function.

a. $f(x) = \sqrt{x - 1}$ **b.** $f(x) = \dfrac{1}{x^2 - 4}$ **c.** $f(x) = x^2 + 3$

Solution

a. Since the square root of a negative number is not a real number, it is necessary that $x - 1 \geq 0$. The inequality is satisfied by the set of real numbers $x \geq 1$. Thus, the domain of f is the interval $[1, \infty)$.
b. The only restriction on x is that $x^2 - 4$ be different from zero, since division by zero is not allowed. But $(x^2 - 4) = (x + 2)(x - 2) = 0$ if $x = -2$ or $x = 2$. Thus, the domain of f in this case consists of the intervals $(-\infty, -2), (-2, 2)$, and $(2, \infty)$.
c. Here, any real number satisfies the equation, so the domain of f is the set of all real numbers.

In many practical applications, the domain of a function is dictated by the nature of the problem, as illustrated in Example 4.

APPLIED EXAMPLE 4 Packaging An open box is to be made from a rectangular piece of cardboard 16 inches long and 10 inches wide by cutting away identical squares (x inches by x inches) from each corner and folding up the resulting flaps (Figure 3). Find an expression that gives the volume V of the box as a function of x. What is the domain of the function?

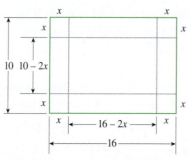
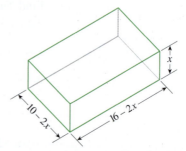

(a) The box is constructed by cutting x in. by x in. squares from each corner.

(b) The dimensions of the resulting box are $(10 - 2x)$ in. by $(16 - 2x)$ in. by x in.

FIGURE 3

Solution The dimensions of the box are $(10 - 2x)$ inches by $(16 - 2x)$ inches by x inches, so its volume (in cubic inches) is given by

$$V = f(x) = (16 - 2x)(10 - 2x)x \qquad \text{Length · width · height}$$
$$= (160 - 52x + 4x^2)x$$
$$= 4x^3 - 52x^2 + 160x$$

Since the length of each side of the box must be greater than or equal to zero, we see that

$$16 - 2x \geq 0 \qquad 10 - 2x \geq 0 \qquad x \geq 0$$

simultaneously; that is,

$$x \leq 8 \qquad x \leq 5 \qquad x \geq 0$$

All three inequalities are satisfied simultaneously provided that $0 \leq x \leq 5$. Thus, the domain of the function f is the interval $[0, 5]$.

Graphs of Functions

If f is a function with domain A, then corresponding to each real number x in A, there is precisely one real number $f(x)$. We can also express this fact by using **ordered pairs** of real numbers. Write each number x in A as the first member of an ordered pair and each number $f(x)$ corresponding to x as the second member of the ordered pair. This gives exactly one ordered pair $(x, f(x))$ for each x in A.

Observe that the condition that there be one and only one number $f(x)$ corresponding to each number x in A translates into the requirement that *no two distinct ordered pairs have the same first number.*

Since ordered pairs of real numbers correspond to points in the plane, we have found a way to exhibit a function graphically.

> ### Graph of a Function of One Variable
>
> The **graph of a function** f is the set of all points (x, y) in the xy-plane such that x is in the domain of f and $y = f(x)$.

Figure 4 shows the graph of a function f. Observe that the y-coordinate of the point (x, y) on the graph of f gives the height of that point (the distance above the x-axis), if $f(x)$ is positive. If $f(x)$ is negative, then $-f(x)$ gives the depth of the point (x, y) (the distance below the x-axis). Also, observe that the domain of f is a set of real numbers lying on the x-axis, whereas the range of f lies on the y-axis.

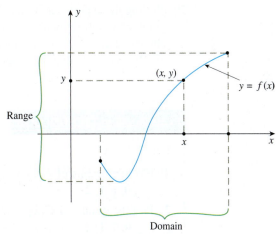

FIGURE 4
The graph of f

EXAMPLE 5 The graph of a function f is shown in Figure 5.

a. What is the value of $f(3)$? The value of $f(5)$?
b. What is the height or depth of the point $(3, f(3))$ from the x-axis? The point $(5, f(5))$ from the x-axis?
c. What is the domain of f? The range of f?

Solution

a. From the graph of f, we see that $y = -2$ when $x = 3$, and we conclude that $f(3) = -2$. Similarly, we see that $f(5) = 3$.
b. Since the point $(3, -2)$ lies below the x-axis, we see that the depth of the point $(3, f(3))$ is $-f(3) = -(-2) = 2$ units below the x-axis. The point

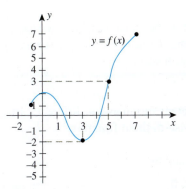

FIGURE 5
The graph of f

$(5, f(5))$ lies above the x-axis and is located at a height of $f(5)$, or 3 units above the x-axis.

c. Observe that x may take on all values between $x = -1$ and $x = 7$, inclusive, so the domain of f is $[-1, 7]$. Next, observe that as x takes on all values in the domain of f, $f(x)$ takes on all values between -2 and 7, inclusive. (You can easily see this by running your index finger along the x-axis from $x = -1$ to $x = 7$ and observing the corresponding values assumed by the y-coordinate of each point of the graph of f.) Therefore, the range of f is $[-2, 7]$.

We can gain much information about the graph of a function by plotting a few points on its graph. Later on, we will develop more systematic and sophisticated techniques for graphing functions.

EXAMPLE 6 Sketch the graph of the function defined by the equation $y = x^2 + 1$. What is the range of f?

Solution   The domain of the function is the set of all real numbers. By assigning several values to the variable x and computing the corresponding values for y, we obtain the following solutions to the equation $y = x^2 + 1$:

x	-3	-2	-1	0	1	2	3
y	10	5	2	1	2	5	10

By plotting these points and then connecting them with a smooth curve, we obtain the graph of $y = f(x)$, which is a parabola (Figure 6). To determine the range of f, we observe that $x^2 \geq 0$ if x is any real number, and so $x^2 + 1 \geq 1$ for all real numbers x. We conclude that the range of f is $[1, \infty)$. The graph of f confirms this result visually.

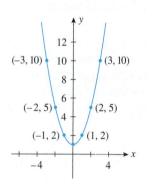

FIGURE 6
The graph of $y = x^2 + 1$ is a parabola.

Exploring with TECHNOLOGY

Let $f(x) = x^2$.

1. Plot the graphs of $F(x) = x^2 + c$ on the same set of axes for $c = -2, -1, -\frac{1}{2}, 0, \frac{1}{2}, 1, 2$.

2. Plot the graphs of $G(x) = (x + c)^2$ on the same set of axes for $c = -2, -1, -\frac{1}{2}, 0, \frac{1}{2}, 1, 2$.

3. Plot the graphs of $H(x) = cx^2$ on the same set of axes for $c = -2, -1, -\frac{1}{2}, -\frac{1}{4}, 0, \frac{1}{4}, \frac{1}{2}, 1, 2$.

4. Study the family of graphs in parts 1–3, and describe the relationship between the graph of a function f and the graphs of the functions defined by (a) $y = f(x) + c$, (b) $y = f(x + c)$, and (c) $y = cf(x)$, where c is a constant.

Sometimes a function is defined by giving different formulas for different parts of its domain. Such a function is said to be a **piecewise-defined function.**

EXAMPLE 7 Sketch the graph of the function f defined by

$$f(x) = \begin{cases} -x & \text{if } x < 0 \\ \sqrt{x} & \text{if } x \geq 0 \end{cases}$$

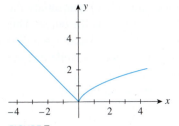

FIGURE 7
The graph of $y = f(x)$ is obtained by graphing $y = -x$ over $(-\infty, 0)$ and $y = \sqrt{x}$ over $[0, \infty)$.

Solution   The function f is defined in a piecewise fashion on the set of all real numbers. In the subdomain $(-\infty, 0)$, the rule for f is given by $f(x) = -x$. The equation $y = -x$ is a linear equation in the slope-intercept form (with slope -1 and intercept 0). Therefore, the graph of f corresponding to the subdomain $(-\infty, 0)$ is the half-line shown in Figure 7. Next, in the subdomain $[0, \infty)$, the rule for f is given by $f(x) = \sqrt{x}$. The values of $f(x)$ corresponding to $x = 0, 1, 2, 3,$ and 4 are shown in the following table:

x	0	1	2	3	4
$f(x)$	0	1	$\sqrt{2}$	$\sqrt{3}$	2

Using these values, we sketch the graph of the function f as shown in Figure 7.

$\boxed{\$}$ APPLIED EXAMPLE 8 Bank Deposits Madison Finance Company plans to open two branch offices 2 years from now in two separate locations: an industrial complex and a newly developed commercial center in the city. As a result of these expansion plans, Madison's total deposits during the next 5 years are expected to grow in accordance with the rule

$$f(x) = \begin{cases} \sqrt{2x} + 20 & \text{if } 0 \le x \le 2 \\ \dfrac{1}{2}x^2 + 20 & \text{if } 2 < x \le 5 \end{cases}$$

where $y = f(x)$ gives the total amount of money (in millions of dollars) on deposit with Madison in year x ($x = 0$ corresponds to the present). Sketch the graph of the function f.

Solution   The function f is defined in a piecewise fashion on the interval $[0, 5]$. In the subdomain $[0, 2]$, the rule for f is given by $f(x) = \sqrt{2x} + 20$. The values of $f(x)$ corresponding to $x = 0, 1,$ and 2 may be tabulated as follows:

x	0	1	2
$f(x)$	20	21.4	22

Next, in the subdomain $(2, 5]$, the rule for f is given by $f(x) = \frac{1}{2}x^2 + 20$. The values of $f(x)$ corresponding to $x = 3, 4,$ and 5 are shown in the following table:

x	3	4	5
$f(x)$	24.5	28	32.5

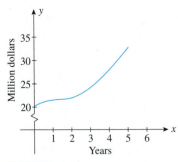

FIGURE 8
We obtain the graph of the function $y = f(x)$ by graphing $y = \sqrt{2x} + 20$ over $[0, 2]$ and $y = \frac{1}{2}x^2 + 20$ over $(2, 5]$.

Using the values of $f(x)$ in this table, we sketch the graph of the function f as shown in Figure 8.

The Vertical Line Test

Although it is true that every function f of a variable x has a graph in the xy-plane, it is not true that every curve in the xy-plane is the graph of a function. For example, consider the curve depicted in Figure 9. This is the graph of the equation $y^2 = x$. In

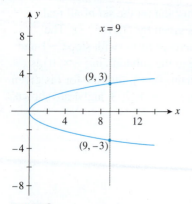

FIGURE 9
Since a vertical line passes through the curve at more than one point, we deduce that the curve is *not* the graph of a function.

general, the **graph of an equation** is the set of all ordered pairs (x, y) that satisfy the given equation. Observe that the points $(9, -3)$ and $(9, 3)$ both lie on the curve. This implies that the number $x = 9$ is associated with *two* numbers: $y = -3$ and $y = 3$. But this clearly violates the uniqueness property of a function. Thus, we conclude that the curve under consideration cannot be the graph of a function.

This example suggests the following **Vertical Line Test** for determining whether a curve is the graph of a function.

> **Vertical Line Test**
>
> A curve in the xy-plane is the graph of a function $y = f(x)$ if and only if each vertical line intersects it in at most one point.

EXAMPLE 9 Determine which of the curves shown in Figure 10 are the graphs of functions of x.

Solution The curves depicted in Figure 10a, c, and d are graphs of functions because each curve satisfies the requirement that each vertical line intersects the curve in at most one point. Note that the vertical line shown in Figure 10c does *not* intersect the graph because the point on the x-axis through which this line passes does not lie in the domain of the function. The curve depicted in Figure 10b is *not* the graph of a function of x because the vertical line shown there intersects the graph at three points.

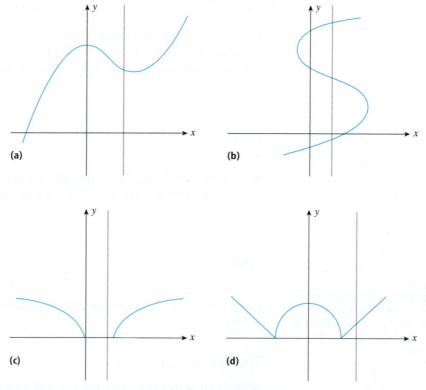

FIGURE 10
The Vertical Line Test can be used to determine which of these curves are graphs of functions.

2.1 Self-Check Exercises

1. Let f be the function defined by

$$f(x) = \frac{\sqrt{x+1}}{x}$$

  a. Find the domain of f.   **b.** Compute $f(3)$.
  c. Compute $f(a+h)$.

2. Let

$$f(x) = \begin{cases} -x+1 & \text{if } -1 \leq x < 1 \\ \sqrt{x-1} & \text{if } 1 \leq x \leq 5 \end{cases}$$

  a. Find $f(0)$ and $f(2)$.
  b. Sketch the graph of f.

3. Let $f(x) = \sqrt{2x+1} + 2$. Determine whether the point $(4, 6)$ lies on the graph of f.

Solutions to Self-Check Exercises 2.1 can be found on page 65.

2.1 Concept Questions

1. **a.** What is a function?
  b. What is the domain of a function? The range of a function?
  c. What is an independent variable? A dependent variable?

2. **a.** What is the graph of a function? Use a drawing to illustrate the graph, the domain, and the range of a function.
  b. If you are given a curve in the xy-plane, how can you tell whether the graph is that of a function f defined by $y = f(x)$?

3. Are the following graphs of functions? Explain.

a.   **b.**

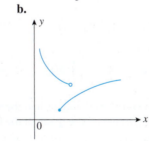

c.   **d.**

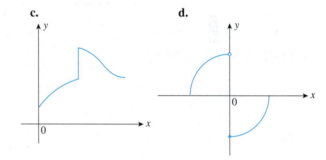

4. What are the domain and range of the function f with the following graph?

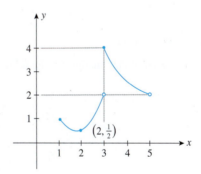

2.1 Exercises

1. Let f be the function defined by $f(x) = 5x + 6$. Find $f(3), f(-3), f(a), f(-a)$, and $f(a+3)$.

2. Let f be the function defined by $f(x) = 4x - 3$. Find $f(4), f(\frac{1}{4}), f(0), f(a)$, and $f(a+1)$.

3. Let g be the function defined by $g(x) = 3x^2 - 6x - 3$. Find $g(0), g(-1), g(a), g(-a)$, and $g(x+1)$.

4. Let h be the function defined by $h(x) = x^3 - x^2 + x + 1$. Find $h(-5), h(0), h(a)$, and $h(-a)$.

5. Let f be the function defined by $f(x) = 2x + 5$. Find $f(a+h), f(-a), f(a^2), f(a-2h)$, and $f(2a-h)$.

6. Let g be the function defined by $g(x) = -x^2 + 2x$. Find $g(a+h), g(-a), g(\sqrt{a}), a + g(a)$, and $\dfrac{1}{g(a)}$.

7. Let s be the function defined by $s(t) = \dfrac{2t}{t^2 - 1}$.
Find $s(4), s(0), s(a), s(2+a)$, and $s(t+1)$.

8. Let g be the function defined by $g(u) = (3u - 2)^{3/2}$. Find $g(1), g(6), g(\frac{11}{3})$, and $g(u + 1)$.

9. Let f be the function defined by $f(t) = \dfrac{2t^2}{\sqrt{t - 1}}$. Find $f(2), f(a), f(x + 1)$, and $f(x - 1)$.

10. Let f be the function defined by $f(x) = 2 + 2\sqrt{5 - x}$. Find $f(-4), f(1), f(\frac{11}{4})$, and $f(x + 5)$.

11. Let f be the function defined by

$$f(x) = \begin{cases} x^2 + 1 & \text{if } x \le 0 \\ \sqrt{x} & \text{if } x > 0 \end{cases}$$

Find $f(-2), f(0)$, and $f(1)$.

12. Let g be the function defined by

$$g(x) = \begin{cases} -\dfrac{1}{2}x + 1 & \text{if } x < 2 \\ \sqrt{x - 2} & \text{if } x \ge 2 \end{cases}$$

Find $g(-2), g(0), g(2)$, and $g(4)$.

13. Let f be the function defined by

$$f(x) = \begin{cases} -\dfrac{1}{2}x^2 + 3 & \text{if } x < 1 \\ 2x^2 + 1 & \text{if } x \ge 1 \end{cases}$$

Find $f(-1), f(0), f(1)$, and $f(2)$.

14. Let f be the function defined by

$$f(x) = \begin{cases} 2 + \sqrt{1 - x} & \text{if } x \le 1 \\ \dfrac{1}{1 - x} & \text{if } x > 1 \end{cases}$$

Find $f(0), f(1)$, and $f(2)$.

15. Refer to the graph of the function f in the following figure.

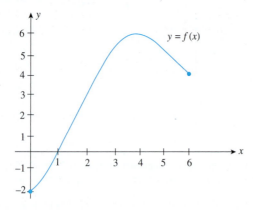

a. Find the value of $f(0)$.
b. Find the value of x for which (i) $f(x) = 3$ and (ii) $f(x) = 0$.
c. Find the domain of f.
d. Find the range of f.

16. Refer to the graph of the function f in the following figure.

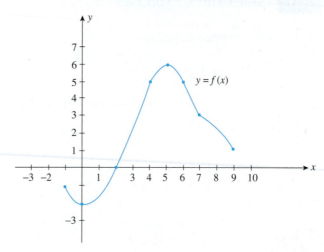

a. Find the value of $f(7)$.
b. Find the values of x corresponding to the point(s) on the graph of f located at a height of 5 units from the x-axis.
c. Find the point on the x-axis at which the graph of f crosses it. What is the value of $f(x)$ at this point?
d. Find the domain and range of f.

In Exercises 17–20, determine whether the point lies on the graph of the function.

17. $(2, \sqrt{3}); g(x) = \sqrt{x^2 - 1}$

18. $(3, 3); f(x) = \dfrac{x + 1}{\sqrt{x^2 + 7}} + 2$

19. $(-2, -3); f(t) = \dfrac{|t - 1|}{t + 1}$

20. $\left(-3, -\dfrac{1}{13}\right); h(t) = \dfrac{|t + 1|}{t^3 + 1}$

In Exercises 21 and 22, find the value of c such that the point $P(a, b)$ lies on the graph of the function f.

21. $f(x) = 2x^2 - 4x + c; P(1, 5)$

22. $f(x) = x\sqrt{9 - x^2} + c; P(2, 4)$

In Exercises 23–36, find the domain of the function.

23. $f(x) = x^2 + 3$

24. $f(x) = 7 - x^2$

25. $f(x) = \dfrac{3x + 1}{x^2}$

26. $g(x) = \dfrac{2x + 1}{x - 1}$

27. $f(x) = \sqrt{x^2 + 1}$

28. $f(x) = \sqrt{x - 5}$

29. $f(x) = \sqrt{5 - x}$

30. $g(x) = \sqrt{2x^2 + 3}$

31. $f(x) = \dfrac{x}{x^2 - 1}$

32. $f(x) = \dfrac{1}{x^2 + x - 2}$

33. $f(x) = (x + 3)^{3/2}$

34. $g(x) = 2(x - 1)^{5/2}$

35. $f(x) = \dfrac{\sqrt{1-x}}{x^2 - 4}$ 36. $f(x) = \dfrac{\sqrt{x-1}}{(x+2)(x-3)}$

37. Let f be the function defined by the rule $f(x) = x^2 - x - 6$.
   a. Find the domain of f.
   b. Compute $f(x)$ for $x = -3, -2, -1, 0, \frac{1}{2}, 1, 2, 3$.
   c. Use the results obtained in parts (a) and (b) to sketch the graph of f.

38. Let f be the function defined by the rule $f(x) = 2x^2 + x - 3$.
   a. Find the domain of f.
   b. Compute $f(x)$ for $x = -3, -2, -1, -\frac{1}{2}, 0, 1, 2, 3$.
   c. Use the results obtained in parts (a) and (b) to sketch the graph of f.

In Exercises 39–50, sketch the graph of the function with the given rule. Find the domain and range of the function.

39. $f(x) = 2x^2 + 1$ 40. $f(x) = 9 - x^2$

41. $f(x) = 2 + \sqrt{x}$ 42. $g(x) = 4 - \sqrt{x}$

43. $f(x) = \sqrt{1-x}$ 44. $f(x) = \sqrt{x-1}$

45. $f(x) = |x| - 1$ 46. $f(x) = |x| + 1$

47. $f(x) = \begin{cases} x & \text{if } x < 0 \\ 2x + 1 & \text{if } x \geq 0 \end{cases}$

48. $f(x) = \begin{cases} 4 - x & \text{if } x < 2 \\ 2x - 2 & \text{if } x \geq 2 \end{cases}$

49. $f(x) = \begin{cases} -x + 1 & \text{if } x \leq 1 \\ x^2 - 1 & \text{if } x > 1 \end{cases}$

50. $f(x) = \begin{cases} -x - 1 & \text{if } x < -1 \\ 0 & \text{if } -1 \leq x \leq 1 \\ x + 1 & \text{if } x > 1 \end{cases}$

In Exercises 51–58, use the Vertical Line Test to determine whether the graph represents y as a function of x.

51.
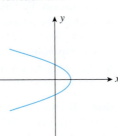

52.
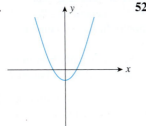

53.
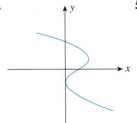

54.
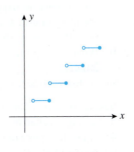

55.
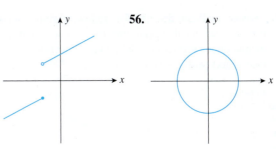

56.

57.
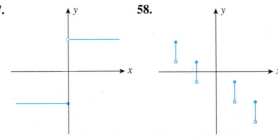

58.

59. The circumference of a circle is given by

$$C(r) = 2\pi r$$

where r is the radius of the circle. What is the circumference of a circle with a 5-in. radius?

60. The volume of a sphere of radius r is given by

$$V(r) = \frac{4}{3}\pi r^3$$

Compute $V(2.1)$ and $V(2)$. What does the quantity $V(2.1) - V(2)$ measure?

61. **CONSUMPTION FUNCTION** The consumption function in a certain economy is given by

$$C(y) = 0.75y + 6$$

where $C(y)$ is the personal consumption expenditure, y is the disposable personal income, and both $C(y)$ and y are measured in billions of dollars. Find $C(0)$, $C(50)$, and $C(100)$.

62. **FRIED'S RULE** Fried's Rule, a method for calculating pediatric drug dosages, is based on a child's age. If a denotes the adult dosage (in milligrams) and if t is the age of the child (in years), then the child's dosage is given by

$$D(t) = \frac{2}{25}ta$$

If the adult dose of a substance is 500 mg, how much should a 4-year-old child receive?

63. **SPENDING ON PHONE SERVICES** The following graphs show the average annual expenditures on cell phone and residential phone services (in dollars) as a function of time t (in years), with $t = 0$ corresponding to 2001.

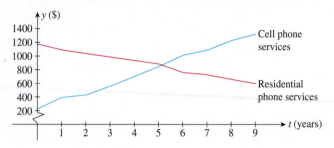

a. In what years were the average annual expenditures on residential phone services greater than those of cell phone services?
b. In what years were the average annual expenditures on cell phone services greater than those of residential phone services?
c. In what year, approximately, were the average annual expenditures on residential phone services the same as those of cell phone services? Estimate the level of expenditure on each service at that time.

Source: Consumer Expenditure Survey.

64. **THE GENDER GAP** The following graph shows the ratio of women's earnings to men's from 1960 through 2000.

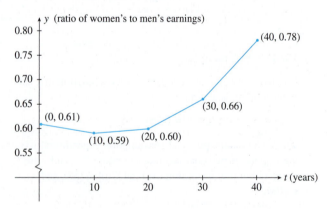

a. Write the rule for the function f giving the ratio of women's earnings to men's in year t, with $t = 0$ corresponding to 1960.
   Hint: The function f is defined piecewise and is linear over each of four subintervals.
b. In what decade(s) was the gender gap expanding? Shrinking?
c. Refer to part (b). How fast was the gender gap expanding or shrinking in each of these decades?

Source: U.S. Bureau of Labor Statistics.

65. **CLOSING THE GENDER GAP IN EDUCATION** The following graph shows the ratio of the number of bachelor's

degrees earned by women to that of men from 1960 through 1990.

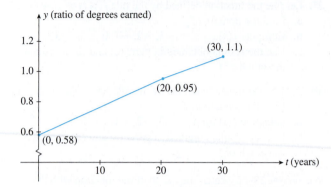

a. Write the rule for the function f giving the ratio of the number of bachelor's degrees earned by women to that of men in year t, with $t = 0$ corresponding to 1960.
   Hint: The function f is defined piecewise and is linear over each of two subintervals.
b. How fast was the ratio changing in the period from 1960 to 1980? From 1980 to 1990?
c. In what year (approximately) was the number of bachelor's degrees earned by women equal for the first time to the number earned by men?

Source: Department of Education.

66. **SALES TAXES** In a certain state, the sales tax T on the amount of taxable goods is 6% of the value of the goods purchased x, where both T and x are measured in dollars.
a. Express T as a function of x.
b. Find $T(200)$ and $T(5.65)$.

67. **COLAS** Social Security recipients receive an automatic cost-of-living adjustment (COLA) once each year. Their monthly benefit is increased by the amount that consumer prices increased during the preceding year. Suppose that consumer prices increased by 5.3% during the preceding year.
a. Express the adjusted monthly benefit of a Social Security recipient as a function of his or her current monthly benefit.
b. If Harrington's monthly Social Security benefit is now $1520, what will be his adjusted monthly benefit?

68. **GLOBAL DEFENSE SPENDING** Global defense spending stood at $1.44 trillion in 2009 and is projected to grow at the rate of $0.058 trillion per year through 2018.
a. Find a function $f(t)$ giving the projected global defense spending in year t, where $t = 0$ corresponds to 2009.
   Hint: The graph of f lies on a straight line.
b. What is the projected global defense spending in 2018?

Source: Homeland Security Research.

69. **SURFACE AREA OF A SINGLE-CELLED ORGANISM** The surface area S of a single-celled organism may be found by multiplying 4π times the square of the radius r of the cell. Express S as a function of r.

70. GROWTH OF A CANCEROUS TUMOR The volume of a spherical cancerous tumor is given by the function

$$V(r) = \frac{4}{3}\pi r^3$$

where r is the radius of the tumor in centimeters. By what factor is the volume of the tumor increased if its radius is doubled?

71. MEDIAN AGE OF EU-27 POPULATION The European Union (EU) was established on November 1, 1993, with 12 member states. As of January 2007, through a series of enlargements, the number of members had grown to 27. Consistently low birth rates and higher life expectancy have resulted in a steady increase in the median age of the population of EU-27, as the Union is now called. The median age of the population of EU-27 (in years) is given by

$$M(t) = 0.3t + 37.9 \qquad (0 \le t \le 11)$$

where t is measured in years, with $t = 0$ corresponding to the year 2000.
a. How fast was the median age of the population of EU-27 changing at any time during the period under consideration?
b. What was the median age in 2011?
c. Assuming that the trend continued, what was the median age in 2015?
Source: Eurostat.

72. COST OF RENTING A TRUCK Ace Truck leases its 10-ft box truck at \$30/day and \$0.45/mi, whereas Acme Truck leases a similar truck at \$25/day and \$0.50/mi.
a. Find the daily cost of leasing from each company as a function of the number of miles driven.
b. Sketch the graphs of the two functions on the same set of axes.
c. Which company should a customer rent a truck from for 1 day if she plans to drive at most 70 mi and wishes to minimize her cost?

73. LINEAR DEPRECIATION A new machine was purchased by National Textile for \$120,000. For income tax purposes, the machine is depreciated linearly over 10 years; that is, the book value of the machine decreases at a constant rate, so that at the end of 10 years the book value is zero.
a. Express the book value of the machine V as a function of the age, in years, of the machine n.
b. Sketch the graph of the function in part (a).
c. Find the book value of the machine at the end of the sixth year.
d. Find the rate at which the machine is being depreciated each year.

74. LINEAR DEPRECIATION Refer to Exercise 73. An office building worth \$1 million when completed in 1997 was depreciated linearly over 50 years. What was the book value of the building in 2012? What will be the book value in 2016? In 2021? (Assume that the book value of the building will be zero at the end of the 50th year.)

75. TAX REFUND FRAUD Incidents of tax refund fraud resulting from identity theft have been rising dramatically since 2010. The number of incidents (in millions) in year t from 2009 ($t = 0$) through 2012 ($t = 3$) is approximated by the function

$$f(t) = 0.2t^2 - 0.14t + 0.46 \qquad (0 \le t \le 3)$$

a. How many incidents occurred in 2009?
b. Assuming that the trend continued, how many incidents occurred in 2013?
Source: IRS.

76. GROWTH OF CRUISE INDUSTRY The number of North American cruise ship passengers has been increasing over the last several decades. The number of cruise ship passengers from 1995 through 2010 is approximated by the function

$$N(t) = 0.011t^2 + 0.521t + 4.6 \qquad (0 \le t \le 15)$$

where $N(t)$ is the number of passengers (in millions) in year t, with $t = 0$ corresponding to 1995. What was the approximate number of passengers in 1995? In 2010?
Source: Cruise Lines International Association.

77. MALE LIFE EXPECTANCY Advances in medical science and healthier lifestyles have resulted in longer life expectancies. The life expectancy of a male whose current age is x years old is

$$f(x) = 0.0069502x^2 - 1.6357x + 93.76 \qquad (60 \le x \le 75)$$

years. What is the life expectancy of a male whose current age is 65? A male whose current age is 75?
Source: Commissioners' Standard Ordinary Mortality Table.

78. CANCER SURVIVORS The number of living Americans who have had a cancer diagnosis has increased drastically since the 1970s. In part, this is due to more testing for cancer and better treatment for some cancers. In part, it is because the population is older, and cancer is largely a disease of the elderly. The number of cancer survivors (in thousands) between 2000 ($t = 0$) and 2012 ($t = 12$) is approximately

$$N(t) = 0.00445t^2 + 0.2903t + 9.564 \qquad (0 \le t \le 12)$$

a. How many living Americans had a cancer diagnosis in 2000? In 2012?
b. Assuming that the trend continued, how many cancer survivors were there in 2014?
Source: National Cancer Institute.

79. ALZHEIMER'S DISEASE The projected number of people aged 65 and over in the U.S. population with Alzheimer's disease (in millions) is given by the function

$$P(t) = -0.0002083t^3 + 0.0157t^2 - 0.093t + 5.2 \\ (4 \le t \le 40)$$

where t is measured in years, with $t = 4$ corresponding to 2014. What is the projected number of people aged 65 and over with Alzheimer's disease in 2030? In 2050?
Source: Alzheimer's Association.

80. WORKER EFFICIENCY An efficiency study conducted for Elektra Electronics showed that the number of Space Commander walkie-talkies assembled by the average worker t hr after starting work at 8 A.M. is given by

$$N(t) = -t^3 + 6t^2 + 15t \qquad (0 \le t \le 4)$$

How many walkie-talkies can an average worker be expected to assemble between 8 and 9 A.M.? Between 9 and 10 A.M.?

81. POLITICS Political scientists have discovered the following empirical rule, known as the "cube rule," which gives the relationship between the proportion of seats in the House of Representatives won by Democratic candidates $s(x)$ and the proportion of popular votes x received by the Democratic presidential candidate:

$$s(x) = \frac{x^3}{x^3 + (1-x)^3} \qquad (0 \le x \le 1)$$

Compute $s(0.6)$ and interpret your result.

82. U.S. HEALTH-CARE INFORMATION TECHNOLOGY SPENDING As health-care costs increase, payers are turning to technology and outsourced services to keep a lid on expenses. The amount of health-care information technology (IT) spending by payer is projected to be

$$S(t) = -0.03t^3 + 0.2t^2 + 0.23t + 5.6 \qquad (0 \le t \le 4)$$

where $S(t)$ is measured in billions of dollars and t is measured in years, with $t = 0$ corresponding to 2004. What was the amount spent by payers on health-care IT in 2004? Assuming that the projection held true, what amount was spent by payers in 2008?
Source: U.S. Department of Commerce.

83. BOYLE'S LAW As a consequence of Boyle's Law, the pressure P of a fixed sample of gas held at a constant temperature is related to the volume V of the gas by the rule

$$P = f(V) = \frac{k}{V}$$

where k is a constant. What is the domain of the function f? Sketch the graph of the function f.

84. POISEUILLE'S LAW According to a law discovered by the nineteenth century physician Poiseuille, the velocity (in centimeters/second) of blood r cm from the central axis of an artery is given by

$$v(r) = k(R^2 - r^2)$$

where k is a constant and R is the radius of the artery. Suppose that for a certain artery, $k = 1000$ and $R = 0.2$, so that $v(r) = 1000(0.04 - r^2)$.
a. What is the domain of the function v?
b. Compute $v(0)$, $v(0.1)$, and $v(0.2)$ and interpret your results.
c. Sketch the graph of the function v on the interval $[0, 0.2]$.

d. What can you say about the velocity of blood as we move away from the central axis toward the artery wall?

85. INVESTMENTS IN HEDGE FUNDS Investments in hedge funds have increased along with their popularity. The assets of hedge funds (in trillions of dollars) from 2002 through 2007 are modeled by the function

$$f(t) = \begin{cases} 0.6 & \text{if } 0 \le t < 1 \\ 0.6t^{0.43} & \text{if } 1 \le t \le 5 \end{cases}$$

where t is measured in years, with $t = 0$ corresponding to the beginning of 2002.
a. What were the assets in hedge funds at the beginning of 2002? At the beginning of 2003?
b. What were the assets in hedge funds at the beginning of 2005? At the beginning of 2007?
Source: Hennessee Group.

86. HOTEL RATES The average daily rate of U.S. hotels from 2006 through 2009 is approximated by the function

$$f(t) = \begin{cases} 0.88t^2 + 3.21t + 96.75 & \text{if } 0 \le t < 2 \\ -5.58t + 117.85 & \text{if } 2 \le t \le 3 \end{cases}$$

where $f(t)$ is measured in dollars and $t = 0$ corresponds to 2006.
a. What was the average daily rate of U.S. hotels in 2006? In 2007? In 2008?
b. Sketch the graph of f.
Source: Smith Travel Research.

87. POSTAL REGULATIONS In 2014, the postage for parcels sent by first-class mail was raised to \$2.32 for any parcel weighing less than 3 oz or fraction thereof and 18¢ for each additional ounce or fraction thereof. Any parcel not exceeding 13 oz may be sent by first-class mail. Letting x denote the weight of a parcel in ounces and $f(x)$ the postage in dollars, complete the following description of the "postage function" f:

$$f(x) = \begin{cases} \$2.32 & \text{if } 0 < x < 4 \\ \$2.50 & \text{if } 4 \le x < 5 \\ \vdots & \\ ? & \text{if } x = 13 \end{cases}$$

a. What is the domain of f?
b. Sketch the graph of f.

88. RISING MEDIAN AGE Increased longevity and the aging of the baby boom generation—those born between 1946 and 1965—are the primary reasons for a rising median age. The median age (in years) of the U.S. population from 1900 through 2011 is approximated by the function

$$f(t) = \begin{cases} 1.3t + 22.9 & \text{if } 0 \le t \le 3 \\ -0.7t^2 + 7.2t + 11.5 & \text{if } 3 < t \le 7 \\ 2.6t + 9.4 & \text{if } 7 < t \le 11 \end{cases}$$

where t is measured in decades, with $t = 0$ corresponding to the beginning of 1900.

a. What was the median age of the U.S. population at the beginning of 1900? At the beginning of 1950? At the beginning of 2000?

b. Sketch the graph of f.

Source: U.S. Census Bureau.

89. DISTANCE BETWEEN TWO SHIPS A passenger ship leaves port sailing east at 14 mph. Two hours later, a cargo ship leaves the same port heading north at 10 mph.

a. Find a function giving the distance between the two ships t hr after the passenger ship leaves port.

b. How far apart are the two ships 3 hr after the cargo ship leaves port?

In Exercises 90–98, determine whether the statement is true or false. If it is true, explain why it is true. If it is false, give an example to show why it is false.

90. If $a = b$, then $f(a) = f(b)$.

91. If $f(a) = f(b)$, then $a = b$.

92. If f is a function, then $f(a + b) = f(a) + f(b)$.

93. A vertical line must intersect the graph of $y = f(x)$ at exactly one point.

94. The domain of $f(x) = \sqrt{x + 2} + \sqrt{2 - x}$ is $[-2, 2]$.

95. If f is a function defined on $(-\infty, \infty)$ and k is a real number, then $f(kx) = kf(x)$.

96. If f is a linear function, then $f(cx + y) = cf(x) + f(y)$, where c is a real number.

97. The functions f and g defined by $f(x) = x + 1$ and $g(x) = \dfrac{x^2 + x}{x}$, respectively, are equal for all values of x.

98. The rule

$$R(x) = \begin{cases} -x & \text{if } -1 \le x \le 0 \\ 1 & \text{if } 0 \le x \le 1 \end{cases}$$

defines a function on the interval $[-1, 1]$.

2.1 Solutions to Self-Check Exercises

1. a. The expression under the radical sign must be nonnegative, so $x + 1 \ge 0$ or $x \ge -1$. Also, $x \ne 0$ because division by zero is not permitted. Therefore, the domain of f is $[-1, 0) \cup (0, \infty)$.

b. $f(3) = \dfrac{\sqrt{3 + 1}}{3} = \dfrac{\sqrt{4}}{3} = \dfrac{2}{3}$

c. $f(a + h) = \dfrac{\sqrt{(a + h) + 1}}{a + h} = \dfrac{\sqrt{a + h + 1}}{a + h}$

2. a. The function f is defined in a piecewise fashion. For $x = 0$, the rule is $f(x) = -x + 1$, and so $f(0) = 1$. For $x = 2$, the rule is $f(x) = \sqrt{x - 1}$, and so $f(2) = \sqrt{2 - 1} = 1$.

b. In the subdomain $[-1, 1)$, the graph of f is the line segment $y = -x + 1$, which is a linear equation with slope -1 and y-intercept 1. In the subdomain $[1, 5]$, the graph of f is given by the rule $f(x) = \sqrt{x - 1}$. From the table below,

x	1	2	3	4	5
$f(x)$	0	1	$\sqrt{2}$	$\sqrt{3}$	2

we obtain the following graph of f.

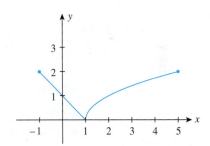

3. A point (x, y) lies on the graph of the function f if and only if the coordinates satisfy the equation $y = f(x)$. Now,

$$f(4) = \sqrt{2(4) + 1} + 2 = \sqrt{9} + 2 = 5 \ne 6$$

and we conclude that the given point does *not* lie on the graph of f.

USING TECHNOLOGY   Graphing a Function

Most of the graphs of functions in this book can be plotted with the help of a graphing utility. Furthermore, a graphing utility can be used to analyze the nature of a function. However, the amount and accuracy of the information obtained by using a graphing utility depend on the experience and sophistication of the user. As you progress through this book, you will see that the more knowledge of calculus you gain, the more effective the graphing utility will prove to be as a tool in problem solving.

Finding a Suitable Viewing Window

The first step in plotting the graph of a function with a graphing utility is to select a suitable viewing window. We usually do this by experimenting. For example, you might first plot the graph using the *standard viewing window* $[-10, 10]$ by $[-10, 10]$. If necessary, you then might adjust the viewing window by enlarging it or reducing it to obtain a sufficiently complete view of the graph or at least the portion of the graph that is of interest.

EXAMPLE 1 Plot the graph of $f(x) = 2x^2 - 4x - 5$ in the standard viewing window.

Solution The graph of f, shown in Figure T1a, is a parabola. From our previous work (Example 6, Section 2.1), we know that the figure does give a good view of the graph.

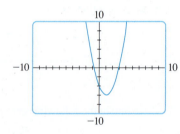

(a)

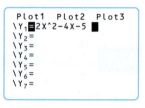

WINDOW
$X_{min} = -10$
$X_{max} = 10$
$X_{scl} = 1$
$Y_{min} = -10$
$Y_{max} = 10$
$Y_{scl} = 1$
$X_{res} = 1$

(b)

Plot1 Plot2 Plot3
$\backslash Y_1 = 2X^2-4X-5$
$\backslash Y_2 =$
$\backslash Y_3 =$
$\backslash Y_4 =$
$\backslash Y_5 =$
$\backslash Y_6 =$
$\backslash Y_7 =$

(c)

FIGURE **T1**
(a) The graph of $f(x) = 2x^2 - 4x - 5$ on $[-10, 10] \times [-10, 10]$; (b) the TI-83/84 window screen for (a); (c) the TI-83/84 equation screen

EXAMPLE 2 Let $f(x) = x^3(x - 3)^4$.

a. Plot the graph of f in the standard viewing window.
b. Plot the graph of f in the window $[-1, 5] \times [-40, 40]$.

Solution

a. The graph of f in the standard viewing window is shown in Figure T2a. Since the graph does not appear to be complete, we need to adjust the viewing window.

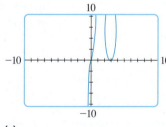

(a)

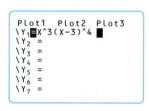

Plot1 Plot2 Plot3
$\backslash Y_1 = X^3(X-3)^4$
$\backslash Y_2 =$
$\backslash Y_3 =$
$\backslash Y_4 =$
$\backslash Y_5 =$
$\backslash Y_6 =$
$\backslash Y_7 =$

(b)

FIGURE **T2**
(a) An incomplete sketch of $f(x) = x^3(x - 3)^4$ on $[-10, 10] \times [-10, 10]$;
(b) the TI-83/84 equation screen

b. The graph of f in the window $[-1, 5] \times [-40, 40]$, shown in Figure T3a, is an improvement over the previous graph. (Later we will be able to show that the figure does in fact give a rather complete view of the graph of f.)

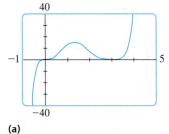

(a)   (b)

FIGURE **T3**
(a) A complete sketch of $f(x) = x^3(x - 3)^4$ is shown using the window
$[-1, 5] \times [-40, 40]$; (b) the TI-83/84 window screen

Evaluating a Function

A graphing utility can be used to find the value of a function with minimal effort, as the next example shows.

EXAMPLE 3 Let $f(x) = x^3 - 4x^2 + 4x + 2$.

a. Plot the graph of f in the standard viewing window.
b. Find $f(3)$ and verify your result by direct computation.
c. Find $f(4.215)$.

Solution

a. The graph of f is shown in Figure T4a.

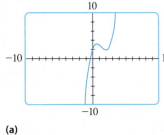   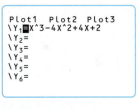

(a)   (b)

FIGURE **T4**
(a) The graph of $f(x) = x^3 - 4x^2 + 4x + 2$ in the standard viewing window;
(b) the TI-83/84 equation screen

b. Using the evaluation function of the graphing utility and the value 3 for x, we find $y = 5$. This result is verified by computing

$$f(3) = 3^3 - 4(3^2) + 4(3) + 2 = 27 - 36 + 12 + 2 = 5$$

c. Using the evaluation function of the graphing utility and the value 4.215 for x, we find $y = 22.679738$. Thus, $f(4.215) = 22.679738$. The efficacy of the graphing utility is clearly demonstrated here!

 APPLIED EXAMPLE 4 Alzheimer's Patients in the United States The number of Alzheimer's patients in the United States is approximated by

$$f(t) = -0.208t^3 + 1.571t^2 - 0.9274t + 5.1 \qquad (0 \le t \le 4)$$

where $f(t)$ is measured in millions and t is measured in decades, with $t = 0$ corresponding to the beginning of 2010.

a. Use a graphing utility to plot the graph of f in the viewing window $[0, 4] \times [0, 14]$.
b. What is the projected number of Alzheimer's patients in the United States at the beginning of 2040 $(t = 3)$?
Source: Alzheimer's Association.

Solution

a. The graph of f in the viewing window $[0, 4] \times [0, 14]$ is shown in Figure T5a.

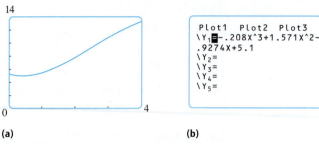

14

0   4

(a)   (b)

FIGURE **T5**
(a) The graph of f in the viewing window $[0, 4] \times [0, 14]$; (b) the TI-83/84 equation screen

b. Using the evaluation function of the graphing utility and the value 3 for x, we see that the anticipated number of Alzheimer's patients at the beginning of 2040 is given by $f(3) \approx 10.84$, or approximately 10.8 million.

TECHNOLOGY EXERCISES

In Exercises 1–4, plot the graph of the function f in (a) the standard viewing window and (b) the indicated window.

1. $f(x) = x^4 - 2x^2 + 8; [-2, 2] \times [6, 10]$

2. $f(x) = x^3 - 20x^2 + 8x - 10; [-20, 20] \times [-1200, 100]$

3. $f(x) = x\sqrt{4 - x^2}; [-3, 3] \times [-2, 2]$

4. $f(x) = \dfrac{4}{x^2 - 8}; [-5, 5] \times [-5, 5]$

In Exercises 5–8, plot the graph of the function f in an appropriate viewing window. (*Note:* The answer is *not* unique.)

5. $f(x) = 2x^4 - 3x^3 + 5x^2 - 20x + 40$

6. $f(x) = -2x^4 + 5x^2 - 4$

7. $f(x) = \dfrac{x^3}{x^3 + 1}$ **8.** $f(x) = \dfrac{2x^4 - 3x}{x^2 - 1}$

In Exercises 9–12, use the evaluation function of your graphing utility to find the value of f at the indicated value of x. Express your answer accurate to four decimal places.

9. $f(x) = 3x^3 - 2x^2 + x - 4; x = 2.145$

10. $f(x) = 5x^4 - 2x^2 + 8x - 3; x = 1.28$

11. $f(x) = \dfrac{2x^3 - 3x + 1}{3x - 2}; x = 2.41$

12. $f(x) = \sqrt{2x^2 + 1} + \sqrt{3x^2 - 1}; x = 0.62$

13. LOBBYISTS' SPENDING Lobbyists try to persuade legislators to propose, pass, or defeat legislation or to change existing laws. The amount (in billions of dollars) spent by lobbyists from 2003 through 2009, where $t = 0$ corresponds to 2003, is given by

$$f(t) = -0.0056t^3 + 0.112t^2 + 0.51t + 8 \quad (0 \le t \le 6)$$

a. Plot the graph of f in the viewing window $[0, 6] \times [0, 15]$.

b. What amount was spent by lobbyists in the year 2005? In 2009?

Source: OpenSecrets.org.

14. SAFE DRIVERS The fatality rate in the United States (per 100 million miles traveled) by age of driver (in years) is given by the function

$$f(x) = 0.00000304x^4 - 0.0005764x^3 + 0.04105x^2 - 1.30366x + 16.579 \quad (18 \le x \le 82)$$

a. Plot the graph of f in the viewing window $[18, 82] \times [0, 8]$.

b. What is the fatality rate for 18-year-old drivers? For 50-year-old drivers? For 80-year-old drivers?

Source: National Highway Traffic Safety Administration.

15. KEEPING WITH THE TRAFFIC FLOW By driving at a speed to match the prevailing traffic speed, you decrease the chances of an accident. According to data obtained in a university study, the number of accidents per 100 million vehicle miles, y, is related to the deviation from the mean speed, x, in miles per hour by

$$y = 1.05x^3 - 21.95x^2 + 155.9x - 327.3 \quad (6 \le x \le 11)$$

a. Plot the graph of y in the viewing window $[6, 11] \times [20, 150]$.

b. What is the number of accidents per 100 million vehicle miles if the deviation from the mean speed is 6 mph, 8 mph, and 11 mph?

Source: University of Virginia School of Engineering and Applied Science.

2.2   The Algebra of Functions

The Sum, Difference, Product, and Quotient of Functions

Let $S(t)$ and $R(t)$ denote the federal government's spending and revenue, respectively, at any time t, measured in billions of dollars. The graphs of these functions for the period between 2006 and 2012 are shown in Figure 11.

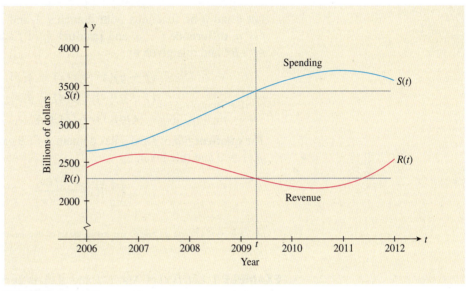

FIGURE 11
$R(t) - S(t)$ gives the federal budget deficit (surplus) at any time t.

Source: Office of Management and Budget.

The difference $R(t) - S(t)$ gives the deficit (surplus) in billions of dollars at any time t if $R(t) - S(t)$ is negative (positive). This observation suggests that we can define a function D whose value at any time t is given by $R(t) - S(t)$. The function D, the *difference* of the two functions R and S, is written $D = R - S$ and may be called the "deficit (surplus) function," since it gives the budget deficit or surplus at any time t. It has the same domain as the functions S and R. The graph of the function D is shown in Figure 12.

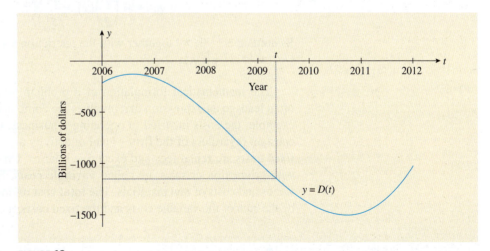

FIGURE 12
The graph of $D(t)$

Source: Office of Management and Budget.

Most functions are built up from other, generally simpler, functions. For example, we may view the function $f(x) = 2x + 4$ as the sum of the two functions $g(x) = 2x$ and $h(x) = 4$. The function $g(x) = 2x$ may in turn be viewed as the product of the functions $p(x) = 2$ and $q(x) = x$.

In general, given the functions f and g, we define the sum $f + g$, the difference $f - g$, the product fg, and the quotient f/g of f and g as follows.

The Sum, Difference, Product, and Quotient of Functions

Let f and g be functions with domains A and B, respectively. Then the **sum** $f + g$, **difference** $f - g$, and **product** fg of f and g are functions with domain $A \cap B$* and rule given by

$$(f + g)(x) = f(x) + g(x) \qquad \text{Sum}$$
$$(f - g)(x) = f(x) - g(x) \qquad \text{Difference}$$
$$(fg)(x) = f(x)g(x) \qquad \text{Product}$$

The **quotient** f/g of f and g has domain $A \cap B$ excluding all numbers x such that $g(x) = 0$ and rule given by

$$\left(\frac{f}{g}\right)(x) = \frac{f(x)}{g(x)} \qquad \text{Quotient}$$

---

*$A \cap B$ is read "A intersected with B" and denotes the set of all points common to both A and B.

EXAMPLE 1 Let $f(x) = \sqrt{x + 1}$ and $g(x) = 2x + 1$. Find the sum s, the difference d, the product p, and the quotient q of the functions f and g.

Solution   Since the domain of f is $A = [-1, \infty)$ and the domain of g is $B = (-\infty, \infty)$, we see that the domain of s, d, and p is $A \cap B = [-1, \infty)$. The rules follow.

$$s(x) = (f + g)(x) = f(x) + g(x) = \sqrt{x + 1} + 2x + 1$$
$$d(x) = (f - g)(x) = f(x) - g(x) = \sqrt{x + 1} - (2x + 1) = \sqrt{x + 1} - 2x - 1$$
$$p(x) = (fg)(x) = f(x)g(x) = \sqrt{x + 1}(2x + 1) = (2x + 1)\sqrt{x + 1}$$

The rule for the quotient function q is

$$q(x) = \left(\frac{f}{g}\right)(x) = \frac{f(x)}{g(x)} = \frac{\sqrt{x + 1}}{2x + 1}$$

Its domain is $[-1, \infty)$ together with the restriction $x \neq -\frac{1}{2}$. We denote this by $\left[-1, -\frac{1}{2}\right) \cup \left(-\frac{1}{2}, \infty\right)$.

The mathematical formulation of a problem arising from a practical situation often leads to an expression that involves the combination of functions. Consider, for example, the costs incurred in operating a business. Costs that remain more or less constant regardless of the firm's level of activity are called **fixed costs.** Examples of fixed costs are rental fees and executive salaries. On the other hand, costs that vary with production or sales are called **variable costs.** Examples of variable costs are wages and costs of raw materials. The **total cost** of operating a business is thus given by the *sum* of the variable costs and the fixed costs, as illustrated in the next example.

 APPLIED EXAMPLE 2 Cost Functions Suppose Puritron, a manufacturer of water filters, has a monthly fixed cost of $10,000 and a variable cost of

$$-0.0001x^2 + 10x \qquad (0 \leq x \leq 40,000)$$

dollars, where x denotes the number of filters manufactured per month. Find a function C that gives the total monthly cost incurred by Puritron in the manufacture of x filters.

Solution   Puritron's monthly fixed cost is always $10,000, regardless of the level of production, and it is described by the constant function $F(x) = 10,000$. Next, the variable cost is described by the function $V(x) = -0.0001x^2 + 10x$. Since the total cost incurred by Puritron at any level of production is the sum of the variable cost and the fixed cost, we see that the required total cost function is given by

$$C(x) = V(x) + F(x)$$
$$= -0.0001x^2 + 10x + 10,000 \qquad (0 \le x \le 40,000)$$

Next, the **total profit** realized by a firm in operating a business is the *difference* between the total revenue realized and the total cost incurred; that is,

$$P(x) = R(x) - C(x)$$

APPLIED EXAMPLE 3 Profit Functions Refer to Example 2. Suppose the total revenue in dollars realized by Puritron from the sale of x water filters per month is given by the total revenue function

$$R(x) = -0.0005x^2 + 20x \qquad (0 \le x \le 40,000)$$

a. Find the total profit function—that is, the function that describes the total profit Puritron realizes in manufacturing and selling x water filters per month.
b. What is the profit when the level of production is 10,000 filters per month?

Solution

a. The total profit realized by Puritron in manufacturing and selling x water filters per month is the difference between the total revenue realized and the total cost incurred. Thus, the required total profit function is given by

$$P(x) = R(x) - C(x)$$
$$= (-0.0005x^2 + 20x) - (-0.0001x^2 + 10x + 10,000)$$
$$= -0.0004x^2 + 10x - 10,000$$

b. The profit realized by Puritron when the level of production is 10,000 filters per month is

$$P(10,000) = -0.0004(10,000)^2 + 10(10,000) - 10,000 = 50,000$$

or $50,000 per month.

Composition of Functions

Another way to build up a function from other functions is through a process known as the *composition of functions*. Consider, for example, the function h, whose rule is given by $h(x) = \sqrt{x^2 - 1}$. Let f and g be functions defined by the rules $f(x) = x^2 - 1$ and $g(x) = \sqrt{x}$. Evaluating the function g at the point $f(x)$ [remember that for each real number x in the domain of f, $f(x)$ is simply a real number], we find that

$$g(f(x)) = \sqrt{f(x)} = \sqrt{x^2 - 1}$$

which is just the rule defining the function h!

In general, the composition of a function g with a function f is defined as follows.

> ### The Composition of Two Functions
>
> Let f and g be functions. Then the composition of g and f is the function $g \circ f$ defined by
>
> $$(g \circ f)(x) = g(f(x))$$
>
> The domain of $g \circ f$ is the set of all x in the domain of f such that $f(x)$ lies in the domain of g.

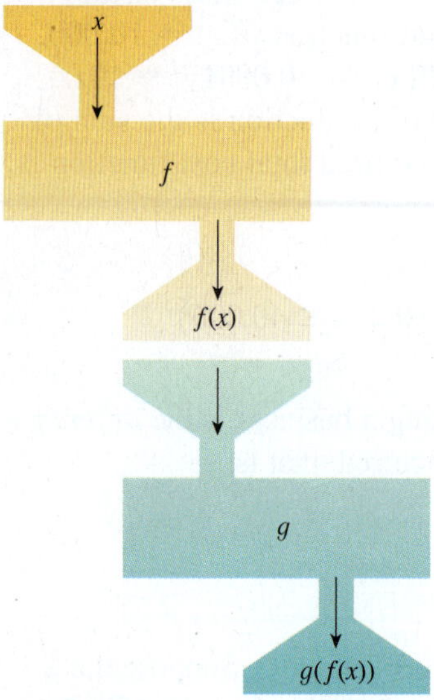

The function $g \circ f$ (read "g circle f") is also called a **composite function.** The interpretation of the function $h = g \circ f$ as a machine is illustrated in Figure 13, and its interpretation as a mapping is shown in Figure 14.

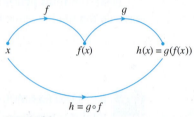

FIGURE 14
The function $h = g \circ f$ viewed as a mapping

FIGURE 13
The composite function $h = g \circ f$ viewed as a machine

EXAMPLE 4 Let $f(x) = x^2 - 1$ and $g(x) = \sqrt{x} + 1$. Find:

a. The rule for the composite function $g \circ f$.
b. The rule for the composite function $f \circ g$.

Solution

a. To find the rule for the composite function $g \circ f$, evaluate the function g at $f(x)$. We obtain

$$(g \circ f)(x) = g(f(x)) = \sqrt{f(x)} + 1 = \sqrt{x^2 - 1} + 1$$

b. To find the rule for the composite function $f \circ g$, evaluate the function f at $g(x)$. Thus,

$$(f \circ g)(x) = f(g(x)) = (g(x))^2 - 1 = (\sqrt{x} + 1)^2 - 1$$
$$= x + 2\sqrt{x} + 1 - 1 = x + 2\sqrt{x}$$

⚠ Example 4 shows us that in general $g \circ f$ is different from $f \circ g$, so care must be taken in finding the rule for a composite function.

> *Explore and Discuss*
>
> Let $f(x) = \sqrt{x} + 1$ for $x \geq 0$, and let $g(x) = (x - 1)^2$ for $x \geq 1$.
>
> **1.** Show that $(g \circ f)(x)$ and $(f \circ g)(x) = x$. (*Note:* The function g is said to be the *inverse* of f and vice versa.)
> **2.** Plot the graphs of f and g together with the straight line $y = x$. Describe the relationship between the graphs of f and g.

 APPLIED EXAMPLE 5 Automobile Pollution An environmental impact study conducted for the city of Oxnard indicates that under existing environmental protection laws, the level of carbon monoxide (CO) present in the air due to pollution from automobile exhaust will be $0.01x^{2/3}$ parts per million when the number of motor vehicles is x thousand. A separate study conducted by a state government agency estimates that t years from now, the number of motor vehicles in Oxnard will be $0.2t^2 + 4t + 64$ thousand.

a. Find an expression for the concentration of CO in the air due to automobile exhaust t years from now.

b. What will be the level of concentration 5 years from now?

Solution

a. The level of CO present in the air due to pollution from automobile exhaust is described by the function $g(x) = 0.01x^{2/3}$, where x is the number (in thousands) of motor vehicles. But the number of motor vehicles x (in thousands) t years from now may be estimated by the rule $f(t) = 0.2t^2 + 4t + 64$. Therefore, the concentration of CO due to automobile exhaust t years from now is given by

$$C(t) = (g \circ f)(t) = g(f(t)) = 0.01(0.2t^2 + 4t + 64)^{2/3}$$

parts per million.

b. The level of concentration 5 years from now will be

$$C(5) = 0.01[0.2(5)^2 + 4(5) + 64]^{2/3}$$
$$= (0.01)89^{2/3} \approx 0.20$$

or approximately 0.20 parts per million.

2.2 Self-Check Exercises

1. Let f and g be functions defined by the rules

$$f(x) = \sqrt{x} + 1 \quad \text{and} \quad g(x) = \frac{x}{1 + x}$$

respectively. Find the rules for
a. The sum s, the difference d, the product p, and the quotient q of f and g.
b. The composite functions $f \circ g$ and $g \circ f$.

2. **HEALTH-CARE SPENDING** Health-care spending per person by the private sector includes payments by individuals, corporations, and their insurance companies and is approximated by the function

$$f(t) = 2.48t^2 + 18.47t + 509 \quad (0 \le t \le 6)$$

where $f(t)$ is measured in dollars and t is measured in years, with $t = 0$ corresponding to the beginning of 1994.

The corresponding government spending—including expenditures for Medicaid, Medicare, and other federal, state, and local government public health care—is

$$g(t) = -1.12t^2 + 29.09t + 429 \quad (0 \le t \le 6)$$

where t has the same meaning as before.
a. Find a function that gives the difference between private and government health-care spending per person at any time t.
b. What was the difference between private and government expenditures per person at the beginning of 1995? At the beginning of 2000?

Source: Health Care Financing Administration.

Solutions to Self-Check Exercises 2.2 can be found on page 77.

2.2 Concept Questions

1. The figure opposite shows the graphs of a total cost function and a total revenue function. Let P, defined by $P(x) = R(x) - C(x)$, denote the total profit function.
a. Find an expression for $P(x_1)$. Explain its significance.
b. Find an expression for $P(x_2)$. Explain its significance.

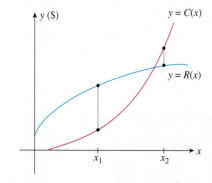

2. a. Explain what is meant by the sum, difference, product, and quotient of the functions f and g with domains A and B, respectively.

  b. If $f(2) = 3$ and $g(2) = -2$, what is $(f + g)(2)$? $(f - g)(2)$? $(fg)(2)$? $(f/g)(2)$?

3. Let f and g be functions, and suppose that (x, y) is a point on the graph of h. What is the value of y for $h = f + g$? $h = f - g$? $h = fg$? $h = f/g$?

4. a. What is the composition of the functions f and g? The functions g and f?

  b. If $f(2) = 3$ and $g(3) = 8$, what is $(g \circ f)(2)$? Can you conclude from the given information what $(f \circ g)(3)$ is? Explain.

5. Let f be a function with domain A, and let g be a function whose domain contains the range of f. If a is any number in A, must $(g \circ f)(a)$ be defined? Explain with an example.

6. The profit P (in dollars) of a one product company is given by $P = g(x)$, where x is the number of units sold. At the same time, the number of units of the product sold is given by $x = f(p)$, where p (in dollars) is the unit price of the product. Write an expression for the profit of the company in terms of the unit price it charges.

2.2 Exercises

In Exercises 1–8, let $f(x) = x^3 + 5$, $g(x) = x^2 - 2$, and $h(x) = 2x + 4$. Find the rule for each function.

1. $f + g$ **2.** $f - g$ **3.** fg **4.** gf

5. $\dfrac{f}{g}$ **6.** $\dfrac{f - g}{h}$ **7.** $\dfrac{fg}{h}$ **8.** fgh

In Exercises 9–18, let $f(x) = x - 1$, $g(x) = \sqrt{x + 1}$, and $h(x) = 2x^3 - 1$. Find the rule for each function.

9. $f + g$ **10.** $g - f$ **11.** fg **12.** gf

13. $\dfrac{g}{h}$ **14.** $\dfrac{h}{g}$ **15.** $\dfrac{fg}{h}$ **16.** $\dfrac{fh}{g}$

17. $\dfrac{f - h}{g}$ **18.** $\dfrac{gh}{g - f}$

In Exercises 19–24, find the functions $f + g$, $f - g$, fg, and f/g.

19. $f(x) = x^2 + 5$; $g(x) = \sqrt{x} - 2$

20. $f(x) = \sqrt{x - 1}$; $g(x) = x^3 + 1$

21. $f(x) = \sqrt{x + 3}$; $g(x) = \dfrac{1}{x - 1}$

22. $f(x) = \dfrac{1}{x^2 + 1}$; $g(x) = \dfrac{1}{x^2 - 1}$

23. $f(x) = \dfrac{x + 1}{x - 1}$; $g(x) = \dfrac{x + 2}{x - 2}$

24. $f(x) = x^2 + 1$; $g(x) = \sqrt{x + 1}$

In Exercises 25–30, find the rules for the composite functions $f \circ g$ and $g \circ f$.

25. $f(x) = x^2 + x + 1$; $g(x) = x^2$

26. $f(x) = 3x^2 + 2x + 1$; $g(x) = x + 3$

27. $f(x) = \sqrt{x} + 1$; $g(x) = x^2 - 1$

28. $f(x) = 2\sqrt{x} + 3$; $g(x) = x^2 + 1$

29. $f(x) = \dfrac{x}{x^2 + 1}$; $g(x) = \dfrac{1}{x}$

30. $f(x) = \sqrt{x + 1}$; $g(x) = \dfrac{1}{x - 1}$

In Exercises 31–34, evaluate $h(2)$, where $h = g \circ f$.

31. $f(x) = x^2 + x + 1$; $g(x) = x^2$

32. $f(x) = \sqrt[3]{x^2 - 1}$; $g(x) = 3x^3 + 1$

33. $f(x) = \dfrac{1}{2x + 1}$; $g(x) = \sqrt{x}$

34. $f(x) = \dfrac{1}{x - 1}$; $g(x) = x^2 + 1$

In Exercises 35–42, find functions f and g such that $h = g \circ f$. (Note: The answer is not unique.)

35. $h(x) = (2x^3 + x^2 + 1)^5$ **36.** $h(x) = (3x^2 - 4)^{-3}$

37. $h(x) = \sqrt{x^2 - 1}$ **38.** $h(x) = (2x - 3)^{3/2}$

39. $h(x) = \dfrac{1}{x^2 - 1}$ **40.** $h(x) = \dfrac{1}{\sqrt{x^2 - 4}}$

41. $h(x) = \dfrac{1}{(3x^2 + 2)^{3/2}}$ **42.** $h(x) = \dfrac{1}{\sqrt{2x + 1}} + \sqrt{2x + 1}$

In Exercises 43–46, find $f(a + h) - f(a)$ for each function. Simplify your answer.

43. $f(x) = 3x + 4$ **44.** $f(x) = -\dfrac{1}{2}x + 3$

45. $f(x) = 4 - x^2$ **46.** $f(x) = x^2 - 2x + 1$

In Exercises 47–52, find and simplify

$$\frac{f(a + h) - f(a)}{h} \qquad (h \neq 0)$$

for each function.

47. $f(x) = x^2 + 1$ **48.** $f(x) = 2x^2 - x + 1$

49. $f(x) = x^3 - x$

50. $f(x) = 2x^3 - x^2 + 1$

51. $f(x) = \dfrac{1}{x}$

52. $f(x) = \sqrt{x}$

53. RESTAURANT REVENUE Nicole owns and operates two restaurants. The revenue of the first restaurant at time t is $f(t)$ dollars, and the revenue of the second restaurant at time t is $g(t)$ dollars. What does the function $F(t) = f(t) + g(t)$ represent?

54. BIRTHRATE OF ENDANGERED SPECIES The birthrate of an endangered species of whales in year t is $f(t)$ whales/year. This species of whales is dying at the rate of $g(t)$ whales/year in year t. What does the function $F(t) = f(t) - g(t)$ represent?

55. VALUE OF AN INVESTMENT The number of IBM shares that Nancy owns is given by $f(t)$. The price per share of the stock of IBM at time t is $g(t)$ dollars. What does the function $f(t)g(t)$ represent?

56. PRODUCTION COSTS The total cost incurred by time t in the production of a certain commodity is $f(t)$ dollars. The number of products produced by time t is $g(t)$ units. What does the function $f(t)/g(t)$ represent?

57. CARBON MONOXIDE POLLUTION The number of cars running in the business district of a town at time t is given by $f(t)$. Carbon monoxide pollution coming from these cars is given by $g(x)$ parts per million, where x is the number of cars being operated in the district. What does the function $g \circ f$ represent?

58. EFFECT OF ADVERTISING ON REVENUE The revenue of Leisure Travel is given by $f(x)$ dollars, where x is the dollar amount spent by the company on advertising. The amount spent by Leisure at time t on advertising is given by $g(t)$ dollars. What does the function $f \circ g$ represent?

59. COST OF PRODUCING DVDs TMI, a manufacturer of blank DVDs, has a monthly fixed cost of $12,100 and a variable cost of $.60/disc. Find a function C that gives the total cost incurred by TMI in the manufacture of x discs/month.

60. BUSINESS EMAIL The average number of email messages sent and received per corporate user per day in year t between 2011 ($t = 1$) and 2015 ($t = 5$) is projected to be

$$f(t) = 3t + 69 \qquad (1 \le t \le 5)$$

The average number of spam emails sent per corporate user per day for the period under consideration is projected to be

$$g(t) = -0.2t + 13.8 \qquad (1 \le t \le 5)$$

a. Find a function, h, giving the projected average number of legitimate (non spam) emails sent and received per corporate user per day in year t.
  Hint: $h(t) = f(t) - g(t)$
b. Compute $f(5), g(5)$, and $h(5)$. Is $h(5) = f(5) - g(5)$?
Source: Radicati Group.

61. PUBLIC TRANSPORTATION BUDGET DEFICIT According to the Massachusetts Bay Transportation Authority (MBTA), the projected cumulative MBTA budget deficit with a $160 million rescue package (in billions of dollars) is given by

$$D_1(t) = 0.0275t^2 + 0.081t + 0.07 \qquad (0 \le t \le 3)$$

and the budget deficit without the rescue package is given by

$$D_2(t) = 0.035t^2 + 0.21t + 0.24 \qquad (0 \le t \le 3)$$

Find the function $D = D_2 - D_1$, and interpret your result.
Source: MBTA Review.

62. MOTORCYCLE DEATHS Suppose the fatality rate (deaths per 100 million miles traveled) of motorcyclists is given by $g(x)$, where x is the percentage of motorcyclists who wear helmets. Next, suppose the percentage of motorcyclists who wear helmets at time t (t measured in years) is $f(t)$, with $t = 0$ corresponding to 2000.
a. If $f(0) = 0.64$ and $g(0.64) = 26$, find $(g \circ f)(0)$ and interpret your result.
b. If $f(6) = 0.51$ and $g(0.51) = 42$, find $(g \circ f)(6)$ and interpret your result.
c. Comment on the results of parts (a) and (b).
Source: National Highway Traffic Safety Administration.

63. FIGHTING CRIME Suppose the reported serious crimes (crimes that include homicide, rape, robbery, aggravated assault, burglary, and car theft) that end in arrests or in the identification of suspects is $g(x)$ percent, where x denotes the total number of detectives. Next, suppose the total number of detectives in year t is $f(t)$, with $t = 0$ corresponding to 2001.
a. If $f(1) = 406$ and $g(406) = 23$, find $(g \circ f)(1)$ and interpret your result.
b. If $f(6) = 326$ and $g(326) = 18$, find $(g \circ f)(6)$ and interpret your result.
c. Comment on the results of parts (a) and (b).
Source: Boston Police Department.

64. PROFIT FROM SALE OF SMARTPHONES Apollo manufactures smartphones at a variable cost of

$$V(x) = 0.000003x^3 - 0.03x^2 + 200x$$

dollars, where x denotes the number of units manufactured per month. The monthly fixed cost attributable to the division that produces them is $100,000. The total revenue realized by Apollo from the sale of x smartphones is given by the total revenue function

$$R(x) = -0.1x^2 + 500x \qquad (0 \le x \le 5000)$$

where $R(x)$ is measured in dollars.
a. Find a function C that gives the total cost incurred by the manufacture of x smartphones.
b. Find the total profit function.
c. What is the profit when 1500 units are produced and sold each month?

65. PROFIT FROM SALE OF PAGERS A division of Chapman Corporation manufactures a pager. The weekly fixed cost for the division is $20,000, and the variable cost for producing x pagers/week is

$$V(x) = 0.000001x^3 - 0.01x^2 + 50x$$

dollars. The company realizes a revenue of

$$R(x) = -0.02x^2 + 150x \qquad (0 \le x \le 7500)$$

dollars from the sale of x pagers/week.
a. Find the total cost function.
b. Find the total profit function.
c. What is the profit for the company if 2000 units are produced and sold each week?

66. FEDERAL DEFICIT The spending by the federal government (in trillions of dollars) in year t from 2006 ($t = 0$) through 2012 ($t = 6$) is approximately

$$S(t) = -0.015278t^3 + 0.11179t^2 + 0.02516t + 2.64$$
$$(0 \le t \le 6)$$

and the revenue realized by the federal government over the same period is approximately

$$R(t) = 0.023611t^3 - 0.19679t^2 + 0.34365t + 2.42$$
$$(0 \le t \le 6)$$

a. Find a function D that gives the approximate deficit of the federal government in year t for $0 \le t \le 6$.
b. Find the spending, revenue, and deficit for the year 2009 ($t = 3$).
c. Is $D(3) = R(3) - S(3)$?
Source: Office of Management and Budget.

67. FAMILY INSURANCE COVERAGE The average annual worker and employer contributions (in dollars) to premiums for family insurance coverage from 2005 ($t = 1$) through 2011 ($t = 7$) are approximated by the functions,

$$f(t) = 4.389t^3 - 47.833t^2 + 374.49t + 2390 \quad (1 \le t \le 7)$$

and

$$g(t) = 13.222t^3 - 132.524t^2 + 757.9t + 7481 \quad (1 \le t \le 7)$$

respectively.
a. Find a function h giving the total premiums for family coverage.
b. Find the average annual worker contribution to premiums in 2010 ($t = 6$), the employer's contribution to premiums in 2010, and the total contributions to premiums.
c. Compute $h(6)$. Compare the result with the total contributions to premiums in 2010 obtained in part (b).
Source: Kaiser/HRET Survey of Employer-Sponsored Health Benefits.

68. EFFECT OF MORTGAGE RATES ON HOUSING STARTS A study prepared for the National Association of Realtors estimated

that the number of housing starts per year over the next 5 years will be

$$N(r) = \frac{7}{1 + 0.02r^2}$$

million units, where r (percent) is the mortgage rate. Suppose the mortgage rate t months from now will be

$$r(t) = \frac{5t + 75}{t + 10} \qquad (0 \le t \le 24)$$

percent/year.
a. Find an expression for the number of housing starts per year as a function of t, t months from now.
b. Using the result from part (a), determine the number of housing starts at present, 12 months from now, and 18 months from now.

69. HOTEL OCCUPANCY RATE The occupancy rate of the all-suite Wonderland Hotel, located near an amusement park, is given by the function

$$r(t) = \frac{10}{81}t^3 - \frac{10}{3}t^2 + \frac{200}{9}t + 55 \qquad (0 \le t \le 11)$$

percent, where t is measured in months and $t = 0$ corresponds to the beginning of January. Management has estimated that the monthly revenue (in thousands of dollars) is approximated by the function

$$R(r) = -\frac{3}{5000}r^3 + \frac{9}{50}r^2 \qquad (0 \le r \le 100)$$

where r (percent) is the occupancy rate.
a. What is the hotel's occupancy rate at the beginning of January? At the beginning of June?
b. What is the hotel's monthly revenue at the beginning of January? At the beginning of June?
Hint: Compute $R(r(0))$ and $R(r(5))$.

70. HOUSING STARTS AND CONSTRUCTION JOBS The president of a major housing construction firm reports that the number of construction jobs (in millions) created is given by

$$N(x) = 1.42x$$

where x denotes the number of housing starts. Suppose the number of housing starts in the next t months is expected to be

$$x(t) = \frac{7(t + 10)^2}{(t + 10)^2 + 2(t + 15)^2}$$

million units. Find an expression for the number of jobs created per year in the next t months. How many jobs per year will have been created 6 months and 12 months from now?

71. a. Let f, g, and h be functions. How would you define the "sum" of f, g, and h?
b. Give a real-life example involving the sum of three functions. (*Note:* The answer is not unique.)

72. a. Let f, g, and h be functions. How would you define the "composition" of h, g, and f, in that order?
    b. Give a real-life example involving the composition of these functions. (*Note:* The answer is not unique.)

In Exercises 73–78, determine whether the statement is true or false. If it is true, explain why it is true. If it is false, give an example to show why it is false.

73. If f and g are functions with domain D, then $f + g = g + f$.

74. If $g \circ f$ is defined at $x = a$, then $f \circ g$ must also be defined at $x = a$.

75. If f and g are functions, then $f \circ g = g \circ f$.

76. If f is a function, then $(f \circ f)(x) = [f(x)]^2$.

77. If f, g, and h are functions, then $h \circ (g \circ f) = (h \circ g) \circ f$.

78. If f, g, and h are functions, then
$$h \circ (g + f) = h \circ g + h \circ f.$$

2.2 Solutions to Self-Check Exercises

1. a. $s(x) = f(x) + g(x) = \sqrt{x} + 1 + \dfrac{x}{1 + x}$

$d(x) = f(x) - g(x) = \sqrt{x} + 1 - \dfrac{x}{1 + x}$

$p(x) = f(x)g(x) = (\sqrt{x} + 1) \cdot \dfrac{x}{1 + x} = \dfrac{x(\sqrt{x} + 1)}{1 + x}$

$q(x) = \dfrac{f(x)}{g(x)} = \dfrac{\sqrt{x} + 1}{\dfrac{x}{1 + x}} = \dfrac{(\sqrt{x} + 1)(1 + x)}{x}$

b. $(f \circ g)(x) = f(g(x)) = \sqrt{\dfrac{x}{1 + x} + 1}$

$(g \circ f)(x) = g(f(x)) = \dfrac{\sqrt{x} + 1}{1 + (\sqrt{x} + 1)} = \dfrac{\sqrt{x} + 1}{\sqrt{x} + 2}$

2. a. The difference between private and government health-care spending per person at any time t is given by the function d with the rule

$d(t) = f(t) - g(t) = (2.48t^2 + 18.47t + 509)$
$\qquad -(-1.12t^2 + 29.09t + 429)$
$\qquad = 3.6t^2 - 10.62t + 80$

b. The difference between private and government expenditures per person at the beginning of 1995 is given by

$$d(1) = 3.6(1)^2 - 10.62(1) + 80$$

or $72.98/person.
    The difference between private and government expenditures per person at the beginning of 2000 is given by

$$d(6) = 3.6(6)^2 - 10.62(6) + 80$$

or $145.88/person.

2.3 Functions and Mathematical Models

Mathematical Models

One of the fundamental goals in this book is to show how mathematics and, in particular, calculus can be used to solve real-world problems such as those arising from the world of business and the social, life, and physical sciences. You have already seen some of these problems earlier. Here are a few more examples of real-world phenomena that we will analyze in this and ensuing chapters.

- The erosion of the middle class (page 79)
- Global warming (page 81)
- The solvency of the U.S. Social Security trust fund (page 82)
- The growth of tablet and smartphone sales (page 170)
- The Case-Shiller Home Price Index (page 286)
- Salaries of married women (page 422)
- Social networks (page 422)

Regardless of the field from which a real-world problem is drawn, the problem is analyzed by using a process called **mathematical modeling.** The four steps in this process are illustrated in Figure 15.

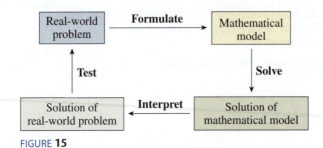

FIGURE **15**

1. **Formulate** Given a real-world problem, our first task is to formulate the problem, using the language of mathematics. The many techniques used in constructing mathematical models range from theoretical consideration of the problem on the one extreme to an interpretation of data associated with the problem on the other. For example, the mathematical model giving the accumulated amount at any time when a certain sum of money is deposited in the bank can be derived theoretically (see Chapter 5). On the other hand, many of the mathematical models in this book are constructed by studying the data associated with the problem (see Using Technology, pages 97–100). In calculus, we are primarily concerned with how one (dependent) variable depends on one or more (independent) variables. Consequently, most of our mathematical models will involve functions of one or more variables or equations defining these functions (implicitly).

2. **Solve** Once a mathematical model has been constructed, we can use the appropriate mathematical techniques, which we will develop throughout the book, to solve the problem.

3. **Interpret** Bearing in mind that the solution obtained in Step 2 is just the solution of the mathematical model, we need to interpret these results in the context of the original real-world problem.

4. **Test** Some mathematical models of real-world applications describe the situations with complete accuracy. For example, the model describing a deposit in a bank account gives the exact accumulated amount in the account at any time. But other mathematical models give, at best, an approximate description of the real-world problem. In this case, we need to test the accuracy of the model by observing how well it describes the original real-world problem and how well it predicts past and/ or future behavior. If the results are unsatisfactory, then we may have to reconsider the assumptions made in the construction of the model or, in the worst case, return to Step 1.

Many real-world phenomena, including those mentioned at the beginning of this section, are modeled by an appropriate function.

In what follows, we will recall some familiar functions and give examples of real-world phenomena that are modeled by using these functions.

Polynomial Functions

A **polynomial function** of degree n is a function of the form

$$f(x) = a_n x^n + a_{n-1} x^{n-1} + \cdots + a_2 x^2 + a_1 x + a_0 \qquad (a_n \neq 0)$$

where n is a nonnegative integer and the numbers $a_0, a_1, \ldots, a_n$ are constants, called the **coefficients** of the polynomial function. For example, the functions

$$f(x) = 2x^5 - 3x^4 + \frac{1}{2}x^3 + \sqrt{2}x^2 - 6$$

and

$$g(x) = 0.001x^3 - 0.2x^2 + 10x + 200$$

are polynomial functions of degrees 5 and 3, respectively. Observe that a polynomial function is defined for every value of x and so its domain is $(-\infty, \infty)$.

A polynomial function of degree 1 $(n = 1)$ has the form

$$y = f(x) = a_1 x + a_0 \qquad (a_1 \neq 0)$$

and is an equation of a straight line in the slope-intercept form with slope $m = a_1$ and y-intercept $b = a_0$ (see Section 1.4). For this reason, a polynomial function of degree 1 is called a **linear function.**

Linear functions are used extensively in mathematical modeling for two important reasons. First, some models are *linear* by nature. For example, the formula for converting temperature from Celsius (°C) to Fahrenheit (°F) is $F = \frac{9}{5}C + 32$, and F is a linear function of C. Second, some natural phenomena exhibit linear characteristics over a small range of values and can therefore be modeled by a linear function restricted to a small interval.

The following example uses a linear function to model the percentage of middle-income adults in the United States from 1971 through 2011. In Section 8.4, we will show how this model is constructed using the *least-squares technique*. In Using Technology on pages 97–100, you will be asked to use a graphing calculator to construct other mathematical models from raw data.

APPLIED EXAMPLE 1 Erosion of the Middle Class The idea of a large, stable middle class (defined as those with annual household incomes in 2010 between $39,000 and $118,000 for a family of three) is central to America's sense of itself. The following table gives the percentage of middle-income adults in the United States from 1971 through 2011:

Year	1971	1981	1991	2001	2011
Percent, y	61	59	56	54	51

A mathematical model giving the percentage of middle-class adults in the United States for the period under consideration is given by

$$f(t) = -2.5t + 61.2 \qquad (0 \leq t \leq 4)$$

where t is measured in decades, with $t = 0$ corresponding to 1971.

a. Plot the data points, and sketch the graph of the function f on the same set of axes.
b. What is the rate of change of the percentage of middle-income adults in the United States over the period from 1971 through 2011?
c. Assuming that the trend continues, what will the percentage of middle-income adults in the United States be in 2021?
Source: Pew Research Center.

Solution

a. The graph of f is shown in Figure 16.

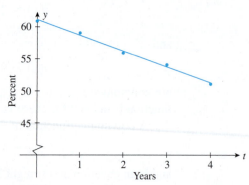

FIGURE 16
The percentage of middle-income adults in the United States from 1971 through 2011

b. The rate of change of the percentage of middle-income adults in the United States over the period from 1971 through 2011 is -2.5% per decade, that is, a decline of 2.5% per decade.

c. The projected percentage of middle-income adults in the United States in 2021 is

$$f(5) = -2.5(5) + 61.2 = 48.7$$

or 48.7%.

A polynomial function of degree 2 has the form

$$y = f(x) = a_2x^2 + a_1x + a_0 \qquad (a_2 \neq 0)$$

or, more simply, $y = ax^2 + bx + c$, and is called a **quadratic function**. The graph of a quadratic function is a parabola (see Figure 17).

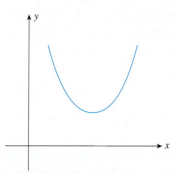

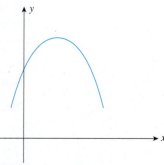

(a) If $a > 0$, the parabola opens upward.

(b) If $a < 0$, the parabola opens downward.

FIGURE 17
The graph of a quadratic function is a parabola.

The parabola opens upward if $a > 0$ and downward if $a < 0$. To see this, we rewrite the equation for y obtaining

$$f(x) = ax^2 + bx + c = x^2\left(a + \frac{b}{x} + \frac{c}{x^2}\right) \qquad (x \neq 0)$$

Observe that if x is large in absolute value, then the expression inside the parentheses is close to a, so $f(x)$ behaves like ax^2 for large values of x. Thus, $y = f(x)$ is large and positive if $a > 0$ (the parabola opens upward) and is large in magnitude and negative if $a < 0$ (the parabola opens downward).

Quadratic functions serve as mathematical models for many phenomena, as Example 2 shows.

APPLIED EXAMPLE 2 Global Warming The increase in carbon dioxide (CO_2) in the atmosphere is a major cause of global warming. The Keeling curve, named after Charles David Keeling, a professor at Scripps Institution of Oceanography, gives the average amount of CO_2, measured in parts per million volume (ppmv), in the atmosphere from 1958 through 2013. Even though data were available for every year in this time interval, we'll construct the curve based only on the following randomly selected data points.

Year	1958	1970	1974	1978	1985	1991	1998	2003	2007	2010	2013
Amount CO_2	315	325	330	335	345	355	365	375	380	390	395

The **scatter plot** associated with these data is shown in Figure 18a. A mathematical model giving the approximate amount of CO_2 in the atmosphere during this period is given by

$$A(t) = 0.012444t^2 + 0.7485t + 313.9 \qquad (1 \leq t \leq 56)$$

where t is measured in years, with $t = 1$ corresponding to 1958. The graph of A is shown in Figure 18b.

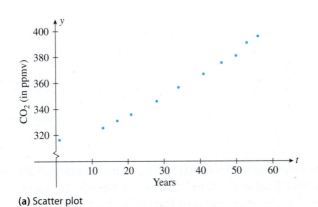

(a) Scatter plot

(b) The graph of A superimposed upon the scatter plot

FIGURE 18

a. Use the model to estimate the average amount of atmospheric CO_2 in 1980 ($t = 23$).
b. Assume that the trend continued, and use the model to predict the average amount of atmospheric CO_2 in 2016 ($t = 59$).
Source: Scripps Institution of Oceanography.

Solution

a. The average amount of atmospheric carbon dioxide in 1980 is given by

$$A(23) = 0.012444(23)^2 + 0.7485(23) + 313.9 \approx 337.70$$

or approximately 338 ppmv.

b. Assuming that the trend continued, the average amount of atmospheric CO_2 in 2016 will be

$$A(59) = 0.012444(59)^2 + 0.7485(59) + 313.9 \approx 401.38$$

or approximately 401 ppmv.

The next example uses a polynomial of degree 4 to help us construct a model that describes the projected assets of the Social Security trust fund.

APPLIED EXAMPLE 3 Social Security Trust Fund Assets The projected assets of the Social Security trust fund (in trillions of dollars) from 2010 through 2033 are given in the following table.

Year	2010	2015	2020	2025	2030	2033
Assets	2.61	2.68	2.44	1.87	0.78	0

The scatter plot associated with these data are shown in Figure 19a, where $t = 0$ corresponds to 2010. A mathematical model giving the approximate value of the assets in the trust fund $A(t)$, in trillions of dollars, in year t is

$$A(t) = 0.000008140t^4 - 0.00043833t^3 - 0.0001305t^2 + 0.02202t + 2.612 \qquad (0 \le t \le 23)$$

The graph of $A(t)$ is shown in Figure 19b. (You will be asked to construct this model in Exercise 20, Using Technology Exercises 2.3.)

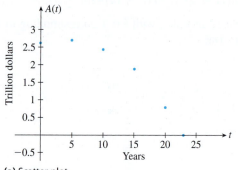

(a) Scatter plot

(b) The graph of A together with the scatter plot

FIGURE **19**

a. The first baby boomers turned 65 in 2011. What were the assets of the Social Security system trust fund at that time? The last of the baby boomers will turn 65 in 2029. What will the assets of the trust fund be at that time?

b. Unless payroll taxes are increased significantly and/or benefits are scaled back dramatically, it is a matter of time before the assets of the current system are depleted. Use the graph of the function $A(t)$ to estimate the year in which the current Social Security system is projected to go broke.

Source: Social Security Administration.

Solution

a. The assets of the Social Security trust fund in 2011 ($t = 1$) were

$$A(1) = 0.000008140(1)^4 - 0.00043833(1)^3 - 0.0001305(1)^2$$
$$+ 0.02202(1) + 2.612 \approx 2.633$$

or approximately \$2.63 trillion. The assets of the trust fund in 2029 ($t = 19$) will be

$$A(19) = 0.000008140(19)^4 - 0.00043833(19)^3 - 0.0001305(19)^2$$
$$+ 0.02202(19) + 2.612 \approx 1.038$$

or approximately \$1.04 trillion.

b. From Figure 19b, we see that the graph of A crosses the t-axis at approximately $t = 23$. So unless the current system is changed, it is projected to go broke in 2033. (At this time, the first of the baby boomers will be 87, and the last of the baby boomers will be 69.) ◼

Rational and Power Functions

Another important class of functions is rational functions. A **rational function** is simply the quotient of two polynomials. Examples of rational functions are

$$F(x) = \frac{3x^3 + x^2 - x + 1}{x - 2}$$

$$G(x) = \frac{x^2 + 1}{x^2 - 1}$$

In general, a rational function has the form

$$R(x) = \frac{f(x)}{g(x)}$$

where $f(x)$ and $g(x)$ are polynomial functions. Since division by zero is not allowed, we conclude that the domain of a rational function is the set of all real numbers except the zeros of g—that is, the roots of the equation $g(x) = 0$. Thus, the domain of the function F is the set of all numbers except $x = 2$, whereas the domain of the function G is the set of all numbers except those that satisfy $x^2 - 1 = 0$, or $x = \pm 1$.

Functions of the form

$$f(x) = x^r$$

where r is any real number, are called **power functions**. We encountered examples of power functions earlier in our work. For example, the functions

$$f(x) = \sqrt{x} = x^{1/2} \quad \text{and} \quad g(x) = \frac{1}{x^2} = x^{-2}$$

are power functions.

Many of the functions that we encounter later will involve combinations of the functions introduced here. For example, the following functions may be viewed as combinations of such functions:

$$f(x) = \sqrt{\frac{1 - x^2}{1 + x^2}}$$

$$g(x) = \sqrt{x^2 - 3x + 4}$$

$$h(x) = (1 + 2x)^{1/2} + \frac{1}{(x^2 + 2)^{3/2}}$$

As with polynomials of degree 3 or greater, analyzing the properties of these functions is facilitated by using the tools of calculus, to be developed later.

In the next example, we use a power function to construct a model that describes the driving costs of a car.

APPLIED EXAMPLE 4 Driving Costs A study of driving costs based on a 2012 medium-sized sedan found the following average costs (car payments, gas, insurance, upkeep, and depreciation), measured in cents per mile.

Miles/year	5000	10,000	15,000	20,000
Cost/mile (¢)	161.2	78.0	61.0	52.3

A mathematical model giving the average cost in cents per mile is

$$C(x) = \frac{1910.5}{x^{1.72}} + 42.9$$

where x (in thousands) denotes the number of miles the car is driven in each year. The scatter plot associated with these data and the graph of C are shown in Figure 20. Using this model, estimate the average cost per mile of driving a 2012 medium-sized sedan 8000 miles per year and 18,000 miles per year.

Source: American Automobile Association.

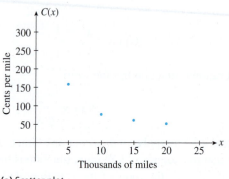

(a) Scatter plot

(b) The graph of the model for driving costs

FIGURE **20**

Solution   The average cost per mile for driving a car 8000 miles per year is

$$C(8) = \frac{1910.5}{8^{1.72}} + 42.9 \approx 96.3$$

or approximately 96.3 cents per mile. The average cost per mile for driving it 18,000 miles per year is

$$C(18) = \frac{1910.5}{18^{1.72}} + 42.9 \approx 56.1$$

or approximately 56.1 cents per mile.

Some Economic Models

In the remainder of this section, we look at some economic models.

In a free-market economy, consumer demand for a particular commodity depends on the commodity's unit price. A **demand equation** expresses the relationship between the unit price and the quantity demanded. The graph of the demand equation is called a **demand curve.** In general, the quantity demanded of a commodity decreases as the commodity's unit price increases, and vice versa. Accordingly, a **demand function** defined by $p = f(x)$, where p measures the unit price and x measures the number of units of the commodity in question, is generally characterized as a decreasing function of x; that is, $p = f(x)$ decreases as x increases. Since both x and p assume only nonnegative values, the demand curve is that part of the graph of $f(x)$ that lies in the first quadrant (Figure 21).

In a competitive market, a relationship also exists between the unit price of a commodity and the commodity's availability in the market. In general, an increase in the commodity's unit price induces the producer to increase the supply of the commodity. Conversely, a decrease in the unit price generally leads to a drop in the supply. The equation that expresses the relation between the unit price and the quantity supplied is called a **supply equation,** and its graph is called a **supply curve.** A **supply function** defined by $p = f(x)$ is generally characterized as an increasing function of x; that is, $p = f(x)$

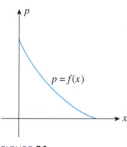

FIGURE **21**
A demand curve

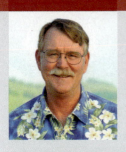

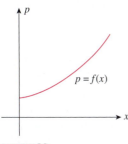

FIGURE 22
A supply curve

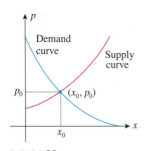

FIGURE 23
Market equilibrium corresponds to (x_0, p_0), the point at which the supply and demand curves intersect.

increases as x increases. Since both x and p assume only nonnegative values, the supply curve is that part of the graph of $f(x)$ that lies in the first quadrant (Figure 22).

Under pure competition, the price of a commodity will eventually settle at a level dictated by the following condition: The supply of the commodity will be equal to the demand for it. If the price is too high, the consumer will not buy; if the price is too low, the supplier will not produce. **Market equilibrium** prevails when the quantity produced is equal to the quantity demanded. The quantity produced at market equilibrium is called the **equilibrium quantity,** and the corresponding price is called the **equilibrium price.**

Market equilibrium corresponds to the point at which the demand curve and the supply curve intersect. In Figure 23, x_0 represents the equilibrium quantity, and p_0 represents the equilibrium price. The point (x_0, p_0) lies on the supply curve and therefore satisfies the supply equation. At the same time, it also lies on the demand curve and therefore satisfies the demand equation. Thus, to find the point (x_0, p_0), and hence the equilibrium quantity and price, we solve the demand and supply equations simultaneously for x and p. For meaningful solutions, x and p must both be positive.

 APPLIED EXAMPLE 5 Supply-Demand for Bluetooth Headsets The demand function for a certain brand of Bluetooth wireless headsets is given by

$$p = d(x) = -0.025x^2 - 0.5x + 60$$

and the corresponding supply function is given by

$$p = s(x) = 0.02x^2 + 0.6x + 20$$

where p is expressed in dollars and x is measured in units of a thousand. Find the equilibrium quantity and price.

Solution   We solve the following system of equations:

$$p = -0.025x^2 - 0.5x + 60$$
$$p = 0.02x^2 + 0.6x + 20$$

Substituting the first equation into the second yields

$$-0.025x^2 - 0.5x + 60 = 0.02x^2 + 0.6x + 20$$

which is equivalent to

$$0.045x^2 + 1.1x - 40 = 0$$
$$45x^2 + 1100x - 40{,}000 = 0 \qquad \text{Multiply by 1000.}$$
$$9x^2 + 220x - 8000 = 0 \qquad \text{Divide by 5.}$$
$$(9x + 400)(x - 20) = 0$$

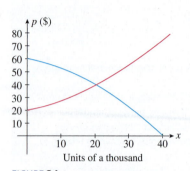

FIGURE 24
The supply curve and the demand curve intersect at the point (20, 40).

Thus, $x = -\frac{400}{9}$ or $x = 20$. Since x must be nonnegative, the root $x = -\frac{400}{9}$ is rejected. Therefore, the equilibrium quantity is 20,000 headsets. The equilibrium price is given by

$$p = 0.02(20)^2 + 0.6(20) + 20 = 40$$

or $40 per headset (Figure 24).

Exploring with TECHNOLOGY

1. **a.** Use a graphing utility to plot the straight lines L_1 and L_2 with equations $y = 2x - 1$ and $y = 2.1x + 3$, respectively, on the same set of axes, using the standard viewing window. Do the lines appear to intersect?
   b. Plot the straight lines L_1 and L_2, using the viewing window $[-100, 100] \times [-100, 100]$. Do the lines appear to intersect? Can you find the point of intersection using **TRACE** and **ZOOM**? Using the "intersection" function of your graphing utility?
   c. Find the point of intersection of L_1 and L_2 algebraically.
   d. Comment on the effectiveness of the methods of solutions in parts (b) and (c).

2. **a.** Use a graphing utility to plot the straight lines L_1 and L_2 with equations $y = 3x - 2$ and $y = -2x + 3$, respectively, on the same set of axes, using the standard viewing window. Then use **TRACE** and **ZOOM** to find the point of intersection of L_1 and L_2. Repeat using the "intersection" function of your graphing utility.
   b. Find the point of intersection of L_1 and L_2 algebraically.
   c. Comment on the effectiveness of the methods.

Constructing Mathematical Models

We close this section by showing how some mathematical models can be constructed by using elementary geometric and algebraic arguments.

The following guidelines can be used to construct mathematical models.

Guidelines for Constructing Mathematical Models

1. Assign a letter to each variable mentioned in the problem. If appropriate, draw and label a figure.
2. Find an expression for the quantity sought.
3. Use the conditions given in the problem to write the quantity sought as a function f of one variable. Note any restrictions to be placed on the domain of f from physical considerations of the problem.

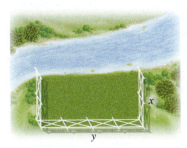

FIGURE 25
The rectangular grazing land has width x and length y.

 APPLIED EXAMPLE 6 Enclosing an Area The owner of Rancho Los Feliz has 3000 yards of fencing with which to enclose a rectangular piece of grazing land along the straight portion of a river. Fencing is not required along the river. Letting x denote the width of the rectangle, find a function f in the variable x giving the area of the grazing land if she uses all of the fencing (Figure 25).

Solution

1. This information was given.
2. The area of the rectangular grazing land is $A = xy$. Next, observe that the amount of fencing is $2x + y$ and this must be equal to 3000, since all the fencing is used; that is,

$$2x + y = 3000$$

3. From the equation, we see that $y = 3000 - 2x$. Substituting this value of y into the expression for A gives

$$A = xy = x(3000 - 2x) = 3000x - 2x^2$$

Finally, observe that both x and y must be nonnegative, since they represent the width and length of a rectangle, respectively. Thus, $x \geq 0$ and $y \geq 0$. But the latter is equivalent to $3000 - 2x \geq 0$, or $x \leq 1500$. So the required function is $f(x) = 3000x - 2x^2$ with domain $0 \leq x \leq 1500$.

Note   Observe that if we view the function $f(x) = 3000x - 2x^2$ strictly as a mathematical entity, then its domain is the set of all real numbers. But physical considerations dictate that its domain should be restricted to the interval $[0, 1500]$.

 APPLIED EXAMPLE 7 Charter-Flight Revenue If exactly 200 people sign up for a charter flight, Leisure World Travel Agency charges $300 per person. However, if more than 200 people sign up for the flight (assume that this is the case), then each fare is reduced by $1 for each additional person. Letting x denote the number of passengers beyond 200, find a function giving the revenue realized by the company.

Solution

1. This information was given.
2. If there are x passengers beyond 200, then the number of passengers signing up for the flight is $200 + x$. Furthermore, the fare will be $(300 - x)$ dollars per passenger.
3. The revenue will be

$$R = (200 + x)(300 - x) \quad \text{Number of passengers} \times$$
$$= -x^2 + 100x + 60{,}000 \quad \text{the fare per passenger}$$

Clearly, x must be nonnegative, and $300 - x \geq 0$, or $x \leq 300$. So the required function is $f(x) = -x^2 + 100x + 60{,}000$ with domain $[0, 300]$.

2.3 Self-Check Exercises

1. **CHILDREN'S DRUG DOSAGE** Thomas Young has suggested the following rule for calculating the dosage of medicine for children from 1 to 12 years of age. If a denotes the adult dosage (in milligrams) and t is the age of the child (in years), then the child's dosage is given by

$$D(t) = \frac{at}{t + 12}$$

If the adult dose of a substance is 500 mg, how much should a 4-year-old child receive?

2. **MARKET EQUILIBRIUM** The demand function for Mrs. Baker's cookies is given by

$$d(x) = -\frac{2}{15}x + 4$$

where $d(x)$ is the wholesale price in dollars per pound and x is the quantity demanded each week, measured in thousands of pounds. The supply function for the cookies is given by

$$s(x) = \frac{1}{75}x^2 + \frac{1}{10}x + \frac{3}{2}$$

where $s(x)$ is the wholesale price in dollars per pound and x is the quantity, in thousands of pounds, that will be made available in the market each week by the supplier.

a. Sketch the graphs of the functions d and s.
b. Find the equilibrium quantity and price.

Solutions to Self-Check Exercises 2.3 can be found on page 95.

2.3 Concept Questions

1. Describe mathematical modeling in your own words.

2. Define (a) a polynomial function and (b) a rational function. Give an example of each.

3. **a.** What is a demand function? A supply function?

   b. What is market equilibrium? Describe how you would go about finding the equilibrium quantity and equilibrium price given the demand and supply equations associated with a commodity.

2.3 Exercises

In Exercises 1–8, determine whether the equation defines y as a linear function of x. If so, write it in the form $y = mx + b$.

1. $2x + 3y = 6$
2. $-2x + 4y = 7$
3. $x = 2y - 4$
4. $2x = 3y + 8$
5. $2x - 4y + 9 = 0$
6. $3x - 6y + 7 = 0$
7. $2x^2 - 8y + 4 = 0$
8. $3\sqrt{x} + 4y = 0$

In Exercises 9–14, determine whether the function is a polynomial function, a rational function, or some other function. State the degree of each polynomial function.

9. $f(x) = 3x^6 - 2x^2 + 1$
10. $f(x) = \dfrac{x^2 - 9}{x - 3}$
11. $G(x) = 2(x^2 - 3)^3$
12. $H(x) = 2x^{-3} + 5x^{-2} + 6$
13. $f(t) = 2t^2 + 3\sqrt{t}$
14. $f(r) = \dfrac{6r}{r^3 - 8}$

15. Find the constants m and b in the linear function $f(x) = mx + b$ such that $f(0) = 2$ and $f(3) = -1$.

16. Find the constants m and b in the linear function $f(x) = mx + b$ such that $f(2) = 4$ and the straight line represented by f has slope -1.

17. A manufacturer has a monthly fixed cost of $40,000 and a production cost of $8 for each unit produced. The product sells for $12/unit.
   a. What is the cost function?
   b. What is the revenue function?
   c. What is the profit function?
   d. Compute the profit (loss) corresponding to production levels of 8000 and 12,000 units.

18. A manufacturer has a monthly fixed cost of $100,000 and a production cost of $14 for each unit produced. The product sells for $20/unit.
   a. What is the cost function?
   b. What is the revenue function?
   c. What is the profit function?
   d. Compute the profit (loss) corresponding to production levels of 12,000 and 20,000 units.

19. **DISPOSABLE INCOME** Economists define the *disposable annual income* for an individual by the equation $D = (1 - r)T$, where T is the individual's total income and r is the net rate at which he or she is taxed. What is the disposable income for an individual whose income is $60,000 and whose net tax rate is 28%?

20. **CHILDREN'S DRUG DOSAGES** A method sometimes used by pediatricians to calculate the dosage of medicine for children is based on the child's surface area. If a denotes the adult dosage (in milligrams) and S is the surface area of the child (in square meters), then the child's dosage is given by

$$D(S) = \frac{Sa}{1.7}$$

If the adult dose of a substance is 500 mg, how much should a child whose surface area is 0.4 m² receive?

21. Cowling's Rule Cowling's Rule is a method for calculating pediatric drug dosages. If a denotes the adult dosage (in milligrams) and t is the age of the child (in years), then the child's dosage is given by

$$D(t) = \left(\frac{t+1}{24}\right)a$$

If the adult dose of a substance is 500 mg, how much should a 4-year-old child receive?

22. Drinking and Driving Among High School Students The percentage of high school students who drink and drive stood at 17.5% at the beginning of 2001 and declined linearly to 10.3% at the beginning of 2011.
   a. Find a linear function $f(t)$ giving the percentage of high school students who drink and drive in year t, where $t = 0$ corresponds to the beginning of 2001.
   b. At what rate was the percentage of students who drink and drive dropping between 2001 and 2011?
   c. Assuming that the trend continued, what was the percentage of high school students who drink and drive at the beginning of 2014?
   Source: Centers for Disease Control and Prevention.

23. California Emissions Caps The California emissions cap is set at 400 million metric tons of carbon dioxide equivalent in 2015 and is expected to drop by 13.2 million metric tons of carbon dioxide equivalent per year through 2020.
   a. Find a linear function f giving the California emissions cap in year t, where $t = 0$ corresponds to 2015.
   b. If the same rate of decline of emissions cap is adopted through 2017, what will the emissions cap be in 2017?
   Source: California Air Resource Board.

24. U.S. Airplane Passenger Projections A report issued by the U.S. Department of Transportation in 2012 predicted that the number of passengers boarding planes in the United States would grow steadily from the current 0.7 billion boardings/year to 1.2 billion boardings/year in 2032.
   a. Find a linear function f giving the projected boardings (in billions) in year t, where $t = 0$ corresponds to 2012.
   b. What is the projected annual rate of growth of boardings between 2012 and 2032?
   c. How many boardings per year are projected in 2022?
   Source: U.S. Department of Transportation.

25. Bounced-Check Charges Overdraft fees have become an important part of a bank's total fee income. The following table gives the bank revenue from overdraft fees (in billions of dollars) from 2004 through 2009.

Year, t	0	1	2	3	4	5
Revenue, y	27.5	29	31	34	36	38

where t is measured in years, with $t = 0$ corresponding to 2004. A mathematical model giving the approximate

projected bank revenue from overdraft fees over the period under consideration is given by

$$f(t) = 2.19t + 27.12 \qquad (0 \le t \le 5)$$

   a. Plot the six data points and sketch the graph of the function f on the same set of axes.
   b. Assuming that the projection held and the trend continued, what was the projected bank revenue from overdraft fees in 2010 ($t = 6$)?
   c. What was the rate of increase of the bank revenue from overdraft fees over the period from 2004 through 2009?
   Source: New York Times.

26. Worker Efficiency An efficiency study showed that the average worker at Delphi Electronics assembled cordless telephones at the rate of

$$f(t) = -\frac{3}{2}t^2 + 6t + 10 \qquad (0 \le t \le 4)$$

phones/hr, t hr after starting work during the morning shift. At what rate does the average worker assemble telephones 2 hr after starting work?

27. Effect of Advertising on Sales The quarterly profit of Cunningham Realty depends on the amount of money x spent on advertising per quarter according to the rule

$$P(x) = -\frac{1}{8}x^2 + 7x + 30 \qquad (0 \le x \le 50)$$

where $P(x)$ and x are measured in thousands of dollars. What is Cunningham's profit when its quarterly advertising budget is $28,000?

28. Baby Boomers and Social Security Benefits Aging baby boomers will put a strain on Social Security benefits unless Congress takes action. The Social Security benefits to be paid out from 2010 through 2040 are projected to be

$$S(t) = 0.1375t^2 + 0.5185t + 0.72 \qquad (0 \le t \le 3)$$

where $S(t)$ is measured in trillions of dollars and t is measured in decades, with $t = 0$ corresponding to 2010.
   a. What was the amount of Social Security benefits paid out in 2010?
   b. What is the amount of Social Security benefits projected to be paid out in 2040?
   Source: Social Security and Medicare Trustees' 2010 report.

29. Mobile Device Usage The average time U.S. adults spent per day on mobile devices (in minutes) for the years 2009 through 2012 is approximated by

$$f(t) = 2.25t^2 + 13.41t + 21.76 \qquad (0 \le t \le 3)$$

where $t = 0$ corresponds to 2009.
   a. What was the average time U.S. adults spent per day on mobile devices in 2009?
   b. If the trend continued through 2013, what was the average time U.S. adults spent per day on mobile devices in 2013?
   Source: eMarketer.

30. U.S. GDP The gross domestic product (GDP) of the United States, in trillions of dollars, from 2011 through 2015 is approximately

$$G(t) = 0.064t^2 + 0.473t + 15.0 \qquad (0 \le t \le 4)$$

where t is measured in years, with $t = 0$ corresponding to 2011. In constructing this model, the government used actual GDP figures from 2011 and 2012 and estimates for the years 2013 through 2015.
a. What was the U.S. GDP in 2011?
b. What is the predicted U.S. GDP for 2015?
Source: World Bank.

31. SUB-SAHARAN AFRICAN GDP The real GDP per capita of sub-Saharan Africa (in 2009 U.S. dollars) from 1990 through 2030 is projected to be

$$f(t) = 1.86251t^2 - 28.08043t + 884 \qquad (0 \le t \le 40)$$

where t is measured in years, with $t = 0$ corresponding to 1990.
a. What was the real GDP per capita of sub-Saharan Africa in 2000?
b. Assuming that the projection holds true, what will be the GDP per capita of sub-Saharan Africa in 2030?
Source: IMF.

32. INSTANT MESSAGING ACCOUNTS The number of enterprise instant messaging (IM) accounts is projected to grow according to the function

$$N(t) = 2.96t^2 + 11.37t + 59.7 \qquad (0 \le t \le 5)$$

where $N(t)$ is measured in millions and t in years, with $t = 0$ corresponding to 2006.
a. How many enterprise IM accounts were there in 2006?
b. What was the expected number of enterprise IM accounts in 2010?
Source: The Radical Group.

33. SOLAR POWER More and more businesses and homeowners are installing solar panels on their roofs to draw energy from the sun's rays. According to the U.S. Department of Energy, the solar cell kilowatt-hour use in the United States (in millions) is projected to be

$$S(t) = 0.73t^2 + 15.8t + 2.7 \qquad (0 \le t \le 8)$$

in year t, with $t = 0$ corresponding to 2000. What was the number of projected solar cell kilowatt-hours used in the United States for 2006? For 2008?
Source: U.S. Department of Energy.

34. U.S. PUBLIC DEBT The U.S. public debt (the outstanding amount owed by the federal government of the United States from the issue of securities by the U.S. Treasury and other federal government agencies) for the years 2005 through 2011 is modeled by the function

$$f(t) = -0.03817t^3 + 0.4571t^2 - 0.1976t + 8.246 \qquad (0 \le t \le 7)$$

where $f(t)$ is measured in trillions of dollars and t is measured in years, with $t = 0$ corresponding to 2005. What was the U.S. public debt in 2005? In 2008?
Source: U.S. Department of the Treasury.

35. WORKERS' EXPECTATIONS The percentage of workers who expect to work past age 65 has more than tripled in 30 years. The function

$$f(t) = 0.004545t^3 - 0.1113t^2 + 1.385t + 11 \qquad (0 \le t \le 22)$$

gives an approximation of the percentage of workers who expect to work past age 65 in year t, where t is measured in years, with $t = 0$ corresponding to 1991. What was the percentage of workers who expected to work past age 65 in 1991? In 2013?
Source: PBS News.

36. AGING DRIVERS The number of fatalities due to car crashes, based on the number of miles driven, begins to climb after the driver is past age 65. Aside from declining ability as one ages, the older driver is more fragile. The number of fatalities per 100 million vehicle miles driven is approximately

$$N(x) = 0.0336x^3 - 0.118x^2 + 0.215x + 0.7 \qquad (0 \le x \le 7)$$

where x denotes the age group of drivers, with $x = 0$ corresponding to those aged 50–54, $x = 1$ corresponding to those aged 55–59, $x = 2$ corresponding to those aged 60–64, . . . , and $x = 7$ corresponding to those aged 85–89. What is the fatality rate per 100 million vehicle miles driven for an average driver in the 50–54 age group? In the 85–89 age group?
Source: U.S. Department of Transportation.

37. TOTAL GLOBAL MOBILE DATA TRAFFIC In a 2009 report, equipment maker Cisco forecast the total global mobile data traffic to be

$$f(t) = 0.021t^3 + 0.015t^2 + 0.12t + 0.06 \qquad (0 \le t \le 5)$$

million terabytes/month in year t, where $t = 0$ corresponds to 2009.
a. What was the total global mobile data traffic in 2009?
b. According to Cisco, what was the total global mobile data traffic in 2014?
Source: Cisco.

38. LEVERAGED RETURN Leanne is contemplating borrowing money from a bank to buy a bond returning 6%/year. The bank requires her to make a down payment of $D\%$ of the loan with the remaining $(100 - D)\%$ borrowed at an interest rate of 5%/year. Then the return on the bond using the borrowed money, called *leveraged return*, is given by

$$L = \frac{1 + 0.05D}{D}$$

If Leanne makes a down payment of 20% to secure the loan on the bond using borrowed money, what is the leveraged return?

Source: Scientific American.

39. **ONLINE VIDEO VIEWERS** As broadband Internet grows more popular, video services such as YouTube will continue to expand. The number of online video viewers (in millions) is projected to grow according to the rule

$$N(t) = 52t^{0.531} \qquad (1 \le t \le 10)$$

where $t = 1$ corresponds to 2003.
a. Sketch the graph of N.
b. How many online video viewers were there in 2012?

Source: eMarketer.com.

40. **INFANT MORTALITY RATES IN MASSACHUSETTS** The deaths of children younger than 1 year old per 1000 live births is modeled by the function

$$R(t) = 162.8t^{-3.025} \qquad (1 \le t \le 3)$$

where t is measured in 50-year intervals, with $t = 1$ corresponding to 1900.
a. Find $R(1)$, $R(2)$, and $R(3)$ and use your result to sketch the graph of the function R over the domain $[1, 3]$.
b. What was the infant mortality rate in 1900? In 1950? In 2000?

Source: Massachusetts Department of Public Health.

41. **OUTSOURCING OF JOBS** According to a study conducted in 2003, the total number of U.S. jobs (in millions) that are projected to leave the country by year t, where $t = 0$ corresponds to 2000, is

$$N(t) = 0.0018425(t + 5)^{2.5} \qquad (0 \le t \le 15)$$

What was the projected number of outsourced jobs for 2005 $(t = 5)$? For 2013 $(t = 13)$?

Source: Forrester Research.

42. **CHIP SALES** The worldwide sales of flash memory chips (in billions of dollars) is approximated by

$$S(t) = 4.3(t + 2)^{0.94} \qquad (0 \le t \le 6)$$

where t is measured in years, with $t = 0$ corresponding to 2002. Flash chips are used in cell phones, digital cameras, and other products.
a. What were the worldwide flash memory chip sales in 2002?
b. What were the estimated sales for 2010?

Source: Web-Feet Research, Inc.

43. **U.S. HEALTH-CARE COSTS** The U.S. health-care costs per capita (in dollars) from 2001 through 2011 can be approximated by the linear function $f(t) = at + b$, where t is measured in years, with $t = 1$ corresponding to 2001, and a and b are constants. The costs per capita in 2001 and 2011 were $5240 and $8680, respectively.
a. Find a and b.

b. Use the model obtained in part (a) to find the approximate per capita costs for 2005.

Source: Centers for Medicare and Medicaid Services.

44. **REACTION OF A FROG TO A DRUG** Experiments conducted by A. J. Clark suggest that the response $R(x)$ of a frog's heart muscle to the injection of x units of acetylcholine (as a percent of the maximum possible effect of the drug) may be approximated by the rational function

$$R(x) = \frac{100x}{b + x} \qquad (x \ge 0)$$

where b is a positive constant that depends on the particular frog.
a. If a concentration of 40 units of acetylcholine produces a response of 50% for a certain frog, find the "response function" for this frog.
b. Using the model found in part (a), find the response of the frog's heart muscle when 60 units of acetylcholine are administered.

45. **DIGITAL VERSUS FILM CAMERAS** The sales of digital cameras (in millions of units) in year t are given by the function

$$f(t) = 3.05t + 6.85 \qquad (0 \le t \le 3)$$

where $t = 0$ corresponds to 2001. Over that same period, the sales of film cameras (in millions of units) are given by

$$g(t) = -1.85t + 16.58 \qquad (0 \le t \le 3)$$

a. Show that more film cameras than digital cameras were sold in 2001.
b. When did the sales of digital cameras first exceed those of film cameras?

Source: Popular Science.

46. **WALKING VERSUS RUNNING** The oxygen consumption (in milliliter per pound per minute) for a person walking at x mph is approximated by the function

$$f(x) = \frac{5}{3}x^2 + \frac{5}{3}x + 10 \qquad (0 \le x \le 9)$$

whereas the oxygen consumption for a runner at x mph is approximated by the function

$$g(x) = 11x + 10 \qquad (4 \le x \le 9)$$

a. Sketch the graphs of f and g.
b. At what speed is the oxygen consumption the same for a walker as it is for a runner? What is the level of oxygen consumption at that speed?
c. What happens to the oxygen consumption of the walker and the runner at speeds beyond that found in part (b)?

Source: William McArdley, Frank Katch, and Victor Katch, Exercise Physiology.

47. **CRICKET CHIRPING AND TEMPERATURE** Entomologists have discovered that a linear relationship exists between the

number of chirps of crickets of a certain species and the air temperature. When the temperature is 70°F, the crickets chirp at the rate of 120 times/min, and when the temperature is 80°F, they chirp at the rate of 160 times/min.
a. Find an equation giving the relationship between the air temperature T and the number of chirps per minute, N, of the crickets.
b. Find N as a function of T, and use this formula to determine the rate at which the crickets chirp when the temperature is 102°F.

48. FARMERS MARKETS Farmers markets have been growing steadily over the years. The number of such markets in the United States from 2006 through 2012 can be modeled by using a quadratic function of the form

$$f(t) = at^2 + bt + c$$

where a, b, and c are constants and t is measured in years, with $t = 0$ corresponding to 2006.
a. Find a, b, and c if $f(0) = 3173$, $f(4) = 6132$, and $f(6) = 7864$.
b. Use the model obtained in part (a) to estimate the number of farmers markets in 2014, assuming that the trend continued.
Source: U.S. Department of Agriculture.

49. SMALL BREWERIES U.S. craft-beer breweries (breweries that make fewer than 6 million barrels annually and are less than 25% owned by big breweries) have been doing booming business. The number of these small breweries from 2008 through 2012 can be modeled by using a quadratic function of the form

$$f(t) = at^2 + bt + c$$

where a, b, and c are constants and t is measured in years, with $t = 0$ corresponding to 2008.
a. Find a, b, and c if $f(0) = 1547$, $f(2) = 1802$, and $f(4) = 2403$.
b. Use the model obtained in part (a) to estimate the number of craft-beer breweries in 2014, assuming that the trend continued.
Source: Breweries Association.

50. LINEAR DEPRECIATION OF AN ASSET In computing income tax, businesses are allowed by law to depreciate certain assets such as buildings, machines, furniture, and automobiles over a period of time. Linear depreciation, or the straight-line method, is often used for this purpose. Suppose an asset has an initial value of $\$C$ and is to be depreciated linearly over n years with a scrap value of $\$S$. Show that the book value of the asset at any time t $(0 \le t \le n)$ is given by the linear function

$$V(t) = C - \frac{C - S}{n}t$$

Hint: Find an equation of the straight line that passes through the points $(0, C)$ and (n, S). Then rewrite the equation in the slope-intercept form.

51. LINEAR DEPRECIATION OF A PRINTING MACHINE Using the linear depreciation model of Exercise 50, find the book value of a printing machine at the end of the second year if its initial value is $100,000 and it is depreciated linearly over 5 years with a scrap value of $30,000.

52. PRICE OF IVORY According to the World Wildlife Fund, a group in the forefront of the fight against illegal ivory trade, the price of ivory (in dollars per kilogram) compiled from a variety of legal and black market sources is approximated by the function

$$f(t) = \begin{cases} 8.37t + 7.44 & \text{if } 0 \le t \le 8 \\ 2.84t + 51.68 & \text{if } 8 < t \le 30 \end{cases}$$

where t is measured in years, with $t = 0$ corresponding to the beginning of 1970.
a. Sketch the graph of the function f.
b. What was the price of ivory at the beginning of 1970? At the beginning of 1990?
Source: World Wildlife Fund.

53. COST OF THE HEALTH-CARE BILL The Congressional Budget Office estimates that the health-care bill passed by the Senate in November 2009, combined with a package of revisions known as the reconciliation bill, will result in a cost by year t (in billions of dollars) of

$$f(t) = \begin{cases} 5 & \text{if } 0 \le t < 2 \\ -0.5278t^3 + 3.012t^2 + 49.23t - 103.29 & \text{if } 2 \le t \le 8 \end{cases}$$

where t is measured in years, with $t = 0$ corresponding to 2010. What will be the cost of the health-care bill by 2011? By 2015?
Source: U.S. Congressional Budget Office.

54. WORKING-AGE POPULATION The ratio of working-age population to the elderly in the United States (including projections after 2000) is given by

$$f(t) = \begin{cases} 4.1 & \text{if } 0 \le t < 5 \\ -0.03t + 4.25 & \text{if } 5 \le t < 15 \\ -0.075t + 4.925 & \text{if } 15 \le t \le 35 \end{cases}$$

with $t = 0$ corresponding to the beginning of 1995.
a. Sketch the graph of f.
b. What was the ratio at the beginning of 2005? What will the ratio be at the beginning of 2020?
c. Over what years is the ratio constant?
d. Over what years is the decline of the ratio greatest?
Source: U.S. Census Bureau.

55. SENIOR CITIZENS' HEALTH CARE According to a study, the out-of-pocket cost to senior citizens for health care, $f(t)$ (as a percentage of income), in year t, where $t = 0$ corresponds to 1977, is given by

$$f(t) = \begin{cases} \dfrac{2}{7}t + 12 & \text{if } 0 \le t \le 7 \\[2mm] t + 7 & \text{if } 7 < t \le 10 \\[2mm] \dfrac{1}{3}t + \dfrac{41}{3} & \text{if } 10 < t \le 25 \end{cases}$$

a. Sketch the graph of f.

b. What was the out-of-pocket cost, as a percentage of income, to senior citizens for health care in 1982? In 2002?

Source: Senate Select Committee on Aging, AARP.

56. SALES OF DVD PLAYERS VERSUS VCRS The sales of DVD players in year t (in millions of units) is given by the function

$$f(t) = 5.6(1 + t) \qquad (0 \le t \le 3)$$

where $t = 0$ corresponds to 2001. Over the same period, the sales of VCRs (in millions of units) is given by

$$g(t) = \begin{cases} -9.6t + 22.5 & \text{if } 0 \le t \le 1 \\ -0.5t + 13.4 & \text{if } 1 < t \le 2 \\ -7.8t + 28 & \text{if } 2 < t \le 3 \end{cases}$$

a. Show that more VCRs than DVD players were sold in 2001.

b. When did the sales of DVD players first exceed those of VCRs?

Source: Popular Science.

For the demand equations in Exercises 57–60, where x represents the quantity demanded in units of a thousand and p is the unit price in dollars, (a) sketch the demand curve and (b) determine the quantity demanded when the unit price is set at $\$p$.

57. $p = -x^2 + 16$; $p = 7$ **58.** $p = -x^2 + 36$; $p = 11$

59. $p = \sqrt{18 - x^2}$; $p = 3$ **60.** $p = \sqrt{9 - x^2}$; $p = 2$

For the supply equations in Exercises 61–64, where x is the quantity supplied in units of a thousand and p is the unit price in dollars, (a) sketch the supply curve and (b) determine the price at which the supplier will make 2000 units of the commodity available in the market.

61. $p = x^2 + 16x + 40$ **62.** $p = 2x^2 + 18$

63. $p = x^3 + 2x + 3$ **64.** $p = x^3 + x + 10$

65. DEMAND FOR SMOKE ALARMS The demand function for the Sentinel smoke alarm is given by

$$p = \frac{30}{0.02x^2 + 1} \qquad (0 \le x \le 10)$$

where x (measured in units of a thousand) is the quantity demanded per week and p is the unit price in dollars.

a. Sketch the graph of the demand function.

b. What is the unit price that corresponds to a quantity demanded of 10,000 units?

66. DEMAND FOR COMMODITIES Assume that the demand function for a certain commodity has the form

$$p = \sqrt{-ax^2 + b} \qquad (a \ge 0, b \ge 0)$$

where x is the quantity demanded, measured in units of a thousand, and p is the unit price in dollars. Suppose the quantity demanded is 6000 ($x = 6$) when the unit price is $\$8$ and 8000 ($x = 8$) when the unit price is $\$6$. Determine the demand equation. What is the quantity demanded when the unit price is set at $\$7.50$?

67. SUPPLY OF DESK LAMPS The supply function for the Luminar desk lamp is given by

$$p = 0.1x^2 + 0.5x + 15$$

where x is the quantity supplied (in thousands) and p is the unit price in dollars.

a. Sketch the graph of the supply function.

b. What unit price will induce the supplier to make 5000 lamps available in the marketplace?

68. SUPPLY OF SATELLITE RADIOS Suppliers of satellite radios will market 10,000 units when the unit price is $\$20$ and 62,500 units when the unit price is $\$35$. Determine the supply function if it is known to have the form

$$p = a\sqrt{x} + b \qquad (a > 0, b > 0)$$

where x is the quantity supplied and p is the unit price in dollars. Sketch the graph of the supply function. What unit price will induce the supplier to make 40,000 satellite radios available in the marketplace?

69. SUPPLY AND DEMAND EQUATIONS Suppose the demand and supply equations for a certain commodity are given by $p = ax + b$ and $p = cx + d$, respectively, where $a < 0$, $c > 0$, and $b > d > 0$ (see the figure below).

a. Find the equilibrium quantity and equilibrium price in terms of a, b, c, and d.

b. Use part (a) to determine what happens to the market equilibrium if c is increased while a, b, and d remain fixed. Interpret your answer in economic terms.

c. Use part (a) to determine what happens to the market equilibrium if b is decreased while a, c, and d remain fixed. Interpret your answer in economic terms.

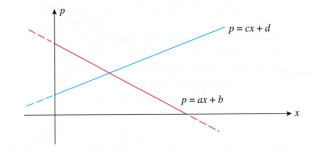

For each pair of supply and demand equations in Exercises 70–73, where x represents the quantity demanded in units of a thousand and p the unit price in dollars, find the equilibrium quantity and the equilibrium price.

70. $p = -x^2 - 2x + 100$ and $p = 8x + 25$

71. $p = -2x^2 + 80$ and $p = 15x + 30$

72. $p = 60 - 2x^2$ and $p = x^2 + 9x + 30$

73. $11p + 3x - 66 = 0$ and $2p^2 + p - x = 10$

74. Market Equilibrium The weekly demand and supply functions for Sportsman 5×7 tents are given by

$$p = -0.1x^2 - x + 40$$
$$p = 0.1x^2 + 2x + 20$$

respectively, where p is measured in dollars and x is measured in units of a hundred. Find the equilibrium quantity and price.

75. Market Equilibrium The management of Titan Tire Company has determined that the weekly demand and supply functions for their Super Titan tires are given by

$$p = 144 - x^2$$
$$p = 48 + \frac{1}{2}x^2$$

respectively, where p is measured in dollars and x is measured in units of a thousand. Find the equilibrium quantity and price.

76. Enclosing an Area Patricia wishes to have a rectangular garden in her backyard. She has 80 ft of fencing with which to enclose her garden. Letting x denote the width of the garden, find a function f in the variable x giving the area of the garden. What is its domain?

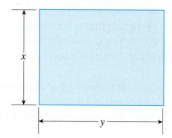

77. Enclosing an Area Juanita wishes to have a rectangular garden in her backyard with an area of 250 ft². Letting x denote the width of the garden, find a function f in the variable x giving the length of the fencing required to construct the garden. What is the domain of the function? **Hint:** Refer to the figure for Exercise 76. The amount of fencing required is equal to the perimeter of the rectangle, which is twice the width plus twice the length of the rectangle.

78. Packaging By cutting away identical squares from each corner of a rectangular piece of cardboard and folding up the resulting flaps, an open box can be made. If the cardboard is 15 in. long and 8 in. wide and the square cutaways have dimensions of x in. by x in., find a function giving the volume of the resulting box.

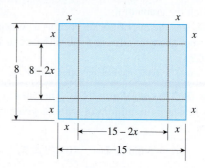

79. Packaging Costs A rectangular box is to have a square base and a volume of 20 ft³. The material for the base costs 30¢/ft², the material for the sides costs 10¢/ft², and the material for the top costs 20¢/ft². Letting x denote the length of one side of the base, find a function in the variable x giving the cost of constructing the box.

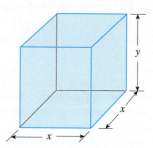

80. Area of a Norman Window A Norman window has the shape of a rectangle surmounted by a semicircle (see the accompanying figure). Suppose a Norman window is to have a perimeter of 28 ft. Find a function in the variable x giving the area of the window.

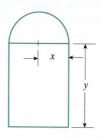

81. Yield of an Apple Orchard An apple orchard has an average yield of 36 bushels of apples per tree if tree density is 22 trees/acre. For each unit increase in tree density, the yield decreases by 2 bushels/tree. Letting x denote the number of trees beyond 22/acre, find a function in x that gives the yield of apples.

82. Book Design A book designer has decided that the pages of a book should have 1-in. margins at the top and bottom and $\frac{1}{2}$-in. margins on the sides. She further stipulated that each page should have a total area of 50 in.^{2}. Find a function in the variable x, giving the area of the printed part of the page. What is the domain of the function?

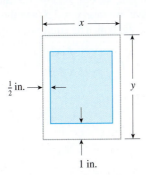

$\frac{1}{2}$ in.

1 in.

83. Profit of a Vineyard Phillip, the proprietor of a vineyard, estimates that if 10,000 bottles of wine were produced this season, then the profit would be $5/bottle. But if more than 10,000 bottles were produced, then the profit per bottle for the entire lot would drop by $0.0002 for each additional bottle sold. Assume that at least 10,000 bottles of wine are produced and sold, and let x denote the number of bottles produced and sold in excess of 10,000.
  a. Find a function P giving the profit in terms of x.
  b. What is the profit Phillip can expect from the sale of 16,000 bottles of wine from his vineyard?

84. Charter Revenue The owner of a luxury motor yacht that sails among the 4000 Greek islands charges $600/person per day if exactly 20 people sign up for the cruise. However, if more than 20 people sign up for the cruise (up to the maximum capacity of 90), the fare for all the passengers is reduced by $4/person for each additional passenger. Assume that at least 20 people sign up for the cruise, and let x denote the number of passengers above 20.
  a. Find a function R giving the revenue per day realized from the charter.
  b. What is the revenue per day if 60 people sign up for the cruise?
  c. What is the revenue per day if 80 people sign up for the cruise?

In Exercises 85–88, determine whether the statement is true or false. If it is true, explain why it is true. If it is false, give an example to show why it is false.

85. A polynomial function is a sum of constant multiples of power functions.

86. A polynomial function is a rational function, but the converse is false.

87. If $r > 0$, then the power function $f(x) = x^r$ is defined for all values of x.

88. The function $f(x) = 2^x$ is a power function.

<h2>2.3 Solutions to Self-Check Exercises</h2>

1. Since the adult dose of the substance is 500 mg, $a = 500$; thus, the rule in this case is

$$D(t) = \frac{500t}{t + 12}$$

A 4-year-old should receive

$$D(4) = \frac{500(4)}{4 + 12}$$

or 125 mg of the substance.

2. a. The graphs of the functions d and s are shown in the following figure:

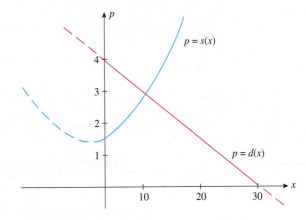

b. Solve the following system of equations:

$$p = -\frac{2}{15}x + 4$$

$$p = \frac{1}{75}x^2 + \frac{1}{10}x + \frac{3}{2}$$

Substituting the first equation into the second yields

$$\frac{1}{75}x^2 + \frac{1}{10}x + \frac{3}{2} = -\frac{2}{15}x + 4$$

$$\frac{1}{75}x^2 + \left(\frac{1}{10} + \frac{2}{15}\right)x - \frac{5}{2} = 0$$

$$\frac{1}{75}x^2 + \frac{7}{30}x - \frac{5}{2} = 0$$

Multiplying both sides of the last equation by 150, we have

$$2x^2 + 35x - 375 = 0$$

$$(2x - 15)(x + 25) = 0$$

Thus, $x = -25$ or $x = 15/2 = 7.5$. Since x must be nonnegative, we take $x = 7.5$, and the equilibrium quantity is 7500 lb. The equilibrium price is given by

$$p = -\frac{2}{15}\left(\frac{15}{2}\right) + 4$$

or $3/lb.

USING TECHNOLOGY

Finding the Points of Intersection of Two Graphs and Modeling

A graphing utility can be used to find the point(s) of intersection of the graphs of two functions.

EXAMPLE 1 Find the points of intersection of the graphs of

$$f(x) = 0.3x^2 - 1.4x - 3 \quad \text{and} \quad g(x) = -0.4x^2 + 0.8x + 6.4$$

Solution   The graphs of both f and g in the standard viewing window are shown in Figure T1a. Using the function for finding the points of intersection of two graphs on a graphing utility, we find the point(s) of intersection, accurate to four decimal places, to be $(-2.4158, 2.1329)$ (Figure T1b) and $(5.5587, -1.5125)$ (Figure T1c). To access this function on the TI-83/84, select **5: intersect** on the Calc menu.

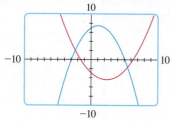
(a)

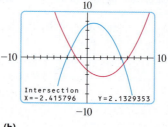
(b)

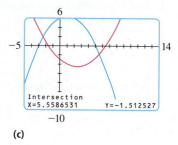
(c)

FIGURE **T1**
(a) The graphs of f and g in the standard viewing window; (b) and (c) the TI-83/84 intersection screens

EXAMPLE 2 Consider the demand and supply functions

$$p = d(x) = -0.01x^2 - 0.2x + 8 \quad \text{and} \quad p = s(x) = 0.01x^2 + 0.1x + 3$$

a. Plot the graphs of d and s in the viewing window $[0, 15] \times [0, 10]$.
b. Verify that the equilibrium point is $(10, 5)$.

Solution

a. The graphs of d and s are shown in Figure T2a.

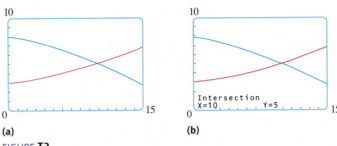

(a)   **(b)**

FIGURE **T2**
(a) The graphs of d and s in the window $[0, 15] \times [0, 10]$; (b) the TI-83/84 intersection screen

b. Using the function for finding the point of intersection of two graphs, we see that $x = 10$ and $y = 5$ (Figure T2b), so the equilibrium point is $(10, 5)$.

Constructing Mathematical Models from Raw Data

A graphing utility can sometimes be used to construct mathematical models from sets of data. For example, if the points corresponding to the given data are scattered about a straight line, then use **LinReg$(ax+b)$** (linear regression) from the statistical calculations menu of the graphing utility to obtain a function (model) that approximates the data at hand. If the points seem to be scattered along a parabola (the graph of a quadratic function), then use **QuadReg** (second-degree polynomial regression), and so on. (These are functions on the TI-83/84 calculator.)

APPLIED EXAMPLE 3 Indian Gaming Industry The following table gives the estimated gross revenues (in billions of dollars) from the Indian gaming industries from 2000 $(t = 0)$ to 2008 $(t = 8)$.

Year	0	1	2	3	4	5	6	7	8
Revenue	11.0	12.8	14.7	16.8	19.5	22.7	25.1	26.4	26.8

a. Use a graphing utility to find a polynomial function f of degree 4 that models the data.
b. Plot the graph of the function f, using the viewing window $[0, 8] \times [0, 30]$.
c. Use the function evaluation capability of the graphing utility to compute $f(0)$, $f(1), \ldots, f(8)$, and compare these values with the original data.
d. If the trend continued, what was the gross revenue for 2009 $(t = 9)$?
Source: National Indian Gaming Association.

Solution

a. First, enter the data using the statistical menu. Then choose **QuartReg** (fourth-degree polynomial regression) from the statistical calculations menu of a graphing utility. We find

$$f(t) = -0.00737t^4 + 0.0655t^3 - 0.008t^2 + 1.61t + 11$$

b. The graph of f is shown in Figure T3.

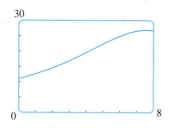

FIGURE T3
The graph of f in the viewing window
$[0, 8] \times [0, 30]$

c. The required values, which compare favorably with the given data, follow:

t	0	1	2	3	4	5	6	7	8
$f(t)$	11.0	12.7	14.6	16.9	19.6	22.4	25.0	26.6	26.7

d. The gross revenue for 2009 ($t = 9$) is given by

$$f(9) = -0.00737(9)^4 + 0.0655(9)^3 - 0.008(9)^2 + 1.61(9) + 11 \approx 24.24$$

or approximately \$24.2 billion.

TECHNOLOGY EXERCISES

In Exercises 1–6, find the points of intersection of the graphs of the functions. Express your answer accurate to four decimal places.

1. $f(x) = 1.2x + 3.8$; $g(x) = -0.4x^2 + 1.2x + 7.5$

2. $f(x) = 0.2x^2 - 1.3x - 3$; $g(x) = -1.3x + 2.8$

3. $f(x) = 0.3x^2 - 1.7x - 3.2$; $g(x) = -0.4x^2 + 0.9x + 6.7$

4. $f(x) = -0.3x^2 + 0.6x + 3.2$; $g(x) = 0.2x^2 - 1.2x - 4.8$

5. $f(x) = 0.3x^3 - 1.8x^2 + 2.1x - 2$; $g(x) = 2.1x - 4.2$

6. $f(x) = -0.2x^3 + 1.2x^2 - 1.2x + 2$; $g(x) = -0.2x^2 + 0.8x + 2.1$

7. MARKET EQUILIBRIUM FOR WALL CLOCKS The monthly demand and supply functions for a certain brand of wall clock are given by

$$p = -0.2x^2 - 1.2x + 50$$
$$p = \;\;\; 0.1x^2 + 3.2x + 25$$

respectively, where p is measured in dollars and x is measured in units of a hundred.
a. Plot the graphs of both functions in an appropriate viewing window.
b. Find the equilibrium quantity and price.

8. MARKET EQUILIBRIUM FOR DIGITAL CAMERAS The quantity demanded x (in units of a hundred) of Mikado digital cameras per week is related to the unit price p (in dollars) by

$$p = -0.2x^2 + 80$$

The quantity x (in units of a hundred) that the supplier is willing to make available in the market is related to the unit price p (in dollars) by

$$p = 0.1x^2 + x + 40$$

a. Plot the graphs of both functions in an appropriate viewing window.
b. Find the equilibrium quantity and price.

In Exercises 9–22, use the statistical calculations menu to construct a mathematical model associated with the given data.

9. CONSUMPTION OF BOTTLED WATER The annual per-capita consumption of bottled water (in gallons) and the scatter plot for these data follow:

Year	2001	2002	2003	2004	2005	2006
Consumption	18.8	20.9	22.4	24	26.1	28.3

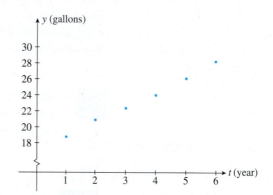

a. Use **LinReg($ax+b$)** to find a first-degree (linear) polynomial regression model for the data. Let $t = 1$ correspond to 2001.
b. Plot the graph of the function f found in part (a), using the viewing window $[1, 6] \times [0, 30]$.
c. Compute the values for $t = 1, 2, 3, 4, 5$, and 6. How do your figures compare with the given data?
d. If the trend continued, what was the annual per-capita consumption of bottled water in 2008?
Source: Beverage Marketing Corporation.

10. WEB CONFERENCING Web conferencing is a big business, and it's growing rapidly. The amount (in billions of dollars) spent on Web conferencing from the beginning of 2003 through 2010, and the scatter diagram for these data follow:

Year	2003	2004	2005	2006	2007	2008	2009	2010
Amount	0.50	0.63	0.78	0.92	1.16	1.38	1.60	1.90

y ($ billion)

a. Let *t* = 0 correspond to the beginning of 2003 and use **QuadReg** to find a second-degree polynomial regression model based on the given data.
b. Plot the graph of the function *f* found in part (a) using the window $[0, 7] \times [0, 2]$.
c. Compute $f(0), f(3), f(6)$, and $f(7)$. Compare these values with the given data.

Source: Gartner Dataquest.

11. **STUDENT POPULATION** The projected total number of students in elementary schools, secondary schools, and colleges (in millions) from 1995 through 2015 is given in the following table:

Year	1995	2000	2005	2010	2015
Number	64.8	68.7	72.6	74.8	78

a. Use **QuadReg** to find a second-degree polynomial regression model for the data. Let *t* be measured in 5-year intervals, with *t* = 0 corresponding to the beginning of 1995.
b. Plot the graph of the function *f* found in part (a), using the viewing window $[0, 4] \times [0, 85]$.
c. Using the model found in part (a), what will be the projected total number of students (all categories) enrolled in 2015?

Source: U.S. National Center for Education Statistics.

12. **MOBILE DEVICE USAGE** The average time U.S. adults spent per day on mobile devices (in minutes) for the years 2009 through 2012 is shown in the following table:

Year	2009	2010	2011	2012
Average Time Spent	22	36.7	58.3	82

a. Let *t* = 0 correspond to the beginning of 2009, and use **QuadReg** to find a second-degree polynomial regression model based on the given data.
b. Obtain the scatter plot and the graph of the function *f* found in part (a), using the viewing window $[0, 3] \times [0, 100]$.

Source: eMarketer.

13. **TIVO OWNERS** The projected number of households (in millions) with digital video recorders that allow viewers to record shows onto a server and skip commercials are given in the following table:

Year	2006	2007	2008	2009	2010
Households	31.2	49.0	71.6	97.0	130.2

a. Let *t* = 0 correspond to the beginning of 2006, and use **QuadReg** to find a second-degree polynomial regression model based on the given data.
b. Obtain the scatter plot and the graph of the function *f* found in part (a), using the viewing window $[0, 4] \times [0, 140]$.

Source: Strategy Analytics.

14. **U.S. PUBLIC DEBT** The U.S. public debt (the outstanding amount owed by the federal government of the U.S. from the issue of securities by the U.S. Treasury and other federal government agencies) for the years 2005 through 2011 (in trillions of dollars) is given in the following table:

Year	2005	2006	2007	2008	2009	2010	2011
Debt	8.170	8.680	9.229	10.700	12.311	14.025	15.223

a. Let *t* = 0 correspond to the beginning of 2005 and use **CubicReg** to find a third-degree polynomial regression model based on the given data.
b. Obtain the scatter plot and the graph of the function *f* found in part (a), using the viewing window $[0, 7] \times [0, 18]$.

Source: U.S. Department of the Treasury.

15. **TELECOMMUNICATIONS INDUSTRY REVENUE** Telecommunications industry revenue is expected to grow in the coming years, fueled by the demand for broadband and high-speed data services. The worldwide revenue for the industry (in trillions of dollars) and the scatter diagram for these data follow:

Year	2000	2002	2004	2006	2008	2010
Revenue	1.7	2.0	2.5	3.0	3.6	4.2

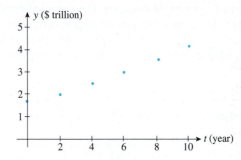

a. Let *t* = 0 correspond to the beginning of 2000 and use **CubicReg** to find a third-degree polynomial regression model based on the given data.
b. Plot the graph of the function *f* found in part (a), using the viewing window $[0, 10] \times [0, 5]$.
c. Find the worldwide revenue for the industry in 2001 and 2005 and find the projected revenue for 2010.

Source: Telecommunication Industry Association.

16. **POPULATION GROWTH IN CLARK COUNTY** Clark County in Nevada—dominated by greater Las Vegas—is one of the fastest-growing metropolitan areas in the United States. The population of the county from 1970 through 2000 is given in the following table:

Year	1970	1980	1990	2000
Population	273,288	463,087	741,459	1,375,765

a. Use **CubicReg** to find a third-degree polynomial regression model for the data. Let t be measured in decades, with $t = 0$ corresponding to the beginning of 1970.

b. Plot the graph of the function f found in part (a), using the viewing window $[0, 3] \times [0, 1,500,000]$.

c. Compare the values of f at $t = 0, 1, 2$, and 3, with the given data.

Source: U.S. Census Bureau.

17. **LOBBYISTS' SPENDING** Lobbyists try to persuade legislators to propose, pass, or defeat legislation or to change existing laws. The amount (in billions of dollars) spent by lobbyists from 2003 through 2009 is shown in the following table:

Year	2003	2004	2005	2006	2007	2008	2009
Amount	8.0	8.5	9.7	10.2	11.3	12.9	13.8

a. Use **CubicReg** to find a third-degree polynomial regression model for the data, letting $t = 0$ correspond to 2003.

b. Plot the scatter diagram and the graph of the function f found in part (a), using the viewing window $[0, 6] \times [0, 15]$.

c. Compare the values of f at $t = 0, 3$, and 6 with the given data.

Source: Center for Public Integrity.

18. **MOBILE ENTERPRISE IM ACCOUNTS** The projected number of mobile enterprise instant messaging (IM) accounts (in millions) from 2006 through 2010 is given in the following table ($t = 0$ corresponds to the beginning of 2006):

Year	0	1	2	3	4
Accounts	2.3	3.6	5.8	8.7	14.9

a. Use **CubicReg** to find a third-degree polynomial regression model based on the given data.

b. Plot the graph of the function f found in part (a), using the viewing window $[0, 5] \times [0, 16]$.

c. Compute $f(0), f(1), f(2), f(3)$, and $f(4)$.

Source: The Radical Group.

19. **NICOTINE CONTENT OF CIGARETTES** Even as measures to discourage smoking have been growing more stringent in recent years, the nicotine content of cigarettes has been rising, making it more difficult for smokers to quit. The following table gives the average amount of nicotine in cigarette smoke (in milligrams) from 1999 through 2004:

Year	1999	2000	2001	2002	2003	2004
Yield per Cigarette	1.71	1.81	1.85	1.84	1.83	1.89

a. Use **QuartReg** to find a fourth-degree polynomial regression model for the data. Let $t = 0$ correspond to the beginning of 1999.

b. Plot the graph of the function f found in part (a), using the viewing window $[0, 5] \times [0, 2]$.

c. Compute the values of $f(t)$ for $t = 0, 1, 2, 3, 4$, and 5.

d. If the trend continued, what was the average amount of nicotine in cigarettes in 2005?

Source: Massachusetts Tobacco Control Program.

20. **SOCIAL SECURITY TRUST FUND ASSETS** The projected assets of the Social Security trust fund (in trillions of dollars) from 2010 through 2033 are given in the following table:

Year	2010	2015	2020	2025	2030	2033
Assets	2.61	2.68	2.44	1.87	0.78	0

Use **QuartReg** to find a fourth-degree polynomial regression model for the data. Let $t = 0$ correspond to 2010.

Source: Social Security Administration.

Introduction to Calculus

Historically, the development of calculus by Isaac Newton (1642–1727) and Gottfried Wilhelm Leibniz (1646–1716) resulted from the investigation of the following problems:

1. Finding the tangent line to a curve at a given point on the curve (Figure 26a)
2. Finding the area of a planar region bounded by an arbitrary curve (Figure 26b)

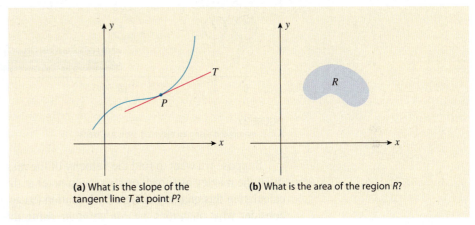

(a) What is the slope of the tangent line *T* at point *P*?

(b) What is the area of the region *R*?

FIGURE **26**

The tangent-line problem might appear to be unrelated to any practical applications of mathematics, but as you will see later, the problem of finding the *rate of change* of one quantity with respect to another is mathematically equivalent to the geometric problem of finding the slope of the *tangent line* to a curve at a given point on the curve. It is precisely the discovery of the relationship between these two problems that spurred the development of calculus in the seventeenth century and made it such an indispensable tool for solving practical problems. The following are a few examples of such problems:

- Finding the velocity of an object
- Finding the rate of change of a bacteria population with respect to time
- Finding the rate of change of a company's profit with respect to time
- Finding the rate of change of a travel agency's revenue with respect to the agency's expenditure for advertising

The study of the tangent-line problem led to the creation of *differential calculus*, which relies on the concept of the *derivative* of a function. The study of the area problem led to the creation of *integral calculus*, which relies on the concept of the *antiderivative*, or *integral*, of a function. (The derivative of a function and the integral of a function are intimately related, as you will see in Section 6.4.) Both the derivative of a function and the integral of a function are defined in terms of a more fundamental concept: the limit, our next topic.

A Real-Life Example

From data obtained in a test run conducted on a prototype of a maglev (magnetic levitation train), which moves along a straight monorail track, engineers have determined that the position of the maglev (in feet) from the origin at time t (in seconds) is given by

$$s = f(t) = 4t^2 \qquad (0 \le t \le 30) \tag{3}$$

where f is called the **position function** of the maglev. The position of the maglev at time $t = 0, 1, 2, 3, \ldots , 10$, measured from its initial position, is

$$f(0) = 0 \qquad f(1) = 4 \qquad f(2) = 16 \qquad f(3) = 36, \ldots \qquad f(10) = 400$$

feet (Figure 27).

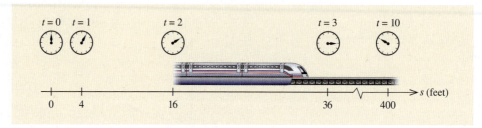

FIGURE 27
A maglev moving along an elevated monorail track

Suppose we want to find the velocity of the maglev at $t = 2$. This is just the velocity of the maglev as shown on its speedometer at that precise instant of time. Offhand, calculating this quantity using only Equation (3) appears to be an impossible task; but consider what quantities we *can* compute using this relationship. Obviously, we can compute the position of the maglev at any time t as we did earlier for some selected values of t. Using these values, we can then compute the *average velocity* of the maglev over an interval of time. For example, the average velocity of the train over the time interval $[2, 4]$ is given by

$$\begin{aligned}\frac{\text{Distance covered}}{\text{Time elapsed}} &= \frac{f(4) - f(2)}{4 - 2} \\ &= \frac{4(4^2) - 4(2^2)}{2} \\ &= \frac{64 - 16}{2} = 24\end{aligned}$$

or 24 feet/second (ft/sec).

Although this is not quite the velocity of the maglev at $t = 2$, it does provide us with an approximation of its velocity at that time.

Can we do better? Intuitively, the smaller the time interval we pick (with $t = 2$ as the left endpoint), the better the average velocity over that time interval will approximate the actual velocity of the maglev at $t = 2$.*

Now, let's describe this process in general terms. Let $t > 2$. Then, the average velocity of the maglev over the time interval $[2, t]$ is given by

$$\frac{f(t) - f(2)}{t - 2} = \frac{4t^2 - 4(2^2)}{t - 2} = \frac{4(t^2 - 4)}{t - 2} \tag{4}$$

By choosing the values of t closer and closer to 2, we obtain a sequence of numbers that give the average velocities of the maglev over smaller and smaller time intervals. As we observed earlier, this sequence of numbers should approach the *instantaneous velocity* of the train at $t = 2$.

---

*Actually, any interval containing $t = 2$ will do.

Let's try some sample calculations. Using Equation (4) and taking the sequence $t = 2.5, 2.1, 2.01, 2.001$, and 2.0001, which approaches 2, we find:

The average velocity over $[2, 2.5]$ is $\dfrac{4(2.5^2 - 4)}{2.5 - 2} = 18$, or 18 ft/sec.

The average velocity over $[2, 2.1]$ is $\dfrac{4(2.1^2 - 4)}{2.1 - 2} = 16.4$, or 16.4 ft/sec.

and so forth. These results are summarized in Table 1.

TABLE 1

			t approaches 2 from the right.		
t	2.5	2.1	2.01	2.001	2.0001
Average Velocity over $[2, t]$	18	16.4	16.04	16.004	16.0004

Average velocity approaches 16 from the right.

From Table 1, we see that the average velocity of the maglev seems to approach the number 16 as it is computed over smaller and smaller time intervals. These computations suggest that the instantaneous velocity of the train at $t = 2$ is 16 ft/sec.

Note Notice that we cannot obtain the instantaneous velocity for the maglev at $t = 2$ by substituting $t = 2$ into Equation (4) because this value of t is not in the domain of the average velocity function.

Intuitive Definition of a Limit

Consider the function g defined by

$$g(t) = \frac{4(t^2 - 4)}{t - 2}$$

which gives the average velocity of the maglev [see Equation (4)]. Suppose we are required to determine the value that $g(t)$ approaches as t approaches the (fixed) number 2. If we take the sequence of values of t approaching 2 from the right-hand side, as we did earlier, we see that $g(t)$ approaches the number 16. Similarly, if we take a sequence of values of t approaching 2 from the left, such as $t = 1.5, 1.9, 1.99, 1.999$, and 1.9999, we obtain the results shown in Table 2.

TABLE 2

			t approaches 2 from the left.		
t	1.5	1.9	1.99	1.999	1.9999
$g(t)$	14	15.6	15.96	15.996	15.9996

Average velocity approaches 16 from the left.

Observe that $g(t)$ approaches the number 16 as t approaches 2—this time from the left-hand side. In other words, as t approaches 2 from *either* side of 2, $g(t)$ approaches 16. In this situation, we say that the limit of $g(t)$ as t approaches 2 is 16, written

$$\lim_{t \to 2} g(t) = \lim_{t \to 2} \frac{4(t^2 - 4)}{t - 2} = 16$$

The graph of the function g, shown in Figure 28, confirms this observation.

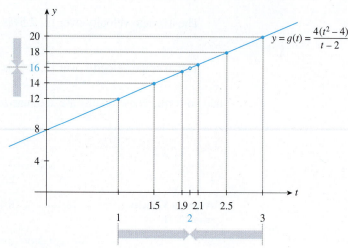

FIGURE 28
As t approaches $t = 2$ from either direction, $g(t)$ approaches $y = 16$.

Observe that the point $t = 2$ is not in the domain of the function g [for this reason, the point $(2, 16)$ is missing from the graph of g]. This, however, is inconsequential because the value, if any, of $g(t)$ at $t = 2$ plays no role in computing the limit.

This example leads to the following informal definition.

Limit of a Function

The function f has the **limit** L as x approaches a, written

$$\lim_{x \to a} f(x) = L$$

if the value of $f(x)$ can be made as close to the number L as we please by taking x sufficiently close to (but not equal to) a.

Exploring with TECHNOLOGY

1. Use a graphing utility to plot the graph of

$$g(x) = \frac{4(x^2 - 4)}{x - 2}$$

in the viewing window $[0, 3] \times [0, 20]$.
2. Use **ZOOM** and **TRACE** to describe what happens to the values of $g(x)$ as x approaches 2, first from the right and then from the left.
3. What happens to the y-value when you try to evaluate $g(x)$ at $x = 2$? Explain.
4. Reconcile your results with those of the preceding example.

Evaluating the Limit of a Function

Let's now consider some examples involving the computation of limits.

EXAMPLE 1 Let $f(x) = x^3$ and evaluate $\lim_{x \to 2} f(x)$.

Solution The graph of f is shown in Figure 29. You can see that $f(x)$ can be made as close to the number 8 as we please by taking x sufficiently close to 2. Therefore,

$$\lim_{x \to 2} x^3 = 8$$

EXAMPLE 2 Let

$$g(x) = \begin{cases} x + 2 & \text{if } x \neq 1 \\ 1 & \text{if } x = 1 \end{cases}$$

Evaluate $\lim_{x \to 1} g(x)$.

Solution The domain of g is the set of all real numbers. From the graph of g shown in Figure 30, we see that $g(x)$ can be made as close to 3 as we please by taking x sufficiently close to 1. Therefore,

$$\lim_{x \to 1} g(x) = 3$$

Observe that $g(1) = 1$, which is not equal to the limit of the function g as x approaches 1. [Once again, the value of $g(x)$ at $x = 1$ has no bearing on the existence or value of the limit of g as x approaches 1.]

EXAMPLE 3 Evaluate the limit of the following functions as x approaches the indicated point.

a. $f(x) = \begin{cases} -1 & \text{if } x < 0 \\ 1 & \text{if } x \geq 0 \end{cases}; x = 0$ **b.** $g(x) = \dfrac{1}{x^2}; x = 0$

Solution The graphs of the functions f and g are shown in Figure 31.

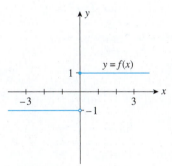

(a) $\lim_{x \to 0} f(x)$ does not exist.

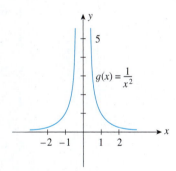

(b) $\lim_{x \to 0} g(x)$ does not exist.

FIGURE 31

a. Referring to Figure 31a, we see that no matter how close x is to zero, $f(x)$ takes on the values 1 or -1, depending on whether x is positive or negative. Thus, there is no *single* real number L that $f(x)$ approaches as x approaches zero. We conclude that the limit of $f(x)$ does *not* exist as x approaches zero.

b. Referring to Figure 31b, we see that as x approaches zero (from either side), $g(x)$ increases without bound and thus does not approach any specific real number. We conclude, accordingly, that the limit of $g(x)$ does *not* exist as x approaches zero.

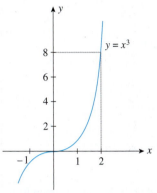

FIGURE 29
$f(x)$ is close to 8 whenever x is close to 2.

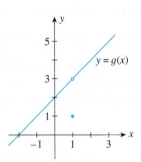

FIGURE 30
$\lim_{x \to 1} g(x) = 3$

Explore and Discuss

Consider the graph of the function h shown in the following figure.

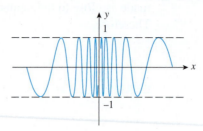

It has the property that as x approaches zero from either the right or the left, the curve oscillates more and more frequently between the lines $y = -1$ and $y = 1$.

1. Explain why $\lim\limits_{x \to 0} h(x)$ does not exist.

2. Compare this function with those in Example 3. More specifically, discuss the different ways the functions fail to have a limit at $x = 0$.

Until now, we have relied on knowing the actual values of a function or the graph of a function near $x = a$ to help us evaluate the limit of the function $f(x)$ as x approaches a. The following properties of limits, which we list without proof, enable us to evaluate limits of functions algebraically.

THEOREM 1

Properties of Limits

Suppose

$$\lim_{x \to a} f(x) = L \quad \text{and} \quad \lim_{x \to a} g(x) = M$$

Then

1. $\lim\limits_{x \to a} [f(x)]^r = \left[\lim\limits_{x \to a} f(x)\right]^r = L^r$ *r, a positive constant*

2. $\lim\limits_{x \to a} cf(x) = c \lim\limits_{x \to a} f(x) = cL$ *c, a real number*

3. $\lim\limits_{x \to a} [f(x) \pm g(x)] = \lim\limits_{x \to a} f(x) \pm \lim\limits_{x \to a} g(x) = L \pm M$

4. $\lim\limits_{x \to a} [f(x)g(x)] = \left[\lim\limits_{x \to a} f(x)\right]\left[\lim\limits_{x \to a} g(x)\right] = LM$

5. $\lim\limits_{x \to a} \dfrac{f(x)}{g(x)} = \dfrac{\lim\limits_{x \to a} f(x)}{\lim\limits_{x \to a} g(x)} = \dfrac{L}{M}$ *Provided that $M \neq 0$*

EXAMPLE 4 Use Theorem 1 to evaluate the following limits.

a. $\lim\limits_{x \to 2} x^3$ **b.** $\lim\limits_{x \to 4} 5x^{3/2}$ **c.** $\lim\limits_{x \to 1} (5x^4 - 2)$

d. $\lim\limits_{x \to 3} 2x^3\sqrt{x^2 + 7}$ **e.** $\lim\limits_{x \to 2} \dfrac{2x^2 + 1}{x + 1}$

Solution

a. $\lim\limits_{x \to 2} x^3 = \left[\lim\limits_{x \to 2} x\right]^3$ Property 1

$\qquad = 2^3 = 8$ $\lim\limits_{x \to 2} x = 2$

b. $\lim\limits_{x\to 4} 5x^{3/2} = 5\big[\lim\limits_{x\to 4} x^{3/2}\big]$ Property 2

$\qquad\qquad = 5(4)^{3/2} = 40$ Property 1

c. $\lim\limits_{x\to 1}(5x^4 - 2) = \lim\limits_{x\to 1} 5x^4 - \lim\limits_{x\to 1} 2$ Property 3

To evaluate $\lim\limits_{x\to 1} 2$, observe that the constant function $g(x) = 2$ has value 2 for all values of x. Therefore, $g(x)$ must approach the limit 2 as x approaches 1 (or any other point for that matter!). Therefore,

$$\lim\limits_{x\to 1}(5x^4 - 2) = 5(1)^4 - 2 = 3$$

d. $\lim\limits_{x\to 3} 2x^3 \sqrt{x^2 + 7} = 2 \lim\limits_{x\to 3} x^3 \sqrt{x^2 + 7}$ Property 2

$\qquad\qquad = 2 \lim\limits_{x\to 3} x^3 \lim\limits_{x\to 3} \sqrt{x^2 + 7}$ Property 4

$\qquad\qquad = 2(3)^3 \sqrt{3^2 + 7}$ Properties 1 and 3

$\qquad\qquad = 2(27)\sqrt{16} = 216$

e. $\lim\limits_{x\to 2} \dfrac{2x^2 + 1}{x + 1} = \dfrac{\lim\limits_{x\to 2}(2x^2 + 1)}{\lim\limits_{x\to 2}(x + 1)}$ Property 5

$\qquad\qquad = \dfrac{2(2)^2 + 1}{2 + 1} = \dfrac{9}{3} = 3$ ◼

Indeterminate Forms

Let's emphasize once again that Property 5 of limits is valid only when the limit of the function that appears in the denominator is not equal to zero at the number in question.

If the numerator has a limit different from zero and the denominator has a limit equal to zero, then the limit of the quotient does not exist at the number in question. This is the case with the function $g(x) = 1/x^2$ in Example 3b. Here, as x approaches zero, the numerator approaches 1 but the denominator approaches zero, so the quotient becomes arbitrarily large. Thus, as was observed earlier, the limit does not exist.

Next, consider

$$\lim\limits_{x\to 2} \frac{4(x^2 - 4)}{x - 2}$$

which we evaluated earlier by looking at the values of the function for x near $x = 2$. If we attempt to evaluate this expression by applying Property 5 of limits, we see that both the numerator and denominator of the function

$$\frac{4(x^2 - 4)}{x - 2}$$

approach zero as x approaches 2; that is, we obtain an expression of the form $0/0$. In this event, we say that the limit of the quotient $f(x)/g(x)$ as x approaches 2 has the **indeterminate form 0/0.**

We need to evaluate limits of this type when we discuss the derivative of a function, a fundamental concept in the study of calculus. As the name suggests, the meaningless expression $0/0$ does not provide us with a solution to our problem. One strategy that can be used to solve this type of problem follows.

> **Strategy for Evaluating Indeterminate Forms**
>
> **1.** Replace the given function with an appropriate one that takes on the same values as the original function everywhere except at $x = a$.
>
> **2.** Evaluate the limit of this function as x approaches a.

Examples 5 and 6 illustrate this strategy.

EXAMPLE 5 Evaluate:

$$\lim_{x \to 2} \frac{4(x^2 - 4)}{x - 2}$$

Solution Since both the numerator and the denominator of this expression approach zero as x approaches 2, we have the indeterminate form 0/0. We rewrite

$$\frac{4(x^2 - 4)}{x - 2} = \frac{4(x - 2)(x + 2)}{(x - 2)}$$

which, upon cancellation of the common factors, is equivalent to $4(x + 2)$, provided that $x \neq 2$. Next, we replace $4(x^2 - 4)/(x - 2)$ with $4(x + 2)$ and find that

$$\lim_{x \to 2} \frac{4(x^2 - 4)}{x - 2} = \lim_{x \to 2} 4(x + 2) = 16$$

The graphs of the functions

$$f(x) = \frac{4(x^2 - 4)}{x - 2} \quad \text{and} \quad g(x) = 4(x + 2)$$

are shown in Figure 32. Observe that the graphs are identical except when $x = 2$. The function g is defined for all values of x and, in particular, its value at $x = 2$ is $g(2) = 4(2 + 2) = 16$. Thus, the point $(2, 16)$ is on the graph of g. However, the function f is not defined at $x = 2$. Since $f(x) = g(x)$ for all values of x except $x = 2$, it follows that the graph of f must look exactly like the graph of g, with the exception that the point $(2, 16)$ is missing from the graph of f. This illustrates graphically why we can evaluate the limit of f by evaluating the limit of the "equivalent" function g.

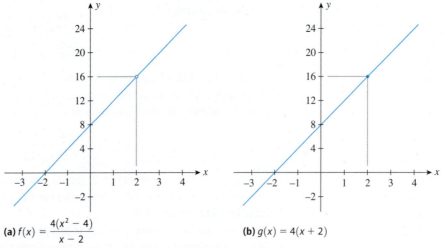

(a) $f(x) = \dfrac{4(x^2 - 4)}{x - 2}$ **(b)** $g(x) = 4(x + 2)$

FIGURE **32**
The graphs of $f(x)$ and $g(x)$ are identical except at the point $(2, 16)$.

Note Notice that the limit in Example 5 is the same limit that we evaluated earlier when we discussed the instantaneous velocity of a maglev at a specified time. ◼

1. Use a graphing utility to plot the graph of

$$f(x) = \frac{4(x^2 - 4)}{x - 2}$$

in the viewing window $[0, 3] \times [0, 20]$. Then use **ZOOM** and **TRACE** to find

$$\lim_{x \to 2} \frac{4(x^2 - 4)}{x - 2}$$

What happens to the y-value when you try to evaluate $f(x)$ at $x = 2$? Explain.

2. Use a graphing utility to plot the graph of $g(x) = 4(x + 2)$ in the viewing window $[0, 3] \times [0, 20]$. Then use **ZOOM** and **TRACE** to find $\lim_{x \to 2} 4(x + 2)$.

3. Can you distinguish between the graphs of f and g?

4. Reconcile your results with those of Example 5.

EXAMPLE 6 Evaluate:

$$\lim_{h \to 0} \frac{\sqrt{1 + h} - 1}{h}$$

(x^2) The algebra icon is used to indicate that the algebraic computation or problem-solving skill used in the example is reviewed on the referenced page. For instance, in Example 6, if you refer to page 19, you will find a review of the process of rationalizing an algebraic fraction. This is followed by a worked example in which the numerator of the expression $\dfrac{\sqrt{1 + h} - 1}{h}$ is rationalized.

Solution Letting h approach zero, we obtain the indeterminate form $0/0$. Next, we rationalize the numerator of the quotient by multiplying both the numerator and the denominator by the expression $(\sqrt{1 + h} + 1)$, obtaining

$$\frac{\sqrt{1 + h} - 1}{h} = \frac{(\sqrt{1 + h} - 1)(\sqrt{1 + h} + 1)}{h(\sqrt{1 + h} + 1)}$$

$(\sqrt{a} - \sqrt{b})(\sqrt{a} + \sqrt{b}) = a - b$

(x^2) See page 19.

$$= \frac{1 + h - 1}{h(\sqrt{1 + h} + 1)}$$

$$= \frac{h}{h(\sqrt{1 + h} + 1)}$$

$$= \frac{1}{\sqrt{1 + h} + 1}$$

Therefore,

$$\lim_{h \to 0} \frac{\sqrt{1 + h} - 1}{h} = \lim_{h \to 0} \frac{1}{\sqrt{1 + h} + 1} = \frac{1}{\sqrt{1} + 1} = \frac{1}{2}$$ ◼

1. Use a graphing utility to plot the graph of

$$g(x) = \frac{\sqrt{1 + x} - 1}{x}$$ in the viewing window

$[-1, 2] \times [0, 1]$. Then use **ZOOM** and **TRACE** to find

$$\lim_{x \to 0} \frac{\sqrt{1 + x} - 1}{x}$$ by observing the values of $g(x)$

as x approaches zero from the left and from the right.

2. Use a graphing utility to plot the graph of

$$f(x) = \frac{1}{\sqrt{1 + x} + 1}$$ in the viewing window

$[-1, 2] \times [0, 1]$. Then use **ZOOM** and **TRACE** to find

$$\lim_{x \to 0} \frac{1}{\sqrt{1 + x} + 1}.$$ What happens to the y-value

when x takes on the value zero? Explain.

3. Can you distinguish between the graphs of f and g?

4. Reconcile your results with those of Example 6.

Limits at Infinity

Up to now, we have studied the limit of a function as x approaches a (finite) number a. There are occasions, however, when we want to know whether $f(x)$ approaches a unique number as x increases without bound. Consider, for example, the function P, giving the number of fruit flies (*Drosophila*) in a container under controlled laboratory conditions, as a function of a time t. The graph of P is shown in Figure 33. You can see from the graph of P that as t increases without bound (gets larger and larger), $P(t)$ approaches the number 400. This number, called the *carrying capacity* of the environment, is determined by the amount of living space and food available, as well as other environmental factors.

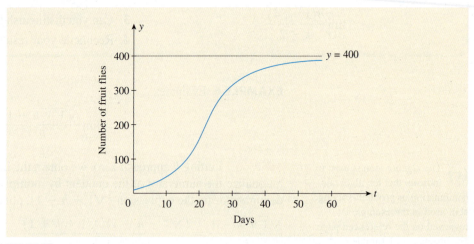

FIGURE 33
The graph of $P(t)$ gives the population of fruit flies in a laboratory experiment.

As another example, suppose we are given the function

$$f(x) = \frac{2x^2}{1 + x^2}$$

and we want to determine what happens to $f(x)$ as x gets larger and larger. Picking the sequence of numbers 1, 2, 5, 10, 100, and 1000 and computing the corresponding values of $f(x)$, we obtain the following table of values:

x	1	2	5	10	100	1000
$f(x)$	1	1.6	1.92	1.98	1.9998	1.999998

From the table, we see that as x gets larger and larger, $f(x)$ gets closer and closer to 2. The graph of the function f shown in Figure 34 confirms this observation. We call the line $y = 2$ a **horizontal asymptote.*** In this situation, we say that the limit of the function $f(x)$ as x increases without bound is 2, written

$$\lim_{x \to \infty} \frac{2x^2}{1 + x^2} = 2$$

In the general case, the following definition for a **limit of a function at infinity** is applicable.

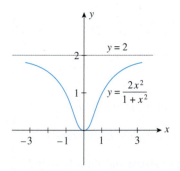

FIGURE 34
The graph of

$$y = \frac{2x^2}{1 + x^2}$$

has a horizontal asymptote at $y = 2$.

---

*We will discuss asymptotes in greater detail in Section 4.3.

Limit of a Function at Infinity

The function f has the limit L as x increases without bound (or as x approaches infinity), written

$$\lim_{x \to \infty} f(x) = L$$

if $f(x)$ can be made arbitrarily close to L by taking x large enough.

Similarly, the function f has the limit M as x decreases without bound (or as x approaches negative infinity), written

$$\lim_{x \to -\infty} f(x) = M$$

if $f(x)$ can be made arbitrarily close to M by taking x to be negative and sufficiently large in absolute value.

EXAMPLE 7 Let f and g be the functions

$$f(x) = \begin{cases} -1 & \text{if } x < 0 \\ 1 & \text{if } x \geq 0 \end{cases} \quad \text{and} \quad g(x) = \frac{1}{x^2}$$

Evaluate:

a. $\lim\limits_{x \to \infty} f(x)$ and $\lim\limits_{x \to -\infty} f(x)$ **b.** $\lim\limits_{x \to \infty} g(x)$ and $\lim\limits_{x \to -\infty} g(x)$

Solution The graphs of $f(x)$ and $g(x)$ are shown in Figure 35. Referring to the graphs of the respective functions, we see that

a. $\lim\limits_{x \to \infty} f(x) = 1$ and $\lim\limits_{x \to -\infty} f(x) = -1$ **b.** $\lim\limits_{x \to \infty} \dfrac{1}{x^2} = 0$ and $\lim\limits_{x \to -\infty} \dfrac{1}{x^2} = 0$

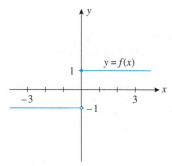

(a) $\lim\limits_{x \to \infty} f(x) = 1$ and $\lim\limits_{x \to -\infty} f(x) = -1$

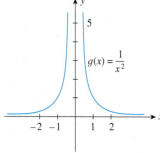

(b) $\lim\limits_{x \to \infty} g(x) = 0$ and $\lim\limits_{x \to -\infty} g(x) = 0$

FIGURE **35**

All the properties of limits listed in Theorem 1 are valid when a is replaced by ∞ or $-\infty$. In addition, we have the following property for the limit at infinity.

THEOREM 2

For all $n > 0$,

$$\lim_{x \to \infty} \frac{1}{x^n} = 0 \quad \text{and} \quad \lim_{x \to -\infty} \frac{1}{x^n} = 0$$

provided that $\dfrac{1}{x^n}$ is defined.

1. Use a graphing utility to plot the graphs of

$$y_1 = \frac{1}{x^{0.5}} \qquad y_2 = \frac{1}{x} \qquad y_3 = \frac{1}{x^{1.5}}$$

in the viewing window $[0, 200] \times [0, 0.5]$. What can you say about $\lim\limits_{x \to \infty} \dfrac{1}{x^n}$ if $n = 0.5$, $n = 1$, and $n = 1.5$? Are these results predicted by Theorem 2?

2. Use a graphing utility to plot the graphs of

$$y_1 = \frac{1}{x} \quad \text{and} \quad y_2 = \frac{1}{x^{5/3}}$$

in the viewing window $[-50, 0] \times [-0.5, 0]$. What can you say about $\lim\limits_{x \to -\infty} \dfrac{1}{x^n}$ if $n = 1$ and $n = \dfrac{5}{3}$? Are these results predicted by Theorem 2?

Hint: To graph y_2, write it in the form $y2 = 1/(x^{(1/3)})^5$.

We often use the following technique to evaluate the limit at infinity of a rational function: *Divide the numerator and denominator of the expression by x^n, where n is the highest power present in the denominator of the expression.*

EXAMPLE 8 Evaluate

$$\lim_{x \to \infty} \frac{x^2 - x + 3}{2x^3 + 1}$$

Solution   Since the limits of both the numerator and the denominator do not exist as x approaches infinity, the property pertaining to the limit of a quotient (Property 5) is not applicable. Let's divide the numerator and denominator of the rational expression by x^3, obtaining

$$\lim_{x \to \infty} \frac{x^2 - x + 3}{2x^3 + 1} = \lim_{x \to \infty} \frac{\dfrac{1}{x} - \dfrac{1}{x^2} + \dfrac{3}{x^3}}{2 + \dfrac{1}{x^3}}$$

$$= \frac{0 - 0 + 0}{2 + 0} = \frac{0}{2} \qquad \text{Use Theorem 2.}$$

$$= 0$$

EXAMPLE 9 Let

$$f(x) = \frac{3x^2 + 8x - 4}{2x^2 + 4x - 5}$$

Compute $\lim\limits_{x \to \infty} f(x)$ if it exists.

Solution   Again, we see that Property 5 is not applicable. Dividing the numerator and the denominator by x^2, we obtain

$$\lim_{x \to \infty} \frac{3x^2 + 8x - 4}{2x^2 + 4x - 5} = \lim_{x \to \infty} \frac{3 + \dfrac{8}{x} - \dfrac{4}{x^2}}{2 + \dfrac{4}{x} - \dfrac{5}{x^2}}$$

$$= \frac{\lim_{x \to \infty} 3 + 8 \lim_{x \to \infty} \dfrac{1}{x} - 4 \lim_{x \to \infty} \dfrac{1}{x^2}}{\lim_{x \to \infty} 2 + 4 \lim_{x \to \infty} \dfrac{1}{x} - 5 \lim_{x \to \infty} \dfrac{1}{x^2}}$$

$$= \frac{3 + 0 - 0}{2 + 0 - 0} \qquad \text{Use Theorem 2.}$$

$$= \frac{3}{2}$$

EXAMPLE 10 Let $f(x) = \dfrac{2x^3 - 3x^2 + 1}{x^2 + 2x + 4}$ and evaluate:

a. $\lim_{x \to \infty} f(x)$ **b.** $\lim_{x \to -\infty} f(x)$

Solution

a. Dividing the numerator and the denominator of the rational expression by x^2, we obtain

$$\lim_{x \to \infty} \frac{2x^3 - 3x^2 + 1}{x^2 + 2x + 4} = \lim_{x \to \infty} \frac{2x - 3 + \dfrac{1}{x^2}}{1 + \dfrac{2}{x} + \dfrac{4}{x^2}}$$

Since the numerator becomes arbitrarily large, whereas the denominator approaches 1 as x approaches infinity, we see that the quotient $f(x)$ gets larger and larger as x approaches infinity. In other words, the limit does not exist. In this case, we indicate this by writing

$$\lim_{x \to \infty} \frac{2x^3 - 3x^2 + 1}{x^2 + 2x + 4} = \infty$$

b. Once again, dividing both the numerator and the denominator by x^2, we obtain

$$\lim_{x \to -\infty} \frac{2x^3 - 3x^2 + 1}{x^2 + 2x + 4} = \lim_{x \to -\infty} \frac{2x - 3 + \dfrac{1}{x^2}}{1 + \dfrac{2}{x} + \dfrac{4}{x^2}}$$

In this case, the numerator becomes arbitrarily large in magnitude but negative in sign, whereas the denominator approaches 1 as x approaches negative infinity. Therefore, the quotient $f(x)$ decreases without bound, and the limit does not exist. In this case, we indicate this by writing

$$\lim_{x \to -\infty} \frac{2x^3 - 3x^2 + 1}{x^2 + 2x + 4} = -\infty$$

Example 11 gives an application of the concept of the limit of a function at infinity.

APPLIED EXAMPLE 11 Average Cost Functions Custom Office makes a line of executive desks. It is estimated that the total cost of making x Senior Executive Model desks is $C(x) = 100x + 200{,}000$ dollars per year, so the average cost of making x desks is given by

$$\overline{C}(x) = \frac{C(x)}{x}$$

$$= \frac{100x + 200{,}000}{x} = 100 + \frac{200{,}000}{x}$$

dollars per desk. Evaluate $\lim\limits_{x \to \infty} \overline{C}(x)$ and interpret your result.

Solution

$$\lim_{x \to \infty} \overline{C}(x) = \lim_{x \to \infty}\left(100 + \frac{200{,}000}{x}\right)$$

$$= \lim_{x \to \infty} 100 + \lim_{x \to \infty}\frac{200{,}000}{x} = 100$$

A sketch of the graph of the function $\overline{C}(x)$ appears in Figure 36. The result we obtained is fully expected if we consider its economic implications. Note that as the level of production increases, the fixed cost per desk produced, represented by the term $(200{,}000/x)$, drops steadily. The average cost should approach a constant unit cost of production—\$100 in this case.

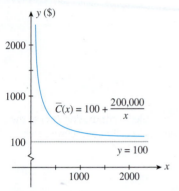

FIGURE 36
As the level of production increases, the average cost approaches $100 per desk.

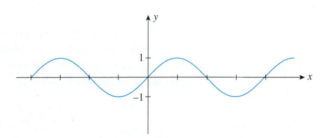

Explore and Discuss

Consider the graph of the function f depicted in the following figure:

It has the property that the curve oscillates between $y = -1$ and $y = 1$ indefinitely in either direction.

1. Explain why $\lim\limits_{x \to -\infty} f(x)$ and $\lim\limits_{x \to \infty} f(x)$ do not exist.
2. Compare this function with those of Example 10. More specifically, discuss the different ways each function fails to have a limit at infinity or minus infinity.

2.4 Self-Check Exercises

1. Find the indicated limit if it exists.

   a. $\lim\limits_{x \to 3} \dfrac{\sqrt{x^2 + 7} + \sqrt{3x - 5}}{x + 2}$

   b. $\lim\limits_{x \to -1} \dfrac{x^2 - x - 2}{2x^2 - x - 3}$

2. **AVERAGE COST OF PRODUCING CDs** The average cost per disc (in dollars) incurred by Herald Records in pressing x CDs

is given by the average cost function

$$\overline{C}(x) = 1.8 + \frac{3000}{x}$$

Evaluate $\lim\limits_{x \to \infty} \overline{C}(x)$ and interpret your result.

Solutions to Self-Check Exercises 2.4 can be found on page 118.

2.4 Concept Questions

1. Explain what is meant by the statement $\lim\limits_{x \to 2} f(x) = 3$.

2. **a.** If $\lim\limits_{x \to 3} f(x) = 5$, what can you say about $f(3)$? Explain.

   b. If $f(2) = 6$, what can you say about $\lim\limits_{x \to 2} f(x)$? Explain.

3. Evaluate the following, and state the property of limits that you use at each step.

   a. $\lim\limits_{x \to 4} \sqrt{x}(2x^2 + 1)$ **b.** $\lim\limits_{x \to 1} \left(\dfrac{2x^2 + x + 5}{x^4 + 1} \right)^{3/2}$

4. What is an indeterminate form? Illustrate with an example.

5. Explain in your own words the meaning of $\lim\limits_{x \to \infty} f(x) = L$ and $\lim\limits_{x \to -\infty} f(x) = M$.

2.4 Exercises

In Exercises 1–8, use the graph of the given function f to determine $\lim\limits_{x \to a} f(x)$ at the indicated value of a, if it exists.

1.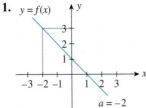

$a = -2$

2.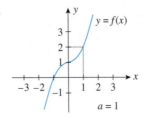

$a = 1$

3.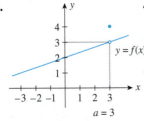

$a = 3$

4.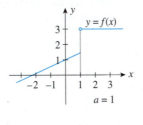

$a = 1$

5.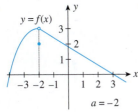

$a = -2$

6.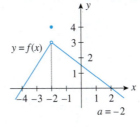

$a = -2$

7.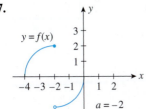

$a = -2$

8.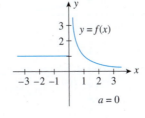

$a = 0$

In Exercises 9–16, complete the table by computing $f(x)$ at the given values of x. Use these results to estimate the indicated limit (if it exists).

9. $f(x) = x^2 + 1$; $\lim\limits_{x \to 2} f(x)$

x	1.9	1.99	1.999	2.001	2.01	2.1
$f(x)$						

10. $f(x) = 2x^2 - 1$; $\lim\limits_{x \to 1} f(x)$

x	0.9	0.99	0.999	1.001	1.01	1.1
$f(x)$						

11. $f(x) = \dfrac{|x|}{x}$; $\lim\limits_{x \to 0} f(x)$

x	-0.1	-0.01	-0.001	0.001	0.01	0.1
$f(x)$						

12. $f(x) = \dfrac{|x - 1|}{x - 1}$; $\lim\limits_{x \to 1} f(x)$

x	0.9	0.99	0.999	1.001	1.01	1.1
$f(x)$						

13. $f(x) = \dfrac{1}{(x - 1)^2}$; $\lim\limits_{x \to 1} f(x)$

x	0.9	0.99	0.999	1.001	1.01	1.1
$f(x)$						

14. $f(x) = \dfrac{1}{x - 2}$; $\lim\limits_{x \to 2} f(x)$

x	1.9	1.99	1.999	2.001	2.01	2.1
$f(x)$						

15. $f(x) = \dfrac{x^2 + x - 2}{x - 1}$; $\lim\limits_{x \to 1} f(x)$

x	0.9	0.99	0.999	1.001	1.01	1.1
$f(x)$						

16. $f(x) = \dfrac{x - 1}{x - 1}$; $\lim\limits_{x \to 1} f(x)$

x	0.9	0.99	0.999	1.001	1.01	1.1
$f(x)$						

In Exercises 17–22, sketch the graph of the function f and evaluate $\lim\limits_{x \to a} f(x)$, if it exists, for the given value of a.

17. $f(x) = \begin{cases} x - 1 & \text{if } x \le 0 \\ -1 & \text{if } x > 0 \end{cases}$ $(a = 0)$

18. $f(x) = \begin{cases} x - 1 & \text{if } x \le 3 \\ -2x + 8 & \text{if } x > 3 \end{cases}$ $(a = 3)$

19. $f(x) = \begin{cases} x & \text{if } x < 1 \\ 0 & \text{if } x = 1 \\ -x + 2 & \text{if } x > 1 \end{cases}$ $(a = 1)$

20. $f(x) = \begin{cases} -2x + 4 & \text{if } x < 1 \\ 4 & \text{if } x = 1 \\ x^2 + 1 & \text{if } x > 1 \end{cases}$ $(a = 1)$

21. $f(x) = \begin{cases} |x| & \text{if } x \ne 0 \\ 1 & \text{if } x = 0 \end{cases}$ $(a = 0)$

22. $f(x) = \begin{cases} |x - 1| & \text{if } x \ne 1 \\ 0 & \text{if } x = 1 \end{cases}$ $(a = 1)$

In Exercises 23–40, find the indicated limit.

23. $\lim\limits_{x \to 2} 3$

24. $\lim\limits_{x \to -2} -3$

25. $\lim\limits_{x \to 3} x$

26. $\lim\limits_{x \to -2} -3x$

27. $\lim\limits_{x \to 1} (1 - 2x^2)$

28. $\lim\limits_{t \to 3} (4t^2 - 2t + 1)$

29. $\lim\limits_{x \to 1} (2x^3 - 3x^2 + x + 2)$

30. $\lim\limits_{x \to 0} (4x^5 - 20x^2 + 2x + 1)$

31. $\lim\limits_{s \to 0} (2s^2 - 1)(2s + 4)$ **32.** $\lim\limits_{x \to 2} (x^2 + 1)(x^2 - 4)$

33. $\lim\limits_{x \to 2} \dfrac{2x + 1}{x + 2}$

34. $\lim\limits_{x \to 1} \dfrac{x^3 + 1}{2x^3 + 2}$

35. $\lim\limits_{x \to 2} \sqrt{x + 2}$

36. $\lim\limits_{x \to -2} \sqrt[3]{5x + 2}$

37. $\lim\limits_{x \to -3} \sqrt{2x^4 + x^2}$

38. $\lim\limits_{x \to 2} \sqrt{\dfrac{2x^3 + 4}{x^2 + 1}}$

39. $\lim\limits_{x \to -1} \dfrac{\sqrt{x^2 + 8}}{2x + 4}$

40. $\lim\limits_{x \to 3} \dfrac{x\sqrt{x^2 + 7}}{2x - \sqrt{2x + 3}}$

In Exercises 41–48, find the indicated limit given that $\lim\limits_{x \to a} f(x) = 3$ and $\lim\limits_{x \to a} g(x) = 4$.

41. $\lim\limits_{x \to a} [f(x) - g(x)]$

42. $\lim\limits_{x \to a} 2f(x)$

43. $\lim\limits_{x \to a} [2f(x) - 3g(x)]$

44. $\lim\limits_{x \to a} [f(x)g(x)]$

45. $\lim\limits_{x \to a} \sqrt{g(x)}$

46. $\lim\limits_{x \to a} \sqrt[3]{5f(x) + 3g(x)}$

47. $\lim\limits_{x \to a} \dfrac{2f(x) - g(x)}{f(x)g(x)}$

48. $\lim\limits_{x \to a} \dfrac{g(x) - f(x)}{f(x) + \sqrt{g(x)}}$

In Exercises 49–62, find the indicated limit, if it exists.

49. $\lim\limits_{x \to 1} \dfrac{x^2 - 1}{x - 1}$

50. $\lim\limits_{x \to -2} \dfrac{x^2 - 4}{x + 2}$

51. $\lim\limits_{x \to 0} \dfrac{x^2 - x}{x}$

52. $\lim\limits_{x \to 0} \dfrac{2x^2 - 3x}{x}$

53. $\lim\limits_{x \to -5} \dfrac{x^2 - 25}{x + 5}$

54. $\lim\limits_{b \to -3} \dfrac{b + 1}{b + 3}$

55. $\lim\limits_{x \to 1} \dfrac{x}{x - 1}$

56. $\lim\limits_{x \to 2} \dfrac{x + 2}{x - 2}$

57. $\lim\limits_{x \to -2} \dfrac{x^2 - x - 6}{x^2 + x - 2}$

58. $\lim\limits_{z \to 2} \dfrac{z^3 - 8}{z - 2}$

59. $\lim\limits_{x \to 1} \dfrac{\sqrt{x} - 1}{x - 1}$

60. $\lim\limits_{x \to 4} \dfrac{x - 4}{\sqrt{x} - 2}$

Hint: Multiply by $\dfrac{\sqrt{x} + 1}{\sqrt{x} + 1}$.   Hint: See Exercise 59.

61. $\lim\limits_{x \to 1} \dfrac{x - 1}{x^3 + x^2 - 2x}$

62. $\lim\limits_{x \to -2} \dfrac{4 - x^2}{2x^2 + x^3}$

In Exercises 63–68, use the graph of the function f to determine $\lim\limits_{x \to \infty} f(x)$ and $\lim\limits_{x \to -\infty} f(x)$, if they exist.

63.

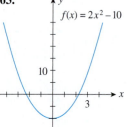

$f(x) = 2x^2 - 10$

64.

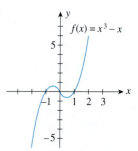

$f(x) = x^3 - x$

65.

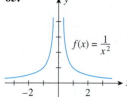

$f(x) = \dfrac{1}{x^2}$

66.

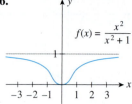

$f(x) = \dfrac{x^2}{x^2 + 1}$

67.

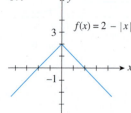

$f(x) = 2 - |x|$

68.

$$f(x) = \begin{cases} \sqrt{-x} & \text{if } x \le 0 \\ \dfrac{x}{x+1} & \text{if } x > 0 \end{cases}$$

In Exercises 69–72, complete the table by computing $f(x)$ at the given values of x. Use the results to guess at the indicated limits, if they exist.

69. $f(x) = \dfrac{1}{x^2 + 1}$; $\lim\limits_{x \to \infty} f(x)$ and $\lim\limits_{x \to -\infty} f(x)$

x	1	10	100	1000
$f(x)$				

x	-1	-10	-100	-1000
$f(x)$				

70. $f(x) = \dfrac{2x}{x + 1}$; $\lim\limits_{x \to \infty} f(x)$ and $\lim\limits_{x \to -\infty} f(x)$

x	1	10	100	1000
$f(x)$				

x	-5	-10	-100	-1000
$f(x)$				

71. $f(x) = 3x^3 - x^2 + 10$; $\lim\limits_{x \to \infty} f(x)$ and $\lim\limits_{x \to -\infty} f(x)$

x	1	5	10	100	1000
$f(x)$					

x	-1	-5	-10	-100	-1000
$f(x)$					

72. $f(x) = \dfrac{|x|}{x}$; $\lim\limits_{x \to \infty} f(x)$ and $\lim\limits_{x \to -\infty} f(x)$

x	1	10	100	-1	-10	-100
$f(x)$						

In Exercises 73–80, find the indicated limits, if they exist.

73. $\lim\limits_{x \to \infty} \dfrac{3x + 2}{x - 5}$

74. $\lim\limits_{x \to -\infty} \dfrac{4x^2 - 1}{x + 2}$

75. $\lim\limits_{x \to -\infty} \dfrac{3x^3 + x^2 + 1}{x^3 + 1}$

76. $\lim\limits_{x \to \infty} \dfrac{2x^2 + 3x + 1}{x^4 - x^2}$

77. $\lim\limits_{x \to -\infty} \dfrac{x^4 + 1}{x^3 - 1}$

78. $\lim\limits_{x \to \infty} \dfrac{4x^4 - 3x^2 + 1}{2x^4 + x^3 + x^2 + x + 1}$

79. $\lim\limits_{x \to \infty} \dfrac{x^5 - x^3 + x - 1}{x^6 + 2x^2 + 1}$

80. $\lim\limits_{x \to \infty} \dfrac{2x^2 - 1}{x^3 + x^2 + 1}$

81. TOXIC WASTE A city's main well was recently found to be contaminated with trichloroethylene, a cancer-causing chemical, as a result of an abandoned chemical dump leaching chemicals into the water. A proposal submitted to city council members indicates that the cost, measured in millions of dollars, of removing $x\%$ of the toxic pollutant is given by

$$C(x) = \frac{0.5x}{100 - x} \qquad (0 < x < 100)$$

a. Find the cost of removing 50%, 60%, 70%, 80%, 90%, and 95% of the pollutant.

b. Evaluate

$$\lim_{x \to 100} \frac{0.5x}{100 - x}$$

and interpret your result.

82. A DOOMSDAY SITUATION The population of a certain breed of rabbits introduced onto an isolated island is given by

$$P(t) = \frac{72}{9 - t} \qquad (0 \le t < 9)$$

where t is measured in months.

a. Find the number of rabbits present on the island initially (at $t = 0$).

b. Show that the population of rabbits is increasing without bound.

c. Sketch the graph of the function P.
(*Comment:* This phenomenon is referred to as a *doomsday situation.*)

83. AVERAGE COST The average cost per disc in dollars incurred by Herald Media in pressing x DVDs is given by the average cost function

$$\overline{C}(x) = 2.2 + \frac{2500}{x}$$

Evaluate $\lim\limits_{x \to \infty} \overline{C}(x)$ and interpret your result.

84. CONCENTRATION OF A DRUG IN THE BLOODSTREAM The concentration of a certain drug in a patient's bloodstream t hr after injection is given by

$$C(t) = \frac{0.2t}{t^2 + 1}$$

mg/cm³. Evaluate $\lim\limits_{t \to \infty} C(t)$ and interpret your result.

85. BOX-OFFICE RECEIPTS The total worldwide box-office receipts for a long-running blockbuster movie are approximated by the function

$$T(x) = \frac{120x^2}{x^2 + 4}$$

where $T(x)$ is measured in millions of dollars and x is the number of months since the movie's release.

a. What are the total box-office receipts after the first month? The second month? The third month?

b. What will the movie gross in the long run (when x is very large)?

86. POPULATION GROWTH A major corporation is building a 4325-acre complex of homes, offices, stores, schools, and churches in the rural community of Glen Cove. As a result of this development, the planners have estimated that Glen Cove's population (in thousands) t years from now will be given by

$$P(t) = \frac{25t^2 + 125t + 200}{t^2 + 5t + 40}$$

a. What is the current population of Glen Cove?
b. What will be the population in the long run?

87. DRIVING COSTS A study of the costs of driving 2012 small-sized sedans found that the average cost per mile (car payments, gas, insurance, upkeep, and depreciation), measured in cents per mile, is approximated by the function

$$C(x) = \frac{2410}{x^{1.95}} + 32.8$$

where x denotes the number of miles (in thousands) the car is driven in a year.

a. What is the average cost per mile of driving a small-sized sedan 5000 mi/year? 10,000 mi/year? 15,000 mi/year? 20,000 mi/year? 25,000 mi/year?
b. Use part (a) to sketch the graph of the function C.
c. What happens to the average cost per mile as the number of miles driven increases without bound?

Source: American Automobile Association.

88. PHOTOSYNTHESIS The rate of production R in photosynthesis is related to the light intensity I by the function

$$R(I) = \frac{aI}{b + I^2}$$

where a and b are positive constants.

a. Taking $a = b = 1$, compute $R(I)$ for $I = 0, 1, 2, 3, 4,$ and 5.
b. Evaluate $\lim_{I \to \infty} R(I)$.
c. Use the results of parts (a) and (b) to sketch the graph of R. Interpret your results.

In Exercises 89–94, determine whether the statement is true or false. If it is true, explain why it is true. If it is false, give an example to show why it is false.

89. If $\lim_{x \to a} f(x)$ exists, then f is defined at $x = a$.

90. If $\lim_{x \to 3} g(x) = 0$ and if $\lim_{x \to 3} f(x)/g(x) = 0$ exists, then $\lim_{x \to 3} f(x) = 0$.

91. If $\lim_{x \to 2} f(x) = 3$ and $\lim_{x \to 2} g(x) = 0$, then $\lim_{x \to 2} [f(x)]/[g(x)]$ does not exist.

92. If $\lim_{x \to 3} f(x) = 0$ and $\lim_{x \to 3} g(x) = 0$, then $\lim_{x \to 3} [f(x)]/[g(x)]$ does not exist.

93. $\lim_{x \to 2} \left(\frac{x}{x+1} + \frac{3}{x-1} \right) = \lim_{x \to 2} \frac{x}{x+1} + \lim_{x \to 2} \frac{3}{x-1}$

94. $\lim_{x \to 1} \left(\frac{2x}{x-1} - \frac{2}{x-1} \right) = \lim_{x \to 1} \frac{2x}{x-1} - \lim_{x \to 1} \frac{2}{x-1}$

95. SPEED OF A CHEMICAL REACTION Certain proteins, known as enzymes, serve as catalysts for chemical reactions in living things. In 1913 Leonor Michaelis and L. M. Menten discovered the following formula giving the initial speed V (in moles per liter per second) at which the reaction begins in terms of the amount of substrate x (the substance being acted upon, measured in moles per liters) present:

$$V = \frac{ax}{x + b}$$

where a and b are positive constants. Evaluate

$$\lim_{x \to \infty} \frac{ax}{x + b}$$

and interpret your result.

96. Show by means of an example that $\lim_{x \to a} [f(x) + g(x)]$ may exist even though neither $\lim_{x \to a} f(x)$ nor $\lim_{x \to a} g(x)$ exists. Does this example contradict Theorem 1?

97. Show by means of an example that $\lim_{x \to a} [f(x)g(x)]$ may exist even though neither $\lim_{x \to a} f(x)$ nor $\lim_{x \to a} g(x)$ exists. Does this example contradict Theorem 1?

98. Show by means of an example that $\lim_{x \to a} f(x)/g(x)$ may exist even though neither $\lim_{x \to a} f(x)$ nor $\lim_{x \to a} g(x)$ exists. Does this example contradict Theorem 1?

2.4 Solutions to Self-Check Exercises

1. a. $\lim_{x \to 3} \frac{\sqrt{x^2 + 7} + \sqrt{3x - 5}}{x + 2} = \frac{\sqrt{9 + 7} + \sqrt{3(3) - 5}}{3 + 2}$

$= \frac{\sqrt{16} + \sqrt{4}}{5}$

$= \frac{6}{5}$

b. Letting x approach -1 leads to the indeterminate form $0/0$. Thus, we proceed as follows:

$\lim_{x \to -1} \frac{x^2 - x - 2}{2x^2 - x - 3} = \lim_{x \to -1} \frac{(x+1)(x-2)}{(x+1)(2x-3)}$

$= \lim_{x \to -1} \frac{x-2}{2x-3}$ Cancel the common factors.

$= \frac{-1-2}{2(-1)-3}$

$= \frac{3}{5}$

2. $\displaystyle \lim_{x \to \infty} \overline{C}(x) = \lim_{x \to \infty} \left(1.8 + \frac{3000}{x} \right)$

$\displaystyle \hspace{2cm} = \lim_{x \to \infty} 1.8 + \lim_{x \to \infty} \frac{3000}{x}$

$\displaystyle \hspace{2cm} = 1.8$

Our computation reveals that as the production of CDs increases without bound, the average cost drops and approaches a unit cost of $1.80/disc.

USING TECHNOLOGY   **Finding the Limit of a Function**

A graphing utility can be used to help us find the limit of a function, if it exists, as illustrated in the following examples.

EXAMPLE 1   Let $f(x) = \dfrac{x^3 - 1}{x - 1}$.

a. Plot the graph of f in the viewing window $[-2, 2] \times [0, 4]$.

b. Use **ZOOM** to find $\displaystyle \lim_{x \to 1} \frac{x^3 - 1}{x - 1}$.

c. Verify your result by evaluating the limit algebraically.

Solution

a. The graph of f in the viewing window $[-2, 2] \times [0, 4]$ is shown in Figure T1a.

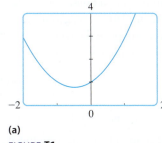
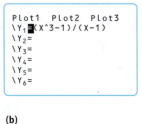

(a)   (b)

FIGURE T1
(a) The graph of $f(x) = (x^3 - 1)/(x - 1)$ in the viewing window $[-2, 2] \times [0, 4]$;
(b) the TI-83/84 equation screen

b. Using **ZOOM-IN** repeatedly, we see that the y-value approaches 3 as the x-value approaches 1. We conclude, accordingly, that

$$\lim_{x \to 1} \frac{x^3 - 1}{x - 1} = 3$$

c. We compute

$$\lim_{x \to 1} \frac{x^3 - 1}{x - 1} = \lim_{x \to 1} \frac{(x - 1)(x^2 + x + 1)}{x - 1}$$

$$= \lim_{x \to 1} (x^2 + x + 1) = 3$$

Note   If you attempt to find the limit in Example 1 by using the evaluation function of your graphing utility to find the value of $f(x)$ when $x = 1$, you will see that the graphing utility does not display the y-value. This happens because $x = 1$ is not in the domain of f.

EXAMPLE 2 Use **ZOOM** to find $\lim_{x \to 0} (1 + x)^{1/x}$.

Solution   We first plot the graph of $f(x) = (1 + x)^{1/x}$ in a suitable viewing window. Figure T2a shows a plot of f in the window $[-1, 1] \times [0, 4]$. Using **ZOOM-IN** repeatedly, we see that $\lim_{x \to 0} (1 + x)^{1/x} \approx 2.71828$.

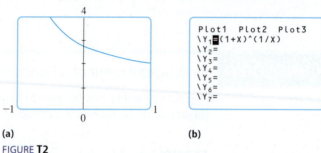

(a)   **(b)**

FIGURE **T2**
(a) The graph of $f(x) = (1 + x)^{1/x}$ in the viewing window $[-1, 1] \times [0, 4]$;
(b) the T1-83/84 equation screen

The limit of $f(x) = (1 + x)^{1/x}$ as x approaches zero, denoted by the letter e, plays a very important role in the study of mathematics and its applications (see Section 5.6). Thus,

$$\lim_{x \to 0} (1 + x)^{1/x} = e$$

where, as we have just seen, $e \approx 2.71828$.

APPLIED EXAMPLE 3 Oxygen Content of a Pond   When organic waste is dumped into a pond, the oxidation process that takes place reduces the pond's oxygen content. However, given time, nature will restore the oxygen content to its natural level. Suppose the oxygen content t days after the organic waste has been dumped into the pond is given by

$$f(t) = 100 \left(\frac{t^2 + 10t + 100}{t^2 + 20t + 100} \right)$$

percent of its normal level.

a. Plot the graph of f in the viewing window $[0, 200] \times [70, 100]$.
b. What can you say about $f(t)$ when t is very large?
c. Verify your observation in part (b) by evaluating $\lim_{t \to \infty} f(t)$.

Solution

a. The graph of f is shown in Figure T3a.
b. From the graph of f, it appears that $f(t)$ approaches 100 steadily as t gets larger and larger. This observation tells us that eventually the oxygen content of the pond will be restored to its natural level.

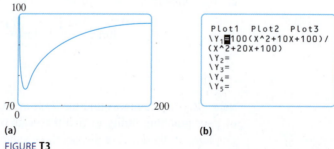

(a)   **(b)**

FIGURE **T3**
(a) The graph of f in the viewing window $[0, 200] \times [70, 100]$; (b) the TI-83/84 equation screen

c. To verify the observation made in part (b), we compute

$$\lim_{t\to\infty} f(t) = \lim_{t\to\infty} 100\left(\frac{t^2 + 10t + 100}{t^2 + 20t + 100}\right)$$

$$= 100 \lim_{t\to\infty}\left(\frac{1 + \dfrac{10}{t} + \dfrac{100}{t^2}}{1 + \dfrac{20}{t} + \dfrac{100}{t^2}}\right) = 100$$

TECHNOLOGY EXERCISES

In Exercises 1–8, find the indicated limit by first plotting the graph of the function in a suitable viewing window and then using the ZOOM-IN feature of the calculator.

1. $\displaystyle\lim_{x\to1}\frac{2x^3 - 2x^2 + 3x - 3}{x - 1}$

2. $\displaystyle\lim_{x\to-2}\frac{2x^3 + 3x^2 - x + 2}{x + 2}$

3. $\displaystyle\lim_{x\to-1}\frac{x^3 + 1}{x + 1}$

4. $\displaystyle\lim_{x\to-1}\frac{x^4 - 1}{x - 1}$

5. $\displaystyle\lim_{x\to1}\frac{x^3 - x^2 - x + 1}{x^3 - 3x + 2}$

6. $\displaystyle\lim_{x\to0}\frac{\sqrt{x + 1} - 1}{x}$

7. $\displaystyle\lim_{x\to0}(1 + 2x)^{1/x}$

8. $\displaystyle\lim_{x\to0}\frac{2^x - 1}{x}$

9. Show that $\displaystyle\lim_{x\to3}\frac{2}{x - 3}$ does not exist.

10. Show that $\displaystyle\lim_{x\to2}\frac{x^3 - 2x + 1}{x - 2}$ does not exist.

11. CITY PLANNING A major developer is building a 5000-acre complex of homes, offices, stores, schools, and churches in the rural community of Marlboro. As a result of this development, the planners have estimated that Marlboro's population (in thousands) t years from now will be given by

$$P(t) = \frac{25t^2 + 125t + 200}{t^2 + 5t + 40}$$

a. Plot the graph of P in the viewing window $[0, 50] \times [0, 30]$.

b. What will be the population of Marlboro in the long run? Hint: Find $\displaystyle\lim_{t\to\infty} P(t)$.

12. AMOUNT OF RAINFALL The total amount of rain (in inches) after t hr during a rainfall is given by

$$T(t) = \frac{0.8t}{t + 4.1}$$

a. Plot the graph of T in the viewing window $[0, 30] \times [0, 0.8]$.

b. What is the total amount of rain during this rainfall? Hint: Find $\displaystyle\lim_{t\to\infty} T(t)$.

2.5 One-Sided Limits and Continuity

One-Sided Limits

Consider the function f defined by

$$f(x) = \begin{cases} x - 1 & \text{if } x < 0 \\ x + 1 & \text{if } x \geq 0 \end{cases}$$

From the graph of f shown in Figure 37, we see that the function f does not have a limit as x approaches zero because, no matter how close x is to zero, $f(x)$ takes on values that are close to 1 if x is positive and values that are close to -1 if x is negative. Therefore, $f(x)$ cannot be close to a single number L—no matter how close x is to zero. Now, if we restrict x to be greater than zero (to the right of zero), then we see that $f(x)$ can be made as close to 1 as we please by taking x sufficiently close to zero. In this situation, we say that the right-hand limit of f as x approaches zero (from the right) is 1, written

$$\lim_{x\to0^+} f(x) = 1$$

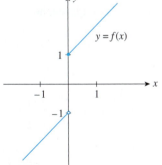

FIGURE 37
The function f does not have a limit as x approaches zero.

Similarly, we see that $f(x)$ can be made as close to -1 as we please by taking x sufficiently close to, but to the left of, zero. In this situation, we say that the left-hand limit of f as x approaches zero (from the left) is -1, written

$$\lim_{x \to 0^-} f(x) = -1$$

These limits are called **one-sided limits.** More generally, we have the following informal definitions.

One-Sided Limits

The function f has the **right-hand limit** L as x approaches a from the right, written

$$\lim_{x \to a^+} f(x) = L$$

if the values of $f(x)$ can be made as close to L as we please by taking x sufficiently close to (but not equal to) a and to the right of a.

Similarly, the function f has the **left-hand limit** M as x approaches a from the left, written

$$\lim_{x \to a^-} f(x) = M$$

if the values of $f(x)$ can be made as close to M as we please by taking x sufficiently close to (but not equal to) a and to the left of a.

The connection between one-sided limits and the two-sided limit defined earlier is given by the following theorem.

THEOREM 3

Let f be a function that is defined for all values of x close to $x = a$ with the possible exception of a itself. Then

$$\lim_{x \to a} f(x) = L \quad \text{if and only if} \quad \lim_{x \to a^+} f(x) = \lim_{x \to a^-} f(x) = L$$

Thus, the two-sided limit exists if and only if the one-sided limits exist and are equal.

EXAMPLE 1 Let

$$f(x) = \begin{cases} -x & \text{if } x \le 0 \\ \sqrt{x} & \text{if } x > 0 \end{cases} \quad \text{and} \quad g(x) = \begin{cases} -1 & \text{if } x < 0 \\ 1 & \text{if } x \ge 0 \end{cases}$$

a. Show that $\lim_{x \to 0} f(x)$ exists by studying the one-sided limits of f as x approaches $x = 0$.

b. Show that $\lim_{x \to 0} g(x)$ does not exist.

Solution

a. For $x \le 0$,

$$\lim_{x \to 0^-} f(x) = \lim_{x \to 0^-} (-x) = 0$$

and for $x > 0$, we find

$$\lim_{x \to 0^+} f(x) = \lim_{x \to 0^+} \sqrt{x} = 0$$

Thus,

$$\lim_{x \to 0} f(x) = 0$$

(Figure 38a).

b. We have

$$\lim_{x \to 0^-} g(x) = -1 \quad \text{and} \quad \lim_{x \to 0^+} g(x) = 1$$

and since these one-sided limits are not equal, we conclude that $\lim_{x \to 0} g(x)$ does not exist (Figure 38b).

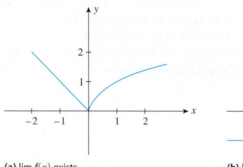
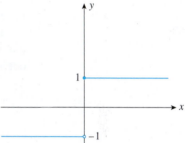

(a) $\lim_{x \to 0} f(x)$ exists.
(b) $\lim_{x \to 0} g(x)$ does not exist.

FIGURE **38**

Continuous Functions

Continuous functions will play an important role throughout most of our study of calculus. Loosely speaking, a function is continuous at a point if the graph of the function at that point is devoid of holes, gaps, jumps, or breaks. Consider, for example, the graph of the function f depicted in Figure 39.

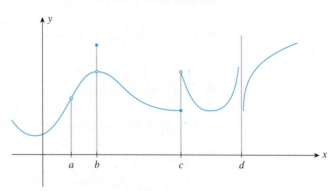

FIGURE **39**
The graph of this function is not continuous at $x = a$, $x = b$, $x = c$, and $x = d$.

Let's take a closer look at the behavior of f at or near $x = a$, $x = b$, $x = c$, and $x = d$. First, note that f is not defined at $x = a$; that is, $x = a$ is not in the domain of f, thereby resulting in a "hole" in the graph of f. Next, observe that the value of f at b, $f(b)$, is not equal to the limit of $f(x)$ as x approaches b, resulting in a "jump" in the graph of f at $x = b$. The function f does not have a limit at $x = c$ since the left-hand and right-hand limits of $f(x)$ are not equal, also resulting in a jump in the graph of f at $x = c$. Finally, the limit of f does not exist at $x = d$, resulting in a break in the graph of f. The function f is *discontinuous* at each of these numbers. It is *continuous* everywhere else.

> **Continuity of a Function at a Number**
>
> A function f is **continuous at a number** $x = a$ if the following conditions are satisfied.
>
> **1.** $f(a)$ is defined.   **2.** $\lim\limits_{x \to a} f(x)$ exists.   **3.** $\lim\limits_{x \to a} f(x) = f(a)$

Thus, a function f is continuous at $x = a$ if the limit of f at $x = a$ exists and has the value $f(a)$. Geometrically, f is continuous at $x = a$ if the proximity of x to a implies the proximity of $f(x)$ to $f(a)$.

If f is not continuous at $x = a$, then f is said to be **discontinuous** at $x = a$. Also, f is **continuous on an interval** if f is continuous at every number in the interval.

Figure 40 depicts the graph of a continuous function on the interval (a, b). Notice that the graph of the function over the stated interval can be sketched without lifting one's pencil from the paper.

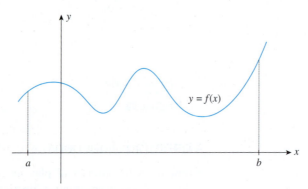

FIGURE 40
The graph of f is continuous on the interval (a, b).

EXAMPLE 2 Find the values of x for which each function is continuous.

a. $f(x) = x + 2$ **b.** $g(x) = \dfrac{x^2 - 4}{x - 2}$ **c.** $h(x) = \begin{cases} x + 2 & \text{if } x \neq 2 \\ 1 & \text{if } x = 2 \end{cases}$

d. $F(x) = \begin{cases} -1 & \text{if } x < 0 \\ 1 & \text{if } x \geq 0 \end{cases}$ **e.** $G(x) = \begin{cases} \dfrac{1}{x} & \text{if } x > 0 \\ -1 & \text{if } x \leq 0 \end{cases}$

The graph of each function is shown in Figure 41 on the next page.

Solution

a. The function f is continuous everywhere because the three conditions for continuity are satisfied for all values of x.

b. The function g is discontinuous at $x = 2$ because g is not defined at that number. It is continuous everywhere else.

c. The function h is discontinuous at $x = 2$ because the third condition for continuity is violated; the limit of $h(x)$ as x approaches 2 exists and has the value 4, but this limit is not equal to $h(2) = 1$. It is continuous for all other values of x.

d. The function F is continuous everywhere except at $x = 0$, where the limit of $F(x)$ fails to exist as x approaches zero (see Example 3a, Section 2.4).

e. Since the limit of $G(x)$ does not exist as x approaches zero, we conclude that G fails to be continuous at $x = 0$. The function G is continuous everywhere else.

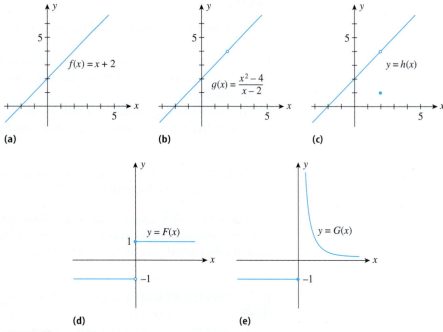

FIGURE **41**

Properties of Continuous Functions

The following properties of continuous functions follow directly from the definition of continuity and the corresponding properties of limits. They are stated without proof.

> **Properties of Continuous Functions**
> 1. The constant function $f(x) = c$ is continuous everywhere.
> 2. The identity function $f(x) = x$ is continuous everywhere.
> *If f and g are continuous at $x = a$, then*
> 3. $[f(x)]^n$, where n is a real number, is continuous at $x = a$ whenever it is defined at that number.
> 4. $f \pm g$ is continuous at $x = a$.
> 5. fg is continuous at $x = a$.
> 6. f/g is continuous at $x = a$ provided that $g(a) \neq 0$.

Using these properties of continuous functions, we can prove the following results. (A proof is sketched in Exercise 101, page 134.)

> **Continuity of Polynomial and Rational Functions**
> 1. A polynomial function $y = P(x)$ is continuous at every value of x.
> 2. A rational function $R(x) = p(x)/q(x)$ is continuous at every value of x where $q(x) \neq 0$.

EXAMPLE 3 Find the values of x for which each function is continuous.

a. $f(x) = 3x^3 + 2x^2 - x + 10$ **b.** $g(x) = \dfrac{8x^{10} - 4x + 1}{x^2 + 1}$

c. $h(x) = \dfrac{4x^3 - 3x^2 + 1}{x^2 - 3x + 2}$

Solution

a. The function f is a polynomial function of degree 3, so $f(x)$ is continuous for all values of x.
b. The function g is a rational function. Observe that the denominator of g—namely, $x^2 + 1$—is never equal to zero. Therefore, we conclude that g is continuous for all values of x.
c. The function h is a rational function. In this case, however, the denominator of h is equal to zero at $x = 1$ and $x = 2$, which can be seen by factoring it. Thus,

$$x^2 - 3x + 2 = (x - 2)(x - 1)$$

We therefore conclude that h is continuous everywhere except at $x = 1$ and $x = 2$, where it is discontinuous.

Up to this point, most of the applications we have discussed involved functions that are continuous everywhere. In Example 4, we consider an application from the field of educational psychology that involves a discontinuous function.

APPLIED EXAMPLE 4 Learning Curves Figure 42 depicts the learning curve associated with a certain individual. Beginning with no knowledge of the subject being taught, the individual makes steady progress toward understanding it over the time interval $0 \le t < t_1$. In this instance, the individual's progress slows as we approach time t_1 because he fails to grasp a particularly difficult concept. All of a sudden, a breakthrough occurs at time t_1, propelling his knowledge of the subject to a higher level. The curve is discontinuous at t_1.

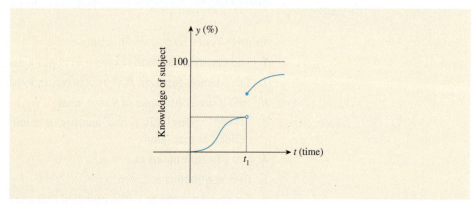

FIGURE 42
A learning curve that is discontinuous at $t = t_1$

Intermediate Value Theorem

Let's look again at our model of the motion of the maglev on a straight stretch of track. We know that the train cannot vanish at any instant of time and it cannot skip portions of the track and reappear someplace else. To put it another way, the train cannot occupy the positions s_1 and s_2 without at least, at some time, occupying an intermediate position (Figure 43).

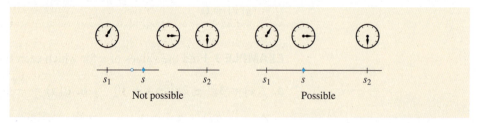

FIGURE 43
The position of the maglev

To state this fact mathematically, recall that the position of the maglev as a function of time is described by

$$f(t) = 4t^2 \qquad (0 \le t \le 10)$$

Suppose the position of the maglev is s_1 at some time t_1 and its position is s_2 at some time t_2 (Figure 44). Then, if s_3 is any number between s_1 and s_2 giving an intermediate position of the maglev, there must be at least one t_3 between t_1 and t_2 giving the time at which the train is at s_3—that is, $f(t_3) = s_3$.

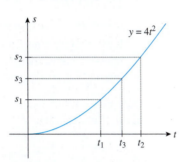

FIGURE 44
If $s_1 \le s_3 \le s_2$, then there must be at least
one t_3 $(t_1 \le t_3 \le t_2)$ such that $f(t_3) = s_3$.

This discussion carries the gist of the Intermediate Value Theorem. The proof of this theorem can be found in most advanced calculus texts.

THEOREM 4

The Intermediate Value Theorem

If f is a continuous function on a closed interval $[a, b]$ and M is any number between $f(a)$ and $f(b)$, then there is at least one number c in $[a, b]$ such that $f(c) = M$ (Figure 45).

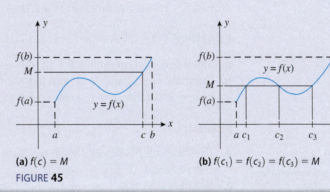

(a) $f(c) = M$

(b) $f(c_1) = f(c_2) = f(c_3) = M$

FIGURE 45

To illustrate the Intermediate Value Theorem, let's look at the example involving the motion of the maglev again (see Figure 27, page 102). Notice that the initial position of the train is $f(0) = 0$ and the position at the end of its test run is $f(10) = 400$. Furthermore, the function f is continuous on $[0, 10]$. So the Intermediate Value Theorem guarantees that if we arbitrarily pick a number between 0 and 400—say, 100—giving the position of the maglev, there must be a $\bar{t}$ (read "t bar") between 0 and 10 at which time the train is at the position $s = 100$. To find the value of $\bar{t}$, we solve the equation $f(\bar{t}) = s$, or

$$4\bar{t}^2 = 100$$

giving $\bar{t} = 5$ (t must lie between 0 and 10).

⚠️ It is important to remember when we use Theorem 4 that the function f must be continuous. The conclusion of the Intermediate Value Theorem may not hold if f is not continuous (see Exercise 102, page 134).

The next theorem is an immediate consequence of the Intermediate Value Theorem. It not only tells us when a **zero of a function** f [root of the equation $f(x) = 0$] exists but also provides the basis for a method of approximating it.

THEOREM 5

Existence of Zeros of a Continuous Function

If f is a continuous function on a closed interval $[a, b]$, and if $f(a)$ and $f(b)$ have opposite signs, then there is at least one solution of the equation $f(x) = 0$ in the interval (a, b) (Figure 46).

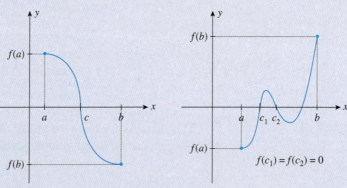

FIGURE 46
If $f(a)$ and $f(b)$ have opposite signs, there must be at least one number c $(a < c < b)$ such that $f(c) = 0$.

Geometrically, this property states that if the graph of a continuous function goes from above the x-axis to below the x-axis or vice versa, it must *cross* the x-axis. This is not necessarily true if the function is discontinuous (Figure 47).

EXAMPLE 5 Let $f(x) = x^3 + x + 1$.

a. Show that f is continuous for all values of x.
b. Compute $f(-1)$ and $f(1)$, and use the results to deduce that there must be at least one number $x = c$, where c lies in the interval $(-1, 1)$ and $f(c) = 0$.

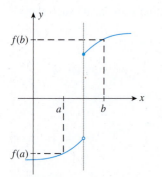

FIGURE 47
$f(a) < 0$ and $f(b) > 0$, but the graph of f does not cross the x-axis between a and b because f is discontinuous.

Solution

a. The function f is a polynomial function of degree 3 and is therefore continuous everywhere.

b. $f(-1) = (-1)^3 + (-1) + 1 = -1$ and $f(1) = 1^3 + 1 + 1 = 3$

Since $f(-1)$ and $f(1)$ have opposite signs, Theorem 5 tells us that there must be at least one number $x = c$ with $-1 < c < 1$ such that $f(c) = 0$. ▪

The next example shows how the Intermediate Value Theorem can be used to help us find a zero of a function.

EXAMPLE 6 Let $f(x) = x^3 + x - 1$. Since f is a polynomial function, it is continuous everywhere. Observe that $f(0) = -1$ and $f(1) = 1$, so Theorem 5 guarantees the existence of at least one root of the equation $f(x) = 0$ in $(0, 1)$.*

---

*It can be shown that f has precisely one zero in $(0, 1)$ (see Exercise 110, Section 4.1).

We can locate the root more precisely by using Theorem 5 once again as follows: Evaluate $f(x)$ at the midpoint of $[0, 1]$, obtaining

$$f(0.5) = -0.375$$

Because $f(0.5) < 0$ and $f(1) > 0$, Theorem 5 now tells us that the root must lie in $(0.5, 1)$.

Repeat the process: Evaluate $f(x)$ at the midpoint of $[0.5, 1]$, which is

$$\frac{0.5 + 1}{2} = 0.75$$

Thus,

$$f(0.75) \approx 0.1719$$

Because $f(0.5) < 0$ and $f(0.75) > 0$, Theorem 5 tells us that the root is in $(0.5, 0.75)$. This process can be continued. Table 3 summarizes the results of our computations through nine steps.

From Table 3, we see that the root is approximately 0.68, accurate to two decimal places. By continuing the process through a sufficient number of steps, we can obtain as accurate an approximation to the root as we please.

TABLE 3

Step	Root of $f(x) = 0$ Lies in
1	$(0, 1)$
2	$(0.5, 1)$
3	$(0.5, 0.75)$
4	$(0.625, 0.75)$
5	$(0.625, 0.6875)$
6	$(0.65625, 0.6875)$
7	$(0.671875, 0.6875)$
8	$(0.6796875, 0.6875)$
9	$(0.6796875, 0.6835937)$

Note The process of finding the root of $f(x) = 0$ used in Example 6 is called the **method of bisection.** It is crude but effective.

2.5 Self-Check Exercises

1. Evaluate $\lim\limits_{x \to -1^-} f(x)$ and $\lim\limits_{x \to -1^+} f(x)$, where

$$f(x) = \begin{cases} 1 & \text{if } x < -1 \\ 1 + \sqrt{x + 1} & \text{if } x \geq -1 \end{cases}$$

Does $\lim\limits_{x \to -1} f(x)$ exist?

2. Determine the values of x for which the given function is discontinuous. At each number where f is discontinuous, indicate which condition(s) for continuity are violated. Sketch the graph of the function.

   a. $f(x) = \begin{cases} -x^2 + 1 & \text{if } x \leq 1 \\ x - 1 & \text{if } x > 1 \end{cases}$

   b. $g(x) = \begin{cases} -x + 1 & \text{if } x < -1 \\ 2 & \text{if } -1 < x \leq 1 \\ -x + 3 & \text{if } x > 1 \end{cases}$

   Solutions to Self-Check Exercises 2.5 can be found on page 134.

2.5 Concept Questions

1. Explain what is meant by the statement $\lim\limits_{x \to 3^-} f(x) = 2$ and $\lim\limits_{x \to 3^+} f(x) = 4$.

2. Suppose $\lim\limits_{x \to 1^-} f(x) = 3$ and $\lim\limits_{x \to 1^+} f(x) = 4$.

   a. What can you say about $\lim\limits_{x \to 1} f(x)$? Explain.

   b. What can you say about $f(1)$? Explain.

3. Explain what it means for a function f to be continuous (a) at a number a and (b) on an interval I.

4. Suppose $\lim\limits_{x \to a^-} f(x) = L$ and $\lim\limits_{x \to a^+} f(x) = M$, where L and M are real numbers. What conditions on L, M, and $f(a)$ will guarantee that f is continuous at $x = a$?

5. Determine whether each function f is continuous or discontinuous. Explain your answer.

   a. $f(t)$ gives the altitude of an airplane at time t.

   b. $f(t)$ measures the total amount of rainfall at time t at the Municipal Airport.

   c. $f(s)$ measures the fare as a function of the distance s for taking a cab from Kennedy Airport to downtown Manhattan.

   d. $f(t)$ gives the interest rate charged by a financial institution at time t.

6. Explain the Intermediate Value Theorem in your own words.

2.5   Exercises

In Exercises 1–8, use the graph of the function f to find $\lim_{x \to a^-} f(x)$, $\lim_{x \to a^+} f(x)$, and $\lim_{x \to a} f(x)$ at the indicated value of a, if the limit exists.

1.

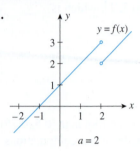

$a = 2$

2.

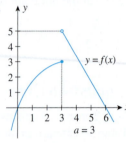

$a = 3$

3.

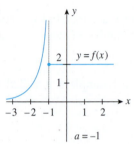

$a = -1$

4.

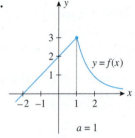

$a = 1$

5.

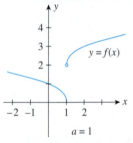

$a = 1$

6.

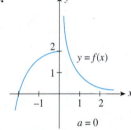

$a = 0$

7.

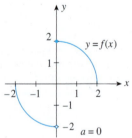

$a = 0$

8.

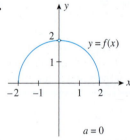

$a = 0$

In Exercises 9–14, refer to the graph of the function f and determine whether each statement is true or false.

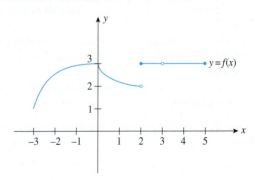

9. $\lim_{x \to -3^+} f(x) = 1$

10. $\lim_{x \to 0} f(x) = f(0)$

11. $\lim_{x \to 2^-} f(x) = 2$

12. $\lim_{x \to 2^+} f(x) = 3$

13. $\lim_{x \to 3} f(x)$ does not exist.

14. $\lim_{x \to 5^-} f(x) = 3$

In Exercises 15–20, refer to the graph of the function f and determine whether each statement is true or false.

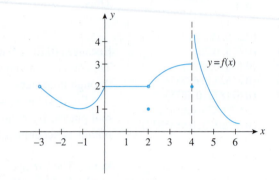

15. $\lim_{x \to -3^+} f(x) = 2$

16. $\lim_{x \to 0} f(x) = 2$

17. $\lim_{x \to 2} f(x) = 1$

18. $\lim_{x \to 4^-} f(x) = 3$

19. $\lim_{x \to 4^+} f(x)$ does not exist.

20. $\lim_{x \to 4} f(x) = 2$

In Exercises 21–38, find the indicated one-sided limit, if it exists.

21. $\lim_{x \to 1^+} (2x + 4)$

22. $\lim_{x \to 1^-} (3x - 4)$

23. $\lim_{x \to 2^-} \dfrac{x - 3}{x + 2}$

24. $\lim_{x \to 1^+} \dfrac{x + 2}{x + 1}$

25. $\lim_{x \to 0^+} \dfrac{1}{x}$

26. $\lim_{x \to 0^-} \dfrac{1}{x}$

27. $\lim_{x \to 0^+} \dfrac{x - 1}{x^2 + 1}$

28. $\lim_{x \to 2^+} \dfrac{x + 1}{x^2 - 2x + 3}$

29. $\lim_{x \to 0^+} \sqrt{x}$

30. $\lim_{x \to 2^+} 2\sqrt{x - 2}$

31. $\lim_{x \to -2^+} (2x + \sqrt{2 + x})$

32. $\lim_{x \to -5^+} x(1 + \sqrt{5 + x})$

33. $\lim_{x \to 1^-} \dfrac{1 + x}{1 - x}$

34. $\lim_{x \to 1^+} \dfrac{1 + x}{1 - x}$

35. $\lim_{x \to 2^-} \dfrac{x^2 - 4}{x - 2}$

36. $\lim_{x \to -3^+} \dfrac{\sqrt{x + 3}}{x^2 + 1}$

37. $\lim_{x \to 0^+} f(x)$ and $\lim_{x \to 0^-} f(x)$, where

$$f(x) = \begin{cases} 2x & \text{if } x < 0 \\ x^2 & \text{if } x \geq 0 \end{cases}$$

38. $\lim_{x \to 0^+} f(x)$ and $\lim_{x \to 0^-} f(x)$, where

$$f(x) = \begin{cases} -x + 1 & \text{if } x \leq 0 \\ 2x + 3 & \text{if } x > 0 \end{cases}$$

In Exercises 39–44, determine the values of x, if any, at which each function is discontinuous. At each number where f is discontinuous, state the condition(s) for continuity that are violated.

39.

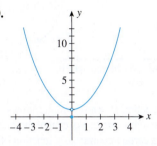

$$f(x) = \begin{cases} 2x - 4 & \text{if } x \le 0 \\ 1 & \text{if } x > 0 \end{cases}$$

40.

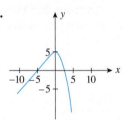

$$f(x) = \begin{cases} x^2 + 1 & \text{if } x \ne 0 \\ 0 & \text{if } x = 0 \end{cases}$$

41.

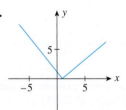

$$f(x) = \begin{cases} x + 5 & \text{if } x \le 0 \\ -x^2 + 5 & \text{if } x > 0 \end{cases}$$

42.

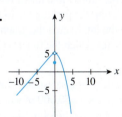

$$f(x) = |x - 1|$$

43.

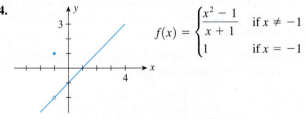

$$f(x) = \begin{cases} x + 5 & \text{if } x < 0 \\ 2 & \text{if } x = 0 \\ -x^2 + 5 & \text{if } x > 0 \end{cases}$$

44.

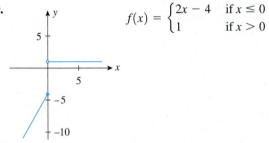

$$f(x) = \begin{cases} \dfrac{x^2 - 1}{x + 1} & \text{if } x \ne -1 \\ 1 & \text{if } x = -1 \end{cases}$$

In Exercises 45–56, find the values of x for which each function is continuous.

45. $f(x) = 2x^2 + x - 1$

46. $f(x) = x^3 - 2x^2 + x - 1$

47. $f(x) = \dfrac{2}{x^2 + 1}$

48. $f(x) = \dfrac{x}{2x^2 + 1}$

49. $f(x) = \dfrac{2}{2x - 1}$

50. $f(x) = \dfrac{x + 1}{x - 1}$

51. $f(x) = \dfrac{2x + 1}{x^2 + x - 2}$

52. $f(x) = \dfrac{x - 1}{x^2 + 2x - 3}$

53. $f(x) = \begin{cases} x & \text{if } x \le 1 \\ 2x - 1 & \text{if } x > 1 \end{cases}$

54. $f(x) = \begin{cases} -2x + 1 & \text{if } x < 0 \\ x^2 + 1 & \text{if } x \ge 0 \end{cases}$

55. $f(x) = |x + 1|$

56. $f(x) = \dfrac{|x - 1|}{x - 1}$

In Exercises 57–60, determine all values of x at which the function is discontinuous.

57. $f(x) = \dfrac{2x}{x^2 - 1}$

58. $f(x) = \dfrac{1}{(x - 1)(x - 2)}$

59. $f(x) = \dfrac{x^2 - 2x}{x^2 - 3x + 2}$

60. $f(x) = \dfrac{x^2 - 3x + 2}{x^2 - 2x}$

61. THE POSTAGE FUNCTION The graph of the "postage function" for 2015,

$$f(x) = \begin{cases} 232 & \text{if } 0 < x < 4 \\ 250 & \text{if } 4 \le x < 5 \\ \vdots \\ 394 & \text{if } 12 \le x < 13 \\ 412 & \text{if } x = 13 \end{cases}$$

where x denotes the weight of a package in ounces and $f(x)$ the postage in cents, is shown in the accompanying figure. Determine the values of x for which f is discontinuous.

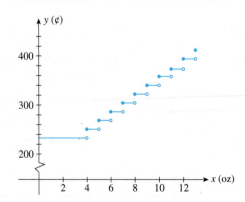

62. INVENTORY CONTROL As part of an optimal inventory policy, the manager of an office supply company orders 500 reams of photocopy paper every 20 days. The accompanying graph shows the *actual* inventory level of paper in an office supply store during the first 60 business days of 2014. Determine the values of t for which the "inventory function" is discontinuous, and give an interpretation of the graph.

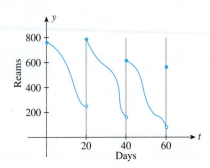

63. LEARNING CURVES The following graph describes the progress Michael made in solving a problem correctly during a mathematics quiz. Here, y denotes the percentage of work completed, and x is measured in minutes. Give an interpretation of the graph.

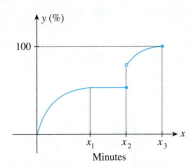

64. AILING FINANCIAL INSTITUTIONS Franklin Savings and Loan acquired two ailing financial institutions in 2010. One of them was acquired at time $t = T_1$, and the other was acquired at time $t = T_2$ ($t = 0$ corresponds to the beginning of 2010). The following graph shows the total amount of money on deposit with Franklin. Explain the significance of the discontinuities of the function at T_1 and T_2.

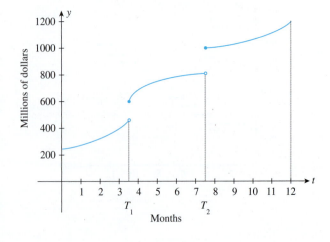

65. ENERGY CONSUMPTION The following graph shows the amount of home heating oil remaining in a 200-gal tank over a 120-day period ($t = 0$ corresponds to October 1). Explain why the function is discontinuous at $t = 40$, 70, 95, and 110.

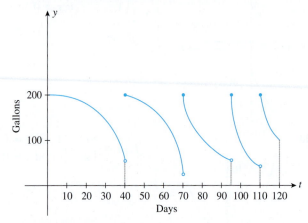

66. PRIME INTEREST RATE The function P, whose graph follows, gives the prime rate (the interest rate banks charge their best corporate customers) for a certain country as a function of time for the first 32 weeks in 2014. Determine the values of t for which P is discontinuous, and interpret your results.

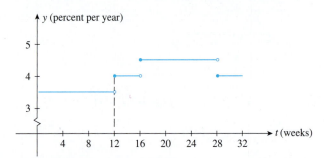

67. PARKING FEES The fee charged per car in a downtown parking lot is $2.00 for the first half hour and $1.00 for each additional half hour or part thereof, subject to a maximum of $10.00. Derive a function f relating the parking fee to the length of time a car is left in the lot. Sketch the graph of f and determine the values of x for which the function f is discontinuous.

68. COMMISSIONS The base monthly salary of a salesman working on commission is $22,000. For each $50,000 of sales beyond $100,000, he is paid a $1000 commission. Sketch a graph showing his earnings as a function of the level of his sales x. Determine the values of x for which the function f is discontinuous.

69. COMMODITY PRICES The function that gives the cost of a certain commodity is defined by

$$C(x) = \begin{cases} 5x & \text{if } 0 < x < 10 \\ 4x & \text{if } 10 \le x < 30 \\ 3.5x & \text{if } 30 \le x < 60 \\ 3.25x & \text{if } x \ge 60 \end{cases}$$

where x is the number of pounds of a certain commodity sold and $C(x)$ is measured in dollars. Sketch the graph of the function C and determine the values of x for which the function C is discontinuous.

70. **ENERGY EXPENDED BY A FISH** Suppose a fish swimming a distance of L ft at a speed of v ft/sec relative to the water and against a current flowing at the rate of u ft/sec $(u < v)$ expends a total energy given by

$$E(v) = \frac{aLv^3}{v - u}$$

where E is measured in foot-pounds (ft-lb) and a is a constant.

a. Evaluate $\lim\limits_{v \to u^+} E(v)$, and interpret your result.

b. Evaluate $\lim\limits_{v \to \infty} E(v)$, and interpret your result.

71. **WEISS'S LAW** According to Weiss's law of excitation of tissue, the strength S of an electric current is related to the time t the current takes to excite tissue by the formula

$$S(t) = \frac{a}{t} + b \qquad (t > 0)$$

where a and b are positive constants.

a. Evaluate $\lim\limits_{t \to 0^+} S(t)$ and interpret your result.

b. Evaluate $\lim\limits_{t \to \infty} S(t)$ and interpret your result.

(*Note:* The limit in part (b) is called the threshold strength of the current. Why?)

72. **LEVERAGED RETURN** The return on assets using borrowed money, called *leveraged return*, is given by

$$L = \frac{Y - (1 - D)R}{D}$$

where Y is the return of the asset, R is the cost of borrowed money, and D is the percentage of money the investor must put down to secure the loan. Suppose that both Y and R are constant.

a. Find $\lim\limits_{D \to 0^+} L$ and interpret your result.

b. Find $\lim\limits_{D \to 1^-} L$ and interpret your result.

Source: Scientific American.

73. Let

$$f(x) = \begin{cases} x + 2 & \text{if } x \le 1 \\ kx^2 & \text{if } x > 1 \end{cases}$$

Find the value of k that will make f continuous on $(-\infty, \infty)$.

74. Let

$$f(x) = \begin{cases} \dfrac{x^2 - 4}{x + 2} & \text{if } x \ne -2 \\ k & \text{if } x = -2 \end{cases}$$

For what value of k will f be continuous on $(-\infty, \infty)$?

In Exercises 75–78, (a) show that the function f is continuous for all values of x in the interval $[a, b]$ and (b) prove that f must have at least one zero in the interval (a, b) by showing that $f(a)$ and $f(b)$ have opposite signs.

75. $f(x) = x^2 - 6x + 8$; $a = 1$, $b = 3$

76. $f(x) = 2x^3 - 3x^2 - 36x + 14$; $a = 0$, $b = 1$

77. $f(x) = x^3 - 2x^2 + 3x + 2$; $a = -1$, $b = 1$

78. $f(x) = 2x^{5/3} - 5x^{4/3}$; $a = 14$, $b = 16$

In Exercises 79 and 80, use the Intermediate Value Theorem to show that there exists a number c in the given interval such that $f(c) = M$. Then find its value.

79. $f(x) = x^2 - 4x + 6$ on $[0, 3]$; $M = 4$

80. $f(x) = x^2 - x + 1$ on $[-1, 4]$; $M = 7$

81. Use the method of bisection (see Example 6) to find the root of the equation $x^5 + 2x - 7 = 0$ accurate to two decimal places.

82. Use the method of bisection (see Example 6) to find the root of the equation $x^3 - x + 1 = 0$ accurate to two decimal places.

83. **FALLING OBJECT** Joan is looking straight out a window of an apartment building at a height of 32 ft from the ground. A boy on the ground throws a tennis ball straight up by the side of the building where the window is located. Suppose the height of the ball (measured in feet) from the ground at time t is $h(t) = 4 + 64t - 16t^2$.

a. Show that $h(0) = 4$ and $h(2) = 68$.

b. Use the Intermediate Value Theorem to conclude that the ball must cross Joan's line of sight at least once.

c. At what time(s) does the ball cross Joan's line of sight? Interpret your results.

84. **OXYGEN CONTENT OF A POND** The oxygen content t days after organic waste has been dumped into a pond is given by

$$f(t) = 100 \left(\frac{t^2 + 10t + 100}{t^2 + 20t + 100} \right)$$

percent of its normal level.

a. Show that $f(0) = 100$ and $f(10) = 75$.

b. Use the Intermediate Value Theorem to conclude that the oxygen content of the pond must have been at a level of 80% at some time.

c. At what time(s) is the oxygen content at the 80% level?

Hint: Use the quadratic formula.

In Exercises 85–97, determine whether the statement is true or false. If it is true, explain why it is true. If it is false, give an example to show why it is false.

85. If $f(2) = 4$, then $\lim\limits_{x \to 2} f(x) = 4$.

86. If $\lim\limits_{x \to 0} f(x) = 3$, then $f(0) = 3$.

87. If $\lim\limits_{x \to 2^+} f(x) = 3$ and $f(2) = 3$ then $\lim\limits_{x \to 2^-} f(x) = 3$.

88. If $\lim\limits_{x\to3^-} f(x)$ and $\lim\limits_{x\to3^+} f(x)$ both exist, then $\lim\limits_{x\to3} f(x)$ exists.

89. If $f(5)$ is not defined, then $\lim\limits_{x\to5^-} f(x)$ does not exist.

90. Suppose the function f is defined on the interval $[a, b]$. If $f(a)$ and $f(b)$ have the same sign, then f has no zero in $[a, b]$.

91. If $\lim\limits_{x\to a^-} f(x) = L$ and $\lim\limits_{x\to a^+} f(x) = L$, then $f(a) = L$.

92. If $\lim\limits_{x\to a} f(x) = L$ and $g(a) = M$, then $\lim\limits_{x\to a} f(x)g(x) = LM$.

93. If f is continuous for all $x \neq 0$ and $f(0) = 0$, then $\lim\limits_{x\to0} f(x) = 0$.

94. If f and g are both discontinuous at a, then $f + g$ must be discontinuous at a.

95. If f and g are both discontinuous at a, then fg must be discontinuous at a.

96. If $g \circ f$ is continuous at a, then f must be continuous at a and g must be continuous at $f(a)$.

97. If f is discontinuous at a and g is discontinuous at $f(a)$, then $g \circ f$ must be discontinuous at a.

98. Is the following statement true or false? Suppose f is continuous on $[a, b]$ and $f(a) < f(b)$. If M is a number that lies outside the interval $[f(a), f(b)]$, then there does not exist a number $a < c < b$ such that $f(c) = M$. Does this contradict the Intermediate Value Theorem?

99. Let $f(x) = \dfrac{x^2}{x^2 + 1}$.
   a. Show that f is continuous for all values of x.
   b. Show that $f(x)$ is nonnegative for all values of x.
   c. Show that f has a zero at $x = 0$. Does this contradict Theorem 5?

100. Let $f(x) = x - \sqrt{1 - x^2}$.
   a. Show that f is continuous for all values of x in the interval $[-1, 1]$.
   b. Show that f has at least one zero in $[-1, 1]$.
   c. Find the zeros of f in $[-1, 1]$ by solving the equation $f(x) = 0$.

101. a. Prove that a polynomial function $y = P(x)$ is continuous at every number x. Follow these steps:
   (i) Use Properties 2 and 3 of continuous functions to establish that the function $g(x) = x^n$, where n is a positive integer, is continuous everywhere.
   (ii) Use Properties 1 and 5 to show that $f(x) = cx^n$, where c is a constant and n is a positive integer, is continuous everywhere.
   (iii) Use Property 4 to complete the proof of the result.
   b. Prove that a rational function $R(x) = p(x)/q(x)$ is continuous at every point x where $q(x) \neq 0$.
   Hint: Use the result of part (a) and Property 6.

102. Show that the conclusion of the Intermediate Value Theorem does not necessarily hold if f is discontinuous on $[a, b]$.

2.5   Solutions to Self-Check Exercises

1. For $x < -1$, $f(x) = 1$, and so

$$\lim\limits_{x\to-1^-} f(x) = \lim\limits_{x\to-1^-} 1 = 1$$

For $x \geq -1$, $f(x) = 1 + \sqrt{x + 1}$, and so

$$\lim\limits_{x\to-1^+} f(x) = \lim\limits_{x\to-1^+} (1 + \sqrt{x + 1}) = 1$$

Since the left-hand and right-hand limits of f exist as x approaches -1 and both are equal to 1, we conclude that

$$\lim\limits_{x\to-1} f(x) = 1$$

2. a. The graph of f follows:

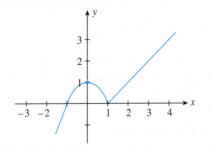

We see that f is continuous everywhere.

b. The graph of g follows:

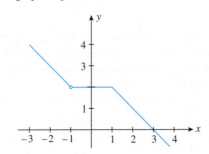

Since g is not defined at $x = -1$, it is discontinuous there. It is continuous everywhere else.

USING TECHNOLOGY

Finding the Points of Discontinuity of a Function

You can very often recognize the points of discontinuity of a function f by examining its graph. For example, Figure T1a shows the graph of $f(x) = x/(x^2 - 1)$ obtained using a graphing utility. It is evident that f is discontinuous at $x = -1$ and $x = 1$. This observation is also borne out by the fact that both these points are not in the domain of f.

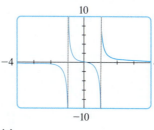
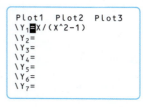

(a)   (b)

FIGURE **T1**
(a) The graph of $f(x) = x/(x^2 - 1)$ in the viewing window $[-4, 4] \times [-10, 10]$; (b) the TI-83/84 equation screen

Consider the function

$$g(x) = \frac{2x^3 + x^2 - 7x - 6}{x^2 - x - 2}$$

Using a graphing utility, we obtain the graph of g shown in Figure T2a.

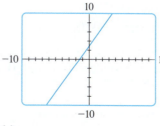
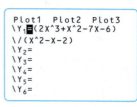

(a)   (b)

FIGURE **T2**
(a) The graph of $g(x) = (2x^3 + x^2 - 7x - 6)/(x^2 - x - 2)$ in the standard viewing window; (b) the TI-83/84 equation screen

An examination of this graph does not reveal any points of discontinuity. However, if we factor both the numerator and the denominator of the rational expression, we see that

$$g(x) = \frac{(x + 1)(x - 2)(2x + 3)}{(x + 1)(x - 2)}$$

$$= 2x + 3$$

provided that $x \neq -1$ and $x \neq 2$, so its graph in fact looks like that shown in Figure T3.

This example shows the limitation of the graphing utility and reminds us of the importance of studying functions analytically!

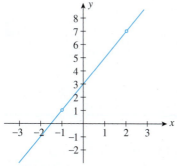

FIGURE **T3**
The graph of g has holes at $(-1, 1)$ and $(2, 7)$.

Graphing Functions Defined Piecewise

The following example illustrates how to plot the graphs of functions defined in a piecewise manner on a graphing utility.

EXAMPLE 1 Plot the graph of

$$f(x) = \begin{cases} x + 1 & \text{if } x \leq 1 \\ \dfrac{2}{x} & \text{if } x > 1 \end{cases}$$

Solution   We enter the function

$$y1 = (x + 1)(x \leq 1) + (2/x)(x > 1)$$

Figure T4a shows the graph of the function in the viewing window $[-5, 5] \times [-2, 4]$.

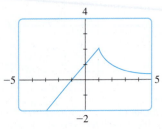

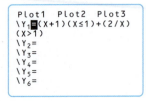

(a)   (b)

FIGURE T4
(a) The graph of *f* in the viewing window $[-5, 5] \times [-2, 4]$; (b) the TI-83/84 equation screen

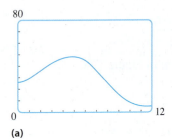

(a)

Plot1 Plot2 Plot3
\Y₁◼(.01354X^4-
.49375X^3+2.58333
X^2+3.8X+31.60704)
(X≥0)(X≤8)+
(1.35X^2-33.05X+208)
(X>8)(X≤12)
\Y₂=

(b)

FIGURE T5
(a) The graph of *P* in the viewing window
$[0, 12] \times [0, 80]$; (b) the TI-83/84 equation
screen

APPLIED EXAMPLE 2 TV Viewing Patterns   The percent of U.S. households, $P(t)$, watching television during weekdays between the hours of 4 P.M. and 4 A.M. is given by

$$P(t) = \begin{cases} 0.01354t^4 - 0.49375t^3 + 2.58333t^2 + 3.8t + 31.60704 & \text{if } 0 \leq t \leq 8 \\ 1.35t^2 - 33.05t + 208 & \text{if } 8 < t \leq 12 \end{cases}$$

where *t* is measured in hours, with $t = 0$ corresponding to 4 P.M. Plot the graph of *P* in the viewing window $[0, 12] \times [0, 80]$.
Source: A. C. Nielsen Co.

Solution   We enter the function

$$y1 = (.01354x^4 - .49375x^3 + 2.58333x^2 + 3.8x + 31.60704)(x \geq 0)(x \leq 8)$$
$$+ (1.35x^2 - 33.05x + 208)(x > 8)(x \leq 12)$$

Figure T5a shows the graph of *P*.

TECHNOLOGY EXERCISES

In Exercises 1–8, plot the graph of *f* and find the points of discontinuity of *f*. Then use analytical means to verify your observation and find all numbers where *f* is discontinuous.

1. $f(x) = \dfrac{2}{x^2 - x}$

2. $f(x) = \dfrac{3}{\sqrt{x}(x + 1)}$

3. $f(x) = \dfrac{6x^3 + x^2 - 2x}{2x^2 - x}$

4. $f(x) = \dfrac{2x^3 - x^2 - 13x - 6}{2x^2 - 5x - 3}$

5. $f(x) = \dfrac{2x^4 - 3x^3 - 2x^2}{2x^2 - 3x - 2}$

6. $f(x) = \dfrac{6x^4 - x^3 + 5x^2 - 1}{6x^2 - x - 1}$

7. $f(x) = \dfrac{x^3 + x^2 - 2x}{x^4 + 2x^3 - x - 2}$
   Hint: $x^4 + 2x^3 - x - 2 = (x^3 - 1)(x + 2)$

8. $f(x) = \dfrac{x^3 - x}{x^{4/3} - x + x^{1/3} - 1}$

Hint: $x^{4/3} - x + x^{1/3} - 1 = (x^{1/3} - 1)(x + 1)$

In Exercises 9 and 10, plot the graph of f in the indicated viewing window.

9. $f(x) = \begin{cases} 2 & \text{if } x \le 0 \\ \sqrt{4 - x^2} & \text{if } x > 0; \end{cases} [-2, 2] \times [-4, 4]$

10. $f(x) = \begin{cases} -x^2 + x + 2 & \text{if } x \le 1 \\ 2x^3 - x^2 - 4 & \text{if } x > 1; \end{cases} [-4, 4] \times [-5, 5]$

11. FLIGHT PATH OF A PLANE The function

$$f(x) = \begin{cases} 0 & \text{if } 0 \le x < 1 \\ -0.00411523x^3 + 0.0679012x^2 \\ \quad -0.123457x + 0.0596708 & \text{if } 1 \le x < 10 \\ 1.5 & \text{if } 10 \le x \le 100 \end{cases}$$

where both x and $f(x)$ are measured in units of 1000 ft, describes the flight path of a plane taking off from the origin and climbing to an altitude of 15,000 ft. Plot the graph of f to visualize the trajectory of the plane.

12. OBESE CHILDREN IN THE UNITED STATES The percentage of obese children aged 12–19 in the United States is approximately

$$P(t) = \begin{cases} 0.04t + 4.6 & \text{if } 0 \le t < 10 \\ -0.01005t^2 + 0.945t - 3.4 & \text{if } 10 \le t \le 30 \end{cases}$$

where t is measured in years, with $t = 0$ corresponding to the beginning of 1970.
  a. Plot the graph of $P(t)$.
  b. What was the percentage of obese children aged 12–19 at the beginning of 1970? At the beginning of 1985? At the beginning of 2000?

Source: Centers for Disease Control and Prevention.

2.6 The Derivative

An Intuitive Example

We mentioned in Section 2.4 that the problem of finding the *rate of change* of one quantity with respect to another is mathematically equivalent to the problem of finding the *slope of the tangent line* to a curve at a given point on the curve. Before going on to establish this relationship, let's show its plausibility by looking at it from an intuitive point of view.

Consider the motion of the maglev discussed in Section 2.4. Recall that the position of the maglev at any time t is given by

$$s = f(t) = 4t^2 \qquad (0 \le t \le 30)$$

where s is measured in feet and t in seconds. The graph of the function f is sketched in Figure 48.

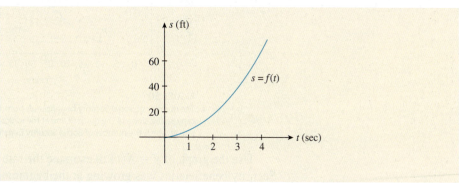

FIGURE 48
Graph showing the position s of a maglev at time t

Observe that the graph of f rises slowly at first but more rapidly as t increases, reflecting the fact that the speed of the maglev is increasing with time. This observation suggests a relationship between the speed of the maglev at any time t and the *steepness* of the curve at the point corresponding to this value of t. Thus, it would appear that we can solve the problem of finding the speed of the maglev at any time if we can find a way to measure the steepness of the curve at any point on the curve.

To discover a yardstick that will measure the steepness of a curve, consider the graph of a function f such as the one shown in Figure 49a. Think of the curve as representing a stretch of roller coaster track (Figure 49b). When the car is at the point P on the curve, a passenger sitting erect in the car and looking straight ahead will have a line of sight that is parallel to the line T, the tangent to the curve at P.

As Figure 49a suggests, the steepness of the curve—that is, the rate at which y is increasing or decreasing with respect to x—is given by the slope of the tangent line to the graph of f at the point $P(x, f(x))$. But for now we will show how this relationship can be used to estimate the rate of change of a function from its graph.

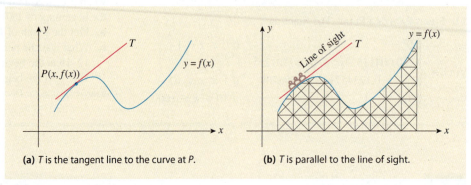

(a) *T* is the tangent line to the curve at *P*.   (b) *T* is parallel to the line of sight.

FIGURE **49**

![$] **APPLIED EXAMPLE 1** Increasing Number of Social Security Beneficiaries The graph of the function $y = N(t)$, shown in Figure 50, gives the number of Social Security beneficiaries from the beginning of 1990 $(t = 0)$ through the year 2045 $(t = 55)$.

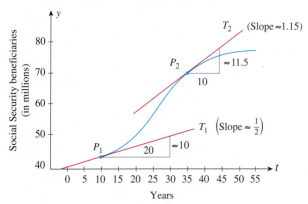

FIGURE **50**
The number of Social Security beneficiaries from 1990 through 2045. We can use the slope of the tangent line at the indicated points to estimate the rate at which the number of Social Security beneficiaries will be changing.

Use the graph of $y = N(t)$ to estimate the rate at which the number of Social Security beneficiaries was growing at the beginning of the year 2000 $(t = 10)$. How fast will the number be growing at the beginning of 2025 $(t = 35)$? [Assume that the rate of change of the function N at any value of t is given by the slope of the tangent line at the point $P(t, N(t))$.]
Source: Social Security Administration.

Solution   From the figure, we see that the slope of the tangent line T_1 to the graph of $y = N(t)$ at $P_1(10, 44.7)$ is approximately 0.5. This tells us that the quantity y is increasing at the rate of $\frac{1}{2}$ unit per unit increase in t, when $t = 10$. In other words, at the beginning of the year 2000, the number of Social Security beneficiaries was increasing at the rate of approximately 0.5 million, or 500,000, per year.

The slope of the tangent line T_2 at $P_2(35, 71.9)$ is approximately 1.15. This tells us that at the beginning of 2025 the number of Social Security beneficiaries will be growing at the rate of approximately 1.15 million, or 1,150,000, per year. ■

Slope of a Tangent Line

In Example 1, we answered the questions raised by drawing the graph of the function N and estimating the position of the tangent lines. Ideally, however, we would like to solve a problem analytically whenever possible. To do this, we need a precise definition of the slope of a tangent line to a curve.

To define the tangent line to a curve C at a point P on the curve, fix P and let Q be any point on C distinct from P (Figure 51). The straight line passing through P and Q is called a **secant line**.

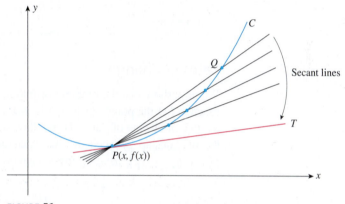

FIGURE **51**
As Q approaches P along the curve C, the secant lines approach the tangent line T.

Now, as the point Q is allowed to move toward P along the curve, the secant line through P and Q rotates about the fixed point P and approaches a fixed line through P. This fixed line, which is the limiting position of the secant lines through P and Q as Q approaches P, is the **tangent line to the graph of f** at the point P.

We can describe the process more precisely as follows. Suppose the curve C is the graph of a function f defined by $y = f(x)$. Then the point P is described by $P(x, f(x))$ and the point Q by $Q(x + h, f(x + h))$, where h is some appropriate nonzero number (Figure 52a). Observe that we can make Q approach P along the curve C by letting h approach zero (Figure 52b).

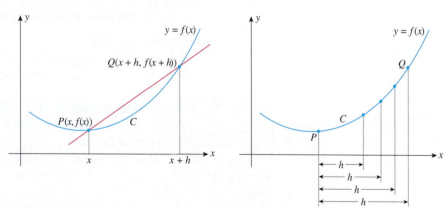

(a) The points $P(x, f(x))$ and $Q(x + h, f(x + h))$

(b) As h approaches zero, Q approaches P.

FIGURE **52**

Next, using the formula for the slope of a line, we can write the slope of the secant line passing through $P(x, f(x))$ and $Q(x + h, f(x + h))$ as

$$\frac{f(x + h) - f(x)}{(x + h) - x} = \frac{f(x + h) - f(x)}{h} \tag{5}$$

As we observed earlier, Q approaches P, and therefore the secant line through P and Q approaches the tangent line T as h approaches zero. Consequently, we might expect that the slope of the secant line would approach the slope of the tangent line T as h approaches zero. This leads to the following definition.

> **Slope of a Tangent Line**
>
> The slope of the tangent line to the graph of f at the point $P(x, f(x))$ is given by
>
> $$\lim_{h \to 0} \frac{f(x + h) - f(x)}{h} \tag{6}$$
>
> if it exists.

Rates of Change

We now show that the problem of finding the slope of the tangent line to the graph of a function f at the point $P(x, f(x))$ is mathematically equivalent to the problem of finding the rate of change of f at x. To see this, suppose we are given a function f that describes the relationship between the two quantities x and y—that is, $y = f(x)$. The number $f(x + h) - f(x)$ measures the change in y that corresponds to a change h in x (Figure 53).

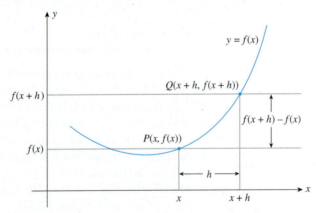

FIGURE 53
$f(x + h) - f(x)$ is the change in y that corresponds to a change h in x.

Then, the **difference quotient**

$$\frac{f(x + h) - f(x)}{h} \tag{7}$$

measures the **average rate of change of y with respect to x** over the interval $[x, x + h]$. For example, if y measures the position of a car at time x, then quotient (7) gives the average velocity of the car over the time interval $[x, x + h]$.

Observe that the difference quotient (7) is the same as (5). We conclude that the difference quotient (7) also measures the slope of the secant line that passes through the two points $P(x, f(x))$ and $Q(x + h, f(x + h))$ lying on the graph of $y = f(x)$. Next, by taking the limit of the difference quotient (7) as h goes to zero—that is, by evaluating

$$\lim_{h \to 0} \frac{f(x + h) - f(x)}{h} \tag{8}$$

we obtain the **rate of change of f at x.** For example, if y measures the position of a car at time x, then the limit (8) gives the velocity of the car at time x. For emphasis, the rate of change of a function f at x is often called the **instantaneous rate of change of f at x.** This distinguishes it from the average rate of change of f, which is computed over an *interval* $[x, x + h]$ rather than at a *number x*.

Observe that the limit (8) is the same as (6). Therefore, the limit of the difference quotient also measures the slope of the tangent line to the graph of $y = f(x)$ at the point $(x, f(x))$. The following summarizes this discussion.

Average and Instantaneous Rates of Change

The **average rate of change** of f over the interval $[x, x + h]$ or **slope of the secant line** to the graph of f through the points $(x, f(x))$ and $(x + h, f(x + h))$ is

$$\frac{f(x + h) - f(x)}{h} \tag{9}$$

The **instantaneous rate of change** of f at x or **slope of the tangent line** to the graph of f at $(x, f(x))$ is

$$\lim_{h \to 0} \frac{f(x + h) - f(x)}{h} \tag{10}$$

Explore and Discuss

Explain the difference between the average rate of change of a function and the instantaneous rate of change of a function.

The Derivative

The limit (6) or (10), which measures both the slope of the tangent line to the graph of $y = f(x)$ at the point $P(x, f(x))$ and the (instantaneous) rate of change of f at x, is given a special name: the **derivative of f at x.**

Derivative of a Function

The derivative of a function f with respect to x is the function f' (read "f prime"),

$$f'(x) = \lim_{h \to 0} \frac{f(x + h) - f(x)}{h} \tag{11}$$

The domain of f' is the set of all x for which the limit exists.

Thus, the derivative of a function f is a function f' that gives the slope of the tangent line to the graph of f at *any* point $(x, f(x))$ and also the rate of change of f at x (Figure 54).

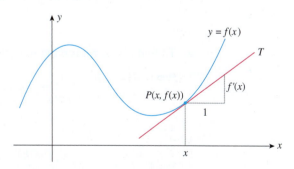

FIGURE 54
The slope of the tangent line at $P(x, f(x))$ is $f'(x)$; f changes at the rate of $f'(x)$ units per unit change in x at x.

Other notations for the derivative of f include the following:

$$D_x f(x)$$ Read "*d* sub *x* of *f* of *x*"

$$\frac{dy}{dx}$$ Read "*d y d x*"

$$y'$$ Read "*y* prime"

The last two are used when the rule for f is written in the form $y = f(x)$.

The calculation of the derivative of f is facilitated by using the following four-step process.

Four-Step Process for Finding $f'(x)$

1. Compute $f(x + h)$.
2. Form the difference $f(x + h) - f(x)$.
3. Form the quotient $\dfrac{f(x + h) - f(x)}{h}$.
4. Compute the limit $f'(x) = \lim\limits_{h \to 0} \dfrac{f(x + h) - f(x)}{h}$.

EXAMPLE 2 Find the slope of the tangent line to the graph of $f(x) = 3x + 5$ at any point $(x, f(x))$.

Solution   The slope of the tangent line at any point on the graph of f is given by the derivative of f at x. To find the derivative, we use the four-step process:

Step 1 $f(x + h) = 3(x + h) + 5 = 3x + 3h + 5$

Step 2 $f(x + h) - f(x) = (3x + 3h + 5) - (3x + 5) = 3h$

Step 3 $\dfrac{f(x + h) - f(x)}{h} = \dfrac{3h}{h} = 3$

Step 4 $f'(x) = \lim\limits_{h \to 0} \dfrac{f(x + h) - f(x)}{h} = \lim\limits_{h \to 0} 3 = 3$

We expect this result, since the tangent line to any point on a straight line must coincide with the line itself and therefore must have the same slope as the line. In this case, the graph of f is a straight line with slope 3.

EXAMPLE 3 Let $f(x) = x^2$.

a. Find $f'(x)$.
b. Compute $f'(2)$ and interpret your result.

Solution

a. To find $f'(x)$, we use the four-step process:

Step 1 $f(x + h) = (x + h)^2 = x^2 + 2xh + h^2$

Step 2 $f(x + h) - f(x) = x^2 + 2xh + h^2 - x^2 = 2xh + h^2 = h(2x + h)$

Step 3 $\dfrac{f(x + h) - f(x)}{h} = \dfrac{h(2x + h)}{h} = 2x + h$

Step 4 $f'(x) = \lim\limits_{h \to 0} \dfrac{f(x + h) - f(x)}{h} = \lim\limits_{h \to 0} (2x + h) = 2x$

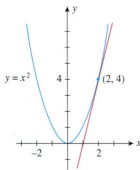

FIGURE 55
The tangent line to the graph of
$f(x) = x^2$ at $(2, 4)$

b. $f'(2) = 2(2) = 4$. This result tells us that the slope of the tangent line to the graph of f at the point $(2, 4)$ is 4. It also tells us that the function f is changing at the rate of 4 units per unit change in x at $x = 2$. The graph of f and the tangent line at $(2, 4)$ are shown in Figure 55. ■

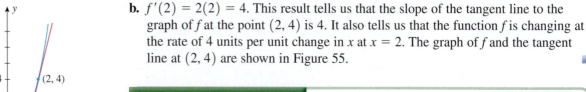

Exploring with TECHNOLOGY

1. Consider the function $f(x) = x^2$ of Example 3. Suppose we want to compute $f'(2)$, using Equation (11). Thus,

$$f'(2) = \lim_{h \to 0} \frac{f(2 + h) - f(2)}{h} = \lim_{h \to 0} \frac{(2 + h)^2 - 2^2}{h}$$

Use a graphing utility to plot the graph of

$$g(x) = \frac{(2 + x)^2 - 4}{x}$$

in the viewing window $[-3, 3] \times [-2, 6]$.

2. Use **ZOOM** and **TRACE** to find $\lim_{x \to 0} g(x)$.

3. Explain why the limit found in part 2 is $f'(2)$.

EXAMPLE 4 Let $f(x) = x^2 - 4x$.

a. Find $f'(x)$.
b. Find the point on the graph of f where the tangent line to the curve is horizontal.
c. Sketch the graph of f and the tangent line to the curve at the point found in part (b).
d. What is the rate of change of f at this point?

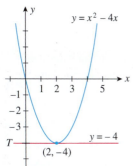

FIGURE 56
The tangent line to the graph of
$y = x^2 - 4x$ at $(2, -4)$ is $y = -4$.

Solution

a. To find $f'(x)$, we use the four-step process:

Step 1 $f(x + h) = (x + h)^2 - 4(x + h) = x^2 + 2xh + h^2 - 4x - 4h$

Step 2 $f(x + h) - f(x) = x^2 + 2xh + h^2 - 4x - 4h - (x^2 - 4x)$
$$= 2xh + h^2 - 4h = h(2x + h - 4)$$

Step 3 $\dfrac{f(x + h) - f(x)}{h} = \dfrac{h(2x + h - 4)}{h} = 2x + h - 4$

Step 4 $f'(x) = \lim_{h \to 0} \dfrac{f(x + h) - f(x)}{h} = \lim_{h \to 0} (2x + h - 4) = 2x - 4$

b. At a point on the graph of f where the tangent line to the curve is horizontal and hence has slope zero, the derivative f' of f is zero. Accordingly, to find such point(s), we set $f'(x) = 0$, which gives $2x - 4 = 0$, or $x = 2$. The corresponding value of y is given by $y = f(2) = -4$, and the required point is $(2, -4)$.

c. The graph of f and the tangent line are shown in Figure 56.

d. The rate of change of f at $x = 2$ is zero. ■

EXAMPLE 5 Let $f(x) = \dfrac{1}{x}$.

a. Find $f'(x)$.
b. Find the slope of the tangent line T to the graph of f at the point where $x = 1$.
c. Find an equation of the tangent line T in part (b).

Solution

a. To find $f'(x)$, we use the four-step process:

Step 1 $f(x + h) = \dfrac{1}{x + h}$

Step 2 $f(x + h) - f(x) = \dfrac{1}{x + h} - \dfrac{1}{x} = \dfrac{x - (x + h)}{x(x + h)} = -\dfrac{h}{x(x + h)}$

Step 3 $\dfrac{f(x + h) - f(x)}{h} = -\dfrac{h}{x(x + h)} \cdot \dfrac{1}{h} = -\dfrac{1}{x(x + h)}$ (x^2) See page 18.

Step 4 $f'(x) = \displaystyle\lim_{h \to 0} \dfrac{f(x + h) - f(x)}{h} = \lim_{h \to 0} -\dfrac{1}{x(x + h)} = -\dfrac{1}{x^2}$

b. The slope of the tangent line T to the graph of f where $x = 1$ is given by $f'(1) = -1$.

c. When $x = 1$, $y = f(1) = 1$ and T is tangent to the graph of f at the point $(1, 1)$. From part (b), we know that the slope of T is -1. Thus, an equation of T is

$$y - 1 = -1(x - 1)$$
$$y = -x + 2$$

(Figure 57).

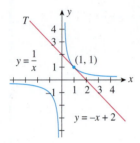

FIGURE 57
The tangent line to the graph of $f(x) = 1/x$ at $(1, 1)$

Exploring with TECHNOLOGY

1. Use the results of Example 5 to draw the graph of $f(x) = 1/x$ and its tangent line at the point $(1, 1)$ by plotting the graphs of $y_1 = 1/x$ and $y_2 = -x + 2$ in the viewing window $[-4, 4] \times [-4, 4]$.

2. Some graphing utilities draw the tangent line to the graph of a function at a given point automatically—you need only specify the function and give the x-coordinate of the point of tangency. If your graphing utility has this feature, verify the result of part 1 without finding an equation of the tangent line.

Explore and Discuss

Consider the following alternative approach to the definition of the derivative of a function: Let h be a positive number and suppose $P(x - h, f(x - h))$ and $Q(x + h, f(x + h))$ are two points on the graph of f.

1. Give a geometric and a physical interpretation of the quotient

$$\dfrac{f(x + h) - f(x - h)}{2h}$$

   Make a sketch to illustrate your answer.

2. Give a geometric and a physical interpretation of the limit

$$\lim_{h \to 0} \dfrac{f(x + h) - f(x - h)}{2h}$$

   Make a sketch to illustrate your answer.

3. Explain why it makes sense to define

$$f'(x) = \lim_{h \to 0} \dfrac{f(x + h) - f(x - h)}{2h}$$

4. Using the definition given in part 3, formulate a four-step process for finding $f'(x)$ similar to that given on page 142, and use it to find the derivative of $f(x) = x^2$. Compare your answer with that obtained in Example 3 on page 142.

APPLIED EXAMPLE 6 Velocity of a Car Suppose the distance (in feet) covered by a car moving along a straight road t seconds after starting from rest is given by the function $f(t) = 2t^2$ $(0 \leq t \leq 30)$.

a. Calculate the average velocity of the car over the time intervals $[22, 23]$, $[22, 22.1]$, and $[22, 22.01]$.
b. Calculate the (instantaneous) velocity of the car when $t = 22$.
c. Compare the results obtained in part (a) with that obtained in part (b).

Solution

a. We first compute the average velocity (average rate of change of f) over the interval $[t, t + h]$ using Formula (9). We find

$$\frac{f(t + h) - f(t)}{h} = \frac{2(t + h)^2 - 2t^2}{h}$$

$$= \frac{2t^2 + 4th + 2h^2 - 2t^2}{h}$$

$$= 4t + 2h$$

Next, using $t = 22$ and $h = 1$, we find that the average velocity of the car over the time interval $[22, 23]$ is

$$4(22) + 2(1) = 90$$

or 90 ft/sec. Similarly, using $t = 22$, $h = 0.1$, and $h = 0.01$, we find that its average velocities over the time intervals $[22, 22.1]$ and $[22, 22.01]$ are 88.2 and 88.02 ft/sec, respectively.

b. Using the limit (10), we see that the instantaneous velocity of the car at any time t is given by

$$\lim_{h \to 0} \frac{f(t + h) - f(t)}{h} = \lim_{h \to 0} (4t + 2h) \qquad \text{Use the results from part (a).}$$

$$= 4t$$

In particular, the velocity of the car 22 seconds from rest ($t = 22$) is given by

$$v = 4(22)$$

or 88 ft/sec.

c. The computations in part (a) show that, as the time intervals over which the average velocity of the car are computed become smaller and smaller, the average velocities over these intervals do approach 88 ft/sec, the instantaneous velocity of the car at $t = 22$.

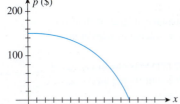

FIGURE **58**
The graph of the demand function $p = 144 - x^2$

APPLIED EXAMPLE 7 Demand for Tires The management of Titan Tire Company has determined that the weekly demand function of their Super Titan tires is given by

$$p = f(x) = 144 - x^2$$

where p, the price per tire, is measured in dollars and x is measured in units of a thousand (Figure 58).

a. Find the average rate of change in the unit price of a tire if the quantity demanded is between 5000 and 6000 tires, between 5000 and 5100 tires, and between 5000 and 5010 tires.
b. What is the instantaneous rate of change of the unit price when the quantity demanded is 5000 units?

Solution

a. The average rate of change of the unit price of a tire if the quantity demanded (in thousands) is between x and $x + h$ is

$$\frac{f(x + h) - f(x)}{h} = \frac{[144 - (x + h)^2] - (144 - x^2)}{h}$$

$$= \frac{144 - x^2 - 2xh - h^2 - 144 + x^2}{h}$$

$$= -2x - h$$

To find the average rate of change of the unit price of a tire when the quantity demanded is between 5000 and 6000 tires (that is, over the interval $[5, 6]$), we take $x = 5$ and $h = 1$, obtaining

$$-2(5) - 1 = -11$$

or $-\$11$ per 1000 tires. (Remember, x is measured in units of a thousand.) Similarly, taking $h = 0.1$ and $h = 0.01$ with $x = 5$, we find that the average rates of change of the unit price when the quantities demanded are between 5000 and 5100 and between 5000 and 5010 are $-\$10.10$ and $-\$10.01$ per 1000 tires, respectively.

b. The instantaneous rate of change of the unit price of a tire when the quantity demanded is x units is given by

$$\lim_{h \to 0} \frac{f(x + h) - f(x)}{h} = \lim_{h \to 0} (-2x - h) \qquad \text{Use the results from part (a).}$$

$$= -2x$$

In particular, the instantaneous rate of change of the unit price per tire when the quantity demanded is 5000 is given by $-2(5)$, or $-\$10$ per 1000 tires.

The derivative of a function provides us with a tool for measuring the rate of change of one quantity with respect to another. Table 4 lists several other applications involving this limit.

TABLE 4

Applications Involving Rate of Change

x stands for	y stands for	$\dfrac{f(a + h) - f(a)}{h}$ measures	$\displaystyle\lim_{h \to 0} \dfrac{f(a + h) - f(a)}{h}$ measures
Time	**Concentration of a drug** in the bloodstream at time x	Average rate of change in the concentration of the drug over the time interval $[a, a + h]$	Instantaneous rate of change in the concentration of the drug in the bloodstream at time $x = a$
Number of items sold	**Revenue** at a sales level of x units	Average rate of change in the revenue when the sales level is between $x = a$ and $x = a + h$	Instantaneous rate of change in the revenue when the sales level is a units
Time	**Volume of sales** at time x	Average rate of change in the volume of sales over the time interval $[a, a + h]$	Instantaneous rate of change in the volume of sales at time $x = a$
Time	**Population** of *Drosophila* (fruit flies) at time x	Average rate of growth of the fruit fly population over the time interval $[a, a + h]$	Instantaneous rate of change of the fruit fly population at time $x = a$
Temperature in a chemical reaction	**Amount of product formed in the chemical reaction** when the temperature is x degrees	Average rate of formation of chemical product over the temperature range $[a, a + h]$	Instantaneous rate of formation of chemical product when the temperature is a degrees

Differentiability and Continuity

In practical applications, one encounters continuous functions that fail to be **differentiable**—that is, do not have a derivative—at certain values in the domain of the function f. It can be shown that a continuous function f fails to be differentiable at $x = a$ when the graph of f makes an abrupt change of direction at $(a, f(a))$. We call such a point a "corner." A function also fails to be differentiable at a point where the tangent line is vertical, since the slope of a vertical line is undefined. These cases are illustrated in Figure 59.

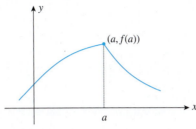

(a) The graph makes an abrupt change of direction at $x = a$.

(b) The slope at $x = a$ is undefined.

FIGURE **59**

The next example illustrates a function that is not differentiable at a point.

APPLIED EXAMPLE 8 Wages Mary works at the B&O department store, where, on a weekday, she is paid $10 an hour for the first 8 hours and $15 an hour for overtime. The function

$$f(x) = \begin{cases} 10x & \text{if } 0 \le x \le 8 \\ 15x - 40 & \text{if } 8 < x \end{cases}$$

gives Mary's earnings on a weekday in which she worked x hours. Sketch the graph of the function f, and explain why it is not differentiable at $x = 8$.

Solution   The graph of f is shown in Figure 60. Observe that the graph of f has a corner at $x = 8$ and consequently is not differentiable at $x = 8$.

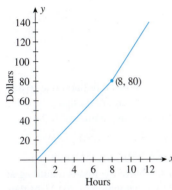

FIGURE **60**
The function f is not differentiable at $(8, 80)$.

We close this section by mentioning the connection between the continuity and the differentiability of a function at a given value $x = a$ in the domain of f. By reexamining the function of Example 8, it becomes clear that f is continuous everywhere and, in particular, when $x = 8$. This shows that in general the continuity of a function at $x = a$ does not necessarily imply the differentiability of the function at that number. The converse, however, is true: If a function f is differentiable at $x = a$, then it is continuous there.

> **Differentiability and Continuity**
>
> If a function is differentiable at $x = a$, then it is continuous at $x = a$.

For a proof of this result, see Exercise 62, page 153.

Explore and Discuss

Suppose a function f is differentiable at $x = a$. Can there be two tangent lines to the graphs of f at the point $(a, f(a))$? Explain your answer.

Exploring with TECHNOLOGY

1. Use a graphing utility to plot the graph of $f(x) = x^{1/3}$ in the viewing window $[-2, 2] \times [-2, 2]$.

2. Use a graphing utility to draw the tangent line to the graph of f at the point $(0, 0)$. Can you explain why the process breaks down?

EXAMPLE 9 Figure 61 depicts a portion of the graph of a function. Explain why the function fails to be differentiable at each of the numbers $x = a, b, c, d, e, f,$ and g.

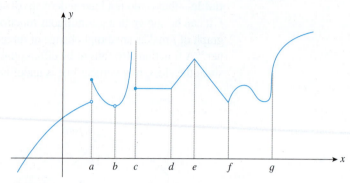

FIGURE 61
The graph of this function is not differentiable at the numbers a–g.

Solution   The function fails to be differentiable at $x = a, b,$ and c because it is discontinuous at each of these numbers. The derivative of the function does not exist at $x = d, e,$ and f because it has a kink at each point on the graph corresponding to these numbers. Finally, the function is not differentiable at $x = g$ because the tangent line is vertical at the corresponding point on the graph.

2.6   Self-Check Exercises

1. Let $f(x) = -x^2 - 2x + 3$.
   a. Find the derivative f' of f, using the definition of the derivative.
   b. Find the slope of the tangent line to the graph of f at the point $(0, 3)$.
   c. Find the rate of change of f when $x = 0$.
   d. Find an equation of the tangent line to the graph of f at the point $(0, 3)$.
   e. Sketch the graph of f and the tangent line to the curve at the point $(0, 3)$.

2. **BANK LOSSES** The losses (in millions of dollars) due to bad loans extended chiefly in agriculture, real estate, shipping, and energy by the Franklin Bank are estimated to be
   $$A = f(t) = -t^2 + 10t + 30 \qquad (0 \leq t \leq 10)$$
   where t is the time in years ($t = 0$ corresponds to the beginning of 2007). How fast were the losses mounting at the beginning of 2010? At the beginning of 2012? At the beginning of 2014?

   Solutions to Self-Check Exercises 2.6 can be found on page 153.

2.6   Concept Questions

For Questions 1 and 2, refer to the following figure.

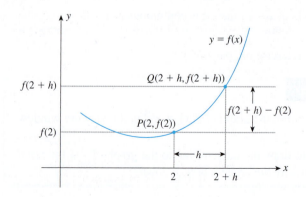

1. Let $P(2, f(2))$ and $Q(2 + h, f(2 + h))$ be points on the graph of a function f.
   a. Find an expression for the slope of the secant line passing through P and Q.
   b. Find an expression for the slope of the tangent line passing through P.

2. Refer to Question 1.
   a. Find an expression for the average rate of change of f over the interval $[2, 2 + h]$.
   b. Find an expression for the instantaneous rate of change of f at 2.
   c. Compare your answers for parts (a) and (b) with those of Question 1.

3. a. Give a geometric and a physical interpretation of the expression

$$\frac{f(x + h) - f(x)}{h}$$

b. Give a geometric and a physical interpretation of the expression

$$\lim_{h \to 0} \frac{f(x + h) - f(x)}{h}$$

4. Under what conditions does a function fail to have a derivative at a number? Illustrate your answer with sketches.

5. The total cost (in dollars) incurred in producing x units of a product is $C(x)$, where C is a differentiable function. Interpret the following:
   a. $C(500)$ **b.** $C'(500)$

6. The population of a city (in thousands) at any time t (in years) is given by $P(t)$, where P is a differentiable function. Interpret the following:
   a. $P(5)$ **b.** $P'(5)$

2.6 Exercises

1. AVERAGE WEIGHT OF AN INFANT The following graph shows the weight measurements of the average infant from the time of birth ($t = 0$) through age 2 ($t = 24$). By computing the slopes of the respective tangent lines, estimate the rate of change of the average infant's weight when $t = 3$ and when $t = 18$. What is the average rate of change in the average infant's weight over the first year of life?

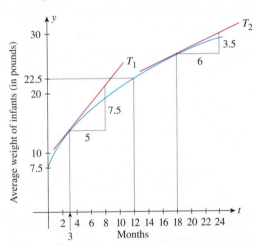

2. FORESTRY The following graph shows the volume of wood produced in a single-species forest. Here, $f(t)$ is measured in cubic meters per hectare, and t is measured in years. By computing the slopes of the respective tangent lines, estimate the rate at which the wood grown is changing at the beginning of year 10 and at the beginning of year 30.
Source: The Random House Encyclopedia.

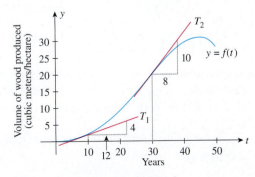

3. TV-VIEWING PATTERNS The following graph shows the percentage of U.S. households watching television during a 24-hr period on a weekday ($t = 0$ corresponds to 6 A.M.). By computing the slopes of the respective tangent lines, estimate the rate of change of the percent of households watching television at 4 P.M. and 11 P.M.
Source: A. C. Nielsen Company.

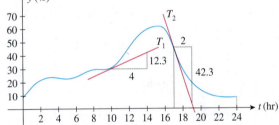

4. CROP YIELD Productivity and yield of cultivated crops are often reduced by insect pests. The following graph shows the relationship between the yield of a certain crop, $f(x)$, as a function of the density of aphids x. (Aphids are small insects that suck plant juices.) Here, $f(x)$ is measured in kilograms per 4000 square meters, and x is measured in hundreds of aphids per bean stem. By computing the slopes of the respective tangent lines, estimate the rate of change of the crop yield with respect to the density of aphids when that density is 200 aphids/bean stem and when it is 800 aphids/bean stem.
Source: The Random House Encyclopedia.

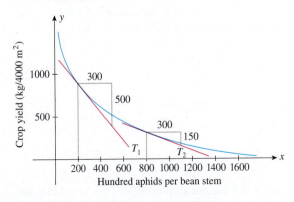

5. The positions of Car A and Car B, starting out side by side and traveling along a straight road, are given by $s = f(t)$ and $s = g(t)$, respectively, where s is measured in feet and t is measured in seconds (see the accompanying figure).

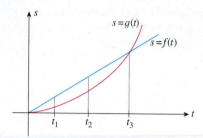

a. Which car is traveling faster at t_1?
b. What can you say about the speed of the cars at t_2?
   Hint: Compare tangent lines.
c. Which car is traveling faster at t_3?
d. What can you say about the positions of the cars at t_3?

6. The velocities of Car A and Car B, which start out side by side and travel along a straight road, are given by $v = f(t)$ and $v = g(t)$, respectively, where v is measured in feet per second and t is measured in seconds (see the accompanying figure).

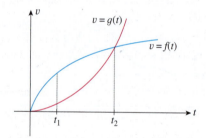

a. What can you say about the velocity and acceleration of the two cars at t_1? (Acceleration is the rate of change of velocity.)
b. What can you say about the velocity and acceleration of the two cars at t_2?

7. EFFECT OF A BACTERICIDE ON BACTERIA In the following figure, $f(t)$ gives the population P_1 of a certain bacteria culture at time t after a portion of Bactericide A was introduced into the population at $t = 0$. The graph of g gives the population P_2 of a similar bacteria culture at time t after a portion of Bactericide B was introduced into the population at $t = 0$.
a. Which population is decreasing faster at t_1?
b. Which population is decreasing faster at t_2?

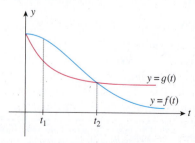

c. Which bactericide is more effective in reducing the population of bacteria in the short run? In the long run?

8. MARKET SHARE The following figure shows the devastating effect the opening of a new discount department store had on an established department store in a small town. The revenue of the discount store at time t (in months) is given by $f(t)$ million dollars, whereas the revenue of the established department store at time t is given by $g(t)$ million dollars. Answer the following questions by giving the value of t at which the specified event took place.

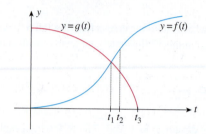

a. The revenue of the established department store is decreasing at the slowest rate.
b. The revenue of the established department store is decreasing at the fastest rate.
c. The revenue of the discount store first overtakes that of the established store.
d. The revenue of the discount store is increasing at the fastest rate.

In Exercises 9–16, use the four-step process to find the slope of the tangent line to the graph of the given function at any point.

9. $f(x) = 13$ **10.** $f(x) = -6$

11. $f(x) = 2x + 7$ **12.** $f(x) = 8 - 4x$

13. $f(x) = 3x^2$ **14.** $f(x) = -\dfrac{1}{2}x^2$

15. $f(x) = -x^2 + 3x$ **16.** $f(x) = 2x^2 + 5x$

In Exercises 17–22, find the slope of the tangent line to the graph of the function at the given point, and determine an equation of the tangent line.

17. $f(x) = 2x + 7$ at $(2, 11)$

18. $f(x) = -3x + 4$ at $(-1, 7)$

19. $f(x) = 3x^2$ at $(1, 3)$

20. $f(x) = 3x - x^2$ at $(-2, -10)$

21. $f(x) = -\dfrac{1}{x}$ at $\left(3, -\dfrac{1}{3}\right)$

22. $f(x) = \dfrac{3}{2x}$ at $\left(1, \dfrac{3}{2}\right)$

23. Let $f(x) = 2x^2 + 1$.
a. Find the derivative f' of f.
b. Find an equation of the tangent line to the curve at the point $(1, 3)$.
c. Sketch the graph of f and its tangent line at $(1, 3)$.

24. Let $f(x) = x^2 + 6x$.
    a. Find the derivative f' of f.
    b. Find the point on the graph of f where the tangent line to the curve is horizontal.
       Hint: Find the value of x for which $f'(x) = 0$.
    c. Sketch the graph of f and the tangent line to the curve at the point found in part (b).

25. Let $f(x) = x^2 - 2x + 1$.
    a. Find the derivative f' of f.
    b. Find the point on the graph of f where the tangent line to the curve is horizontal.
    c. Sketch the graph of f and the tangent line to the curve at the point found in part (b).
    d. What is the rate of change of f at this point?

26. Let $f(x) = \dfrac{1}{x - 1}$.
    a. Find the derivative f' of f.
    b. Find an equation of the tangent line to the curve at the point $\left(-1, -\frac{1}{2}\right)$.
    c. Sketch the graph of f and the tangent line to the curve at $\left(-1, -\frac{1}{2}\right)$.

27. Let $y = f(x) = x^2 + x$.
    a. Find the average rate of change of y with respect to x in the interval from $x = 2$ to $x = 3$, from $x = 2$ to $x = 2.5$, and from $x = 2$ to $x = 2.1$.
    b. Find the (instantaneous) rate of change of y at $x = 2$.
    c. Compare the results obtained in part (a) with the result of part (b).

28. Let $y = f(x) = x^2 - 4x$.
    a. Find the average rate of change of y with respect to x in the interval from $x = 3$ to $x = 4$, from $x = 3$ to $x = 3.5$, and from $x = 3$ to $x = 3.1$.
    b. Find the (instantaneous) rate of change of y at $x = 3$.
    c. Compare the results obtained in part (a) with the result of part (b).

29. **VELOCITY OF A CAR** Suppose the distance s (in feet) covered by a car moving along a straight road after t sec is given by the function $s = f(t) = 2t^2 + 48t$.
    a. Calculate the average velocity of the car over the time intervals $[20, 21]$, $[20, 20.1]$, and $[20, 20.01]$.
    b. Calculate the (instantaneous) velocity of the car when $t = 20$.
    c. Compare the results of part (a) with the result of part (b).

30. **VELOCITY OF A BALL THROWN INTO THE AIR** A ball is thrown straight up with an initial velocity of 128 ft/sec, so that its height (in feet) after t sec is given by $s(t) = 128t - 16t^2$.
    a. What is the average velocity of the ball over the time intervals $[2, 3]$, $[2, 2.5]$, and $[2, 2.1]$?
    b. What is the instantaneous velocity at time $t = 2$?
    c. What is the instantaneous velocity at time $t = 5$? Is the ball rising or falling at this time?
    d. When will the ball hit the ground?

31. **VELOCITY OF A FALLING OBJECT** During the construction of a high-rise building, a worker accidentally dropped his portable electric screwdriver from a height of 400 ft. After t sec, the screwdriver had fallen a distance of $s = 16t^2$ ft.
    a. How long did it take the screwdriver to reach the ground?
    b. What was the average velocity of the screwdriver between the time it was dropped and the time it hit the ground?
    c. What was the velocity of the screwdriver at the time it hit the ground?

32. **VELOCITY OF A HOT-AIR BALLOON** A hot-air balloon rises vertically from the ground so that its height after t sec is $h = \frac{1}{2}t^2 + \frac{1}{2}t$ ft $(0 \le t \le 60)$.
    a. What is the height of the balloon at the end of 40 sec?
    b. What is the average velocity of the balloon between $t = 0$ and $t = 40$?
    c. What is the velocity of the balloon at the end of 40 sec?

33. At a temperature of 20°C, the volume V (in liters) of 1.33 g of O_2 is related to its pressure p (in atmospheres) by the formula $V = 1/p$.
    a. What is the average rate of change of V with respect to p as p increases from $p = 2$ to $p = 3$?
    b. What is the rate of change of V with respect to p when $p = 2$?

34. **COST OF PRODUCING SURFBOARDS** The total cost $C(x)$ (in dollars) incurred by Aloha Company in manufacturing x surfboards a day is given by
    $$C(x) = -10x^2 + 300x + 130 \qquad (0 \le x \le 15)$$
    a. Find $C'(x)$.
    b. What is the rate of change of the total cost when the level of production is ten surfboards a day?

35. **EFFECT OF ADVERTISING ON PROFIT** The quarterly profit (in thousands of dollars) of Cunningham Realty is given by
    $$P(x) = -\frac{1}{3}x^2 + 7x + 30 \qquad (0 \le x \le 50)$$
    where x (in thousands of dollars) is the amount of money Cunningham spends on advertising per quarter.
    a. Find $P'(x)$.
    b. What is the rate of change of Cunningham's quarterly profit if the amount it spends on advertising is $10,000/quarter ($x = 10$) and $30,000/quarter ($x = 30$)?

36. **DEMAND FOR TENTS** The demand function for Sportsman 5×7 tents is given by
    $$p = f(x) = -0.1x^2 - x + 40$$
    where p is measured in dollars and x is measured in units of a thousand.
    a. Find the average rate of change in the unit price of a tent if the quantity demanded is between 5000 and 5050 tents; between 5000 and 5010 tents.
    b. What is the rate of change of the unit price if the quantity demanded is 5000?

37. A Country's GDP The gross domestic product (GDP) of a certain country is projected to be

$$N(t) = t^2 + 2t + 50 \qquad (0 \le t \le 5)$$

billion dollars t years from now. What will be the rate of change of the country's GDP 2 years and 4 years from now?

38. Growth of Bacteria Under a set of controlled laboratory conditions, the size of the population of a certain bacteria culture at time t (in minutes) is described by the function

$$P = f(t) = 3t^2 + 2t + 1$$

Find the rate of population growth at $t = 10$ min.

39. Air Temperature The air temperature at a height of h ft from the surface of the earth is $T = f(h)$ degrees Fahrenheit.
  a. Give a physical interpretation of $f'(h)$. Give units.
  b. Generally speaking, what do you expect the sign of $f'(h)$ to be?
  c. If you know that $f'(1000) = -0.05$, estimate the change in the air temperature if the altitude changes from 1000 ft to 1001 ft.

40. Revenue of a Travel Agency Suppose that the total revenue realized by the Odyssey Travel Agency is $R = f(x)$ thousand dollars if x thousand dollars are spent on advertising.
  a. What does

$$\frac{f(b) - f(a)}{b - a} \qquad (0 < a < b)$$

measure? What are the units?
  b. What does $f'(x)$ measure? Give units.
  c. Given that $f'(20) = 3$, what is the approximate change in the revenue if Odyssey increases its advertising budget from $20,000 to $21,000?

In Exercises 41–46, let x and $f(x)$ represent the given quantities. Fix $x = a$ and let h be a small positive number. Give an interpretation of the quantities

$$\frac{f(a + h) - f(a)}{h} \quad \text{and} \quad \lim_{h \to 0} \frac{f(a + h) - f(a)}{h}$$

41. x denotes time, and $f(x)$ denotes the population of seals at time x.

42. x denotes time, and $f(x)$ denotes the prime interest rate at time x.

43. x denotes time, and $f(x)$ denotes a country's industrial production.

44. x denotes the level of production of a certain commodity, and $f(x)$ denotes the total cost incurred in producing x units of the commodity.

45. x denotes altitude, and $f(x)$ denotes atmospheric pressure.

46. x denotes the speed of a car in miles per hour (mph), and $f(x)$ denotes the fuel consumption of the car measured in miles per gallon (mpg).

In each of Exercises 47–52, the graph of a function is shown. For each function, state whether or not (a) $f(x)$ has a limit at $x = a$, as x approaches a, (b) $f(x)$ is continuous at $x = a$, and (c) $f(x)$ is differentiable at $x = a$. Justify your answers.

47.
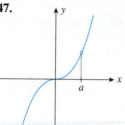

48.
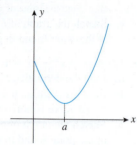

49.
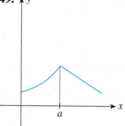

50.
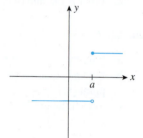

51.
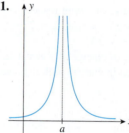

52.
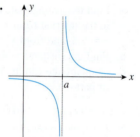

53. Velocity of a Motorcycle The distance s (in feet) covered by a motorcycle traveling in a straight line and starting from rest in t sec is given by the function

$$s(t) = -0.1t^3 + 2t^2 + 24t \qquad (0 \le t \le 3)$$

Calculate the motorcycle's average velocity over the time interval $[2, 2 + h]$ for $h = 1, 0.1, 0.01, 0.001, 0.0001$, and 0.00001, and use your results to guess at the motorcycle's instantaneous velocity at $t = 2$.

54. Rate of Change of Production Costs The daily total cost $C(x)$ incurred by Trappee and Sons for producing x cases of TexaPep hot sauce is given by

$$C(x) = 0.000002x^3 + 5x + 400$$

Calculate

$$\frac{C(100 + h) - C(100)}{h}$$

for $h = 1, 0.1, 0.01, 0.001$, and 0.0001, and use your results to estimate the rate of change of the total cost function when the level of production is 100 cases/day.

In Exercises 55 and 56, determine whether the statement is true or false. If it is true, explain why it is true. If it is false, give an example to show why it is false.

55. If f is continuous at $x = a$, then f is differentiable at $x = a$.

56. If f is continuous at $x = a$ and g is differentiable at $x = a$, then $\lim_{x \to a} f(x)g(x) = f(a)g(a)$.

57. Sketch the graph of the function $f(x) = |x + 1|$, and show that the function does not have a derivative at $x = -1$.

58. Sketch the graph of the function $f(x) = 1/(x - 1)$, and show that the function does not have a derivative at $x = 1$.

59. Let

$$f(x) = \begin{cases} x^2 & \text{if } x \leq 1 \\ ax + b & \text{if } x > 1 \end{cases}$$

Find the values of a and b such that f is continuous and has a derivative at $x = 1$. Sketch the graph of f.

60. Sketch the graph of the function $f(x) = x^{2/3}$. Is the function continuous at $x = 0$? Does $f'(0)$ exist? Why or why not?

61. Prove that the derivative of the function $f(x) = |x|$ for $x \neq 0$ is given by

$$f'(x) = \begin{cases} 1 & \text{if } x > 0 \\ -1 & \text{if } x < 0 \end{cases}$$

Hint: Recall the definition of the absolute value of a number.

62. Show that if a function f is differentiable at $x = a$, then f must be continuous at $x = a$.
Hint: Write

$$f(x) - f(a) = \left[\frac{f(x) - f(a)}{x - a} \right] (x - a)$$

Use the product rule for limits and the definition of the derivative to show that

$$\lim_{x \to a} [f(x) - f(a)] = 0$$

2.6 Solutions to Self-Check Exercises

1. a.

$$f'(x) = \lim_{h \to 0} \frac{f(x + h) - f(x)}{h}$$

$$= \lim_{h \to 0} \frac{[-(x + h)^2 - 2(x + h) + 3] - (-x^2 - 2x + 3)}{h}$$

$$= \lim_{h \to 0} \frac{-x^2 - 2xh - h^2 - 2x - 2h + 3 + x^2 + 2x - 3}{h}$$

$$= \lim_{h \to 0} \frac{h(-2x - h - 2)}{h}$$

$$= \lim_{h \to 0} (-2x - h - 2) = -2x - 2$$

b. From the result of part (a), we see that the slope of the tangent line to the graph of f at any point $(x, f(x))$ is given by

$$f'(x) = -2x - 2$$

In particular, the slope of the tangent line to the graph of f at $(0, 3)$ is

$$f'(0) = -2$$

c. The rate of change of f when $x = 0$ is given by $f'(0) = -2$, or -2 units/unit change in x.

d. Using the result from part (b), we see that an equation of the required tangent line is

$$y - 3 = -2(x - 0)$$

$$y = -2x + 3$$

e.

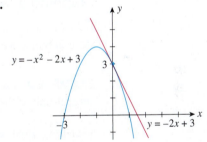

$y = -x^2 - 2x + 3$

$y = -2x + 3$

2. The rate of change of the losses at any time t is given by

$$f'(t) = \lim_{h \to 0} \frac{f(t + h) - f(t)}{h}$$

$$= \lim_{h \to 0} \frac{[-(t + h)^2 + 10(t + h) + 30] - (-t^2 + 10t + 30)}{h}$$

$$= \lim_{h \to 0} \frac{-t^2 - 2th - h^2 + 10t + 10h + 30 + t^2 - 10t - 30}{h}$$

$$= \lim_{h \to 0} \frac{h(-2t - h + 10)}{h}$$

$$= \lim_{h \to 0} (-2t - h + 10)$$

$$= -2t + 10$$

Therefore, the rate of change of the losses suffered by the bank at the beginning of 2010 ($t = 3$) was

$$f'(3) = -2(3) + 10 = 4$$

In other words, the losses were increasing at the rate of $4 million/year. At the beginning of 2012 ($t = 5$),

$$f'(5) = -2(5) + 10 = 0$$

and we see that the growth in losses due to bad loans was zero at this point. At the beginning of 2014 ($t = 7$),

$$f'(7) = -2(7) + 10 = -4$$

and we conclude that the losses were decreasing at the rate of $4 million/year.

USING TECHNOLOGY

Finding the Derivative of a Function for a Given Value of x

The numerical derivative operation of a graphing utility can be used to find an approximate value of the derivative of a function for a given value of x.

```
nDeriv(X^.5,X,2)
         .3535534017
```

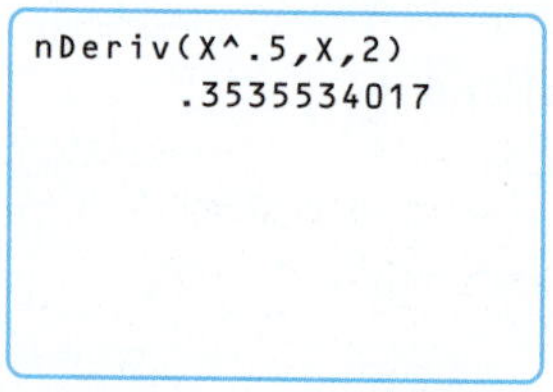

FIGURE T1
The TI-83/84 numerical derivative screen

EXAMPLE 1 Use the numerical derivative operation of a graphing calculator to find the derivative of $f(x) = \sqrt{x}$ when $x = 2$.

Solution   Access the numerical derivative operation, **nDeriv**, by pressing MATH 8 . The display **nDeriv(** appears on the screen. Next, we enter the function whose derivative we want to find, the variable x, followed by the value of x at which the derivative is to be evaluated, and then press ENTER . The approximate derivative of f at $x = 2$ is shown on the screen (Figure T1).

Graphing a Tangent Line

We can use a graphing utility to plot the graph of a function f and the tangent line to the graph at a specified point on the graph.

EXAMPLE 2 Use a graphing utility to plot the graph of $f(x) = x^2 - 4x$ and the tangent line to the graph of f at the point $(3, -3)$.

Solution   First, we plot the graph of $f(x) = x^2 - 4x$ using the standard viewing window according to the instructions starting on page 66. To draw the desired tangent line, we continue by pressing 2nd DRAW 5 to call the operation **Tangent.** Then, press 3 to obtain the x-coordinate of the point of tangency. Press ENTER , and the window will display both the graph of f and the tangent line at the point $(3, -3)$ together with its equation $y = 2x - 9$ (Figure T2).

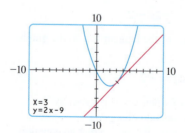

FIGURE T2
The TI-83/84 screen showing the graph of f, the tangent line, and its equation at $(3, -3)$

TECHNOLOGY EXERCISES

In Exercises 1–8, (a) use the numerical derivative operation to find the derivative of f for the given value of x (to two decimal places of accuracy), (b) plot the graph of f and the tangent line on the same set of axes, and (c) find an equation of the tangent line to the graph of f at the indicated point. Use a suitable viewing window.

1. $f(x) = 2x^2 + x - 3; x = 2; (2, 7)$

2. $f(x) = x + \dfrac{1}{x}; x = 1; (1, 2)$

3. $f(x) = x^{1/3}; x = 8; (8, 2)$

4. $f(x) = \dfrac{1}{\sqrt{x}}; x = 4; \left(4, \dfrac{1}{2}\right)$

5. $f(x) = x^3 + x + 1; x = 1; (1, 3)$

6. $f(x) = \dfrac{1}{x + 1}; x = 1; \left(1, \dfrac{1}{2}\right)$

7. $f(x) = x\sqrt{x^2 + 1}; x = 2; (2, 2\sqrt{5})$

8. $f(x) = \dfrac{x}{\sqrt{x^2 + 1}}; x = 1; \left(1, \dfrac{\sqrt{2}}{2}\right)$

9. ALZHEIMER'S PATIENTS IN THE UNITED STATES The number of patients with Alzheimer's disease in the United States is approximated by

$$f(t) = -0.208t^3 + 1.571t^2 - 0.9274t + 5.1 \qquad (0 \le t \le 4)$$

where $f(t)$ is measured in millions and t is measured in decades, with $t = 0$ corresponding to the beginning of 1990.
a. Plot the graph of $f(t)$ in the window $[0, 4] \times [0, 14]$.
b. How fast will the number of Alzheimer's patients in the United States be increasing at the beginning of 2020 ($t = 3$)?

Source: Alzheimer's Association.

10. MODELING WITH DATA Annual retail sales in the United States from 2000 through the year 2008 (in billions of dollars) are given in the following table:

Year	2000	2001	2002	2003	2004
Sales	2.988	3.068	3.134	3.267	3.480

Year	2005	2006	2007	2008
Sales	3.698	3.882	4.005	3.959

a. Let $t = 0$ correspond to 2000, and use **QuadReg** to find a second-degree polynomial regression model based on the given data.
b. Plot the graph of the function found in part (a) in the viewing window $[0, 8] \times [0, 5]$.
c. What were the annual retail sales in the United States in 2006 ($t = 6$)?
d. Approximately how fast were the retail sales changing in 2006 ($t = 6$)?

Source: U.S. Census Bureau.

CHAPTER 2   Summary of Principal Formulas and Terms

FORMULAS

1. Average rate of change of f over $[x, x + h]$ or Slope of the secant line to the graph of f through $(x, f(x))$ and $(x + h, f(x + h))$ or Difference quotient	$\dfrac{f(x + h) - f(x)}{h}$
2. Instantaneous rate of change of f at $(x, f(x))$ or Slope of tangent line to the graph of f at $(x, f(x))$ at x or Derivative of f	$\displaystyle\lim_{h \to 0} \dfrac{f(x + h) - f(x)}{h}$

TERMS

function (52)	independent variable (53)	graph of a function (55)
domain (52)	dependent variable (53)	graph of an equation (58)
range (52)	ordered pairs (55)	Vertical Line Test (58)

composite function (72)

polynomial function (78)

linear function (79)

quadratic function (80)

rational function (83)

power function (83)

demand function (84)

supply function (84)

market equilibrium (85)

equilibrium quantity (85)

equilibrium price (85)

limit of a function (104)

indeterminate form (107)

limit of a function at infinity (110)

right-hand limit of a function (122)

left-hand limit of a function (122)

continuity of a function at a number (124)

zero of a function (128)

secant line (139)

tangent line to the graph of f (139)

differentiable function (147)

CHAPTER 2   Concept Review Questions

Fill in the blanks.

1. If f is a function from the set A to the set B, then A is called the _____ of f, and the set of all values of $f(x)$ as x takes on all possible values in A is called the _____ of f. The range of f is contained in the set _____.

2. The graph of a function is the set of all points (x, y) in the xy-plane such that x is in the _____ of f and $y = $ _____. The Vertical Line Test states that a curve in the xy-plane is the graph of a function $y = f(x)$ if and only if each _____ line intersects it in at most one _____.

3. If f and g are functions with domains A and B, respectively, then (a) $(f \pm g)(x) = $ _____, (b) $(fg)(x) = $ _____, and (c) $(f/g)(x) = $ _____. The domain of $f + g$ is _____. The domain of f/g is _____ with the additional condition that $g(x)$ is never _____.

4. The composition of g and f is the function with rule $(g \circ f)(x) = $ _____. Its domain is the set of all x in the domain of _____ such that _____ lies in the domain of _____.

5. a. A polynomial function of degree n is a function of the form _____.
   b. A polynomial function of degree 1 is called a/an _____ function; one of degree 2 is called a/an _____ function; one of degree 3 is called a/an _____ function.
   c. A rational function is a/an _____ of two _____.
   d. A power function has the form $f(x) = $ _____.

6. The statement $\lim_{x \to a} f(x) = L$ means that the values of _____ can be made as close to _____ as we please by taking x sufficiently close to _____.

7. If $\lim_{x \to a} f(x) = L$ and $\lim_{x \to a} g(x) = M$, then
   a. $\lim_{x \to a} [f(x)]^r = $ _____, where r is a positive constant.
   b. $\lim_{x \to a} [f(x) \pm g(x)] = $ _____.
   c. $\lim_{x \to a} [f(x)g(x)] = $ _____.
   d. $\lim_{x \to a} \dfrac{f(x)}{g(x)} = $ _____ provided that _____.

8. a. The statement $\lim_{x \to \infty} f(x) = L$ means that $f(x)$ can be made arbitrarily close to _____ by taking _____ large enough.

   b. The statement $\lim_{x \to -\infty} f(x) = M$ means that $f(x)$ can be made arbitrarily close to _____ by taking x to be _____ and sufficiently large in _____ value.

9. a. The statement $\lim_{x \to a^+} f(x) = L$ is similar to the statement $\lim_{x \to a} f(x) = L$, but here x is required to lie to the _____ of a.
   b. The statement $\lim_{x \to a^-} f(x) = L$ is similar to the statement $\lim_{x \to a} f(x) = L$, but here x is required to lie to the _____ of a.
   c. $\lim_{x \to a} f(x) = L$ if and only if both $\lim_{x \to a^-} f(x) = $ _____ and $\lim_{x \to a^+} f(x) = $ _____.

10. a. If $f(a)$ is defined, $\lim_{x \to a} f(x)$ exists and $\lim_{x \to a} f(x) = f(a)$, then f is _____ at a.
    b. If f is not continuous at a, then it is _____ at a.
    c. f is continuous on an interval I if f is continuous at _____ number in the interval.

11. a. If f and g are continuous at a, then $f \pm g$ and fg are continuous at _____. Also, f/g is continuous at _____, provided _____ $\neq 0$.
    b. A polynomial function is continuous _____.
    c. A rational function $R = P/Q$ is continuous everywhere except at values of x for which _____ $= 0$.

12. a. Suppose f is continuous on $[a, b]$ and $f(a) < M < f(b)$. Then the Intermediate Value Theorem guarantees the existence of at least one number c in _____ such that _____.
    b. If f is continuous on $[a, b]$ and $f(a)f(b) < 0$, then there must be at least one solution of the equation _____ in the interval _____.

13. a. The tangent line at $P(a, f(a))$ to the graph of f is the line passing through P and having slope _____.
    b. If the slope of the tangent line at $P(a, f(a))$ is m, then an equation of the tangent line at P is _____.

14. a. The slope of the secant line passing through $P(a, f(a))$ and $Q(a + h, f(a + h))$ and the average rate of change of f over the interval $[a, a + h]$ are both given by _____.
    b. The slope of the tangent line at $P(a, f(a))$ and the instantaneous rate of change of f at a are both given by _____.

CHAPTER 2 Review Exercises

1. Find the domain of the function.

a. $f(x) = \sqrt{9 - x}$ **b.** $f(x) = \dfrac{x + 3}{2x^2 - x - 3}$

2. Find the domain of the function.

a. $f(x) = \dfrac{\sqrt{2 - x}}{x + 3}$ **b.** $f(x) = \dfrac{x^2 + 3x + 4}{\sqrt{x^2 + 1}}$

3. Let $f(x) = 3x^2 + 5x - 2$. Find:

a. $f(-2)$ **b.** $f(a + 2)$
c. $f(2a)$ **d.** $f(a + h)$

4. Let $f(x) = 2x^2 - x + 1$. Find:

a. $f(x - 1) + f(x + 1)$
b. $f(x + 2h)$

5. Let $y^2 = 2x + 1$.

a. Sketch the graph of this equation.
b. Is y a function of x? Why?
c. Is x a function of y? Why?

6. Sketch the graph of the function defined by

$$f(x) = \begin{cases} x + 1 & \text{if } x < 1 \\ -x^2 + 4x - 1 & \text{if } x \geq 1 \end{cases}$$

7. Let $f(x) = 1/x$ and $g(x) = 2x + 3$. Find:

a. $f(x)g(x)$ **b.** $f(x)/g(x)$
c. $f(g(x))$ **d.** $g(f(x))$

8. Find the rules for the composite functions $f \circ g$ and $g \circ f$.

a. $f(x) = 2x - 1; g(x) = x^2 + 4$

b. $f(x) = 1 - x; g(x) = \dfrac{1}{3x + 4}$

c. $f(x) = x - 3; g(x) = \dfrac{1}{\sqrt{x + 1}}$

9. Find functions f and g such that $h = g \circ f$. (*Note:* The answer is *not* unique.)

a. $h(x) = \dfrac{1}{(2x^2 + x + 1)^3}$ **b.** $h(x) = \sqrt{x^2 + x + 4}$

10. Find the value of c such that the point $(4, 2)$ lies on the graph of $f(x) = cx^2 + 3x - 4$.

In Exercises 11–24, find the indicated limits, if they exist.

11. $\lim\limits_{x \to 0} (5x - 3)$ **12.** $\lim\limits_{x \to 1} (x^2 + 1)$

13. $\lim\limits_{x \to -1} (3x^2 + 4)(2x - 1)$

14. $\lim\limits_{x \to 3} \dfrac{x - 3}{x + 4}$ **15.** $\lim\limits_{x \to 2} \dfrac{x + 3}{x^2 - 9}$

16. $\lim\limits_{x \to -2} \dfrac{x^2 - 2x - 3}{x^2 + 5x + 6}$ **17.** $\lim\limits_{x \to 3} \sqrt{2x^3 - 5}$

18. $\lim\limits_{x \to 3} \dfrac{4x - 3}{\sqrt{x + 1}}$ **19.** $\lim\limits_{x \to 1^+} \dfrac{x - 1}{x(x - 1)}$

20. $\lim\limits_{x \to 1^-} \dfrac{\sqrt{x} - 1}{x - 1}$ **21.** $\lim\limits_{x \to \infty} \dfrac{x^2}{x^2 - 1}$

22. $\lim\limits_{x \to -\infty} \dfrac{x + 1}{x}$ **23.** $\lim\limits_{x \to \infty} \dfrac{3x^2 + 2x + 4}{2x^2 - 3x + 1}$

24. $\lim\limits_{x \to -\infty} \dfrac{x^2}{x + 1}$

25. Sketch the graph of the function

$$f(x) = \begin{cases} 2x - 3 & \text{if } x \leq 2 \\ -x + 3 & \text{if } x > 2 \end{cases}$$

and evaluate $\lim\limits_{x \to a^+} f(x)$, $\lim\limits_{x \to a^-} f(x)$, and $\lim\limits_{x \to a} f(x)$ at the point $a = 2$, if the limits exist.

26. Sketch the graph of the function

$$f(x) = \begin{cases} 4 - x & \text{if } x \leq 2 \\ x + 2 & \text{if } x > 2 \end{cases}$$

and evaluate $\lim\limits_{x \to a^+} f(x)$, $\lim\limits_{x \to a^-} f(x)$, and $\lim\limits_{x \to a} f(x)$ at the point $a = 2$, if the limits exist.

In Exercises 27–30, determine all values of x for which each function is discontinuous.

27. $g(x) = \begin{cases} x + 3 & \text{if } x \neq 2 \\ 0 & \text{if } x = 2 \end{cases}$

28. $f(x) = \dfrac{3x + 4}{4x^2 - 2x - 2}$

29. $f(x) = \begin{cases} \dfrac{1}{(x + 1)^2} & \text{if } x \neq -1 \\ 2 & \text{if } x = -1 \end{cases}$

30. $f(x) = \dfrac{|2x|}{x}$

31. Let $y = x^2 + 2$.

a. Find the average rate of change of y with respect to x over the intervals $[1, 2]$, $[1, 1.5]$, and $[1, 1.1]$.
b. Find the (instantaneous) rate of change of y at $x = 1$.

32. Use the definition of the derivative to find the slope of the tangent line to the graph of the function $f(x) = 4x + 5$ at any point $P(x, f(x))$ on the graph.

33. Use the definition of the derivative to find the slope of the tangent line to the graph of the function $f(x) = \frac{3}{2}x + 5$ at the point $(-2, 2)$ and determine an equation of the tangent line.

34. Use the definition of the derivative to find the slope of the tangent line to the graph of the function $f(x) = -x^2$ at the point $(2, -4)$ and determine an equation of the tangent line.

35. Use the definition of the derivative to find the slope of the tangent line to the graph of the function $f(x) = -1/x$ at any point $P(x, f(x))$ on the graph.

36. The graph of the function f is shown in the accompanying figure.
   a. Is f continuous at $x = a$? Why or why not?
   b. Is f differentiable at $x = a$? Justify your answers.

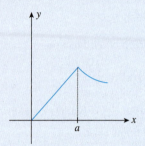

37. **SALES OF MP3 CLOCK RADIOS** Sales of a certain MP3 clock radio are approximated by the relationship $S(x) = 6000x + 30,000$ $(0 \leq x \leq 5)$, where $S(x)$ denotes the number of clock radios sold in year x ($x = 0$ corresponds to the year 2010). Find the number of clock radios expected to be sold in 2014.

38. **SALES OF A COMPANY** A company's total sales (in millions of dollars) are approximately linear as a function of t in years ($t = 0$ corresponds to the year 2008). Sales in 2008 were $2.4 million, whereas sales in 2013 amounted to $7.4 million.
   a. Find an equation that gives the company's sales as a function of time.
   b. What were the sales in 2011?

39. **PROFIT FUNCTIONS** A company has a fixed cost of $30,000 and a production cost of $6 for each unit it manufactures. A unit sells for $10.
   a. What is the cost function?
   b. What is the revenue function?
   c. What is the profit function?
   d. Compute the profit (loss) corresponding to production levels of 6000, 8000, and 12,000 units, respectively.

40. Find the point of intersection of the two straight lines having the equations $y = \frac{3}{4}x + 6$ and $3x - 2y + 3 = 0$.

41. The cost and revenue functions for a certain firm are given by $C(x) = 12x + 20,000$ and $R(x) = 20x$, respectively. Find the company's profit function.

42. **MARKET EQUILIBRIUM** Given the demand equation $3x + p - 40 = 0$ and the supply equation $2x - p + 10 = 0$, where p is the unit price in dollars and x represents the quantity in units of a thousand, determine the equilibrium quantity and the equilibrium price.

43. **CLARK'S RULE** Clark's Rule is a method for calculating pediatric drug dosages based on a child's weight. If a denotes the adult dosage (in milligrams) and w is the weight of the child (in pounds), then the child's dosage is given by

$$D(w) = \frac{aw}{150}$$

If the adult dose of a substance is 500 mg, how much should a child who weighs 35 lb receive?

44. **REVENUE FUNCTIONS** The revenue (in dollars) realized by Apollo from the sale of its ink-jet printers is given by

$$R(x) = -0.1x^2 + 500x$$

where x denotes the number of units manufactured each month. What is Apollo's revenue when 1000 units are produced?

45. **REVENUE FUNCTIONS** The monthly revenue R (in hundreds of dollars) realized from the sale of Royal electric shavers is related to the unit price p (in dollars) by the equation

$$R(p) = -\frac{1}{2}p^2 + 30p$$

Find the revenue when an electric shaver is priced at $30.

46. **HEALTH CLUB MEMBERSHIP** The membership of the newly opened Venus Health Club is approximated by the function

$$N(x) = 200(4 + x)^{1/2} \quad (1 \leq x \leq 24)$$

where $N(x)$ denotes the number of members x months after the club's grand opening. Find $N(0)$ and $N(12)$, and interpret your results.

47. **POPULATION GROWTH** A study prepared for a Sunbelt town's Chamber of Commerce projected that the population of the town in the next 3 years will grow according to the rule

$$P(x) = 50,000 + 30x^{3/2} + 20x$$

where $P(x)$ denotes the population x months from now. By how much will the population increase during the next 9 months? During the next 16 months?

48. **THURSTONE LEARNING CURVE** Psychologist L. L. Thurstone discovered the following model for the relationship between the learning time T and the length of a list n:

$$T = f(n) = An\sqrt{n - b}$$

where A and b are constants that depend on the person and the task. Suppose that, for a certain person and a certain task, $A = 4$ and $b = 4$. Compute $f(4), f(5), \ldots, f(12)$, and use this information to sketch the graph of the function f. Interpret your results.

49. **FORECASTING SALES** The annual sales of Crimson Drug Store are expected to be given by

$$S_1(t) = 2.3 + 0.4t$$

million dollars t years from now, whereas the annual sales of Cambridge Drug Store are expected to be given by

$$S_2(t) = 1.2 + 0.6t$$

million dollars t years from now. When will the annual sales of Cambridge first surpass the annual sales of Crimson?

50. **MARKET EQUILIBRIUM** The monthly demand and supply functions for the Luminar desk lamp are given by

$$p = d(x) = -1.1x^2 + 1.5x + 40$$
$$p = s(x) = 0.1x^2 + 0.5x + 15$$

respectively, where p is measured in dollars and x is measured in units of a thousand. Find the equilibrium quantity and price.

51. **FEMALE LIFE EXPECTANCY** Advances in medical science and healthier lifestyles have resulted in longer life expectancies. The life expectancy of a female whose current age is x years old is

$$f(x) = 0.0053694x^2 - 1.4663x + 92.74 \qquad (60 \le x \le 75)$$

years. What is the life expectancy of a female whose current age is 65? Whose current age is 75?
Source: Commissioners' Standard Ordinary Mortality Table.

52. **BABY BOOMERS AND MEDICARE BENEFITS** Aging baby boomers will put a strain on Medicare benefits unless Congress takes action. The Medicare benefits to be paid out from 2010 through 2040 are projected to be

$$B(t) = 0.09t^2 + 0.102t + 0.25 \qquad (0 \le t \le 3)$$

where $B(t)$ is measured in trillions of dollars and t is measured in decades, with $t = 0$ corresponding to 2010.
   a. What was the amount of Medicare benefits paid out in 2010?
   b. What is the amount of Medicare benefits projected to be paid out in 2040?
Source: Social Security and Medicare Trustees' 2010 report.

53. **TESTOSTERONE USE** Fueled by the promotion of testosterone as an antiaging elixir, use of the hormone by middle-age and older men grew dramatically. The total number of prescriptions for testosterone from 1999 through 2002 is given by

$$N(t) = -35.8t^3 + 202t^2 + 87.8t + 648 \qquad (0 \le t \le 3)$$

where $N(t)$ is measured in thousands and t is measured in years, with $t = 0$ corresponding to the beginning of 1999. Find the total number of prescriptions for testosterone in 1999, 2000, 2001, and 2002.
Source: IMS Health.

54. **U.S. NUTRITIONAL SUPPLEMENTS MARKET** The size of the U.S. nutritional supplements market from 1999 through 2003 is approximated by the function

$$A(t) = 16.4(t + 1)^{0.1} \qquad (0 \le t \le 4)$$

where $A(t)$ is measured in billions of dollars and t is measured in years, with $t = 0$ corresponding to the beginning of 1999.
   a. Compute $A(0)$, $A(1)$, $A(2)$, $A(3)$, and $A(4)$. Interpret your results.
   b. Use the results of part (a) to sketch the graph of A.
Source: Nutrition Business Journal.

55. **GLOBAL SUPPLY OF PLUTONIUM** The global stockpile of plutonium for military applications between 1990 ($t = 0$) and 2003 ($t = 13$) stood at a constant 267 tons. On the other hand, the global stockpile of plutonium for civilian use was

$$2t^2 + 46t + 733$$

tons in year t over the same period.
   a. Find the function f giving the global stockpile of plutonium for military use from 1990 through 2003 and the function g giving the global stockpile of plutonium for civilian use over the same period.
   b. Find the function h giving the total global stockpile of plutonium between 1990 and 2003.
   c. What was the total global stockpile of plutonium in 2003?
Source: Institute for Science and International Security.

56. **HOTEL OCCUPANCY RATE** A forecast released by PricewaterhouseCoopers in June of 2004 predicted the occupancy rate of U.S. hotels between 2001 ($t = 0$) and 2005 ($t = 4$) to be

$$P(t) = \begin{cases} -0.9t + 59.8 & \text{if } 0 \le t < 1 \\ 0.3t + 58.6 & \text{if } 1 \le t < 2 \\ 56.79t^{0.06} & \text{if } 2 \le t \le 4 \end{cases}$$

percent.
   a. Compute $P(0)$, $P(1)$, $P(2)$, $P(3)$, and $P(4)$.
   b. Sketch the graph of P.
   c. What was the predicted occupancy rate of hotels for 2004?
Source: PricewaterhouseCoopers LLP Hospitality & Leisure Research.

57. **OIL SPILLS** The oil spilling from the ruptured hull of a grounded tanker spreads in all directions in calm waters. Suppose the area polluted is a circle of radius r and the radius is increasing at the rate of 2 ft/sec.
   a. Find a function f giving the area polluted in terms of r.
   b. Find a function g giving the radius of the polluted area in terms of t.
   c. Find a function h giving the area polluted in terms of t.
   d. What is the size of the polluted area 30 sec after the hull was ruptured?

58. **PACKAGING** By cutting away identical squares from each corner of a 20-in. × 20-in. piece of cardboard and folding up the resulting flaps, an open box can be made. Denoting the length of a side of a cutaway by x, find a function of x giving the volume of the resulting box.

59. CONSTRUCTION COSTS The length of a rectangular box is to be twice its width, and its volume is to be 30 ft³. The material for the base costs 30¢/ft², the material for the sides costs 15¢/ft², and the material for the top costs 20¢/ft². Letting x denote the width of the box, find a function in the variable x giving the cost of constructing the box.

60. FILM CONVERSION PRICES PhotoMart transfers movie films to DVDs. The fees charged for this service are shown in the following table. Find a function C relating the cost $C(x)$ to the number of feet x of film transferred. Sketch the graph of the function C and discuss its continuity.

Length of Film in Feet, x	Price for Conversion ($)
$1 \leq x \leq 100$	5.00
$100 < x \leq 200$	9.00
$200 < x \leq 300$	12.50
$300 < x \leq 400$	15.00
$x > 400$	$7 + 0.02x$

61. AVERAGE PRICE OF A COMMODITY The average cost (in dollars) of producing x units of a certain commodity is given by

$$\overline{C}(x) = 20 + \frac{400}{x}$$

Evaluate $\lim_{x \to \infty} \overline{C}(x)$, and interpret your results.

62. MANUFACTURING COSTS Suppose that the total cost of manufacturing x units of a certain product is $C(x)$ dollars.
a. What does $C'(x)$ measure? Give units.
b. What can you say about the sign of C'?
c. Given that $C'(1000) = 20$, estimate the additional cost to be incurred by the company in producing the 1001st unit of the product.

In Exercises 63 and 64, determine whether the statement is true or false. If it is true, explain why it is true. If it is false, give an example to show why it is false.

63. There does not exist a function f whose rule is
$$f(x) = \sqrt{-x} + \sqrt{x} \ (x \neq 0).$$

64. A tangent line to the graph of a function cannot intersect the graph of the function at any point other than the point of tangency of the tangent line.

The problem-solving skills that you learn in each chapter are building blocks for the rest of the course. Therefore, it is a good idea to make sure that you have mastered these skills before moving on to the next chapter. The Before Moving On exercises that follow are designed for that purpose. After completing these exercises, you can identify the skills that you should review before starting the next chapter.

CHAPTER 2 Before Moving On . . .

1. Let
$$f(x) = \begin{cases} -2x + 1 & \text{if } -1 \leq x < 0 \\ x^2 + 2 & \text{if } 0 \leq x \leq 2 \end{cases}$$
Find (a) $f(-1)$, (b) $f(0)$, and (c) $f(\frac{3}{2})$.

2. Let $f(x) = \dfrac{1}{x + 1}$ and $g(x) = x^2 + 1$. Find the rules for
(a) $f + g$, (b) fg, (c) $f \circ g$, and (d) $g \circ f$.

3. POSTAL REGULATIONS Postal regulations specify that a parcel sent by priority mail may have a combined length and girth of no more than 108 in. Suppose a rectangular package that has a square cross section of x in. × x in. is to have a combined length and girth of exactly 108 in. Find a function in terms of x giving the volume of the package.
Hint: The length plus the girth is $4x + h$ (see the accompanying figure).

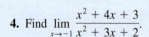

4. Find $\lim_{x \to -1} \dfrac{x^2 + 4x + 3}{x^2 + 3x + 2}$.

5. Let
$$f(x) = \begin{cases} x^2 - 1 & \text{if } -2 \leq x < 1 \\ x^3 & \text{if } 1 \leq x \leq 2 \end{cases}$$
Find (a) $\lim_{x \to 1^-} f(x)$ and (b) $\lim_{x \to 1^+} f(x)$. Is f continuous at $x = 1$? Explain.

6. Find the slope of the tangent line to the graph of $f(x) = x^2 - 3x + 1$ at the point $(1, -1)$. What is an equation of the tangent line?

3 Differentiation

THIS CHAPTER GIVES several rules that will greatly simplify the task of finding the derivative of a function, thus enabling us to study how fast one quantity is changing with respect to another in many real-world situations. For example, we will be able to find how fast the population of an endangered species of whales grows after certain conservation measures have been implemented, how fast an economy's consumer price index (CPI) is changing at any time, and how fast the time taken to learn the items on a list changes with respect to the length of a list. We also see how these rules of differentiation facilitate the study of the rate of change of economic quantities—that is, the study of marginal analysis. Finally, we introduce the notion of the differential of a function. Differentials are used to approximate the change in one quantity due to a small change in a related quantity.

What happens to the sales of a DVD recording of a certain hit movie over a 10-year period after it is first released into the market? In Example 6, page 178, you will see how to find the rate of change of sales for the DVD over the first 10 years after its release.

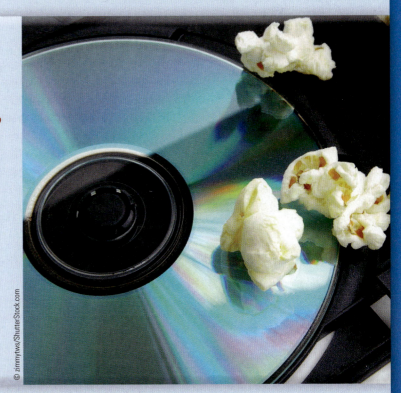

© zimmytws/ShutterStock.com

3.1   Basic Rules of Differentiation

Four Basic Rules

The method used in Chapter 2 for computing the derivative of a function is based on a faithful interpretation of the definition of the derivative as the limit of a quotient. To find the rule for the derivative f' of a function f, we first computed the difference quotient

$$\frac{f(x + h) - f(x)}{h}$$

and then evaluated its limit as h approached zero. As you have probably observed, this method is tedious even for relatively simple functions.

The main purpose of this chapter is to derive certain rules that will simplify the process of finding the derivative of a function. We will use the notation

$$\frac{d}{dx}[f(x)] \qquad \text{Read "} d, d\,x \text{ of } f \text{ of } x\text{"}$$

to mean "the derivative of f with respect to x at x."

Rule 1: Derivative of a Constant

$$\frac{d}{dx}(c) = 0 \qquad (c, \text{ a constant})$$

The derivative of a constant function is equal to zero.

We can see this from a geometric viewpoint by recalling that the graph of a constant function is a straight line parallel to the x-axis (Figure 1). Since the tangent line to a straight line at any point on the line coincides with the straight line itself, its slope [as given by the derivative of $f(x) = c$] must be zero. We can also use the definition of the derivative to prove this result by computing

$$f'(x) = \lim_{h \to 0} \frac{f(x + h) - f(x)}{h}$$

$$= \lim_{h \to 0} \frac{c - c}{h}$$

$$= \lim_{h \to 0} 0 = 0$$

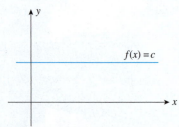

FIGURE 1
The slope of the tangent line to the graph of $f(x) = c$, where c is a constant, is zero.

EXAMPLE 1

a. If $f(x) = 28$, then

$$f'(x) = \frac{d}{dx}(28) = 0$$

b. If $f(x) = \pi^2$, then

$$f'(x) = \frac{d}{dx}(\pi^2) = 0$$

> **Rule 2: The Power Rule**
>
> If n is any real number, then $\dfrac{d}{dx}(x^n) = nx^{n-1}$.

Let's verify the Power Rule for the special case $n = 2$. If $f(x) = x^2$, then

$$
\begin{aligned}
f'(x) = \frac{d}{dx}(x^2) &= \lim_{h \to 0} \frac{f(x+h) - f(x)}{h} \\
&= \lim_{h \to 0} \frac{(x+h)^2 - x^2}{h} \\
&= \lim_{h \to 0} \frac{x^2 + 2xh + h^2 - x^2}{h} \\
&= \lim_{h \to 0} \frac{2xh + h^2}{h} = \lim_{h \to 0} \frac{h(2x + h)}{h} \\
&= \lim_{h \to 0} (2x + h) = 2x
\end{aligned}
$$

as we set out to show.

The Power Rule for the general case is not easy to prove, and its proof will be omitted. However, you will be asked to prove the rule for the special case $n = 3$ in Exercise 78, page 173.

EXAMPLE 2

a. If $f(x) = x$, then

$$
f'(x) = \frac{d}{dx}(x) = 1 \cdot x^{1-1} = x^0 = 1
$$

b. If $f(x) = x^8$, then

$$
f'(x) = \frac{d}{dx}(x^8) = 8x^7
$$

c. If $f(x) = x^{5/2}$, then

$$
f'(x) = \frac{d}{dx}(x^{5/2}) = \frac{5}{2}x^{3/2}
$$

◼

To differentiate a function whose rule involves a radical, we first rewrite the radical using fractional powers. The resulting expression can then be differentiated by using the Power Rule.

EXAMPLE 3 Find the derivative of the following functions:

a. $f(x) = \sqrt{x}$ **b.** $g(x) = \dfrac{1}{\sqrt[3]{x}}$

Solution

a. Rewriting $\sqrt{x}$ in the form $x^{1/2}$, we obtain (x^z) See page 6.

$$
\begin{aligned}
f'(x) &= \frac{d}{dx}(x^{1/2}) \\
&= \frac{1}{2}x^{-1/2} = \frac{1}{2x^{1/2}} = \frac{1}{2\sqrt{x}}
\end{aligned}
$$

b. Rewriting $\dfrac{1}{\sqrt[3]{x}}$ in the form $x^{-1/3}$, we obtain

$$g'(x) = \frac{d}{dx}(x^{-1/3})$$

$$= -\frac{1}{3}x^{-4/3} = -\frac{1}{3x^{4/3}}$$

In stating the remaining rules of differentiation, we assume that the functions f and g are differentiable.

Rule 3: Derivative of a Constant Multiple of a Function

$$\frac{d}{dx}[cf(x)] = c\frac{d}{dx}[f(x)] \qquad \text{(c, a constant)}$$

The derivative of a constant times a differentiable function is equal to the constant times the derivative of the function.

This result follows from the following computations:

If $g(x) = cf(x)$, then

$$g'(x) = \lim_{h \to 0} \frac{g(x+h) - g(x)}{h} = \lim_{h \to 0} \frac{cf(x+h) - cf(x)}{h}$$

$$= c \lim_{h \to 0} \frac{f(x+h) - f(x)}{h}$$

$$= cf'(x)$$

EXAMPLE 4

a. If $f(x) = 5x^3$, then

$$f'(x) = \frac{d}{dx}(5x^3) = 5\frac{d}{dx}(x^3)$$

$$= 5(3x^2) = 15x^2$$

b. If $f(x) = \dfrac{3}{\sqrt{x}}$, then

$$f'(x) = \frac{d}{dx}(3x^{-1/2}) \qquad \text{Rewrite } \frac{3}{\sqrt{x}} \text{ as } \frac{3}{x^{1/2}} = 3x^{-1/2}$$

$$= 3\left(-\frac{1}{2}x^{-3/2}\right) = -\frac{3}{2x^{3/2}}$$

Rule 4: The Sum Rule

$$\frac{d}{dx}[f(x) \pm g(x)] = \frac{d}{dx}[f(x)] \pm \frac{d}{dx}[g(x)]$$

The derivative of the sum (difference) of two differentiable functions is equal to the sum (difference) of their derivatives.

This result may be extended to the sum and difference of any finite number of differentiable functions. Let's verify the rule for a sum of two functions.

If $s(x) = f(x) + g(x)$, then

$$s'(x) = \lim_{h \to 0} \frac{s(x+h) - s(x)}{h}$$

$$= \lim_{h \to 0} \frac{[f(x+h) + g(x+h)] - [f(x) + g(x)]}{h}$$

$$= \lim_{h \to 0} \frac{[f(x+h) - f(x)] + [g(x+h) - g(x)]}{h}$$

$$= \lim_{h \to 0} \frac{f(x+h) - f(x)}{h} + \lim_{h \to 0} \frac{g(x+h) - g(x)}{h}$$

$$= f'(x) + g'(x)$$

EXAMPLE 5 Find the derivatives of the following functions:

a. $f(x) = 4x^5 + 3x^4 - 8x^2 + x + 3$ **b.** $g(t) = \dfrac{t^2}{5} + \dfrac{5}{t^3}$

Solution

a. $f'(x) = \dfrac{d}{dx}(4x^5 + 3x^4 - 8x^2 + x + 3)$

$$= \frac{d}{dx}(4x^5) + \frac{d}{dx}(3x^4) - \frac{d}{dx}(8x^2) + \frac{d}{dx}(x) + \frac{d}{dx}(3)$$

$$= 20x^4 + 12x^3 - 16x + 1$$

b. Here, the independent variable is t instead of x, so we differentiate with respect to t. Thus,

$$g'(t) = \frac{d}{dt}\left(\frac{1}{5}t^2 + 5t^{-3}\right) \qquad \text{Rewrite } \frac{1}{t^3} \text{ as } t^{-3}.$$

$$= \frac{2}{5}t - 15t^{-4} = \frac{2}{5}t - \frac{15}{t^4} \qquad \text{Rewrite } t^{-4} \text{ as } \frac{1}{t^4}.$$

$$= \frac{2t^5 - 75}{5t^4}$$

EXAMPLE 6 Find the slope and an equation of the tangent line to the graph of $f(x) = 2x + 1/\sqrt{x}$ at the point $(1, 3)$.

Solution   The slope of the tangent line at any point on the graph of f is given by

$$f'(x) = \frac{d}{dx}\left(2x + \frac{1}{\sqrt{x}}\right)$$

$$= \frac{d}{dx}(2x + x^{-1/2}) \qquad \text{Rewrite } \frac{1}{\sqrt{x}} \text{ as } \frac{1}{x^{1/2}} = x^{-1/2}.$$

$$= 2 - \frac{1}{2}x^{-3/2} \qquad \text{Use the Sum Rule.}$$

$$= 2 - \frac{1}{2x^{3/2}} \qquad \text{Rewrite } \frac{1}{2}x^{-3/2} \text{ as } \frac{1}{2x^{3/2}}.$$

In particular, the slope of the tangent line to the graph of f at $(1, 3)$ (where $x = 1$) is

$$f'(1) = 2 - \frac{1}{2(1^{3/2})} = 2 - \frac{1}{2} = \frac{3}{2}$$

Using the point-slope form of an equation of a line with slope $\frac{3}{2}$ and the point $(1, 3)$, we see that an equation of the tangent line is

$$y - 3 = \frac{3}{2}(x - 1) \qquad \textcolor{red}{y - y_1 = m(x - x_1)} \qquad (x^2) \text{ See page 37.}$$

or, upon simplification,

$$y = \frac{3}{2}x + \frac{3}{2}$$

(see Figure 2).

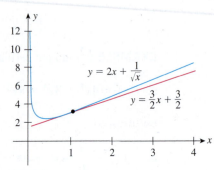

FIGURE 2
The tangent line to the graph of
$f(x) = 2x + 1/\sqrt{x}$ at $(1, 3)$.

APPLIED EXAMPLE 7 Conservation of a Species A group of marine biologists at the Neptune Institute of Oceanography recommended that a series of conservation measures be carried out over the next decade to save a certain species of whale from extinction. After the conservation measures are implemented, the population of this species is expected to be

$$N(t) = 3t^3 + 2t^2 - 10t + 600 \qquad (0 \le t \le 10)$$

where $N(t)$ denotes the population at the end of year t. Find the rate of growth of the whale population when $t = 2$ and $t = 6$. How large will the whale population be 8 years after the conservation measures are implemented?

Solution   The rate of growth of the whale population at any time t is given by

$$N'(t) = 9t^2 + 4t - 10$$

In particular, when $t = 2$ and $t = 6$, we have

$$N'(2) = 9(2)^2 + 4(2) - 10$$
$$= 34$$
$$N'(6) = 9(6)^2 + 4(6) - 10$$
$$= 338$$

Thus, the whale population's rate of growth will be 34 whales per year after 2 years and 338 per year after 6 years.

The whale population at the end of the eighth year will be

$$N(8) = 3(8)^3 + 2(8)^2 - 10(8) + 600$$
$$= 2184$$

The graph of the function N appears in Figure 3. Note the rapid growth of the population in later years, as the conservation measures begin to pay off, compared with the growth in the early years.

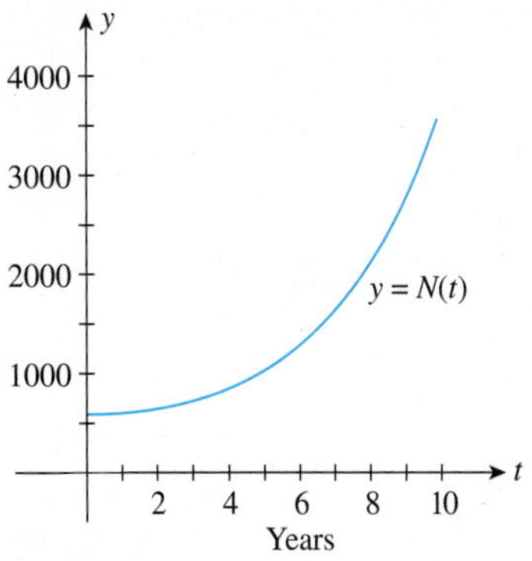

$$4000$$
$$3000$$
$$2000$$
$$y = N(t)$$
$$1000$$

$$2 \quad 4 \quad 6 \quad 8 \quad 10 \quad t$$
Years

FIGURE **3**
The whale population at the end of year t is given by $N(t)$.

APPLIED EXAMPLE 8 Altitude of a Rocket An experimental rocket lifts off vertically. Its altitude (in feet) t seconds into flight is given by

$$s = f(t) = -t^3 + 96t^2 + 5 \qquad (t \geq 0)$$

a. Find an expression v for the rocket's velocity at any time t.
b. Compute the rocket's velocity when $t = 0, 30, 50, 64$, and 70. Interpret your results.
c. Using the results from the solution to part (b) and the observation that at the highest point in its trajectory the rocket's velocity is zero, find the maximum altitude attained by the rocket.

Solution

a. The rocket's velocity at any time t is given by

$$v = f'(t) = -3t^2 + 192t$$

b. The rocket's velocity when $t = 0, 30, 50, 64$, and 70 is given by

$$f'(0) = -3(0)^2 + 192(0) = 0$$
$$f'(30) = -3(30)^2 + 192(30) = 3060$$
$$f'(50) = -3(50)^2 + 192(50) = 2100$$
$$f'(64) = -3(64)^2 + 192(64) = 0$$
$$f'(70) = -3(70)^2 + 192(70) = -1260$$

or 0, 3060, 2100, 0, and -1260 feet per second (ft/sec).

Thus, the rocket has an initial velocity of 0 ft/sec at $t = 0$ and accelerates to a velocity of 3060 ft/sec at $t = 30$. Fifty seconds into the flight, the rocket's velocity is 2100 ft/sec, which is less than the velocity at $t = 30$. This means that the rocket begins to decelerate after an initial period of acceleration. (Later on, we will learn how to determine the rocket's maximum velocity.)

The deceleration continues: The velocity is 0 ft/sec at $t = 64$ and -1260 ft/sec when $t = 70$. This result tells us that 70 seconds into flight, the rocket is heading back to the earth with a speed of 1260 ft/sec.

c. The results of part (b) show that the rocket's velocity is zero when $t = 64$. At this instant, the rocket's maximum altitude is

$$s = f(64) = -(64)^3 + 96(64)^2 + 5$$
$$= 131,077$$

or 131,077 feet. A sketch of the graph of f appears in Figure 4 (see next page).

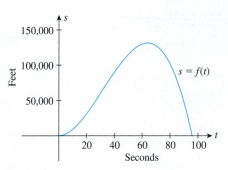

FIGURE 4
The rocket's altitude t seconds into flight is given by $f(t)$.

Note   You may have observed that the domain of the function f in Example 8 is restricted, for practical reasons, to the interval $[0, \infty)$. Since the definition of the derivative of a function f at a number a requires that f be defined in an open interval containing a, the derivative of f is not, strictly speaking, defined at 0. But notice that the function f can, in fact, be defined for all values of t, and hence it makes sense to calculate $f'(0)$. You will encounter situations such as this throughout the book, especially in exercises pertaining to real-world applications. The nature of the functions appearing in these applications obviates the necessity to consider "one-sided" derivatives.

Exploring with TECHNOLOGY

Refer to Example 8.

1. Use a graphing utility to plot the graph of the velocity function

$$v = f'(t) = -3t^2 + 192t$$

using the viewing window $[0, 120] \times [-5000, 5000]$. Then, using **ZOOM** and **TRACE** or the root-finding capability of your graphing utility, verify that $f'(64) = 0$.

2. Plot the graph of the position function of the rocket

$$s = f(t) = -t^3 + 96t^2 + 5$$

using the viewing window $[0, 120] \times [0, 150{,}000]$. Then, using **ZOOM** and **TRACE** repeatedly, verify that the maximum altitude of the rocket is 131,077 feet.

3. Use **ZOOM** and **TRACE** or the root-finding capability of your graphing utility to find when the rocket returns to the earth.

3.1   Self-Check Exercises

1. Find the derivative of each function using the rules of differentiation.

   a. $f(x) = 1.5x^2 + 2x^{1.5}$

   b. $g(x) = 2\sqrt{x} + \dfrac{3}{\sqrt{x}}$

2. Let $f(x) = 2x^3 - 3x^2 + 2x - 1$.

   a. Compute $f'(x)$.

   b. What is the slope of the tangent line to the graph of f when $x = 2$?

   c. What is the rate of change of the function f at $x = 2$?

3. **GDP OF A COUNTRY**   A certain country's gross domestic product (GDP) (in billions of dollars) is described by the function

$$G(t) = -2t^3 + 45t^2 + 20t + 6000 \qquad (0 \le t \le 11)$$

where $t = 0$ corresponds to the beginning of 2005.

   a. At what rate was the GDP changing at the beginning of 2010? At the beginning of 2012? At the beginning of 2015?

   b. What was the average rate of growth of the GDP from the beginning of 2010 to the beginning of 2015?

Solutions to Self-Check Exercises 3.1 can be found on page 173.

3.1 Concept Questions

1. State the following rules of differentiation in your own words.
   a. The rule for differentiating a constant function
   b. The Power Rule
   c. The Constant Multiple Rule
   d. The Sum Rule

2. If $f'(2) = 3$ and $g'(2) = -2$, find
   a. $h'(2)$ if $h(x) = 2f(x)$
   b. $F'(2)$ if $F(x) = 3f(x) - 4g(x)$

3. Suppose f and g are differentiable functions and a and b are nonzero numbers. Find $F'(x)$ if
   a. $F(x) = af(x) + bg(x)$
   b. $F(x) = \dfrac{f(x)}{a}$

4. If f is differentiable at $x = a$, is $[f'(x)](a) = \dfrac{d}{dx}[f(a)]$? Explain.

3.1 Exercises

In Exercises 1–34, find the derivative of the function f by using the rules of differentiation.

1. $f(x) = -3$
2. $f(x) = 365$
3. $f(x) = x^5$
4. $f(x) = x^7$
5. $f(x) = x^{3.1}$
6. $f(x) = x^{0.8}$
7. $f(x) = 3x^2$
8. $f(x) = -2x^3$
9. $f(r) = \pi r^2$
10. $f(r) = \dfrac{4}{3}\pi r^3$
11. $f(x) = 9x^{1/3}$
12. $f(x) = \dfrac{5}{4}x^{4/5}$
13. $f(x) = 3\sqrt{x}$
14. $f(u) = \dfrac{2}{\sqrt{u}}$
15. $f(x) = 7x^{-12}$
16. $f(x) = 0.3x^{-1.2}$
17. $f(x) = 5x^2 - 3x + 7$
18. $f(x) = x^3 - 3x^2 + 1$
19. $f(x) = -x^3 + 2x^2 - 6$
20. $f(x) = (1 + 2x^2)^2 + 2x^3$
21. $f(x) = 0.03x^2 - 0.4x + 10$
22. $f(x) = 0.002x^3 - 0.05x^2 + 0.1x - 20$
23. $f(x) = \dfrac{2x^3 - 4x^2 + 3}{x}$
24. $f(x) = \dfrac{x^3 + 2x^2 + x - 1}{x}$
25. $f(x) = 4x^4 - 3x^{5/2} + 2$
26. $f(x) = 5x^{4/3} - \dfrac{2}{3}x^{3/2} + x^2 - 3x + 1$
27. $f(x) = 5x^{-1} + 4x^{-2}$
28. $f(x) = -\dfrac{1}{3}(x^{-3} - x^6)$
29. $f(t) = \dfrac{4}{t^4} - \dfrac{3}{t^3} + \dfrac{2}{t}$
30. $f(x) = \dfrac{5}{x^3} - \dfrac{2}{x^2} - \dfrac{1}{x} + 200$

31. $f(x) = 3x - 5\sqrt{x}$
32. $f(t) = 2t^2 + \sqrt{t^3}$
33. $f(x) = \dfrac{2}{x^2} - \dfrac{3}{x^{1/3}}$
34. $f(x) = \dfrac{3}{x^3} + \dfrac{4}{\sqrt{x}} + 1$

35. Let $f(x) = 2x^3 - 4x$. Find:
   a. $f'(-2)$ b. $f'(0)$ c. $f'(2)$

36. Let $f(x) = 4x^{5/4} + 2x^{3/2} + x$. Find:
   a. $f'(4)$ b. $f'(16)$

In Exercises 37–40, find each limit by evaluating the derivative of a suitable function at an appropriate point.
Hint: Look at the definition of the derivative.

37. $\displaystyle\lim_{h \to 0} \dfrac{(1 + h)^3 - 1}{h}$
38. $\displaystyle\lim_{x \to 1} \dfrac{x^5 - 1}{x - 1}$
   Hint: Let $h = x - 1$.

39. $\displaystyle\lim_{h \to 0} \dfrac{3(2 + h)^2 - (2 + h) - 10}{h}$

40. $\displaystyle\lim_{t \to 0} \dfrac{1 - (1 + t)^2}{t(1 + t)^2}$

In Exercises 41–44, find the slope and an equation of the tangent line to the graph of the function f at the specified point.

41. $f(x) = 2x^2 - 3x + 4$; $(2, 6)$
42. $f(x) = -\dfrac{5}{3}x^2 + 2x + 2$; $\left(-1, -\dfrac{5}{3}\right)$
43. $f(x) = x^4 - 3x^3 + 2x^2 - x + 1$; $(2, -1)$
44. $f(x) = \sqrt{x} + \dfrac{1}{\sqrt{x}}$; $\left(4, \dfrac{5}{2}\right)$

45. Let $f(x) = x^3$.
   a. Find the point on the graph of f where the tangent line is horizontal.
   b. Sketch the graph of f and draw the horizontal tangent line.

46. Let $f(x) = x^3 - 4x^2$. Find the points on the graph of f where the tangent line is horizontal.

47. Let $f(x) = x^3 + 1$.
   a. Find the points on the graph of f where the slope of the tangent line is equal to 12.
   b. Find the equation(s) of the tangent line(s) of part (a).
   c. Sketch the graph of f showing the tangent line(s).

48. Let $f(x) = \frac{2}{3}x^3 + x^2 - 12x + 6$. Find the values of x for which:
   a. $f'(x) = -12$ **b.** $f'(x) = 0$
   c. $f'(x) = 12$

49. Let $f(x) = \frac{1}{4}x^4 - \frac{1}{3}x^3 - x^2$. Find the points on the graph of f where the slope of the tangent line is equal to:
   a. $-2x$ **b.** 0 **c.** $10x$

50. A straight line perpendicular to and passing through the point of tangency of the tangent line is called the *normal* to the curve at that point. Find an equation of the tangent line and the normal to the curve $y = x^3 - 3x + 1$ at the point $(2, 3)$.

51. GROWTH OF A CANCEROUS TUMOR The volume of a spherical cancerous tumor is given by the function

$$V(r) = \frac{4}{3}\pi r^3$$

where r is the radius of the tumor in centimeters. Find the rate of change in the volume of the tumor with respect to its radius when
   a. $r = \frac{2}{3}$ cm   **b.** $r = \frac{5}{4}$ cm

52. VELOCITY OF BLOOD IN AN ARTERY The velocity (in centimeters per second) of blood r cm from the central axis of an artery is given by

$$v(r) = k(R^2 - r^2)$$

where k is a constant and R is the radius of the artery (see the accompanying figure). Suppose $k = 1000$ and $R = 0.2$ cm. Find $v(0.1)$ and $v'(0.1)$, and interpret your results.

Blood vessel

53. GROWTH OF TABLET AND SMARTPHONE USERS The number of tablets and smartphones in use worldwide (in millions) in year t from 2010 through 2012 is approximately

$$f(t) = 128.1t^{1.94} \qquad (1 \le t \le 3)$$

where $t = 1$ corresponds to 2010.
   a. How many tablets and smartphones were in use in 2011?
   b. How fast was the number of tablets and smartphones changing in 2011?
Source: MIT Technology Review.

54. U.K. MOBILE PHONE VIDEO VIEWERS As mobile phones continue to proliferate, more and more people will watch video content on them through a mobile browser, subscription, download, or application at least once a month. In a study released in 2013, eMarketer estimated that the number of mobile phone video viewers in the United Kingdom in year t from 2011 ($t = 0$) through 2017 ($t = 6$) will be

$$P(t) = 13.86t^{0.535}$$

percent of the population.
   a. What percentage of the U.K. population is expected to watch video content on mobile phones in 2015?
   b. How fast is the percentage of mobile phone video viewers in the United Kingdom in 2015 expected to change?
Source: eMarketer.

55. MARRIED COUPLES WITH CHILDREN The percentage of families that were married couples with children between 1970 and 2010 is approximately

$$P(t) = \frac{49.6}{t^{0.27}} \qquad (1 \le t \le 5)$$

where t is measured in decades, with $t = 1$ corresponding to 1970.
   a. What percentage of families were married couples with children in 1970? In 1990? In 2010?
   b. How fast was the percentage of families that were married couples with children changing in 1990? In 2000?
Source: American Community Survey.

56. EFFECT OF STOPPING ON AVERAGE SPEED According to data from a General Motors study, the average speed of your trip A (in miles per hour) is related to the number of stops per mile you make on the trip x by the equation

$$A = \frac{26.5}{x^{0.45}}$$

Compute dA/dx for $x = 0.25$ and $x = 2$. How is the rate of change with respect to x of the average speed of your trip affected by the number of stops per mile?
Source: General Motors.

57. DECLINE OF MIDDLE-CLASS INCOME The share of households in the United States that are earning middle-class incomes has been in decline since the 1970s. The percentage of households with annual incomes within 50% of the median income is given by

$$P(t) = 50.3t^{-0.09} \qquad (1 \le t \le 6)$$

where t is measured in decades, with $t = 1$ corresponding to 1980.
   a. What percentage of households had annual incomes within 50% of the median income in 2010?
   b. How fast was the percentage of households with annual incomes within 50% of the median income decreasing in 2010?
Source: Alan Krueger.

58. DEMAND FUNCTION FOR DESK LAMPS The demand function for the Luminar desk lamp is given by

$$p = f(x) = -0.1x^2 - 0.4x + 35$$

where x is the quantity demanded in thousands and p is the unit price in dollars.
  a. Find $f'(x)$.
  b. What is the rate of change of the unit price when the quantity demanded is 10,000 units ($x = 10$)? What is the unit price at that level of demand?

59. STOPPING DISTANCE OF A RACING CAR During a test by the editors of an auto magazine, the distance s (in feet) traveled by the MacPherson X-2 racing car t sec after the brakes were applied conformed to the rule

$$s = f(t) = 120t - 15t^2 \qquad (t \geq 0)$$

  a. Find an expression for the car's velocity v at any time t.
  b. What was the car's velocity when the brakes were first applied?
  c. What was the car's stopping distance for that particular test?
  Hint: The stopping time is found by setting $v = 0$.

60. MOBILE INSTANT MESSAGING ACCOUNTS Mobile instant messaging (IM) is a small portion of total IM usage, but it is growing sharply. The function

$$P(t) = 0.257t^2 + 0.57t + 3.9 \qquad (0 \leq t \leq 4)$$

gives the mobile IM accounts in year t as a percentage of total enterprise IM accounts from 2006 ($t = 0$) through 2010 ($t = 4$).
  a. What percentage of total enterprise IM accounts were the mobile accounts in 2008?
  b. How fast was this percentage changing in 2008?
  Source: The Radical Group.

61. MEDICAL COSTS FOR A FAMILY OF FOUR The average annual medical costs for a family of four in the United States from 2001 through 2011 is approximated by the function

$$C(t) = 22.9883t^2 + 830.358t + 7513 \qquad (1 \leq t \leq 11)$$

where t is measured in years, with $t = 1$ corresponding to 2001, and $C(t)$ is measured in dollars.
  a. What was the approximate average annual medical costs for a family of four in 2010?
  b. How fast was the approximate average medical costs for a family of four increasing in 2010?
  Source: Milliman Medical Index.

62. SPENDING ON MEDICARE Based on the current eligibility requirement, a study conducted in 2004 showed that federal spending on entitlement programs, particularly Medicare, would grow enormously in the future. The study predicted that spending on Medicare, as a percentage of the gross domestic product (GDP), will be

$$P(t) = 0.27t^2 + 1.4t + 2.2 \qquad (0 \leq t \leq 5)$$

percent in year t, where t is measured in decades, with $t = 0$ corresponding to 2000.
  a. How fast was the spending on Medicare, as a percentage of the GDP, growing in 2010? How fast will it be growing in 2020?
  b. What was the predicted spending on Medicare in 2010? What will it be in 2020?
  Source: Congressional Budget Office.

63. CONSUMER PRICE INDEX An economy's consumer price index (CPI) is described by the function

$$I(t) = -0.2t^3 + 3t^2 + 100 \qquad (0 \leq t \leq 11)$$

in year t, where $t = 0$ corresponds to 2003.
  a. At what rate was the CPI changing in 2008? In 2010? In 2013?
  b. What was the average rate of increase in the CPI over the period from 2008 to 2013?

64. WORKER EFFICIENCY An efficiency study conducted for Elektra Electronics showed that the number of Space Commander walkie-talkies assembled by the average worker during the morning shift t hr after starting work at 8 A.M. is given by

$$N(t) = -t^3 + 6t^2 + 15t \qquad (0 \leq t \leq 4)$$

  a. Find the rate at which the average worker will be assembling walkie-talkies t hr after starting work.
  b. At what rate will the average worker be assembling walkie-talkies at 10 A.M.? At 11 A.M.?
  c. How many walkie-talkies will the average worker assemble between 10 A.M. and 11 A.M.?

65. CURBING POPULATION GROWTH Five years ago, the government of a Pacific Island country launched an extensive propaganda campaign aimed toward curbing the country's population growth. According to the Census Department, the population (measured in thousands of people) for the following 4 years was

$$P(t) = -\frac{1}{3}t^3 + 64t + 3000$$

where t is measured in years and $t = 0$ corresponds to the start of the campaign. Find the rate of change of the population at the end of years 1, 2, 3, and 4. Was the plan working?

66. CONSERVATION OF SPECIES A certain species of turtle faces extinction because dealers collect truckloads of turtle eggs to be sold as aphrodisiacs. After severe conservation measures are implemented, it is hoped that the turtle population will grow according to the rule

$$N(t) = 2t^3 + 3t^2 - 4t + 1000 \qquad (0 \leq t \leq 10)$$

where $N(t)$ denotes the population at the end of year t. Find the rate of growth of the turtle population when $t = 2$ and $t = 8$. What will be the population 10 years after the conservation measures are implemented?

67. FLIGHT OF A MODEL ROCKET The altitude (in feet) of a model rocket t sec into a trial flight is given by

$$s = f(t) = -2t^3 + 12t^2 + 5 \qquad (t \ge 0)$$

a. Find an expression v for the rocket's velocity at any time t.

b. Compute the rocket's vertical velocity when $t = 0, 2, 4,$ and 6. Interpret your results.

c. Using the results from the solution to part (b), find the maximum altitude attained by the rocket.

Hint: At its highest point, the velocity of the rocket is zero.

68. SUPPLY FUNCTION FOR SATELLITE RADIOS The supply function for a certain make of satellite radio is given by

$$p = f(x) = 0.0001x^{5/4} + 10$$

where x is the quantity supplied and p is the unit price in dollars.

a. Find $f'(x)$.

b. What is the rate of change of the unit price if the quantity supplied is 10,000 satellite radios?

69. POPULATION GROWTH FOR A RESORT TOWN A study prepared for a Sunbelt town's chamber of commerce projected that the town's population in the next 3 years will grow according to the rule

$$P(t) = 50,000 + 30t^{3/2} + 20t$$

where $P(t)$ denotes the population t months from now. How fast will the population be increasing 9 months and 16 months from now?

70. AVERAGE SPEED OF A VEHICLE ON A HIGHWAY The average speed of a vehicle on a stretch of Route 134 between 6 A.M. and 10 A.M. on a typical weekday is approximated by the function

$$f(t) = 20t - 40\sqrt{t} + 50 \qquad (0 \le t \le 4)$$

where $f(t)$ is measured in miles per hour and t is measured in hours, with $t = 0$ corresponding to 6 A.M.

a. Compute $f'(t)$.

b. What is the average speed of a vehicle on that stretch of Route 134 at 6 A.M.? At 7 A.M.? At 8 A.M.?

c. How fast is the average speed of a vehicle on that stretch of Route 134 changing at 6:30 A.M.? At 7 A.M.? At 8 A.M.?

71. EFFECT OF ADVERTISING ON SALES The relationship between the amount of money x that Cannon Precision Instruments spends on advertising and the company's total sales $S(x)$ is given by the function

$$S(x) = -0.002x^3 + 0.6x^2 + x + 500 \qquad (0 \le x \le 200)$$

where x is measured in thousands of dollars. Find the rate of change of the sales with respect to the amount of money spent on advertising. Are Cannon's total sales increasing at a faster rate when the amount of money spent on advertising is (a) $100,000 or (b) $150,000?

72. NATIONAL HEALTH CARE EXPENDITURE The per capita health spending in the United States in year t for the years 2000 through 2011 is approximately

$$C(t) = -1.1708t^3 + 7.029t^2 + 389.69t + 4780$$
$$(0 \le t \le 11)$$

dollars in year t, with $t = 0$ corresponding to 2000.

a. What was the per capita health spending in 2010?

b. How fast was the per capita health spending changing in 2010?

Source: NHCM Foundation.

73. OBESITY IN AMERICA The body mass index (BMI) measures body weight in relation to height. A BMI of 25 to 29.9 is considered overweight, a BMI of 30 or more is considered obese, and a BMI of 40 or more is morbidly obese. The percentage of the U.S. population that is obese is approximated by the function

$$P(t) = 0.0004t^3 + 0.0036t^2 + 0.8t + 12 \qquad (0 \le t \le 20)$$

where t is measured in years, with $t = 0$ corresponding to the beginning of 1991.

a. What percentage of the U.S. population was deemed obese at the beginning of 1991? At the beginning of 2010?

b. How fast was the percentage of the U.S. population that is deemed obese changing at the beginning of 1991? At the beginning of 2010?

(Note: A formula for calculating the BMI of a person is given in Exercise 35, page 563.)

Source: Centers for Disease Control and Prevention.

74. AGING POPULATION The population age 65 and over (in millions) of developed countries from 2005 through 2034 is projected to be

$$f(t) = 3.567t + 175.2 \qquad (5 \le t \le 35)$$

where t is measured in years and $t = 5$ corresponds to 2005. On the other hand, the population age 65 and over of underdeveloped/emerging countries over the same period is projected to be

$$g(t) = 0.46t^2 + 0.16t + 287.8 \qquad (5 \le t \le 35)$$

a. What does the function $D = g + f$ represent?

b. Find D' and $D'(10)$, and interpret your results.

Source: U.S. Census Bureau, United Nations.

75. SHORTAGE OF NURSES The projected number of nurses (in millions) in year t from 2000 through 2015 is given by

$$N(t) = \begin{cases} 1.9 & \text{if } 0 \le t < 5 \\ -0.0004t^2 + 0.038t + 1.72 & \text{if } 5 \le t \le 15 \end{cases}$$

where $t = 0$ corresponds to 2000. The projected number of nursing jobs (in millions) over the same period is

$$J(t) = \begin{cases} -0.0002t^2 + 0.032t + 2 & \text{if } 0 \le t < 10 \\ -0.0016t^2 + 0.12t + 1.26 & \text{if } 10 \le t \le 15 \end{cases}$$

a. Find the rule for the function $G = J - N$ giving the gap between the supply and the demand of nurses from 2000 through 2015.

b. How fast was the gap between the supply and the demand of nurses changing in 2008? In 2012?

Source: U.S. Department of Health and Human Services.

In Exercises 76 and 77, determine whether the statement is true or false. If it is true, explain why it is true. If it is false, give an example to show why it is false.

76. If f and g are differentiable, then

$$\frac{d}{dx}[2f(x) - 5g(x)] = 2f'(x) - 5g'(x)$$

77. If $f(x) = \pi^x$, then $f'(x) = x\pi^{x-1}$.

78. Prove the Power Rule (Rule 2) for the special case $n = 3$.

Hint: Compute $\lim\limits_{h \to 0}\left[\dfrac{(x + h)^3 - x^3}{h}\right]$.

3.1 Solutions to Self-Check Exercises

1. a. $f'(x) = \dfrac{d}{dx}(1.5x^2) + \dfrac{d}{dx}(2x^{1.5})$

$= (1.5)(2x) + (2)(1.5x^{0.5})$

$= 3x + 3x^{0.5}$

b. $g'(x) = \dfrac{d}{dx}(2x^{1/2}) + \dfrac{d}{dx}(3x^{-1/2})$

$= (2)\left(\dfrac{1}{2}x^{-1/2}\right) + (3)\left(-\dfrac{1}{2}x^{-3/2}\right)$

$= x^{-1/2} - \dfrac{3}{2}x^{-3/2} = \dfrac{1}{\sqrt{x}} - \dfrac{3}{2\sqrt{x^3}}$

2. a. $f'(x) = \dfrac{d}{dx}(2x^3) - \dfrac{d}{dx}(3x^2) + \dfrac{d}{dx}(2x) - \dfrac{d}{dx}(1)$

$= (2)(3x^2) - (3)(2x) + 2$

$= 6x^2 - 6x + 2$

b. The slope of the tangent line to the graph of f when $x = 2$ is given by

$$f'(2) = 6(2)^2 - 6(2) + 2 = 14$$

c. The rate of change of f at $x = 2$ is given by $f'(2)$. Using the results of part (b), we see that the required rate of change is 14 units/unit change in x.

3. a. The rate at which the GDP was changing at any time t $(0 < t < 11)$ is given by

$$G'(t) = -6t^2 + 90t + 20$$

In particular, the rates of change of the GDP at the beginning of the years 2010 ($t = 5$), 2012 ($t = 7$), and 2015 ($t = 10$) are given by

$$G'(5) = 320 \qquad G'(7) = 356 \qquad G'(10) = 320$$

respectively—that is, by \$320 billion/year, \$356 billion/year, and \$320 billion/year, respectively.

b. The average rate of growth of the GDP over the period from the beginning of 2010 ($t = 5$) to the beginning of 2015 ($t = 10$) is given by

$$\frac{G(10) - G(5)}{10 - 5} = \frac{[-2(10)^3 + 45(10)^2 + 20(10) + 6000]}{5}$$

$$- \frac{[-2(5)^3 + 45(5)^2 + 20(5) + 6000]}{5}$$

$$= \frac{8700 - 6975}{5}$$

or \$345 billion/year.

USING TECHNOLOGY   Finding the Rate of Change of a Function

We can use the numerical derivative operation of a graphing utility to obtain the value of the derivative at a given value of x. Since the derivative of a function f measures the rate of change of the function with respect to x, the numerical derivative operation can be used to answer questions pertaining to the rate of change of one quantity y with respect to another quantity x, where $y = f(x)$, for a specific value of x.

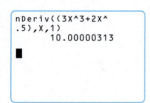

FIGURE **T1**
The TI-83/84 numerical derivative screen
for computing $f'(1)$

EXAMPLE 1 Let $y = 3t^3 + 2\sqrt{t}$.

a. Use the numerical derivative operation of a graphing utility to find how fast y is changing with respect to t when $t = 1$.

b. Verify the result of part (a), using the rules of differentiation of this section.

Solution

a. Let $f(t) = 3t^3 + 2\sqrt{t}$. Using the numerical derivative operation of a graphing utility, we find that the rate of change of y with respect to t when $t = 1$ is given by $f'(1) = 10$ (Figure T1).

b. Here, $f(t) = 3t^3 + 2t^{1/2}$, and

$$f'(t) = 9t^2 + 2\left(\frac{1}{2}t^{-1/2}\right) = 9t^2 + \frac{1}{\sqrt{t}}$$

Using this result, we see that when $t = 1$, y is changing at the rate of

$$f'(1) = 9(1^2) + \frac{1}{\sqrt{1}} = 10$$

units per unit change in t, as obtained earlier.

APPLIED EXAMPLE 2 Fuel Economy of Cars According to data obtained from the U.S. Department of Energy and the Shell Development Company, a typical car's fuel economy depends on the speed at which it is driven and is approximated by the function

$$f(x) = 0.00000310315x^4 - 0.000455174x^3$$
$$+ 0.00287869x^2 + 1.25986x \quad (0 \le x \le 75)$$

where x is measured in miles per hour and $f(x)$ is measured in miles per gallon (mpg).

a. Use a graphing utility to graph the function f on the interval $[0, 75]$.

b. Find the rate of change of f when $x = 20$ and when $x = 50$.

c. Interpret your results.

Source: U.S. Department of Energy and the Shell Development Company.

FIGURE **T2**
The graph of the function f on the interval $[0, 75]$

Solution

a. The graph is shown in Figure T2.

b. Using the numerical derivative operation of a graphing utility, we see that $f'(20) = 0.9280996$. The rate of change of f when $x = 50$ is given by $f'(50) = -0.3145009995$. (See Figure T3a and T3b.)

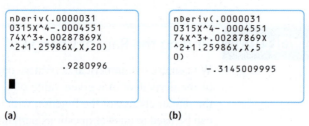

(a)   (b)

FIGURE **T3**
The TI-83/84 numerical derivative screen for computing (a) $f'(20)$ and (b) $f'(50)$

c. The results of part (b) tell us that when a typical car is being driven at 20 mph, its fuel economy increases at the rate of approximately 0.9 mpg per 1 mph increase in its speed. At a speed of 50 mph, its fuel economy decreases at the rate of approximately 0.3 mpg per 1 mph increase in its speed.

TECHNOLOGY EXERCISES

In Exercises 1–6, use the numerical derivative operation to find the rate of change of f at the given value of x. Give your answer accurate to four decimal places.

1. $f(x) = 4x^5 - 3x^3 + 2x^2 + 1; x = 0.5$

2. $f(x) = -x^5 + 4x^2 + 3; x = 0.4$

3. $f(x) = x - 2\sqrt{x}; x = 3$

4. $f(x) = \dfrac{\sqrt{x} - 1}{x}; x = 2$

5. $f(x) = x^{1/2} - x^{1/3}; x = 1.2$

6. $f(x) = 2x^{5/4} + x; x = 2$

7. CARBON MONOXIDE IN THE ATMOSPHERE The projected average global atmospheric concentration of carbon monoxide is approximated by the function

$$f(t) = 0.881443t^4 - 1.45533t^3 + 0.695876t^2 + 2.87801t + 293 \quad (0 \le t \le 4)$$

where t is measured in 40-year intervals, with $t = 0$ corresponding to the beginning of 1860, and $f(t)$ is measured in parts per million by volume.

a. Plot the graph of f in the viewing window $[0, 4] \times [280, 500]$.

b. Use a graphing utility to estimate how fast the projected average global atmospheric concentration of carbon monoxide was changing at the beginning of 1900 ($t = 1$) and at the beginning of 2010 ($t = 4$).

Source: Meadows et al., *Beyond the Limits.*

8. SPREAD OF HIV TO CHILDREN The estimated number of children newly infected with HIV through mother-to-child contact worldwide is given by

$$f(t) = -0.2083t^3 + 3.0357t^2 + 44.0476t + 200.2857 \quad (0 \le t \le 12)$$

where $f(t)$ is measured in thousands and t is measured in years, with $t = 0$ corresponding to the beginning of 1990.

a. Plot the graph of f in the viewing window $[0, 12] \times [0, 850]$.

b. How fast was the estimated number of children newly infected with HIV through mother-to-child contact worldwide increasing at the beginning of the year 2000?

Source: United Nations.

3.2 The Product and Quotient Rules

In this section, we study two more rules of differentiation: the **Product Rule** and the **Quotient Rule.**

The Product Rule

The derivative of the product of two differentiable functions is given by the following rule:

> **Rule 5: The Product Rule**
>
> $$\frac{d}{dx}[f(x)g(x)] = f(x)g'(x) + g(x)f'(x)$$

The derivative of the product of two functions is the first function times the derivative of the second plus the second function times the derivative of the first.

The Product Rule may be extended to the case involving the product of any finite number of functions (see Exercise 73, page 184). We prove the Product Rule at the end of this section.

The derivative of the product of two functions is *not* given by the product of the derivatives of the functions; that is, in general

$$\frac{d}{dx}[f(x)g(x)] \ne f'(x)g'(x)$$

For example, if $f(x) = x$ and $g(x) = 2x^2$. Then

$$\frac{d}{dx}[f(x)g(x)] = \frac{d}{dx}[x(2x^2)] = \frac{d}{dx}(2x^3) = 6x^2$$

On the other hand, $f'(x)g'(x) = (1)(4x) = 4x$. So

$$\frac{d}{dx}[f(x)g(x)] \neq f'(x)g'(x)$$

EXAMPLE 1 Find the derivative of the function

$$f(x) = (2x^2 - 1)(x^3 + 3)$$

Solution By the Product Rule,

$$\begin{aligned}
f'(x) &= (2x^2 - 1)\frac{d}{dx}(x^3 + 3) + (x^3 + 3)\frac{d}{dx}(2x^2 - 1) \\
&= (2x^2 - 1)(3x^2) + (x^3 + 3)(4x) && (x^2)\ \text{See page 8.} \\
&= 6x^4 - 3x^2 + 4x^4 + 12x \\
&= 10x^4 - 3x^2 + 12x && \text{Combine like terms.} \\
&= x(10x^3 - 3x + 12) && \text{Factor out } x.
\end{aligned}$$

EXAMPLE 2 Differentiate (that is, find the derivative of) the function

$$f(x) = x^3(\sqrt{x} + 1)$$

Solution First, we express the function in exponential form, obtaining

$$f(x) = x^3(x^{1/2} + 1)$$

By the Product Rule,

$$\begin{aligned}
f'(x) &= x^3\frac{d}{dx}(x^{1/2} + 1) + (x^{1/2} + 1)\frac{d}{dx}x^3 \\
&= x^3\left(\frac{1}{2}x^{-1/2}\right) + (x^{1/2} + 1)(3x^2) \\
&= \frac{1}{2}x^{5/2} + 3x^{5/2} + 3x^2 \\
&= \frac{7}{2}x^{5/2} + 3x^2
\end{aligned}$$

Note We can also solve the problem by first expanding the product before differentiating f. Examples for which this is not possible will be considered in Section 3.3, where the true value of the Product Rule will be appreciated.

The Quotient Rule

The derivative of the quotient of two differentiable functions is given by the following rule:

Rule 6: The Quotient Rule

$$\frac{d}{dx}\left[\frac{f(x)}{g(x)}\right] = \frac{g(x)f'(x) - f(x)g'(x)}{[g(x)]^2} \qquad (g(x) \neq 0)$$

As an aid to remembering this expression, observe that it has the following form:

$$\frac{d}{dx}\left[\frac{f(x)}{g(x)}\right] = \frac{(\text{Denominator})\begin{pmatrix}\text{Derivative of}\\\text{numerator}\end{pmatrix} - (\text{Numerator})\begin{pmatrix}\text{Derivative of}\\\text{denominator}\end{pmatrix}}{(\text{Square of denominator})}$$

For a proof of the Quotient Rule, see Exercise 74, page 184.

The derivative of a quotient is *not* equal to the quotient of the derivatives; that is,

$$\frac{d}{dx}\left[\frac{f(x)}{g(x)}\right] \neq \frac{f'(x)}{g'(x)}$$

For example, if $f(x) = x^3$ and $g(x) = x^2$, then

$$\frac{d}{dx}\left[\frac{f(x)}{g(x)}\right] = \frac{d}{dx}\left(\frac{x^3}{x^2}\right) = \frac{d}{dx}(x) = 1$$

which is *not* equal to

$$\frac{f'(x)}{g'(x)} = \frac{\dfrac{d}{dx}(x^3)}{\dfrac{d}{dx}(x^2)} = \frac{3x^2}{2x} = \frac{3}{2}x$$

EXAMPLE 3 Find $f'(x)$ if $f(x) = \dfrac{x}{2x - 4}$.

Solution Using the Quotient Rule, we obtain

$$\begin{aligned}
f'(x) &= \frac{(2x - 4)\dfrac{d}{dx}(x) - x\dfrac{d}{dx}(2x - 4)}{(2x - 4)^2} \\[2mm]
&= \frac{(2x - 4)(1) - x(2)}{(2x - 4)^2} \\[2mm]
&= \frac{2x - 4 - 2x}{(2x - 4)^2} = -\frac{4}{(2x - 4)^2}
\end{aligned}$$

EXAMPLE 4 Find $f'(x)$ if $f(x) = \dfrac{x^2 + 1}{x^2 - 1}$.

Solution By the Quotient Rule,

$$\begin{aligned}
f'(x) &= \frac{(x^2 - 1)\dfrac{d}{dx}(x^2 + 1) - (x^2 + 1)\dfrac{d}{dx}(x^2 - 1)}{(x^2 - 1)^2} \\[2mm]
&= \frac{(x^2 - 1)(2x) - (x^2 + 1)(2x)}{(x^2 - 1)^2} \\[2mm]
&= \frac{2x^3 - 2x - 2x^3 - 2x}{(x^2 - 1)^2} \\[2mm]
&= -\frac{4x}{(x^2 - 1)^2}
\end{aligned}$$

EXAMPLE 5 Find $h'(x)$ if $h(x) = \dfrac{\sqrt{x}}{x^2 + 1}$.

Solution Rewrite $h(x)$ in the form $h(x) = \dfrac{x^{1/2}}{x^2 + 1}$. By the Quotient Rule, we find

$$
\begin{aligned}
h'(x) &= \frac{(x^2 + 1)\dfrac{d}{dx}(x^{1/2}) - x^{1/2}\dfrac{d}{dx}(x^2 + 1)}{(x^2 + 1)^2} \\[2mm]
&= \frac{(x^2 + 1)(\tfrac{1}{2}x^{-1/2}) - x^{1/2}(2x)}{(x^2 + 1)^2} \\[2mm]
&= \frac{\tfrac{1}{2}x^{-1/2}(x^2 + 1 - 4x^2)}{(x^2 + 1)^2} \qquad \text{Factor out } \tfrac{1}{2}x^{-1/2} \text{ from the numerator.} \qquad (x^2)\ \text{See page 10.} \\[2mm]
&= \frac{1 - 3x^2}{2\sqrt{x}(x^2 + 1)^2}
\end{aligned}
$$

$ APPLIED EXAMPLE 6 Rate of Change of DVD Sales The annual sales (in millions of dollars per year) of a DVD recording of a hit movie t years from the date of release is given by

$$ S(t) = \frac{5t}{t^2 + 1} $$

a. Find the rate at which the annual sales are changing at time t.
b. How fast are the annual sales changing at the time the DVDs are released $(t = 0)$? Two years from the date of release?

Solution

a. The rate at which the annual sales are changing at time t is given by $S'(t)$. Using the Quotient Rule, we obtain

$$
\begin{aligned}
S'(t) &= \frac{d}{dt}\left[\frac{5t}{t^2 + 1}\right] = 5\,\frac{d}{dt}\left[\frac{t}{t^2 + 1}\right] \\[2mm]
&= 5\left[\frac{(t^2 + 1)(1) - t(2t)}{(t^2 + 1)^2}\right] \qquad (x^2)\ \text{See page 17.} \\[2mm]
&= 5\left[\frac{t^2 + 1 - 2t^2}{(t^2 + 1)^2}\right] = \frac{5(1 - t^2)}{(t^2 + 1)^2}
\end{aligned}
$$

b. The rate at which the annual sales are changing at the time the DVDs are released is given by

$$ S'(0) = \frac{5(1 - 0)}{(0 + 1)^2} = 5 $$

That is, they are increasing at the rate of $5 million per year per year.

Two years from the date of release, the annual sales are changing at the rate of

$$ S'(2) = \frac{5(1 - 4)}{(4 + 1)^2} = -\frac{3}{5} = -0.6 $$

That is, they are decreasing at the rate of $600,000 per year per year.

The graph of the function S is shown in Figure 5.

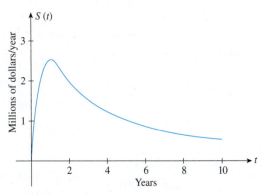

FIGURE 5
After a spectacular rise, the annual sales begin to taper off.

Exploring with TECHNOLOGY

Refer to Example 6.

1. Use a graphing utility to plot the graph of the function S, using the viewing window $[0, 10] \times [0, 3]$.

2. Use **TRACE** and **ZOOM** to determine the coordinates of the highest point on the graph of S in the interval $[0, 10]$. Interpret your results.

Explore and Discuss

Suppose the revenue of a company is given by $R(x) = xp(x)$, where x is the number of units of the product sold at a unit price of $p(x)$ dollars.

1. Compute $R'(x)$ and explain, in words, the relationship between $R'(x)$ and $p(x)$ and/or its derivative.

2. What can you say about $R'(x)$ if $p(x)$ is constant? Is this expected?

APPLIED EXAMPLE 7 Oxygen-Restoration Rate in a Pond When organic waste is dumped into a pond, the oxidation process that takes place reduces the pond's oxygen content. However, given time, nature will restore the oxygen content to its natural level. Suppose the oxygen content t days after organic waste has been dumped into the pond is given by

$$f(t) = 100\left[\frac{t^2 + 10t + 100}{t^2 + 20t + 100}\right] \qquad (0 < t < \infty)$$

percent of its normal level.

a. Derive a general expression that gives the rate of change of the pond's oxygen level at any time t.
b. How fast is the pond's oxygen content changing 1 day, 10 days, and 20 days after the organic waste has been dumped?

Solution

a. The rate of change of the pond's oxygen level at any time t is given by the derivative of the function f. Thus, the required expression is

$$f'(t) = 100 \frac{d}{dt}\left[\frac{t^2 + 10t + 100}{t^2 + 20t + 100}\right]$$

$$= 100\left[\frac{(t^2 + 20t + 100)\frac{d}{dt}(t^2 + 10t + 100) - (t^2 + 10t + 100)\frac{d}{dt}(t^2 + 20t + 100)}{(t^2 + 20t + 100)^2}\right]$$

$$= 100\left[\frac{(t^2 + 20t + 100)(2t + 10) - (t^2 + 10t + 100)(2t + 20)}{(t^2 + 20t + 100)^2}\right] \qquad (x^2) \text{ See page 17.}$$

$$= 100\left[\frac{2t^3 + 10t^2 + 40t^2 + 200t + 200t + 1000 - 2t^3 - 20t^2 - 20t^2 - 200t - 200t - 2000}{(t^2 + 20t + 100)^2}\right]$$

$$= 100\left[\frac{10t^2 - 1000}{(t^2 + 20t + 100)^2}\right] \qquad \text{Combine like terms in the numerator.}$$

b. The rate at which the pond's oxygen content is changing 1 day after the organic waste has been dumped is given by

$$f'(1) = 100\left[\frac{10 - 1000}{(1 + 20 + 100)^2}\right] \approx -6.76$$

That is, it is dropping at the rate of 6.8% per day. After 10 days, the rate is

$$f'(10) = 100\left[\frac{10(10)^2 - 1000}{(10^2 + 20(10) + 100)^2}\right] = 0$$

That is, it is neither increasing nor decreasing. After 20 days, the rate is

$$f'(20) = 100\left[\frac{10(20)^2 - 1000}{(20^2 + 20(20) + 100)^2}\right] \approx 0.37$$

That is, the oxygen content is increasing at the rate of 0.37% per day, and the restoration process has indeed begun.

Verification of the Product Rule

We will now verify the Product Rule. If $p(x) = f(x)g(x)$, then

$$p'(x) = \lim_{h \to 0} \frac{p(x + h) - p(x)}{h} = \lim_{h \to 0} \frac{f(x + h)g(x + h) - f(x)g(x)}{h}$$

By adding $-f(x + h)g(x) + f(x + h)g(x)$ (which is zero!) to the numerator and factoring, we have

$$p'(x) = \lim_{h \to 0} \frac{f(x + h)[g(x + h) - g(x)] + g(x)[f(x + h) - f(x)]}{h}$$

$$= \lim_{h \to 0}\left\{ f(x + h)\left[\frac{g(x + h) - g(x)}{h}\right] + g(x)\left[\frac{f(x + h) - f(x)}{h}\right]\right\}$$

$$= \lim_{h \to 0} f(x + h)\left[\frac{g(x + h) - g(x)}{h}\right] + \lim_{h \to 0} g(x)\left[\frac{f(x + h) - f(x)}{h}\right] \qquad \text{By Property 3 of limits}$$

$$= \lim_{h \to 0} f(x + h) \cdot \lim_{h \to 0} \frac{g(x + h) - g(x)}{h}$$

$$+ \lim_{h \to 0} g(x) \cdot \lim_{h \to 0} \frac{f(x + h) - f(x)}{h} \qquad \text{By Property 4 of limits}$$

$$= f(x)g'(x) + g(x)f'(x)$$

Observe that in the last link in the chain of equalities, we have used the fact that $\lim_{h \to 0} f(x + h) = f(x)$ because f is continuous at x.

3.2 Self-Check Exercises

1. Find the derivative of $f(x) = \dfrac{2x+1}{x^2-1}$.

2. What is the slope of the tangent line to the graph of

$$f(x) = (x^2 + 1)(2x^3 - 3x^2 + 1)$$

at the point $(2, 25)$? How fast is the function f changing when $x = 2$?

3. **SALES OF ADS SECURITY SYSTEMS** The total sales of ADS Security Systems in its first 2 years of operation are given by

$$S = f(t) = \frac{0.3t^3}{1 + 0.4t^2} \qquad (0 \le t \le 2)$$

where S is measured in millions of dollars and $t = 0$ corresponds to the date ADS Security Systems began operations. How fast were the sales increasing at the beginning of the company's second year of operation?

Solutions to Self-Check Exercises 3.2 can be found on page 184.

3.2 Concept Questions

1. State the rule of differentiation in your own words.
   a. Product Rule   b. Quotient Rule

2. If $f(1) = 3, g(1) = 2, f'(1) = -1$, and $g'(1) = 4$, find
   a. $h'(1)$ if $h(x) = f(x)g(x)$ b. $F'(1)$ if $F(x) = \dfrac{f(x)}{g(x)}$

3.2 Exercises

In Exercises 1–30, find the derivative of each function.

1. $f(x) = 2x(x^2 + 1)$

2. $f(x) = 3x^2(x - 1)$

3. $f(t) = (t - 1)(2t + 1)$

4. $f(x) = (2x + 3)(3x - 4)$

5. $f(x) = (3x + 1)(x^2 - 2)$

6. $f(x) = (x + 1)(2x^2 - 3x + 1)$

7. $f(x) = (x^3 - 1)(x + 1)$

8. $f(x) = (x^3 - 12x)(3x^2 + 2x)$

9. $f(w) = (w^3 - w^2 + w - 1)(w^2 + 2)$

10. $f(x) = \dfrac{1}{5}x^5 + (x^2 + 1)(x^2 - x - 1) + 28$

11. $f(x) = (5x^2 + 1)(2\sqrt{x} - 1)$

12. $f(t) = (1 + \sqrt{t})(2t^2 - 3)$

13. $f(x) = (x^2 - 5x + 2)\left(x - \dfrac{2}{x}\right)$

14. $f(x) = (x^3 + 2x + 1)\left(2 + \dfrac{1}{x^2}\right)$

15. $f(x) = \dfrac{1}{x - 2}$

16. $g(x) = \dfrac{3}{2x + 4} + 2x^2$

17. $f(x) = \dfrac{2x - 1}{2x + 1}$

18. $f(t) = \dfrac{1 - 2t}{1 + 3t}$

19. $f(x) = \dfrac{1}{x^2 + x + 2}$

20. $f(u) = \dfrac{u}{u^2 + 1}$

21. $f(s) = \dfrac{s^2 - 4}{s + 1}$

22. $f(x) = \dfrac{x^3 - 2}{x^2 + 1}$

23. $f(x) = \dfrac{\sqrt{x} + 1}{x^2 + 1}$

24. $f(x) = \dfrac{x}{\sqrt{x} + 2}$

25. $f(x) = \dfrac{x^2 + 2}{x^2 + x + 1}$

26. $f(x) = \dfrac{x + 1}{2x^2 + 2x + 3}$

27. $f(x) = \dfrac{(x + 1)(x^2 + 1)}{x - 2}$

28. $f(x) = (3x^2 - 1)\left(x^2 - \dfrac{1}{x}\right)$

29. $f(x) = \dfrac{x}{x^2 - 4} - \dfrac{x - 1}{x^2 + 4}$

30. $f(x) = \dfrac{x + \sqrt{3x}}{3x - 1}$

In Exercises 31–34, suppose f and g are functions that are differentiable at $x = 1$ and that $f(1) = 2, f'(1) = -1, g(1) = -2$, and $g'(1) = 3$. Find the value of $h'(1)$.

31. $h(x) = f(x)g(x)$

32. $h(x) = (x^2 + 1)g(x)$

33. $h(x) = \dfrac{xf(x)}{x + g(x)}$

34. $h(x) = \dfrac{f(x)g(x)}{f(x) - g(x)}$

In Exercises 35–38, find the derivative of each function and evaluate $f'(x)$ at the given value of x.

35. $f(x) = (2x - 1)(x^2 + 3)$; $x = 1$

36. $f(x) = \dfrac{2x + 1}{2x - 1}$; $x = 2$

37. $f(x) = \dfrac{x}{x^4 - 2x^2 - 1}$; $x = -1$

38. $f(x) = (\sqrt{x} + 2x)(x^{3/2} - x)$; $x = 4$

In Exercises 39–42, find the slope and an equation of the tangent line to the graph of the function f at the specified point.

39. $f(x) = (x^3 + 1)(x^2 - 2)$; $(2, 18)$

40. $f(x) = \dfrac{x^2}{x + 1}$; $\left(2, \dfrac{4}{3}\right)$ **41.** $f(x) = \dfrac{x + 1}{x^2 + 1}$; $(1, 1)$

42. $f(x) = \dfrac{1 + 2x^{1/2}}{1 + x^{3/2}}$; $\left(4, \dfrac{5}{9}\right)$

43. Suppose $g(x) = x^2 f(x)$ and it is known that $f(2) = 3$ and $f'(2) = -1$. Evaluate $g'(2)$.

44. Suppose $g(x) = (x^2 + 1)f(x)$ and it is known that $f(2) = 3$ and $f'(2) = -1$. Evaluate $g'(2)$.

45. Find an equation of the tangent line to the graph of the function $f(x) = (x^3 + 1)(3x^2 - 4x + 2)$ at the point $(1, 2)$.

46. Find an equation of the tangent line to the graph of the function $f(x) = \dfrac{3x}{x^2 - 2}$ at the point $(2, 3)$.

47. Let $f(x) = (x^2 + 1)(2 - x)$. Find the point(s) on the graph of f where the tangent line is horizontal.

48. Let $f(x) = \dfrac{x}{x^2 + 1}$. Find the point(s) on the graph of f where the tangent line is horizontal.

49. Find the point(s) on the graph of the function $f(x) = (x^2 + 6)(x - 5)$ where the slope of the tangent line is equal to -2.

50. Find the point(s) on the graph of the function $f(x) = \dfrac{x + 1}{x - 1}$ where the slope of the tangent line is equal to $-\frac{1}{2}$.

51. A straight line perpendicular to and passing through the point of tangency of the tangent line is called the *normal* to the curve at that point. Find the equation of the tangent line and the normal to the curve

$$y = \frac{1}{1 + x^2}$$

at the point $\left(1, \frac{1}{2}\right)$.

52. CONCENTRATION OF A DRUG IN THE BLOODSTREAM The concentration of a certain drug in a patient's bloodstream t hr after injection is given by

$$C(t) = \frac{0.2t}{t^2 + 1}$$

a. Find the rate at which the concentration of the drug is changing with respect to time.
b. How fast is the concentration changing $\frac{1}{2}$ hr, 1 hr, and 2 hr after the injection?

53. COST OF REMOVING TOXIC WASTE A city's main water reservoir was recently found to be contaminated with trichloroethylene, a cancer-causing chemical, as a result of an abandoned chemical dump leaching chemicals into the water. A proposal submitted to the city's council members indicates that the cost, measured in millions of dollars, of removing $x\%$ of the toxic pollutant is given by

$$C(x) = \frac{0.5x}{100 - x}$$

Find $C'(80)$, $C'(90)$, $C'(95)$, and $C'(99)$. What does your result tell you about the cost of removing *all* of the pollutant?

54. DRUG DOSAGES Thomas Young has suggested the following rule for calculating the dosage of medicine for children 1 to 12 years old. If a denotes the adult dosage (in milligrams) and if t is the child's age (in years), then the child's dosage is given by

$$D(t) = \frac{at}{t + 12}$$

Suppose the adult dosage of a substance is 500 mg. Find an expression that gives the rate of change of a child's dosage with respect to the child's age. What is the rate of change of a child's dosage with respect to his or her age for a 6-year-old child? A 10-year-old child?

55. EFFECT OF BACTERICIDE The number of bacteria $N(t)$ in a certain culture t min after an experimental bactericide is introduced is given by

$$N(t) = \frac{10,000}{1 + t^2} + 2000$$

Find the rate of change of the number of bacteria in the culture 1 min and 2 min after the bactericide is introduced. What is the population of the bacteria in the culture 1 min and 2 min after the bactericide is introduced?

56. DEMAND FUNCTION FOR SPORTS WATCHES The demand function for the Sicard sports watch is given by

$$d(x) = \frac{50}{0.01x^2 + 1} \qquad (0 \le x \le 20)$$

where x (measured in units of a thousand) is the quantity demanded per week and $d(x)$ is the unit price in dollars.
a. Find $d'(x)$.
b. Find $d'(5)$, $d'(10)$, and $d'(15)$, and interpret your results.

57. REVENUE FUNCTION FOR SPORTS WATCHES Refer to Exercise 56.
   a. Find an expression for the revenue function R for the Sicard sports watch.
      Hint: $R(x) = xd(x)$
   b. Find $R'(x)$.
   c. Find $R'(8)$, $R'(10)$, and $R'(12)$, and interpret your results.

58. PROFIT FUNCTION FOR SPORTS WATCHES Refer to Exercise 56. The total profit function P (in thousands of dollars) for the Sicard sports watch is given by

$$P(x) = \frac{50x}{0.01x^2 + 1} - 0.025x^3 + 0.35x^2 - 10x - 30$$

$$(0 \le x \le 20)$$

where x is measured in units of a thousand.
   a. Find $P(0)$, and interpret your result.
   b. Find $P'(5)$ and $P'(10)$, and interpret your results.

59. MORTGAGE RATES The average 30-year fixed mortgage rate in the United States in the first week of May in 2010 through 2012 is approximated by

$$M(t) = \frac{55.9}{t^2 - 0.31t + 11.2}$$

percent/year. Here t is measured in years, with $t = 0$ corresponding to the first week of May in 2010.
   a. What was the average 30-year fixed mortgage rate in the first week of May in 2011 ($t = 1$)?
   b. How fast was the 30-year fixed mortgage rate changing in the first week of May in 2011 ($t = 1$)?
   Source: Mortgage Bankers Association.

60. BOX-OFFICE RECEIPTS The total worldwide box-office receipts for a long-running movie are approximated by the function

$$T(x) = \frac{120x^2}{x^2 + 4}$$

where $T(x)$ is measured in millions of dollars and x is the number of years since the movie's release. How fast are the total receipts changing 1 year, 3 years, and 5 years after its release?

61. LEARNING CURVES From experience, Emory Secretarial School knows that the average student taking Advanced Computer Typing will progress according to the rule

$$N(t) = \frac{60t + 180}{t + 6} \qquad (t \ge 0)$$

where $N(t)$ measures the number of words per minute the student can type after t weeks in the course.
   a. Find an expression for $N'(t)$.
   b. Compute $N'(t)$ for $t = 1, 3, 4$, and 7, and interpret your results.
   c. Sketch the graph of the function N. Does it confirm the results obtained in part (b)?
   d. What will be the average student's typing speed at the end of the 12-week course?

62. UNEMPLOYMENT RATE The unemployment rate of a certain country shortly after the Great Recession was approximately

$$f(t) = \frac{5t + 300}{t^2 + 25} \qquad (0 \le t \le 4)$$

percent in year t, where $t = 0$ corresponds to the beginning of 2010. How fast was the unemployment rate of the country changing at the beginning of 2013?

63. HOME FORMALDEHYDE LEVELS A study on formaldehyde levels in 900 homes indicates that emissions of various chemicals can decrease over time. The formaldehyde level (parts per million) in an average home in the study is given by

$$f(t) = \frac{0.055t + 0.26}{t + 2} \qquad (0 \le t \le 12)$$

where t is the age of the house in years. How fast is the formaldehyde level of the average house dropping when it is new? At the beginning of its fourth year? (See Note on page 168.)
Source: Bonneville Power Administration.

64. POPULATION GROWTH A major corporation is building a 4325-acre complex of homes, offices, stores, schools, and churches in the rural community of Glen Cove. As a result of this development, the planners have estimated that Glen Cove's population (in thousands) t years from now will be given by

$$P(t) = \frac{25t^2 + 125t + 200}{t^2 + 5t + 40}$$

   a. Find the rate at which Glen Cove's population is changing with respect to time.
   b. What will be the population after 10 years? At what rate will the population be increasing when $t = 10$?

65. CHANGE IN REVENUE Suppose that the demand equation for a product is $p = D(x)$, where x is the quantity demanded. If x units of the product are sold at a price of p dollars/unit, then the revenue realized from the sales is $R(x) = xD(x)$ (number of units sold times the price per unit).
   a. Use the Product Rule to find $R'(x)$.
   b. Use the result of part (a) to find $R'(x)$ for the case of a linear demand equation $p = a - bx$, where a and b are positive constants.
   c. Verify your result by first finding an expression for $R(x)$ and then differentiating $R(x)$ directly.

66. CHANGE IN PER CAPITA INCOME The gross domestic product (GDP) of a country t years from now is $g(t) = 60 + 4t$ billion dollars. The population of the country at that time is $f(t) = 3 + 0.06t$ million. How fast will the per-capita income of the country be changing 2 years from now?
   Hint: The per-capita income at time t is $g(t)/f(t)$.

67. **MONOD'S GROWTH MODEL** The relative growth rate of a biomass at time t, R, is related to the concentration of a substrate s at time t by the equation

$$R(s) = \frac{cs}{k + s}$$

where c and k are positive constants.
a. What is the relative growth rate of the biomass if there is no substrate present?
b. Show that the relative rate of growth of the biomass approaches an upper limit when the substrate is in great excess. What is this limit?
Hint: Find $\lim\limits_{s \to \infty} R$.

c. How fast is the relative growth rate changing with respect to s?

68. **OPTICS** The equation

$$\frac{1}{f} = \frac{1}{p} + \frac{1}{q}$$

sometimes called a **lens-maker's equation,** gives the relationship between the focal length f of a thin lens, the distance p of the object from the lens, and the distance q of its image from the lens. We can think of the eye as an optical system in which the ciliary muscle constantly adjusts the curvature of the cornea-lens system to focus the image on the retina. Assume that the distance from the cornea to the retina is 2.5 cm, as shown in the figure below.

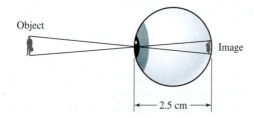

a. Find the focal length of the cornea-lens system if an object located 50 cm away is to be focused on the retina.
b. What is the rate of change of the focal length with respect to the distance of the object when the object is 50 cm away?

In Exercises 69–72, determine whether the statement is true or false. If it is true, explain why it is true. If it is false, give an example to show why it is false.

69. If f and g are differentiable, then

$$\frac{d}{dx}[f(x)g(x)] = f'(x)g'(x)$$

70. If f is differentiable, then

$$\frac{d}{dx}[xf(x)] = f(x) + xf'(x)$$

71. If f is differentiable, then

$$\frac{d}{dx}\left[\frac{f(x)}{x^2}\right] = \frac{f'(x)}{2x}$$

72. If f, g, and h are differentiable, then

$$\frac{d}{dx}\left[\frac{f(x)g(x)}{h(x)}\right] = \frac{f'(x)g(x)h(x) + f(x)g'(x)h(x) - f(x)g(x)h'(x)}{[h(x)]^2}$$

73. Extend the Product Rule for differentiation to the following case involving the product of three differentiable functions: Let $h(x) = u(x)v(x)w(x)$, and show that $h'(x) = u(x)v(x)w'(x) + u(x)v'(x)w(x) + u'(x)v(x)w(x)$.
Hint: Let $f(x) = u(x)v(x)$, $g(x) = w(x)$, and $h(x) = f(x)g(x)$, and apply the Product Rule to the function h.

74. Prove the Quotient Rule for differentiation (Rule 6).
Hint: Let $k(x) = f(x)/g(x)$, and verify the following steps:

a. $\dfrac{k(x + h) - k(x)}{h} = \dfrac{f(x + h)g(x) - f(x)g(x + h)}{hg(x + h)g(x)}$

b. By adding $[-f(x)g(x) + f(x)g(x)]$ to the numerator and simplifying, show that

$$\frac{k(x + h) - k(x)}{h} = \frac{1}{g(x + h)g(x)}$$

$$\times \left\{ \left[\frac{f(x + h) - f(x)}{h}\right] \cdot g(x) \right.$$

$$\left. - \left[\frac{g(x + h) - g(x)}{h}\right] \cdot f(x) \right\}$$

c. $k'(x) = \lim\limits_{h \to 0} \dfrac{k(x + h) - k(x)}{h}$

$$= \frac{g(x)f'(x) - f(x)g'(x)}{[g(x)]^2}$$

3.2 Solutions to Self-Check Exercises

1. We use the Quotient Rule to obtain

$$f'(x) = \frac{(x^2 - 1)\dfrac{d}{dx}(2x + 1) - (2x + 1)\dfrac{d}{dx}(x^2 - 1)}{(x^2 - 1)^2}$$

$$= \frac{(x^2 - 1)(2) - (2x + 1)(2x)}{(x^2 - 1)^2}$$

$$= \frac{2x^2 - 2 - 4x^2 - 2x}{(x^2 - 1)^2}$$

$$= \frac{-2x^2 - 2x - 2}{(x^2 - 1)^2}$$

$$= \frac{-2(x^2 + x + 1)}{(x^2 - 1)^2}$$

2. The slope of the tangent line to the graph of f at any point is given by

$$f'(x) = (x^2 + 1)\frac{d}{dx}(2x^3 - 3x^2 + 1)$$

$$+ (2x^3 - 3x^2 + 1)\frac{d}{dx}(x^2 + 1)$$

$$= (x^2 + 1)(6x^2 - 6x) + (2x^3 - 3x^2 + 1)(2x)$$

In particular, the slope of the tangent line to the graph of f when $x = 2$ is

$$f'(2) = (2^2 + 1)[6(2)^2 - 6(2)]$$
$$+ [2(2)^3 - 3(2)^2 + 1][2(2)]$$
$$= 60 + 20 = 80$$

Note that it is not necessary to simplify the expression for $f'(x)$, since we are required only to evaluate the expression at $x = 2$. We also conclude, from this result, that the function f is changing at the rate of 80 units/unit change in x when $x = 2$.

3. The rate at which the company's total sales are changing at any time t is given by

$$S'(t) = \frac{(1 + 0.4t^2)\dfrac{d}{dt}(0.3t^3) - (0.3t^3)\dfrac{d}{dt}(1 + 0.4t^2)}{(1 + 0.4t^2)^2}$$

$$= \frac{(1 + 0.4t^2)(0.9t^2) - (0.3t^3)(0.8t)}{(1 + 0.4t^2)^2}$$

Therefore, at the beginning of the second year of operation, ADS Security Systems's sales were increasing at the rate of

$$S'(1) = \frac{(1 + 0.4)(0.9) - (0.3)(0.8)}{(1 + 0.4)^2} \approx 0.52$$

or \$520,000/year.

USING TECHNOLOGY   The Product and Quotient Rules

EXAMPLE 1 Let $f(x) = (2\sqrt{x} + 0.5x)(0.3x^3 + 2x - \frac{0.3}{x})$. Find $f'(0.2)$.

Solution   Using the numerical derivative operation of a graphing utility, we find

$$f'(0.2) = 6.4797499802$$

See Figure T1.

```
nDeriv((2X^.5+.5
X)(.3X^3+2X-.3/X),
X,.2)
         6.4797499802
■
```

FIGURE **T1**
The TI-83/84 numerical derivative screen for computing $f'(0.2)$

APPLIED EXAMPLE 2 Importance of Time in Treating Heart Attacks
According to the American Heart Association, the treatment benefit for heart attacks depends on the time until treatment and is described by the function

$$f(t) = \frac{0.44t^4 + 700}{0.1t^4 + 7} \qquad (0 \le t \le 24)$$

where t is measured in hours and $f(t)$ is expressed as a percent.

a. Use a graphing utility to graph the function f using the viewing window $[0, 24] \times [0, 100]$.
b. Use a graphing utility to find the derivative of f when $t = 0$ and $t = 2$.
c. Interpret the results obtained in part (b).
Source: American Heart Association.

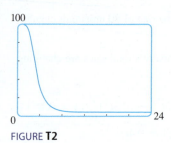

FIGURE T2

Solution

a. The graph of f is shown in Figure T2.

b. Using the numerical derivative operation of a graphing utility, we find

$$f'(0) \approx 0 \quad \text{and} \quad f'(2) \approx -28.95402429$$

(see Figure T3).

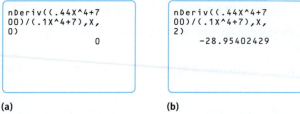

(a)   (b)

FIGURE T3
TI-83/84 numerical derivative screens (a) for computing $f'(0)$ and (b) for computing $f'(2)$

c. The results of part (b) show that there is no drop in the treatment benefit when the heart attack is treated immediately. But the treatment benefit drops off at the rate of approximately 29% per hour when the time to treatment is 2 hours. Thus, it is extremely urgent that a patient suffering a heart attack receive medical attention as soon as possible.

TECHNOLOGY EXERCISES

In Exercises 1–6, use the numerical derivative operation to find the rate of change of $f(x)$ at the given value of x. Give your answer accurate to four decimal places.

1. $f(x) = (2x^2 + 1)(x^3 + 3x + 4); x = -0.5$

2. $f(x) = (\sqrt{x} + 1)(2x^2 + x - 3); x = 1.5$

3. $f(x) = \dfrac{\sqrt{x} - 1}{\sqrt{x} + 1}; x = 3$

4. $f(x) = \dfrac{\sqrt{x}(x^2 + 4)}{x^3 + 1}; x = 4$

5. $f(x) = \dfrac{\sqrt{x}(1 + x^{-1})}{x + 1}; x = 1$

6. $f(x) = \dfrac{x^2(2 + \sqrt{x})}{1 + \sqrt{x}}; x = 1$

7. NEW CONSTRUCTION JOBS The president of a major housing construction company claims that the number of construc-

tion jobs created in the next t months is given by

$$f(t) = 1.42\left(\frac{7t^2 + 140t + 700}{3t^2 + 80t + 550}\right)$$

where $f(t)$ is measured in millions of jobs per year. At what rate will construction jobs be created 1 year from now, assuming that her projection is correct?

8. POPULATION GROWTH A major corporation is building a 4325-acre complex of homes, offices, stores, schools, and churches in the rural community of Glen Cove. As a result of this development, the planners have estimated that Glen Cove's population (in thousands) t years from now will be given by

$$P(t) = \frac{25t^2 + 125t + 200}{t^2 + 5t + 40}$$

a. What will be the population 10 years from now?
b. At what rate will the population be increasing 10 years from now?

3.3   The Chain Rule

The population of Americans age 55 years and older as a percentage of the total population is approximated by the function

$$f(t) = 10.72(0.9t + 10)^{0.3} \qquad (0 \le t \le 20)$$

where t is measured in years with $t = 0$ corresponding to the year 2000 (Figure 6).

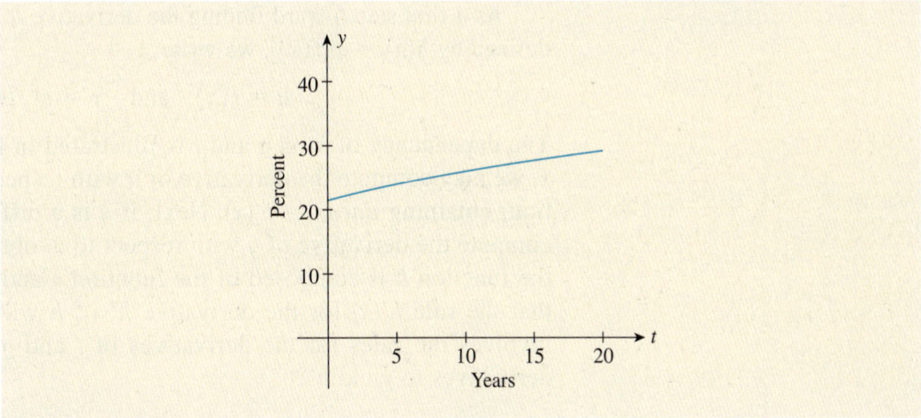

FIGURE 6
Percentage of population of Americans age 55 years and older
Source: U.S. Census Bureau.

How fast will the population age 55 years and older be increasing at the beginning of 2015? To answer this question, we have to evaluate $f'(15)$, where f' is the derivative of f. But the rules of differentiation that we have developed up to now will not help us find the derivative of f.

In this section, we will introduce another rule of differentiation called the **Chain Rule.** When used in conjunction with the rules of differentiation developed in the last two sections, the Chain Rule enables us to greatly enlarge the class of functions that we are able to differentiate. (In Exercise 72, page 195, we will use the Chain Rule to answer the question posed in the introductory example.)

The Chain Rule

Consider the function $h(x) = (x^2 + x + 1)^2$. If we were to compute $h'(x)$ using only the rules of differentiation from the previous sections, then our approach might be to expand $h(x)$. Thus,

$$h(x) = (x^2 + x + 1)^2 = (x^2 + x + 1)(x^2 + x + 1)$$
$$= x^4 + 2x^3 + 3x^2 + 2x + 1$$

from which we find

$$h'(x) = 4x^3 + 6x^2 + 6x + 2$$

But what about the function $H(x) = (x^2 + x + 1)^{100}$? The same technique may be used to find the derivative of the function H, but the amount of work involved in this case would be prodigious! Consider, also, the function $G(x) = \sqrt{x^2 + 1}$. For each of the two functions H and G, the rules of differentiation of the previous sections cannot be applied directly to compute the derivatives H' and G'.

Observe that both H and G are **composite functions;** that is, each is composed of, or built up from, simpler functions. For example, the function H is composed of the two simpler functions $f(x) = x^2 + x + 1$ and $g(x) = x^{100}$ as follows:

$$H(x) = g[f(x)] = [f(x)]^{100}$$
$$= (x^2 + x + 1)^{100}$$

In a similar manner, we see that the function G is composed of the two simpler functions $f(x) = x^2 + 1$ and $g(x) = \sqrt{x}$. Thus,

$$G(x) = g[f(x)] = \sqrt{f(x)}$$
$$= \sqrt{x^2 + 1}$$

As a first step toward finding the derivative h' of a composite function $h = g \circ f$ defined by $h(x) = g[f(x)]$, we write

$$u = f(x) \quad \text{and} \quad y = g[f(x)] = g(u)$$

The dependency of h on g and f is illustrated in Figure 7. Since u is a function of x, we may compute the derivative of u with respect to x if f is a differentiable function, obtaining $du/dx = f'(x)$. Next, if g is a differentiable function of u, we may compute the derivative of g with respect to u, obtaining $dy/du = g'(u)$. Now, since the function h is composed of the function g and the function f, we might suspect that the rule $h'(x)$ for the derivative h' of h will be given by an expression that involves the rules for the derivatives of f and g. But how do we combine these derivatives to yield h'?

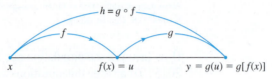

FIGURE 7
The composite function $h(x) = g[f(x)]$

This question can be answered by interpreting the derivative of each function as the rate of change of that function. For example, suppose $u = f(x)$ changes three times as fast as x—that is,

$$f'(x) = \frac{du}{dx} = 3$$

And suppose $y = g(u)$ changes twice as fast as u—that is,

$$g'(u) = \frac{dy}{du} = 2$$

Then we would expect $y = h(x)$ to change six times as fast as x—that is,

$$h'(x) = g'(u)f'(x) = (2)(3) = 6$$

or, equivalently,

$$\frac{dy}{dx} = \frac{dy}{du} \cdot \frac{du}{dx} = (2)(3) = 6$$

This observation suggests the following result, which we state without proof.

Rule 7: The Chain Rule

If $h(x) = g[f(x)]$, then

$$h'(x) = \frac{d}{dx} g[f(x)] = g'[f(x)]f'(x) \tag{1}$$

Equivalently, if we write $y = h(x) = g(u)$, where $u = f(x)$, then

$$\frac{dy}{dx} = \frac{dy}{du} \cdot \frac{du}{dx} \tag{2}$$

Notes

1. If we label the composite function h in the following manner:

$$\overset{\text{Inside function}}{\underset{\uparrow}{\downarrow}}$$
$$h(x) = g[f(x)]$$

Outside function

then $h'(x)$ is just the *derivative* of the "outside function" *evaluated at* the "inside function" times the *derivative* of the "inside function."

2. Equation (2) can be remembered by observing that if we "cancel" the du's, then

$$\frac{dy}{dx} = \frac{dy}{d\bcancel{u}} \cdot \frac{d\bcancel{u}}{dx} = \frac{dy}{dx}$$

The Chain Rule for Powers of Functions

Many composite functions have the special form $h(x) = g(f(x))$ where g is defined by the rule $g(x) = x^n$ (n, a real number)—that is,

$$h(x) = [f(x)]^n$$

In other words, the function h is given by the power of a function f. The functions

$$h(x) = (x^2 + x + 1)^2 \qquad H(x) = (x^2 + x + 1)^{100} \qquad G(x) = \sqrt{x^2 + 1}$$

discussed earlier are examples of this type of composite function. By using the following corollary of the Chain Rule, the General Power Rule, we can find the derivative of this type of function much more easily than by using the Chain Rule directly.

> **The General Power Rule**
>
> If the function f is differentiable and $h(x) = [f(x)]^n$ (n, a real number), then
>
> $$h'(x) = \frac{d}{dx}[f(x)]^n = n[f(x)]^{n-1}f'(x) \qquad (3)$$

To see this, we observe that $h(x) = g(f(x))$ where $g(x) = x^n$, so by virtue of the Chain Rule, we have

$$h'(x) = g'[f(x)]f'(x)$$
$$= n[f(x)]^{n-1}f'(x)$$

since $g'(x) = nx^{n-1}$.

EXAMPLE 1 Let $F(x) = (3x + 1)^2$.

a. Find $F'(x)$, using the General Power Rule.
b. Verify your result without the benefit of the Chain Rule or the General Power Rule.

Solution

a. Using the General Power Rule, we obtain

$$F'(x) = 2(3x + 1)^1 \frac{d}{dx}(3x + 1)$$
$$= 2(3x + 1)(3)$$
$$= 6(3x + 1)$$

b. We first expand $F(x)$. Thus,

$$F(x) = (3x + 1)^2 = 9x^2 + 6x + 1$$

Next, differentiating, we have

$$F'(x) = \frac{d}{dx}(9x^2 + 6x + 1)$$
$$= 18x + 6$$
$$= 6(3x + 1)$$

as before.

EXAMPLE 2 Differentiate the function $G(x) = \sqrt{x^2 + 1}$.

Solution   We rewrite the function $G(x)$ as

$$G(x) = (x^2 + 1)^{1/2}$$

and apply the General Power Rule, obtaining

$$G'(x) = \frac{1}{2}(x^2 + 1)^{-1/2}\frac{d}{dx}(x^2 + 1)$$
$$= \frac{1}{2}(x^2 + 1)^{-1/2} \cdot 2x = \frac{x}{\sqrt{x^2 + 1}}$$

EXAMPLE 3 Differentiate the function $f(x) = x^2(2x + 3)^5$.

Solution   Applying the Product Rule followed by the General Power Rule, we obtain

$$f'(x) = x^2\frac{d}{dx}(2x + 3)^5 + (2x + 3)^5\frac{d}{dx}(x^2)$$
$$= (x^2)5(2x + 3)^4 \cdot \frac{d}{dx}(2x + 3) + (2x + 3)^5(2x)$$
$$= 5x^2(2x + 3)^4(2) + 2x(2x + 3)^5$$
$$= 2x(2x + 3)^4(5x + 2x + 3) = 2x(7x + 3)(2x + 3)^4$$

EXAMPLE 4 Find $f'(x)$ if $f(x) = (2x^2 + 3)^4(3x - 1)^5$.

Solution   Applying the Product Rule, we have

$$f'(x) = (2x^2 + 3)^4\frac{d}{dx}(3x - 1)^5 + (3x - 1)^5\frac{d}{dx}(2x^2 + 3)^4$$

Next, we apply the General Power Rule to each term, obtaining

$$f'(x) = (2x^2 + 3)^4 \cdot 5(3x - 1)^4\frac{d}{dx}(3x - 1) + (3x - 1)^5 \cdot 4(2x^2 + 3)^3\frac{d}{dx}(2x^2 + 3)$$
$$= 5(2x^2 + 3)^4(3x - 1)^4 \cdot 3 + 4(3x - 1)^5(2x^2 + 3)^3(4x)$$

Finally, observing that $(2x^2 + 3)^3(3x - 1)^4$ is common to both terms, we can factor and simplify as follows:

$$f'(x) = (2x^2 + 3)^3(3x - 1)^4[15(2x^2 + 3) + 16x(3x - 1)]$$
$$= (2x^2 + 3)^3(3x - 1)^4(30x^2 + 45 + 48x^2 - 16x)$$
$$= (2x^2 + 3)^3(3x - 1)^4(78x^2 - 16x + 45)$$

EXAMPLE 5 Find $f'(x)$ if $f(x) = \dfrac{1}{(4x^2 - 7)^2}$.

Solution Rewriting $f(x)$ and then applying the General Power Rule, we obtain

$$f'(x) = \frac{d}{dx}\left[\frac{1}{(4x^2 - 7)^2}\right] = \frac{d}{dx}(4x^2 - 7)^{-2}$$

$$= -2(4x^2 - 7)^{-3}\frac{d}{dx}(4x^2 - 7)$$

$$= -2(4x^2 - 7)^{-3}(8x) = -\frac{16x}{(4x^2 - 7)^3}$$

EXAMPLE 6 Find the slope of the tangent line to the graph of the function

$$f(x) = \left(\frac{2x + 1}{3x + 2}\right)^3$$

at the point $\left(0, \frac{1}{8}\right)$.

Solution The slope of the tangent line to the graph of f at any point x is given by $f'(x)$. To compute $f'(x)$, we use the General Power Rule followed by the Quotient Rule, obtaining

$$f'(x) = 3\left(\frac{2x + 1}{3x + 2}\right)^2 \frac{d}{dx}\left(\frac{2x + 1}{3x + 2}\right)$$

$$= 3\left(\frac{2x + 1}{3x + 2}\right)^2 \left[\frac{(3x + 2)(2) - (2x + 1)(3)}{(3x + 2)^2}\right] \qquad \text{(x^2) See page 17.}$$

$$= 3\left(\frac{2x + 1}{3x + 2}\right)^2 \left[\frac{6x + 4 - 6x - 3}{(3x + 2)^2}\right]$$

$$= \frac{3(2x + 1)^2}{(3x + 2)^4} \qquad \text{Combine like terms, and simplify.}$$

In particular, the slope of the tangent line to the graph of f at $\left(0, \frac{1}{8}\right)$ is given by

$$f'(0) = \frac{3(0 + 1)^2}{(0 + 2)^4} = \frac{3}{16}$$

Exploring with **TECHNOLOGY**

Refer to Example 6.

1. Use a graphing utility to plot the graph of the function f, using the viewing window $[-2, 1] \times [-1, 2]$. Then draw the tangent line to the graph of f at the point $\left(0, \frac{1}{8}\right)$.

2. For a better picture, repeat part 1 using the viewing window $[-1, 1] \times [-0.1, 0.3]$.

3. Use the numerical differentiation capability of the graphing utility to verify that the slope of the tangent line at $\left(0, \frac{1}{8}\right)$ is $\frac{3}{16}$.

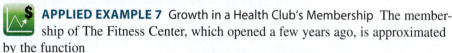 **APPLIED EXAMPLE 7** Growth in a Health Club's Membership The membership of The Fitness Center, which opened a few years ago, is approximated by the function

$$N(t) = 100(64 + 4t)^{2/3} \qquad (0 \le t \le 52)$$

where $N(t)$ gives the number of members at the beginning of week t.

a. Find $N'(t)$.
b. How fast was the center's membership increasing initially ($t = 0$)? (See Note on page 168.)
c. How fast was the membership increasing at the beginning of the 40th week?
d. What was the membership when the center first opened? At the beginning of the 40th week?

Solution

a. Using the General Power Rule, we obtain

$$N'(t) = \frac{d}{dt}\left[100(64 + 4t)^{2/3}\right]$$

$$= 100\frac{d}{dt}(64 + 4t)^{2/3}$$

$$= 100\left(\frac{2}{3}\right)(64 + 4t)^{-1/3}\frac{d}{dt}(64 + 4t)$$

$$= \frac{200}{3}(64 + 4t)^{-1/3}(4)$$

$$= \frac{800}{3(64 + 4t)^{1/3}}$$

b. The rate at which the membership was increasing when the center first opened is given by

$$N'(0) = \frac{800}{3(64)^{1/3}} \approx 66.7$$

or approximately 67 people per week.

c. The rate at which the membership was increasing at the beginning of the 40th week is given by

$$N'(40) = \frac{800}{3(64 + 160)^{1/3}} \approx 43.9$$

or approximately 44 people per week.

d. The membership when the center first opened is given by

$$N(0) = 100(64)^{2/3} = 100(16)$$

or approximately 1600 people. The membership at the beginning of the 40th week is given by

$$N(40) = 100(64 + 160)^{2/3} \approx 3688.3$$

or approximately 3688 people.

Explore and Discuss

The profit P of a one-product software manufacturer depends on the number of units of its product sold. The manufacturer estimates that it will sell x units of its product per week. Suppose $P = g(x)$ and $x = f(t)$, where g and f are differentiable functions and t is measured in weeks.

1. Write an expression giving the rate of change of the profit with respect to the number of units sold.
2. Write an expression giving the rate of change of the number of units sold per week.
3. Write an expression giving the rate of change of the profit per week.

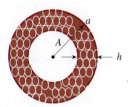

FIGURE 8
Cross section of the aorta

APPLIED EXAMPLE 8 Arteriosclerosis Arteriosclerosis begins during childhood when plaque (soft masses of fatty material) forms in the arterial walls, blocking the flow of blood through the arteries and leading to heart attacks, strokes, and gangrene. Suppose the idealized cross section of the aorta is circular with radius a cm and by year t, the thickness of the plaque (assume that it is uniform) is $h = g(t)$ cm (Figure 8). Then the area of the opening is given by $A = \pi(a - h)^2$ square centimeters (cm^2).

Suppose the radius of an individual's artery is 1 cm ($a = 1$) and the thickness of the plaque in cm in year t is given by

$$h = g(t) = 1 - 0.01(10,000 - t^2)^{1/2} - 0.001t$$

Since the area of the arterial opening is given by

$$A = f(h) = \pi(1 - h)^2$$

the rate at which A is changing with respect to time is given by

$$\frac{dA}{dt} = \frac{dA}{dh} \cdot \frac{dh}{dt} = f'(h) \cdot g'(t) \qquad \text{By the Chain Rule}$$

$$= 2\pi(1 - h)(-1)\left[-0.01\left(\frac{1}{2}\right)(10,000 - t^2)^{-1/2}(-2t) - 0.001\right] \qquad \text{Use the Chain Rule three times.}$$

$$= -2\pi(1 - h)\left[\frac{0.01t}{(10,000 - t^2)^{1/2}} - 0.001\right]$$

For example, when $t = 50$,

$$h = g(50) = 1 - 0.01(10,000 - 2500)^{1/2} - (0.001)(50) \approx 0.08397$$

so that

$$\frac{dA}{dt} = -2\pi(1 - 0.08397)\left[\frac{(0.01)(50)}{\sqrt{10,000 - (50)^2}} - 0.001\right] \approx -0.027$$

That is, the area of the arterial opening is decreasing at the rate of 0.03 cm^2 per year.

Explore and Discuss

Suppose the population P of a certain bacteria culture is given by $P = f(T)$, where T is the temperature of the medium. Further, suppose the temperature T is a function of time t in seconds—that is, $T = g(t)$. Give an interpretation of each of the following quantities:

1. $\dfrac{dP}{dT}$ 2. $\dfrac{dT}{dt}$ 3. $\dfrac{dP}{dt}$ 4. $(f \circ g)(t)$ 5. $f'(g(t))g'(t)$

3.3   Self-Check Exercises

1. Find the derivative of

$$f(x) = -\frac{1}{\sqrt{2x^2 - 1}}$$

2. **FEMALE LIFE EXPECTANCY** Suppose the life expectancy at birth (in years) of a female in a certain country is described by the function

$$g(t) = 50.02(1 + 1.09t)^{0.1} \qquad (0 \le t \le 150)$$

where t is measured in years, with $t = 0$ corresponding to the beginning of 1900.

a. What is the life expectancy at birth of a female born at the beginning of 1980? At the beginning of 2010?

b. How fast is the life expectancy at birth of a female born at any time t changing?

Solutions to Self-Check Exercises 3.3 can be found on page 197.

3.3 Concept Questions

1. In your own words, state the Chain Rule for differentiating the composite function $h(x) = g[f(x)]$.

2. In your own words, state the General Power Rule for differentiating the function $h(x) = [f(x)]^n$, where n is a real number.

3. If $f(t)$ gives the number of units of a certain product sold by a company after t days and $g(x)$ gives the revenue (in dollars) realized from the sale of x units of the company's products, what does $(g \circ f)'(t)$ describe?

4. Suppose $f(x)$ gives the air temperature in the gondola of a hot-air balloon when it is at an altitude of x ft from the ground and $g(t)$ gives the altitude of the balloon t min after lifting off from the ground. Find a function giving the rate of change of the air temperature in the gondola at time t.

3.3 Exercises

In Exercises 1–48, find the derivative of each function.

1. $f(x) = (2x - 1)^3$

2. $f(x) = (1 - x)^4$

3. $f(x) = (x^2 + 2)^5$

4. $f(t) = 2(t^3 - 1)^5$

5. $f(x) = (2x - x^2)^3$

6. $f(x) = 3(x^3 - x)^4$

7. $f(x) = (2x + 1)^{-2}$

8. $f(t) = \frac{1}{2}(2t^2 + t)^{-3}$

9. $f(x) = (x^2 - 4)^{5/2}$

10. $f(t) = (3t^2 - 2t + 1)^{3/2}$

11. $f(x) = \sqrt{3x - 2}$

12. $f(t) = \sqrt{3t^2 - t}$

13. $f(x) = \sqrt[3]{1 - x^2}$

14. $f(x) = \sqrt{2x^2 - 2x + 3}$

15. $f(x) = \frac{1}{(2x + 3)^3}$

16. $f(x) = \frac{2}{(x^2 - 1)^4}$

17. $f(t) = \frac{1}{\sqrt{2t - 4}}$

18. $f(x) = \frac{1}{\sqrt{2x^2 - 1}}$

19. $y = \frac{1}{(4x^4 + x)^{3/2}}$

20. $f(t) = \frac{4}{\sqrt[3]{2t^2 + t}}$

21. $f(x) = (3x^2 + 2x + 1)^{-2}$

22. $f(t) = (5t^3 + 2t^2 - t + 4)^{-3}$

23. $f(x) = (x^2 + 1)^3 - (x^3 + 1)^2$

24. $f(t) = (2t - 1)^4 + (2t + 1)^4$

25. $f(t) = (t^{-1} - t^{-2})^3$

26. $f(v) = (v^{-3} + 4v^{-2})^3$

27. $f(x) = \sqrt{x + 1} + \sqrt{x - 1}$

28. $f(u) = (2u + 1)^{3/2} + (u^2 - 1)^{-3/2}$

29. $f(x) = 2x^2(3 - 4x)^4$

30. $h(t) = t^2(3t + 4)^3$

31. $f(x) = (x - 1)^2(2x + 1)^4$

32. $g(u) = \sqrt{u + 1}(1 - 2u^2)^8$

33. $f(x) = \left(\dfrac{x + 3}{x - 2}\right)^3$

34. $f(x) = \left(\dfrac{x + 1}{x - 1}\right)^5$

35. $s(t) = \left(\dfrac{t}{2t + 1}\right)^{3/2}$

36. $g(s) = \left(s^2 + \dfrac{1}{s}\right)^{3/2}$

37. $g(u) = \sqrt{\dfrac{u + 1}{3u + 2}}$

38. $g(x) = \sqrt{\dfrac{2x + 1}{2x - 1}}$

39. $f(x) = \dfrac{x^2}{(x^2 - 1)^4}$

40. $g(u) = \dfrac{2u^2}{(u^2 + u)^3}$

41. $h(x) = \dfrac{(3x^2 + 1)^3}{(x^2 - 1)^4}$

42. $g(t) = \dfrac{(2t - 1)^2}{(3t + 2)^4}$

43. $f(x) = \dfrac{\sqrt{2x + 1}}{x^2 - 1}$

44. $f(t) = \dfrac{4t^2}{\sqrt{2t^2 + 2t - 1}}$

45. $g(t) = \dfrac{\sqrt{t + 1}}{\sqrt{t^2 + 1}}$

46. $f(x) = \dfrac{\sqrt{x^2 + 1}}{\sqrt{x^2 - 1}}$

47. $f(x) = (3x + 1)^4(x^2 - x + 1)^3$

48. $g(t) = (2t + 3)^2(3t^2 - 1)^{-3}$

In Exercises 49–54, find $\dfrac{dy}{du}, \dfrac{du}{dx},$ and $\dfrac{dy}{dx}$.

49. $y = u^{4/3}$ and $u = 3x^2 - 1$

50. $y = \sqrt{u}$ and $u = 7x - 2x^2$

51. $y = u^{-2/3}$ and $u = 2x^3 - x + 1$

52. $y = 2u^2 + 1$ and $u = x^2 + 1$

53. $y = \sqrt{u} + \dfrac{1}{\sqrt{u}}$ and $u = x^3 - x$

54. $y = \dfrac{1}{u}$ and $u = \sqrt{x} + 1$

55. If $g(x) = f(2x + 1)$, what is $g'(x)$?

56. If $h(x) = f(-x^3)$, what is $h'(x)$?

57. Suppose $F(x) = g(f(x))$ and $f(2) = 3, f'(2) = -3$, $g(3) = 5$, and $g'(3) = 4$. Find $F'(2)$.

58. Suppose $h = f \circ g$. Find $h'(0)$ given that $f(0) = 6$, $f'(5) = -2, g(0) = 5$, and $g'(0) = 3$.

59. Suppose $F(x) = f(x^2 + 1)$. Find $F'(1)$ if $f'(2) = 3$.

60. Let $F(x) = f(f(x))$. Does it follow that $F'(x) = [f'(x)]^2$?
   Hint: Let $f(x) = x^2$.

61. Suppose $h = g \circ f$. Does it follow that $h' = g' \circ f'$?
Hint: Let $f(x) = x$ and $g(x) = x^2$.

62. Suppose $h = f \circ g$. Show that $h' = (f' \circ g)g'$.

In Exercises 63–66, find an equation of the tangent line to the graph of the function at the given point.

63. $f(x) = (1 - x)(x^2 - 1)^2; (2, -9)$

64. $f(x) = \left(\dfrac{x + 1}{x - 1}\right)^2; (3, 4)$

65. $f(x) = x\sqrt{2x^2 + 7}; (3, 15)$

66. $f(x) = \dfrac{8}{\sqrt{x^2 + 6x}}; (2, 2)$

67. TELEVISION VIEWERSHIP The number of viewers of a television series introduced several years ago is approximated by the function

$$N(t) = (60 + 2t)^{2/3} \qquad (1 \le t \le 26)$$

where $N(t)$ (measured in millions) denotes the number of weekly viewers of the series in the tth week. Find the rate of increase of the weekly audience at the end of week 2 and at the end of week 12. How many viewers were there in week 2? In week 24?

68. DIGITAL INFORMATION CREATION The amount of digital information created each month globally, t months after the beginning of 2008, is approximately

$$f(t) = 400\left(\dfrac{t}{12} + 1\right)^{1.09} \qquad (0 \le t \le 36)$$

billion gigabytes.
a. How many billion gigabytes of information were created at the beginning of 2008?
b. How fast was digital information being created at the beginning of 2010 $(t = 24)$?
Source: MIT Technology Review.

69. OUTSOURCING OF JOBS The cumulative number of jobs outsourced overseas by U.S.-based multinational companies in year t from 2005 $(t = 0)$ through 2009 is approximated by

$$N(t) = -0.05(t + 1.1)^{2.2} + 0.7t + 0.9 \qquad (0 \le t \le 4)$$

where $N(t)$ is measured in millions. How fast was the number of jobs that were outsourced changing in 2008 $(t = 3)$?
Source: Forrester Research.

70. SELLING PRICE OF DVD RECORDERS The rise of digital video and the improvement to the DVD format are some of the reasons why the average selling price of stand-alone DVD recorders has dropped in recent years. The function

$$A(t) = \dfrac{699}{(t + 1)^{0.94}} \qquad (0 \le t \le 13)$$

gives the projected average selling price (in dollars) of stand-alone DVD recorders in year t, where $t = 0$

corresponds to the beginning of 2002. How fast was the average selling price of stand-alone DVD recorders falling at the beginning of 2002? How fast was it falling at the beginning of 2012?
Source: Consumer Electronics Association.

71. BRAIN CANCER SURVIVAL RATE Glioblastoma is the most common and most deadly of brain tumors, and it kills most patients in a little over a year. The probability of survival for patients with a glioblastoma t years after diagnosis is approximated by the function

$$P(t) = \dfrac{100}{(1 + 0.14t)^{9.2}} \qquad (0 \le t \le 10)$$

where P is the percent of surviving patients.
a. Compute $P(1)$ and $P(2)$, and interpret your results.
b. Compute $P'(1)$ and $P'(2)$, and interpret your results.
Source: National Cancer Institute.

72. AGING POPULATION The population of Americans age 55 and older as a percentage of the total population is approximated by the function

$$f(t) = 10.72(0.9t + 10)^{0.3} \qquad (0 \le t \le 20)$$

where t is measured in years, with $t = 0$ corresponding to the year 2000. At what rate was the percentage of Americans age 55 and older changing at the beginning of 2000? At what rate was the percentage of Americans age 55 and older changing in 2015? What was the percentage of the population of Americans age 55 and older in 2015? (See Note on page 168.)
Source: U.S. Census Bureau.

73. CONCENTRATION OF CARBON MONOXIDE (CO) IN THE AIR According to a joint study conducted by Oxnard's Environmental Management Department and a state government agency, the concentration of CO in the air due to automobile exhaust t years from now is given by

$$C(t) = 0.01(0.2t^2 + 4t + 64)^{2/3}$$

parts per million.
a. Find the rate at which the level of CO is changing with respect to time.
b. Find the rate at which the level of CO will be changing 5 years from now.

74. CONTINUING EDUCATION ENROLLMENT The registrar of Kellogg University estimates that the total student enrollment in the Continuing Education division will be given by

$$N(t) = -\dfrac{20,000}{\sqrt{1 + 0.2t}} + 21,000$$

where $N(t)$ denotes the number of students enrolled in the division t years from now. Find an expression for $N'(t)$. How fast will the student enrollment be increasing 1 year from now? How fast will it be increasing 5 years from now?

75. AIR POLLUTION According to the South Coast Air Quality Management District, the level of nitrogen dioxide, a brown gas that impairs breathing, present in the atmosphere on a certain May day in downtown Los Angeles is approximated by

$$A(t) = 0.03t^3(t - 7)^4 + 60.2 \quad (0 \leq t \leq 7)$$

where $A(t)$ is measured in pollutant standard index and t is measured in hours, with $t = 0$ corresponding to 7 A.M.
a. Find $A'(t)$.
b. Find $A'(1), A'(3)$, and $A'(4)$, and interpret your results.
Source: Los Angeles Times.

76. EFFECT OF LUXURY TAX ON CONSUMPTION Government economists of a developing country determined that the purchase of imported perfume is related to a proposed "luxury tax" by the formula

$$N(x) = \sqrt{10,000 - 40x - 0.02x^2} \quad (0 \leq x \leq 200)$$

where $N(x)$ measures the percentage of normal consumption of perfume when a "luxury tax" of $x\%$ is imposed on it. Find the rate of change of $N(x)$ for taxes of 10%, 100%, and 150%.

77. PULSE RATE OF AN ATHLETE The pulse rate (the number of heartbeats per minute) of a long-distance runner t sec after leaving the starting line is given by

$$P(t) = \frac{300\sqrt{\frac{1}{2}t^2 + 2t + 25}}{t + 25} \quad (t \geq 0)$$

Compute $P'(t)$. How fast is the athlete's pulse rate increasing 10 sec, 60 sec, and 2 min into the run? What is her pulse rate 2 min into the run?

78. THURSTONE LEARNING MODEL Psychologist L. L. Thurstone suggested the following relationship between learning time T and the length of a list n:

$$T = f(n) = An\sqrt{n - b}$$

where A and b are constants that depend on the person and the task.
a. Compute dT/dn, and interpret your result.
b. For a certain person and a certain task, suppose $A = 4$ and $b = 4$. Compute $f'(13)$ and $f'(29)$, and interpret your results.

79. OIL SPILLS In calm waters, the oil spilling from the ruptured hull of a grounded tanker spreads in all directions. Assuming that the area polluted is a circle and that its radius is increasing at a rate of 2 ft/sec, determine how fast the area is increasing when the radius of the circle is 40 ft.

80. ARTERIOSCLEROSIS Refer to Example 8, page 193. Suppose the radius of an individual's artery is 1 cm and the thickness of the plaque (in centimeters) t years from now is given by

$$h = g(t) = \frac{0.5t^2}{t^2 + 10} \quad (0 \leq t \leq 10)$$

How fast will the arterial opening be decreasing 5 years from now?

81. EFFECT OF HOUSING STARTS ON JOBS The president of a major housing construction firm claims that the number of construction jobs created is given by

$$N(x) = 1.42x$$

where x denotes the number of housing starts. Suppose the number of housing starts in the next t months is expected to be

$$x(t) = \frac{7t^2 + 140t + 700}{3t^2 + 80t + 550}$$

million units/year. Find an expression that gives the rate at which the number of construction jobs will be created t months from now. At what rate will construction jobs be created 1 year from now?

82. HOTEL OCCUPANCY RATES The occupancy rate of the all-suite Wonderland Hotel, located near an amusement park, is given by the function

$$r(t) = \frac{10}{81}t^3 - \frac{10}{3}t^2 + \frac{200}{9}t + 56.2 \quad (0 \leq t \leq 12)$$

where t is measured in months, with $t = 0$ corresponding to the beginning of January. Management has estimated that the monthly revenue (in thousands of dollars per month) is approximated by the function

$$R(r) = -\frac{3}{5000}r^3 + \frac{9}{50}r^2 \quad (0 \leq r \leq 100)$$

where r is the occupancy rate.
a. Find an expression that gives the rate of change of Wonderland's occupancy rate with respect to time.
b. Find an expression that gives the rate of change of Wonderland's monthly revenue with respect to the occupancy rate.
c. What is the rate of change of Wonderland's monthly revenue with respect to time at the beginning of January? At the beginning of July?
Hint: Use the Chain Rule to find $R'(r(0))r'(0)$ and $R'(r(6))r'(6)$.

83. DEMAND FOR TABLET COMPUTERS The quantity demanded per month, x, of the Zephyr tablet computer is related to the average unit price, p (in dollars), of tablet computers by the equation

$$x = f(p) = \frac{100}{9}\sqrt{810,000 - p^2}$$

It is estimated that t months from now, the average price of a tablet computer will be given by

$$p(t) = \frac{400}{1 + \frac{1}{8}\sqrt{t}} + 200 \quad (0 \leq t \leq 60)$$

dollars. Find the rate at which the quantity demanded per month of the tablet computers will be changing 16 months from now.

84. CRUISE SHIP BOOKINGS The management of Cruise World, operators of Caribbean luxury cruises, expects that the percentage of young adults booking passage on their cruises in the years ahead will rise dramatically. They have constructed the following model, which gives the percentage of young adult passengers in year t:

$$p = f(t) = 50\left(\frac{t^2 + 2t + 4}{t^2 + 4t + 8}\right) \qquad (0 \le t \le 5)$$

Young adults normally pick shorter cruises and generally spend less on their passage. The following model gives an approximation of the average amount of money R (in dollars) spent per passenger on a cruise when the percentage of young adults is p:

$$R(p) = 1000\left(\frac{p + 4}{p + 2}\right)$$

Find the rate at which the amount of money spent per passenger on a cruise will be changing 2 years from now.

In Exercises 85–88, determine whether the statement is true or false. If it is true, explain why it is true. If it is false, give an example to show why it is false.

85. If f and g are differentiable and $h = f \circ g$, then $h'(x) = f'[g(x)]g'(x)$.

86. If f is differentiable and c is a constant, then

$$\frac{d}{dx}[f(cx)] = cf'(cx)$$

87. If f is differentiable and $f(x) > 0$, then

$$\frac{d}{dx}\sqrt{f(x)} = \frac{f'(x)}{2\sqrt{f(x)}}$$

88. If f is differentiable, then

$$\frac{d}{dx}\left[f\left(\frac{1}{x}\right)\right] = f'\left(\frac{1}{x}\right)$$

89. In Section 3.1, we proved that

$$\frac{d}{dx}(x^n) = nx^{n-1}$$

for the special case when $n = 2$. Use the Chain Rule to show that

$$\frac{d}{dx}(x^{1/n}) = \frac{1}{n}x^{1/n-1}$$

for any nonzero integer n, assuming that $f(x) = x^{1/n}$ is differentiable.
Hint: Let $f(x) = x^{1/n}$, so that $[f(x)]^n = x$. Differentiate both sides with respect to x.

90. With the aid of Exercise 89, prove that

$$\frac{d}{dx}(x^r) = rx^{r-1}$$

for every rational number r.
Hint: Let $r = m/n$, where m and n are integers, with $n \ne 0$, and write $x^r = (x^m)^{1/n}$.

3.3 Solutions to Self-Check Exercises

1. Rewriting, we have

$$f(x) = -(2x^2 - 1)^{-1/2}$$

Using the General Power Rule, we find

$$f'(x) = -\frac{d}{dx}(2x^2 - 1)^{-1/2}$$

$$= -\left(-\frac{1}{2}\right)(2x^2 - 1)^{-3/2}\frac{d}{dx}(2x^2 - 1)$$

$$= \frac{1}{2}(2x^2 - 1)^{-3/2}(4x)$$

$$= \frac{2x}{(2x^2 - 1)^{3/2}}$$

2. a. The life expectancy at birth of a female born at the beginning of 1980 is given by

$$g(80) = 50.02[1 + 1.09(80)]^{0.1} \approx 78.29$$

or approximately 78 years. Similarly, the life expectancy at birth of a female born at the beginning of the year 2010 is given by

$$g(110) = 50.02[1 + 1.09(110)]^{0.1} \approx 80.80$$

or approximately 81 years.

b. The rate of change of the life expectancy at birth of a female born at any time t is given by $g'(t)$. Using the General Power Rule, we have

$$g'(t) = 50.02\frac{d}{dt}(1 + 1.09t)^{0.1}$$

$$= (50.02)(0.1)(1 + 1.09t)^{-0.9}\frac{d}{dt}(1 + 1.09t)$$

$$= (50.02)(0.1)(1.09)(1 + 1.09t)^{-0.9}$$

$$= 5.45218(1 + 1.09t)^{-0.9}$$

$$= \frac{5.45218}{(1 + 1.09t)^{0.9}}$$

USING TECHNOLOGY Finding the Derivative of a Composite Function

EXAMPLE 1 Find the rate of change of $f(x) = \sqrt{x}(1 + 0.02x^2)^{3/2}$ when $x = 2.1$.

Solution   Using the numerical derivative operation of a graphing utility, we find

$$f'(2.1) = 0.5821463392$$

or approximately 0.58 unit per unit change in x. (See Figure T1.)

```
nDeriv(X^.5(1+.0
2X^2)^1.5,X,2.1)

      .5821463392
■
```

FIGURE **T1**
The TI-83/84 numerical derivative
screen for computing $f'(2.1)$

APPLIED EXAMPLE 2 Amusement Park Attendance The management of AstroWorld ("The Amusement Park of the Future") estimates that the total number of visitors (in thousands) to the amusement park t hours after opening time at 9 A.M. is given by

$$N(t) = \frac{30t}{\sqrt{2 + t^2}}$$

What is the rate at which visitors are admitted to the amusement park at 10:30 A.M.?

Solution   Using the numerical derivative operation of a graphing utility, we find

$$N'(1.5) \approx 6.8481$$

or approximately 6848 visitors per hour. (See Figure T2.)

```
nDeriv((30X)/(2+
X^2)^.5,X,1.5)
          6.848066034
■
```

FIGURE **T2**
The TI-83/84 numerical derivative
screen for computing $N'(1.5)$

TECHNOLOGY EXERCISES

In Exercises 1–6, use the numerical derivative operation to find the rate of change of f at the given value of x. Give your answer accurate to four decimal places.

1. $f(x) = \sqrt{x^2 - x^4}; x = 0.5$

2. $f(x) = x - \sqrt{1 - x^2}; x = 0.4$

3. $f(x) = x\sqrt{1 - x^2}; x = 0.2$

4. $f(x) = (x + \sqrt{x^2 + 4})^{3/2}; x = 1$

5. $f(x) = \dfrac{\sqrt{1 + x^2}}{x^3 + 2}; x = -1$

6. $f(x) = \dfrac{x^3}{1 + (1 + x^2)^{3/2}}; x = 3$

7. **WATCHING TV ON SMARTPHONES** The number of people watching TV on smartphones (in millions) is approximated by

$$N(t) = 11.9\sqrt{1 + 0.91t} \qquad (0 \le t \le 4)$$

where t is measured in years, with $t = 0$ corresponding to the beginning of 2007.
  a. What was the rate of change of the number of people watching TV on smartphones at the beginning of 2007?
  b. What was the rate of change of the number of people watching TV on smartphones at the beginning of 2011?
Source: IDC, U.S. forecast.

8. ACCUMULATION YEARS Demographic studies pertaining to investors are of particular importance to financial institutions. People from their mid-40s to their mid-50s are in the prime investing years. The function

$$N(t) = 34.4(1 + 0.32125t)^{0.15} \quad (0 \le t \le 12)$$

gives the projected number of people in this age group in the United States (in millions) in year t, where $t = 0$ corresponds to the beginning of 1996.

a. How large was this segment of the population projected to be at the beginning of 2008?
b. How fast was this segment of the population growing at the beginning of 2008?

Source: U.S. Census Bureau.

3.4 Marginal Functions in Economics

Marginal analysis is the study of the rate of change of economic quantities. For example, an economist is not merely concerned with the value of an economy's gross domestic product (GDP) at a given time but is equally concerned with the rate at which it is growing or declining. In the same vein, a manufacturer is not only interested in the total cost corresponding to a certain level of production of a commodity but also is interested in the rate of change of the total cost with respect to the level of production, and so on. Let's begin with an example to explain the meaning of the adjective *marginal*, as used by economists.

Cost Functions

APPLIED EXAMPLE 1 Rate of Change of Cost Functions Suppose the total cost in dollars incurred each week by Polaraire for manufacturing x refrigerators is given by the total cost function

$$C(x) = 8000 + 200x - 0.2x^2 \quad (0 \le x \le 400)$$

a. What is the actual cost incurred for manufacturing the 251st refrigerator?
b. Find the rate of change of the total cost function with respect to x when $x = 250$.
c. Compare the results obtained in parts (a) and (b).

Solution

a. The actual cost incurred in producing the 251st refrigerator is the difference between the total cost incurred in producing the first 251 refrigerators and the total cost of producing the first 250 refrigerators:

$$
\begin{aligned}
C(251) - C(250) &= [8000 + 200(251) - 0.2(251)^2] \\
&\quad - [8000 + 200(250) - 0.2(250)^2] \\
&= 45{,}599.8 - 45{,}500 \\
&= 99.8
\end{aligned}
$$

or $99.80.

b. The rate of change of the total cost function C with respect to x is given by the derivative of C—that is, $C'(x) = 200 - 0.4x$. Thus, when the level of production is 250 refrigerators, the rate of change of the total cost with respect to x is given by

$$
\begin{aligned}
C'(250) &= 200 - 0.4(250) \\
&= 100
\end{aligned}
$$

or $100.

c. From the solution to part (a), we know that the actual cost for producing the 251st refrigerator is $99.80. This answer is very closely approximated by the answer to part (b), $100. To see why this is so, observe that the difference $C(251) - C(250)$ may be written in the form

$$\frac{C(251) - C(250)}{1} = \frac{C(250 + 1) - C(250)}{1} = \frac{C(250 + h) - C(250)}{h}$$

where $h = 1$. In other words, the difference $C(251) - C(250)$ is precisely the average rate of change of the total cost function C over the interval $[250, 251]$ or, equivalently, the slope of the secant line through the points $(250, 45,500)$ and $(251, 45,599.8)$. However, the number $C'(250) = 100$ is the instantaneous rate of change of the total cost function C at $x = 250$ or, equivalently, the slope of the tangent line to the graph of C at $x = 250$.

Now when h is small, the average rate of change of the function C is a good approximation to the instantaneous rate of change of the function C, or, equivalently, the slope of the secant line through the points in question is a good approximation to the slope of the tangent line through the point in question. Therefore, we may expect

$$C(251) - C(250) = \frac{C(251) - C(250)}{1} \approx \frac{C(250 + h) - C(250)}{h} \quad (h \text{ small})$$

$$\approx \lim_{h \to 0} \frac{C(250 + h) - C(250)}{h} = C'(250)$$

which is precisely the case in this example.

The actual cost incurred in producing an additional unit of a certain commodity given that a plant is already at a certain level of operation is called the **marginal cost**. Knowing this cost is very important to management. As we saw in Example 1, the marginal cost is approximated by the rate of change of the total cost function evaluated at the appropriate point. For this reason, economists have defined the **marginal cost function** to be the derivative of the corresponding total cost function. In other words, if C is a total cost function, then the marginal cost function is defined to be its derivative C'. Thus, the adjective *marginal* is synonymous with *derivative of*.

$ APPLIED EXAMPLE 2 Marginal Cost Functions A subsidiary of Elektra Electronics manufactures a portable DVD player. Management determined that the daily total cost of producing these DVD players (in dollars) is given by

$$C(x) = 0.0001x^3 - 0.08x^2 + 40x + 5000$$

where x stands for the number of DVD players produced.

a. Find the marginal cost function.
b. What is the marginal cost when $x = 200, 300, 400,$ and 600?
c. Interpret your results.

Solution

a. The marginal cost function C' is given by the derivative of the total cost function C. Thus,

$$C'(x) = 0.0003x^2 - 0.16x + 40$$

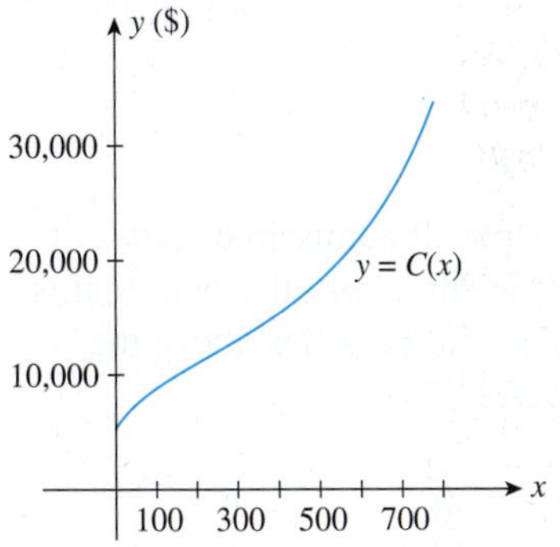

y ($)

30,000

20,000

y = C(x)

10,000

100   300   500   700   x

FIGURE **9**
The cost of producing x DVD players is
given by C(x).

b. The marginal cost when $x = 200$, 300, 400, and 600 is given by

$$C'(200) = 0.0003(200)^2 - 0.16(200) + 40 = 20$$
$$C'(300) = 0.0003(300)^2 - 0.16(300) + 40 = 19$$
$$C'(400) = 0.0003(400)^2 - 0.16(400) + 40 = 24$$
$$C'(600) = 0.0003(600)^2 - 0.16(600) + 40 = 52$$

or $20, $19, $24, and $52 per unit, respectively.

c. From the results of part (b), we see that Elektra's actual cost for producing the 201st DVD player is approximately $20. The actual cost incurred for producing one additional DVD player when the level of production is already 300 players is approximately $19, and so on. Observe that when the level of production is already 600 units, the actual cost of producing one additional unit is approximately $52. The higher cost for producing this additional unit when the level of production is 600 units may be the result of several factors, among them excessive costs incurred because of overtime or higher maintenance, production breakdown caused by greater stress and strain on the equipment, and so on. The graph of the total cost function appears in Figure 9. ◼

Average Cost Functions

Let's now introduce another marginal concept that is closely related to the marginal cost. Let $C(x)$ denote the total cost incurred in producing x units of a certain commodity. Then the **average cost** of producing x units of the commodity is obtained by dividing the total production cost by the number of units produced. This leads to the following definition:

> **Average Cost Function**
>
> Suppose $C(x)$ is a total cost function. Then the **average cost function,** denoted by $\overline{C}(x)$ (read "C bar of x"), is
>
> $$\frac{C(x)}{x} \qquad\qquad (4)$$

The derivative $\overline{C}'(x)$ of the average cost function, called the **marginal average cost function,** measures the rate of change of the average cost function with respect to the number of units produced.

APPLIED EXAMPLE 3 Marginal Average Cost Functions The total cost of producing x units of a certain commodity is given by $C(x) = 400 + 20x$ dollars.

a. Find the average cost function $\overline{C}$.
b. Find the marginal average cost function $\overline{C}'$.
c. What are the economic implications of your results?

Solution

a. The average cost function is given by

$$\overline{C}(x) = \frac{C(x)}{x} = \frac{400 + 20x}{x}$$

$$= 20 + \frac{400}{x}$$

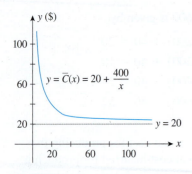

FIGURE 10
As the level of production increases, the average cost approaches $20.

b. The marginal average cost function is

$$\overline{C}'(x) = -\frac{400}{x^2}$$

c. Since the marginal average cost function is negative for all admissible values of x, the rate of change of the average cost function is negative for all $x > 0$; that is, $\overline{C}(x)$ decreases as x increases. However, the graph of $\overline{C}$ always lies above the horizontal line $y = 20$, but it approaches the line, since

$$\lim_{x \to \infty} \overline{C}(x) = \lim_{x \to \infty} \left(20 + \frac{400}{x} \right) = 20$$

A sketch of the graph of the function $\overline{C}(x)$ appears in Figure 10. This result is fully expected if we consider the economic implications. Note that as the level of production increases, the fixed cost per unit of production, represented by the term $(400/x)$, drops steadily. The average cost approaches the constant unit cost of production, which is $20 in this case.

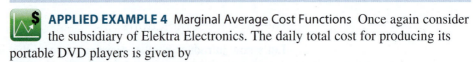

APPLIED EXAMPLE 4 Marginal Average Cost Functions Once again consider the subsidiary of Elektra Electronics. The daily total cost for producing its portable DVD players is given by

$$C(x) = 0.0001x^3 - 0.08x^2 + 40x + 5000$$

dollars, where x stands for the number of DVD players produced (see Example 2).

a. Find the average cost function $\overline{C}$.
b. Find the marginal average cost function $\overline{C}'$. Compute $\overline{C}'(500)$.
c. Sketch the graph of the function $\overline{C}$ and interpret the results obtained in parts (a) and (b).

Solution

a. The average cost function is given by

$$\overline{C}(x) = \frac{C(x)}{x} = 0.0001x^2 - 0.08x + 40 + \frac{5000}{x}$$

b. The marginal average cost function is given by

$$\overline{C}'(x) = 0.0002x - 0.08 - \frac{5000}{x^2}$$

Also,

$$\overline{C}'(500) = 0.0002(500) - 0.08 - \frac{5000}{(500)^2} = 0$$

c. To sketch the graph of the function $\overline{C}$, observe that if x is a small positive number, then $\overline{C}(x) > 0$. Furthermore, $\overline{C}(x)$ becomes arbitrarily large as x approaches zero from the right, since the term $(5000/x)$ becomes arbitrarily large as x approaches zero. Next, the result $\overline{C}'(500) = 0$ obtained in part (b) tells us that the tangent line to the graph of the function $\overline{C}$ is horizontal at the point $(500, 35)$ on the graph. Finally, plotting the points on the graph corresponding to, say, $x = 100, 200, 300, \ldots, 900$, we obtain the sketch in Figure 11. As expected, the average cost drops as the level of production increases. But in this case, in contrast to the case in Example 3, the average cost reaches a minimum value of $35, corresponding to a production level of 500, and *increases* thereafter.

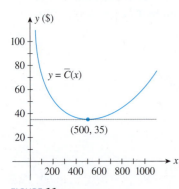

FIGURE 11
The average cost reaches a minimum of $35 when 500 DVD players are produced.

This phenomenon is typical in situations in which the marginal cost increases from some point on as production increases, as in Example 2. This situation is in contrast to that of Example 3, in which the marginal cost remains constant at any level of production. ∎

Exploring with TECHNOLOGY

Refer to Example 4.

1. Use a graphing utility to plot the graph of the average cost function

$$\overline{C}(x) = 0.0001x^2 - 0.08x + 40 + \frac{5000}{x}$$

using the viewing window $[0, 1000] \times [0, 100]$. Then, using ZOOM and TRACE, show that the lowest point on the graph of $\overline{C}$ is (500, 35).

2. Draw the tangent line to the graph of $\overline{C}$ (500, 35). What is its slope? Is this expected?

3. Plot the graph of the marginal average cost function

$$\overline{C}'(x) = 0.0002x - 0.08 - \frac{5000}{x^2}$$

using the viewing window $[0, 2000] \times [-1, 1]$. Then use ZOOM and TRACE to show that the zero of the function $\overline{C}'$ occurs at $x = 500$. Verify this result using the root-finding capability of your graphing utility. Is this result compatible with that obtained in part 2? Explain your answer.

Revenue Functions

Recall that a revenue function $R(x)$ gives the revenue realized by a company from the sale of x units of a certain commodity. If the company charges p dollars per unit, then

$$R(x) = px \tag{5}$$

However, the price that a company can command for the product depends on the market in which the company operates. If the company is one of many—none of which is able to dictate the price of the commodity—then in this competitive market environment, the price is determined by market equilibrium (see Section 2.3). On the other hand, if the company is the sole supplier of the product, then under this monopolistic situation, it can manipulate the price of the commodity by controlling the supply. The unit selling price p of the commodity is related to the quantity x of the commodity demanded. This relationship between p and x is called a *demand equation* (see Section 2.3). Solving the demand equation for p in terms of x, we obtain the unit price function f. Thus,

$$p = f(x)$$

and the revenue function R is given by

$$R(x) = px = xf(x)$$

The **marginal revenue** gives the actual revenue realized from the sale of an additional unit of the commodity given that sales are already at a certain level. Following an argument parallel to that applied to the cost function in Example 1, you can convince yourself that the marginal revenue is approximated by $R'(x)$. Thus, we define the **marginal revenue function** to be $R'(x)$, where R is the revenue function. The derivative R' of the function R measures the rate of change of the revenue function.

 APPLIED EXAMPLE 5 Marginal Revenue Functions Suppose the relationship between the unit price p in dollars and the quantity demanded x of the Acrosonic model F loudspeaker system is given by the equation

$$p = -0.02x + 400 \qquad (0 \le x \le 20{,}000)$$

a. Find the revenue function R.
b. Find the marginal revenue function R'.
c. Compute $R'(2000)$, and interpret your result.

Solution

a. The revenue function R is given by

$$
\begin{aligned}
R(x) &= px \\
&= x(-0.02x + 400) \\
&= -0.02x^2 + 400x \qquad (0 \le x \le 20{,}000)
\end{aligned}
$$

b. The marginal revenue function R' is given by

$$R'(x) = -0.04x + 400$$

c.
$$R'(2000) = -0.04(2000) + 400 = 320$$

Thus, the actual revenue to be realized from the sale of the 2001st loudspeaker system is approximately \$320.

Profit Functions

Our final example of a marginal function involves the profit function. The profit function P is given by

$$P(x) = R(x) - C(x) \tag{6}$$

where R and C are the revenue and cost functions and x is the number of units of a commodity produced and sold. The **marginal profit function** $P'(x)$ measures the rate of change of the profit function P and provides us with a good approximation of the actual profit or loss realized from the sale of the $(x + 1)$st unit of the commodity (assuming that the xth unit has been sold).

 APPLIED EXAMPLE 6 Marginal Profit Functions Refer to Example 5. Suppose the cost of producing x units of the Acrosonic model F loudspeaker is

$$C(x) = 100x + 200{,}000$$

dollars.

a. Find the profit function P.
b. Find the marginal profit function P'.
c. Compute $P'(2000)$, and interpret your result.
d. Sketch the graph of the profit function P.

Solution

a. From the solution to Example 5a, we have

$$R(x) = -0.02x^2 + 400x$$

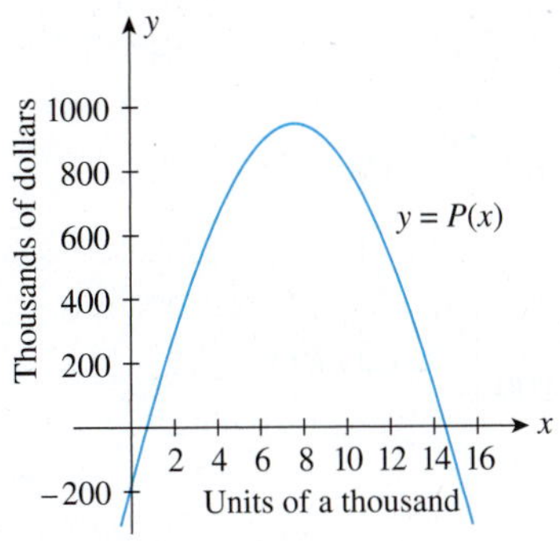

FIGURE 12
The total profit made when x loud-speakers are produced and sold is given by $P(x)$.

Thus, the required profit function P is given by

$$\begin{aligned} P(x) &= R(x) - C(x) \\ &= (-0.02x^2 + 400x) - (100x + 200{,}000) \\ &= -0.02x^2 + 300x - 200{,}000 \end{aligned}$$

b. The marginal profit function P' is given by

$$P'(x) = -0.04x + 300$$

c.

$$P'(2000) = -0.04(2000) + 300 = 220$$

Thus, the actual profit realized from the sale of the 2001st loudspeaker system is approximately $220.

d. The graph of the profit function P appears in Figure 12.   ▪

Relative Rate of Change

The *relative change* of the size of a quantity is defined to be

$$\frac{\text{Change in the size of the quantity}}{\text{Size of the quantity}}$$

For example, suppose that the mortgage rate increases from the current rate of 10% per year to a rate of 11% per year. Then the relative change in the mortgage rate is

$$\frac{1}{10} = 0.1 \quad \text{or} \quad 10\%$$

But if the current mortgage rate is 5% per year, then a change of 1% per year, to 6% per year, in the mortgage rate would yield a relative change in the mortgage rate of

$$\frac{1}{5} = 0.2 \quad \text{or} \quad 20\%$$

This example shows that the relative change sometimes conveys a better sense of what is going on than does a simple look at the change in the quantity itself.

Similarly, it is often more meaningful to look at the *relative rate of change* of a function at a given value of x than at the rate of change of f at x. We have the following definition.

> **Relative Rate of Change**
>
> The **relative rate of change** of a differentiable function f at x per unit change in x is
>
> $$\frac{f'(x)}{f(x)} \quad \text{or} \quad \frac{100f'(x)}{f(x)}\%$$
>
> The second expression is called the **percentage rate of change of f at x.**

APPLIED EXAMPLE 7 Inflation An economy's consumer price index (CPI) is described by the function

$$I(t) = -0.05t^3 + 0.5t^2 + 100 \qquad (0 \le t \le 4)$$

where t is measured in years, with $t = 0$ corresponding to the beginning of 2012. Find the annual percentage rate of inflation in the CPI of the country (defined as the percentage relative rate of change of I) at the beginning of 2014 ($t = 2$).

Solution The inflation rate in year t is

$$R(t) = \frac{100I'(t)}{I(t)} = 100\left[\frac{-0.15t^2 + t}{-0.05t^3 + 0.5t^2 + 100}\right]$$

So the inflation rate at the beginning of 2014 is

$$R(2) = 100\left[\frac{-0.15(2^2) + 2}{-0.05(2^3) + 0.5(2^2) + 100}\right] \approx 1.37795$$

or approximately 1.4% per year.

Elasticity of Demand

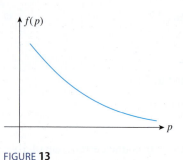

FIGURE 13
The graph of a demand function

Finally, let's use the marginal concepts introduced in this section to derive an important criterion that economists use to analyze a demand function: elasticity of demand. This measurement enables a businessperson to see how a small percentage change in the price of a commodity affects the percentage change in the quantity demanded of the commodity.

In what follows, it will be convenient to write the demand function f in the form $x = f(p)$; that is, we will think of the quantity demanded of a certain commodity as a function of its unit price. Since the quantity demanded of a commodity usually decreases as its unit price increases, the function f is typically a decreasing function of p (Figure 13).

Since the rate of change of the demand function with respect to p is $f'(p)$, we see that the relative rate of change of f with respect to p at p is

$$\frac{f'(p)}{f(p)} \quad \text{or} \quad \frac{100 f'(p)}{f(p)} \%$$

Next, we see that the relative rate of change of the price p of the commodity is

$$\frac{\dfrac{d}{dp}(p)}{p} = \frac{1}{p} \quad \text{or} \quad \frac{100}{p} \%$$

Therefore,

$$\frac{\text{Percentage rate of change of } f}{\text{Percentage rate of change of } p} = \frac{\dfrac{100f'(p)}{f(p)}}{\dfrac{100}{p}} = \frac{pf'(p)}{f(p)}$$

Economists call the *negative* of this quantity the elasticity of demand.

> **Elasticity of Demand**
>
> If f is a differentiable demand function defined by $x = f(p)$, then the **elasticity of demand** at price p is given by
>
> $$E(p) = -\frac{pf'(p)}{f(p)} \tag{7}$$

Note It will be shown later (in Section 4.1) that if f is decreasing on an interval, then $f'(p) < 0$ for p in that interval. In light of this, we see that since both p and $f(p)$ are positive, the quantity $\dfrac{pf'(p)}{f(p)}$ is negative. Because economists would rather work with a positive value, the elasticity of demand $E(p)$ is defined to be the negative of this quantity.

APPLIED EXAMPLE 8 Elasticity of Demand for Loudspeakers Consider the demand equation

$$p = -0.02x + 400 \qquad (0 \le x \le 20{,}000)$$

which describes the relationship between the unit price in dollars and the quantity demanded x of the Acrosonic model F loudspeaker systems.

a. Find the elasticity of demand $E(p)$.
b. Compute $E(100)$, and interpret your result.
c. Compute $E(300)$, and interpret your result.

Solution

a. Solving the given demand equation for x in terms of p, we find

$$x = f(p) = -50p + 20{,}000$$

from which we see that

$$f'(p) = -50$$

Therefore,

$$E(p) = -\frac{pf'(p)}{f(p)} = -\frac{p(-50)}{-50p + 20{,}000}$$

$$= \frac{p}{400 - p}$$

b.
$$E(100) = \frac{100}{400 - 100} = \frac{1}{3}$$

which is the elasticity of demand when $p = 100$. To interpret this result, recall that $E(100)$ is the negative of the ratio of the percentage change in the quantity demanded to the percentage change in the unit price when $p = 100$. Therefore, our result tells us that when the unit price p is set at $100 per speaker, an increase of 1% in the unit price will cause a decrease of approximately 0.33% in the quantity demanded.

c.
$$E(300) = \frac{300}{400 - 300} = 3$$

which is the elasticity of demand when $p = 300$. It tells us that when the unit price is set at $300 per speaker, an increase of 1% in the unit price will cause a decrease of approximately 3% in the quantity demanded.

Economists often use the following terminology to describe demand in terms of elasticity.

Elasticity of Demand

The demand is said to be **elastic** if $E(p) > 1$.
The demand is said to be **unitary** if $E(p) = 1$.
The demand is said to be **inelastic** if $E(p) < 1$.

As an illustration, our computations in Example 8 revealed that demand for Acrosonic loudspeakers is elastic when $p = 300$ but inelastic when $p = 100$. These computations confirm that when demand is elastic, a small percentage change in the unit price will result in a greater percentage change in the quantity demanded; and when demand is inelastic, a small percentage change in the unit price will cause a smaller percentage change in the quantity demanded. Finally, when demand is unitary, a small percentage change in the unit price will result in approximately the same percentage change in the quantity demanded.

We can describe the way revenue responds to changes in the unit price using the notion of elasticity. If the quantity demanded of a certain commodity is related to its unit price by the equation $x = f(p)$, then the revenue realized through the sale of x units of the commodity at a price of p dollars each is

$$R(p) = px = pf(p)$$

The rate of change of the revenue with respect to the unit price p is given by

$$R'(p) = f(p) + pf'(p)$$
$$= f(p)\left[1 + \frac{pf'(p)}{f(p)}\right]$$
$$= f(p)[1 - E(p)]$$

Now, suppose demand is elastic when the unit price is set at a dollars. Then $E(a) > 1$, and so $1 - E(a) < 0$. Since $f(p)$ is positive for all values of p, we see that

$$R'(a) = f(a)[1 - E(a)] < 0$$

and so $R(p)$ is decreasing at $p = a$. This implies that a small increase in the unit price when $p = a$ results in a decrease in the revenue, whereas a small decrease in the unit price will result in an increase in the revenue. Similarly, you can show that if the demand is inelastic when the unit price is set at a dollars, then a small increase in the unit price will cause the revenue to increase, and a small decrease in the unit price will cause the revenue to decrease. Finally, if the demand is unitary when the unit price is set at a dollars, then $E(a) = 1$ and $R'(a) = 0$. This implies that a small increase or decrease in the unit price will not result in a change in the revenue. The following statements summarize this discussion.

---

1. If the demand is elastic at $p[E(p) > 1]$, then a small increase in the unit price will cause the revenue to decrease, whereas a small decrease in the unit price will cause the revenue to increase.

2. If the demand is inelastic at $p[E(p) < 1]$, then a small increase in the unit price will cause the revenue to increase, and a small decrease in the unit price will cause the revenue to decrease.

3. If the demand is unitary at $p[E(p) = 1]$, then a small increase or decrease in the unit price will cause the revenue to stay about the same.

These results are illustrated in Figure 14.

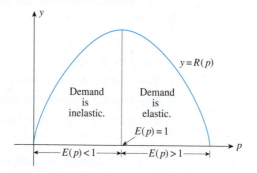

FIGURE 14
The revenue is increasing on an interval where the demand is inelastic, decreasing on an interval where the demand is elastic, and stationary at the point where the demand is unitary.

Note   As an aid to remembering, note the following:

1. If demand is elastic, then the change in revenue and the change in the unit price move in opposite directions.
2. If demand is inelastic, then they move in the same direction.

 APPLIED EXAMPLE 9 Elasticity of Demand for Loudspeakers Refer to Example 8.

a. Is demand elastic, unitary, or inelastic when $p = 100$? When $p = 300$?
b. If the price is \$100, will raising the unit price slightly cause the revenue to increase or decrease?

Solution

a. From the results of Example 8, we see that $E(100) = \frac{1}{3} < 1$ and $E(300) = 3 > 1$. We conclude accordingly that demand is inelastic when $p = 100$ and elastic when $p = 300$.
b. Since demand is inelastic when $p = 100$, raising the unit price slightly will cause the revenue to increase.

3.4 Self-Check Exercises

1. **DEMAND FOR DVRs** The weekly demand for Pulsar DVRs is given by the demand equation

$$p = -0.02x + 300 \qquad (0 \le x \le 15{,}000)$$

where p denotes the wholesale unit price in dollars and x denotes the quantity demanded. The weekly total cost function associated with manufacturing these recorders is

$$C(x) = 0.000003x^3 - 0.04x^2 + 200x + 70{,}000$$

dollars.

a. Find the revenue function R and the profit function P.

b. Find the marginal cost function C', the marginal revenue function R', and the marginal profit function P'.
c. Find the marginal average cost function $\overline{C}'$.
d. Compute $C'(3000)$, $R'(3000)$, and $P'(3000)$, and interpret your results.

2. **ELASTICITY OF DEMAND FOR DVRs** Refer to the preceding exercise. Determine whether the demand is elastic, unitary, or inelastic when $p = 100$ and when $p = 200$.

Solutions to Self-Check Exercises 3.4 can be found on page 213.

3.4 Concept Questions

1. Explain each term in your own words:
   a. Marginal cost function
   b. Average cost function
   c. Marginal average cost function
   d. Marginal revenue function
   e. Marginal profit function

2. a. Define the elasticity of demand.

   b. When is the elasticity of demand elastic? Unitary? Inelastic? Explain the meaning of each term.

3. The proprietor of a company finds that his marginal revenue and marginal cost are \$3/unit and \$2.80/unit, respectively, at a production level of 500 units. Should the proprietor increase production in order to increase the profit of the company? Explain.

3.4 Exercises

1. **PRODUCTION COSTS** The graph of a typical total cost function $C(x)$ associated with the manufacture of x units of a certain commodity is shown in the figure on the next page.

a. Explain why the function C is always increasing.
b. As the level of production x increases, the cost per unit drops so that $C(x)$ increases but at a slower pace.

However, a level of production is soon reached at which the cost per unit begins to increase dramatically (owing to a shortage of raw material, overtime, breakdown of machinery due to excessive stress and strain), so $C(x)$ continues to increase at a faster pace. Use the graph of C to find the approximate level of production x_0 where this occurs.

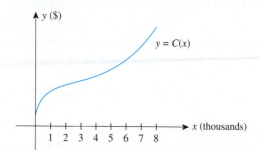

2. PRODUCTION COSTS The graph of a typical average cost function $A(x) = C(x)/x$, where $C(x)$ is a total cost function associated with the manufacture of x units of a certain commodity, is shown in the following figure.
  a. Explain in economic terms why $A(x)$ is large if x is small and why $A(x)$ is large if x is large.
  b. What is the significance of the numbers x_0 and y_0, the x- and y-coordinates of the lowest point on the graph of the function A?

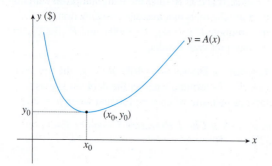

3. MARGINAL COST The total weekly cost (in dollars) incurred by Lincoln Records in pressing x compact discs is

$$C(x) = 2000 + 2x - 0.0001x^2 \qquad (0 \le x \le 6000)$$

  a. What is the actual cost incurred in producing the 1001st disc and the 2001st disc?
  b. What is the marginal cost when $x = 1000$ and 2000?

4. MARGINAL COST A division of Ditton Industries manufactures the Futura model microwave oven. The daily cost (in dollars) of producing these microwave ovens is

$$C(x) = 0.0002x^3 - 0.06x^2 + 120x + 5000$$

where x stands for the number of units produced.
  a. What is the actual cost incurred in manufacturing the 101st oven? The 201st oven? The 301st oven?
  b. What is the marginal cost when $x = 100$, 200, and 300?

5. MARGINAL AVERAGE COST FOR PRODUCING DESKS Custom Office makes a line of executive desks. It is estimated that the total cost for making x units of their Senior Executive model is

$$C(x) = 100x + 200{,}000$$

dollars/year.
  a. Find the average cost function $\overline{C}$.
  b. Find the marginal average cost function $\overline{C}'$.
  c. What happens to $\overline{C}(x)$ when x is very large? Interpret your results.

6. MARGINAL AVERAGE COST FOR PRODUCING THERMOMETERS The management of ThermoMaster Company, whose Mexican subsidiary manufactures an indoor–outdoor thermometer, has estimated that the total weekly cost (in dollars) for producing x thermometers is

$$C(x) = 5000 + 2x$$

  a. Find the average cost function $\overline{C}$.
  b. Find the marginal average cost function $\overline{C}'$.
  c. Interpret your results.

7. AVERAGE COST FOR PRODUCING CDS Find the average cost function $\overline{C}$ and the marginal average cost function $\overline{C}'$ associated with the total cost function C of Exercise 3.

8. AVERAGE COST FOR PRODUCING MICROWAVES Find the average cost function $\overline{C}$ and the marginal average cost function $\overline{C}'$ associated with the total cost function C of Exercise 4.

9. MARGINAL REVENUE OF A COMMUTER AIR SERVICE Williams Commuter Air Service realizes a monthly revenue of

$$R(x) = 8000x - 100x^2$$

dollars when the price charged per passenger is x dollars.
  a. Find the marginal revenue R'.
  b. Compute $R'(39)$, $R'(40)$, and $R'(41)$.
  c. Based on the results of part (b), what price should the airline charge in order to maximize their revenue?

10. MARGINAL REVENUE FOR PRODUCING LOUDSPEAKERS The management of Acrosonic plans to market the ElectroStat, an electrostatic speaker system. The marketing department has determined that the demand function for these speakers is

$$p = -0.04x + 800 \qquad (0 \le x \le 20{,}000)$$

where p denotes the speaker's unit price (in dollars) and x denotes the quantity demanded.
  a. Find the revenue function R.
  b. Find the marginal revenue function R'.
  c. Compute $R'(5000)$, and interpret your results.

11. MARGINAL PROFIT FOR PRODUCING LOUDSPEAKERS Refer to Exercise 10. Acrosonic's production department estimates that the total cost (in dollars) incurred in manufacturing x ElectroStat speaker systems in the first year of production will be

$$C(x) = 200x + 300{,}000$$

a. Find the profit function P.
b. Find the marginal profit function P'.
c. Compute $P'(5000)$ and $P'(8000)$.
d. Sketch the graph of the profit function, and interpret your results.

12. **MARGINAL PROFIT FOR AN APARTMENT COMPLEX** Lynbrook West, an apartment complex, has 100 two-bedroom units. The monthly profit (in dollars) realized from renting x apartments is

$$P(x) = -10x^2 + 1760x - 50,000$$

a. What is the actual profit realized from renting the 51st unit, assuming that 50 units have already been rented?
b. Compute the marginal profit when $x = 50$, and compare your results with that obtained in part (a).

13. **MARGINAL COST, REVENUE, AND PROFIT FOR PRODUCING LED TVs** The weekly demand for the Pulsar 25 color LED television is

$$p = 600 - 0.05x \qquad (0 \le x \le 12,000)$$

where p denotes the wholesale unit price in dollars and x denotes the quantity demanded. The weekly total cost function associated with manufacturing the Pulsar 25 is given by

$$C(x) = 0.000002x^3 - 0.03x^2 + 400x + 80,000$$

where $C(x)$ denotes the total cost incurred in producing x sets.
a. Find the revenue function R and the profit function P.
b. Find the marginal cost function C', the marginal revenue function R', and the marginal profit function P'.
c. Compute $C'(2000)$, $R'(2000)$, and $P'(2000)$, and interpret your results.
d. Sketch the graphs of the functions C, R, and P, and interpret parts (b) and (c), using the graphs obtained.

14. **MARGINAL COST, REVENUE, AND PROFIT FOR PRODUCING LCD TVs** Pulsar manufactures a series of 20-in. flat-tube LCD televisions. The quantity x of these sets demanded each week is related to the wholesale unit price p by the equation

$$p = -0.006x + 180$$

The weekly total cost incurred by Pulsar for producing x sets is

$$C(x) = 0.000002x^3 - 0.02x^2 + 120x + 60,000$$

dollars. Answer the questions in Exercise 13 for these data.

15. **MARGINAL AVERAGE COST FOR PRODUCING LED TVs** Refer to Exercise 13.
a. Find the average cost function $\overline{C}$ associated with the total cost function C of Exercise 13.
b. What is the marginal average cost function $\overline{C}'$?
c. Compute $\overline{C}'(5000)$ and $\overline{C}'(10,000)$, and interpret your results.
d. Sketch the graph of $\overline{C}$.

16. **MARGINAL AVERAGE COST FOR PRODUCING LCD TVs** Refer to Exercise 14.
a. Find the average cost function $\overline{C}$ associated with the total cost function C of Exercise 14.
b. What is the marginal average cost function $\overline{C}'$?
c. Compute $\overline{C}'(5000)$ and $\overline{C}'(10,000)$, and interpret your results.
d. Sketch the graph of $\overline{C}$.

17. **MARGINAL REVENUE FOR GAMING MICE** The quantity of Sensitech laser gaming mice demanded each month is related to the unit price by the equation

$$p = \frac{50}{0.01x^2 + 1} \qquad (0 \le x \le 20)$$

where p is measured in thousands of dollars and x in units of a thousand.
a. Find the revenue function R.
b. Find the marginal revenue function R'.
c. Compute $R'(2)$, and interpret your result.

18. **MARGINAL PROPENSITY TO CONSUME** The consumption function of the U.S. economy from 1929 to 1941 is

$$C(x) = 0.712x + 95.05$$

where $C(x)$ is the personal consumption expenditure and x is the personal income, both measured in billions of dollars. Find the rate of change of consumption with respect to income, dC/dx. This quantity is called the *marginal propensity to consume*.

19. **MARGINAL PROPENSITY TO CONSUME** Refer to Exercise 18. Suppose a certain economy's consumption function is

$$C(x) = 0.873x^{1.1} + 20.34$$

where $C(x)$ and x are measured in billions of dollars. Find the marginal propensity to consume when $x = 10$.

20. **MARGINAL PROPENSITY TO SAVE** Suppose $C(x)$ measures an economy's personal consumption expenditure and x measures the personal income, both in billions of dollars. Then

$$S(x) = x - C(x) \qquad \text{Income} - \text{consumption}$$

measures the economy's savings corresponding to an income of x billion dollars. Show that

$$\frac{dS}{dx} = 1 - \frac{dC}{dx}$$

The quantity dS/dx is called the *marginal propensity to save*.

21. **MARGINAL PROPENSITY TO SAVE** Refer to Exercise 20. For the consumption function of Exercise 18, find the marginal propensity to save.

22. MARGINAL PROPENSITY TO SAVE Refer to Exercise 20. For the consumption function of Exercise 19, find the marginal propensity to save when $x = 10$.

In Exercises 23–26, find the percentage rate of change of f at the given value of x.

23. $f(x) = 2x^2 + x + 1; x = 2$

24. $f(x) = \sqrt{2x^2 + 7}; x = 3$

25. $f(x) = \dfrac{x + 1}{x^3 + x + 1}; x = 2$

26. $f(x) = \left(\dfrac{x}{x^2 + 3x + 12}\right)^{3/2}; x = 1$

For each demand equation in Exercises 27–31, compute the elasticity of demand and determine whether the demand is elastic, unitary, or inelastic at the indicated price.

27. $x = -\dfrac{5}{4}p + 20; p = 10$

28. $x = -\dfrac{3}{2}p + 9; p = 2$

29. $x + \dfrac{1}{3}p - 20 = 0; p = 30$

30. $p = 144 - x^2; p = 96$ **31.** $p = 169 - x^2; p = 29$

32. INFLATION The consumer price index (CPI) of a certain country is given by

$$I(t) = -0.02t^3 + 0.4t^2 + 120 \qquad (0 \le t \le 4)$$

where $t = 0$ corresponds to the beginning of 2013. Find the annual percentage rate of inflation in the CPI of the country at the beginning of 2014.

33. GROWTH IN INCOME The per capita income of a country t years from now is defined to be

$$C(t) = \dfrac{I(t)}{P(t)}$$

where $I(t)$ and $P(t)$ are the income and the population of the country in year t, respectively.

  a. Find an expression for the percentage rate of change of the per capita income of the country in year t.
  b. Use your result to find the percentage rate of change in the per capita income of the country whose income in year t is $I(t) = 10^9(300 + 12t)$ dollars and whose population in year t is $P(t) = 2 \times 10^7 e^{0.02t}$.
  c. What is the percentage rate of change of the per capita income 2 years from now?

34. PERCENTAGE RATE OF CHANGE IN REVENUE Marianne Designs operates two boutiques in two locations in the Bay Area. The revenues for Boutique I and Boutique II at time $t = a$, where t is measured in years, are $3.2 million and $2.6 million, respectively. Management estimates that the revenue for Boutique I is growing at the rate of

$0.24 million/year, while that for Boutique II is growing at the rate of $0.30 million/year, at that time. What is the percentage rate of growth of the revenue of Marianne Designs at time $t = a$?
Hint: Marianne Designs's revenue is $R = f + g$, where f and g are the revenue of Boutique I and Boutique II, respectively. Find $100R'/R$ at $t = a$.

35. ELASTICITY OF DEMAND FOR HAIR DRYERS The demand equation for the Roland portable hair dryer is given by

$$x = \dfrac{1}{5}(225 - p^2) \qquad (0 \le p \le 15)$$

where x (measured in units of a hundred) is the quantity demanded per week and p is the unit price in dollars.
  a. Is the demand elastic or inelastic when $p = 8$ and when $p = 10$?
  b. When is the demand unitary?
    Hint: Solve $E(p) = 1$ for p.
  c. If the unit price is lowered slightly from $10, will the revenue increase or decrease?
  d. If the unit price is increased slightly from $8, will the revenue increase or decrease?

36. ELASTICITY OF DEMAND FOR TIRES The management of Titan Tire Company has determined that the quantity demanded x of their Super Titan tires per week is related to the unit price p by the equation

$$x = \sqrt{144 - p} \qquad (0 \le p \le 144)$$

where p is measured in dollars and x in units of a thousand.
  a. Compute the elasticity of demand when $p = 63$, 96, and 108.
  b. Interpret the results obtained in part (a).
  c. Is the demand elastic, unitary, or inelastic when $p = 63$, 96, and 108?

37. ELASTICITY OF DEMAND FOR DVD RENTALS The proprietor of Showplace, a video store, has estimated that the rental price p (in dollars) of prerecorded DVDs is related to the quantity x (in thousands) rented per day by the demand equation

$$x = \dfrac{2}{3}\sqrt{36 - p^2} \qquad (0 \le p \le 6)$$

Currently, the rental price is $2/disc.
  a. Is the demand elastic or inelastic at this rental price?
  b. If the rental price is increased, will the revenue increase or decrease?

38. ELASTICITY OF DEMAND FOR DIGITAL CAMERAS The quantity demanded each week x (in units of a hundred) of the Mikado digital camera is related to the unit price p (in dollars) by the demand equation

$$x = \sqrt{400 - 5p} \qquad (0 \le p \le 80)$$

  a. Is the demand elastic or inelastic when $p = 40$? When $p = 60$?
  b. When is the demand unitary?
    Hint: Solve $E(p) = 1$ for p.

c. If the unit price is lowered slightly from $60, will the revenue increase or decrease?

d. If the unit price is increased slightly from $40, will the revenue increase or decrease?

39. **ELASTICITY OF DEMAND FOR EXERCISE BICYCLES** The demand function for a certain make of exercise bicycle sold exclusively through cable television is

$$p = \sqrt{9 - 0.02x} \qquad (0 \le x \le 450)$$

where p is the unit price in hundreds of dollars and x is the quantity demanded per week. Compute the elasticity of demand and determine the range of prices corresponding to inelastic, unitary, and elastic demand.
Hint: Solve the equation $E(p) = 1$.

40. **ELASTICITY OF DEMAND FOR SPORTS WATCHES** The demand equation for the Sicard sports watch is given by

$$x = 10\sqrt{\frac{50 - p}{p}} \qquad (0 < p \le 50)$$

where x (measured in units of a thousand) is the quantity demanded per week and p is the unit price in dollars. Compute the elasticity of demand and determine the range of prices corresponding to inelastic, unitary, and elastic demand.

In Exercises 41 and 42, determine whether the statement is true or false. If it is true, explain why it is true. If it is false, give an example to show why it is false.

41. If C is a differentiable total cost function, then the marginal average cost function is

$$\overline{C}'(x) = \frac{xC'(x) - C(x)}{x^2}$$

42. If the marginal profit function is positive at $x = a$, then it makes sense to decrease the level of production.

3.4 Solutions to Self-Check Exercises

1. **a.** $R(x) = px$

$$= x(-0.02x + 300)$$
$$= -0.02x^2 + 300x \qquad (0 \le x \le 15,000)$$

$$P(x) = R(x) - C(x)$$
$$= -0.02x^2 + 300x$$
$$\quad - (0.000003x^3 - 0.04x^2 + 200x + 70,000)$$
$$= -0.000003x^3 + 0.02x^2 + 100x - 70,000$$

b. $C'(x) = 0.000009x^2 - 0.08x + 200$
$R'(x) = -0.04x + 300$
$P'(x) = -0.000009x^2 + 0.04x + 100$

c. The average cost function is

$$\overline{C}(x) = \frac{C(x)}{x}$$

$$= \frac{0.000003x^3 - 0.04x^2 + 200x + 70,000}{x}$$

$$= 0.000003x^2 - 0.04x + 200 + \frac{70,000}{x}$$

Therefore, the marginal average cost function is

$$\overline{C}'(x) = 0.000006x - 0.04 - \frac{70,000}{x^2}$$

d. Using the results from part (b), we find

$$C'(3000) = 0.000009(3000)^2 - 0.08(3000) + 200$$
$$= 41$$

That is, when the level of production is already 3000 recorders, the actual cost of producing one additional recorder is approximately $41. Next,

$$R'(3000) = -0.04(3000) + 300 = 180$$

That is, the actual revenue to be realized from selling the 3001st recorder is approximately $180. Finally,

$$P'(3000) = -0.000009(3000)^2 + 0.04(3000) + 100$$
$$= 139$$

That is, the actual profit realized from selling the 3001st DVR is approximately $139.

2. We first solve the given demand equation for x in terms of p, obtaining

$$x = f(p) = -50p + 15,000$$
$$f'(p) = -50$$

Therefore,

$$E(p) = -\frac{pf'(p)}{f(p)} = -\frac{p(-50)}{-50p + 15,000}$$

$$= \frac{p}{300 - p} \qquad (0 \le p < 300)$$

Next, we compute

$$E(100) = \frac{100}{300 - 100} = \frac{1}{2} < 1$$

and we conclude that demand is inelastic when $p = 100$. Also,

$$E(200) = \frac{200}{300 - 200} = 2 > 1$$

and we see that demand is elastic when $p = 200$.

3.5   Higher-Order Derivatives

Higher-Order Derivatives

The derivative f' of a function f is also a function. As such, the differentiability of f' may be considered. Thus, the function f' has a derivative f'' at a point x in the domain of f' if the limit of the quotient

$$\frac{f'(x + h) - f'(x)}{h}$$

exists as h approaches zero. In other words, it is the derivative of the first derivative.

The function f'' obtained in this manner is called the **second derivative of the function f,** just as the derivative f' of f is often called the first derivative of f. Continuing in this fashion, we are led to considering the third, fourth, and higher-order derivatives of f whenever they exist. Notations for the first, second, third, and, in general, nth derivatives of a function f at a point x are

$$f'(x), f''(x), f'''(x), \ldots, f^{(n)}(x)$$

or

$$D^1 f(x), D^2 f(x), D^3 f(x), \ldots, D^n f(x)$$

If f is written in the form $y = f(x)$, then the notations for its derivatives are

$$y', y'', y''', \ldots, y^{(n)}$$

$$\frac{dy}{dx}, \frac{d^2 y}{dx^2}, \frac{d^3 y}{dx^3}, \ldots, \frac{d^n y}{dx^n}$$

or

$$D^1 y, D^2 y, D^3 y, \ldots, D^n y$$

EXAMPLE 1 Find the derivatives of all orders of the polynomial function

$$f(x) = x^5 - 3x^4 + 4x^3 - 2x^2 + x - 8$$

Solution   We have

$$f'(x) = 5x^4 - 12x^3 + 12x^2 - 4x + 1$$

$$f''(x) = \frac{d}{dx} f'(x) = 20x^3 - 36x^2 + 24x - 4$$

$$f'''(x) = \frac{d}{dx} f''(x) = 60x^2 - 72x + 24$$

$$f^{(4)}(x) = \frac{d}{dx} f'''(x) = 120x - 72$$

$$f^{(5)}(x) = \frac{d}{dx} f^{(4)}(x) = 120$$

and

$$f^{(n)}(x) = 0 \quad \text{for } n > 5$$

EXAMPLE 2 Find the third derivative of the function f defined by $y = x^{2/3}$. What is its domain?

Solution We have

$$y' = \frac{2}{3}x^{-1/3}$$

$$y'' = \left(\frac{2}{3}\right)\left(-\frac{1}{3}\right)x^{-4/3} = -\frac{2}{9}x^{-4/3}$$

so the required derivative is

$$y''' = \left(-\frac{2}{9}\right)\left(-\frac{4}{3}\right)x^{-7/3} = \frac{8}{27}x^{-7/3} = \frac{8}{27x^{7/3}}$$

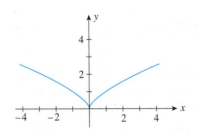

FIGURE 15
The graph of the function $y = x^{2/3}$

The common domain of the functions f', f'', and f''' is the set of all real numbers except $x = 0$. The domain of $y = x^{2/3}$ is the set of all real numbers. The graph of the function $y = x^{2/3}$ appears in Figure 15. ◼

Note Always simplify an expression before differentiating it to obtain the next order derivative. ◼

EXAMPLE 3 Find the second derivative of the function $y = (2x^2 + 3)^{3/2}$.

Solution We have, using the General Power Rule,

$$y' = \frac{3}{2}(2x^2 + 3)^{1/2}(4x) = 6x(2x^2 + 3)^{1/2}$$

Next, using the Product Rule and then the General Power Rule, we find

$$y'' = (6x) \cdot \frac{d}{dx}(2x^2 + 3)^{1/2} + \left[\frac{d}{dx}(6x)\right](2x^2 + 3)^{1/2}$$

$$= (6x)\left(\frac{1}{2}\right)(2x^2 + 3)^{-1/2}(4x) + 6(2x^2 + 3)^{1/2} \quad \text{See page 10.}$$

$$= 12x^2(2x^2 + 3)^{-1/2} + 6(2x^2 + 3)^{1/2}$$

$$= 6(2x^2 + 3)^{-1/2}[2x^2 + (2x^2 + 3)] \quad \text{Factor out } 6(2x^2 + 3)^{-1/2}.$$

$$= \frac{6(4x^2 + 3)}{\sqrt{2x^2 + 3}}$$

◼

Just as the derivative of a function f at a point x measures the rate of change of the function f at that point, the second derivative of f (the derivative of f') measures the rate of change of the derivative f' of the function f. The third derivative of the function f, f''', measures the rate of change of f'', and so on.

In Chapter 4, we will discuss applications involving the geometric interpretation of the second derivative of a function. The following example gives an interpretation of the second derivative in a familiar role.

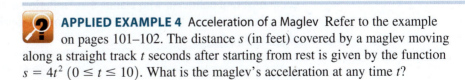

APPLIED EXAMPLE 4 Acceleration of a Maglev Refer to the example on pages 101–102. The distance s (in feet) covered by a maglev moving along a straight track t seconds after starting from rest is given by the function $s = 4t^2 \; (0 \le t \le 10)$. What is the maglev's acceleration at any time t?

Solution The velocity of the maglev t seconds from rest is given by

$$v = \frac{ds}{dt} = \frac{d}{dt}(4t^2) = 8t$$

The acceleration of the maglev t seconds from rest is given by the rate of change of the velocity of t—that is,

$$a = \frac{d}{dt}v = \frac{d}{dt}\left(\frac{ds}{dt}\right) = \frac{d^2s}{dt^2} = \frac{d}{dt}(8t) = 8$$

or 8 feet per second per second, normally abbreviated 8 ft/sec^2.

APPLIED EXAMPLE 5 Acceleration and Velocity of a Falling Object A ball is thrown straight up into the air from the roof of a building. The height of the ball as measured from the ground is given by

$$s = -16t^2 + 24t + 120$$

where s is measured in feet and t in seconds. Find the velocity and acceleration of the ball 3 seconds after it is thrown into the air.

Solution   The velocity v and acceleration a of the ball at any time t are given by

$$v = \frac{ds}{dt} = \frac{d}{dt}(-16t^2 + 24t + 120) = -32t + 24$$

and

$$a = \frac{d^2s}{dt^2} = \frac{d}{dt}\left(\frac{ds}{dt}\right) = \frac{d}{dt}(-32t + 24) = -32$$

Therefore, the velocity of the ball 3 seconds after it is thrown into the air is

$$v = -32(3) + 24 = -72$$

That is, the ball is falling downward at a speed of 72 ft/sec. The acceleration of the ball is 32 ft/sec^2 downward at any time during the motion.

Another interpretation of the second derivative of a function—this time from the field of economics—follows. Suppose the consumer price index (CPI) of an economy between the years a and b is described by the function $I(t)$ $(a \le t \le b)$ (Figure 16). Then the first derivative of I at $t = c$, $I'(c)$, where $a < c < b$, gives the rate of change of I at c. The quantity

$$\frac{I'(c)}{I(c)}$$

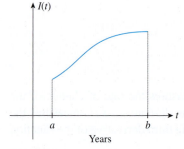

FIGURE 16
The CPI of a certain economy from year a to year b is given by $I(t)$.

measures the *inflation rate* of the economy at $t = c$. The second derivative of I at $t = c$, $I''(c)$, gives the rate of change of I' at $t = c$. Now, it is possible for $I'(t)$ to be positive and $I''(t)$ to be negative at $t = c$ (see Example 6). This tells us that at $t = c$, the economy is experiencing inflation (the CPI is increasing) but the rate at which the CPI is growing is in fact decreasing. This is precisely the situation described by an economist or a politician who claims that "inflation is slowing." One may not jump to the conclusion from the aforementioned quote that prices of goods and services are about to drop!

APPLIED EXAMPLE 6 Inflation Rate of an Economy The function

$$I(t) = -0.2t^3 + 3t^2 + 100 \qquad (0 \le t \le 9)$$

gives the CPI of an economy, where t is measured in years, with $t = 0$ corresponding to the beginning of 2004.

a. Find the inflation rate at the beginning of 2010 ($t = 6$).

b. Show that inflation was moderating at that time.

Solution

a. We find $I'(t) = -0.6t^2 + 6t$. Next, we compute

$$I'(6) = -0.6(6)^2 + 6(6) = 14.4 \quad \text{and} \quad I(6) = -0.2(6)^3 + 3(6)^2 + 100 = 164.8$$

from which we see that the inflation rate is

$$\frac{I'(6)}{I(6)} = \frac{14.4}{164.8} \approx 0.0874$$

or approximately 8.7%.

b. We find

$$I''(t) = \frac{d}{dt}(-0.6t^2 + 6t) = -1.2t + 6$$

Since

$$I''(6) = -1.2(6) + 6 = -1.2$$

we see that I' is indeed decreasing at $t = 6$, and we conclude that inflation was moderating at that time (Figure 17).

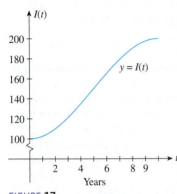

FIGURE 17
The CPI of an economy is given by $I(t)$.

3.5 Self-Check Exercises

1. Find the third derivative of

$$f(x) = 2x^5 - 3x^3 + x^2 - 6x + 10$$

2. Let

$$f(x) = \frac{1}{1 + x}$$

Find $f'(x)$, $f''(x)$, and $f'''(x)$.

3. CONSERVATION OF SPECIES A certain species of turtles faces extinction because dealers collect truckloads of turtle eggs

to be sold as aphrodisiacs. After severe conservation measures are implemented, it is hoped that the turtle population will grow according to the rule

$$N(t) = 2t^3 + 3t^2 - 4t + 1000 \qquad (0 \le t \le 10)$$

where $N(t)$ denotes the population at the end of year t. Compute $N''(2)$ and $N''(8)$. What do your results tell you about the effectiveness of the program?

Solutions to Self-Check Exercises 3.5 can be found on page 220.

3.5 Concept Questions

1. a. What is the second derivative of a function f?

b. How do you find the second derivative of a function f, assuming that it exists?

2. If $s = f(t)$ gives the position of an object moving on the coordinate line, what do $f'(t)$ and $f''(t)$ measure?

3. Suppose $f(t)$ measures the population of a country at time t. What can you say about the signs of $f'(t)$ and $f''(t)$ over the time interval (a, b)

a. If the population is increasing at an increasing rate over (a, b)?

b. If the population is increasing at a decreasing rate over (a, b)?

c. If the population is decreasing at an increasing rate over (a, b)?

d. If the population is decreasing at a decreasing rate over (a, b)?

4. Suppose $f(t)$ measures the population of a country at time t. What can you say about the signs of $f'(t)$ and $f''(t)$ over the time interval

a. If the population is increasing at a constant rate over (a, b)?

b. If the population is decreasing at a constant rate over (a, b)?

c. If the population is constant over (a, b)?

3.5 Exercises

In Exercises 1–20, find the first and second derivatives of the function.

1. $f(x) = 4x^2 - 2x + 1$

2. $f(x) = -0.2x^2 + 0.3x + 4$

3. $f(x) = 2x^3 - 3x^2 + 1$

4. $g(x) = -3x^3 + 24x^2 + 6x - 64$

5. $h(t) = t^4 - 2t^3 + 6t^2 - 3t + 10$

6. $f(x) = x^5 - x^4 + x^3 - x^2 + x - 1$

7. $f(x) = (x^2 + 2)^5$ **8.** $g(t) = t^2(3t + 1)^4$

9. $g(t) = (2t^2 - 1)^2(3t^2)$

10. $h(x) = (x^2 + 1)^2(x - 1)$

11. $f(x) = (2x^2 + 2)^{7/2}$

12. $h(w) = (w^2 + 2w + 4)^{5/2}$

13. $f(x) = x(x^2 + 1)^2$ **14.** $g(u) = u(2u - 1)^3$

15. $f(x) = \dfrac{x}{2x + 1}$ **16.** $g(t) = \dfrac{t^2}{t - 1}$

17. $f(s) = \dfrac{s - 1}{s + 1}$ **18.** $f(u) = \dfrac{u}{u^2 + 1}$

19. $f(u) = \sqrt{4 - 3u}$ **20.** $f(x) = \sqrt{2x - 1}$

In Exercises 21–28, find the third derivative of the given function.

21. $f(x) = 3x^4 - 4x^3$

22. $f(x) = 3x^5 - 6x^4 + 2x^2 - 8x + 12$

23. $f(x) = \dfrac{1}{x}$ **24.** $f(x) = \dfrac{2}{x^2}$

25. $g(s) = \sqrt{3s - 2}$ **26.** $g(t) = \sqrt{2t + 3}$

27. $f(x) = (2x - 3)^4$ **28.** $g(t) = \left(\dfrac{1}{2}t^2 - 1\right)^5$

29. ACCELERATION OF A FALLING OBJECT During the construction of an office building, a hammer is accidentally dropped from a height of 256 ft. The distance (in feet) the hammer falls in t sec is $s = 16t^2$. What is the hammer's velocity when it strikes the ground? What is its acceleration?

30. ACCELERATION OF A CAR The distance s (in feet) covered by a car after t sec is given by

$$s = -t^3 + 8t^2 + 20t \qquad (0 \le t \le 6)$$

Find a general expression for the car's acceleration at any time t $(0 \le t \le 6)$. Show that the car is decelerating after $2\frac{2}{3}$ sec.

31. AUTO TRANSMISSIONS AND FUEL ECONOMY In trying to extract maximum efficiency out of every subsystem of a vehicle, auto transmission engineers are developing transmissions with up to ten gears. This is one of the ways in which manufacturers are trying to meet stricter federal fuel economy and pollution standards. The projected percentage of new cars equipped with transmissions that have seven speeds or more is given by the function

$$P(t) = 0.38t^2 + 1.3t + 3 \qquad (0 \le t \le 20)$$

where t is measured in years, with $t = 0$ corresponding to 2010.

a. What will the percentage of vehicles equipped with transmissions that have seven speeds or more be in 2015 according to the projection?

b. How fast will the percentage of vehicles equipped with transmissions that have seven speeds or more be changing in 2015?

c. What is $P''(5)$? Interpret your result.

Source: IHS Automotive.

32. ALZHEIMER'S DISEASE As baby boomers enter their golden years, the number of people afflicted with Alzheimer's disease is expected to rise dramatically. In a study published in the *Journal of Neurology*, the number of people with Alzheimer's disease in the United States age 65 years and over is projected to be

$$N(t) = 0.00525t^2 + 0.075t + 4.7 \qquad (0 \le t \le 4)$$

million in decade t, where $t = 0$ corresponds to 2010.

a. What is the projected number of people with Alzheimer's disease in the United States age 65 years and over in 2030?

b. How fast is the number of people with Alzheimer's disease in the United States age 65 years and over projected to grow in 2030?

c. How fast is the rate of growth of people with Alzheimer's disease in the United States age 65 years and over projected to change in the period covered by the study?

Source: American Academy of Neurology.

33. CRIME RATES The number of major crimes committed in Bronxville between 2006 and 2014 is approximated by the function

$$N(t) = -0.1t^3 + 1.5t^2 + 100 \qquad (0 \le t \le 8)$$

where $N(t)$ denotes the number of crimes committed in year t, with $t = 0$ corresponding to 2006. Enraged by the dramatic increase in the crime rate, Bronxville's citizens, with the help of the local police, organized Neighborhood Crime Watch groups in early 2010 to combat this menace.

a. Verify that the crime rate was increasing from 2006 through 2014.

 Hint: Compute $N'(0)$, $N'(1)$, . . . , $N'(8)$. (See Note on page 168.)

b. Show that the Neighborhood Crime Watch program was working by computing $N''(4)$, $N''(5)$, $N''(6)$, $N''(7)$, and $N''(8)$.

34. GDP OF A DEVELOPING COUNTRY A developing country's gross domestic product (GDP) from 2006 to 2015 is approximated by the function

$$G(t) = -0.2t^3 + 2.4t^2 + 60 \qquad (0 \le t \le 9)$$

where $G(t)$ is measured in billions of dollars, with $t = 0$ corresponding to 2006.
a. Compute $G'(1), \ldots, G'(8)$.
b. Compute $G''(1), \ldots, G''(8)$.
c. Using the results obtained in parts (a) and (b), show that after a slow start, the GDP increases quickly and then cools off.

35. SALES OF VINYL RECORDS Vinyl records (LPs) almost disappeared after CDs were introduced in the early 1980s, but they have been making a comeback. Wagner's Records estimates that t months after the grand opening of the store, the sales of LPs will be

$$S(t) = 4t^3 + 2t^2 + 300t$$

units. Compute $S(6)$, $S'(6)$, and $S''(6)$, and interpret your results.

36. MEDIAN AGE OF U.S. POPULATION The median age (in years) of the U.S. population over the decades from 1960 through 2010 is given by

$$f(t) = -0.2176t^3 + 1.962t^2 - 2.833t + 29.4 \qquad (0 \le t \le 5)$$

where t is measured in decades, with $t = 0$ corresponding to 1960.
a. What was the median age of the population in the year 2000?
b. At what rate was the median age of the population changing in the year 2000?
c. Calculate $f''(4)$ and interpret your result.
Source: U.S. Census Bureau.

37. OBESITY IN AMERICA The body mass index (BMI) measures body weight in relation to height. A BMI of 25 to 29.9 is considered overweight, a BMI of 30 or more is considered obese, and a BMI of 40 or more is morbidly obese. The percent of the U.S. population that is obese is approximated by the function

$$P(t) = 0.0004t^3 + 0.0036t^2 + 0.8t + 12 \qquad (0 \le t \le 13)$$

where t is measured in years, with $t = 0$ corresponding to the beginning of 1991. Show that the rate of the rate of change of the percent of the U.S. population that is deemed obese was positive from 1991 to 2004. What does this mean?
Source: Centers for Disease Control and Prevention.

38. TEST FLIGHT OF A VTOL In a test flight of the McCord Terrier, McCord Aviation's experimental VTOL (vertical takeoff and landing) aircraft, it was determined that t sec after liftoff, when the craft was operated in the vertical takeoff mode, its altitude (in feet) was

$$h(t) = \frac{1}{16}t^4 - t^3 + 4t^2 \qquad (0 \le t \le 8)$$

a. Find an expression for the aircraft's velocity at time t.
b. Find the aircraft's velocity when $t = 0$ (the initial velocity), $t = 4$, and $t = 8$.
c. Find an expression for the aircraft's acceleration at time t.
d. Find the aircraft's acceleration when $t = 0$, 4, and 8.
e. Find the aircraft's height when $t = 0$, 4, and 8.

39. AIR PURIFICATION During testing of a certain brand of air purifier, the amount of smoke remaining t min after the start of the test was

$$A(t) = -0.00006t^5 + 0.00468t^4 - 0.1316t^3 + 1.915t^2 - 17.63t + 100$$

percent of the original amount. Compute $A'(10)$ and $A''(10)$, and interpret your results.
Source: Consumer Reports.

40. WORKING MOTHERS The percentage of mothers who work outside the home and have children younger than 6 years old is approximated by the function

$$P(t) = 33.55(t + 5)^{0.205} \qquad (0 \le t \le 32)$$

where t is measured in years, with $t = 0$ corresponding to the beginning of 1980. Compute $P''(20)$, and interpret your result.
Source: U.S. Bureau of Labor Statistics.

41. AGING POPULATION The population of Americans age 55 and older as a percentage of the total population is approximated by the function

$$f(t) = 10.72(0.9t + 10)^{0.3} \qquad (0 \le t \le 20)$$

where t is measured in years, with $t = 0$ corresponding to 2000. Compute $f''(10)$, and interpret your result.
Source: U.S. Census Bureau.

In Exercises 42–46, determine whether the statement is true or false. If it is true, explain why it is true. If it is false, give an example to show why it is false.

42. If the second derivative of f exists at $x = a$, then $f''(a) = [f'(a)]^2$.

43. If $h = fg$ where f and g have second-order derivatives, then

$$h''(x) = f''(x)g(x) + 2f'(x)g'(x) + f(x)g''(x)$$

44. If f is a polynomial of degree n, then $f^{(n+1)}(x) = 0$.

45. Suppose $P(t)$ represents the population of bacteria at time t, and suppose $P'(t) > 0$ and $P''(t) < 0$; then the population is increasing at time t but at a decreasing rate.

46. If $h(x) = f(2x)$, then $h''(x) = 4f''(2x)$.

47. Let $\overline{C}(x)$ be the average cost function associated with a total cost function $C(x)$. Show that

$$\overline{C}''(x) = \frac{C''(x)}{x} - \frac{2C'(x)}{x^2} + \frac{2C(x)}{x^3}$$

48. Let f be the function defined by the rule $f(x) = x^{7/3}$. Show that f has first- and second-order derivatives at all points x, and in particular at $x = 0$. Also show that the third derivative of f does *not* exist at $x = 0$.

49. Construct a function f that has derivatives of order up through and including n at a point a but fails to have the $(n + 1)$st derivative there.
Hint: See Exercise 48.

50. Show that a polynomial function has derivatives of all orders.
Hint: Let $P(x) = a_n x^n + a_{n-1} x^{n-1} + a_{n-2} x^{n-2} + \cdots + a_0$ be a polynomial of degree n, where n is a positive integer and $a_0, a_1, \ldots, a_n$ are constants with $a_n \neq 0$. Compute $P'(x)$, $P''(x), \ldots$.

3.5 Solutions to Self-Check Exercises

1. $f'(x) = \dfrac{d}{dx}(2x^5 - 3x^3 + x^2 - 6x + 10)$

$\qquad = 10x^4 - 9x^2 + 2x - 6$

$f''(x) = \dfrac{d}{dx}(10x^4 - 9x^2 + 2x - 6)$

$\qquad = 40x^3 - 18x + 2$

$f'''(x) = \dfrac{d}{dx}(40x^3 - 18x + 2) = 120x^2 - 18$

2. We write $f(x) = (1 + x)^{-1}$ and use the General Power Rule, obtaining

$f'(x) = (-1)(1 + x)^{-2} \dfrac{d}{dx}(1 + x) = -(1 + x)^{-2}(1)$

$\qquad = -(1 + x)^{-2} = -\dfrac{1}{(1 + x)^2}$

Continuing, we find

$f''(x) = -(-2)(1 + x)^{-3}$

$\qquad = 2(1 + x)^{-3} = \dfrac{2}{(1 + x)^3}$

$f'''(x) = 2(-3)(1 + x)^{-4}$

$\qquad = -6(1 + x)^{-4} = -\dfrac{6}{(1 + x)^4}$

3. We first find the second derivative of $N(t) = 2t^3 + 3t^2 - 4t + 1000$, obtaining

$$N'(t) = 6t^2 + 6t - 4 \quad \text{and} \quad N''(t) = 12t + 6 = 6(2t + 1)$$

Therefore, $N''(2) = 30$ and $N''(8) = 102$. The results of our computations reveal that at the end of year 2, the *rate* of growth of the turtle population is increasing at the rate of 30 turtles/year/year. At the end of year 8, the rate is increasing at the rate of 102 turtles/year/year. Clearly, the conservation measures are paying off handsomely.

USING TECHNOLOGY   Finding the Second Derivative of a Function at a Given Point

Some graphing utilities have the capability of numerically computing the second derivative of a function at a point. If your graphing utility has this capability, use it to work through the examples and exercises of this section.

EXAMPLE 1 Use the (second) numerical derivative operation of a graphing utility to find the second derivative of $f(x) = \sqrt{x}$ when $x = 4$.

Solution   Using the (second) numerical derivative operation, we find

$$f''(4) = \text{der2}(x^{\wedge}.5, x, 4) = -.03125$$

(Figure T1).

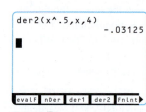

FIGURE T1
The TI-86 second derivative screen for computing $f''(4)$

APPLIED EXAMPLE 2 Prevalence of Alzheimer's Patients The number of Alzheimer's patients in the United States is given by

$$f(t) = -0.02765t^4 + 0.3346t^3 - 1.1261t^2$$
$$+ 1.7575t + 3.7745 \qquad (0 \le t \le 5)$$

where $f(t)$ is measured in millions and t is measured in decades, with $t = 0$ corresponding to the beginning of 1990.

a. How fast is the number of Alzheimer's patients in the United States anticipated to be changing at the beginning of 2030?
b. How fast is the rate of change of the number of Alzheimer's patients in the United States anticipated to be changing at the beginning of 2030?
c. Plot the graph of f in the viewing window $[0, 5] \times [0, 12]$.
Source: Alzheimer's Association.

Solution

a. Using the numerical derivative operation of a graphing utility, we find that the number of Alzheimer's patients at the beginning of 2030 can be anticipated to be changing at the rate of

$$f'(4) = 1.7311$$

That is, the number is increasing at the rate of approximately 1.7 million patients per decade.

b. Using the (second) numerical derivative operation of a graphing utility, we find that

$$f''(4) = 0.4694$$

(Figure T2); that is, the rate of change of the number of Alzheimer's patients is increasing at the rate of approximately 0.5 million patients per decade per decade.

c. Figure T3 shows the graph.

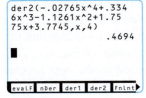

FIGURE T2
The TI-86 second derivative screen for computing $f''(4)$

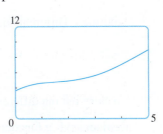

FIGURE T3
The graph of f in the viewing window $[0, 5] \times [0, 12]$

TECHNOLOGY EXERCISES

In Exercises 1–8, find the value of the second derivative of f at the given value of x. Express your answer correct to four decimal places.

1. $f(x) = 2x^3 - 3x^2 + 1; x = -1$

2. $f(x) = 2.5x^5 - 3x^3 + 1.5x + 4; x = 2.1$

3. $f(x) = 2.1x^{3.1} - 4.2x^{1.7} + 4.2; x = 1.4$

4. $f(x) = 1.7x^{4.2} - 3.2x^{1.3} + 4.2x - 3.2; x = 2.2$

5. $f(x) = \dfrac{x^2 + 2x - 5}{x^3 + 1}; x = 2.1$

6. $f(x) = \dfrac{x^3 + x + 2}{2x^2 - 5x + 4}; x = 1.2$

7. $f(x) = \dfrac{x^{1/2} + 2x^{3/2} + 1}{2x^{1/2} + 3}; x = 0.5$

8. $f(x) = \dfrac{\sqrt{x} - 1}{2x + \sqrt{x} + 4}; x = 2.3$

3.6   Implicit Differentiation and Related Rates

Differentiating Implicitly

Up to now, we have dealt with functions expressed in the form $y = f(x)$, that is, functions in which the dependent variable y is expressed *explicitly* in terms of the independent variable x. However, not all functions are expressed in this form. Consider, for example, the equation

$$x^2y + y - x^2 + 1 = 0 \qquad (8)$$

This equation does express y *implicitly* as a function of x. In fact, solving (8) for y in terms of x, we obtain

$$(x^2 + 1)y = x^2 - 1 \qquad \text{Implicit equation}$$

$$y = f(x) = \frac{x^2 - 1}{x^2 + 1} \qquad \text{Explicit equation}$$

which gives an explicit representation of f.

Next, consider the equation

$$y^4 - y^3 - y + 2x^3 - x = 8$$

If certain restrictions are placed on x and y, this equation defines y as a function of x. But in this instance, we would be hard pressed to find y explicitly in terms of x. The following question arises naturally: How do we compute dy/dx in this case?

As it turns out, thanks to the Chain Rule, a method *does* exist for computing the derivative of a function directly from the implicit equation defining the function. This method is called **implicit differentiation** and is demonstrated in the next several examples.

EXAMPLE 1 Given the equation $y^2 = x$, find $\dfrac{dy}{dx}$.

Solution   Differentiating both sides of the equation with respect to x, we obtain

$$\frac{d}{dx}(y^2) = \frac{d}{dx}(x)$$

To carry out the differentiation of the term $\dfrac{d}{dx}(y^2)$, we note that y (with suitable restrictions) is a function of x. Writing $y = f(x)$ to remind us of this fact, we find that

$$\frac{d}{dx}(y^2) = \frac{d}{dx}[f(x)]^2 \qquad \text{Write } y = f(x).$$

$$= 2f(x)f'(x) \qquad \text{Use the Chain Rule.}$$

$$= 2y\frac{dy}{dx} \qquad \text{Replace } f(x) \text{ with } y.$$

Therefore, the equation

$$\frac{d}{dx}(y^2) = \frac{d}{dx}(x)$$

is equivalent to

$$2y\frac{dy}{dx} = 1$$

Solving for $\dfrac{dy}{dx}$ yields

$$\frac{dy}{dx} = \frac{1}{2y}$$

Before considering other examples, let's summarize the important steps involved in implicit differentiation. (Here, we assume that dy/dx exists.)

> **Finding $\dfrac{dy}{dx}$ by Implicit Differentiation**
>
> **1.** Differentiate both sides of the equation *with respect to x.* (Make sure that the derivative of any term involving y includes the factor dy/dx.)
> **2.** Solve the resulting equation for dy/dx in terms of x and y.

EXAMPLE 2 Find $\dfrac{dy}{dx}$ given the equation

$$y^3 - y + 2x^3 - x = 8$$

Solution Differentiating both sides of the given equation with respect to x, we obtain

$$\frac{d}{dx}(y^3 - y + 2x^3 - x) = \frac{d}{dx}(8)$$

$$\frac{d}{dx}(y^3) - \frac{d}{dx}(y) + \frac{d}{dx}(2x^3) - \frac{d}{dx}(x) = 0$$

Now, recalling that y is a function of x, we apply the Chain Rule to the first two terms on the left. Thus,

$$3y^2\frac{dy}{dx} - \frac{dy}{dx} + 6x^2 - 1 = 0$$

$$(3y^2 - 1)\frac{dy}{dx} = 1 - 6x^2$$

$$\frac{dy}{dx} = \frac{1 - 6x^2}{3y^2 - 1}$$

Explore and Discuss

Refer to Example 2. Suppose we think of the equation $y^3 - y + 2x^3 - x = 8$ as defining x implicitly as a function of y. Find dx/dy, and justify your method of solution.

EXAMPLE 3 Consider the equation $x^2 + y^2 = 4$.

a. Find dy/dx by implicit differentiation.
b. Find the slope of the tangent line to the graph of the function $y = f(x)$ at the point $(1, \sqrt{3})$.
c. Find an equation of the tangent line of part (b).

Solution

a. Differentiating both sides of the equation with respect to x, we obtain

$$\frac{d}{dx}(x^2 + y^2) = \frac{d}{dx}(4)$$

$$\frac{d}{dx}(x^2) + \frac{d}{dx}(y^2) = 0$$

$$2x + 2y\frac{dy}{dx} = 0$$

$$2y\frac{dy}{dx} = -2x$$

$$\frac{dy}{dx} = -\frac{x}{y} \qquad (y \neq 0)$$

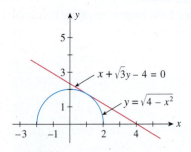

FIGURE 18
The line $x + \sqrt{3}y - 4 = 0$ is tangent to the graph of the function $y = f(x)$.

b. The slope of the tangent line to the graph of the function at the point $(1, \sqrt{3})$ is given by

$$\left.\frac{dy}{dx}\right|_{(1,\sqrt{3})} = \left.-\frac{x}{y}\right|_{(1,\sqrt{3})} = -\frac{1}{\sqrt{3}}$$

(*Note:* This notation is read "dy/dx evaluated at the point $(1, \sqrt{3})$.")

c. An equation of the tangent line in question is found by using the point-slope form of the equation of a line with the slope $m = -1/\sqrt{3}$ and the point $(1, \sqrt{3})$. Thus,

$$y - \sqrt{3} = -\frac{1}{\sqrt{3}}(x - 1) \quad \text{(x^2) See page 37.}$$
$$\sqrt{3}y - 3 = -x + 1$$
$$x + \sqrt{3}y - 4 = 0$$

A sketch of this tangent line and the graph of $y = f(x)$ are shown in Figure 18.
We can also solve the equation $x^2 + y^2 = 4$ explicitly for y in terms of x. If we do this, we obtain

$$y = \pm\sqrt{4 - x^2}$$

From this, we see that the equation $x^2 + y^2 = 4$ defines the two functions

$$y = f(x) = \sqrt{4 - x^2}$$
$$y = g(x) = -\sqrt{4 - x^2}$$

Since the point $(1, \sqrt{3})$ does not lie on the graph of $y = g(x)$, we conclude that

$$y = f(x) = \sqrt{4 - x^2}$$

is the required function. The graph of f is the semicircle shown in Figure 18. ◼

Note The notation

$$\left.\frac{dy}{dx}\right|_{(a,b)}$$

is used to denote the value of dy/dx at the point (a, b). ◼

Explore and Discuss

Refer to Example 3. Yet another function defined implicitly by the equation $x^2 + y^2 = 4$ is the function

$$y = h(x) = \begin{cases} \sqrt{4 - x^2} & \text{if } -2 \le x < 0 \\ -\sqrt{4 - x^2} & \text{if } 0 \le x \le 2 \end{cases}$$

1. Sketch the graph of h.
2. Show that $h'(x) = -x/y$ if $x \neq 0$ and $y \neq 0$.
3. Find an equation of the tangent line to the graph of h at the point $(1, -\sqrt{3})$.

To find dy/dx at a *specific point* (a, b), differentiate the given equation implicitly with respect to x and then replace x and y by a and b, respectively, *before* solving the equation for dy/dx. This often simplifies the amount of algebra involved.

EXAMPLE 4 Find $\dfrac{dy}{dx}$ given that x and y are related by the equation

$$x^2y^3 + 6x^2 = y + 12$$

and that $y = 2$ when $x = 1$.

Solution Differentiating both sides of the given equation with respect to x, we obtain

$$\frac{d}{dx}(x^2y^3) + \frac{d}{dx}(6x^2) = \frac{d}{dx}(y) + \frac{d}{dx}(12)$$

$$x^2 \cdot \frac{d}{dx}(y^3) + y^3 \cdot \frac{d}{dx}(x^2) + 12x = \frac{dy}{dx} \qquad \text{Use the Product Rule on } \frac{d}{dx}(x^2y^3).$$

$$3x^2y^2\frac{dy}{dx} + 2xy^3 + 12x = \frac{dy}{dx}$$

Substituting $x = 1$ and $y = 2$ into this equation gives

$$3(1)^2(2)^2\frac{dy}{dx} + 2(1)(2)^3 + 12(1) = \frac{dy}{dx}$$

$$12\frac{dy}{dx} + 16 + 12 = \frac{dy}{dx}$$

Solving for $\dfrac{dy}{dx}$, we have

$$\frac{dy}{dx} = -\frac{28}{11}$$

Note that it is not necessary to find an explicit expression for dy/dx. ▪

Note In Examples 3 and 4, you can verify that the points at which we evaluated dy/dx actually lie on the curve in question by showing that the coordinates of the points satisfy the given equations. ▪

EXAMPLE 5 Find $\dfrac{d^2y}{dx^2}$ if $xy - y^3 = 4$.

Solution Differentiating both sides of the given equation with respect to x, we obtain

$$\frac{d}{dx}(xy) - \frac{d}{dx}(y^3) = \frac{d}{dx}(4)$$

$$x\frac{dy}{dx} + y - 3y^2\frac{dy}{dx} = 0$$

$$y - (3y^2 - x)\frac{dy}{dx} = 0$$

or

$$\frac{dy}{dx} = \frac{y}{3y^2 - x}$$

Next, we differentiate both sides of the last equation with respect to x again, obtaining

$$\frac{d}{dx}\left(\frac{dy}{dx}\right) = \frac{d}{dx}\left(\frac{y}{3y^2 - x}\right)$$

$$\frac{d^2y}{dx^2} = \frac{(3y^2 - x)\dfrac{d}{dx}(y) - y\dfrac{d}{dx}(3y^2 - x)}{(3y^2 - x)^2}$$

$$= \frac{(3y^2 - x)\dfrac{dy}{dx} - y\left(6y\dfrac{dy}{dx} - 1\right)}{(3y^2 - x)^2} = \frac{y - (3y^2 + x)\dfrac{dy}{dx}}{(3y^2 - x)^2}$$

$$= \frac{y - (3y^2 + x)\left(\dfrac{y}{3y^2 - x}\right)}{(3y^2 - x)^2} = \frac{3y^3 - xy - 3y^3 - xy}{(3y^2 - x)^3}$$

$$= -\frac{2xy}{(3y^2 - x)^3}$$

APPLIED EXAMPLE 6 Output of a Country The chief economist of a country estimates that the output of the country is given by $Q = 10x^{3/4}y^{1/4}$, where x is the amount of money spent on labor and y is the amount spent on capital. Here, x, y, and Q are measured in billions of dollars.

a. Find the output of the country if $625 billion is spent on labor and $81 billion is spent on capital.

b. Suppose that the output of the country is to be maintained at the level found in part (a). By how much should the amount spent on capital be changed if the amount spent on labor is to be increased by $1 billion?

Solution

a. Replacing x and y by 625 and 81, respectively, in the expression for Q, we see that the required output is

$$Q = 10(625)^{3/4}(81)^{1/4} = 3750$$

or $3750 billion.

b. We want to find the change in y per unit change in x when $x = 625$ and $y = 81$ given that $10x^{3/4}y^{1/4} = 3750$. But as you saw in Section 3.4, this quantity is approximated by dy/dx (evaluated at the specified values of x and y). To find dy/dx, we differentiate the equation

$$10x^{3/4}y^{1/4} = 3750$$

or, equivalently, the equation

$$x^{3/4}y^{1/4} = 375$$

implicitly. Thus,

$$\frac{d}{dx}(x^{3/4}y^{1/4}) = \frac{d}{dx}(375)$$

$$\frac{3}{4}x^{-1/4}y^{1/4} + x^{3/4}\left(\frac{1}{4}y^{-3/4}\frac{dy}{dx}\right) = 0$$

$$\frac{1}{4}x^{3/4}y^{-3/4}\frac{dy}{dx} = -\frac{3}{4}x^{-1/4}y^{1/4}$$

or

$$\frac{dy}{dx} = (4x^{-3/4}y^{3/4})\left(-\frac{3}{4}x^{-1/4}y^{1/4}\right) = -3x^{-1}y = -3\left(\frac{y}{x}\right)$$

Therefore, when $x = 625$ and $y = 81$, we find

$$\frac{dy}{dx} = -3\left(\frac{81}{625}\right) = -0.3888$$

So to keep the output at the constant level of $3750 billion, the amount spent on capital should decrease by $0.3888 billion if the amount spent on labor were to be increased by $1 billion.

Note The negative of dy/dx found in Example 6 is called the **marginal rate of technical substitution** (MRTS). In general, the MRTS measures the rate at which a producer is technically able to reduce one input (capital) for a unit increase of another input (labor) while maintaining a constant level of output. Thus, the MRTS for the country in Example 6 is $0.3888 billion per $1 billion.

Related Rates

Implicit differentiation is a useful technique for solving a class of problems known as **related-rates** problems. The following is a typical related-rates problem: Suppose x and y are two quantities that depend on a third quantity t and we know the relationship between x and y in the form of an equation. Can we find a relationship between dx/dt and dy/dt? In particular, if we know one of the rates of change at a specific value of t—say, dx/dt—can we find the other rate, dy/dt, at that value of t?

APPLIED EXAMPLE 7 Rate of Change of Housing Starts A study prepared for the National Association of Realtors estimates that the number of housing starts in the southwest, $N(t)$ (in units of a million), over the next 5 years is related to the mortgage rate $r(t)$ (percent per year) by the equation

$$9N^2 + r = 36$$

where t is measured in years. What is the rate of change of the number of housing starts with respect to time when the mortgage rate is 11% per year and is increasing at the rate of 1.5% per year?

Solution We are given that

$$r = 11 \quad \text{and} \quad \frac{dr}{dt} = 1.5$$

at a certain instant of time, and we are required to find dN/dt. First, by substituting $r = 11$ into the given equation, we find

$$9N^2 + 11 = 36$$
$$N^2 = \frac{25}{9}$$

or $N = 5/3$ (we reject the negative root). Next, differentiating the given equation implicitly on both sides with respect to t, we obtain

$$\frac{d}{dt}(9N^2) + \frac{d}{dt}(r) = \frac{d}{dt}(36)$$

$$18N\frac{dN}{dt} + \frac{dr}{dt} = 0 \qquad \text{\color{red}{Use the Chain Rule on the first term.}}$$

Then, substituting $N = 5/3$ and $dr/dt = 1.5$ into this equation gives

$$18\left(\frac{5}{3}\right)\frac{dN}{dt} + 1.5 = 0$$

Solving this equation for dN/dt then gives

$$\frac{dN}{dt} = -\frac{1.5}{30} = -0.05$$

Thus, at the instant of time under consideration, the number of housing starts is decreasing at the rate of 50,000 units per year.

APPLIED EXAMPLE 8 Effect of Price on Supply of Flash Drives Texar Inc., a manufacturer of disk drives is willing to make x thousand IGB USB flash drives available in the marketplace each week when the wholesale price is $\$p$ per drive. It is known that the relationship between x and p is governed by the supply equation

$$x^2 - 3xp + p^2 = 5$$

How fast is the supply of drives changing when the price per drive is \$11, the quantity supplied is 4000 drives, and the wholesale price per drive is increasing at the rate of \$0.10 per drive each week?

Solution   We are given that

$$p = 11 \qquad x = 4 \qquad \frac{dp}{dt} = 0.1$$

at a certain instant of time, and we are required to find dx/dt. Differentiating the given equation on both sides with respect to t, we obtain

$$\frac{d}{dt}(x^2) - \frac{d}{dt}(3xp) + \frac{d}{dt}(p^2) = \frac{d}{dt}(5)$$

$$2x\frac{dx}{dt} - 3\left(p\frac{dx}{dt} + x\frac{dp}{dt}\right) + 2p\frac{dp}{dt} = 0 \qquad \text{\textcolor{red}{Use the Product Rule on the second term.}}$$

Substituting the given values of p, x, and dp/dt into the last equation, we have

$$2(4)\frac{dx}{dt} - 3\left[(11)\frac{dx}{dt} + 4(0.1)\right] + 2(11)(0.1) = 0$$

$$8\frac{dx}{dt} - 33\frac{dx}{dt} - 1.2 + 2.2 = 0$$

$$25\frac{dx}{dt} = 1$$

$$\frac{dx}{dt} = 0.04$$

Thus, at the instant of time under consideration, the supply of drives is increasing at the rate of $(0.04)(1000)$, or 40, drives per week.

In certain related-rates problems, we need to formulate the problem mathematically before analyzing it. The following guidelines can be used to help solve problems of this type.

> **Solving Related-Rates Problems**
> 1. Assign a variable to each quantity. Draw a diagram if needed.
> 2. Write the *given* values of the variables and their rates of change with respect to t.
> 3. Find an equation giving the relationship between the variables.
> 4. Differentiate both sides of this equation implicitly with respect to t.
> 5. Replace the variables and their derivatives by the numerical data found in Step 2, and solve the equation for the required rate of change.

APPLIED EXAMPLE 9 Watching a Rocket Launch   At a distance of 4000 feet from the launch site, a spectator is observing a rocket being launched. If the rocket lifts off vertically and is rising at a speed of 600 feet per

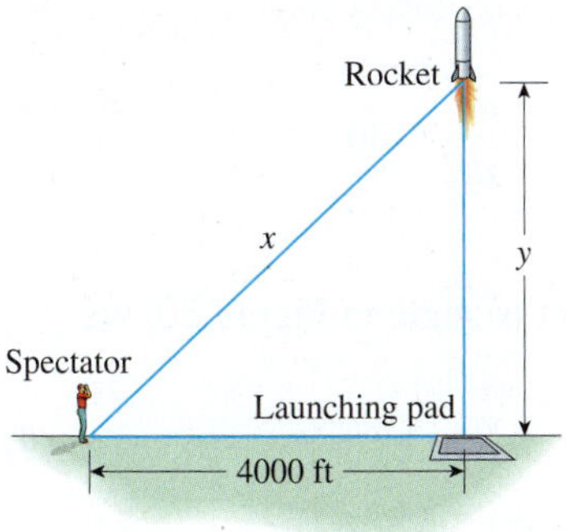

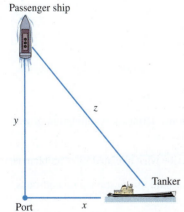

FIGURE 19
The rate at which x is changing with respect to time is related to the rate of change of y with respect to time.

second (ft/sec) when it is at an altitude of 3000 feet, how fast is the distance between the rocket and the spectator changing at that instant?

Solution

Step 1   Let

$$y = \text{altitude of the rocket}$$
$$x = \text{distance between the rocket and the spectator}$$

at any time t (Figure 19).

Step 2   We are given that at a certain instant of time

$$y = 3000 \quad \text{and} \quad \frac{dy}{dt} = 600$$

and are asked to find dx/dt at that instant.

Step 3   Applying the Pythagorean Theorem to the right triangle in Figure 19, we find that

$$x^2 = y^2 + 4000^2$$

Therefore, when $y = 3000$,

$$x = \sqrt{3000^2 + 4000^2} = 5000$$

Step 4   Next, we differentiate the equation $x^2 = y^2 + 4000^2$ with respect to t, obtaining

$$2x\frac{dx}{dt} = 2y\frac{dy}{dt}$$

(Remember, both x and y are functions of t.)

Step 5   Substituting $x = 5000$, $y = 3000$, and $dy/dt = 600$, we find

$$2(5000)\frac{dx}{dt} = 2(3000)(600)$$

$$\frac{dx}{dt} = 360$$

Therefore, the distance between the rocket and the spectator is changing at a rate of 360 ft/sec.

⚠ Be sure that you do *not* replace the variables in the equation found in Step 3 by their numerical values before differentiating the equation.

 APPLIED EXAMPLE 10 Distance Between Two Ships A passenger ship and an oil tanker left port together sometime in the morning; the former headed north, and the latter headed east. At noon, the passenger ship was 40 miles from port and sailing at 30 mph, while the oil tanker was 30 miles from port and sailing at 20 mph. How fast was the distance between the two ships changing at that time?

Solution

Step 1   Let

$$x = \text{distance of the oil tanker from port}$$
$$y = \text{distance of the passenger ship from port}$$
$$z = \text{distance between the two ships}$$

See Figure 20.

FIGURE 20
We want to find dz/dt, the rate at which the distance between the two ships is changing at a certain instant of time.

Step 2   We are given that at noon,

$$x = 30 \qquad y = 40 \qquad \frac{dx}{dt} = 20 \qquad \frac{dy}{dt} = 30$$

and we are required to find dz/dt at that time.

Step 3   Applying the Pythagorean Theorem to the right triangle in Figure 20, we find that

$$z^2 = x^2 + y^2 \tag{9}$$

In particular, when $x = 30$ and $y = 40$, we have

$$z^2 = 30^2 + 40^2 = 2500 \quad \text{or} \quad z = 50$$

Step 4   Differentiating (9) implicitly with respect to t, we obtain

$$2z \frac{dz}{dt} = 2x \frac{dx}{dt} + 2y \frac{dy}{dt}$$

$$z \frac{dz}{dt} = x \frac{dx}{dt} + y \frac{dy}{dt}$$

Step 5   Finally, substituting $x = 30$, $y = 40$, $z = 50$, $dx/dt = 20$, and $dy/dt = 30$ into the last equation, we find

$$50 \frac{dz}{dt} = (30)(20) + (40)(30) \quad \text{and} \quad \frac{dz}{dt} = 36$$

Therefore, at noon on the day in question, the ships are moving apart at the rate of 36 mph.

3.6   Self-Check Exercises

1. Given the equation $x^3 + 3xy + y^3 = 4$, find dy/dx by implicit differentiation.

2. Find an equation of the tangent line to the graph of

$$16x^2 + 9y^2 = 144$$

at the point

$$\left(2, -\frac{4\sqrt{5}}{3}\right)$$

Solutions to Self-Check Exercises 3.6 can be found on page 234.

3.6   Concept Questions

1. **a.** Suppose the equation $F(x, y) = 0$ defines y as a function of x. Explain how implicit differentiation can be used to find dy/dx.
   b. What is the role of the Chain Rule in implicit differentiation?

2. Suppose the equation $xg(y) + yf(x) = 0$, where f and g

are differentiable functions, defines y as a function of x. Find an expression for dy/dx.

3. In your own words, describe what a related-rates problem is.

4. Give the steps that you would use to solve a related-rates problem.

3.6 Exercises

In Exercises 1–8, find the derivative dy/dx (a) by solving each of the implicit equations for y explicitly in terms of x and (b) by differentiating each of the equations implicitly. Show that in each case, the results are equivalent.

1. $x + 2y = 5$

2. $3x + 4y = 6$

3. $xy = 1$

4. $xy - y - 1 = 0$

5. $x^3 - x^2 - xy = 4$

6. $x^2y - x^2 + y - 1 = 0$

7. $\dfrac{x}{y} - x^2 = 1$

8. $\dfrac{y}{x} - 2x^3 = 4$

In Exercises 9–30, find dy/dx by implicit differentiation.

9. $x^2 + y^2 = 16$

10. $2x^2 + y^2 = 16$

11. $x^2 - 2y^2 = 16$

12. $x^3 + y^3 + y - 4 = 0$

13. $x^2 - 2xy = 6$

14. $x^2 + 5xy + y^2 = 10$

15. $x^2y^2 - xy = 8$

16. $x^2y^3 - 2xy^2 = 5$

17. $x^{1/2} + y^{1/2} = 1$

18. $x^{1/3} + y^{1/3} = 1$

19. $\sqrt{x + y} = x$

20. $(2x + 3y)^{1/3} = x^2$

21. $\dfrac{1}{x^2} + \dfrac{1}{y^2} = 1$

22. $\dfrac{1}{x^3} + \dfrac{1}{y^3} = 5$

23. $\sqrt{xy} = x + y$

24. $\sqrt{xy} = 2x + y^2$

25. $\dfrac{x + y}{x - y} = 3x$

26. $\dfrac{x - y}{2x + 3y} = 2x$

27. $xy^{3/2} = x^2 + y^2$

28. $x^2y^{1/2} = x + 2y^3$

29. $(x + y)^3 + x^3 + y^3 = 0$ **30.** $(x + y^2)^{10} = x^2 + 25$

In Exercises 31–34, find an equation of the tangent line to the graph of the function f defined by the equation at the indicated point.

31. $4x^2 + 9y^2 = 36; (0, 2)$

32. $y^2 - x^2 = 16; (2, 2\sqrt{5})$

33. $x^2y^3 - y^2 + xy - 1 = 0; (1, 1)$

34. $(x - y - 1)^3 = x; (1, -1)$

In Exercises 35–38, find the second derivative d^2y/dx^2 of each function defined implicitly by the equation.

35. $xy = 1$

36. $x^3 + y^3 = 28$

37. $y^2 - xy = 8$

38. $x^{1/3} + y^{1/3} = 1$

39. VOLUME OF A CYLINDER The volume of a right-circular cylinder of radius r and height h is $V = \pi r^2 h$. Suppose the radius and height of the cylinder are changing with respect to time t.
  a. Find a relationship between dV/dt, dr/dt, and dh/dt.
  b. At a certain instant of time, the radius and height of the cylinder are 2 and 6 in. and are increasing at the

rate of 0.1 in./sec and 0.3 in./sec, respectively. How fast is the volume of the cylinder increasing?

40. DISTANCE BETWEEN TWO CARS A car leaves an intersection traveling west. Its position 4 sec later is 20 ft from the intersection. At the same time, another car leaves the same intersection heading north so that its position 4 sec later is 28 ft from the intersection. If the speeds of the cars at that instant of time are 9 ft/sec and 11 ft/sec, respectively, find the rate at which the distance between the two cars is changing.

41. EFFECT OF PRICE ON DEMAND FOR TIRES Suppose the quantity demanded weekly of the Super Titan radial tires is related to its unit price by the equation

$$p + x^2 = 144$$

where p is measured in dollars and x is measured in units of a thousand. How fast is the quantity demanded weekly changing when $x = 9$, $p = 63$, and the price per tire is increasing at the rate of $2/week?

42. EFFECT OF PRICE ON SUPPLY OF TIRES Suppose the quantity x of Super Titan radial tires made available each week in the marketplace is related to the unit selling price by the equation

$$p - \frac{1}{2}x^2 = 48$$

where x is measured in units of a thousand and p is in dollars. How fast is the weekly supply of Super Titan radial tires being introduced into the marketplace changing when $x = 6$, $p = 66$, and the price per tire is decreasing at the rate of $3/week?

43. EFFECT OF PRICE ON DEMAND FOR HEADPHONES The demand equation for a certain brand of two-way headphones is

$$100x^2 + 9p^2 = 3600$$

where x represents the number (in thousands) of headphones demanded each week when the unit price is $p. How fast is the quantity demanded increasing when the unit price per headphone is $14 and the price is dropping at the rate of $0.15/headphone/week?
Hint: To find the value of x when $p = 14$, solve the equation $100x^2 + 9p^2 = 3600$ for x when $p = 14$.

44. EFFECT OF PRICE ON SUPPLY OF EGGS Suppose the wholesale price of a certain brand of medium-sized eggs p (in dollars per carton) is related to the weekly supply x (in thousands of cartons) by the equation

$$625p^2 - x^2 = 100$$

If 25,000 cartons of eggs are available at the beginning of a certain week and the price is falling at the rate of 2¢/carton/week, at what rate is the weekly supply falling?
Hint: To find the value of p when $x = 25$, solve the supply equation for p when $x = 25$.

45. **EFFECT OF SUPPLY ON PRICE OF EGGS** Refer to Exercise 44. If 25,000 cartons of eggs are available at the beginning of a certain week and the weekly supply is falling at the rate of 1000 cartons/week, at what rate is the wholesale price changing?

46. **ELASTICITY OF DEMAND FOR INK-JET CARTRIDGES** The demand function for a certain make of ink-jet cartridge is

$$p = -0.01x^2 - 0.1x + 6$$

where p is the unit price in dollars and x is the quantity demanded each week, measured in units of a thousand. Compute the elasticity of demand and determine whether the demand is inelastic, unitary, or elastic when $x = 10$.

47. **ELASTICITY OF DEMAND FOR CDs** The demand function for a certain brand of compact disc is

$$p = -0.01x^2 - 0.2x + 8$$

where p is the wholesale unit price in dollars and x is the quantity demanded each week, measured in units of a thousand. Compute the elasticity of demand and determine whether the demand is inelastic, unitary, or elastic when $x = 15$.

48. **PRODUCTION FUNCTION FOR A FURNITURE COMPANY** The manager of Dixie Furniture Company estimates that the daily output of her factory (in thousands of dollars) Q is given by

$$Q = 5x^{1/4}y^{3/4}$$

where x is the amount spent on labor and y is the amount spent on capital (both measured in thousands of dollars).
   a. Find the daily output of the factory if $16,000 is spent on labor and $81,000 is spent on capital each day.
   b. Suppose that the output of the factory is to be maintained at the level found in part (a). By how much should the amount spent on capital be changed if the amount on labor is increased by $1000? What is the MRTS?

49. **OUTPUT OF A COUNTRY** Suppose that the output Q of a certain country is given by $Q = 20x^{3/5}y^{2/5}$ billion dollars if x billion dollars are spent on labor and y billion dollars are spent on capital.
   a. Find the output of the country if it spends $32 billion on labor and $243 billion on capital.
   b. Suppose that the output of the country is to be maintained at the level found in part (a). By how much should the amount spent on capital be changed if the amount spent on labor is increased by $1 billion? What is the MRTS?

50. **VOLUME OF A CUBE** The volume V of a cube with sides of length x in. is changing with respect to time. At a certain instant of time, the sides of the cube are 5 in. long and increasing at the rate of 0.1 in./sec. How fast is the volume of the cube changing at that instant of time?

51. **OIL SPILLS** In calm waters, the oil spilling from the ruptured hull of a grounded tanker spreads in all directions. Assuming that the area polluted is circular, determine how fast the area is increasing when the radius of the circle is 60 ft and is increasing at the rate of $\frac{1}{2}$ ft/sec?

52. **DISTANCE BETWEEN TWO SHIPS** Two ships leave the same port at noon. Ship A sails north at 15 mph, and Ship B sails east at 12 mph. How fast is the distance between them changing at 1 P.M.?

53. **OIL SPILLS** In calm waters, the oil spilling from the ruptured hull of a grounded tanker spreads in all directions. Assuming that the area polluted is circular, determine how fast the radius of the circle is changing when the area of the circle is 1600π ft^2 and increasing at the rate of 80π ft^2/sec.

54. **DISTANCE BETWEEN TWO CARS** A car leaves an intersection traveling east. Its position t sec later is given by $x = t^2 + t$ ft. At the same time, another car leaves the same intersection heading north, traveling $y = t^2 + 3t$ ft in t sec. Find the rate at which the distance between the two cars will be changing 5 sec later.

55. **DISTANCE BETWEEN TWO CARS** A car leaves an intersection traveling west. Its position 4 sec later is 20 ft from the intersection. At the same time, another car leaves the same intersection heading north so that its position t sec later is $t^2 + 2t$ ft from the intersection. If the speed of the first car 4 sec after leaving the intersection is 9 ft/sec, find the rate at which the distance between the two cars is changing at that instant of time.

56. **WATCHING A HELICOPTER TAKE OFF** At a distance of 50 ft from the pad, a man observes a helicopter taking off from a heliport. If the helicopter lifts off vertically and is rising at a speed of 44 ft/sec when it is at an altitude of 120 ft, how fast is the distance between the helicopter and the man changing at that instant?

57. **WATCHING A ROWING RACE** A spectator watches a rowing race from the edge of a river bank. The lead boat is moving in a straight line that is 120 ft from the river bank. If the boat is moving at a constant speed of 20 ft/sec, how fast is the boat moving away from the spectator when it is 50 ft past her?

58. **DOCKING A BOAT** A boat is pulled toward a dock by means of a rope wound on a drum that is located 4 ft above the bow of the boat. If the rope is being pulled in at the rate of 3 ft/sec, how fast is the boat approaching the dock when it is 25 ft from the dock?

59. **A MELTING SNOWBALL** Assume that a snowball is in the shape of a sphere. If the snowball melts at a rate that is proportional to its surface area, show that its radius decreases at a constant rate.
   Hint: Its volume is $V = (4/3)\pi r^3$, and its surface area is $S = 4\pi r^2$.

60. **BLOWING SOAP BUBBLES** Carlos is blowing air into a soap bubble at the rate of 8 cm³/sec. Assuming that the bubble is spherical, how fast is its radius changing at the instant of time when the radius is 10 cm? How fast is the surface area of the bubble changing at that instant of time?

61. **COAST GUARD PATROL SEARCH MISSION** The pilot of a Coast Guard patrol aircraft on a search mission had just spotted a disabled fishing trawler and decided to go in for a closer look. Flying in a straight line at a constant altitude of 1000 ft and at a steady speed of 264 ft/sec, the aircraft passed directly over the trawler. How fast was the aircraft receding from the trawler when it was 1500 ft from the trawler?

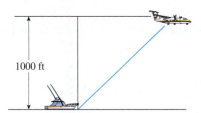

1000 ft

62. **FILLING A COFFEE POT** A coffee pot in the form of a circular cylinder of radius 4 in. is being filled with water flowing at a constant rate. If the water level is rising at the rate of 0.4 in./sec, what is the rate at which water is flowing into the coffee pot?

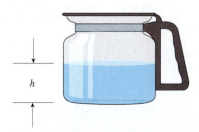

h

63. **MOVEMENT OF A SHADOW** A 6-ft tall man is walking away from a street light 18 ft high at a speed of 6 ft/sec. How fast is the tip of his shadow moving along the ground?

64. **A SLIDING LADDER** A 20-ft ladder leaning against a wall begins to slide. How fast is the top of the ladder sliding down the wall at the instant of time when the bottom of the ladder is 12 ft from the wall and sliding away from the wall at the rate of 5 ft/sec?
Hint: Refer to the accompanying figure. By the Pythagorean Theorem, $x^2 + y^2 = 400$. Find dy/dt when $x = 12$ and $dx/dt = 5$.

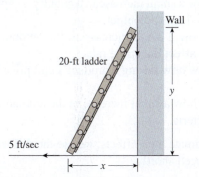

Wall

20-ft ladder

y

5 ft/sec

x

65. **A SLIDING LADDER** The base of a 13-ft ladder leaning against a wall begins to slide away from the wall. At the instant of time when the base is 12 ft from the wall, the base is moving at the rate of 8 ft/sec. How fast is the top of the ladder sliding down the wall at that instant of time?
Hint: Refer to the hint in Problem 64.

66. **FLOW OF WATER FROM A TANK** Water flows from a tank of constant cross-sectional area 50 ft² through an orifice of constant cross-sectional area 1.4 ft² located at the bottom of the tank (see the figure).

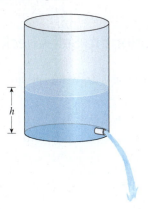

h

Initially, the height of the water in the tank was 20 ft, and its height t sec later is given by the equation

$$2\sqrt{h} + \frac{1}{25}t - 2\sqrt{20} = 0 \qquad (0 \le t \le 50\sqrt{20})$$

How fast was the height of the water decreasing when its height was 8 ft?

67. **VOLUME OF A GAS** In an adiabatic process (one in which no heat transfer takes place), the pressure P and volume V of an ideal gas such as oxygen satisfy the equation $P^5 V^7 = C$, where C is a constant. Suppose that at a certain instant of time, the volume of the gas is 4 L, the pressure is 100 kPa, and the pressure is decreasing at the rate of 5 kPa/sec. Find the rate at which the volume is changing.

68. **MASS OF A MOVING PARTICLE** The mass m of a particle moving at a velocity v is related to its rest mass m_0 by the equation

$$m = \frac{m_0}{\sqrt{1 - \dfrac{v^2}{c^2}}}$$

where c (2.98 × 10⁸ m/sec) is the speed of light. Suppose an electron of mass 9.11 × 10⁻³¹ kg is being accelerated in a particle accelerator. When its velocity is 2.92 × 10⁸ m/sec and its acceleration is 2.42 × 10⁵ m/sec², how fast is the mass of the electron changing?

In Exercises 69–74, determine whether the statement is true or false. If it is true, explain why it is true. If it is false, give an example to show why it is false.

69. The equation $x^2 + y^2 + 1 = 0$ defines y as a function of x.

70. The function

$$f(x) = \begin{cases} \sqrt{1 - x^2} & \text{if } -1 \le x < 0 \\ -\sqrt{1 - x^2} & \text{if } 0 \le x \le 1 \end{cases}$$

may be defined implicitly by the equation $x^2 + y^2 = 1$.

71. If f and g are differentiable and $f(x)g(y) = 0$, then

$$\frac{dy}{dx} = -\frac{f'(x)g(y)}{f(x)g'(y)} \qquad f(x) \ne 0 \text{ and } g'(y) \ne 0$$

72. If f and g are differentiable, $f(x) + g(y) = 0$, and $g'(y) \ne 0$, then

$$\frac{dy}{dx} = -\frac{f'(x)}{g'(y)}$$

73. If f is differentiable, then $f(x + \Delta x) \approx f(x) + f'(x)\,\Delta x$.

74. If h is small in comparison to a, then

$$(a + h)^{1/3} \approx a^{1/3} + \frac{h}{3a^{2/3}}$$

3.6 Solutions to Self-Check Exercises

1. Differentiating both sides of the equation with respect to x, we have

$$3x^2 + 3y + 3xy' + 3y^2y' = 0$$
$$(x^2 + y) + (x + y^2)y' = 0$$
$$(x + y^2)y' = -(x^2 + y)$$
$$y' = -\frac{x^2 + y}{x + y^2}$$

2. To find the slope of the tangent line to the graph of the function at any point, we differentiate the equation implicitly with respect to x, obtaining

$$32x + 18yy' = 0$$
$$y' = -\frac{16x}{9y}$$

In particular, the slope of the tangent line at $\left(2, -\dfrac{4\sqrt{5}}{3}\right)$ is

$$m = -\frac{16(2)}{9\left(-\dfrac{4\sqrt{5}}{3}\right)} = \frac{8}{3\sqrt{5}}$$

Using the point-slope form of the equation of a line, we find

$$y - \left(-\frac{4\sqrt{5}}{3}\right) = \frac{8}{3\sqrt{5}}(x - 2)$$
$$y = \frac{8\sqrt{5}}{15}x - \frac{36\sqrt{5}}{15} = \frac{8\sqrt{5}}{15}x - \frac{12\sqrt{5}}{5}$$

3.7 Differentials

The Millers are planning to buy a house in the near future and estimate that they will need a 30-year fixed-rate mortgage of $240,000. If the interest rate increases from the present rate of 5% per year to 5.4% per year between now and the time the Millers decide to secure the loan, approximately how much more per month will their mortgage be? (You will be asked to answer this question in Exercise 46, page 242.)

Questions such as this, in which one wishes to *estimate* the change in the dependent variable (monthly mortgage payment) corresponding to a small change in the independent variable (interest rate per year), occur in many real-life applications. For example:

- An economist would like to know how a small increase in a country's capital expenditure will affect the country's gross domestic output.
- A sociologist would like to know how a small increase in the amount of capital investment in a housing project will affect the crime rate.
- A businesswoman would like to know how raising a product's unit price by a small amount will affect her profit.
- A bacteriologist would like to know how a small increase in the amount of a bactericide will affect a population of bacteria.

To calculate these changes and estimate their effects, we use the **differential** of a function, a concept that will be introduced shortly.

Increments

Let x denote a variable quantity and suppose x changes from x_1 to x_2. This change in x is called the **increment in x** and is denoted by the symbol Δx (read "delta x"). Thus,

$$\Delta x = x_2 - x_1 \qquad \text{Final value} - \text{initial value} \tag{10}$$

EXAMPLE 1 Find the increment in x as x changes (a) from 3 to 3.2 and (b) from 3 to 2.7.

Solution

a. Here, $x_1 = 3$ and $x_2 = 3.2$, so

$$\Delta x = x_2 - x_1 = 3.2 - 3 = 0.2$$

b. Here, $x_1 = 3$ and $x_2 = 2.7$. Therefore,

$$\Delta x = x_2 - x_1 = 2.7 - 3 = -0.3$$

Observe that Δx plays the same role that h played in Section 2.4.

Now, suppose two quantities, x and y, are related by an equation $y = f(x)$, where f is a function. If x changes from x to $x + \Delta x$, then the corresponding change in y is called the **increment in y.** It is denoted by Δy and is defined by

$$\Delta y = f(x + \Delta x) - f(x) \tag{11}$$

(see Figure 21).

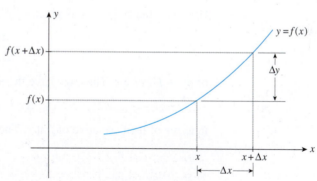

FIGURE 21
An increment of Δx in x induces an increment of $\Delta y = f(x + \Delta x) - f(x)$ in y.

EXAMPLE 2 Let $y = x^3$. Find Δx and Δy when x changes (a) from 2 to 2.01 and (b) from 2 to 1.98.

Solution Let $f(x) = x^3$.

a. Here, $\Delta x = 2.01 - 2 = 0.01$. Next,

$$\begin{aligned}
\Delta y &= f(x + \Delta x) - f(x) = f(2.01) - f(2) \\
&= (2.01)^3 - 2^3 = 8.120601 - 8 = 0.120601
\end{aligned}$$

b. Here, $\Delta x = 1.98 - 2 = -0.02$. Next,

$$\begin{aligned}
\Delta y &= f(x + \Delta x) - f(x) = f(1.98) - f(2) \\
&= (1.98)^3 - 2^3 = 7.762392 - 8 = -0.237608
\end{aligned}$$

Differentials

We can obtain a relatively quick and simple way of approximating Δy, the change in y due to a small change Δx, by examining the graph of the function f shown in Figure 22.

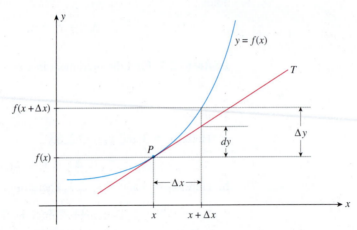

FIGURE 22
If Δx is small, dy is a good approximation of Δy.

Observe that near the point of tangency P, the tangent line T is close to the graph of f. Therefore, if Δx is small, then dy is a good approximation of Δy. We can find an expression for dy as follows. Notice that the slope of T is given by

$$\frac{dy}{\Delta x} \qquad \text{Rise} \div \text{run}$$

However, the slope of T is given by $f'(x)$. Therefore, we have

$$\frac{dy}{\Delta x} = f'(x)$$

or $dy = f'(x)\,\Delta x$. Thus, we have the approximation

$$\Delta y \approx dy = f'(x)\,\Delta x$$

in terms of the derivative of f at x. The quantity dy is called the *differential of y*.

The Differential

Let $y = f(x)$ define a differentiable function of x. Then,

1. The **differential dx** of the independent variable x is $dx = \Delta x$.
2. The **differential dy** of the dependent variable y is

$$dy = f'(x)\,\Delta x = f'(x)\,dx \qquad\qquad (12)$$

Notes

1. For the independent variable x: There is no difference between Δx and dx—both measure the change in x from x to $x + \Delta x$.
2. For the dependent variable y: Δy measures the *actual* change in y as x changes from x to $x + \Delta x$, whereas dy measures the *approximate* change in y corresponding to the same change in x.
3. The differential dy depends on both x and dx, but for fixed x, dy is a linear function of dx.

EXAMPLE 3 Let $y = x^3$.

a. Find the differential dy of y.
b. Use dy to approximate Δy when x changes from 2 to 2.01.
c. Use dy to approximate Δy when x changes from 2 to 1.98.
d. Compare the results of part (b) with those of Example 2.

Solution

a. Let $f(x) = x^3$. Then,

$$dy = f'(x)\,dx = 3x^2\,dx$$

b. Here, $x = 2$ and $dx = 2.01 - 2 = 0.01$. Therefore,

$$dy = 3x^2\,dx = 3(2)^2(0.01) = 0.12$$

c. Here, $x = 2$ and $dx = 1.98 - 2 = -0.02$. Therefore,

$$dy = 3x^2\,dx = 3(2)^2(-0.02) = -0.24$$

d. As you can see, both approximations 0.12 and -0.24 are quite close to the actual changes Δy obtained in Example 2: 0.120601 and -0.237608.

Observe how much easier it is to find an approximation to the exact change in a function with the help of the differential, rather than calculating the exact change in the function itself. In the following examples, we take advantage of this fact.

EXAMPLE 4 Approximate the value of $\sqrt{26.5}$ using differentials. Verify your result using the $\boxed{\sqrt{}}$ key on your calculator.

Solution Since we want to compute the square root of a number, let's consider the function $y = f(x) = \sqrt{x}$. Since 25 is the number nearest 26.5 whose square root is readily recognized, let's take $x = 25$. We want to know the change in y, Δy, as x changes from $x = 25$ to $x = 26.5$, an increase of $\Delta x = 1.5$ units. Using Equation (12), we find

$$\Delta y \approx dy = f'(x)\,\Delta x$$

$$= \left(\left. \frac{1}{2\sqrt{x}} \right|_{x=25} \right) \cdot (1.5) = \left(\frac{1}{10} \right)(1.5) = 0.15$$

Therefore,

$$\sqrt{26.5} - \sqrt{25} = \Delta y \approx 0.15$$
$$\sqrt{26.5} \approx \sqrt{25} + 0.15 = 5.15$$

The exact value of $\sqrt{26.5}$, rounded off to five decimal places, is 5.14782. Thus, the error incurred in the approximation is 0.00218.

APPLIED EXAMPLE 5 The Effect of Speed on Vehicular Operating Cost The total cost incurred in operating a certain type of truck on a 500-mile trip, traveling at an average speed of v mph, is estimated to be

$$C(v) = 125 + v + \frac{4500}{v}$$

dollars. Find the approximate change in the total operating cost when the average speed is increased from 55 mph to 58 mph.

Solution   With $v = 55$ and $\Delta v = dv = 3$, we find

$$\Delta C \approx dC = C'(v)\, dv = \left(1 - \frac{4500}{v^2}\right)\Bigg|_{v=55} \cdot (3)$$

$$= \left(1 - \frac{4500}{3025}\right)(3) \approx -1.46$$

so the total operating cost is found to decrease by $1.46. This might explain why so many independent truckers often exceed the speed limit where it is 55 mph.

 APPLIED EXAMPLE 6 The Effect of Advertising on Sales The relationship between the amount of money x spent by Cannon Precision Instruments on advertising and Cannon's total sales $S(x)$ is given by the function

$$S(x) = -0.002x^3 + 0.6x^2 + x + 500 \qquad (0 \le x \le 200)$$

where x is measured in thousands of dollars. Use differentials to estimate the change in Cannon's total sales if advertising expenditures are increased from $100,000 ($x = 100$) to $105,000 ($x = 105$).

Solution   The required change in sales is given by

$$\Delta S \approx dS = S'(100)\, dx$$

$$= -0.006x^2 + 1.2x + 1\,\Bigg|_{x=100} \cdot (5) \qquad {\color{red} dx = 105 - 100 = 5}$$

$$= (-60 + 120 + 1)(5) = 305$$

that is, an increase of $305,000.

APPLIED EXAMPLE 7 The Rings of Neptune

a. A ring has an inner radius of r units and an outer radius of R units, where $(R - r)$ is small in comparison to r (Figure 23a). Use differentials to estimate the area of the ring.

b. Recent observations, including those of *Voyager I* and *II*, showed that Neptune's ring system is considerably more complex than had been believed. For one thing, it is made up of a large number of distinguishable rings rather than one continuous great ring as was previously thought (Figure 23b). The outermost ring, 1989N1R, has an inner radius of approximately 62,900 kilometers (measured from the center of the planet) and a radial width of approximately 50 kilometers. Using these data, estimate the area of the ring.

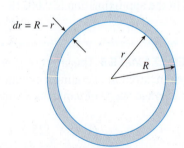

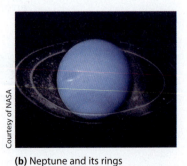

(a) The area of the ring is approximately equal to the circumference of the inner circle times the thickness.

(b) Neptune and its rings

Courtesy of NASA

FIGURE 23

Solution

a. Since the area of a circle of radius x is $A = f(x) = \pi x^2$, we find

$$\pi R^2 - \pi r^2 = f(R) - f(r)$$
$$= \Delta A$$
$$\approx dA$$
$$= f'(r)\, dr$$

Remember, ΔA = change in f when x changes from $x = r$ to $x = R$.

where $dr = R - r$. So we see that the area of the ring is approximately $2\pi r(R - r)$ square units. In words, the area of the ring is approximately equal to

Circumference of the inner circle $\times$ Thickness of the ring

b. Applying the results of part (a) with $r = 62{,}900$ and $dr = 50$, we find that the area of the ring is approximately $2\pi(62{,}900)(50)$, or $19{,}760{,}000$ square kilometers, which is roughly 4% of the earth's surface.

Before looking at the next example, we need to familiarize ourselves with some terminology. If a quantity with exact value q is measured or calculated with an error of Δq, then the quantity $\Delta q/q$ is called the *relative error* in the measurement or calculation of q. If the quantity $\Delta q/q$ is expressed as a percentage, it is then called the *percentage error*. Because Δq is approximated by dq, we normally approximate the relative error $\Delta q/q$ by dq/q.

 APPLIED EXAMPLE 8 Estimating Errors in Measurement Suppose the radius of a ball bearing is measured to be 0.5 inch, with a maximum error of ±0.0002 inch. Then, the relative error in r is

$$\frac{dr}{r} = \frac{\pm0.0002}{0.5} = \pm0.0004 \qquad \text{Relative error}$$

and the percentage error is $\pm0.04\%$.

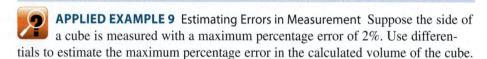 **APPLIED EXAMPLE 9** Estimating Errors in Measurement Suppose the side of a cube is measured with a maximum percentage error of 2%. Use differentials to estimate the maximum percentage error in the calculated volume of the cube.

Solution Suppose the side of the cube is x, so that its volume is

$$V = x^3$$

We are given that $\left|\dfrac{dx}{x}\right| \le 0.02$. Now,

$$dV = 3x^2\, dx$$

and so

$$\frac{dV}{V} = \frac{3x^2\, dx}{x^3} = 3\,\frac{dx}{x}$$

Therefore,

$$\left|\frac{dV}{V}\right| = 3\left|\frac{dx}{x}\right| \le 3(0.02) = 0.06$$

and we see that the maximum percentage error in the measurement of the volume of the cube is approximately 6%.

Finally, if at some point in reading this section you have a sense of déjà vu, do not be surprised, because the notion of the differential was first used in Section 3.4 (see Example 1). There, we took $\Delta x = 1$, since we were interested in finding the marginal cost when the level of production was increased from $x = 250$ to $x = 251$. If we had used differentials, we would have found

$$C(251) - C(250) \approx C'(250)\,dx$$

so taking $dx = \Delta x = 1$, we have $C(251) - C(250) \approx C'(250)$, which agrees with the result obtained in Example 1. Thus, in Section 3.4, we touched upon the notion of the differential, albeit in the special case in which $dx = 1$.

3.7 Self-Check Exercises

1. Find the differential of $f(x) = \sqrt{x} + 1$.

2. FORECASTING CORN PRICES A certain country's government economists have determined that the demand equation for corn in that country is given by

$$p = f(x) = \frac{125}{x^2 + 1}$$

where p is expressed in dollars per bushel and x, the quantity demanded each year, is measured in billions of

bushels. The economists are forecasting a harvest of 6 billion bushels for the year. If the actual production of corn were 6.2 billion bushels for the year instead, what would be the approximate drop in the predicted price of corn per bushel?

Solutions to Self-Check Exercises 3.7 can be found on page 243.

3.7 Concept Questions

1. If $y = f(x)$, what is the differential of x? Write an expression for the differential dy.

2. Refer to the following figure.

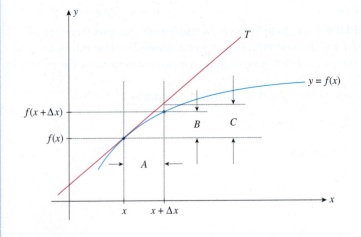

    a. Express A, B, and C in terms of Δx, Δy, and dy.
    b. Express $f'(x)$ (the slope of the tangent line T at x) in terms of Δx and dy.
    c. Observing that $B \approx C$, explain why $\Delta y \approx f'(x)\,\Delta x = f'(x)\,dx$.

3. Suppose the population of a city at time t_0 is $P(t_0)$. If $\Delta t = t - t_0$, what quantity does the expression $P'(t_0)\,\Delta t$ approximate?

4. Suppose that the total profit of a company at time t_0 is $P(t_0)$ dollars and the profit is growing at the rate of $P'(t_0)$ dollars per month at that instant of time. Find an expression giving the approximate profit of the company at time $t = t_0 + \Delta t$, where Δt is a small increment in t.

3.7 Exercises

In Exercises 1–14, find the differential of the function.

1. $f(x) = 2x^2$

2. $f(x) = 3x^2 + 1$

3. $f(x) = x^3 - x$

4. $f(x) = 2x^3 + x$

5. $f(x) = \sqrt{x + 1}$

6. $f(x) = \dfrac{3}{\sqrt{x}}$

7. $f(x) = 2x^{3/2} + x^{1/2}$

8. $f(x) = 3x^{5/6} + 7x^{2/3}$

9. $f(x) = x + \dfrac{2}{x}$

10. $f(x) = \dfrac{3}{x-1}$

11. $f(x) = \dfrac{x-1}{x^2+1}$

12. $f(x) = \dfrac{2x^2+1}{x+1}$

13. $f(x) = \sqrt{3x^2 - x}$

14. $f(x) = (2x^2 + 3)^{1/3}$

15. Let f be the function defined by
$$y = f(x) = x^2 - 1$$

   a. Find the differential of f.
   b. Use your result from part (a) to find the approximate change in y if x changes from 1 to 1.02.
   c. Find the actual change in y if x changes from 1 to 1.02, and compare your result with that obtained in part (b).

16. Let f be the function defined by
$$y = f(x) = 3x^2 - 2x + 6$$

   a. Find the differential of f.
   b. Use your result from part (a) to find the approximate change in y if x changes from 2 to 1.97.
   c. Find the actual change in y if x changes from 2 to 1.97, and compare your result with that obtained in part (b).

17. Let f be the function defined by
$$y = f(x) = \dfrac{1}{x}$$

   a. Find the differential of f.
   b. Use your result from part (a) to find the approximate change in y if x changes from -1 to -0.95.
   c. Find the actual change in y if x changes from -1 to -0.95, and compare your result with that obtained in part (b).

18. Let f be the function defined by
$$y = f(x) = \sqrt{2x + 1}$$

   a. Find the differential of f.
   b. Use your result from part (a) to find the approximate change in y if x changes from 4 to 4.1.
   c. Find the actual change in y if x changes from 4 to 4.1, and compare your result with that obtained in part (b).

In Exercises 19–26, use differentials to approximate the quantity.

19. $\sqrt{10}$

20. $\sqrt{17}$

21. $\sqrt{49.5}$

22. $\sqrt{99.7}$

23. $\sqrt[3]{7.8}$

24. $\sqrt[4]{81.6}$

25. $\sqrt{0.089}$

26. $\sqrt[3]{0.00096}$

27. Use a differential to approximate $\sqrt{4.02} + \dfrac{1}{\sqrt{4.02}}$.

   Hint: Let $f(x) = \sqrt{x} + \dfrac{1}{\sqrt{x}}$ and compute dy with $x = 4$ and $dx = 0.02$.

28. Use a differential to approximate $\dfrac{2(4.98)}{(4.98)^2 + 1}$.

   Hint: Study the hint for Exercise 27.

29. **ERROR ESTIMATION** The length of each edge of a cube is 12 cm, with a possible error in measurement of 0.02 cm. Use differentials to estimate the error that might occur when the volume of the cube is calculated.

30. **ESTIMATING THE AMOUNT OF PAINT REQUIRED** A coat of paint of thickness 0.05 cm is to be applied uniformly to the faces of a cube of edge 30 cm. Use differentials to find the approximate amount of paint required for the job.

31. **ESTIMATING THE AMOUNT OF RUST-PROOFER REQUIRED** A hemisphere-shaped dome of radius 60 ft is to be coated with a layer of rust-proofer before painting. Use differentials to estimate the amount of rust-proofer needed if the coat is to be 0.01 in. thick.

   Hint: The volume of a hemisphere of radius r is $V = \frac{2}{3}\pi r^3$.

32. **GROWTH OF A CANCEROUS TUMOR** The volume of a spherical cancerous tumor is given by
$$V(r) = \dfrac{4}{3}\pi r^3$$

   If the radius of a tumor is estimated at 1.1 cm, with a maximum error in measurement of 0.005 cm, determine the error that might occur when the volume of the tumor is calculated.

33. **UNCLOGGING ARTERIES** Research done in the 1930s by the French physiologist Jean Poiseuille showed that the resistance R of a blood vessel of length l and radius r is $R = kl/r^4$, where k is a constant. Suppose a dose of the drug TPA increases r by 10%. How will this affect the resistance R? Assume that l is constant.

34. **EFFECT OF CAPITAL EXPENDITURES ON GDP** An economist has determined that a certain country's gross domestic product (GDP) is approximated by the function $f(x) = 640x^{1/5}$, where $f(x)$ is measured in billions of dollars and x is the capital outlay in billions of dollars. Use differentials to estimate the change in the country's GDP if the country's capital expenditure changes from $243 billion to $248 billion.

35. **LEARNING CURVES** The length of time (in seconds) a certain individual takes to learn a list of n items is approximated by
$$f(n) = 4n\sqrt{n - 4}$$

   Use differentials to approximate the additional time it takes the individual to learn the items on a list when n is increased from 85 to 90 items.

36. **EFFECT OF ADVERTISING ON PROFITS** The relationship between Cunningham Realty's quarterly profits, $P(x)$, and the amount of money x spent on advertising per quarter is described by the function
$$P(x) = -\dfrac{1}{8}x^2 + 7x + 30 \qquad (0 \le x \le 50)$$

where both $P(x)$ and x are measured in thousands of dollars. Use differentials to estimate the increase in profits when advertising expenditure each quarter is increased from \$24,000 to \$26,000.

37. **EFFECT OF MORTGAGE RATES ON HOUSING STARTS** A study prepared for the National Association of Realtors estimates that the number of housing starts per year over the next 5 years will be

$$N(r) = \frac{7}{1 + 0.02r^2}$$

million units, where r (percent) is the mortgage rate. Use differentials to estimate the decrease in the number of annual housing starts when the mortgage rate is increased from 6% to 6.5%.

38. **EFFECT OF SUPPLY OF RADIOS ON PRICE** The supply equation for a certain brand of radio is given by

$$p = s(x) = 0.3\sqrt{x} + 10$$

where x is the quantity supplied and p is the unit price in dollars. Use differentials to approximate the change in price when the quantity supplied is increased from 10,000 units to 10,500 units.

39. **EFFECT OF DEMAND FOR SMOKE ALARMS ON PRICE** The demand function for the Sentinel smoke alarm is given by

$$p = d(x) = \frac{30}{0.02x^2 + 1}$$

where x is the quantity demanded (in units of a thousand) and p is the unit price in dollars. Use differentials to estimate the change in the price p when the quantity demanded changes from 5000 to 5500 units per week.

40. **SURFACE AREA OF AN ANIMAL** Animal physiologists use the formula

$$S = kW^{2/3}$$

to calculate an animal's surface area (in square meters) from its weight W (in kilograms), where k is a constant that depends on the animal under consideration. Suppose a physiologist calculates the surface area of a horse ($k = 0.1$). If the horse's weight is estimated at 300 kg, with a maximum error in measurement of 0.6 kg, determine the percentage error in the calculation of the horse's surface area.

41. **FORECASTING PROFITS** The management of Trappee and Sons forecast that they will sell 200,000 cases of their TexaPep hot sauce next year. Their annual profit is described by

$$P(x) = -0.000032x^3 + 6x - 100$$

thousand dollars, where x is measured in thousands of cases. If the maximum error in the forecast is 15%, determine the corresponding error in Trappee's profits.

42. **FORECASTING COMMODITY PRICES** A certain country's government economists have determined that the demand equation for soybeans in that country is given by

$$p = f(x) = \frac{55}{2x^2 + 1}$$

where p is expressed in dollars per bushel and x, the quantity demanded each year, is measured in billions of bushels. The economists are forecasting a harvest of 1.8 billion bushels for the year, with a maximum error of 15% in their forecast. Determine the corresponding maximum error in the predicted price per bushel of soybeans.

43. **DEMAND FOR TIRES** The management of Titan Tire Company has determined that the quantity demanded x of the Super Titan tires per week is related to the unit price p by the equation

$$x = \sqrt{144 - p} \qquad (0 \le p \le 144)$$

where p is measured in dollars and x is measured in units of a thousand. Use differentials to find the approximate change in the quantity of the tires demanded per week if the unit price of the tires is increased from \$108 per tire to \$110 per tire.

44. **AIR POLLUTION** The amount of nitrogen dioxide, a brown gas that impairs breathing, present in the atmosphere on a certain May day in the city of Long Beach is approximated by

$$A(t) = \frac{136}{1 + 0.25(t - 4.5)^2} + 28 \qquad (0 \le t \le 11)$$

where $A(t)$ is measured in pollutant standard index (PSI) and t is measured in hours, with $t = 0$ corresponding to 7 A.M. Use differentials to find the approximate change in the PSI from 8 A.M. to 8:05 A.M.

45. **EFFECT OF COMMUNITY INVESTMENT ON CRIME** A sociologist has found that the number of serious crimes in a certain city each year is described by the function

$$N(x) = \frac{500(400 + 20x)^{1/2}}{(5 + 0.2x)^2}$$

where x (in cents per dollar deposited) is the level of reinvestment in the area in conventional mortgages by the city's ten largest banks. Use differentials to estimate the change in the number of crimes if the level of reinvestment changes from 20¢/dollar deposited to 22¢/dollar deposited.

46. **FINANCING A HOME** The Millers are planning to buy a home in the near future and estimate that they will need a 30-year fixed-rate mortgage for \$240,000. Their monthly payment P (in dollars) can be computed by using the formula

$$P = \frac{20,000r}{1 - \left(1 + \frac{r}{12}\right)^{-360}}$$

where r is the interest rate per year.

a. Find the differential of P.

b. If the interest rate increases from the present rate of 5%/year to 5.2%/year between now and the time the Millers decide to secure the loan, approximately how much more will their monthly mortgage payment be? How much more will it be if the interest rate increases to 5.3%/year? To 5.4%/year? To 5.5%/year?

47. **INVESTMENTS** Lupé deposits a sum of $10,000 into an account that pays interest at the rate of r/year compounded monthly. Her investment at the end of 10 years is given by

$$A = 10,000\left(1 + \frac{r}{12}\right)^{120}$$

a. Find the differential of A.

b. Approximately how much more would Lupé's account be worth at the end of the term if her account paid 3.1%/year instead of 3%/year? 3.2%/year instead of 3%/year? 3.3%/year instead of 3%/year?

48. **DEFINED BENEFIT ACCOUNTS** Ian, who is self-employed, contributes $2000 a month into a defined benefit retirement account earning interest at the rate of r/year compounded monthly. At the end of 25 years, his account will be worth

$$S = \frac{24,000\left[\left(1 + \frac{r}{12}\right)^{300} - 1\right]}{r}$$

dollars.

a. Find the differential of S.

b. Approximately how much more would Ian's account be worth at the end of 25 years if his account earned 4.1%/year instead of 4%/year? 4.2%/year instead of 4%/year? 4.3%/year instead of 4%/year?

In Exercises 49 and 50, determine whether the statement is true or false. If it is true, explain why it is true. If it is false, give an example to show why it is false.

49. If $y = ax + b$ where a and b are constants, then $\Delta y = dy$.

50. If $A = f(x)$, then the percentage change in A is approximately

$$\frac{100f'(x)}{f(x)}\,dx$$

3.7 Solutions to Self-Check Exercises

1. We find

$$f'(x) = \frac{1}{2}x^{-1/2} = \frac{1}{2\sqrt{x}}$$

Therefore, the required differential of f is

$$dy = \frac{1}{2\sqrt{x}}\,dx$$

2. We first compute the differential

$$dp = -\frac{250x}{(x^2 + 1)^2}\,dx$$

Next, using Equation (12) with $x = 6$ and $dx = 0.2$, we find

$$\Delta p \approx dp = -\frac{250(6)}{(36 + 1)^2}(0.2) \approx -0.22$$

or a drop in price of 22¢/bushel.

USING TECHNOLOGY   Finding the Differential of a Function

The calculation of the differential of f at a given value of x involves the evaluation of the derivative of f at that point and can be facilitated through the use of the numerical derivative function.

EXAMPLE 1 Use dy to approximate Δy if $y = x^2(2x^2 + x + 1)^{2/3}$ and x changes from 2 to 1.98.

Solution   Let $f(x) = x^2(2x^2 + x + 1)^{2/3}$. Since $dx = 1.98 - 2 = -0.02$, we find the required approximation to be

$$dy = f'(2)(-0.02)$$

But using the numerical derivative operation, we find

$$f'(2) = 30.57581679$$

(see Figure T1). Thus,

$$dy = (-0.02)(30.57581679) = -0.6115163358$$

```
nDeriv(X^2(2X^2+
X+1)^(2/3),X,2)
        30.57581679
```

FIGURE **T1**
The TI-83/84 numerical derivative
screen for computing $f'(2)$

APPLIED EXAMPLE 2 Financing a Home The Meyers are considering the purchase of a house in the near future and estimate that they will need a loan of $240,000. Based on a 30-year conventional mortgage with an annual interest rate of r, their monthly repayment will be

$$P = \frac{20,000r}{1 - \left(1 + \dfrac{r}{12}\right)^{-360}}$$

dollars. If the interest rate increases from 5% per year to 5.2% per year between now and the time the Meyers decide to secure the loan, approximately how much more will their monthly mortgage payment be?

Solution Let's write

$$P = f(r) = \frac{20,000r}{1 - \left(1 + \dfrac{r}{12}\right)^{-360}}$$

Then the increase in the mortgage payment will be approximately

$$dP = f'(0.05)\ dr = f'(0.05)(0.002) \qquad \text{Since } dr = 0.052 - 0.05$$
$$\approx 29.34 \qquad\qquad\qquad\qquad \text{Use the numerical derivative operation.}$$

or approximately $29.34 per month. (See Figure T2.)

```
nDeriv((20000X)/
(1-(1+X/12)^-360
),X,.05)
        14667.66333
```

FIGURE **T2**
The TI-83/84 numerical derivative
screen for computing $f'(0.05)$

TECHNOLOGY EXERCISES

In Exercises 1–6, use dy to approximate Δy for the function $y = f(x)$ when x changes from $x = a$ to $x = b$.

1. $f(x) = 0.21x^7 - 3.22x^4 + 5.43x^2 + 1.42x + 12.42$;
$a = 3, b = 3.01$

2. $f(x) = \dfrac{0.2x^2 + 3.1}{1.2x + 1.3}; a = 2, b = 1.96$

3. $f(x) = \sqrt{2.2x^2 + 1.3x + 4}; a = 1, b = 1.03$

4. $f(x) = x\sqrt{2x^3 - x + 4};\ a = 2,\ b = 1.98$

5. $f(x) = \dfrac{\sqrt{x^2 + 4}}{x - 1};\ a = 4,\ b = 4.1$

6. $f(x) = 2.1x^2 + \dfrac{3}{\sqrt{x}} + 5;\ a = 3,\ b = 2.95$

7. CALCULATING MORTGAGE PAYMENTS Refer to Example 2. How much more will the Meyers's mortgage payment be each month if the interest rate increases from 5% to 5.3%/year? To 5.4%/year? To 5.5%/year?

8. ESTIMATING THE AREA OF A RING OF NEPTUNE The ring 1989N2R of the planet Neptune has an inner radius of approximately 53,200 km (measured from the center of the planet) and a radial width of 15 km. Use differentials to estimate the area of the ring.

9. EFFECT OF PRICE ON DEMAND FOR WATCHES The quantity demanded each week of the Alpha Sports Watch, x (in thousands), is related to its unit price of p dollars by the equation

$$x = f(p) = 10\sqrt{\dfrac{50 - p}{p}} \qquad (0 \le p \le 50)$$

Use differentials to find the decrease in the quantity of the watches demanded each week if the unit price is increased from $40 to $42.

10. PERIOD OF A COMMUNICATIONS SATELLITE According to Kepler's Third Law of Planetary Motion, the period T (in days) of a satellite moving in a circular orbit d mi above the surface of the earth is given by

$$T = 0.0588\left(1 + \dfrac{d}{3959}\right)^{3/2}$$

Suppose a communications satellite that was moving in a circular orbit 22,000 mi above the earth's surface at one time has, because of friction, dropped down to a new orbit that is 21,500 mi above the earth's surface. Estimate the decrease in the period of the satellite to the nearest $\frac{1}{100}$th hr.

CHAPTER 3   Summary of Principal Formulas and Terms

FORMULAS

1. Derivative of a constant	$\dfrac{d}{dx}(c) = 0 \qquad (c,\ \text{a constant})$
2. Power Rule	$\dfrac{d}{dx}(x^n) = nx^{n-1}$
3. Constant Multiple Rule	$\dfrac{d}{dx}[cf(x)] = cf'(x)$
4. Sum Rule	$\dfrac{d}{dx}[f(x) \pm g(x)] = f'(x) \pm g'(x)$
5. Product Rule	$\dfrac{d}{dx}[f(x)g(x)] = f(x)g'(x) + g(x)f'(x)$
6. Quotient Rule	$\dfrac{d}{dx}\left[\dfrac{f(x)}{g(x)}\right] = \dfrac{g(x)f'(x) - f(x)g'(x)}{[g(x)]^2}$
7. Chain Rule	$\dfrac{d}{dx}g(f(x)) = g'(f(x))f'(x)$
8. General Power Rule	$\dfrac{d}{dx}[f(x)]^n = n[f(x)]^{n-1}f'(x)$
9. Average cost function	$\bar{C}(x) = \dfrac{C(x)}{x}$

10. Revenue function	$R(x) = px$
11. Profit function	$P(x) = R(x) - C(x)$
12. Elasticity of demand	$E(p) = -\dfrac{pf'(p)}{f(p)}$
13. Differential of y	$dy = f'(x)\,dx$

TERMS

marginal cost (200)

marginal cost function (200)

average cost (201)

marginal average cost function (201)

marginal revenue (203)

marginal revenue function (203)

marginal profit function (204)

relative rate of change (205)

elasticity of demand (206)

elastic demand (207)

unitary demand (207)

inelastic demand (207)

second derivative of f (214)

implicit differentiation (222)

marginal rate of technical substitution (226)

related rates (227)

differential (234)

CHAPTER 3 Concept Review Questions

Fill in the blanks.

1. a. If c is a constant, then $\dfrac{d}{dx}(c) =$ _____.

  b. The Power Rule states that if n is any real number, then $\dfrac{d}{dx}(x^n) =$ _____.

  c. The Constant Multiple Rule states that if c is a constant, then $\dfrac{d}{dx}[cf(x)] =$ _____.

  d. The Sum Rule states that $\dfrac{d}{dx}[f(x) \pm g(x)] =$ _____.

2. a. The Product Rule states that $\dfrac{d}{dx}[f(x)g(x)] =$ _____.

  b. The Quotient Rule states that $\dfrac{d}{dx}[f(x)/g(x)] =$ _____.

3. a. The Chain Rule states that if $h(x) = g[f(x)]$, then $h'(x) =$ _____.

  b. The General Power Rule states that if $h(x) = [f(x)]^n$, then $h'(x) =$ _____.

4. If $C, R, P,$ and $\overline{C}$ denote the total cost function, the total revenue function, the profit function, and the average cost function, respectively, then C' denotes the _____ _____ function, R' denotes the _____ _____ function, P' denotes the _____ _____ function, and $\overline{C}'$ denotes the _____ _____ _____ function.

5. a. If f is a differentiable demand function defined by $x = f(p)$, then the elasticity of demand at price p is given by $E(p) =$ _____.

  b. The demand is _____ if $E(p) > 1$; it is _____ if $E(p) = 1$; it is _____ if $E(p) < 1$.

6. Suppose a function $y = f(x)$ is defined implicitly by an equation in x and y. To find $\dfrac{dy}{dx}$, we differentiate _____ _____ of the equation with respect to x and then solve the resulting equation for $\dfrac{dy}{dx}$. The derivative of a term involving y includes _____ as a factor.

7. In a related-rates problem, we are given a relationship between x and _____ that depends on a third variable t. Knowing the values of $x, y,$ and $\dfrac{dx}{dt}$ at a, we want to find _____ at _____.

8. Let $y = f(t)$ and $x = g(t)$. If $x^2 + y^2 = 4$, then $\dfrac{dx}{dt} =$ _____. If $xy = 1$, then $\dfrac{dy}{dt} =$ _____.

9. a. If a variable quantity x changes from x_1 to x_2, then the increment in x is $\Delta x =$ _____.

  b. If $y = f(x)$ and x changes from x to $x + \Delta x$, then the increment in y is $\Delta y =$ _____.

10. If $y = f(x)$, where f is a differentiable function, then the differential dx of x is $dx =$ _____, where _____ is an increment in _____, and the differential dy of y is $dy =$ _____.

Review Exercises

In Exercises 1–30, find the derivative of the function.

1. $f(x) = 3x^5 - 2x^4 + 3x^2 - 2x + 1$

2. $f(x) = 4x^6 + 2x^4 + 3x^2 - 2$

3. $g(x) = -2x^{-3} + 3x^{-1} + 2$

4. $f(t) = 2t^2 - 3t^3 - t^{-1/2}$

5. $g(t) = 2t^{-1/2} + 4t^{-3/2} + 2$

6. $h(x) = x^2 + \dfrac{2}{x}$ **7.** $f(t) = t + \dfrac{2}{t} + \dfrac{3}{t^2}$

8. $g(s) = 2s^2 - \dfrac{4}{s} + \dfrac{2}{\sqrt{s}}$ **9.** $h(x) = x^2 - \dfrac{2}{x^{3/2}}$

10. $f(x) = \dfrac{x+1}{2x-1}$ **11.** $g(t) = \dfrac{t^2}{2t^2+1}$

12. $h(t) = \dfrac{\sqrt{t}}{\sqrt{t}+1}$ **13.** $f(x) = \dfrac{\sqrt{x}-1}{\sqrt{x}+1}$

14. $f(t) = \dfrac{t}{2t^2+1}$ **15.** $f(x) = \dfrac{x^2(x^2+1)}{x^2-1}$

16. $f(x) = (2x^2 + x)^3$ **17.** $f(x) = (3x^3 - 2)^8$

18. $h(x) = (\sqrt{x} + 2)^5$ **19.** $f(t) = \sqrt{2t^2 + 1}$

20. $g(t) = \sqrt[3]{1 - 2t^3}$ **21.** $s(t) = (3t^2 - 2t + 5)^{-2}$

22. $f(x) = (2x^3 - 3x^2 + 1)^{-3/2}$

23. $h(x) = \left(x + \dfrac{1}{x}\right)^2$ **24.** $h(x) = \dfrac{1+x}{(2x^2+1)^2}$

25. $h(t) = (t^2 + t)^4(2t^2)$ **26.** $f(x) = (2x+1)^3(x^2+x)^2$

27. $g(x) = \sqrt{x}(x^2 - 1)^3$ **28.** $f(x) = \dfrac{x}{\sqrt{x^3 + 2}}$

29. $h(x) = \dfrac{\sqrt{3x+2}}{4x-3}$ **30.** $f(t) = \dfrac{\sqrt{2t+1}}{(t+1)^3}$

In Exercises 31–36, find the second derivative of the function.

31. $f(x) = 2x^4 - 3x^3 + 2x^2 + x + 4$

32. $g(x) = \sqrt{x} + \dfrac{1}{\sqrt{x}}$ **33.** $h(t) = \dfrac{t}{t^2 + 4}$

34. $f(x) = (x^3 + x + 1)^2$ **35.** $f(x) = \sqrt{2x^2 + 1}$

36. $f(t) = t(t^2 + 1)^3$

In Exercises 37–42, find dy/dx by implicit differentiation.

37. $6x^2 - 3y^2 = 9$ **38.** $2x^3 - 3xy = 4$

39. $y^3 + 3x^2 = 3y$

40. $x^2 + 2x^2y^2 + y^2 = 10$

41. $x^2 - 4xy - y^2 = 12$

42. $3x^2y - 4xy + x - 2y = 6$

43. Find the differential of $f(x) = x^2 + \dfrac{1}{x^2}$.

44. Find the differential of $f(x) = \dfrac{1}{\sqrt{x^3 + 1}}$.

45. Let f be the function defined by $f(x) = \sqrt{2x^2 + 4}$.
   a. Find the differential of f.
   b. Use your result from part (a) to find the approximate change in $y = f(x)$ if x changes from 4 to 4.1.
   c. Find the actual change in y if x changes from 4 to 4.1 and compare your result with that obtained in part (b).

46. Use a differential to approximate $\sqrt[3]{26.8}$.

47. Let $f(x) = 2x^3 - 3x^2 - 16x + 3$.
   a. Find the points on the graph of f at which the slope of the tangent line is equal to -4.
   b. Find the equation(s) of the tangent line(s) of part (a).

48. Let $f(x) = \frac{1}{3}x^3 + \frac{1}{2}x^2 - 4x + 1$.
   a. Find the points on the graph of f at which the slope of the tangent line is equal to -2.
   b. Find the equation(s) of the tangent line(s) of part (a).

49. Find an equation of the tangent line to the graph of $y = \sqrt{4 - x^2}$ at the point $(1, \sqrt{3})$.

50. Find an equation of the tangent line to the graph of $y = x(x + 1)^5$ at the point $(1, 32)$.

51. Find the third derivative of the function

$$f(x) = \dfrac{1}{2x - 1}$$

What is its domain?

52. **SALES OF DSPs** The annual sales of digital signal processors (DSPs) in billions of dollars is approximated by

$$S(t) = 0.14t^2 + 0.68t + 3.1 \qquad (0 \le t \le 6)$$

where t is measured in years, with $t = 0$ corresponding to the beginning of 1997.
   a. What were the sales of DSPs at the beginning of 1997? What were the sales at the beginning of 2002?
   b. How fast was the level of sales increasing at the beginning of 1997? How fast were sales increasing at the beginning of 2002?
Source: World Semiconductor Trade Statistics.

53. **U.K. DIGITAL VIDEO VIEWERS** Digital video viewing is one of the top online activities among U.K. Internet users. It is expected that between 2012 and 2017, the U.K. digital video audience will be given by

$$N(t) = 65.71t^{0.085} \qquad (2 \le t \le 7)$$

where $N(t)$ is measured in millions and t is measured in years, with $t = 2$ corresponding to 2012.

a. How many U.K. digital video viewers will there be in 2015?

b. How fast was the U.K. digital video audience expected to be changing in 2015?

Source: eMarketer.

54. **ADULT OBESITY** In the United States, the percentage of adults (age 20–74) classified as obese held steady through the 1960s and 1970s at around 14% but began to rise rapidly during the 1980s and 1990s. This rise in adult obesity coincided with the period when an increasing number of Americans began eating more sugar and fats. The function

$$P(t) = 0.01484t^2 + 0.446t + 15 \qquad (0 \le t \le 22)$$

where t is measured in years, gives the percentage of obese adults from 1978 ($t = 0$) through 2000 ($t = 22$).

a. What percentage of adults were obese in 1978? In 2000?

b. How fast was the percentage of obese adults increasing in 1980 ($t = 2$)? In 1998 ($t = 20$)?

Source: Journal of the American Medical Association.

55. **SALES OF CAMERAS** The shipments of Lica digital single-lens reflex cameras (SLRs) are projected to be

$$N(t) = 6t^2 + 200t + 4\sqrt{t} + 20,000 \qquad (0 \le t \le 4)$$

units t years from now.

a. How many Lica SLRs will be shipped after 2 years?

b. At what rate will the number of Lica SLRs shipped be changing after 2 years?

56. **GDP OF A COUNTRY** The gross domestic product (GDP) of a certain country is

$$f(t) = 0.1t^3 + 0.5t^2 + 2t + 20 \qquad (0 \le t \le 4)$$

billion dollars in year t, where t is measured in years with $t = 0$ corresponding to 2010.

a. What was the GDP of the country in 2013?

b. How fast was the GDP of the country changing in 2013?

57. **POPULATION GROWTH IN A SUBURB** The population of a certain suburb is expected to be

$$P(t) = 30 - \frac{20}{2t + 3} \qquad (0 \le t \le 5)$$

thousand t years from now.

a. By how much will the population have grown after 3 years?

b. How fast is the population changing after 3 years?

58. **BEST-SELLING NOVEL** The number of copies of a best-selling novel sold t weeks after it was introduced is given by

$$N(t) = (4 + 5t)^{5/3} \qquad (1 \le t \le 30)$$

where $N(t)$ is measured in thousands.

a. How many copies of the novel were sold after 12 weeks?

b. How fast were the sales of the novel changing after 12 weeks?

59. **CABLE TV SUBSCRIBERS** The number of subscribers to CNC Cable Television in the town of Randolph is approximated by the function

$$N(x) = 1000(1 + 2x)^{1/2} \qquad (1 \le x \le 30)$$

where $N(x)$ denotes the number of subscribers to the service in the xth week. Find the rate of increase in the number of subscribers at the end of the 12th week.

60. **COST OF WIRELESS PHONE CALLS** As cellular phone usage continues to soar, the airtime costs have dropped. The average price per minute of use (in cents) is approximated by

$$f(t) = 31.88(1 + t)^{-0.45} \qquad (0 \le t \le 6)$$

where t is measured in years and $t = 0$ corresponds to the beginning of 1998. Compute $f'(t)$. How fast was the average price per minute of use changing at the beginning of 2000? What was the average price/minute of use at the beginning of 2000?

Source: Cellular Telecommunications Industry Association.

61. **MALE LIFE EXPECTANCY** Suppose the life expectancy of a male at birth in a certain country is described by the function

$$f(t) = 46.9(1 + 1.09t)^{0.1} \qquad (0 \le t \le 150)$$

where t is measured in years, with $t = 0$ corresponding to the beginning of 1900. How long can a male born at the beginning of 2000 in that country expect to live? What is the rate of change of the life expectancy of a male born in that country at the beginning of 2000?

62. **COST OF PRODUCING DVDs** The total weekly cost in dollars incurred by Herald Media Corp. in producing x DVDs is given by the total cost function

$$C(x) = 2500 + 2.2x \qquad (0 \le x \le 8000)$$

a. What is the marginal cost when $x = 1000$ and 2000?

b. Find the average cost function $\overline{C}$ and the marginal average cost function $\overline{C}'$.

c. Using the results from part (b), show that the average cost incurred by Herald in producing a DVD approaches $2.20/disc when the level of production is high enough.

63. **SUPPLY FUNCTION FOR SATELLITE RADIOS** The supply function for a certain brand of satellite radios is given by

$$p = \frac{1}{10}x^{3/2} + 10 \qquad (0 \le x \le 50)$$

where x is the quantity demanded (in thousands) if the unit price is $\$p$. Find $p'(40)$, and interpret your result.

64. Demand Function for Electric Shavers The demand for a certain brand of electric shavers is given by

$$p = 20\sqrt{-x^2 + 100} \qquad (0 \le x \le 10)$$

where x (in thousands) is the quantity demanded if the unit price is $\$p$. Find $p'(6)$, and interpret your result.

65. Marginal Cost The total daily cost (in dollars) incurred by Delta Electronics in producing x MP3 players is

$$C(x) = 0.0001x^3 - 0.02x^2 + 24x + 2000 \qquad (0 \le x \le 500)$$

where x stands for the number of units produced.
a. What is the actual cost incurred in manufacturing the 301st MP3 player, assuming that the 300th player was manufactured?
b. What is the marginal cost when $x = 300$?

66. Demand for Smartphones The marketing department of Telecon has determined that the demand for their smartphones obeys the relationship

$$p = -0.02x + 600 \qquad (0 \le x \le 30,000)$$

where p denotes the phone's unit price (in dollars) and x denotes the quantity demanded.
a. Find the revenue function R.
b. Find the marginal revenue function R'.
c. Compute $R'(10,000)$, and interpret your result.

67. Demand for Photocopying Machines The weekly demand for the LectroCopy photocopying machine is given by the demand equation

$$p = 2000 - 0.04x \qquad (0 \le x \le 50,000)$$

where p denotes the wholesale unit price in dollars and x denotes the quantity demanded. The weekly total cost function for manufacturing these copiers is given by

$$C(x) = 0.000002x^3 - 0.02x^2 + 1000x + 120,000$$

where $C(x)$ denotes the total cost incurred in producing x units.
a. Find the revenue function R, the profit function P, and the average cost function $\overline{C}$.
b. Find the marginal cost function C', the marginal revenue function R', the marginal profit function P', and the marginal average cost function $\overline{C}'$.
c. Compute $C'(3000)$, $R'(3000)$, and $P'(3000)$.
d. Compute $\overline{C}'(5000)$ and $\overline{C}'(8000)$, and interpret your results.

68. Marginal Average Cost The Custom Office makes a line of executive desks. It is estimated that the total cost for making x units of the Junior Executive model is

$$C(x) = 80x + 150,000 \qquad (0 \le x \le 20,000)$$

dollars/year.

a. Find the average cost function $\overline{C}$.
b. Find the marginal average cost function $\overline{C}'$.
c. What happens to $\overline{C}(x)$ when x is very large? Interpret your result.

69. Elasticity of Demand The demand equation for a certain product is $2x + 5p - 60 = 0$, where p is the unit price and x is the quantity demanded of the product. Find the elasticity of demand and determine whether the demand is elastic or inelastic, at the indicated prices.
a. $p = 3$ b. $p = 6$ c. $p = 9$

70. Elasticity of Demand The demand equation for a certain product is

$$x = \frac{25}{\sqrt{p}} - 1$$

where p is the unit price and x is the quantity demanded for the product. Compute the elasticity of demand and determine the range of prices corresponding to inelastic, unitary, and elastic demand.

71. Elasticity of Demand The demand equation for a certain product is $x = 100 - 0.01p^2$.
a. Is the demand elastic, unitary, or inelastic when $p = 40$?
b. If the price is $\$40$, will raising the price slightly cause the revenue to increase or decrease?

72. Elasticity of Demand The demand equation for a certain product is

$$p = 9\sqrt[3]{1000 - x}$$

a. Is the demand elastic, unitary, or inelastic when $p = 60$?
b. If the price is $\$60$, will raising the price slightly cause the revenue to increase or decrease?

73. GDP of a Country The GDP of a country from the years 2006 to 2013 is approximated by the function

$$G(t) = -0.3t^3 + 1.2t^2 + 500 \qquad (0 \le t \le 7)$$

where $G(t)$ is measured in billions of dollars and t is measured in years, with $t = 0$ corresponding to 2006. Find $G'(2)$ and $G''(2)$, and interpret your results.

74. Motion of an Object The position of an object moving along a straight line is given by

$$s = t\sqrt{2t^2 + 1} \qquad (0 \le t \le 5)$$

where s is measured in feet and t in seconds. Find the velocity and acceleration of the object after 2 sec.

The problem-solving skills that you learn in each chapter are building blocks for the rest of the course. Therefore, it is a good idea to make sure that you have mastered these skills before moving on to the next chapter. The Before Moving On exercises that follow are designed for that purpose. After completing these exercises, you can identify the skills that you should review before starting the next chapter.

CHAPTER 3 Before Moving On . . .

1. Find the derivative of $f(x) = 2x^3 - 3x^{1/3} + 5x^{-2/3}$.

2. Differentiate $g(x) = x\sqrt{2x^2 - 1}$.

3. Find $\dfrac{dy}{dx}$ if $y = \dfrac{2x + 1}{x^2 + x + 1}$.

4. Find the first three derivatives of $f(x) = \dfrac{1}{\sqrt{x + 1}}$.

5. Find $\dfrac{dy}{dx}$ given that $xy^2 - x^2y + x^3 = 4$.

6. Let $y = x\sqrt{x^2 + 5}$.
   a. Find the differential of y.
   b. If x changes from $x = 2$ to $x = 2.01$, what is the approximate change in y?

4

Applications of the Derivative

THIS CHAPTER FURTHER EXPLORES the power of the derivative as a tool to help analyze the properties of functions. We can use this information to accurately sketch graphs of functions. We also see how the derivative is used in solving a large class of optimization problems, including finding what level of production will yield a maximum profit for a company, finding what level of production will result in minimal cost to a company, finding the maximum height attained by a rocket, finding the maximum velocity at which air is expelled when a person coughs, and a host of other problems.

How many loudspeaker systems should the Acrosonic company produce to maximize its profit? In Example 4, page 309, you will see how the techniques of calculus can be used to help answer this question.

© rakenroll/ShutterStock.com

4.1 | Applications of the First Derivative

Determining the Intervals Where a Function Is Increasing or Decreasing

According to a study by the U.S. Department of Energy and the Shell Development Company, a typical car's fuel economy as a function of its speed is described by the graph shown in Figure 1. Observe that the fuel economy $f(x)$ in miles per gallon (mpg) improves as x, the vehicle's speed in miles per hour (mph), increases from 0 to 42 and then drops as the speed increases beyond 42 mph. We use the terms *increasing* and *decreasing* to describe the behavior of a function as we move from left to right along its graph.

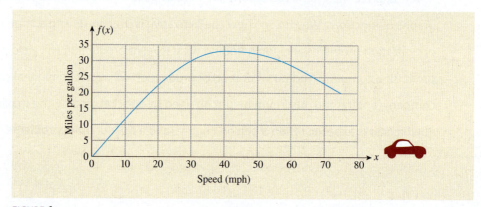

FIGURE 1
A typical car's fuel economy improves as the speed at which it is driven increases from 0 mph to 42 mph and drops at speeds greater than 42 mph.

Source: U.S. Department of Energy and Shell Development Co.

More precisely, we have the following definitions:

Increasing and Decreasing Functions

A function f is **increasing** on an interval (a, b) if for every two numbers x_1 and x_2 in (a, b), $f(x_1) < f(x_2)$ whenever $x_1 < x_2$ (Figure 2a).

A function f is **decreasing** on an interval (a, b) if for every two numbers x_1 and x_2 in (a, b), $f(x_1) > f(x_2)$ whenever $x_1 < x_2$ (Figure 2b).

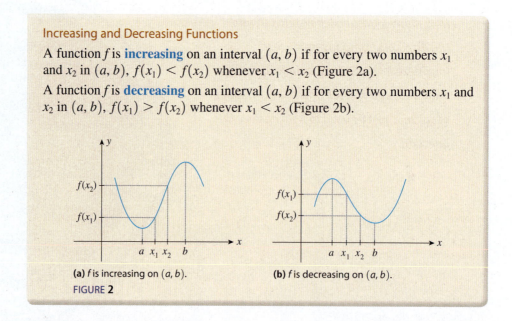

(a) f is increasing on (a, b).

(b) f is decreasing on (a, b).

FIGURE 2

We say that f is *increasing at a number c* if there exists an interval (a, b) containing c such that f is increasing on (a, b). Similarly, we say that f is *decreasing at a number c* if there exists an interval (a, b) containing c such that f is decreasing on (a, b).

Since the rate of change of a function at $x = c$ is given by the derivative of the function at that number, the derivative lends itself naturally to being a tool for determining the intervals where a differentiable function is increasing or decreasing. Indeed, as we saw in Chapter 2, the derivative of a function at a number measures both the slope of the tangent line to the graph of the function at the point on the graph of f corresponding to that number and the rate of change of the function at that number. In fact, at a number where the derivative is positive, the slope of the tangent line to the graph is positive, and the function is increasing. At a number where the derivative is negative, the slope of the tangent line to the graph is negative, and the function is decreasing (Figure 3).

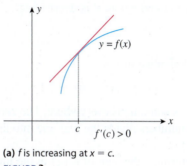
(a) f is increasing at $x = c$.

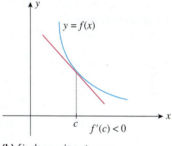
(b) f is decreasing at $x = c$.

FIGURE 3

These observations lead to the following important theorem, which we state without proof.

THEOREM 1

a. If $f'(x) > 0$ for every value of x in an interval (a, b), then f is increasing on (a, b).

b. If $f'(x) < 0$ for every value of x in an interval (a, b), then f is decreasing on (a, b).

c. If $f'(x) = 0$ for every value of x in an interval (a, b), then f is constant on (a, b).

EXAMPLE 1 Find the interval where the function $f(x) = x^2$ is increasing and the interval where it is decreasing.

Solution The derivative of $f(x) = x^2$ is $f'(x) = 2x$. Since

$$f'(x) = 2x > 0 \quad \text{if } x > 0 \qquad \text{and} \qquad f'(x) = 2x < 0 \quad \text{if } x < 0$$

f is increasing on the interval $(0, \infty)$ and decreasing on the interval $(-\infty, 0)$ (Figure 4).

FIGURE 4
The graph of f falls on $(-\infty, 0)$ where $f'(x) < 0$ and rises on $(0, \infty)$ where $f'(x) > 0$.

Recall that the graph of a continuous function cannot have any breaks. As a consequence, a continuous function cannot change sign unless it equals zero for some value of x. (See Theorem 5, page 128.) This observation suggests the following procedure for determining the sign of the derivative f' of a function f and hence the intervals where the function f is increasing and where it is decreasing.

> **Determining the Intervals Where a Function Is Increasing or Decreasing**
> 1. Find all values of x for which $f'(x) = 0$ or f' is discontinuous, and identify the open intervals determined by these numbers.
> 2. Select a test number c in each interval found in step 1, and determine the sign of $f'(c)$ in that interval.
>   **a.** If $f'(c) > 0$, f is increasing on that interval.
>   **b.** If $f'(c) < 0$, f is decreasing on that interval.

Explore and Discuss

True or false? If f is continuous at c and f is increasing at c, then $f'(c) > 0$. Explain your answer.
Hint: Consider $f(x) = x^3$ and $c = 0$.

EXAMPLE 2 Determine the intervals where the function

$$f(x) = x^3 - 3x^2 - 24x + 32$$

is increasing and where it is decreasing.

Solution

1. The derivative of f is

$$f'(x) = 3x^2 - 6x - 24 = 3(x + 2)(x - 4) \quad \text{(x²)} \text{ See page 11.}$$

and it is continuous everywhere. The zeros of $f'(x)$ are $x = -2$ and $x = 4$, and these numbers divide the real line into the intervals $(-\infty, -2)$, $(-2, 4)$, and $(4, \infty)$.

2. To determine the sign of $f'(x)$ in the intervals $(-\infty, -2)$, $(-2, 4)$, and $(4, \infty)$, compute $f'(x)$ at a convenient test point in each interval. The results are shown in the following table:

Interval	Test Point c	$f'(c)$	Sign of $f'(x)$
$(-\infty, -2)$	-3	21	$+$
$(-2, 4)$	0	-24	$-$
$(4, \infty)$	5	21	$+$

(x²) See page 21.

Using these results, we obtain the sign diagram shown in Figure 5. We conclude that f is increasing on the intervals $(-\infty, -2)$ and $(4, \infty)$ and is decreasing on the interval $(-2, 4)$. Figure 6 shows the graph of f.

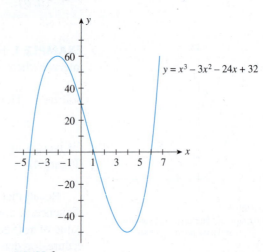

$y = x^3 - 3x^2 - 24x + 32$

+ + + 0 − − − − − 0 + + + +
        −2 0   4

FIGURE 5
Sign diagram for f'

FIGURE 6
The graph of f rises on $(-\infty, -2)$, falls on $(-2, 4)$, and rises again on $(4, \infty)$.

Note We will learn how to sketch these graphs later. However, if you are familiar with the use of a graphing utility, you may go ahead and verify each graph. ■

Exploring with **TECHNOLOGY**

Refer to Example 2.

1. Use a graphing utility to plot the graphs of $f(x) = x^3 - 3x^2 - 24x + 32$ and its derivative function $f'(x) = 3x^2 - 6x - 24$ using the viewing window $[-10, 10] \times [-50, 70]$.

2. By looking at the graph of f', determine the intervals where $f'(x) > 0$ and the intervals where $f'(x) < 0$. Next, look at the graph of f and determine the intervals where it is increasing and the intervals where it is decreasing. Describe the relationship. Is it what you expected?

EXAMPLE 3 Find the interval where the function $f(x) = x^{2/3}$ is increasing and the interval where it is decreasing.

Solution

1. The derivative of f is

$$f'(x) = \frac{2}{3}x^{-1/3} = \frac{2}{3x^{1/3}}$$

The function f' is not defined at $x = 0$, so f' is discontinuous there. It is continuous everywhere else. Furthermore, f' is not equal to zero anywhere. The number 0 divides the real line (the domain of f) into the intervals $(-\infty, 0)$ and $(0, \infty)$.

2. Pick a test point (say, $x = -1$) in the interval $(-\infty, 0)$, and compute

$$f'(-1) = -\frac{2}{3}$$

Since $f'(-1) < 0$, we know that $f'(x) < 0$ on $(-\infty, 0)$. Next, we pick a test point (say, $x = 1$) in the interval $(0, \infty)$ and compute

$$f'(1) = \frac{2}{3}$$

Since $f'(1) > 0$, we know that $f'(x) > 0$ on $(0, \infty)$. Figure 7 shows these results in the form of a sign diagram.

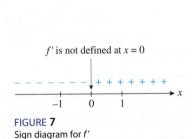

FIGURE **7**
Sign diagram for f'

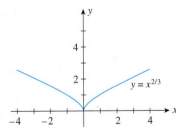

FIGURE **8**
f decreases on $(-\infty, 0)$ and increases on $(0, \infty)$.

We conclude that f is decreasing on the interval $(-\infty, 0)$ and increasing on the interval $(0, \infty)$. The graph of f, shown in Figure 8, confirms these results. ■

EXAMPLE 4 Find the intervals where the function $f(x) = x + \dfrac{1}{x}$ is increasing and where it is decreasing.

Solution

1. The derivative of f is

$$f'(x) = 1 - \frac{1}{x^2} = \frac{x^2 - 1}{x^2} = \frac{(x + 1)(x - 1)}{x^2} \qquad (x^2) \ \text{See page 17.}$$

Since f' is not defined at $x = 0$, it is discontinuous there. Furthermore, $f'(x)$ is equal to zero when $(x + 1)(x - 1) = 0$ or $x = \pm 1$. Note that the value of f' is different from zero in the open intervals $(-\infty, -1)$, $(-1, 0)$, $(0, 1)$, and $(1, \infty)$.

2. To determine the sign of f' in each of these intervals, we compute $f'(x)$ at the test points $x = -2, -\frac{1}{2}, \frac{1}{2},$ and 2, respectively, obtaining $f'(-2) = \frac{3}{4}, f'(-\frac{1}{2}) = -3,$ $f'(\frac{1}{2}) = -3,$ and $f'(2) = \frac{3}{4}$. From the sign diagram for f' (Figure 9), we conclude that f is increasing on $(-\infty, -1)$ and $(1, \infty)$ and decreasing on $(-1, 0)$ and $(0, 1)$.

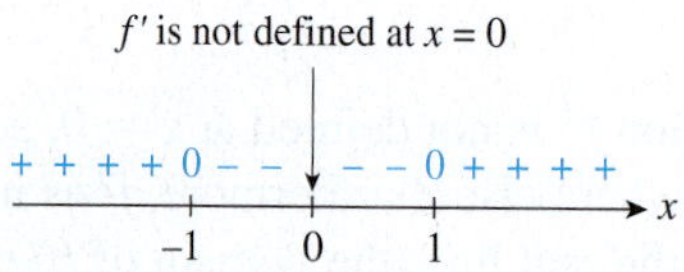

f' is not defined at $x = 0$

$+ + + + 0 - - \ - - 0 + + + +$

$\hphantom{}-1 \qquad 0 \qquad 1$ $\qquad x$

FIGURE 9
f' does not change sign as we move across $x = 0$.

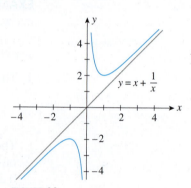

FIGURE 10
The graph of f rises on $(-\infty, -1)$, falls on $(-1, 0)$ and $(0, 1)$, and rises again on $(1, \infty)$.

The graph of f appears in Figure 10. Note that f' does not change sign as we move across $x = 0$. (Compare this with Example 3.)

⚠️ Example 4 reminds us that we must *not* automatically conclude that the derivative f' must change sign when we move across a number where f' is discontinous or a zero of f'.

Explore and Discuss

Consider the profit function P associated with a certain commodity defined by

$$P(x) = R(x) - C(x) \qquad (x \geq 0)$$

where R is the revenue function, C is the total cost function, and x is the number of units of the product produced and sold.

1. Find an expression for $P'(x)$.
2. Find relationships in terms of the derivatives of R and C such that
   a. P is increasing at $x = a$.
   b. P is decreasing at $x = a$.
   c. P is neither increasing nor decreasing at $x = a$.
   Hint: Recall that the derivative of a function at $x = a$ measures the rate of change of the function at that number.
3. Explain the results of part 2 in economic terms.

1. Use a graphing utility to sketch the graphs of $f(x) = x^3 - ax$ for $a = -2$, $-1, 0, 1$, and 2, using the viewing window $[-2, 2] \times [-2, 2]$.

2. Use the results of part 1 to guess at the values of a for which f is increasing on $(-\infty, \infty)$.

3. Prove your conjecture analytically.

Relative Extrema

Besides helping us determine where the graph of a function is increasing and decreasing, the first derivative may be used to help us locate certain "high points" and "low points" on the graph of f. Knowing these points is invaluable in sketching the graphs of functions and solving optimization problems. These "high points" and "low points" correspond to the *relative (local) maxima* and *relative minima* of a function. They are so called because they are the highest or the lowest points when compared with points nearby.

The graph shown in Figure 11 gives the U.S. budget surplus (deficit) from 1996 ($t = 0$) to 2010 ($t = 14$). The relative maxima and the relative minima of the function f are indicated on the graph.

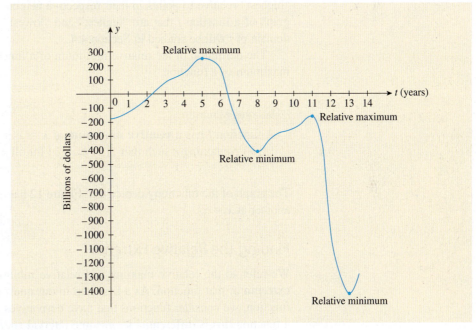

FIGURE 11
U.S. budget surplus (deficit) from 1996 to 2010
Source: Office of Management and Budget.

More generally, we have the following definition:

Relative Maximum

A function f has a **relative maximum** at $x = c$ if there exists an open interval (a, b) containing c such that $f(x) \leq f(c)$ for all x in (a, b).

Geometrically, this means that there is *some* interval containing $x = c$ such that no point on the graph of f with its x-coordinate in that interval can lie above the point

$(c, f(c))$; that is, $f(c)$ is the largest value of $f(x)$ in some interval around $x = c$. Figure 12 depicts the graph of a function f that has a relative maximum at $x = x_1$ and another at $x = x_3$.

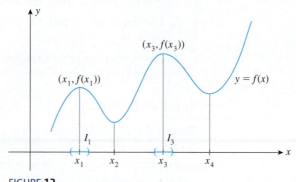

FIGURE 12
f has a relative maximum at $x = x_1$ and at $x = x_3$.

Observe that all the points on the graph of f with x-coordinates in the interval I_1 containing x_1 (shown in blue) lie on or below the point $(x_1, f(x_1))$. This is also true for the point $(x_3, f(x_3))$ and the interval I_3. Thus, even though there are points on the graph of f that are "higher" than the points $(x_1, f(x_1))$ and $(x_3, f(x_3))$, the latter points are "highest" relative to points in their respective neighborhoods (intervals). Points on the graph of a function f that are "highest" and "lowest" with respect to *all* points in the domain of f will be studied in Section 4.4.

The definition of the relative minimum of a function parallels that of the relative maximum of a function.

> **Relative Minimum**
>
> A function f has a **relative minimum** at $x = c$ if there exists an open interval (a, b) containing c such that $f(x) \geq f(c)$ for all x in (a, b).

The graph of the function f depicted in Figure 12 has a relative minimum at $x = x_2$ and another at $x = x_4$.

Finding the Relative Extrema

We refer to the relative maxima and relative minima of a function as the **relative extrema** of that function. As a first step in our quest to find the relative extrema of a function, we consider functions that have derivatives at such points. Suppose that f is a function that is differentiable on some interval (a, b) that contains a number c and that f has a relative maximum at $x = c$ (Figure 13a).

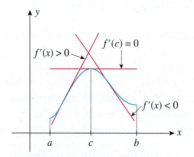

(a) f has a relative maximum at $x = c$.

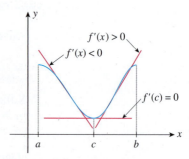

(b) f has a relative minimum at $x = c$.

FIGURE 13

Observe that the slope of the tangent line to the graph of f must change from positive to negative as we move across $x = c$ from left to right. Therefore, the tangent line to the graph of f at the point $(c, f(c))$ must be horizontal; that is, $f'(c) = 0$ (Figure 13a).

Using a similar argument, it may be shown that the derivative f' of a differentiable function f must also be equal to zero at $x = c$ if f has a relative minimum at $x = c$ (Figure 13b).

This analysis reveals an important characteristic of the relative extrema of a differentiable function f: *At any number c where f has a relative extremum, $f'(c) = 0$.*

Before we develop a procedure for finding such numbers, a few words of caution are in order. First, this result tells us that if a differentiable function f has a relative extremum at a number $x = c$, then $f'(c) = 0$. The converse of this statement—if $f'(c) = 0$ at $x = c$, then f must have a relative extremum at that number—is *not* true. Consider, for example, the function $f(x) = x^3$. Here, $f'(x) = 3x^2$, so $f'(0) = 0$. However, f has neither a relative maximum nor a relative minimum at $x = 0$ (Figure 14).

Second, our result assumes that the function is differentiable and therefore has a derivative at a number that gives rise to a relative extremum. The functions $f(x) = |x|$ and $g(x) = x^{2/3}$ demonstrate that a relative extremum of a function may exist at a number at which the derivative does not exist. Both these functions fail to be differentiable at $x = 0$, but each has a relative minimum there. Figure 15 shows the graphs of these functions. Note that the slopes of the tangent lines change from negative to positive as we move across $x = 0$, just as in the case of a function that is differentiable at a value of x that gives rise to a relative minimum.

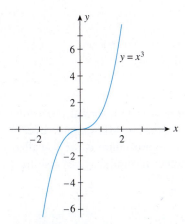

FIGURE 14
$f'(0) = 0$, but f does not have a relative extremum at $(0, 0)$.

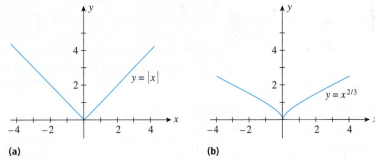

(a)   (b)

FIGURE 15
Each of these functions has a relative extremum at $(0, 0)$, but the derivative does not exist there.

We refer to a number in the domain of f that *might* give rise to a relative extremum as a critical number.

> **Critical Number of f**
>
> A **critical number** of a function f is any number x in the domain of f such that $f'(x) = 0$ or $f'(x)$ does not exist.

Figure 16 (see next page) depicts the graph of a function that has critical numbers at $x = a, b, c, d,$ and e. Observe that $f'(x) = 0$ at $x = a, b,$ and c. Next, since there is a corner at $x = d$, $f'(x)$ does not exist there. Finally, $f'(x)$ does not exist at $x = e$ because the tangent line there is vertical. Also, observe that the critical numbers

$x = a$, b, and d give rise to relative extrema of f, whereas the critical numbers $x = c$ and $x = e$ do not.

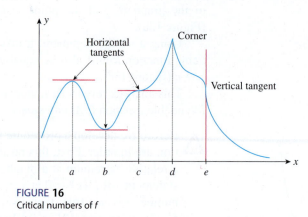

FIGURE 16
Critical numbers of f

Having defined what a critical number is, we can now state a formal procedure for finding the relative extrema of a continuous function that is differentiable everywhere except at isolated values of x. Incorporated into the procedure is the so-called **First Derivative Test,** which helps us determine whether a number gives rise to a relative maximum or a relative minimum of the function f.

The First Derivative Test

> **Procedure for Finding the Relative Extrema of a Continuous Function f**
>
> **1.** Determine the critical numbers of f.
> **2.** Determine the sign of $f'(x)$ to the left and right of each critical number.
>   **a.** If $f'(x)$ changes sign from *positive* to *negative* as we move across a critical number c, then f has a relative maximum at $x = c$.
>   **b.** If $f'(x)$ changes sign from *negative* to *positive* as we move across a critical number c, then f has a relative minimum at $x = c$.
>   **c.** If $f'(x)$ does not change sign as we move across a critical number c, then f does not have a relative extremum at $x = c$.

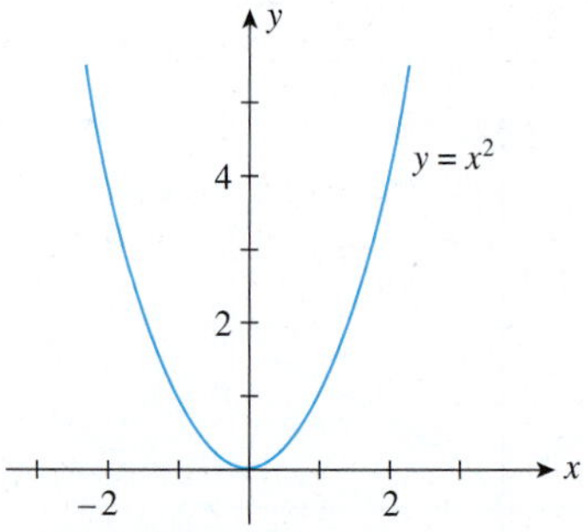

FIGURE 17
f has a relative minimum at $x = 0$.

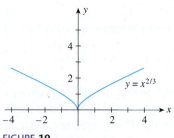

f' is not defined at $x = 0$

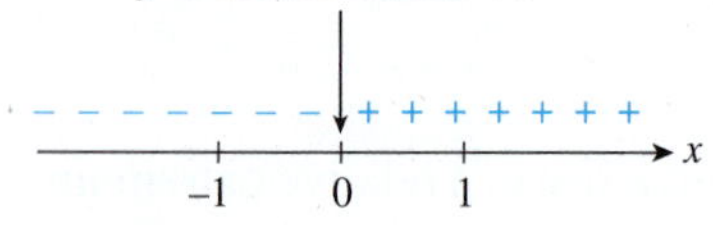

FIGURE 18
Sign diagram for f'

FIGURE 19
f has a relative minimum at $x = 0$.

EXAMPLE 5 Find the relative maxima and relative minima of the function $f(x) = x^2$.

Solution The derivative of $f(x) = x^2$ is given by $f'(x) = 2x$. Setting $f'(x) = 0$ yields $x = 0$ as the only critical number of f. Since

$$f'(x) < 0 \quad \text{if } x < 0 \qquad \text{and} \qquad f'(x) > 0 \quad \text{if } x > 0$$

we see that $f'(x)$ changes sign from negative to positive as we move across the critical number 0. Thus, we conclude that $f(0) = 0$ is a relative minimum of f (Figure 17).

EXAMPLE 6 Find the relative maxima and relative minima of the function $f(x) = x^{2/3}$ (see Example 3).

Solution The derivative of f is $f'(x) = \frac{2}{3}x^{-1/3}$. As was noted in Example 3, f' is not defined at $x = 0$, is continuous everywhere else, and is not equal to zero in its domain. Thus, $x = 0$ is the only critical number of the function f.

The sign diagram obtained in Example 3 is reproduced in Figure 18. We can see that the sign of $f'(x)$ changes from negative to positive as we move across $x = 0$ from left to right. Thus, an application of the First Derivative Test tells us that $f(0) = 0$ is a relative minimum of f (Figure 19).

Recall that the average cost function $\overline{C}$ is defined by

$$\overline{C} = \frac{C(x)}{x}$$

where $C(x)$ is the total cost function and x is the number of units of a commodity manufactured (see Section 3.4).

1. Show that

$$\overline{C}'(x) = \frac{C'(x) - \overline{C}(x)}{x} \qquad (x > 0)$$

2. Use the result of part 1 to conclude that $\overline{C}$ is decreasing for values of x at which $C'(x) < \overline{C}(x)$. Find similar conditions for which $\overline{C}$ is increasing and for which $\overline{C}$ is constant.

3. Explain the results of part 2 in economic terms.

EXAMPLE 7 Find the relative maxima and relative minima of the function

$$f(x) = x^3 - 3x^2 - 24x + 32$$

Solution   The derivative of f is

$$f'(x) = 3x^2 - 6x - 24 = 3(x + 2)(x - 4) \qquad (x^2) \text{ See page 11.}$$

and it is continuous everywhere. The zeros of $f'(x)$, $x = -2$ and $x = 4$, are the only critical numbers of the function f. The sign diagram for f' is shown in Figure 20. Examine the two critical numbers $x = -2$ and $x = 4$ for a relative extremum using the First Derivative Test and the sign diagram for f'.

1. *The critical number* -2: Since the function $f'(x)$ changes sign from positive to negative as we move across $x = -2$ from left to right, we conclude that a relative maximum of f occurs at $x = -2$. The value of $f(x)$ when $x = -2$ is

$$f(-2) = (-2)^3 - 3(-2)^2 - 24(-2) + 32 = 60$$

2. *The critical number* 4: $f'(x)$ changes sign from negative to positive as we move across $x = 4$ from left to right, so $f(4) = -48$ is a relative minimum of f. The graph of f appears in Figure 21.

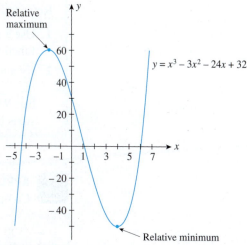

FIGURE 20
Sign diagram for f'

FIGURE 21
f has a relative maximum at $x = -2$ and a relative minimum at $x = 4$.

EXAMPLE 8 Find the relative maxima and the relative minima of the function

$$f(x) = x + \frac{1}{x}$$

Solution   The derivative of f is

$$f'(x) = 1 - \frac{1}{x^2} = \frac{x^2 - 1}{x^2} = \frac{(x + 1)(x - 1)}{x^2}$$

Since f' is equal to zero at $x = -1$ and $x = 1$, these are critical numbers for the function f. Next, observe that f' is discontinuous at $x = 0$. However, because f *is not defined at that number*, $x = 0$ does not qualify as a critical number of f. Figure 22 shows the sign diagram for f'.

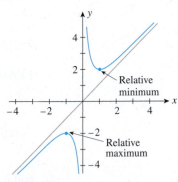

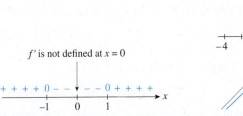

FIGURE **22**
$x = 0$ is not a critical number because f is not defined at $x = 0$.

FIGURE **23**
$f(x) = x + \dfrac{1}{x}$

Since $f'(x)$ changes sign from positive to negative as we move across $x = -1$ from left to right, the First Derivative Test implies that $f(-1) = -2$ is a relative maximum of the function f. Next, $f'(x)$ changes sign from negative to positive as we move across $x = 1$ from left to right, so $f(1) = 2$ is a relative minimum of the function f. The graph of f appears in Figure 23. Note that this function has a relative maximum that lies below its relative minimum.

Exploring with TECHNOLOGY

Refer to Example 8.

1. Use a graphing utility to plot the graphs of $f(x) = x + 1/x$ and its derivative function $f'(x) = 1 - 1/x^2$, using the viewing window $[-4, 4] \times [-8, 8]$.

2. By studying the graph of f', determine the critical numbers of f. Next, note the sign of $f'(x)$ immediately to the left and to the right of each critical number. What can you conclude about each critical number? Are your conclusions borne out by the graph of f?

 APPLIED EXAMPLE 9 Profit Functions   The profit function of Acrosonic Company is given by

$$P(x) = -0.02x^2 + 300x - 200,000$$

dollars, where x is the number of Acrosonic model F loudspeaker systems produced. Find where the function P is increasing and where it is decreasing.

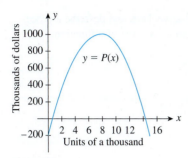

FIGURE 24
The profit function is increasing on
$(0, 7500)$ and decreasing on $(7500, \infty)$.

Solution The derivative P' of the function P is

$$P'(x) = -0.04x + 300 = -0.04(x - 7500)$$

Thus, $P'(x) = 0$ when $x = 7500$. Furthermore, $P'(x) > 0$ for x in the interval $(0, 7500)$, and $P'(x) < 0$ for x in the interval $(7500, \infty)$. This means that the profit function P is increasing on $(0, 7500)$ and decreasing on $(7500, \infty)$ (Figure 24).

 APPLIED EXAMPLE 10 Crime Rates The number of major crimes committed in the city of Bronxville from 2006 to 2013 is approximated by the function

$$N(t) = -0.1t^3 + 1.5t^2 + 100 \qquad (0 \le t \le 7)$$

where $N(t)$ denotes the number of crimes committed in year t, with $t = 0$ corresponding to the beginning of 2006. Find where the function N is increasing and where it is decreasing.

Solution The derivative N' of the function N is

$$N'(t) = -0.3t^2 + 3t = -0.3t(t - 10)$$

Since $N'(t) > 0$ for t in the interval $(0, 7)$, the function N is increasing throughout that interval (Figure 25).

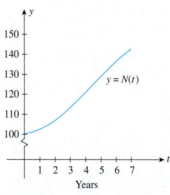

FIGURE 25
The number of crimes, $N(t)$, is increasing over the 7-year interval.

4.1 Self-Check Exercises

1. Find the intervals where the function

$$f(x) = \frac{2}{3}x^3 - x^2 - 12x + 3$$

is increasing and the intervals where it is decreasing.

2. Find the relative extrema of $f(x) = \dfrac{x^2}{1 - x^2}$.

Solutions to Self-Check Exercises 4.1 can be found on page 269.

4.1 Concept Questions

1. Explain each of the following:
   a. f is increasing on an interval I.
   b. f is decreasing on an interval I.

2. Describe a procedure for determining where a function is increasing and where it is decreasing.

3. Explain each term: (a) relative maximum and (b) relative minimum.

4. **a.** What is a critical number of a function f?
   b. Explain the role of critical numbers in determining the relative extrema of a function.

5. Describe the First Derivative Test, and describe a procedure for finding the relative extrema of a function.

4.1 Exercises

In Exercises 1–8, you are given the graph of a function f. Determine the intervals where f is increasing, constant, or decreasing.

1.
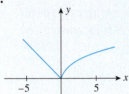

2.
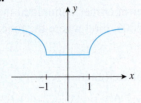

3.
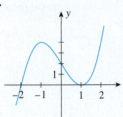

4.
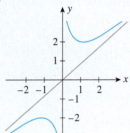

5.
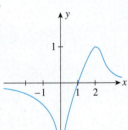

6.
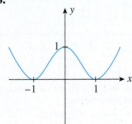

7.
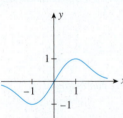

8.
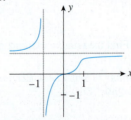

9. **THE BOSTON MARATHON** The graph of the function f shown in the following figure gives the elevation of the part of the Boston Marathon course that includes the notorious Heartbreak Hill. Determine the intervals (stretches of the course) where the function f is increasing (the runner is laboring), where it is constant (the runner is taking a breather), and where it is decreasing (the runner is coasting).

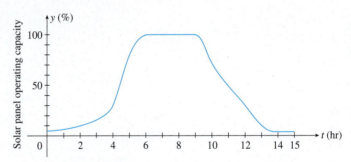

10. **SOLAR PANEL POWER OUTPUT** The graph of the function f shown in the accompanying figure gives the average "fixed" solar panel power output over a 15-hr period on a typical day. Determine the interval(s) where f is increasing, the interval(s) where f is constant, and the interval(s) where f is decreasing. Here, $t = 0$ corresponds to 5 A.M. Interpret your result.

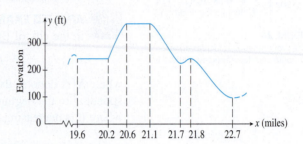

Source: Solarcity.com/California.

11. **AIRCRAFT STRUCTURAL INTEGRITY** Among the important factors in determining the structural integrity of an aircraft is its age. Advancing age makes planes more likely to crack. The graph of the function f, shown in the accompanying figure, is referred to as a "bathtub curve" in the airline industry. It gives the fleet damage rate (damage due to corrosion, accident, and metal fatigue) of a typical fleet of commercial aircraft as a function of the number of years of service.

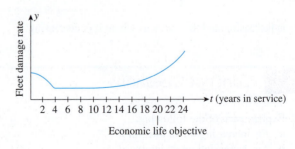

a. Determine the interval where f is decreasing. This corresponds to the time period when the fleet damage rate is dropping as problems are found and corrected during the initial "shakedown" period.

b. Determine the interval where f is constant. After the initial shakedown period, planes have few structural problems, and this is reflected by the fact that the function is constant on this interval.

c. Determine the interval where f is increasing. Beyond the time period mentioned in part (b), the function is increasing—reflecting an increase in structural defects due mainly to metal fatigue.

12. Refer to the following figure:

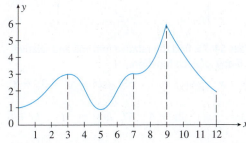

What is the sign of the following?
a. $f'(2)$ b. $f'(x)$ in the interval $(1, 3)$
c. $f'(4)$ d. $f'(x)$ in the interval $(3, 6)$
e. $f'(7)$ f. $f'(x)$ in the interval $(6, 9)$
g. $f'(x)$ in the interval $(9, 12)$

13. Refer to the following figure:

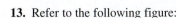

a. What are the critical numbers of f? Give reasons for your answers.
b. Draw the sign diagram for f'.
c. Find the relative extrema of f.

In Exercises 14–37, find the interval(s) where the function is increasing and the interval(s) where it is decreasing.

14. $f(x) = 4 - 5x$ **15.** $f(x) = 3x + 5$

16. $f(x) = x^2 - 3x$ **17.** $f(x) = 2x^2 + x + 1$

18. $f(x) = x^3 - 3x^2$ **19.** $g(x) = x - x^3$

20. $f(x) = x^3 - 3x + 4$ **21.** $g(x) = x^3 + 3x^2 + 1$

22. $f(x) = \dfrac{1}{3}x^3 - 3x^2 + 9x + 20$

23. $f(x) = \dfrac{2}{3}x^3 - 2x^2 - 6x - 2$

24. $g(x) = x^4 - 2x^2 + 4$ **25.** $h(x) = x^4 - 4x^3 + 10$

26. $f(x) = \dfrac{1}{x - 2}$ **27.** $h(x) = \dfrac{1}{2x + 3}$

28. $h(t) = \dfrac{t}{t - 1}$ **29.** $g(t) = \dfrac{2t}{t^2 + 1}$

30. $f(x) = x^{3/5}$ **31.** $f(x) = x^{2/3} + 5$

32. $f(x) = \sqrt{x + 1}$ **33.** $f(x) = (x - 5)^{2/3}$

34. $f(x) = \sqrt{16 - x^2}$ **35.** $g(x) = x\sqrt{x + 1}$

36. $f(x) = \dfrac{1 - x^2}{x}$ **37.** $h(x) = \dfrac{x^2}{x - 1}$

In Exercises 38–45, you are given the graph of a function f. Determine the relative maxima and relative minima, if any.

38.
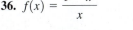

39.
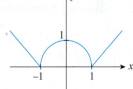

40.
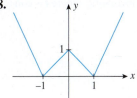

41.
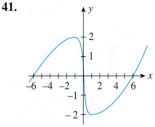

42.
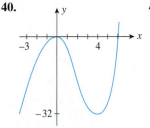

43.
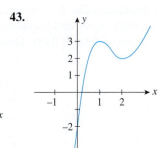

44.
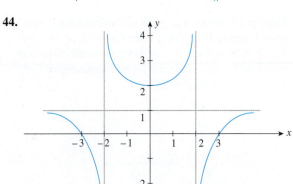

45.

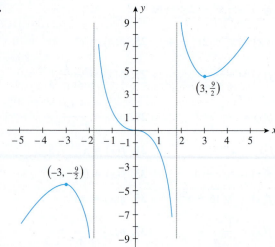

$\left(3, \frac{9}{2}\right)$

$\left(-3, -\frac{9}{2}\right)$

In Exercises 46 and 47, you are given the graph of the derivative, f', of the function f on an interval (a, b). Use this graph to determine where the graph of f is increasing, where f is constant, and where f is decreasing.

46.

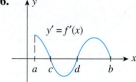

$y' = f'(x)$

47.

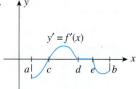

$y' = f'(x)$

48. MOTION OF A CAR A car travels along a straight road. Its velocity at any time t in the time interval (a, b) is shown in the accompanying figure. Describe the motion of the car.

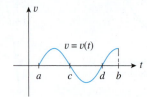

$v = v(t)$

49. PROFIT OF A COMPANY The graph of the derivative P' of a profit function, P, is shown in the following figure. Find the levels of production, x, at which the profit of the company is increasing, stationary (neither increasing or decreasing), and decreasing.

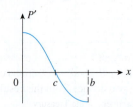

P'

In Exercises 50–53, match the graph of the function with the graph of its derivative in (a)–(d).

(a)   **(b)**

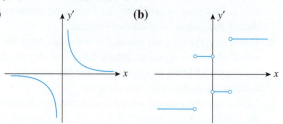

(c)   **(d)**

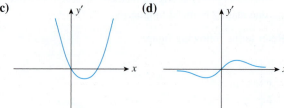

50.   **51.**

52.   **53.**

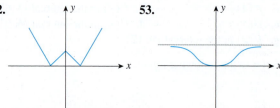

In Exercises 54–75, find the relative maxima and relative minima, if any, of each function.

54. $f(x) = x^2 - 4x$

55. $g(x) = x^2 + 3x + 8$

56. $f(x) = \frac{1}{2}x^2 - 2x + 4$

57. $h(t) = -t^2 + 6t + 6$

58. $f(x) = x^{5/3}$

59. $f(x) = x^{2/3} + 2$

60. $g(x) = x^3 - 3x^2 + 5$

61. $f(x) = x^3 - 3x + 6$

62. $F(x) = \frac{1}{3}x^3 - x^2 - 3x + 4$

63. $f(x) = \frac{1}{2}x^4 - x^2$

64. $h(x) = \frac{1}{2}x^4 - 3x^2 + 4x - 8$

65. $g(x) = x^4 - 4x^3 + 12$

66. $f(x) = 3x^4 - 2x^3 + 4$

67. $F(t) = 3t^5 - 20t^3 + 20$

68. $h(x) = \dfrac{x}{x + 1}$

69. $g(x) = \dfrac{x + 1}{x}$

70. $f(x) = x + \dfrac{9}{x} + 2$

71. $g(x) = 2x^2 + \dfrac{4000}{x} + 10$

72. $f(x) = \dfrac{x}{1 + x^2}$

73. $g(x) = \dfrac{x}{x^2 - 1}$

74. $f(x) = (x - 1)^{2/3}$

75. $g(x) = x\sqrt{x - 4}$

76. A stone is thrown straight up from the roof of an 80-ft building. The distance (in feet) of the stone from the ground at any time t (in seconds) is given by

$$h(t) = -16t^2 + 64t + 80$$

When is the stone rising, and when is it falling? If the stone were to miss the building, when would it hit the ground? Sketch the graph of h.
Hint: The stone is on the ground when $h(t) = 0$.

77. SMART METERS The percentage of U.S. homes and businesses with digital electrical meters between 2008 and 2012 is approximated by

$$N(t) = 1.1375t^2 + 0.25t + 4.6 \qquad (0 \le t \le 4)$$

Here, t is measured in years, with $t = 0$ corresponding to 2008. Show that the percentage of U.S. homes and businesses with digital electrical meters was always increasing between 2008 and 2012.
Source: Federal Energy Regulatory Commission.

78. PROFIT FUNCTION FOR PRODUCING THERMOMETERS The Mexican subsidiary of ThermoMaster manufactures an indoor–outdoor thermometer. Management estimates that the profit (in dollars) realizable by the company for the manufacture and sale of x units of thermometers each week is

$$P(x) = -0.001x^2 + 8x - 5000$$

Find the intervals where the profit function P is increasing and the intervals where P is decreasing.

79. GROWTH OF MANAGED SERVICES Almost half of companies let other firms manage some of their Web operations—a practice called Web hosting. Managed services—monitoring a customer's technology services—is the fastest-growing part of Web hosting. Managed services sales are approximated by the function

$$f(t) = 0.469t^2 + 0.758t + 0.44 \qquad (0 \le t \le 10)$$

where $f(t)$ is measured in billions of dollars and t is measured in years, with $t = 0$ corresponding to 1999.
a. Find the interval where f is increasing.
b. What does your result tell you about sales in managed services from 1999 through 2009?
Source: International Data Corp.

80. VOTER TURNOUT According to a study conducted by the Pew Research Center in 2013, the number of Hispanic voters in presidential elections in each of the years from 2000 through 2016 is approximated by the function

$$N(t) = 0.0046875t^2 + 0.38125t + 6 \qquad (t = 0, 4, 8, 12, 16)$$

where $N(t)$ is measured in millions with t taking on the discrete values, 0, 4, 8, 12, and 16. (The number for 2016 is estimated.)

a. Find the approximate number of Hispanic voters in each of the presidential elections from the year 2000 through 2016.
b. Thinking of the function N as a function defined for all values of t on the interval $(0, 16)$, find $N'(t)$.
c. Using the result of part (b), find $N'(4)$, $N'(8)$, and $N'(12)$. What do these numbers tell you about the growth in the number of Hispanic voters in succeeding elections since the year 2000?
Source: Pew Research Center.

81. FLIGHT OF A MODEL ROCKET The height (in feet) attained by a model rocket t sec into flight is given by the function

$$h(t) = -\frac{1}{3}t^3 + 16t^2 + 33t + 10 \qquad (t \ge 0)$$

When is the rocket rising, and when is it descending?

82. ENVIRONMENT OF FORESTS Following the lead of the National Wildlife Federation, the Department of the Interior of a South American country began to record an index of environmental quality that measured progress and decline in the environmental quality of its forests. The index for the years 2004 through 2014 is approximated by the function

$$I(t) = \frac{1}{3}t^3 - \frac{5}{2}t^2 + 80 \qquad (0 \le t \le 10)$$

where $t = 0$ corresponds to 2004. Find the intervals where the function I is increasing and the intervals where it is decreasing. Interpret your results.

83. AVERAGE SPEED OF A HIGHWAY VEHICLE The average speed of a vehicle on a stretch of Route 134 between 6 A.M. and 10 A.M. on a typical weekday is approximated by the function

$$f(t) = 20t - 40\sqrt{t} + 50 \qquad (0 \le t \le 4)$$

where $f(t)$ is measured in miles per hour and t is measured in hours, with $t = 0$ corresponding to 6 A.M. Find the interval where f is increasing and the interval where f is decreasing and interpret your results.

84. AVERAGE COST FOR PRODUCING CDs The average cost (in dollars) incurred by Lincoln Media each week in pressing x compact discs is given by

$$\overline{C}(x) = -0.0001x + 2 + \frac{2000}{x} \qquad (0 < x \le 6000)$$

Show that $\overline{C}(x)$ is always decreasing over the interval $(0, 6000)$.

85. PUBLIC DEBT The U.S. public debt outstanding from 1990 through 2011 is approximated by

$$D(t) = 0.0032t^3 - 0.0698t^2 + 0.6048t + 3.22 \qquad (0 \le t \le 21)$$

trillion dollars in year t, with $t = 0$ corresponding to 1990. Show that $D'(t) > 0$ for all t in the interval $(0, 21)$. What conclusion can you deduce from this result?
Source: U.S. Department of the Treasury.

86. **HAWAII TOURISM** Numbers of visitors to the Aloha State dropped dramatically during the recession. The number of visitors to Hawaii between 2007 ($t = 0$) and 2011 ($t = 4$) is approximately

$$N(t) = -0.062t^3 + 0.617t^2 - 1.557t + 7.7 \qquad (0 \le t \le 4)$$

Show that the number of visitors to Hawaii was approximately 7.7 million in 2007, then dropped to a low sometime short of 2009, before recovering to approximately 7.4 million in 2011.
Hint: Use the quadratic formula.
Source: Los Angeles Times.

87. **WEB HOSTING** Refer to Exercise 79. Sales in the Web-hosting industry are projected to grow in accordance with the function

$$f(t) = -0.05t^3 + 0.56t^2 + 5.47t + 7.5 \qquad (0 \le t \le 10)$$

where $f(t)$ is measured in billions of dollars and t is measured in years, with $t = 0$ corresponding to 1999.
a. Find the interval where f is increasing.
   Hint: Use the quadratic formula.
b. What does your result tell you about sales in the Web-hosting industry from 1999 through 2009?
Source: International Data Corp.

88. **DRIVING COSTS** A study of driving costs based on 2012 medium-sized sedans found that the average cost (car payments, gas, insurance, upkeep, and depreciation) in cents per mile is approximately

$$C(x) = \frac{1910.5}{x^{1.72}} + 42.9 \qquad (5 \le x \le 20)$$

where x (in thousands) denotes the number of miles the car is driven each year. Show that C is a decreasing function of x on the interval $(5, 20)$. What does your result tell you about the average cost of driving a 2012 medium-sized sedan in terms of the number of miles driven?
Source: American Automobile Association.

89. **FIRST-CLASS MAIL** The volume of first-class mail deliveries (in billions of pieces) from 2008 through 2012 is approximated by

$$f(t) = \frac{90.7}{0.01t^2 + 0.01t + 1} \qquad (0 \le t \le 4)$$

where t is measured in years, with $t = 0$ corresponding to 2008. Show that the volume of first-class mail deliveries has been declining throughout the period under consideration.
Source: U.S. Postal Service.

90. **AIR POLLUTION** According to the South Coast Air Quality Management District, the level of nitrogen dioxide, a brown gas that impairs breathing, present in the atmosphere on a certain May day in downtown Los Angeles is approximated by

$$A(t) = 0.03t^3(t - 7)^4 + 60.2 \qquad (0 \le t \le 7)$$

where $A(t)$ is measured in pollutant standard index (PSI) and t is measured in hours, with $t = 0$ corresponding to 7 A.M. At what time of day is the air pollution increasing, and at what time is it decreasing?

91. **PROJECTED RETIREMENT FUNDS** Based on data from the Central Provident Fund of a certain country (a government agency similar to the Social Security Administration), the estimated cash in the fund in 2010 is given by

$$A(t) = -96.6t^4 + 403.6t^3 + 660.9t^2 + 250 \qquad (0 \le t \le 5)$$

where $A(t)$ is measured in billions of dollars and t is measured in decades, with $t = 0$ corresponding to 2010. Find the interval where A is increasing and the interval where A is decreasing, and interpret your results.
Hint: Use the quadratic formula.

92. **DRUG CONCENTRATION IN THE BLOOD** The concentration (in milligrams per cubic centimeter) of a certain drug in a patient's body t hr after injection is given by

$$C(t) = \frac{t^2}{2t^3 + 1} \qquad (0 \le t \le 4)$$

When is the concentration of the drug increasing, and when is it decreasing?

93. **SMALL CAR MARKET SHARE** Owing in part to an aging population and the squeeze from carbon-cutting regulations, the percentage of small and lower-midsize vehicles is expected to increase in the near future. The function

$$f(t) = \frac{5.3\sqrt{t} - 300}{\sqrt{t} - 10} \qquad (0 \le t \le 10)$$

gives the projected percentage of small and lower-midsize vehicles t years after 2005.
a. What was the percentage of small and lower-midsize vehicles in 2005? What is the projected percentage of small and lower-midsize vehicles in 2015?
b. Show that f is increasing on the interval $(0, 10)$, and interpret your results.
Source: J.D. Power Automotive.

94. **AIR POLLUTION** The amount of nitrogen dioxide, a brown gas that impairs breathing, present in the atmosphere on a certain May day in the city of Long Beach is approximated by

$$A(t) = \frac{136}{1 + 0.25(t - 4.5)^2} + 28 \qquad (0 \le t \le 11)$$

where $A(t)$ is measured in pollutant standard index (PSI) and t is measured in hours, with $t = 0$ corresponding to 7 A.M. Find the intervals where A is increasing and where A is decreasing, and interpret your results.
Source: Los Angeles Times.

95. **REVENUE FUNCTIONS** Suppose that the unit price of a product, p, is related to the quantity demanded, x, by the linear

demand equation $p = a - bx$, where a and b are positive constants.

a. Write an expression, R, giving the revenue realized in the sale of x units of the product.

b. Find the levels of production at which the revenue is increasing, at which the revenue is decreasing, and at which the revenue is stationary (neither increasing nor decreasing).

c. Find the level of production that will result in a maximum revenue. What is the maximum revenue?

96. **U.S. NURSING SHORTAGE** The demand for nurses between 2000 and 2015 is estimated to be

$$D(t) = 0.0007t^2 + 0.0265t + 2 \qquad (0 \le t \le 15)$$

where $D(t)$ is measured in millions and $t = 0$ corresponds to the year 2000. The supply of nurses over the same time period is estimated to be

$$S(t) = -0.0014t^2 + 0.0326t + 1.9 \qquad (0 \le t \le 15)$$

where $S(t)$ is also measured in millions.

a. Find an expression $G(t)$ giving the gap between the demand and supply of nurses over the period in question.

b. Find the interval where G is decreasing and where it is increasing. Interpret your result.

c. Find the relative extrema of G. Interpret your result.

Source: U.S. Department of Health and Human Services.

In Exercises 97–104, determine whether the statement is true or false. If it is true, explain why it is true. If it is false, give an example to show why it is false.

97. If f is decreasing on (a, b), then $f'(x) < 0$ for each x in (a, b).

98. If f is increasing on (a, b), then $-f$ is decreasing on (a, b).

99. If f and g are both increasing on (a, b), then $f + g$ is increasing on (a, b).

100. If $f(x)$ and $g(x)$ are positive on (a, b) and both f and g are increasing on (a, b), then fg is increasing on (a, b).

101. If f and g are decreasing on (a, b), then fg must be decreasing on (a, b).

102. If $f'(c) = 0$, then f has a relative maximum or a relative minimum at $x = c$.

103. A polynomial function of degree n $(n \ge 2)$ can have at most $(n - 1)$ relative extrema.

104. If f is decreasing on (a, b) and g is decreasing on (a, b), then the composite function $h = g \circ f$ is also decreasing on (a, b).

105. Using Theorem 1, verify that the linear function $f(x) = mx + b$ is (a) increasing everywhere if $m > 0$, (b) decreasing everywhere if $m < 0$, and (c) constant if $m = 0$.

106. Let $f(x) = -x^2 + ax + b$. Determine the constants a and b such that f has a relative maximum at $x = 2$ and the relative maximum value is 7.

107. Let $f(x) = ax^3 + 6x^2 + bx + 4$. Determine the constants a and b such that f has a relative minimum at $x = -1$ and a relative maximum at $x = 2$.

108. Let

$$f(x) = \begin{cases} -3x & \text{if } x < 0 \\ 2x + 4 & \text{if } x \ge 0 \end{cases}$$

a. Compute $f'(x)$, and show that it changes sign from negative to positive as we move across $x = 0$.

b. Show that f does not have a relative minimum at $x = 0$. Does this contradict the First Derivative Test? Explain your answer.

109. Let

$$f(x) = \begin{cases} \dfrac{1}{x^2} & \text{if } x > 0 \\ x^2 & \text{if } x \le 0 \end{cases}$$

a. Compute $f'(x)$, and show that it does not change sign as we move across $x = 0$.

b. Show that f has a relative minimum at $x = 0$. Does this contradict the First Derivative Test? Explain your answer.

110. Refer to Example 6, page 128.

a. Show that f is increasing on the interval $(0, 1)$.

b. Show that $f(0) = -1$ and $f(1) = 1$, and use the result of part (a) together with the Intermediate Value Theorem to conclude that there is exactly one root of $f(x) = 0$ in $(0, 1)$.

4.1 Solutions to Self-Check Exercises

1. The derivative of f is

$$f'(x) = 2x^2 - 2x - 12 = 2(x + 2)(x - 3)$$

and it is continuous everywhere. The zeros of $f'(x)$ are $x = -2$ and $x = 3$. The sign diagram of f' is shown in the accompanying figure. We conclude that f is increasing

on the intervals $(-\infty, -2)$ and $(3, \infty)$ and decreasing on the interval $(-2, 3)$.

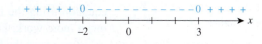

2. The derivative of f is

$$f'(x) = \frac{(1-x^2)\dfrac{d}{dx}(x^2) - x^2\dfrac{d}{dx}(1-x^2)}{(1-x^2)^2}$$

$$= \frac{(1-x^2)(2x) - x^2(-2x)}{(1-x^2)^2} = \frac{2x}{(1-x^2)^2}$$

and it is continuous everywhere except at $x = \pm 1$. Since $f'(x)$ is equal to zero at $x = 0$, $x = 0$ is a critical number of f. Next, observe that $f'(x)$ is discontinuous at $x = \pm 1$,

but since these numbers are not in the domain of f, they do not qualify as critical numbers of f. Finally, from the sign diagram of f' shown in the accompanying figure, we conclude that $f(0) = 0$ is a relative minimum of f.

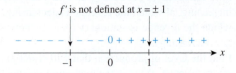

f' is not defined at $x = \pm 1$

USING **TECHNOLOGY**

Using the First Derivative to Analyze a Function

A graphing utility is an effective tool for analyzing the properties of functions. This is especially true when we also bring into play the power of calculus, as the following examples show.

EXAMPLE 1 Let $f(x) = 2.4x^4 - 8.2x^3 + 2.7x^2 + 4x + 1$.

a. Use a graphing utility to plot the graph of f.
b. Find the intervals where f is increasing and the intervals where f is decreasing.
c. Find the relative extrema of f.

Solution

a. The graph of f in the viewing window $[-2, 4] \times [-10, 10]$ is shown in Figure T1.
b. We compute

$$f'(x) = 9.6x^3 - 24.6x^2 + 5.4x + 4$$

and observe that f' is continuous everywhere, so the critical numbers of f occur at values of x where $f'(x) = 0$. To solve this last equation, observe that $f'(x)$ is a *polynomial function* of degree 3. The easiest way to solve the polynomial equation

$$9.6x^3 - 24.6x^2 + 5.4x + 4 = 0$$

is to use the function on a graphing utility for solving polynomial equations. (Not all graphing utilities have this function.) You can also use **TRACE** and **ZOOM**, but this will not give the same accuracy without a much greater effort.

We find

$$x_1 \approx 2.22564943249 \qquad x_2 \approx 0.63272944121 \qquad x_3 \approx -0.295878873696$$

Referring to Figure T1, we conclude that f is decreasing on $(-\infty, -0.2959)$ and $(0.6327, 2.2256)$ (correct to four decimal places) and f is increasing on $(-0.2959, 0.6327)$ and $(2.2256, \infty)$.

c. Using the evaluation function of a graphing utility, we find the value of f at each of the critical numbers found in part (b). Upon referring to Figure T1 once again, we see that $f(x_3) \approx 0.2836$ and $f(x_1) \approx -8.2366$ are relative minimum values of f and $f(x_2) \approx 2.9194$ is a relative maximum value of f. ∎

Note The equation $f'(x) = 0$ in Example 1 is a polynomial equation, so it is easily solved by using the function for solving polynomial equations. We could also solve the equation using the function for finding the roots of equations, but that would require much more work. For equations that are *not* polynomial equations, however, our only choice is to use the function for finding the roots of equations. ∎

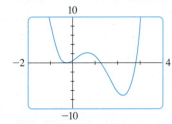

FIGURE T1
The graph of f in the viewing window $[-2, 4] \times [-10, 10]$

If the derivative of a function is difficult to compute or simplify and we do not require great precision in the solution, we can find the relative extrema of the function using a combination of **ZOOM** and **TRACE**. This technique, which does not require the use of the derivative of f, is illustrated in the following example.

EXAMPLE 2 Let $f(x) = x^{1/3}(x^2 + 1)^{-3/2}3^{-x}$.

a. Use a graphing utility to plot the graph of f.*
b. Find the relative extrema of f.

Solution

a. The graph of f in the viewing window $[-4, 2] \times [-2, 1]$ is shown in Figure T2.
b. From the graph of f in Figure T2, we see that f has relative maxima when $x \approx -2$ and $x \approx 0.25$ and a relative minimum when $x \approx -0.75$. To obtain a better approximation of the first relative maximum, we zoom in with the cursor at approximately the point on the graph corresponding to $x \approx -2$. Then, using **TRACE,** we see that a relative maximum occurs when $x \approx -1.76$ with value $y \approx -1.01$. Similarly, we find the other relative maximum where $x \approx 0.20$ with value $y \approx 0.44$. Repeating the procedure, we find the relative minimum at $x \approx -0.86$ and $y \approx -1.07$.

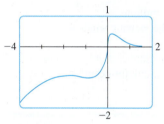

FIGURE **T2**
The graph of f in the viewing window
$[-4, 2] \times [-2, 1]$

You can also use the "minimum" and "maximum" functions of a graphing utility to find the relative extrema of the function. See the website for the procedure.

Finally, we comment that if you have access to a computer and software such as Derive, Maple, or Mathematica, then symbolic differentiation will yield the derivative $f'(x)$ of a differentiable function. This software will also solve the equation $f'(x) = 0$ with ease. Thus, the use of a computer will simplify even more greatly the analysis of functions.

---

*Functions of the form $f(x) = 3^{-x}$ are called *exponential functions*, and we will study them in greater detail in Chapter 5.

TECHNOLOGY EXERCISES

In Exercises 1–4, find (a) the intervals where f is increasing and the intervals where f is decreasing and (b) the relative extrema of f. Express your answers accurate to four decimal places.

1. $f(x) = 3.4x^4 - 6.2x^3 + 1.8x^2 + 3x - 2$

2. $f(x) = 1.8x^4 - 9.1x^3 + 5x - 4$

3. $f(x) = 2x^5 - 5x^3 + 8x^2 - 3x + 2$

4. $f(x) = 3x^5 - 4x^2 + 3x - 1$

In Exercises 5–8, use the ZOOM and TRACE features to find (a) the intervals where f is increasing and the intervals where f is decreasing and (b) the relative extrema of f. Express your answers accurate to two decimal places.

5. $f(x) = (2x + 1)^{1/3}(x^2 + 1)^{-2/3}$

6. $f(x) = [x^2(x^3 - 1)]^{1/3} + \dfrac{1}{x}$

7. $f(x) = x - \sqrt{1 - x^2}$

8. $f(x) = \dfrac{\sqrt{x}(x^2 - 1)^2}{x - 2}$

9. WORKERS' EXPECTATIONS The percentage of workers who expect to work past age 65 years has more than tripled in 30 years. The function

$$f(t) = 0.004545t^3 - 0.1113t^2 + 1.385t + 11 \qquad (0 \le t \le 22)$$

gives an approximation of the percentage of workers who expect to work past age 65 in year t, where t is measured in years, with $t = 0$ corresponding to 1991.
   a. Plot the graph of f in the viewing window $[0, 22] \times [10, 38]$.
   b. Prove that the percentage of workers who expect to work past age 65 was increasing over this interval.

10. OUTSOURCING OF JOBS The cumulative number of jobs outsourced overseas by U.S.-based multinational companies from 2005 ($t = 0$) through 2009 is approximated by

$$N(t) = -0.05(t + 1.1)^{2.2} + 0.7t + 0.9 \qquad (0 \le t \le 4)$$

where $N(t)$ is measured in millions.
   a. Plot the graph of N in the viewing window $[0, 4] \times [0, 2]$.
   b. What was the maximum number of jobs outsourced during this period? When did this occur?
   Source: Forrester Research.

11. AIR POLLUTION The amount of nitrogen dioxide, a brown gas that impairs breathing, present in the atmosphere on a certain May day in the city of Long Beach, is approximated by

$$A(t) = \dfrac{136}{1 + 0.25(t - 4.5)^2} + 28 \qquad (0 \le t \le 11)$$

where $A(t)$ is measured in pollutant standard index (PSI) and t is measured in hours, with $t = 0$ corresponding to 7 A.M. When is the PSI increasing, and when is it decreasing? At what time is the PSI highest, and what is its value at that time?

12. MEDIAN AGE OF WOMEN AT FIRST MARRIAGE The following data gives the median age of women in the United States at first marriage from 1960 through 2010.

Decade	1960	1970	1980	1990	2000	2010
Median Age	20.1	21.0	22.0	24.0	25.1	26.1

   a. Use **CubicReg** to find a third-degree polynomial regression model for these data. Let t be measured in decades, with $t = 0$ corresponding to 1960.
   b. Plot the graph of the polynomial function f found in part (a) using the viewing window $[0, 5] \times [18, 28]$.
   c. Where is f increasing? What does this tell us?
   d. Verify the result of part (c) analytically.
   Source: The National Marriage Project, Rutgers University.

4.2 Applications of the Second Derivative

Determining the Intervals of Concavity

Consider the graphs shown in Figure 26, which give the estimated population of the world and of the United States through the year 2000. Both graphs are rising, indicating that both the U.S. population and the world population continued to increase through the

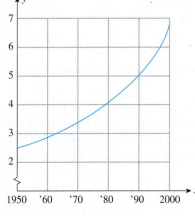

(a) World population in billions

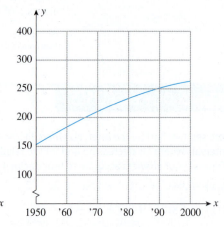

(b) U.S. population in millions

FIGURE 26
Source: U.S. Department of Commerce and Worldwatch Institute.

year 2000. But observe that the graph in Figure 26a opens upward, whereas the graph in Figure 26b opens downward. What is the significance of this? To answer this question, let's look at the slopes of the tangent lines to various points on each graph (Figure 27).

In Figure 27a, we see that the slopes of the tangent lines to the graph are increasing as we move from left to right. Since the slope of the tangent line to the graph at a point on the graph measures the rate of change of the function at that point, we conclude that the world population not only was increasing through the year 2000 but also was increasing at an *increasing* pace. A similar analysis of Figure 27b reveals that the U.S. population was increasing, but at a *decreasing* pace.

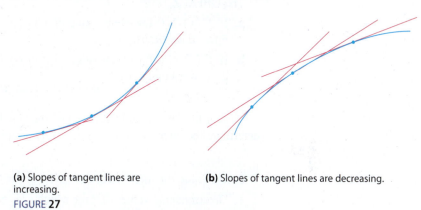

(a) Slopes of tangent lines are increasing.

(b) Slopes of tangent lines are decreasing.

FIGURE **27**

The shape of a curve can be described by using the notion of concavity.

> ### Concavity of a Function *f*
>
> Let the function *f* be differentiable on an interval (a, b). Then,
>
> **1.** The graph of *f* is **concave upward** on (a, b) if f' is increasing on (a, b).
>
> **2.** The graph of *f* is **concave downward** on (a, b) if f' is decreasing on (a, b).

Geometrically, a curve is concave upward if it lies above its tangent lines (Figure 28a). Similarly, a curve is concave downward if it lies below its tangent lines (Figure 28b).

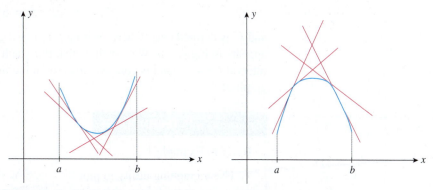

(a) The graph of *f* is concave upward on (a, b).

(b) The graph of *f* is concave downward on (a, b).

FIGURE **28**

We also say that the graph of *f* is *concave upward at a number c* if there exists an interval (a, b) containing *c* on which the graph of *f* is concave upward. Similarly, we say that the graph of *f* is *concave downward at a number c* if there exists an interval (a, b) containing *c* on which the graph of *f* is concave downward.

If a function f has a second derivative f'', we can use f'' to determine the intervals of concavity of the graph of the function. Recall that $f''(x)$ measures the rate of change of the slope $f'(x)$ of the tangent line to the graph of f at the point $(x, f(x))$. Thus, if $f''(x) > 0$ on an interval (a, b), then the slopes of the tangent lines to the graph of f are increasing on (a, b), so the graph of f is concave upward on (a, b). Similarly, if $f''(x) < 0$ on (a, b), then the graph of f is concave downward on (a, b). These observations suggest the following theorem.

THEOREM 2

a. If $f''(x) > 0$ for every value of x in (a, b), then the graph of f is concave upward on (a, b).

b. If $f''(x) < 0$ for every value of x in (a, b), then the graph of f is concave downward on (a, b).

The following procedure, based on the conclusions of Theorem 2, may be used to determine the intervals of concavity of the graph of a function.

Determining the Intervals of Concavity of the Graph of f

1. Determine the values of x for which f'' is zero or where f'' is not defined, and identify the open intervals determined by these numbers.

2. Determine the sign of f'' in each interval found in Step 1. To do this, compute $f''(c)$, where c is any conveniently chosen test number in the interval.
  a. If $f''(c) > 0$, then the graph of f is concave upward on that interval.
  b. If $f''(c) < 0$, then the graph of f is concave downward on that interval.

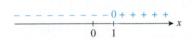

FIGURE 29
Sign diagram for f''

EXAMPLE 1 Determine where the graph of the function

$$f(x) = x^3 - 3x^2 - 24x + 32$$

is concave upward and where it is concave downward.

Solution   Here,

$$f'(x) = 3x^2 - 6x - 24$$
$$f''(x) = 6x - 6 = 6(x - 1)$$

and f'' is defined everywhere. Setting $f''(x) = 0$ gives $x = 1$. The sign diagram of f'' appears in Figure 29. We conclude that the graph of f is concave downward on the interval $(-\infty, 1)$ and is concave upward on the interval $(1, \infty)$. Figure 30 shows the graph of f.

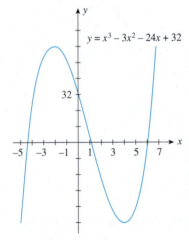

$y = x^3 - 3x^2 - 24x + 32$

FIGURE 30
The graph of f is concave downward on $(-\infty, 1)$ and concave upward on $(1, \infty)$.

Exploring with TECHNOLOGY

Refer to Example 1.

1. Use a graphing utility to plot the graph of $f(x) = x^3 - 3x^2 - 24x + 32$ and its second derivative $f''(x) = 6x - 6$ using the viewing window $[-10, 10] \times [-80, 90]$.

2. By studying the graph of f'', determine the intervals where $f''(x) > 0$ and the intervals where $f''(x) < 0$. Next, look at the graph of f, and determine the intervals where the graph of f is concave upward and the intervals where it is concave downward. Are these observations what you might have expected?

EXAMPLE 2 Determine the intervals where the graph of the function $f(x) = x + \dfrac{1}{x}$ is concave upward and where it is concave downward.

Solution   We have

$$f'(x) = 1 - \frac{1}{x^2}$$

$$f''(x) = \frac{2}{x^3}$$

We deduce from the sign diagram for f'' (Figure 31) that the graph of the function f is concave downward on the interval $(-\infty, 0)$ and concave upward on the interval $(0, \infty)$. The graph of f is sketched in Figure 32.

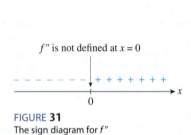

f'' is not defined at $x = 0$

FIGURE 31
The sign diagram for f''

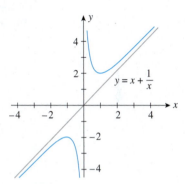

$y = x + \dfrac{1}{x}$

FIGURE 32
The graph of f is concave downward on $(-\infty, 0)$ and concave upward on $(0, \infty)$.

Inflection Points

Figure 33 shows the total sales S of a manufacturer of automobile air conditioners versus the amount of money x that the company spends on advertising its product.

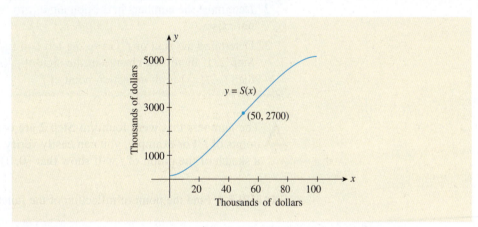

$y = S(x)$

(50, 2700)

Thousands of dollars

Thousands of dollars

FIGURE 33
The graph of S has a point of inflection at (50, 2700).

Notice that the graph of the continuous function $y = S(x)$ changes concavity—from upward to downward—at the point $(50, 2700)$. This point is called an *inflection point* of S. To understand the significance of this inflection point, observe that the total sales increase rather slowly at first, but as more money is spent on advertising, the total sales increase rapidly. This rapid increase reflects the effectiveness of the company's ads. However, a point is soon reached after which any additional advertising expenditure results in increased sales but at a slower rate of increase. This point, commonly known as the *point of diminishing returns,* is the point of inflection of the function *S*. We will return to this example later.

Let's now state formally the definition of an inflection point.

> **Inflection Point**
>
> A point on the graph of a continuous function *f* where the tangent line exists and where the concavity changes is called a **point of inflection** or an **inflection point.**

Observe that the graph of a function crosses its tangent line at a point of inflection (Figure 34).

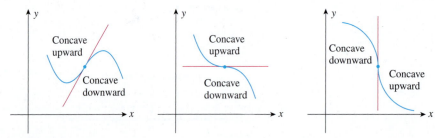

FIGURE 34
At each point of inflection, the graph of a function crosses its tangent line.

The following procedure may be used to find inflection points:

> **Finding Inflection Points**
>
> 1. Compute $f''(x)$.
> 2. Determine the numbers in the domain of *f* for which $f''(x) = 0$ or $f''(x)$ does not exist.
> 3. Determine the sign of $f''(x)$ to the left and right of each number *c* found in Step 2. If there is a change in the sign of $f''(x)$ as we move across $x = c$, then $(c, f(c))$ is an inflection point of *f*.

⚠️ The numbers that were found in Step 2 are only *candidates* for the inflection points of *f*. For example, you can easily verify that $f''(0) = 0$ if $f(x) = x^4$, but a sketch of the graph of *f* will show that $(0, 0)$ is *not* an inflection point.

EXAMPLE 3 Find the point of inflection of the function $f(x) = x^3$.

Solution

$$f'(x) = 3x^2$$
$$f''(x) = 6x$$

Observe that f'' is continuous everywhere and is zero if $x = 0$. The sign diagram of f'' is shown in Figure 35 (see next page). From this diagram, we see that $f''(x)$ changes sign as we move across $x = 0$. Thus, the point $(0, 0)$ is an inflection point (Figure 36).

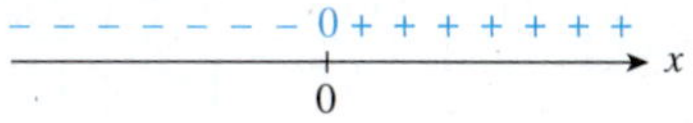

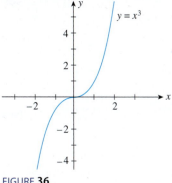

FIGURE **36**
f has an inflection point at (0, 0).

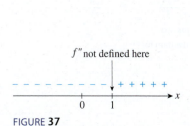

FIGURE **35**
Sign diagram for *f*″

EXAMPLE 4 Determine the intervals where the graph of the function $f(x) = (x - 1)^{5/3}$ is concave upward and where it is concave downward, and find the inflection points of *f*.

Solution   The first derivative of *f* is

$$f'(x) = \frac{5}{3}(x - 1)^{2/3}$$

and the second derivative of *f* is

$$f''(x) = \frac{10}{9}(x - 1)^{-1/3} = \frac{10}{9(x - 1)^{1/3}}$$

We see that *f*″ is not defined at *x* = 1. Furthermore, *f*″(*x*) is not equal to zero anywhere. The sign diagram of *f*″ is shown in Figure 37. From the sign diagram, we see that the graph of *f* is concave downward on (−∞, 1) and concave upward on (1, ∞). Next, since *x* = 1 does lie in the domain of *f*, we see that the point (1, 0) is an inflection point (Figure 38).

f″ not defined here

FIGURE **37**
Sign diagram for *f*″

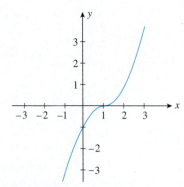

FIGURE **38**
f has an inflection point at (1, 0).

EXAMPLE 5 Determine the intervals where the graph of the function

$$f(x) = \frac{1}{x^2 + 1}$$

is concave upward and where it is concave downward, and find the inflection points of *f*.

Solution   The first derivative of f is

$$f'(x) = \frac{d}{dx}(x^2 + 1)^{-1} = -2x(x^2 + 1)^{-2}$$ Rewrite the original function, and use the General Power Rule.

$$= -\frac{2x}{(x^2 + 1)^2}$$

Next, using the Quotient Rule, we find

$$f''(x) = \frac{(x^2 + 1)^2(-2) + (2x)2(x^2 + 1)(2x)}{(x^2 + 1)^4}$$ (x^2) See page 16.

$$= \frac{(x^2 + 1)[-2(x^2 + 1) + 8x^2]}{(x^2 + 1)^4} = \frac{(x^2 + 1)(6x^2 - 2)}{(x^2 + 1)^4}$$

$$= \frac{2(3x^2 - 1)}{(x^2 + 1)^3}$$ Cancel the common factors.

Observe that f'' is continuous everywhere and is zero if

$$3x^2 - 1 = 0$$

$$x^2 = \frac{1}{3}$$

or $x = \pm\sqrt{3}/3$. The sign diagram for f'' is shown in Figure 39. From the sign diagram for f'', we see that the graph of f is concave upward on $(-\infty, -\sqrt{3}/3)$ and $(\sqrt{3}/3, \infty)$, and is concave downward on $(-\sqrt{3}/3, \sqrt{3}/3)$. Also, observe that $f''(x)$ changes sign as we move across the numbers $x = -\sqrt{3}/3$ and $x = \sqrt{3}/3$. Since

$$f\left(-\frac{\sqrt{3}}{3}\right) = \frac{1}{\frac{1}{3} + 1} = \frac{3}{4} \quad \text{and} \quad f\left(\frac{\sqrt{3}}{3}\right) = \frac{3}{4}$$

we see that the points $(-\sqrt{3}/3, 3/4)$, and $(\sqrt{3}/3, 3/4)$, are inflection points. The graph of f is shown in Figure 40.

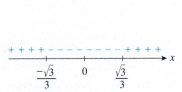

FIGURE 39
Sign diagram for f''

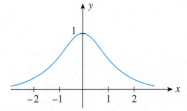

FIGURE 40
The graph of $f(x) = \dfrac{1}{x^2 + 1}$ is concave upward on $(-\infty, -\sqrt{3}/3)$ and $(\sqrt{3}/3, \infty)$ and is concave downward on $(-\sqrt{3}/3, \sqrt{3}/3)$.

Explore and Discuss

1. Suppose $(c, f(c))$ is an inflection point of f. Can you conclude that f does not have a relative extremum at $x = c$? Explain your answer.

2. True or false: A polynomial function of degree 3 has exactly one inflection point.
   Hint: Study the function $f(x) = ax^3 + bx^2 + cx + d \ (a \neq 0)$.

The next example uses an interpretation of the first and second derivatives to help us sketch the graph of a function.

EXAMPLE 6 Sketch the graph of a function having the following properties:

$$f(-1) = 4$$
$$f(0) = 2$$
$$f(1) = 0$$
$$f'(-1) = 0$$
$$f'(1) = 0$$
$$f'(x) > 0 \quad \text{on } (-\infty, -1) \text{ and } (1, \infty)$$
$$f'(x) < 0 \quad \text{on } (-1, 1)$$
$$f''(x) < 0 \quad \text{on } (-\infty, 0)$$
$$f''(x) > 0 \quad \text{on } (0, \infty)$$

Solution First, we plot the points $(-1, 4)$, $(0, 2)$, and $(1, 0)$ that lie on the graph of f. Since $f'(-1) = 0$ and $f'(1) = 0$, the tangent lines at the points $(-1, 4)$ and $(1, 0)$ are horizontal. Since $f'(x) > 0$ on $(-\infty, -1)$ and $f'(x) < 0$ on $(-1, 1)$, we see that f has a relative maximum at the point $(-1, 4)$. Also, $f'(x) < 0$ on $(-1, 1)$ and $f'(x) > 0$ on $(1, \infty)$ implies that f has a relative minimum at the point $(1, 0)$ (Figure 41a).

Since $f''(x) < 0$ on $(-\infty, 0)$ and $f''(x) > 0$ on $(0, \infty)$, we see that the point $(0, 2)$ is an inflection point. Finally, we complete the graph, making use of the fact that f is increasing on $(-\infty, -1)$ and $(1, \infty)$, where it is given that $f'(x) > 0$, and f is decreasing on $(-1, 1)$, where $f'(x) < 0$. Also, make sure that the graph of f is concave downward on $(-\infty, 0)$ and concave upward on $(0, \infty)$ (Figure 41b).

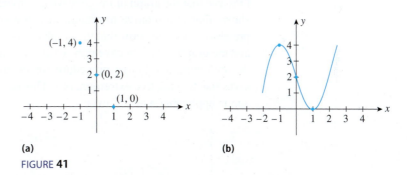

(a)   (b)

FIGURE **41**

Example 7 illustrates a familiar interpretation of the significance of an inflection point of a function.

APPLIED EXAMPLE 7 Effect of Advertising on Sales The total sales S (in thousands of dollars) of Arctic Air Corporation, a manufacturer of automobile air conditioners, is related to the amount of money x (in thousands of dollars) the company spends on advertising its products by the formula

$$S(x) = -0.01x^3 + 1.5x^2 + 200 \qquad (0 \le x \le 100)$$

Find the inflection point of the function S.

Solution The first two derivatives of S are given by

$$S'(x) = -0.03x^2 + 3x$$
$$S''(x) = -0.06x + 3$$

Setting $S''(x) = 0$ gives $x = 50$. So $(50, S(50))$ is the only candidate for an inflection point of S. Moreover, since

$$S''(x) > 0 \quad \text{for } x < 50$$

and

$$S''(x) < 0 \quad \text{for } x > 50$$

the point $(50, 2700)$ is an inflection point of the function S. The graph of S appears in Figure 42. Notice that this is the same graph as that shown in Figure 33.

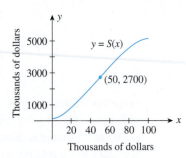

FIGURE 42
The graph of S has a point of inflection at (50, 2700).

The Second Derivative Test

We now show how the second derivative f'' of a function f can be used to help us determine whether a critical number of f gives rise to a relative extremum of f. Figure 43a shows the graph of a function that has a relative maximum at $x = c$. Observe that the graph of f is concave downward at that number. Similarly, Figure 43b shows that at a relative minimum of f, the graph is concave upward. But from our previous work, we know that the graph of f is concave downward at $x = c$ if $f''(c) < 0$ and the graph of f is concave upward at $x = c$ if $f''(c) > 0$. These observations suggest the following alternative procedure for determining whether a critical number of f gives rise to a relative extremum of f. This result is called the **Second Derivative Test** and is applicable when f'' exists.

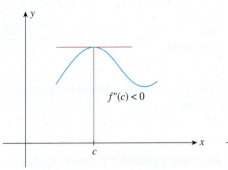   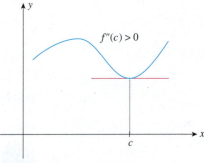

(a) *f* has a relative maximum at *x* = *c*.   **(b)** *f* has a relative minimum at *x* = *c*.

FIGURE 43

> ### The Second Derivative Test
> 1. Compute $f'(x)$ and $f''(x)$.
> 2. Find all the critical numbers of f at which $f'(x) = 0$.
> 3. Compute $f''(c)$ for each such critical number c.
>   a. If $f''(c) < 0$, then f has a relative maximum at c.
>   b. If $f''(c) > 0$, then f has a relative minimum at c.
>   c. If $f''(c) = 0$ or $f''(c)$ does not exist, the test fails; that is, it is inconclusive.

Note   The Second Derivative Test does not yield a conclusion if $f''(c) = 0$ or if $f''(c)$ does not exist. In other words, $x = c$ may or may not give rise to a relative extremum (see Exercise 110, page 290). In such cases, you should revert to the First Derivative Test.   ▪

EXAMPLE 8 Determine the relative extrema of the function

$$f(x) = x^3 - 3x^2 - 24x + 32$$

using the Second Derivative Test. (See Example 7, Section 4.1.)

Solution   We have

$$f'(x) = 3x^2 - 6x - 24 = 3(x + 2)(x - 4)$$

so $f'(x) = 0$ when $x = -2$ and $x = 4$, the critical numbers of f, as in Example 7, Section 4.1. Next, we compute

$$f''(x) = 6x - 6 = 6(x - 1)$$

Since

$$f''(-2) = 6(-2 - 1) = -18 < 0$$

the Second Derivative Test implies that $f(-2) = 60$ is a relative maximum of f. Also,

$$f''(4) = 6(4 - 1) = 18 > 0$$

and the Second Derivative Test implies that $f(4) = -48$ is a relative minimum of f, which confirms the results obtained earlier.   ▪

Explore and Discuss

Suppose a function f has the following properties:

1. $f''(x) > 0$ for all x in an interval (a, b).
2. There is a number c between a and b such that $f'(c) = 0$.

What special property can you ascribe to the point $(c, f(c))$? Answer the question if Property 1 is replaced by the property that $f''(x) < 0$ for all x in (a, b).

Comparing the First and Second Derivative Tests

Notice that both the First Derivative Test and the Second Derivative Test are used to classify the critical numbers of f. What are the pros and cons of the two tests? Since the Second Derivative Test is applicable only when f'' exists, it is less versatile than the First Derivative Test. For example, it cannot be used to locate the relative minimum $f(0) = 0$ of the function $f(x) = x^{2/3}$.

Furthermore, the Second Derivative Test is inconclusive when f'' is equal to zero at a critical number of f, whereas the First Derivative Test always yields a conclusion; that is, it tells us if f has a relative maximum, relative minimum, or neither. The Second Derivative Test is also inconvenient to use when f'' is difficult to compute. On the plus side, if f'' is computed easily, then we use the Second Derivative Test, since it involves just the evaluation of f'' at the critical number(s) of f. Also, the conclusions of the Second Derivative Test are important in theoretical work.

We close this section by summarizing the different roles played by the first derivative f' and the second derivative f'' of a function f in determining the properties of the graph of f. The first derivative f' tells us where f is increasing and where f is decreasing, whereas the second derivative f'' tells us where the graph of f is concave upward and where it is concave downward. These different properties of f are reflected by the signs of f' and f'' in the interval of interest. The following table shows the general characteristics of the function f for various possible combinations of the signs of f' and f'' in an interval (a, b).

Signs of f' and f''	Properties of f	General Shape of the Graph of f
$f'(x) > 0$ $f''(x) > 0$	f is increasing. The graph of f is concave upward.	
$f'(x) > 0$ $f''(x) < 0$	f is increasing. The graph of f is concave downward.	
$f'(x) < 0$ $f''(x) > 0$	f is decreasing. The graph of f is concave upward.	
$f'(x) < 0$ $f''(x) < 0$	f is decreasing. The graph of f is concave downward.	

4.2 Self-Check Exercises

1. Determine where the graph of the function
   $f(x) = 4x^3 - 3x^2 + 6$ is concave upward and where it is concave downward.

2. Using the Second Derivative Test, if applicable, find the relative extrema of the function
   $f(x) = 2x^3 - \frac{1}{2}x^2 - 12x - 10$.

3. **GDP OF A COUNTRY** A certain country's gross domestic product (GDP) (in millions of dollars) in year t is described by the function

$$G(t) = -2t^3 + 45t^2 + 20t + 6000 \qquad (0 \le t \le 11)$$

where $t = 0$ corresponds to the beginning of 2004. Find the inflection point of the function G, and discuss its significance.

Solutions to Self-Check Exercises 4.2 can be found on page 290.

4.2 Concept Questions

1. Explain what it means for the graph of a function f to be (a) concave upward and (b) concave downward on an open interval I. Given that f has a second derivative on I (except at isolated numbers), how do you determine where the graph of f is concave upward and where it is concave downward?

2. What is an inflection point of a function f? How do you find the inflection point(s) of a function f whose rule is given?

3. State the Second Derivative Test. What are the pros and cons of using the First Derivative Test and the Second Derivative Test?

4.2 Exercises

In Exercises 1–8, you are given the graph of a function f. Determine the intervals where the graph of f is concave upward and where it is concave downward. Also, find all inflection points of f, if any.

1.

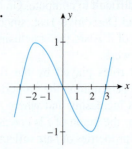

2.

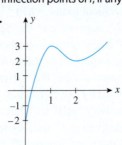

3.

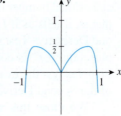

4.

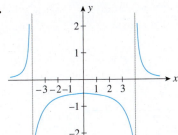

5.

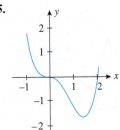

6.

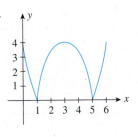

7.

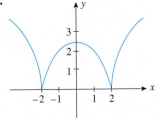

8.

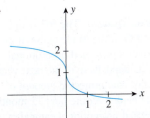

9. Refer to the graph of f shown in the following figure:

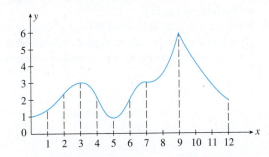

    a. Find the intervals where f is concave upward and the intervals where f is concave downward.

    b. Find the inflection points of f.

10. Refer to the figure for Exercise 9.

    a. Explain how the Second Derivative Test can be used to show that the critical number 3 gives rise to a relative maximum of f and the critical number 5 gives rise to a relative minimum of f.

    b. Explain why the Second Derivative Test cannot be used to show that the critical number 7 does not give rise to a relative extremum of f nor can it be used to show that the critical number 9 gives rise to a relative maximum of f.

In Exercises 11–14, determine which graph—(a), (b), or (c)—is the graph of the function f with the specified properties.

11. $f(2) = 1, f'(2) > 0$, and $f''(2) < 0$

    (a)

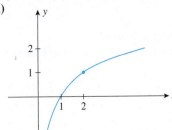

    (b)

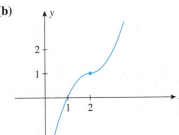

    (c)

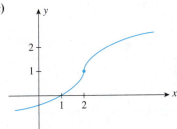

12. $f(1) = 2, f'(x) > 0$ on $(-\infty, 1)$ and $(1, \infty)$, and $f''(1) = 0$

    (a)

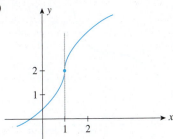

    (b)

(c)

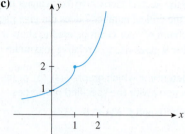

(a)

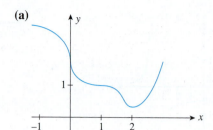

13. $f'(0)$ is undefined, f is decreasing on $(-\infty, 0)$, f is concave downward on $(0, 3)$, and f has an inflection point at $x = 3$.

(b)

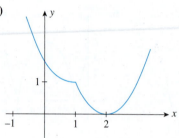

(a)

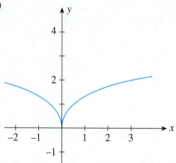

(c)

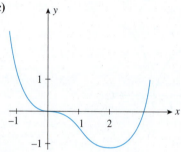

(b)

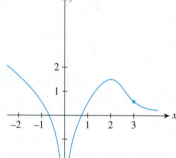

15. EFFECT OF ADVERTISING ON BANK DEPOSITS The following graphs were used by the CEO of the Madison Savings Bank to illustrate what effect a projected promotional campaign would have on its deposits over the next year. The functions D_1 and D_2 give the projected amount of money on deposit with the bank over the next 12 months with and without the proposed promotional campaign, respectively.

a. Determine the signs of $D_1'(t)$, $D_2'(t)$, $D_1''(t)$, and $D_2''(t)$ on the interval $(0, 12)$.

b. What can you conclude about the rate of change of the growth rate of the money on deposit with the bank with and without the proposed promotional campaign?

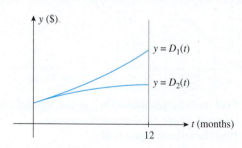

(c)

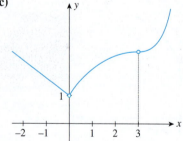

14. f is decreasing on $(-\infty, 2)$ and increasing on $(2, \infty)$, the graph of f is concave upward on $(1, \infty)$, and f has inflection points at $x = 0$ and $x = 1$.

16. MOTION OF CARS Two cars start out side by side and travel along a straight road. The velocity of Car A is $f(t)$ ft/sec, and the velocity of Car B is $g(t)$ ft/sec over the interval $[0, t_2]$. Furthermore, suppose the graphs of f and g are as depicted in the accompanying figure.

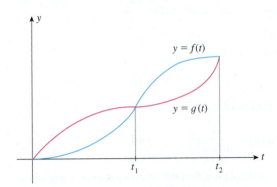

a. What can you say about the acceleration of Car A on the interval $(0, t_1)$? The acceleration of Car B on the interval $(0, t_1)$?
b. What can you say about the acceleration of Car A on the interval (t_1, t_2)? The acceleration of Car B over (t_1, t_2)?
c. What can you say about the acceleration of Car A at t_1? The acceleration of Car B at t_1?
d. At what time do both cars have the same velocity?

In Exercises 17–20, match the graphs (a), (b), (c), or (d) with the corresponding statement.

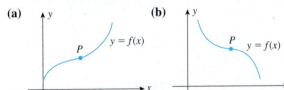

17. The function f is increasing most rapidly at P.

18. The function f is increasing least rapidly at P.

19. The function f is decreasing most rapidly at P.

20. The function f is decreasing least rapidly at P.

21. ASSEMBLY TIME OF A WORKER In the following graph, $N(t)$ gives the number of smartphones assembled by the average worker by the tth hr, where $t = 0$ corresponds to 8 A.M. and $0 \le t \le 4$. The point P is an inflection point.

a. What can you say about the rate of change of the rate of change of the number of smartphones assembled by the average worker between 8 A.M. and 10 A.M.? Between 10 A.M. and 12 noon?
b. At what time is the rate at which the smartphones are being assembled by the average worker greatest?

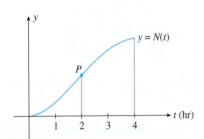

22. RUMORS OF A RUN ON A BANK The graph of the function f shows the total deposits with a bank t days after rumors abounded that there was a run on the bank due to heavy loan losses incurred by the bank.

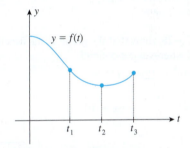

a. Determine the signs of $f'(t)$ on the intervals $(0, t_2)$ and (t_2, t_3), and determine the signs of $f''(t)$ on the intervals $(0, t_1)$ and (t_1, t_3).
b. Find where the inflection point(s) of f occur.
c. Interpret the results of parts (a) and (b).

23. WATER POLLUTION When organic waste is dumped into a pond, the oxidation process that takes place reduces the pond's oxygen content. However, given time, nature will restore the oxygen content to its natural level. In the following graph, $P(t)$ gives the oxygen content (as a percent of its normal level) t days after organic waste has been dumped into the pond. Explain the significance of the inflection point Q.

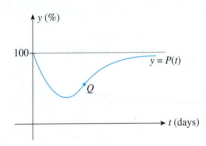

24. CASE-SHILLER HOME PRICE INDEX The following graph shows the change in the S&P/Case-Shiller Home Price Index based on a 20-city average from June 2001 $\left(t = \frac{1}{2}\right)$ through June 2008 $\left(t = 7\frac{1}{2}\right)$, adjusted for inflation.

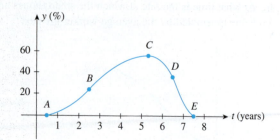

a. What do the points $A\left(\frac{1}{2}, 0\right)$ and $E\left(7\frac{1}{2}, 0\right)$ tell you about the change in the Case-Shiller Home Price Index over the period under consideration?

b. What does the point $C\left(5\frac{1}{3}, 56\right)$ tell you about the Case-Shiller Home Price Index?

c. Give an interpretation of the inflection points $B\left(2\frac{1}{2}, 24\right)$ and $D\left(6\frac{1}{2}, 36\right)$.

Source: New York Times.

In Exercises 25–28, show that the graph of the function is concave upward wherever it is defined.

25. $f(x) = 4x^2 - 12x + 7$

26. $g(x) = x^4 + \frac{1}{2}x^2 + 6x + 10$

27. $f(x) = \frac{1}{x^4}$ **28.** $g(x) = -\sqrt{4 - x^2}$

In Exercises 29–48, determine where the graph of the function is concave upward and where it is concave downward.

29. $f(x) = 2x^2 - 3x + 4$ **30.** $g(x) = -x^2 + 3x + 4$

31. $f(x) = 1 - x^3$ **32.** $g(x) = x^3 - x$

33. $f(x) = x^4 - 6x^3 + 2x + 8$

34. $f(x) = 3x^4 - 6x^3 + x - 8$

35. $f(x) = x^{4/7}$ **36.** $f(x) = \sqrt[3]{x}$

37. $f(x) = \sqrt{4 - x}$ **38.** $g(x) = \sqrt{x - 2}$

39. $f(x) = \frac{1}{x - 2}$ **40.** $g(x) = \frac{x}{x + 1}$

41. $f(x) = \frac{1}{2 + x^2}$ **42.** $g(x) = \frac{x}{1 + x^2}$

43. $h(t) = \frac{t^2}{t - 1}$ **44.** $f(x) = \frac{x + 1}{x - 1}$

45. $g(x) = x + \frac{1}{x^2}$ **46.** $h(r) = -\frac{1}{(r - 2)^2}$

47. $g(t) = (2t - 4)^{1/3}$ **48.** $f(x) = (x - 2)^{2/3}$

In Exercises 49–60, find the inflection point(s), if any, of each function.

49. $f(x) = x^3 - 2$ **50.** $g(x) = x^3 - 6x$

51. $f(x) = 6x^3 - 18x^2 + 12x - 20$

52. $g(x) = 2x^3 - 3x^2 + 18x - 8$

53. $f(x) = 3x^4 - 4x^3 + 1$

54. $f(x) = x^4 - 2x^3 + 6$

55. $g(t) = \sqrt[3]{t}$ **56.** $f(x) = \sqrt[5]{x}$

57. $f(x) = (x - 1)^3 + 2$ **58.** $f(x) = (x - 2)^{4/3}$

59. $f(x) = \frac{2}{1 + x^2}$ **60.** $f(x) = 2 + \frac{3}{x}$

In Exercises 61–76, find the relative extrema, if any, of each function. Use the Second Derivative Test if applicable.

61. $f(x) = -x^2 + 2x + 4$ **62.** $g(x) = 2x^2 + 3x + 7$

63. $f(x) = 2x^3 + 1$ **64.** $g(x) = x^3 - 6x$

65. $f(x) = \frac{1}{3}x^3 - 2x^2 - 5x - 5$

66. $f(x) = 2x^3 + 3x^2 - 12x - 4$

67. $g(t) = t + \frac{9}{t}$ **68.** $f(t) = 2t + \frac{3}{t}$

69. $f(x) = \frac{x}{1 - x}$ **70.** $f(x) = \frac{2x}{x^2 + 1}$

71. $f(t) = t^2 - \frac{16}{t}$ **72.** $g(x) = x^2 + \frac{2}{x}$

73. $g(s) = \frac{s}{1 + s^2}$ **74.** $g(x) = \frac{1}{1 + x^2}$

75. $f(x) = \frac{x^4}{x - 1}$ **76.** $f(x) = \frac{x^2}{x^2 + 1}$

In Exercises 77–82, sketch the graph of a function having the given properties.

77. $f(2) = 4, f'(2) = 0, f''(x) < 0$ on $(-\infty, \infty)$

78. $f(2) = 2, f'(2) = 0, f'(x) > 0$ on $(-\infty, 2), f'(x) > 0$ on $(2, \infty), f''(x) < 0$ on $(-\infty, 2), f''(x) > 0$ on $(2, \infty)$

79. $f(-2) = 4, f(3) = -2, f'(-2) = 0, f'(3) = 0,$ $f'(x) > 0$ on $(-\infty, -2)$ and $(3, \infty), f'(x) < 0$ on $(-2, 3)$, inflection point at $(1, 1)$

80. $f(0) = 0, f'(0)$ does not exist, $f''(x) < 0$ if $x \neq 0$

81. $f(0) = 1, f'(0) = 0, f(x) > 0$ on $(-\infty, \infty), f''(x) < 0$ on $(-\sqrt{2}/2, \sqrt{2}/2), f''(x) > 0$ on $(-\infty, -\sqrt{2}/2)$ and $(\sqrt{2}/2, \infty)$

82. f has domain $[-1, 1], f(-1) = -1, f\left(-\frac{1}{2}\right) = -2,$ $f'\left(-\frac{1}{2}\right) = 0, f''(x) > 0$ on $(-1, 1)$

83. DEMAND FOR RNs The following graph gives the total number of help-wanted ads for RNs (registered nurses) in 22 cities over the last 12 months as a function of time t (t measured in months).
   a. Explain why $N'(t)$ is positive on the interval $(0, 12)$.
   b. Determine the signs of $N''(t)$ on the interval $(0, 6)$ and the interval $(6, 12)$.
   c. Interpret the results of part (b).

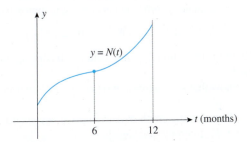

84. EFFECT OF BUDGET CUTS ON DRUG-RELATED CRIMES The graphs below were used by a police commissioner to illustrate what effect a budget cut would have on crime in the city. The number $N_1(t)$ gives the projected number of drug-related crimes in the next 12 months. The number $N_2(t)$ gives the projected number of drug-related crimes in the same time frame if next year's budget is cut.
   a. Explain why $N_1'(t)$ and $N_2'(t)$ are both positive on the interval $(0, 12)$.
   b. What are the signs of $N_1''(t)$ and $N_2''(t)$ on the interval $(0, 12)$?
   c. Interpret the results of part (b).

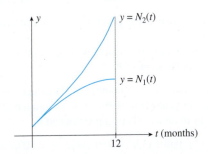

85. In the following figure, water is poured into the vase at a constant rate (in appropriate units), and the water level rises to a height of $f(t)$ units at time t as measured from the base of the vase. The graph of f follows. Explain the shape of the curve in terms of its concavity. What is the significance of the inflection point?

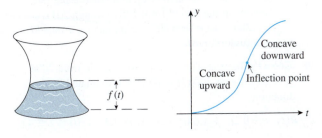

86. In the following figure, water is poured into an urn at a constant rate (in appropriate units), and the water level rises to a height of $f(t)$ units at time t as measured from the base of the urn. Sketch the graph of f and explain its shape, indicating where it is concave upward and concave downward. Indicate the inflection point on the graph, and explain its significance.
Hint: Study Exercise 85.

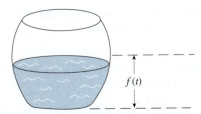

87. STATE CIGARETTE TAXES The average state cigarette tax per pack (in dollars) from 2001 through 2007 is approximated by the function

$$T(t) = 0.43t^{0.43} \qquad (1 \le t \le 7)$$

where t is measured in years, with $t = 1$ corresponding to 2001.
   a. Show that the average state cigarette tax per pack was increasing throughout the period in question.
   b. What can you say about the rate at which the average state cigarette tax per pack was increasing over the period in question?
Source: Campaign for Tobacco-Free Kids.

88. DRIVING COSTS A study of driving costs based on 2012 medium-sized sedans found that the average cost (car payments, gas, insurance, upkeep, and depreciation) in cents per mile is approximately

$$C(x) = \frac{1910.5}{x^{1.72}} + 42.9 \qquad (5 \le x \le 20)$$

where x (in thousands) denotes the number of miles the car is driven each year. Show that the graph of C is concave upward on the interval $(5, 20)$. What does your result tell you about the average cost of driving a 2012 medium-sized sedan in terms of the number of miles driven?
Source: American Automobile Association.

89. GLOBAL WARMING The increase in carbon dioxide (CO_2) in the atmosphere is a major cause of global warming. Using data obtained by Charles David Keeling, professor at Scripps Institution of Oceanography, the average amount of CO_2 in the atmosphere from 1958 through 2013 is approximated by

$$A(t) = 0.012414t^2 + 0.7485t + 313.9 \qquad (1 \le t \le 58)$$

where $A(t)$ is measured in parts per million volume (ppmv) and t in years, with $t = 1$ corresponding to 1958.
   a. What can you say about the rate of change of the average amount of atmospheric CO_2 from 1958 through 2013?

b. What can you say about the rate of change of the rate of change of the average amount of atmospheric CO_2 from 1958 through 2013?
Source: Scripps Institution of Oceanography.

90. FLIGHT OF A ROCKET The altitude (in feet) of a rocket t sec into flight is given by

$$s = f(t) = -t^3 + 54t^2 + 480t + 6 \quad (t \geq 0)$$

Find the point of inflection of the graph of f, and interpret your result. What is the maximum velocity attained by the rocket?

91. PUBLIC DEBT The U.S. public debt outstanding from 1990 to 2011 is approximated by

$$D(t) = 0.0032t^3 - 0.0698t^2 + 0.6048t + 3.22$$
$$(0 \leq t \leq 21)$$

trillion dollars in year t, where $t = 0$ is measured in years, with $t = 0$ corresponding to 1990.
a. Find the interval where the graph of D is concave downward and the interval where it is concave upward.
b. What is the inflection point of D? Interpret your result.
Source: U.S. Department of the Treasury.

92. DEATH RATE FROM AIDS The estimated number of deaths from AIDS worldwide is given by

$$f(t) = -0.0004401t^3 + 0.007t^2 + 0.112t + 0.28$$
$$(0 \leq t \leq 21)$$

million/year, where t is measured in years, with $t = 0$ corresponding to 1990. At what time was the death rate from AIDS worldwide increasing most rapidly? What was the rate?
Source: UNAIDS.

93. AGE AT FIRST MARRIAGE According to a study conducted by Rutgers University, the median age of women in the U.S. at first marriage is approximated by the function

$$f(t) = -0.083t^3 + 0.6t^2 + 0.18t + 20.1 \quad (0 \leq t \leq 5)$$

where t is measured in decades and $t = 0$ corresponds to the beginning of 1960.
a. What was the median age of women at first marriage at the beginning of 1960? At the beginning of the year 2000? At the beginning of 2010?
b. When was the median age of women at first marriage changing most rapidly over the time period under consideration?
Source: The National Marriage Project, Rutgers University.

94. EFFECT OF ADVERTISING ON HOTEL REVENUE The total annual revenue R of the Miramar Resorts Hotel is related to the amount of money x the hotel spends on advertising its services by the function

$$R(x) = -0.003x^3 + 1.35x^2 + 2x + 8000 \quad (0 \leq x \leq 400)$$

where both R and x are measured in thousands of dollars.

a. Find the interval where the graph of R is concave upward and the interval where the graph of R is concave downward. What is the inflection point of the graph of R?
b. Would it be more beneficial for the hotel to increase its advertising budget slightly when the budget is $140,000 or when it is $160,000?

95. OFFSHORE OUTSOURCING The amount (in billions of dollars) spent by the top 15 U.S. financial institutions on IT (information technology) offshore outsourcing is approximated by

$$A(t) = 0.92(t + 1)^{0.61} \quad (0 \leq t \leq 4)$$

where t is measured in years, with $t = 0$ corresponding to 2004.
a. Show that A is increasing on $(0, 4)$, and interpret your result.
b. Show that the graph of A is concave downward on $(0, 4)$. Interpret your result.
Source: Tower Group.

96. FORECASTING PROFITS As a result of increasing energy costs, the growth rate of the profit of the 4-year old Venice Glassblowing Company has begun to decline. Venice's management, after consulting with energy experts, decides to implement certain energy-conservation measures aimed at cutting energy bills. The general manager reports that, according to his calculations, the growth rate of Venice's profit should be on the increase again within 4 years. If Venice's profit (in hundreds of dollars) t years from now is given by the function

$$P(t) = t^3 - 9t^2 + 40t + 50 \quad (0 \leq t \leq 8)$$

determine whether the general manager's forecast will be accurate.
Hint: Find the inflection point of the graph of P, and study the concavity of the graph of P.

97. SENIORS IN CANADA The number of Canadians aged over 80 years in year t from 1981 ($t = 1$) through 2011 ($t = 31$) in thousands is approximated by

$$P(t) = -0.007333t^3 + 0.91343t^2 + 8.507t + 439$$
$$(0 \leq t \leq 31)$$

a. Verify that the number of Canadians aged over 80 years has been increasing throughout the period from 1981 through 2011.
Hint: Use the quadratic formula.
b. Verify that the number of Canadians aged over 80 years has been increasing at an increasing rate throughout the period from 1981 through 2011.
Hint: Show that P'' is positive on the interval in question.
Source: Statistics Canada.

98. GOOGLE'S REVENUE The revenue for Google from 2004 ($t = 0$) through 2008 ($t = 4$) is approximated by the function

$$R(t) = -0.2t^3 + 1.64t^2 + 1.31t + 3.2 \quad (0 \leq t \leq 4)$$

where $R(t)$ is measured in billions of dollars.

a. Find $R'(t)$ and $R''(t)$.

b. Show that $R'(t) > 0$ for all t in the interval $(0, 4)$ and interpret your result.
   Hint: Use the quadratic formula.

c. Find the inflection point of the graph of R, and interpret your result.

Source: Google company report.

99. POPULATION GROWTH IN CLARK COUNTY Clark County in Nevada—dominated by greater Las Vegas—is one of the fastest-growing metropolitan areas in the United States. The population of the county from 1970 through 2010 is approximated by the function

$$P(t) = 44{,}560t^3 - 89{,}394t^2 + 234{,}633t + 273{,}288$$
$$(0 \le t \le 4)$$

where t is measured in decades, with $t = 0$ corresponding to the beginning of 1970.

a. Show that the population of Clark County was always increasing over the time period in question.
   Hint: Show that $P'(t) > 0$ for all t in the interval $(0, 4)$.

b. Show that the population of Clark County was increasing at the slowest pace some time in August 1976.
   Hint: Find the inflection point of the graph of P in the interval $(0, 4)$.

Source: U.S. Census Bureau.

100. PUBLIC TRANSPORTATION BUDGET DEFICIT According to the Massachusetts Bay Transportation Authority (MBTA), the projected cumulative MBTA budget deficit with the $160 million rescue package (in billions of dollars) is given by

$$D_1(t) = 0.0275t^2 + 0.081t + 0.07 \qquad (0 \le t \le 3)$$

and the budget deficit without the rescue package is given by

$$D_2(t) = 0.035t^2 + 0.211t + 0.24 \qquad (0 \le t \le 3)$$

Here, t is measured in years, with $t = 0$ corresponding to 2011. Let $D = D_2 - D_1$.

a. Show that D is increasing on $(0, 3)$, and interpret your result.

b. Show that the graph of D is concave upward on $(0, 3)$, and interpret your result.

Source: MBTA Review.

101. AIR POLLUTION The level of ozone, an invisible gas that irritates and impairs breathing, present in the atmosphere on a certain May day in the city of Riverside was approximated by

$$A(t) = 1.0974t^3 - 0.0915t^4 \qquad (0 \le t \le 11)$$

where $A(t)$ is measured in pollutant standard index (PSI) and t is measured in hours, with $t = 0$ corresponding to 7 A.M. Use the Second Derivative Test to show that the function A has a relative maximum at approximately $t = 9$. Interpret your results.

102. WOMEN'S SOCCER Starting with the youth movement that took hold in the 1970s and buoyed by the success of the U.S. national women's team in international competition in recent years, girls and women have taken to soccer in ever-growing numbers. The function

$$N(t) = -0.9307t^3 + 74.04t^2 + 46.8667t + 3967$$
$$(0 \le t \le 16)$$

gives the number of participants in women's soccer in year t, with $t = 0$ corresponding to the beginning of 1985.

a. Verify that the number of participants per year in women's soccer had been increasing from 1985 through 2000.
   Hint: Use the quadratic formula.

b. Show that the number of participants per year in women's soccer had been increasing at an increasing rate from 1985 through 2000.
   Hint: Show that the sign of N'' is positive on the interval in question.

Source: NCCA News.

103. DEPENDENCY RATIO The share of the world population that is over 60 years of age compared to the rest of the working population in the world is of concern to economists. An increasing dependency ratio means that there will be fewer workers to support an aging population. The dependency ratio over the next century is forecast to be

$$R(t) = 0.00731t^4 - 0.174t^3 + 1.528t^2 + 0.48t + 19.3$$
$$(0 \le t \le 6)$$

in year t, where t is measured in decades with $t = 0$ corresponding to 2000.

a. Show that the dependency ratio will be increasing at the fastest pace around 2052.
   Hint: Use the quadratic formula.

b. What will the dependency ratio be at that time?

Source: International Institute for Applied Systems Analysis.

In Exercises 104–108, determine whether the statement is true or false. If it is true, explain why it is true. If it is false, give an example to show why it is false.

104. If the graph of f is concave upward on (a, b), then the graph of $-f$ is concave downward on (a, b).

105. If the graph of f is concave upward on (a, c) and concave downward on (c, b), where $a < c < b$, then f has an inflection point at $(c, f(c))$.

106. Suppose the second derivative of f exists on (a, b). If the graph of f is concave upward on (a, b), then the graph of $g = f^2$ defined by $g(x) = [f(x)]^2$ is also concave upward on (a, b).

107. If a function f is defined on (a, b) and the graph of f has an inflection point at $(c, f(c))$, where $a < c < b$, then $f''(c) = 0$.

108. If c is a critical number of f where $a < c < b$ and $f''(x) < 0$ on (a, b), then f has a relative maximum at $x = c$.

109. Show that the graph of the quadratic function

$$f(x) = ax^2 + bx + c \qquad (a \neq 0)$$

is concave upward if $a > 0$ and concave downward if $a < 0$. Thus, by examining the sign of the coefficient of x^2, one can tell immediately whether the parabola opens upward or downward.

110. Consider the functions $f(x) = x^3$, $g(x) = x^4$, and $h(x) = -x^4$.

a. Show that $x = 0$ is a critical number of each of the functions f, g, and h.
b. Show that the second derivative of each of the functions f, g, and h equals zero at $x = 0$.
c. Show that f has neither a relative maximum nor a relative minimum at $x = 0$, that g has a relative minimum at $x = 0$, and that h has a relative maximum at $x = 0$.

4.2 Solutions to Self-Check Exercises

1. We first compute

$$f'(x) = 12x^2 - 6x$$
$$f''(x) = 24x - 6 = 6(4x - 1)$$

Observe that f'' is continuous everywhere and has a zero at $x = \frac{1}{4}$. The sign diagram of f'' is shown in the accompanying figure.

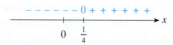

From the sign diagram for f'', we see that the graph of f is concave upward on $\left(\frac{1}{4}, \infty\right)$ and concave downward on $\left(-\infty, \frac{1}{4}\right)$.

2. First, we find the critical numbers of f by solving the equation

$$f'(x) = 6x^2 - x - 12 = 0$$

That is,

$$(3x + 4)(2x - 3) = 0$$

giving $x = -\frac{4}{3}$ and $x = \frac{3}{2}$. Next, we compute

$$f''(x) = 12x - 1$$

Since

$$f''\left(-\frac{4}{3}\right) = 12\left(-\frac{4}{3}\right) - 1 = -17 < 0$$

the Second Derivative Test implies that $f\left(-\frac{4}{3}\right) = \frac{10}{27}$ is a relative maximum of f. Also,

$$f''\left(\frac{3}{2}\right) = 12\left(\frac{3}{2}\right) - 1 = 17 > 0$$

and we see that $f\left(\frac{3}{2}\right) = -\frac{179}{8}$ is a relative minimum.

3. We compute the second derivative of G. Thus,

$$G'(t) = -6t^2 + 90t + 20$$
$$G''(t) = -12t + 90$$

Now, G'' is continuous everywhere, and $G''(t) = 0$, when $t = \frac{15}{2}$, giving $t = \frac{15}{2}$ as the only candidate for an inflection point of G. Since $G''(t) > 0$ for $t < \frac{15}{2}$ and $G''(t) < 0$ for $t > \frac{15}{2}$, we see that $\left(\frac{15}{2}, \frac{15,675}{2}\right)$ is an inflection point of G. The results of our computations tell us that the country's GDP was increasing most rapidly at the beginning of July 2011.

4.3 Curve Sketching

A Real-Life Example

As we have seen on numerous occasions, the graph of a function is a useful aid for visualizing the function's properties. From a practical point of view, the graph of a function also gives, at one glance, a complete summary of all the information captured by the function.

Consider, for example, the graph of the function giving the Dow-Jones Industrial Average (DJIA) on Black Monday, October 19, 1987 (Figure 44). Here, $t = 0$ corresponds to 9:30 A.M., when the market was open for business, and $t = 6.5$ corresponds to 4 P.M., the closing time. The following information may be gleaned from studying the graph.

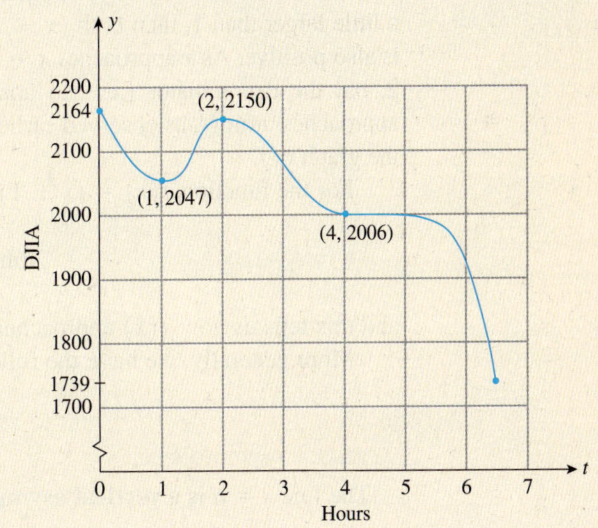

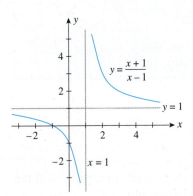

FIGURE **44**

The Dow-Jones Industrial Average on Black Monday

Source: Wall Street Journal.

The graph is *falling* rapidly from $t = 0$ to $t = 1$, reflecting the sharp drop in the index in the first hour of trading. The point $(1, 2047)$ is a *relative minimum* point of the function, and this turning point coincides with the start of an aborted recovery. The short-lived rally, represented by the portion of the graph that is *rising* on the interval $(1, 2)$, quickly fizzled out at $t = 2$ (11:30 A.M.). The *relative maximum* point $(2, 2150)$ marks the highest point of the recovery. The function is decreasing in the rest of the interval. The point $(4, 2006)$ is an *inflection point* of the function; it shows that there was a temporary respite at $t = 4$ (1:30 P.M.). However, selling pressure continued unabated, and the DJIA continued to fall until the closing bell. Finally, the graph also shows that the index opened at the high of the day [$f(0) = 2164$ is the *absolute maximum* of the function] and closed at the low of the day [$f(\frac{13}{2}) = 1739$ is the *absolute minimum* of the function], a drop of 508 points from the previous close!*

Before we turn our attention to the actual task of sketching the graph of a function, let's look at some properties of graphs that will be helpful in this connection.

Vertical Asymptotes

Before going on, you might want to review the material on one-sided limits and the limit at infinity of a function (Sections 2.4 and 2.5).

Consider the graph of the function

$$f(x) = \frac{x + 1}{x - 1}$$

shown in Figure 45. Observe that $f(x)$ increases without bound (tends to infinity) as x approaches $x = 1$ from the right; that is,

$$\lim_{x \to 1^+} \frac{x + 1}{x - 1} = \infty$$

You can verify this by taking a sequence of values of x approaching $x = 1$ from the right and looking at the corresponding values of $f(x)$.

FIGURE **45**

The graph of *f* has a vertical asymptote at $x = 1$.

*Absolute maxima and absolute minima of functions are covered in Section 4.4.

Here is another way of looking at the situation: Observe that if x is a number that is a little larger than 1, then both $(x + 1)$ and $(x - 1)$ are positive, so $(x + 1)/(x - 1)$ is also positive. As x approaches $x = 1$, the numerator $(x + 1)$ approaches the number 2, but the denominator $(x - 1)$ approaches zero, so the quotient $(x + 1)/(x - 1)$ approaches infinity, as observed earlier. The line $x = 1$ is called a vertical asymptote of the graph of f.

For the function $f(x) = (x + 1)/(x - 1)$, you can show that

$$\lim_{x \to 1^-} \frac{x + 1}{x - 1} = -\infty$$

and this tells us how $f(x)$ approaches the asymptote $x = 1$ from the left.

More generally, we have the following definition:

Vertical Asymptote

The line $x = a$ is a **vertical asymptote** of the graph of a function f if

$$\lim_{x \to a^+} f(x) = \infty \quad \text{or} \quad -\infty$$

or

$$\lim_{x \to a^-} f(x) = \infty \quad \text{or} \quad -\infty$$

Note Although a vertical asymptote of a graph is not part of the graph, it serves as a useful aid for sketching the graph. ∎

For rational functions

$$f(x) = \frac{P(x)}{Q(x)}$$

there is a simple criterion for determining whether the graph of f has any vertical asymptotes.

Finding Vertical Asymptotes of Rational Functions

Suppose f is a rational function

$$f(x) = \frac{P(x)}{Q(x)}$$

where P and Q are polynomial functions. Then, the line $x = a$ is a vertical asymptote of the graph of f if $Q(a) = 0$ but $P(a) \neq 0$.

For the function

$$f(x) = \frac{x + 1}{x - 1}$$

considered earlier, $P(x) = x + 1$ and $Q(x) = x - 1$. Observe that $Q(1) = 0$ but $P(1) = 2 \neq 0$, so $x = 1$ is a vertical asymptote of the graph of f.

EXAMPLE 1 Find the vertical asymptotes of the graph of the function

$$f(x) = \frac{x^2}{4 - x^2}$$

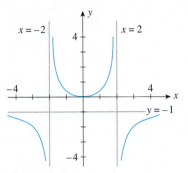

FIGURE 46
$x = -2$ and $x = 2$ are vertical asymptotes of the graph of f.

Solution The function f is a rational function with $P(x) = x^2$ and $Q(x) = 4 - x^2$. The zeros of Q are found by solving

$$4 - x^2 = 0$$

that is,

$$(2 + x)(2 - x) = 0$$

giving $x = -2$ and $x = 2$. These are candidates for the vertical asymptotes of the graph of f. Examining $x = -2$, we compute $P(-2) = (-2)^2 = 4 \neq 0$, and we see that $x = -2$ is indeed a vertical asymptote of the graph of f. Similarly, we find $P(2) = 2^2 = 4 \neq 0$, so $x = 2$ is also a vertical asymptote of the graph of f. The graph of f sketched in Figure 46 confirms these results.

 Recall that if the line $x = a$ is a vertical asymptote of the graph of a rational function f, then *only* the denominator of $f(x)$ is equal to zero at $x = a$. If *both* $P(a)$ and $Q(a)$ are equal to zero, then $x = a$ need *not* be a vertical asymptote. For example, look at the function

$$f(x) = \frac{4(x^2 - 4)}{x - 2}$$

whose graph appears in Figure 32a in Chapter 2 (page 108).

Horizontal Asymptotes

Let's return to the function f defined by

$$f(x) = \frac{x + 1}{x - 1}$$

(Figure 47).

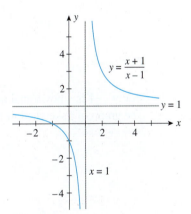

FIGURE 47
The graph of f has a horizontal asymptote at $y = 1$.

Observe that the graph of f approaches the horizontal line $y = 1$ as x approaches infinity, and in this case, the graph of f approaches $y = 1$ as x approaches minus infinity as well. The line $y = 1$ is called a horizontal asymptote of the graph of f. More generally, we have the following definition:

> **Horizontal Asymptote**
> The line $y = b$ is a **horizontal asymptote** of the graph of a function f if either
> $$\lim_{x \to \infty} f(x) = b \quad \text{or} \quad \lim_{x \to -\infty} f(x) = b$$

For the function

$$f(x) = \frac{x + 1}{x - 1}$$

we see that

$$\lim_{x \to \infty} \frac{x + 1}{x - 1} = \lim_{x \to \infty} \frac{1 + \dfrac{1}{x}}{1 - \dfrac{1}{x}} \qquad \textcolor{red}{\text{Divide numerator and denominator by } x.}$$

$$= 1$$

Also,

$$\lim_{x \to -\infty} \frac{x + 1}{x - 1} = \lim_{x \to -\infty} \frac{1 + \dfrac{1}{x}}{1 - \dfrac{1}{x}}$$

$$= 1$$

In either case, we conclude that $y = 1$ is a horizontal asymptote of the graph of f, as observed earlier.

EXAMPLE 2 Find the horizontal asymptotes of the graph of the function

$$f(x) = \frac{x^2}{4 - x^2}$$

Solution   We compute

$$\lim_{x \to \infty} \frac{x^2}{4 - x^2} = \lim_{x \to \infty} \frac{1}{\dfrac{4}{x^2} - 1} \qquad \textcolor{red}{\text{Divide numerator and denominator by } x^2.}$$

$$= -1$$

so $y = -1$ is a horizontal asymptote, as before. (Similarly, $\lim\limits_{x \to -\infty} f(x) = -1$ as well.) The graph of f sketched in Figure 48 confirms this result.

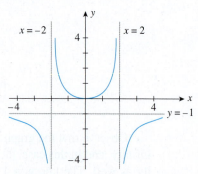

FIGURE 48
The graph of f has a horizontal asymptote at $y = -1$.

We next state an important property of polynomial functions.

> The graph of a polynomial function has no vertical or horizontal asymptotes.

To see this, note that a polynomial function $P(x)$ can be written as a rational function with denominator equal to 1. Thus,

$$P(x) = \frac{P(x)}{1}$$

Since the denominator is never equal to zero, P has no vertical asymptotes. Next, if P is a polynomial of degree greater than or equal to 1, then

$$\lim_{x \to \infty} P(x) \quad \text{and} \quad \lim_{x \to -\infty} P(x)$$

are either infinity or minus infinity; that is, they do not exist. Therefore, P has no horizontal asymptotes.

  In Sections 4.1 and 4.2, we saw how the first and second derivatives of a function are used to reveal various properties of the graph of a function f. We now show how this information can be used to help us sketch the graph of f. We begin by giving a general procedure for curve sketching.

A Guide to Curve Sketching

1. Determine the domain of f.
2. Find the x- and y-intercepts of f.*
3. Determine the behavior of f for large absolute values of x.
4. Find all horizontal and vertical asymptotes of the graph of f.
5. Determine the intervals where f is increasing and where f is decreasing.
6. Find the relative extrema of f.
7. Determine the concavity of the graph of f.
8. Find the inflection points of f.
9. Plot a few additional points to help further identify the shape of the graph of f and sketch the graph.

---

*The equation $f(x) = 0$ may be difficult to solve, in which case one may decide against finding the x-intercepts or to use technology, if available, for assistance.

  We illustrate the techniques of curve sketching in the next two examples.

Two Step-by-Step Examples

EXAMPLE 3 Sketch the graph of the function

$$y = f(x) = x^3 - 6x^2 + 9x + 2$$

Solution   Obtain the following information on the graph of f.

Step 1   The domain of f is the interval $(-\infty, \infty)$.

Step 2   By setting $x = 0$, we find that the y-intercept is 2. The x-intercept is found by setting $y = 0$, which in this case leads to a cubic equation. Since the solution is not readily found, we will not use this information.

Step 3   Since

$$\lim_{x \to -\infty} f(x) = \lim_{x \to -\infty} (x^3 - 6x^2 + 9x + 2) = -\infty$$

$$\lim_{x \to \infty} f(x) = \lim_{x \to \infty} (x^3 - 6x^2 + 9x + 2) = \infty$$

we see that f decreases without bound as x decreases without bound and that f increases without bound as x increases without bound.

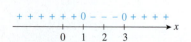

FIGURE 49
Sign diagram for f'

FIGURE 50
Sign diagram for f''

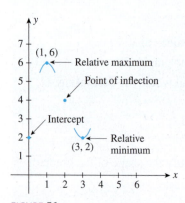

FIGURE 51
We first plot the intercept, the relative extrema, and the inflection point.

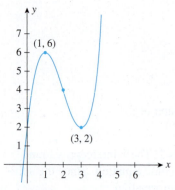

FIGURE 52
The graph of $y = x^3 - 6x^2 + 9x + 2$

Step 4   Since f is a polynomial function, there are no asymptotes.

Step 5
$$f'(x) = 3x^2 - 12x + 9 = 3(x^2 - 4x + 3)$$
$$= 3(x - 1)(x - 3)$$

Setting $f'(x) = 0$ gives $x = 1$ or $x = 3$. The sign diagram for f' shows that f is increasing on the intervals $(-\infty, 1)$ and $(3, \infty)$ and decreasing on the interval $(1, 3)$ (Figure 49).

Step 6   From the results of Step 5, we see that $x = 1$ and $x = 3$ are critical numbers of f. Furthermore, f' changes sign from positive to negative as we move across $x = 1$, so a relative maximum of f occurs at $x = 1$. Similarly, we see that a relative minimum of f occurs at $x = 3$. Now,

$$f(1) = 1 - 6 + 9 + 2 = 6$$
$$f(3) = 3^3 - 6(3)^2 + 9(3) + 2 = 2$$

so $f(1) = 6$ is a relative maximum of f and $f(3) = 2$ is a relative minimum of f.

Step 7
$$f''(x) = 6x - 12 = 6(x - 2)$$

which is equal to zero when $x = 2$. The sign diagram of f'' shows that the graph of f is concave downward on the interval $(-\infty, 2)$ and concave upward on the interval $(2, \infty)$ (Figure 50).

Step 8   From the results of Step 7, we see that f'' changes sign as we move across $x = 2$. Next,

$$f(2) = 2^3 - 6(2)^2 + 9(2) + 2 = 4$$

so the required inflection point of f is $(2, 4)$.

Step 9   Summarizing, we have the following:

> Domain: $(-\infty, \infty)$
> Intercept: $(0, 2)$
> $\lim\limits_{x \to -\infty} f(x)$; $\lim\limits_{x \to \infty} f(x)$: $-\infty$; ∞
> Asymptotes: None
> Intervals where f is $\nearrow$ or $\searrow$: $\nearrow$ on $(-\infty, 1)$ and $(3, \infty)$; $\searrow$ on $(1, 3)$
> Relative extrema: Relative maximum at $(1, 6)$; relative minimum at $(3, 2)$
> Concavity: Downward on $(-\infty, 2)$; upward on $(2, \infty)$
> Point of inflection: $(2, 4)$

In general, it is a good idea to start graphing by plotting the intercept(s), relative extrema, and inflection point(s) (Figure 51). Then, using the rest of the information, we complete the graph of f, as sketched in Figure 52.

Explore and Discuss

The average price of gasoline at the pump over a 3-month period, during which there was a temporary shortage of oil, is described by the function f defined on the interval $[0, 3]$. During the first month, the price was increasing at an increasing rate. Starting with the second month, the good news was that the rate of increase was slowing down, although the price of gas was still increasing. This pattern continued until the end of the second month. The price of gas peaked at $t = 2$ and began to fall at an increasing rate until $t = 3$.

1. Describe the signs of $f'(t)$ and $f''(t)$ over each of the intervals $(0, 1)$, $(1, 2)$, and $(2, 3)$.
2. Make a sketch showing a plausible graph of f over $[0, 3]$.

EXAMPLE 4 Sketch the graph of the function

$$y = f(x) = \frac{x+1}{x-1}$$

Solution Obtain the following information:

Step 1 f is undefined when $x = 1$, so the domain of f is the set of all real numbers other than $x = 1$.

Step 2 Setting $y = 0$ gives $x = -1$, the x-intercept of f. Next, setting $x = 0$ gives $y = -1$ as the y-intercept of f.

Step 3 Earlier, we found that

$$\lim_{x \to \infty} \frac{x+1}{x-1} = 1 \quad \text{and} \quad \lim_{x \to -\infty} \frac{x+1}{x-1} = 1$$

(see page 294). Consequently, we see that the graph of f approaches the line $y = 1$ as $|x|$ becomes arbitrarily large. For $x > 1$, $f(x) > 1$ and the graph of f approaches the line $y = 1$ from above. For $x < 1$, $f(x) < 1$, so the graph of f approaches the line $y = 1$ from below.

Step 4 The straight line $x = 1$ is a vertical asymptote of the graph of f. Also, from the results of Step 3, we conclude that $y = 1$ is a horizontal asymptote of the graph of f.

Step 5
$$f'(x) = \frac{(x-1)(1) - (x+1)(1)}{(x-1)^2} = -\frac{2}{(x-1)^2}$$

and is discontinuous at $x = 1$. The sign diagram of f' shows that $f'(x) < 0$ whenever it is defined. Thus, f is decreasing on the intervals $(-\infty, 1)$ and $(1, \infty)$ (Figure 53).

Step 6 From the results of Step 5, we see that there are no critical numbers of f, since $f'(x)$ is never equal to zero for any value of x in the domain of f.

Step 7
$$f''(x) = \frac{d}{dx}[-2(x-1)^{-2}] = 4(x-1)^{-3} = \frac{4}{(x-1)^3}$$

The sign diagram of f'' shows immediately that the graph of f is concave downward on the interval $(-\infty, 1)$ and concave upward on the interval $(1, \infty)$ (Figure 54).

Step 8 From the results of Step 7, we see that there are no candidates for inflection points of f, since $f''(x)$ is never equal to zero for any value of x in the domain of f. Hence, f has no inflection points.

Step 9 Summarizing, we have the following:

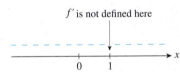

FIGURE 53
The sign diagram for f'

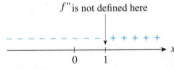

FIGURE 54
The sign diagram for f''

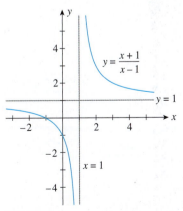

FIGURE 55
The graph of f has a horizontal asymptote at $y = 1$ and a vertical asymptote at $x = 1$.

Domain: $(-\infty, 1) \cup (1, \infty)$

Intercepts: $(-1, 0); (0, -1)$

$\lim_{x \to -\infty} f(x); \lim_{x \to \infty} f(x)$: 1; 1

Asymptotes: $x = 1$ is a vertical asymptote
$y = 1$ is a horizontal asymptote

Intervals where f is ↗ or ↘: ↘ on $(-\infty, 1)$ and $(1, \infty)$

Relative extrema: None

Concavity: Downward on $(-\infty, 1)$; upward on $(1, \infty)$

Points of inflection: None

The graph of f is sketched in Figure 55.

4.3   Self-Check Exercises

1. Find the horizontal and vertical asymptotes of the graph of the function

$$f(x) = \frac{2x^2}{x^2 - 1}$$

2. Sketch the graph of the function

$$f(x) = \frac{2}{3}x^3 - 2x^2 - 6x + 4$$

Solutions to Self-Check Exercises 4.3 can be found on page 302.

4.3   Concept Questions

1. Explain the following terms in your own words:
   a. Vertical asymptote   **b.** Horizontal asymptote

2. **a.** How many vertical asymptotes can the graph of a function *f* have? Explain using graphs.
   b. How many horizontal asymptotes can the graph of a function *f* have? Explain using graphs.

3. How do you find the vertical asymptotes of the graph of a rational function?

4. Give a procedure for sketching the graph of a function.

4.3   Exercises

In Exercises 1–10, find the horizontal and vertical asymptotes of the graph of the function.

1.

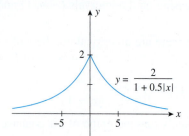

$$y = \frac{2}{1 + 0.5|x|}$$

2.

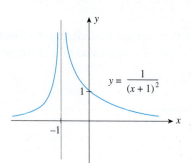

$$y = \frac{1}{(x + 1)^2}$$

3.

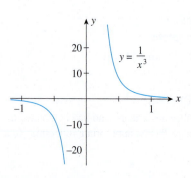

$$y = \frac{1}{x^3}$$

4.

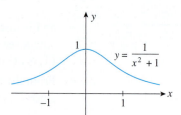

$$y = \frac{1}{x^2 + 1}$$

5.

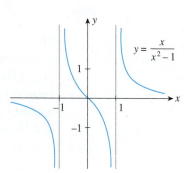

$$y = \frac{x}{x^2 - 1}$$

6.

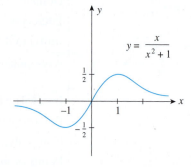

$$y = \frac{x}{x^2 + 1}$$

7.

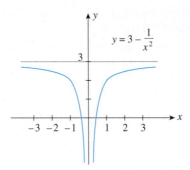

$y = 3 - \dfrac{1}{x^2}$

8.

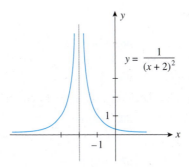

$y = \dfrac{1}{(x+2)^2}$

9.

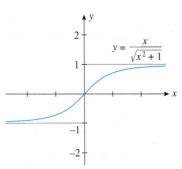

$y = \dfrac{x}{\sqrt{x^2+1}}$

10.

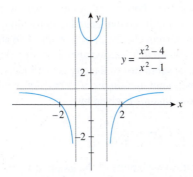

$y = \dfrac{x^2-4}{x^2-1}$

In Exercises 11–28, find the horizontal and vertical asymptotes of the graph of the function. (You need not sketch the graph.)

11. $f(x) = \dfrac{1}{x}$ **12.** $f(x) = \dfrac{1}{x+2}$

13. $f(x) = -\dfrac{2}{x^2}$ **14.** $g(x) = \dfrac{1}{1+2x^2}$

15. $f(x) = \dfrac{x-2}{x+2}$ **16.** $g(t) = \dfrac{t+1}{2t-1}$

17. $h(x) = x^3 - 3x^2 + x + 1$

18. $g(x) = 2x^3 + x^2 + 1$

19. $f(t) = \dfrac{t^2}{t^2-16}$ **20.** $g(x) = \dfrac{x^3}{x^2-4}$

21. $f(x) = \dfrac{3x}{x^2-x-6}$ **22.** $g(x) = \dfrac{2x}{x^2+x-2}$

23. $g(t) = 2 + \dfrac{5}{(t-2)^2}$ **24.** $f(x) = 1 + \dfrac{2}{x-3}$

25. $f(x) = \dfrac{x^2-2}{x^2-4}$ **26.** $h(x) = \dfrac{2-x^2}{x^2+x}$

27. $g(x) = \dfrac{x^3-x}{x(x+1)}$ **28.** $f(x) = \dfrac{x^4-x^2}{x(x-1)(x+2)}$

In Exercises 29 and 30, you are given the graphs of two functions f and g. One function is the derivative function of the other. Identify each of them.

29.

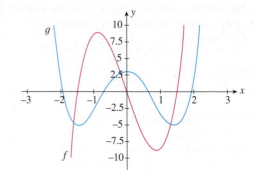

30.

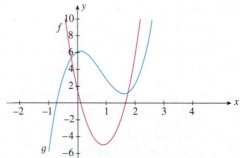

31. TERMINAL VELOCITY A skydiver leaps from the gondola of a hot-air balloon. As she free-falls, air resistance, which is proportional to her velocity, builds up to a point at which it balances the force due to gravity. The resulting motion may be described in terms of her velocity as follows: Starting at rest (zero velocity), her velocity increases and approaches a constant velocity, called the *terminal velocity*. Sketch a graph of her velocity v versus time t.

32. SPREAD OF A FLU EPIDEMIC Initially, 10 students at a junior high school contracted influenza. The flu spread over time, and the total number of students who eventually contracted the flu approached but never exceeded 200. Let $P(t)$ denote the number of students who had contracted the flu after t days, where P is an appropriate function.

a. Make a sketch of the graph of P. (Your answer will *not* be unique.)

b. Where is the function increasing?

c. Does the graph of P have a horizontal asymptote? If so, what is it?

d. Discuss the concavity of the graph of P. Explain its significance.

e. Is there an inflection point on the graph of P? If so, explain its significance.

In Exercises 33–36, use the information summarized in the table to sketch the graph of f.

33. $f(x) = x^3 - 3x^2 + 1$

Domain: $(-\infty, \infty)$
Intercept: y-intercept: 1
Asymptotes: None
Intervals where f is $\nearrow$ and $\searrow$: $\nearrow$ on $(-\infty, 0)$ and $(2, \infty)$; $\searrow$ on $(0, 2)$
Relative extrema: Rel. max. at $(0, 1)$; rel. min. at $(2, -3)$
Concavity: Downward on $(-\infty, 1)$; upward on $(1, \infty)$
Point of inflection: $(1, -1)$

34. $f(x) = \frac{1}{9}(x^4 - 4x^3)$

Domain: $(-\infty, \infty)$
Intercepts: x-intercepts: 0, 4; y-intercept: 0
Asymptotes: None
Intervals where f is $\nearrow$ and $\searrow$: $\nearrow$ on $(3, \infty)$; $\searrow$ on $(-\infty, 3)$
Relative extrema: Rel. min. at $(3, -3)$
Concavity: Downward on $(0, 2)$; upward on $(-\infty, 0)$ and $(2, \infty)$
Points of inflection: $(0, 0)$ and $\left(2, -\frac{16}{9}\right)$

35. $f(x) = \frac{4x - 4}{x^2}$

Domain: $(-\infty, 0) \cup (0, \infty)$
Intercept: x-intercept: 1
Asymptotes: x-axis and y-axis
Intervals where f is $\nearrow$ and $\searrow$: $\nearrow$ on $(0, 2)$; $\searrow$ on $(-\infty, 0)$ and $(2, \infty)$
Relative extrema: Rel. max. at $(2, 1)$
Concavity: Downward on $(-\infty, 0)$ and $(0, 3)$; upward on $(3, \infty)$
Point of inflection: $\left(3, \frac{8}{9}\right)$

36. $f(x) = x - 3x^{1/3}$

Domain: $(-\infty, \infty)$
Intercepts: x-intercepts: $\pm 3\sqrt{3}$, 0; y-intercept: 0
Asymptotes: None
Intervals where f is $\nearrow$ and $\searrow$: $\nearrow$ on $(-\infty, -1)$ and $(1, \infty)$; $\searrow$ on $(-1, 1)$
Relative extrema: Rel. max. at $(-1, 2)$; rel. min. at $(1, -2)$
Concavity: Downward on $(-\infty, 0)$; upward on $(0, \infty)$
Point of inflection: $(0, 0)$

In Exercises 37–60, sketch the graph of the function, using the curve-sketching guide of this section.

37. $g(x) = 4 - 3x - 2x^3$ **38.** $f(x) = x^2 - 2x + 3$

39. $h(x) = x^3 - 3x + 1$ **40.** $f(x) = 2x^3 + 1$

41. $f(x) = -2x^3 + 3x^2 + 12x + 2$

42. $f(t) = 2t^3 - 15t^2 + 36t - 20$

43. $h(x) = \frac{3}{2}x^4 - 2x^3 - 6x^2 + 8$

44. $f(t) = 3t^4 + 4t^3$

45. $f(t) = \sqrt{t^2 - 4}$ **46.** $f(x) = \sqrt{x^2 + 5}$

47. $g(x) = \frac{1}{2}x - \sqrt{x}$ **48.** $f(x) = \sqrt[3]{x^2}$

49. $g(x) = \frac{2}{x - 1}$ **50.** $f(x) = \frac{1}{x + 1}$

51. $h(x) = \frac{x + 2}{x - 2}$ **52.** $g(x) = \frac{x}{x - 1}$

53. $f(t) = \frac{t^2}{1 + t^2}$ **54.** $g(x) = \frac{x}{x^2 - 4}$

55. $g(t) = -\frac{t^2 - 2}{t - 1}$ **56.** $f(x) = \frac{x^2 - 9}{x^2 - 4}$

57. $g(t) = \frac{t^2}{t^2 - 1}$ **58.** $h(x) = \frac{1}{x^2 - x - 2}$

59. $h(x) = (x - 1)^{2/3} + 1$ **60.** $g(x) = (x + 2)^{3/2} + 1$

61. COST OF REMOVING TOXIC POLLUTANTS A city's main well was recently found to be contaminated with trichloroethylene (a cancer-causing chemical) as a result of an abandoned chemical dump that leached chemicals into the water. A proposal submitted to the city council indicated that the cost, measured in millions of dollars, of removing $x\%$ of the toxic pollutants is given by

$$C(x) = \frac{0.5x}{100 - x}$$

a. Find the vertical asymptote of the graph of C.

b. Is it possible to remove 100% of the toxic pollutant from the water?

62. AVERAGE COST OF PRODUCING DVDS The average cost per disc (in dollars) incurred by Herald Media Corporation in pressing x DVDs is given by the average cost function

$$\overline{C}(x) = 2.2 + \frac{2500}{x}$$

a. Find the horizontal asymptote of the graph of $\overline{C}$.

b. What is the limiting value of the average cost?

63. Concentration of a Drug in the Bloodstream The concentration (in milligrams per cubic centimeter) of a certain drug in a patient's bloodstream t hr after injection is given by

$$C(t) = \frac{0.2t}{t^2 + 1}$$

a. Find the horizontal asymptote of the graph of C.
b. Interpret your result.

64. Effect of Enzymes on Chemical Reactions Certain proteins, known as enzymes, serve as catalysts for chemical reactions in living things. In 1913, Leonor Michaelis and L. M. Menten discovered the following formula giving the initial speed V (in moles per liter per second) at which the reaction begins in terms of the amount of substrate x (the substance that is being acted upon, measured in moles per liter):

$$V = \frac{ax}{x + b}$$

where a and b are positive constants.
a. Find the horizontal asymptote of the graph of V.
b. What does the result of part (a) tell you about the initial speed at which the reaction begins, if the amount of substrate is very large?

65. GDP of a Developing Country A developing country's gross domestic product (GDP) from 2002 to 2010 is approximated by the function

$$G(t) = -0.2t^3 + 2.4t^2 + 60 \qquad (0 \le t \le 8)$$

where $G(t)$ is measured in billions of dollars, and t is measured in years, with $t = 0$ corresponding to 2002. Sketch the graph of the function G, and interpret your results.

66. Crime Rate The number of major crimes per 100,000 committed in a city between 2006 and 2013 is approximated by the function

$$N(t) = -0.1t^3 + 1.5t^2 + 80 \qquad (0 \le t \le 7)$$

where $N(t)$ denotes the number of crimes per 100,000 committed in year t, with $t = 0$ corresponding to 2006. Enraged by the dramatic increase in the crime rate, the citizens, with the help of the local police, organized Neighborhood Crime Watch groups in early 2010 to combat this menace. Sketch the graph of the function N, and interpret your results. Is the Neighborhood Crime Watch program working?

67. Worker Efficiency An efficiency study showed that the total number of smartphones assembled by an average worker at Delphi Electronics t hr after starting work at 8 A.M. is given by

$$N(t) = -\frac{1}{2}t^3 + 3t^2 + 10t \qquad (0 \le t \le 4)$$

Sketch the graph of the function N, and interpret your results.

68. Concentration of a Drug in the Bloodstream The concentration (in milligrams per cubic centimeter) of a certain drug in a patient's bloodstream t hr after injection is given by

$$C(t) = \frac{0.2t}{t^2 + 1}$$

Sketch the graph of the function C, and interpret your results.

69. Box-Office Receipts The total worldwide box-office receipts for a long-running movie are approximated by the function

$$T(x) = \frac{120x^2}{x^2 + 4}$$

where $T(x)$ is measured in millions of dollars and x is the number of years since the movie's release. Sketch the graph of the function T, and interpret your results.

70. Traffic Flow Analysis The speed of traffic flow in miles per hour on a stretch of Route 123 between 6 A.M. and 10 A.M. on a typical workday is approximated by the function

$$f(t) = 20t - 40\sqrt{t} + 52 \qquad (0 \le t \le 4)$$

where t is measured in hours, with $t = 0$ corresponding to 6 A.M. Sketch the graph of f, and interpret your results.

71. Cost of Removing Toxic Pollutants Refer to Exercise 61. The cost, measured in millions of dollars, of removing $x\%$ of a toxic pollutant is given by

$$C(x) = \frac{0.5x}{100 - x}$$

Sketch the graph of the function C, and interpret your results.

In Exercises 72–74, determine whether the statement is true or false. If it is true, explain why it is true. If it is false, give an example to show why it is false.

72. If the graph of a function f has a vertical asymptote at $x = a$, then $\lim_{x \to a^-} f(x) = \infty$ or $-\infty$ and $\lim_{x \to a^+} f(x) = \infty$ or $-\infty$.

73. The graph of a function f cannot intersect its vertical asymptote.

74. The graph of a function f cannot intersect its horizontal asymptote at more than one point.

4.3   Solutions to Self-Check Exercises

1. Since

$$\lim_{x \to \infty} \frac{2x^2}{x^2 - 1} = \lim_{x \to \infty} \frac{2}{1 - \frac{1}{x^2}}$$ Divide the numerator and denominator by x^2.

$$= 2$$

we see that $y = 2$ is a horizontal asymptote. Next, since

$$x^2 - 1 = (x + 1)(x - 1) = 0$$

implies $x = -1$ or $x = 1$, these are candidates for the vertical asymptotes of the graph of f. Since the numerator of f is not equal to zero for $x = -1$ or $x = 1$, we conclude that $x = -1$ and $x = 1$ are vertical asymptotes of the graph of f.

2. We obtain the following information on the graph of f.
  (1) The domain of f is the interval $(-\infty, \infty)$.
  (2) By setting $x = 0$, we find that the y-intercept is 4.
  (3) Since

$$\lim_{x \to -\infty} f(x) = \lim_{x \to -\infty} \left(\frac{2}{3}x^3 - 2x^2 - 6x + 4 \right) = -\infty$$

$$\lim_{x \to \infty} f(x) = \lim_{x \to \infty} \left(\frac{2}{3}x^3 - 2x^2 - 6x + 4 \right) = \infty$$

we see that $f(x)$ decreases without bound as x decreases without bound and that $f(x)$ increases without bound as x increases without bound.
  (4) Since f is a polynomial function, the graph of f has no asymptotes.

  (5) $f'(x) = 2x^2 - 4x - 6 = 2(x^2 - 2x - 3)$
$$= 2(x + 1)(x - 3)$$

Setting $f'(x) = 0$ gives $x = -1$ or $x = 3$. The accompanying sign diagram for f' shows that f is increasing on the intervals $(-\infty, -1)$ and $(3, \infty)$ and decreasing on $(-1, 3)$.

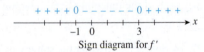

$$+ + + + 0 - - - - - - 0 + + + +$$
$\quad\quad -1 \;\; 0 \quad\quad\quad 3$

Sign diagram for f'

  (6) From the results of Step 5, we see that $x = -1$ and $x = 3$ are critical numbers of f. Furthermore, the sign diagram of f' tells us that $x = -1$ gives rise to a relative maximum of f and $x = 3$ gives rise to a relative minimum of f. Now,

$$f(-1) = \frac{2}{3}(-1)^3 - 2(-1)^2 - 6(-1) + 4 = \frac{22}{3}$$

$$f(3) = \frac{2}{3}(3)^3 - 2(3)^2 - 6(3) + 4 = -14$$

so $f(-1) = \frac{22}{3}$ is a relative maximum of f and $f(3) = -14$ is a relative minimum of f.

  (7) $f''(x) = 4x - 4 = 4(x - 1)$

which is equal to zero when $x = 1$. The accompanying sign diagram of f'' shows that the graph of f is concave downward on the interval $(-\infty, 1)$ and concave upward on the interval $(1, \infty)$.

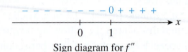

$$- - - - - - - - - 0 + + + + \quad\quad\to x$$
$\quad\quad\quad\quad\quad 0 \quad 1$

Sign diagram for f''

  (8) From the results of Step 7, we see that $x = 1$ is the only candidate for an inflection point of f. Since $f''(x)$ changes sign as we move across the point $x = 1$ and

$$f(1) = \frac{2}{3}(1)^3 - 2(1)^2 - 6(1) + 4 = -\frac{10}{3}$$

we see that the required inflection point is $\left(1, -\frac{10}{3}\right)$.
  (9) Summarizing this information, we have the following:

Domain: $(-\infty, \infty)$
Intercept: $(0, 4)$
$\lim_{x \to -\infty} f(x); \lim_{x \to \infty} f(x): -\infty; \infty$
Asymptotes: None
Intervals where f is $\nearrow$ or $\searrow$: $\nearrow$ on $(-\infty, -1)$ and $(3, \infty)$; $\searrow$ on $(-1, 3)$
Relative extrema: Rel. max. at $\left(-1, \frac{22}{3}\right)$; rel. min. at $(3, -14)$
Concavity: Downward on $(-\infty, 1)$; upward on $(1, \infty)$
Point of inflection: $\left(1, -\frac{10}{3}\right)$

The graph of f is sketched in the accompanying figure.

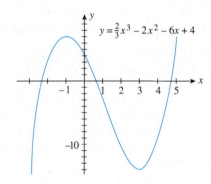

$$y = \frac{2}{3}x^3 - 2x^2 - 6x + 4$$

Analyzing the Properties of a Function

One of the main purposes of studying Section 4.3 is to see how the many concepts of calculus come together to paint a picture of a function. The techniques of graphing also play a very practical role. For example, using the techniques of graphing developed in Section 4.3, you can tell whether the graph of a function generated by a graphing utility is reasonably complete. Furthermore, these techniques can often reveal details that are missing from a graph.

EXAMPLE 1 Consider the function $f(x) = 2x^3 - 3.5x^2 + x - 10$. A plot of the graph of f in the standard viewing window is shown in Figure T1. Since the domain of f is the interval $(-\infty, \infty)$, we see that Figure T1 does not reveal the part of the graph to the left of the y-axis. This suggests that we enlarge the viewing window accordingly. Figure T2 shows the graph of f in the viewing window $[-10, 10] \times [-20, 10]$.

The behavior of f for large values of x

$$\lim_{x \to -\infty} f(x) = -\infty \quad \text{and} \quad \lim_{x \to \infty} f(x) = \infty$$

suggests that this viewing window has captured a sufficiently complete picture of f.

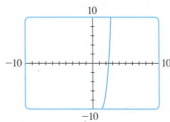

FIGURE T1
The graph of f in the standard viewing window

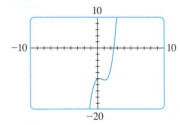

FIGURE T2
The graph of f in the viewing window $[-10, 10] \times [-20, 10]$

Next, an analysis of the first derivative of x,

$$f'(x) = 6x^2 - 7x + 1 = (6x - 1)(x - 1)$$

reveals that f has critical values at $x = \frac{1}{6}$ and $x = 1$. In fact, a sign diagram of f' shows that f has a relative maximum at $x = \frac{1}{6}$ and a relative minimum at $x = 1$, details that are not revealed in the graph of f shown in Figure T2. To examine this portion of the graph of f, we use, say, the viewing window $[-1, 2] \times [-11, -8]$. The resulting graph of f is shown in Figure T3, which certainly reveals the hitherto missing details! Thus, through an interaction of calculus and a graphing utility, we are able to obtain a good picture of the properties of f.

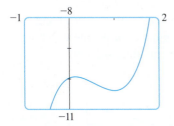

FIGURE T3
The graph of f in the viewing window $[-1, 2] \times [-11, -8]$

Finding x-Intercepts

As was noted in Section 4.3, it is not always easy to find the x-intercepts of the graph of a function. But this information is very important in applications. By using the function for solving polynomial equations or the function for finding the roots of an equation, we can solve the equation $f(x) = 0$ quite easily and hence yield the x-intercepts of the graph of a function.

EXAMPLE 2 Let $f(x) = x^3 - 3x^2 + x + 1.5$.

a. Use the function for solving polynomial equations on a graphing utility to find the x-intercepts of the graph of f.
b. Use the function for finding the roots of an equation on a graphing utility to find the x-intercepts of the graph of f.

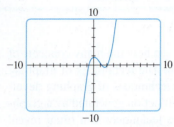

FIGURE **T4**
The graph of $f(x) = x^3 - 3x^2 + x + 1.5$

Solution

a. Observe that f is a polynomial function of degree 3, and so we may use the function for solving polynomial equations to solve the equation $x^3 - 3x^2 + x + 1.5 = 0$ $[f(x) = 0]$. We find that the solutions (x-intercepts) are

$$x_1 \approx -0.525687120865 \qquad x_2 \approx 1.2586520225 \qquad x_3 \approx 2.26703509836$$

b. Using the graph of f (Figure T4), we see that $x_1 \approx -0.5$, $x_2 \approx 1$, and $x_3 \approx 2$. Using the function for finding the roots of an equation on a graphing utility and these values of x as initial guesses, we find

$$x_1 \approx -0.5256871209 \qquad x_2 \approx 1.2586520225 \qquad x_3 \approx 2.2670350984$$

Note   The function for solving polynomial equations on a graphing utility will solve a polynomial equation $f(x) = 0$, where f is a polynomial function. The function for finding the roots of an equation, however, will solve equations $f(x) = 0$ even if f is not a polynomial.

APPLIED EXAMPLE 3 TV on Smartphones   The number of people watching TV on smartphones (in millions) is approximated by

$$N(t) = 11.9\sqrt{1 + 0.91t} \qquad (0 \le t \le 4)$$

where t is measured in years, with $t = 0$ corresponding to the beginning of 2007.

a. Use a graphing calculator to plot the graph of N.
b. Based on this model, when did the number of people watching TV on smartphones first exceed 20 million?
Source: IDC, U.S. forecast.

Solution

a. The graph of N in the window $[0, 4] \times [0, 30]$ is shown in Figure T5a.
b. Using the function for finding the intersection of the graphs of $y_1 = N(t)$ and $y_2 = 20$, we find $t \approx 2.005$ (see Figure T5b). So the number of people watching TV on smartphones first exceeded 20 million at the beginning of 2009.

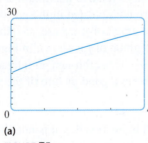
(a)

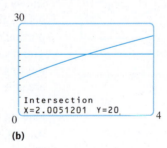
(b)

FIGURE **T5**
(a) The graph of N in the viewing window $[0, 4] \times [0, 30]$; (b) the graph showing the intersection of $y_1 = N(t)$ and $y_2 = 20$ on the TI-83/84.

TECHNOLOGY EXERCISES

In Exercises 1–4, use the method of Example 1 to analyze the function. (*Note:* Your answers will *not* be unique.)

1. $f(x) = 4x^3 - 4x^2 + x + 10$

2. $f(x) = x^3 + 2x^2 + x - 12$

3. $f(x) = \dfrac{1}{2}x^4 + x^3 + \dfrac{1}{2}x^2 - 10$

4. $f(x) = 2.25x^4 - 4x^3 + 2x^2 + 2$

In Exercises 5–8, find the *x*-intercepts of the graph of *f*. Give your answers accurate to four decimal places.

5. $f(x) = 0.2x^3 - 1.2x^2 + 0.8x + 2.1$

6. $f(x) = -0.2x^4 + 0.8x^3 - 2.1x + 1.2$

7. $f(x) = 2x^2 - \sqrt{x + 1} - 3$

8. $f(x) = x - \sqrt{1 - x^2}$

9. AIR POLLUTION The level of ozone, an invisible gas that irritates and impairs breathing, present in the atmosphere on a certain day in June in the city of Riverside is approximated by

$$S(t) = 1.0974t^3 - 0.0915t^4 \qquad (0 \le t \le 11)$$

where $S(t)$ is measured in pollutant standard index (PSI) and *t* is measured in hours, with $t = 0$ corresponding to 7 A.M. Sketch the graph of *S*, and interpret your results.
Source: Los Angeles Times.

10. FLIGHT PATH OF A PLANE The function

$$f(x) = \begin{cases} 0 & \text{if } 0 \le x < 1 \\ -0.0411523x^3 + 0.679012x^2 \\ \qquad -1.23457x + 0.596708 & \text{if } 1 \le x < 10 \\ 15 & \text{if } 10 \le x \le 11 \end{cases}$$

where both *x* and $f(x)$ are measured in units of 1000 ft, describes the flight path of a plane taking off from the origin and climbing to an altitude of 15,000 ft. Sketch the graph of *f* to visualize the trajectory of the plane.

4.4 Optimization I

Absolute Extrema

The graph of the function *f* in Figure 56 shows the average age of cars in use in the United States from the beginning of 1946 ($t = 0$) to the beginning of 2009 ($t = 63$). Observe that the highest average age of cars in use during this period is 9.3 years, whereas the lowest average age of cars in use during the same period is 5.5 years. The number 9.3, the largest value of $f(t)$ for all values of *t* in the interval $[0, 63]$ (the domain of *f*), is called the *absolute maximum value of f* on that interval. The number 5.5, the smallest value of $f(t)$ for all values of *t* in $[0, 63]$, is called the *absolute minimum value of f* on that interval. Notice, too, that the absolute maximum value of *f* is attained at the endpoint $t = 63$ of the interval, whereas the absolute minimum value of *f* is attained at the points $t = 12$ (corresponding to 1958) and $t = 23$ (corresponding to 1969) that lie within the interval $(0, 63)$.

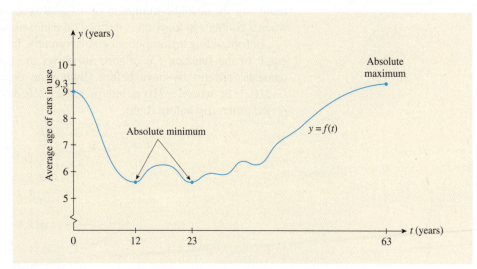

FIGURE 56
$f(t)$ gives the average age of cars in use in year *t*, *t* in [0, 63].
Source: American Automobile Association.

(Incidentally, it is interesting to note that 1946 marked the first year of peace following World War II and the two years 1958 and 1969 marked the end of two periods of prosperity in recent U.S. history.)

A precise definition of the **absolute extrema** (absolute maximum or absolute minimum) of a function follows.

> ### The Absolute Extrema of a Function f
>
> If $f(x) \leq f(c)$ for all x in the domain of f, then $f(c)$ is called the **absolute maximum value** of f.
>
> If $f(x) \geq f(c)$ for all x in the domain of f, then $f(c)$ is called the **absolute minimum value** of f.

Figure 57 shows the graphs of several functions and gives the absolute maximum and absolute minimum of each function, if they exist.

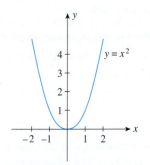
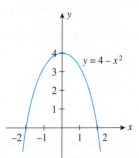
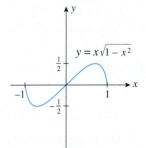
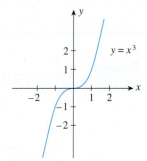

(a) $f(0) = 0$ is the absolute minimum of f; f has no absolute maximum.

(b) $f(0) = 4$ is the absolute maximum of f; f has no absolute minimum.

(c) $f(\sqrt{2}/2) = 1/2$ is the absolute maximum of f; $f(-\sqrt{2}/2) = -1/2$ is the absolute minimum of f.

(d) f has no absolute extrema.

FIGURE 57

Absolute Extrema on a Closed Interval

As the preceding examples show, a continuous function defined on an arbitrary interval does not always have an absolute maximum or an absolute minimum. But an important case arises often in practical applications in which both the absolute maximum and the absolute minimum of a function are guaranteed to exist. This occurs when a continuous function is defined on a *closed* interval.

Before stating this important result formally, let's look at a real-life example. The graph of the function f in Figure 58 shows the average price, $f(t)$, in dollars, of domestic airfares by days before flight. The domain of f is the closed interval $[-210, -1]$, where -210 is interpreted as 210 days before flight and -1 is interpreted as the day before flight.

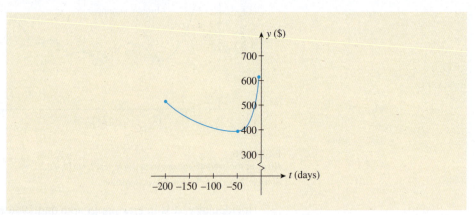

FIGURE 58
Average price before flight

Source: Cheapair.com.

Observe that f attains the minimum value of 395 when $t = -49$ and the maximum value of 614 when $t = -1$. This result tells us that the best time to book a domestic flight is seven weeks in advance and the worst day to book a domestic flight is the day before the flight. Probably most surprising of all, booking too early can be almost as expensive as booking too late. Note that the function f is continuous on a closed interval. For such functions, we have the following theorem.

THEOREM 3

If a function f is continuous on a closed interval $[a, b]$, then f has both an absolute maximum value and an absolute minimum value on $[a, b]$.

Observe that if an absolute extremum of a continuous function f occurs at a point in an open interval (a, b), then it must be a relative extremum of f, and hence its x-coordinate must be a critical number of f. Otherwise, the absolute extremum of f must occur at one or both of the endpoints of the interval $[a, b]$. A typical situation is illustrated in Figure 59.

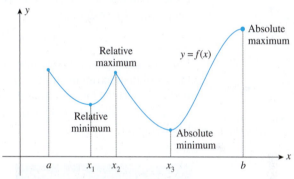

FIGURE 59
The relative minimum of f at x_3 is the absolute minimum of f. The right endpoint b of the interval $[a, b]$ gives rise to the absolute maximum value $f(b)$ of f.

Here, x_1, x_2, and x_3 are critical numbers of f. The absolute minimum of f occurs at x_3, which lies in the open interval (a, b) and is a critical number of f. The absolute maximum of f occurs at b, an endpoint. This observation suggests the following procedure for finding the absolute extrema of a continuous function on a closed interval.

Finding the Absolute Extrema of f on a Closed Interval

1. Find the critical numbers of f that lie in (a, b).
2. Compute the value of f at each critical number found in Step 1 and compute $f(a)$ and $f(b)$.
3. The absolute maximum value and absolute minimum value of f will correspond to the largest and smallest numbers, respectively, found in Step 2.

EXAMPLE 1 Find the absolute extrema of the function $F(x) = x^2$ defined on the interval $[-1, 2]$.

Solution The function F is continuous on the closed interval $[-1, 2]$ and differentiable on the open interval $(-1, 2)$. The derivative of F is

$$F'(x) = 2x$$

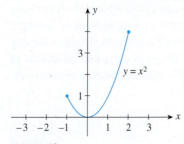

FIGURE 60
F has an absolute minimum value of 0 and an absolute maximum value of 4.

so 0 is the only critical number of *F*. Next, evaluate *F*(*x*) at *x* = −1, *x* = 0, and *x* = 2. Thus,

$$F(-1) = 1 \qquad F(0) = 0 \qquad F(2) = 4$$

It follows that 0 is the absolute minimum value of *F* and 4 is the absolute maximum value of *F*. The graph of *F*, in Figure 60, confirms our results.

EXAMPLE 2 Find the absolute extrema of the function

$$f(x) = x^3 - 2x^2 - 4x + 4$$

defined on the interval $[0, 3]$.

Solution The function *f* is continuous on the closed interval $[0, 3]$ and differentiable on the open interval $(0, 3)$. The derivative of *f* is

$$f'(x) = 3x^2 - 4x - 4 = (3x + 2)(x - 2)$$

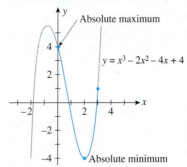

FIGURE 61
f has an absolute maximum value of 4 and an absolute minimum value of −4.

and it is equal to zero when $x = -\frac{2}{3}$ and $x = 2$. Since $x = -\frac{2}{3}$ lies outside the interval $[0, 3]$, it is dropped from further consideration, and $x = 2$ is seen to be the sole critical number of *f*. Next, we evaluate $f(x)$ at the critical number of *f* as well as the endpoints of *f*, obtaining

$$f(0) = 4 \qquad f(2) = -4 \qquad f(3) = 1$$

From these results, we conclude that −4 is the absolute minimum value of *f* and 4 is the absolute maximum value of *f*. The graph of *f*, which appears in Figure 61, confirms our results. Observe that the absolute maximum of *f* occurs at the endpoint $x = 0$ of the interval $[0, 3]$, while the absolute minimum of *f* occurs at $x = 2$, which lies in the interval $(0, 3)$.

Exploring with TECHNOLOGY

Let $f(x) = x^3 - 2x^2 - 4x + 4$. (This is the function of Example 2.)

1. Use a graphing utility to plot the graph of *f*, using the viewing window $[0, 3] \times [-5, 5]$. Use **ZOOM** and **TRACE** to find the absolute extrema of *f* on the interval $[0, 3]$ and thus verify the results obtained analytically in Example 2.
2. Plot the graph of *f*, using the viewing window $[-2, 1] \times [-5, 6]$. Use **ZOOM** and **TRACE** to find the absolute extrema of *f* on the interval $[-2, 1]$. Verify your results analytically.

EXAMPLE 3 Find the absolute maximum and absolute minimum values of the function $f(x) = x^{2/3}$ on the interval $[-1, 8]$.

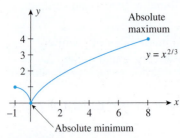

FIGURE 62
f has an absolute minimum value of $f(0) = 0$ and an absolute maximum value of $f(8) = 4$.

Solution The derivative of *f* is

$$f'(x) = \frac{2}{3}x^{-1/3} = \frac{2}{3x^{1/3}}$$

Note that *f'* is not defined at $x = 0$ and does not equal zero for any *x*. Therefore, 0 is the only critical number of *f*. Evaluating $f(x)$ at $x = -1$, 0, and 8, we obtain

$$f(-1) = 1 \qquad f(0) = 0 \qquad f(8) = 4$$

We conclude that the absolute minimum value of *f* is 0, attained at $x = 0$, and the absolute maximum value of *f* is 4, attained at $x = 8$ (Figure 62).

Many real-world applications call for finding the absolute maximum value or the absolute minimum value of a given function. For example, management is interested in finding what level of production will yield the maximum profit for a company; a farmer is interested in finding the right amount of fertilizer to maximize crop yield; a doctor is interested in finding the maximum concentration of a drug in a patient's body and the time at which it occurs; and an engineer is interested in finding the dimensions of a container with a specified shape and volume that can be constructed at a minimum cost.

APPLIED EXAMPLE 4 Maximizing Profits Acrosonic's total profit (in dollars) from manufacturing and selling x units of their model F loudspeaker systems is given by

$$P(x) = -0.02x^2 + 300x - 200{,}000 \qquad (0 \le x \le 20{,}000)$$

How many units of the loudspeaker system must Acrosonic produce to maximize its profits?

Solution To find the absolute maximum of P on $[0, 20{,}000]$, first find the critical points of P on the interval $(0, 20{,}000)$. To do this, compute

$$P'(x) = -0.04x + 300$$

Solving the equation $P'(x) = 0$ gives $x = 7500$. Next, evaluate $P(x)$ at $x = 7500$ as well as the endpoints $x = 0$ and $x = 20{,}000$ of the interval $[0, 20{,}000]$, obtaining

$$P(0) = -200{,}000$$
$$P(7500) = 925{,}000$$
$$P(20{,}000) = -2{,}200{,}000$$

From these computations, we see that the absolute maximum value of the function P is 925,000. Thus, by producing 7500 units, Acrosonic will realize a maximum profit of \$925,000. The graph of P is sketched in Figure 63.

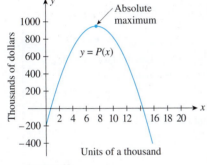

FIGURE 63
P has an absolute maximum at (7500, 925,000).

> **Explore and Discuss**
>
> Recall that the total profit function P is defined as $P(x) = R(x) - C(x)$, where R is the total revenue function, C is the total cost function, and x is the number of units of a product produced and sold. (Assume that all derivatives exist.)
>
> 1. Show that at the level of production x_0 that yields the maximum profit for the company, the following two conditions are satisfied:
>
> $$R'(x_0) = C'(x_0) \quad \text{and} \quad R''(x_0) < C''(x_0)$$
>
> 2. Interpret the two conditions in part 1 in economic terms, and explain why they make sense.

APPLIED EXAMPLE 5 Trachea Contraction During a Cough When a person coughs, the trachea (windpipe) contracts, allowing air to be expelled at a maximum velocity. It can be shown that during a cough, the velocity v of airflow is given by the function

$$v = f(r) = kr^2(R - r)$$

where r is the trachea's radius (in centimeters) during a cough, R is the trachea's normal radius (in centimeters), and k is a positive constant that depends on the length of the trachea. Find the radius r for which the velocity of airflow is greatest.

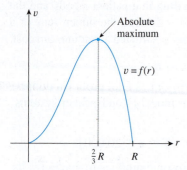

FIGURE 64
The velocity of airflow is greatest when the radius of the contracted trachea is $\frac{2}{3}R$.

Solution   To find the absolute maximum of f on $[0, R]$, first find the critical numbers of f on the interval $(0, R)$. We compute

$$f'(r) = 2kr(R - r) - kr^2 \qquad \text{Use the Product Rule.}$$
$$= -3kr^2 + 2kRr = kr(-3r + 2R)$$

Setting $f'(r) = 0$ gives $r = 0$ or $r = \frac{2}{3}R$, so $\frac{2}{3}R$ is the sole critical number of f ($r = 0$ is an endpoint). Evaluating $f(r)$ at $r = \frac{2}{3}R$, as well as at the endpoints $r = 0$ and $r = R$, we obtain

$$f(0) = 0$$
$$f\left(\frac{2}{3}R\right) = \frac{4k}{27}R^3$$
$$f(R) = 0$$

from which we deduce that the velocity of airflow is greatest when the radius of the contracted trachea is $\frac{2}{3}R$—that is, when the radius is contracted by approximately 33%. The graph of the function f is shown in Figure 64.

Explore and Discuss

Prove that if a cost function $C(x)$ is concave upward $[C''(x) > 0]$, then the level of production that will result in the smallest average production cost occurs when

$$\overline{C}(x) = C'(x)$$

that is, when the average cost $\overline{C}(x)$ is equal to the marginal cost $C'(x)$.
Hints:

1. Show that

$$\overline{C}'(x) = \frac{xC'(x) - C(x)}{x^2}$$

   so the critical number of the function $\overline{C}$ occurs when

$$xC'(x) - C(x) = 0$$

2. Show that at a critical number of $\overline{C}$

$$\overline{C}''(x) = \frac{C''(x)}{x}$$

Use the Second Derivative Test to reach the desired conclusion.

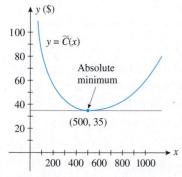

FIGURE 65
The minimum average cost is $35 per unit.

 APPLIED EXAMPLE 6 Minimizing Average Cost   The daily average cost function (in dollars per unit) of Elektra Electronics is given by

$$\overline{C}(x) = 0.0001x^2 - 0.08x + 40 + \frac{5000}{x} \qquad (x > 0)$$

where x stands for the number of graphing calculators that Elektra produces. Show that a production level of 500 units per day results in a minimum average cost for the company.

Solution   The domain of the function $\overline{C}$ is the interval $(0, \infty)$, which is not closed. To solve the problem, we resort to the graphical method. Using the techniques of graphing from the last section, we sketch the graph of $\overline{C}$ (Figure 65).

Now,

$$\overline{C}'(x) = 0.0002x - 0.08 - \frac{5000}{x^2}$$

Substituting the given value of x, 500, into $\overline{C}'(x)$ gives $\overline{C}'(500) = 0$, so 500 is a critical number of $\overline{C}$. Next,

$$\overline{C}''(x) = 0.0002 + \frac{10,000}{x^3}$$

Thus,

$$\overline{C}''(500) = 0.0002 + \frac{10,000}{(500)^3} > 0$$

and by the Second Derivative Test, a relative minimum of the function $\overline{C}$ occurs at 500. Furthermore, $\overline{C}''(x) > 0$ for $x > 0$, which implies that the graph of $\overline{C}$ is concave upward everywhere, so the relative minimum of $\overline{C}$ must be the absolute minimum of $\overline{C}$. The minimum average cost is given by

$$\overline{C}(500) = 0.0001(500)^2 - 0.08(500) + 40 + \frac{5000}{500}$$

$$= 35$$

or $35 per unit.

Exploring with **TECHNOLOGY**

Refer to the preceding Explore and Discuss and Example 6.

1. Using a graphing utility, plot the graphs of

$$\overline{C}(x) = 0.0001x^2 - 0.08x + 40 + \frac{5000}{x}$$

$$C'(x) = 0.0003x^2 - 0.16x + 40$$

using the viewing window $[0, 1000] \times [0, 150]$.
   Note: $C(x) = 0.0001x^3 - 0.08x^2 + 40x + 5000$. (Why?)

2. Find the point of intersection of the graphs of $\overline{C}$ and C' and thus verify the assertion in the Explore and Discuss for the special case studied in Example 6.

 APPLIED EXAMPLE 7 Flight of a Rocket The altitude (in feet) of a rocket t seconds into flight is given by

$$s = f(t) = -t^3 + 96t^2 + 5 \qquad (t \geq 0)$$

a. Find the maximum altitude attained by the rocket.
b. Find the maximum velocity attained by the rocket.

Solution

a. The maximum altitude attained by the rocket is given by the largest value of the function f in the closed interval $[0, T]$, where T denotes the time the rocket touches the earth. We know that such a number exists because the dominant term in the expression for the continuous function f is $-t^3$. So for t large enough, the value of $f(t)$ must change from positive to negative and, in particular, it must attain the value 0 for some T.

To find the absolute maximum of f, compute

$$f'(t) = -3t^2 + 192t$$
$$= -3t(t - 64) \qquad (x^2) \text{ See page 9.}$$

and solve the equation $f'(t) = 0$, obtaining $t = 0$ and $t = 64$. Evaluating f at the critical number $t = 64$ and the endpoints of f, we have

$$f(0) = 5 \qquad f(64) = 131{,}077$$

and we conclude, accordingly, that the absolute maximum value of f is 131,077. Thus, the maximum altitude of the rocket is 131,077 feet, attained 64 seconds into flight. The graph of f is sketched in Figure 66.

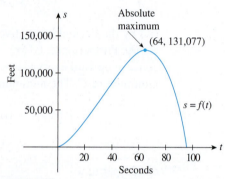

FIGURE 66
The maximum altitude of the rocket is 131,077 feet.

b. To find the maximum velocity attained by the rocket, find the largest value of the function that describes the rocket's velocity at any time t—namely,

$$v = f'(t) = -3t^2 + 192t \qquad (t \geq 0)$$

We find the critical number of v by setting $v' = 0$. But

$$v' = -6t + 192$$

and the critical number of v is 32. Since

$$v'' = -6 < 0$$

the Second Derivative Test implies that a relative maximum of v occurs at $t = 32$. Our computation has in fact clarified the property of the "velocity curve." Since $v'' < 0$ everywhere, the velocity curve is concave downward everywhere. With this observation, we assert that the relative maximum must in fact be the absolute maximum of v. The maximum velocity of the rocket is given by evaluating v at $t = 32$:

$$f'(32) = -3(32)^2 + 192(32)$$

or 3072 feet per second. The graph of the velocity function v is sketched in Figure 67.

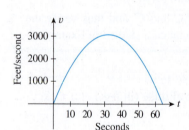

FIGURE 67
The maximum velocity of the rocket is 3072 feet per second.

4.4 Self-Check Exercises

1. Let $f(x) = x - 2\sqrt{x}$.
  a. Find the absolute extrema of f on the interval $[0, 9]$.
  b. Find the absolute extrema of f.

2. Find the absolute extrema of $f(x) = 3x^4 + 4x^3 + 1$ on $[-2, 1]$.

3. **FACTORY OPERATING RATE** The operating rate (expressed as a percent) of factories, mines, and utilities in a certain region of the country on the tth day of 2014 is given by the function

$$f(t) = 80 + \frac{1200t}{t^2 + 40,000} \qquad (0 \le t \le 250)$$

On which of the first 250 days of 2014 was the operating rate highest?

Solutions to Self-Check Exercises 4.4 can be found on page 318.

4.4 Concept Questions

1. Explain the following terms: (a) absolute maximum and (b) absolute minimum.

2. Describe the procedure for finding the absolute extrema of a continuous function on a closed interval.

4.4 Exercises

In Exercises 1–8, you are given the graph of a function f defined on the indicated interval. Find the absolute maximum and the absolute minimum of f, if they exist.

1.

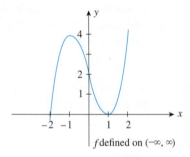

f defined on $(-\infty, \infty)$

2.

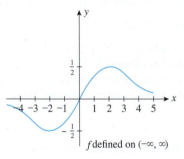

f defined on $(-\infty, \infty)$

3.

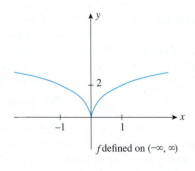

f defined on $(-\infty, \infty)$

4.

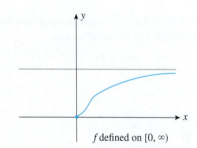

f defined on $[0, \infty)$

5.

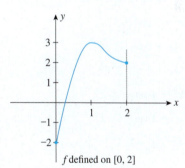

f defined on $[0, 2]$

6.

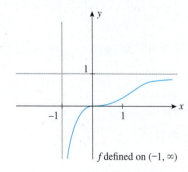

f defined on $(-1, \infty)$

7.

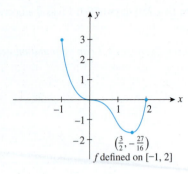

$\left(\frac{3}{2}, -\frac{27}{16}\right)$

f defined on $[-1, 2]$

8.

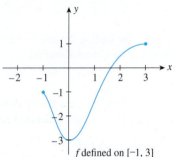

f defined on $[-1, 3]$

In Exercises 9–38, find the absolute maximum value and the absolute minimum value, if any, of each function.

9. $f(x) = 2x^2 + 3x - 4$ **10.** $g(x) = -x^2 + 4x + 3$

11. $h(x) = x^{1/3}$ **12.** $f(x) = x^{2/3}$

13. $f(x) = \dfrac{1}{1 + x^2}$ **14.** $f(x) = \dfrac{x}{1 + x^2}$

15. $f(x) = x^2 - 2x - 3$ on $[-2, 3]$

16. $g(x) = x^2 - 2x - 3$ on $[0, 4]$

17. $f(x) = -x^2 + 4x + 6$ on $[0, 5]$

18. $f(x) = -x^2 + 4x + 6$ on $[3, 6]$

19. $f(x) = x^3 + 3x^2 - 1$ on $[-3, 2]$

20. $g(x) = x^3 + 3x^2 - 1$ on $[-3, 1]$

21. $g(x) = 3x^4 + 4x^3$ on $[-2, 1]$

22. $f(x) = \dfrac{1}{2}x^4 - \dfrac{2}{3}x^3 - 2x^2 + 3$ on $[-2, 3]$

23. $f(x) = \dfrac{x + 1}{x - 1}$ on $[2, 4]$ **24.** $g(t) = \dfrac{t}{t - 1}$ on $[2, 4]$

25. $f(x) = 4x + \dfrac{1}{x}$ on $[1, 4]$

26. $f(x) = 9x - \dfrac{1}{x}$ on $[1, 3]$

27. $f(x) = \dfrac{1}{2}x^2 - 2\sqrt{x}$ on $[0, 3]$

28. $g(x) = \dfrac{1}{8}x^2 - 4\sqrt{x}$ on $[0, 9]$

29. $f(x) = \dfrac{1}{x}$ on $(0, \infty)$ **30.** $g(x) = \dfrac{1}{x + 1}$ on $(0, \infty)$

31. $f(x) = 3x^{2/3} - 2x$ on $[0, 3]$

32. $g(x) = x^2 + 2x^{2/3}$ on $[-2, 2]$

33. $f(x) = x^{2/3}(x^2 - 4)$ on $[-1, 2]$

34. $f(x) = x^{2/3}(x^2 - 4)$ on $[-1, 3]$

35. $f(x) = \dfrac{x}{x^2 + 2}$ on $[-1, 2]$

36. $f(x) = \dfrac{1}{x^2 + 2x + 5}$ on $[-2, 1]$

37. $f(x) = \dfrac{x}{\sqrt{x^2 + 1}}$ on $[-1, 1]$

38. $g(x) = x\sqrt{4 - x^2}$ on $[0, 2]$

39. A stone is thrown straight up from the roof of an 80-ft building. The height (in feet) of the stone at any time *t* (in seconds), measured from the ground, is given by

$$h(t) = -16t^2 + 64t + 80$$

What is the maximum height the stone reaches?

40. MAXIMIZING PROFITS Lynbrook West, an apartment complex, has 100 two-bedroom units. The monthly profit (in dollars) realized from renting out *x* apartments is given by

$$P(x) = -10x^2 + 1760x - 50,000$$

To maximize the monthly rental profit, how many units should be rented out? What is the maximum monthly profit realizable?

41. STRIKE OUTS The rate at which major league players were striking out in the years 2009 through 2013 is approximately

$$f(t) = 0.136t^2 + 0.127t + 18.1 \qquad (0 \le t \le 4)$$

percent in year *t*, where *t* = 0 corresponds to 2009.
a. What was the lowest rate of strikeouts over the years under consideration? When did it occur?
b. What was the highest rate of strikeouts. When did it occur?
Source: USA Today.

42. END OF THE IPOD ERA Apple introduced the first iPod in October 2001. Sales of the portable music player grew slowly in the early years but began to grow rapidly after 2005. But the iPod era is coming to a close. Smartphones with music and video players are replacing the iPod, along with the category of device it helped to create. Sales of the iPod worldwide from 2007 through 2011 (in millions) were approximately

$$N(t) = -2.65t^2 + 13.13t + 39.9 \qquad (0 \le t \le 4)$$

in year *t*, where *t* = 0 corresponds to 2007. Show that the worldwide sales of the iPod peaked sometime in 2009. What was the approximate largest number of iPods sold worldwide from 2007 through 2011?
Source: Popular Mechanics.

43. **AVERAGE SPEED OF A VEHICLE** The average speed of a vehicle on a stretch of Route 134 between 6 A.M. and 10 A.M. on a typical weekday is approximated by the function

$$f(t) = 20t - 40\sqrt{t} + 50 \qquad (0 \le t \le 4)$$

where $f(t)$ is measured in miles per hour and t is measured in hours, with $t = 0$ corresponding to 6 A.M. At what time of the morning commute is the traffic moving at the slowest rate? What is the average speed of a vehicle at that time?

44. **MAXIMIZING REVENUE** The quantity demanded of a certain brand of handbags per day, x, is related to the unit price, p, in dollars, by the equation

$$p = \frac{100,000}{250 + x} - 100 \qquad (0 \le x \le 750)$$

Find the level of sales and the corresponding unit price of the handbags that will result in a maximum revenue per day for the company. What is that revenue?

45. **FLIGHT OF A ROCKET** The altitude (in feet) attained by a model rocket t sec into flight is given by the function

$$h(t) = -\frac{1}{3}t^3 + 4t^2 + 20t + 2 \qquad (t \ge 0)$$

Find the maximum altitude attained by the rocket.

46. **MAXIMIZING PROFITS** The management of Trappee and Sons, producers of the famous TexaPep hot sauce, estimate that their profit (in dollars) from the daily production and sale of x cases (each case consisting of 24 bottles) of the hot sauce is given by

$$P(x) = -0.000002x^3 + 6x - 400$$

What is the largest possible profit Trappee can make in 1 day?

47. **MAXIMIZING PROFITS** The quantity demanded each month of the Walter Serkin recording of Beethoven's *Moonlight Sonata*, produced by Phonola Media, is related to the price per compact disc. The equation

$$p = -0.00042x + 6 \qquad (0 \le x \le 12,000)$$

where p denotes the unit price in dollars and x is the number of discs demanded, relates the demand to the price. The total monthly cost (in dollars) for pressing and packaging x copies of this classical recording is given by

$$C(x) = 600 + 2x - 0.00002x^2 \qquad (0 \le x \le 20,000)$$

To maximize its profits, how many copies should Phonola produce each month?
Hint: The revenue is $R(x) = px$, and the profit is $P(x) = R(x) - C(x)$.

48. **MAXIMIZING PROFIT** A manufacturer of tennis rackets finds that the total cost $C(x)$ (in dollars) of manufacturing x rackets/day is given by $C(x) = 400 + 4x + 0.0001x^2$. Each racket can be sold at a price of p dollars, where p is related to x by the demand equation $p = 10 - 0.0004x$. If

all rackets that are manufactured can be sold, find the daily level of production that will yield a maximum profit for the manufacturer.

49. **MAXIMIZING PROFIT** A division of Chapman Corporation manufactures a pager. The weekly fixed cost for the division is $20,000, and the variable cost for producing x pagers per week is

$$V(x) = 0.000001x^3 - 0.01x^2 + 50x$$

dollars. The company realizes a revenue of

$$R(x) = -0.02x^2 + 150x \qquad (0 \le x \le 7500)$$

dollars from the sale of x pagers/week. Find the level of production that will yield a maximum profit for the manufacturer.
Hint: Use the quadratic formula.

50. **MAXIMIZING PROFIT** The weekly demand for the Pulsar 40-in. high-definition television is given by the demand equation

$$p = -0.05x + 600 \qquad (0 \le x \le 12,000)$$

where p denotes the wholesale unit price in dollars and x denotes the quantity demanded. The weekly total cost function associated with manufacturing these sets is given by

$$C(x) = 0.000002x^3 - 0.03x^2 + 400x + 80,000$$

where $C(x)$ denotes the total cost incurred in producing x sets. Find the level of production that will yield a maximum profit for the manufacturer.
Hint: Use the quadratic formula.

51. **MINIMIZING AVERAGE COST** Suppose the total cost function for manufacturing a certain product is $C(x) = 0.2(0.01x^2 + 120)$ dollars, where x represents the number of units produced. Find the level of production that will minimize the average cost.

52. **MINIMIZING PRODUCTION COSTS** The total monthly cost (in dollars) incurred by Cannon Precision Instruments for manufacturing x units of the model M1 digital camera is given by the function

$$C(x) = 0.0025x^2 + 80x + 10,000$$

a. Find the average cost function $\bar{C}$.
b. Find the level of production that results in the smallest average production cost.
c. Find the level of production for which the average cost is equal to the marginal cost.
d. Compare the result of part (c) with that of part (b).

53. **MINIMIZING PRODUCTION COSTS** The daily total cost (in dollars) incurred by Trappee and Sons for producing x cases of TexaPep hot sauce is given by the function

$$C(x) = 0.000002x^3 + 5x + 400$$

Using this function, answer the questions posed in Exercise 52.

54. MINIMIZING AVERAGE COST Suppose that the total cost incurred in manufacturing x units of a certain product is given by $C(x)$, where C is a differentiable cost function. Show that the average cost is minimized at the level of production where the average cost is equal to the marginal cost.

55. MINIMIZING PRODUCTION COSTS Re-solve Exercise 52 using the result of Exercise 54.

56. MAXIMIZING REVENUE Suppose the quantity demanded per week of a certain dress is related to the unit price p by the demand equation $p = \sqrt{800 - x}$, where p is in dollars and x is the number of dresses made. To maximize the revenue, how many dresses should be made and sold each week?
Hint: $R(x) = px$

57. MAXIMIZING REVENUE The quantity demanded each month of the Sicard sports watch is related to the unit price by the equation

$$p = \frac{50}{0.01x^2 + 1} \qquad (0 \le x \le 20)$$

where p is measured in dollars and x is measured in units of a thousand. To yield a maximum revenue, how many watches must be sold?

58. AIR POLLUTION The amount of nitrogen dioxide, a brown gas that impairs breathing, present in the atmosphere on a certain May day in the city of Long Beach is approximated by

$$A(t) = \frac{136}{1 + 0.25(t - 4.5)^2} + 28 \qquad (0 \le t \le 11)$$

where $A(t)$ is measured in pollutant standard index (PSI) and t is measured in hours, with $t = 0$ corresponding to 7 A.M. Determine the time of day when the pollution is at its highest level.

59. OXYGEN CONTENT OF A POND When organic waste is dumped into a pond, the oxidation process that takes place reduces the pond's oxygen content. However, given time, nature will restore the oxygen content to its natural level. Suppose the oxygen content t days after organic waste has been dumped into the pond is given by

$$f(t) = 100\left(\frac{t^2 - 4t + 4}{t^2 + 4}\right) \qquad (0 \le t < \infty)$$

percent of its normal level.
a. When is the level of oxygen content lowest?
b. When is the rate of oxygen regeneration greatest?

60. VELOCITY OF BLOOD According to a law discovered by the 19th-century physician Jean Louis Marie Poiseuille, the velocity (in centimeters per second) of blood r cm from the central axis of an artery is given by

$$v(r) = k(R^2 - r^2)$$

where k is a constant and R is the radius of the artery. Show that the velocity of blood is greatest along the central axis.

61. MAXIMIZING REVENUE The average revenue is defined as the function

$$\overline{R}(x) = \frac{R(x)}{x} \qquad (x > 0)$$

Prove that if a revenue function $R(x)$ is concave downward $[R''(x) < 0]$, then the level of sales that will result in the largest average revenue occurs when $\overline{R}(x) = R'(x)$.

62. CRIME RATES The number of major crimes committed in the city of Bronxville between 2007 and 2014 is approximated by the function

$$N(t) = -0.1t^3 + 1.5t^2 + 100 \qquad (0 \le t \le 7)$$

where $N(t)$ denotes the number of crimes committed in year t ($t = 0$ corresponds to 2007). Enraged by the dramatic increase in the crime rate, the citizens of Bronxville, with the help of the local police, organized Neighborhood Crime Watch groups in early 2011 to combat this menace. Show that the growth in the crime rate was maximal in 2012, giving credence to the claim that the Neighborhood Crime Watch program was working.

63. GDP OF A DEVELOPING COUNTRY A developing country's gross domestic product (GDP) from 2006 to 2014 is approximated by the function

$$G(t) = -0.2t^3 + 2.4t^2 + 60 \qquad (0 \le t \le 8)$$

where $G(t)$ is measured in billions of dollars and $t = 0$ corresponds to 2006. Show that the growth rate of the country's GDP was maximal in 2010.

64. FEDERAL DEFICIT The deficit of the federal government (in trillions of dollars) in year t from 2006 ($t = 0$) through 2012 ($t = 6$) is approximately

$$D(t) = -0.038898t^3 + 0.30858t^2 - 0.31849t + 0.22$$
$$(0 \le t \le 6)$$

What was the largest federal deficit over the period under consideration?
Hint: Use the quadratic formula.
Source: Office of Management and Budget.

65. NEW INMATES The number of new prison admissions (into state or federal facilities) between 2002 and 2011 is given by the function

$$N(t) = -87.244444t^3 - 2482.35t^2 + 46009.26t + 579185$$
$$(2 \le t \le 11)$$

where t is measured in years, with $t = 2$ corresponding to 2002. Show that the number of new prison admissions peaked in 2006. What was the highest number of new inmates that year?
Hint: Use the quadratic formula.
Source: Bureau of Justice Statistics.

66. DEATH RATE FROM AIDS The estimated number of deaths from AIDS worldwide is given by

$$f(t) = -0.0004401t^3 + 0.007t^2 + 0.112t + 0.28$$
$$(0 \le t \le 21)$$

million per year, where t is measured in years, with $t = 0$ corresponding to 1990. What was the highest rate of deaths from AIDS worldwide over the period from 1990 through 2011?

Hint: Use the quadratic formula.

Source: UNAIDS.

67. **BRAIN GROWTH AND IQS** In a study conducted at the National Institute of Mental Health, researchers followed the development of the cortex, the thinking part of the brain, in 307 children. Using repeated magnetic resonance imaging scans from childhood to the latter teens, they measured the thickness (in millimeters) of the cortex of children of age t years with the highest IQs—121 to 149. These data lead to the model

$$S(t) = 0.000989t^3 - 0.0486t^2 + 0.7116t + 1.46$$
$$(5 \le t \le 19)$$

Show that the cortex of children with superior intelligence reaches maximum thickness around age 11 years.

Hint: Use the quadratic formula.

Source: Nature.

68. **BRAIN GROWTH AND IQS** Refer to Exercise 67. The researchers at the Institute also measured the thickness (also in millimeters) of the cortex of children of age t years who were of average intelligence. These data lead to the model

$$A(t) = -0.00005t^3 - 0.000826t^2 + 0.0153t + 4.55$$
$$(5 \le t \le 19)$$

Show that the cortex of children with average intelligence reaches maximum thickness at age 6 years.

Source: Nature.

69. **WORLD POPULATION** The total world population is forecast to be

$$P(t) = 0.00074t^3 - 0.0704t^2 + 0.89t + 6.04$$
$$(0 \le t \le 10)$$

in year t, where t is measured in decades, with $t = 0$ corresponding to 2000 and $P(t)$ is measured in billions.

a. Show that the world population is forecast to peak around 2071.

Hint: Use the quadratic formula.

b. At what number will the population peak?

Source: International Institute for Applied Systems Analysis.

70. **VENTURE-CAPITAL INVESTMENT** Venture-capital investment increased dramatically in the late 1990s but came to a screeching halt after the dot-com bust. The venture-capital investment (in billions of dollars) from 1995 ($t = 0$) through 2003 ($t = 8$) is approximated by the function

$$C(t) = \begin{cases} 0.6t^2 + 2.4t + 7.6 & \text{if } 0 \le t < 3 \\ 3t^2 + 18.8t - 63.2 & \text{if } 3 \le t < 5 \\ -3.3167t^3 + 80.1t^2 \\ \quad - 642.583t + 1730.8025 & \text{if } 5 \le t < 8 \end{cases}$$

a. In what year did venture-capital investment peak over the period under consideration? What was the amount of that investment?

b. In what year was the venture-capital investment lowest over this period? What was the amount of that investment?

Hint: Find the absolute extrema of C on each of the closed intervals $[0, 3]$, $[3, 5]$, and $[5, 8]$.

Sources: Venture One; Ernst & Young.

71. **ENERGY EXPENDED BY A FISH** It has been conjectured that a fish swimming a distance of L ft at a speed of v ft/sec relative to the water and against a current flowing at the rate of u ft/sec ($u < v$) expends a total energy given by

$$E(v) = \frac{aLv^3}{v - u}$$

where E is measured in foot-pounds (ft-lb) and a is a constant. Find the speed v at which the fish must swim in order to minimize the total energy expended. (*Note:* This result has been verified by biologists.)

72. **REACTION TO A DRUG** The strength of a human body's reaction R to a dosage D of a certain drug is given by

$$R = D^2 \left(\frac{k}{2} - \frac{D}{3} \right)$$

where k is a positive constant. Show that the maximum reaction is achieved if the dosage is k units.

73. Refer to Exercise 72. Show that the rate of change in the reaction R with respect to the dosage D is maximal if $D = k/2$.

74. **CHEMICAL REACTION** In an autocatalytic chemical reaction, the product formed acts as a catalyst for the reaction. If Q is the amount of the original substrate present initially and x is the amount of catalyst formed, then the rate of change of the chemical reaction with respect to the amount of catalyst present in the reaction is

$$R(x) = kx(Q - x) \qquad (0 \le x \le Q)$$

where k is a constant. Show that the rate of the chemical reaction is greatest at the point when exactly half of the original substrate has been transformed.

75. **MAXIMUM POWER OUTPUT** Suppose the source of current in an electric circuit is a battery. Then the power output P (in watts) obtained if the circuit has a resistance of R ohms is given by

$$P = \frac{E^2R}{(R + r)^2}$$

where E is the electromotive force in volts and r is the internal resistance of the battery in ohms. If E and r are constant, find the value of R that will result in the greatest power output. What is the maximum power output?

76. **VELOCITY OF A WAVE** In deep water, a wave of length L travels with a velocity

$$v = k\sqrt{\frac{L}{C} + \frac{C}{L}}$$

where k and C are positive constants. Find the length of the wave that has a minimum velocity.

77. **A MIXTURE PROBLEM** A tank initially contains 10 gal of brine with 2 lb of salt. Brine with 1.5 lb of salt per gallon enters the tank at the rate of 3 gal/min, and the well-stirred mixture leaves the tank at the rate of 4 gal/min. It can be shown that the amount of salt in the tank after t min is x lb, where

$$x = f(t) = 1.5(10 - t) - 0.0013(10 - t)^4 \qquad (0 \le t \le 10)$$

What is the maximum amount of salt present in the tank at any time?

In Exercises 78–84, determine whether the statement is true or false. If it is true, explain why it is true. If it is false, give an example to show why it is false.

78. If f is not continuous on the closed interval $[a, b]$, then f cannot have an absolute maximum value.

79. If f is defined on a closed interval $[a, b]$, then f has an absolute maximum value.

80. If $f''(x) < 0$ on (a, b) and $f'(c) = 0$, where $a < c < b$, then $f(c)$ is the absolute maximum value of f on $[a, b]$.

81. If f is continuous on $[a, b]$, f is differentiable on (a, b), and $f'(x) \ne 0$ for all x in (a, b), then the absolute maximum value of f on $[a, b]$ is $f(a)$ or $f(b)$.

82. If f is continuous on $[a, b]$, f is differentiable on (a, b), and $f'(x) > 0$ for all x in (a, b), then the absolute minimum value of f on $[a, b]$ is $f(a)$.

83. If the level of production, x, is constrained to be between a and b units, inclusive, then the maximum profit is $P(a)$, $P(b)$, or $P(c)$, where c is any number in (a, b) for which $R'(c) = C'(c)$ and $R''(c) < C''(c)$.

84. If $f''(x) > 0$ for all x in an interval I, then f must have an absolute minimum value at some number c in that interval.

85. Let f be a constant function—that is, let $f(x) = c$, where c is some real number. Show that every number a gives rise to an absolute maximum and, at the same time, an absolute minimum of f.

86. Show that a polynomial function defined on the interval $(-\infty, \infty)$ cannot have both an absolute maximum and an absolute minimum unless it is a constant function.

87. One condition that must be satisfied before Theorem 3 (page 307) is applicable is that the function f must be continuous on the closed interval $[a, b]$. Define a function f on the closed interval $[-1, 1]$ by

$$f(x) = \begin{cases} \dfrac{1}{x} & \text{if } x \in [-1, 1] \quad (x \ne 0) \\ 0 & \text{if } x = 0 \end{cases}$$

a. Show that f is not continuous at $x = 0$.
b. Show that f does not attain an absolute maximum or an absolute minimum on the interval $[-1, 1]$.
c. Confirm your results by sketching the function f.

88. One condition that must be satisfied before Theorem 3 (page 307) is applicable is that the interval on which f is defined must be a closed interval $[a, b]$. Define a function f on the *open* interval $(-1, 1)$ by $f(x) = x$. Show that f does not attain an absolute maximum or an absolute minimum on the interval $(-1, 1)$.
Hint: What happens to $f(x)$ if x is close to but not equal to $x = -1$? If x is close to but not equal to $x = 1$?

4.4   Solutions to Self-Check Exercises

1. **a.** The function f is continuous in its domain and differentiable on the interval $(0, 9)$. The derivative of f is

$$f'(x) = 1 - x^{-1/2} = \frac{x^{1/2} - 1}{x^{1/2}}$$

and it is equal to zero when $x = 1$. Evaluating $f(x)$ at the endpoints $x = 0$ and $x = 9$ and at the critical number 1 of f, we have

$$f(0) = 0 \qquad f(1) = -1 \qquad f(9) = 3$$

From these results, we see that -1 is the absolute minimum value of f and 3 is the absolute maximum value of f.

b. In this case, the domain of f is the interval $[0, \infty)$, which is not closed. Therefore, we resort to the graphical method. Using the techniques of graphing, we sketch the graph of f in the accompanying figure.

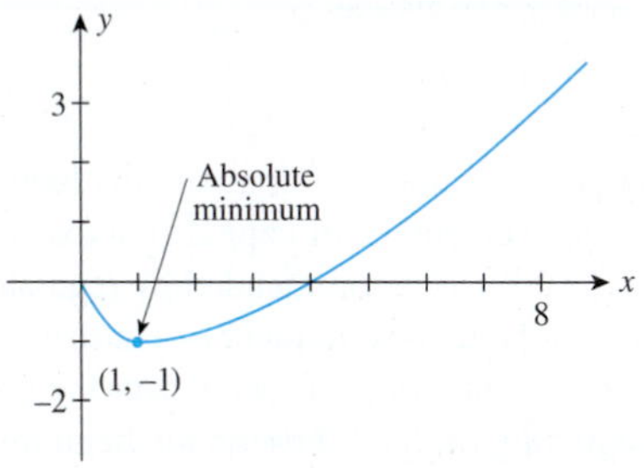

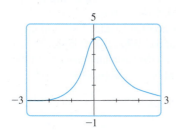

The graph of f shows that -1 is the absolute minimum value of f but f has no absolute maximum, since $f(x)$ increases without bound as x increases without bound.

2. The function f is continuous on the interval $[-2, 1]$. It is also differentiable on the open interval $(-2, 1)$. The derivative of f is

$$f'(x) = 12x^3 + 12x^2 = 12x^2(x + 1)$$

and it is continuous on $(-2, 1)$. Setting $f'(x) = 0$ gives -1 and 0 as critical numbers of f. Evaluating $f(x)$ at these critical numbers of f as well as at the endpoints of the interval $[-2, 1]$, we obtain

$$f(-2) = 17 \qquad f(-1) = 0 \qquad f(0) = 1 \qquad f(1) = 8$$

From these results, we see that 0 is the absolute minimum value of f and 17 is the absolute maximum value of f.

3. The problem is solved by finding the absolute maximum of the function f on $[0, 250]$. Differentiating $f(t)$, we obtain

$$f'(t) = \frac{(t^2 + 40{,}000)(1200) - 1200t(2t)}{(t^2 + 40{,}000)^2}$$

$$= \frac{-1200(t^2 - 40{,}000)}{(t^2 + 40{,}000)^2}$$

Upon setting $f'(t) = 0$ and solving the resulting equation, we obtain $t = -200$ or 200. Since -200 lies outside the interval $[0, 250]$, we are interested only in the critical number 200 of f. Evaluating $f(t)$ at $t = 0$, $t = 200$, and $t = 250$, we find

$$f(0) = 80 \qquad f(200) = 83 \qquad f(250) \approx 82.93$$

We conclude that the operating rate was the highest on the 200th day of 2014—that is, a little past the middle of July 2014.

USING TECHNOLOGY

Finding the Absolute Extrema of a Function

Some graphing utilities have a function for finding the absolute maximum and the absolute minimum values of a continuous function on a closed interval. If your graphing utility has this capability, use it to work through the example and exercises of this section.

EXAMPLE 1 Let $f(x) = \dfrac{2x + 4}{(x^2 + 1)^{3/2}}$.

a. Use a graphing utility to plot the graph of f in the viewing window $[-3, 3] \times [-1, 5]$.

b. Find the absolute maximum and absolute minimum values of f on the interval $[-3, 3]$. Express your answers accurate to four decimal places.

Solution

a. The graph of f is shown in Figure T1.

b. Using the function on a graphing utility for finding the absolute minimum value of a continuous function on a closed interval, we find the absolute minimum value of f to be -0.0632. Similarly, using the function for finding the absolute maximum value, we find the absolute maximum value to be 4.1593.

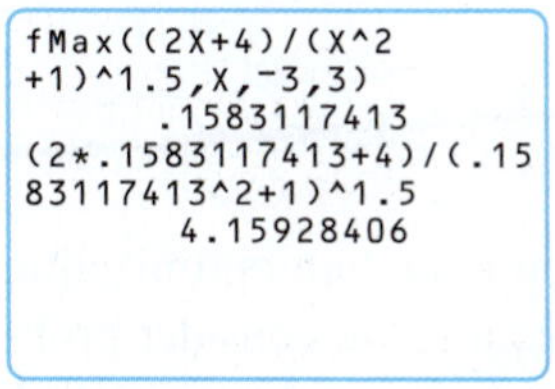

Note Some graphing utilities will enable you to find the absolute minimum and absolute maximum values of a continuous function on a closed interval without having to graph the function. For example, using **fMax** on the TI-83/84 will yield the x-coordinate of the absolute maximum of f. The absolute maximum value can then be found by evaluating f at that value of x. Figure T2 shows the work involved in finding the absolute maximum of the function of Example 1.

FIGURE T1
The graph of f in the viewing window $[-3, 3] \times [-1, 5]$

```
fMax((2X+4)/(X^2
+1)^1.5,X,-3,3)
         .1583117413
(2*.1583117413+4)/(.15
83117413^2+1)^1.5
         4.15928406
```

FIGURE T2
The TI-83/84 screen for Example 1

TECHNOLOGY EXERCISES

In Exercises 1–6, find the absolute maximum and the absolute minimum values of f in the given interval using the method of Example 1. Express your answers accurate to four decimal places.

1. $f(x) = 3x^4 - 4.2x^3 + 6.1x - 2; [-2, 3]$

2. $f(x) = 2.1x^4 - 3.2x^3 + 4.1x^2 + 3x - 4; [-1, 2]$

3. $f(x) = \dfrac{2x^3 - 3x^2 + 1}{x^2 + 2x - 8}; [-3, 1]$

4. $f(x) = \sqrt{x}(x^3 - 4)^2; [0.5, 1]$

5. $f(x) = \dfrac{x^3 - 1}{x^2}; [1, 3]$

6. $f(x) = \dfrac{x^3 - x^2 + 1}{x - 2}; [1, 3]$

7. BANK FAILURE The Haven Trust Bank of Duluth, Ga., founded in 2000, quickly increased its risky commercial real estate portfolio, despite many red flags from regulators. The bank failed in December 2008. The amount of construction loans of the bank as a percentage of its capital is approximated by the function

$$f(t) = -5.92t^4 + 58.89t^3 - 165.75t^2 + 56.21t + 629$$
$$(0 \le t \le 5)$$

where $t = 0$ corresponds to the beginning of 2003.
a. Plot the graph of f using the viewing window $[0, 5] \times [0, 650]$.
b. Show that at no time during the period from the beginning of 2003 through the beginning of 2008 did the amount of construction loans of the bank as a percentage of its capital fall below 415%. Note: The maximum percentage recommended by regulators in 2008 was 100%.
Source: FDIC Office of Inspector General.

8. CONSTRUCTION LOANS Refer to Exercise 7. The amount of construction loans of peer banks as a percentage of capital from the beginning of 2003 ($t = 0$) through the beginning of 2008 is approximated by the function

$$g(t) = -0.656t^4 + 5.693t^3 - 16.798t^2 + 36.083t^2 + 51.9$$
$$(0 \le t \le 5)$$

a. When did the amount of construction loans of peer banks as a percentage of capital first exceed the maximum of 100% as recommended by regulators in 2006?
b. What was the highest amount of construction loans as a share of capital of peer banks over the period from the beginning of 2003 through the beginning of 2008?
Source: FDIC Office of Inspector General.

9. SICKOUTS In a sickout by pilots of American Airlines in February 1999, the number of canceled flights from February 6 ($t = 0$) through February 14 ($t = 8$) is approximated by the function

$$N(t) = 1.2576t^4 - 26.357t^3 + 127.98t^2 + 82.3t + 43$$
$$(0 \le t \le 8)$$

where t is measured in days. The sickout ended after the union was threatened with millions of dollars in fines.
a. Show that the number of canceled flights was increasing at the fastest rate on February 8.
b. Estimate the maximum number of canceled flights in a day during the sickout.
Source: Associated Press.

10. AVERAGE 401(K) ACCOUNT BALANCE The following data give the average account balance (in thousands of dollars) of a 401(k) investor in year t in the United States during a certain 6-year period.

Year, t	0	1	2	3	4	5	6
Account Balance	37.5	40.8	47.3	55.5	49.4	43	40

a. Use **QuartReg** to find a fourth-degree polynomial regression model for the data. Let $t = 0$ correspond to the beginning of the first year.
b. Plot the graph of the function found in part (a), using the viewing window $[0, 6] \times [0, 60]$.
c. When was the average account balance lowest in the period under consideration? When was it highest?
d. What were the lowest average account balance and the highest average account balance during the period under consideration?
Source: Investment Company Institute.

4.5 Optimization II

In Section 4.4, we outlined a method for finding the solution to certain optimization problems in which the objective function is given. In this section, we consider problems in which we are required to first find the appropriate function to be optimized. The following guidelines will be useful for solving these problems.

Note   In carrying out Step 4, remember that if the function f to be optimized is continuous on a closed interval, then the absolute maximum and absolute minimum of f are, respectively, the largest and smallest values of $f(x)$ on the set composed of the critical numbers of f and the endpoints of the interval. If the domain of f is not a closed interval, then we resort to the graphical or some other method. ◾

Maximization Problems

APPLIED EXAMPLE 1 Fencing a Garden A man wishes to have a rectangular-shaped garden in his backyard. He has 50 feet of fencing with which to enclose his garden. Find the dimensions for the largest garden he can have if he uses all of the fencing.

Solution

Step 1   Let x and y denote the dimensions (in feet) of two adjacent sides of the garden (Figure 68), and let A denote its area. (x^2) See pages 86–87.

Step 2   The area of the garden

$$A = xy \qquad (1)$$

is the quantity to be maximized.

Step 3   The perimeter of the rectangle, $(2x + 2y)$ feet, must equal 50 feet. Therefore, we have the equation

$$2x + 2y = 50$$

Next, solving this equation for y in terms of x yields

$$y = 25 - x \qquad (2)$$

which, when substituted into Equation (1), gives

$$A = x(25 - x)$$
$$= -x^2 + 25x$$

(Remember, the function to be optimized must involve just one variable.) Since the sides of the rectangle must be nonnegative, we must have $x \geq 0$ and $y = 25 - x \geq 0$; that is, we must have $0 \leq x \leq 25$. Thus, the problem is reduced to that of finding the absolute maximum of $A = f(x) = -x^2 + 25x$ on the closed interval $[0, 25]$.

Step 4   Observe that f is continuous on $[0, 25]$, so the absolute maximum value of f must occur at the endpoint(s) of the interval or at the critical number(s) of f. The derivative of the function A is given by

$$A' = f'(x) = -2x + 25$$

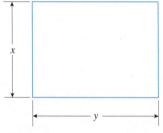

FIGURE 68
What is the maximum rectangular area that can be enclosed with 50 feet of fencing?

Setting $A' = 0$ gives

$$-2x + 25 = 0$$

or 12.5, as the critical number of A. Next, we evaluate the function $A = f(x)$ at $x = 12.5$ and at the endpoints $x = 0$ and $x = 25$ of the interval $[0, 25]$, obtaining

$$f(0) = 0 \qquad f(12.5) = 156.25 \qquad f(25) = 0$$

We see that the absolute maximum value of the function f is 156.25. From Equation (2), we see that $y = 12.5$ when $x = 12.5$. Thus, the garden of maximum area (156.25 square feet) is a square with sides of length 12.5 feet.

APPLIED EXAMPLE 2 Packaging By cutting away identical squares from each corner of a rectangular piece of cardboard and folding up the resulting flaps, the cardboard may be turned into an open box. If the cardboard is 16 inches long and 10 inches wide, find the dimensions of the box that will yield the maximum volume.

Solution

Step 1   Let x denote the length (in inches) of one side of each of the identical squares to be cut out of the cardboard (Figure 69), and let V denote the volume of the resulting box.

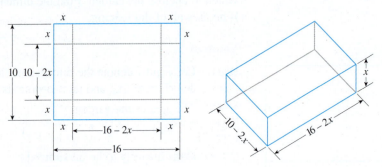

FIGURE **69**
The dimensions of the open box are $(16 - 2x)$ inches by $(10 - 2x)$ inches by x inches.

Step 2   The dimensions of the box are $(16 - 2x)$ inches by $(10 - 2x)$ inches by x inches. Therefore, its volume (in cubic inches),

$$V = (16 - 2x)(10 - 2x)x$$
$$= 4(x^3 - 13x^2 + 40x) \qquad \text{Expand the expression.}$$

is the quantity to be maximized.

Step 3   Since each side of the box must be nonnegative, x must satisfy the inequalities $x \geq 0$, $16 - 2x \geq 0$, and $10 - 2x \geq 0$. This set of inequalities is equivalent to $0 \leq x \leq 5$. Thus, the problem at hand is equivalent to that of finding the absolute maximum of

$$V = f(x) = 4(x^3 - 13x^2 + 40x)$$

on the closed interval $[0, 5]$.

Step 4   Observe that f is continuous on $[0, 5]$, so the absolute maximum value of f must be attained at the endpoint(s) or at the critical number(s) of f.
Differentiating $f(x)$, we obtain

$$f'(x) = 4(3x^2 - 26x + 40)$$
$$= 4(3x - 20)(x - 2)$$

Upon setting $f'(x) = 0$ and solving the resulting equation for x, we obtain $x = \frac{20}{3}$ or $x = 2$. Since $\frac{20}{3}$ lies outside the interval $[0, 5]$, it is no longer considered, and we are interested only in the critical number 2 of f. Next, evaluating $f(x)$ at $x = 0$, $x = 5$ (the endpoints of the interval $[0, 5]$), and $x = 2$, we obtain

$$f(0) = 0 \qquad f(2) = 144 \qquad f(5) = 0$$

Thus, the volume of the box is maximized by taking $x = 2$. The dimensions of the box are 12 in. $\times$ 6 in. $\times$ 2 in., and the volume is 144 cubic inches. ∎

Exploring with TECHNOLOGY

Refer to Example 2.

1. Use a graphing utility to plot the graph of

$$f(x) = 4(x^3 - 13x^2 + 40x)$$

using the viewing window $[0, 5] \times [0, 150]$. Explain what happens to $f(x)$ as x increases from $x = 0$ to $x = 5$ and give a physical interpretation.

2. Using ZOOM and TRACE, find the absolute maximum of f on the interval $[0, 5]$ and thus verify the solution for Example 2 obtained analytically.

APPLIED EXAMPLE 3 Optimal Subway Fare A city's Metropolitan Transit Authority (MTA) operates a subway line for commuters from a certain suburb to the downtown metropolitan area. Currently, an average of 6000 passengers a day take the trains, paying a fare of $3.00 per ride. The board of the MTA, contemplating raising the fare to $3.50 per ride in order to generate a larger revenue, engages the services of a consulting firm. The firm's study reveals that for each $0.50 increase in fare, the ridership will be reduced by an average of 1000 passengers a day. Therefore, the consulting firm recommends that MTA stick to the current fare of $3.00 per ride, which already yields a maximum revenue. Show that the consultants are correct.

Solution

Step 1 Let x denote the number of passengers per day, p denote the fare per ride, and R be MTA's revenue. (x^2) See pages 86–87.

Step 2 To find a relationship between x and p, observe that the given data imply that when $x = 6000$, $p = 3$, and when $x = 5000$, $p = 3.50$. Therefore, the points $(6000, 3)$ and $(5000, 3.50)$ lie on a straight line. (Why?) To find the linear relationship between p and x, use the point-slope form of the equation of a straight line. Now, the slope of the line is

$$m = \frac{3.50 - 3}{5000 - 6000} = -0.0005$$

Therefore, the required equation is

$$p - 3 = -0.0005(x - 6000)$$
$$= -0.0005x + 3$$
$$p = -0.0005x + 6$$

Therefore, the revenue

$$R = f(x) = xp = -0.0005x^2 + 6x \qquad \text{Number of riders} \times \text{unit fare}$$

is the quantity to be maximized.

Step 3 Since both p and x must be nonnegative, we see that $0 \leq x \leq 12{,}000$, and the problem is that of finding the absolute maximum of the function f on the closed interval $[0, 12{,}000]$.

Step 4 Observe that f is continuous on $[0, 12{,}000]$. To find the critical number of R, we compute

$$f'(x) = -0.001x + 6$$

and set it equal to zero, giving $x = 6000$. Evaluating the function f at $x = 6000$, as well as at the endpoints $x = 0$ and $x = 12{,}000$, yields

$$f(0) = 0$$
$$f(6000) = 18{,}000$$
$$f(12{,}000) = 0$$

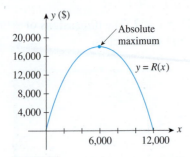

FIGURE 70
f has an absolute maximum of 18,000 when $x = 6000$.

We conclude that a maximum revenue of \$18,000 per day is realized when the ridership is 6000 per day. The optimum price of the fare per ride is therefore \$3.00, as recommended by the consultants. The graph of the revenue function R is shown in Figure 70.

Minimization Problems

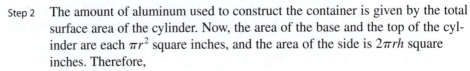

APPLIED EXAMPLE 4 Packaging Betty Moore Company requires that its corned beef hash containers have a capacity of 54 cubic inches, have the shape of a right circular cylinder, and be made of aluminum. Determine the radius and height of the container that requires the least amount of metal.

Solution

Step 1 Let the radius and height of the container be r and h inches, respectively, and let S denote the surface area of the container (Figure 71).

Step 2 The amount of aluminum used to construct the container is given by the total surface area of the cylinder. Now, the area of the base and the top of the cylinder are each πr^2 square inches, and the area of the side is $2\pi rh$ square inches. Therefore,

FIGURE 71
We want to minimize the amount of material used to construct the container.

$$S = 2\pi r^2 + 2\pi rh \qquad (3)$$

is the quantity to be minimized.

Step 3 The requirement that the volume of a container be 54 cubic inches implies that

$$\pi r^2 h = 54 \qquad (4)$$

Solving Equation (4) for h, we obtain

$$h = \frac{54}{\pi r^2} \qquad (5)$$

which, when substituted into Equation (3), yields

$$S = 2\pi r^2 + 2\pi r \left(\frac{54}{\pi r^2}\right)$$

$$= 2\pi r^2 + \frac{108}{r}$$

Clearly, the radius r of the container must satisfy the inequality $r > 0$. The problem now is reduced to finding the absolute minimum of the function $S = f(r)$ on the interval $(0, \infty)$.

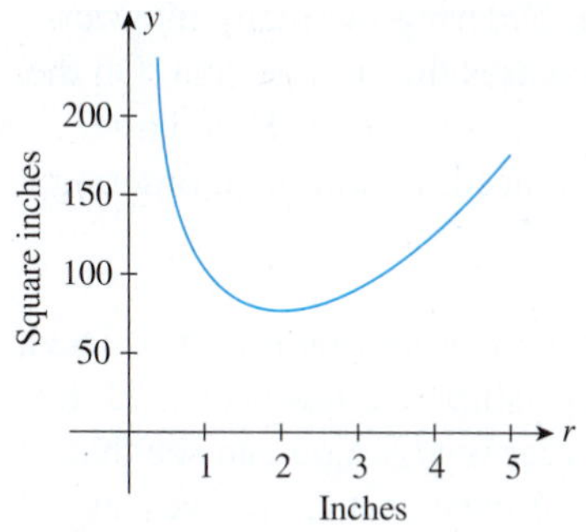

FIGURE 72
The total surface area of the right cylindrical container is graphed as a function of r.

Step 4 Using the curve-sketching techniques of Section 4.3, we obtain the graph of f in Figure 72.

To find the critical number of f, we compute

$$S' = 4\pi r - \frac{108}{r^2}$$

and solve the equation $S' = 0$ for r:

$$4\pi r - \frac{108}{r^2} = 0$$

$$4\pi r^3 - 108 = 0$$

$$r^3 = \frac{27}{\pi}$$

$$r = \frac{3}{\sqrt[3]{\pi}} \approx 2 \tag{6}$$

Next, let's show that this value of r gives rise to the absolute minimum of f. To show this, we first compute

$$S'' = 4\pi + \frac{216}{r^3}$$

Since $S'' > 0$ for $r = 3/\sqrt[3]{\pi}$, the Second Derivative Test implies that the value of r in Equation (6) gives rise to a relative minimum of f. Finally, this relative minimum of f is also the absolute minimum of f, since the graph of f is always concave upward ($S'' > 0$ for all $r > 0$). To find the height of the given container, we substitute the value of r given in Equation (6) into Equation (5). Thus,

$$h = \frac{54}{\pi r^2} = \frac{54}{\pi \left(\dfrac{3}{\pi^{1/3}}\right)^2}$$

$$= \frac{54\pi^{2/3}}{(\pi)9}$$

$$= \frac{6}{\pi^{1/3}} = \frac{6}{\sqrt[3]{\pi}}$$

$$= 2r$$

We conclude that the required container has a radius of approximately 2 inches and a height of approximately 4 inches, or twice the size of the radius. ∎

An Inventory Problem

One problem faced by many companies is that of controlling the inventory of goods carried. Ideally, the manager must ensure that the company has sufficient stock to meet customer demand at all times. At the same time, she must make sure that this is accomplished without overstocking (incurring unnecessary storage costs) and also without having to place orders too frequently (incurring reordering costs).

APPLIED EXAMPLE 5 Inventory Control and Planning Dixie Import-Export is the sole agent for the Excalibur 250-cc motorcycle. Management estimates that the demand for these motorcycles is 10,000 per year and that they will sell at a uniform rate throughout the year. The cost incurred in ordering each shipment of motorcycles is $10,000, and the cost per year of storing each motorcycle is $200.

Dixie's management faces the following problem: Ordering too many motorcycles at one time ties up valuable storage space and increases the storage cost. On the other hand, placing orders too frequently increases the ordering costs. How large should each order be, and how often should orders be placed, to minimize ordering and storage costs?

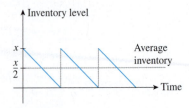

FIGURE 73
As each lot is depleted, the new lot arrives. The average inventory level is $x/2$ if x is the lot size.

Solution   Let x denote the number of motorcycles in each order (the lot size). Then, assuming that each shipment arrives just as the previous shipment has been sold, the average number of motorcycles in storage during the year is $x/2$. You can see that this is the case by examining Figure 73. Thus, Dixie's storage cost for the year is given by $200(x/2)$, or $100x$ dollars.

Next, since the company requires 10,000 motorcycles for the year and since each order is for x motorcycles, the number of orders required is

$$\frac{10,000}{x}$$

This gives an ordering cost of

$$10,000\left(\frac{10,000}{x}\right) = \frac{100,000,000}{x}$$

dollars for the year. Thus, the total yearly cost incurred by Dixie, which includes the ordering and storage costs attributed to the sale of these motorcycles, is given by

$$C(x) = 100x + \frac{100,000,000}{x}$$

The problem is reduced to finding the absolute minimum of the function C on the interval $(0, 10,000]$. To accomplish this, we compute

$$C'(x) = 100 - \frac{100,000,000}{x^2}$$

Setting $C'(x) = 0$ and solving the resulting equation, we obtain $x = \pm 1000$. Since the number -1000 is outside the domain of the function C, it is rejected, leaving 1000 as the only critical number of C. Next, we find

$$C''(x) = \frac{200,000,000}{x^3}$$

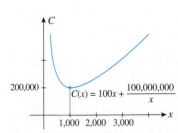

FIGURE 74
C has an absolute minimum at $(1000, 200,000)$.

Since $C''(1000) > 0$, the Second Derivative Test implies that the critical number 1000 is a relative minimum of the function C (Figure 74). Also, since $C''(x) > 0$ for all x in $(0, 10,000)$, the graph of C is concave upward everywhere, so $x = 1000$ also gives the absolute minimum of C. Thus, to minimize the ordering and storage costs, Dixie should place 10,000/1000, or 10, orders a year, each for a shipment of 1000 motorcycles. ▪

4.5   Self-Check Exercises

1. **FENCING A GARDEN** A man wishes to have an enclosed vegetable garden in his backyard. If the garden is to be a rectangular area of 300 ft², find the dimensions of the garden that will minimize the amount of fencing needed.

2. **INVENTORY CONTROL AND PLANNING** The demand for Super Titan tires is 1,000,000/year. The setup cost for each production run is $4000, and the manufacturing cost is

$20/tire. The cost of storing each tire over the year is $2. Assuming uniformity of demand throughout the year and instantaneous production, determine how many tires should be manufactured per production run to keep the total cost to a minimum.

Solutions to Self-Check Exercises 4.5 can be found on page 331.

4.5 Concept Questions

1. If the domain of a function f is not a closed interval, how would you find the absolute extrema of f, if they exist?

2. Refer to Example 4 (page 324). In the solution given in the example, we solved for h in terms of r, resulting in a function of r, which we then optimized with respect to r. Write S in terms of h and re-solve the problem. Which choice is better?

4.5 Exercises

1. Find the dimensions of a rectangle with a perimeter of 100 ft that has the largest possible area.

2. Find the dimensions of a rectangle of area 144 ft² that has the smallest possible perimeter.

3. **ENCLOSING THE LARGEST AREA** The owner of the Rancho Los Feliz has 3000 yd of fencing with which to enclose a rectangular piece of grazing land along the straight portion of a river. If fencing is not required along the river, what are the dimensions of the largest area that he can enclose? What is this area?

4. **ENCLOSING THE LARGEST AREA** Refer to Exercise 3. As an alternative plan, the owner of the Rancho Los Feliz might use the 3000 yd of fencing to enclose the rectangular piece of grazing land along the straight portion of the river and then subdivide it by means of a fence running parallel to the sides. Again, no fencing is required along the river. What are the dimensions of the largest area that can be enclosed? What is this area? (See the accompanying figure.)

5. **MINIMIZING CONSTRUCTION COSTS** The management of the UNICO department store has decided to enclose an 800-ft² area outside the building for displaying potted plants and flowers. One side will be formed by the external wall of the store, two sides will be constructed of pine boards, and the fourth side will be made of galvanized steel fencing. If the pine board fencing costs $6/running foot and the steel fencing costs $3/running foot,

determine the dimensions of the enclosure that can be erected at minimum cost.

6. **PACKAGING** By cutting away identical squares from each corner of a rectangular piece of cardboard and folding up the resulting flaps, an open box may be made. If the cardboard is 15 in. long and 8 in. wide, find the dimensions of the box that will yield the maximum volume.

7. **METAL FABRICATION** If an open box is made from a tin sheet 8 in. square by cutting out identical squares from each corner and bending up the resulting flaps, determine the dimensions of the largest box that can be made.

8. **MINIMIZING PACKAGING COSTS** If an open box has a square base and a volume of 108 in.³ and is constructed from a tin sheet, find the dimensions of the box, assuming that a minimum amount of material is used in its construction.

9. **MINIMIZING COSTS** A pencil cup with a capacity of 36 in.³ is to be constructed in the shape of a rectangular box with a square base and an open top. If the material for the sides costs 15¢/in.² and the material for the base costs 40¢/in.², what should the dimensions of the cup be to minimize the construction cost?

10. MINIMIZING PACKAGING COSTS What are the dimensions of a closed rectangular box that has a square cross section and a capacity of 128 in.³ and is constructed using the least amount of material?

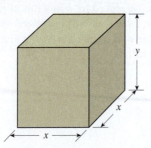

11. MINIMIZING PACKAGING COSTS A rectangular box is to have a square base and a volume of 20 ft³. If the material for the base costs 30¢/ft², the material for the sides costs 10¢/ft², and the material for the top costs 20¢/ft², determine the dimensions of the box that can be constructed at minimum cost. (Refer to the figure for Exercise 10.)

12. PARCEL POST REGULATIONS Postal regulations specify that a parcel sent by priority mail may have a combined length and girth of no more than 108 in. Find the dimensions of a rectangular package that has a square cross section and the largest volume that may be sent via priority mail. What is the volume of such a package?
Hint: The length plus the girth is $4x + h$ (see the accompanying figure).

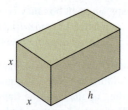

13. BOOK DESIGN A book designer has decided that the pages of a book should have 1-in. margins at the top and bottom and $\frac{1}{2}$-in. margins on the sides. She further stipulated that each page should have an area of 50 in.² (see the accompanying figure). Determine the page dimensions that will result in the maximum printed area on the page.

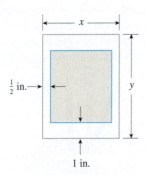

14. PARCEL POST REGULATIONS Postal regulations specify that a parcel sent by priority mail may have a combined length and girth of no more than 108 in. Find the dimensions of the cylindrical package of greatest volume that may be sent via priority mail. What is the volume of such a package? Compare with Exercise 12.
Hint: The length plus the girth is $2\pi r + l$.

15. MINIMIZING COSTS A pencil cup with a capacity of 9π in.³ is to be constructed in the shape of a right circular cylinder with an open top. If the material for the side costs $\frac{3}{8}$ of the cost of the material for the base, what dimensions should the cup have to minimize the construction cost?

16. MINIMIZING COSTS For its beef stew, Betty Moore Company uses aluminum containers that have the form of right circular cylinders. Find the radius and height of a container if it has a capacity of 36 in.³ and is constructed using the least amount of metal.

17. PRODUCT DESIGN The cabinet that will enclose the Acrosonic model D loudspeaker system will be rectangular and will have an internal volume of 2.4 ft³. For aesthetic reasons, the design team has decided that the height of the cabinet is to be 1.5 times its width. If the top, bottom, and sides of the cabinet are constructed of veneer costing 40¢/ft² and the front (ignore the cutouts in the baffle) and rear are constructed of particle board costing 20¢/ ft², what are the dimensions of the enclosure that can be constructed at a minimum cost?

18. DESIGNING A NORMAN WINDOW A Norman window has the shape of a rectangle surmounted by a semicircle (see the accompanying figure). If a Norman window is to have a perimeter of 28 ft, what should its dimensions be in order to allow the maximum amount of light through the window?

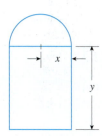

19. OPTIMAL CHARTER-FLIGHT FARE If exactly 200 people sign up for a charter flight, Leisure World Travel Agency charges $300/person. However, if more than 200 people sign up for the flight (assume that this is the case), then each fare is reduced by $1 for each additional person. Determine how many passengers will result in a maximum revenue for the travel agency. What is the maximum revenue? What would be the fare per passenger in this case?
Hint: Let x denote the number of passengers above 200. Show that the revenue function R is given by $R(x) = (200 + x)(300 - x)$.

20. MAXIMIZING YIELD An apple orchard has an average yield of 36 bushels of apples/tree if tree density is 22 trees/acre. For each unit increase in tree density, the yield decreases by 2 bushels/tree. How many trees should be planted to maximize the yield?

21. CHARTER REVENUE The owner of a luxury motor yacht that sails among the 4000 Greek islands charges $600/person/day if exactly 20 people sign up for the cruise. However, if more than 20 people sign up (up to the maximum capacity of 90) for the cruise, then every fare is reduced by $4 for each additional passenger. Assuming that at least 20 people sign up for the cruise, determine how many passengers will result in the maximum revenue for the owner of the yacht. What is the maximum revenue? What would be the fare per passenger in this case?

22. PROFIT OF A VINEYARD Phillip, the proprietor of a vineyard, estimates that the first 10,000 bottles of wine produced this season will fetch a profit of $5/bottle. But if more than 10,000 bottles are produced, then the profit per bottle for the entire lot will drop by $0.0002 for each additional bottle sold. Assuming that at least 10,000 bottles of wine are produced and sold, what is the maximum profit?

23. OPTIMAL SPEED OF A TRUCK A truck gets $600/x$ mpg when driven at a constant speed of x mph (between 50 and 70 mph). If the price of fuel is $3/gal and the driver is paid $18/hr, at what speed between 50 and 70 mph is it most economical to drive?

24. MINIMIZING COSTS Suppose the cost incurred in operating a cruise ship for one hour is $a + bv^3$ dollars, where a and b are positive constants and v is the ship's speed in miles per hour. At what speed should the ship be operated between two ports to minimize the cost?

25. STRENGTH OF A BEAM A wooden beam has a rectangular cross section of height h in. and width w in. (see the accompanying figure). The strength S of the beam is directly proportional to its width and the square of its height. What are the dimensions of the cross section of the strongest beam that can be cut from a round log of diameter 24 in.?
Hint: $S = kh^2w$, where k is a constant of proportionality.

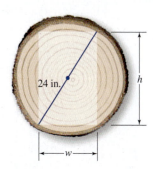

26. DESIGNING A GRAIN SILO A grain silo has the shape of a right circular cylinder surmounted by a hemisphere (see the accompanying figure). If the silo is to have a capacity of 504π ft^3, find the radius and height of the silo that requires the least amount of material to construct.
Hint: The volume of the silo is $\pi r^2h + \frac{2}{3}\pi r^3$, and the surface area (including the floor) is $\pi(3r^2 + 2rh)$.

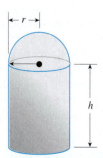

27. MINIMIZING COST OF LAYING CABLE In the following diagram, S represents the position of a power relay station located on a straight coast, and E shows the location of a marine biology experimental station on an island. A cable is to be laid connecting the relay station with the experimental station. If the cost of running the cable on land is $1.50/running foot and the cost of running the cable under water is $2.50/running foot, locate the point P that will result in a minimum cost (solve for x).

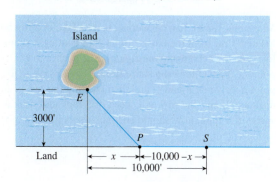

28. **STORING RADIOACTIVE WASTE** A cylindrical container for storing radioactive waste is to be constructed from lead and have a thickness of 6 in. (see the accompanying figure). If the volume of the outside cylinder is to be 16π ft^3, find the radius and the height of the inside cylinder that will result in a container of maximum storage capacity.

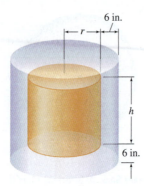

6 in.

r

h

6 in.

Hint: Show that the storage capacity (inside volume) is given by

$$V(r) = \pi r^2 \left[\frac{16}{(r + \frac{1}{2})^2} - 1 \right] \qquad \left(0 \le r \le \frac{7}{2} \right)$$

29. **FLIGHTS OF BIRDS** During daylight hours, some birds fly more slowly over water than over land because some of their energy is expended in overcoming the downdrafts of air over open bodies of water. Suppose a bird that flies at a constant speed of 4 mph over water and 6 mph over land starts its journey at the point E on an island and ends at its nest N on the shore of the mainland, as shown in the accompanying figure. Find the location of the point P that allows the bird to complete its journey in the minimum time (solve for x).

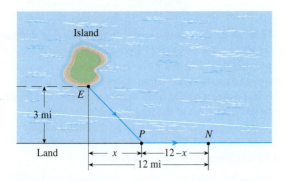

Island

E

3 mi

Land

P

N

x

$12 - x$

12 mi

30. **MINIMIZING TRAVEL TIME** A woman is on a lake in a rowboat located 1 mi from the closest point P of a straight shoreline (see the accompanying figure). She wishes to get to point Q, 10 mi along the shore from P, by rowing to a point R between P and Q and then walking the rest of the distance. If she can row at a speed of 3 mph and walk at a speed of 4 mph, how should she pick the point R in order

to get to Q as quickly as possible? How much time does she require?

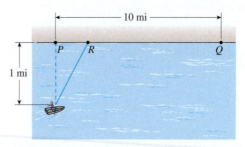

10 mi

P R Q

1 mi

31. **RACETRACK DESIGN** The accompanying figure depicts a racetrack with ends that are semicircular in shape. The length of the track is 1760 ft ($\frac{1}{3}$ mi). Find l and r such that the area enclosed by the rectangular region of the racetrack is as large as possible. What is the area enclosed by the track in this case?

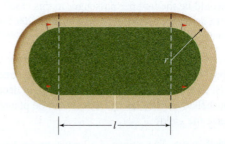

r

l

32. **INVENTORY CONTROL AND PLANNING** The demand for motorcycle tires imported by Dixie Import-Export is 40,000/year and may be assumed to be uniform throughout the year. The cost of ordering a shipment of tires is $400, and the cost of storing each tire for a year is $2. Determine how many tires should be in each shipment if the ordering and storage costs are to be minimized. (Assume that each shipment arrives just as the previous one has been sold.)

33. **INVENTORY CONTROL AND PLANNING** McDuff Preserves expects to bottle and sell 2,000,000 32-oz jars of jam at a uniform rate throughout the year. The company orders its containers from Consolidated Bottle Company. The cost of ordering a shipment of bottles is $200, and the cost of storing each empty bottle for a year is $0.40. How many orders should McDuff place per year, and how many bottles should be in each shipment if the ordering and storage costs are to be minimized? (Assume that each shipment of bottles is used up before the next shipment arrives.)

34. **INVENTORY CONTROL AND PLANNING** Neilsen Cookie Company sells its assorted butter cookies in containers that have a net content of 1 lb. The estimated demand for the cookies is 1,000,000 1-lb containers per year. The setup cost for each production run is $500, and the manufacturing cost is $0.50 for each container of cookies. The cost of storing each container of cookies over the year is $0.40. Assuming

uniformity of demand throughout the year and instantaneous production, how many containers of cookies should Neilsen produce per production run in order to minimize the production cost?

Hint: Following the method of Example 5, show that the total production cost is given by the function

$$C(x) = \frac{500,000,000}{x} + 0.2x + 500,000$$

Then minimize the function C on the interval $(0, 1,000,000)$.

35. **INVENTORY CONTROL AND PLANNING** A company expects to sell D units of a certain product per year. Sales are assumed to be at a steady rate with no shortages allowed. Each time an order for the product is placed, an ordering cost of K dollars is incurred. Each item costs p dollars, and the holding cost is h dollars per item per year.

   a. Show that the inventory cost (the combined ordering cost, purchasing cost, and holding cost) is

   $$C(x) = \frac{KD}{x} + pD + \frac{hx}{2} \qquad (x > 0)$$

   where x is the order quantity (the number of items in each order).

   b. Use the result of part (a) to show that the inventory cost is minimized if

   $$x = \sqrt{\frac{2KD}{h}}$$

   This quantity is called the *economic order quantity* (EOQ).

36. **INVENTORY CONTROL AND PLANNING** Refer to Exercise 35. The Camera Store sells 960 Yamaha A35 digital cameras per year. Each time an order for cameras is placed with the manufacturer, an ordering cost of $10 is incurred. The store pays $80 for each camera, and the cost for holding a camera (mainly due to the opportunity cost incurred in tying up capital in inventory) is $12/year. Assume that the cameras sell at a uniform rate and no shortages are allowed.

   a. What is the EOQ?
   b. How many orders will be placed each year?
   c. What is the interval between orders?

4.5 Solutions to Self-Check Exercises

1. Let x and y (measured in feet) denote the length and width of the rectangular garden.

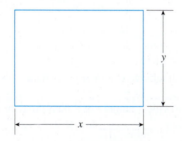

Since the area is to be 300 ft², we have

$$xy = 300$$

Next, the amount of fencing to be used is given by the perimeter, and this quantity is to be minimized. Thus, we want to minimize

$$2x + 2y$$

Since $y = 300/x$ (obtained by solving for y in the first equation), we see that the expression to be minimized is

$$f(x) = 2x + 2\left(\frac{300}{x}\right)$$

$$= 2x + \frac{600}{x}$$

for positive values of x. Now,

$$f'(x) = 2 - \frac{600}{x^2}$$

Setting $f'(x) = 0$ yields $x = -\sqrt{300}$ or $x = \sqrt{300}$. We consider only the critical number $\sqrt{300}$, since $-\sqrt{300}$ lies outside the interval $(0, \infty)$. We then compute

$$f''(x) = \frac{1200}{x^3}$$

Since

$$f''(300) > 0$$

the Second Derivative Test implies that a relative minimum of f occurs at $x = \sqrt{300}$. In fact, since $f''(x) > 0$ for all x in $(0, \infty)$, we conclude that $x = \sqrt{300}$ gives rise to the absolute minimum of f. The corresponding value of y, obtained by substituting this value of x into the equation $xy = 300$, is $y = \sqrt{300}$. Therefore, the required dimensions of the vegetable garden are approximately 17.3 ft × 17.3 ft.

2. Let x denote the number of tires in each production run. Then, the average number of tires in storage is $x/2$, so the storage cost incurred by the company is $2(x/2)$, or x dollars. Next, since the company needs to manufacture 1,000,000 tires for the year to meet the demand, the

number of production runs is $1{,}000{,}000/x$. This gives setup costs amounting to

$$4000\left(\frac{1{,}000{,}000}{x}\right) = \frac{4{,}000{,}000{,}000}{x}$$

dollars for the year. The total manufacturing cost is $20,000,000. Thus, the total yearly cost incurred by the company is given by

$$C(x) = x + \frac{4{,}000{,}000{,}000}{x} + 20{,}000{,}000$$

Differentiating $C(x)$, we find

$$C'(x) = 1 - \frac{4{,}000{,}000{,}000}{x^2}$$

Setting $C'(x) = 0$ gives 63,246 as the critical number in the interval $(0, 1{,}000{,}000)$. Next, we find

$$C''(x) = \frac{8{,}000{,}000{,}000}{x^3}$$

Since $C''(x) > 0$ for all $x > 0$, we see that the graph of C is concave upward for all $x > 0$. Furthermore, $C''(63{,}246) > 0$ implies that $x = 63{,}246$ gives rise to a relative minimum of C (by the Second Derivative Test). Since the graph of C is always concave upward for $x > 0$, $x = 63{,}246$ gives the absolute minimum of C. Therefore, the company should manufacture 63,246 tires in each production run.

CHAPTER 4 Summary of Principal Terms

TERMS

increasing function (252)

decreasing function (252)

relative maximum (257)

relative minimum (258)

relative extrema (258)

critical number (259)

First Derivative Test (260)

concave upward (273)

concave downward (273)

point of inflection (276)

inflection point (276)

Second Derivative Test (280)

vertical asymptote (292)

horizontal asymptote (293)

absolute extrema (306)

absolute maximum value (306)

absolute minimum value (306)

CHAPTER 4 Concept Review Questions

Fill in the blanks.

1. **a.** A function f is increasing on an interval I if for any two numbers x_1 and x_2 in I, $x_1 < x_2$ implies that _____.
   b. A function f is decreasing on an interval I if for any two numbers x_1 and x_2 in I, $x_1 < x_2$ implies that _____.

2. **a.** If f is differentiable on an open interval (a, b) and $f'(x) > 0$ on (a, b), then f is _____ on (a, b).
   b. If f is differentiable on an open interval (a, b) and _____ on (a, b), then f is decreasing on (a, b).
   c. If $f'(x) = 0$ for each value of x in the interval (a, b), then f is _____ on (a, b).

3. **a.** A function f has a relative maximum at c if there exists an open interval (a, b) containing c such that _____ for all x in (a, b).
   b. A function f has a relative minimum at c if there exists an open interval (a, b) containing c such that _____ for all x in (a, b).

4. **a.** A critical number of a function f is any number in the _____ of f at which $f'(c)$ _____ or $f'(c)$ does not _____.

   b. If f has a relative extremum at c, then c must be a/an _____ _____ of f.
   c. If c is a critical number of f, then f may or may not have a/an _____ _____ at c.

5. **a.** The graph of a differentiable function f is concave upward on an interval I if _____ is increasing on I.
   b. If f has a second derivative on an open interval I and $f''(x)$ _____ on I, then the graph of f is concave upward on I.
   c. If the graph of a continuous function f has a tangent line at $P(c, f(c))$ and the graph of f changes _____ at P, then P is called an inflection point of f.
   d. Suppose f has a second derivative on an interval (a, b), containing a critical number c of f. If $f''(c) < 0$, then f has a/an _____ _____ at c. If $f''(c) = 0$, then f may or may not have a/an _____ _____ at c.

6. The line $x = a$ is a vertical asymptote of the graph f if at least one of the following is true: $\displaystyle\lim_{x \to a^+} f(x) = $ _____ or $\displaystyle\lim_{x \to a^-} f(x) = $ _____.

7. For a rational function $f(x) = \dfrac{P(x)}{Q(x)}$, the line $x = a$ is a vertical asymptote of the graph of f if $Q(a) = $ _____ but $P(a) \neq $ _____.

8. The line $y = b$ is a horizontal asymptote of the graph of a function f if either $\lim\limits_{x \to \infty} f(x) = $ _____ or $\lim\limits_{x \to -\infty} f(x) = $ _____.

9. a. A function f has an absolute maximum at c if _____ for all x in the domain D of f. The number $f(c)$ is called the _____ _____ _____ of f on D.

b. A function f has a relative minimum at c if _____ for all values of x in some _____ _____ containing c.

10. The extreme value theorem states that if f is _____ on the closed interval $[a, b]$, then f has both a/an _____ maximum value and a/an _____ minimum value on $[a, b]$.

CHAPTER 4 Review Exercises

In Exercises 1–10, (a) find the intervals where the function f is increasing and where it is decreasing, (b) find the relative extrema of f, (c) find the intervals where the graph of f is concave upward and where it is concave downward, and (d) find the inflection points, if any, of f.

1. $f(x) = \dfrac{1}{3}x^3 - x^2 + x - 6$

2. $f(x) = (x - 2)^3$

3. $f(x) = x^4 - 2x^2$

4. $f(x) = x + \dfrac{4}{x}$

5. $f(x) = \dfrac{x^2}{x - 1}$

6. $f(x) = \sqrt{x - 1}$

7. $f(x) = (1 - x)^{1/3}$

8. $f(x) = x\sqrt{x - 1}$

9. $f(x) = \dfrac{2x}{x + 1}$

10. $f(x) = -\dfrac{1}{1 + x^2}$

In Exercises 11–18, use the curve-sketching guide on page 295 to sketch the graph of the function.

11. $f(x) = x^2 - 5x + 5$

12. $f(x) = -2x^2 - x + 1$

13. $g(x) = 2x^3 - 6x^2 + 6x + 1$

14. $g(x) = \dfrac{1}{3}x^3 - x^2 + x - 3$

15. $h(x) = x\sqrt{x - 2}$

16. $h(x) = \dfrac{2x}{1 + x^2}$

17. $f(x) = \dfrac{x - 2}{x + 2}$

18. $f(x) = x - \dfrac{1}{x}$

In Exercises 19–22, find the horizontal and vertical asymptotes of the graph of each function. Do not sketch the graph.

19. $f(x) = \dfrac{1}{2x + 3}$

20. $f(x) = \dfrac{2x}{x + 1}$

21. $f(x) = \dfrac{5x}{x^2 - 2x - 8}$

22. $f(x) = \dfrac{x^2 + x}{x(x - 1)}$

In Exercises 23–32, find the absolute maximum value and the absolute minimum value, if any, of the function.

23. $f(x) = 2x^2 + 3x - 2$

24. $g(x) = x^{2/3}$

25. $g(t) = \sqrt{25 - t^2}$

26. $f(x) = \dfrac{1}{3}x^3 - x^2 + x + 1$ on $[0, 2]$

27. $h(t) = t^3 - 6t^2$ on $[2, 5]$

28. $g(x) = \dfrac{x}{x^2 + 1}$ on $[0, 5]$

29. $f(x) = x - \dfrac{1}{x}$ on $[1, 3]$

30. $h(t) = 8t - \dfrac{1}{t^2}$ on $[1, 3]$

31. $f(s) = s\sqrt{1 - s^2}$ on $[-1, 1]$

32. $f(x) = \dfrac{x^2}{x - 1}$ on $[-1, 3]$

33. REVENUE OF COMPETING BOOKSTORES The graphs of R_1 and R_2 that follow show the revenue of a neighborhood bookstore and that of a branch of a national bookstore, three months after the opening of the latter ($t = 0$) until sometime later T.

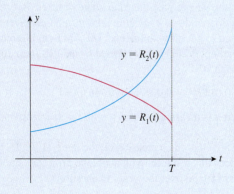

a. Find the signs of the first and second derivatives of R_1 and R_2 on the interval $(0, T)$.

b. Give an interpretation of the results obtained in part (a) in terms of the revenues of the two bookstores.

34. SPREAD OF A RUMOR Initially, a handful of students heard a rumor on campus. The rumor spread, and after t hr, the number of students who had heard it had grown to $N(t)$. The graph of the function N is shown in the figure below. Describe the spread of the rumor in terms of the speed at which it was spread. In particular, explain the significance of the inflection point P of the graph of N.

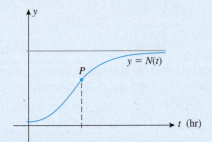

35. MAXIMIZING PROFITS Odyssey Travel Agency's monthly profit (in thousands of dollars) depends on the amount of money x (in thousands of dollars) spent on advertising each month according to the rule

$$P(x) = -x^2 + 8x + 20$$

To maximize its monthly profits, what should be Odyssey's monthly advertising budget?

36. EFFECT OF SMOKING BANS The sales (in billions of dollars) in restaurants and bars in California from 1993 ($t = 0$) through 2000 ($t = 7$) are approximated by the function

$$S(t) = 0.195t^2 + 0.32t + 23.7 \qquad (0 \le t \le 7)$$

a. Show that the sales in restaurants and bars continued to rise after smoking bans were implemented in restaurants in 1995 and in bars in 1998.
   Hint: Show that S is increasing on the interval $(2, 7)$.

b. What can you say about the rate at which the sales were rising after smoking bans were implemented?

Source: California Board of Equalization.

37. ALTERNATIVE MINIMUM TAX Congress created the alternative minimum tax (AMT) in the late 1970s to ensure that wealthy people paid their fair share of taxes. But because of quirks in the law, even middle-income taxpayers have started to get hit with the tax. The AMT (in billions of dollars) projected to be collected by the IRS from 2001 through 2010 is

$$f(t) = 0.0117t^3 + 0.0037t^2 + 0.7563t + 4.1 \qquad (0 \le t \le 9)$$

where t is measured in years, with $t = 0$ corresponding to 2001.

a. Show that f is increasing on the interval $(0, 9)$. What does this result tell you about the projected amount of AMT paid over the years in question?

b. Show that f' is increasing on the interval $(0, 9)$. What conclusion can you draw from this result concerning the rate of growth of the amount of the AMT collected over the years in question?

Source: U.S. Congress Joint Economic Committee.

38. MEASLES DEATHS Measles is still a leading cause of vaccine-preventable death among children, but because of improvements in immunizations, measles deaths have dropped globally. The function

$$N(t) = -2.42t^3 + 24.5t^2 - 123.3t + 506 \qquad (0 \le t \le 6)$$

gives the number of measles deaths (in thousands) in sub-Saharan Africa in year t, with $t = 0$ corresponding to the beginning of 1999.

a. How many measles deaths were there in 1999? In 2005?

b. Show that $N'(t) < 0$ on $(0, 6)$. What does this say about the number of measles deaths from 1999 through 2005?

c. When was the number of measles deaths decreasing most rapidly? What was the rate of decline of measles deaths at that instant of time?

Source: Centers for Disease Control and Prevention and World Health Organization.

39. EFFECT OF ADVERTISING ON SALES The total sales S of Cannon Precision Instruments is related to the amount of money x that Cannon spends on advertising its products by the function

$$S(x) = -0.002x^3 + 0.6x^2 + x + 500 \qquad (0 \le x \le 200)$$

where S and x are measured in thousands of dollars. Find the inflection point of the function S, and discuss its significance.

40. ELDERLY WORKFORCE The percentage of men 65 years and older in the workforce from 1970 through 2000 is approximated by the function

$$P(t) = 0.00093t^3 - 0.018t^2 - 0.51t + 25 \qquad (0 \le t \le 30)$$

where t is measured in years, with $t = 0$ corresponding to the beginning of 1970.

a. Find the interval where P is decreasing and the interval where P is increasing.

b. Interpret the results of part (a).

Source: U.S. Census Bureau.

41. COST OF PRODUCING CALCULATORS A subsidiary of Elektra Electronics manufactures graphing calculators. Management determines that the daily cost $C(x)$ (in dollars) of producing these calculators is

$$C(x) = 0.0001x^3 - 0.08x^2 + 40x + 5000$$

where x is the number of calculators produced. Find the inflection point of the function C, and interpret your result.

42. SALES OF MOBILE PROCESSORS The rising popularity of notebook computers is fueling the sales of mobile PC processors. In a study conducted in 2003, the sales of these chips (in billions of dollars) was projected to be

$$S(t) = 6.8(t + 1.03)^{0.49} \qquad (0 \le t \le 4)$$

where t is measured in years, with $t = 0$ corresponding to 2003.
a. Show that S is increasing on the interval $(0, 4)$, and interpret your result.
b. Show that the graph of S is concave downward on the interval $(0, 4)$. Interpret your result.
Source: International Data Corp.

43. **SMALL CAR MARKET SHARE IN EUROPE** Owing in part to an aging population and the squeeze from carbon-cutting regulations, the percentage of small and lower-midsize vehicles in Europe is expected to increase in the near future. The function

$$f(t) = \frac{150\sqrt{t} + 766}{59 - \sqrt{t}} \qquad (0 \le t \le 10)$$

gives the projected percentage of small and lower-midsize vehicles t years after 2005.
a. What was the percentage of small and lower-midsize vehicles in 2005? What is the projected percentage of small and lower-midsize vehicles in 2015?
b. Show that f is increasing on the interval $(0, 10)$, and interpret your results.
Source: J.D. Power Automotive.

44. **INDEX OF ENVIRONMENTAL QUALITY** The Department of the Interior of an African country began to record an index of environmental quality to measure progress or decline in the environmental quality of its wildlife. The index for the years 2002 through 2012 is approximated by the function

$$I(t) = \frac{50t^2 + 600}{t^2 + 10} \qquad (0 \le t \le 10)$$

a. Compute $I'(t)$ and show that $I(t)$ is decreasing on the interval $(0, 10)$.
b. Compute $I''(t)$. Study the concavity of the graph of I.
c. Sketch the graph of I.
d. Interpret your results.

45. **MAXIMIZING PROFITS** The weekly demand for DVDs manufactured by Herald Media Corporation is given by

$$p = -0.0005x^2 + 60$$

where p denotes the unit price in dollars and x denotes the quantity demanded. The weekly total cost function associated with producing these discs is given by

$$C(x) = -0.001x^2 + 18x + 4000$$

where $C(x)$ denotes the total cost (in dollars) incurred in pressing x discs. Find the production level that will yield a maximum profit for the manufacturer.
Hint: Use the quadratic formula.

46. **MAXIMIZING PROFITS** The estimated monthly profit (in dollars) realizable by Cannon Precision Instruments for manufacturing and selling x units of its model M1 digital camera is

$$P(x) = -0.04x^2 + 240x - 10,000$$

To maximize its profits, how many cameras should Cannon produce each month?

47. **MINIMIZING AVERAGE COST** The total monthly cost (in dollars) incurred by Carlota Music in manufacturing x units of its Professional Series guitars is given by the function

$$C(x) = 0.001x^2 + 100x + 4000$$

a. Find the average cost function $\overline{C}$.
b. Determine the production level that will result in the smallest average production cost.

48. **WORKER EFFICIENCY** The average worker at Wakefield Avionics will have assembled

$$N(t) = -2t^3 + 12t^2 + 2t \qquad (0 \le t \le 4)$$

ready-to-fly radio-controlled model airplanes t hr into the 8 A.M. to 12 noon shift. At what time during this shift is the average worker performing at peak efficiency?

49. **SENIOR WORKFORCE** The percentage of women 65 years and older in the workforce from 1970 through 2000 is approximated by the function

$$P(t) = -0.0002t^3 + 0.018t^2 - 0.36t + 10 \qquad (0 \le t \le 30)$$

where t is measured in years, with $t = 0$ corresponding to the beginning of 1970.
a. Find the interval where P is decreasing and the interval where P is increasing.
b. Find the absolute minimum of P.
c. Interpret the results of parts (a) and (b).
Source: U.S. Census Bureau.

50. **AGE OF DRIVERS IN CRASH FATALITIES** The number of crash fatalities per 100,000 vehicle miles of travel in a certain year is approximated by the model

$$f(x) = \frac{15}{0.08333x^2 + 1.91667x + 1} \qquad (0 \le x \le 11)$$

where x is the age of the driver in years, with $x = 0$ corresponding to age 16. Show that f is decreasing on $(0, 11)$, and interpret your result.
Source: National Highway Traffic Safety Administration.

51. **SPREAD OF A CONTAGIOUS DISEASE** The incidence (number of new cases per day) of a contagious disease spreading in a population of M people is given by

$$R(x) = kx(M - x)$$

where k is a positive constant and x denotes the number of people already infected. Show that the incidence R is greatest when half the population is infected.

52. MEDICAL SCHOOL APPLICANTS According to a study from the American Medical Association, the number of medical school applicants from academic year 1997–1998 ($t = 0$) through the academic year 2008–2009 is approximated by the function

$$N(t) = \begin{cases} 0.36t^2 - 3.10t + 41.2 & \text{if } 0 \le t < 9 \\ 42.46 & \text{if } 9 \le t \le 11 \end{cases}$$

Find the years when the number of medical school applicants was increasing, when it was decreasing, and when it was approximately constant.

Source: Journal of the American Medical Association.

53. Let

$$f(x) = \begin{cases} -x^2 + 3 & \text{if } x \ne 0 \\ 2 & \text{if } x = 0 \end{cases}$$

a. Compute $f'(x)$, and show that it changes sign from positive to negative as we move across $x = 0$.

b. Show that f does not have a relative maximum at $x = 0$. Does this contradict the First Derivative Test? Explain your answer.

54. MAXIMIZING THE VOLUME OF A BOX A box with an open top is to be constructed from a square piece of cardboard, 10 in. wide, by cutting out a square from each of the four corners and bending up the sides. What is the maximum volume of such a box?

55. MINIMIZING CONSTRUCTION COSTS A man wishes to construct a cylindrical barrel with a capacity of 32π ft^3. The cost per square foot of the material for the side of the barrel is half that of the cost per square foot for the top and bottom. Help him find the dimensions of the barrel that can be constructed at a minimum cost in terms of material used.

56. PACKAGING You wish to construct a closed rectangular box that has a volume of 4 ft^3. The length of the base of the box will be twice as long as its width. The material for the top and bottom of the box costs 30¢/ft^2. The material

for the sides of the box costs 20¢/ft^2. Find the dimensions of the least expensive box that can be constructed.

57. INVENTORY CONTROL AND PLANNING Lehen Vinters imports a certain brand of beer. The demand, which may be assumed to be uniform, is 800,000 cases/year. The cost of ordering a shipment of beer is $500, and the cost of storing each case of beer for a year is $2. Determine how many cases of beer should be in each shipment if the ordering and storage costs are to be kept at a minimum. (Assume that each shipment of beer arrives just as the previous one has been sold.)

58. In what interval is the quadratic function

$$f(x) = ax^2 + bx + c \qquad (a \ne 0)$$

increasing? In what interval is f decreasing?

59. Let $f(x) = x^2 + ax + b$. Determine the constants a and b such that f has a relative minimum at $x = 2$ and the relative minimum value is 7.

60. Find the values of c such that the graph of

$$f(x) = x^4 + 2x^3 + cx^2 + 2x + 2$$

is concave upward everywhere.

61. Suppose that the point $(a, f(a))$ is an inflection point of the graph of $y = f(x)$. Show that the number a gives rise to a relative extremum of the function f'.

62. Let

$$f(x) = \begin{cases} x^3 + 1 & \text{if } x \ne 0 \\ 2 & \text{if } x = 0 \end{cases}$$

a. Compute $f'(x)$, and show that it does not change sign as we move across $x = 0$.

b. Show that f has a relative maximum at $x = 0$. Does this contradict the First Derivative Test? Explain your answer.

CHAPTER 4 | Before Moving On . . .

1. Find the interval(s) where $f(x) = \dfrac{x^2}{1 - x}$ is increasing and where it is decreasing.

2. Find the relative maxima and relative minima, if any, of $f(x) = 2x^2 - 12x^{1/3}$.

3. Find the intervals where the graph of

$$f(x) = \frac{1}{3}x^3 - \frac{1}{4}x^2 - \frac{1}{2}x + 1$$

is concave upward, the intervals where the graph of f is concave downward, and the inflection point(s) of f.

4. Sketch the graph of $f(x) = 2x^3 - 9x^2 + 12x - 1$.

5. Find the absolute maximum and absolute minimum values of $f(x) = 2x^3 + 3x^2 - 1$ on the interval $[-2, 3]$.

6. An open bucket in the form of a right circular cylinder is to be constructed with a capacity of 1 ft^3. Find the radius and height of the cylinder if the amount of material used is minimal.

5 Exponential and Logarithmic Functions

THE EXPONENTIAL FUNCTION is, without doubt, the most important function in mathematics and its applications. After a brief introduction to the exponential function and its *inverse,* the logarithmic function, we learn how to differentiate such functions. This lays the foundation for exploring the many applications involving exponential functions. For example, we look at the role played by exponential functions in computing earned interest on a bank account and in studying the growth of a bacteria population in the laboratory, the rate at which radioactive matter decays, the rate at which a factory worker learns a certain process, and the rate at which a communicable disease is spread over time.

How many cameras can a new employee at Eastman Optical assemble after completing the basic training program, and how many cameras can he assemble after being on the job for 6 months? In Example 5, page 395, you will see how to answer these questions.

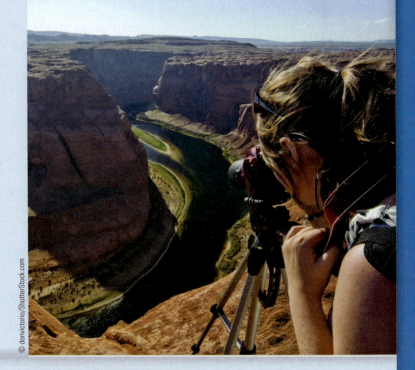

© donvictorio/ShutterStock.com

337

5.1 Exponential Functions

Exponential Functions and Their Graphs

Suppose you deposit a sum of $1000 in an account earning interest at the rate of 10% per year *compounded continuously* (the way most financial institutions compute interest). Then, the accumulated amount at the end of t years ($0 \le t \le 20$) is described by the function f, whose graph appears in Figure 1.* This function is called an *exponential function*. Observe that the graph of f rises rather slowly at first but very rapidly as time goes by. For purposes of comparison, we have also shown the graph of the function $y = g(t) = 1000(1 + 0.10t)$, giving the accumulated amount for the same principal ($1000) but earning *simple* interest at the rate of 10% per year. The moral of the story: It is never too early to save.

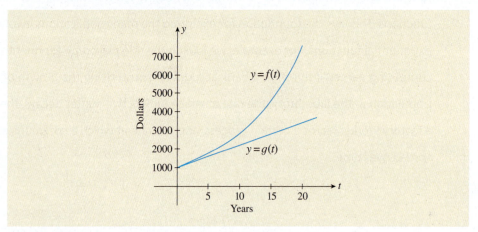

FIGURE 1
Under continuous compounding, a sum of money grows exponentially.

Exponential functions play an important role in many real-world applications, as you will see throughout this chapter.

Recall that if b is a positive number and n is any real number, the expression b^n is a real number. This enables us to define an exponential function as follows:

> **Exponential Function**
>
> The function defined by
>
> $$f(x) = b^x \qquad (b > 0, b \ne 1)$$
>
> is called an **exponential function with base b and exponent x.** The domain of f is the set of all real numbers.

For example, the exponential function with base 2 is the function

$$f(x) = 2^x$$

with domain $(-\infty, \infty)$. The values of $f(x)$ for selected values of x follow:

$$f(3) = 2^3 = 8 \qquad f\left(\frac{3}{2}\right) = 2^{3/2} = 2 \cdot 2^{1/2} = 2\sqrt{2} \qquad f(0) = 2^0 = 1$$

$$f(-1) = 2^{-1} = \frac{1}{2} \qquad f\left(-\frac{2}{3}\right) = 2^{-2/3} = \frac{1}{2^{2/3}} = \frac{1}{\sqrt[3]{4}}$$

---

*We will derive the rule for f in Section 5.3.

Computations involving exponentials are facilitated by the laws of exponents. These laws were stated in Section 1.1, and you might want to review the material there. For convenience, however, we will restate these laws.

Laws of Exponents

Let a and b be positive numbers and let x and y be real numbers. Then,

1. $b^x \cdot b^y = b^{x+y}$ **4.** $(ab)^x = a^x b^x$

2. $\dfrac{b^x}{b^y} = b^{x-y}$ **5.** $\left(\dfrac{a}{b}\right)^x = \dfrac{a^x}{b^x}$

3. $(b^x)^y = b^{xy}$

The use of the laws of exponents is illustrated in the next two examples.

EXAMPLE 1

a. $16^{7/4} \cdot 16^{-1/2} = 16^{7/4-1/2} = 16^{5/4} = 2^5 = 32$ Law 1

b. $\dfrac{8^{5/3}}{8^{-1/3}} = 8^{5/3-(-1/3)} = 8^2 = 64$ Law 2

c. $(64^{4/3})^{-1/2} = 64^{(4/3)(-1/2)} = 64^{-2/3}$

$$= \frac{1}{64^{2/3}} = \frac{1}{(64^{1/3})^2} = \frac{1}{4^2} = \frac{1}{16} \quad \text{Law 3}$$

d. $(16 \cdot 81)^{-1/4} = 16^{-1/4} \cdot 81^{-1/4} = \dfrac{1}{16^{1/4}} \cdot \dfrac{1}{81^{1/4}} = \dfrac{1}{2} \cdot \dfrac{1}{3} = \dfrac{1}{6}$ Law 4

e. $\left(\dfrac{3^{1/2}}{2^{1/3}}\right)^4 = \dfrac{3^{4/2}}{2^{4/3}} = \dfrac{9}{2^{4/3}}$ Law 5

EXAMPLE 2 Let $f(x) = 2^{2x-1}$. Find the value of x for which $f(x) = 16$.

Solution We want to solve the equation

$$2^{2x-1} = 16 = 2^4$$

But this equation holds if and only if

$$2x - 1 = 4 \qquad b^m = b^n \Rightarrow m = n$$

giving $x = \frac{5}{2}$.

Exponential functions play an important role in mathematical analysis. Because of their special characteristics, they are some of the most useful functions and are found in virtually every field in which mathematics is applied. To mention a few examples: Under ideal conditions, the number of bacteria present at any time t in a culture may be described by an exponential function of t; radioactive substances decay over time in accordance with an "exponential" law of decay; money left on fixed deposit and earning compound interest grows exponentially; and some of the most important distribution functions encountered in statistics are exponential.

Let's begin our investigation into the properties of exponential functions by studying their graphs.

EXAMPLE 3 Sketch the graph of the exponential function $y = 2^x$.

Solution First, as was discussed earlier, the domain of the exponential function $y = f(x) = 2^x$ is the set of real numbers. Next, putting $x = 0$ gives $y = 2^0 = 1$, the

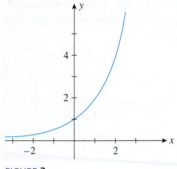

FIGURE 2
The graph of $y = 2^x$

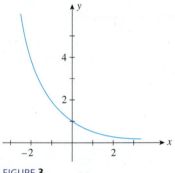

FIGURE 3
The graph of $y = \left(\dfrac{1}{2}\right)^x$

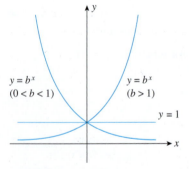

FIGURE 4
$y = b^x$ is an increasing function of x if $b > 1$, a constant function if $b = 1$, and a decreasing function if $0 < b < 1$.

y-intercept of f. There is no x-intercept, since there is no value of x for which $y = 0$. To find the range of f, consider the following table of values:

x	-5	-4	-3	-2	-1	0	1	2	3	4	5
y	$\frac{1}{32}$	$\frac{1}{16}$	$\frac{1}{8}$	$\frac{1}{4}$	$\frac{1}{2}$	1	2	4	8	16	32

We see from these computations that 2^x decreases and approaches zero as x decreases without bound and that 2^x increases without bound as x increases without bound. Thus, the range of f is the interval $(0, \infty)$—that is, the set of positive real numbers. Finally, we sketch the graph of $y = f(x) = 2^x$ in Figure 2.

EXAMPLE 4 Sketch the graph of the exponential function $y = (1/2)^x$.

Solution The domain of the exponential function $y = (1/2)^x$ is the set of all real numbers. The y-intercept is $(1/2)^0 = 1$; there is no x-intercept, since there is no value of x for which $y = 0$. From the following table of values

x	-5	-4	-3	-2	-1	0	1	2	3	4	5
y	32	16	8	4	2	1	$\frac{1}{2}$	$\frac{1}{4}$	$\frac{1}{8}$	$\frac{1}{16}$	$\frac{1}{32}$

we deduce that $(1/2)^x = 1/2^x$ increases without bound as x decreases without bound and that $(1/2)^x$ decreases and approaches zero as x increases without bound. Thus, the range of f is the interval $(0, \infty)$. The graph of $y = f(x) = (1/2)^x$ is sketched in Figure 3.

The functions $y = 2^x$ and $y = (1/2)^x$, whose graphs you studied in Examples 3 and 4, are special cases of the exponential function $y = f(x) = b^x$, obtained by setting $b = 2$ and $b = 1/2$, respectively. In general, the exponential function $y = b^x$ with $b > 1$ has a graph similar to that of $y = 2^x$, whereas the graph of $y = b^x$ for $0 < b < 1$ is similar to that of $y = (1/2)^x$ (Exercises 27 and 28 on page 342). When $b = 1$, the function $y = b^x$ reduces to the constant function $y = 1$. For comparison, the graphs of all three functions are sketched in Figure 4.

Properties of the Exponential Function

The exponential function $y = b^x$ ($b > 0$, $b \neq 1$) has the following properties:

1. Its domain is $(-\infty, \infty)$.
2. Its range is $(0, \infty)$.
3. Its graph passes through the point $(0, 1)$.
4. It is continuous on $(-\infty, \infty)$.
5. It is increasing on $(-\infty, \infty)$ if $b > 1$ and decreasing on $(-\infty, \infty)$ if $b < 1$.

The Base e

Exponential functions to the base e, where e is an irrational number whose value is $2.7182818\ldots$, play an important role in both theoretical and applied problems. It can be shown, although we will not do so here, that

$$e = \lim_{m \to \infty} \left(1 + \frac{1}{m}\right)^m \tag{1}$$

TABLE 1	
m	$\left(1 + \dfrac{1}{m}\right)^m$
10	2.59374
100	2.70481
1000	2.71692
10,000	2.71815
100,000	2.71827
1,000,000	2.71828

However, you may convince yourself of the plausibility of this definition of the number e by examining Table 1, which can be constructed with the help of a calculator.

Exploring with TECHNOLOGY

To obtain a visual confirmation of the fact that the expression $(1 + 1/m)^m$ approaches the number $e = 2.71828.\ldots$ as m increases without bound, plot the graph of $f(x) = (1 + 1/x)^x$ in a suitable viewing window and observe that $f(x)$ approaches $2.71828.\ldots$ as x increases without bound. Use **ZOOM** and **TRACE** to find the value of $f(x)$ for large values of x.

EXAMPLE 5 Sketch the graph of the function $y = e^x$.

Solution   Since $e > 1$, it follows from our previous discussion that the graph of $y = e^x$ is similar to the graph of $y = 2^x$ (see Figure 2). With the aid of a calculator, we obtain the following table:

x	-3	-2	-1	0	1	2	3
y	0.05	0.14	0.37	1	2.72	7.39	20.09

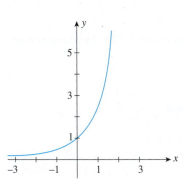

FIGURE 5
The graph of $y = e^x$

The graph of $y = e^x$ is sketched in Figure 5.

Next, we consider another exponential function to the base e that is closely related to the previous function and is particularly useful in constructing models that describe "exponential decay."

EXAMPLE 6 Sketch the graph of the function $y = e^{-x}$.

Solution   Since $e > 1$, it follows that $0 < 1/e < 1$, so $f(x) = e^{-x} = 1/e^x = (1/e)^x$ is an exponential function with base less than 1. Therefore, it has a graph similar to that of the exponential function $y = (1/2)^x$. As before, we construct the following table of values of $y = e^{-x}$ for selected values of x:

x	-3	-2	-1	0	1	2	3
y	20.09	7.39	2.72	1	0.37	0.14	0.05

FIGURE 6
The graph of $y = e^{-x}$

Using this table, we sketch the graph of $y = e^{-x}$ in Figure 6.

5.1 Self-Check Exercises

1. Solve the equation $2^{2x+1} \cdot 2^{-3} = 2^{x-1}$.

2. Sketch the graph of $y = e^{0.4x}$.

Solutions to Self-Check Exercises 5.1 can be found on page 344.

5.1 Concept Questions

1. Define the exponential function f with base b and exponent x. What restrictions, if any, are placed on b?

2. For the exponential function $y = b^x$ $(b > 0, b \neq 1)$, state (a) its domain and range, (b) its y-intercept, (c) where it is continuous, and (d) where it is increasing and where it is decreasing for the case $b > 1$ and the case $b < 1$.

5.1 Exercises

In Exercises 1–8, evaluate the expression.

1. **a.** $4^{-3} \cdot 4^5$ **b.** $3^{-3} \cdot 3^6$

2. **a.** $(2^{-1})^3$ **b.** $(3^{-2})^3$

3. **a.** $9(9)^{-1/2}$ **b.** $5(5)^{-1/2}$

4. **a.** $\left[\left(-\dfrac{1}{2}\right)^3\right]^{-2}$ **b.** $\left[\left(-\dfrac{1}{3}\right)^2\right]^{-3}$

5. **a.** $\dfrac{(-3)^4(-3)^5}{(-3)^8}$ **b.** $\dfrac{(2^{-4})(2^6)}{2^{-1}}$

6. **a.** $3^{1/4} \cdot 9^{-5/8}$ **b.** $2^{3/4} \cdot 4^{-3/2}$

7. **a.** $\dfrac{5^{3.3} \cdot 5^{-1.6}}{5^{-0.3}}$ **b.** $\dfrac{4^{2.7} \cdot 4^{-1.3}}{4^{-0.4}}$

8. **a.** $\left(\dfrac{1}{16}\right)^{-1/4}\left(\dfrac{27}{64}\right)^{-1/3}$ **b.** $\left(\dfrac{8}{27}\right)^{-1/3}\left(\dfrac{81}{256}\right)^{-1/4}$

In Exercises 9–16, simplify the expression.

9. **a.** $(64x^9)^{1/3}$ **b.** $(25x^3y^4)^{1/2}$

10. **a.** $(2x^3)(-4x^{-2})$ **b.** $(4x^{-2})(-3x^5)$

11. **a.** $\dfrac{6a^{-4}}{3a^{-3}}$ **b.** $\dfrac{4b^{-4}}{12b^{-6}}$

12. **a.** $y^{-3/2}y^{5/3}$ **b.** $x^{-3/5}x^{8/3}$

13. **a.** $(2x^3y^2)^3$ **b.** $(4x^2y^2z^3)^2$

14. **a.** $(x^{r/s})^{s/r}$ **b.** $(x^{-b/a})^{-a/b}$

15. **a.** $\dfrac{5^0}{(2^{-3}x^{-3}y^2)^2}$ **b.** $\dfrac{(x+y)(x-y)}{(x-y)^0}$

16. **a.** $\dfrac{(a^m \cdot a^{-n})^{-2}}{(a^{m+n})^2}$ **b.** $\left(\dfrac{x^{2n-2}y^{2n}}{x^{5n+1}y^{-n}}\right)^{1/3}$

In Exercises 17–26, solve the equation for x.

17. $6^{2x} = 6^6$

18. $5^{-x} = 5^3$

19. $3^{3x-4} = 3^5$

20. $10^{2x-1} = 10^{x+3}$

21. $(2.1)^{x+2} = (2.1)^5$

22. $(1.3)^{x-2} = (1.3)^{2x+1}$

23. $8^x = \left(\dfrac{1}{32}\right)^{x-2}$

24. $3^{x-x^2} = \dfrac{1}{9^x}$

25. $3^{2x} - 12 \cdot 3^x + 27 = 0$

26. $2^{2x} - 4 \cdot 2^x + 4 = 0$

In Exercises 27–36, sketch the graphs of the given functions on the same axes.

27. $y = 2^x$, $y = 3^x$, and $y = 4^x$

28. $y = \left(\dfrac{1}{2}\right)^x$, $y = \left(\dfrac{1}{3}\right)^x$, and $y = \left(\dfrac{1}{4}\right)^x$

29. $y = 2^{-x}$, $y = 3^{-x}$, and $y = 4^{-x}$

30. $y = 4^{0.5x}$ and $y = 4^{-0.5x}$

31. $y = 4^{0.5x}$, $y = 4^x$, and $y = 4^{2x}$

32. $y = e^x$, $y = 2e^x$, and $y = 3e^x$

33. $y = e^{0.5x}$, $y = e^x$, and $y = e^{1.5x}$

34. $y = e^{-0.5x}$, $y = e^{-x}$, and $y = e^{-1.5x}$

35. $y = 0.5e^{-x}$, $y = e^{-x}$, and $y = 2e^{-x}$

36. $y = 1 - e^{-x}$ and $y = 1 - e^{-0.5x}$

37. A function f has the form $f(x) = Ae^{kx}$. Find f if it is known that $f(0) = 100$ and $f(1) = 120$.
Hint: $e^{kx} = (e^k)^x$

38. If $f(x) = Axe^{-kx}$, find $f(3)$ if $f(1) = 5$ and $f(2) = 7$.
Hint: $e^{kx} = (e^k)^x$

39. If
$$f(t) = \frac{1000}{1 + Be^{-kt}}$$
find $f(5)$ given that $f(0) = 20$ and $f(2) = 30$.
Hint: $e^{kx} = (e^k)^x$

40. **DECLINE OF AMERICAN IDOL** After having been on the air for more than a decade, Fox's *American Idol* seemed to be suffering from viewer fatigue. The average number of viewers from the 2011 season through the 2013 season is approximated by
$$f(t) = 32.744e^{-0.252t} \qquad (1 \leq t \leq 3)$$
where $f(t)$ is measured in millions, with $t = 1$ corresponding to the 2011 season.

a. What was the average number of viewers in the 2011 season?
b. What was the average number of viewers in the 2014 season, assuming that the trend continued into that season?
Source: Nielsen Ratings.

41. RENEWABLE ENERGY Developing countries are accelerating the pace of their investment in renewable energy. According to a report by the Frankfurt School of Finance and Management, the amount of investment in renewable energy (in billions of dollars) by developing countries between 2009 ($t = 0$) and 2012 is given by

$$f(t) = 64e^{0.188t} \qquad (0 \le t \le 3)$$

a. Find the amount of investment in renewable energy by developing countries in each of the years 2009 through 2012 by completing the following table:

t	0	1	2	3
$f(t)$				

b. Use the information from part (a) to sketch the graph of f.
Source: Frankfurt School of Finance and Management.

42. ONLINE SHOPPERS According to a study conducted by Forrester Research, Inc., the amount of money spent by online shoppers in the United States is projected to be

$$f(t) = 105e^{0.095t} \qquad (1 \le t \le 6)$$

billion dollars in year t, where $t = 1$ corresponds to 2011.
a. Find the projected spending in each of the years 2011 through 2016 by completing the following table:

t	1	2	3	4	5	6
$f(t)$						

b. Use the information from part (a) to sketch the graph of f.
Source: Forrester Research, Inc.

43. INTERNET USERS IN CHINA The number of Internet users in China is approximated by

$$N(t) = 94.5e^{0.2t} \qquad (1 \le t \le 6)$$

where $N(t)$ is measured in millions and t is measured in years, with $t = 1$ corresponding to 2005.
a. How many Internet users were there in 2005? In 2006? In 2010?
b. Sketch the graph of N.
Source: C. E. Unterberg.

44. U.S. CELL PHONE SUBSCRIBERS The number of cell phone subscribers in the United States between the years 2000 and 2010 is approximated by the function

$$N(t) = \frac{385.474}{1 + 2.521e^{-0.214t}} \qquad (0 \le t \le 10)$$

where $N(t)$ is measured in millions and t is measured in years, with $t = 0$ corresponding to the year 2000. How many cell phone subscribers were there in the United States in 2000? If the trend continued, how many subscribers were there in 2012?
Source: CTIA—The Wireless Association.

45. ALTERNATIVE MINIMUM TAX The alternative minimum tax was created in 1969 to prevent the very wealthy from using creative deductions and shelters to avoid having to pay anything to the Internal Revenue Service. But it has increasingly hit the middle class. The number of taxpayers subjected to an alternative minimum tax is projected to be

$$N(t) = \frac{35.5}{1 + 6.89e^{-0.8674t}} \qquad (0 \le t \le 7)$$

where $N(t)$ is measured in millions and t is measured in years, with $t = 0$ corresponding to 2004. What was the projected number of taxpayers subjected to an alternative minimum tax in 2010?
Source: Brookings Institution.

46. ABSORPTION OF DRUGS The concentration of a drug in an organ at any time t (in seconds) is given by

$$x(t) = 0.08 + 0.12(1 - e^{-0.02t})$$

where $x(t)$ is measured in grams per cubic centimeter (g/cm³).
a. What is the initial concentration of the drug in the organ?
b. What is the concentration of the drug in the organ after 20 sec?
c. What will be the concentration of the drug in the organ in the long run?
Hint: Evaluate $\lim_{t \to \infty} x(t)$.
d. Sketch the graph of x.

47. ABSORPTION OF DRUGS The concentration of a drug in an organ at any time t (in seconds) is given by

$$C(t) = \begin{cases} 0.3t - 18(1 - e^{-t/60}) & \text{if } 0 \le t \le 20 \\ 18e^{-t/60} - 12e^{-(t-20)/60} & \text{if } t > 20 \end{cases}$$

where $C(t)$ is measured in grams per cubic centimeter (g/cm³).
a. What is the initial concentration of the drug in the organ?
b. What is the concentration of the drug in the organ after 10 sec?
c. What is the concentration of the drug in the organ after 30 sec?
d. What will be the concentration of the drug in the long run?
Hint: Evaluate $\lim_{t \to \infty} C(t)$.

48. ABSORPTION OF DRUGS Jane took 100 mg of a drug in the morning and another 100 mg of the same drug at the same time the following morning. The amount of the drug (in milligrams) in her body t days after the first dose was taken is given by

$$A(t) = \begin{cases} 100e^{-1.4t} & \text{if } 0 \le t < 1 \\ 100(1 + e^{1.4})e^{-1.4t} & \text{if } t \ge 1 \end{cases}$$

a. What was the amount of drug in Jane's body immediately after taking the second dose? After 2 days? In the long run?

b. Sketch the graph of A.

In Exercises 49–54, determine whether the statement is true or false. If it is true, explain why it is true. If it is false, give an example to show why it is false.

49. If a and b are positive numbers, then $(a + b)^x = a^x + b^x$.

50. If $x < y$, then $e^x < e^y$.

51. If $0 < b < 1$ and $x < y$, then $b^x > b^y$.

52. If $e^{kx} > 1$, then $k > 0$ and $x > 0$.

53. $f(x) = e^{kx}$ is an increasing function if $k > 0$ and a decreasing function if $k < 0$.

54. $f(x) = \dfrac{x}{1 + e^x}$ is continuous on $(-\infty, \infty)$.

5.1 Solutions to Self-Check Exercises

1.
$$2^{2x+1} \cdot 2^{-3} = 2^{x-1}$$

$$\frac{2^{2x+1}}{2^{x-1}} \cdot 2^{-3} = 1 \qquad \text{Divide both sides by } 2^{x-1}.$$

$$2^{(2x+1)-(x-1)-3} = 1$$

$$2^{x-1} = 1$$

This is true if and only if $x - 1 = 0$ or $x = 1$.

2. We first construct the following table of values:

x	-3	-2	-1	0	1	2	3	4
$y = e^{0.4x}$	0.3	0.4	0.7	1	1.5	2.2	3.3	5

Next, we plot these points and join them by a smooth curve to obtain the graph of f shown in the accompanying figure.

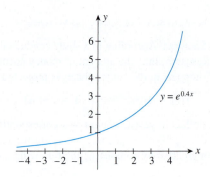

USING TECHNOLOGY

Although the proof is outside the scope of this book, it can be proved that an exponential function of the form $f(x) = b^x$, where $b > 1$, will ultimately grow faster than the power function $g(x) = x^n$ for *any* positive real number n. To give a visual demonstration of this result for the special case of the exponential function $f(x) = e^x$, we can use a graphing utility to plot the graphs of both f and g (for selected values of n) on the same set of axes in an appropriate viewing window and observe that the graph of f ultimately lies above that of g.

EXAMPLE 1 Use a graphing utility to plot the graphs of (a) $f(x) = e^x$ and $g(x) = x^3$ on the same set of axes in the viewing window $[0, 6] \times [0, 250]$ and (b) $f(x) = e^x$ and $g(x) = x^5$ in the viewing window $[0, 20] \times [0, 1{,}000{,}000]$.

Solution

a. The graphs of $f(x) = e^x$ and $g(x) = x^3$ in the viewing window $[0, 6] \times [0, 250]$ are shown in Figure T1a.

b. The graphs of $f(x) = e^x$ and $g(x) = x^5$ in the viewing window $[0, 20] \times [0, 1{,}000{,}000]$ are shown in Figure T1b.

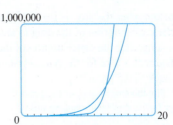

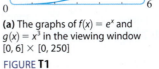

(a) The graphs of $f(x) = e^x$ and $g(x) = x^3$ in the viewing window $[0, 6] \times [0, 250]$

(b) The graphs of $f(x) = e^x$ and $g(x) = x^5$ in the viewing window $[0, 20] \times [0, 1,000,000]$

FIGURE **T1**

In the exercises that follow, you are asked to use a graphing utility to reveal the properties of exponential functions.

TECHNOLOGY EXERCISES

In Exercises 1 and 2, plot the graphs of the functions f and g on the same set of axes in the specified viewing window.

1. $f(x) = e^x$ and $g(x) = x^2$; $[0, 4] \times [0, 30]$

2. $f(x) = e^x$ and $g(x) = x^4$; $[0, 15] \times [0, 20,000]$

In Exercises 3 and 4, plot the graphs of the functions f and g on the same set of axes in an appropriate viewing window to demonstrate that f ultimately grows faster than g. (*Note:* Your answer will *not* be unique.)

3. $f(x) = 2^x$ and $g(x) = x^{2.5}$

4. $f(x) = 3^x$ and $g(x) = x^3$

5. Plot the graphs of $f(x) = 2^x$, $g(x) = 3^x$, and $h(x) = 4^x$ on the same set of axes in the viewing window $[0, 5] \times [0, 100]$. Comment on the relationship between the base b and the growth of the function $f(x) = b^x$.

6. Plot the graphs of $f(x) = (1/2)^x$, $g(x) = (1/3)^x$, and $h(x) = (1/4)^x$ on the same set of axes in the viewing window $[0, 4] \times [0, 1]$. Comment on the relationship between the base b and the growth of the function $f(x) = b^x$.

7. Plot the graphs of $f(x) = e^x$, $g(x) = 2e^x$, and $h(x) = 3e^x$ on the same set of axes in the viewing window $[-3, 3] \times [0, 10]$. Comment on the role played by the constant k in the graph of $f(x) = ke^x$.

8. Plot the graphs of $f(x) = -e^x$, $g(x) = -2e^x$, and $h(x) = -3e^x$ on the same set of axes in the viewing window $[-3, 3] \times [-10, 0]$. Comment on the role played by the constant k in the graph of $f(x) = ke^x$.

9. Plot the graphs of $f(x) = e^{0.5x}$, $g(x) = e^x$, and $h(x) = e^{1.5x}$ on the same set of axes in the viewing window $[-2, 2] \times [0, 4]$. Comment on the role played by the constant k in the graph of $f(x) = e^{kx}$.

10. Plot the graphs of $f(x) = e^{-0.5x}$, $g(x) = e^{-x}$, and $h(x) = e^{-1.5x}$ on the same set of axes in the viewing

window $[-2, 2] \times [0, 4]$. Comment on the role played by the constant k in the graph of $f(x) = e^{kx}$.

11. ABSORPTION OF DRUGS The concentration of a drug in an organ at any time t (in seconds) is given by

$$x(t) = 0.08 + 0.12(1 - e^{-0.02t})$$

where $x(t)$ is measured in grams per cubic centimeter (g/cm³).
a. Plot the graph of the function x in the viewing window $[0, 200] \times [0, 0.2]$.
b. What is the initial concentration of the drug in the organ?
c. What is the concentration of the drug in the organ after 20 sec?
d. What will be the concentration of the drug in the organ in the long run?
Hint: Evaluate $\lim_{t \to \infty} x(t)$.

12. ABSORPTION OF DRUGS Jane took 100 mg of a drug in the morning and another 100 mg of the same drug at the same time the following morning. The amount of the drug in her body t days after the first dosage was taken is given by

$$A(t) = \begin{cases} 100e^{-1.4t} & \text{if } 0 \le t < 1 \\ 100(1 + e^{1.4})e^{-1.4t} & \text{if } t \ge 1 \end{cases}$$

a. Plot the graph of the function A in the viewing window $[0, 5] \times [0, 140]$.
b. Verify the results of Exercise 48, page 344.

13. ABSORPTION OF DRUGS The concentration of a drug in an organ at any time t (in seconds) is given by

$$C(t) = \begin{cases} 0.3t - 18(1 - e^{-t/60}) & \text{if } 0 \le t \le 20 \\ 18e^{-t/60} - 12e^{-(t-20)/60} & \text{if } t > 20 \end{cases}$$

where $C(t)$ is measured in grams per cubic centimeter (g/cm³).
a. Plot the graph of the function C in the viewing window $[0, 120] \times [0, 1]$.

b. How long after the drug is first introduced will it take for the concentration of the drug to reach a peak?

c. How long after the concentration of the drug has peaked will it take for the concentration of the drug to fall back to 0.5 g/cm³?

Hint: Plot the graphs of $y_1 = C(x)$ and $y_2 = 0.5$, and use the **ISECT** function of your graphing utility.

14. MODELING WITH DATA The estimated number of Internet users in China (in millions) from 2005 through 2010 are shown in the following table:

Year	2005	2006	2007	2008	2009	2010
Number	116.1	141.9	169.0	209.0	258.1	314.8

a. Use **ExpReg** to find an exponential regression model for the data. Let $t = 1$ correspond to 2005.

Hint: $a^x = e^{x \ln a}$

b. Plot the scatter diagram and the graph of the function f found in part (a).

5.2   Logarithmic Functions

Logarithms

You are already familiar with exponential equations of the form

$$b^y = x \qquad (b > 0, b \neq 1)$$

where the variable x is expressed in terms of a real number b and a variable y. But what about solving this same equation for y? You may recall from your study of algebra that the number y is called the **logarithm of x to the base b** and is denoted by **$\log_b x$.** It is the power to which the base b must be raised to obtain the number x.

> **Logarithm of x to the Base b**
>
> $y = \log_b x$ if and only if $x = b^y$ $(b > 0, b \neq 1, \text{ and } x > 0)$

⚠️ Observe that the logarithm $\log_b x$ is defined only for positive values of x.

EXAMPLE 1

a. $\log_{10} 100 = 2$ since $100 = 10^2$

b. $\log_5 125 = 3$ since $125 = 5^3$

c. $\log_3 \dfrac{1}{27} = -3$ since $\dfrac{1}{27} = \dfrac{1}{3^3} = 3^{-3}$

d. $\log_{20} 20 = 1$ since $20 = 20^1$

EXAMPLE 2   Solve each of the following equations for x.

a. $\log_3 x = 4$ **b.** $\log_{16} 4 = x$ **c.** $\log_x 8 = 3$

Solution

a. By definition, $\log_3 x = 4$ implies $x = 3^4 = 81$.

b. $\log_{16} 4 = x$ is equivalent to $4 = 16^x = (4^2)^x = 4^{2x}$, or $4^1 = 4^{2x}$, from which we deduce that

$$2x = 1 \qquad \color{red}{b^m = b^n \Rightarrow m = n}$$

$$x = \frac{1}{2}$$

c. Referring once again to the definition, we see that the equation $\log_x 8 = 3$ is equivalent to

$$8 = 2^3 = x^3$$

$$x = 2 \qquad a^m = b^m \Rightarrow a = b$$

The two most widely used systems of logarithms are the system of **common logarithms,** which uses the number 10 as its base, and the system of **natural logarithms,** which uses the irrational number $e = 2.71828\ldots$ as its base. Also, it is standard practice to write **log** for $\log_{10}$ and **ln** for $\log_e$.

Logarithmic Notation

$$\log x = \log_{10} x \qquad \text{Common logarithm}$$

$$\ln x = \log_e x \qquad \text{Natural logarithm}$$

The system of natural logarithms is widely used in theoretical work. Using natural logarithms rather than logarithms to other bases often leads to simpler expressions.

Laws of Logarithms

Computations involving logarithms are facilitated by the following **laws of logarithms.**

Laws of Logarithms

If m and n are positive numbers and $b > 0$, $b \neq 1$, then

1. $\log_b mn = \log_b m + \log_b n$

2. $\log_b \dfrac{m}{n} = \log_b m - \log_b n$

3. $\log_b m^n = n \log_b m$

4. $\log_b 1 = 0$

5. $\log_b b = 1$

⚠️ Do not confuse the expression $\log m/n$ (Law 2) with the expression $\log m/\log n$. For example,

$$\log \frac{100}{10} = \log 100 - \log 10 = 2 - 1 = 1 \neq \frac{\log 100}{\log 10} = \frac{2}{1} = 2$$

You will be asked to prove these laws in Exercises 66–68 on page 353. Their derivations are based on the definition of a logarithm and the corresponding laws of exponents. The following examples illustrate the properties of logarithms.

EXAMPLE 3

a. $\log(2 \cdot 3) = \log 2 + \log 3$ **b.** $\ln \dfrac{5}{3} = \ln 5 - \ln 3$

c. $\log \sqrt{7} = \log 7^{1/2} = \dfrac{1}{2} \log 7$ **d.** $\log_5 1 = 0$

e. $\log_{45} 45 = 1$

EXAMPLE 4 Given that $\log 2 \approx 0.3010$, $\log 3 \approx 0.4771$, and $\log 5 \approx 0.6990$, use the laws of logarithms to find

a. $\log 15$ **b.** $\log 7.5$ **c.** $\log 81$ **d.** $\log 50$

Solution

a. Note that $15 = 3 \cdot 5$, so by Law 1 for logarithms,

$$\log 15 = \log 3 \cdot 5$$
$$= \log 3 + \log 5$$
$$\approx 0.4771 + 0.6990$$
$$= 1.1761$$

b. Observing that $7.5 = 15/2 = (3 \cdot 5)/2$, we apply Laws 1 and 2, obtaining

$$\log 7.5 = \log \frac{(3)(5)}{2}$$
$$= \log 3 + \log 5 - \log 2$$
$$\approx 0.4771 + 0.6990 - 0.3010$$
$$= 0.8751$$

c. Since $81 = 3^4$, we apply Law 3 to obtain

$$\log 81 = \log 3^4$$
$$= 4 \log 3$$
$$\approx 4(0.4771)$$
$$= 1.9084$$

d. We write $50 = 5 \cdot 10$ and find

$$\log 50 = \log(5)(10)$$
$$= \log 5 + \log 10$$
$$\log 50 \approx 0.6990 + 1 \qquad \text{Use Law 5}$$
$$= 1.6990$$

EXAMPLE 5 Expand and simplify the following expressions:

a. $\log_3 x^2 y^3$ **b.** $\log_2 \dfrac{x^2 + 1}{2^x}$ **c.** $\ln \dfrac{x^2 \sqrt{x^2 - 1}}{e^x}$

Solution

a. $\log_3 x^2 y^3 = \log_3 x^2 + \log_3 y^3$ Law 1
$$= 2 \log_3 x + 3 \log_3 y \qquad \text{Law 3}$$

b. $\log_2 \dfrac{x^2 + 1}{2^x} = \log_2(x^2 + 1) - \log_2 2^x$ Law 2

$$= \log_2(x^2 + 1) - x \log_2 2 \qquad \text{Law 3}$$
$$= \log_2(x^2 + 1) - x \qquad \text{Law 5}$$

c. $\ln \dfrac{x^2 \sqrt{x^2 - 1}}{e^x} = \ln \dfrac{x^2(x^2 - 1)^{1/2}}{e^x}$ Rewrite

$$= \ln x^2 + \ln(x^2 - 1)^{1/2} - \ln e^x \qquad \text{Laws 1 and 2}$$
$$= 2 \ln x + \frac{1}{2} \ln(x^2 - 1) - x \ln e \qquad \text{Law 3}$$
$$= 2 \ln x + \frac{1}{2} \ln(x^2 - 1) - x \qquad \text{Law 5}$$

Logarithmic Functions and Their Graphs

The definition of a logarithm implies that if b and n are positive numbers and b is different from 1, then the expression $\log_b n$ is a real number. This enables us to define a logarithmic function as follows.

> **Logarithmic Function**
>
> The function defined by
>
> $$f(x) = \log_b x \qquad (b > 0, b \neq 1)$$
>
> is called the **logarithmic function with base b.** The domain of f is the set of all positive numbers.

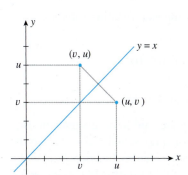

FIGURE 7
The points (u, v) and (v, u) are mirror reflections of each other.

One easy way to obtain the graph of the logarithmic function $y = \log_b x$ is to construct a table of values of the logarithm (base b). However, another method—and a more instructive one—is based on exploiting the intimate relationship between logarithmic and exponential functions.

If a point (u, v) lies on the graph of $y = \log_b x$, then

$$v = \log_b u$$

But we can also write this equation in exponential form as

$$u = b^v$$

So the point (v, u) also lies on the graph of the function $y = b^x$. Let's look at the relationship between the points (u, v) and (v, u) and the line $y = x$ (Figure 7). If we think of the line $y = x$ as a mirror, then the point (v, u) is the mirror reflection of the point (u, v). Similarly, the point (u, v) is the mirror reflection of the point (v, u). We can take advantage of this relationship to help us draw the graph of logarithmic functions. For example, if we wish to draw the graph of $y = \log_b x$, where $b > 1$, then we need only draw the mirror reflection of the graph of $y = b^x$ with respect to the line $y = x$ (Figure 8).

You may discover the following properties of the logarithmic function by taking the reflection of the graph of an appropriate exponential function (Exercises 33 and 34 on page 352).

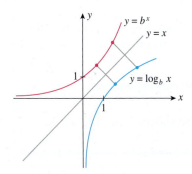

FIGURE 8
The graphs of $y = b^x$ and $y = \log_b x$ are mirror reflections of each other.

> **Properties of the Logarithmic Function**
>
> The logarithmic function $y = \log_b x$ $(b > 0, b \neq 1)$ has the following properties:
>
> 1. Its domain is $(0, \infty)$.
> 2. Its range is $(-\infty, \infty)$.
> 3. Its graph passes through the point $(1, 0)$.
> 4. It is continuous on $(0, \infty)$.
> 5. It is increasing on $(0, \infty)$ if $b > 1$ and decreasing on $(0, \infty)$ if $b < 1$.

EXAMPLE 6 Sketch the graph of the function $y = \ln x$.

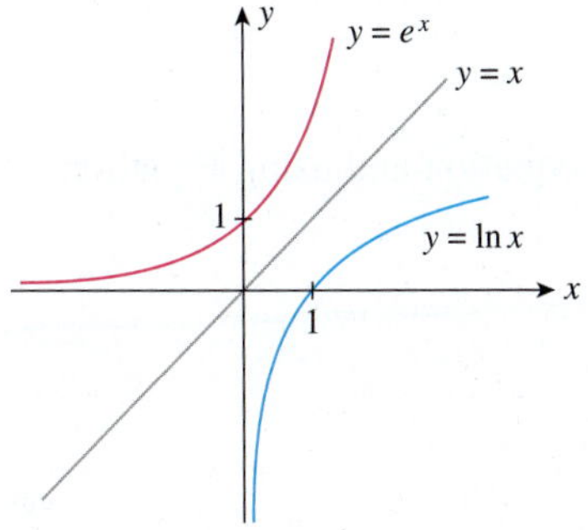

FIGURE 9
The graph of $y = \ln x$ is the mirror reflection of the graph of $y = e^x$.

Solution We first sketch the graph of $y = e^x$. Then, the required graph is obtained by tracing the mirror reflection of the graph of $y = e^x$ with respect to the line $y = x$ (Figure 9). ▪

Properties Relating the Exponential and Logarithmic Functions

We made use of the relationship that exists between the exponential function $f(x) = e^x$ and the logarithmic function $g(x) = \ln x$ when we sketched the graph of g in Example 6. This relationship is further described by the following properties, which are an immediate consequence of the definition of the logarithm of a number.

Properties Relating e^x and $\ln x$

$$e^{\ln x} = x \qquad \text{(for } x > 0\text{)} \tag{2}$$
$$\ln e^x = x \qquad \text{(for any real number } x\text{)} \tag{3}$$

(Try to verify these properties.)

From Properties 2 and 3, we conclude that the composite function satisfies

$$(f \circ g)(x) = f[g(x)]$$
$$= e^{\ln x} = x \qquad \text{(for all } x > 0\text{)}$$
$$(g \circ f)(x) = g[f(x)]$$
$$= \ln e^x = x \qquad \text{(for all } x > 0\text{)}$$

Any two functions f and g that satisfy this relationship are said to be **inverses** of each other. Note that the function f undoes what the function g does, and vice versa, so the composition of the two functions in any order results in the identity function $F(x) = x$.*

The relationships expressed in Equations (2) and (3) are useful in solving equations that involve exponentials and logarithms.

Exploring with TECHNOLOGY

You can demonstrate the validity of Properties 2 and 3, which state that the exponential function $f(x) = e^x$ and the logarithmic function $g(x) = \ln x$ are inverses of each other, as follows:

1. Sketch the graph of $(f \circ g)(x) = e^{\ln x}$, using the viewing window $[0, 10] \times [0, 10]$. Interpret the result.
2. Sketch the graph of $(g \circ f)(x) = \ln e^x$, using the standard viewing window. Interpret the result.

EXAMPLE 7 Solve the equation $2e^{x+2} = 5$.

Solution We first divide both sides of the equation by 2 to obtain

$$e^{x+2} = \frac{5}{2} = 2.5$$

Next, taking the natural logarithm of each side of the equation and using Equation (3), we have

$$\ln e^{x+2} = \ln 2.5$$
$$x + 2 = \ln 2.5$$
$$x = -2 + \ln 2.5$$
$$\approx -1.08$$

---

*For a more extensive treatment of inverse functions, see Appendix A.

Explore and Discuss

Consider the equation $y = y_0 b^{kx}$, where y_0 and k are positive constants and $b > 0$, $b \neq 1$. Suppose we want to express y in the form $y = y_0 e^{px}$. Use the laws of logarithms to show that $p = k \ln b$ and hence that $y = y_0 e^{(k \ln b)x}$ is an alternative form of $y = y_0 b^{kx}$ using the base e.

EXAMPLE 8 Solve the equation $5 \ln x + 3 = 0$.

Solution Adding -3 to both sides of the equation leads to

$$5 \ln x = -3$$

$$\ln x = -\frac{3}{5} = -0.6$$

and so

$$e^{\ln x} = e^{-0.6}$$

Using Equation (2), we conclude that

$$x = e^{-0.6}$$

$$\approx 0.55$$

5.2 Self-Check Exercises

1. Sketch the graph of $y = 3^x$ and $y = \log_3 x$ on the same set of axes.

2. Solve the equation $3e^{x+1} - 2 = 4$.

Solutions to Self-Check Exercises 5.2 can be found on page 354.

5.2 Concept Questions

1. **a.** Define $y = \log_b x$.
   b. Define the logarithmic function f with base b. What restrictions, if any, are placed on b?

2. For the logarithmic function $y = \log_b x$ $(b > 0, b \neq 1)$, state (a) its domain and range, (b) its x-intercept, (c) where it is continuous, and (d) where it is increasing and where it is decreasing for the case $b > 1$ and the case $b < 1$.

3. **a.** If $x > 0$, what is $e^{\ln x}$?
   b. If x is any real number, what is $\ln e^x$?

4. Let $f(x) = \ln x^2$ and $g(x) = 2 \ln x$. Are f and g identical?
   Hint: Look at their domains.

5.2 Exercises

In Exercises 1–10, express each equation in logarithmic form.

1. $2^6 = 64$

2. $3^5 = 243$

3. $4^{-2} = \dfrac{1}{16}$

4. $5^{-3} = \dfrac{1}{125}$

5. $\left(\dfrac{1}{3}\right)^1 = \dfrac{1}{3}$

6. $\left(\dfrac{1}{2}\right)^{-4} = 16$

7. $32^{4/5} = 16$

8. $81^{3/4} = 27$

9. $10^{-3} = 0.001$

10. $16^{-1/4} = 0.5$

In Exercises 11–16, given that $\log 3 \approx 0.4771$ and $\log 4 \approx 0.6021$, find the value of each logarithm.

11. $\log 12$

12. $\log \dfrac{3}{4}$

13. $\log 16$

14. $\log \sqrt{3}$

15. $\log 48$

16. $\log \dfrac{1}{300}$

In Exercises 17–20, write the expression as the logarithm of a single quantity.

17. $2 \ln a + 3 \ln b$

18. $\dfrac{1}{2} \ln x + 2 \ln y - 3 \ln z$

19. $\ln 3 + \dfrac{1}{2} \ln x + \ln y - \dfrac{1}{3} \ln z$

20. $\ln 2 + \dfrac{1}{2} \ln(x + 1) - 2 \ln(1 + \sqrt{x})$

In Exercises 21–28, use the laws of logarithms to expand and simplify the expression.

21. $\log x(x + 1)^4$

22. $\log x(x^2 + 1)^{-1/2}$

23. $\log \dfrac{\sqrt{x + 1}}{x^2 + 1}$

24. $\ln \dfrac{e^x}{1 + e^x}$

25. $\ln xe^{-x^2}$

26. $\ln x(x + 1)(x + 2)$

27. $\ln \dfrac{x^{1/2}}{x^2\sqrt{1 + x^2}}$

28. $\ln \dfrac{x^2}{\sqrt{x}(1 + x)^2}$

In Exercises 29–32, sketch the graph of the equation.

29. $y = \log_3 x$

30. $y = \log_{1/3} x$

31. $y = \ln 2x$

32. $y = \ln \dfrac{1}{2} x$

In Exercises 33 and 34, sketch the graphs of the equations on the same coordinate axes.

33. $y = 2^x$ and $y = \log_2 x$

34. $y = e^{3x}$ and $y = \dfrac{1}{3} \ln x$

In Exercises 35–44, use logarithms to solve the equation for t.

35. $e^{0.4t} = 8$

36. $\dfrac{1}{3} e^{-3t} = 0.9$

37. $5e^{-2t} = 6$

38. $4e^{t-1} = 4$

39. $2e^{-0.2t} - 4 = 6$

40. $12 - e^{0.4t} = 3$

41. $\dfrac{50}{1 + 4e^{0.2t}} = 20$

42. $\dfrac{200}{1 + 3e^{-0.3t}} = 100$

43. $A = Be^{-t/2}$

44. $\dfrac{A}{1 + Be^{t/2}} = C$

45. A function f has the form $f(x) = a + b \ln x$. Find f if it is known that $f(1) = 2$ and $f(2) = 4$.

46. AVERAGE LIFE SPAN One reason for the increase in human life span over the years has been the advances in medical technology. The average life span for American women from 1907 through 2007 is given by

$$W(t) = 49.9 + 17.1 \ln t \qquad (1 \le t \le 6)$$

where $W(t)$ is measured in years and t is measured in 20-year intervals, with $t = 1$ corresponding to 1907.
  a. What was the average life expectancy for women in 1907?
  b. If the trend continues, what will be the average life expectancy for women in 2027 ($t = 7$)?
  Source: American Association of Retired Persons (AARP).

47. BLOOD PRESSURE A normal child's systolic blood pressure may be approximated by the function

$$p(x) = m(\ln x) + b$$

where $p(x)$ is measured in millimeters of mercury, x is measured in pounds, and m and b are constants. Given

that $m = 19.4$ and $b = 18$, determine the systolic blood pressure of a child who weighs 92 lb.

48. MAGNITUDE OF EARTHQUAKES On the Richter scale, the magnitude R of an earthquake is given by the formula

$$R = \log \dfrac{I}{I_0}$$

where I is the intensity of the earthquake being measured and I_0 is the standard reference intensity.
  a. Express the intensity I of an earthquake of magnitude $R = 5$ in terms of the standard intensity I_0.
  b. Express the intensity I of an earthquake of magnitude $R = 8$ in terms of the standard intensity I_0. How many times greater is the intensity of an earthquake of magnitude 8 than one of magnitude 5?
  c. In modern times, the greatest loss of life attributable to an earthquake occurred in Haiti in 2010. Known as the Haiti earthquake, it registered 7.0 on the Richter scale. How does the intensity of this earthquake compare with the intensity of an earthquake of magnitude $R = 5$?

49. SOUND INTENSITY The relative loudness of a sound D of intensity I is measured in decibels (db), where

$$D = 10 \log \dfrac{I}{I_0}$$

and I_0 is the standard threshold of audibility.
  a. Express the intensity I of a 30-db sound (the sound level of normal conversation) in terms of I_0.
  b. Determine how many times greater the intensity of an 80-db sound (rock music) is than that of a 30-db sound.
  c. Prolonged noise above 150 db causes permanent deafness. How does the intensity of a 150-db sound compare with the intensity of an 80-db sound?

50. BAROMETRIC PRESSURE Halley's Law states that the barometric pressure (in inches of mercury) at an altitude of x mi above sea level is approximated by the equation

$$p(x) = 29.92e^{-0.2x} \qquad (x \ge 0)$$

If the barometric pressure as measured by a hot-air balloonist is 20 in. of mercury, what is the balloonist's altitude?

51. NEWTON'S LAW OF COOLING The temperature of a cup of coffee t min after it is poured is given by

$$T = 70 + 100e^{-0.0446t}$$

where T is measured in degrees Fahrenheit.
  a. What was the temperature of the coffee when it was poured?
  b. When will the coffee be cool enough to drink (say, 120°F)?

52. **HEIGHT OF TREES** The height (in feet) of a certain kind of tree is approximated by

$$h(t) = \frac{160}{1 + 240e^{-0.2t}}$$

where t is the age of the tree in years. Estimate the age of an 80-ft tree.

53. **OBESITY OVER TIME** The percentage of obese adults, ages 20 through 74 years, is projected to be

$$f(t) = \frac{46.5}{1 + 2.324e^{-0.05113t}} \qquad (0 \le t \le 60)$$

in year t, with $t = 0$ corresponding to 1970. Estimate the year when the percentage reached 40%.

Source: Centers for Disease Control and Prevention.

54. **LENGTHS OF FISH** The length (in centimeters) of a typical Pacific halibut t years old is approximately

$$f(t) = 200(1 - 0.956e^{-0.18t})$$

Suppose a Pacific halibut caught by Mike measures 140 cm. What is its approximate age?

55. **ABSORPTION OF DRUGS** The concentration of a drug in an organ t seconds after it has been administered is given by

$$x(t) = 0.08 + 0.12e^{-0.02t}$$

where $x(t)$ is measured in grams per cubic centimeter (g/cm³).

   a. How long would it take for the concentration of the drug in the organ to reach 0.18 g/cm³?

   b. How long would it take for the concentration of the drug in the organ to reach 0.16 g/cm³?

56. **ABSORPTION OF DRUGS** The concentration of a drug in an organ t seconds after it has been administered is given by

$$x(t) = 0.08(1 - e^{-0.02t})$$

where $x(t)$ is measured in grams per cubic centimeter (g/cm³).

   a. How long would it take for the concentration of the drug in the organ to reach 0.02 g/cm³?

   b. How long would it take for the concentration of the drug in the organ to reach 0.04 g/cm³?

57. **FORENSIC SCIENCE** Forensic scientists use the following law to determine the time of death of accident or murder victims. If T denotes the temperature of a body t hr after death, then

$$T = T_0 + (T_1 - T_0)(0.97)^t$$

where T_0 is the air temperature and T_1 is the body temperature at the time of death. John Doe was found murdered at midnight in his house; the room temperature was 70°F, and his body temperature was 80°F when he was found. When was he killed? Assume that the normal body temperature is 98.6°F.

58. **ELASTICITY OF DEMAND** The demand function for a certain brand of backpacks is

$$p = 50 \ln \frac{50}{x} \qquad (0 < x \le 50)$$

where p is the unit price in dollars and x is the quantity (in hundreds) demanded per month.

   a. Find the elasticity of demand, and determine the range of prices corresponding to inelastic, unitary, and elastic demand.

   b. If the unit price is increased slightly from $50, will the revenue increase or decrease?

In Exercises 59–62, determine whether the statement is true or false. If it is true, explain why it is true. If it is false, give an example to show why it is false.

59. $(\ln x)^3 = 3 \ln x$ for all x in $(0, \infty)$.

60. If $a > 0$ and $b > 0$, then $\ln(a + b) = \ln a + \ln b$.

61. If $b > 0$, then $e^{\ln b} = \ln e^b$.

62. $\ln a - \ln b = \ln(a - b)$ for all positive real numbers a and b.

63. The function $f(x) = \dfrac{1}{\ln x}$ is continuous on $(1, \infty)$.

64. $(\log_2 3)(\log_3 2) = 1$

65. a. Given that $2^x = e^{kx}$, find k.

   b. Show that, in general, if b is a positive real number, then any equation of the form $y = b^x$ may be written in the form $y = e^{kx}$, for some real number k.

66. Use the definition of a logarithm to prove

   a. $\log_b mn = \log_b m + \log_b n$

   b. $\log_b \dfrac{m}{n} = \log_b m - \log_b n$

   Hint: Let $\log_b m = p$ and $\log_b n = q$. Then, $b^p = m$ and $b^q = n$.

67. Use the definition of a logarithm to prove

$$\log_b m^n = n \log_b m$$

68. Use the definition of a logarithm to prove

   a. $\log_b 1 = 0$

   b. $\log_b b = 1$

5.2 Solutions to Self-Check Exercises

1. First, sketch the graph of $y = 3^x$ with the help of the following table of values:

x	-3	-2	-1	0	1	2	3
$y = 3^x$	$\frac{1}{27}$	$\frac{1}{9}$	$\frac{1}{3}$	1	3	9	27

Next, take the mirror reflection of this graph with respect to the line $y = x$ to obtain the graph of $y = \log_3 x$.

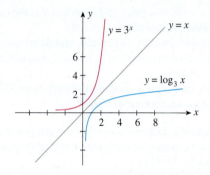

2.
$$3e^{x+1} - 2 = 4$$
$$3e^{x+1} = 6$$
$$e^{x+1} = 2$$
$$\ln e^{x+1} = \ln 2 \qquad \text{Take the logarithm of both sides.}$$
$$(x + 1)\ln e = \ln 2 \qquad \text{Law 3}$$
$$x + 1 = \ln 2 \qquad \text{Law 5}$$
$$x = \ln 2 - 1$$
$$\approx -0.3069$$

5.3 Compound Interest

Compound Interest

Compound interest is a natural application of the exponential function to the business world. We begin by recalling that simple interest is interest that is computed only on the original principal. Thus, if I denotes the interest on a principal P (in dollars) at an interest rate of r per year for t years, then we have

$$I = Prt$$

The **accumulated amount** A, the sum of the principal and interest after t years, is given by

$$A = P + I = P + Prt$$
$$= P(1 + rt) \qquad \text{Simple interest formula} \qquad (4)$$

Frequently, interest earned is periodically added to the principal and thereafter earns interest itself at the same rate. This is called **compound interest.** To find a formula for the accumulated amount, let's consider a numerical example. Suppose $1000 (the principal) is deposited in a bank for a **term** of 3 years, earning interest at the rate of 8% per year (called the **nominal,** or **stated, rate**) compounded annually. Then, using Formula (4) with $P = 1000$, $r = 0.08$, and $t = 1$, we see that the accumulated amount at the end of the first year is

$$A_1 = P(1 + rt)$$
$$= 1000[1 + 0.08(1)] = 1000(1.08) = 1080$$

or $1080.

To find the accumulated amount A_2 at the end of the second year, we use Equation (4) once again, this time with $P = A_1$. (Remember, the principal *and* interest now earn interest over the second year.) We obtain

$$\begin{aligned} A_2 &= P(1 + rt) = A_1(1 + rt) \\ &= 1000[1 + 0.08(1)][1 + 0.08(1)] \\ &= 1000(1 + 0.08)^2 = 1000(1.08)^2 = 1166.40 \end{aligned}$$

or $1166.40.

Finally, the accumulated amount A_3 at the end of the third year is found by using Equation (4) with $P = A_2$, giving

$$\begin{aligned} A_3 &= P(1 + rt) = A_2(1 + rt) \\ &= 1000[1 + 0.08(1)]^2[1 + 0.08(1)] \\ &= 1000(1 + 0.08)^3 = 1000(1.08)^3 \approx 1259.71 \end{aligned}$$

or approximately $1259.71.

If you reexamine our calculations in this example, you will see that the accumulated amounts at the end of each year have the following form:

First year: $A_1 = 1000(1 + 0.08)$ or $A_1 = P(1 + r)$

Second year: $A_2 = 1000(1 + 0.08)^2$ or $A_2 = P(1 + r)^2$

Third year: $A_3 = 1000(1 + 0.08)^3$ or $A_3 = P(1 + r)^3$

These observations suggest the following general result: If P dollars are invested over a term of t years earning interest at the rate of r per year compounded annually, then the accumulated amount is

$$A = P(1 + r)^t \tag{5}$$

Formula (5) was derived under the assumption that interest was compounded *annually*. In practice, however, interest is usually compounded more than once a year. The interval of time between successive interest calculations is called the **conversion period.**

If interest at a nominal rate of r per year is compounded m times a year on a principal of P dollars, then the simple interest rate per conversion period is

$$i = \frac{r}{m} \quad \frac{\text{Annual interest rate}}{\text{Periods per year}}$$

For example, if the nominal interest rate is 8% per year ($r = 0.08$) and interest is compounded quarterly ($m = 4$), then

$$i = \frac{r}{m} = \frac{0.08}{4} = 0.02$$

or 2% per period.

To find a general formula for the accumulated amount if a principal of P dollars is deposited in a bank for a term of t years and earns interest at the (nominal) rate of r per year compounded m times per year, we proceed as before, using Formula (5) repeatedly with the interest rate $i = r/m$. We see that the accumulated amount at the end of each period is as follows:

First period: $A_1 = P(1 + i)$

Second period: $A_2 = A_1(1 + i) = [P(1 + i)](1 + i) = P(1 + i)^2$

Third period: $A_3 = A_2(1 + i) = [P(1 + i)^2](1 + i) = P(1 + i)^3$
$$\vdots \qquad\qquad \vdots$$

nth period: $A_n = A_{n-1}(1 + i) = [P(1 + i)^{n-1}](1 + i) = P(1 + i)^n$

But there are $n = mt$ periods in t years (number of conversion periods per year times the term). Therefore, the accumulated amount at the end of t years is given by

Compound Interest Formula

$$A = P\left(1 + \frac{r}{m}\right)^{mt} \tag{6}$$

where

A = Accumulated amount at the end of t years
P = Principal
r = Nominal interest rate per year
m = Number of conversion periods per year
t = Term (number of years)

EXAMPLE 1 Find the accumulated amount after 3 years if $1000 is invested at 8% per year compounded (a) annually, (b) semiannually, (c) quarterly, (d) monthly, and (e) daily.

Solution

a. Here, $P = 1000$, $r = 0.08$, $m = 1$, and $t = 3$, so Formula (6) gives

$$A = 1000(1 + 0.08)^3$$
$$\approx 1259.71$$

or $1259.71.

b. Here, $P = 1000$, $r = 0.08$, $m = 2$, and $t = 3$, so Formula (6) gives

$$A = 1000\left(1 + \frac{0.08}{2}\right)^{(2)(3)}$$
$$\approx 1265.32$$

or $1265.32.

c. In this case, $P = 1000$, $r = 0.08$, $m = 4$, and $t = 3$, so Formula (6) gives

$$A = 1000\left(1 + \frac{0.08}{4}\right)^{(4)(3)}$$
$$\approx 1268.24$$

or $1268.24.

d. Here, $P = 1000$, $r = 0.08$, $m = 12$, and $t = 3$, so Formula (6) gives

$$A = 1000\left(1 + \frac{0.08}{12}\right)^{(12)(3)}$$
$$\approx 1270.24$$

or $1270.24.

e. Here, $P = 1000$, $r = 0.08$, $m = 365$, and $t = 3$, so Formula (6) gives

$$A = 1000\left(1 + \frac{0.08}{365}\right)^{(365)(3)}$$
$$\approx 1271.22$$

or $1271.22. These results are summarized in Table 2.

TABLE 2				
Nominal Rate, r	Conversion Period	Term in Years	Initial Investment	Accumulated Amount
8%	Annual $(m = 1)$	3	$1000	$1259.71
8	Semiannual $(m = 2)$	3	1000	1265.32
8	Quarterly $(m = 4)$	3	1000	1268.24
8	Monthly $(m = 12)$	3	1000	1270.24
8	Daily $(m = 365)$	3	1000	1271.22

Exploring with TECHNOLOGY

Investments that are allowed to grow over time can increase in value surprisingly fast. Consider the potential growth of $10,000 if earnings are reinvested. More specifically, suppose $A_1(t), A_2(t), A_3(t), A_4(t)$, and $A_5(t)$ denote the accumulated values of an investment of $10,000 over a term of t years and earning interest at the rate of 4%, 6%, 8%, 10%, and 12% per year compounded annually.

1. Find expressions for $A_1(t), A_2(t), \ldots, A_5(t)$.
2. Use a graphing utility to plot the graphs of $A_1, A_2, \ldots, A_5$ on the same set of axes, using the viewing window $[0, 20] \times [0, 100{,}000]$.
3. Use **TRACE** to find $A_1(20), A_2(20), \ldots, A_5(20)$, and then interpret your results.

Effective Rate of Interest

In Example 1, we saw that the interest earned on an investment depends on the frequency with which the interest is compounded. Thus, the stated, or nominal, rate of 8% per year does not reflect the actual rate at which interest is earned. This suggests that we need to find a common basis for comparing interest rates. One way of comparing interest rates is provided by using the effective rate. The **effective rate** is the *simple* interest rate that would produce the same accumulated amount in 1 year as the nominal rate compounded m times a year. The effective rate is also called the **true rate.**

To derive a relation between the nominal interest rate, r per year compounded m times and its corresponding effective rate, r_{eff} per year, let's assume an initial investment of P dollars. Then, the accumulated amount after 1 year at a simple interest rate of r_{eff} per year is

$$A = P(1 + r_{eff})$$

Also, the accumulated amount after 1 year at an interest rate of r per year compounded m times a year is

$$A = P\left(1 + \frac{r}{m}\right)^m \qquad \text{Since } t = 1$$

Equating the two expressions gives

$$P(1 + r_{eff}) = P\left(1 + \frac{r}{m}\right)^m$$

$$1 + r_{eff} = \left(1 + \frac{r}{m}\right)^m \qquad \text{Divide both sides by } P.$$

or, upon solving for r_{eff}, we obtain the formula for computing the effective rate of interest:

Effective Rate of Interest Formula

$$r_{\text{eff}} = \left(1 + \frac{r}{m}\right)^m - 1 \qquad (7)$$

where

r_{eff} = Effective rate of interest

$\quad r$ = Nominal interest rate per year

$\quad m$ = Number of conversion periods per year

EXAMPLE 2 Find the effective rate of interest corresponding to a nominal rate of 8% per year compounded (a) annually, (b) semiannually, (c) quarterly, (d) monthly, and (e) daily.

Solution

a. The effective rate of interest corresponding to a nominal rate of 8% per year compounded annually is of course given by 8% per year. This result is also confirmed by using Formula (7) with $r = 0.08$ and $m = 1$. Thus,

$$r_{\text{eff}} = (1 + 0.08) - 1 = 0.08$$

b. Let $r = 0.08$ and $m = 2$. Then, Formula (7) yields

$$r_{\text{eff}} = \left(1 + \frac{0.08}{2}\right)^2 - 1$$

$$= 0.0816$$

so the required effective rate is 8.16% per year.

c. Let $r = 0.08$ and $m = 4$. Then, Formula (7) yields

$$r_{\text{eff}} = \left(1 + \frac{0.08}{4}\right)^4 - 1$$

$$\approx 0.0824$$

so the corresponding effective rate in this case is 8.24% per year.

d. Let $r = 0.08$ and $m = 12$. Then, Formula (7) yields

$$r_{\text{eff}} = \left(1 + \frac{0.08}{12}\right)^{12} - 1$$

$$\approx 0.0830$$

so the corresponding effective rate in this case is 8.30% per year.

e. Let $r = 0.08$ and $m = 365$. Then, Formula (7) yields

$$r_{\text{eff}} = \left(1 + \frac{0.08}{365}\right)^{365} - 1$$

$$\approx 0.0833$$

so the corresponding effective rate in this case is 8.33% per year.

Now, if the effective rate of interest r_{eff} is known, then the accumulated amount after t years on an investment of P dollars may be more readily computed by using the formula

$$A = P(1 + r_{\text{eff}})^t$$

The 1968 Truth in Lending Act passed by Congress requires that the effective rate of interest be disclosed in all contracts involving interest charges. The passage of this act has benefited consumers because they now have a common basis for comparing the various nominal rates quoted by different financial institutions. Furthermore, knowing the effective rate enables consumers to compute the actual charges involved in a transaction. The effective rate is also called the **annual percentage yield** (APY). Thus, if the effective rates of interest found in Example 2 were known, the accumulated values of Example 1, shown in Table 3, could have been readily found.

TABLE 3

Nominal Rate, r	Frequency of Interest Payment	Effective Rate	Initial Investment	Accumulated Amount after 3 Years	
8%	Annually	8%	$1000	$1000(1 + 0.08)^3$	$= \$1259.71$
8	Semiannually	8.16	1000	$1000(1 + 0.0816)^3$	$= 1265.32$
8	Quarterly	8.243	1000	$1000(1 + 0.08243)^3$	$= 1268.23$
8	Monthly	8.300	1000	$1000(1 + 0.08300)^3$	$= 1270.24$
8	Daily	8.328	1000	$1000(1 + 0.08328)^3$	$= 1271.22$

 APPLIED EXAMPLE 3 Investment Options Jane has narrowed her investment options down to two:

1. Purchase a CD that matures in 24 years and pays interest upon maturity at the rate of 5% per year compounded daily (assume 365 days in a year).
2. Purchase a zero coupon CD that will triple her investment in the same period.

Which option will optimize Jane's investment?

Solution Let's compute the accumulated amount under option 1. Here,

$$r = 0.05 \qquad m = 365 \qquad t = 24$$

so $n = 24(365) = 8760$ and $i = \frac{0.05}{365}$. The accumulated amount at the end of 24 years (after 8760 conversion periods) is

$$A = P\left(1 + \frac{0.05}{365}\right)^{8760} \approx 3.32P$$

or $3.32P$. If Jane chooses option 2, the accumulated amount of her investment after 24 years will be $3P$. Therefore, she should choose option 1.

APPLIED EXAMPLE 4 IRAs Moesha has an Individual Retirement Account (IRA) with a brokerage firm. Her money is invested in a money market mutual fund that pays interest on a daily basis. Over a 2-year period in which no deposits or withdrawals were made, her account grew from $4500 to $4792.61. Find the effective rate at which Moesha's account was earning interest over that period (assume 365 days in a year).

Solution Let r_{eff} denote the required effective rate of interest. We have

$$4792.61 = 4500(1 + r_{\text{eff}})^2$$
$$(1 + r_{\text{eff}})^2 \approx 1.06502$$
$$1 + r_{\text{eff}} \approx 1.031998 \qquad \text{Take the square root on both sides.}$$

or $r_{\text{eff}} \approx 0.031998$. Therefore, the effective rate was approximately 3.20% per year.

Present Value

Let's return to the compound interest Formula (6), which expresses the accumulated amount at the end of t years when interest at the rate of r is compounded m times a year. The principal P in Formula (6) is often referred to as the **present value,** and the accumulated value A is called the **future value,** since it is realized at a future date. In certain instances, an investor may wish to determine how much money he or she should invest now, at a fixed rate of interest, to realize a certain sum at some future date. This problem may be solved by expressing P in terms of A. Thus, from Formula (6), we find

> **Present Value Formula for Compound Interest**
>
> $$P = A\left(1 + \frac{r}{m}\right)^{-mt} \tag{8}$$

EXAMPLE 5 How much money should be deposited in a bank paying interest at the rate of 6% per year compounded monthly so that at the end of 3 years, the accumulated amount will be $20,000?

Solution   Here, $A = 20{,}000$, $r = 0.06$, $m = 12$, and $t = 3$. Using Formula (8), we obtain

$$P = 20{,}000\left(1 + \frac{0.06}{12}\right)^{-(12)(3)}$$

$$\approx 16{,}713$$

or $16,713.

EXAMPLE 6 Find the present value of $36,605.70 due in 5 years at an interest rate of 4% per year compounded quarterly.

Solution   Using Formula (8) with $A = 36{,}605.70$, $r = 0.4$, $m = 4$, and $t = 5$, we obtain

$$P = 36{,}605.70\left(1 + \frac{0.04}{4}\right)^{-(4)(5)} \approx 30{,}000$$

or $30,000.

Continuous Compounding of Interest

One question that arises naturally in the study of compound interest is: What happens to the accumulated amount over a fixed period of time if the interest is computed more and more frequently?

Intuition suggests that the more often interest is compounded, the larger the accumulated amount will be. This is confirmed by the results of Example 1, in which we found that the accumulated amounts did in fact increase when we increased the number of conversion periods per year.

This leads us to another question: Does the accumulated amount approach a limit when the interest is computed more and more frequently over a fixed period of time?

To answer this question, let's look again at the compound interest formula:

$$A = P\left(1 + \frac{r}{m}\right)^{mt} \tag{9}$$

Recall that m is the number of conversion periods per year. So to find an answer to our question, we should let m get larger and larger (approach infinity) in Equation (9). But first we will rewrite this equation in the form

$$A = P\left[\left(1 + \frac{r}{m}\right)^m\right]^t \qquad \text{Since } b^{xy} = (b^x)^y$$

Now, letting $m \to \infty$, we find that

$$\lim_{m\to\infty}\left[P\left(1 + \frac{r}{m}\right)^m\right]^t = P\left[\lim_{m\to\infty}\left(1 + \frac{r}{m}\right)^m\right]^t \qquad \text{Why?}$$

Next, upon making the substitution $u = m/r$ and observing that $u \to \infty$ as $m \to \infty$, the foregoing expression reduces to

$$P\left[\lim_{u\to\infty}\left(1 + \frac{1}{u}\right)^{ur}\right]^t = P\left[\lim_{u\to\infty}\left(1 + \frac{1}{u}\right)^u\right]^{rt}$$

But

$$\lim_{u\to\infty}\left(1 + \frac{1}{u}\right)^u = e \qquad \text{Use Equation (1).}$$

so

$$\lim_{m\to\infty} P\left[\left(1 + \frac{r}{m}\right)^m\right]^t = Pe^{rt}$$

Our computations tell us that as the frequency with which interest is compounded increases without bound, the accumulated amount approaches Pe^{rt}. In this situation, we say that interest is *compounded continuously*. Let's summarize this important result.

Continuous Compound Interest Formula

$$A = Pe^{rt} \qquad\qquad (10)$$

where

P = Principal

r = Annual interest rate compounded continuously

t = Time in years

A = Accumulated amount at the end of t years

EXAMPLE 7 Find the accumulated amount after 3 years if $1000 is invested at 8% per year compounded (a) daily (assume a 365-day year) and (b) continuously.

Solution

a. Using Formula (6) with $P = 1000$, $r = 0.08$, $m = 365$, and $t = 3$, we find

$$A = 1000\left(1 + \frac{0.08}{365}\right)^{(365)(3)} \approx 1271.22$$

or $1271.22.

b. Here, we use Formula (10) with $P = 1000$, $r = 0.08$, and $t = 3$, obtaining

$$A = 1000e^{(0.08)(3)}$$
$$\approx 1271.25$$

or $1271.25.

Observe that the accumulated amounts corresponding to interest compounded daily and interest compounded continuously differ by very little. The continuous compound interest formula is a very important tool in theoretical work in financial analysis.

Exploring with TECHNOLOGY

In the opening paragraph of Section 5.1, we pointed out that the accumulated amount of an account earning interest *compounded continuously* will eventually outgrow by far the accumulated amount of an account earning interest at the same nominal rate but earning simple interest. Illustrate this fact using the following example.

Suppose you deposit $1000 in Account I, earning interest at the rate of 10% per year compounded continuously so that the accumulated amount at the end of t years is $A_1(t) = 1000e^{0.1t}$. Suppose you also deposit $1000 in Account II, earning simple interest at the rate of 10% per year so that the accumulated amount at the end of t years is $A_2(t) = 1000(1 + 0.1t)$. Use a graphing utility to sketch the graphs of the functions A_1 and A_2 in the viewing window $[0, 20] \times [0, 10,000]$ to see the accumulated amounts $A_1(t)$ and $A_2(t)$ over a 20-year period.

If we solve Formula (10) for P, we obtain

$$P = Ae^{-rt} \tag{11}$$

which gives the present value in terms of the future (accumulated) value for the case of continuous compounding.

APPLIED EXAMPLE 8 Real Estate Investment Blakely Investment Company owns an office building located in the commercial district of a city. As a result of the continued success of an urban renewal program, local business is enjoying a miniboom. The market value of Blakely's property is

$$V(t) = 300,000e^{\sqrt{t}/2}$$

where $V(t)$ is measured in dollars and t is the time in years from the present. If the expected rate of appreciation is 9% compounded continuously for the next 10 years, find an expression for the present value $P(t)$ of the market price of the property valid for the next 10 years. Compute $P(7)$, $P(8)$, and $P(9)$, and interpret your results.

Solution Using Formula (11) with $A = V(t)$ and $r = 0.09$, we find that the present value of the market price of the property t years from now is

$$\begin{aligned} P(t) &= V(t)e^{-0.09t} \\ &= 300,000e^{-0.09t + \sqrt{t}/2} \qquad (0 \le t \le 10) \end{aligned}$$

Letting $t = 7, 8$, and 9, respectively, we find that

$$\begin{aligned} P(7) &= 300,000e^{-0.09(7) + \sqrt{7}/2} \approx 599,837 \quad \text{or} \quad \$599,837 \\ P(8) &= 300,000e^{-0.09(8) + \sqrt{8}/2} \approx 600,640 \quad \text{or} \quad \$600,640 \\ P(9) &= 300,000e^{-0.09(9) + \sqrt{9}/2} \approx 598,115 \quad \text{or} \quad \$598,115 \end{aligned}$$

From the results of these computations, we see that the present value of the property's market price seems to decrease after a certain period of growth. This suggests that there is an optimal time for the owners to sell. Later we will show that the highest present value of the property's market price is $600,779, which occurs at time $t = 7.72$ years.

Exploring with TECHNOLOGY

The effective rate of interest is given by

$$r_{\text{eff}} = \left(1 + \frac{r}{m}\right)^m - 1$$

where the number of conversion periods per year is m. In Exercise 55 on page 367 you will be asked to show that the effective rate of interest r_{eff} corresponding to a nominal interest rate r per year compounded continuously is given by

$$\hat{r}_{\text{eff}} = e^r - 1$$

To obtain a visual confirmation of this result, consider the special case in which $r = 0.1$ (10% per year).

1. Use a graphing utility to plot the graph of both

$$y_1 = \left(1 + \frac{0.1}{x}\right)^x - 1 \quad \text{and} \quad y_2 = e^{0.1} - 1$$

in the viewing window $[0, 3] \times [0, 0.12]$.
2. Does your result seem to imply that

$$\left(1 + \frac{r}{m}\right)^m - 1$$

approaches

$$\hat{r}_{\text{eff}} = e^r - 1$$

as m increases without bound for the special case $r = 0.1$?

The next two examples show how logarithms can be used to solve problems involving compound interest.

EXAMPLE 9 How long will it take $10,000 to grow to $15,000 if the investment earns interest at the rate of 6% per year compounded quarterly?

Solution Using Formula (6) with $A = 15,000$, $P = 10,000$, $r = 0.06$, and $m = 4$, we obtain

$$15,000 = 10,000\left(1 + \frac{0.06}{4}\right)^{4t}$$

$$(1.015)^{4t} = \frac{15,000}{10,000} = 1.5$$

Taking the logarithm on each side of the equation gives

$$\ln(1.015)^{4t} = \ln 1.5$$
$$4t \ln 1.015 = \ln 1.5 \qquad \log_b m^n = n \log_b m$$
$$4t = \frac{\ln 1.5}{\ln 1.015}$$
$$t = \frac{\ln 1.5}{4 \ln 1.015} \approx 6.808$$

So it will take approximately 6.8 years for the investment to grow from $10,000 to $15,000.

EXAMPLE 10 Find the interest rate needed for an investment of $10,000 to grow to an amount of $18,000 in 10 years if the interest is compounded monthly.

Solution   Using Formula (6) with $A = 18{,}000$, $P = 10{,}000$, $m = 12$, and $t = 10$, we obtain

$$18{,}000 = 10{,}000\left(1 + \frac{r}{12}\right)^{12(10)}$$

Dividing both sides of the equation by 10,000 gives

$$\frac{18{,}000}{10{,}000} = \left(1 + \frac{r}{12}\right)^{120}$$

or, upon simplification,

$$\left(1 + \frac{r}{12}\right)^{120} = 1.8$$

Now, we take the logarithm on each side of the equation, obtaining

$$\ln\left(1 + \frac{r}{12}\right)^{120} = \ln 1.8$$

$$120 \ln\left(1 + \frac{r}{12}\right) = \ln 1.8$$

$$\ln\left(1 + \frac{r}{12}\right) = \frac{\ln 1.8}{120} \approx 0.004898$$

$$\left(1 + \frac{r}{12}\right) \approx e^{0.004898} \qquad \text{By Property 2}$$

$$\approx 1.00491$$

and

$$\frac{r}{12} \approx 1.00491 - 1$$

$$r \approx 0.05892$$

or approximately 5.89% per year.

5.3   Self-Check Exercises

1. Find the present value of $20,000 due in 3 years at an interest rate of 6%/year compounded monthly.

2. **INVESTMENT INCOME** Glen is a retiree living on Social Security and the income from his investment. Currently, his $100,000 investment in a 1-year CD is yielding 4.6% interest compounded daily. If he reinvests the principal ($100,000) on the due date of the CD in another 1-year CD paying 3.2% interest compounded daily, find the net decrease in his yearly income from his investment.

3. **a.** What is the accumulated amount after 5 years if $10,000 is invested at 5%/year compounded continuously?

   b. Find the present value of $10,000 due in 5 years at an interest rate of 5%/year compounded continuously.

   Solutions to Self-Check Exercises 5.3 can be found on page 368.

5.3 Concept Questions

1. **a.** What is the difference between simple interest and compound interest?
   b. State the simple interest formula and the compound interest formula.

2. **a.** What is the effective rate of interest?
   b. State the formula for computing the effective rate of interest.

3. What is the present value formula for compound interest?

4. State the continuous compound interest formula.

5.3 Exercises

In Exercises 1–4, find the accumulated amount A if the principal P is invested at an interest rate of r per year for t years.

1. $P = \$2500$, $r = 4\%$, $t = 10$, compounded semiannually

2. $P = \$12,000$, $r = 5\%$, $t = 10$, compounded quarterly

3. $P = \$150,000$, $r = 6\%$, $t = 4$, compounded monthly

4. $P = \$150,000$, $r = 4\%$, $t = 3$, compounded daily

In Exercises 5 and 6, find the effective rate corresponding to the given nominal rate.

5. **a.** 6%/year compounded semiannually
   b. 5%/year compounded quarterly

6. **a.** 4.5%/year compounded monthly
   b. 4.5%/year compounded daily

In Exercises 7 and 8, find the present value of $40,000 due in 4 years at the given rate of interest.

7. **a.** 5%/year compounded semiannually
   b. 5%/year compounded quarterly

8. **a.** 4%/year compounded monthly
   b. 6%/year compounded daily

9. Find the accumulated amount after 4 years if $5000 is invested at 5%/year compounded continuously.

10. An amount of $25,000 is deposited in a bank that pays interest at the rate of 4%/year, compounded annually. What is the total amount on deposit at the end of 6 years, assuming that there are no deposits or withdrawals during those 6 years? What is the interest earned in that period of time?

11. How much money should be deposited in a bank paying interest at the rate of 4%/year compounded daily (assume a 365-day year) so that at the end of 2 years, the accumulated amount will be $10,000?

12. **PRESENT VALUE OF AN INVESTMENT** Jada deposited an amount of money in a bank 3 years ago. If the bank had been paying interest at the rate of 5%/year compounded daily (assume a 365-day year) and she has $15,000 on deposit today, what was her initial deposit?

13. **PRESENT VALUE OF AN INVESTMENT** How much money should Jack deposit in a bank paying interest at the rate of 6%/year compounded continuously so that at the end of 3 years, the accumulated amount will be $20,000?

14. **PRESENT VALUE OF AN INVESTMENT** Diego deposited a certain sum of money in a bank 2 years ago. If the bank had been paying interest at the rate of 6% compounded continuously and he has $12,000 on deposit today, what was his initial deposit?

15. Find the present value of $59,673 due in 5 years at an interest rate of 6%/year compounded continuously.

16. Find the interest rate needed for an investment of $5000 to grow to an amount of $7500 in 3 years if interest is compounded quarterly.

17. Find the interest rate needed for an investment of $5000 to grow to an amount of $7500 in 3 years if interest is compounded monthly.

18. Find the interest rate needed for an investment of $5000 to grow to an amount of $8000 in 4 years if interest is compounded semiannually.

19. Find the interest rate needed for an investment of $5000 to grow to an amount of $5500 in 6 months if interest is compounded monthly.

20. Find the interest rate needed for an investment of $2000 to double in 5 years if interest is compounded annually.

21. Find the interest rate needed for an investment of $2000 to triple in 5 years if interest is compounded monthly.

22. How long will it take $12,000 to grow to $15,000 if the investment earns interest at the rate of 5%/year compounded monthly?

23. How long will it take $5000 to grow to $6500 if the investment earns interest at the rate of 6%/year compounded monthly?

24. How long will it take an investment of $2000 to double if the investment earns interest at the rate of 6%/year compounded monthly?

25. How long will it take an investment of $5000 to triple if the investment earns interest at the rate of 4%/year compounded daily?

26. Find the interest rate needed for an investment of $5000 to grow to an amount of $6000 in 3 years if interest is compounded continuously.

27. Find the interest rate needed for an investment of $4000 to double in 5 years if interest is compounded continuously.

28. How long will it take an investment of $6000 to grow to $7000 if the investment earns interest at the rate of $7\frac{1}{2}$%/year compounded continuously?

29. How long will it take an investment of $8000 to double if the investment earns interest at the rate of 5%/year compounded continuously?

30. **HOUSING PRICES** The Estradas are planning to buy a house 4 years from now. Housing experts in their area have estimated that the cost of a home will increase at a rate of 4%/year during that 4-year period. If this economic prediction holds true, how much can they expect to pay for a house that currently costs $180,000?

31. **ENERGY CONSUMPTION** A metropolitan utility company in a western city of the United States expects the consumption of electricity to increase by 8%/year during the next decade, owing mainly to an expected population increase. If consumption does increase at this rate, find the amount by which the utility company will have to increase its generating capacity to meet the area's needs at the end of the decade.

32. **PENSION FUNDS** The managers of a pension fund have invested $1.5 million in U.S. government certificates of deposit (CDs) that pay interest at the rate of 2.5%/year compounded semiannually over a period of 10 years. At the end of this period, how much will the investment be worth?

33. **COMPARING INVESTMENT RETURNS** The value of Maria's investments increased by 20% in the first year and by a further 10% in the second year. The value of Laura's investments grew 10% in the first year, followed by a gain of 20% in the second year. Both Maria and Laura started out with a $10,000 investment. Whose investment increased the most in the 2-year period? Explain.

34. **INVESTMENTS** The value of Alan's stock portfolio grew by 20% in the first year, followed by a growth of 10% in the second year. It dropped 10% and 20% in the third and fourth years, respectively. Is the value of Alan's stock portfolio after 4 years the same as that when he started out? Explain.

35. **INVESTMENTS** The value of Jack's investment portfolio fell by 20% in the first year but rebounded by 20% in the second year. Did Jack regain all of the money that he lost in the first year at the end of the second year? Explain.

36. **INVESTMENTS** The value of Arabella's stock portfolio dropped 20% in the first year. Find the annual rate of growth, compounded yearly, that she must achieve in the next 2 years to bring the value of her stock portfolio back to its initial value (its value at the beginning of the first year).

37. **MUTUAL FUNDS** Jodie invested $15,000 in a mutual fund 4 years ago. If the fund grew at the rate of 7.8%/year compounded monthly, what would Jodie's account be worth today?

38. **TRUST FUNDS** A young man is the beneficiary of a trust fund established for him 21 years ago at his birth. If the original amount placed in trust was $10,000, how much will he receive if the money has earned interest at the rate of 6%/year compounded annually? Compounded quarterly? Compounded monthly?

39. **RETIREMENT FUNDS** Five and a half years ago, Chris invested $10,000 in a retirement fund that grew at the rate of 6.82%/year compounded quarterly. What is his account worth today?

40. **TAX-DEFERRED ANNUITIES** Kate is in the 28% tax bracket and has $25,000 available for investment during her current tax year. Assume that she remains in the same tax bracket over the next 10 years, and determine the accumulated amount of her investment after taxes if she puts the $25,000 into a
   a. Tax-deferred annuity that pays 6%/year, tax deferred for 10 years.
   b. Taxable instrument that pays 6%/year for 10 years. Hint: In this case the yield after taxes is 4.32%/year.

41. **REVENUE GROWTH OF A HOME THEATER BUSINESS** Maxwell started a home theater business in 2013. The revenue of his company for that year was $240,000. The revenue grew by 20% in 2014 and 30% in 2015. Maxwell projects that the revenue growth for his company in the next 3 years will be at least 25%/year. How much does Maxwell expect his minimum revenue to be for 2018?

42. **ONLINE RETAIL SALES** Online retail sales stood at $141.4 billion for 2004. For the next 2 years, they grew by 24.3% and 14% per year, respectively. For the next 3 years, online retail sales grew at the rate of approximately 30.5%, 17.6%, and 10.5% per year, respectively. What were the online sales for 2009?
   Source: Jupiter Research.

43. **PURCHASING POWER** The year-end inflation rates in the U.S. economy in 2009–2012 were 2.7%, 1.5%, 3.0%, and

1.7%, respectively. What was the purchasing power of a dollar at the beginning of 2013 compared to that at the beginning of 2009?

Source: U.S. Census Bureau.

44. SAVING FOR COLLEGE Having received a large inheritance, a child's parents wish to establish a trust for the child's college education. If they need an estimated $120,000 7 years from now, how much should they set aside in trust now if they invest the money at 4.6% compounded quarterly? Continuously?

45. EFFECT OF INFLATION ON SALARIES Omar's current annual salary is $65,000. How much will he need to earn 10 years from now to retain his present purchasing power if the rate of inflation over that period is 3%/year? Assume that inflation is continuously compounded.

46. SAVINGS ACCOUNTS Bernie invested a sum of money 5 years ago in a savings account, which has since paid interest at the rate of 3%/year compounded quarterly. His investment is now worth $22,289.22. How much did he originally invest?

47. LOAN CONSOLIDATION The proprietors of the Coachmen Inn secured two loans from the Union Bank: one for $8000 due in 3 years and one for $15,000 due in 6 years, both at an interest rate of 8%/year compounded semiannually. The bank agreed to allow the two loans to be consolidated into one loan payable in 5 years at the same interest rate. How much will the proprietors have to pay the bank at the end of 5 years?

48. PENSIONS Eleni, who is now 50 years old, is employed by a firm that guarantees her a pension of $40,000/year at age 65. What is the present value of her first year's pension if inflation over the next 15 years is (a) 3%? (b) 4%? (c) 6%? Assume that inflation is continuously compounded.

49. REAL ESTATE INVESTMENTS An investor purchased a piece of waterfront property. Because of the development of a marina in the vicinity, the market value of the property is expected to increase according to the rule

$$V(t) = 80,000e^{\sqrt{t}/2}$$

where $V(t)$ is measured in dollars and t is the time in years from the present. If the rate of appreciation is expected to be 9% compounded continuously for the next 8 years, find an expression for the present value $P(t)$ of the property's market price valid for the next 8 years. What is $P(t)$ expected to be in 4 years?

50. REAL ESTATE INVESTMENTS Tower Investments owns a shopping mall located just outside the city. The market value of the mall is

$$V(t) = 500,000e^{0.5t^{0.4}}$$

dollars, where t is measured in years from the present. The rate of appreciation of the mall is expected to be 8%/year compounded continuously for the next 8 years.

a. Find an expression for the present value $P(t)$ of the market price of the mall valid for the next 8 years.

b. Compute $P(4)$, $P(5)$, and $P(6)$, and interpret your results.

51. INVESTMENT OPTIONS Investment A offers an 8% return compounded semiannually, and Investment B offers a 7.75% return compounded continuously. Which investment has a higher rate of return over a 4-year period?

52. INVESTMENT RETURNS Zoe purchased a house in 2008 for $300,000. In 2014, she sold the house and made a net profit of $66,000. Find the effective annual rate of return on her investment over the 6-year period.

53. INVESTMENT RETURNS Julio purchased 1000 shares of a certain stock for $25,250 (including commissions). He sold the shares 2 years later and received $32,100 after deducting commissions. Find the effective annual rate of return on his investment over the 2-year period.

54. REAL ESTATE INVESTMENTS A condominium complex was purchased by a group of private investors for $1.4 million and sold 6 years later for $3.6 million. Find the annual rate of return (compounded continuously) on their investment.

55. Show that the effective interest rate $\hat{r}_{eff}$ that corresponds to a nominal interest rate r per year compounded continuously is given by

$$\hat{r}_{eff} = e^r - 1$$

Hint: From Formula (7), we see that the effective rate $\hat{r}_{eff}$ corresponding to a nominal interest rate r per year compounded m times a year is given by

$$\hat{r}_{eff} = \left(1 + \frac{r}{m}\right)^m - 1$$

Let m tend to infinity in this expression.

56. Refer to Exercise 55. Find the effective interest rate that corresponds to a nominal rate of 10%/year compounded (a) quarterly, (b) monthly, and (c) continuously.

57. INVESTMENT ANALYSIS Refer to Exercise 55. Bank A pays interest on deposits at a 7% annual rate compounded quarterly, and Bank B pays interest on deposits at a $7\frac{1}{8}\%$ annual rate compounded continuously. Which bank has the higher effective rate of interest?

58. INVESTMENT ANALYSIS Find the nominal interest rate that, when compounded monthly, yields an effective interest rate of 6%/year.
Hint: Use Equation (7).

59. EFFECTIVE RATE OF INTEREST Suppose an initial investment of $P grows to an accumulated amount of $A in t years. Show that the effective rate (annual effective yield) is

$$r_{eff} = \left(\frac{A}{P}\right)^{1/t} - 1$$

Use the formula given in Exercise 59 to solve Exercises 60–61.

60. EFFECTIVE RATE OF INTEREST Martha invested \$40,000 in a boutique 5 years ago. Her investment is worth \$60,000 today. What is the effective rate (annual effective yield) of her investment?

61. MONEY MARKET MUTUAL FUNDS Carlos invested \$5000 in a money market mutual fund that pays interest on a daily basis. The balance in his account at the end of 8 months (245 days) was \$5070.42. Find the effective rate at which Carlos's account earned interest over this period (assume a 365-day year).

62. ANNUITIES An annuity is a sequence of payments made at regular time intervals. The future value of an annuity of n payments of R dollars each paid at the end of each investment period into an account that earns an interest rate of i/period is

$$S = R \left[\frac{(1 + i)^n - 1}{i} \right]$$

Determine

$$\lim_{i \to 0} R \left[\frac{(1 + i)^n - 1}{i} \right]$$

and interpret your result.

Hint: Use the definition of the derivative.

In Exercises 63 and 64, determine whether the statement is true or false. If it is true, explain why it is true. If it is false, give an example to show why it is false.

63. If interest is compounded annually, then the effective rate is the same as the nominal rate.

64. Susan's salary increased from \$50,000/year to \$60,000/year over a 4-year period. Therefore, Susan received annual increases of 5% over that period.

5.3 Solutions to Self-Check Exercises

1. Using Formula (8) with $A = 20,000$, $r = 0.06$, $m = 12$, and $t = 3$, we find the required present value to be

$$P = 20,000 \left(1 + \frac{0.06}{12} \right)^{-(12)(3)}$$

$$\approx 16,712.90$$

or \$16,712.90.

2. The accumulated amount of Glen's current investment is found by using Formula (6) with $P = 100,000$, $r = 0.046$, and $m = 365$. Thus, the required accumulated amount is

$$A = 100,000 \left(1 + \frac{0.046}{365} \right)^{365}$$

$$\approx 104,707.14$$

or \$104,707.14. Next, we compute the accumulated amount of Glen's reinvestment. Once again, using Formula (6) with $P = 100,000$, $r = 0.032$, and $m = 365$, we find the required accumulated amount in this case to be

$$\bar{A} = 100,000 \left(1 + \frac{0.032}{365} \right)^{365}$$

or approximately \$103,251.61. Therefore, Glen can expect to experience a net decrease in yearly income of

$$104,707.14 - 103,251.61$$

or \$1455.53.

3. a. Using Formula (10) with $P = 10,000$, $r = 0.05$, and $t = 5$, we find that the required accumulated amount is given by

$$A = 10,000 e^{(0.05)(5)}$$

$$\approx 12,840.25$$

or \$12,840.25.

b. Using Formula (11) with $A = 10,000$, $r = 0.05$, and $t = 5$, we see that the required present value is given by

$$P = 10,000 e^{-(0.05)(5)}$$

$$\approx 7788.01$$

or \$7788.01.

USING TECHNOLOGY   Finding the Accumulated Amount of an Investment, the Effective Rate of Interest, and the Present Value of an Investment

Graphing Utility

Some graphing utilities have built-in routines for solving problems involving the mathematics of finance. For example, the TI-83/84 Solver function incorporates several functions that can be used to solve the problems that are encountered in this section. To access the **TVM SOLVER** on the TI-83 press 2ND , press FINANCE , and then select 1: TVM Solver . To access the **TVM SOLVER** on the TI-83 Plus and the TI-84, press APPS , press 1: Finance , and then select 1: TVM Solver .

EXAMPLE 1 Finding the Accumulated Amount of an Investment Find the accumulated amount after 10 years if $5000 is invested at a rate of 10% per year compounded monthly.

Solution Using the TI-83/84 **TVM SOLVER** with the following inputs,

$$N = 120 \qquad \text{(10)(12)}$$
$$I\% = 10$$
$$PV = -5000 \qquad \text{We use a minus sign because an investment is an outflow.}$$
$$PMT = 0$$
$$FV = 0$$
$$P/Y = 12 \qquad \text{The number of payments each year}$$
$$C/Y = 12 \qquad \text{The number of conversion periods each year}$$
$$\text{PMT:} \boxed{\text{END}} \text{ BEGIN}$$

Move the cursor up to the FV line and press ALPHA SOLVE . We obtain the display shown in Figure T1. We conclude that the required accumulated amount is $13,535.21.

```
N=120
I%=10
PV=-5000
PMT=0
■ FV=13535.20745
P/Y=12
C/Y=12
PMT:END  BEGIN
```

FIGURE T1
The TI-83/84 screen showing the future value (FV) of an investment

EXAMPLE 2 Finding the Effective Rate of Interest Find the effective rate of interest corresponding to a nominal rate of 10% per year compounded quarterly.

Solution Here we use the **Eff** function of the TI-83/84 calculator to obtain the result shown in Figure T2. The required effective rate is approximately 10.38% per year.

```
►Eff(10, 4)
       10.38128906
```

FIGURE T2
The TI-83/84 screen showing the effective rate of interest (Eff)

EXAMPLE 3 Finding the Present Value of an Investment Find the present value of $20,000 due in 5 years if the interest rate is 7.5% per year compounded daily.

Solution Using the TI-83/84 **TVM SOLVER** with the following inputs,

$$N = 1825 \qquad \text{(5)(365)}$$
$$I\% = 7.5$$
$$PV = 0$$
$$PMT = 0$$
$$FV = 20000$$
$$P/Y = 365 \qquad \text{The number of payments each year}$$
$$C/Y = 365 \qquad \text{The number of conversions each year}$$
$$\text{PMT:} \boxed{\text{END}} \text{ BEGIN}$$

```
N=1825
I%=7.5
■ PV=-13746.3151
PMT=0
FV=20000
P/Y=365
C/Y=365
PMT:END  BEGIN
```

FIGURE T3
The TI-83/84 screen showing the present value (PV) of an investment

By moving the cursor up to the PV line and pressing ALPHA SOLVE , we obtain the display shown in Figure T3. We see that the required present value is approximately $13,746.32. Note that PV is negative because an investment is an outflow (money is paid out).

Note: Boldfaced words/characters enclosed in a box (for example, Enter) indicate that an action (click, select, or press) is required. Words/characters printed blue (for example **Chart sub-type:**) indicate words/characters appearing on the screen.

1. Find the accumulated amount A if $5000 is invested at the interest rate of $5\frac{3}{8}\%$/year compounded monthly for 3 years.

2. Find the accumulated amount A if $327.35 is invested at the interest rate of $5\frac{1}{3}\%$/year compounded daily for 7 years.

3. Find the effective rate corresponding to $8\frac{2}{3}\%$/year compounded quarterly.

4. Find the effective rate corresponding to $10\frac{5}{8}\%$/year compounded monthly.

5. Find the present value of $38,000 due in 3 years at $8\frac{1}{4}\%$/year compounded quarterly.

6. Find the present value of $150,000 due in 5 years at $9\frac{3}{8}\%$/year compounded monthly.

5.4 Differentiation of Exponential Functions

The Derivative of the Exponential Function

To study the effects of budget deficit-reduction plans at different income levels, it is important to know the income distribution of American households. Based on data from the U.S. Census Bureau, the graph of f shown in Figure 10 gives the percentage of American households y as a function of their annual income x (in thousands of dollars) in 2010.

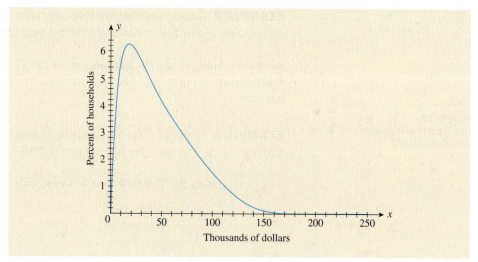

FIGURE **10**
The graph of f shows the percentage of households versus their annual income.
Source: U.S. Census Bureau.

Observe that the graph of f rises very quickly and then tapers off. From the graph of f, you can see that the bulk of American households earned less than $150,000 per year. (We will refer to this model again in Using Technology at the end of this section.)

To analyze mathematical models involving exponential and logarithmic functions in greater detail, we need to develop rules for computing the derivative of these functions. We begin by looking at the rule for computing the derivative of the exponential function.

> **Rule 1: Derivative of the Exponential Function**
>
> $$\frac{d}{dx}(e^x) = e^x$$

Thus, the derivative of the exponential function with base e is equal to the function itself. To demonstrate the validity of this rule, we compute

$$f'(x) = \lim_{h \to 0} \frac{f(x + h) - f(x)}{h}$$

$$= \lim_{h \to 0} \frac{e^{x+h} - e^x}{h}$$

$$= \lim_{h \to 0} \frac{e^x(e^h - 1)}{h} \qquad \text{Write } e^{x+h} = e^x e^h \text{ and factor.}$$

$$= e^x \lim_{h \to 0} \frac{e^h - 1}{h} \qquad \text{Why?}$$

To evaluate

$$\lim_{h \to 0} \frac{e^h - 1}{h}$$

let's refer to Table 4, which is constructed with the aid of a calculator. From the table, we see that

$$\lim_{h \to 0} \frac{e^h - 1}{h} = 1$$

(Although a rigorous proof of this fact is possible, it is beyond the scope of this book. Also see Example 1, Using Technology, page 380.) Using this result, we conclude that

$$f'(x) = e^x \cdot 1 = e^x$$

as we set out to show.

TABLE 4

h	$\dfrac{e^h - 1}{h}$
0.1	1.0517
0.01	1.0050
0.001	1.0005
−0.1	0.9516
−0.01	0.9950
−0.001	0.9995

EXAMPLE 1 Find the derivative of each of the following functions:

a. $f(x) = x^2 e^x$ **b.** $g(t) = (e^t + 2)^{3/2}$

Solution

a. The Product Rule gives

$$f'(x) = \frac{d}{dx}(x^2 e^x) = x^2 \frac{d}{dx}(e^x) + e^x \frac{d}{dx}(x^2)$$

$$= x^2 e^x + e^x(2x)$$

$$= xe^x(x + 2) \qquad \qquad (x^2) \text{ See page 9.}$$

b. Using the General Power Rule, we find

$$g'(t) = \frac{3}{2}(e^t + 2)^{1/2} \frac{d}{dt}(e^t + 2)$$

$$= \frac{3}{2}(e^t + 2)^{1/2} e^t = \frac{3}{2} e^t (e^t + 2)^{1/2}$$

Exploring with TECHNOLOGY

Consider the exponential function $f(x) = b^x (b > 0, b \neq 1)$.

1. Use the definition of the derivative of a function to show that

$$f'(x) = b^x \cdot \lim_{h \to 0} \frac{b^h - 1}{h}$$

2. Use the result of part 1 to show that

$$\frac{d}{dx}(2^x) = 2^x \cdot \lim_{h \to 0} \frac{2^h - 1}{h}$$

$$\frac{d}{dx}(3^x) = 3^x \cdot \lim_{h \to 0} \frac{3^h - 1}{h}$$

3. Use the technique in Using Technology, page 380, to show that (to two decimal places)

$$\lim_{h \to 0} \frac{2^h - 1}{h} = 0.69$$

and

$$\lim_{h \to 0} \frac{3^h - 1}{h} = 1.10$$

4. Conclude from the results of parts 2 and 3 that

$$\frac{d}{dx}(2^x) \approx (0.69)2^x$$

and

$$\frac{d}{dx}(3^x) \approx (1.10)3^x$$

Thus,

$$\frac{d}{dx}(b^x) = k \cdot b^x$$

where k is an appropriate constant.

5. The results of part 4 suggest that, for convenience, we pick the base b, where $2 < b < 3$, so that $k = 1$. This value of b is $e \approx 2.718281828. \ldots$ Thus,

$$\frac{d}{dx}(e^x) = e^x$$

This is why we prefer to work with the exponential function $f(x) = e^x$.

Applying the Chain Rule to Exponential Functions

To enlarge the class of exponential functions to be differentiated, we appeal to the Chain Rule to obtain the following rule for differentiating composite functions of the form $h(x) = e^{f(x)}$. An example of such a function is $h(x) = e^{x^2 - 2x}$. Here, $f(x) = x^2 - 2x$.

Rule 2: The Chain Rule for Exponential Functions

If $f(x)$ is a differentiable function, then

$$\frac{d}{dx}(e^{f(x)}) = e^{f(x)}f'(x)$$

To see this, observe that if $h(x) = g[f(x)]$, where $g(x) = e^x$, then by virtue of the Chain Rule,

$$h'(x) = g'[f(x)]f'(x) = e^{f(x)}f'(x)$$

since $g'(x) = e^x$.

As an aid to remembering the Chain Rule for Exponential Functions, observe that it has the following form:

$$\frac{d}{dx}\left(e^{f(x)}\right) = e^{f(x)} \cdot \text{derivative of the exponent}$$
$$\underset{\text{Same}}{\underbrace{\qquad\qquad}}$$

EXAMPLE 2 Find the derivative of each of the following functions:

a. $f(x) = e^{2x}$ **b.** $y = e^{-3x}$ **c.** $g(t) = e^{2t^2+t}$

Solution

a. $f'(x) = e^{2x}\dfrac{d}{dx}(2x) = e^{2x} \cdot 2 = 2e^{2x}$

b. $\dfrac{dy}{dx} = e^{-3x}\dfrac{d}{dx}(-3x) = -3e^{-3x}$

c. $g'(t) = e^{2t^2+t} \cdot \dfrac{d}{dt}(2t^2 + t) = (4t + 1)e^{2t^2+t}$

EXAMPLE 3 Differentiate the function $y = xe^{-2x}$.

Solution Using the Product Rule, followed by the Chain Rule, we find

$$\frac{dy}{dx} = x\frac{d}{dx}\left(e^{-2x}\right) + e^{-2x}\frac{d}{dx}(x)$$

$$= xe^{-2x}\frac{d}{dx}(-2x) + e^{-2x} \qquad \textcolor{red}{\text{Use the Chain Rule on the first term.}}$$

$$= -2xe^{-2x} + e^{-2x}$$

$$= e^{-2x}(1 - 2x)$$

EXAMPLE 4 Differentiate the function $g(t) = \dfrac{e^t}{e^t + e^{-t}}$.

Solution Using the Quotient Rule, followed by the Chain Rule, we find

$$g'(t) = \frac{(e^t + e^{-t})\dfrac{d}{dt}(e^t) - e^t\dfrac{d}{dt}(e^t + e^{-t})}{(e^t + e^{-t})^2}$$

$$= \frac{(e^t + e^{-t})e^t - e^t(e^t - e^{-t})}{(e^t + e^{-t})^2} \qquad \textcolor{red}{(x^2)} \text{ See page 9.}$$

$$= \frac{e^{2t} + 1 - e^{2t} + 1}{(e^t + e^{-t})^2} \qquad \textcolor{red}{e^0 = 1}$$

$$= \frac{2}{(e^t + e^{-t})^2}$$

EXAMPLE 5 In Section 5.6, we will discuss some practical applications of the exponential function

$$Q(t) = Q_0 e^{kt}$$

where Q_0 and k are positive constants and $t \in [0, \infty)$. A quantity $Q(t)$ growing according to this law experiences *exponential growth*. Show that for a quantity $Q(t)$ experiencing exponential growth, the rate of growth of the quantity, $Q'(t)$, at any time t is directly proportional to the amount of the quantity present.

Solution Using the Chain Rule for Exponential Functions, we compute the derivative Q' of the function Q. Thus,

$$Q'(t) = Q_0 e^{kt} \frac{d}{dt}(kt)$$
$$= Q_0 e^{kt}(k)$$
$$= kQ_0 e^{kt}$$
$$= kQ(t) \qquad \textcolor{red}{Q(t) = Q_0 e^{kt}}$$

which is the desired conclusion.

EXAMPLE 6 Find the inflection points of the function $f(x) = e^{-x^2}$.

Solution The first derivative of f is

$$f'(x) = -2xe^{-x^2}$$

Differentiating $f'(x)$ with respect to x yields

$$f''(x) = (-2x)(-2xe^{-x^2}) - 2e^{-x^2}$$
$$= 2e^{-x^2}(2x^2 - 1)$$

Setting $f''(x) = 0$ gives

$$2e^{-x^2}(2x^2 - 1) = 0$$

Since e^{-x^2} never equals zero for any real value of x, we see that $x = \pm 1/\sqrt{2}$ are the only candidates for inflection points of f. The sign diagram of f'', shown in Figure 11, tells us that both $x = -1/\sqrt{2}$ and $x = 1/\sqrt{2}$ give rise to inflection points of f. Next,

$$f\left(-\frac{1}{\sqrt{2}}\right) = f\left(\frac{1}{\sqrt{2}}\right) = e^{-1/2}$$

and the inflection points of f are $(-1/\sqrt{2}, e^{-1/2})$ and $(1/\sqrt{2}, e^{-1/2})$. The graph of f appears in Figure 12.

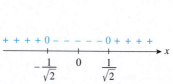

FIGURE 11
Sign diagram for f''

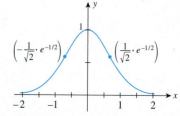

FIGURE 12
The graph of $y = e^{-x^2}$ has two inflection points.

Our final example involves finding the absolute maximum of an exponential function.

APPLIED EXAMPLE 7 Optimal Market Price Refer to Example 8, Section 5.3. The present value of the market price of the Blakely Office Building is given by

$$P(t) = 300{,}000e^{-0.09t + \sqrt{t}/2} \qquad (0 \le t \le 10)$$

Find the optimal present value of the building's market price.

Solution To find the maximum value of P over $[0, 10]$, we compute

$$P'(t) = 300{,}000e^{-0.09t + \sqrt{t}/2} \frac{d}{dt}\left(-0.09t + \frac{1}{2}t^{1/2}\right)$$

$$= 300{,}000e^{-0.09t + \sqrt{t}/2}\left(-0.09 + \frac{1}{4}t^{-1/2}\right)$$

Setting $P'(t) = 0$ gives

$$-0.09 + \frac{1}{4t^{1/2}} = 0$$

since $e^{-0.09t + \sqrt{t}/2}$ is never zero for any value of t. Solving this equation, we find

$$\frac{1}{4t^{1/2}} = 0.09$$

$$t^{1/2} = \frac{1}{4(0.09)}$$

$$= \frac{1}{0.36}$$

$$t = \left(\frac{1}{0.36}\right)^2 \approx 7.72$$

the sole critical number of the function P. Finally, evaluating $P(t)$ at the critical number as well as at the endpoints of $[0, 10]$, we have

t	0	7.72	10
$P(t)$	300,000	600,779	592,838

We conclude, accordingly, that the optimal present value of the property's market price is $600,779 and that this will occur 7.72 years from now.

5.4 Self-Check Exercises

1. Let $f(x) = xe^{-x}$.
   a. Find the first and second derivatives of f.
   b. Find the relative extrema of f.
   c. Find the inflection points of f.

2. **DEPRECIATION OF AN INDUSTRIAL ASSET** An industrial asset is being depreciated at a rate such that its book value t years from now will be

$$V(t) = 50{,}000e^{-0.4t}$$

dollars. How fast will the book value of the asset be changing 3 years from now?

Solutions to Self-Check Exercises 5.4 can be found on page 380.

5.4 Concept Questions

1. State the rule for differentiating (a) $f(x) = e^x$ and (b) $g(x) = e^{f(x)}$, where f is a differentiable function.

2. Let $f(x) = e^{kx}$.
   a. Compute $f'(x)$.
   b. Use the result to deduce the behavior of f for the case $k > 0$ and the case $k < 0$.

5.4 Exercises

In Exercises 1–28, find the derivative of the function.

1. $f(x) = e^{3x}$

2. $f(x) = 3e^x$

3. $g(t) = e^{-t}$

4. $f(x) = e^{-2x}$

5. $f(x) = e^x + x^2$

6. $f(x) = 2e^x - x^2$

7. $f(x) = x^3 e^x$

8. $f(u) = u^2 e^{-u}$

9. $f(x) = \dfrac{e^x}{x}$

10. $f(x) = \dfrac{x}{e^x}$

11. $f(x) = 3(e^x + e^{-x})$

12. $f(x) = \dfrac{e^x + e^{-x}}{2}$

13. $f(w) = \dfrac{e^w + 2}{e^w}$

14. $f(x) = \dfrac{e^x}{e^x + 1}$

15. $f(x) = 2e^{3x-1}$

16. $f(t) = 4e^{3t+2}$

17. $h(x) = e^{-x^2}$

18. $f(x) = e^{x^2-1}$

19. $f(x) = 3e^{-1/x}$

20. $f(x) = e^{1/(2x)}$

21. $f(x) = (e^x + 1)^{25}$

22. $f(x) = (4 - e^{-3x})^3$

23. $f(x) = e^{\sqrt{x}}$

24. $f(t) = -e^{-\sqrt{2t}}$

25. $f(x) = (x - 1)e^{3x+2}$

26. $f(s) = (s^2 + 1)e^{-s^2}$

27. $f(x) = \dfrac{e^x - 1}{e^x + 1}$

28. $g(t) = \dfrac{e^{-t}}{1 + t^2}$

In Exercises 29–32, find the second derivative of the function.

29. $f(x) = e^{-4x} + e^{3x}$

30. $f(t) = 3e^{-2t} - 5e^{-t}$

31. $f(x) = 2xe^{3x}$

32. $f(t) = t^2 e^{-2t}$

33. Find an equation of the tangent line to the graph of $y = e^{2x-3}$ at the point $(\frac{3}{2}, 1)$.

34. Find an equation of the tangent line to the graph of $y = e^{-x^2}$ at the point $(1, 1/e)$.

35. Determine the intervals where the function $f(x) = e^{-x^2/2}$ is increasing and where it is decreasing.

36. Determine the intervals where the function $f(x) = x^2 e^{-x}$ is increasing and where it is decreasing.

37. Determine the intervals of concavity for the graph of the function $f(x) = \dfrac{e^x - e^{-x}}{2}$.

38. Determine the intervals of concavity for the graph of the function $f(x) = xe^x$.

39. Find the inflection point of the function $f(x) = xe^{-2x}$.

40. Find the inflection point(s) of the function $f(x) = 2e^{-x^2}$.

41. Find the equations of the tangent lines to the graph of $f(x) = e^{-x^2}$ at its inflection points.

42. Find an equation of the tangent line to the graph of $f(x) = xe^{-x}$ at its inflection point.

In Exercises 43–46, find the absolute extrema of the function.

43. $f(x) = e^{-x^2}$ on $[-1, 1]$

44. $h(x) = e^{x^2-4}$ on $[-2, 2]$

45. $g(x) = (2x - 1)e^{-x}$ on $[0, \infty)$

46. $f(x) = xe^{-x^2}$ on $[0, 2]$

In Exercises 47–50, use the curve-sketching guidelines of Chapter 4, page 295, to sketch the graph of the function.

47. $f(t) = e^t - t$

48. $h(x) = \dfrac{e^x + e^{-x}}{2}$

49. $f(x) = 2 - e^{-x}$

50. $f(x) = \dfrac{3}{1 + e^{-x}}$

In Exercises 51 and 52, find dy/dx by implicit differentiation.

51. $x^2 + y^3 = 2e^{2y}$

52. $xy^2 + \sqrt{x}e^y = 10$

In Exercises 53 and 54, find d^2y/dx^2 by implicit differentiation.

53. $x = y + e^y$

54. $e^x - e^y = y - x$

55. Find dy/dx at the point $(0, 1)$ using implicit differentiation given that $xy + e^y = e$.

56. Find an equation of the tangent line to the graph of the equation $x + y - e^{x-y} = 1$ at the point $(1, 1)$.

57. ONLINE VIDEO VIEWERS As broadband Internet grows more popular, video services such as YouTube will continue to expand. The number of online video viewers (in millions) is projected to grow from 2008 through 2013 according to the rule

$$N(t) = 135e^{0.067t} \qquad (1 \le t \le 6)$$

where $t = 1$ corresponds to 2008.

a. How many online video viewers were there in 2012?
b. How fast was the number of online video viewers changing in 2012?

Source: eMarketer.com.

58. OUTPUT PER WORKER In leading advanced manufacturing countries, output per worker has grown impressively as factories have become more automated. The index for the United States (100 in 2002) from 1990 through 2011 is approximately

$$I(t) = 50e^{0.05t} \qquad (0 \le t \le 21)$$

in year t, where $t = 0$ corresponds to 1990.
a. What was the index in 1990? In 2011?
b. Sketch the graph of I.
c. How fast was the index changing in the year 2000?

Source: MIT Technology Review.

59. PHARMACEUTICAL THEFT Pharmaceutical theft has been rising rapidly in recent years. Experts believe that pharmaceuticals are the new "street gold." The value of stolen drugs (in millions of dollars per year) from 2007 through 2009 is approximated by the function

$$f(t) = 20.5e^{0.74t} \qquad (1 \le t \le 3)$$

where t is the number of years since 2007. Find the rate of change of the value of stolen drugs at $t = 2$, and interpret your result.

Source: New York Times.

60. WORLD POPULATION GROWTH After its fastest rate of growth ever during the 1980s and 1990s, the rate of growth of world population is expected to slow dramatically in the twenty-first century. The function

$$G(t) = 1.58e^{-0.213t}$$

gives the projected annual average percent population growth per decade in the tth decade, with $t = 1$ corresponding to 2000.
a. What will the projected annual average population growth rate be in 2020 ($t = 3$)?
b. How fast will the projected annual average population growth rate be changing in 2020?

Source: U.S. Census Bureau.

61. SALES PROMOTION The Lady Bug, a women's clothing chain store, found that t days after the end of a sales promotion the volume of sales was given by

$$S(t) = 20{,}000(1 + e^{-0.5t}) \qquad (0 \le t \le 5)$$

dollars.
a. Find the rate of change of The Lady Bug's sales volume when $t = 1$, $t = 2$, $t = 3$, and $t = 4$.
b. In how many days will the sales volume drop below $27,400?

62. BLOOD ALCOHOL LEVEL The percentage of alcohol in a person's bloodstream t hr after drinking 8 fluid oz of whiskey is given by

$$A(t) = 0.23te^{-0.4t} \qquad (0 \le t \le 12)$$

a. What is the percentage of alcohol in a person's bloodstream after $\frac{1}{2}$ hr? After 8 hr?
b. How fast is the percentage of alcohol in a person's bloodstream changing after $\frac{1}{2}$ hr? After 8 hr?

Source: Encyclopedia Britannica.

63. POLIO IMMUNIZATION Polio, a once-feared killer, declined markedly in the United States in the 1950s after Jonas Salk developed the inactivated polio vaccine and mass immunization of children took place. The number of polio cases in the United States from the beginning of 1959 to the beginning of 1963 is approximated by the function

$$N(t) = 5.3e^{0.095t^2 - 0.85t} \qquad (0 \le t \le 4)$$

where $N(t)$ gives the number of polio cases (in thousands) and t is measured in years, with $t = 0$ corresponding to the beginning of 1959.
a. Show that the function N is decreasing over the time interval under consideration.
b. How fast was the number of polio cases decreasing at the beginning of 1959? At the beginning of 1962? (*Comment:* Since the introduction of the oral vaccine developed by Dr. Albert B. Sabin in 1963, polio in the United States has, for all practical purposes, been eliminated.)

Source: Centers for Disease Control and Prevention.

64. AUTISTIC BRAIN At birth, the autistic brain is similar in size to a healthy child's brain. Between birth and 2 years, it grows to be abnormally large, reaching its maximum size between 3 and 6 years of age. The percentage difference in size between the autistic brain and the normal brain to age 40 is approximated by

$$D(t) = 6.9te^{-0.24t} \qquad (0 \le t \le 40)$$

where t is measured in years.
a. At what ages is the difference in the size between the autistic brain and the normal brain increasing? Decreasing?
b. At what age is the difference in the size between the autistic brain and the normal brain the greatest? What is the maximum difference?
c. At what age is the difference in the size between the autistic brain and the normal brain decreasing at the fastest rate?
d. Sketch the graph of D on the interval $[0, 40]$.

Source: Newsweek.

65. ALZHEIMER'S DISEASE Alzheimer's disease can occur at any age, even as young as 40 years old, but its occurrence is much more common as the years go by. The frequency of occurrence of the disease (as a percentage) is given by

$$f(t) = 0.71e^{0.7t} \qquad (1 \le t \le 5)$$

where t is measured in 5-year intervals, with $t = 1$ corresponding to an age of 70 years.
a. What is the frequency of occurrence of Alzheimer's disease for 70-year-old people? For 90-year-old people?

b. Show that f is increasing on the interval $(1, 5)$. Interpret your results.

c. Show that f is concave upward on the interval $(1, 5)$. Interpret your results.

Source: World Health Organization.

66. **AGING POPULATION** According to information obtained from the U.S. Census, the population of the United States aged 75 years and above is estimated to be

$$f(t) = \frac{72.15}{1 + 2.7975e^{-0.02145t}} \qquad (0 \le t \le 80)$$

in year t, where $t = 0$ corresponds to 2010 and $f(t)$ is measured in millions.

a. What was the population aged 75 years and over in 2010?

b. How fast is the population aged 75 years and over expected to grow in 2030?

c. What is the population aged 75 years and over expected to be in 2030?

Source: U.S. Census Bureau.

67. **MARGINAL REVENUE** The relationship between the unit selling price p (in dollars) and the quantity demanded x (in pairs) of a certain brand of women's gloves are given by the demand equation

$$p = 100e^{-0.0001x} \qquad (0 \le x \le 20{,}000)$$

a. Find the revenue function R.
    Hint: $R(x) = px$

b. Find the marginal revenue function R'.

c. What is the marginal revenue when $x = 10{,}000$?

68. **DEATH DUE TO STROKES** Before 1950, little was known about strokes. By 1960, however, risk factors such as hypertension were identified. In recent years, CAT scans used as a diagnostic tool have helped to prevent strokes. As a result, the number of deaths due to strokes has fallen dramatically. The function

$$N(t) = 130.7e^{-0.1155t^2} + 50 \qquad (0 \le t \le 6)$$

gives the number of deaths due to stroke per 100,000 people from 1950 through 2010, where t is measured in decades, with $t = 0$ corresponding to 1950.

a. How many deaths due to strokes per 100,000 people were there in 1950?

b. How fast was the number of deaths due to strokes per 100,000 people changing in 1950? In 1960? In 1970? In 1980?

c. When was the rate of decline in the number of deaths due to strokes per 100,000 people greatest?

d. How many deaths due to strokes per 100,000 people were there in 2010?

Source: American Heart Association, Centers for Disease Control, and National Institutes of Health.

69. **PRICE OF PERFUME** The monthly demand for a certain brand of perfume is given by the demand equation

$$p = 100e^{-0.0002x} + 150$$

where p denotes the retail unit price (in dollars) and x denotes the quantity (in 1-oz bottles) demanded.

a. Find the rate of change of the price per bottle when $x = 1000$ and when $x = 2000$.

b. What is the price per bottle when $x = 1000$? When $x = 2000$?

70. **PRICE OF WINE** The monthly demand for a certain brand of table wine is given by the demand equation

$$p = 240\left(1 - \frac{3}{3 + e^{-0.0005x}}\right)$$

where p denotes the wholesale price per case (in dollars) and x denotes the number of cases demanded.

a. Find the rate of change of the price per case when $x = 1000$.

b. What is the price per case when $x = 1000$?

71. **SPREAD OF AN EPIDEMIC** During a flu epidemic, the total number of students on a state university campus who had contracted influenza by the xth day was given by

$$N(x) = \frac{3000}{1 + 99e^{-x}} \qquad (x \ge 0)$$

a. How many students had influenza initially?

b. Derive an expression for the rate at which the disease was being spread, and prove that the function N is increasing on the interval $(0, \infty)$.

c. Sketch the graph of N. What was the total number of students who contracted influenza during that particular epidemic?

72. **SALES OF A PRODUCT** The total number of units of a new product sold t months after its introduction is given by

$$N(t) = 20{,}000(1 - e^{-0.05t})^2 \qquad (0 \le t \le 36)$$

a. How many units of the product were sold 24 months after its introduction?

b. How fast was the product selling 24 months after its introduction?

73. **ELASTICITY OF DEMAND** The quantity demanded each month x (in units of a hundred) of the Soundex model A alarm clock radio/CD player is related to the unit price p (in dollars) by the demand equation

$$x = 50e^{-0.02p} \qquad (p > 0)$$

a. Find the elasticity of demand for the model A players.

b. Find the values of p for which the demand is inelastic, unitary, or elastic.

c. If the unit price of the player is decreased slightly from \$40, will the revenue increase or decrease?

d. If the unit price of the player is increased slightly from \$60, will the revenue increase or decrease?
    Hint: Refer to Section 3.4.

74. **ELASTICITY OF DEMAND** Suppose that the demand equation for a certain commodity has the form $x = ae^{-bp}$, where a and b are positive constants.

a. Find the elasticity of demand $E(p)$.

b. Find the values of p for which the demand is inelastic, unitary, or elastic.

75. WEIGHTS OF CHILDREN The Ehrenberg equation

$$W = 2.4e^{1.84h}$$

gives the relationship between the height h (in meters) and the average weight W (in kilograms) for children between 5 and 13 years of age.

a. What is the average weight of a 10-year-old child who stands 1.6 m tall?

b. Use differentials to estimate the change in the average weight of a 10-year-old child whose height increases from 1.6 m to 1.65 m.

76. POPULATION DISTRIBUTION The number of people living x mi from the center of town is given by

$$P(x) = 50,000(1 - e^{-0.01x^2}) \qquad (0 < x < 25)$$

Use differentials to estimate the number of people living between 10 and 10.1 mi from the center of town.

77. OPTIMAL SELLING TIME Refer to Exercise 49, page 367. The present value of a piece of waterfront property purchased by an investor is given by the function

$$P(t) = 80,000e^{\sqrt{t/2} - 0.09t} \qquad (0 \le t \le 8)$$

Determine the optimal time (based on present value) for the investor to sell the property. What is the property's optimal present value?

78. MAXIMUM OIL PRODUCTION It has been estimated that the total production of oil from a certain oil well is given by

$$T(t) = -1000(t + 10)e^{-0.1t} + 10,000$$

thousand barrels t years after production has begun. Determine the year when the oil well will be producing at maximum capacity.

79. BLOOD ALCOHOL LEVEL Refer to Exercise 62, page 377. At what time after drinking the alcohol is the percentage of alcohol in the person's bloodstream at its highest level? What is that level?

80. PRICE OF A COMMODITY The price of a certain commodity in dollars per unit at time t (measured in weeks) is given by $p = 8 + 4e^{-2t} + te^{-2t}$.

a. What is the price of the commodity at $t = 0$?

b. How fast is the price of the commodity changing at $t = 0$?

c. Find the equilibrium price of the commodity.
Hint: It's given by $\lim_{t \to \infty} p$. Also, use the fact that $\lim_{t \to \infty} te^{-2t} = 0$.

81. THERMOMETER READINESS A thermometer is moved from inside a house out to the deck. Its temperature t min after it has been moved is given by

$$T(t) = 30 + 40e^{-0.98t}$$

a. What is the temperature inside the house?

b. How fast is the reading on the thermometer changing 1 min after it has been taken out of the house?

c. What is the outdoor temperature?
Hint: Evaluate $\lim_{t \to \infty} T(t)$.

82. DIFFUSION Suppose a cell of volume V is surrounded by a homogeneous chemical solution of concentration C g/cc. Let y denote the concentration of the solute inside the cell at any time t, and suppose that the concentration initially is y_0. Then

$$y = (y_0 - C)e^{-kt/V} + C$$

where k is a constant. Find the rate of diffusion, dy/dt, of the solute across the cell membrane.

83. CHEMICAL REACTION Two chemicals, A and B, interact to form a Chemical C. Suppose the amount (in grams) of Chemical C formed t min after the interaction begins is

$$A(t) = \frac{150(1 - e^{0.022662t})}{1 - 2.5e^{0.022662t}}$$

a. How fast is Chemical C being formed 1 min after the interaction first began?

b. How much Chemical C will there be eventually?
Hint: Evaluate $\lim_{t \to \infty} A(t)$.

84. CONCENTRATION OF A DRUG IN THE BLOODSTREAM The concentration of a drug in the bloodstream t sec after injection into a muscle is given by

$$y = c(e^{-bt} - e^{-at}) \qquad (t \ge 0)$$

where a, b, and c are positive constants, with $a > b$.

a. Find the time at which the concentration is maximal.

b. Find the time at which the concentration of the drug in the bloodstream is decreasing most rapidly.

85. ABSORPTION OF DRUGS A liquid carries a drug into an organ of volume V cm^3 at the rate of a cm^3/sec and leaves at the same rate. The concentration of the drug in the entering liquid is c g/cm^3. Letting $x(t)$ denote the concentration of the drug in the organ at any time t, we have $x(t) = c(1 - e^{-at/V})$.

a. Show that x is an increasing function on $(0, \infty)$.

b. Sketch the graph of x.

86. ABSORPTION OF DRUGS Refer to Exercise 85. Suppose the maximum concentration of the drug in the organ must *not* exceed m g/cm^3, where $m < c$. Show that the liquid must not be allowed to enter the organ for a time longer than

$$T = \left(\frac{V}{a}\right) \ln\left(\frac{c}{c - m}\right)$$

minutes.

87. ABSORPTION OF DRUGS Jane took 100 mg of a drug one morning and another 100 mg of the same drug at the same time the following morning. The amount of the drug in her body t days after the first dose was taken is given by

$$A(t) = \begin{cases} 100e^{-1.4t} & \text{if } 0 \le t < 1 \\ 100(1 + e^{1.4})e^{-1.4t} & \text{if } t \ge 1 \end{cases}$$

a. How fast was the amount of drug in Jane's body changing after 12 hr $\left(t = \frac{1}{2}\right)$? After 2 days?
b. When was the amount of drug in Jane's body a maximum?
c. What was the maximum amount of drug in Jane's body?

In Exercises 88–92, determine whether the statement is true or false. If it is true, explain why it is true. If it is false, give an example to show why it is false.

88. If $f(x) = 3^x$, then $f'(x) = x \cdot 3^{x-1}$.

89. If $f(x) = e^\pi$, then $f'(x) = e^\pi$.

90. If $f(x) = \pi^x$, then $f'(x) = \pi^x$.

91. If $f(x) = e^{x^2 + x}$, then $f'(x) = e^{2x+1}$.

92. If $x^2 + e^y = 10$, then $y' = \dfrac{-2x}{e^y}$.

5.4   Solutions to Self-Check Exercises

1. a. Using the Product Rule, we obtain

$$f'(x) = x \frac{d}{dx}(e^{-x}) + e^{-x}\frac{d}{dx}(x)$$

$$= -xe^{-x} + e^{-x} = (1 - x)e^{-x}$$

Using the Product Rule once again, we obtain

$$f''(x) = (1 - x)\frac{d}{dx}(e^{-x}) + e^{-x}\frac{d}{dx}(1 - x)$$

$$= (1 - x)(-e^{-x}) + e^{-x}(-1)$$

$$= -e^{-x} + xe^{-x} - e^{-x} = (x - 2)e^{-x}$$

b. Setting $f'(x) = 0$ gives

$$(1 - x)e^{-x} = 0$$

Since $e^{-x} \ne 0$, we see that $1 - x = 0$, and this gives 1 as the only critical number of f. The sign diagram of f' shown in the accompanying figure tells us that the point $(1, e^{-1})$ is a relative maximum of f.

c. Setting $f''(x) = 0$ gives $x - 2 = 0$, so $x = 2$ is a candidate for an inflection point of f. The sign diagram of f'' (see the accompanying figure) shows that $(2, 2e^{-2})$ is an inflection point of f.

2. The rate change of the book value of the asset t years from now is

$$V'(t) = 50{,}000\frac{d}{dt}(e^{-0.4t})$$

$$= 50{,}000(-0.4)e^{-0.4t} = -20{,}000e^{-0.4t}$$

Therefore, 3 years from now, the book value of the asset will be changing at the rate of

$$V'(3) = -20{,}000e^{-0.4(3)} = -20{,}000e^{-1.2} \approx -6023.88$$

—that is, decreasing at the rate of approximately $6024/year.

USING TECHNOLOGY

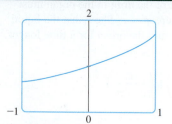

FIGURE T1
The graph of f in the viewing window $[-1, 1] \times [0, 2]$

EXAMPLE 1 At the beginning of Section 5.4, we demonstrated via a table of values of $(e^h - 1)/h$ for selected values of h the plausibility of the result

$$\lim_{h \to 0} \frac{e^h - 1}{h} = 1$$

To obtain a visual confirmation of this result, we plot the graph of

$$f(x) = \frac{e^x - 1}{x}$$

in the viewing window $[-1, 1] \times [0, 2]$ (Figure T1). From the graph of f, we see that $f(x)$ appears to approach 1 as x approaches 0.

The numerical derivative function of a graphing utility will yield the derivative of an exponential or logarithmic function for any value of x, just as it did for algebraic functions.*

---

*The rules for differentiating logarithmic functions will be covered in Section 5.5. However, the exercises given here can be done without using these rules.

TECHNOLOGY EXERCISES

In Exercises 1–6, use the numerical derivative operation to find the rate of change of $f(x)$ at the given value of x. Give your answer accurate to four decimal places.

1. $f(x) = x^3 e^{-1/x}; x = -1$

2. $f(x) = (\sqrt{x} + 1)^{3/2} e^{-x}; x = 0.5$

3. $f(x) = x^3 \sqrt{\ln x}; x = 2$

4. $f(x) = \dfrac{\sqrt{x} \ln x}{x + 1}; x = 3.2$

5. $f(x) = e^{-x} \ln(2x + 1); x = 0.5$

6. $f(x) = \dfrac{e^{-\sqrt{x}}}{\ln(x^2 + 1)}; x = 1$

7. AN EXTINCTION SITUATION The number of saltwater crocodiles in a certain area of northern Australia in year t is given by

$$P(t) = \frac{300e^{-0.024t}}{5e^{-0.024t} + 1}$$

a. How many crocodiles were in the population initially?

b. Show that $\lim\limits_{t \to \infty} P(t) = 0$.

c. Plot the graph of P in the viewing window $[0, 200] \times [0, 70]$.

(*Comment:* This phenomenon is referred to as an *extinction situation.*)

8. INCOME OF AMERICAN HOUSEHOLDS On the basis of government data, it is estimated that the percentage of American households y who earned x thousand dollars in 2010 is given by the equation

$$y = 1.168xe^{-0.00000312x^3 + 0.000659x^2 - 0.0783x} \quad (x > 0)$$

a. Plot the graph of the equation in the viewing window $[0, 150] \times [0, 2]$.

b. How fast is y changing with respect to x when $x = 10$? When $x = 50$? Interpret your results.

Source: U.S. Census Bureau.

9. LOAN AMORTIZATION The Sotos plan to secure a loan of $160,000 to purchase a house. They are considering a conventional 30-year home mortgage at 9%/year on the unpaid balance. It can be shown that the Sotos will have an outstanding principal of

$$B(x) = \frac{160{,}000(1.0075^{360} - 1.0075^x)}{1.0075^{360} - 1}$$

dollars after making x monthly payments of $1287.40.

a. Plot the graph of $B(x)$, using the viewing window $[0, 360] \times [0, 160{,}000]$.

b. Compute $B(0)$ and $B'(0)$, and interpret your results; compute $B(180)$ and $B'(180)$, and interpret your results.

10. INCREASING CROP YIELDS If left untreated on bean stems, aphids (small insects that suck plant juices) will multiply at an increasing rate during the summer months and reduce productivity and crop yield of cultivated crops. But if the aphids are treated in mid-June, the numbers decrease sharply to fewer than 100/bean stem, allowing for steep rises in crop yield. The function

$$F(t) = \begin{cases} 62e^{1.152t} & \text{if } 0 \le t < 1.5 \\ 349e^{-1.324(t-1.5)} & \text{if } 1.5 \le t \le 3 \end{cases}$$

gives the number of aphids after treatment on a typical bean stem at time t, where t is measured in months, with $t = 0$ corresponding to the beginning of May.

a. How many aphids are there on a typical bean stem at the beginning of June $(t = 1)$? At the beginning of July $(t = 2)$?

b. How fast is the population of aphids changing at the beginning of June? At the beginning of July?

Source: The Random House Encyclopedia.

5.5 Differentiation of Logarithmic Functions

The Derivative of ln x

Let's now turn our attention to the differentiation of logarithmic functions.

> **Rule 3: Derivative of ln x**
>
> $$\frac{d}{dx} \ln|x| = \frac{1}{x} \qquad (x \neq 0)$$

To derive Rule 3, suppose $x > 0$ and write $f(x) = \ln x$ in the equivalent form

$$x = e^{f(x)}$$

Differentiating both sides of the equation with respect to x, we find, using the Chain Rule,

$$1 = e^{f(x)} \cdot f'(x)$$

from which we see that

$$f'(x) = \frac{1}{e^{f(x)}}$$

or, since $e^{f(x)} = x$,

$$f'(x) = \frac{1}{x}$$

as we set out to show. You are asked to prove the rule for the case $x < 0$ in Exercise 99, page 390.

EXAMPLE 1 Find the derivative of each function:

a. $f(x) = x \ln x$ **b.** $g(x) = \dfrac{\ln x}{x}$

Solution

a. Using the Product Rule, we obtain

$$f'(x) = \frac{d}{dx}(x \ln x) = x\frac{d}{dx}(\ln x) + (\ln x)\frac{d}{dx}(x)$$

$$= x\left(\frac{1}{x}\right) + \ln x = 1 + \ln x$$

b. Using the Quotient Rule, we obtain

$$g'(x) = \frac{x\dfrac{d}{dx}(\ln x) - (\ln x)\dfrac{d}{dx}(x)}{x^2} = \frac{x\left(\dfrac{1}{x}\right) - \ln x}{x^2} = \frac{1 - \ln x}{x^2}$$

> **Explore and Discuss**
>
> You can derive the formula for the derivative of $f(x) = \ln x$ directly from the definition of the derivative, as follows.

1. Show that

$$f'(x) = \lim_{h \to 0} \frac{f(x+h) - f(x)}{h} = \lim_{h \to 0} \ln\left(1 + \frac{h}{x}\right)^{1/h}$$

2. Put $m = x/h$ and note that $m \to \infty$ as $h \to 0$. Then, $f'(x)$ can be written in the form

$$f'(x) = \lim_{m \to \infty} \ln\left(1 + \frac{1}{m}\right)^{m/x}$$

3. Finally, use both the fact that the natural logarithmic function is continuous and the definition of the number e to show that

$$f'(x) = \frac{1}{x} \ln\left[\lim_{m \to \infty}\left(1 + \frac{1}{m}\right)^m\right] = \frac{1}{x}$$

The Chain Rule and Logarithmic Functions

To enlarge the class of logarithmic functions to be differentiated, we appeal once again to the Chain Rule to obtain the following rule for differentiating composite functions of the form $h(x) = \ln f(x)$, where $f(x)$ is assumed to be a positive differentiable function.

> **Rule 4: Derivative of ln $f(x)$**
>
> If $f(x)$ is a differentiable function, then
>
> $$\frac{d}{dx}[\ln f(x)] = \frac{f'(x)}{f(x)} \qquad (f(x) > 0)$$

To see this, observe that $h(x) = g[f(x)]$, where $g(x) = \ln x$ $(x > 0)$. Since $g'(x) = 1/x$, we have, using the Chain Rule,

$$h'(x) = g'[f(x)]f'(x)$$
$$= \frac{1}{f(x)}f'(x) = \frac{f'(x)}{f(x)}$$

Observe that in the special case $f(x) = x$, $h(x) = \ln x$, so the derivative of h is, by Rule 3, given by $h'(x) = 1/x$.

EXAMPLE 2 Find the derivative of the function $f(x) = \ln(x^2 + 1)$.

Solution Using Rule 4, we see immediately that

$$f'(x) = \frac{\dfrac{d}{dx}(x^2 + 1)}{x^2 + 1} = \frac{2x}{x^2 + 1}$$

When we differentiate functions involving logarithms, the rules of logarithms may be used to advantage, as shown in Examples 3 and 4.

EXAMPLE 3 Differentiate the function $y = \ln[(x^2 + 1)(x^3 + 2)^6]$.

Solution We first rewrite the given function using the properties of logarithms:

$$y = \ln[(x^2 + 1)(x^3 + 2)^6]$$
$$= \ln(x^2 + 1) + \ln(x^3 + 2)^6 \qquad \ln mn = \ln m + \ln n$$
$$= \ln(x^2 + 1) + 6\ln(x^3 + 2) \qquad \ln m^n = n\ln m$$

Differentiating and using Rule 4, we obtain

$$y' = \frac{\frac{d}{dx}(x^2 + 1)}{x^2 + 1} + \frac{6\frac{d}{dx}(x^3 + 2)}{x^3 + 2}$$

$$= \frac{2x}{x^2 + 1} + \frac{6(3x^2)}{x^3 + 2} = \frac{2x}{x^2 + 1} + \frac{18x^2}{x^3 + 2}$$

Exploring with TECHNOLOGY

Use a graphing utility to plot the graphs of $f(x) = \ln x$; its first derivative function, $f'(x) = 1/x$; and its second derivative function $f''(x) = -1/x^2$, using the same viewing window $[0, 4] \times [-3, 3]$.

1. Describe the properties of the graph of f revealed by studying the graph of $f'(x)$. What can you say about the rate of increase of f?

2. Describe the properties of the graph of f revealed by studying the graph of $f''(x)$. What can you say about the concavity of f?

EXAMPLE 4 Find the derivative of the function $g(t) = \ln(t^2 e^{-t^2})$.

Solution   Here again, to save a lot of work, we first simplify the given expression using the properties of logarithms. We have

$$g(t) = \ln(t^2 e^{-t^2})$$

$$= \ln t^2 + \ln e^{-t^2} \qquad \text{\textcolor{red}{ln } mn = \text{ln } m + \text{ln } n}$$

$$= 2 \ln t - t^2 \qquad \text{\textcolor{red}{ln } m^n = n \text{ ln } m \quad \text{and} \quad \text{ln } e = 1}$$

Therefore,

$$g'(t) = \frac{2}{t} - 2t = \frac{2(1 - t^2)}{t}$$

Logarithmic Differentiation

As we saw in Examples 3 and 4, the task of finding the derivative of a given function can sometimes be made easier by first applying the laws of logarithms to simplify the function. We now illustrate a process called **logarithmic differentiation**, which not only simplifies the calculation of the derivatives of certain functions but also enables us to compute the derivatives of functions that we could not otherwise differentiate using the techniques developed thus far.

EXAMPLE 5 Differentiate $y = x(x + 1)(x^2 + 1)$, using logarithmic differentiation.

Solution   First, we take the natural logarithm on both sides of the given equation, obtaining

$$\ln y = \ln[x(x + 1)(x^2 + 1)]$$

Next, we use the properties of logarithms to rewrite the right-hand side of this equation, obtaining

$$\ln y = \ln x + \ln(x + 1) + \ln(x^2 + 1)$$

If we differentiate both sides of this equation, we have

$$\frac{d}{dx} \ln y = \frac{d}{dx} \left[\ln x + \ln(x + 1) + \ln(x^2 + 1) \right]$$

$$= \frac{1}{x} + \frac{1}{x + 1} + \frac{2x}{x^2 + 1} \qquad \text{Use Rule 4.}$$

To evaluate the expression on the left-hand side, note that y is a function of x. Therefore, writing $y = f(x)$ to remind us of this fact, we have

$$\frac{d}{dx} \ln y = \frac{d}{dx} \ln[f(x)] \qquad \text{Write } y = f(x).$$

$$= \frac{f'(x)}{f(x)} \qquad \text{Use Rule 4.}$$

$$= \frac{y'}{y} \qquad \text{Return to using } y \text{ instead of } f(x).$$

Therefore, we have

$$\frac{y'}{y} = \frac{1}{x} + \frac{1}{x + 1} + \frac{2x}{x^2 + 1}$$

Finally, solving for y', we have

$$y' = y \left(\frac{1}{x} + \frac{1}{x + 1} + \frac{2x}{x^2 + 1} \right)$$

$$= x(x + 1)(x^2 + 1) \left(\frac{1}{x} + \frac{1}{x + 1} + \frac{2x}{x^2 + 1} \right)$$

Before considering other examples, let's summarize the important steps involved in logarithmic differentiation.

Finding $\dfrac{dy}{dx}$ by Logarithmic Differentiation

1. Take the natural logarithm on both sides of the equation, and use the properties of logarithms to write any "complicated expression" as a sum of simpler terms.
2. Differentiate both sides of the equation with respect to x.
3. Solve the resulting equation for $\dfrac{dy}{dx}$.

EXAMPLE 6 Differentiate $y = x^2(x - 1)(x^2 + 4)^3$.

Solution Taking the natural logarithm on both sides of the given equation and using the laws of logarithms, we obtain

$$\ln y = \ln[x^2(x - 1)(x^2 + 4)^3]$$

$$= \ln x^2 + \ln(x - 1) + \ln(x^2 + 4)^3$$

$$= 2 \ln x + \ln(x - 1) + 3 \ln(x^2 + 4)$$

Differentiating both sides of the equation with respect to x, we have

$$\frac{d}{dx} \ln y = \frac{y'}{y} = \frac{2}{x} + \frac{1}{x - 1} + 3 \cdot \frac{2x}{x^2 + 4}$$

Finally, solving for y', we have

$$y' = y\left(\frac{2}{x} + \frac{1}{x-1} + \frac{6x}{x^2+4}\right)$$

$$= x^2(x-1)(x^2+4)^3\left(\frac{2}{x} + \frac{1}{x-1} + \frac{6x}{x^2+4}\right)$$

Recall from Section 3.4 that the relative rate of change of a differentiable function Q of x is $Q'(x)/Q(x)$. In view of Rule 4, we see that the relative rate of change of Q at x can also be obtained by finding the derivative of $\ln Q$. We exploit this fact in Example 7.

 APPLIED EXAMPLE 7 Population Growth The population of a town t months after the opening of an auto assembly plant in the surrounding area is given by the function

$$P(t) = 18000e^{-(\ln 9)e^{-0.1t}}$$

What is the relative rate of growth of the population 6 months after the opening of the auto assembly plant?

Solution We could find the required relative rate by computing $P'(t)/P(t)$ directly. Alternatively, we can proceed as follows:

$$\ln P(t) = \ln 18000e^{-(\ln 9)e^{-0.1t}}$$

$$= \ln 18000 + \ln e^{-(\ln 9)e^{-0.1t}}$$

$$= \ln 18000 - (\ln 9)e^{-0.1t} \qquad \ln e^x = x$$

So

$$\frac{P'(t)}{P(t)} = \frac{d}{dt}[\ln P(t)] = \frac{d}{dt}\ln 18000 - \frac{d}{dt}(\ln 9)e^{-0.1t}$$

$$= 0 - (\ln 9)(-0.1)e^{-0.1t} = (0.1)(\ln 9)e^{-0.1t}$$

Therefore,

$$\frac{P'(t)}{P(t)}\bigg|_{t=6} = (0.1)(\ln 9)e^{-(0.1)(6)} \approx 0.121$$

This tells us that 6 months after the opening of the auto assembly plant, the relative rate of growth of the population is approximately 12.1% per month.

5.5 Self-Check Exercises

1. Find an equation of the tangent line to the graph of $f(x) = x\ln(2x+3)$ at the point $(-1, 0)$.

2. Use logarithmic differentiation to compute y', given $y = (2x+1)^3(3x+4)^5$.

Solutions to Self-Check Exercises 5.5 can be found on page 390.

5.5 Concept Questions

1. State the rule for differentiating (a) $f(x) = \ln|x|$ $(x \neq 0)$, and (b) $g(x) = \ln f(x)$ $[f(x) > 0]$, where f is a differentiable function.

2. Explain the technique of logarithmic differentiation.

5.5 Exercises

In Exercises 1–34, find the derivative of the function.

1. $f(x) = 5 \ln x$

2. $f(x) = \ln 5x$

3. $f(x) = \ln(x + 1)$

4. $g(x) = \ln(2x + 1)$

5. $f(x) = \ln x^8$

6. $h(t) = 2 \ln t^5$

7. $f(x) = \ln \sqrt{x}$

8. $f(x) = \ln(\sqrt{x} + 1)$

9. $f(x) = \ln \dfrac{1}{x^2}$

10. $f(x) = \ln \dfrac{1}{2x^3}$

11. $f(x) = \ln(4x^2 - 5x + 3)$

12. $f(x) = \ln(3x^2 - 2x + 1)$

13. $f(x) = \ln \dfrac{2x}{x + 1}$

14. $f(x) = \ln \dfrac{x + 1}{x - 1}$

15. $f(x) = x^2 \ln x$

16. $f(x) = 3x^2 \ln 2x$

17. $f(x) = \dfrac{2 \ln x}{x}$

18. $f(x) = \dfrac{3 \ln x}{x^2}$

19. $f(u) = \ln(u - 2)^3$

20. $f(x) = \ln(x^3 - 3)^4$

21. $f(x) = \sqrt{\ln x}$

22. $f(x) = \sqrt{\ln x + x}$

23. $f(x) = (\ln x)^2$

24. $f(x) = 2(\ln x)^{3/2}$

25. $f(x) = \ln(x^3 + 1)$

26. $f(x) = \ln\sqrt{x^2 - 4}$

27. $f(x) = e^x \ln x$

28. $f(x) = e^x \ln\sqrt{x + 3}$

29. $f(t) = e^{2t} \ln(t + 1)$

30. $g(t) = t^2 \ln(e^{2t} + 1)$

31. $f(x) = \dfrac{\ln x}{x^2}$

32. $g(t) = \dfrac{t}{\ln t}$

33. $f(x) = \ln(\ln x)$

34. $g(x) = \ln(e^x + \ln x)$

In Exercises 35–40, find the second derivative of the function.

35. $f(x) = \ln 2x$

36. $f(x) = \ln(x + 5)$

37. $f(x) = \ln(x^2 + 2)$

38. $f(x) = (\ln x)^2$

39. $f(x) = x^2 \ln x$

40. $g(x) = e^{2x} \ln x$

In Exercises 41–50, use logarithmic differentiation to find the derivative of the function.

41. $y = (x + 1)^2(x + 2)^3$

42. $y = (3x + 2)^4(5x - 1)^2$

43. $y = (x - 1)^2(x + 1)^3(x + 3)^4$

44. $y = \sqrt{3x + 5}(2x - 3)^4$

45. $y = \dfrac{(2x^2 - 1)^5}{\sqrt{x + 1}}$

46. $y = \dfrac{\sqrt{4 + 3x^2}}{\sqrt[3]{x^2 + 1}}$

47. $y = 3^x$

48. $y = x^{x+2}$

49. $y = (x^2 + 1)^x$

50. $y = x^{\ln x}$

In Exercises 51 and 52, use implicit differentiation to find dy/dx.

51. $\ln y - x \ln x = -1$

52. $\ln xy - y^2 = 5$

53. Find an equation of the tangent line to the graph of $y = x \ln x$ at the point $(1, 0)$.

54. Find an equation of the tangent line to the graph of $y = \ln x^2$ at the point $(2, \ln 4)$.

55. Determine the intervals where the function $f(x) = \ln x^2$ is increasing and where it is decreasing.

56. Determine the intervals where the function $f(x) = \dfrac{\ln x}{x}$ is increasing and where it is decreasing.

57. Determine the intervals of concavity for the graph of the function $f(x) = x^2 + \ln x^2$.

58. Determine the intervals of concavity for the graph of the function $f(x) = \dfrac{\ln x}{x}$.

59. Find the inflection points of the function
$f(x) = \ln(x^2 + 1)$.

60. Find the inflection points of the function $f(x) = x^2 \ln x$.

61. Find an equation of the tangent line to the graph of $f(x) = x^2 + 2 \ln x$ at its inflection point.

62. Find an equation of the tangent line to the graph of $f(x) = e^{x/2} \ln x$ at its inflection point.
Hint: Show that $(1, 0)$ is the only inflection point of f.

63. Find the absolute extrema of the function $f(x) = x - \ln x$ on $\left[\frac{1}{2}, 3\right]$.

64. Find the absolute extrema of the function $g(x) = \dfrac{x}{\ln x}$ on $[2, 5]$.

In Exercises 65 and 66, find dy/dx by implicit differentiation.

65. $\ln(xy) = x + y$

66. $\ln x + e^{-y/x} = 10$

In Exercises 67 and 68, find d^2y/dx^2 by implicit differentiation.

67. $\ln x + xy = 5$

68. $\ln y + y = x$

69. Find dy/dx at the point $(1, 1)$ using implicit differentiation if $\ln y + xy = 1$.

70. Find an equation of the tangent line to the graph of the equation $\ln x + xe^y = 1$ at the point $(1, 0)$.

71. STRAIN ON VERTEBRAE The strain (percentage of compression) on the lumbar vertebral disks in an adult human as a function of the load x (in kilograms) is given by

$$f(x) = 7.2956 \ln(0.0645012x^{0.95} + 1)$$

What is the rate of change of the strain with respect to the load when the load is 100 kg? When the load is 500 kg?
Source: Benedek and Villars, Physics with Illustrative Examples from Medicine and Biology.

72. **HEIGHTS OF CHILDREN** For children between the ages of 5 and 13 years, the Ehrenberg equation

$$\ln W = \ln 2.4 + 1.84h$$

gives the relationship between the weight W (in kilograms) and the height h (in meters) of the child. Use differentials to estimate the change in the weight of a child who grows from 1 m to 1.1 m.

73. **AVERAGE LIFE SPAN** One reason for the increase in the life span over the years has been the advances in medical technology. The average life span for American women from 1907 through 2007 is given by

$$W(t) = 49.9 + 17.1 \ln t \qquad (1 \le t \le 6)$$

where $W(t)$ is measured in years and t is measured in 20-year intervals, with $t = 1$ corresponding to the beginning of 1907.
a. Show that W is increasing on $(1, 6)$.
b. What can you say about the concavity of the graph of W on the interval $(1, 6)$?

74. **STREAMING MUSIC** Consumers are spending more on music from digital sources than they do on CDs. Purchases of streamed music from services such as Pandora and Spotify are projected to grow in accordance with the model

$$f(t) = 3.7 + 0.84 \ln(t + 1) \qquad (0 \le t \le 5)$$

where $f(t)$ is measured in billions of dollars and t is measured in years, with $t = 0$ corresponding to 2012.
a. What was the amount spent by consumers on streamed music in 2012? The amount spent in 2014?
b. How fast was the amount spent by consumers on streamed music growing in 2014?
Source: Strategy Analytics.

75. **DEPRECIATION OF EQUIPMENT** For assets such as machines, whose market values drop rapidly in the early years of usage, businesses often use the double declining–balance method. In practice, a business firm normally employs the double declining–balance method for depreciating such assets for a certain number of years and then switches over to the linear method (see Exercise 50, page 92). The double declining–balance formula is

$$V(n) = C\left(1 - \frac{2}{N}\right)^n$$

where C is the initial value of the asset in dollars, $V(n)$ denotes the book value of the assets at the end of n years, and N is the number of years over which the asset is depreciated.
a. Find $V'(n)$.
   Hint: Use logarithmic differentiation.
b. What is the relative rate of change of $V(n)$?
   Hint: Find $[V'(n)]/[V(n)]$. See Section 3.4.

76. **DEPRECIATION OF EQUIPMENT** Refer to Exercise 75. A tractor purchased at a cost of $60,000 is to be depreciated by the double declining–balance method over 10 years.

a. What is the book value of the tractor at the end of 2 years?
b. What is the relative rate of change of the book value of the tractor at the end of 2 years?

77. **POPULATION GROWTH** The population of a town t months after the establishment of a biotech research center nearby is given by

$$P(t) = \frac{40 + 80e^{0.06t}}{20 + e^{0.06t}}$$

where $P(t)$ is measured in thousands. Find the relative rate of growth of the population 5 years after the establishment of the biotech research center.

78. **MARGINAL REVENUE** The demand function for the Viking Boat's 34-ft *Sundancer* yacht is

$$p = 200 - 0.01x \ln x$$

where x denotes the number of yachts and p is the price per yacht in hundreds of dollars.
a. Find the revenue and the marginal revenue function for this model of yacht.
b. Use the result of part (a) to estimate the revenue to be realized from the sale of the 500th 34-ft *Sundancer* yacht.

79. **MAXIMIZING PROFIT** The manager of Seko, an information technology (IT) consulting company, estimates that the annual profit of the company, in millions of dollars, is given by

$$P(x) = 2 \ln(2x + 1) + 2x - x^2 - 0.3$$

where x is the number of IT consultants (in hundreds) in its employ. Find the number of consultants the firm should hire so that its profit is maximized. What is the maximum profit?

80. **LAMBERT'S LAW OF ABSORPTION** Lambert's law of absorption states that the light intensity $I(x)$ (in calories per square centimeter per second) at a depth of x m as measured from the surface of a material is given by $I = I_0 a^x$, where I_0 and a are positive constants.
a. Find the rate of change of the light intensity with respect to x at a depth of x m from the surface of the material.
b. Using the result of part (a), conclude that the rate of change $I'(x)$ at a depth of x m is proportional to $I(x)$. What is the constant of proportion?

81. **PERCENTAGE RATE OF GROWTH OF A COMPANY**
a. Show that the percentage rate of change of a function is

$$100 \frac{d}{dx}[\ln f(x)]$$

b. The revenue of a newly formed company t years from now is predicted to be

$$R(t) = 0.1t^{1.5}e^{0.2t} \qquad (0 \le t \le 5)$$

million dollars. Find the percentage rate of growth of the company 3 years from now.

82. ELASTICITY OF DEMAND

a. In Section 3.4, we defined the elasticity of demand at price p as

$$E(p) = -\frac{pf'(p)}{f(p)}$$

where f is a differentiable demand function $x = f(p)$. Show that

$$E(p) = -\frac{\dfrac{d}{dp}[\ln f(p)]}{\dfrac{d}{dp}(\ln p)}$$

b. The demand function of a product is

$$x = f(p) = \frac{e^{-0.1\sqrt{p}}}{\sqrt{2p+1}} \qquad (0 \le p \le 10)$$

where p is the unit price in dollars and x is the quantity demanded, measured in units of a thousand. Use the result of part (a) to find the elasticity of demand.

83. GOMPERTZ FUNCTION The Gompertz function defined by

$$P(t) = Le^{-\ln(L/P_0)e^{-ct}}$$

provides us with a model for population growth. Here, L is an upper bound for the population (called the carrying capacity for the environment), c is a positive constant, and P_0 is the population at $t = 0$. Show that the relative rate of change of P is

$$c \ln \frac{L}{P_0} e^{-ct}$$

84. ABSORPTION OF LIGHT When light passes through a window glass, some of it is absorbed. It can be shown that if $r\%$ of the light is absorbed by a glass of thickness w, then the percentage of light absorbed by a piece of glass of thickness nw is

$$A(n) = 100\left[1 - \left(1 - \frac{r}{100}\right)^n\right] \qquad (0 \le r \le 100)$$

a. Show that A is an increasing function of n on $(0, \infty)$ if $0 < r < 100$.
Hint: Use logarithmic differentiation.

b. Sketch the graph of A for the special case in which $r = 10$.

c. Evaluate $\lim\limits_{n \to \infty} A(n)$ and interpret your result.

85. MAGNITUDE OF EARTHQUAKES On the Richter scale, the magnitude R of an earthquake is given by the formula

$$R = \log \frac{I}{I_0}$$

where I is the intensity of the earthquake being measured and I_0 is the standard reference intensity.

a. What is the magnitude of an earthquake that has intensity 1 million times that of I_0?

b. Suppose an earthquake is measured with a magnitude of 6 on the Richter scale with an error of at most 2%. Use differentials to find the error in the intensity of the earthquake.
Hint: Observe that $I = I_0 10^R$, and use logarithmic differentiation.

86. WEBER–FECHNER LAW The Weber–Fechner Law

$$R = k \ln \frac{S}{S_0}$$

where k is a positive constant, describes the relationship between a stimulus S and the resulting response R. Here, S_0, a positive constant, is the threshold level.

a. Show that $R = 0$ if the stimulus is at the threshold level S_0.

b. The derivative dR/dS is the *sensitivity* corresponding to the stimulus level S and measures the capability to detect small changes in the stimulus level. Show that dR/dS is inversely proportional to S, and interpret your result.

87. PREDATOR–PREY MODEL The relationship between the number of rabbits $y(t)$ and the number of foxes $x(t)$ at any time t is given by

$$-C \ln y + Dy = A \ln x - Bx + E$$

where A, B, C, D, and E are constants. This relationship is based on a model by Lotka (1880–1949) and Volterra (1860–1940) for analyzing the ecological balance between two species of animals, one of which is a prey and the other a predator. Use implicit differentiation to find the relationship between the rate of change of the rabbit population in terms of the rate of change of the fox population.

88. RATE OF A CATALYTIC CHEMICAL REACTION A catalyst is a substance that either accelerates a chemical reaction or is necessary for the reaction to occur. Suppose an enzyme E (a catalyst) combines with a substrate S (a reacting chemical) to form an intermediate product X that then produces a product P and releases the enzyme. If initially there are x_0 moles/liter of S and there is no P, then based on the theory of Michaelis and Menten, the concentration of P, $p(t)$, after t hr is given by the equation

$$Vt = p - k \ln\left(1 - \frac{p}{x_0}\right)$$

where the constant V is the maximum possible speed of the reaction and the constant k is called the **Michaelis constant** for the reaction. Find the rate of change of the formation of the product P in this reaction.

In Exercises 89 and 90, use the guidelines on page 295 to sketch the graph of the given function.

89. $f(x) = 2x - \ln x$

90. $f(x) = \ln(x - 1)$

91. DERIVATIVE OF b^x

a. Let $f(x) = b^x$ $(b > 0, b \neq 1)$. Show that $f'(x) = (\ln b)b^x$.

b. Use the result of part (a) to find the derivative of $f(x) = 3^x$.

92. DERIVATIVE OF $\log_b x$

a. Let $f(x) = \log_b x$ $(b > 0, b \neq 1)$. Use the result of Exercise 91 to show that $f'(x) = \dfrac{1}{\ln b} \cdot \dfrac{1}{x}$.

b. Use the result of part (a) to find the derivative of $f(x) = \log_{10} x$.

In Exercises 93–96, use the results of Exercises 91 and 92 to find the derivative of the given function.

93. $f(x) = x^3 2^x$

94. $g(x) = \dfrac{10^x}{x + 1}$

95. $h(x) = x^2 \log_{10} x$

96. $f(x) = 3^{x^2} + \log_2(x^2 + 1)$

In Exercises 97 and 98, determine whether the statement is true or false. If it is true, explain why it is true. If it is false, give an example to show why it is false.

97. If $f(x) = \ln 5$, then $f'(x) = \frac{1}{5}$.

98. If $f(x) = \ln a^x$, then $f'(x) = \ln a$.

99. Prove that $\dfrac{d}{dx} \ln|x| = \dfrac{1}{x}$ $(x \neq 0)$ for the case $x < 0$.

100. Use the definition of the derivative to show that

$$\lim_{x \to 0} \frac{\ln(x + 1)}{x} = 1$$

5.5   Solutions to Self-Check Exercises

1. The slope of the tangent line to the graph of f at any point $(x, f(x))$ lying on the graph of f is given by $f'(x)$. Using the Product Rule, we find

$$f'(x) = \frac{d}{dx}[x \ln(2x + 3)]$$

$$= x \frac{d}{dx} \ln(2x + 3) + \ln(2x + 3) \cdot \frac{d}{dx}(x)$$

$$= x\left(\frac{2}{2x + 3}\right) + \ln(2x + 3) \cdot 1$$

$$= \frac{2x}{2x + 3} + \ln(2x + 3)$$

In particular, the slope of the tangent line to the graph of f at the point $(-1, 0)$ is

$$f'(-1) = \frac{-2}{-2 + 3} + \ln 1 = -2$$

Therefore, using the point-slope form of the equation of a line, we see that a required equation is

$$y - 0 = -2(x + 1)$$

$$y = -2x - 2$$

2. Taking the logarithm on both sides of the equation gives

$$\ln y = \ln[(2x + 1)^3(3x + 4)^5]$$
$$= \ln(2x + 1)^3 + \ln(3x + 4)^5$$
$$= 3 \ln(2x + 1) + 5 \ln(3x + 4)$$

Differentiating both sides of the equation with respect to x, keeping in mind that y is a function of x, we obtain

$$\frac{d}{dx}(\ln y) = \frac{y'}{y} = 3 \cdot \frac{2}{2x + 1} + 5 \cdot \frac{3}{3x + 4}$$

$$= 3\left(\frac{2}{2x + 1} + \frac{5}{3x + 4}\right)$$

and

$$y' = 3(2x + 1)^3(3x + 4)^5\left(\frac{2}{2x + 1} + \frac{5}{3x + 4}\right)$$

5.6   Exponential Functions as Mathematical Models

Exponential Growth

Many problems arising from practical situations can be described mathematically in terms of exponential functions or functions closely related to the exponential function. In this section, we look at some applications involving exponential functions from the fields of the life and social sciences.

In Section 5.1, we saw that the exponential function $f(x) = b^x$ is an increasing function when $b > 1$. In particular, the function $f(x) = e^x$ has this property. Suppose

PORTFOLIO   Carol A. Reeb, Ph.D.

TITLE Research Associate
INSTITUTION Hopkins Marine Station, Stanford University

Historically, the world's oceans were thought to provide an unlimited source of inexpensive seafood. However, in a world in which the human population now exceeds six billion people, overfishing has pushed one third of all marine fishery stocks toward a state of collapse.

As a fishery geneticist at Hopkins Marine Station, I study commercially harvested marine populations and use exponential models in my work. The equation for determining the size of a population that grows or declines exponentially is $x_t = x_0 e^{rt}$, where x_0 is the initial population, t is time, and r is the growth or decay constant (positive for growth, negative for decay).

This equation can be used to estimate the population in the past as well as in the future. We know that the demand for seafood increased as the human population grew, eventually causing fish populations to decline. Because genetic diversity is linked to population size, the exponential function is useful to model change in fishery populations and their gene pools over time.

Interestingly, exponential functions can also be used to model the increase in the market value of seafood in the United States over the past 60 years. In general, the price of seafood has increased exponentially, although the price did stabilize briefly in 1995.

Although exponential curves are important to my work, they are not always the best fit. Exponential curves are best applied across short time frames when environments or markets are unlimited. Over longer periods, the logistic growth function is more suitable. In my research, selecting the most accurate model requires examining many possibilities.

that $Q(t)$ represents a quantity at time t, then one may deduce that the function $Q(t) = Q_0 e^{kt}$, where Q_0 and k are positive constants, has the following properties:

1. $Q(0) = Q_0$
2. $Q(t)$ increases "rapidly" without bound as t increases without bound (Figure 13).

Property 1 follows from the computation

$$Q(0) = Q_0 e^0 = Q_0$$

Next, to study the rate of change of the function $Q(t)$, we differentiate it with respect to t, obtaining

$$Q'(t) = \frac{d}{dt}\left(Q_0 e^{kt}\right)$$

$$= Q_0 \frac{d}{dt}\left(e^{kt}\right)$$

$$= kQ_0 e^{kt}$$

$$= kQ(t) \tag{12}$$

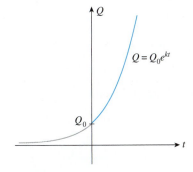

FIGURE 13
Exponential growth

Since $Q(t) > 0$ (because Q_0 is assumed to be positive) and $k > 0$, we see that $Q'(t) > 0$, so $Q(t)$ is an increasing function of t. Our computation has in fact shed more light on an important property of the function $Q(t)$. Equation (12) says that the rate of increase of the function $Q(t)$ is proportional to the amount $Q(t)$ of the quantity present at time t. The implication is that as $Q(t)$ increases, so does the *rate of increase* of $Q(t)$, resulting in a very rapid increase in $Q(t)$ as t increases without bound.

Thus, the exponential function

$$Q(t) = Q_0 e^{kt} \qquad (0 \le t < \infty) \tag{13}$$

provides us with a mathematical model of a quantity $Q(t)$ that is initially present in the amount of $Q(0) = Q_0$ and whose rate of growth at any time t is directly proportional to the amount of the quantity present at time t. Such a quantity is said to exhibit unrestricted **exponential growth**, and the constant k of proportionality is called the **growth constant**.

Interest earned on a fixed deposit when compounded continuously exhibits exponential growth. Other examples of unrestricted exponential growth follow.

 APPLIED EXAMPLE 1 Growth of Bacteria Under ideal laboratory conditions, the number of bacteria in a culture grows in accordance with the law $Q(t) = Q_0 e^{kt}$, where Q_0 denotes the number of bacteria initially present in the culture, k is a constant determined by the strain of bacteria under consideration and other factors, and t is the elapsed time measured in hours. Suppose 10,000 bacteria are present initially in the culture and 60,000 present 2 hours later.

a. How many bacteria will there be in the culture at the end of 4 hours?
b. What is the rate of growth of the population after 4 hours?

Solution

a. We are given that $Q(0) = Q_0 = 10,000$, so $Q(t) = 10,000e^{kt}$. Next, the fact that 60,000 bacteria are present 2 hours later translates into $Q(2) = 60,000$. Thus,

$$60,000 = 10,000e^{2k}$$
$$e^{2k} = 6$$

Taking the natural logarithm on both sides of the equation, we obtain

$$\ln e^{2k} = \ln 6$$
$$2k = \ln 6 \qquad \text{Since } \ln e = 1$$
$$k = \frac{\ln 6}{2}$$
$$k \approx 0.8959$$

Thus, the number of bacteria present at any time t is given by

$$Q(t) \approx 10,000e^{0.8959t}$$

In particular, the number of bacteria present in the culture at the end of 4 hours is given by

$$Q(4) \approx 10,000e^{0.8959(4)}$$
$$\approx 360,000$$

b. The rate of growth of the bacteria population at any time t is given by

$$Q'(t) = kQ(t)$$

Thus, using the result from part (a), we find that the rate at which the population is growing at the end of 4 hours is

$$Q'(4) = kQ(4)$$
$$\approx (0.8959)(360,000)$$
$$\approx 322,500$$

or approximately 322,500 bacteria per hour.

Exponential Decay

In contrast to exponential growth, a quantity exhibits **exponential decay** if it decreases at a rate that is directly proportional to its size. Such a quantity may be described by the exponential function

$$Q(t) = Q_0 e^{-kt} \qquad (0 \le t < \infty) \tag{14}$$

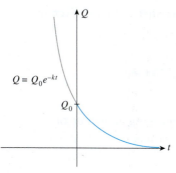

$Q = Q_0 e^{-kt}$

FIGURE 14
Exponential decay

where the positive constant Q_0 measures the amount present initially $(t = 0)$ and k is some suitable positive number, called the **decay constant.** The choice of this number is determined by the nature of the substance under consideration and other factors. The graph of this function is sketched in Figure 14.

To verify the properties ascribed to the function $Q(t)$, we simply compute

$$Q(0) = Q_0 e^0 = Q_0$$

$$Q'(t) = \frac{d}{dt}(Q_0 e^{-kt})$$

$$= Q_0 \frac{d}{dt}(e^{-kt})$$

$$= -kQ_0 e^{-kt} = -kQ(t)$$

APPLIED EXAMPLE 2 Radioactive Decay Radioactive substances decay exponentially. For example, the amount of radium present at any time t obeys the law $Q(t) = Q_0 e^{-kt}$, where Q_0 is the initial amount present and k is a specific positive constant. The **half-life of a radioactive substance** is the time required for a given amount to be reduced by one-half. It is known that the half-life of radium is approximately 1600 years. Suppose initially there are 200 milligrams of pure radium.

a. Find the amount left after t years.
b. What is the amount left after 800 years?
c. How fast is the amount of radium present changing after t years?
d. How fast is the amount of radium present changing after 800 years?

Solution

a. The initial amount of radium present is 200 milligrams, so $Q(0) = Q_0 = 200$. Thus, $Q(t) = 200e^{-kt}$. Next, the datum concerning the half-life of radium implies that $Q(1600) = 100$, and this gives

$$100 = 200e^{-1600k}$$

$$e^{-1600k} = \frac{1}{2}$$

Taking the natural logarithm on both sides of this equation yields

$$-1600k \ln e = \ln \frac{1}{2}$$

$$-1600k = \ln \frac{1}{2} \qquad \text{\color{red}{ln } } e = 1$$

$$k = -\frac{1}{1600} \ln\left(\frac{1}{2}\right) \approx 0.0004332$$

Therefore, the amount of radium left after t years is

$$Q(t) = 200e^{-0.0004332t}$$

b. The amount of radium left after 800 years is

$$Q(800) = 200e^{-0.0004332(800)} \approx 141.42$$

or approximately 141 milligrams.

c. The rate of change of the amount of radium present after t years is given by

$$Q'(t) = \frac{d}{dt}\left[200e^{-0.0004332t}\right]$$

$$= (200)(-0.0004332)e^{-0.0004332t}$$

$$= -0.08664e^{-0.0004332t}$$

d. After 800 years, the amount of radium present is changing at the rate of

$$Q'(800) = -0.08664e^{(-0.0004332)(800)}$$

$$\approx -0.06126$$

or decreasing at the rate of approximately 0.06126 milligrams per year.

 APPLIED EXAMPLE 3 Carbon-14 Decay Carbon 14, a radioactive isotope of carbon, has a half-life of 5730 years. What is its decay constant?

Solution   We have $Q(t) = Q_0 e^{-kt}$. Since the half-life of the element is 5730 years, half of the substance is left at the end of that period; that is,

$$Q(5730) = Q_0 e^{-5730k} = \frac{1}{2}Q_0$$

$$e^{-5730k} = \frac{1}{2}$$

Taking the natural logarithm on both sides of this equation, we have

$$\ln e^{-5730k} = \ln\frac{1}{2}$$

$$-5730k = -0.693147$$

$$k \approx 0.000121$$

Carbon-14 dating is a well-known method used by anthropologists to establish the age of animal and plant fossils. This method assumes that the proportion of carbon 14 (C-14) present in the atmosphere has remained fairly constant over the past 50,000 years. Professor Willard Libby, recipient of the Nobel Prize in chemistry in 1960, proposed this theory.

The amount of C-14 in the tissues of a living plant or animal is fairly constant. However, when an organism dies, it stops absorbing new quantities of C-14, and the amount of C-14 in the remains diminishes because of the natural decay of the radioactive substance. Therefore, the approximate age of a plant or animal fossil can be determined by measuring the amount of C-14 present in the remains.

APPLIED EXAMPLE 4 Carbon-14 Dating A skull from an archeological site has one tenth the amount of C-14 that it originally contained. Determine the approximate age of the skull.

Solution   Here,

$$Q(t) = Q_0 e^{-kt}$$

$$= Q_0 e^{-0.000121t}$$

where Q_0 is the amount of C-14 present originally and k, the decay constant, is equal to 0.000121 (see Example 3). Since $Q(t) = (1/10)Q_0$, we have

$$\frac{1}{10}Q_0 = Q_0e^{-0.000121t}$$

$$\ln\frac{1}{10} = -0.000121t \qquad \text{Take the natural logarithm on both sides.}$$

$$t = \frac{\ln\frac{1}{10}}{-0.000121}$$

$$\approx 19{,}030$$

or approximately 19,030 years.

Learning Curves

The next example shows how the exponential function may be applied to describe certain types of learning processes. Consider the function

$$Q(t) = C - Ae^{-kt}$$

where C, A, and k are positive constants. To sketch the graph of the function Q, observe that its y-intercept is given by $Q(0) = C - A$. Next, we compute

$$Q'(t) = kAe^{-kt}$$

Since both k and A are positive, we see that $Q'(t) > 0$ for all values of t. Thus, $Q(t)$ is an increasing function of t. Also,

$$\lim_{t\to\infty} Q(t) = \lim_{t\to\infty} (C - Ae^{-kt})$$

$$= \lim_{t\to\infty} C - \lim_{t\to\infty} Ae^{-kt}$$

$$= C$$

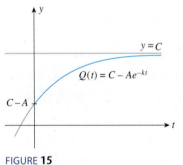

FIGURE 15
A learning curve

so $y = C$ is a horizontal asymptote of Q. Thus, $Q(t)$ increases and approaches the number C as t increases without bound. The graph of the function Q is shown in Figure 15, where that part of the graph corresponding to the negative values of t is drawn with a gray line since, in practice, one normally restricts the domain of the function to the interval $[0, \infty)$.

Observe that starting at $t = 0$, $Q(t)$ increases rather rapidly but then the rate of increase slows down considerably after a while. To see this, we compute

$$\lim_{t\to\infty} Q'(t) = \lim_{t\to\infty} kAe^{-kt} = 0$$

This behavior of the graph of the function Q closely resembles the learning pattern experienced by workers engaged in highly repetitive work. For example, the productivity of an assembly-line worker increases very rapidly in the early stages of the training period. This productivity increase is a direct result of the worker's training and accumulated experience. But the rate of increase of productivity slows as time goes by, and the worker's productivity level approaches some fixed level due to the limitations of the worker and the machine. Because of this characteristic, the graph of the function $Q(t) = C - Ae^{-kt}$ is often called a **learning curve.**

APPLIED EXAMPLE 5 Assembly Time The Camera Division of Eastman Optical produces a 35-mm single-lens reflex camera. Eastman's training department determines that after completing the basic training program, a new, previously inexperienced employee will be able to assemble

$$Q(t) = 50 - 30e^{-0.5t}$$

model F cameras per day t months after the employee starts work on the assembly line.

a. How many model F cameras can a new employee assemble per day after basic training?
b. How many model F cameras can an employee with 1 month of experience assemble per day? An employee with 2 months of experience? An employee with 6 months of experience?
c. How many model F cameras can the average experienced employee ultimately be expected to assemble per day?

Solution

a. The number of model F cameras a new employee can assemble is given by

$$Q(0) = 50 - 30 = 20$$

b. The number of model F cameras that an employee with 1 month of experience, 2 months of experience, and 6 months of experience can assemble per day is given by

$$Q(1) = 50 - 30e^{-0.5} \approx 31.80$$
$$Q(2) = 50 - 30e^{-1} \approx 38.96$$
$$Q(6) = 50 - 30e^{-3} \approx 48.51$$

or approximately 32, 39, and 49, respectively.

c. To find how many model F cameras the average experienced employee can ultimately be expected to assemble per day, we evaluate

$$\lim_{t \to \infty} Q(t) = \lim_{t \to \infty} (50 - 30e^{-0.5t})$$

$$= \lim_{t \to \infty} 50 - \lim_{t \to \infty} 30e^{-0.5t}$$

$$= 50$$

So the average experienced employee can ultimately be expected to assemble 50 model F cameras per day.

Other applications of the learning curve are found in models that describe the dissemination of information about a product or the velocity of an object dropped into a viscous medium.

Logistic Growth Functions

Our last example of an application of exponential functions to the description of natural phenomena involves the **logistic** (also called the **S-shaped**, or **sigmoidal**) **curve,** which is the graph of the function

$$Q(t) = \frac{A}{1 + Be^{-kt}}$$

where A, B, and k are positive constants. The function Q is called a **logistic growth function.** The graph of the function Q is sketched in Figure 16.

Observe that $Q(t)$ increases slowly at first but more rapidly as t increases. In fact, for small positive values of t, the logistic curve resembles an exponential growth curve. However, the *rate of growth* of $Q(t)$ decreases quite rapidly as t increases and $Q(t)$ approaches the number A as t increases without bound.

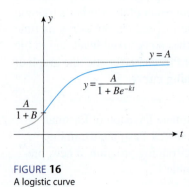

FIGURE 16
A logistic curve

Thus, the logistic curve exhibits both the property of rapid growth of the exponential growth curve as well as the "saturation" property of the learning curve. Because of these characteristics, the logistic curve serves as a suitable mathematical model for describing many natural phenomena. For example, if a small number of rabbits were introduced to a tiny island in the South Pacific, the rabbit population might be expected to grow very rapidly at first, but the growth rate would decrease quickly as overcrowding, scarcity of food, and other environmental factors affected it. The population would eventually stabilize at a level compatible with the life-support capacity of the environment. This level, given by A, is called the *carrying capacity* of the environment. Models describing the spread of rumors and epidemics are other examples of the application of the logistic curve.

 APPLIED EXAMPLE 6 Spread of Flu The number of soldiers at Fort MacArthur who contracted influenza after t days during a flu epidemic is approximated by the exponential model

$$Q(t) = \frac{5000}{1 + 1249e^{-kt}}$$

a. If 40 soldiers contracted the flu by day 7, find how many soldiers contracted the flu by day 15.

b. At what rate is the number of soldiers contracting the flu changing on day 15?

Solution

a. The given information implies that

$$Q(7) = \frac{5000}{1 + 1249e^{-7k}} = 40$$

Thus,

$$40(1 + 1249e^{-7k}) = 5000$$

$$1 + 1249e^{-7k} = \frac{5000}{40} = 125$$

$$e^{-7k} = \frac{124}{1249}$$

$$-7k = \ln\frac{124}{1249}$$

$$k = -\frac{\ln\frac{124}{1249}}{7} \approx 0.33$$

Therefore, the number of soldiers who contracted the flu after t days is given by

$$Q(t) = \frac{5000}{1 + 1249e^{-0.33t}}$$

In particular, the number of soldiers who contracted the flu by day 15 is given by

$$Q(15) = \frac{5000}{1 + 1249e^{-15(0.33)}}$$

$$\approx 508$$

or approximately 508 soldiers.

b. The rate at which the number of soldiers contracting the flu on day t is changing is given by

$$Q'(t) = \frac{d}{dt}\left[\frac{5000}{1 + 1249e^{-0.33t}}\right] = 5000\frac{d}{dt}[(1 + 1249e^{-0.33t})^{-1}]$$

$$= (5000)(-1)(1 + 1249e^{-0.33t})^{-2} \cdot \frac{d}{dt}(1 + 1249e^{-0.33t})$$

$$= \frac{(-5000)(1249)(-0.33)e^{-0.33t}}{(1 + 1249e^{-0.33t})^2}$$

$$= \frac{2,060,850e^{-0.33t}}{(1 + 1249e^{-0.33t})^2}$$

So the rate at which the number of soldiers contracting the flu is changing on day 15 is

$$Q'(15) = \frac{2,060,850e^{(-0.33)(15)}}{(1 + 1249e^{(-0.33)(15)})^2} \approx 150.54$$

or approximately 151 soldiers per day.

Exploring with TECHNOLOGY

Refer to Example 6.

1. Use a graphing utility to plot the graph of the function Q, using the viewing window $[0, 40] \times [0, 5000]$.

2. How long will it take for the first 1000 soldiers to contract the flu?
   Hint: Plot the graphs of $y_1 = Q(t)$ and $y_2 = 1000$, and find the point of intersection of the two graphs.

5.6 Self-Check Exercise

IMMIGRATION INTO THE UNITED STATES Suppose the population (in millions) of a country at any time t grows in accordance with the rule

$$P = \left(P_0 + \frac{I}{k}\right)e^{kt} - \frac{I}{k}$$

where P denotes the population at any time t, k is a constant reflecting the natural growth rate of the population, I is a constant giving the (constant) rate of immigration into the country, and P_0 is the total population of the country at time $t = 0$. The population of the United States in 1980 ($t = 0$) was 226.5 million. If the natural growth rate is 0.8% annually ($k = 0.008$) and net immigration is allowed at the rate of half a million people per year ($I = 0.5$), what is the projected population of the United States in 2015?

The solution to Self-Check Exercise 5.6 can be found on page 403.

5.6 Concept Questions

1. Give the model for unrestricted exponential growth and the model for exponential decay. What effect does the magnitude of the growth (decay) constant have on the growth (decay) of a quantity?

2. What is the half-life of a radioactive substance?

3. What is the logistic growth function? What are its characteristics?

5.6 Exercises

1. EXPONENTIAL GROWTH Given that a quantity $Q(t)$ is described by the exponential growth function

$$Q(t) = 300e^{0.02t}$$

where t is measured in minutes, answer the following questions:

a. What is the growth constant?

b. What quantity is present initially?

c. Complete the following table of values:

t	0	10	20	100	1000
Q					

2. EXPONENTIAL DECAY Given that a quantity $Q(t)$ exhibiting exponential decay is described by the function

$$Q(t) = 2000e^{-0.06t}$$

where t is measured in years, answer the following questions:

a. What is the decay constant?

b. What quantity is present initially?

c. Complete the following table of values:

t	0	5	10	20	100
Q					

3. GROWTH OF BACTERIA The growth rate of the bacterium *Escherichia coli*, a common bacterium found in the human intestine, is proportional to its size. Under ideal laboratory conditions, when this bacterium is grown in a nutrient broth medium, the number of cells in a culture doubles approximately every 20 min.

a. If the initial cell population is 100, determine the function $Q(t)$ that expresses the exponential growth of the number of cells of this bacterium as a function of time t (in minutes).

b. How long will it take for a colony of 100 cells to increase to a population of 1 million?

c. If the initial cell population were 1000, how would this alter our model?

4. WORLD POPULATION The world population at the beginning of 1990 was 5.3 billion. Assume that the population continues to grow at the rate of approximately 2%/year and find the function $Q(t)$ that expresses the world population (in billions) as a function of time t (in years), with $t = 0$ corresponding to the beginning of 1990.

a. Using this function, complete the following table of values and sketch the graph of the function Q.

Year	1990	1995	2000	2005
World Population				

Year	2010	2015	2020	2025
World Population				

b. Find the estimated rate of growth in 2010.

5. WORLD POPULATION Refer to Exercise 4.

a. If the world population continues to grow at the rate of approximately 2%/year, find the length of time t_0 required for the world population to triple in size.

b. Using the time t_0 found in part (a), what would be the world population if the growth rate were reduced to 1.8%/year?

6. RESALE VALUE Garland Mills purchased a certain piece of machinery 3 years ago for \$500,000. Its present resale value is \$320,000. Assuming that the machine's resale value decreases exponentially, what will it be 4 years from now?

7. ATMOSPHERIC PRESSURE If the temperature is constant, then the atmospheric pressure P (in pounds per square inch) varies with the altitude above sea level h in accordance with the law

$$P = p_0 e^{-kh}$$

where p_0 is the atmospheric pressure at sea level and k is a constant. If the atmospheric pressure is 15 lb/in.^{2} at sea level and 12.5 lb/in.^{2} at 4000 ft, find the atmospheric pressure at an altitude of 12,000 ft. How fast is the atmospheric pressure changing with respect to altitude at an altitude of 12,000 ft?

8. RADIOACTIVE DECAY The radioactive element polonium decays according to the law

$$Q(t) = Q_0 \cdot 2^{-(t/140)}$$

where Q_0 is the initial amount and the time t is measured in days. If the amount of polonium left after 280 days is 20 mg, what was the initial amount present?

9. RADIOACTIVE DECAY Phosphorus 32 (P-32) has a half-life of 14.2 days. If 100 g of this substance are present initially, find the amount present after t days. What amount will be left after 7.1 days? How fast is P-32 decaying when $t = 7.1$?

10. NUCLEAR FALLOUT Strontium 90 (Sr-90), a radioactive isotope of strontium, is present in the fallout resulting from nuclear explosions. It is especially hazardous to animal life, including humans, because, upon ingestion of contaminated food, it is absorbed into the bone structure. Its half-life is 27 years. If the amount of Sr-90 in a certain area is found to be four times the "safe" level, find how much time must elapse before the safe level is reached.

11. **CARBON-14 DATING** Wood deposits recovered from an archeological site contain 20% of the C-14 they originally contained. How long ago did the tree from which the wood was obtained die?

12. **CARBON-14 DATING** The skeletal remains of the so-called Pittsburgh Man, unearthed in Pennsylvania, had lost 82% of the C-14 they originally contained. Determine the approximate age of the bones.

13. **BANK FAILURES** In the wake of the 2008 financial crisis, bank failures started spiraling upward. The number of bank failures peaked in 2010 at 157. Thereafter, the number of bank failures began to fall sharply. The number of bank failures from 2010 through 2012 is described by the function

$$f(t) = 157e^{-0.55t} \qquad (0 \le t \le 2)$$

where t is measured in years, with $t = 0$ corresponding to 2010.
   a. How fast was the number of bank failures dropping in 2011?
   b. If the trend continued, how many bank failures were there in 2013?
   Source: FDIC.

14. **ONLINE SHOPPERS** The number of consumers researching products or shopping online is growing steadily. According to the research firm, eMarketer, the number of online shoppers is projected to be

$$f(t) = 172.2e^{0.031t} \qquad (0 \le t \le 5)$$

million in year t, where $t = 0$ corresponds to 2010.
   a. Find the projected number of online shoppers in each of the years 2010 through 2015 by completing the following table:

t	0	1	2	3	4	5
$f(t)$						

   b. Use the information from part (a) to sketch the graph of f.
   Source: eMarketer.

15. **EXPONENTIAL SALES GROWTH** The sales of the Garland Corporation are projected to grow exponentially for the years between 2010 and 2015 from $100 million to $150 million.
   a. Find a model giving the sales of Garland Corporation in year t between 2010 ($t = 0$) and 2015 ($t = 5$). Hint: The sales in year t are $S = S_0 e^{kt}$.
   b. What were the sales of Garland Corporation in 2013?

16. **LEARNING CURVES** The American Court Reporting Institute finds that the average student taking Advanced Machine Shorthand, an intensive 20-week course, progresses according to the function

$$Q(t) = 120(1 - e^{-0.05t}) + 60 \qquad (0 \le t \le 20)$$

where $Q(t)$ measures the number of words (per minute) of dictation that the student can take in machine shorthand after t weeks in the course. Sketch the graph of the function Q and answer the following questions:

a. What is the beginning shorthand speed for the average student in this course?
b. What shorthand speed does the average student attain halfway through the course?
c. How many words per minute can the average student take after completing this course?

17. **DRIVER'S LICENSES** The percentage of teenagers and young adults with a driver's license in 2010 is approximated by

$$P(x) = 90(1 - e^{-0.37(x-15)}) \qquad (16 \le x \le 39)$$

where x is the age of the person. Sketch the graph of the function P, and answer the following questions:
   a. What percentage of 16-year-olds had a driver's license in 2010?
   b. What percentage of 20-year-olds had a driver's license in 2010?
   c. What percentage of 39-year-olds had a driver's license in 2010?
   Source: University of Michigan.

18. **EFFECT OF ADVERTISING ON SALES** Metro Department Store found that t weeks after the end of a sales promotion the volume of sales was given by

$$S(t) = B + Ae^{-kt} \qquad (0 \le t \le 4)$$

where $B = 50,000$ and is equal to the average weekly volume of sales before the promotion. The sales volumes at the end of the first and third weeks were $83,515 and $65,055, respectively. Assume that the sales volume is decreasing exponentially.
   a. Find the decay constant k.
   b. Find the sales volume at the end of the fourth week.
   c. How fast is the sales volume dropping at the end of the fourth week?

19. **DEMAND FOR COMPUTERS** Universal Instruments found that the monthly demand for its new line of Galaxy tablet computers t months after the line was placed on the market was given by

$$D(t) = 2000 - 1500e^{-0.05t} \qquad (t > 0)$$

Graph this function and answer the following questions:
   a. What is the demand after 1 month? After 1 year? After 2 years? After 5 years?
   b. At what level is the demand expected to stabilize?
   c. Find the rate of growth of the demand after the tenth month.

20. **RELIABILITY OF COMPUTER CHIPS** The percentage of a certain brand of computer chips that will fail after t years of use is estimated to be

$$P(t) = 100(1 - e^{-0.1t})$$

   a. What percentage of this brand of computer chips are expected to be usable after 3 years?
   b. Evaluate $\lim_{t \to \infty} P(t)$. Did you expect this result?

21. Lengths of Fish The length (in centimeters) of a typical Pacific halibut t years old is approximately

$$f(t) = 200(1 - 0.956e^{-0.18t})$$

a. What is the length of a typical 5-year-old Pacific halibut?

b. How fast is the length of a typical 5-year-old Pacific halibut increasing?

c. What is the maximum length a typical Pacific halibut can attain?

22. Spread of an Epidemic During a flu epidemic, the number of children in the Woodbridge Community School System who contracted influenza after t days was given by

$$Q(t) = \frac{1000}{1 + 199e^{-0.8t}}$$

a. How many children were stricken by the flu after the first day?

b. How many children had the flu after 10 days?

c. How many children eventually contracted the disease?

23. Growth of a Fruit Fly Population On the basis of data collected during an experiment, a biologist found that the growth of a fruit fly (*Drosophila*) with a limited food supply could be approximated by the exponential model

$$N(t) = \frac{400}{1 + 39e^{-0.16t}}$$

where t denotes the number of days since the beginning of the experiment.

a. What was the initial fruit fly population in the experiment?

b. What was the maximum fruit fly population that could be expected under this laboratory condition?

c. What was the population of the fruit fly colony on the 20th day?

d. How fast was the population changing on the 20th day?

24. Demographics The number of citizens aged 45–64 years is approximated by

$$P(t) = \frac{197.9}{1 + 3.274e^{-0.0361t}} \qquad (0 \le t \le 25)$$

where $P(t)$ is measured in millions and t is measured in years, with $t = 0$ corresponding to the beginning of 1990. People belonging to this age group are the targets of insurance companies that want to sell them annuities. What was the expected population of citizens aged 45–64 years in 2010? In 2015?

Source: K. G. Securities.

25. U.S. Crude Oil Production The rate of production of crude oil in the United States in year t (in millions of barrels per day) from 1900 through 2010 is approximated by the function

$$f(t) = \frac{40e^{-(t-1975)/20}}{(1 + e^{-(t-1975)/20})^2} \qquad (1900 \le t \le 2010)$$

What was the maximum rate of production of crude oil in the United States for the period under consideration?
Hint: Let $u = (t - 1975)/20$.
Source: U.S. Energy Information Administration.

26. Population Growth in the Twenty-First Century The U.S. population is approximated by the function

$$P(t) = \frac{616.5}{1 + 4.02e^{-0.5t}}$$

where $P(t)$ is measured in millions of people and t is measured in 30-year intervals, with $t = 0$ corresponding to 1930. What is the expected population of the United States in 2020 ($t = 3$)?

27. Dissemination of Information Three hundred students attended the dedication ceremony of a new building on a college campus. The president of the traditionally female college announced a new expansion program, which included plans to make the college coeducational. The number of students who learned of the new program t hr later is given by the function

$$f(t) = \frac{3000}{1 + Be^{-kt}}$$

If 600 students on campus had heard about the new program 2 hr after the ceremony, how many students had heard about the policy after 4 hr? How fast was the news spreading 4 hr after the ceremony?

28. Price of a Commodity The unit price of a certain commodity is given by

$$p = f(t) = 6 + 4e^{-2t}$$

where p is measured in dollars and t is measured in months.

a. Show that f is decreasing on $(0, \infty)$.

b. Show that the graph of f is concave upward on $(0, \infty)$.

c. Evaluate $\lim_{t \to \infty} f(t)$. (*Note:* This value is called the *equilibrium price* of the commodity, and in this case, we have *price stability*.)

d. Sketch the graph of f.

29. Chemical Mixtures Two chemicals react to form a compound. Suppose the amount of the compound formed in time t (in hours) is given by

$$x(t) = \frac{15\left[1 - \left(\frac{2}{3}\right)^{3t}\right]}{1 - \frac{1}{4}\left(\frac{2}{3}\right)^{3t}}$$

where $x(t)$ is measured in pounds. How many pounds of the compound are formed eventually?
Hint: You need to evaluate $\lim_{t \to \infty} x(t)$.

30. Von Bertalanffy Growth Function The length (in centimeters) of a common commercial fish is approximated by the von Bertalanffy growth function

$$f(t) = a(1 - be^{-kt})$$

where a, b, and k are positive constants.

a. Show that f is increasing on the interval $(0, \infty)$.

b. Show that the graph of f is concave downward on $(0, \infty)$.

c. Show that $\lim_{t \to \infty} f(t) = a$.

d. Use the results of parts (a)–(c) to sketch the graph of f.

31. Absorption of Drugs The concentration of a drug in grams per cubic centimeter (g/cm^3) t min after it has been injected into the bloodstream is given by

$$C(t) = \frac{k}{b - a}(e^{-at} - e^{-bt})$$

where a, b, and k are positive constants, with $b > a$.

a. At what time is the concentration of the drug the greatest?

b. What will be the concentration of the drug in the long run?

32. Concentration of Glucose in the Bloodstream A glucose solution is administered intravenously into the bloodstream at a constant rate of r mg/hr. As the glucose is being administered, it is converted into other substances and removed from the bloodstream. Suppose the concentration of the glucose solution at time t is given by

$$C(t) = \frac{r}{k} - \left[\frac{r}{k} - C_0\right]e^{-kt}$$

where C_0 is the concentration at time $t = 0$ and k is a positive constant. Assuming that $C_0 < r/k$, evaluate $\lim_{t \to \infty} C(t)$.

a. What does your result say about the concentration of the glucose solution in the long run?

b. Show that the function C is increasing on $(0, \infty)$.

c. Show that the graph of C is concave downward on $(0, \infty)$.

d. Sketch the graph of the function C.

33. Radioactive Decay A radioactive substance decays according to the formula

$$Q(t) = Q_0 e^{-kt}$$

where $Q(t)$ denotes the amount of the substance present at time t (measured in years), Q_0 denotes the amount of the substance present initially, and k (a positive constant) is the decay constant.

a. Show that half-life of the substance is $\bar{t} = (\ln 2)/k$.

b. Suppose a radioactive substance decays according to the formula

$$Q(t) = 20e^{-0.0001238t}$$

How long will it take for the substance to decay to half the original amount?

34. Logistic Growth Function Consider the logistic growth function

$$Q(t) = \frac{A}{1 + Be^{-kt}}$$

where A, B, and k are positive constants.

a. Show that Q satisfies the equation

$$Q'(t) = kQ\left(1 - \frac{Q}{A}\right)$$

b. Show that $Q(t)$ is increasing on $(0, \infty)$.

35. a. Use the results of Exercise 34 to show that the graph of Q has an inflection point when $Q = A/2$ and that this occurs when $t = (\ln B)/k$.

b. Interpret your results.

36. Logistic Growth Function Consider the logistic growth function

$$Q(t) = \frac{A}{1 + Be^{-kt}}$$

Suppose the population is Q_1 when $t = t_1$ and Q_2 when $t = t_2$. Show that the value of k is

$$k = \frac{1}{t_2 - t_1} \ln\left[\frac{Q_2(A - Q_1)}{Q_1(A - Q_2)}\right]$$

37. Logistic Growth Function The carrying capacity of a colony of fruit flies (*Drosophila*) is 600. The population of fruit flies after 14 days is 76, and the population after 21 days is 167. What is the value of the growth constant k? Hint: Use the result of Exercise 36.

38. Gompertz Growth Curve Consider the function

$$Q(t) = Ce^{-Ae^{-kt}}$$

where $Q(t)$ is the size of a quantity at time t, and A, C, and k are positive constants. The graph of this function, called the *Gompertz growth curve*, is used by biologists to describe restricted population growth.

a. Show that the function Q is always increasing.

b. Find the time t at which the growth rate $Q'(t)$ is increasing most rapidly. Hint: Find the inflection point of Q.

c. Show that $\lim_{t \to \infty} Q(t) = C$, and interpret your result.

5.6 Solution to Self-Check Exercise

We are given that $P_0 = 226.5$, $k = 0.008$, and $I = 0.5$. So

$$P = \left(226.5 + \frac{0.5}{0.008}\right)e^{0.008t} - \frac{0.5}{0.008}$$

$$= 289e^{0.008t} - 62.5$$

Therefore, the expected population in 2015 is given by

$$P(35) = 289e^{0.28} - 62.5$$

$$\approx 319.9$$

or approximately 319.9 million.

USING TECHNOLOGY | Analyzing Mathematical Models

We can use a graphing utility to analyze the mathematical models encountered in this section.

APPLIED EXAMPLE 1 Internet-Gaming Sales The estimated growth in global Internet-gaming revenue (in billions of dollars), as predicted by industry analysts, is given in the following table:

Year	2001	2002	2003	2004	2005	2006	2007	2008	2009	2010
Revenue	3.1	3.9	5.6	8.0	11.8	15.2	18.2	20.4	22.7	24.5

a. Use **Logistic** to find a regression model for the data. Let $t = 0$ correspond to 2001.
b. Plot the scatter diagram and the graph of the function f found in part (a) using the viewing window $[0, 9] \times [0, 35]$.
c. How fast was the revenue from global gaming on the Internet changing in 2001? In 2007?

Source: Christiansen Capital/Advisors.

Solution

a. Using **Logistic**, we find

$$f(t) = \frac{27.11}{1 + 9.64e^{-0.49t}} \qquad (0 \le t \le 9)$$

b. The scatter plot for the data, and the graph of f in the viewing window $[0, 9] \times [0, 35]$, are shown in Figure T1.
c. Using the numerical derivative operation, we find $f'(0) \approx 1.13$ and $f'(6) \approx 2.971$. We conclude that the revenue was increasing at the rate of $1.13 billion per year in 2001 and at the rate of $2.97 billion per year in 2007.

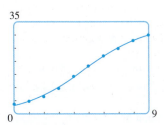

FIGURE T1
The graph of f in the viewing window $[0, 9] \times [0, 35]$

TECHNOLOGY EXERCISES

1. **AIR TRAVEL** Air travel has been rising dramatically in the past 30 years. In a study conducted in 2000, the FAA projected further exponential growth for air travel through 2010. The function

$$f(t) = 666e^{0.0413t} \qquad (0 \le t \le 10)$$

gives the number of passengers (in millions) in year t, with $t = 0$ corresponding to 2000.

a. Plot the graph of f, using the viewing window $[0, 10] \times [0, 1000]$.
b. How many air passengers were there in 2000? What was the projected number of air passengers for 2008?
c. What was the rate of change of the number of air passengers in 2008?

Source: Federal Aviation Administration.

2. NEWTON'S LAW OF COOLING The temperature of a cup of coffee t min after it is poured is given by

$$T = 70 + 100e^{-0.0446t}$$

where T is measured in degrees Fahrenheit.
   a. Plot the graph of T, using the viewing window $[0, 30] \times [0, 200]$.
   b. When will the coffee be cool enough to drink (say, $120°$)?
   Hint: Use the **ISECT** function.

3. COMPUTER GAME SALES The total number of Starr Communication's newest game, Laser Beams, sold t months after its release is given by

$$N(t) = -20(t + 20)e^{-0.05t} + 400$$

thousand units.
   a. Plot the graph of N, using the viewing window $[0, 500] \times [0, 500]$.
   b. Use the result of part (a) to find $\lim_{t \to \infty} N(t)$, and interpret this result.

4. POPULATION GROWTH IN THE TWENTY-FIRST CENTURY The U.S. population is approximated by the function

$$P(t) = \frac{616.5}{1 + 4.02e^{-0.5t}}$$

where $P(t)$ is measured in millions of people and t is measured in 30-year intervals, with $t = 0$ corresponding to 1930.
   a. Plot the graph of P, using the viewing window $[0, 4] \times [0, 650]$.
   b. What is the projected population of the United States in 2020 $(t = 3)$?
   c. What is the projected rate of growth of the U.S. population in 2020?

5. TIME RATE OF GROWTH OF A TUMOR The rate at which a tumor grows, with respect to time, is given by

$$R = Ax \ln \frac{B}{x} \qquad (0 < x < B)$$

where A and B are positive constants and x is the radius of the tumor in centimeters.
   a. Plot the graph of R for the case $A = B = 10$, using the viewing window $[0, 10] \times [0, 45]$.
   b. Find the radius of the tumor when the tumor is growing most rapidly with respect to time for the case $A = B = 10$.

6. ANNUITIES At the time of retirement, Christine expects to have a sum of $500,000 in her retirement account. Assuming that the account pays interest at the rate of 5%/year compounded continuously, her accountant pointed out to her that if she made withdrawals amounting to x dollars per year ($x > 25,000$), then the time required to deplete her savings would be T years, where

$$T = f(x) = 20 \ln\left(\frac{x}{x - 25,000}\right) \qquad (x > 25,000)$$

   a. Plot the graph of f, using the viewing window $[25,000, 50,000] \times [0, 100]$.
   b. How much should Christine plan to withdraw from her retirement account each year if she wants it to last for 25 years?
   c. Evaluate $\lim_{x \to 25,000^+} f(x)$. Is the result expected? Explain.
   d. Evaluate $\lim_{x \to \infty} f(x)$. Is the result expected? Explain.

7. ABSORPTION OF DRUGS The concentration of a drug in an organ at any time t (in seconds) is given by

$$C(t) = \begin{cases} 0.3t - 18(1 - e^{-t/60}) & \text{if } 0 \le t \le 20 \\ 18e^{-t/60} - 12e^{-(t-20)/60} & \text{if } t > 20 \end{cases}$$

where $C(t)$ is measured in grams per cubic centimeter (g/cm^3).
   a. Plot the graph of C, using the viewing window $[0, 120] \times [0, 1]$.
   b. What is the initial concentration of the drug in the organ?
   c. What is the concentration of the drug in the organ after 10 sec?
   d. What is the concentration of the drug in the organ after 30 sec?
   e. What will be the concentration of the drug in the long run?

8. MODELING WITH DATA The snowfall accumulation at Logan Airport (in inches), t hr after the beginning of a 33-hr snowstorm in Boston on a certain day, follows:

Hour	0	3	6	9	12	15	18	21	24	27	30	33
Inches	0.1	0.4	3.6	6.5	9.1	14.4	19.5	22	23.6	24.8	26.6	27

Here, $t = 0$ corresponds to noon of February 6.
   a. Use **Logistic** to find a regression model for the data.
   b. Plot the scatter diagram and the graph of the function f found in part (a), using the viewing window $[0, 33] \times [0, 30]$.
   c. How fast was the snowfall accumulating at midnight on February 6? At noon on February 7?
   d. At what time during the storm was the snowfall accumulating at the greatest rate? What was the rate of accumulation?

Source: Boston Globe.

CHAPTER 5 Summary of Principal Formulas and Terms

FORMULAS

1. Exponential function with base b	$y = b^x$
2. The number e	$e = \lim\limits_{m \to \infty} \left(1 + \dfrac{1}{m}\right)^m = 2.71828\ldots$
3. Exponential function with base e	$y = e^x$
4. Logarithmic function with base b	$y = \log_b x$
5. Logarithmic function with base e	$y = \ln x$
6. Inverse properties of $\ln x$ and e^x	$\ln e^x = x$ and $e^{\ln x} = x$
7. Compound interest (accumulated amount)	$A = P\left(1 + \dfrac{r}{m}\right)^{mt}$
8. Effective rate of interest	$r_{\text{eff}} = \left(1 + \dfrac{r}{m}\right)^m - 1$
9. Compound interest (present value)	$P = A\left(1 + \dfrac{r}{m}\right)^{-mt}$
10. Continuous compound interest	$A = Pe^{rt}$
11. Derivative of the exponential function	$\dfrac{d}{dx}(e^x) = e^x$
12. Chain Rule for Exponential Functions	$\dfrac{d}{dx}(e^u) = e^u \dfrac{du}{dx}$
13. Derivative of the logarithmic function	$\dfrac{d}{dx}\ln|x| = \dfrac{1}{x}$
14. Chain Rule for Logarithmic Functions	$\dfrac{d}{dx}\ln|u| = \dfrac{1}{u}\dfrac{du}{dx}$

TERMS

common logarithm (347)

natural logarithm (347)

compound interest (354)

logarithmic differentiation (384)

exponential growth (391)

growth constant (391)

exponential decay (392)

decay constant (393)

half-life of a radioactive substance (393)

logistic growth function (396)

CHAPTER 5 Concept Review Questions

Fill in the blanks.

1. The function $f(x) = x^b$ (b, a real number) is called a/an _____ function, whereas the function $g(x) = b^x$, where $b >$ _____ and $b \ne$ _____, is called a/an _____ function.

2. a. The domain of the function $y = 3^x$ is _____, and its range is _____.

b. The graph of the function $y = 0.3^x$ passes through the point _____ and is decreasing on _____.

3. a. If $b > 0$ and $b \ne 1$, then the logarithmic function $y = \log_b x$ has domain _____ and range _____; its graph passes through the point _____.

b. The graph of $y = \log_b x$ is decreasing if b _____ and increasing if b _____.

4. a. If $x > 0$, then $e^{\ln x} =$ _____.

b. If x is any real number, then $\ln e^x =$ _____.

5. In the compound interest formula $A = P\left(1 + \frac{r}{m}\right)^{mt}$, A stands for the _____ _____, P stands for the _____, r stands for the _____ _____ _____ per year, m stands for the _____ _____ _____ _____ per year, and t stands for the _____.

6. The effective rate r_{eff} is related to the nominal interest rate r per year and the number of conversion periods per year by $r_{eff} =$ _____.

7. If interest earned at the rate of r per year is compounded continuously over t years, then a principal of P dollars will have an accumulated value of $A =$ _____ dollars.

8. **a.** If $g(x) = e^{f(x)}$, where f is a differentiable function, then $g'(x) =$ _____.
   b. If $g(x) = \ln f(x)$, where $f(x) > 0$ is differentiable, then $g'(x) =$ _____.

9. **a.** In the unrestricted exponential growth model $Q = Q_0 e^{kt}$, Q_0 represents the quantity present _____, and k is called the _____ constant.

b. In the exponential decay model $Q = Q_0 e^{-kt}$, k is called the _____ constant.

c. The half-life of a radioactive substance is the _____ required for a substance to decay to _____ _____ of its original amount.

10. **a.** For the model $Q(t) = C - Ae^{-kt}$ describing a learning curve, $y = C$ is a/an _____ _____ of the graph of Q. The value of $Q(t)$ never exceeds _____.

b. For the logistic growth model $Q(t) = \dfrac{A}{1 + Be^{-kt}}$, $y = A$ is a/an _____ _____ of the graph of Q. If the quantity $Q(t)$ is initially smaller than A, then $Q(t)$ will eventually approach _____ as t increases; the number A represents the life-support capacity of the environment and is called the _____ _____ of the environment.

CHAPTER 5 Review Exercises

1. Sketch the graphs of the exponential functions defined by the equations on the same set of coordinate axes.

   a. $y = 2^{-x}$ **b.** $y = \left(\dfrac{1}{2}\right)^x$

In Exercises 2 and 3, express each equation in logarithmic form.

2. $\left(\dfrac{2}{3}\right)^{-3} = \dfrac{27}{8}$ 3. $16^{-3/4} = 0.125$

In Exercises 4 and 5, solve each equation for x.

4. $\log_4(2x + 1) = 2$

5. $\ln(x - 1) + \ln 4 = \ln(2x + 4) - \ln 2$

In Exercises 6–8, given that $\ln 2 = x$, $\ln 3 = y$, and $\ln 5 = z$, express each of the given logarithmic values in terms of x, y, and z.

6. $\ln 30$ 7. $\ln 3.6$ 8. $\ln 75$

9. Sketch the graph of the function $y = \log_2(x + 3)$.

10. Sketch the graph of the function $y = \log_3(x + 1)$.

11. A sum of $10,000 is deposited in a bank. Find the amount on deposit after 2 years if the bank pays interest at the rate of 6%/year compounded (a) daily (assume a 365-day year) and (b) continuously.

12. Find the interest rate needed for an investment of $10,000 to grow to an amount of $12,000 in 3 years if interest is compounded quarterly.

13. How long will it take for an investment of $10,000 to grow to $15,000 if the investment earns interest at the rate of 6%/year compounded quarterly?

14. Find the nominal interest rate that yields an effective interest rate of 8%/year compounded quarterly.

In Exercises 15–32, find the derivative of the function.

15. $f(x) = xe^{2x}$ 16. $f(t) = \sqrt{t}e^t + t$

17. $g(t) = \sqrt{t}e^{-2t}$ 18. $g(x) = e^x\sqrt{1 + x^2}$

19. $y = \dfrac{e^{2x}}{1 + e^{-2x}}$ 20. $f(x) = e^{2x^2 - 1}$

21. $f(x) = xe^{-x^2}$ 22. $g(x) = (1 + e^{2x})^{3/2}$

23. $f(x) = x^2e^x + e^x$ 24. $g(t) = t \ln t$

25. $f(x) = \ln(e^{x^2} + 1)$ 26. $f(x) = \dfrac{x}{\ln x}$

27. $f(x) = \dfrac{\ln x}{x + 1}$ 28. $y = (x + 1)e^x$

29. $y = \ln(e^{4x} + 3)$ 30. $f(r) = \dfrac{re^r}{1 + r^2}$

31. $f(x) = \dfrac{\ln x}{1 + e^x}$ 32. $g(x) = \dfrac{e^{x^2}}{1 + \ln x}$

33. Find the second derivative of the function $y = \ln(3x + 1)$.

34. Find the second derivative of the function $y = x \ln x$.

35. Find $h'(0)$ if $h(x) = g(f(x))$, $g(x) = x + \dfrac{1}{x}$, and $f(x) = e^x$.

36. Find $h'(1)$ if $h(x) = g(f(x))$, $g(x) = \dfrac{x + 1}{x - 1}$, and $f(x) = \ln x$.

37. Use logarithmic differentiation to find the derivative of $f(x) = (2x^3 + 1)(x^2 + 2)^3$.

38. Use logarithmic differentiation to find the derivative of

$$f(x) = \frac{x(x^2 - 2)^2}{x - 1}$$

39. Find an equation of the tangent line to the graph of $y = e^{-2x}$ at the point $(1, e^{-2})$.

40. Find an equation of the tangent line to the graph of $y = xe^{-x}$ at the point $(1, e^{-1})$.

41. Sketch the graph of the function $f(x) = xe^{-2x}$.

42. Sketch the graph of the function $f(x) = x^2 - \ln x$.

43. Find the absolute extrema of the function $f(t) = te^{-t}$.

44. Find the absolute extrema of the function

$$g(t) = \frac{\ln t}{t}$$

on $[1, 2]$.

45. **INVESTMENT RETURN** A hotel was purchased by a conglomerate for \$4.5 million and sold 5 years later for \$8.2 million. Find the annual rate of return (compounded continuously).

46. Find the present value of \$30,000 due in 5 years at an interest rate of 5%/year compounded monthly.

47. Find the present value of \$119,346 due in 4 years at an interest rate of 6%/year compounded continuously.

48. **CONSUMER PRICE INDEX** At an annual inflation rate of 7.5%, how long will it take the Consumer Price Index (CPI) to double?

49. **GROWTH OF BACTERIA** A culture of bacteria that initially contained 2000 bacteria has a count of 18,000 bacteria after 2 hr.
    a. Determine the function $Q(t)$ that expresses the exponential growth of the number of cells of this bacterium as a function of time t (in minutes).
    b. Find the number of bacteria present after 4 hr.

50. **RADIOACTIVE DECAY** The radioactive element radium has a half-life of 1600 years. What is its decay constant?

51. **DEMAND FOR DVD PLAYERS** VCA Television found that the monthly demand for its new line of DVD players t months after placing the players on the market is given by

$$D(t) = 4000 - 3000e^{-0.06t} \qquad (t \geq 0)$$

Graph this function and answer the following questions:
    a. What was the demand after 1 month? After 1 year? After 2 years?
    b. At what level is the demand expected to stabilize?

52. **RADIOACTIVITY** The mass of a radioactive isotope at time t (in years) is $M(t) = 200e^{-0.14t}$ g. What is the mass of the isotope initially? How fast is the mass of the isotope changing 2 years later?

53. **ENERGY CONSUMPTION OF APPLIANCES** The average energy consumption of the typical refrigerator/freezer manufactured by York Industries is approximately

$$C(t) = 1486e^{-0.073t} + 500 \qquad (0 \leq t \leq 20)$$

kilowatt-hours (kWh) per year, where t is measured in years, with $t = 0$ corresponding to 1972.
    a. What was the average energy consumption of the York refrigerator/freezer at the beginning of 1972?
    b. Prove that the average energy consumption of the York refrigerator/freezer is decreasing over the years in question.
    c. All refrigerator/freezers manufactured since January 1, 1990, must meet the 950-kWh/year maximum energy-consumption standard set by the National Appliance Conservation Act. Show that the York refrigerator/freezer satisfies this requirement.

54. **OIL USED TO FUEL PRODUCTIVITY** A study on worldwide oil use was prepared for a major oil company. The study predicted that the amount of oil used to fuel productivity in a certain country is given by

$$f(t) = 1.5 + 1.8te^{-1.2t} \qquad (0 \leq t \leq 4)$$

where $f(t)$ denotes the number of barrels per \$1000 of economic output and t is measured in decades ($t = 0$ corresponds to 1965). Compute $f'(0), f'(1), f'(2)$, and $f'(3)$ and interpret your results.

55. **PRICE OF A COMMODITY** The price of a certain commodity in dollars per unit at time t (measured in weeks) is given by $p = 18 - 3e^{-2t} - 6e^{-t/3}$.
    a. What is the price of the commodity at $t = 0$?
    b. How fast is the price of the commodity changing at $t = 0$?
    c. Find the equilibrium price of the commodity.
    Hint: It is given by $\lim\limits_{t \to \infty} p$.

56. **FLU EPIDEMIC** During a flu epidemic, the number of students at a certain university who contracted influenza after t days could be approximated by the exponential model

$$Q(t) = \frac{3000}{1 + 499e^{-kt}}$$

If 90 students contracted the flu by day 10, how many students contracted the flu by day 20?

57. **WORLD POPULATION** The world population is projected to grow according to the model

$$P(t) = \frac{12}{1 + 3e^{-0.2747t}} \qquad (0 \leq t \leq 8)$$

where $P(t)$ is measured in billions and t in decades, with $t = 0$ corresponding to 1960.

a. What was the world population in 1960?

b. What is the world population expected to be in 2040?

c. At what rate will the world population be growing in 2030?

Source: U.S. Census Bureau.

58. **U.S. INFANT MORTALITY RATE** The U.S. infant mortality rate (per 1000 live births) is approximated by the function

$$N(t) = 12.5e^{-0.0294t} \qquad (0 \le t \le 21)$$

where t is measured in years, with $t = 0$ corresponding to 1980.

a. What was the mortality rate in 1980? In 1990? In 2000?

b. Sketch the graph of N.

Source: U.S. Department of Health and Human Services.

59. **MAXIMIZING REVENUE** The unit selling price p (in dollars) and the quantity demanded x (in pairs) of a certain brand of men's socks is given by the demand equation

$$p = 20e^{-0.0002x} \qquad (0 \le x \le 10,000)$$

How many pairs of socks must be sold to yield a maximum revenue? What will the maximum revenue be?

60. **ABSORPTION OF DRUGS** The concentration of a drug in an organ at any time t (in seconds) is given by

$$x(t) = 0.08(1 - e^{-0.02t})$$

where $x(t)$ is measured in grams per cubic centimeter (g/cm^3).

a. What is the initial concentration of the drug in the organ?

b. What is the concentration of the drug in the organ after 30 sec?

c. What will be the concentration of the drug in the organ in the long run?

d. Sketch the graph of x.

CHAPTER 5 **Before Moving On . . .**

1. Solve the equation $\dfrac{100}{1 + 2e^{0.3t}} = 40$ for t.

2. Find the accumulated amount after 4 years if $3000 is invested at 8%/year compounded weekly.

3. Find the slope of the tangent line to the graph of $f(x) = e^{\sqrt{x}}$.

4. Find the rate at which $y = x \ln(x^2 + 1)$ is changing at $x = 1$.

5. Find the second derivative of $y = e^{2x} \ln 3x$.

6. The temperature of a cup of coffee at time t (in minutes) is

$$T(t) = 70 + ce^{-kt}$$

Initially, the temperature of the coffee was 200°F. Three minutes later, it was 180°. When will the temperature of the coffee be 150°F?

6

Integration

DIFFERENTIAL CALCULUS focuses on the problem of finding the rate of change of one quantity with respect to another. In this chapter, we begin the study of the other branch of calculus, known as integral calculus. Here we are interested in precisely the opposite problem: If we know the rate of change of one quantity with respect to another, can we find the relationship between the two quantities? The principal tool used in the study of integral calculus is the *antiderivative* of a function, and we develop rules for antidifferentiation, or *integration*, as the process of finding the antiderivative is called. We also show that a link is established between differential and integral calculus—via the Fundamental Theorem of Calculus.

How much electricity should be produced over the next 3 years to meet the projected demand? In Example 9, page 451, you will see how the current rate of consumption can be used to answer this question.

© Songquan Deng/ShutterStock.com

6.1   Antiderivatives and the Rules of Integration

Antiderivatives

Let's return once again to the example involving the motion of the maglev (Figure 1).

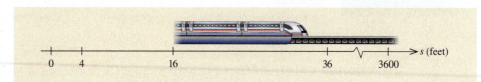

FIGURE 1
A maglev moving along an elevated monorail track

In Chapter 2, we discussed the following problem:

If we know the position of the maglev at any time t, can we find its velocity at time t?

As it turns out, if the position of the maglev is described by the position function f, then its velocity at any time t is given by $f'(t)$. Here, f'—the velocity function of the maglev—is just the derivative of f.

Now, in Chapters 6 and 7, we will consider precisely the opposite problem:

If we know the velocity of the maglev at any time t, can we find its position at time t?

Stated another way, if we know the velocity function f' of the maglev, can we find its position function f?

To solve this problem, we need the concept of an antiderivative of a function.

> **Antiderivative**
>
> A function F is an **antiderivative** of f on an interval I if $F'(x) = f(x)$ for all x in I.

Thus, an antiderivative of a function f is a function F whose derivative is f. For example, $F(x) = x^2$ is an antiderivative of $f(x) = 2x$ because

$$F'(x) = \frac{d}{dx}(x^2) = 2x = f(x)$$

and $F(x) = x^3 + 2x + 1$ is an antiderivative of $f(x) = 3x^2 + 2$ because

$$F'(x) = \frac{d}{dx}(x^3 + 2x + 1) = 3x^2 + 2 = f(x)$$

EXAMPLE 1 Let $F(x) = \frac{1}{3}x^3 - 2x^2 + x - 1$. Show that F is an antiderivative of $f(x) = x^2 - 4x + 1$.

Solution   Differentiating the function F, we obtain

$$F'(x) = x^2 - 4x + 1 = f(x)$$

and the desired result follows.

EXAMPLE 2 Let $F(x) = x$, $G(x) = x + 2$, and $H(x) = x + C$, where C is a constant. Show that F, G, and H are all antiderivatives of the function f defined by $f(x) = 1$.

Solution   Since

$$F'(x) = \frac{d}{dx}(x) = 1 = f(x)$$

$$G'(x) = \frac{d}{dx}(x + 2) = 1 = f(x)$$

$$H'(x) = \frac{d}{dx}(x + C) = 1 = f(x)$$

we see that F, G, and H are indeed antiderivatives of f.

Example 2 shows that once an antiderivative G of a function f is known, then another antiderivative of f may be found by adding an arbitrary constant to the function G. The following theorem states that no function other than one obtained in this manner can be an antiderivative of f. (We omit the proof.)

> **THEOREM 1**
>
> Let G be an antiderivative of a function f on an interval I. Then, every antiderivative F of f on I must be of the form $F(x) = G(x) + C$, where C is a constant.

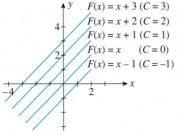

FIGURE 2
The graphs of some antiderivatives of $f(x) = 1$

Returning to Example 2, we see that there are infinitely many antiderivatives of the function $f(x) = 1$. We obtain each one by specifying the constant C in the function $F(x) = x + C$. Figure 2 shows the graphs of some of these antiderivatives for selected values of C. These graphs constitute part of a family of infinitely many parallel straight lines, each having a slope equal to 1. This result is expected, since there are infinitely many curves (straight lines) with a given slope equal to 1. The antiderivatives $F(x) = x + C$ (C, a constant) are precisely the functions representing this family of straight lines.

EXAMPLE 3 Prove that the function $G(x) = x^2$ is an antiderivative of the function $f(x) = 2x$. Write a general expression for the antiderivatives of f.

Solution   Since $G'(x) = 2x = f(x)$, we have shown that $G(x) = x^2$ is an antiderivative of $f(x) = 2x$. By Theorem 1, every antiderivative of the function $f(x) = 2x$ has the form $F(x) = x^2 + C$, where C is some constant. The graphs of a few of the antiderivatives of f are shown in Figure 3.

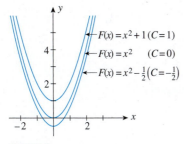

FIGURE 3
The graphs of some antiderivatives of $f(x) = 2x$

> **Exploring with TECHNOLOGY**
>
> Let $f(x) = x^2 - 1$.
>
> 1. Show that $F(x) = \frac{1}{3}x^3 - x + C$, where C is an arbitrary constant, is an antiderivative of f.
>
> 2. Use a graphing utility to plot the graphs of the antiderivatives of f corresponding to $C = -2$, $C = -1$, $C = 0$, $C = 1$, and $C = 2$ on the same set of axes, using the viewing window $[-4, 4] \times [-4, 4]$.
>
> 3. If your graphing utility has the capability, draw the tangent line to each of the graphs in part 2 at the point whose x-coordinate is 2. What can you say about this family of tangent lines?
>
> 4. What is the slope of a tangent line in this family? Explain how you obtained your answer.

The Indefinite Integral

The process of finding all antiderivatives of a function is called **antidifferentiation,** or **integration.** We use the symbol ∫, called an **integral sign,** to indicate that the operation of integration is to be performed on some function f. Thus,

$$\int f(x)\, dx = F(x) + C$$

[read "the indefinite integral of f of x with respect to x equals F of x plus C"] tells us that the **indefinite integral** of f is the family of functions given by $F(x) + C$, where $F'(x) = f(x)$. The function f to be integrated is called the **integrand,** and the constant C is called a **constant of integration.** The expression dx following the integrand $f(x)$ reminds us that the operation is performed with respect to x. If the independent variable is t, we write $\int f(t)\, dt$ instead. In this sense, both t and x are "dummy variables."

Using this notation, we can write the results of Examples 2 and 3 as

$$\int 1\, dx = x + C \quad \text{and} \quad \int 2x\, dx = x^2 + K$$

where C and K are arbitrary constants.

Basic Integration Rules

Our next task is to develop some rules for finding the indefinite integral of a given function f. Because integration and differentiation are reverse operations, we discover many of the rules of integration by first making an "educated guess" at the antiderivative F of the function f to be integrated. Then this result is verified by demonstrating that $F' = f$.

> **Rule 1: The Indefinite Integral of a Constant**
> $$\int k\, dx = kx + C \qquad (k, \text{ a constant})$$

To prove this result, observe that

$$F'(x) = \frac{d}{dx}(kx + C) = k$$

EXAMPLE 4 Find each of the following indefinite integrals:

a. $\displaystyle\int 2\, dx$ **b.** $\displaystyle\int \pi^2\, dx$

Solution Each of the integrands has the form $f(x) = k$, where k is a constant. Applying Rule 1 in each case yields

a. $\displaystyle\int 2\, dx = 2x + C$

b. $\displaystyle\int \pi^2\, dx = \pi^2 x + C$

Next, from the rule of differentiation,

$$\frac{d}{dx} x^n = nx^{n-1}$$

we obtain the following rule of integration.

Rule 2: The Power Rule

$$\int x^n \, dx = \frac{1}{n+1} x^{n+1} + C \qquad (n \neq -1)$$

An antiderivative of a power function is another power function obtained from the integrand by increasing its power by 1 and dividing the resulting expression by the new power.

To prove this result, observe that

$$
\begin{aligned}
F'(x) &= \frac{d}{dx}\left(\frac{1}{n+1} x^{n+1} + C\right) \\
&= \frac{n+1}{n+1} x^n \\
&= x^n \\
&= f(x)
\end{aligned}
$$

EXAMPLE 5 Find each of the following indefinite integrals:

a. $\displaystyle\int x^3 \, dx$ **b.** $\displaystyle\int x^{3/2} \, dx$ **c.** $\displaystyle\int \frac{1}{x^{3/2}} \, dx$

Solution   Each integrand is a power function with exponent $n \neq -1$. Applying Rule 2 in each case yields the following results:

a. $\displaystyle\int x^3 \, dx = \frac{1}{4} x^4 + C$

b. $\displaystyle\int x^{3/2} \, dx = \frac{1}{\frac{5}{2}} x^{5/2} + C = \frac{2}{5} x^{5/2} + C$

c. $\displaystyle\int \frac{1}{x^{3/2}} \, dx = \int x^{-3/2} \, dx = \frac{1}{-\frac{1}{2}} x^{-1/2} + C = -2x^{-1/2} + C = -\frac{2}{x^{1/2}} + C$

These results may be verified by differentiating each of the antiderivatives and showing that the result is equal to the corresponding integrand. ◼

The next rule tells us that a constant factor may be moved through an integral sign.

Rule 3: The Indefinite Integral of a Constant Multiple of a Function

$$\int cf(x) \, dx = c \int f(x) \, dx \qquad (c, \text{ a constant})$$

The indefinite integral of a constant multiple of a function is equal to the constant multiple of the indefinite integral of the function.

This result follows from the corresponding rule of differentiation (see Rule 3, Section 3.1).

⚠️ Only a constant can be "moved out" of an integral sign. For example, it would be incorrect to write

$$\int x^2 \, dx = x^2 \int 1 \, dx$$

In fact, $\int x^2 \, dx = \frac{1}{3} x^3 + C$, whereas $x^2 \int 1 \, dx = x^2(x + C) = x^3 + Cx^2$.

EXAMPLE 6 Find each of the following indefinite integrals:

a. $\displaystyle\int 2t^3\,dt$ **b.** $\displaystyle\int -3x^{-2}\,dx$

Solution   Each integrand has the form $cf(x)$, where c is a constant. Applying Rule 3, we obtain

a. $\displaystyle\int 2t^3\,dt = 2\int t^3\,dt = 2\left(\frac{1}{4}t^4 + K\right) = \frac{1}{2}t^4 + 2K = \frac{1}{2}t^4 + C$

where $C = 2K$. From now on, we will write the constant of integration as C, since any nonzero multiple of an arbitrary constant is an arbitrary constant.

b. $\displaystyle\int -3x^{-2}\,dx = -3\int x^{-2}\,dx = (-3)(-1)x^{-1} + C = \frac{3}{x} + C$

Rule 4: The Sum Rule

$$\int [f(x) + g(x)]\,dx = \int f(x)\,dx + \int g(x)\,dx$$

$$\int [f(x) - g(x)]\,dx = \int f(x)\,dx - \int g(x)\,dx$$

The indefinite integral of a sum (difference) of two functions is equal to the sum (difference) of their indefinite integrals.

This result is easily extended to the case involving the sum and difference of any finite number of functions. As in Rule 3, the proof of Rule 4 follows from the corresponding rule of differentiation (see Rule 4, Section 3.1).

EXAMPLE 7 Find the indefinite integral

$$\int (3x^5 + 4x^{3/2} - 2x^{-1/2})\,dx$$

Solution   Applying the extended version of Rule 4, we find that

$$\int (3x^5 + 4x^{3/2} - 2x^{-1/2})\,dx$$

$$= \int 3x^5\,dx + \int 4x^{3/2}\,dx - \int 2x^{-1/2}\,dx$$

$$= 3\int x^5\,dx + 4\int x^{3/2}\,dx - 2\int x^{-1/2}\,dx \qquad \text{Rule 3}$$

$$= (3)\left(\frac{1}{6}\right)x^6 + (4)\left(\frac{2}{5}\right)x^{5/2} - (2)(2)x^{1/2} + C \qquad \text{Rule 2}$$

$$= \frac{1}{2}x^6 + \frac{8}{5}x^{5/2} - 4x^{1/2} + C$$

Observe that we have combined the three constants of integration, which arise from evaluating the three indefinite integrals, to obtain one constant C. After all, the sum of three arbitrary constants is also an arbitrary constant.

Rule 5: The Indefinite Integral of the Exponential Function

$$\int e^x\,dx = e^x + C$$

The indefinite integral of the exponential function with base e is equal to the function itself (except, of course, for the constant of integration).

EXAMPLE 8 Find the indefinite integral

$$\int (2e^x - x^3)\, dx$$

Solution We have

$$\int (2e^x - x^3)\, dx = \int 2e^x\, dx - \int x^3\, dx$$

$$= 2\int e^x\, dx - \int x^3\, dx$$

$$= 2e^x - \frac{1}{4}x^4 + C \qquad \blacksquare$$

The last rule of integration in this section covers the integration of the function $f(x) = x^{-1}$. Remember that this function constituted the only exceptional case in the integration of the power function $f(x) = x^n$ (see Rule 2).

> **Rule 6: The Indefinite Integral of the Function $f(x) = x^{-1}$**
>
> $$\int x^{-1}\, dx = \int \frac{1}{x}\, dx = \ln|x| + C \qquad (x \neq 0)$$

To prove Rule 6, observe that

$$\frac{d}{dx}\ln|x| = \frac{1}{x} \qquad \text{See Rule 3, Section 5.5.}$$

EXAMPLE 9 Find the indefinite integral

$$\int \left(2x + \frac{3}{x} + \frac{4}{x^2}\right) dx$$

Solution

$$\int \left(2x + \frac{3}{x} + \frac{4}{x^2}\right) dx = \int 2x\, dx + \int \frac{3}{x}\, dx + \int \frac{4}{x^2}\, dx$$

$$= 2\int x\, dx + 3\int \frac{1}{x}\, dx + 4\int x^{-2}\, dx$$

$$= 2\left(\frac{1}{2}\right)x^2 + 3\ln|x| + 4(-1)x^{-1} + C$$

$$= x^2 + 3\ln|x| - \frac{4}{x} + C \qquad \blacksquare$$

Differential Equations

Let's return to the problem posed at the beginning of the section: *Given the derivative of a function, f', can we find the function f?* As an example, suppose we are given the function

$$f'(x) = 2x - 1 \tag{1}$$

and we wish to find $f(x)$. From what we now know, we can find f by integrating both sides of Equation (1). Thus,

$$f(x) = \int f'(x)\,dx = \int (2x - 1)\,dx = x^2 - x + C \qquad \text{(2)}$$

where C is an arbitrary constant. Thus, infinitely many functions have the derivative f', each differing from the other by a constant.

Equation (1) is called a differential equation. In general, a **differential equation** is an equation that involves the derivative or differential of an unknown function. [In the case of Equation (1), the unknown function is f.] A **solution** of a differential equation is any function that satisfies the differential equation. Thus, Equation (2) gives *all* the solutions of the differential Equation (1), and it is, accordingly, called the **general solution** of the differential equation $f'(x) = 2x - 1$.

The graphs of $f(x) = x^2 - x + C$ for selected values of C are shown in Figure 4. These graphs have one property in common: For any fixed value of x, the tangent lines to these graphs have the same slope. This follows because any member of the family $f(x) = x^2 - x + C$ must have the same slope at x—namely, $2x - 1$!

Although there are infinitely many solutions to the differential equation $f'(x) = 2x - 1$, we can obtain a **particular solution** by specifying the value the function must assume at a certain value of x. For example, suppose we stipulate that the function f under consideration must satisfy the condition $f(1) = 3$ or, equivalently, the graph of f must pass through the point $(1, 3)$. Then, using the condition on the general solution $f(x) = x^2 - x + C$, we find that

$$f(1) = 1 - 1 + C = 3$$

and $C = 3$. Thus, the particular solution is $f(x) = x^2 - x + 3$ (see Figure 4).

The condition $f(1) = 3$ is an example of an initial condition. More generally, an **initial condition** is a condition imposed on the value of f at $x = a$.

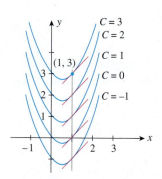

FIGURE 4
The graphs of some of the functions having the derivative $f'(x) = 2x - 1$. Observe that the slopes of the tangent lines to the graphs are the same for a fixed value of x.

Initial-Value Problems

An **initial-value problem** is one in which we are required to find a function satisfying (1) a differential equation and (2) one or more initial conditions. The following are examples of initial-value problems.

EXAMPLE 10 Find the function f if it is known that

$$f'(x) = 3x^2 - 4x + 8 \quad \text{and} \quad f(1) = 9$$

Solution   We are required to solve the initial-value problem

$$\begin{cases} f'(x) = 3x^2 - 4x + 8 \\ f(1) = \qquad\quad 9 \end{cases}$$

Integrating the function f', we find

$$\begin{aligned} f(x) &= \int f'(x)\,dx \\ &= \int (3x^2 - 4x + 8)\,dx \\ &= x^3 - 2x^2 + 8x + C \end{aligned}$$

Using the condition $f(1) = 9$, we have

$$9 = f(1) = 1^3 - 2(1)^2 + 8(1) + C = 7 + C \quad \text{or} \quad C = 2$$

Therefore, the required function f is given by $f(x) = x^3 - 2x^2 + 8x + 2$.   ◼

APPLIED EXAMPLE 11 Velocity of a Maglev In a test run of a maglev along a straight elevated monorail track, data obtained from reading its speedometer indicate that the velocity of the maglev at time t can be described by the velocity function

$$v(t) = 8t \qquad (0 \le t \le 30)$$

Find the position function of the maglev. Assume that initially the maglev is located at the origin of a coordinate line.

Solution Let $s(t)$ denote the position of the maglev at any time t $(0 \le t \le 30)$. Then, $s'(t) = v(t)$. So we have the initial-value problem

$$\begin{cases} s'(t) = 8t \\ s(0) = 0 \end{cases}$$

Integrating both sides of the differential equation $s'(t) = 8t$, we obtain

$$s(t) = \int s'(t)\, dt = \int 8t\, dt = 4t^2 + C$$

where C is an arbitrary constant. To evaluate C, we use the initial condition $s(0) = 0$ to write

$$s(0) = 4(0)^2 + C = 0 \quad \text{or} \quad C = 0$$

Therefore, the required position function is $s(t) = 4t^2$ $(0 \le t \le 30)$.

APPLIED EXAMPLE 12 Magazine Circulation The current circulation of the *Investor's Digest* is 3000 copies per week. The managing editor of this weekly projects a growth rate of

$$4 + 5t^{2/3}$$

copies per week, t weeks from now, for the next 3 years. On the basis of her projection, what will be the circulation of the digest 125 weeks from now?

Solution Let $S(t)$ denote the circulation of the digest t weeks from now. Then $S'(t)$ is the rate of change in the circulation in the tth week and is given by

$$S'(t) = 4 + 5t^{2/3}$$

Furthermore, the current circulation of 3000 copies per week translates into the initial condition $S(0) = 3000$. Integrating both sides of the differential equation with respect to t gives

$$S(t) = \int S'(t)\, dt = \int (4 + 5t^{2/3})\, dt$$
$$= 4t + 5\left(\frac{t^{5/3}}{\frac{5}{3}}\right) + C = 4t + 3t^{5/3} + C$$

To determine the value of C, we use the condition $S(0) = 3000$ to write

$$S(0) = 4(0) + 3(0)^{5/3} + C = 3000$$

which gives $C = 3000$. Therefore, the circulation of the digest t weeks from now will be

$$S(t) = 4t + 3t^{5/3} + 3000$$

In particular, the circulation 125 weeks from now will be

$$S(125) = 4(125) + 3(125)^{5/3} + 3000 = 12,875$$

copies per week.

6.1 Self-Check Exercises

1. Evaluate $\int \left(\dfrac{1}{\sqrt{x}} - \dfrac{2}{x} + 3e^x \right) dx$.

2. Find the function f given that (1) the slope of the tangent line to the graph of f at any point $P(x, f(x))$ is given by the expression $3x^2 - 6x + 3$ and (2) the graph of f passes through the point $(2, 9)$.

3. **MARKET SHARE OF AN AUTO COMPANY** Suppose United Motors' share of the new cars sold in a certain country is changing at the rate of

$$f(t) = -0.01875t^2 + 0.15t - 1.2 \qquad (0 \le t \le 12)$$

percent/year at year t ($t = 0$ corresponds to the beginning of 2004). The company's market share at the beginning of 2004 was 48.4%. What was United Motors' market share at the beginning of 2016?

Solutions to Self-Check Exercises 6.1 can be found on page 423.

6.1 Concept Questions

1. What is an antiderivative? Give an example.

2. If $f'(x) = g'(x)$ for all x in an interval I, what is the relationship between f and g?

3. What is the difference between an antiderivative of f and the indefinite integral of f?

4. Can the Power Rule be used to integrate $\int \dfrac{1}{x}\, dx$? Explain your answer.

6.1 Exercises

In Exercises 1–4, verify directly that F is an antiderivative of f.

1. $F(x) = \dfrac{1}{3}x^3 + 2x^2 - x + 2; f(x) = x^2 + 4x - 1$

2. $F(x) = xe^x + \pi; f(x) = e^x(1 + x)$

3. $F(x) = \sqrt{2x^2 - 1}; f(x) = \dfrac{2x}{\sqrt{2x^2 - 1}}$

4. $F(x) = x \ln x - x; f(x) = \ln x$

In Exercises 5–8, (a) verify that G is an antiderivative of f, (b) find all antiderivatives of f, and (c) sketch the graphs of a few members of the family of antiderivatives found in part (b).

5. $G(x) = 2x; f(x) = 2$ 　　6. $G(x) = 2x^2; f(x) = 4x$

7. $G(x) = \dfrac{1}{3}x^3; f(x) = x^2$ 　8. $G(x) = e^x; f(x) = e^x$

In Exercises 9–50, find the indefinite integral.

9. $\displaystyle\int 6\, dx$

10. $\displaystyle\int \sqrt{2}\, dx$

11. $\displaystyle\int x^3\, dx$

12. $\displaystyle\int 2x^5\, dx$

13. $\displaystyle\int x^{-4}\, dx$

14. $\displaystyle\int 3t^{-7}\, dt$

15. $\displaystyle\int x^{2/3}\, dx$

16. $\displaystyle\int 2u^{3/4}\, du$

17. $\displaystyle\int x^{-5/4}\, dx$

18. $\displaystyle\int 3x^{-2/3}\, dx$

19. $\displaystyle\int \dfrac{2}{x^3}\, dx$

20. $\displaystyle\int \dfrac{1}{3x^5}\, dx$

21. $\displaystyle\int \pi\sqrt{t}\, dt$

22. $\displaystyle\int \dfrac{3}{\sqrt{t}}\, dt$

23. $\displaystyle\int (3 - 4x)\, dx$

24. $\displaystyle\int (1 + u + u^2)\, du$

25. $\displaystyle\int (x^2 + x + x^{-3})\, dx$

26. $\displaystyle\int (0.3t^2 + 0.02t + 2)\, dt$

27. $\displaystyle\int 5e^x\, dx$

28. $\displaystyle\int (1 + e^x)\, dx$

29. $\displaystyle\int (1 + x + e^x)\, dx$

30. $\displaystyle\int (2 + x + 2x^2 + e^x)\, dx$

31. $\displaystyle\int \left(4x^3 - \dfrac{2}{x^2} - 1 \right) dx$

32. $\displaystyle\int \left(6x^3 + \dfrac{3}{x^2} - x \right) dx$

33. $\int (x^{5/2} + 2x^{3/2} - x)\, dx$ **34.** $\int (t^{3/2} + 2t^{1/2} - 4t^{-1/2})\, dt$

35. $\int \left(\sqrt{x} + \dfrac{2}{\sqrt{x}} \right) dx$ **36.** $\int \left(\sqrt[3]{x^2} - \dfrac{1}{x^2} \right) dx$

37. $\int \left(\dfrac{u^3 + 2u^2 - u}{3u} \right) du$

Hint: $\dfrac{u^3 + 2u^2 - u}{3u} = \dfrac{1}{3}u^2 + \dfrac{2}{3}u - \dfrac{1}{3}$.

38. $\int \dfrac{x^4 - 1}{x^2}\, dx$

Hint: $\dfrac{x^4 - 1}{x^2} = x^2 - x^{-2}$

39. $\int (2t + 1)(t - 2)\, dt$ **40.** $\int u^{-2}(1 - u^2 + u^4)\, du$

41. $\int \dfrac{1}{x^2}(x^4 - 2x^2 + 1)\, dx$ **42.** $\int \sqrt{t}\,(t^2 + t - 1)\, dt$

43. $\int \dfrac{ds}{(s + 1)^{-2}}$ **44.** $\int \left(\sqrt{x} + \dfrac{3}{x} - 2e^x \right) dx$

45. $\int (e^t + t^e)\, dt$ **46.** $\int \left(\dfrac{1}{x^2} - \dfrac{1}{\sqrt[3]{x^2}} + \dfrac{1}{\sqrt{x}} \right) dx$

47. $\int \dfrac{x^3 + x^2 - x + 1}{x^2}\, dx$

Hint: Simplify the integrand first.

48. $\int \dfrac{t^3 + \sqrt[3]{t}}{t^2}\, dt$

Hint: Simplify the integrand first.

49. $\int \dfrac{(\sqrt{x} - 1)^2}{x^2}\, dx$

Hint: Simplify the integrand first.

50. $\int (x + 1)^2 \left(1 - \dfrac{1}{x} \right) dx$

Hint: Simplify the integrand first.

In Exercises 51–58, find $f(x)$ by solving the initial-value problem.

51. $f'(x) = 3x + 1; f(1) = 3$

52. $f'(x) = 3x^2 - 6x; f(2) = 4$

53. $f'(x) = 3x^2 + 4x - 1; f(2) = 9$

54. $f'(x) = \dfrac{1}{\sqrt{x}}; f(4) = 2$

55. $f'(x) = 1 + \dfrac{1}{x^2}; f(1) = 3$

56. $f'(x) = e^x - 2x; f(0) = 2$

57. $f'(x) = \dfrac{x + 1}{x}; f(1) = 1$

58. $f'(x) = 1 + e^x + \dfrac{1}{x}; f(1) = 3 + e$

In Exercises 59–62, find the function f given that the slope of the tangent line to the graph of f at any point $(x, f(x))$ is $f'(x)$ and that the graph of f passes through the given point.

59. $f'(x) = \dfrac{1}{2}x^{-1/2}; (2, \sqrt{2})$

60. $f'(t) = t^2 - 2t + 3; (1, 2)$

61. $f'(x) = e^x + x; (0, 3)$ **62.** $f'(x) = \dfrac{2}{x} + 1; (1, 2)$

63. BANK DEPOSITS Madison Finance opened two branches on September 1 $(t = 0)$. Branch A is located in an established industrial park, and Branch B is located in a fast-growing new development. The net rate at which money was deposited into Branch A and Branch B in the first 180 business days is given by the graphs of f and g, respectively (see the figure). Which branch has the larger amount on deposit at the end of 180 business days? Justify your answer.

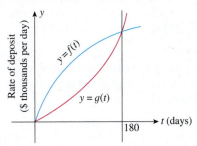

64. VELOCITY OF A CAR Two cars, side by side, start from rest and travel along a straight road. The velocity of Car A is given by $v = f(t)$, and the velocity of Car B is given by $v = g(t)$. The graphs of f and g are shown in the figure below. Are the cars still side by side after T sec? If not, which car is ahead of the other? Justify your answer.

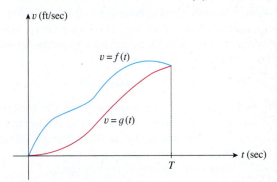

65. U.S. SMARTPHONE USERS The number of smartphone users and penetration in the United States continues to grow steadily. The number of users (in millions) from 2011 through 2015 is projected to grow at the rate of

$$R(t) = 14.3 \qquad (0 \le t \le 4)$$

million/year. The number of users in 2011 $(t = 0)$ was 90.1 million. Find an expression giving the projected number of smartphone users in year t. What is the estimated number of smartphone users in 2015?
Source: eMarketer.

66. HOUSEHOLDS OWNING MORE THAN ONE TV Television is the most popular mass medium in the United States. The percentage of households with multiple TV sets from the year 2000 through 2010 grew at the approximate rate of

$$R(t) = 0.7 \qquad (0 \le t \le 10)$$

percent/year in year t. The percentage of American households owning multiple TV sets stood at 75% in the year 2000 ($t = 0$). Find an expression giving the approximate percentage of households owning multiple TV sets in year t. What percentage of households owned multiple TV sets in 2010?

67. VELOCITY OF A MAGLEV The velocity (in feet per second) at time t of a maglev is

$$v(t) = 0.2t + 3 \qquad (0 \le t \le 120)$$

At $t = 0$, it is at the station. Find the function giving the position of the maglev at time t, assuming that the motion takes place along a straight stretch of track.

68. REVENUE FUNCTIONS The management of Lorimar Watch Company has determined that the daily marginal revenue function associated with producing and selling their travel clocks is given by

$$R'(x) = -0.009x + 12$$

where x denotes the number of units produced and sold and $R'(x)$ is measured in dollars per unit.
a. Determine the revenue function $R(x)$ associated with producing and selling these clocks.
b. What is the demand equation that relates the wholesale unit price with the quantity of travel clocks demanded?

69. PROFIT FUNCTIONS Cannon Precision Instruments makes an automatic electronic flash with Thyrister circuitry. The estimated marginal profit associated with producing and selling these electronic flashes is

$$P'(x) = -0.004x + 20$$

dollars per unit per month when the production level is x units per month. Cannon's fixed cost for producing and selling these electronic flashes is $16,000/month. At what level of production does Cannon realize a maximum profit? What is the maximum monthly profit?

70. COST OF PRODUCING GUITARS Carlota Music Company estimates that the marginal cost of manufacturing its Professional Series guitars is

$$C'(x) = 0.002x + 100$$

dollars/month when the level of production is x guitars/month. The fixed costs incurred by Carlota are $4000/month. Find the total monthly cost incurred by Carlota in manufacturing x guitars/month.

71. WIND ENERGY IN CHINA China's push to install more wind energy capacity has started paying off. As of 2012, wind energy had become the country's third largest source of energy, after coal and hydroelectric power. The amount of wind energy generated in China from 2005 through 2012 grew at the rate of

$$r(t) = 5.018t - 3.204 \qquad (0 \le t \le 7)$$

terawatt-hours/year, in year t, where $t = 0$ corresponds to 2005. The amount of wind energy generated in 2005 stood at 1.8 terawatt-hours.
a. Find an expression giving the amount of wind energy generated in year t.
b. How much wind energy was generated in 2012?
c. If the trend continued into 2013, how much wind energy was generated in China in that year?
Source: earth-policy.org.

72. GROWTH OF NATIONAL HEALTH COSTS National health expenditures are projected to grow at the rate of

$$r(t) = 0.0058t + 0.159 \qquad (0 \le t \le 13)$$

trillion dollars/year from 2002 through 2015. Here, $t = 0$ corresponds to 2002. The expenditure in 2002 was $1.60 trillion.
a. Find a function f giving the projected national health expenditures in year t.
b. What does your model project the national health expenditure to be in 2015?
Source: National Health Expenditures.

73. GENETICALLY MODIFIED CROPS The total number of acres of genetically modified crops grown in developing countries from 2006 through 2012 was changing at the rate of

$$R(t) = 150t + 14.82 \qquad (0 \le t \le 6)$$

million acres/year. The total number of acres of such crops grown in 2006 ($t = 0$) was 27.2 million acres. How many acres of genetically modified crops were grown in developing countries in 2012?
Source: Clive James/ISAAA.

74. VELOCITY OF A CAR The velocity of a car (in feet per second) t sec after starting from rest is given by the function

$$f(t) = 2\sqrt{t} \qquad (0 \le t \le 30)$$

Find the car's position, $s(t)$, at any time t. Assume that $s(0) = 0$.

75. BALLAST DROPPED FROM A BALLOON A ballast is dropped from a stationary hot-air balloon that is hovering at an altitude of 400 ft. The velocity of the ballast after t sec is $(-32t + 4)$ ft/sec.
a. Find the height $h(t)$ of the ballast from the ground at time t.
    Hint: $h'(t) = -32t + 4$ and $h(0) = 400$.
b. When will the ballast strike the ground?

c. Find the velocity of the ballast when it hits the ground.

Ballast

76. POPULATION GROWTH The development of AstroWorld ("The Amusement Park of the Future") on the outskirts of a city will increase the city's population at the rate of

$$4500\sqrt{t} + 1000$$

people/year, t years from the start of construction. The population before construction is 30,000. Determine the projected population 9 years after construction of the park has begun.

77. CABLE TV SUBSCRIBERS A study conducted by TeleCable estimates that the number of cable TV subscribers will grow at the rate of

$$100 + 210t^{3/4}$$

new subscribers/month, t months from the start date of the service. If 5000 subscribers signed up for the service before the starting date, how many subscribers will there be 16 months from that date?

78. BLOOD FLOW IN AN ARTERY Nineteenth-century physician Jean Louis Marie Poiseuille discovered that the rate of change of the velocity of blood r cm from the central axis of an artery (in centimeters per second per centimeter) is given by

$$a(r) = -kr$$

where k is a constant. If the radius of an artery is R cm, find an expression for the velocity of blood as a function of r (see the accompanying figure).
Hint: $v'(r) = a(r)$ and $v(R) = 0$. (Why?)

Blood vessel

79. FLIGHT OF A ROCKET The velocity, in feet per second, of a rocket t sec into vertical flight is given by

$$v(t) = -3t^2 + 192t$$

Find an expression $h(t)$ that gives the rocket's altitude, in feet, t sec after liftoff. What is the altitude of the rocket 30 sec after liftoff?
Hint: $h'(t) = v(t)$; $h(0) = 0$.

80. QUALITY CONTROL As part of a quality-control program, the chess sets manufactured by Jones Brothers are subjected to a final inspection before packing. The rate of increase in the number of sets checked per hour by an inspector t hr into the 8 A.M. to 12 noon shift is approximately

$$N'(t) = -3t^2 + 12t + 45 \qquad (0 \le t \le 4)$$

a. Find an expression $N(t)$ that approximates the number of sets inspected at the end of t hours.
Hint: $N(0) = 0$
b. How many sets does the average inspector check during a morning shift?

81. COST OF PRODUCING CLOCKS Lorimar Watch Company manufactures travel clocks. The daily marginal cost function associated with producing these clocks is

$$C'(x) = 0.000009x^2 - 0.009x + 8$$

where $C'(x)$ is measured in dollars per unit and x denotes the number of units produced. Management has determined that the daily fixed cost incurred in producing these clocks is $120. Find the total cost incurred by Lorimar in producing the first 500 travel clocks per day.

82. RISK OF DOWN SYNDROME The rate at which the risk of Down syndrome is changing is approximated by the function

$$r(x) = 0.004641x^2 - 0.3012x + 4.9 \qquad (20 \le x \le 45)$$

where $r(x)$ is measured in percentage of all births per year and x is the maternal age at delivery.
a. Find a function f giving the risk as a percentage of all births when the maternal age at delivery is x years, given that the risk of Down syndrome at age 30 is 0.14% of all births.
b. On the basis of this model, what is the risk of Down syndrome when the maternal age at delivery is 40 years? 45 years?
Source: New England Journal of Medicine.

83. AMOUNT OF RAINFALL During a thunderstorm, rain was falling at the rate of

$$8\sqrt{2t} - 32t^3 \qquad (0 \le t \le 0.6)$$

in./hr.
a. Find an expression giving the total amount of rainfall after t hr.
Hint: The total amount of rainfall at $t = 0$ is zero.
b. How much rain had fallen after $\frac{1}{2}$ hr?

84. **SALARIES OF MARRIED WOMEN** The percentage of married women who earned more than their husbands over the period from 1960 ($t = 0$) through 2011 ($t = 51$) was growing at the rate of approximately

$$r(t) = -0.00025142t^2 + 0.02116t + 0.0328$$
$$(0 < t \leq 51)$$

percent/year in year t. The percentage of married women earning more than their husbands in 1960 stood at 6.2%.

a. Find an expression giving the approximate percentage of married women who earned more than their husbands in year t.

b. Use the result of part (a) to estimate the percentage of married women who earned more than their husbands in 2013, assuming that the trend continued.

Source: Pew Research Center.

85. **SOCIAL NETWORKS** The percentage of people age 12 years and older using social network sites and/or services "several" times a day from 2009 through 2013 grew at the rate of

$$R(t) = 5.92t^{-0.158} \qquad (1 \leq t \leq 5)$$

percent/year in year t, with $t = 1$ corresponding to 2009. The percentage of people age 12 years and older using social network sites and/or services in 2009 was 7%.

a. Find an expression giving the percentage of people age 12 years and older using social network sites and/or services in year t ($1 \leq t \leq 5$).

b. According to this model, what percentage of people age 12 years and older used social network sites and/or services in 2013?

Source: Arbitron and Edison Research.

86. **COAL EXPORTS** The U.S. coal industry, under increasing pressure from tougher antipollution rules, is ratcheting up its export business. The rate of change of coal exports from 2010 through 2012 is given by

$$f(t) = 31.863t^{-0.61} \qquad (0 \leq t \leq 2)$$

million short tons (2000 lb) per year in year t, where t is measured in years, with $t = 0$ corresponding to 2010. U.S. coal exports in 2010 were 81.7 million short tons.

a. Find an expression for U.S. coal exports in year t.

b. Assuming that the trend continued through 2013, what were U.S. coal exports for that year?

Source: U.S. Department of Energy.

87. **SURFACE AREA OF A HUMAN** Empirical data suggest that the surface area of a 180-cm-tall human body changes at the rate of

$$S'(W) = 0.131773W^{-0.575}$$

m²/kg, where W is the mass of the body in kilograms. If the surface area of a 180-cm-tall human body weighing

70 kg is 1.886277 m², what is the surface area of a human body of the same height weighing 75 kg?

88. **OZONE POLLUTION** The rate of change of the level of ozone, an invisible gas that is an irritant and impairs breathing, present in the atmosphere on a certain May day in the city of Riverside is given by

$$R(t) = 3.2922t^2 - 0.366t^3 \qquad (0 \leq t \leq 11)$$

(measured in pollutant standard index per hour). Here, t is measured in hours, with $t = 0$ corresponding to 7 A.M. Find the ozone level $A(t)$ at any time t, assuming that at 7 A.M. it is zero.
Hint: $A'(t) = R(t)$ and $A(0) = 0$.
Source: Los Angeles Times.

89. **HEIGHTS OF CHILDREN** According to the Jenss model for predicting the height of preschool children, the rate of growth of a typical infant is

$$R(t) = 1.0490t^4 - 4.2255t^3 + 12.7659t^2 - 25.7119t + 32.28$$
$$(\tfrac{1}{4} \leq t \leq 1)$$

cm/year, where t is measured in years. The height of a typical 3-month-old infant is 60.30 cm.

a. What is the height of a typical infant at age t?

b. Use the result of part (a) to estimate the height of a typical 1-year-old child.

90. **ACCELERATION OF A CAR** A car traveling along a straight road at 66 ft/sec accelerated to a speed of 88 ft/sec over a distance of 440 ft. What was the acceleration of the car, assuming that it was constant?

91. **DECELERATION OF A CAR** What constant deceleration would a car moving along a straight road have to be subjected to if it were brought to rest from a speed of 88 ft/sec in 9 sec? What would be the stopping distance?

92. **CARRIER LANDING** A pilot lands a fighter aircraft on an aircraft carrier. At the moment of touchdown, the speed of the aircraft is 160 mph. If the aircraft is brought to a complete stop in 1 sec and the deceleration is assumed to be constant, find the number of g's the pilot is subjected to during landing (1 g = 32 ft/sec²).

93. **CROSSING THE FINISH LINE** After rounding the final turn in the bell lap, two runners emerge ahead of the pack. When Runner A is 200 ft from the finish line, his speed is 22 ft/sec, a speed that he maintains until he crosses the line. At that instant of time, Runner B, who is 20 ft behind Runner A and running at a speed of 20 ft/sec, begins to sprint. Assuming that Runner B sprints with a constant acceleration, what minimum acceleration will enable him to cross the finish line ahead of Runner A?

94. **DRAINING A TANK** A tank has a constant cross-sectional area of 50 ft² and an orifice of constant cross-sectional

area of $\frac{1}{2}$ ft^2 located at the bottom of the tank (see the accompanying figure).

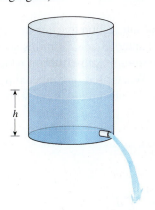

If the tank is filled with water to a height of h ft and allowed to drain, then the height of the water decreases at a rate (in feet per second) that is described by the equation

$$\frac{dh}{dt} = -\frac{1}{25}\left(\sqrt{20} - \frac{t}{50}\right) \qquad (0 \le t \le 50\sqrt{20})$$

Find an expression for the height of the water at any time t (in seconds) if its height initially is 20 ft.

95. LAUNCHING A FIGHTER AIRCRAFT A fighter aircraft is launched from the deck of a Nimitz-class aircraft carrier with the help of a steam catapult. If the aircraft is to attain a takeoff speed of at least 240 ft/sec after traveling 800 ft along the flight deck, find the minimum acceleration to which it must be subjected, assuming that it is constant.

In Exercises 96–100, determine whether the statement is true or false. If it is true, explain why it is true. If it is false, give an example to show why it is false.

96. If F and G are antiderivatives of f on an interval I, then $F(x) = G(x) + C$ on I.

97. If F is an antiderivative of f on an interval I, then $\int f(x)\, dx = F(x)$.

98. If f and g are integrable, then

$$\int [2f(x) - 3g(x)]\, dx = 2\int f(x)\, dx - 3\int g(x)\, dx$$

99. $\int \dfrac{d}{dx}[f(x)]\, dx = f(x)$

100. If f and g are integrable, then

$$\int f(x)g(x)\, dx = \left[\int f(x)\, dx\right]\left[\int g(x)\, dx\right]$$

6.1 Solutions to Self-Check Exercises

1. $\displaystyle\int \left(\frac{1}{\sqrt{x}} - \frac{2}{x} + 3e^x\right) dx = \int \left(x^{-1/2} - \frac{2}{x} + 3e^x\right) dx$

$$= \int x^{-1/2}\, dx - 2\int \frac{1}{x}\, dx + 3\int e^x\, dx$$

$$= 2x^{1/2} - 2\ln|x| + 3e^x + C$$

$$= 2\sqrt{x} - 2\ln|x| + 3e^x + C$$

2. The slope of the tangent line to the graph of the function f at any point $P(x, f(x))$ is given by the derivative f' of f. Thus, the first condition implies that

$$f'(x) = 3x^2 - 6x + 3$$

which, upon integration, yields

$$f(x) = \int (3x^2 - 6x + 3)\, dx$$

$$= x^3 - 3x^2 + 3x + k$$

where k is the constant of integration.

To evaluate k, we use the initial condition (2), which implies that $f(2) = 9$, or

$$9 = f(2) = 2^3 - 3(2)^2 + 3(2) + k$$

or $k = 7$. Hence, the required function f is

$$f(x) = x^3 - 3x^2 + 3x + 7$$

3. Let $M(t)$ denote United Motors' market share at year t. Then,

$$M(t) = \int f(t)\, dt$$

$$= \int (-0.01875t^2 + 0.15t - 1.2)\, dt$$

$$= -0.00625t^3 + 0.075t^2 - 1.2t + C$$

To determine the value of C, we use the initial condition $M(0) = 48.4$, obtaining $C = 48.4$. Therefore,

$$M(t) = -0.00625t^3 + 0.075t^2 - 1.2t + 48.4$$

In particular, United Motors' market share of new cars at the beginning of 2016 is given by

$$M(12) = -0.00625(12)^3 + 0.075(12)^2$$
$$-1.2(12) + 48.4$$
$$= 34$$

or 34%.

6.2 Integration by Substitution

In Section 6.1, we developed certain rules of integration that are closely related to the corresponding rules of differentiation in Chapters 3 and 5. In this section, we introduce a method of integration called the **method of substitution,** which is related to the chain rule for differentiating functions. When used in conjunction with the rules of integration developed earlier, the method of substitution is a powerful tool for integrating a large class of functions.

How the Method of Substitution Works

Consider the indefinite integral

$$\int 2(2x + 4)^5 \, dx \tag{3}$$

One way of evaluating this integral is to expand the expression $2(2x + 4)^5$ and then integrate the resulting integrand term by term. As an alternative approach, let's see whether we can simplify the integral by making a change of variable. Write

$$u = 2x + 4$$

with differential

$$du = 2 \, dx$$

If we formally substitute these quantities into the integral (3), we obtain

$$\int 2(2x + 4)^5 \, dx = \int (2x + 4)^5 (2 \, dx) = \int u^5 \, du$$

↑
Rewrite

$$\begin{cases} u = 2x + 4 \\ du = 2 \, dx \end{cases}$$

Now, the last integral involves a power function and is easily evaluated using Rule 2 of Section 6.1. Thus,

$$\int u^5 \, du = \frac{1}{6} u^6 + C$$

Therefore, using this result and replacing u by $u = 2x + 4$, we obtain

$$\int 2(2x + 4)^5 \, dx = \frac{1}{6} (2x + 4)^6 + C$$

We can verify that the foregoing result is indeed correct by computing

$$\frac{d}{dx}\left[\frac{1}{6}(2x + 4)^6 + C\right] = \frac{1}{6} \cdot 6(2x + 4)^5(2) \qquad \text{Use the Chain Rule.}$$
$$= 2(2x + 4)^5$$

and observing that the last expression is just the integrand of the integral (3).

The Method of Integration by Substitution

To see why the approach used in evaluating the integral in (3) is successful, write

$$f(x) = x^5 \quad \text{and} \quad g(x) = 2x + 4$$

Then, $g'(x) = 2$. Furthermore, the integrand of (3) is just 2 times the composition of f and g, that is,

$$2(f \circ g)(x) = 2f(g(x))$$
$$= 2[g(x)]^5 = 2(2x + 4)^5$$

Therefore, the integral (3) can be written as

$$\int 2f(g(x))g'(x)\,dx \tag{4}$$

Next, let's show that an integral having the form (4) can always be written as

$$\int f(u)\,du \tag{5}$$

Suppose F is an antiderivative of f. By the Chain Rule, we have

$$\frac{d}{dx}[F(g(x))] = F'(g(x))g'(x)$$

Therefore,

$$\int F'(g(x))g'(x)\,dx = F(g(x)) + C$$

Letting $F' = f$ and making the substitution $u = g(x)$, we have

$$\int f(g(x))g'(x)\,dx = F(u) + C = \int F'(u)\,du = \int f(u)\,du$$

as we wished to show. Thus, if the transformed integral is readily evaluated, as is the case with the integral (3), then the method of substitution will prove successful.

Before we look at more examples, let's summarize the steps involved in integration by substitution.

Integration by Substitution

Step 1   Let $u = g(x)$, where $g(x)$ is part of the integrand, usually the "inside function" of a composite function $f(g(x))$.

Step 2   Find $du = g'(x)\,dx$.

Step 3   Use the substitution $u = g(x)$ and $du = g'(x)\,dx$ to convert the *entire* integral into one involving *only u*.

Step 4   Find the resulting integral.

Step 5   Replace u by $g(x)$ to obtain the final solution as a function of x.

Note   Sometimes we need to consider different choices of g for the substitution $u = g(x)$ in order to carry out Step 3 and/or Step 4.

EXAMPLE 1 Find $\displaystyle\int 2x(x^2 + 3)^4\,dx$.

Solution

Step 1   Observe that the integrand involves the composite function $(x^2 + 3)^4$ with "inside function" $g(x) = x^2 + 3$. So we choose $u = x^2 + 3$.

Step 2   Find $du = 2x\,dx$.

Step 3   Use the substitution $u = x^2 + 3$ and $du = 2x\,dx$ to obtain

$$\int 2x(x^2 + 3)^4\,dx = \int \underset{\substack{\uparrow \\ \text{Rewrite}}}{(x^2 + 3)^4(2x\,dx)} = \int u^4\,du$$

an integral involving only the variable u.

Step 4   Find the resulting integral:

$$\int u^4\,du = \frac{1}{5}u^5 + C$$

Step 5   Replacing u by $x^2 + 3$, we obtain

$$\int 2x(x^2 + 3)^4 \, dx = \frac{1}{5}(x^2 + 3)^5 + C$$

EXAMPLE 2 Find $\int 3\sqrt{3x + 1} \, dx$.

Solution

Step 1   The integrand involves the composite function $\sqrt{3x + 1}$ with "inside function" $g(x) = 3x + 1$. So let $u = 3x + 1$.

Step 2   Find $du = 3 \, dx$.

Step 3   Use the substitution $u = 3x + 1$ and $du = 3 \, dx$ to obtain

$$\int 3\sqrt{3x + 1} \, dx = \int \sqrt{3x + 1}(3 \, dx) = \int \sqrt{u} \, du$$

an integral involving only the variable u.

Step 4   Find the resulting integral:

$$\int \sqrt{u} \, du = \int u^{1/2} \, du = \frac{2}{3} u^{3/2} + C$$

Step 5   Replacing u by $3x + 1$, we obtain

$$\int 3\sqrt{3x + 1} \, dx = \frac{2}{3}(3x + 1)^{3/2} + C$$

EXAMPLE 3 Find $\int x^2(x^3 + 1)^{3/2} \, dx$.

Solution

Step 1   The integrand contains the composite function $(x^3 + 1)^{3/2}$ with "inside function" $g(x) = x^3 + 1$. So let $u = x^3 + 1$.

Step 2   Find $du = 3x^2 \, dx$.

Step 3   Use the substitution $u = x^3 + 1$ and $du = 3x^2 \, dx$, or $x^2 \, dx = \frac{1}{3} \, du$ to obtain

$$\int x^2(x^3 + 1)^{3/2} \, dx = \int (x^3 + 1)^{3/2}(x^2 \, dx)$$

$$= \int u^{3/2}\left(\frac{1}{3} \, du\right) = \frac{1}{3}\int u^{3/2} \, du$$

an integral involving only the variable u.

Step 4   Find the resulting integral:

$$\frac{1}{3}\int u^{3/2} \, du = \frac{1}{3} \cdot \frac{2}{5} u^{5/2} + C = \frac{2}{15} u^{5/2} + C$$

Step 5   Replacing u by $x^3 + 1$, we obtain

$$\int x^2(x^3 + 1)^{3/2} \, dx = \frac{2}{15}(x^3 + 1)^{5/2} + C$$

Explore and Discuss

Let $f(x) = x^2(x^3 + 1)^{3/2}$. Using the result of Example 3, we see that an antiderivative of f is $F(x) = \frac{2}{15}(x^3 + 1)^{5/2}$. However, in terms of u (where $u = x^3 + 1$), an antiderivative of f is $G(u) = \frac{2}{15} u^{5/2}$. Compute $F(2)$. Next, suppose we want to compute $F(2)$ using the function G instead. At what value of u should you evaluate $G(u)$ to obtain the desired result? Explain your answer.

In the remaining examples, we drop the practice of labeling the steps involved in evaluating each integral.

EXAMPLE 4 Find $\int e^{-3x}\, dx$.

Solution   Let $u = -3x$, so that $du = -3\, dx$, or $dx = -\frac{1}{3}\, du$. Then,

$$\int e^{-3x}\, dx = \int e^u\left(-\frac{1}{3}\, du\right) = -\frac{1}{3}\int e^u\, du$$

$$= -\frac{1}{3}\, e^u + C = -\frac{1}{3}\, e^{-3x} + C \qquad \blacksquare$$

EXAMPLE 5 Find $\int \dfrac{x}{3x^2 + 1}\, dx$.

Solution   Let $u = 3x^2 + 1$. Then, $du = 6x\, dx$, or $x\, dx = \frac{1}{6}\, du$. Making the appropriate substitutions, we have

$$\int \frac{x}{3x^2 + 1}\, dx = \int \frac{\frac{1}{6}}{u}\, du$$

$$= \frac{1}{6}\int \frac{1}{u}\, du$$

$$= \frac{1}{6}\ln|u| + C$$

$$= \frac{1}{6}\ln(3x^2 + 1) + C \qquad \text{\color{red}Since } 3x^2 + 1 > 0 \qquad \blacksquare$$

EXAMPLE 6 Find $\int \dfrac{(\ln x)^2}{2x}\, dx$.

Solution   Let $u = \ln x$. Then,

$$du = \frac{d}{dx}\,(\ln x)\, dx = \frac{1}{x}\, dx$$

$$\int \frac{(\ln x)^2}{2x}\, dx = \frac{1}{2}\int \frac{(\ln x)^2}{x}\, dx$$

$$= \frac{1}{2}\int u^2\, du$$

$$= \frac{1}{6}\, u^3 + C$$

$$= \frac{1}{6}\,(\ln x)^3 + C \qquad \blacksquare$$

Explore and Discuss

Suppose $\int f(u)\, du = F(u) + C$.

1. Show that $\int f(ax + b)\, dx = \dfrac{1}{a}\, F(ax + b) + C$.

2. How can you use this result to facilitate the evaluation of integrals such as $\int (2x + 3)^5\, dx$ and $\int e^{3x-2}\, dx$? Explain your answer.

Examples 7 and 8 show how the method of substitution can be used in practical situations.

APPLIED EXAMPLE 7 Cost of Producing Solar Cell Panels In 2000, the head of the research and development department of Soloron Corporation claimed that the cost of producing solar cell panels would drop at the rate of

$$\frac{105}{2(3t + 5)^2} \qquad (0 \le t \le 15)$$

dollars per peak watt for the next t years, with $t = 0$ corresponding to the beginning of 2000. (A peak watt is the power produced at noon on a sunny day.) In 2000, the panels, which are used for photovoltaic power systems, cost \$4 per peak watt. Find an expression giving the cost per peak watt of producing solar cell panels at the beginning of year t. What was the cost at the beginning of 2012?

Solution Let $C(t)$ denote the cost per peak watt for producing solar cell panels at the beginning of year t. Then,

$$C'(t) = -\frac{105}{2(3t + 5)^2}$$

Integrating, we find that

$$C(t) = \int \frac{-105}{2(3t + 5)^2}\, dt$$

$$= -\frac{105}{2} \int (3t + 5)^{-2}\, dt$$

Let $u = 3t + 5$ so that

$$du = 3\, dt \quad \text{or} \quad dt = \frac{1}{3}\, du$$

Then,

$$C(t) = -\frac{105}{2}\left(\frac{1}{3}\right) \int u^{-2}\, du$$

$$= -\frac{35}{2}(-1)u^{-1} + k$$

$$= \frac{35}{2(3t + 5)} + k$$

where k is an arbitrary constant. To determine the value of k, note that the cost per peak watt of producing solar cell panels at the beginning of 2000 ($t = 0$) was 4, or $C(0) = 4$. This gives

$$C(0) = \frac{35}{2(5)} + k = 4$$

or $k = \frac{1}{2}$. Therefore, the required expression is given by

$$C(t) = \frac{35}{2(3t + 5)} + \frac{1}{2}$$

$$= \frac{35 + (3t + 5)}{2(3t + 5)}$$

$$= \frac{3t + 40}{2(3t + 5)}$$

The cost per peak watt for producing solar cell panels at the beginning of 2012 is given by

$$C(12) = \frac{3(12) + 40}{2[3(12) + 5]} \approx 0.93$$

or approximately $0.93 per peak watt.

Exploring with TECHNOLOGY

Refer to Example 7.

1. Use a graphing utility to plot the graph of

$$C(t) = \frac{3t + 40}{2(3t + 5)}$$

using the viewing window $[0, 15] \times [0, 5]$. Then, use the numerical differentiation capability of the graphing utility to compute $C'(12)$.

2. Plot the graph of

$$C'(t) = -\frac{105}{2(3t + 5)^2}$$

using the viewing window $[0, 15] \times [-2, 0]$. Then, use the evaluation capability of the graphing utility to find $C'(12)$. Is this value of $C'(12)$ the same as that obtained in part 1? Explain your answer.

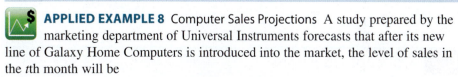 **APPLIED EXAMPLE 8** Computer Sales Projections A study prepared by the marketing department of Universal Instruments forecasts that after its new line of Galaxy Home Computers is introduced into the market, the level of sales in the tth month will be

$$2000 - 1500e^{-0.05t} \qquad (0 \le t \le 60)$$

units per month. Find an expression that gives the total number of computers that will have been sold in the first t months after they become available on the market. How many computers will Universal sell in the first year they are on the market?

Solution   Let $N(t)$ denote the total number of computers that may be expected to be sold t months after their introduction in the market. Then, the rate of growth of sales is given by $N'(t)$ units per month. Thus,

$$N'(t) = 2000 - 1500e^{-0.05t}$$

so that

$$N(t) = \int (2000 - 1500e^{-0.05t})\, dt$$
$$= \int 2000\, dt - 1500 \int e^{-0.05t}\, dt$$

Upon integrating the second integral by the method of substitution, we obtain

$$N(t) = 2000t + \frac{1500}{0.05} e^{-0.05t} + C \qquad \text{Let } u = -0.05t;$$
$$\text{then } du = -0.05\, dt.$$
$$= 2000t + 30{,}000e^{-0.05t} + C$$

To determine the value of C, note that the number of computers sold at the end of month 0 is nil, so $N(0) = 0$. This gives

$$N(0) = 30{,}000 + C = 0 \quad \text{Since } e^0 = 1$$

or $C = -30{,}000$. Therefore, the required expression is given by

$$N(t) = 2000t + 30{,}000e^{-0.05t} - 30{,}000$$
$$= 2000t + 30{,}000(e^{-0.05t} - 1)$$

The number of computers that Universal can expect to sell in the first year is given by

$$N(12) = 2000(12) + 30{,}000(e^{-0.05(12)} - 1)$$
$$\approx 10{,}464$$

6.2 Self-Check Exercises

1. Find $\displaystyle\int \sqrt{2x + 5}\, dx$.

2. Find $\displaystyle\int \frac{x^2}{(2x^3 + 1)^{3/2}}\, dx$.

3. Find $\displaystyle\int xe^{2x^2-1}\, dx$.

4. **Automobile Pollution** According to a joint study conducted by Oxnard's Environmental Management Department and a state government agency, the concentration of carbon monoxide (CO) in the air due to automobile exhaust is increasing at the rate given by

$$f(t) = \frac{8(0.1t + 1)}{300(0.2t^2 + 4t + 64)^{1/3}}$$

parts per million (ppm) per year t. Currently, the CO concentration due to automobile exhaust is 0.16 ppm. Find an expression giving the CO concentration t years from now.

Solutions to Self-Check Exercises 6.2 can be found on page 433.

6.2 Concept Questions

1. Explain how the method of substitution works by showing the steps used to find $\int f(g(x))g'(x)\, dx$.

2. Explain why the method of substitution works for the integral $\int xe^{-x^2}\, dx$ but not for the integral $\int e^{-x^2}\, dx$.

6.2 Exercises

In Exercises 1–50, find the indefinite integral.

1. $\displaystyle\int 4(4x + 3)^4\, dx$

2. $\displaystyle\int 4x(2x^2 + 1)^7\, dx$

3. $\displaystyle\int (x^3 - 2x)^2(3x^2 - 2)\, dx$

4. $\displaystyle\int (3x^2 - 2x + 1)(x^3 - x^2 + x)^4\, dx$

5. $\displaystyle\int \frac{4x}{(2x^2 + 3)^3}\, dx$

6. $\displaystyle\int \frac{3x^2 + 2}{(x^3 + 2x)^2}\, dx$

7. $\displaystyle\int 3t^2 \sqrt{t^3 + 2}\, dt$

8. $\displaystyle\int 3t^2 (t^3 + 2)^{3/2}\, dt$

9. $\displaystyle\int 2(x^2 - 1)^9 x\, dx$

10. $\displaystyle\int x^2(2x^3 + 3)^4\, dx$

11. $\displaystyle\int \frac{x^4}{1 - x^5}\, dx$

12. $\displaystyle\int \frac{x^2}{\sqrt{x^3 - 1}}\, dx$

13. $\displaystyle\int \frac{2}{x - 2}\, dx$

14. $\displaystyle\int \frac{x^2}{x^3 - 3}\, dx$

15. $\displaystyle\int \frac{0.3x - 0.2}{0.3x^2 - 0.4x + 2}\, dx$

16. $\displaystyle\int \frac{2x^2 + 1}{0.2x^3 + 0.3x}\, dx$

17. $\displaystyle\int \frac{2x}{3x^2 - 1}\, dx$

18. $\displaystyle\int \frac{x^2 - 1}{x^3 - 3x + 1}\, dx$

19. $\displaystyle\int e^{-2x}\, dx$

20. $\displaystyle\int e^{-0.02x}\, dx$

21. $\displaystyle\int e^{2-x}\, dx$

22. $\displaystyle\int e^{2t+3}\, dt$

23. $\displaystyle\int xe^{-x^2}\, dx$

24. $\displaystyle\int x^2 e^{x^3-1}\, dx$

25. $\displaystyle\int (e^x - e^{-x})\, dx$

26. $\displaystyle\int (e^{2x} + e^{-3x})\, dx$

27. $\displaystyle\int \frac{2e^x}{1 + e^x}\,dx$

28. $\displaystyle\int \frac{e^{2x}}{1 + e^{2x}}\,dx$

29. $\displaystyle\int \frac{e^{\sqrt{x}}}{\sqrt{x}}\,dx$

30. $\displaystyle\int \frac{e^{-1/x}}{x^2}\,dx$

31. $\displaystyle\int \frac{e^{3x} + x^2}{(e^{3x} + x^3)^3}\,dx$

32. $\displaystyle\int \frac{e^x - e^{-x}}{(e^x + e^{-x})^{3/2}}\,dx$

33. $\displaystyle\int e^{2x}(e^{2x} + 1)^3\,dx$

34. $\displaystyle\int e^{-x}(1 + e^{-x})\,dx$

35. $\displaystyle\int \frac{\ln 5x}{x}\,dx$

36. $\displaystyle\int \frac{(\ln u)^3}{u}\,du$

37. $\displaystyle\int \frac{3}{x \ln x}\,dx$

38. $\displaystyle\int \frac{1}{x(\ln x)^2}\,dx$

39. $\displaystyle\int \frac{\sqrt{\ln x}}{x}\,dx$

40. $\displaystyle\int \frac{(\ln x)^{7/2}}{x}\,dx$

41. $\displaystyle\int \left(xe^{x^2} - \frac{x}{x^2 + 2}\right) dx$

42. $\displaystyle\int \left(xe^{-x^2} + \frac{e^x}{e^x + 3}\right) dx$

43. $\displaystyle\int \frac{x + 1}{\sqrt{x} - 1}\,dx$
Hint: Let $u = \sqrt{x} - 1$.

44. $\displaystyle\int \frac{e^{-u} - 1}{e^{-u} + u}\,du$
Hint: Let $v = e^{-u} + u$.

45. $\displaystyle\int x(x - 1)^5\,dx$
Hint: $u = x - 1$ implies $x = u + 1$.

46. $\displaystyle\int \frac{t}{t + 1}\,dt$
Hint: $\dfrac{t}{t + 1} = 1 - \dfrac{1}{t + 1}$.

47. $\displaystyle\int \frac{1 - \sqrt{x}}{1 + \sqrt{x}}\,dx$
Hint: Let $u = 1 + \sqrt{x}$.

48. $\displaystyle\int \frac{1 + \sqrt{x}}{1 - \sqrt{x}}\,dx$
Hint: Let $u = 1 - \sqrt{x}$.

49. $\displaystyle\int v^2(1 - v)^6\,dv$
Hint: Let $u = 1 - v$.

50. $\displaystyle\int x^3(x^2 + 1)^{3/2}\,dx$
Hint: Let $u = x^2 + 1$.

In Exercises 51–54, find the function f given that the slope of the tangent line to the graph of f at any point $(x, f(x))$ is $f'(x)$ and that the graph of f passes through the given point.

51. $f'(x) = 5(2x - 1)^4;\ (1, 3)$

52. $f'(x) = \dfrac{3x^2}{2\sqrt{x^3 - 1}};\ (1, 1)$

53. $f'(x) = -2xe^{-x^2+1};\ (1, 0)$

54. $f'(x) = 1 - \dfrac{2x}{x^2 + 1};\ (0, 2)$

55. STUDENT ENROLLMENT The registrar of Kellogg University estimates that the student enrollment in the Continuing Education division will grow at the rate of

$$N'(t) = 2000(1 + 0.2t)^{-3/2}$$

students/year, t years from now. If the current student enrollment is 1000, find an expression giving the enrollment t years from now. What will be the enrollment 5 years from now?

56. TV VIEWERS: NEWSMAGAZINE SHOWS The number of viewers of a weekly TV newsmagazine show, introduced in the 2009 season, has been increasing at the rate of

$$3\left(2 + \frac{1}{2}t\right)^{-1/3} \qquad (1 \le t \le 6)$$

million viewers/year in its tth year on the air, where $t = 1$ corresponds to 2009. The number of viewers of the program during its first year on the air is given by $9(5/2)^{2/3}$ million. Find how many viewers were expected in the 2014 season.

57. CREDIT CARD LOSSES IN THE UNITED KINGDOM The mail and the Internet are major routes for fraud against merchants who sell and ship products. These merchants must often accept credit cards that are not physically present (called CNPs, credit cards not present). After peaking at 328.9 million British pounds (GBP) in 2008 ($t = 0$), CNP fraud in the United Kingdom began to decline at the rate of

$$R(t) = \frac{92.07}{t + 1} \qquad (0 \le t \le 4)$$

million GBP/year in year t through 2012. What was the amount of CNP fraud on U.K.-issued credit cards in 2012?
Source: Financial Fraud Action UK.

58. DEMAND: WOMEN'S BOOTS The rate of change of the unit price p (in dollars) of Apex women's boots is given by

$$p'(x) = \frac{-250x}{(16 + x^2)^{3/2}}$$

where x is the quantity demanded daily in units of a hundred. Find the demand function for these boots if the quantity demanded daily is 300 pairs ($x = 3$) when the unit price is $50/pair.

59. POPULATION GROWTH The population of a certain city is projected to grow at the rate of

$$r(t) = 400\left(1 + \frac{2t}{24 + t^2}\right) \qquad (0 \le t \le 5)$$

people/year, t years from now. The current population is 60,000. What will be the population 5 years from now?

60. OIL SPILL In calm waters, the oil spilling from the ruptured hull of a grounded tanker forms an oil slick that is circular in shape. If the radius r of the circle is increasing at the rate of

$$r'(t) = \frac{30}{\sqrt{2t + 4}}$$

ft/min t min after the rupture occurs, find an expression for the radius at any time t. How large is the polluted area 16 min after the rupture occurred?
Hint: $r(0) = 0$

61. LIFE EXPECTANCY OF A FEMALE Suppose that in a certain country, the life expectancy at birth of a female is changing at the rate of

$$g'(t) = \frac{5.45218}{(1 + 1.09t)^{0.9}}$$

years/year. Here, t is measured in years, with $t = 0$ corresponding to the beginning of 1900. Find an expression $g(t)$ giving the life expectancy at birth (in years) of a female in that country if the life expectancy at the beginning of 1900 is 50.02 years. What is the life expectancy at birth of a female born in 2000 in that country?

62. LEARNING CURVES The average student enrolled in the 20-week Court Reporting I course at the American Institute of Court Reporting progresses according to the rule

$$N'(t) = 6e^{-0.05t} \qquad (0 \le t \le 20)$$

where $N'(t)$ measures the rate of change in the number of words per minute of dictation the student takes in machine shorthand after t weeks in the course. Assuming that the average student enrolled in this course begins with a dictation speed of 60 words/min, find an expression $N'(t)$ that gives the dictation speed of the student after t weeks in the course.

63. U.S. ONLINE VIDEO VIEWERS The number of online video viewers in the United States was growing at the rate of

$$r(t) = 9.045e^{0.067t} \qquad (0 \le t \le 5)$$

million viewers/year between 2008 ($t = 0$) and 2013 ($t = 5$). The number of viewers stood at 135 million in 2008.
a. Find an expression giving the number of online video viewers in year t.
b. How many viewers were there in 2012? In 2013?
Source: eMarketer.

64. VIDEO ADS In recent years, video ads have become the most popular rich media format for ad buyers as advertisers have begun to realize returns on online video advertising. According to research and forecast by Forrester, media spending on ads for the years 2012 through 2017 is projected to grow at the rate of

$$R(t) = 0.538434e^{0.234t} \qquad (0 \le t \le 5)$$

billion dollars/year in year t, with $t = 0$ corresponding to 2012. The expenditure on media spending on ads in 2012 stood at $2.9 billion.
a. Find an expression for media spending on ads in year t.
b. Assuming that the forecast was accurate, what would the media spending on ads be in 2016?
Source: Forrester.

65. U.S. ONLINE AD REVENUES Online advertisement revenues took a slight dip in 2009, but since then, they have continued to grow spectacularly. In fact, the rate of growth of online advertisement revenues from 2009 ($t = 0$) through 2012 ($t = 3$) was approximately

$$r(t) = 3.1182e^{0.163(t+1)} \qquad (0 \le t \le 3)$$

billion dollars/year in year t. Online advertisement revenues in 2009 were $22.7 billion.
a. Find an expression for the online advertisement revenue in year t.
b. What were the online advertisement revenues in 2012?
Source: IAB/PricewaterhouseCoopers.

66. AVERAGE BIRTH HEIGHT OF BOYS Using data collected at Kaiser Hospital, pediatricians estimate that the average height of male children changes at the rate of

$$h'(t) = \frac{52.8706e^{-0.3277t}}{(1 + 2.449e^{-0.3277t})^2}$$

in./year, where the child's height $h(t)$ is measured in inches and t, the child's age, is measured in years, with $t = 0$ corresponding to birth. Find an expression $h(t)$ for the average height of a boy at age t if the height at birth of an average child is 19.4 in. What is the height of an average 8-year-old boy?

67. AMOUNT OF GLUCOSE IN THE BLOODSTREAM Suppose a patient is given a continuous intravenous infusion of glucose at a constant rate of r mg/min. Then, the rate at which the amount of glucose in the bloodstream is changing at time t (in minutes) because of this infusion is given by

$$A'(t) = re^{-at}$$

mg/min, where a is a positive constant associated with the rate at which excess glucose is eliminated from the bloodstream and is dependent on the patient's metabolism rate. Derive an expression for the amount of glucose in the bloodstream at time t.
Hint: $A(0) = 0$

68. CONCENTRATION OF A DRUG IN AN ORGAN A drug is carried into an organ of volume V cm^3 by a liquid that enters the organ at the rate of a cm^3/sec and leaves it at the rate of b cm^3/sec. The concentration of the drug in the liquid entering the organ is c g/cm^3. If the concentration of the drug in the organ at time t (in seconds) is increasing at the rate of

$$x'(t) = \frac{1}{V}(ac - bx_0)e^{-bt/V}$$

g/cm^3/sec and the concentration of the drug in the organ initially is x_0 g/cm^3, show that the concentration of the drug in the organ at time t is given by

$$x(t) = \frac{ac}{b} + \left(x_0 - \frac{ac}{b}\right)e^{-bt/V}$$

In Exercises 69–72, determine whether the statement is true or false. If it is true, explain why it is true. If it is false, give an example to show why it is false.

69. If f is integrable, then $\int xf(x^2)\, dx = \frac{1}{2}\int f(x)\, dx$.

70. If f is integrable, then $\int f(ax + b)\, dx = a\int f(x)\, dx$.

71. If f is integrable, then $\int e^{kx} f(e^{kx}) \, dx = \dfrac{1}{k} \int f(x) \, dx$. $(k \neq 0)$.

72. If f is integrable, then $\int \dfrac{f(\ln x)}{\ln x} \, dx = \int f(x) \, dx$.

6.2 Solutions to Self-Check Exercises

1. Let $u = 2x + 5$. Then, $du = 2 \, dx$, or $dx = \frac{1}{2} du$. Making the appropriate substitutions, we have

$$\int \sqrt{2x + 5} \, dx = \int \sqrt{u} \left(\frac{1}{2} du \right) = \frac{1}{2} \int u^{1/2} \, du$$

$$= \frac{1}{2} \left(\frac{2}{3} \right) u^{3/2} + C$$

$$= \frac{1}{3} (2x + 5)^{3/2} + C$$

2. Let $u = 2x^3 + 1$, so that $du = 6x^2 \, dx$, or $x^2 \, dx = \frac{1}{6} du$. Making the appropriate substitutions, we have

$$\int \frac{x^2}{(2x^3 + 1)^{3/2}} \, dx = \int \frac{\left(\frac{1}{6} \right) du}{u^{3/2}} = \frac{1}{6} \int u^{-3/2} \, du$$

$$= \left(\frac{1}{6} \right) (-2) u^{-1/2} + C$$

$$= -\frac{1}{3} (2x^3 + 1)^{-1/2} + C$$

$$= -\frac{1}{3\sqrt{2x^3 + 1}} + C$$

3. Let $u = 2x^2 - 1$, so that $du = 4x \, dx$, or $x \, dx = \frac{1}{4} du$. Then,

$$\int x e^{2x^2 - 1} \, dx = \frac{1}{4} \int e^u \, du$$

$$= \frac{1}{4} e^u + C$$

$$= \frac{1}{4} e^{2x^2 - 1} + C$$

4. Let $C(t)$ denote the CO concentration in the air due to automobile exhaust t years from now. Then,

$$C'(t) = f(t) = \frac{8(0.1t + 1)}{300(0.2t^2 + 4t + 64)^{1/3}}$$

$$= \frac{8}{300} (0.1t + 1)(0.2t^2 + 4t + 64)^{-1/3}$$

Integrating, we find

$$C(t) = \int \frac{8}{300} (0.1t + 1)(0.2t^2 + 4t + 64)^{-1/3} \, dt$$

$$= \frac{8}{300} \int (0.1t + 1)(0.2t^2 + 4t + 64)^{-1/3} \, dt$$

Let $u = 0.2t^2 + 4t + 64$, so that

$$du = (0.4t + 4) \, dt = 4(0.1t + 1) \, dt$$

or

$$(0.1t + 1) \, dt = \frac{1}{4} du$$

Then,

$$C(t) = \frac{8}{300} \left(\frac{1}{4} \right) \int u^{-1/3} \, du$$

$$= \frac{1}{150} \left(\frac{3}{2} u^{2/3} \right) + k$$

$$= 0.01(0.2t^2 + 4t + 64)^{2/3} + k$$

where k is an arbitrary constant. To determine the value of k, we use the condition $C(0) = 0.16$, obtaining

$$C(0) = 0.16 = 0.01(64)^{2/3} + k$$

$$0.16 = 0.16 + k$$

$$k = 0$$

Therefore,

$$C(t) = 0.01(0.2t^2 + 4t + 64)^{2/3}$$

6.3 Area and the Definite Integral

An Intuitive Look

Suppose a certain state's annual rate of petroleum consumption over a 4-year period is constant and is given by the function

$$f(t) = 1.2 \qquad (0 \leq t \leq 4)$$

where t is measured in years and $f(t)$ in millions of barrels per year. Then, the state's total petroleum consumption over the period of time in question is

$$(1.2)(4 - 0) \qquad \text{Rate of consumption} \times \text{time elapsed}$$

or 4.8 million barrels. If you examine the graph of f shown in Figure 5, you will see that this total is just the area of the rectangular region bounded above by the graph of f, below by the t-axis, and to the left and right by the vertical lines $t = 0$ (the y-axis) and $t = 4$, respectively.

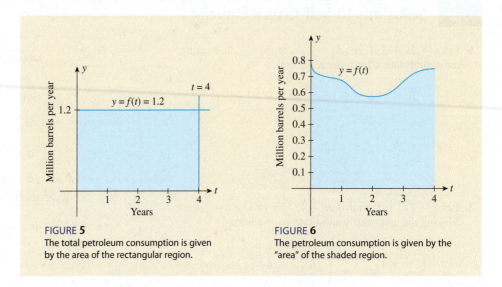

FIGURE 5
The total petroleum consumption is given by the area of the rectangular region.

FIGURE 6
The petroleum consumption is given by the "area" of the shaded region.

Figure 6 shows the actual petroleum consumption of a certain New England state over a 4-year period from 2008 ($t = 0$) to 2012 ($t = 4$). Observe that the rate of consumption is not constant; that is, the function f is not a constant function. What is the state's total petroleum consumption over this 4-year period? It seems reasonable to conjecture that it is given by the "area" of the region bounded above by the graph of f, below by the t-axis, and to the left and right by the vertical lines $t = 0$ and $t = 4$, respectively. We will show that this conjecture is justified at the end of this section.

This example raises two questions:

1. What is the "area" of the region shown in Figure 6?
2. How do we compute this area?

The Area Problem

The preceding example touches on the second fundamental problem in calculus: Calculate the area of the region bounded by the graph of a nonnegative function f, the x-axis, and the vertical lines $x = a$ and $x = b$ where $a < b$ (Figure 7). This area is called the **area of the region under the graph of** f on the interval $[a, b]$ or from a to b.

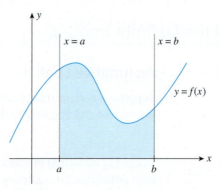

FIGURE 7
The area of the region under the graph of f on $[a, b]$

Defining Area—Two Examples

Just as we used the slopes of secant lines (quantities that we could compute) to help us define the slope of the tangent line to a point on the graph of a function, we now adopt a parallel approach and use the areas of rectangles (quantities that we can compute) to help us define the area under the graph of a function. We begin by looking at a specific example.

EXAMPLE 1 Let $f(x) = x^2$ and consider the region R under the graph of f on the interval $[0, 1]$ (Figure 8a). To obtain an approximation of the area of R, let's construct four nonoverlapping rectangles, except for their edges, as follows: Divide the interval $[0, 1]$ into four subintervals

$$\left[0, \frac{1}{4}\right], \quad \left[\frac{1}{4}, \frac{1}{2}\right], \quad \left[\frac{1}{2}, \frac{3}{4}\right], \quad \left[\frac{3}{4}, 1\right]$$

of equal length $\frac{1}{4}$. Next, construct four rectangles with these subintervals as bases and with heights given by the values of the function at the midpoints

$$\frac{1}{8}, \quad \frac{3}{8}, \quad \frac{5}{8}, \quad \frac{7}{8}$$

of each subinterval. Then, each of these rectangles has width $\frac{1}{4}$ and height

$$f\left(\frac{1}{8}\right), \quad f\left(\frac{3}{8}\right), \quad f\left(\frac{5}{8}\right), \quad f\left(\frac{7}{8}\right)$$

respectively (Figure 8b).

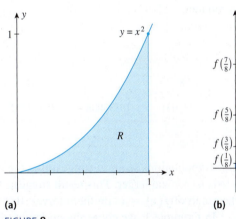 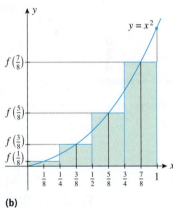

(a) **(b)**

FIGURE **8**
The area of the region under the graph of f on [0, 1] in (a) is approximated by the sum of the areas of the four rectangles in (b).

If we approximate the area A of R by the sum of the areas of the four rectangles, we obtain

$$A \approx \frac{1}{4}f\left(\frac{1}{8}\right) + \frac{1}{4}f\left(\frac{3}{8}\right) + \frac{1}{4}f\left(\frac{5}{8}\right) + \frac{1}{4}f\left(\frac{7}{8}\right)$$

$$= \frac{1}{4}\left[f\left(\frac{1}{8}\right) + f\left(\frac{3}{8}\right) + f\left(\frac{5}{8}\right) + f\left(\frac{7}{8}\right)\right]$$

$$= \frac{1}{4}\left[\left(\frac{1}{8}\right)^2 + \left(\frac{3}{8}\right)^2 + \left(\frac{5}{8}\right)^2 + \left(\frac{7}{8}\right)^2\right] \qquad \text{Recall that } f(x) = x^2.$$

$$= \frac{1}{4}\left(\frac{1}{64} + \frac{9}{64} + \frac{25}{64} + \frac{49}{64}\right) = \frac{21}{64}$$

or approximately 0.328125.

Following the procedure of Example 1, we can obtain approximations of the area of the region R using any number n of rectangles ($n = 4$ in Example 1). Figure 9a shows the approximation of the area A of R using 8 rectangles ($n = 8$), and Figure 9b shows the approximation of the area A of R using 16 rectangles.

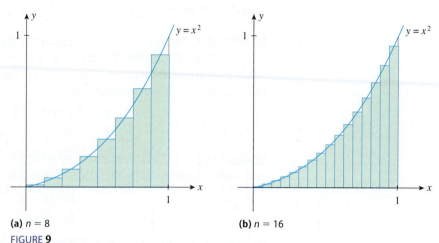

(a) $n = 8$ **(b)** $n = 16$

FIGURE **9**
As n increases, the number of rectangles increases, and the approximation improves.

These figures suggest that the approximations seem to get better as n increases. This is borne out by the results given in Table 1, which were obtained by using a computer.

TABLE 1

Number of Rectangles, n	4	8	16	32	64	100	200
Approximation of A	0.328125	0.332031	0.333008	0.333252	0.333313	0.333325	0.333331

Our computations seem to suggest that the approximations approach the number $\frac{1}{3}$ as n gets larger and larger. This result suggests that we *define* the area of the region under the graph of $f(x) = x^2$ on the interval $[0, 1]$ to be $\frac{1}{3}$.

In Example 1, we chose the *midpoint* of each subinterval as the point at which to evaluate $f(x)$ to obtain the height of the approximating rectangle. Let's consider another example, this time choosing the *left endpoint* of each subinterval.

EXAMPLE 2 Let R be the region under the graph of $f(x) = 16 - x^2$ on the interval $[1, 3]$. Find an approximation of the area A of R using four subintervals of $[1, 3]$ of equal length and picking the left endpoint of each subinterval to evaluate $f(x)$ to obtain the height of the approximating rectangle.

Solution The graph of f is sketched in Figure 10a. Since the length of $[1, 3]$ is 2, we see that the length of each subinterval is $\frac{2}{4}$, or $\frac{1}{2}$. Therefore, the four subintervals are

$$\left[1, \frac{3}{2}\right], \quad \left[\frac{3}{2}, 2\right], \quad \left[2, \frac{5}{2}\right], \quad \left[\frac{5}{2}, 3\right]$$

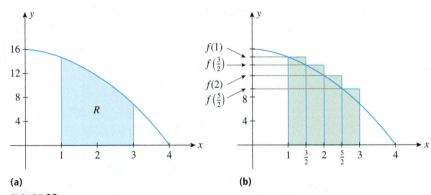

FIGURE 10
The area of R in (a) is approximated by the sum of the areas of the four rectangles in (b).

The left endpoints of these subintervals are $1, \frac{3}{2}, 2$, and $\frac{5}{2}$, respectively, so the heights of the approximating rectangles are $f(1), f(\frac{3}{2}), f(2)$, and $f(\frac{5}{2})$, respectively (Figure 10b). Therefore, the required approximation is

$$A \approx \frac{1}{2}f(1) + \frac{1}{2}f\left(\frac{3}{2}\right) + \frac{1}{2}f(2) + \frac{1}{2}f\left(\frac{5}{2}\right)$$

$$= \frac{1}{2}\left[f(1) + f\left(\frac{3}{2}\right) + f(2) + f\left(\frac{5}{2}\right)\right]$$

$$= \frac{1}{2}\left\{[16 - (1)^2] + \left[16 - \left(\frac{3}{2}\right)^2\right]\right.$$

$$\left. + [16 - (2)^2] + \left[16 - \left(\frac{5}{2}\right)^2\right]\right\} \qquad \text{Recall that } f(x) = 16 - x^2.$$

$$= \frac{1}{2}\left(15 + \frac{55}{4} + 12 + \frac{39}{4}\right) = \frac{101}{4}$$

or approximately 25.25.

Table 2 shows the approximations of the area A of the region R of Example 2 when n rectangles are used for the approximation and the heights of the approximating rectangles are found by evaluating $f(x)$ at the left endpoints.

TABLE 2

Number of Rectangles, n	4	10	100	1,000	10,000	50,000	100,000
Approximation of A	25.2500	24.1200	23.4132	23.3413	23.3341	23.3335	23.3334

Once again, we see that the approximations seem to approach a unique number as n gets larger and larger. This time, the number is $23\frac{1}{3}$. This result suggests that we *define* the area of the region under the graph of $f(x) = 16 - x^2$ on the interval $[1, 3]$ to be $23\frac{1}{3}$.

Defining Area—The General Case

Examples 1 and 2 point the way to defining the area A of the region R under the graph of an arbitrary but continuous, nonnegative function f on an interval $[a, b]$ (Figure 11a).

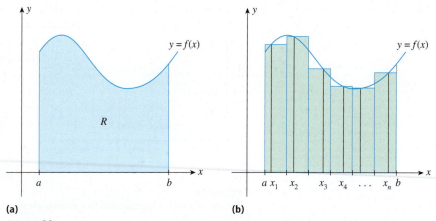

(a)   **(b)**

FIGURE **11**
The area of the region under the graph of f on [a, b] in (a) is approximated by the sum of the areas of the n rectangles shown in (b).

Divide the interval $[a, b]$ into n subintervals of equal length $\Delta x = (b - a)/n$. Next, pick n arbitrary points $x_1, x_2, \ldots,$ and x_n, called *representative points*, from the first, second, $\ldots,$ and nth subintervals, respectively (Figure 11b). Then, approximating the area A of the region R by the n rectangles of width Δx and heights $f(x_1), f(x_2), \ldots, f(x_n)$, so that the areas of the rectangles are $f(x_1)\,\Delta x, f(x_2)\,\Delta x, \ldots, f(x_n)\,\Delta x$, we have

$$A \approx f(x_1)\,\Delta x + f(x_2)\,\Delta x + \cdots + f(x_n)\,\Delta x$$

The sum on the right-hand side of this expression is called a **Riemann sum** in honor of the German mathematician Bernhard Riemann (1826–1866). Now, as the earlier examples seem to suggest, the Riemann sum will approach a unique number as n becomes arbitrarily large.* We define this number to be the area A of the region R.

> **The Area Under the Graph of a Function**
>
> Let f be a nonnegative continuous function on $[a, b]$. Then, the area of the region under the graph of f is
>
> $$A = \lim_{n \to \infty} [f(x_1) + f(x_2) + \cdots + f(x_n)]\Delta x \qquad (6)$$
>
> where $x_1, x_2, \ldots, x_n$ are arbitrary points in the n subintervals of $[a, b]$ of equal width $\Delta x = (b - a)/n$.

The Definite Integral

As we have just seen, the area under the graph of a continuous *nonnegative* function f on an interval $[a, b]$ is defined by the limit of the Riemann sum

$$\lim_{n \to \infty} [f(x_1)\,\Delta x + f(x_2)\,\Delta x + \cdots + f(x_n)\,\Delta x]$$

We now turn our attention to the study of limits of Riemann sums involving functions that are not necessarily nonnegative. Such limits arise in many applications of calculus.

---

*Even though we chose the representative points to be the midpoints of the subintervals in Example 1 and the left endpoints in Example 2, it can be shown that each of the respective sums will always approach the same unique number as n approaches infinity.

For example, the calculation of the distance covered by a body traveling along a straight line involves evaluating a limit of this form. The computation of the total revenue realized by a company over a certain time period, the calculation of the total amount of electricity consumed in a typical home over a 24-hour period, the average concentration of a drug in a body over a certain interval of time, and the volume of a solid—all involve limits of this type.

We begin with the following definition.

The Definite Integral

Let f be a function defined on $[a, b]$. If

$$\lim_{n \to \infty} [f(x_1) \, \Delta x + f(x_2) \, \Delta x + \cdots + f(x_n) \, \Delta x]$$

exists and is the same for all choices of representative points $x_1, x_2, \ldots, x_n$ in the n subintervals of $[a, b]$ of equal width $\Delta x = (b - a)/n$, then this limit is called the **definite integral** of f from a to b and is denoted by $\int_a^b f(x) \, dx$. Thus,

$$\int_a^b f(x) \, dx = \lim_{n \to \infty} [f(x_1) \, \Delta x + f(x_2) \, \Delta x + \cdots + f(x_n) \, \Delta x] \qquad \textbf{(7)}$$

The number a is called the **lower limit of integration**, and the number b is called the **upper limit of integration**.

Notes

1. If f is nonnegative, then the limit in Equation (7) is the same as the limit in (6); therefore, the definite integral gives the area under the graph of f on $[a, b]$.

2. The limit in Equation (7) is denoted by the integral sign $\int$ because, as we will see later, the definite integral and the antiderivative of a function f are related.

3. It is important to realize that the definite integral $\int_a^b f(x) \, dx$ is a *number*, whereas the indefinite integral $\int f(x) \, dx$ represents a *family of functions* (the antiderivatives of f).

4. If the limit in Equation (7) exists, we say that f is **integrable** on the interval $[a, b]$. ◼

When Is a Function Integrable?

The following theorem, which we state without proof, guarantees that a continuous function is integrable.

Integrability of a Function

Let f be continuous on $[a, b]$. Then, f is integrable on $[a, b]$; that is, the definite integral $\int_a^b f(x) \, dx$ exists.

Geometric Interpretation of the Definite Integral

If f is nonnegative and integrable on $[a, b]$, then we have the following geometric interpretation of the definite integral $\int_a^b f(x) \, dx$.

Geometric Interpretation of $\int_a^b f(x)\, dx$ for $f(x) \geq 0$ on $[a, b]$

If f is nonnegative and continuous on $[a, b]$, then

$$\int_a^b f(x)\, dx \tag{8}$$

is equal to the area of the region under the graph of f on $[a, b]$ (Figure 12).

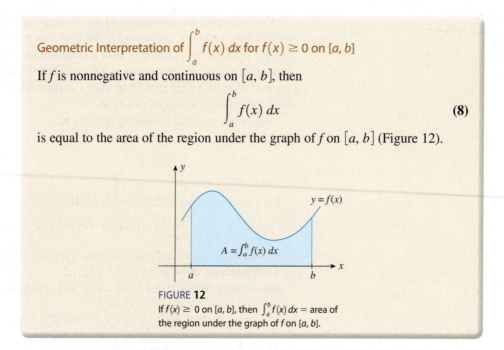

FIGURE 12
If $f(x) \geq 0$ on $[a, b]$, then $\int_a^b f(x)\, dx =$ area of the region under the graph of f on $[a, b]$.

Explore and Discuss

Suppose f is nonpositive [that is, $f(x) \leq 0$] and continuous on $[a, b]$. Explain why the area of the region below the x-axis and above the graph of f is given by $-\int_a^b f(x)\, dx$.

Next, let's extend our geometric interpretation of the definite integral to include the case where f assumes both positive and negative values on $[a, b]$. Consider a typical Riemann sum of the function f,

$$f(x_1)\, \Delta x + f(x_2)\, \Delta x + \cdots + f(x_n)\, \Delta x$$

corresponding to a partition of $[a, b]$ into n subintervals of equal width $(b - a)/n$, where $x_1, x_2, \ldots, x_n$ are representative points in the subintervals. The sum consists of n terms in which a positive term corresponds to the area of a rectangle of height $f(x_k)$ (for some positive integer k) lying above the x-axis and a negative term corresponds to the negative of the area of a rectangle of height $-f(x_k)$ lying below the x-axis. (See Figure 13, which depicts a situation with $n = 6$.)

As n gets larger and larger, the sums of the areas of the rectangles lying above the x-axis seem to give a better and better approximation of the area of the region lying above the x-axis (Figure 14). Similarly, the sums of the areas of those rectangles lying below the x-axis seem to give a better and better approximation of the area of the region lying below the x-axis.

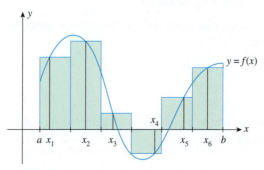

FIGURE 13
The positive terms in the Riemann sum are associated with the areas of the rectangles that lie above the x-axis, and the negative terms are associated with the areas of those that lie below the x-axis.

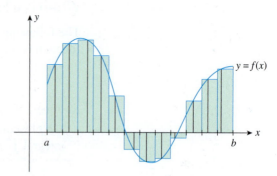

FIGURE 14
As n gets larger, the approximations get better. Here, $n = 12$, and we are approximating with twice as many rectangles as in Figure 13.

These observations suggest the following geometric interpretation of the definite integral for an arbitrary continuous function on an interval $[a, b]$.

Geometric Interpretation of $\int_a^b f(x)\,dx$ on $[a, b]$

If f is continuous on $[a, b]$, then

$$\int_a^b f(x)\,dx$$

is equal to the area of the region above $[a, b]$ minus the area of the region below $[a, b]$ (Figure 15).

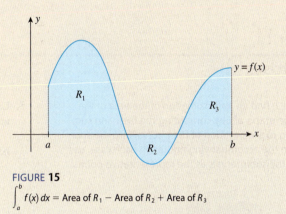

FIGURE **15**

$\int_a^b f(x)\,dx$ = Area of R_1 − Area of R_2 + Area of R_3

Finally, as promised at the beginning of the section, we will demonstrate that the total daily petroleum consumption of the New England state between 2008 and 2012 was given by the area under the graph of f between $t = 0$ and $t = 4$.

We begin by dividing the time interval $[0, 4]$ into n subintervals of equal length $\Delta t = 4/n$, with endpoints $t_0 = 0, t_1 = \Delta t, t_2 = 2\,\Delta t, \dots, t_n = n\,\Delta t$. In the first subinterval, $[t_0, t_1]$, we pick an arbitrary point t_1^*. Observe that because of the continuity of f, the values of f in $[t_0, t_1]$ do not vary appreciably from the constant value $f(t_1^*)$. So the consumption of petroleum over the period from $t = 0$ to $t = \Delta t$ is approximately $f(t_1^*)\,\Delta t$ million barrels. Similarly, we see that the petroleum consumption over the period from $t = t_1$ to $t = t_2$ is approximately $f(t_2^*)\,\Delta t$, where t_2^* is an arbitrary point in the subinterval $[t_1, t_2]$. Continuing, we see that the total petroleum consumption over the interval $[0, 4]$ is approximately

$$f(t_1^*)\,\Delta t + f(t_2^*)\,\Delta t + \cdots + f(t_n^*)\,\Delta t$$

where t_1^* is an arbitrary point in the interval $[t_{i-1}, t_i]$ $(1 \leq i \leq n)$. Intuitively, we see that as n gets larger and larger and the length of each subinterval gets smaller and smaller, the approximation gets better and better. Thus, as n becomes arbitrarily large, the sum approaches the limit that we take to be the total petroleum consumption on $[0, 4]$. But the sum is just the Riemann sum of the nonnegative function on $[0, 4]$, so its limit gives the area of the graph of f under $[0, 4]$.

6.3 Self-Check Exercise

Find an approximation of the area of the region R under the graph of $f(x) = 2x^2 + 1$ on the interval $[0, 3]$, using four subintervals of $[0, 3]$ of equal length and picking the midpoint of each subinterval as a representative point.

The solution to Self-Check Exercise 6.3 can be found on page 444.

6.3 Concept Questions

1. Explain how you would define the area of the region under the graph of a nonnegative continuous function f on the interval $[a, b]$.

2. Define the definite integral of a continuous function on the interval $[a, b]$. Give a geometric interpretation of

$$\int_a^b f(x)\, dx$$

for the case in which (a) f is nonnegative on $[a, b]$ and (b) f assumes both positive and negative values on $[a, b]$. Illustrate your answers graphically.

6.3 Exercises

In Exercises 1 and 2, find an approximation of the area of the region R under the graph of f by computing the Riemann sum of f corresponding to the partition of the interval into the subintervals shown in the accompanying figures. In each case, use the midpoints of the subintervals as the representative points.

1.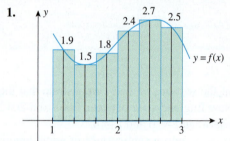

2.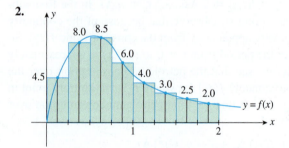

3. Let $f(x) = 3x$.
   a. Sketch the region R under the graph of f on the interval $[0, 2]$, and find its exact area using geometry.
   b. Use a Riemann sum with four subintervals of equal length ($n = 4$) to approximate the area of R. Choose the representative points to be the left endpoints of the subintervals.
   c. Repeat part (b) with eight subintervals of equal length ($n = 8$).
   d. Compare the approximations obtained in parts (b) and (c) with the exact area found in part (a). Do the approximations improve with larger n?

4. Repeat Exercise 3, choosing the representative points to be the right endpoints of the subintervals.

5. Let $f(x) = 4 - 2x$.
   a. Sketch the region R under the graph of f on the interval $[0, 2]$, and find its exact area using geometry.
   b. Use a Riemann sum with five subintervals of equal length ($n = 5$) to approximate the area of R. Choose the representative points to be the left endpoints of the subintervals.
   c. Repeat part (b) with ten subintervals of equal length ($n = 10$).
   d. Compare the approximations obtained in parts (b) and (c) with the exact area found in part (a). Do the approximations improve with larger n?

6. Repeat Exercise 5, choosing the representative points to be the right endpoints of the subintervals.

7. Let $f(x) = x^2$, and compute the Riemann sum of f over the interval $[2, 4]$, choosing the representative points to be the midpoints of the subintervals and using:
   a. Two subintervals of equal length ($n = 2$).
   b. Five subintervals of equal length ($n = 5$).
   c. Ten subintervals of equal length ($n = 10$).
   d. Can you guess at the area of the region under the graph of f on the interval $[2, 4]$?

8. Repeat Exercise 7, choosing the representative points to be the left endpoints of the subintervals.

9. Repeat Exercise 7, choosing the representative points to be the right endpoints of the subintervals.

10. Let $f(x) = x^3$, and compute the Riemann sum of f over the interval $[0, 1]$, choosing the representative points to be the midpoints of the subintervals and using:
    a. Two subintervals of equal length ($n = 2$).
    b. Five subintervals of equal length ($n = 5$).
    c. Ten subintervals of equal length ($n = 10$).
    d. Can you guess at the area of the region under the graph of f on the interval $[0, 1]$?

11. Repeat Exercise 10, choosing the representative points to be the left endpoints of the subintervals.

12. Repeat Exercise 10, choosing the representative points to be the right endpoints of the subintervals.

In Exercises 13–16, find an approximation of the area of the region *R* under the graph of the function *f* on the interval [*a*, *b*]. In each case, use *n* subintervals and choose the representative points as indicated.

13. $f(x) = x^2 + 1$; [0, 2]; *n* = 5; midpoints

14. $f(x) = 4 - x^2$; [−1, 2]; *n* = 6; left endpoints

15. $f(x) = \dfrac{1}{x}$; [1, 3]; *n* = 4; right endpoints

16. $f(x) = e^x$; [0, 3]; *n* = 5; midpoints

17. REAL ESTATE Figure (a) shows a vacant lot with a 100-ft frontage in a development. To estimate its area, we introduce a coordinate system so that the *x*-axis coincides with the edge of the straight road forming the lower boundary of the property, as shown in Figure (b). Then, thinking of the upper boundary of the property as the graph of a continuous function *f* over the interval [0, 100], we see that the problem is mathematically equivalent to that of finding the area under the graph of *f* on [0, 100]. To estimate the area of the lot using a Riemann sum, we divide the interval [0, 100] into five equal subintervals of length 20 ft. Then, using surveyor's equipment, we measure the distance from the midpoint of each of these subintervals to the upper boundary of the property. These measurements give the values of *f*(*x*) at *x* = 10, 30, 50, 70, and 90. What is the approximate area of the lot?

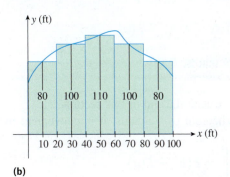

(a)

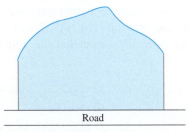

(b)

18. REAL ESTATE Use the technique of Exercise 17 to obtain an estimate of the area of the vacant lot shown in the accompanying figures.

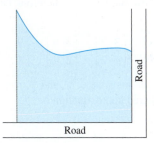

(a)

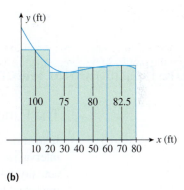

(b)

In Exercises 19 and 20, determine whether the statement is true or false. If it is true, explain why it is true. If it is false, give an example to show why it is false.

19. If *f* is continuous on [*a*, *b*] and $\int_a^b f(x)\,dx > 0$, then $f(x) \geq 0$ for all *x* in [*a*, *b*].

20. If *f* is continuous on [*a*, *b*] and $f(x) \neq 0$ for all *x* in [*a*, *b*], then $\int [f(x)]^2\,dx > 0$.

6.3 Solution to Self-Check Exercise

The length of each subinterval is $\frac{3}{4}$. Therefore, the four subintervals are

$$\left[0, \frac{3}{4}\right], \quad \left[\frac{3}{4}, \frac{3}{2}\right], \quad \left[\frac{3}{2}, \frac{9}{4}\right], \quad \left[\frac{9}{4}, 3\right]$$

The representative points are $\frac{3}{8}, \frac{9}{8}, \frac{15}{8}$, and $\frac{21}{8}$, respectively. Therefore, the required approximation is

$$
\begin{aligned}
A &= \frac{3}{4}f\left(\frac{3}{8}\right) + \frac{3}{4}f\left(\frac{9}{8}\right) + \frac{3}{4}f\left(\frac{15}{8}\right) + \frac{3}{4}f\left(\frac{21}{8}\right) \\
&= \frac{3}{4}\left[f\left(\frac{3}{8}\right) + f\left(\frac{9}{8}\right) + f\left(\frac{15}{8}\right) + f\left(\frac{21}{8}\right)\right] \\
&= \frac{3}{4}\left\{\left[2\left(\frac{3}{8}\right)^2 + 1\right] + \left[2\left(\frac{9}{8}\right)^2 + 1\right] + \left[2\left(\frac{15}{8}\right)^2 + 1\right] \right. \\
&\quad \left. + \left[2\left(\frac{21}{8}\right)^2 + 1\right]\right\} \\
&= \frac{3}{4}\left(\frac{41}{32} + \frac{113}{32} + \frac{257}{32} + \frac{473}{32}\right) = \frac{663}{32}
\end{aligned}
$$

or approximately 20.72.

6.4 The Fundamental Theorem of Calculus

The Fundamental Theorem of Calculus

In Section 6.3, we defined the definite integral of an arbitrary continuous function on an interval $[a, b]$ as a limit of Riemann sums. Calculating the value of a definite integral by actually taking the limit of such sums is tedious and in most cases impractical. It is important to realize that the numerical results we obtained in Examples 1 and 2 of Section 6.3 were *approximations* of the respective areas of the regions in question, even though these results enabled us to *conjecture* what the actual areas might be. Fortunately, there is a much better way of finding the exact value of a definite integral.

The following theorem shows how to evaluate the definite integral of a continuous function provided that we can find an antiderivative of that function. Because of its importance in establishing the relationship between differentiation and integration, this theorem—discovered independently by Sir Isaac Newton (1642–1727) in England and Gottfried Wilhelm Leibniz (1646–1716) in Germany—is called the **Fundamental Theorem of Calculus.**

THEOREM 2

The Fundamental Theorem of Calculus

Let f be continuous on $[a, b]$. Then,

$$\int_a^b f(x)\, dx = F(b) - F(a) \tag{9}$$

where F is any antiderivative of f; that is, $F'(x) = f(x)$.

We will explain why this theorem is true at the end of this section.

In applying the Fundamental Theorem of Calculus, it is convenient to use the notation

$$F(x)\Big|_a^b = F(b) - F(a)$$

For example, in this notation, Equation (9) is written

$$\int_a^b f(x)\, dx = F(x)\Big|_a^b = F(b) - F(a)$$

Molly H. Fisher, David C. Royster, and Diandra Leslie-Pelecky

TITLE Professors of Mathematics Education, Mathematics, and Physics

INSTITUTION University of Kentucky (Fisher and Royster); Author, *The Physics of NASCAR®* (Leslie-Pelecky)

The last thing a fan thinks about at a NASCAR Sprint Cup race is calculus. The Sprint Cup series is NASCAR's most popular and most profitable series. During a ten-month season, drivers compete in 36 races across the United States and Mexico.

Typically, a racing team is made up not only of a driver, who is the most visible, and mechanics, who are often seen at the pit stops, but also of many people who work behind the scenes, away from the track. The last group includes highly trained engineers and scientists (some of whom hold Ph.D.s in engineering, mathematics, or physics) who study how each component of a race car performs under different conditions.

According to Dr. Andrew Randolph, who is a chemical engineer and the Engine Technical Director of Earnhardt Childress Racing Engines, "a calculation we do every week is obtaining the average power over a given speed range to provide a single engine performance metric for a given track. The calculation is a simple integral of power versus speed divided by the speed delta over which the integration is performed."

Much of the data available for this purpose is in graphical form. So recognizing the relationship between the area under a curve and the corresponding definite integral is essential to the solution of the problem. Even though there are many parameters that determine how an engine in a race car performs, the problem ultimately reduces to one of finding the average value of a function as given by the integral:

$$\frac{1}{\Delta v} \int_{v_0}^{v_1} P(v)\, dv$$

This is just one of many examples of how calculus can be used in a sport that some people think is reserved for speed-hungry daredevils.

EXAMPLE 1 Let R be the region under the graph of $f(x) = x$ on the interval $[1, 3]$. Use the Fundamental Theorem of Calculus to find the area A of R, and verify your result by elementary means.

Solution The region R is shown in Figure 16a. Since f is nonnegative on $[1, 3]$, the area of R is given by the definite integral of f from 1 to 3; that is,

$$A = \int_{1}^{3} x\, dx$$

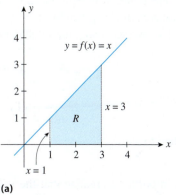

(a)

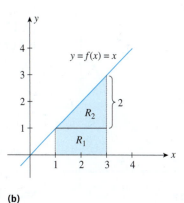

(b)

FIGURE 16
The area of R can be computed in two different ways.

To evaluate the definite integral, observe that an antiderivative of $f(x) = x$ is $F(x) = \frac{1}{2}x^2 + C$, where C is an arbitrary constant. Therefore, by the Fundamental Theorem of Calculus, we have

$$A = \int_1^3 x\,dx = \frac{1}{2}x^2 + C\,\Big|_1^3$$

$$= \left(\frac{9}{2} + C\right) - \left(\frac{1}{2} + C\right) = 4$$

To verify this result by elementary means, refer to Figure 16b. Observe that the area A is the area of the rectangle R_1 (width $\times$ height) plus the area of the triangle R_2 ($\frac{1}{2}$ base $\times$ height); that is,

$$2(1) + \frac{1}{2}(2)(2) = 2 + 2 = 4$$

which agrees with the result obtained earlier.

Observe that in evaluating the definite integral in Example 1, the constant of integration "dropped out." This is true in general, for if $F(x) + C$ denotes an antiderivative of some function f, then

$$F(x) + C\,\Big|_a^b = [F(b) + C] - [F(a) + C]$$

$$= F(b) + C - F(a) - C$$

$$= F(b) - F(a)$$

With this fact in mind, we may, in all future computations involving the evaluation of a definite integral, drop the constant of integration from our calculations.

Finding the Area of a Region Under a Curve

Having seen how effective the Fundamental Theorem of Calculus is in helping us find the area of simple regions, we now use it to find the area of more complicated regions.

EXAMPLE 2 In Section 6.3, we conjectured that the area of the region R under the graph of $f(x) = x^2$ on the interval $[0, 1]$ was $\frac{1}{3}$. Use the Fundamental Theorem of Calculus to verify this conjecture.

Solution   The region R is reproduced in Figure 17. Observe that f is nonnegative on $[0, 1]$, so the area of R is given by $A = \int_0^1 x^2\,dx$. Since an antiderivative of $f(x) = x^2$ is $F(x) = \frac{1}{3}x^3$, we see, using the Fundamental Theorem of Calculus, that

$$A = \int_0^1 x^2\,dx = \frac{1}{3}x^3\,\Big|_0^1$$

$$= \frac{1}{3}(1) - \frac{1}{3}(0) = \frac{1}{3}$$

as we wished to show.

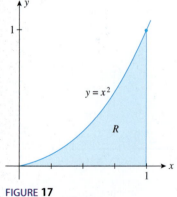

FIGURE 17
The area of R is $\int_0^1 x^2\,dx = \frac{1}{3}$.

Note   It is important to realize that the value $\frac{1}{3}$ is by definition the exact value of the area of R.

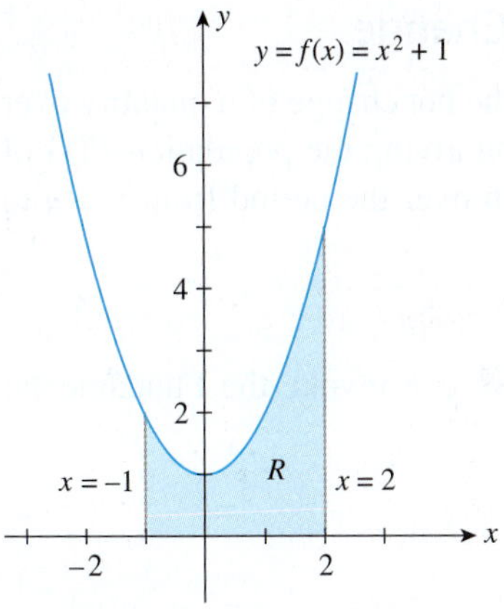

FIGURE **18**
The area of R is $\int_{-1}^{2} (x^2 + 1)\, dx$.

EXAMPLE 3 Find the area of the region R under the graph of $y = x^2 + 1$ from $x = -1$ to $x = 2$.

Solution   The region R is shown in Figure 18. Using the Fundamental Theorem of Calculus, we find that the required area is

$$\int_{-1}^{2} (x^2 + 1)\, dx = \left(\frac{1}{3}x^3 + x\right)\Big|_{-1}^{2}$$

$$= \left[\frac{1}{3}(2)^3 + 2\right] - \left[\frac{1}{3}(-1)^3 + (-1)\right] = 6$$

Evaluating Definite Integrals

In Examples 4 and 5, we use the rules of integration of Section 6.1 to help us evaluate the definite integrals.

EXAMPLE 4 Evaluate $\int_{1}^{3} (3x^2 + e^x)\, dx$.

Solution

$$\int_{1}^{3} (3x^2 + e^x)\, dx = x^3 + e^x \Big|_{1}^{3}$$

$$= (27 + e^3) - (1 + e) = 26 + e^3 - e$$

EXAMPLE 5 Evaluate $\int_{1}^{2}\left(\frac{1}{x} - \frac{1}{x^2}\right) dx$.

Solution

$$\int_{1}^{2}\left(\frac{1}{x} - \frac{1}{x^2}\right) dx = \int_{1}^{2}\left(\frac{1}{x} - x^{-2}\right) dx$$

$$= \ln|x| + \frac{1}{x}\Big|_{1}^{2}$$

$$= \left(\ln 2 + \frac{1}{2}\right) - (\ln 1 + 1)$$

$$= \ln 2 - \frac{1}{2} \qquad \text{Recall that } \ln 1 = 0.$$

Explore and Discuss

Consider the definite integral $\int_{-1}^{1} \frac{1}{x^2}\, dx$.

1. Show that a formal application of Equation (9) leads to

$$\int_{-1}^{1} \frac{1}{x^2}\, dx = -\frac{1}{x}\Big|_{-1}^{1} = -1 - 1 = -2$$

2. Observe that $f(x) = 1/x^2$ is positive at each value of x in $[-1, 1]$ where it is defined. So one might expect that the definite integral with integrand f has a positive value, if it exists.

3. Resolve this apparent contradiction in the result (1) and the observation (2).

The Definite Integral as a Measure of Net Change

In real-world applications, we are often interested in the net change of a quantity over a period of time. For example, suppose P is a function giving the population $P(t)$ of a city at time t. Then the *net change* in the population over the period from $t = a$ to $t = b$ is given by

$$P(b) - P(a) \qquad \text{Population at } t = b \text{ minus population at } t = a$$

If P has a continuous derivative P' in $[a, b]$, then we can invoke the Fundamental Theorem of Calculus to write

$$P(b) - P(a) = \int_a^b P'(t) \, dt \qquad P \text{ is an antiderivative of } P'.$$

Thus, if we know the *rate of change* of the population at any time t, then we can calculate the net change in the population from $t = a$ to $t = b$ by evaluating an appropriate definite integral.

APPLIED EXAMPLE 6 Population Growth in Clark County As the twentieth century ended, Clark County in Nevada—dominated by Las Vegas—was one of the fastest-growing metropolitan areas in the United States. From 1970 through 2000, the population was growing at the rate of

$$R(t) = 133{,}680t^2 - 178{,}788t + 234{,}633 \qquad (0 \le t \le 3)$$

people per decade, where $t = 0$ corresponds to the beginning of 1970. What was the net change in the population over the decade from 1980 to 1990?
Source: U.S. Census Bureau.

Solution The net change in the population over the decade from 1980 ($t = 1$) to 1990 ($t = 2$) is given by $P(2) - P(1)$, where P denotes the population in the county at time t. But $P' = R$, so

$$
\begin{aligned}
P(2) - P(1) &= \int_1^2 P'(t) \, dt = \int_1^2 R(t) \, dt \\
&= \int_1^2 (133{,}680t^2 - 178{,}788t + 234{,}633) \, dt \\
&= 44{,}560t^3 - 89{,}394t^2 + 234{,}633t \Big|_1^2 \\
&= [44{,}560(2)^3 - 89{,}394(2)^2 + 234{,}633(2)] \\
&\quad - [44{,}560 - 89{,}394 + 234{,}633] \\
&= 278{,}371
\end{aligned}
$$

and so the net change is 278,371.

More generally, we have the following result. We assume that f has a continuous derivative, even though the integrability of f' is sufficient.

Net Change Formula

The net change in a function f over an interval $[a, b]$ is given by

$$f(b) - f(a) = \int_a^b f'(x) \, dx \tag{10}$$

provided that f' is continuous on $[a, b]$.

Additional examples of the net change of a function follow.

APPLIED EXAMPLE 7 Production Costs The management of Staedtler Office Equipment has determined that the daily marginal cost function associated with producing battery-operated pencil sharpeners is given by

$$C'(x) = 0.000006x^2 - 0.006x + 4$$

where $C'(x)$ is measured in dollars per unit and x denotes the number of units produced. Management has also determined that the daily fixed cost incurred in producing these pencil sharpeners is $100. Find Staedtler's daily total cost for producing (a) the first 500 units and (b) the 201st through 400th units.

Solution

a. Since $C'(x)$ is the marginal cost function, its antiderivative $C(x)$ is the total cost function. The daily fixed cost incurred in producing the pencil sharpeners is $C(0)$ dollars. Since the daily fixed cost is given as $100, we have $C(0) = 100$. We are required to find $C(500)$. Let's compute $C(500) - C(0)$, the net change in the total cost function $C(x)$ over the interval $[0, 500]$. Using the Fundamental Theorem of Calculus, we find

$$C(500) - C(0) = \int_0^{500} C'(x)\, dx$$

$$= \int_0^{500} (0.000006x^2 - 0.006x + 4)\, dx$$

$$= 0.000002x^3 - 0.003x^2 + 4x \Big|_0^{500}$$

$$= [0.000002(500)^3 - 0.003(500)^2 + 4(500)]$$
$$\quad - [0.000002(0)^3 - 0.003(0)^2 + 4(0)]$$

$$= 1500$$

Therefore, $C(500) = 1500 + C(0) = 1500 + 100 = 1600$, so the total cost incurred daily by Staedtler in producing the first 500 pencil sharpeners is $1600.

b. The daily total cost incurred by Staedtler in producing the 201st through 400th units of battery-operated pencil sharpeners is given by

$$C(400) - C(200) = \int_{200}^{400} C'(x)\, dx$$

$$= \int_{200}^{400} (0.000006x^2 - 0.006x + 4)\, dx$$

$$= 0.000002x^3 - 0.003x^2 + 4x \Big|_{200}^{400}$$

$$= [0.000002(400)^3 - 0.003(400)^2 + 4(400)]$$
$$\quad - [0.000002(200)^3 - 0.003(200)^2 + 4(200)]$$

$$= 552$$

or $552.

Since $C'(x)$ is nonnegative for x in the interval $[0, \infty)$, we have the following geometric interpretation of the two definite integrals in Example 7: $\int_0^{500} C'(x)\,dx$ is the area of the region under the graph of the function C' from $x = 0$ to $x = 500$, shown in Figure 19a, and $\int_{200}^{400} C'(x)\,dx$ is the area of the region from $x = 200$ to $x = 400$, shown in Figure 19b.

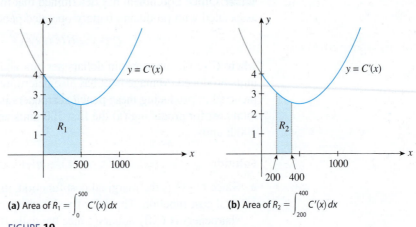

(a) Area of $R_1 = \displaystyle\int_0^{500} C'(x)\,dx$ **(b)** Area of $R_2 = \displaystyle\int_{200}^{400} C'(x)\,dx$

FIGURE **19**

APPLIED EXAMPLE 8 Assembly Time of Workers An efficiency study conducted for Elektra Electronics showed that the rate at which Space Commander walkie-talkies are assembled by the average worker t hours after starting work at 8 A.M. is given by the function

$$f(t) = -3t^2 + 12t + 15 \qquad (0 \le t \le 4)$$

Determine how many walkie-talkies can be assembled by the average worker in the first hour of the morning shift.

Solution Let $N(t)$ denote the number of walkie-talkies assembled by the average worker during the first t hours after starting work in the morning shift. Then we have

$$N'(t) = f(t) = -3t^2 + 12t + 15$$

Therefore, the number of units assembled by the average worker in the first hour of the morning shift is

$$N(1) - N(0) = \int_0^1 N'(t)\,dt = \int_0^1 (-3t^2 + 12t + 15)\,dt$$

$$= -t^3 + 6t^2 + 15t \Big|_0^1 = -1 + 6 + 15$$

$$= 20$$

or 20 units.

Exploring with TECHNOLOGY

You can demonstrate graphically that $\int_0^x t\,dt = \frac{1}{2}x^2$ as follows:

1. Plot the graphs of $y1 = \text{fnInt}\,(t, t, 0, x) = \int_0^x t\,dt$ and $y2 = \frac{1}{2}x^2$ on the same set of axes, using the viewing window $[-5, 5] \times [0, 10]$.
2. Compare the graphs of $y1$ and $y2$, and draw the desired conclusion.

APPLIED EXAMPLE 9 Projected Demand for Electricity A certain city's rate of electricity consumption is expected to grow exponentially with a growth constant of $k = 0.04$. If the present rate of consumption is 40 million kilowatt-hours (kWh) per year, what should be the total production of electricity over the next 3 years in order to meet the projected demand?

Solution If $R(t)$ denotes the expected rate of consumption of electricity t years from now, then

$$R(t) = 40e^{0.04t}$$

million kWh per year. Next, if $C(t)$ denotes the expected total consumption of electricity over the next t years, then

$$C'(t) = R(t)$$

Therefore, the total consumption of electricity expected over the next 3 years is given by

$$
\begin{aligned}
\int_{0}^{3} C'(t)\,dt &= \int_{0}^{3} 40e^{0.04t}\,dt \\
&= \frac{40}{0.04}\, e^{0.04t}\Big|_{0}^{3} \\
&= 1000(e^{0.12} - 1) \\
&\approx 127.5
\end{aligned}
$$

or 127.5 million kWh, the amount that must be produced over the next 3 years to meet the demand.

Validity of the Fundamental Theorem of Calculus

To demonstrate the plausibility of the Fundamental Theorem of Calculus for the case in which f is nonnegative on an interval $[a, b]$, let's define an "area function" A as follows. Let $A(t)$ denote the area of the region R under the graph of $y = f(x)$ from $x = a$ to $x = t$, where $a \le t \le b$ (Figure 20).

If h is a small positive number, then $A(t + h)$ is the area of the region under the graph of $y = f(x)$ from $x = a$ to $x = t + h$. Therefore, the difference

$$A(t + h) - A(t)$$

is the area of the region under the graph of $y = f(x)$ from $x = t$ to $x = t + h$ (Figure 21).

Now, the area of this last region can be approximated by the area of the rectangle of width h and height $f(t)$—that is, by the expression $h \cdot f(t)$ (see Figure 22 on the next page). Thus,

$$A(t + h) - A(t) \approx h \cdot f(t)$$

where the approximations improve as h is taken to be smaller and smaller.

Dividing both sides of the foregoing relationship by h, we obtain

$$\frac{A(t + h) - A(t)}{h} \approx f(t)$$

Taking the limit as h approaches zero, we find, by the definition of the derivative, that the left-hand side is

$$\lim_{h \to 0} \frac{A(t + h) - A(t)}{h} = A'(t)$$

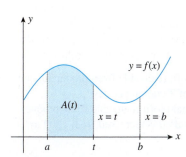

FIGURE 20
$A(t)$ = area under the graph of f from $x = a$ to $x = t$

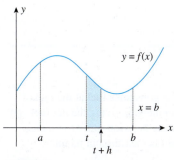

FIGURE 21
$A(t + h) - A(t)$ = area under the graph of f from $x = t$ to $x = t + h$

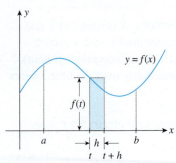

FIGURE 22
The area of the rectangle is $h \cdot f(t)$.

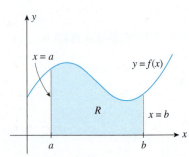

FIGURE 23
The area of R is given by $A(b)$.

The right-hand side, which is independent of h, remains constant throughout the limiting process. Because the approximation becomes exact as h approaches zero, we find that

$$A'(t) = f(t)$$

Since this equation holds for all values of t in the interval $[a, b]$, we have shown that the *area function A* is an antiderivative of the function f. By Theorem 1 of Section 6.1, we conclude that $A(x)$ must have the form

$$A(x) = F(x) + C$$

where F is any antiderivative of f and C is a constant. To determine the value of C, observe that $A(a) = 0$. This condition implies that

$$A(a) = F(a) + C = 0$$

or $C = -F(a)$. Next, since the area of the region R is $A(b)$ (Figure 23), we see that the required area is

$$A(b) = F(b) + C$$
$$= F(b) - F(a)$$

Since the area of the region R is

$$\int_a^b f(x)\, dx$$

we have

$$\int_a^b f(x)\, dx = F(b) - F(a)$$

as we set out to show.

6.4 Self-Check Exercises

1. Evaluate $\int_0^2 (x + e^x)\, dx$.

2. **MARGINAL PROFIT OF A COMPANY** The daily marginal profit function associated with producing and selling TexaPep hot sauce is

$$P'(x) = -0.000006x^2 + 6$$

where x denotes the number of cases (each case contains 24 bottles) produced and sold daily and $P'(x)$ is measured in dollars per unit. The fixed cost is $400.

a. What is the total profit realizable from producing and selling 1000 cases of TexaPep per day?

b. What is the additional profit realizable if the production and sale of TexaPep is increased from 1000 to 1200 cases/day?

Solutions to Self-Check Exercises 6.4 can be found on page 455.

6.4 Concept Questions

1. State the Fundamental Theorem of Calculus.

2. State the net change formula, and use it to answer the following questions:
   a. If a company generates income at the rate of R dollars/day, explain what $\int_a^b R(t)\, dt$ measures, where a and b are measured in days with $a < b$.

   b. If a private jet airplane consumes fuel at the rate of $R(t)$ gal/min, write an integral giving the net fuel consumption by the airplane between times $t = a$ and $t = b$ $(a < b)$, where t is measured in minutes.

6.4 Exercises

In Exercises 1–4, find the area of the region under the graph of the function f on the interval $[a, b]$, using the Fundamental Theorem of Calculus. Then verify your result using geometry.

1. $f(x) = 2; [1, 4]$

2. $f(x) = 4; [-1, 2]$

3. $f(x) = 2x; [1, 3]$

4. $f(x) = -\dfrac{1}{4}x + 1; [1, 4]$

In Exercises 5–16, find the area of the region under the graph of the function f on the interval $[a, b]$.

5. $f(x) = 2x + 3; [-1, 2]$

6. $f(x) = 4x - 1; [2, 4]$

7. $f(x) = -x^2 + 4; [-1, 2]$

8. $f(x) = 4x - x^2; [0, 4]$

9. $f(x) = \dfrac{1}{x}; [1, 2]$

10. $f(x) = \dfrac{1}{x^2}; [2, 4]$

11. $f(x) = \sqrt{x}; [1, 9]$

12. $f(x) = x^3; [1, 3]$

13. $f(x) = 1 - \sqrt[3]{x}; [-8, -1]$

14. $f(x) = \dfrac{1}{\sqrt{x}}; [1, 9]$

15. $f(x) = e^x; [0, 2]$

16. $f(x) = e^x - x; [1, 2]$

In Exercises 17–40, evaluate the definite integral.

17. $\displaystyle\int_2^4 3\, dx$

18. $\displaystyle\int_{-1}^2 -2\, dx$

19. $\displaystyle\int_1^4 (2x + 3)\, dx$

20. $\displaystyle\int_{-1}^0 (4 - x)\, dx$

21. $\displaystyle\int_{-1}^3 2x^2\, dx$

22. $\displaystyle\int_0^2 8x^3\, dx$

23. $\displaystyle\int_{-2}^2 (x^2 - 1)\, dx$

24. $\displaystyle\int_1^4 \sqrt{u}\, du$

25. $\displaystyle\int_1^8 2x^{1/3}\, dx$

26. $\displaystyle\int_1^4 2x^{-3/2}\, dx$

27. $\displaystyle\int_0^1 (x^3 - 2x^2 + 1)\, dx$

28. $\displaystyle\int_1^2 (t^5 - t^3 + 1)\, dt$

29. $\displaystyle\int_1^4 \dfrac{1}{x}\, dx$

30. $\displaystyle\int_1^3 \dfrac{2}{x}\, dx$

31. $\displaystyle\int_0^4 x(x^2 - 1)\, dx$

32. $\displaystyle\int_0^2 (x - 4)(x - 1)\, dx$

33. $\displaystyle\int_1^3 (t^2 - t)^2\, dt$

34. $\displaystyle\int_{-1}^1 (x^2 - 1)^2\, dx$

35. $\displaystyle\int_{-3}^{-1} \dfrac{1}{x^2}\, dx$

36. $\displaystyle\int_1^2 \dfrac{2}{x^3}\, dx$

37. $\displaystyle\int_1^4 \left(\sqrt{x} - \dfrac{1}{\sqrt{x}}\right) dx$

38. $\displaystyle\int_0^1 \sqrt{2x}(\sqrt{x} + \sqrt{2})\, dx$

39. $\displaystyle\int_1^4 \dfrac{3x^3 - 2x^2 + 4}{x^2}\, dx$

40. $\displaystyle\int_1^2 \left(1 + \dfrac{1}{u} + \dfrac{1}{u^2}\right) du$

41. Personal Bankruptcy The number of personal bankruptcy filings by fiscal years ending September 30 between 2010 and 2012 was declining at the rate of

$$R(t) = 0.077t + 0.0825 \qquad (0 \le t \le 2)$$

million cases/year, t years after September 30, 2010. The number of filings as of September 30, 2010, stood at approximately 1.538 million cases.

a. Estimate the change in the number of personal bankruptcy cases filed between September 30, 2010, and September 30, 2012.

b. What was the approximate number of personal bankruptcy cases filed in 2012?

Hint: If $N(t)$ denotes the number of bankruptcy filings in year t, then $N'(t) = -R(t)$.

Source: Administrative Office of the U.S. Courts.

42. Health Care Costs According to a study conducted by the Centers for Medicare & Medicaid Services in 2010, the national spending for out-of-pocket health-care costs is projected to increase over the next several years. The amount spent annually from 2010 $(t = 0)$ is expected to grow at the rate of

$$R(t) = 1.0952t + 17.357 \qquad (0 \le t \le 6)$$

billion dollars/year in year t. The national spending in 2010 was \$317 billion. What is the projected national spending in 2016?

Source: Centers for Medicare & Medicaid Services.

43. Marginal Cost A division of Ditton Industries manufactures a deluxe toaster oven. Management has determined that the daily marginal cost function associated with producing these toaster ovens is given by

$$C'(x) = 0.0003x^2 - 0.12x + 20$$

where $C'(x)$ is measured in dollars per unit and x denotes the number of units produced. Management has also determined that the daily fixed cost incurred in the production is \$800.

a. Find the total cost incurred by Ditton in producing the first 300 units of these toaster ovens per day.

b. What is the total cost incurred by Ditton in producing the 201st through 300th units/day?

44. Marginal Revenue The management of Ditton Industries has determined that the daily marginal revenue function associated with selling x units of their deluxe toaster ovens is given by

$$R'(x) = -0.1x + 40$$

where $R'(x)$ is measured in dollars per unit.

a. Find the daily total revenue realized from the sale of 200 units of the toaster oven.

b. Find the additional revenue realized when the production (and sales) level is increased from 200 to 300 units.

45. **MARGINAL PROFIT** Refer to Exercise 43. The daily marginal profit function associated with the production and sales of the deluxe toaster ovens is known to be

$$P'(x) = -0.0003x^2 + 0.02x + 20$$

where x denotes the number of units manufactured and sold daily and $P'(x)$ is measured in dollars per unit.
a. Find the total profit realizable from the manufacture and sale of 200 units of the toaster ovens per day.
   Hint: $P(200) - P(0) = \int_0^{200} P'(x)\, dx$, $P(0) = -800$.
b. What is the additional daily profit realizable if the production and sales of the toaster ovens are increased from 200 to 220 units/day?

46. **EFFICIENCY STUDIES** Tempco Electronics, a division of Tempco Toys, manufactures an electronic football game. An efficiency study showed that the rate at which the games are assembled by the average worker t hr after starting work at 8 A.M. is

$$-\frac{3}{2}t^2 + 6t + 20 \qquad (0 \le t \le 4)$$

units/hr.
a. Find the total number of games the average worker can be expected to assemble in the 4-hr morning shift.
b. How many units can the average worker be expected to assemble in the first hour of the morning shift? In the second hour of the morning shift?

47. **SPEEDBOAT RACING** In a recent pretrial run for the world water speed record, the velocity of the *Sea Falcon II* t sec after firing the booster rocket was given by

$$v(t) = -t^2 + 20t + 440 \qquad (0 \le t \le 20)$$

ft/sec. Find the distance covered by the boat over the 20-sec period after the booster rocket was activated.
   Hint: The distance is given by $\int_0^{20} v(t)\, dt$.

48. **TABLET COMPUTERS** Annual sales (in millions of units) of a certain brand of tablet computers are expected to grow in accordance with the function

$$f(t) = 0.18t^2 + 0.16t + 2.64 \qquad (0 \le t \le 6)$$

per year, where t is measured in years. How many tablet computers will be sold over the next 6 years?

49. **CREDIT CARD DEBT** The average U.S. household credit card debt stood at $8382 per household at the beginning of 2008 ($t = 0$) and was changing at the rate of

$$f(t) = 258t^2 - 680t - 316 \qquad (0 \le t \le 4)$$

dollars/year in year t from the beginning of 2008 until the beginning of 2012. What was the average U.S. household credit card debt at the beginning of 2012?
   Source: TIM.

50. **U.S. NATIONAL DEBT** The U.S. national debt stood at $10.025 trillion in 2008 ($t = 0$) and was growing at the rate of

$$R(t) = 0.070251t^2 - 0.51548t + 2.1667 \qquad (0 \le t \le 4)$$

trillion dollars/year in year t between 2008 and 2012. What was the U.S. national debt in 2012?
   Source: U.S. Treasury Direct.

51. **AIR PURIFICATION** To test air purifiers, engineers ran a purifier in a smoke-filled 10-ft × 20-ft room. While conducting a test for a certain brand of air purifier, they determined that the amount of smoke in the room was decreasing at the rate of

$$R(t) = 0.00032t^4 - 0.01872t^3 + 0.3948t^2$$
$$- 3.83t + 17.63 \qquad (0 \le t \le 20)$$

percent of the (original) amount of the smoke per minute, t min after the start of the test. How much smoke was left in the room 5 min after the start of the test? 10 min after the start of the test?
   Source: Consumer Reports.

52. **SOLAR PANEL PRODUCTION** The manager of Sodex Corporation estimates that t months after it first began production, it was manufacturing the company's 140-watt 12-volt nominal solar panels at the rate of

$$R(t) = \frac{4t}{1 + t^2} + 3\sqrt{t}$$

hundred panels per month. What is the manager's estimate of the number of solar panels manufactured by the company during the second year?

53. **CELL PHONE AD SPENDING** Cell phone advertising spending is expected to grow at the rate of

$$R(t) = 0.8256t^{-0.04} \qquad (1 \le t \le 5)$$

billion dollars/year between 2007 ($t = 1$) and 2011 ($t = 5$). The cell phone ad spending in 2007 was $0.9 billion.
a. Find an expression giving the cell phone ad spending in year t.
b. If the trend continued, what was the cell phone ad spending in 2014?
   Source: Interactive Advertising Bureau.

54. **CREDIT CARD DELINQUENCY RATE** The credit card delinquency rate stood at 5.7% at the beginning of 2010. It was declining at the rate of

$$R(t) = 0.150975e^{-0.0275t} \qquad (0 \le t \le 24)$$

percent t months later. What was the credit card delinquency rate at the beginning of 2012?
   Source: Federal Reserve.

55. **AMAZON'S GROWTH** Amazon's revenue was growing at the rate of

$$R'(t) = 0.545043e^{0.291t} \qquad (0 \le t \le 6)$$

billion dollars/year in year t with $t = 0$ corresponding to 2006. The revenue of the company for 2006 was $10.71 billion.

a. By how much did the revenue of Amazon increase over the period from 2006 through 2012?

b. What was the revenue of the company in 2012?

Source: Amazon.

56. **CANADIAN OIL-SANDS PRODUCTION** The production of oil (in millions of barrels per day) extracted from oil sands in Canada is projected to grow according to the function

$$P(t) = \frac{4.76}{1 + 4.11e^{-0.22t}} \qquad (0 \le t \le 15)$$

where t is measured in years, with $t = 0$ corresponding to 2005. What is the expected total production of oil from oil sands over the years from 2005 until 2020 ($t = 15$)?

Hint: Multiply the integrand by $\dfrac{e^{0.22t}}{e^{0.22t}}$.

Source: Canadian Association of Petroleum Producers.

57. **SENIOR CITIZENS** The U.S. population aged 65 years and older (in millions) from 2000 to 2050 is projected to grow at the rate of

$$f(t) = \frac{85}{1 + 1.859e^{-0.66t}} \qquad (0 \le t \le 5)$$

where t is measured in decades, with $t = 0$ corresponding to 2000. By how much will the population aged 65 years and older increase from the beginning of 2000 until the beginning of 2030?

Hint: Multiply the integrand by $\dfrac{e^{0.66t}}{e^{0.66t}}$.

Source: U.S. Census Bureau.

58. **BLOOD FLOW** Consider an artery of length L cm and radius R cm. By using Poiseuille's Law (page 316), it can be shown that the rate at which blood flows through the artery (measured in cubic centimeters per second) is given by

$$V = \int_0^R \frac{k}{L} x(R^2 - x^2) \, dx$$

where k is a constant. Find an expression for V that does *not* involve an integral.

Hint: Use the substitution $u = R^2 - x^2$.

59. Find the area of the region bounded by the graph of the function $f(x) = x^4 - 2x^2 + 2$, the x-axis, and the lines $x = a$ and $x = b$, where $a < b$ and a and b are the x-coordinates of the relative maximum point and a relative minimum point of f, respectively.

60. Find the area of the region bounded by the graph of the function $f(x) = (x + 1)/\sqrt{x}$, the x-axis, and the lines $x = a$ and $x = b$ where a and b are, respectively, the x-coordinates of the relative minimum point and the inflection point of f.

In Exercises 61–64, determine whether the statement is true or false. Give a reason for your answer.

61. $\displaystyle\int_{-1}^{1} \frac{1}{x^3} \, dx = -\frac{1}{2x^2} \Big|_{-1}^{1} = -\frac{1}{2} - \left(-\frac{1}{2}\right) = 0$

62. $\displaystyle\int_{-1}^{1} \frac{1}{x} \, dx = \ln|x| \Big|_{-1}^{1} = \ln|1| - \ln|-1| = \ln 1 - \ln 1 = 0$

63. $\int_0^2 (1 - x) \, dx$ gives the area of the region under the graph of $f(x) = 1 - x$ on the interval $[0, 2]$.

64. The total revenue realized in selling the first 500 units of a product is given by

$$\int_0^{500} R'(x) \, dx = R(500) - R(0)$$

where R is the total revenue.

6.4 Solutions to Self-Check Exercises

1. $\displaystyle\int_0^2 (x + e^x) \, dx = \frac{1}{2}x^2 + e^x \Big|_0^2$

$$= \left[\frac{1}{2}(2)^2 + e^2\right] - \left[\frac{1}{2}(0) + e^0\right]$$

$$= 2 + e^2 - 1$$

$$= e^2 + 1$$

2. **a.** We want $P(1000)$, but

$$P(1000) - P(0) = \int_0^{1000} P'(x) \, dx = \int_0^{1000} (-0.000006x^2 + 6) \, dx$$

$$= -0.000002x^3 + 6x \Big|_0^{1000}$$

$$= -0.000002(1000)^3 + 6(1000)$$

$$= 4000$$

So, $P(1000) = 4000 + P(0) = 4000 - 400$, or 3600/day [$P(0) = -$fixed cost].

b. The additional profit realizable is given by

$$\int_{1000}^{1200} P'(x) \, dx = -0.000002x^3 + 6x \Big|_{1000}^{1200}$$

$$= [-0.000002(1200)^3 + 6(1200)]$$

$$- [-0.000002(1000)^3 + 6(1000)]$$

$$= 3744 - 4000$$

$$= -256$$

That is, the company's profit is reduced by $256/day if production is increased from 1000 to 1200 cases/day.

USING TECHNOLOGY Evaluating Definite Integrals

Some graphing utilities have an operation for finding the definite integral of a function. If your graphing utility has this capability, use it to work through the example and exercises of this section.

EXAMPLE 1 Use the numerical integral operation of a graphing utility to evaluate

$$\int_{-1}^{2} \frac{2x + 4}{(x^2 + 1)^{3/2}} \, dx$$

Solution   Using the numerical integral operation of a graphing utility, we find

$$\int_{-1}^{2} \frac{2x + 4}{(x^2 + 1)^{3/2}} \, dx = \text{fnInt}((2x + 4)/(x^2 + 1)\wedge 1.5, x, -1, 2) \approx 6.92592226 \quad \blacksquare$$

TECHNOLOGY EXERCISES

In Exercises 1–4, find the area of the region under the graph of f on the interval $[a, b]$. Express your answer to four decimal places.

1. $f(x) = 0.002x^5 + 0.032x^4 - 0.2x^2 + 2; [-1.1, 2.2]$

2. $f(x) = x\sqrt{x^3 + 1}; [1, 2]$

3. $f(x) = \sqrt{x}e^{-x}; [0, 3]$

4. $f(x) = \dfrac{\ln x}{\sqrt{1 + x^2}}; [1, 2]$

In Exercises 5–10, evaluate the definite integral.

5. $\displaystyle\int_{-1.2}^{2.3} (0.2x^4 - 0.32x^3 + 1.2x - 1) \, dx$

6. $\displaystyle\int_{1}^{3} x(x^4 - 1)^{3.2} \, dx$

7. $\displaystyle\int_{0}^{2} \frac{3x^3 + 2x^2 + 1}{2x^2 + 3} \, dx$

8. $\displaystyle\int_{1}^{2} \frac{\sqrt{x} + 1}{2x^2 + 1} \, dx$

9. $\displaystyle\int_{0}^{2} \frac{e^x}{\sqrt{x^2 + 1}} \, dx$

10. $\displaystyle\int_{1}^{3} e^{-x} \ln(x^2 + 1) \, dx$

11. Rework Exercise 51, Exercises 6.4.

12. Rework Exercise 57, Exercises 6.4.

13. **MARIJUANA ARRESTS** The number of arrests for marijuana sales and possession in New York City grew at the rate of approximately

$$f(t) = 0.0125t^4 - 0.01389t^3 + 0.55417t^2$$
$$+ 0.53294t + 4.95238 \qquad (0 \le t \le 5)$$

thousand/year, where t is measured in years, with $t = 0$ corresponding to the beginning of 1992. Find the approximate number of marijuana arrests in the city from the beginning of 1992 to the end of 1997.
Source: State Division of Criminal Justice Services.

14. **POPULATION GROWTH** The population of a certain city is projected to grow at the rate of $9 \sqrt{t + 1} \ln \sqrt{t + 1}$ thousand people/year t years from now. If the current population is 800,000, what will be the population 45 years from now?

6.5 Evaluating Definite Integrals

This section continues our discussion of the applications of the Fundamental Theorem of Calculus.

Properties of the Definite Integral

Before going on, we list the following useful properties of the definite integral, some of which parallel the rules of integration of Section 6.1.

> **Properties of the Definite Integral**
>
> Let f and g be integrable functions; then,
>
> **1.** $\displaystyle\int_a^a f(x)\,dx = 0$
>
> **2.** $\displaystyle\int_a^b f(x)\,dx = -\int_b^a f(x)\,dx$
>
> **3.** $\displaystyle\int_a^b cf(x)\,dx = c\int_a^b f(x)\,dx \qquad (c,\text{ a constant})$
>
> **4.** $\displaystyle\int_a^b \left[f(x) \pm g(x)\right]dx = \int_a^b f(x)\,dx \pm \int_a^b g(x)\,dx$
>
> **5.** $\displaystyle\int_a^b f(x)\,dx = \int_a^c f(x)\,dx + \int_c^b f(x)\,dx \qquad (a < c < b)$

Property 5 states that if c is a number lying between a and b, so that the interval $[a, b]$ is divided into the intervals $[a, c]$ and $[c, b]$, then the integral of f over $[a, b]$ may be expressed as the sum of the integral of f over $[a, c]$ and the integral of f over $[c, b]$.

Property 5 has the following geometric interpretation when f is nonnegative. By definition,

$$\int_a^b f(x)\,dx$$

is the area of the region under the graph of $y = f(x)$ from $x = a$ to $x = b$ (Figure 24). Similarly, we interpret the definite integrals

$$\int_a^c f(x)\,dx \quad \text{and} \quad \int_c^b f(x)\,dx$$

as the areas of the regions under the graph of $y = f(x)$ from $x = a$ to $x = c$ and from $x = c$ to $x = b$, respectively. Since the two regions do not overlap, we see that

$$\int_a^b f(x)\,dx = \int_a^c f(x)\,dx + \int_c^b f(x)\,dx$$

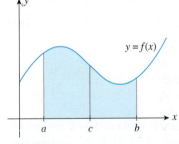

FIGURE 24

$\displaystyle\int_a^b f(x)\,dx = \int_a^c f(x)\,dx + \int_c^b f(x)\,dx$

The Method of Substitution for Definite Integrals

Our first example shows two approaches that are generally used in evaluating a definite integral using the method of substitution.

EXAMPLE 1 Evaluate $\displaystyle\int_0^4 x\sqrt{9 + x^2}\,dx$.

Solution

Method 1 We first find the corresponding indefinite integral:

$$I = \int x\sqrt{9 + x^2}\,dx$$

Make the substitution $u = 9 + x^2$, so that

$$du = \frac{d}{dx}(9 + x^2)\,dx$$

$$= 2x\,dx$$

$$x\,dx = \frac{1}{2}\,du \qquad \text{Divide both sides by 2.}$$

Then

$$I = \int \frac{1}{2}\sqrt{u}\,du = \frac{1}{2}\int u^{1/2}\,du$$

$$= \frac{1}{3}u^{3/2} + C = \frac{1}{3}(9 + x^2)^{3/2} + C \qquad \begin{array}{l}\text{Substitute}\\ 9 + x^2 \text{ for } u.\end{array}$$

Using this result, we now evaluate the given definite integral:

$$\int_0^4 x\sqrt{9 + x^2}\,dx = \frac{1}{3}(9 + x^2)^{3/2}\Big|_0^4$$

$$= \frac{1}{3}\left[(9 + 16)^{3/2} - 9^{3/2}\right]$$

$$= \frac{1}{3}(125 - 27) = \frac{98}{3} = 32\tfrac{2}{3}$$

Method 2 *Changing the Limits of Integration.* As before, we make the substitution

$$u = 9 + x^2 \tag{11}$$

so that

$$du = 2x\,dx$$

$$x\,dx = \frac{1}{2}\,du$$

Next, observe that the given definite integral is evaluated *with respect to* x with the range of integration given by the interval $[0, 4]$. If we perform the integration *with respect to* u via the substitution (11), then we must adjust the range of integration to reflect the fact that the integration is being performed with respect to the new variable u. To determine the proper range of integration, note that when $x = 0$, Equation (11) implies that

$$u = 9 + 0^2 = 9$$

which gives the required lower limit of integration with respect to u. Similarly, when $x = 4$,

$$u = 9 + 16 = 25$$

is the required upper limit of integration with respect to u. Thus, the range of integration when the integration is performed with respect to u is given by the interval $[9, 25]$. Therefore, we have

$$\int_0^4 x\sqrt{9 + x^2}\,dx = \int_9^{25} \frac{1}{2}\sqrt{u}\,du = \frac{1}{2}\int_9^{25} u^{1/2}\,du$$

$$= \frac{1}{3}u^{3/2}\Big|_9^{25} = \frac{1}{3}(25^{3/2} - 9^{3/2})$$

$$= \frac{1}{3}(125 - 27) = \frac{98}{3} = 32\tfrac{2}{3}$$

which agrees with the result obtained by using Method 1.

⚠️ When you use Method 2, make sure you adjust the limits of integration to reflect integrating with respect to the new variable u.

Exploring with TECHNOLOGY

Refer to Example 1. You can confirm the results obtained there by using a graphing utility as follows:

1. Use the numerical integration operation of the graphing utility to evaluate
$$\int_0^4 x\sqrt{9 + x^2}\, dx$$

2. Evaluate $\dfrac{1}{2}\displaystyle\int_9^{25} \sqrt{u}\, du$.

3. Conclude that $\displaystyle\int_0^4 x\sqrt{9 + x^2}\, dx = \dfrac{1}{2}\int_9^{25} \sqrt{u}\, du$.

EXAMPLE 2 Evaluate $\displaystyle\int_0^2 xe^{2x^2}\, dx$.

Solution Let $u = 2x^2$, so that $du = 4x\, dx$, or $x\, dx = \frac{1}{4}\, du$. When $x = 0$, $u = 0$; and when $x = 2$, $u = 8$. These give the lower and upper limits of integration with respect to u. Making the indicated substitutions, we find

$$\int_0^2 xe^{2x^2}\, dx = \int_0^8 \frac{1}{4} e^u\, du = \frac{1}{4} e^u \Big|_0^8 = \frac{1}{4}\left(e^8 - 1\right)$$ ■

EXAMPLE 3 Evaluate $\displaystyle\int_0^1 \frac{x^2}{x^3 + 1}\, dx$.

Solution Let $u = x^3 + 1$, so that $du = 3x^2\, dx$, or $x^2\, dx = \frac{1}{3}\, du$. When $x = 0$, $u = 1$; and when $x = 1$, $u = 2$. These give the lower and upper limits of integration with respect to u. Making the indicated substitutions, we find

$$\int_0^1 \frac{x^2}{x^3 + 1}\, dx = \frac{1}{3}\int_1^2 \frac{du}{u} = \frac{1}{3}\ln|u| \Big|_1^2$$

$$= \frac{1}{3}\left(\ln 2 - \ln 1\right) = \frac{1}{3}\ln 2$$ ■

Finding the Area Under a Curve

EXAMPLE 4 Find the area of the region R under the graph of $f(x) = e^{(1/2)x}$ from $x = -1$ to $x = 1$.

Solution The region R is shown in Figure 25. Its area is given by

$$A = \int_{-1}^1 e^{(1/2)x}\, dx$$

To evaluate this integral, we make the substitution

$$u = \frac{1}{2} x$$

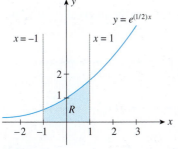

FIGURE 25

Area of $R = \displaystyle\int_{-1}^1 e^{(1/2)x}\, dx$

so that

$$du = \frac{1}{2} dx$$

$$dx = 2 \, du$$

When $x = -1$, $u = -\frac{1}{2}$, and when $x = 1$, $u = \frac{1}{2}$. Making the indicated substitutions, we obtain

$$A = \int_{-1}^{1} e^{(1/2)x} \, dx = 2 \int_{-1/2}^{1/2} e^u \, du$$

$$= 2e^u \Big|_{-1/2}^{1/2} = 2(e^{1/2} - e^{-1/2})$$

or approximately 2.08.

Explore and Discuss

Let f be the function defined piecewise by the rule

$$f(x) = \begin{cases} \sqrt{x} & \text{if } 0 \le x \le 1 \\ \dfrac{1}{x} & \text{if } 1 < x \le 2 \end{cases}$$

How would you use Property 5 of definite integrals to find the area of the region under the graph of f on $[0, 2]$? What is the area?

Average Value of a Function

The *average value* of a function over an interval provides us with an application of the definite integral. Recall that the average value of a set of n numbers $y_1, y_2, \ldots, y_n$ is the number

$$\frac{y_1 + y_2 + \cdots + y_n}{n}$$

Now, suppose f is a continuous function defined on $[a, b]$. Let's divide the interval $[a, b]$ into n subintervals of equal length $(b - a)/n$. Choose points $x_1, x_2, \ldots, x_n$ in the first, second, . . . , nth subintervals, respectively. Then, the average value of the numbers $f(x_1), f(x_2), \ldots, f(x_n)$, given by

$$\frac{f(x_1) + f(x_2) + \cdots + f(x_n)}{n}$$

is an approximation of the average of all the values of $f(x)$ on the interval $[a, b]$. This expression can be written in the form

$$\frac{b-a}{b-a} \left[f(x_1) \cdot \frac{1}{n} + f(x_2) \cdot \frac{1}{n} + \cdots + f(x_n) \cdot \frac{1}{n} \right]$$

$$= \frac{1}{b-a} \left[f(x_1) \cdot \frac{b-a}{n} + f(x_2) \cdot \frac{b-a}{n} + \cdots + f(x_n) \cdot \frac{b-a}{n} \right]$$

$$= \frac{1}{b-a} \left[f(x_1) \, \Delta x + f(x_2) \, \Delta x + \cdots + f(x_n) \, \Delta x \right] \tag{12}$$

As n gets larger and larger, the expression (12) approximates the average value of $f(x)$ over $[a, b]$ with increasing accuracy. But the sum inside the brackets in (12) is a Riemann sum of the function f over $[a, b]$. In view of this, we have

$$\lim_{n \to \infty} \left[\frac{f(x_1) + f(x_2) + \cdots + f(x_n)}{n} \right]$$

$$= \frac{1}{b - a} \lim_{n \to \infty} \left[f(x_1)\,\Delta x + f(x_2)\,\Delta x + \cdots + f(x_n)\,\Delta x \right]$$

$$= \frac{1}{b - a} \int_a^b f(x)\,dx$$

This discussion motivates the following definition.

> **The Average Value of a Function**
>
> Suppose f is integrable on $[a, b]$. Then the **average value** of f over $[a, b]$ is
>
> $$\frac{1}{b - a} \int_a^b f(x)\,dx$$

EXAMPLE 5 Find the average value of the function $f(x) = \sqrt{x}$ over the interval $[0, 4]$.

Solution The required average value is given by

$$\frac{1}{4 - 0} \int_0^4 \sqrt{x}\,dx = \frac{1}{4} \int_0^4 x^{1/2}\,dx$$

$$= \frac{1}{6} x^{3/2} \Big|_0^4 = \frac{1}{6} (4^{3/2})$$

$$= \frac{4}{3}$$

$\boxed{\sim\!\!\!\$}$ APPLIED EXAMPLE 6 Automobile Financing The interest rates charged by Madison Finance on auto loans for used cars over a certain 6-month period in 2016 are approximated by the function

$$r(t) = -\frac{1}{12} t^3 + \frac{7}{8} t^2 - 3t + 6 \qquad (0 \le t \le 6)$$

where t is measured in months and $r(t)$ is the annual percentage rate. What is the average rate on auto loans extended by Madison over the 6-month period?

Solution The average rate over the 6-month period is given by

$$\frac{1}{6 - 0} \int_0^6 \left(-\frac{1}{12} t^3 + \frac{7}{8} t^2 - 3t + 6 \right) dt$$

$$= \frac{1}{6} \left(-\frac{1}{48} t^4 + \frac{7}{24} t^3 - \frac{3}{2} t^2 + 6t \right) \Big|_0^6$$

$$= \frac{1}{6} \left[-\frac{1}{48} (6^4) + \frac{7}{24} (6^3) - \frac{3}{2} (6^2) + 6(6) \right]$$

$$= 3$$

or 3% per year.

APPLIED EXAMPLE 7 Drug Concentration in a Body The amount of a certain drug in a patient's body t days after the drug has been administered is

$$C(t) = 5e^{-0.2t}$$

units. Determine the average amount of the drug present in the patient's body for the first 4 days after the drug has been administered.

Solution   The average amount of the drug present in the patient's body for the first 4 days after it has been administered is given by

$$\frac{1}{4-0} \int_0^4 5e^{-0.2t}\, dt = \frac{5}{4} \int_0^4 e^{-0.2t}\, dt$$

$$= \frac{5}{4} \left[\left(-\frac{1}{0.2} \right) e^{-0.2t} \Big|_0^4 \right]$$

$$= \frac{5}{4} (-5e^{-0.8} + 5)$$

$$\approx 3.44$$

or approximately 3.44 units.

We now give a geometric interpretation of the average value of a function f over an interval $[a, b]$. Suppose $f(x)$ is nonnegative, so that the definite integral

$$\int_a^b f(x)\, dx$$

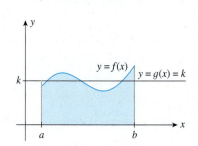

FIGURE 26
The average value of f over $[a, b]$ is k.

gives the area under the graph of f from $x = a$ to $x = b$ (Figure 26). Observe that, in general, the "height" $f(x)$ varies from point to point. Can we replace $f(x)$ by a constant function $g(x) = k$ (which has constant height) such that the areas under each of the two functions f and g are the same? If so, since the area under the graph of g from $x = a$ to $x = b$ is $k(b - a)$, we have

$$k(b - a) = \int_a^b f(x)\, dx$$

$$k = \frac{1}{b - a} \int_a^b f(x)\, dx$$

so k is the average value of f over $[a, b]$. Thus, the average value of a function f over an interval $[a, b]$ is the height of a rectangle with base of length $(b - a)$ that has the same area as that of the region under the graph of f from $x = a$ to $x = b$.

6.5 Self-Check Exercises

1. Evaluate $\displaystyle\int_0^2 \sqrt{2x + 5}\, dx$.

2. Find the average value of the function $f(x) = 1 - x^2$ over the interval $[-1, 2]$.

3. **MEDIAN HOME PRICE** The median price of a house in a southwestern state between January 1, 2009, and January 1, 2014, is approximated by the function

$$f(t) = t^3 - 7t^2 + 17t + 280 \qquad (0 \le t \le 5)$$

where $f(t)$ is measured in thousands of dollars and t is expressed in years, with $t = 0$ corresponding to the beginning of 2009. Determine the average median price of a house over that time interval.

Solutions to Self-Check Exercises 6.5 can be found on page 467.

6.5 Concept Questions

1. Describe two approaches that are used to evaluate a definite integral using the method of substitution. Illustrate with the integral $\int_0^1 x^2(x^3 + 1)^2 \, dx$.

2. Define the average value of a function f over an interval $[a, b]$. Give a geometric interpretation.

6.5 Exercises

In Exercises 1–28, evaluate the definite integral.

1. $\displaystyle\int_0^2 x(x^2 - 1)^3 \, dx$

2. $\displaystyle\int_0^1 x^2(2x^3 - 1)^4 \, dx$

3. $\displaystyle\int_0^1 x\sqrt{5x^2 + 4} \, dx$

4. $\displaystyle\int_1^3 x\sqrt{3x^2 - 2} \, dx$

5. $\displaystyle\int_0^2 x^2(x^3 + 1)^{3/2} \, dx$

6. $\displaystyle\int_1^5 (2x - 1)^{5/2} \, dx$

7. $\displaystyle\int_0^1 \frac{1}{\sqrt{2x + 1}} \, dx$

8. $\displaystyle\int_0^2 \frac{x}{\sqrt{x^2 + 5}} \, dx$

9. $\displaystyle\int_1^3 (2x - 1)^4 \, dx$

10. $\displaystyle\int_1^2 (2x + 4)(x^2 + 4x - 8)^3 \, dx$

11. $\displaystyle\int_{-1}^1 x^2(x^3 + 1)^4 \, dx$

12. $\displaystyle\int_1^2 \left(x^3 + \frac{3}{4}\right)(x^4 + 3x)^{-2} \, dx$

13. $\displaystyle\int_1^5 x\sqrt{x - 1} \, dx$

14. $\displaystyle\int_1^4 x\sqrt{x + 1} \, dx$

Hint: Let $u = x + 1$.

15. $\displaystyle\int_0^2 2xe^{x^2} \, dx$

16. $\displaystyle\int_0^1 e^{-x} \, dx$

17. $\displaystyle\int_0^1 (e^{2x} + x^2 + 1) \, dx$

18. $\displaystyle\int_0^2 (e^t - e^{-t}) \, dt$

19. $\displaystyle\int_{-1}^1 xe^{x^2 + 1} \, dx$

20. $\displaystyle\int_1^4 \frac{e^{\sqrt{x}}}{\sqrt{x}} \, dx$

21. $\displaystyle\int_3^6 \frac{1}{x - 2} \, dx$

22. $\displaystyle\int_0^1 \frac{x}{1 + 2x^2} \, dx$

23. $\displaystyle\int_1^2 \frac{x^2 + 2x}{x^3 + 3x^2 - 1} \, dx$

24. $\displaystyle\int_0^1 \frac{e^x}{1 + e^x} \, dx$

25. $\displaystyle\int_1^2 \left(4e^{2u} - \frac{1}{u}\right) du$

26. $\displaystyle\int_1^2 \left(1 + \frac{1}{x} + e^x\right) dx$

27. $\displaystyle\int_1^2 \left(2e^{-4x} - \frac{1}{x^2}\right) dx$

28. $\displaystyle\int_1^2 \frac{\ln x}{x} \, dx$

In Exercises 29–34, find the area of the region under the graph of f on $[a, b]$.

29. $f(x) = x^2 - 2x + 2; [-1, 2]$

30. $f(x) = x^3 + x; [0, 1]$ **31.** $f(x) = \dfrac{1}{x^2}; [1, 2]$

32. $f(x) = 2 + \sqrt{x + 1}; [0, 3]$

33. $f(x) = e^{-x/2}; [-1, 2]$ **34.** $f(x) = \dfrac{\ln x}{4x}; [1, 2]$

In Exercises 35–44, find the average value of the function f over the indicated interval $[a, b]$.

35. $f(x) = 2x + 3; [0, 2]$ **36.** $f(x) = 8 - x; [1, 4]$

37. $f(x) = 2x^2 - 3; [1, 3]$ **38.** $f(x) = 4 - x^2; [-2, 3]$

39. $f(x) = x^2 + 2x - 3; [-1, 2]$

40. $f(x) = x^3; [-1, 1]$ **41.** $f(x) = \sqrt{2x + 1}; [0, 4]$

42. $f(x) = e^{-x}; [0, 4]$ **43.** $f(x) = xe^{x^2}; [0, 2]$

44. $f(x) = \dfrac{1}{x + 1}; [0, 2]$

45. VELOCITY OF A CAR A car moves along a straight road in such a way that its velocity (in feet per second) at any time t (in seconds) is given by

$$v(t) = 3t\sqrt{16 - t^2} \qquad (0 \le t \le 4)$$

Find the distance traveled by the car in the 4 sec from $t = 0$ to $t = 4$.

46. OIL PRODUCTION On the basis of a preliminary report by a geological survey team, it is estimated that a newly discovered oil field can be expected to produce oil at the rate of

$$R(t) = \frac{600t^2}{t^3 + 32} + 5 \qquad (0 \le t \le 20)$$

thousand barrels/year, t years after production begins. Find the amount of oil that the field can be expected to yield during the first 5 years of production, assuming that the projection holds true.

47. NET INVESTMENT FLOW The net investment flow (rate of capital formation) of the giant conglomerate LTF incorporated is projected to be

$$t\sqrt{\frac{1}{2}t^2 + 1}$$

million dollars/year in year t. Find the accruement on the company's capital stock in the second year.

Hint: The amount is given by

$$\int_1^2 t\sqrt{\frac{1}{2}t^2 + 1}\, dt$$

48. NEWTON'S LAW OF COOLING A bottle of white wine at room temperature (68°F) is placed in a refrigerator at 4 P.M. Its temperature after t hr is changing at the rate of

$$-18e^{-0.6t}$$

°F/hr. By how many degrees will the temperature of the wine have dropped by 7 P.M.? What will the temperature of the wine be at 7 P.M.?

49. WORLD PRODUCTION OF COAL A study proposed in 1980 by researchers from the major producers and consumers of the world's coal concluded that coal could and must play an important role in fueling global economic growth over the next 20 years. The world production of coal in 1980 was 3.5 billion metric tons. If output increased at the rate of $3.5e^{0.05t}$ billion metric tons/year in year t ($t = 0$ corresponding to 1980), determine how much coal was produced worldwide between 1980 and the end of the twentieth century.

50. DEPRECIATION: DOUBLE DECLINING–BALANCE METHOD Suppose a tractor purchased at a price of $60,000 is to be depreciated by the *double declining–balance method* over a 10-year period. It can be shown that the rate at which the book value will be decreasing is given by

$$R(t) = 13388.61e^{-0.22314t} \qquad (0 \le t \le 10)$$

dollars/year at year t. Find the amount by which the book value of the tractor will depreciate over the first 5 years of its life.

51. CELL PHONE AD SPENDING Cell phone ad spending between 2005 ($t = 1$) and 2011 ($t = 7$) is given by

$$S(t) = 0.86t^{0.96} \qquad (1 \le t \le 7)$$

where $S(t)$ is measured in billions of dollars and t is measured in years. What was the average spending per year on cell phone ads between 2005 and 2011?

Source: Interactive Advertising Bureau.

52. GLOBAL WARMING The increase in carbon dioxide (CO_2) in the atmosphere is a major cause of global warming. Using data obtained by Charles David Keeling, professor at Scripps Institution of Oceanography, the average amount of CO_2 in the atmosphere from 1958 through 2016 is approximated by

$$A(t) = 0.012414t^2 + 0.7485t + 313.9 \qquad (1 \le t \le 59)$$

where $A(t)$ is measured in parts per million volume (ppmv) and t in years, with $t = 1$ corresponding to 1958. Find the average rate of increase of the average amount of CO_2 in the atmosphere from 1958 through 2016.

Source: Scripps Institution of Oceanography.

53. PROJECTED U.S. GASOLINE USAGE The White House wants to cut gasoline usage from 140 billion gallons per year in 2007 to 128 billion gallons per year in 2017. But estimates by the Department of Energy's Energy Information Agency suggest that won't happen. In fact, the agency's projection of gasoline usage from the beginning of 2007 until the beginning of 2017 is given by

$$A(t) = 0.014t^2 + 1.93t + 140 \qquad (0 \le t \le 10)$$

where $A(t)$ is measured in billions of gallons/year and t is in years, with $t = 0$ corresponding to 2007.

a. According to the agency's projection, what will be gasoline consumption at the beginning of 2017?

b. What will be the average consumption/year over the period from the beginning of 2007 until the beginning of 2017?

Source: U.S. Department of Energy, Energy Information Agency.

54. U.S. CITIZENS 65 YEARS AND OLDER The number of U.S. citizens aged 65 years and older from 1900 through 2050 is estimated to be growing at the rate of

$$R(t) = 0.063t^2 - 0.48t + 3.87 \qquad (0 \le t \le 15)$$

million people/decade, where t is measured in decades, with $t = 0$ corresponding to 1900. Show that the average rate of growth of the number of U.S. citizens aged 65 years and older between 2000 and 2050 will be about twice the average rate between 1950 and 2000.

Source: American Heart Association.

55. BASEBALL A batted baseball is subjected to a force whose direction is constant but whose magnitude is given by

$$F(t) = -9,400,000(t^2 - 0.02t) \qquad (0 \le t \le 0.02)$$

where $F(t)$ is measured in newtons and t in seconds. The force acts only over the time interval $[0, 0.02]$. Find the impulse I exerted by the bat on the baseball where

$$I = \int_0^{0.02} F(t)\, dt$$

N-sec. What is the average force acting on the baseball?

Hint: Average force $= \dfrac{\displaystyle\int_a^b F(t)\, dt}{b - a}$

56. COMMERCIAL VEHICLE REGISTRATIONS In a report issued by the Economist Intelligence Unit in 2013, the number of commercial vehicle registrations in the United States (in thousands) by year from 2010 through 2015 is approximated by the function

$$N(t) = 1.3926t^3 - 9.2873t^2 + 74.719t + 228.3$$
$$(0 \le t \le 5)$$

where t is measured in years with $t = 0$ corresponding to 2010. If the projection holds up, what would the approximate average commercial vehicle registration per year be in the period from 2010 through 2015?

Source: Economist Intelligence Unit.

57. FLOW OF BLOOD IN AN ARTERY According to a law discovered by nineteenth century physician Jean Louis Marie Poiseuille, the velocity of blood (in centimeters per second) r cm from the central axis of an artery is given by

$$v(r) = k(R^2 - r^2)$$

where k is a constant and R is the radius of the artery. Find the average velocity of blood along a radius of the artery (see the accompanying figure).

Hint: Evaluate $\dfrac{1}{R} \displaystyle\int_0^R v(r)\, dr$.

Blood vessel

58. TOTAL KNEE REPLACEMENT PROCEDURES The number of total knee replacement procedures (in millions) performed in the United States from the beginning of 1991 through the end of 2030 is projected to be

$$f(t) = 0.0000846t^3 - 0.002116t^2 + 0.03897t + 0.16$$
$$(0 \le t \le 40)$$

where $t = 0$ corresponds to 1991. What is the projected average number of total knee replacement procedures performed in the United States in the period from 1991 through 2030?

Source: American Academy of Orthopaedic Surgeons.

59. CONCENTRATION OF A DRUG IN THE BLOODSTREAM The concentration of a certain drug in a patient's bloodstream t hr after injection is

$$C(t) = \frac{0.2t}{t^2 + 1}$$

mg/cm³. Determine the average concentration of the drug in the patient's bloodstream over the first 4 hr after the drug is injected.

60. SEAT BELT USE According to the U.S. Department of Transportation, the percentage of drivers using seat belts from 2001 through 2009 is modeled by the function

$$f(t) = 72.9(t + 1)^{0.057}$$

where t is measured in years, with $t = 0$ corresponding to the beginning of 2001. What was the average percentage use of seat belts over the period from the beginning of 2001 through the end of 2009 ($t = 9$)?

Source: U.S. Department of Transportation.

61. AVERAGE YEARLY SALES The sales of Universal Instruments in the first t years of its operation are approximated by the function

$$S(t) = t\sqrt{0.2t^2 + 4}$$

where $S(t)$ is measured in millions of dollars. What were Universal's average yearly sales over its first 5 years of operation?

62. CABLE TV SUBSCRIBERS The manager of TeleStar Cable Service estimates that the total number of subscribers to the service in a certain city t years from now will be

$$N(t) = -\frac{40{,}000}{\sqrt{1 + 0.2t}} + 50{,}000$$

Find the average number of cable television subscribers over the next 5 years if this prediction holds true.

63. CREDIT CARDS IN CHINA The total number of credit cards issued in China from 2009 through 2012 grew at the rate of

$$r(t) = 153e^{0.21t} \qquad (1 \le t \le 4)$$

million/year, where ($t = 1$) corresponds to 2009. What was the average growth in the number of credit cards issued in China for the period under consideration?

Source: Chinese Credit Card Industry Development Blue Book.

64. U.S. RETAIL E-COMMERCE SALES In a 2010 report, U.S. retail e-commerce sales (excluding travel, digital downloads, and event tickets) from 2011 through 2014 are projected to be

$$f(t) = 157.6e^{0.09t} \qquad (1 \le t \le 4)$$

billion dollars in year t, where ($t = 1$) corresponds to 2011. What is the average number of e-commerce sales in the United States over the period under consideration?

Source: eMarketer.

65. INTERNATIONAL AIR TRAFFIC International air traffic in and out of the United States is projected to grow at the rate of

$$R(t) = 6.69e^{0.0456t} \qquad (2 \le t \le 20)$$

million passengers per year between 2012 and 2030. Here, $t = 2$ corresponds to 2012. The number of passengers in 2012 was 160 million.

a. Find an expression for the projected number of passengers in year t.

b. What is the projected number of passengers in 2030 ($t = 20$)?

c. What is the estimated average growth rate of international air traffic out of the United States from 2012 to 2030?

Source: Federal Aviation Administration.

66. AVERAGE PRICE OF A COMMODITY The price of a certain commodity in dollars per unit at time t (measured in weeks) is given by

$$p = 18 - 3e^{-2t} - 6e^{-t/3}$$

What is the average price of the commodity over the 5-week period from ($t = 0$) to ($t = 5$)?

67. VELOCITY OF A FALLING HAMMER During the construction of a high-rise apartment building, a construction worker accidentally drops a hammer that falls vertically a distance of h ft. The velocity of the hammer after falling a distance of x ft is $v = \sqrt{2gx}$ ft/sec ($0 \le x \le h$). Show that the average velocity of the hammer over this path is $\bar{v} = \frac{2}{3}\sqrt{2gh}$ ft/sec.

68. WASTE DISPOSAL When organic waste is dumped into a pond, the oxidization process that takes place reduces the pond's oxygen content. However, in time, nature will restore the oxygen content to its natural level. Suppose that the oxygen content t days after organic waste has been dumped into a pond is given by

$$f(t) = 100\left(\frac{t^2 + 10t + 100}{t^2 + 20t + 100}\right)$$

percent of its normal level. Find the average content of oxygen in the pond over the first 10 days after organic waste has been dumped into it.
Hint: Show that

$$\frac{t^2 + 10t + 100}{t^2 + 20t + 100} = 1 - \frac{10}{t + 10} + \frac{100}{(t + 10)^2}$$

69. Prove Property 1 of the definite integral.
Hint: Let F be an antiderivative of f, and use the definition of the definite integral.

70. Prove Property 2 of the definite integral.
Hint: See Exercise 69.

71. Verify by direct computation that

$$\int_1^3 x^2\, dx = -\int_3^1 x^2\, dx$$

72. Prove Property 3 of the definite integral.
Hint: See Exercise 69.

73. Verify by direct computation that

$$\int_1^9 2\sqrt{x}\, dx = 2\int_1^9 \sqrt{x}\, dx$$

74. Verify by direct computation that

$$\int_0^1 (1 + x - e^x)\, dx = \int_0^1 dx + \int_0^1 x\, dx - \int_0^1 e^x\, dx$$

What property of the definite integral is illustrated in this exercise?

75. Verify by direct computation that

$$\int_0^3 (1 + x^3)\, dx = \int_0^1 (1 + x^3)\, dx + \int_1^3 (1 + x^3)\, dx$$

What property of the definite integral is illustrated here?

76. Verify by direct computation that

$$\int_0^3 (1 + x^3)\, dx$$
$$= \int_0^1 (1 + x^3)\, dx + \int_1^2 (1 + x^3)\, dx + \int_2^3 (1 + x^3)\, dx$$

hence illustrating that Property 5 may be extended.

77. Evaluate $\displaystyle\int_3^3 (1 + \sqrt{x})e^{-x}\, dx$.

78. Evaluate $\displaystyle\int_3^0 f(x)\, dx$, given that $\displaystyle\int_0^3 f(x)\, dx = 4$.

79. Given that $\displaystyle\int_{-1}^2 f(x)\, dx = -2$ and $\displaystyle\int_{-1}^2 g(x)\, dx = 3$, evaluate

a. $\displaystyle\int_{-1}^2 [2f(x) + g(x)]\, dx$

b. $\displaystyle\int_{-1}^2 [g(x) - f(x)]\, dx$

c. $\displaystyle\int_{-1}^2 [2f(x) - 3g(x)]\, dx$

80. Given that $\displaystyle\int_{-1}^2 f(x)\, dx = 2$ and $\displaystyle\int_0^2 f(x)\, dx = 3$, evaluate

a. $\displaystyle\int_{-1}^0 f(x)\, dx$

b. $\displaystyle\int_0^2 f(x)\, dx - \int_{-1}^0 f(x)\, dx$

In Exercises 81–86, determine whether the statement is true or false. If it is true, explain why it is true. If it is false, explain why or give an example to show why it is false.

81. $\displaystyle\int_2^2 \frac{e^x}{\sqrt{1 + x}}\, dx = 0$

82. $\displaystyle\int_1^3 \frac{dx}{x - 2} = -\int_3^1 \frac{dx}{x - 2}$

83. $\displaystyle\int_0^1 x\sqrt{x + 1}\, dx = \sqrt{x + 1}\int_0^1 x\, dx$
$$= \frac{1}{2}x^2\sqrt{x + 1}\Big|_0^1 = \frac{\sqrt{2}}{2}$$

84. If f' is continuous on $[0, 2]$, then

$$\int_0^2 f'(x)\, dx = f(2) - f(0)$$

85. If f and g are continuous on $[a, b]$ and k is a constant, then

$$\int_a^b [kf(x) + g(x)]\, dx = k\int_a^b f(x)\, dx + \int_a^b g(x)\, dx$$

86. If f is continuous on $[a, b]$ and $a < c < b$, then

$$\int_b^c f(x)\, dx = \int_a^c f(x)\, dx - \int_a^b f(x)\, dx$$

6.5 Solutions to Self-Check Exercises

1. Let $u = 2x + 5$. Then, $du = 2\,dx$, or $dx = \frac{1}{2}\,du$. Also, when $x = 0$, $u = 5$, and when $x = 2$, $u = 9$. Therefore,

$$\int_0^2 \sqrt{2x + 5}\,dx = \int_0^2 (2x + 5)^{1/2}\,dx$$

$$= \frac{1}{2}\int_5^9 u^{1/2}\,du$$

$$= \left(\frac{1}{2}\right)\left(\frac{2}{3}u^{3/2}\right)\Big|_5^9$$

$$= \frac{1}{3}\left[9^{3/2} - 5^{3/2}\right]$$

$$= \frac{1}{3}(27 - 5\sqrt{5})$$

2. The required average value is given by

$$\frac{1}{2 - (-1)}\int_{-1}^2 (1 - x^2)\,dx = \frac{1}{3}\int_{-1}^2 (1 - x^2)\,dx$$

$$= \frac{1}{3}\left(x - \frac{1}{3}x^3\right)\Big|_{-1}^2$$

$$= \frac{1}{3}\left[\left(2 - \frac{8}{3}\right) - \left(-1 + \frac{1}{3}\right)\right] = 0$$

3. The average median price of a house over the stated time interval is given by

$$\frac{1}{5 - 0}\int_0^5 (t^3 - 7t^2 + 17t + 280)\,dt$$

$$= \frac{1}{5}\left(\frac{1}{4}t^4 - \frac{7}{3}t^3 + \frac{17}{2}t^2 + 280t\right)\Big|_0^5$$

$$= \frac{1}{5}\left[\frac{1}{4}(5)^4 - \frac{7}{3}(5)^3 + \frac{17}{2}(5)^2 + 280(5)\right]$$

$$\approx 295.417$$

or approximately \$295,417.

USING TECHNOLOGY

Evaluating Definite Integrals for Piecewise-Defined Functions

We continue using graphing utilities to find the definite integral of a function. But here we will make use of Property 5 of the properties of the definite integral (page 457).

EXAMPLE 1 Use the numerical integral operation of a graphing utility to evaluate

$$\int_{-1}^2 f(x)\,dx$$

where

$$f(x) = \begin{cases} -x^2 & \text{if } x < 0 \\ \sqrt{x} & \text{if } x \geq 0 \end{cases}$$

Solution Using Property 5 of the definite integral, we can write

$$\int_{-1}^2 f(x)\,dx = \int_{-1}^0 -x^2\,dx + \int_0^2 x^{1/2}\,dx$$

Using a graphing utility, we find

$$\int_{-1}^2 f(x)\,dx = \text{fnInt}(-x\wedge2, x, -1, 0) + \text{fnInt}(x\wedge0.5, x, 0, 2)$$

$$\approx -0.333333 + 1.885618$$

$$= 1.552285$$

TECHNOLOGY EXERCISES

In Exercises 1–4, use Property 5 of the definite integral (page 457) to evaluate the definite integral accurate to five decimal places.

1. $\int_{-1}^{2} f(x)\, dx$, where

$$f(x) = \begin{cases} 2.3x^3 - 3.1x^2 + 2.7x + 3 & \text{if } x < 1 \\ -1.7x^2 + 2.3x + 4.3 & \text{if } x \ge 1 \end{cases}$$

2. $\int_{0}^{3} f(x)\, dx$, where $f(x) = \begin{cases} \dfrac{\sqrt{x}}{1 + x^2} & \text{if } 0 \le x < 1 \\ 0.5e^{-0.1x^2} & \text{if } x \ge 1 \end{cases}$

3. $\int_{-2}^{2} f(x)\, dx$, where $f(x) = \begin{cases} x^4 - 2x^2 + 4 & \text{if } x < 0 \\ 2\ln(x + e^2) & \text{if } x \ge 0 \end{cases}$

4. $\int_{-2}^{6} f(x)\, dx$, where

$$f(x) = \begin{cases} 2x^3 - 3x^2 + x + 2 & \text{if } x < -1 \\ \sqrt{3x + 4} - 5 & \text{if } -1 \le x \le 4 \\ x^2 - 3x - 5 & \text{if } x > 4 \end{cases}$$

5. CROP YIELD If left untreated on bean stems, aphids (small insects that suck plant juices) will multiply at an increasing rate during the summer months and reduce productivity and crop yield of cultivated crops. But if the aphids are treated in mid-June, the numbers decrease sharply to less than 100/bean stem, allowing for steep rises in crop yield. The function

$$F(t) = \begin{cases} 62e^{1.152t} & \text{if } 0 \le t < 1.5 \\ 349e^{-1.324(t - 1.5)} & \text{if } 1.5 \le t \le 3 \end{cases}$$

gives the number of aphids on a typical bean stem at time t, where t is measured in months, $t = 0$ corresponding to the beginning of May. Find the average number of aphids on a typical bean stem over the period from the beginning of May to the beginning of August.

6. ABSORPTION OF DRUGS Jane took 100 mg of a drug in the morning and another 100 mg of the same drug at the same time the following morning. The amount of the drug (in mg) in her body t days after the first dose was taken is given by

$$A(t) = \begin{cases} 100e^{-1.4t} & \text{if } 0 \le t < 1 \\ 100(1 + e^{1.4})e^{-1.4t} & \text{if } t \ge 1 \end{cases}$$

Find the average amount of the drug in Jane's body over the first 2 days.

7. ABSORPTION OF DRUGS The concentration of a drug in an organ t seconds after it is administered is given by

$$C(t) = \begin{cases} 0.3t - 18(1 - e^{-t/60}) & \text{if } 0 \le t \le 20 \\ 18e^{-t/60} - 12e^{-(t - 20)/60} & \text{if } t > 20 \end{cases}$$

where $C(t)$ is measured in grams per cubic centimeter (g/cm^3). Find the average concentration of the drug in the organ over the first 30 sec after it is administered.

6.6   Area Between Two Curves

Suppose a certain country's petroleum consumption is expected to be $f(t)$ million barrels per year, t years from now, for the next 5 years. Then the country's total petroleum consumption over the period of time in question is given by the area of the region under the graph of f on the interval $[0, 5]$ (Figure 27).

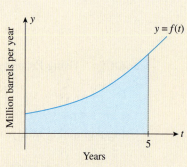

FIGURE 27
At a rate of consumption of $f(t)$ million barrels per year, the total petroleum consumption is given by the area of the region under the graph of f.

Next, suppose that because of the implementation of certain energy-conservation measures, the petroleum consumption is expected to be $g(t)$ million barrels per year instead. Then, the country's projected total petroleum consumption over the 5-year period is given by the area of the region under the graph of g on the interval $[0, 5]$ (Figure 28).

Therefore, the area of the shaded region S lying between the graphs of f and g on the interval $[0, 5]$ (Figure 29) gives the amount of petroleum that would be saved over the 5-year period because of the conservation measures.

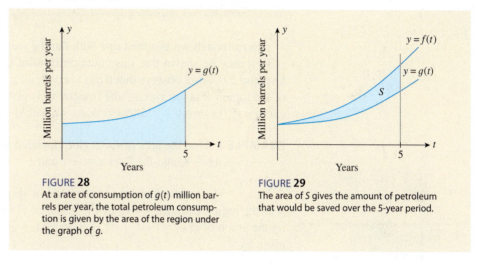

FIGURE 28
At a rate of consumption of $g(t)$ million barrels per year, the total petroleum consumption is given by the area of the region under the graph of g.

FIGURE 29
The area of S gives the amount of petroleum that would be saved over the 5-year period.

But the area of S is given by

Area of the region under the graph of f on $[a, b]$
$-$ Area of the region under the graph of g on $[a, b]$

$$= \int_0^5 f(t)\, dt - \int_0^5 g(t)\, dt$$

$$= \int_0^5 [f(t) - g(t)]\, dt \qquad \text{By Property 4, Section 6.5}$$

This example shows that some practical problems can be solved by finding the area of a region between two curves, which in turn can be found by evaluating an appropriate definite integral.

Finding the Area of the Region Between Two Curves

We now turn our attention to the general problem of finding the area of a plane region bounded both above and below by the graphs of functions. First, consider the situation in which the graph of one function lies above that of another. More specifically, let R be the region in the xy-plane (Figure 30) that is bounded above by the graph of a continuous function f, below by the graph of a continuous function g, where $f(x) \geq g(x) \geq 0$ on $[a, b]$, and to the left and right by the vertical lines $x = a$ and $x = b$, respectively. From the figure, we see that

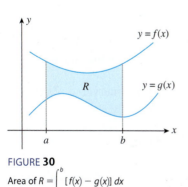

FIGURE 30
Area of $R = \int_a^b [f(x) - g(x)]\, dx$

Area of $R =$ Area of the region under the graph of $f(x)$
$-$ Area of the region under the graph of $g(x)$

$$= \int_a^b f(x)\, dx - \int_a^b g(x)\, dx$$

$$= \int_a^b [f(x) - g(x)]\, dx$$

upon using Property 4 of the definite integral.

The Area of the Region Between Two Curves

Let f and g be continuous functions such that $f(x) \geq g(x)$ on the interval $[a, b]$. Then the area of the region bounded above by $y = f(x)$ and below by $y = g(x)$ on $[a, b]$ is given by

$$\int_a^b [f(x) - g(x)] \, dx \qquad (13)$$

Even though we assumed that both f and g were nonnegative in the derivation of (13), it may be shown that this equation is valid if f and g are not nonnegative (see Exercise 59). Also, observe that if $g(x)$ is 0 for all x—that is, when the lower boundary of the region R is the x-axis—the integral (13) gives the area of the region under the curve $y = f(x)$ from $x = a$ to $x = b$, as we would expect.

EXAMPLE 1 Find the area of the region bounded by the x-axis, the graph of $y = -x^2 + 4x - 8$, and the lines $x = -1$ and $x = 4$.

Solution The region R under consideration is shown in Figure 31. We can view R as the region bounded above by the graph of $f(x) = 0$ (the x-axis) and below by the graph of $g(x) = -x^2 + 4x - 8$ on $[-1, 4]$. Therefore, the area of R is given by

$$\int_a^b [f(x) - g(x)] \, dx = \int_{-1}^4 [0 - (-x^2 + 4x - 8)] \, dx$$

$$= \int_{-1}^4 (x^2 - 4x + 8) \, dx$$

$$= \frac{1}{3} x^3 - 2x^2 + 8x \Big|_{-1}^4$$

$$= \left[\frac{1}{3}(64) - 2(16) + 8(4) \right] - \left[\frac{1}{3}(-1) - 2(1) + 8(-1) \right]$$

$$= 31\tfrac{2}{3}$$

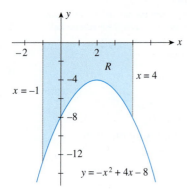

FIGURE 31

Area of $R = -\int_{-1}^4 g(x) \, dx$

EXAMPLE 2 Find the area of the region R bounded by the graphs of

$$f(x) = 2x - 1 \quad \text{and} \quad g(x) = x^2 - 4$$

and the vertical lines $x = 1$ and $x = 2$.

Solution We first sketch the graphs of the functions $f(x) = 2x - 1$ and $g(x) = x^2 - 4$ and the vertical lines $x = 1$ and $x = 2$, and then we identify the region R whose area is to be calculated (Figure 32).

Since the graph of f always lies above that of g for x in the interval $[1, 2]$, we see by integral (13) that the required area is given by

$$\int_1^2 [f(x) - g(x)] \, dx = \int_1^2 [(2x - 1) - (x^2 - 4)] \, dx$$

$$= \int_1^2 (-x^2 + 2x + 3) \, dx$$

$$= -\frac{1}{3} x^3 + x^2 + 3x \Big|_1^2$$

$$= \left(-\frac{8}{3} + 4 + 6 \right) - \left(-\frac{1}{3} + 1 + 3 \right) = \frac{11}{3}$$

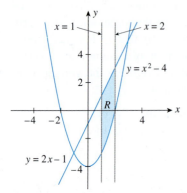

FIGURE 32

Area of $R = \int_1^2 [f(x) - g(x)] \, dx$

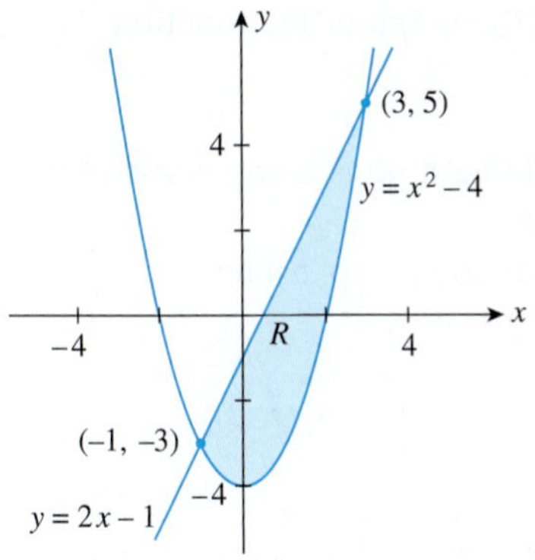

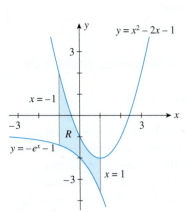

FIGURE 33

Area of $R = -\int_{-1}^{3} [f(x) - g(x)]\, dx$

EXAMPLE 3 Find the area of the region R that is completely enclosed by the graphs of the functions

$$f(x) = 2x - 1 \quad \text{and} \quad g(x) = x^2 - 4$$

Solution The region R is shown in Figure 33. First, we find the points of intersection of the two curves. To do this, we solve the system that consists of the two equations $y = 2x - 1$ and $y = x^2 - 4$. Equating the two values of y gives

$$x^2 - 4 = 2x - 1$$
$$x^2 - 2x - 3 = 0$$
$$(x + 1)(x - 3) = 0$$

so $x = -1$ or $x = 3$. That is, the two curves intersect when $x = -1$ and $x = 3$.

Observe that we could also view the region R as the region bounded above by the graph of the function $f(x) = 2x - 1$, below by the graph of the function $g(x) = x^2 - 4$, and to the left and right by the vertical lines $x = -1$ and $x = 3$, respectively.

Next, since the graph of the function f always lies above that of the function g on $[-1, 3]$, we can use integral (13) to compute the desired area:

$$\int_a^b [f(x) - g(x)]\, dx = \int_{-1}^{3} [(2x - 1) - (x^2 - 4)]\, dx$$

$$= \int_{-1}^{3} (-x^2 + 2x + 3)\, dx$$

$$= -\frac{1}{3}x^3 + x^2 + 3x \Big|_{-1}^{3}$$

$$= (-9 + 9 + 9) - \left(\frac{1}{3} + 1 - 3\right) = \frac{32}{3}$$

$$= 10\tfrac{2}{3}$$

EXAMPLE 4 Find the area of the region R bounded by the graphs of the functions

$$f(x) = x^2 - 2x - 1 \quad \text{and} \quad g(x) = -e^x - 1$$

and the vertical lines $x = -1$ and $x = 1$.

Solution The region R is shown in Figure 34. Since the graph of the function f always lies above that of the function g, the area of the region R is given by

$$\int_a^b [f(x) - g(x)]\, dx = \int_{-1}^{1} [(x^2 - 2x - 1) - (-e^x - 1)]\, dx$$

$$= \int_{-1}^{1} (x^2 - 2x + e^x)\, dx$$

$$= \frac{1}{3}x^3 - x^2 + e^x \Big|_{-1}^{1}$$

$$= \left(\frac{1}{3} - 1 + e\right) - \left(-\frac{1}{3} - 1 + e^{-1}\right)$$

$$= \frac{2}{3} + e - \frac{1}{e} \approx 3.02$$

FIGURE 34

Area of $R = \int_{-1}^{1} [f(x) - g(x)]\, dx$

Integral (13), which gives the area of the region between the curves $y = f(x)$ and $y = g(x)$ for $a \le x \le b$, is valid when the graph of the function f lies above that of the function g over the interval $[a, b]$. Example 5 shows how to use Equation (13) to find the area of a region when this condition does not hold.

EXAMPLE 5 Find the area of the region bounded by the graph of the function $f(x) = x^3$, the x-axis, and the lines $x = -1$ and $x = 1$.

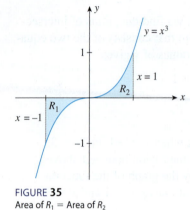

FIGURE 35
Area of R_1 = Area of R_2

Solution   The region R under consideration can be thought of as being composed of the two subregions R_1 and R_2, as shown in Figure 35.

Recall that the x-axis is represented by the function $g(x) = 0$. Since $g(x) \geq f(x)$ on $[-1, 0]$, we see that the area of R_1 is given by

$$\int_a^b [g(x) - f(x)]\, dx = \int_{-1}^0 (0 - x^3)\, dx = -\int_{-1}^0 x^3\, dx$$

$$= -\frac{1}{4} x^4 \Big|_{-1}^0 = 0 - \left(-\frac{1}{4}\right) = \frac{1}{4}$$

To find the area of R_2, we observe that $f(x) \geq g(x)$ on $[0, 1]$, so it is given by

$$\int_a^b [f(x) - g(x)]\, dx = \int_0^1 (x^3 - 0)\, dx = \int_0^1 x^3\, dx$$

$$= \frac{1}{4} x^4 \Big|_0^1 = \left(\frac{1}{4}\right) - 0 = \frac{1}{4}$$

Therefore, the area of R is $\frac{1}{4} + \frac{1}{4}$, or $\frac{1}{2}$.

By making use of symmetry, we could have obtained the same result by computing

$$-2 \int_{-1}^0 x^3\, dx \quad \text{or} \quad 2 \int_0^1 x^3\, dx$$

as you may verify.

Explore and Discuss

A function is *even* if it satisfies the condition $f(-x) = f(x)$, and it is *odd* if it satisfies the condition $f(-x) = -f(x)$. Show that the graph of an even function is symmetric with respect to the y-axis while the graph of an odd function is symmetric with respect to the origin. Explain why

$$\int_{-a}^a f(x)\, dx = 2 \int_0^a f(x)\, dx \quad \text{if } f \text{ is even}$$

and

$$\int_{-a}^a f(x)\, dx = 0 \quad \text{if } f \text{ is odd}$$

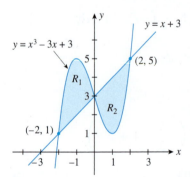

FIGURE 36
Area of R_1 + Area of R_2

$$= \int_{-2}^0 [f(x) - g(x)]\, dx$$

$$+ \int_0^2 [g(x) - f(x)]\, dx$$

EXAMPLE 6 Find the area of the region completely enclosed by the graphs of the functions

$$f(x) = x^3 - 3x + 3 \quad \text{and} \quad g(x) = x + 3$$

Solution   First, we sketch the graphs of $y = x^3 - 3x + 3$ and $y = x + 3$, and we then identify the required region R. We can view the region R as being composed of the two subregions R_1 and R_2, as shown in Figure 36. By solving the equations $y = x + 3$ and $y = x^3 - 3x + 3$ simultaneously, we find the points of intersection of the two curves. Equating the two values of y, we have

$$x^3 - 3x + 3 = x + 3$$
$$x^3 - 4x = 0$$
$$x(x^2 - 4) = 0$$
$$x(x + 2)(x - 2) = 0$$
$$x = 0, -2, 2$$

Hence the points of intersection of the two curves are $(-2, 1)$, $(0, 3)$, and $(2, 5)$.

For $-2 \leq x \leq 0$, we see that the graph of the function f lies above that of the function g, so the area of the region R_1 is, by Equation (13),

$$\int_{-2}^{0} [(x^3 - 3x + 3) - (x + 3)] \, dx = \int_{-2}^{0} (x^3 - 4x) \, dx$$
$$= \frac{1}{4}x^4 - 2x^2 \Big|_{-2}^{0}$$
$$= -(4 - 8)$$
$$= 4$$

For $0 \leq x \leq 2$, the graph of the function g lies above that of the function f, and the area of R_2 is given by

$$\int_{0}^{2} [(x + 3) - (x^3 - 3x + 3)] \, dx = \int_{0}^{2} (-x^3 + 4x) \, dx$$
$$= -\frac{1}{4}x^4 + 2x^2 \Big|_{0}^{2}$$
$$= -4 + 8$$
$$= 4$$

Therefore, the required area is the sum of the areas of the two regions R_1 and R_2—that is, $4 + 4$, or 8. ◾

APPLIED EXAMPLE 7 Conservation of Oil In a 2006 study for a developing country's Economic Development Board, government economists and energy experts concluded that if the Energy Conservation Bill were implemented in 2009, the country's oil consumption for the next 5 years is expected to grow in accordance with the model

$$R(t) = 20e^{0.05t}$$

where t is measured in years ($t = 0$ corresponding to the year 2009) and $R(t)$ in millions of barrels per year. Without the government-imposed conservation measures, however, the expected rate of growth of oil consumption would be given by

$$R_1(t) = 20e^{0.08t}$$

million barrels per year. Using these models, determine how much oil would have been saved from 2009 through 2014 if the bill had been implemented.

Solution Under the Energy Conservation Bill, the total amount of oil that would have been consumed between 2009 and 2014 is given by

$$\int_{0}^{5} R(t) \, dt = \int_{0}^{5} 20e^{0.05t} \, dt \tag{14}$$

Without the bill, the total amount of oil that would have been consumed between 2009 and 2014 is given by

$$\int_{0}^{5} R_1(t) \, dt = \int_{0}^{5} 20e^{0.08t} \, dt \tag{15}$$

The first integral in (14) may be interpreted as the area of the region under the curve $y = R(t)$ from $t = 0$ to $t = 5$. Similarly, we interpret the integrals in (15) as giving the area of the region under the curve $y = R_1(t)$ from $t = 0$ to $t = 5$. Furthermore, note that the graph of $y = R_1(t) = 20e^{0.08t}$ always lies on or above the graph of

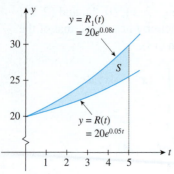

FIGURE 37

Area of $S = \int_0^5 [R_1(t) - R(t)] \, dt$

$y = R(t) = 20e^{0.05t}$ $(t \geq 0)$. Thus, the area of the shaded region S in Figure 37 shows the amount of oil that would have been saved from 2009 to 2014 if the Energy Conservation Bill had been implemented. But the area of the region S is given by

$$\int_0^5 [R_1(t) - R(t)] \, dt = \int_0^5 (20e^{0.08t} - 20e^{0.05t}) \, dt$$

$$= 20 \int_0^5 (e^{0.08t} - e^{0.05t}) \, dt$$

$$= 20 \left(\frac{e^{0.08t}}{0.08} - \frac{e^{0.05t}}{0.05} \right) \Big|_0^5$$

$$= 20 \left[\left(\frac{e^{0.4}}{0.08} - \frac{e^{0.25}}{0.05} \right) - \left(\frac{1}{0.08} - \frac{1}{0.05} \right) \right]$$

$$\approx 9.3$$

Thus, the amount of oil that would have been saved is 9.3 million barrels.

Exploring with TECHNOLOGY

Refer to Example 7. Suppose we want to construct a mathematical model giving the amount of oil saved from 2009 through the year $(2009 + x)$ where $x \geq 0$. In Example 7, for instance, we take $x = 5$.

1. Show that this model is given by

$$F(x) = \int_0^x [R_1(t) - R(t)] \, dt$$
$$= 250e^{0.08x} - 400e^{0.05x} + 150$$

Hint: You may find it helpful to use some of the results of Example 7.

2. Use a graphing utility to plot the graph of F, using the viewing window $[0, 10] \times [0, 50]$.

3. Find $F(5)$ and thus confirm the result of Example 7.

4. What is the main advantage of this model?

6.6 **Self-Check Exercises**

1. Find the area of the region bounded by the graphs of $f(x) = x^2 + 2$ and $g(x) = 1 - x$ and the vertical lines $x = 0$ and $x = 1$.

2. Find the area of the region completely enclosed by the graphs of $f(x) = -x^2 + 6x + 5$ and $g(x) = x^2 + 5$.

3. **PROFIT OF A HOTEL CHAIN** The management of Kane Corporation, which operates a chain of hotels, expects its profits to grow at the rate of $(1 + t^{2/3})$ million dollars/year

t years from now. However, with renovations and improvements of existing hotels and proposed acquisitions of new hotels, Kane's profits are expected to grow at the rate of $(t - 2\sqrt{t} + 4)$ million dollars/year in the next decade. What additional profits are expected over the next 10 years if the group implements the proposed plans?

Solutions to Self-Check Exercises 6.6 can be found on page 479.

6.6 Concept Questions

1. Suppose f and g are continuous functions such that $f(x) \geq g(x)$ on the interval $[a, b]$. Write an integral giving the area of the region bounded above by the graph of f, below by the graph of g, and on the left and right by the lines $x = a$ and $x = b$.

2. Write an expression in terms of definite integrals giving the area of the shaded region in the figure.

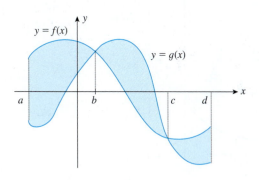

6.6 Exercises

In Exercises 1–8, find the area of the shaded region.

1.

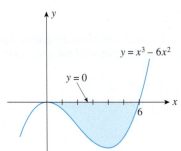

2.

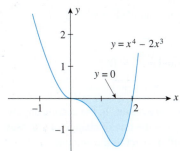

3.

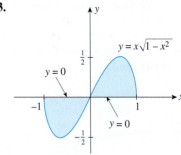

4.

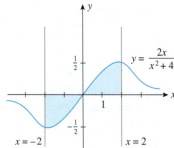

5.

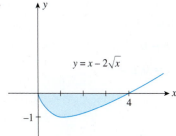

6.

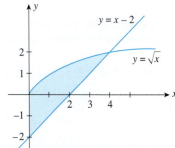

7.

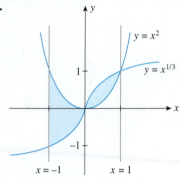

8.

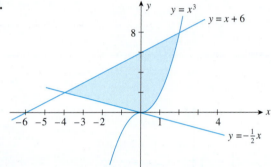

In Exercises 9–16, sketch the graph, and find the area of the region bounded below by the graph of each function and above by the x-axis from $x = a$ to $x = b$.

9. $f(x) = -x^2; a = -1, b = 2$

10. $f(x) = x^2 - 4; a = -2, b = 2$

11. $f(x) = x^2 - 5x + 4; a = 1, b = 3$

12. $f(x) = x^3; a = -1, b = 0$

13. $f(x) = -1 - \sqrt{x}; a = 0, b = 9$

14. $f(x) = \frac{1}{2}x - \sqrt{x}; a = 0, b = 4$

15. $f(x) = -e^{(1/2)x}; a = -2, b = 4$

16. $f(x) = -xe^{-x^2}; a = 0, b = 1$

In Exercises 17–26, sketch the graphs of the functions f and g, and find the area of the region enclosed by these graphs and the vertical lines $x = a$ and $x = b$.

17. $f(x) = x^2 + 3, g(x) = 1; a = 1, b = 3$

18. $f(x) = x + 2, g(x) = x^2 - 4; a = -1, b = 2$

19. $f(x) = -x^2 + 2x + 3, g(x) = -x + 3; a = 0, b = 2$

20. $f(x) = 9 - x^2, g(x) = 2x + 3; a = -1, b = 1$

21. $f(x) = x^2 + 1, g(x) = \frac{1}{3}x^3; a = -1, b = 2$

22. $f(x) = \sqrt{x}, g(x) = -\frac{1}{2}x - 1; a = 1, b = 4$

23. $f(x) = \frac{1}{x}, g(x) = 2x - 1; a = 1, b = 4$

24. $f(x) = x^2, g(x) = \frac{1}{x^2}; a = 1, b = 3$

25. $f(x) = e^x, g(x) = \frac{1}{x}; a = 1, b = 2$

26. $f(x) = x, g(x) = e^{2x}; a = 1, b = 3$

In Exercises 27–34, sketch the graph, and find the area of the region bounded by the graph of the function f and the lines $y = 0, x = a$, and $x = b$.

27. $f(x) = x; a = -1, b = 2$

28. $f(x) = x^2 - 2x; a = -1, b = 1$

29. $f(x) = -x^2 + 4x - 3; a = -1, b = 2$

30. $f(x) = x^3 - x^2; a = -1, b = 1$

31. $f(x) = x^3 - 4x^2 + 3x; a = 0, b = 2$

32. $f(x) = 4x^{1/3} + x^{4/3}; a = -1, b = 8$

33. $f(x) = e^x - 1; a = -1, b = 3$

34. $f(x) = xe^{x^2}; a = 0, b = 2$

In Exercises 35–42, sketch the graph, and find the area of the region completely enclosed by the graphs of the given functions f and g.

35. $f(x) = x + 2$ and $g(x) = x^2 - 4$

36. $f(x) = -x^2 + 4x$ and $g(x) = 2x - 3$

37. $f(x) = x^2$ and $g(x) = x^3$

38. $f(x) = x^3 + 2x^2 - 3x$ and $g(x) = 0$

39. $f(x) = x^3 - 6x^2 + 9x$ and $g(x) = x^2 - 3x$

40. $f(x) = \sqrt{x}$ and $g(x) = x^2$

41. $f(x) = x\sqrt{9 - x^2}$ and $g(x) = 0$

42. $f(x) = 2x$ and $g(x) = x\sqrt{x + 1}$

43. EFFECT OF ADVERTISING ON REVENUE In the accompanying figure, the function f gives the rate of change of Odyssey Travel's revenue with respect to the amount x it spends on advertising with their current advertising agency. By engaging the services of a different advertising agency, it is expected that Odyssey's revenue will grow at the rate given by the function g. Give an interpretation of the area A of the region S and find an expression for A in terms of a definite integral involving f and g.

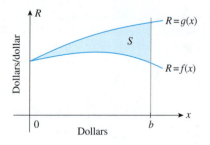

44. PULSE RATE DURING EXERCISE In the accompanying figure, the function *f* gives the rate of increase of an individual's pulse rate when he walked a prescribed course on a treadmill 6 months ago. The function *g* gives the rate of increase of his pulse rate when he recently walked the same prescribed course. Give an interpretation of the area *A* of the region *S*, and find an expression for *A* in terms of a definite integral involving *f* and *g*.

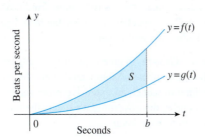

45. OIL PRODUCTION SHORTFALL Energy experts disagree about when global oil production will begin to decline. In the following figure, the function *f* gives the annual world oil production in billions of barrels from 1980 to 2050, according to the Department of Energy projection. The function *g* gives the world oil production in billions of barrels per year over the same period, according to longtime petroleum geologist Colin Campbell. Find an expression in terms of the definite integrals involving *f* and *g* giving the shortfall in the total oil production over the period in question heeding Campbell's dire warnings.
Source: U.S. Department of Energy; Colin Campbell.

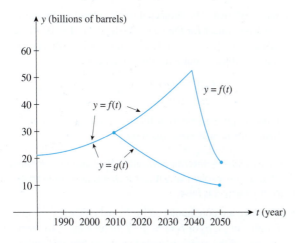

46. AIR PURIFICATION To study the effectiveness of air purifiers in removing smoke, engineers ran each purifier in a smoke-filled 10-ft × 20-ft room. In the accompanying figure, the function *f* gives the rate of change of the smoke level per minute, *t* min after the start of the test, when a Brand *A* purifier is used. The function *g* gives the rate of change of the smoke level per minute when a Brand *B* purifier is used.
  a. Give an interpretation of the area of the region *S*.

b. Find an expression for the area of *S* in terms of a definite integral involving *f* and *g*.

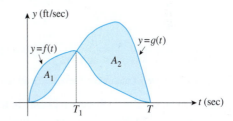

47. MOTION OF TWO CARS Two cars start out side by side and travel along a straight road. The velocity of Car 1 is *f*(*t*) ft/sec over the interval [0, *T*], the velocity of Car 2 is *g*(*t*) ft/sec over the interval [0, *T*], and 0 < T_1 < *T*. Furthermore, suppose the graphs of *f* and *g* are as depicted in the accompanying figure. Let A_1 and A_2 denote the areas of the regions (shown shaded).

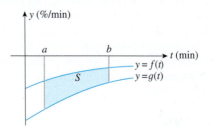

  a. Write the number
$$\int_{T_1}^{T} [g(t) - f(t)]\, dt - \int_{0}^{T_1} [f(t) - g(t)]\, dt$$
  in terms of A_1 and A_2.
  b. What does the number obtained in part (a) represent?

48. COMPARISON OF COMPANIES' REVENUES The rate of change of the revenue of Company *A* over the (time) interval [0, *T*] is *f*(*t*) dollars/week, whereas the rate of change of the revenue of Company *B* over the same period is *g*(*t*) dollars/week. The graphs of *f* and *g* are depicted in the accompanying figure. Find an expression in terms of definite integrals involving *f* and *g* giving the additional revenue that Company *B* will have over Company *A* in the period [0, *T*].

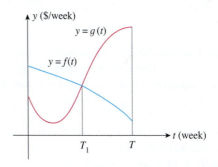

49. **MEXICO'S HEDGING TACTIC** Agustin Carstens, the Mexican finance minister, hedged all his country's oil sales for 2009 at $70/barrel, in effect gambling that prices would stay low that year. The figure below shows the U.S. crude oil price, $f(t)$, and Mexico's hedge level, $g(t) = 70$, (in dollars per barrel) from the beginning of January 2009 ($t = 0$) through the end of the year. Write, in terms of definite integrals, the oil profits that Mexico realized in 2009 using this tactic.

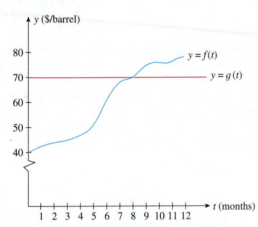

Sources: Thomson Reuters Datastream; *Financial Times* research.

50. **TURBO-CHARGED ENGINE VERSUS STANDARD ENGINE** In tests conducted by *Auto Test Magazine* on two identical models of the Phoenix Elite—one equipped with a standard engine and the other with a turbo-charger—it was found that the acceleration of the former is given by

$$a = f(t) = 4 + 0.8t \qquad (0 \le t \le 12)$$

ft/sec/sec, t sec after starting from rest at full throttle, whereas the acceleration of the latter is given by

$$a = g(t) = 4 + 1.2t + 0.03t^2 \qquad (0 \le t \le 12)$$

ft/sec/sec. How much faster is the turbo-charged model moving than the model with the standard engine at the end of a 10-sec test run at full throttle?

51. **ALTERNATIVE ENERGY SOURCES** Because of the increasingly important role played by coal as a viable alternative energy source, the production of coal has been growing at the rate of

$$3.5e^{0.05t}$$

billion metric tons/year, t years from 1980 (which corresponds to $t = 0$). Had it not been for the energy crisis, the rate of production of coal since 1980 might have been only

$$3.5e^{0.01t}$$

billion metric tons/year, t years from 1980. Determine how much additional coal was produced between 1980 and 2000 as an alternative energy source.

52. **EFFECT OF TV ADVERTISING ON CAR SALES** Carl Williams, the proprietor of Carl Williams Auto Sales, estimates that

with extensive television advertising, car sales over the next several years could be increasing at the rate of

$$5e^{0.3t}$$

thousand cars/year, t years from now, instead of at the current rate of

$$5 + 0.5t^{3/2}$$

thousand cars/year, t years from now. Find how many more cars Carl expects to sell over the next 5 years by implementing his advertising plans.

53. **POPULATION GROWTH** In an endeavor to curb population growth in a Southeast Asian island state, the government has decided to launch an extensive propaganda campaign. Without curbs, the government expects the rate of population growth to have been

$$60e^{0.02t}$$

thousand people/year, t years from now, over the next 5 years. However, successful implementation of the proposed campaign is expected to result in a population growth rate of

$$-t^2 + 60$$

thousand people/year, t years from now, over the next 5 years. Assuming that the campaign is mounted, how many fewer people will there be in that country 5 years from now than there would have been if no curbs had been imposed?

54. **BRITISH DEFICIT** The projected public spending by the British government for the years 2010–2011 ($t = 0$) through 2014–2015 ($t = 4$) is 692 billion pounds/year. The projected public revenue for the same period is

$$f(t) = \frac{292.6t + 134.4}{t + 6.9} + 500 \qquad (0 \le t \le 4)$$

Find the projected deficit for the period in question.
Source: Thomson Reuters Datastream.

In Exercises 55–58, determine whether the statement is true or false. If it is true, explain why it is true. If it is false, give an example to show why it is false.

55. If f and g are continuous on $[a, b]$ and either $f(x) \ge g(x)$ for all x in $[a, b]$ or $f(x) \le g(x)$ for all x in $[a, b]$, then the area of the region bounded by the graphs of f and g and the vertical lines $x = a$ and $x = b$ is given by $\int_a^b |f(x) - g(x)| \, dx$.

56. The area of the region bounded by the graphs of $f(x) = 2 - x$ and $g(x) = 4 - x^2$ and the vertical lines $x = 0$ and $x = 2$ is given by $\int_0^2 [f(x) - g(x)] \, dx$.

57. If A denotes the area bounded by the graphs of f and g on $[a, b]$, then

$$A^2 = \int_a^b [f(x) - g(x)]^2 \, dx$$

58. If f and g are continuous on $[a, b]$ and
$\int_a^b [f(t) - g(t)] \, dt > 0$, then $f(t) \geq g(t)$ for all t in $[a, b]$.

59. Show that the area of a region R bounded above by the graph of a function f and below by the graph of a function g from $x = a$ to $x = b$ is given by

$$\int_a^b [f(x) - g(x)] \, dx$$

where f and g are continuous functions.

Hint: The validity of the formula was verified earlier for the case in which both f and g were nonnegative. Now, let f and g be two functions such that $f(x) \geq g(x)$ for $a \leq x \leq b$. Then there exists some nonnegative constant c such that the translated curves

$y = f(x) + c$ and $y = g(x) + c$ lie entirely above the x-axis. The region R' has the same area as the region R (see the accompanying figures). Show that the area of R' is given by

$$\int_a^b \{[f(x) + c] - [g(x) + c]\} \, dx = \int_a^b [f(x) - g(x)] \, dx$$

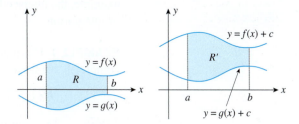

6.6 Solutions to Self-Check Exercises

1. The region in question is shown in the accompanying figure. Since the graph of the function f lies above that of the function g for $0 \leq x \leq 1$, we see that the required area is given by

$$\int_0^1 [(x^2 + 2) - (1 - x)] \, dx = \int_0^1 (x^2 + x + 1) \, dx$$

$$= \frac{1}{3} x^3 + \frac{1}{2} x^2 + x \Big|_0^1$$

$$= \frac{1}{3} + \frac{1}{2} + 1$$

$$= \frac{11}{6}$$

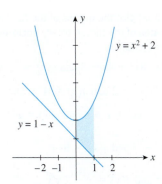

2. The region in question is shown in the accompanying figure. To find the points of intersection of the two curves, we solve the equations

$$-x^2 + 6x + 5 = x^2 + 5$$
$$2x^2 - 6x = 0$$
$$2x(x - 3) = 0$$

giving $x = 0$ or $x = 3$. Therefore, the points of intersection are $(0, 5)$ and $(3, 14)$.

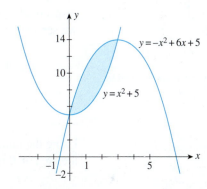

Since the graph of f always lies above that of g for $0 \leq x \leq 3$, we see that the required area is given by

$$\int_0^3 [(-x^2 + 6x + 5) - (x^2 + 5)] \, dx = \int_0^3 (-2x^2 + 6x) \, dx$$

$$= -\frac{2}{3} x^3 + 3x^2 \Big|_0^3$$

$$= -18 + 27$$

$$= 9$$

3. The additional profits realizable over the next 10 years are given by

$$\int_0^{10} [(t - 2\sqrt{t} + 4) - (1 + t^{2/3})] \, dt$$

$$= \int_0^{10} (t - 2t^{1/2} + 3 - t^{2/3}) \, dt$$

$$= \frac{1}{2} t^2 - \frac{4}{3} t^{3/2} + 3t - \frac{3}{5} t^{5/3} \Big|_0^{10}$$

$$= \frac{1}{2} (10)^2 - \frac{4}{3} (10)^{3/2} + 3(10) - \frac{3}{5} (10)^{5/3}$$

$$\approx 9.99$$

or approximately \$10 million.

USING TECHNOLOGY   Finding the Area Between Two Curves

The numerical integral operation can be used to find the area between two curves. We do this by using the numerical integral operation to evaluate an appropriate definite integral or the sum (difference) of appropriate definite integrals. In the following example, the intersection operation is also used to advantage to help us find the limits of integration.

EXAMPLE 1 Use a graphing utility to find the area of the smaller region R that is completely enclosed by the graphs of the functions

$$f(x) = 2x^3 - 8x^2 + 4x - 3 \quad \text{and} \quad g(x) = 3x^2 + 10x - 11$$

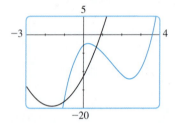

FIGURE T1
The region R is completely enclosed by the graphs of f and g.

Solution   The graphs of f and g in the viewing window $[-3, 4] \times [-20, 5]$ are shown in Figure T1.

Using the intersection operation of a graphing utility, we find the x-coordinates of the points of intersection of the two graphs to be approximately -1.04 and 0.65, respectively. Since the graph of f lies above that of g on the interval $[-1.04, 0.65]$, we see that the area of R is given by

$$A \approx \int_{-1.04}^{0.65} \left[(2x^3 - 8x^2 + 4x - 3) - (3x^2 + 10x - 11) \right] dx$$

$$= \int_{-1.04}^{0.65} (2x^3 - 11x^2 - 6x + 8) \, dx$$

Using the numerical integration function of a graphing utility, we find $A \approx 9.87$, so the area of R is approximately 9.87.

TECHNOLOGY EXERCISES

In Exercises 1–6, (a) plot the graphs of the functions f and g, and (b) find the area of the region enclosed by these graphs and the vertical lines $x = a$ and $x = b$. Express your answers accurate to four decimal places.

1. $f(x) = x^3(x - 5)^4$, $g(x) = 0$; $a = 1$, $b = 3$

2. $f(x) = x - \sqrt{1 - x^2}$, $g(x) = 0$; $a = -\dfrac{1}{2}$, $b = \dfrac{1}{2}$

3. $f(x) = x^{1/3}(x + 1)^{1/2}$, $g(x) = x^{-1}$; $a = 1.2$, $b = 2$

4. $f(x) = 2$, $g(x) = \ln(1 + x^2)$; $a = -1$, $b = 1$

5. $f(x) = \sqrt{x}$, $g(x) = \dfrac{x^2 - 3}{x^2 + 1}$; $a = 0$, $b = 3$

6. $f(x) = \dfrac{4}{x^2 + 1}$, $g(x) = x^4$; $a = -1$, $b = 1$

In Exercises 7–12, (a) plot the graphs of the functions f and g, and (b) find the area of the region totally enclosed by the graphs of these functions.

7. $f(x) = 2x^3 - 8x^2 + 4x - 3$ and $g(x) = -3x^2 + 10x - 10$

8. $f(x) = x^4 - 2x^2 + 2$ and $g(x) = 4 - 2x^2$

9. $f(x) = 2x^3 - 3x^2 + x + 5$ and $g(x) = e^{2x} - 3$

10. $f(x) = \dfrac{1}{2}x^2 - 3$ and $g(x) = \ln x$

11. $f(x) = xe^{-x}$ and $g(x) = x - 2\sqrt{x}$

12. $f(x) = e^{-x^2}$ and $g(x) = x^4$

13. Refer to Example 1. Find the area of the larger region that is completely enclosed by the graphs of the functions f and g.

6.7   Applications of the Definite Integral to Business and Economics

In this section, we consider several applications of the definite integral in the fields of business and economics.

Consumers' and Producers' Surplus

We begin by deriving a formula for computing the consumers' surplus. Suppose $p = D(x)$ is the demand function that relates the unit price p of a commodity to the quantity x demanded of it. Furthermore, suppose a fixed unit market price $\bar{p}$ has been established for the commodity and corresponding to this unit price the quantity demanded is $\bar{x}$ units (Figure 38). Then those consumers who would be willing to pay a unit price higher than $\bar{p}$ for the commodity would in effect experience a savings. This difference between what the consumers *would* be willing to pay for $\bar{x}$ units of the commodity and what they *actually* pay for them is called the **consumers' surplus.**

To derive a formula for computing the consumers' surplus, divide the interval $[0, \bar{x}]$ into n subintervals, each of length $\Delta x = \bar{x}/n$, and denote the right endpoints of these subintervals by $x_1, x_2, \ldots, x_n = \bar{x}$ (Figure 39).

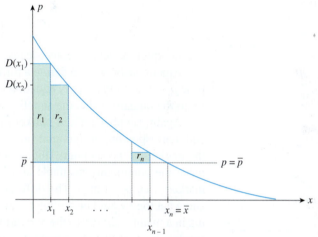

FIGURE 38

$D(x)$ is a demand function.

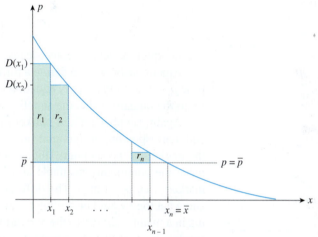

FIGURE 39

Approximating consumers' surplus by the sum of the areas of the rectangles $r_1, r_2, \ldots, r_n$

We observe in Figure 39 that there are consumers who would pay a unit price of at least $D(x_1)$ dollars for the first Δx units of the commodity instead of the market price of $\bar{p}$ dollars per unit. The savings to these consumers is approximated by

$$D(x_1)\, \Delta x - \bar{p}\, \Delta x = [D(x_1) - \bar{p}]\, \Delta x$$

which is the area of the rectangle r_1. Pursuing the same line of reasoning, we find that the savings to the consumers who would be willing to pay a unit price of at least $D(x_2)$ dollars for the next Δx units (from x_1 through x_2) of the commodity, instead of the market price of $\bar{p}$ dollars per unit, is approximated by

$$D(x_2)\, \Delta x - \bar{p}\, \Delta x = [D(x_2) - \bar{p}]\, \Delta x$$

Continuing, we approximate the total savings to the consumers in purchasing $\bar{x}$ units of the commodity by the sum

$$
\begin{aligned}
[D(x_1) - \bar{p}]\, \Delta x &+ [D(x_2) - \bar{p}]\, \Delta x + \cdots + [D(x_n) - \bar{p}]\, \Delta x \\
&= [D(x_1) + D(x_2) + \cdots + D(x_n)]\, \Delta x - \underbrace{[\bar{p}\, \Delta x + \bar{p}\, \Delta x + \cdots + \bar{p}\, \Delta x]}_{n \text{ terms}}
\end{aligned}
$$

$$
\begin{aligned}
&= [D(x_1) + D(x_2) + \cdots + D(x_n)]\, \Delta x - n\bar{p}\, \Delta x \\
&= [D(x_1) + D(x_2) + \cdots + D(x_n)]\, \Delta x - \bar{p}\, \bar{x}
\end{aligned}
$$

Now, the first term in the last expression is the Riemann sum of the demand function $p = D(x)$ over the interval $[0, \bar{x}]$ with representative points $x_1, x_2, \ldots, x_n$. Letting n approach infinity, we obtain the following formula for the consumers' surplus CS.

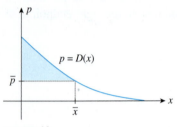

FIGURE 40
Consumers' surplus

The consumer's surplus is given by the area of the region bounded above by the demand curve $p = D(x)$ and below by the horizontal line $p = \bar{p}$ from $x = 0$ to $x = \bar{x}$ (Figure 40). We can also see this if we rewrite Equation (16) in the form

$$CS = \int_0^{\bar{x}} [D(x) - \bar{p}] \, dx$$

and interpret the result geometrically.

Analogously, we can derive a formula for computing the producers' surplus. Suppose $p = S(x)$ is the supply equation that relates the unit price p of a certain commodity to the quantity x that the supplier will make available in the market at that price.

Again, suppose a fixed market price $\bar{p}$ has been established for the commodity, and, corresponding to this unit price, a quantity of $\bar{x}$ units will be made available in the market by the suppliers (Figure 41). Then the suppliers who would be willing to make the commodity available at a lower price stand to gain from the fact that the market price is set at $\bar{p}$. The difference between what the suppliers actually receive and what they would be willing to receive is called the **producers' surplus.** Proceeding in a manner similar to the derivation of the equation for computing the consumers' surplus, we find that the producers' surplus PS is defined as follows:

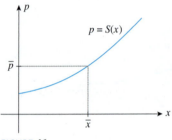

FIGURE 41
$S(x)$ is a supply function.

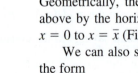

FIGURE 42
Producers' surplus

Geometrically, the producers' surplus is given by the area of the region bounded above by the horizontal line $p = \bar{p}$ and below by the supply curve $p = S(x)$ from $x = 0$ to $x = \bar{x}$ (Figure 42).

We can also show that the last statement is true by converting Equation (17) to the form

$$PS = \int_0^{\bar{x}} [\bar{p} - S(x)] \, dx$$

and interpreting the definite integral geometrically.

EXAMPLE 1 The demand function for a certain make of ten-speed bicycle is given by

$$p = D(x) = -0.001x^2 + 250$$

where p is the unit price in dollars and x is the quantity demanded in thousands of units. The supply function for these bicycles is given by

$$p = S(x) = 0.0006x^2 + 0.02x + 100$$

where p stands for the unit price in dollars and x stands for the number of bicycles that the supplier will put on the market, in units of a thousand. Determine the consumers' surplus and the producers' surplus if the market price of a bicycle is set at the equilibrium price.

Solution   Recall that the equilibrium price is the unit price of the commodity when market equilibrium occurs. We determine the equilibrium price by solving for the point of intersection of the demand curve and the supply curve (Figure 43). To solve the system of equations

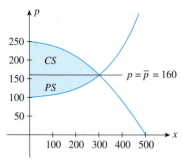

FIGURE 43
Consumers' surplus and producers' surplus when market price = equilibrium price

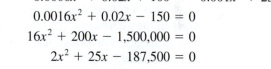

$$p = -0.001x^2 \qquad + 250$$
$$p = 0.0006x^2 + 0.02x + 100$$

we simply substitute the second equation into the first, obtaining

$$0.0006x^2 + 0.02x + 100 = -0.001x^2 + 250$$
$$0.0016x^2 + 0.02x - 150 = 0$$
$$16x^2 + 200x - 1,500,000 = 0$$
$$2x^2 + 25x - 187,500 = 0$$

Factoring this last equation, we obtain

$$(2x + 625)(x - 300) = 0$$

Thus, $x = -625/2$ or $x = 300$. The first number lies outside the interval of interest, so we are left with the solution $x = 300$, with a corresponding value of

$$p = -0.001(300)^2 + 250 = 160$$

Thus, the equilibrium point is $(300, 160)$; that is, the equilibrium quantity is 300,000, and the equilibrium price is \$160. Setting the market price at \$160 per unit and using Formula (16) with $\bar{p} = 160$ and $\bar{x} = 300$, we find that the consumers' surplus is given by

$$CS = \int_0^{300} (-0.001x^2 + 250)\, dx - (160)(300)$$

$$= \left(-\frac{1}{3000}x^3 + 250x \right)\Bigg|_0^{300} - 48,000$$

$$= -\frac{300^3}{3000} + (250)(300) - 48,000$$

$$= 18,000$$

or \$18,000,000. (Recall that x is measured in units of a thousand.)

Next, using Equation (17), we find that the producers' surplus is given by

$$PS = (160)(300) - \int_0^{300} (0.0006x^2 + 0.02x + 100)\, dx$$

$$= 48,000 - (0.0002x^3 + 0.01x^2 + 100x)\Bigg|_0^{300}$$

$$= 48,000 - [(0.0002)(300)^3 + (0.01)(300)^2 + 100(300)]$$

$$= 11,700$$

or \$11,700,000.

The Future and Present Value of an Income Stream

Suppose a firm generates a stream of income over a period of time—for example, the revenue generated by a large chain of retail stores over a 5-year period. As the income is realized, it is reinvested and earns interest at a fixed rate. The **accumulated future income stream** over the 5-year period is the amount of money the firm ends up with at the end of that period.

The definite integral can be used to determine this accumulated, or total, future income stream over a period of time. The total future value of an income stream gives us a way to measure the value of such a stream. To find the **total future value of an income stream,** suppose that

$$R(t) = \text{Rate of income generation at any time } t \qquad \textcolor{red}{\text{In dollars per year}}$$

$$r = \text{Interest rate compounded continuously}$$

$$T = \text{Term} \qquad\qquad\qquad\qquad\qquad\qquad \textcolor{red}{\text{In years}}$$

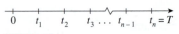

FIGURE 44
The time interval [0, T] is partitioned into n subintervals.

Let's divide the time interval $[0, T]$ into n subintervals of equal length $\Delta t = T/n$ and denote the right endpoints of these intervals by $t_1, t_2, \ldots, t_n = T$, as shown in Figure 44.

If R is a continuous function on $[0, T]$, then $R(t)$ will not differ by much from $R(t_1)$ in the subinterval $[0, t_1]$ provided that the subinterval is small (which is true if n is large). Therefore, the income generated over the time interval $[0, t_1]$ is approximately

$$R(t_1)\,\Delta t \qquad \textcolor{red}{\text{Constant rate of income} \times \text{length of time}}$$

dollars. The future value of this amount, T years from now, calculated as if it were earned at time t_1, is

$$\left[R(t_1)\,\Delta t\right]e^{r(T-t_1)} \qquad \textcolor{red}{\text{Equation (10), Section 5.3}}$$

dollars. Similarly, the income generated over the time interval $[t_1, t_2]$ is approximately $R(t_2)\,\Delta t$ dollars and has a future value, T years from now, of approximately

$$\left[R(t_2)\,\Delta t\right]e^{r(T-t_2)}$$

dollars. Therefore, the sum of the future values of the income stream generated over the time interval $[0, T]$ is approximately

$$R(t_1)e^{r(T-t_1)}\,\Delta t + R(t_2)e^{r(T-t_2)}\,\Delta t + \cdots + R(t_n)e^{r(T-t_n)}\,\Delta t$$
$$= e^{rT}\left[R(t_1)e^{-rt_1}\,\Delta t + R(t_2)e^{-rt_2}\,\Delta t + \cdots + R(t_n)e^{-rt_n}\,\Delta t\right]$$

dollars. But this sum is just the Riemann sum of the function $e^{rT}R(t)e^{-rt}$ over the interval $[0, T]$ with representative points $t_1, t_2, \ldots, t_n$. Letting n approach infinity, we obtain the following result.

Accumulated or Total Future Value of an Income Stream

The accumulated, or total, future value after T years of an income stream of $R(t)$ dollars per year, earning interest at the rate of r per year compounded continuously, is given by

$$A = e^{rT}\int_0^T R(t)e^{-rt}\,dt \qquad\qquad \textbf{(18)}$$

APPLIED EXAMPLE 2 Income Stream Crystal Car Wash recently bought an automatic car-washing machine that is expected to generate $40,000 in revenue per year for the next 5 years. If the income is reinvested in a business earning interest at the rate of 12% per year compounded continuously, find the total accumulated value of this income stream at the end of 5 years.

Solution We are required to find the total future value of the given income stream after 5 years. Using Equation (18) with $R(t) = 40{,}000$, $r = 0.12$, and $T = 5$, we see that the required value is given by

$$e^{0.12(5)} \int_0^5 40{,}000 e^{-0.12t}\, dt$$

$$= e^{0.6}\left(-\frac{40{,}000}{0.12} e^{-0.12t}\right)\Bigg|_0^5 \qquad \text{Integrate using the substitution } u = -0.12t.$$

$$= -\frac{40{,}000 e^{0.6}}{0.12}\left(e^{-0.6} - 1\right) \approx 274{,}039.60$$

or approximately $274,040.   ■

Another way of measuring the value of an income stream is by considering its present value. The **present value of an income stream** of $R(t)$ dollars per year over a term of T years, earning interest at the rate of r per year compounded continuously, is the principal P that will yield the same accumulated value as the income stream itself when P is invested today for a period of T years at the same rate of interest. In other words,

$$Pe^{rT} = e^{rT} \int_0^T R(t) e^{-rt}\, dt$$

Dividing both sides of the equation by e^{rT} gives the following result.

Present Value of an Income Stream

The present value of an income stream of $R(t)$ dollars per year, earning interest at the rate of r per year compounded continuously, is given by

$$PV = \int_0^T R(t) e^{-rt}\, dt \qquad\qquad \text{(19)}$$

APPLIED EXAMPLE 3 Investment Analysis The owner of a local cinema is considering two alternative plans for renovating and improving the theater. Plan A calls for an immediate cash outlay of $250,000, whereas Plan B requires an immediate cash outlay of $180,000. It has been estimated that adopting Plan A would result in a net income stream generated at the rate of

$$f(t) = 630{,}000$$

dollars per year, whereas adopting Plan B would result in a net income stream generated at the rate of

$$g(t) = 580{,}000$$

dollars per year for the next 3 years. If the prevailing interest rate for the next 3 years is 10% per year, which plan will generate the higher net income by the end of 3 years?

Solution   Since the initial outlay is $250,000, we find—using Equation (19) with $R(t) = 630,000$, $r = 0.1$, and $T = 3$—that the present value of the net income under Plan A is given by

$$\int_0^3 630,000e^{-0.1t}\,dt - 250,000$$

$$= \frac{630,000}{-0.1}e^{-0.1t}\Big|_0^3 - 250,000 \qquad \text{\color{red}{Integrate using the substitution } } u = -0.1t.$$

$$= -6,300,000e^{-0.3} + 6,300,000 - 250,000$$

$$\approx 1,382,845$$

or approximately $1,382,845.

To find the present value of the net income under Plan B, we use Equation (19) with $R(t) = 580,000$, $r = 0.1$, and $T = 3$, obtaining

$$\int_0^3 580,000e^{-0.1t}\,dt - 180,000$$

dollars. Proceeding as in the previous computation, we see that the required value is $1,323,254 (see Exercise 12, page 491).

Comparing the present value of each plan, we conclude that Plan A would generate the higher total net income over the 3 years.

Note   The function R in Example 3 is a constant function. If R is not a constant function, then we may need more sophisticated techniques of integration to evaluate the integral in (19). Exercises 7.1 and 7.2 contain problems of this type.

The Amount and Present Value of an Annuity

An annuity is a sequence of payments made at regular time intervals. The time period in which these payments are made is called the *term* of the annuity. Although the payments need not be equal in size, they are equal in many important applications, and we will assume that they are equal in our discussion. Examples of annuities are regular deposits to a savings account, monthly home mortgage payments, and monthly insurance payments.

The **amount of an annuity** is the sum of the payments plus the interest earned. A formula for computing the amount of an annuity A can be derived with the help of Equation (18). Let

$$P = \text{Size of each payment in the annuity}$$
$$r = \text{Interest rate compounded continuously}$$
$$T = \text{Term of the annuity (in years)}$$
$$m = \text{Number of payments per year}$$

Assume that the payments into the annuity constitute a constant income stream of $R(t) = mP$ dollars per year. With this value of $R(t)$, Equation (18) yields

$$A = e^{rT}\int_0^T R(t)e^{-rt}\,dt = e^{rT}\int_0^T mPe^{-rt}\,dt$$

$$= mPe^{rT}\left[-\frac{e^{-rt}}{r}\right]\Big|_0^T = mPe^{rT}\left[-\frac{e^{-rT}}{r} + \frac{1}{r}\right]$$

$$= \frac{mP}{r}\left(e^{rT} - 1\right) \qquad \text{\color{red}{Since } } e^{rT}\cdot e^{-rT} = 1$$

This leads us to the following formula.

> **Amount of an Annuity**
> The amount of an annuity is
> $$A = \frac{mP}{r}(e^{rT} - 1) \tag{20}$$
> where P, r, T, and m are as defined earlier.

APPLIED EXAMPLE 4 IRAs On January 1, 1998, Marcus Chapman deposited $2000 into an Individual Retirement Account (IRA) paying interest at the rate of 5% per year compounded continuously. Assuming that he deposited $2000 annually into the account, how much did he have in his IRA at the beginning of 2014?

Solution We use Equation (20), with $P = 2000$, $r = 0.05$, $T = 16$, and $m = 1$, obtaining

$$A = \frac{2000}{0.05}(e^{0.8} - 1)$$
$$\approx 49{,}021.64$$

Thus, Marcus had approximately $49,022 in his account at the beginning of 2014.

> **Exploring with TECHNOLOGY**
>
> Refer to Example 4. Suppose Marcus wished to know how much he would have in his IRA at any time in the future, not just at the beginning of 2014, as you were asked to compute in the example.
>
> 1. Using Formula (18) and the relevant data from Example 4, show that the required amount at any time x (x measured in years, $x > 0$) is given by
> $$A = f(x) = 40{,}000(e^{0.05x} - 1)$$
> 2. Use a graphing utility to plot the graph of f, using the viewing window $[0, 30] \times [0, 200{,}000]$.
> 3. Using ZOOM and TRACE, or using the function evaluation capability of your graphing utility, use the result of part 2 to verify the result obtained in Example 4. Comment on the advantage of the mathematical model found in part 1.

Using Equation (19), we can derive the following formula for the present value of an annuity.

> **Present Value of an Annuity**
> The present value of an annuity is given by
> $$PV = \frac{mP}{r}(1 - e^{-rT}) \tag{21}$$
> where P, r, T, and m are as defined earlier.

APPLIED EXAMPLE 5 Sinking Funds Tomas Perez, the proprietor of a hardware store, wants to establish a fund from which he will withdraw $1000 per month for the next 10 years. If the fund earns interest at the rate of 6% per year compounded continuously, how much money does he need to establish the fund?

Solution We want to find the present value of an annuity with $P = 1000$, $r = 0.06$, $T = 10$, and $m = 12$. Using Equation (21), we find

$$PV = \frac{12{,}000}{0.06}\left(1 - e^{-(0.06)(10)}\right)$$

$$\approx 90{,}237.70$$

Therefore, Tomas needs approximately $90,238 to establish the fund. ∎

Lorenz Curves and Income Distributions

One method used by economists to study the distribution of income in a society is based on the **Lorenz curve,** named after American statistician M.O. Lorenz. To describe the Lorenz curve, let $f(x)$ denote the proportion of the total income received by the poorest $100x\%$ of the population for $0 \le x \le 1$. Using this terminology, $f(0.3) = 0.1$ simply states that the lowest 30% of the income recipients receive 10% of the total income.

The function f has the following properties:

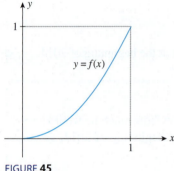

FIGURE 45
A Lorenz curve

1. The domain of f is $[0, 1]$.
2. The range of f is $[0, 1]$.
3. $f(0) = 0$ and $f(1) = 1$.
4. $f(x) \le x$ for every x in $[0, 1]$.
5. f is increasing on $[0, 1]$.

The first two properties follow from the fact that both x and $f(x)$ are fractions of a whole. Property 3 is a statement that 0% of the income recipients receive 0% of the total income and 100% of the income recipients receive 100% of the total income. Property 4 follows from the fact that the lowest $100x\%$ of the income recipients cannot receive more than $100x\%$ of the total income. A typical Lorenz curve is shown in Figure 45.

 APPLIED EXAMPLE 6 Lorenz Curves A developing country's income distribution is described by the function

$$f(x) = \frac{19}{20}x^2 + \frac{1}{20}x$$

a. Sketch the Lorenz curve for the given function.
b. Compute $f(0.2)$ and $f(0.8)$, and interpret your results.

Solution

a. The Lorenz curve is shown in Figure 46.

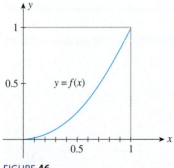

FIGURE 46
The Lorenz curve $f(x) = \dfrac{19}{20}x^2 + \dfrac{1}{20}x$

b.
$$f(0.2) = \frac{19}{20}(0.2)^2 + \frac{1}{20}(0.2) = 0.048$$

Thus, the lowest 20% of the people receive 4.8% of the total income.

$$f(0.8) = \frac{19}{20}(0.8)^2 + \frac{1}{20}(0.8) = 0.648$$

Thus, the lowest 80% of the people receive 64.8% of the total income. ∎

Next, let's consider the Lorenz curve described by the function $y = f(x) = x$. Since exactly $100x\%$ of the total income is received by the lowest $100x\%$ of income

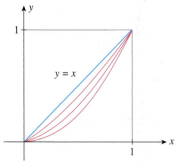

recipients, the line $y = x$ is called the **line of complete equality.** For example, 10% of the total income is received by the lowest 10% of income recipients, 20% of the total income is received by the lowest 20% of income recipients, and so on. Now, it is evident that the closer a Lorenz curve is to this line, the more equitable the income distribution is among the income recipients. But the proximity of a Lorenz curve to the line of complete equality is reflected by the area between the Lorenz curve and the line $y = x$ (Figure 47). The closer the curve is to the line, the smaller the enclosed area.

This observation suggests that we may define a number, called the coefficient of inequality of a Lorenz curve, as the ratio of the area between the line of complete equality and the Lorenz curve to the area under the line of complete equality. Since the area under the line of complete equality is $\frac{1}{2}$, we see that the coefficient of inequality is given by the following formula.

Coefficient of Inequality of a Lorenz Curve

The coefficient of inequality, or **Gini Index,** of a Lorenz curve is

$$L = 2 \int_0^1 [x - f(x)]\, dx \qquad (22)$$

The coefficient of inequality is a number between 0 and 1. For example, a coefficient of zero implies that the income distribution is perfectly uniform.

APPLIED EXAMPLE 7 Income Distributions In a study conducted by a certain country's Economic Development Board with regard to the income distribution of certain segments of the country's workforce, it was found that the Lorenz curves for the distributions of incomes of medical doctors and of movie actors are described by the functions

$$f(x) = \frac{14}{15}x^2 + \frac{1}{15}x \quad \text{and} \quad g(x) = \frac{5}{8}x^4 + \frac{3}{8}x$$

respectively. Compute the coefficient of inequality for each Lorenz curve. Which profession has the more equitable income distribution?

Solution The required coefficients of inequality are, respectively,

$$L_1 = 2 \int_0^1 \left[x - \left(\frac{14}{15}x^2 + \frac{1}{15}x \right) \right] dx = 2 \int_0^1 \left(\frac{14}{15}x - \frac{14}{15}x^2 \right) dx$$

$$= \frac{28}{15} \int_0^1 (x - x^2)\, dx = \frac{28}{15} \left(\frac{1}{2}x^2 - \frac{1}{3}x^3 \right) \Big|_0^1$$

$$= \frac{14}{45} \approx 0.311$$

$$L_2 = 2 \int_0^1 \left[x - \left(\frac{5}{8}x^4 + \frac{3}{8}x \right) \right] dx = 2 \int_0^1 \left(\frac{5}{8}x - \frac{5}{8}x^4 \right) dx$$

$$= \frac{5}{4} \int_0^1 (x - x^4)\, dx = \frac{5}{4} \left(\frac{1}{2}x^2 - \frac{1}{5}x^5 \right) \Big|_0^1$$

$$= \frac{15}{40} = 0.375$$

We conclude that in this country, the incomes of medical doctors are more evenly distributed than the incomes of movie actors.

6.7 Self-Check Exercise

CONSUMERS' AND PRODUCERS' SURPLUS The demand function for a certain make of exercise bicycle that is sold exclusively through cable television is

$$p = d(x) = \sqrt{9 - 0.02x}$$

where p is the unit price in hundreds of dollars and x is the quantity demanded per week. The corresponding supply function is given by

$$p = s(x) = \sqrt{1 + 0.02x}$$

where p has the same meaning as before and x is the number of exercise bicycles the supplier will make available at price p. Determine the consumers' surplus and the producers' surplus if the unit price is set at the equilibrium price.

The solution to Self-Check Exercise 6.7 can be found on page 493.

6.7 Concept Questions

1. a. Define consumers' surplus. Give a formula for computing it.
   b. Define producers' surplus. Give a formula for computing it.

2. a. Define the accumulated (future) value of an income stream. Give a formula for computing it.
   b. Define the present value of an income stream. Give a formula for computing it.

3. Define the amount of an annuity. Give a formula for computing it.

4. Explain the following terms: (a) Lorenz curve and (b) coefficient of inequality of a Lorenz curve.

6.7 Exercises

1. CONSUMERS' SURPLUS The demand function for a certain make of replacement cartridges for a water purifier is given by

$$p = -0.01x^2 - 0.1x + 6$$

where p is the unit price in dollars and x is the quantity demanded each week, measured in units of a thousand. Determine the consumers' surplus if the market price is set at $4/cartridge.

2. CONSUMERS' SURPLUS The demand function for a certain brand of CD is given by

$$p = -0.01x^2 - 0.2x + 8$$

where p is the unit price in dollars and x is the quantity demanded each week, measured in units of a thousand. Determine the consumers' surplus if the market price is set at $5/disc.

3. CONSUMERS' SURPLUS It is known that the quantity demanded of a certain make of portable hair dryer is x hundred units/week and the corresponding unit price is

$$p = \sqrt{225 - 5x}$$

dollars. Determine the consumers' surplus if the market price is set at $10/unit.

4. PRODUCERS' SURPLUS The supplier of the portable hair dryers in Exercise 3 will make x hundred units of hair dryers available in the market when the unit price is

$$p = \sqrt{36 + 1.8x}$$

dollars. Determine the producers' surplus if the market price is set at $9/unit.

5. PRODUCERS' SURPLUS The supply function for the CDs of Exercise 2 is given by

$$p = 0.01x^2 + 0.1x + 3$$

where p is the unit price in dollars and x stands for the quantity that will be made available in the market by the supplier, measured in units of a thousand. Determine the producers' surplus if the market price is set at the equilibrium price.

6. CONSUMERS' SURPLUS The demand function for a certain brand of mattress is given by

$$p = 600e^{-0.04x}$$

where p is the unit price in dollars and x (in units of a hundred) is the quantity demanded per month.
   a. Find the number of mattresses demanded per month if the unit price is set at $400/mattress.
   b. Use the results of part (a) to find the consumers' surplus if the selling price is set at $400 per mattress.

7. **PRODUCERS' SURPLUS** The manufacturer of the brand of mattresses of Exercise 6 will make x hundred units available in the market when the unit price is

$$p = 100 + 80e^{0.05x}$$

dollars.
  a. Find the number of mattresses the manufacturer will make available in the market place if the unit price is set at \$250/mattress.
  b. Use the result of part (a) to find the producers' surplus if the unit price is set at \$250/mattress.

8. **CONSUMERS' SURPLUS** The demand function for a certain model of Blu-ray player is given by

$$p = \frac{600}{0.5x + 2}$$

where p is the unit price in dollars and x (in units of a thousand) is the quantity demanded per week. What is the consumers' surplus if the selling price is set at \$200/unit?

9. **PRODUCERS' SURPLUS** The manufacturer of the Blu-ray players in Exercise 8 will make x thousand units available in the market per week when the unit price is

$$p = 100\left(0.5x + \frac{0.4}{1 + x}\right)$$

dollars. What is the producers' surplus if the selling price is set at \$160/unit?

10. **CONSUMERS' AND PRODUCERS' SURPLUS** The management of the Titan Tire Company has determined that the quantity demanded x of their Super Titan tires per week is related to the unit price p by the relation

$$p = 144 - x^2$$

where p is measured in dollars and x is measured in units of a thousand. Titan will make x units of the tires available in the market if the unit price is

$$p = 48 + \frac{1}{2}x^2$$

dollars. Determine the consumers' surplus and the producers' surplus when the market unit price is set at the equilibrium price.

11. **CONSUMERS' AND PRODUCERS' SURPLUS** The quantity demanded x (in units of a hundred) of the Mikado miniature cameras per week is related to the unit price p (in dollars) by

$$p = -0.2x^2 + 80$$

and the quantity x (in units of a hundred) that the supplier is willing to make available in the market is related to the unit price p (in dollars) by

$$p = 0.1x^2 + x + 40$$

If the market price is set at the equilibrium price, find the consumers' surplus and the producers' surplus.

12. Refer to Example 3, page 486. Verify that

$$\int_0^3 580{,}000e^{-0.1t}\, dt - 180{,}000 \approx 1{,}323{,}254$$

13. **FUTURE VALUE OF AN INVESTMENT** The newly opened Mario's Trattoria is expected to produce a continuous income stream at the rate of

$$R(t) = 120{,}000$$

dollars/year for the next 4 years. If the prevailing interest rate is 3.5%/year compounded continuously, find the future value of this income stream.

14. **FUTURE VALUE OF AN INVESTMENT** An investment is projected to generate a continuous revenue stream at the rate of

$$R(t) = 30{,}000e^{0.03t}$$

dollars/year for the next 3 years. If the income stream is invested in a bank that pays interest at the rate of 4.5%/year compounded continuously, find the total accumulated value of this income stream at the end of 3 years.

15. **PRESENT VALUE OF AN INVESTMENT** Suppose an investment is expected to generate income at the rate of

$$R(t) = 200{,}000$$

dollars/year for the next 5 years. Find the present value of this investment if the prevailing interest rate is 8%/year compounded continuously.

16. **FRANCHISES** Camille purchased a 15-year franchise for a fitness club that is expected to generate income at the rate of

$$R(t) = 400{,}000$$

dollars/year. If the prevailing interest rate is 8%/year compounded continuously, find the present value of the franchise.

17. **THE AMOUNT OF AN ANNUITY** Find the amount of an annuity if \$250/month is paid into it for a period of 20 years, earning interest at the rate of 4%/year compounded continuously.

18. **THE AMOUNT OF AN ANNUITY** Find the amount of an annuity if \$400/month is paid into it for a period of 20 years, earning interest at the rate of 5%/year compounded continuously.

19. **VALUE OF AN INVESTMENT** An investment generates a continuous income stream of

$$R(t) = 20e^{0.08t}$$

thousand dollars/year at time t for the next 5 years. The prevailing rate of interest is 3%/year compounded continuously for this 5-year period.
  a. What is the future value of the investment at the end of 5 years?
  b. What is the present value of the investment over the term of 5 years?

20. **BUSINESS DECISIONS** Sharon, the owner of the Brentwood Motel, is planning to renovate all the rooms in her motel. There are two plans before her. Plan A calls for an immediate cash outlay of $600,000, whereas plan B calls for an immediate outlay of $400,000. Sharon estimates that adopting plan A would yield an income stream of

$$f(t) = 3,050,000e^{0.02t}$$

dollars/year for the next 5 years, whereas adopting plan B would yield an income stream of

$$g(t) = 3,200,000$$

dollars/year for the next 5 years. If the prevailing rate of interest is 3%/year compounded continuously, which plan will yield the higher net income at the end of 5 years?

21. **THE AMOUNT OF AN ANNUITY** Aiso deposits $150/month in a savings account paying 5%/year compounded continuously. Estimate the amount that will be in his account after 15 years.

22. **CUSTODIAL ACCOUNTS** The Armstrongs wish to establish a custodial account to finance their children's education. If they deposit $200 monthly for 10 years in a savings account paying 4%/year compounded continuously, how much will their savings account be worth at the end of this period?

23. **IRA ACCOUNTS** Refer to Example 4, page 487. Suppose Marcus made his IRA payment on April 1, 1998, and annually thereafter. If interest is paid at the same initial rate, approximately how much did Marcus have in his account at the beginning of 2014?

24. **PRESENT VALUE OF AN ANNUITY** Estimate the present value of an annuity if payments are $800 monthly for 12 years and the account earns interest at the rate of 5%/year compounded continuously.

25. **PRESENT VALUE OF AN ANNUITY** Estimate the present value of an annuity if payments are $1200 monthly for 15 years and the account earns interest at the rate of 6%/year compounded continuously.

26. **LOTTERY PAYMENTS** A state lottery commission pays the winner of the "Million Dollar" lottery 20 annual installments of $50,000 each. If the prevailing interest rate is 6%/year compounded continuously, find the present value of the winning ticket.

27. **REVERSE ANNUITY MORTGAGES** Sinclair wishes to supplement his retirement income by $300/month for the next 10 years. He plans to obtain a reverse annuity mortgage (RAM) on his home to meet this need. Estimate the amount of the mortgage he will require if the prevailing interest rate is 5%/year compounded continuously.

28. **REVERSE ANNUITY MORTGAGE** Refer to Exercise 27. Leah wishes to supplement her retirement income by $400/month for the next 15 years by obtaining a RAM. Estimate the amount of the mortgage she will require if the prevailing interest rate is 6%/year compounded continuously.

29. **GINI INDEX** The income distribution of a certain country is described by the function

$$f(x) = 0.3x^{1.5} + 0.7x^{2.5}$$

Find the Gini Index for the country.

30. **GINI INDEX** A certain country's income distribution is described by the function

$$f(x) = \frac{e^{0.1x} - 1}{e^{0.1} - 1}$$

Find the Gini Index for the country.

31. **LORENZ CURVES** A certain country's income distribution is described by the function

$$f(x) = \frac{15}{16}x^2 + \frac{1}{16}x$$

   a. Sketch the Lorenz curve for this function.
   b. Compute $f(0.4)$ and $f(0.9)$ and interpret your results.

32. **LORENZ CURVES** A certain country's income distribution is described by the function

$$f(x) = \frac{14}{15}x^2 + \frac{1}{15}x$$

   a. Sketch the Lorenz curve for this function.
   b. Compute $f(0.3)$ and $f(0.7)$.

33. **LORENZ CURVES** In a study conducted by a certain country's Economic Development Board, it was found that the Lorenz curve for the distribution of income of college teachers was described by the function

$$f(x) = \frac{13}{14}x^2 + \frac{1}{14}x$$

and that of lawyers by the function

$$g(x) = \frac{9}{11}x^4 + \frac{2}{11}x$$

   a. Compute the coefficient of inequality for each Lorenz curve.
   b. Which profession has a more equitable income distribution?

34. LORENZ CURVES In a study conducted by a certain country's Economic Development Board, it was found that the Lorenz curve for the distribution of income of stockbrokers was described by the function

$$f(x) = \frac{11}{12}x^2 + \frac{1}{12}x$$

and that of high school teachers by the function

$$g(x) = \frac{5}{6}x^2 + \frac{1}{6}x$$

a. Compute the coefficient of inequality for each Lorenz curve.

b. Which profession has a more equitable income distribution?

6.7 Solution to Self-Check Exercise

We find the equilibrium price and equilibrium quantity by solving the system of equations

$$p = \sqrt{9 - 0.02x}$$
$$p = \sqrt{1 + 0.02x}$$

simultaneously. Substituting the first equation into the second, we have

$$\sqrt{9 - 0.02x} = \sqrt{1 + 0.02x}$$

Squaring both sides of the equation then leads to

$$9 - 0.02x = 1 + 0.02x$$
$$x = 200$$

Therefore,

$$p = \sqrt{9 - 0.02(200)}$$
$$= \sqrt{5} \approx 2.24$$

The equilibrium price is $224, and the equilibrium quantity is 200. The consumers' surplus is given by

$$CS \approx \int_0^{200} \sqrt{9 - 0.02x}\, dx - (2.24)(200)$$

$$= \int_0^{200} (9 - 0.02x)^{1/2}\, dx - 448$$

$$= -\frac{1}{0.02}\left(\frac{2}{3}\right)(9 - 0.02x)^{3/2}\Big|_0^{200} - 448 \qquad \text{Integrate by substitution.}$$

$$= -\frac{1}{0.03}(5^{3/2} - 9^{3/2}) - 448$$

$$\approx 79.32$$

or approximately $7932.

Next, the producers' surplus is given by

$$PS = (2.24)(200) - \int_0^{200} \sqrt{1 + 0.02x}\, dx$$

$$= 448 - \int_0^{200} (1 + 0.02x)^{1/2}\, dx$$

$$= 448 - \frac{1}{0.02}\left(\frac{2}{3}\right)(1 + 0.02x)^{3/2}\Big|_0^{200}$$

$$= 448 - \frac{1}{0.03}(5^{3/2} - 1)$$

$$\approx 108.66$$

or approximately $10,866.

USING TECHNOLOGY

Business and Economic Applications: Technology Exercises

1. Re-solve Example 1, Section 6.7, using a graphing utility.

Hint: Use the intersection operation to find the equilibrium quantity and the equilibrium price. Use the numerical integral operation to evaluate the definite integral.

2. Re-solve Exercise 11, Section 6.7, using a graphing utility.

Hint: See Exercise 1.

3. CONSUMERS' AND PRODUCERS' SURPLUS The demand function for a certain brand of travel alarm clocks is given by

$$p = -0.01x^2 - 0.3x + 10$$

where p is the unit price in dollars and x is the quantity demanded each month, measured in units of a thousand. The supply function for this brand of clocks is given by

$$p = -0.01x^2 + 0.2x + 4$$

where p has the same meaning as before and x is the quantity, in thousands, the supplier will make available in the marketplace per month. Determine the consumers' surplus and the producers' surplus when the market unit price is set at the equilibrium price.

4. **CONSUMERS' AND PRODUCERS' SURPLUS** The quantity demanded of a certain make of compact disc organizer is x thousand units per week when the corresponding unit price is

$$p = \sqrt{400 - 8x}$$

dollars. The supplier of the organizers will make x thousand units available in the market when the unit price is

$$p = 0.02x^2 + 0.04x + 5$$

dollars. Determine the consumers' surplus and the producers' surplus when the market unit price is set at the equilibrium price.

5. **RETURN ON INVESTMENTS** Investment A is expected to generate income at the rate of

$$R_1(t) = 50,000 + 10,000\sqrt{t}$$

dollars/year over 5 years, and Investment B is expected to generate income at the rate of

$$R_2(t) = 50,000 + 6000t$$

dollars/year over the same period of time. If the prevailing interest rate for the next 5 years is 10%/year, which investment will generate the higher net income over of 5 years?

6. **RETURN ON INVESTMENTS** Investment A is expected to generate income at the rate of

$$R_1(t) = 40,000 + 5000t + 100t^2$$

dollars/year for the next 10 years, and Investment B is expected to generate income at the rate of

$$R_2(t) = 60,000 + 2000t$$

dollars/year over the same period of time. If the prevailing interest rate for the next 10 years is 8%/year, which investment will generate the higher net income over the 10 years?

CHAPTER 6 Summary of Principal Formulas and Terms

FORMULAS

1. Indefinite integral of a constant	$\int k\, du = ku + C$
2. Power Rule	$\int u^n\, du = \dfrac{u^{n+1}}{n+1} + C \qquad (n \neq -1)$
3. Constant Multiple Rule	$\int k f(u)\, du = k \int f(u)\, du$ (k, a constant)

4. Sum Rule	$\int [f(u) \pm g(u)] \, du$ $= \int f(u) \, du \pm \int g(u) \, du$
5. Indefinite integral of the exponential function	$\int e^u \, du = e^u + C$
6. Indefinite integral of $f(u) = \dfrac{1}{u}$	$\int \dfrac{du}{u} = \ln|u| + C$
7. Method of substitution	$\int F'(g(x))g'(x) \, dx = \int F'(u) \, du$
8. Definite integral as the limit of a sum	$\int_a^b f(x) \, dx = \lim_{n \to \infty} S_n,$ where S_n is a Riemann sum
9. Fundamental Theorem of Calculus	$\int_a^b f(x) \, dx = F(b) - F(a),$ where $F'(x) = f(x)$
10. Average value of f over $[a, b]$	$\dfrac{1}{b-a} \int_a^b f(x) \, dx$
11. Area between two curves	$\int_a^b [f(x) - g(x)] \, dx,$ where $f(x) \geq g(x)$
12. Consumers' surplus	$CS = \int_0^{\bar{x}} D(x) \, dx - \bar{p}\,\bar{x}$
13. Producers' surplus	$PS = \bar{p}\,\bar{x} - \int_0^{\bar{x}} S(x) \, dx$
14. Accumulated (future) value of an income stream	$A = e^{rT} \int_0^T R(t)e^{-rt} \, dt$
15. Present value of an income stream	$PV = \int_0^T R(t)e^{-rt} \, dt$
16. Amount of an annuity	$A = \dfrac{mP}{r}\left(e^{rT} - 1\right)$
17. Present value of an annuity	$PV = \dfrac{mP}{r}\left(1 - e^{-rT}\right)$
18. Coefficient of inequality of a Lorenz curve	$L = 2 \int_0^1 [x - f(x)] \, dx$

TERMS

antiderivative (410)

antidifferentiation (412)

integration (412)

indefinite integral (412)

integrand (412)

constant of integration (412)

differential equation (416)

initial-value problem (416)

Riemann sum (438)

definite integral (439)

lower limit of integration (439)

upper limit of integration (439)

Lorenz curve (488)

line of complete equality (489)

CHAPTER 6   Concept Review Questions

Fill in the blanks.

1. a. A function F is an antiderivative of f on an interval I if _____ for all x in I.

b. If F is an antiderivative of f on an interval I, then every antiderivative of f on I has the form _____.

2. a. $\int cf(x)\, dx =$ _____

b. $\int [f(x) \pm g(x)]\, dx =$ _____

3. a. A differential equation is an equation that involves the derivative or differential of a/an _____ function.

b. A solution of a differential equation on an interval I is any _____ that satisfies the differential equation.

4. If we let $u = g(x)$, then $du =$ _____, and the substitution transforms the integral $\int f(g(x))g'(x)\, dx$ into the integral _____ involving only u.

5. a. If f is continuous and nonnegative on an interval $[a, b]$, then the area of the region under the graph of f on $[a, b]$ is given by _____.

b. If f is continuous on an interval $[a, b]$, then $\int_a^b f(x)\, dx$ is equal to the areas of the regions lying above the x-axis and bounded by the graph of f on $[a, b]$ _____ the areas of the regions lying below the x-axis and bounded by the graph of f on $[a, b]$.

6. a. The Fundamental Theorem of Calculus states that if f is continuous on $[a, b]$, then $\int_a^b f(x)\, dx =$ _____, where F is a/an _____ of f.

b. The net change in a function f over an interval $[a, b]$ is given by $f(b) - f(a) =$ _____, provided that f' is continuous on $[a, b]$.

7. a. If f is continuous on $[a, b]$, then the average value of f over $[a, b]$ is the number _____.

b. If f is a continuous and nonnegative function on $[a, b]$, then the average value of f over $[a, b]$ may be thought of as the _____ of the rectangle with base lying on the interval $[a, b]$ and having the same _____ as the region under the graph of f on $[a, b]$.

8. If f and g are continuous on $[a, b]$ and $f(x) \geq g(x)$ for all x in $[a, b]$, then the area of the region between the graphs of f and g and the vertical lines $x = a$ and $x = b$ is $A =$ _____.

9. a. The consumers' surplus is given by $CS =$ _____.

b. The producers' surplus is given by $PS =$ _____.

10. a. The accumulated value after T years of an income stream of $R(t)$ dollars/year, earning interest of r/year compounded continuously, is given by $A =$ _____.

b. The present value of an income stream is given by $PV =$ _____.

11. The amount of an annuity is $A =$ _____.

12. The coefficient of inequality of a Lorenz curve is $L =$ _____.

CHAPTER 6   Review Exercises

In Exercises 1–20, find each indefinite integral.

1. $\int (x^3 + 2x^2 - x)\, dx$ **2.** $\int \left(\frac{1}{3}x^3 - 2x^2 + 8\right) dx$

3. $\int \left(x^4 - 2x^3 + \frac{1}{x^2}\right) dx$ **4.** $\int (x^{1/3} - \sqrt{x} + 4)\, dx$

5. $\int x(2x^2 + x^{1/2})\, dx$ **6.** $\int (x^2 + 1)(\sqrt{x} - 1)\, dx$

7. $\int \left(x^2 - x + \frac{2}{x} + 5\right) dx$ **8.** $\int \sqrt{2x + 1}\, dx$

9. $\int (3x - 1)(3x^2 - 2x + 1)^{1/3}\, dx$

10. $\int x^2(x^3 + 2)^{10}\, dx$ **11.** $\int \frac{x - 1}{x^2 - 2x + 5}\, dx$

12. $\int 2e^{-2x}\, dx$ **13.** $\int \left(x + \frac{1}{2}\right) e^{x^2 + x + 1}\, dx$

14. $\int \frac{e^{-x} - 1}{(e^{-x} + x)^2}\, dx$ **15.** $\int \frac{(\ln x)^5}{x}\, dx$

16. $\int \frac{\ln x^2}{x}\, dx$ **17.** $\int x^3(x^2 + 1)^{10}\, dx$

18. $\int x\sqrt{x + 1}\, dx$ **19.** $\int \frac{x}{\sqrt{x - 2}}\, dx$

20. $\int \frac{3x}{\sqrt{x + 1}}\, dx$

In Exercises 21–32, evaluate each definite integral.

21. $\int_0^1 (2x^3 - 3x^2 + 1)\, dx$

22. $\int_0^2 (4x^3 - 9x^2 + 2x - 1)\, dx$

23. $\int_1^4 (\sqrt{x} + x^{-3/2})\, dx$ **24.** $\int_0^1 20x(2x^2 + 1)^4\, dx$

25. $\int_{-1}^0 12(x^2 - 2x)(x^3 - 3x^2 + 1)^3\, dx$

26. $\int_4^7 x\sqrt{x - 3}\, dx$ **27.** $\int_0^2 \frac{x}{x^2 + 1}\, dx$

28. $\int_0^1 \dfrac{dx}{(5-2x)^2}$

29. $\int_0^2 \dfrac{4x}{\sqrt{1+2x^2}}\, dx$

30. $\int_0^2 xe^{(-1/2)x^2}\, dx$

31. $\int_{-1}^0 \dfrac{e^{-x}}{(1+e^{-x})^2}\, dx$

32. $\int_1^e \dfrac{\ln x}{x}\, dx$

In Exercises 33–36, find the function f given that the slope of the tangent line to the graph at any point $(x, f(x))$ is $f'(x)$ and that the graph of f passes through the given point.

33. $f'(x) = 3x^2 - 4x + 1; (1, 1)$

34. $f'(x) = \dfrac{x}{\sqrt{x^2 + 1}}; (0, 1)$

35. $f'(x) = 1 - e^{-x}; (0, 2)$

36. $f'(x) = \dfrac{\ln x}{x}; (1, -2)$

In Exercises 37 and 38, refer to the following figure.

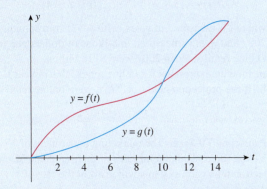

37. Motion of Two Cars Suppose $f(t)$ and $g(t)$ (both measured in feet per second) give the velocities of Car A and Car B, respectively, t seconds after starting side by side on a straight road.
  a. Give a physical interpretation of the integral

$$\int_0^T [f(t) - g(t)]\, dt \qquad (0 < t < 14)$$

  b. At what time is the distance between the two cars greatest? Write an integral giving this distance.

38. Revenues of Companies Suppose $f(t)$ and $g(t)$ give the rate of change of the revenues (measured in dollars per day) of two new branches, A and B, of an office supply company t weeks after their simultaneous grand opening.
  a. Give an interpretation of the integral

$$\int_0^T [f(t) - g(t)]\, dt \qquad (0 < t < 14)$$

  b. At what time over the first two weeks since opening is the difference between the revenues of the two branches greatest? Write an integral giving this difference.

39. Let $f(x) = -2x^2 + 1$, and compute the Riemann sum of f over the interval $[1, 2]$ by partitioning the interval into five subintervals of the same length ($n = 5$), where the points p_i ($1 \le i \le 5$) are taken to be the *right* endpoints of the respective subintervals.

40. Growth Rate of Smartphone Users The rate of growth of smartphone users, as a percentage of mobile phone users, is projected to be

$$R(t) = 10.8 \qquad (0 \le t \le 2)$$

percent/year. The percentage in October 2011 was 38.5. Find an expression giving the percentage of mobile phone users that were smartphone users in year t. What was the projected percentage of mobile phone users that were smartphone users in October 2013?
Source: eMarketer.

41. Computer Resale Value Franklin National Life Insurance Company purchased new computers for $200,000. If the rate at which the computers' resale value changes is given by the function

$$V'(t) = 3800(t - 10)$$

where t is the length of time since the purchase date and $V'(t)$ is measured in dollars per year, find an expression $V(t)$ that gives the resale value of the computers after t years. How much would the computers sell for after 6 years?

42. Marginal Cost Functions The management of National Electric has determined that the daily marginal cost function associated with producing their automatic drip coffeemakers is given by

$$C'(x) = 0.00003x^2 - 0.03x + 20$$

where $C'(x)$ is measured in dollars per unit and x denotes the number of units produced. Management has also determined that the daily fixed cost incurred in producing these coffeemakers is $500. What is the total cost incurred by National in producing the first 400 coffeemakers/day?

43. Marginal Revenue Functions Refer to Exercise 42. Management has also determined that the daily marginal revenue function associated with producing and selling their coffeemakers is given by

$$R'(x) = -0.03x + 60$$

where x denotes the number of units produced and sold and $R'(x)$ is measured in dollars per unit.
  a. Determine the revenue function $R(x)$ associated with producing and selling these coffeemakers.
  b. What is the demand equation relating the unit price to the quantity of coffeemakers demanded?

44. Measuring Temperature The temperature on a certain day as measured at the airport of a city is changing at the rate of

$$T'(t) = 0.15t^2 - 3.6t + 14.4 \qquad (0 \le t \le 4)$$

°F/hr, where t is measured in hours, with $t = 0$ corresponding to 6 A.M. The temperature at 6 A.M. was 24°F.
a. Find an expression giving the temperature T at the airport at any time between 6 A.M. and 10 A.M.
b. What was the temperature at 10 A.M.?

45. DVD SALES The total number of DVDs sold to U.S. dealers for rental and sale from 1999 through 2003 grew at the rate of approximately

$$R(t) = -0.03t^2 + 0.218t - 0.032 \qquad (0 \le t \le 4)$$

billion discs/year, where t is measured in years, with $t = 0$ corresponding to 1999. The total number of DVDs sold as of 1999 was 0.1 billion.
a. Find an expression giving the total number of DVDs sold by year t.
b. How many DVDs were sold in 2003?
Source: Adams Media.

46. AIR POLLUTION On an average summer day, the level of carbon monoxide (CO) in a city's air is 2 parts per million (ppm). An environmental protection agency's study predicts that, unless more stringent measures are taken to protect the city's atmosphere, the CO concentration present in the air will increase at the rate of

$$0.003t^2 + 0.06t + 0.1$$

ppm/year, t years from now. If no further pollution-control efforts are made, what will be the CO concentration on an average summer day 5 years from now?

47. MARGINAL COST FUNCTIONS The management of a division of Ditton Industries has determined that the daily marginal cost function associated with producing their hot-air corn poppers is given by

$$C'(x) = 0.00003x^2 - 0.03x + 10$$

where $C'(x)$ is measured in dollars/unit and x denotes the number of units manufactured. Management has also determined that the daily fixed cost incurred in producing these corn poppers is $600. Find the total cost incurred by Ditton in producing the first 500 corn poppers.

48. U.S. CENSUS The number of Americans aged 45–54 years (which stood at 25 million at the beginning of 1990) grew at the rate of

$$R(t) = 0.00933t^3 + 0.019t^2 - 0.10833t + 1.3467$$

million people/year, t years from the beginning of 1990. How many Americans aged 45 to 54 were added to the population between 1990 and 2000?
Source: U.S. Census Bureau.

49. COMMUTER TRENDS Because of the increasing cost of fuel, the manager of the City Transit Authority estimates that the number of commuters using the city subway system will increase at the rate of

$$3000(1 + 0.4t)^{-1/2} \qquad (0 \le t \le 36)$$

commuters per month, t months from now. If 100,000 commuters are currently using the system, find an

expression giving the total number of commuters who will be using the subway t months from now. How many commuters will be using the subway 6 months from now?

50. SUPPLY: WOMEN'S BOOTS The rate of change of the unit price p (in dollars) of Apex women's boots is given by

$$p'(x) = \frac{240}{(5 - x)^2}$$

where x is the number of pairs in units of a hundred that the supplier will make available in the market daily when the unit price is $\$p$/pair. Find the supply equation for these boots if the quantity the supplier is willing to make available is 200 pairs daily ($x = 2$) when the unit price is $50/pair.

51. ONLINE RETAIL SALES Since the inception of the Web, online commerce has enjoyed phenomenal growth. But growth, led by such major sectors as books, tickets, and office supplies, is expected to slow in the coming years. The projected growth of online retail sales is given by

$$R(t) = 15.82e^{-0.176t} \qquad (0 \le t \le 4)$$

where t is measured in years, with $t = 0$ corresponding to 2007 and $R(t)$ is measured in billions of dollars per year. Online retail sales in 2007 were $116 billion.
a. Find an expression for online retail sales in year t.
b. If the projection held true, what were online retail sales in 2011?
Source: Jupiter Research.

52. PROJECTION TV SALES The marketing department of Vista Vision forecasts that total sales of their new line of projection television systems will grow at the rate of

$$3000 - 2000e^{-0.04t} \qquad (0 \le t \le 24)$$

systems/month once they are introduced into the market. Find an expression giving the total number of the projection television systems that Vista may expect to sell in the t months after they are put on the market. How many of these systems can Vista expect to sell during the first year?

53. SALES: LOUDSPEAKERS Sales of the Acrosonic model F loudspeaker systems have been growing at the rate of

$$f'(t) = 2000(3 - 2e^{-t})$$

systems/year, where t denotes the number of years these loudspeaker systems have been on the market. Determine the number of loudspeaker systems that were sold in the first 5 years after they appeared on the market.

54. Find the area of the region under the curve $y = 3x^2 + 2x + 1$ from $x = -1$ to $x = 2$.

55. Find the area of the region under the curve $y = e^{2x}$ from $x = 0$ to $x = 2$.

56. Find the area of the region bounded by the graph of the function $y = 1/x^2$, the x-axis, and the lines $x = 1$ and $x = 3$.

57. Find the area of the region bounded by the curve $y = -x^2 - x + 2$ and the x-axis.

58. Find the area of the region bounded by the graphs of the functions $f(x) = e^x$ and $g(x) = x$ and the vertical lines $x = 0$ and $x = 2$.

59. Find the area of the region that is completely enclosed by the graphs of $f(x) = x^4$ and $g(x) = x$.

60. Find the area of the region between the curve $y = x(x - 1)(x - 2)$ and the x-axis.

61. OIL PRODUCTION On the basis of current production techniques, the rate of oil production from a certain oil well t years from now is estimated to be

$$R_1(t) = 100e^{0.05t}$$

thousand barrels/year. On the basis of a new production technique, however, it is estimated that the rate of oil production from that oil well t years from now will be

$$R_2(t) = 100e^{0.08t}$$

thousand barrels/year. Determine how much additional oil will be produced over the next 10 years if the new technique is adopted.

62. Find the average value of the function

$$f(x) = \frac{x}{\sqrt{x^2 + 16}}$$

over the interval $[0, 3]$.

63. AVERAGE TEMPERATURE The temperature (in degrees Fahrenheit) in Boston over a 12-hr period on a certain December day was given by

$$T = -0.05t^3 + 0.4t^2 + 3.8t + 5.6 \qquad (0 \le t \le 12)$$

where t is measured in hours, with $t = 0$ corresponding to 6 A.M. Determine the average temperature on that day over the 12-hr period from 6 A.M. to 6 P.M.

64. AVERAGE VELOCITY OF A TRUCK A truck traveling along a straight road has a velocity (in feet per second) at time t (in seconds) given by

$$v(t) = \frac{1}{12}t^2 + 2t + 44 \qquad (0 \le t \le 5)$$

What is the average velocity of the truck over the time interval from $t = 0$ to $t = 5$?

65. MEMBERSHIP IN CREDIT UNIONS The membership in Massachusetts credit unions grew at the rate of

$$R(t) = -0.0039t^2 + 0.0374t + 0.0046 \qquad (0 \le t \le 9)$$

million members/year in year t between 1994 ($t = 0$) and 2003 ($t = 9$). Find the average rate of growth of membership in Massachusetts credit unions over the period in question.

Source: Massachusetts Credit Union League.

66. DEMAND FOR DIGITAL CAMCORDER TAPES The demand function for a brand of blank digital camcorder tapes is given by

$$p = -0.01x^2 - 0.2x + 23$$

where p is the unit price in dollars and x is the quantity demanded each week, measured in units of a thousand. Determine the consumers' surplus if the unit price is $8/tape.

67. CONSUMERS' AND PRODUCERS' SURPLUS The quantity demanded x (in units of a hundred) of the Sportsman 5×7 tents, per week, is related to the unit price p (in dollars) by the relation

$$p = -0.1x^2 - x + 40$$

The quantity x (in units of a hundred) that the supplier is willing to make available in the market is related to the unit price by the relation

$$p = 0.1x^2 + 2x + 20$$

If the market price is set at the equilibrium price, find the consumers' surplus and the producers' surplus.

68. RETIREMENT ACCOUNT SAVINGS Chi-Tai plans to deposit $4000/year in his defined benefit retirement account. If interest is compounded continuously at the rate of 8%/year, how much will he have in his retirement account after 20 years?

69. INSTALLMENT CONTRACTS Glenda sold her house under an installment contract whereby the buyer gave her a down payment of $20,000 and agreed to make monthly payments of $925/month for 30 years. If the prevailing interest rate is 6%/year compounded continuously, find the present value of the purchase price of the house.

70. PRESENT VALUE OF A FRANCHISE Alicia purchased a 10-year franchise for a health spa that is expected to generate income at the rate of

$$R(t) = 80,000$$

dollars/year. If the prevailing interest rate is 10%/year compounded continuously, find the present value of the franchise.

71. INCOME DISTRIBUTION OF A COUNTRY A certain country's income distribution is described by the function

$$f(x) = \frac{17}{18}x^2 + \frac{1}{18}x$$

a. Sketch the Lorenz curve for this function.
b. Compute $f(0.3)$ and $f(0.6)$, and interpret your results.
c. Compute the coefficient of inequality for this Lorenz curve.

72. POPULATION GROWTH The population of a certain Sunbelt city, currently 80,000, is expected to grow exponentially in the next 5 years with a growth constant of 0.05. If the prediction comes true, what will be the average population of the city over the next 5 years?

CHAPTER 6 Before Moving On . . .

1. Find $\int \left(2x^3 + \sqrt{x} + \dfrac{2}{x} - \dfrac{2}{\sqrt{x}} \right) dx$.

2. Find f if $f'(x) = e^x + x$ and $f(0) = 2$.

3. Find $\int \dfrac{x}{\sqrt{x^2 + 1}} \, dx$.

4. Evaluate $\int_0^1 x\sqrt{2 - x^2} \, dx$.

5. Find the area of the region completely enclosed by the graphs of $y = x^2 - 1$ and $y = 1 - x$.

7

Additional Topics in Integration

BESIDES THE BASIC rules of integration developed in Chapter 6, there are more sophisticated techniques for finding the antiderivatives of functions. We begin this chapter by looking at the method of integration by parts. We then look at a technique of integration that involves using tables of integrals. We also look at numerical methods of integration, which enable us to obtain approximations to definite integrals, especially those whose exact value cannot be found. Numerical integration methods are especially useful when the integrand is known only at discrete points. We also learn how to evaluate integrals in which the intervals of integration are unbounded. Such integrals, called *improper integrals,* play an important role in the study of probability. Finally, we see how the definite integral can be used to help us find the volume of a solid of revolution.

How much money is needed to establish an endowment for a scholarship fund that will pay $5000 annually? In Example 6, page 538, you will learn how to determine the amount required.

© Monkey Business Images/Shutterstock.com

7.1　Integration by Parts

The Method of Integration by Parts

Integration by parts is another technique of integration that, like the method of substitution discussed in Chapter 6, is based on a corresponding rule of differentiation. In this case, the rule of differentiation is the Product Rule, which asserts that if f and g are differentiable functions, then

$$\frac{d}{dx}[f(x)g(x)] = f(x)g'(x) + g(x)f'(x) \qquad (1)$$

If we integrate both sides of Equation (1) with respect to x, we obtain

$$\int \frac{d}{dx}f(x)g(x)\,dx = \int f(x)g'(x)\,dx + \int g(x)f'(x)\,dx$$

$$f(x)g(x) = \int f(x)g'(x)\,dx + \int g(x)f'(x)\,dx$$

This last equation, which may be written in the form

$$\int f(x)g'(x)\,dx = f(x)g(x) - \int g(x)f'(x)\,dx \qquad (2)$$

is called the formula for **integration by parts.** This formula is useful because it enables us to express one indefinite integral in terms of another, which may be easier to evaluate. Formula (2) may be simplified by letting

$$u = f(x) \qquad dv = g'(x)\,dx$$
$$du = f'(x)\,dx \qquad v = g(x)$$

giving the following version of the formula for integration by parts.

> **Integration by Parts Formula**
>
> $$\int u\,dv = uv - \int v\,du \qquad (3)$$

EXAMPLE 1 Evaluate $\int xe^x\,dx$.

Solution　No method of integration developed thus far enables us to evaluate the given indefinite integral in its present form. Therefore, we attempt to write it in terms of an indefinite integral that will be easier to evaluate. Let's use the integration by parts Formula (3) by letting

$$u = x \quad \text{and} \quad dv = e^x\,dx$$

so that

$$du = dx \quad \text{and} \quad v = e^x$$

Therefore,

$$\int xe^x\,dx = \int u\,dv$$
$$= uv - \int v\,du$$
$$= xe^x - \int e^x\,dx$$
$$= xe^x - e^x + C$$
$$= (x-1)e^x + C$$

The success of the method of integration by parts depends on the proper choice of u and dv. For example, if we had chosen

$$u = e^x \quad \text{and} \quad dv = x \, dx$$

in Example 1, then

$$du = e^x \, dx \quad \text{and} \quad v = \frac{1}{2} x^2$$

Thus, Formula (3) would have yielded

$$\int xe^x \, dx = \int u \, dv$$
$$= uv - \int v \, du$$
$$= \frac{1}{2} x^2 e^x - \int \frac{1}{2} x^2 e^x \, dx$$

Since the indefinite integral on the right-hand side of this equation is not readily evaluated (it is in fact more complicated than the original integral!), choosing u and dv as shown has not helped us to evaluate the given indefinite integral.

In general, we can use the following guidelines.

Guidelines for Choosing u and dv

Choose u and dv so that

1. du is simpler than u.
2. dv is easy to integrate.

EXAMPLE 2 Evaluate $\int x \ln x \, dx$.

Solution Letting

$$u = \ln x \quad \text{and} \quad dv = x \, dx$$

we have

$$du = \frac{1}{x} \, dx \quad \text{and} \quad v = \frac{1}{2} x^2$$

Therefore,

$$\int x \ln x \, dx = \int u \, dv = uv - \int v \, du$$
$$= \frac{1}{2} x^2 \ln x - \int \frac{1}{2} x^2 \cdot \left(\frac{1}{x} \right) dx$$
$$= \frac{1}{2} x^2 \ln x - \frac{1}{2} \int x \, dx$$
$$= \frac{1}{2} x^2 \ln x - \frac{1}{4} x^2 + C$$
$$= \frac{1}{4} x^2 (2 \ln x - 1) + C$$

EXAMPLE 3 Evaluate $\displaystyle\int \frac{xe^x}{(x+1)^2}\,dx$.

Solution Let

$$u = xe^x \quad \text{and} \quad dv = \frac{1}{(x+1)^2}\,dx$$

Then

$$du = (xe^x + e^x)\,dx = e^x(x+1)\,dx \quad \text{and} \quad v = -\frac{1}{x+1}$$

Therefore,

$$\int \frac{xe^x}{(x+1)^2}\,dx = \int u\,dv = uv - \int v\,du$$

$$= xe^x\left(-\frac{1}{x+1}\right) - \int \left(-\frac{1}{x+1}\right)e^x(x+1)\,dx$$

$$= -\frac{xe^x}{x+1} + \int e^x\,dx$$

$$= -\frac{xe^x}{x+1} + e^x + C$$

$$= \frac{e^x}{x+1} + C \qquad \text{\color{red}{Combine first two terms.}}$$

As illustrated in the next example, we may need to apply the technique of integration by parts more than once to evaluate an integral.

EXAMPLE 4 Find $\displaystyle\int x^2 e^x\,dx$.

Solution Let

$$u = x^2 \quad \text{and} \quad dv = e^x\,dx$$

Then

$$du = 2x\,dx \quad \text{and} \quad v = e^x$$

Therefore,

$$\int x^2 e^x\,dx = \int u\,dv = uv - \int v\,du$$

$$= x^2 e^x - \int e^x(2x)\,dx = x^2 e^x - 2\int xe^x\,dx$$

To complete the solution of the problem, we need to evaluate the integral

$$\int xe^x\,dx$$

But this integral may be found by using integration by parts. In fact, you will recognize this integral as precisely that of Example 1. Using the results obtained there, we now find

$$\int x^2 e^x\,dx = x^2 e^x - 2\left[(x-1)e^x\right] + C = e^x(x^2 - 2x + 2) + C$$

Explore and Discuss

1. Use the method of integration by parts to derive the formula

$$\int x^n e^{ax}\, dx = \frac{1}{a} x^n e^{ax} - \frac{n}{a} \int x^{n-1} e^{ax}\, dx$$

where n is a positive integer and a is a real number.

2. Use the formula of part 1 to evaluate

$$\int x^3 e^x\, dx$$

Hint: You may find the results of Example 4 helpful.

 APPLIED EXAMPLE 5 Oil Production The estimated rate at which oil will be produced from a certain oil well t years after production has begun is given by

$$R(t) = 100te^{-0.1t} \qquad (t \geq 0)$$

thousand barrels per year. Find an expression that describes the total production of oil at the end of year t.

Solution Let $T(t)$ denote the total production of oil from the well at the end of year t. Then the rate of oil production will be given by $T'(t)$ thousand barrels per year. Thus,

$$T'(t) = R(t) = 100te^{-0.1t}$$

so

$$T(t) = \int 100te^{-0.1t}\, dt$$
$$= 100 \int te^{-0.1t}\, dt$$

We use the technique of integration by parts to evaluate this integral. Let

$$u = t \quad \text{and} \quad dv = e^{-0.1t}\, dt$$

so that

$$du = dt \quad \text{and} \quad v = -\frac{1}{0.1} e^{-0.1t} = -10e^{-0.1t}$$

Therefore,

$$T(t) = 100\left[-10te^{-0.1t} + 10 \int e^{-0.1t}\, dt \right]$$
$$= 100(-10te^{-0.1t} - 100e^{-0.1t}) + C$$
$$= -1000e^{-0.1t}(t + 10) + C \qquad (x^2) \text{ See page 10.}$$

To determine the value of C, note that the total quantity of oil produced at the end of year 0 is nil, so $T(0) = 0$. This gives

$$T(0) = -1000(10) + C = 0$$
$$C = 10{,}000$$

Thus, the required production function is given by

$$T(t) = -1000e^{-0.1t}(t + 10) + 10{,}000$$

The integration by parts formula is easily adapted for definite integrals by observing that

$$\int_a^b u \, dv = uv \Big|_a^b - \int_a^b v \, du \tag{4}$$

We illustrate the use of this formula in Example 6.

EXAMPLE 6 Find the area of the region under the graph of $f(x) = \dfrac{x}{\sqrt{x + 5}}$ on $[0, 4]$.

Solution The region is shown in Figure 1. Since $f(x) \geq 0$ on $[0, 4]$, we see that the required area is given by

$$A = \int_0^4 \frac{x}{\sqrt{x + 5}} \, dx = \int_0^4 x(x + 5)^{-1/2} \, dx$$

To evaluate this integral, we use Formula (4). Let

$$u = x \quad \text{and} \quad dv = (x + 5)^{-1/2} \, dx$$

Then

$$du = dx \quad \text{and} \quad v = 2(x + 5)^{1/2}$$

Therefore,

$$A = \int_0^4 x(x + 5)^{-1/2} \, dx = x(2)(x + 5)^{1/2} \Big|_0^4 - \int_0^4 2(x + 5)^{1/2} \, dx$$

$$= 2x\sqrt{x + 5} \Big|_0^4 - \frac{4}{3}(x + 5)^{3/2} \Big|_0^4$$

$$= [2(4)\sqrt{4 + 5} - 0] - \frac{4}{3}[(4 + 5)^{3/2} - 5^{3/2}]$$

$$= 24 - \frac{4}{3}(27 - 5\sqrt{5}) = \frac{4}{3}(5\sqrt{5} - 9)$$

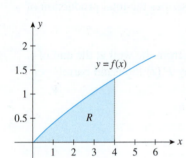

FIGURE 1
The region under the graph of
$y = \dfrac{x}{\sqrt{x + 5}}$ on $[0, 4]$

7.1 Self-Check Exercises

1. Evaluate $\displaystyle\int x^2 \ln x \, dx$.

2. **COMMUTER AIRLINE GROWTH** Since the inauguration of Ryan's Express at the beginning of 2012, the number of passengers (in millions) flying on this commuter airline has been growing at the rate of

$$R(t) = 0.1 + 0.2te^{-0.4t}$$

passengers/year ($t = 0$ corresponds to the beginning of 2012). Assuming that this trend continues through 2016, determine how many passengers will have flown on Ryan's Express by that time.

Solutions to Self-Check Exercises 7.1 can be found on page 509.

7.1 Concept Questions

1. Write the formula for integration by parts.

2. Explain how you would choose u and dv when using the integration by parts formula. Illustrate your answer with $\int x^2 e^{-x}\,dx$. What happens if you reverse your choices of u and dv?

7.1 Exercises

In Exercises 1–28, find each indefinite integral.

1. $\displaystyle\int xe^{2x}\,dx$

2. $\displaystyle\int xe^{-x}\,dx$

3. $\displaystyle\int \frac{1}{2}xe^{x/4}\,dx$

4. $\displaystyle\int 6xe^{3x}\,dx$

5. $\displaystyle\int (e^x - x)^2\,dx$

6. $\displaystyle\int (e^{-x} + x)^2\,dx$

7. $\displaystyle\int (x + 1)e^x\,dx$

8. $\displaystyle\int (x - 3)e^{3x}\,dx$

9. $\displaystyle\int x(x + 1)^{-3/2}\,dx$

10. $\displaystyle\int x(x + 4)^{-2}\,dx$

11. $\displaystyle\int x\sqrt{x - 5}\,dx$

12. $\displaystyle\int \frac{3x}{\sqrt{2x + 3}}\,dx$

13. $\displaystyle\int x \ln 2x\,dx$

14. $\displaystyle\int x^2 \ln 2x\,dx$

15. $\displaystyle\int x^3 \ln x\,dx$

16. $\displaystyle\int \sqrt{x} \ln x\,dx$

17. $\displaystyle\int \sqrt{x} \ln \sqrt{x}\,dx$

18. $\displaystyle\int \frac{\ln x}{\sqrt{x}}\,dx$

19. $\displaystyle\int \frac{\ln x}{x^2}\,dx$

20. $\displaystyle\int \frac{\ln x}{x^3}\,dx$

21. $\displaystyle\int (x + 1)^2 e^x\,dx$

22. $\displaystyle\int \ln(xe^{x^2})\,dx$

23. $\displaystyle\int \ln x\,dx$
Hint: Let $u = \ln x$ and $dv = dx$.

24. $\displaystyle\int \ln(x + 1)\,dx$

25. $\displaystyle\int x^2 e^{-x}\,dx$
Hint: Integrate by parts twice.

26. $\displaystyle\int e^{-\sqrt{x}}\,dx$
Hint: First, make the substitution $u = \sqrt{x}$; then integrate by parts.

27. $\displaystyle\int x(\ln x)^2\,dx$
Hint: Integrate by parts twice.

28. $\displaystyle\int x \ln(x + 1)\,dx$
Hint: First, make the substitution $u = x + 1$; then integrate by parts.

In Exercises 29–34, evaluate each definite integral by using the method of integration by parts.

29. $\displaystyle\int_0^{\ln 3} xe^x\,dx$

30. $\displaystyle\int_0^2 xe^{-x}\,dx$

31. $\displaystyle\int_1^4 3 \ln x\,dx$

32. $\displaystyle\int_1^2 x \ln x\,dx$

33. $\displaystyle\int_0^2 xe^{2x}\,dx$

34. $\displaystyle\int_0^1 x^2 e^{-x}\,dx$

35. Find the function f given that the slope of the tangent line to the graph of f at any point $(x, f(x))$ is xe^{-2x} and that the graph passes through the point $(0, 3)$.

36. Find the function f given that the slope of the tangent line to the graph of f at any point $(x, f(x))$ is $x\sqrt{x + 1}$ and that the graph passes through the point $(3, 6)$.

37. Find the area of the region under the graph of $f(x) = \ln x$ from $x = 1$ to $x = 5$.

38. Find the area of the region under the graph of $f(x) = xe^{-x}$ from $x = 0$ to $x = 3$.

39. VELOCITY OF A DRAGSTER The velocity of a dragster t sec after leaving the starting line is

$$100te^{-0.2t}$$

ft/sec. What is the distance covered by the dragster in the first 10 sec of its run?

40. PRODUCTION OF STEAM COAL In keeping with the projected increase in worldwide demand for steam coal, the boiler-firing fuel used for generating electricity, the management of Consolidated Mining has decided to step up its mining operations. Plans call for increasing the yearly production of steam coal over its current level by

$$2te^{-0.05t}$$

million metric tons/year for the next 20 years. The current yearly production is 20 million metric tons. Find a function that describes Consolidated's total production of steam coal at the end of t years. How much coal will Consolidated have produced over the next 20 years if this plan is carried out?

41. COMPACT DISC SALES Sales of the latest recording by Brittania, a British rock group, are currently $2te^{-0.1t}$ units/week (each unit representing 10,000 CDs), where t denotes the number of weeks since the recording's release. Find an expression that gives the total number of CDs sold as a function of t.

42. CONCENTRATION OF A DRUG IN THE BLOODSTREAM The concentration (in milligrams per milliliter) of a certain drug in a patient's bloodstream t hr after it has been administered is given by $C(t) = 3te^{-t/3}$ mg/mL. Find the average concentration of the drug in the patient's bloodstream over the first 12 hr after administration.

43. **ALCOHOL-RELATED TRAFFIC ACCIDENTS** As a result of increasingly stiff laws aimed at reducing the number of alcohol-related traffic accidents in a certain state, preliminary data indicate that the number of such accidents has been changing at the rate of

$$R(t) = -10 - te^{0.1t}$$

accidents/month t months after the laws took effect. There were 982 alcohol-related accidents for the year before the enactment of the laws. Determine how many alcohol-related accidents were expected during the first year the laws were in effect.

44. **AVERAGE PRICE OF A COMMODITY** The price of a certain commodity in dollars per unit at time t (measured in weeks) is given by

$$p = 8 + 4e^{-2t} + te^{-2t}$$

What is the average price of the commodity over the 4-week period from $t = 0$ to $t = 4$?

45. **GROWTH OF HMOs** The membership of the Cambridge Community Health Plan (a health maintenance organization) is projected to grow at the rate of

$$9\sqrt{t + 1} \ln \sqrt{t + 1}$$

thousand people/year, t years from now. If the HMO's current membership is 50,000, what will be the membership 5 years from now?

46. **DISTANCE COVERED BY A CAR** The velocity of the Zephyr electric sedan traveling along a straight road is

$$v(t) = 8t \ln(t + 1) \qquad (0 \le t \le 6)$$

ft/sec t sec after starting from rest. What is the total distance covered by the car at the end of 6 sec?

47. **CREDIT CARD LOSSES IN THE UNITED KINGDOM** The mail and the Internet are major routes for fraud against merchants who sell and ship products. These merchants must often accept credit cards that are not physically present (called CNPs, "credit cards not present"). The amount of CNP fraud in the United Kingdom from 2008 through 2012 was

$$f(t) = 328.9 - 92.07 \ln(t + 1) \qquad (0 \le t \le 4)$$

million British pounds (GBP) in year t, where $t = 0$ corresponds to 2008. What was the average amount of CNP fraud losses in the United Kingdom over the period from 2008 through 2012?
Source: Financial Fraud Action UK.

48. **U.S. SMARTPHONE USERS** Smartphone users, as a percentage of the population, are projected to be

$$f(t) = 18.952 + 14.088 \ln(t + 1) \qquad (0 \le t \le 5)$$

percent in year t, where t is measured in years, with $t = 0$ corresponding to 2010. What was the average percentage

of the population that were smartphone users over the period from 2010 through 2015?
Source: eMarketer.

49. **FUTURE VALUE OF AN INCOME STREAM** The chef-owner of the Vegan Restaurant expects that in the first 5 years of its operation, revenue will be generated at the rate of $R(t) = 100 + 20t$ thousand dollars/year, where t is the number of years after the grand opening of the restaurant. If the prevailing rate of interest for the period is 5%/year compounded continuously, what will be the value of the restaurant's revenue stream at the end of 5 years?
Hint: Use Formula (18), Section 6.7 (page 484).

50. **FUTURE VALUE OF AN INVESTMENT** Laura, the sole proprietor of Laura's Boutique, plans to invest $R(t) = 10,000e^{-0.02t}$ dollars/year into her savings account for 5 years, starting now. If the bank pays interest at the rate of 3%/year compounded continuously, how much will Laura have in her account at the end of the 5-year period?

51. **RATE OF RETURN ON AN INVESTMENT** Suppose an investment is expected to generate income at the rate of

$$P(t) = 30,000 + 800t$$

dollars/year for the next 5 years. Find the present value of the income from this investment if the prevailing interest rate is 5%/year compounded continuously.
Hint: Use Formula (19), Section 6.7 (page 485).

52. **PRESENT VALUE OF A FRANCHISE** Tracy purchased a 15-year franchise for a computer outlet store that is expected to generate income at the rate of

$$P(t) = 50,000 + 3000t$$

dollars/year. If the prevailing interest rate is 4%/year compounded continuously, find the present value of the income from the franchise.
Hint: Use Formula (19), Section 6.7 (page 485).

53. **CONSUMERS' SURPLUS FOR INFLATABLE CANOES** Refer to Section 6.7. The demand function for Sea Hawk inflatable canoes is given by

$$p = 300 - 2(x + 1)\ln(x + 1) \qquad (x \ge 0)$$

where p is the unit price in dollars and x (in units of a hundred) is the quantity demanded. Estimate the consumers' surplus if the quantity demanded is set at 2000 units.

54. **PRODUCERS' SURPLUS FOR LAPTOP LOCKS** Refer to Section 6.7. The supply function for a certain brand of laptop lock is given by

$$p = x \ln(0.1x^2 + 1) + 5 \qquad (0 \le x \le 10)$$

where p is the unit price in dollars and x is the corresponding quantity that the manufacturers will make available in the market, measured in units of a thousand. Find the producers' surplus if the quantity demanded is 7000.

55. GINI INDEX Refer to Section 6.7. The income distribution of a country is described by the Lorenz function

$$f(x) = xe^{2(x-1)} \qquad (0 \le x \le 1)$$

Find the Gini Index for the income distribution of this country.

56. DIFFUSION A cylindrical membrane with inner radius r_1 cm and outer radius r_2 cm containing a chemical solution is introduced into a salt bath with constant concentration c_2 moles/L (see the accompanying figure).

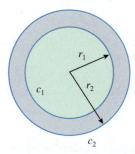

If the concentration of the chemical inside the membrane is kept constant at a different concentration of c_1 moles/L, then the concentration of the chemical across the membrane will be given by

$$c(r) = \left(\frac{c_1 - c_2}{\ln r_1 - \ln r_2}\right)(\ln r - \ln r_2) + c_2 \qquad (r_1 < r < r_2)$$

moles/L. Find the average concentration of the chemical across the membrane from $r = r_1$ to $r = r_2$.

57. A MIXTURE PROBLEM Two tanks are connected in tandem as shown in the following figure. Each tank contains 60 gal of water. Starting at time $t = 0$, brine containing 3 lb/gal of salt flows into tank 1 at the rate of 2 gal/min. The mixture then enters and leaves tank 2 at the same rate. The mixtures in both tanks are stirred uniformly. It can be shown that the amount of salt in tank 2 after t min is given by

$$A(t) = 180(1 - e^{-t/30}) - 6te^{-t/30}$$

where $A(t)$ is measured in pounds.

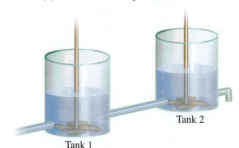

Tank 2

Tank 1

a. What is the initial amount of salt in tank 2?
b. What is the amount of salt in tank 2 after 3 hr (180 min)?
c. What is the average amount of salt in tank 2 over the first 3 hr?

58. Suppose f'' is continuous on $[1, 3]$ and $f(1) = 2$, $f(3) = -1, f'(1) = 2$, and $f'(3) = -1$. Evaluate $\int_1^3 xf''(x)\, dx$.

In Exercises 59–62, determine whether the statement is true or false. If it is true, explain why it is true. If it is false, give an example to show why it is false.

59. $\displaystyle\int u\, dv + \int v\, du = uv$

60. $\displaystyle\int e^x g'(x)\, dx = e^x g(x) - \int e^x g(x)\, dx$

61. $\displaystyle\int uv\, dw = uvw - \int uw\, dv - \int vw\, du$

62. $\displaystyle\int_a^b u\, dv = uv - \int_a^b v\, du$

7.1 Solutions to Self-Check Exercises

1. Let $u = \ln x$ and $dv = x^2\, dx$, so that $du = \dfrac{1}{x}\, dx$ and $v = \dfrac{1}{3}x^3$. Therefore,

$$\int x^2 \ln x\, dx = \int u\, dv = uv - \int v\, du$$

$$= \frac{1}{3}x^3 \ln x - \int \frac{1}{3}x^2\, dx$$

$$= \frac{1}{3}x^3 \ln x - \frac{1}{9}x^3 + C$$

$$= \frac{1}{9}x^3(3 \ln x - 1) + C$$

2. Let $N(t)$ denote the total number of passengers who will have flown on Ryan's Express by the end of year t. Then $N'(t) = R(t)$, so

$$N(t) = \int R(t)\, dt$$

$$= \int (0.1 + 0.2te^{-0.4t})\, dt$$

$$= \int 0.1\, dt + 0.2 \int te^{-0.4t}\, dt$$

We now use the technique of integration by parts on the second integral. Letting $u = t$ and $dv = e^{-0.4t}\, dt$, we have

$$du = dt \quad \text{and} \quad v = -\frac{1}{0.4}e^{-0.4t} = -2.5e^{-0.4t}$$

Therefore,

$$N(t) = 0.1t + 0.2\left(-2.5te^{-0.4t} + 2.5\int e^{-0.4t}\,dt\right)$$

$$= 0.1t - 0.5te^{-0.4t} - \frac{0.5}{0.4}e^{-0.4t} + C$$

$$= 0.1t - 0.5(t + 2.5)e^{-0.4t} + C$$

To determine the value of C, note that $N(0) = 0$, which gives

$$N(0) = -0.5(2.5) + C = 0$$

$$C = 1.25$$

Therefore,

$$N(t) = 0.1t - 0.5(t + 2.5)e^{-0.4t} + 1.25$$

The number of passengers who will have flown on Ryan's Express by the end of 2016 is given by

$$N(5) = 0.1(5) - 0.5(5 + 2.5)e^{-0.4(5)} + 1.25$$

$$= 1.242493$$

that is, 1,242,493 passengers.

7.2 Integration Using Tables of Integrals

A Table of Integrals

We have studied several techniques for finding an antiderivative of a function. However, useful as they are, these techniques are not always applicable. There are numerous other methods for finding an antiderivative of a function. Extensive lists of integration formulas have been compiled based on these methods.

A small sample of the integration formulas that can be found in many mathematical handbooks is given in the following table of integrals. The formulas are grouped according to the basic form of the integrand. Note that it may be necessary to modify the integrand of the integral to be evaluated in order to use one of these formulas.

TABLE OF INTEGRALS

Forms Involving $a + bu$

1. $\displaystyle\int \frac{u\,du}{a + bu} = \frac{1}{b^2}(a + bu - a\ln|a + bu|) + C$

2. $\displaystyle\int \frac{u^2\,du}{a + bu} = \frac{1}{2b^3}[(a + bu)^2 - 4a(a + bu) + 2a^2\ln|a + bu|] + C$

3. $\displaystyle\int \frac{u\,du}{(a + bu)^2} = \frac{1}{b^2}\left(\frac{a}{a + bu} + \ln|a + bu|\right) + C$

4. $\displaystyle\int u\sqrt{a + bu}\,du = \frac{2}{15b^2}(3bu - 2a)(a + bu)^{3/2} + C$

5. $\displaystyle\int \frac{u\,du}{\sqrt{a + bu}} = \frac{2}{3b^2}(bu - 2a)\sqrt{a + bu} + C$

6. $\displaystyle\int \frac{du}{u\sqrt{a + bu}} = \frac{1}{\sqrt{a}}\ln\left|\frac{\sqrt{a + bu} - \sqrt{a}}{\sqrt{a + bu} + \sqrt{a}}\right| + C$ (if $a > 0$)

Forms Involving $\sqrt{a^2 + u^2},\, a > 0$

7. $\displaystyle\int \sqrt{a^2 + u^2}\,du = \frac{u}{2}\sqrt{a^2 + u^2} + \frac{a^2}{2}\ln|u + \sqrt{a^2 + u^2}| + C$

8. $\displaystyle\int u^2\sqrt{a^2 + u^2}\,du = \frac{u}{8}(a^2 + 2u^2)\sqrt{a^2 + u^2} - \frac{a^4}{8}\ln|u + \sqrt{a^2 + u^2}| + C$

9. $\displaystyle\int \frac{du}{\sqrt{a^2 + u^2}} = \ln|u + \sqrt{a^2 + u^2}| + C$

TABLE OF INTEGRALS (*continued*)

Forms Involving $\sqrt{a^2 + u^2}, a > 0$ (*continued*)

10. $\displaystyle\int \frac{du}{u\sqrt{a^2 + u^2}} = -\frac{1}{a}\ln\left|\frac{\sqrt{a^2 + u^2} + a}{u}\right| + C$

11. $\displaystyle\int \frac{du}{u^2\sqrt{a^2 + u^2}} = -\frac{\sqrt{a^2 + u^2}}{a^2 u} + C$

12. $\displaystyle\int \frac{du}{(a^2 + u^2)^{3/2}} = \frac{u}{a^2\sqrt{a^2 + u^2}} + C$

Forms Involving $\sqrt{u^2 - a^2}, a > 0$

13. $\displaystyle\int \sqrt{u^2 - a^2}\, du = \frac{u}{2}\sqrt{u^2 - a^2} - \frac{a^2}{2}\ln|u + \sqrt{u^2 - a^2}| + C$

14. $\displaystyle\int u^2\sqrt{u^2 - a^2}\, du = \frac{u}{8}(2u^2 - a^2)\sqrt{u^2 - a^2} - \frac{a^4}{8}\ln|u + \sqrt{u^2 - a^2}| + C$

15. $\displaystyle\int \frac{\sqrt{u^2 - a^2}}{u^2}\, du = -\frac{\sqrt{u^2 - a^2}}{u} + \ln|u + \sqrt{u^2 - a^2}| + C$

16. $\displaystyle\int \frac{du}{\sqrt{u^2 - a^2}} = \ln|u + \sqrt{u^2 - a^2}| + C$

17. $\displaystyle\int \frac{du}{u^2\sqrt{u^2 - a^2}} = \frac{\sqrt{u^2 - a^2}}{a^2 u} + C$

18. $\displaystyle\int \frac{du}{(u^2 - a^2)^{3/2}} = -\frac{u}{a^2\sqrt{u^2 - a^2}} + C$

Forms Involving $\sqrt{a^2 - u^2}, a > 0$

19. $\displaystyle\int \frac{\sqrt{a^2 - u^2}}{u}\, du = \sqrt{a^2 - u^2} - a\ln\left|\frac{a + \sqrt{a^2 - u^2}}{u}\right| + C$

20. $\displaystyle\int \frac{du}{u\sqrt{a^2 - u^2}} = -\frac{1}{a}\ln\left|\frac{a + \sqrt{a^2 - u^2}}{u}\right| + C$

21. $\displaystyle\int \frac{du}{u^2\sqrt{a^2 - u^2}} = -\frac{\sqrt{a^2 - u^2}}{a^2 u} + C$

22. $\displaystyle\int \frac{du}{(a^2 - u^2)^{3/2}} = \frac{u}{a^2\sqrt{a^2 - u^2}} + C$

Forms Involving e^{au} and $\ln u$

23. $\displaystyle\int u e^{au}\, du = \frac{1}{a^2}(au - 1)e^{au} + C$

24. $\displaystyle\int u^n e^{au}\, du = \frac{1}{a}u^n e^{au} - \frac{n}{a}\int u^{n-1} e^{au}\, du$

25. $\displaystyle\int \frac{du}{1 + be^{au}} = u - \frac{1}{a}\ln(1 + be^{au}) + C$

26. $\displaystyle\int \ln u\, du = u\ln u - u + C$

27. $\displaystyle\int u^n \ln u\, du = \frac{u^{n+1}}{(n+1)^2}[(n+1)\ln u - 1] + C \qquad (n \neq -1)$

28. $\displaystyle\int \frac{du}{u\ln u} = \ln|\ln u| + C$

29. $\displaystyle\int (\ln u)^n\, du = u(\ln u)^n - n\int (\ln u)^{n-1}\, du$

Using a Table of Integrals

We now consider several examples that illustrate how the table of integrals can be used to evaluate an integral.

EXAMPLE 1 Use the table of integrals to find $\int \dfrac{2x}{\sqrt{3+x}}\,dx$.

Solution   We first write

$$\int \frac{2x}{\sqrt{3+x}}\,dx = 2\int \frac{x}{\sqrt{3+x}}\,dx$$

Since $\sqrt{3+x}$ is of the form $\sqrt{a+bu}$, with $a = 3$, $b = 1$, and $u = x$, we use Formula (5),

$$\int \frac{u}{\sqrt{a+bu}}\,du = \frac{2}{3b^2}(bu - 2a)\sqrt{a+bu} + C$$

obtaining

$$2\int \frac{x}{\sqrt{3+x}}\,dx = 2\left[\frac{2}{3(1)}(x-6)\sqrt{3+x}\right] + C$$

$$= \frac{4}{3}(x-6)\sqrt{3+x} + C$$

Explore and Discuss

All formulas given in the table of integrals can be verified by direct computation. Describe a method you would use, and apply it to verify a formula of your choice.

EXAMPLE 2 Use the table of integrals to find $\int x^2\sqrt{3+x^2}\,dx$.

Solution   Observe that if we write 3 as $(\sqrt{3})^2$, then $3 + x^2$ has the form $\sqrt{a^2+u^2}$, with $a = \sqrt{3}$ and $u = x$. Using Formula (8), we have

$$\int u^2\sqrt{a^2+u^2}\,du = \frac{u}{8}(a^2+2u^2)\sqrt{a^2+u^2} - \frac{a^4}{8}\ln\left|u + \sqrt{a^2+u^2}\right| + C$$

and we obtain

$$\int x^2\sqrt{3+x^2}\,dx = \frac{x}{8}(3+2x^2)\sqrt{3+x^2} - \frac{9}{8}\ln\left|x + \sqrt{3+x^2}\right| + C$$

EXAMPLE 3 Use the table of integrals to evaluate

$$\int_3^4 \frac{dx}{x^2\sqrt{50-2x^2}}$$

Solution   We first find the indefinite integral

$$I = \int \frac{dx}{x^2\sqrt{50-2x^2}}$$

Observe that $\sqrt{50-2x^2} = \sqrt{2(25-x^2)} = \sqrt{2}\sqrt{25-x^2}$, so we can write I as

$$I = \frac{1}{\sqrt{2}}\int \frac{dx}{x^2\sqrt{25-x^2}} = \frac{\sqrt{2}}{2}\int \frac{dx}{x^2\sqrt{25-x^2}} \qquad \frac{1}{\sqrt{2}} = \frac{\sqrt{2}}{2}$$

Next, using Formula (21),

$$\int \frac{du}{u^2\sqrt{a^2-u^2}} = -\frac{\sqrt{a^2-u^2}}{a^2u} + C$$

with $a = 5$ and $u = x$, we find

$$I = \frac{\sqrt{2}}{2}\left(-\frac{\sqrt{25 - x^2}}{25x}\right)$$

$$= -\left(\frac{\sqrt{2}}{50}\right)\frac{\sqrt{25 - x^2}}{x}$$

Finally, using this result, we obtain

$$\int_3^4 \frac{dx}{x^2\sqrt{50 - 2x^2}} = -\frac{\sqrt{2}}{50}\frac{\sqrt{25 - x^2}}{x}\Big|_3^4$$

$$= -\frac{\sqrt{2}}{50}\frac{\sqrt{25 - 16}}{4} - \left(-\frac{\sqrt{2}}{50}\frac{\sqrt{25 - 9}}{3}\right)$$

$$= -\frac{3\sqrt{2}}{200} + \frac{2\sqrt{2}}{75} = \frac{7\sqrt{2}}{600}$$

EXAMPLE 4 Use the table of integrals to find $\int e^{2x}\sqrt{5 + 2e^x}\,dx$.

Solution Let $u = e^x$. Then $du = e^x\,dx$. Therefore, the given integral can be written

$$\int e^x\sqrt{5 + 2e^x}\,(e^x\,dx) = \int u\sqrt{5 + 2u}\,du$$

Using Formula (4),

$$\int u\sqrt{a + bu}\,du = \frac{2}{15b^2}(3bu - 2a)(a + bu)^{3/2} + C$$

with $a = 5$ and $b = 2$, we see that

$$\int u\sqrt{5 + 2u}\,du = \frac{2}{15(4)}(6u - 10)(5 + 2u)^{3/2} + C$$

$$= \frac{1}{15}(3u - 5)(5 + 2u)^{3/2} + C$$

Finally, recalling the substitution $u = e^x$, we find

$$\int e^{2x}\sqrt{5 + 2e^x}\,dx = \frac{1}{15}(3e^x - 5)(5 + 2e^x)^{3/2} + C$$

Explore and Discuss

The formulas given in the table of integrals were derived by using various techniques, including the method of substitution and the method of integration by parts studied earlier. For example, Formula (1),

$$\int \frac{u\,du}{a + bu} = \frac{1}{b^2}[a + bu - a\ln|a + bu|] + C$$

can be derived by using the method of substitution. Show how this is done.

As illustrated in the next example, we may need to apply a formula more than once to evaluate an integral.

EXAMPLE 5 Use the table of integrals to find $\int x^2 e^{(-1/2)x} \, dx$.

Solution Scanning the table of integrals for a formula involving e^{au} in the integrand, we are led to Formula (24),

$$\int u^n e^{au} \, du = \frac{1}{a} u^n e^{au} - \frac{n}{a} \int u^{n-1} e^{au} \, du$$

With $n = 2$, $a = -\frac{1}{2}$, and $u = x$, we have

$$\int x^2 e^{(-1/2)x} \, dx = \left(\frac{1}{-\frac{1}{2}}\right) x^2 e^{(-1/2)x} - \frac{2}{\left(-\frac{1}{2}\right)} \int x e^{(-1/2)x} \, dx$$

$$= -2x^2 e^{(-1/2)x} + 4 \int x e^{(-1/2)x} \, dx$$

If we use Formula (24) once again, with $n = 1$, $a = -\frac{1}{2}$, and $u = x$, to evaluate the integral on the right, we obtain

$$\int x^2 e^{(-1/2)x} \, dx = -2x^2 e^{(-1/2)x} + 4\left[\left(\frac{1}{-\frac{1}{2}}\right) x e^{(-1/2)x} - \frac{1}{\left(-\frac{1}{2}\right)} \int e^{(-1/2)x} \, dx\right]$$

$$= -2x^2 e^{(-1/2)x} + 4\left[-2x e^{(-1/2)x} + 2 \cdot \frac{1}{\left(-\frac{1}{2}\right)} e^{(-1/2)x}\right] + C$$

$$= -2e^{(-1/2)x}(x^2 + 4x + 8) + C$$

APPLIED EXAMPLE 6 Mortgage Rates A study prepared for the National Association of Realtors estimated that the mortgage rate over the next t months will be

$$r(t) = \frac{6t + 75}{t + 10} \qquad (0 \le t \le 24)$$

percent per year. If the prediction holds true, what will be the average mortgage rate over the next 12 months?

Solution The average mortgage rate over the next 12 months will be given by

$$A = \frac{1}{12 - 0} \int_0^{12} \frac{6t + 75}{t + 10} \, dt = \frac{1}{12} \left(\int_0^{12} \frac{6t}{t + 10} \, dt + \int_0^{12} \frac{75}{t + 10} \, dt \right)$$

$$= \frac{1}{2} \int_0^{12} \frac{t}{t + 10} \, dt + \frac{25}{4} \int_0^{12} \frac{1}{t + 10} \, dt$$

Using Formula (1),

$$\int \frac{u}{a + bu} \, du = \frac{1}{b^2} (a + bu - a \ln|a + bu|) + C \qquad a = 10, b = 1, u = t$$

to evaluate the first integral, we have

$$A = \left(\frac{1}{2}\right)[10 + t - 10 \ln(t + 10)]\Big|_0^{12} + \left(\frac{25}{4}\right) \ln(t + 10)\Big|_0^{12}$$

$$= \left(\frac{1}{2}\right)[(22 - 10 \ln 22) - (10 - 10 \ln 10)] + \left(\frac{25}{4}\right)[\ln 22 - \ln 10]$$

$$\approx 6.99$$

or approximately 6.99% per year.

7.2 Self-Check Exercises

1. Use the table of integrals to evaluate

$$\int_0^2 \frac{dx}{(5 - x^2)^{3/2}}$$

2. **FLU EPIDEMIC** During a flu epidemic, the number of children in Easton Middle School who had contracted influenza t days after the outbreak began was given by

$$N(t) = \frac{200}{1 + 9e^{-0.8t}}$$

Determine the average number of children who contracted the flu in the first 10 days of the epidemic.

Solutions to Self-Check Exercises 7.2 can be found on page 517.

7.2 Concept Questions

1. Consider the integral $\int \frac{\sqrt{2 - x^2}}{x}\, dx$.

   a. Which formula from the table of integrals would you choose to help you find the integral?
   b. Find the integral showing the appropriate substitutions you need to use to make the given integral conform to the formula.

2. Consider the integral $\int_2^3 \frac{dx}{\sqrt{2x^2 - 5}}$.

   a. Which formula from the table of integrals would you choose to help you evaluate the integral?
   b. Evaluate the integral.

7.2 Exercises

In Exercises 1–32, use the table of integrals in this section to find or evaluate each integral.

1. $\displaystyle\int \frac{2x}{2 + 3x}\, dx$

2. $\displaystyle\int \frac{x}{(1 + 2x)^2}\, dx$

3. $\displaystyle\int \frac{3x^2}{2 + 4x}\, dx$

4. $\displaystyle\int \frac{x^2}{3 + x}\, dx$

5. $\displaystyle\int x^2\sqrt{9 + 4x^2}\, dx$

6. $\displaystyle\int x^2\sqrt{4 + x^2}\, dx$

7. $\displaystyle\int \frac{dx}{x\sqrt{1 + 4x}}$

8. $\displaystyle\int_0^2 \frac{x + 1}{\sqrt{2 + 3x}}\, dx$

9. $\displaystyle\int_0^2 \frac{dx}{\sqrt{9 + 4x^2}}$

10. $\displaystyle\int \frac{dx}{x\sqrt{4 + 8x^2}}$

11. $\displaystyle\int \frac{dx}{(9 - x^2)^{3/2}}$

12. $\displaystyle\int \frac{dx}{(2 - x^2)^{3/2}}$

13. $\displaystyle\int x^2\sqrt{x^2 - 4}\, dx$

14. $\displaystyle\int_4^5 \frac{dx}{x^2\sqrt{x^2 - 9}}$

15. $\displaystyle\int \frac{\sqrt{4 - x^2}}{x}\, dx$

16. $\displaystyle\int_0^1 \frac{dx}{(4 - x^2)^{3/2}}$

17. $\displaystyle\int xe^{2x}\, dx$

18. $\displaystyle\int \frac{dx}{1 + e^{-x}}$

19. $\displaystyle\int \frac{dx}{(x + 1)\ln(1 + x)}$
    Hint: First use the substitution $u = x + 1$.

20. $\displaystyle\int \frac{x}{(x^2 + 1)\ln(x^2 + 1)}\, dx$
    Hint: First use the substitution $u = x^2 + 1$.

21. $\displaystyle\int \frac{3e^{2x}}{(1 + 3e^x)^2}\, dx$

22. $\displaystyle\int \frac{e^{2x}}{\sqrt{1 + 3e^x}}\, dx$

23. $\displaystyle\int \frac{3e^x}{1 + e^{x/2}}\, dx$

24. $\displaystyle\int \frac{dx}{1 - 2e^{-x}}$

25. $\displaystyle\int \frac{4 \ln x}{x(2 + 3 \ln x)}\, dx$

26. $\displaystyle\int_1^e (\ln x)^2\, dx$

27. $\displaystyle\int_0^1 x^2 e^x\, dx$

28. $\displaystyle\int x^3 e^{2x}\, dx$

29. $\displaystyle\int x^2 \ln x\, dx$

30. $\displaystyle\int x^3 \ln x\, dx$

31. $\displaystyle\int (\ln x)^3\, dx$

32. $\displaystyle\int (\ln x)^4\, dx$

33. **AMUSEMENT PARK ATTENDANCE** The management of Astro-World ("The Amusement Park of the Future") estimates that the number of visitors (in thousands) entering the amusement park t hr after opening time at 9 A.M. is given by

$$R(t) = \frac{60}{(2 + t^2)^{3/2}}$$

per hour. Determine the number of visitors admitted by noon.

34. Voter Registration The number of voters in a certain district of a city is expected to grow at the rate of

$$R(t) = \frac{3000}{\sqrt{4 + t^2}}$$

people/year, t years from now. If the number of voters at present is 20,000, how many voters will be in the district 5 years from now?

35. Growth of Fruit Flies On the basis of data collected during an experiment, a biologist found that the number of fruit flies (*Drosophila*) with a limited food supply could be approximated by the logistic model

$$N(t) = \frac{1000}{1 + 24e^{-0.02t}}$$

where t denotes the number of days since the beginning of the experiment. Find the average number of fruit flies in the colony in the first 10 days of the experiment and in the first 20 days.

36. Recycling Programs The commissioner of the City of Newton Department of Public Works estimates that the number of people in the city who have been recycling their magazines in year t following the introduction of the recycling program at the beginning of 2010 is

$$N(t) = \frac{100,000}{2 + 3e^{-0.2t}}$$

Find the average number of people who will have recycled their magazines during the first 5 years since the program was introduced.

37. Average Life Span One reason for the increase in the life span over the years has been the advances in medical technology. The average life span for American women from 1907 through 2007 is given by

$$W(t) = 49.9 + 17.1 \ln t \qquad (1 \le t \le 6)$$

where $W(t)$ is measured in years and t is measured in 20-year intervals, with $t = 1$ corresponding to 1907. What is the *average* average life expectancy for women from 1907 through 2007?
Source: AARP.

38. Consumers' Surplus for Women's Boots Refer to Section 6.7. The demand function for Apex women's boots is

$$p = \frac{250}{\sqrt{16 + x^2}}$$

where p is the unit price in dollars and x is the quantity demanded daily, in units of a hundred. Find the consumers' surplus if the price is set at $50/pair.

39. Producers' Surplus for Women's Boots Refer to Section 6.7. The supplier of Apex women's boots will make x

hundred pairs of the boots available in the market daily when the unit price is

$$p = \frac{30x}{5 - x}$$

dollars. Find the producers' surplus if the price is set at $50/pair.

40. Consumers' Surplus for Bluetooth Speakers Refer to Section 6.7. The demand function for a certain make of wireless portable Bluetooth speaker is given by

$$p = \frac{160}{\sqrt{4 + x^2}}$$

where p is the unit price in dollars and x (in units of a thousand) is the quantity demanded per month. What is the consumers' surplus corresponding to a quantity demanded of 2000 units/month?

41. Producers' Surplus for Multitask Desk Lamps Refer to Section 6.7. The manufacturer of a multitask desk lamp will make x thousand units of the lamps available in the market when the unit price is

$$p = 50 + x\sqrt{1 + x}$$

dollars. Find the producers' surplus if the price is set at $74/unit.
Hint: Show that the corresponding quantity demanded is 8000 units.

42. Accumulated Value of an Income Stream The revenue of Virtual Reality, a video-game arcade, is generated at the rate of $R(t) = 20,000t$ dollars/year. If the revenue is invested t years from now in a business earning income at the rate of 10%/year compounded continuously, find the accumulated value of this stream of income at the end of 5 years.
Hint: Use Formula (18), Section 6.7.

43. Franchises Elaine purchased a 10-year franchise for a fast-food restaurant that is expected to generate income at the rate of $R(t) = 250,000 + 2000t^2$ dollars/year, t years from now. If the prevailing interest rate is 10%/year compounded continuously, find the present value of the franchise.
Hint: Use Formula (19), Section 6.7.

44. Future Value of an Investment Refer to Section 6.7. Joanne expects to deposit

$$R(t) = t^2 e^{-0.02t}$$

thousand dollars/year for 10 years into a bank account that pays interest at the rate of 4%/year, compounded continuously. How much will Joanne's account be worth at the end of the 10-year period?

45. Lorenz Curves In a study conducted by a certain country's Economic Development Board regarding the income distribution of certain segments of the country's work-force, it was found that the Lorenz curve for the distribution of income of college professors is described by the function

$$g(x) = \frac{1}{3}x\sqrt{1 + 8x}$$

Compute the coefficient of inequality of the Lorenz curve. Hint: Use Formula (22), Section 6.7.

46. **PRESENT VALUE OF AN INVESTMENT** Refer to Section 6.7. John has decided to participate as a silent partner in a hobby store. In the first 5 years of operation, the hobby store is projected to generate an income of

$$R(t) = 10t + 4e^{0.02t}$$

thousand dollars/year. If the prevailing rate of interest for the period under consideration is 4%/year, compounded continuously, what is the present value of John's investment over the first 5 years?

47. **GINI INDEX** Refer to Section 6.7. The income distributions of two countries, Country A and Country B, are described by the Lorenz functions

$$f(x) = x^2 e^{x-1} \qquad (0 \le x \le 1)$$

and

$$g(x) = \frac{1}{2}x^2\sqrt{3 + x^2} \qquad (0 \le x \le 1)$$

respectively. Find the Gini Index for each of these income distributions. Which country has the more equitable income distribution?

7.2 Solutions to Self-Check Exercises

1. Using Formula (22) from the Table of Integrals,

$$\int \frac{du}{(a^2 - u^2)^{3/2}} = \frac{u}{a^2\sqrt{a^2 - u^2}} + C$$

with $a^2 = 5$ and $u = x$, we see that

$$\int_0^2 \frac{dx}{(5 - x^2)^{3/2}} = \frac{x}{5\sqrt{5 - x^2}}\Big|_0^2$$

$$= \frac{2}{5\sqrt{5 - 4}}$$

$$= \frac{2}{5}$$

2. The average number of children who contracted the flu in the first 10 days of the epidemic is given by

$$A = \frac{1}{10}\int_0^{10} \frac{200}{1 + 9e^{-0.8t}}\, dt = 20\int_0^{10} \frac{dt}{1 + 9e^{-0.8t}}$$

$$= 20\left[t + \frac{1}{0.8}\ln(1 + 9e^{-0.8t})\right]\Big|_0^{10} \quad \text{Formula (25), } a = -0.8, \; b = 9, u = t$$

$$= 20\left[10 + \frac{1}{0.8}\ln(1 + 9e^{-8})\right] - 20\left(\frac{1}{0.8}\right)\ln 10$$

$$\approx 200.07537 - 57.56463 \approx 143$$

or 143 children.

7.3 Numerical Integration

Approximating Definite Integrals

One method of measuring cardiac output is to inject 5 to 10 milligrams (mg) of a dye into a vein leading to the heart. After making its way through the lungs, the dye returns to the heart and is pumped into the aorta, where its concentration is measured at equal time intervals. The graph of the function c in Figure 2 (see next page) shows the concentration of dye in a person's aorta, measured at 2-second intervals after 5 mg of dye have been injected. The person's cardiac output, measured in liters per minute (L/min), is computed by using the formula

$$R = \frac{60D}{\int_0^{28} c(t)\, dt} \tag{5}$$

where D is the quantity of dye injected (see Exercise 54, on page 531).

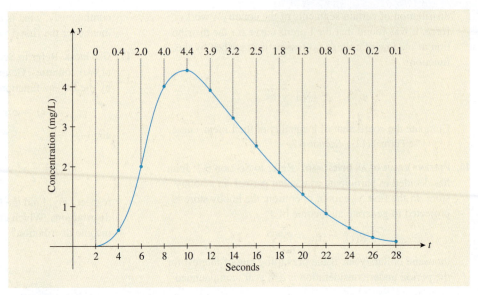

FIGURE 2
The function c gives the concentration of a dye measured at the aorta. The graph is constructed by drawing a smooth curve through a set of discrete points.

Now, to use Formula (5), we need to evaluate the definite integral

$$\int_0^{28} c(t) \, dt$$

But we do not have the algebraic rule defining the integrand c for all values of t in $[0, 28]$. In fact, we are given its values only at a set of discrete points in that interval. In situations such as this, the Fundamental Theorem of Calculus proves useless because we cannot find an antiderivative of c. (We will complete the solution to this problem in Example 4.)

Other situations also arise in which an integrable function has an antiderivative that cannot be found in terms of elementary functions (functions that can be expressed as a finite combination of algebraic, exponential, logarithmic, and trigonometric functions). Examples of such functions are

$$f(x) = e^{x^2} \qquad g(x) = x^{-1/2} e^x \qquad h(x) = \frac{1}{\ln x}$$

Riemann sums provide us with a good approximation of a definite integral, provided that the number of subintervals in the partitions is large enough. But there are better techniques and formulas, called *quadrature formulas*, that give a more efficient way of computing approximate values of definite integrals. In this section, we look at two rather simple but effective ways of approximating definite integrals.

The Trapezoidal Rule

We assume that $f(x) \geq 0$ on $[a, b]$ to simplify the derivation of the Trapezoidal Rule, but the result is valid without this restriction. We begin by subdividing the interval $[a, b]$ into n subintervals of equal length Δx, by means of the $(n + 1)$ points $x_0 = a$, $x_1, x_2, \ldots, x_n = b$, where n is a positive integer (Figure 3).

Then the length of each subinterval is given by

$$\Delta x = \frac{b - a}{n}$$

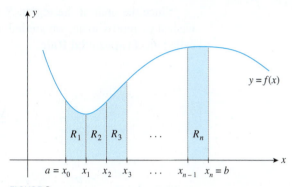

FIGURE **3**
The area under the curve is equal to the sum of the areas of the
n subregions $R_1, R_2, \ldots, R_n$.

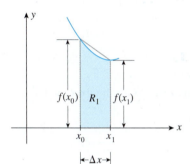

FIGURE **4**
The area of R_1 is approximated by the
area of the trapezoid.

Furthermore, as we saw earlier, we may view the definite integral

$$\int_a^b f(x)\,dx$$

as the area of the region R under the curve $y = f(x)$ between $x = a$ and $x = b$. This area is given by the sum of the areas of the n nonoverlapping subregions $R_1, R_2, \ldots, R_n$, where R_1 represents the region under the curve $y = f(x)$ from $x = x_0$ to $x = x_1$, and so on.

The basis for the Trapezoidal Rule lies in the approximation of each region, $R_1, R_2, \ldots, R_n$ by a suitable trapezoid. This often leads to a much better approximation than one obtained by means of rectangles (a Riemann sum).

Let's consider the subregion R_1, shown magnified for the sake of clarity in Figure 4. Observe that the area of the region R_1 may be approximated by the trapezoid of width Δx whose parallel sides are of lengths $f(x_0)$ and $f(x_1)$. The area of the trapezoid is given by

$$\left[\frac{f(x_0) + f(x_1)}{2}\right]\Delta x \qquad \text{Average of the lengths of the parallel sides} \times \text{width}$$

Similarly, the area of the region R_2 may be approximated by the trapezoid of width Δx and sides of lengths $f(x_1)$ and $f(x_2)$. The area of the trapezoid is given by

$$\left[\frac{f(x_1) + f(x_2)}{2}\right]\Delta x$$

Continuing, we see that the area of the last (nth) approximating trapezoid is given by

$$\left[\frac{f(x_{n-1}) + f(x_n)}{2}\right]\Delta x$$

Then the area of the region R is approximated by the sum of the areas of the n trapezoids—that is,

$$\left[\frac{f(x_0) + f(x_1)}{2}\right]\Delta x + \left[\frac{f(x_1) + f(x_2)}{2}\right]\Delta x + \cdots + \left[\frac{f(x_{n-1}) + f(x_n)}{2}\right]\Delta x$$

$$= \frac{\Delta x}{2}\left[f(x_0) + f(x_1) + f(x_1) + f(x_2) + \cdots + f(x_{n-1}) + f(x_n)\right]$$

$$= \frac{\Delta x}{2}\left[f(x_0) + 2f(x_1) + 2f(x_2) + \cdots + 2f(x_{n-1}) + f(x_n)\right]$$

Since the area of the region R is given by the value of the definite integral we wished to approximate, we are led to the following approximation formula, which is called the **Trapezoidal Rule.**

> **Trapezoidal Rule**
>
> $$\int_a^b f(x)\,dx \approx \frac{\Delta x}{2}\left[f(x_0) + 2f(x_1) + 2f(x_2)\right.$$
> $$\left. + \cdots + 2f(x_{n-1}) + f(x_n)\right] \tag{6}$$
>
> where $\Delta x = \dfrac{b-a}{n}$.

The approximation generally improves with larger values of n.

EXAMPLE 1 Approximate the value of

$$\int_1^2 \frac{1}{x}\,dx$$

using the Trapezoidal Rule with $n = 10$. Compare this result with the exact value of the integral.

Solution   Here, $a = 1$, $b = 2$, and $n = 10$, so

$$\Delta x = \frac{b-a}{n} = \frac{1}{10} = 0.1$$

and

$$x_0 = 1 \qquad x_1 = 1.1 \qquad x_2 = 1.2 \qquad x_3 = 1.3 \quad \cdots \quad x_9 = 1.9 \qquad x_{10} = 2$$

The Trapezoidal Rule yields

$$\int_1^2 \frac{1}{x}\,dx \approx \frac{0.1}{2}\left[1 + 2\left(\frac{1}{1.1}\right) + 2\left(\frac{1}{1.2}\right) + 2\left(\frac{1}{1.3}\right) + \cdots + 2\left(\frac{1}{1.9}\right) + \frac{1}{2}\right]$$
$$\approx 0.693771$$

In this case, we can easily compute the actual value of the definite integral under consideration. In fact,

$$\int_1^2 \frac{1}{x}\,dx = \ln x\,\Big|_1^2 = \ln 2 - \ln 1 = \ln 2$$
$$\approx 0.693147$$

Thus, the Trapezoidal Rule with $n = 10$ yields a result with an error of 0.000624 to six decimal places.

 APPLIED EXAMPLE 2 Consumers' Surplus   The demand function for a certain brand of perfume is given by

$$p = D(x) = \sqrt{10{,}000 - 0.01x^2}$$

where p is the unit price in dollars and x is the quantity demanded each week, measured in ounces. Find the consumers' surplus if the market price is set at \$60 per ounce.

Solution   When $p = 60$, we have

$$\sqrt{10,000 - 0.01x^2} = 60$$
$$10,000 - 0.01x^2 = 3,600$$
$$x^2 = 640,000$$

or $x = 800$ since x must be nonnegative. Next, using the consumers' surplus formula (page 482) with $\bar{p} = 60$ and $\bar{x} = 800$, we see that

$$CS = \int_0^{800} \sqrt{10,000 - 0.01x^2}\, dx - (60)(800)$$

It is not easy to evaluate this definite integral by finding an antiderivative of the integrand. Instead, let's use the Trapezoidal Rule with $n = 10$.

With $a = 0$ and $b = 800$, we find that

$$\Delta x = \frac{b - a}{n} = \frac{800}{10} = 80$$

and

$$x_0 = 0 \qquad x_1 = 80 \qquad x_2 = 160 \qquad x_3 = 240 \qquad \cdots \qquad x_9 = 720 \qquad x_{10} = 800$$

so

$$\int_0^{800} \sqrt{10,000 - 0.01x^2}\, dx$$

$$\approx \frac{80}{2} [100 + 2\sqrt{10,000 - (0.01)(80)^2}$$

$$+ 2\sqrt{10,000 - (0.01)(160)^2} + \cdots + 2\sqrt{10,000 - (0.01)(720)^2}$$

$$+ \sqrt{10,000 - (0.01)(800)^2}]$$

$$= 40(100 + 199.3590 + 197.4234 + 194.1546 + 189.4835$$

$$+ 183.3030 + 175.4537 + 165.6985$$

$$+ 153.6750 + 138.7948 + 60)$$

$$\approx 70,293.82$$

Therefore, the consumers' surplus is approximately $70,294 - 48,000$, or $\$22,294$. ◼

Explore and Discuss

Explain how you would approximate the value of $\int_0^2 f(x)\, dx$ using the Trapezoidal Rule with $n = 10$, where

$$f(x) = \begin{cases} \sqrt{1 + x^2} & \text{if } 0 \le x \le 1 \\ \dfrac{2}{\sqrt{1 + x^2}} & \text{if } 1 < x \le 2 \end{cases}$$

and find the value.

Simpson's Rule

Before stating Simpson's Rule, let's review the two rules we have used in approximating a definite integral. Let f be a continuous, nonnegative function defined on the interval $[a, b]$. Suppose the interval $[a, b]$ is partitioned by means of the $n + 1$

equally spaced points $x_0 = a, x_1, x_2, \ldots, x_n = b$, where n is a positive integer, so that the length of each subinterval is $\Delta x = (b - a)/n$ (Figure 5).

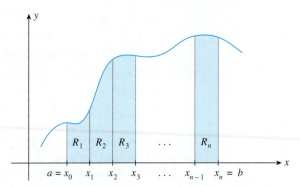

FIGURE 5
The area of the region under the curve is equal to the sum of the areas of the n subregions $R_1, R_2, \ldots, R_n$.

Let's concentrate on the portion of the graph of $y = f(x)$ defined on the interval $[x_0, x_2]$. When we use a Riemann sum to approximate the definite integral, we are in effect approximating the function $f(x)$ on $[x_0, x_1]$ by the *constant* function $y = f(p_1)$, where p_1 is chosen to be a point in $[x_0, x_1]$; and we are approximating the function $f(x)$ on $[x_1, x_2]$ by the constant function $y = f(p_2)$, where p_2 lies in $[x_1, x_2]$. Using a Riemann sum, we see that the area of the region under the curve $y = f(x)$ between $x = x_0$ and $x = x_2$ is approximated by the area of the region under the approximating "step" function (Figure 6a).

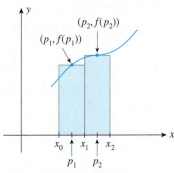
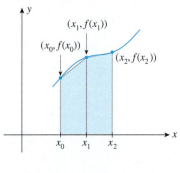

(a) The area of the region under the curve is approximated by the area of the rectangles.

(b) The area of the region under the curve is approximated by the area of the trapezoids.

FIGURE 6

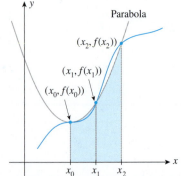

FIGURE 7
Simpson's Rule approximates the area of the region under the curve by the area of the region under the parabola.

When we use the Trapezoidal Rule, we are in effect approximating the function $f(x)$ on the interval $[x_0, x_1]$ by a *linear* function through the two points $(x_0, f(x_0))$ and $(x_1, f(x_1))$ and the function $f(x)$ on $[x_1, x_2]$ by a *linear* function through the two points $(x_1, f(x_1))$ and $(x_2, f(x_2))$. Thus, the Trapezoidal Rule simply approximates the actual area of the region under the curve $y = f(x)$ from $x = x_0$ to $x = x_2$ by the area of the region under the approximating polygonal curve (Figure 6b).

A natural extension of the preceding idea is to approximate portions of the graph of $y = f(x)$ by means of portions of the graphs of second-degree polynomials (parts of parabolas). It can be shown that given any three noncollinear points, there is a unique parabola that passes through the given points. Choose the points $(x_0, f(x_0))$, $(x_1, f(x_1))$, and $(x_2, f(x_2))$ corresponding to the first three points of the partition. Then we can approximate the function $f(x)$ on $[x_0, x_2]$ by means of a quadratic function whose graph contains these three points (Figure 7).

Although we will not do so here, it can be shown that the area of the region under the parabola between $x = x_0$ and $x = x_2$ is given by

$$\frac{\Delta x}{3} [f(x_0) + 4f(x_1) + f(x_2)]$$

Repeating this argument on the interval $[x_2, x_4]$, we see that the area of the region under the curve between $x = x_2$ and $x = x_4$ is approximated by the area of the region under the parabola between x_2 and x_4—that is, by

$$\frac{\Delta x}{3} [f(x_2) + 4f(x_3) + f(x_4)]$$

Proceeding, we conclude that if n is even (Why?), then the area of the region under the curve $y = f(x)$ from $x = a$ to $x = b$ may be approximated by the sum of the areas of the regions under the $n/2$ approximating parabolas—that is,

$$\frac{\Delta x}{3} [f(x_0) + 4f(x_1) + f(x_2)] + \frac{\Delta x}{3} [f(x_2) + 4f(x_3) + f(x_4)] + \cdots$$

$$+ \frac{\Delta x}{3} [f(x_{n-2}) + 4f(x_{n-1}) + f(x_n)]$$

$$= \frac{\Delta x}{3} [f(x_0) + 4f(x_1) + f(x_2) + f(x_2) + 4f(x_3) + f(x_4) + \cdots$$

$$+ f(x_{n-2}) + 4f(x_{n-1}) + f(x_n)]$$

$$= \frac{\Delta x}{3} [f(x_0) + 4f(x_1) + 2f(x_2) + 4f(x_3)$$

$$+ 2f(x_4) + \cdots + 4f(x_{n-1}) + f(x_n)]$$

The preceding is the derivation of the approximation formula known as **Simpson's Rule.**

Simpson's Rule

$$\int_a^b f(x)\, dx \approx \frac{\Delta x}{3} [f(x_0) + 4f(x_1) + 2f(x_2) + 4f(x_3) + 2f(x_4)$$

$$+ \cdots + 4f(x_{n-1}) + f(x_n)] \tag{7}$$

where $\Delta x = \dfrac{b - a}{n}$ and n is even.

In using Simpson's Rule, remember that n must be even.

EXAMPLE 3 Find an approximation of

$$\int_1^2 \frac{1}{x}\, dx$$

using Simpson's Rule with $n = 10$. Compare this result with that of Example 1 and also with the exact value of the integral.

 Solution   We have $a = 1$, $b = 2$, $f(x) = \dfrac{1}{x}$, and $n = 10$, so

$$\Delta x = \frac{b - a}{n} = \frac{1}{10} = 0.1$$

and

$$x_0 = 1 \qquad x_1 = 1.1 \qquad x_2 = 1.2 \qquad x_3 = 1.3 \quad \cdots \quad x_9 = 1.9 \qquad x_{10} = 2$$

Simpson's Rule yields

$$\int_1^2 \frac{1}{x}\, dx \approx \frac{0.1}{3}\left[f(1) + 4f(1.1) + 2f(1.2) + \cdots + 4f(1.9) + f(2) \right]$$

$$= \frac{0.1}{3}\left[1 + 4\left(\frac{1}{1.1}\right) + 2\left(\frac{1}{1.2}\right) + 4\left(\frac{1}{1.3}\right) + 2\left(\frac{1}{1.4}\right) + 4\left(\frac{1}{1.5}\right) \right.$$

$$\left. + 2\left(\frac{1}{1.6}\right) + 4\left(\frac{1}{1.7}\right) + 2\left(\frac{1}{1.8}\right) + 4\left(\frac{1}{1.9}\right) + \frac{1}{2} \right]$$

$$\approx 0.693150$$

The Trapezoidal Rule with $n = 10$ yielded an approximation of 0.693771, which is 0.000624 off the value of ln $2 \approx 0.693147$ to six decimal places. Simpson's Rule yields an approximation with an error of 0.000003, a definite improvement over the Trapezoidal Rule.

 APPLIED EXAMPLE 4 Cardiac Output Solve the problem posed at the beginning of this section. Recall that we wished to find a person's cardiac output by using the formula

$$R = \frac{60\, D}{\displaystyle\int_0^{28} c(t)\, dt}$$

where D (the quantity of dye injected) is equal to 5 milligrams and the function c has the graph shown in Figure 8. Use Simpson's Rule with $n = 14$ to estimate the value of the integral.

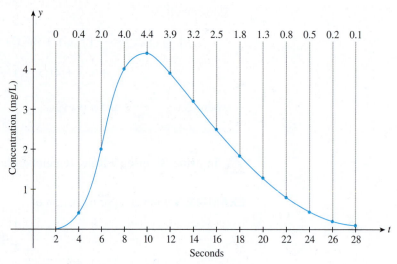

FIGURE 8
The function *c* gives the concentration of a dye measured at the aorta. The graph is constructed by drawing a smooth curve through a set of discrete points.

Solution Using Simpson's Rule with $n = 14$ and $\Delta t = 2$ so that

$$t_0 = 0 \qquad t_1 = 2 \qquad t_2 = 4 \qquad t_3 = 6 \quad \cdots \quad t_{14} = 28$$

we obtain

$$
\int_0^{28} c(t)\, dt \approx \frac{2}{3} \big[c(0) + 4c(2) + 2c(4) + 4c(6) + \cdots
$$
$$
+ 4c(26) + c(28) \big]
$$
$$
= \frac{2}{3} \big[0 + 4(0) + 2(0.4) + 4(2.0) + 2(4.0)
$$
$$
+ 4(4.4) + 2(3.9) + 4(3.2) + 2(2.5) + 4(1.8)
$$
$$
+ 2(1.3) + 4(0.8) + 2(0.5) + 4(0.2) + 0.1 \big]
$$
$$
\approx 49.9
$$

Therefore, the person's cardiac output is

$$
R \approx \frac{60(5)}{49.9} \approx 6.0
$$

or 6.0 liters per minute.

APPLIED EXAMPLE 5 Oil Spill An oil spill off the coastline was caused by a ruptured tank in a grounded oil tanker. Using aerial photographs, the Coast Guard was able to obtain the dimensions of the oil spill (Figure 9). Using Simpson's Rule with $n = 10$, estimate the area of the oil spill.

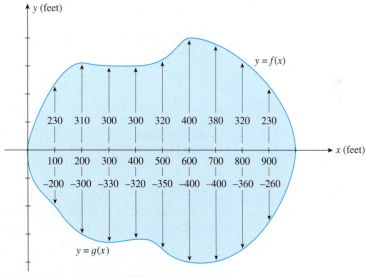

FIGURE 9
Simpson's Rule can be used to calculate the area of the oil spill.

Solution   We may think of the area affected by the oil spill as the area of the plane region bounded above by the graph of the function $f(x)$ and below by the graph of the function $g(x)$ between $x = 0$ and $x = 1000$ (Figure 9). Then the required area is given by

$$
A = \int_0^{1000} [f(x) - g(x)]\, dx
$$

Using Simpson's Rule with $n = 10$ and $\Delta x = 100$, so that

$$
x_0 = 0 \qquad x_1 = 100 \qquad x_2 = 200 \quad \cdots \quad x_{10} = 1000
$$

we have

$$A = \int_0^{1000} \left[f(x) - g(x) \right] dx$$

$$\approx \frac{\Delta x}{3} \{ [f(x_0) - g(x_0)] + 4[f(x_1) - g(x_1)] + 2[f(x_2) - g(x_2)]$$

$$+ \cdots + 4[f(x_9) - g(x_9)] + [f(x_{10}) - g(x_{10})] \}$$

$$= \frac{100}{3} \{ [0 - 0] + 4[230 - (-200)] + 2[310 - (-300)]$$

$$+ 4[300 - (-330)] + 2[300 - (-320)] + 4[320 - (-350)]$$

$$+ 2[400 - (-400)] + 4[380 - (-400)] + 2[320 - (-360)]$$

$$+ 4[230 - (-260)] + [0 - 0] \}$$

$$= \frac{100}{3} \left[0 + 4(430) + 2(610) + 4(630) + 2(620) + 4(670) \right.$$

$$\left. + 2(800) + 4(780) + 2(680) + 4(490) + 0 \right]$$

$$= \frac{100}{3} (17{,}420)$$

$$\approx 580{,}667$$

or approximately 580,667 square feet.

Explore and Discuss

Explain how you would approximate the value of $\int_0^2 f(x)\, dx$ using Simpson's Rule with $n = 10$, where

$$f(x) = \begin{cases} \sqrt{1 + x^2} & \text{if } 0 \le x \le 1 \\ \dfrac{2}{\sqrt{1 + x^2}} & \text{if } 1 < x \le 2 \end{cases}$$

and find the value.

Error Analysis

The following results give the bounds on the errors incurred when the Trapezoidal Rule and Simpson's Rule are used to approximate a definite integral (proof omitted).

Errors in the Trapezoidal and Simpson Approximations

Suppose the definite integral

$$\int_a^b f(x)\, dx$$

is approximated with n subintervals.

1. The *maximum* error incurred in using the Trapezoidal Rule is

$$\frac{M(b - a)^3}{12n^2} \tag{8}$$

where M is a number such that $|f''(x)| \le M$ for all x in $[a, b]$.

2. The *maximum* error incurred in using Simpson's Rule is

$$\frac{M(b-a)^5}{180n^4} \tag{9}$$

where M is a number such that $|f^{(4)}(x)| \le M$ for all x in $[a, b]$.

Note In many instances, the actual error is less than the upper error bounds given.

EXAMPLE 6 Find bounds on the errors incurred when

$$\int_1^2 \frac{1}{x}\, dx$$

is approximated by using (a) the Trapezoidal Rule and (b) Simpson's Rule with $n = 10$. Compare these with the actual errors found in Examples 1 and 3.

Solution

a. Here, $a = 1$, $b = 2$, and $f(x) = 1/x$. Next, to find a value for M, we compute

$$f'(x) = -\frac{1}{x^2} \quad \text{and} \quad f''(x) = \frac{2}{x^3}$$

Since $f''(x)$ is positive and decreasing on $(1, 2)$ (Why?), it attains its maximum value of 2 at $x = 1$, the left endpoint of the interval. Therefore, if we take $M = 2$, then $|f''(x)| \le 2$. Using (8), we see that the maximum error incurred is

$$\frac{2(2-1)^3}{12(10)^2} = \frac{2}{1200} = 0.0016667$$

The actual error found in Example 1, 0.000624, is much less than the upper bound just found.

b. We compute

$$f'''(x) = \frac{-6}{x^4} \quad \text{and} \quad f^{(4)}(x) = \frac{24}{x^5}$$

Since $f^{(4)}(x)$ is positive and decreasing on $(1, 2)$ (just look at $f^{(5)}$ to verify this fact), it attains its maximum at the left endpoint of $[1, 2]$. Now,

$$f^{(4)}(1) = 24$$

so we may take $M = 24$. Using (9), we obtain the maximum error of

$$\frac{24(2-1)^5}{180(10)^4} = 0.0000133$$

The actual error is 0.000003 (see Example 3).

Explore and Discuss

Refer to the Explore and Discuss on page 521. Explain how you would find the maximum error incurred in using the Trapezoidal Rule with $n = 10$ to approximate $\int_0^2 f(x)\, dx$.

7.3   Self-Check Exercises

1. Use the Trapezoidal Rule and Simpson's Rule with $n = 8$ to approximate the value of the definite integral

$$\int_0^2 \frac{1}{\sqrt{1 + x^2}}\, dx$$

2. **U.S. CONSUMPTION OF PETROLEUM** The graph in the accompanying figure shows the consumption of petroleum in the United States in millions of barrels per day, from 1996 to 2006. Using Simpson's Rule with $n = 10$, estimate the average consumption during the 10-year period.

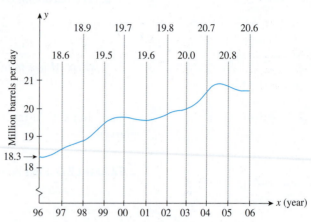

Source: BP Statistical Review of World Energy.

Solutions to Self-Check Exercises 7.3 can be found on page 532.

7.3   Concept Questions

1. Explain why n can be odd or even in the Trapezoidal Rule but must be even in Simpson's Rule.

2. Explain, without alluding to the error formulas, why the Trapezoidal Rule gives the exact value of $\int_a^b f(x)\, dx$ if f is a linear function and why Simpson's Rule gives the exact value of the integral if f is a quadratic function.

3. Refer to Concept Question 2, and answer the questions using the error formulas for the Trapezoidal Rule and Simpson's Rule.

7.3   Exercises

In Exercises 1–14, use the Trapezoidal Rule and Simpson's Rule to approximate the value of each definite integral. Compare your result with the exact value of the integral. Express your answers correct to four decimal places.

1. $\int_0^2 x^2\, dx;\ n = 6$

2. $\int_1^3 (x^2 - 1)\, dx;\ n = 4$

3. $\int_0^1 x^3\, dx;\ n = 4$

4. $\int_1^2 x^3\, dx;\ n = 6$

5. $\int_1^2 \frac{1}{x}\, dx;\ n = 4$

6. $\int_1^2 \frac{1}{x}\, dx;\ n = 8$

7. $\int_1^2 \frac{1}{x^2}\, dx;\ n = 4$

8. $\int_0^1 \frac{1}{1 + x}\, dx;\ n = 4$

9. $\int_0^4 \sqrt{x}\, dx;\ n = 8$

10. $\int_0^2 x\sqrt{2x^2 + 1}\, dx;\ n = 6$

11. $\int_0^1 e^{-x}\, dx;\ n = 6$

12. $\int_0^1 xe^{-x^2}\, dx;\ n = 6$

13. $\int_1^2 \ln x\, dx;\ n = 4$

14. $\int_0^1 x\ln(x^2 + 1)\, dx;\ n = 8$

In Exercises 15–22, use the Trapezoidal Rule and Simpson's Rule to approximate the value of each definite integral. Express your answers correct to four decimal places.

15. $\int_0^1 \sqrt{1 + x^3}\, dx;\ n = 4$

16. $\int_0^2 x\sqrt{1 + x^3}\, dx;\ n = 4$

17. $\int_0^2 \frac{1}{\sqrt{x^3 + 1}}\, dx;\ n = 4$

18. $\int_0^1 \sqrt{1 - x^2}\, dx;\ n = 4$

19. $\int_0^2 e^{-x^2}\, dx;\ n = 4$

20. $\int_0^1 e^{x^2}\, dx;\ n = 6$

21. $\int_1^2 x^{-1/2}e^x\, dx;\ n = 4$

22. $\int_2^4 \frac{dx}{\ln x};\ n = 6$

In Exercises 23–28, find a bound on the error in approximating each definite integral using (a) the Trapezoidal Rule and (b) Simpson's Rule with n intervals.

23. $\int_{-1}^2 x^5\, dx;\ n = 10$

24. $\int_0^1 e^{-x}\, dx;\ n = 8$

25. $\int_1^3 \frac{1}{x}\, dx;\ n = 10$

26. $\int_1^3 \frac{1}{x^2}\, dx;\ n = 8$

27. $\int_0^2 \dfrac{1}{\sqrt{1+x}}\,dx; \; n = 8$ **28.** $\int_1^3 \ln x\,dx; \; n = 10$

29. TRIAL RUN OF AN ATTACK SUBMARINE In a submerged trial run of an attack submarine, a reading of the sub's velocity was made every quarter hour, as shown in the accompanying table. Use the Trapezoidal Rule to estimate the distance traveled by the submarine during the 2-hr period.

Time, t (hr)	0	$\frac{1}{4}$	$\frac{1}{2}$	$\frac{3}{4}$
Velocity, $V(t)$ (mph)	19.5	24.3	34.2	40.5

Time, t (hr)	1	$\frac{5}{4}$	$\frac{3}{2}$	$\frac{7}{4}$	2
Velocity, $V(t)$ (mph)	38.4	26.2	18	16	8

30. CONDOMINIUM RESORT COMPLEX Cooper Realty is considering development of a time-sharing condominium resort complex along the oceanfront property illustrated in the accompanying graph. To obtain an estimate of the area of this property, measurements of the distances from the edge of a straight road, which defines one boundary of the property, to the corresponding points on the shoreline are made at 100-ft intervals. Using Simpson's Rule with $n = 10$, estimate the area of the oceanfront property.

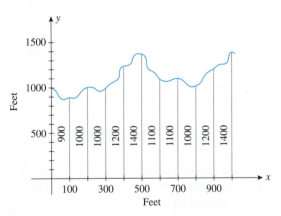

31. AVERAGE TEMPERATURE The graph depicted in the following figure shows the daily mean temperatures recorded during one September in Cameron Highlands. Using (a) the Trapezoidal Rule and (b) Simpson's Rule with $n = 10$, estimate the average temperature during that month.

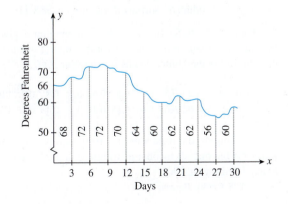

32. SURFACE AREA OF THE JACQUELINE KENNEDY ONASSIS RESERVOIR The reservoir located in Central Park in New York City has the shape depicted in the following figure. The measurements shown were taken at 206-ft intervals. Use Simpson's Rule with $n = 10$ to estimate the surface area of the lake.

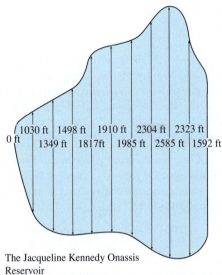

The Jacqueline Kennedy Onassis Reservoir

Source: Boston Globe.

33. FUEL CONSUMPTION OF DOMESTIC CARS Thanks to smaller and more fuel-efficient models, American carmakers doubled their average fuel economy over a 13-year period, from 1974 to 1987. The following figure gives the average fuel consumption in miles per gallon (mpg) of domestic-built cars over the period under consideration ($t = 0$ corresponds to the beginning of 1974). Use the Trapezoidal Rule to estimate the average fuel consumption of the domestic cars built during this period.

Hint: Approximate the integral $\frac{1}{13}\int_0^{13} f(t)\,dt$.

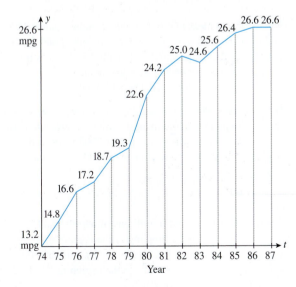

34. WATER FLOW IN A RIVER At a certain point, a river is 78 ft wide, and its depth, measured at 6-ft intervals across the river, is recorded in the following table.

x (ft)	0	6	12	18	24	30	36
y (ft)	0.8	2.6	5.8	6.2	8.2	10.1	10.8

x (ft)	42	48	54	60	66	72	78
y (ft)	9.8	7.6	6.4	5.2	3.9	2.4	1.4

Here, x denotes the distance (in feet) from one bank of the river and y (in feet) is the corresponding depth. If the average rate of flow through this section of the river is 4 ft/sec, use the Trapezoidal Rule with $n = 13$ to find the rate of the volume of flow of water in the river.
Hint: Volume of flow = rate of flow × area of cross section.

35. MEDICAL COSTS FOR VETERANS According to a study by Linda Bilmes of the Kennedy School of Government, the cost of medical treatment of veterans (in billions of dollars) is projected to be

$$C(t) = 0.0048t^3 - 0.0091t^2 + 0.3789t + 0.63$$

$$(1 \leq t \leq 9)$$

in year t, where $t = 1$ corresponds to 2006. Use the Trapezoidal Rule with $n = 8$ to estimate the total projected cost of medical treatment of veterans over the years from 2006 through 2014.
Source: "Soldiers Returning from Iraq and Afghanistan" by Linda Bilmes, Kennedy School of Government.

36. AIR POLLUTION The amount of nitrogen dioxide, a brown gas that impairs breathing, present in the atmosphere on a certain May day in the city of Long Beach has been approximated by

$$A(t) = \frac{136}{1 + 0.25(t - 4.5)^2} + 28 \qquad (0 \leq t \leq 11)$$

where $A(t)$ is measured in pollutant standard index (PSI), and t is measured in hours, with $t = 0$ corresponding to 7 A.M. Use the Trapezoidal Rule with $n = 10$ to estimate the average PSI between 7 A.M. and noon.
Hint: The average value is given by $\frac{1}{5} \int_0^5 A(t)\, dt$.

37. U.S. STRATEGIC PETROLEUM RESERVES According to data from the American Petroleum Institute, the U.S. strategic petroleum reserves from the beginning of 2002 through the beginning of 2012 can be approximated by the function

$$S(t) = \frac{720t^2 + 3480}{t^2 + 6.3} \qquad (0 \leq t \leq 10)$$

where $S(t)$ is measured in millions of barrels and t in years, with $t = 0$ corresponding to the beginning of 2002. Using the Trapezoidal Rule with $n = 10$, estimate the average petroleum reserves from the beginning of 2002 through the beginning of 2012.
Source: American Petroleum Institute.

38. CONSUMERS' SURPLUS Refer to Section 6.7. Before the launch of a new product, the company's research department conducted a study to determine the demand for the product. The following table summarizes the data obtained by the researchers. Here, p denotes the unit selling price (in dollars) of the product, and x denotes the quantity demanded per week (in units of a thousand) at that price.

Quantity Demanded, x	0	1	2	3	4	5	6	7
Unit Price, p		80.0	61.5	44.4	32.0	23.5	17.8	13.8	11.0

Use the Trapezoidal Rule with $n = 7$ to estimate the consumers' surplus if the unit selling price is set at $11.

39. FUTURE VALUE OF AN INVESTMENT Ivan expects to deposit

$$R(t) = t^{1.5}e^{-0.02t}$$

thousand dollars/year for 5 years into a bank that pays interest at the rate of 4%/year, compounded continuously. Use the Trapezoidal Rule with $n = 10$ to estimate the value of Ivan's account at the end of 5 years.

40. CONSUMERS' SURPLUS FOR SPORTS WATCHES Refer to Section 6.7. The demand equation for the Sicard sports watch is given by

$$p = D(x) = \frac{50}{0.01x^2 + 1} \qquad (0 \leq x \leq 20)$$

where x (measured in units of a thousand) is the quantity demanded per week and p is the unit price in dollars. Use (a) the Trapezoidal Rule and (b) Simpson's Rule (take $n = 8$) to estimate the consumers' surplus if the market price is $25/watch.

41. PRODUCERS' SURPLUS FOR CDS Refer to Section 6.7. The supply function for the CD manufactured by Herald Records is given by

$$p = S(x) = \sqrt{0.01x^2 + 0.11x + 38}$$

where p is the unit price in dollars and x stands for the quantity that will be made available in the market by the supplier, measured in units of a thousand. Use (a) the Trapezoidal Rule and (b) Simpson's Rule (take $n = 8$) to estimate the producers' surplus if the price is $8/CD.

42. GINI INDEX Refer to Section 6.7. The following table gives the values of the Lorenz function for the market income distribution in the United States for the year 2007.

x	0.0	0.1	0.2	0.3	0.4	0.5	0.6	0.7	0.8	0.9	1.0
f(x)	0.00	0.01	0.02	0.05	0.09	0.15	0.21	0.31	0.40	0.56	1.00

Use the Trapezoidal Rule with $n = 10$ to estimate the Gini Index for the market income distribution in the United States for the year 2007.
Source: U.S. Census Bureau.

43. GINI INDEX Refer to Section 6.7. The following table gives the values of the Lorenz function for the after-tax income distribution in the United States for the year 2007.

x	0.0	0.1	0.2	0.3	0.4	0.5	0.6	0.7	0.8	0.9	1.0
$f(x)$	0.00	0.02	0.04	0.08	0.13	0.20	0.27	0.36	0.47	0.60	1.00

Use Simpson's Rule with $n = 10$ to estimate the Gini Index for the market income distribution in the United States for the year 2007.
Source: U.S. Census Bureau.

44. GROWTH OF SERVICE INDUSTRIES It has been estimated that service industries, which currently make up 30% of the nonfarm workforce in a certain country, will continue to grow at the rate of

$$R(t) = 5e^{1/(t+1)}$$

percent/decade, t decades from now. Estimate the percentage of the nonfarm workforce in the service industries one decade from now.
Hint: (a) Show that the answer is given by $30 + \int_0^1 5e^{1/(t+1)}\, dt$ and (b) use Simpson's Rule with $n = 10$ to approximate the definite integral.

45. PRODUCERS' SURPLUS Refer to Section 6.7. Researchers at Market Research obtained the following data pertaining to a popular product. Here, x denotes the number of units of the product (in thousands) per week that the manufacturer will make available in the market if the price is p dollars.

Quantity Supplied, x	0	2	4	6	8	10	12
Unit Price, p		20.0	21.5	25.0	31.2	40.3	53.0	69.7

Use Simpson's Rule with $n = 6$ to estimate the producers' surplus if the unit price is set at $69.70.

46. MEASURING CARDIAC OUTPUT Eight milligrams of a dye are injected into a vein leading to an individual's heart. The concentration of the dye in the aorta (in milligrams per liter) measured at 2-sec intervals is shown in the accompanying table. Use Simpson's Rule and the formula of Example 4 to estimate the person's cardiac output.

t	0	2	4	6	8	10	12
$C(t)$	0	0	2.8	6.1	9.7	7.6	4.8

t	14	16	18	20	22	24
$C(t)$	3.7	1.9	0.8	0.3	0.1	0

47. ESTIMATING THE FLOW RATE OF A RIVER A stream is 120 ft wide. The following table gives the depths of the river measured across a section of the river in intervals of 6 ft. Here, x denotes the distance from one bank of the river, and y denotes the corresponding depth (in feet). The average rate of flow of the river across this section of the river

is 4.2 ft/sec. Use Simpson's Rule with $n = 20$ to estimate the rate of the volume of flow of the river.

x (ft)	0	6	12	18	24	30	36	42	48	54	60
y (ft)	0.8	1.2	3.0	4.1	5.8	6.6	6.8	7.0	7.2	7.4	7.8

x (ft)	66	72	78	84	90	96	102	108	114	120
y (ft)	7.6	7.4	7.0	6.6	6.0	5.1	4.3	3.2	2.2	1.1

Hint: Volume of flow = rate of flow $\times$ area of cross section.

48. TREAD LIFE OF TIRES Under normal driving conditions, the percent of Super Titan radial tires expected to have a useful tread life of between 30,000 and 40,000 mi is given by

$$P = 100 \int_{30,000}^{40,000} \frac{1}{2000\sqrt{2\pi}}\, e^{-(1/2)[(x-40,000)/2000]^2}\, dx$$

Use Simpson's Rule with $n = 10$ to estimate P.

49. LENGTH OF INFANTS AT BIRTH Medical records of infants delivered at Kaiser Memorial Hospital show that the percentage of infants whose length at birth is between 19 and 21 in. is given by

$$P = 100 \int_{19}^{21} \frac{1}{2.6\sqrt{2\pi}}\, e^{-(1/2)[(x-20)/2.6]^2}\, dx$$

Use Simpson's Rule with $n = 10$ to estimate P.

In Exercises 50–53, determine whether the statement is true or false. It it is true, explain why it is true. If it is false, give an example to show why it is false.

50. In using the Trapezoidal Rule, the number of subintervals n must be even.

51. In using Simpson's Rule, the number of subintervals n may be chosen to be odd or even.

52. Simpson's Rule is more accurate than the Trapezoidal Rule.

53. If f is a polynomial function of degree less than or equal to 3, then the approximation of $\int_a^b f(x)\, dx$ using Simpson's Rule is exact.

54. Derive the formula

$$R = \frac{60\, D}{\int_0^T c(t)\, dt}$$

for calculating the cardiac output of a person in liters per minute. Here, $c(t)$ is the concentration of dye in the aorta (in milligrams per liter) at time t (in seconds) for t in $[0, T]$, and D is the amount of dye (in milligrams) injected into a vein leading to the heart.
Hint: Partition the interval $[0, T]$ into n subintervals of equal length Δt. The amount of dye that flows past the measuring point in the aorta during the time interval $[0, \Delta t]$ is

approximately $c(t_i)(R \, \Delta t)/60$ (concentration times volume). Therefore, the total amount of dye measured at the aorta is

$$\frac{[c(t_1)R \, \Delta t + c(t_2)R \, \Delta t + \cdots + c(t_n)R \, \Delta t]}{60} = D$$

Take the limit of the Riemann sum to obtain

$$R = \frac{60 \, D}{\displaystyle\int_0^T c(t) \, dt}$$

---

1. We have $a = 0$, $b = 2$, and $n = 8$, so

$$\Delta x = \frac{b - a}{n} = \frac{2}{8} = 0.25$$

and $x_0 = 0$, $x_1 = 0.25$, $x_2 = 0.50$, $x_3 = 0.75$, ..., $x_7 = 1.75$, and $x_8 = 2$. The Trapezoidal Rule gives

$$\int_0^2 \frac{1}{\sqrt{1 + x^2}} \, dx \approx \frac{0.25}{2}\left[1 + \frac{2}{\sqrt{1 + (0.25)^2}} + \frac{2}{\sqrt{1 + (0.5)^2}}\right.$$

$$+ \cdots + \frac{2}{\sqrt{1 + (1.75)^2}} + \left.\frac{1}{\sqrt{5}}\right]$$

$$\approx 0.125(1 + 1.9403 + 1.7889 + 1.6000 + 1.4142$$

$$+ 1.2494 + 1.1094 + 0.9923 + 0.4472)$$

$$\approx 1.4427$$

Using Simpson's Rule with $n = 8$ gives

$$\int_0^2 \frac{1}{\sqrt{1 + x^2}} \, dx \approx \frac{0.25}{3}\left[1 + \frac{4}{\sqrt{1 + (0.25)^2}} + \frac{2}{\sqrt{1 + (0.5)^2}}\right.$$

$$+ \frac{4}{\sqrt{1 + (0.75)^2}} + \cdots + \frac{4}{\sqrt{1 + (1.75)^2}} + \left.\frac{1}{\sqrt{5}}\right]$$

$$\approx \frac{0.25}{3}(1 + 3.8806 + 1.7889 + 3.2000 + 1.4142$$

$$+ 2.4988 + 1.1094 + 1.9846 + 0.4472)$$

$$\approx 1.4436$$

2. The average consumption of petroleum in millions of barrels per day during the 10-year period is given by

$$\frac{1}{10}\int_0^{10} f(x) \, dx$$

where f is the function describing the given graph. Using Simpson's Rule with $a = 0$, $b = 10$, and $n = 10$ so that $\Delta x = 1$ and

$$x_0 = 0 \qquad x_1 = 1 \qquad x_2 = 2 \quad \cdots \quad x_{10} = 10$$

we have

$$\frac{1}{10}\int_0^{10} f(x) \, dx = \left(\frac{1}{10}\right)\left(\frac{1}{3}\right)[f(x_0) + 4f(x_1) + 2f(x_2) + 4f(x_3)$$

$$+ \cdots + 4f(x_9) + f(x_{10})]$$

$$= \frac{1}{30}[18.3 + 4(18.6) + 2(18.9) + 4(19.5) + 2(19.7) + 4(19.6)$$

$$+ 2(19.8) + 4(20.0) + 2(20.7) + 4(20.8) + 20.6]$$

$$\approx 19.7$$

or approximately 19.7 million barrels/day.

Improper Integrals

All the definite integrals that we have encountered have had finite intervals of integration. In many applications, however, we are concerned with integrals that have unbounded intervals of integration. These integrals are called **improper integrals.**

To lead us to the definition of an improper integral of a function f over an infinite interval, consider the problem of finding the area of the region R under the curve $y = f(x) = 1/x^2$ and to the right of the vertical line $x = 1$, as shown in Figure 10. Because the interval over which the integration must be performed is unbounded, the methods of integration presented previously cannot be applied directly to solve this problem. However, we can approximate the area of the region R by the definite integral

$$\int_1^b \frac{1}{x^2} \, dx \tag{10}$$

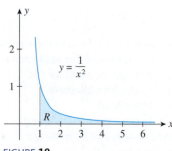

FIGURE 10
The area of the unbounded region R can be approximated by a definite integral.

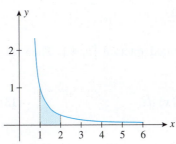

FIGURE 11
Area of shaded region = $\int_1^b \frac{1}{x^2}\,dx$

which gives the area of the region under the curve $y = f(x) = 1/x^2$ from $x = 1$ to $x = b$ (Figure 11). The approximation of the area of the region R by the definite integral (10) seems to improve as the upper limit of integration, b, becomes larger and larger. Figure 12 illustrates the situation for $b = 2$, 3, and 4, respectively.

This observation suggests that if we define a function $I(b)$ by

$$I(b) = \int_1^b \frac{1}{x^2}\,dx \tag{11}$$

then we can find the area of the required region R by evaluating the limit of $I(b)$ as b tends to infinity; that is, the area of R is given by

$$\lim_{b\to\infty} I(b) = \lim_{b\to\infty} \int_1^b \frac{1}{x^2}\,dx \tag{12}$$

if the limit exists.

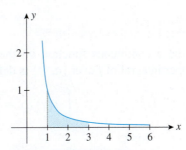

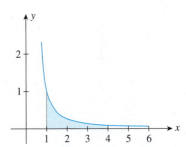

(a) Area of region under the graph of f on [1, 2]

(b) Area of region under the graph of f on [1, 3]

(c) Area of region under the graph of f on [1, 4]

FIGURE 12
As b increases, the approximation of R by the definite integral seems to improve.

EXAMPLE 1

a. Evaluate the definite integral $I(b)$ in Equation (11).
b. Compute $I(b)$ for $b = 10$, 100, 1000, 10,000.
c. Evaluate the limit in Equation (11).
d. Interpret the results of parts (b) and (c).

Solution

a. $I(b) = \displaystyle\int_1^b \frac{1}{x^2}\,dx = -\frac{1}{x}\Big|_1^b = -\frac{1}{b} + 1$

b. From the result of part (a),

$$I(b) = 1 - \frac{1}{b}$$

Therefore,

$$I(10) = 1 - \frac{1}{10} = 0.9$$

$$I(100) = 1 - \frac{1}{100} = 0.99$$

$$I(1000) = 1 - \frac{1}{1000} = 0.999$$

$$I(10{,}000) = 1 - \frac{1}{10{,}000} = 0.9999$$

c. Once again, using the result of part (a), we find

$$\lim_{b \to \infty} I(b) = \lim_{b \to \infty} \int_1^b \frac{1}{x^2}\, dx$$

$$= \lim_{b \to \infty} \left(1 - \frac{1}{b} \right)$$

$$= 1$$

d. The result of part (c) tells us that the area of the region R is 1. The results of the computations performed in part (b) reinforce our expectation that $I(b)$ should approach 1, the area of the region R, as b approaches infinity.

The preceding discussion and the results of Example 1 suggest that we define the improper integral of a continuous function f over the unbounded interval $[a, \infty)$ as follows.

Improper Integral of f over $[a, \infty)$

Let f be a continuous function on the unbounded interval $[a, \infty)$. Then, the improper integral of f over $[a, \infty)$ is defined by

$$\int_a^\infty f(x)\, dx = \lim_{b \to \infty} \int_a^b f(x)\, dx \tag{13}$$

if the limit exists.

If the limit exists, the improper integral is said to be **convergent.** An improper integral for which the limit in Equation (13) fails to exist is said to be **divergent.**

EXAMPLE 2 Evaluate $\int_2^\infty \frac{1}{x}\, dx$ if it converges.

Solution

$$\int_2^\infty \frac{1}{x}\, dx = \lim_{b \to \infty} \int_2^b \frac{1}{x}\, dx$$

$$= \lim_{b \to \infty} \ln x \Big|_2^b$$

$$= \lim_{b \to \infty} (\ln b - \ln 2)$$

Since $\ln b \to \infty$ as $b \to \infty$, the limit does not exist, and we conclude that the given improper integral is divergent.

Explore and Discuss

1. Suppose f is continuous and nonnegative on $[0, \infty)$. Furthermore, suppose $\lim_{x \to \infty} f(x) = L$, where L is a positive number. What can you say about the convergence of the improper integral $\int_0^\infty f(x)\, dx$? Explain your answer, and illustrate with an example.

2. Suppose f is continuous and nonnegative on $[0, \infty)$ and satisfies the condition $\lim_{x \to \infty} f(x) = 0$. Can you conclude that the improper integral $\int_0^\infty f(x)\, dx$ converges? Explain your answer and illustrate with examples.

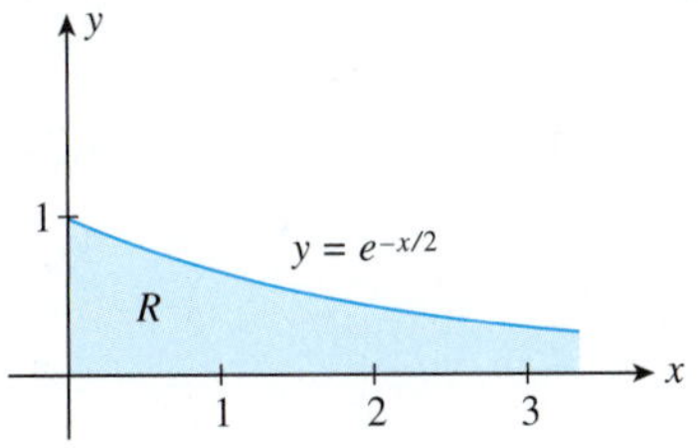

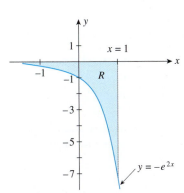

FIGURE 13

Area of $R = \int_0^{\infty} e^{-x/2}\, dx$

EXAMPLE 3 Find the area of the region R under the curve $y = e^{-x/2}$ for $x \geq 0$.

Solution   The region R is shown in Figure 13. Taking $b > 0$, we compute the area of the region under the curve $y = e^{-x/2}$ from $x = 0$ to $x = b$, namely,

$$I(b) = \int_0^b e^{-x/2}\, dx = -2e^{-x/2}\Big|_0^b = -2e^{-b/2} + 2$$

Then the area of the region R is given by

$$\lim_{b \to \infty} I(b) = \lim_{b \to \infty} (2 - 2e^{-b/2}) = 2 - 2 \lim_{b \to \infty} \frac{1}{e^{b/2}}$$
$$= 2$$

Exploring with TECHNOLOGY

You can see how fast the improper integral in Example 3 converges, as follows:

1. Use a graphing utility to plot the graph of $I(b) = 2 - 2e^{-b/2}$, using the viewing window $[0, 50] \times [0, 3]$.

2. Use **TRACE** to follow the values of y for increasing values of x, starting at the origin.

The improper integral defined in Equation (13) has an interval of integration that is unbounded on the right. Improper integrals with intervals of integration that are unbounded on the left also arise in practice and are defined in a similar manner.

Improper Integral of f over $(-\infty, b]$

Let f be a continuous function on the unbounded interval $(-\infty, b]$. Then the improper integral of f over $(-\infty, b]$ is defined by

$$\int_{-\infty}^b f(x)\, dx = \lim_{a \to -\infty} \int_a^b f(x)\, dx \tag{14}$$

if the limit exists.

In this case, the improper integral is said to be convergent. Otherwise, the improper integral is said to be divergent.

EXAMPLE 4 Find the area of the region R bounded above by the x-axis, below by the curve $y = -e^{2x}$, and on the right by the vertical line $x = 1$.

Solution   The region R is shown in Figure 14. Taking $a < 1$, compute the area of the region bounded above by the x-axis ($y = 0$) and below by the curve $y = -e^{2x}$ from $x = a$ to $x = 1$, namely,

$$I(a) = \int_a^1 [0 - (-e^{2x})]\, dx = \int_a^1 e^{2x}\, dx$$
$$= \frac{1}{2} e^{2x}\Big|_a^1 = \frac{1}{2} e^2 - \frac{1}{2} e^{2a}$$

FIGURE 14

Area of $R = -\int_{-\infty}^1 (-e^{2x})\, dx$

Then the area of the required region is given by

$$\lim_{a \to -\infty} I(a) = \lim_{a \to -\infty} \left(\frac{1}{2} e^2 - \frac{1}{2} e^{2a} \right)$$

$$= \frac{1}{2} e^2 - \frac{1}{2} \lim_{a \to -\infty} e^{2a}$$

$$= \frac{1}{2} e^2$$

Another improper integral found in practical applications involves the integration of a function f over the unbounded interval $(-\infty, \infty)$.

> **Improper Integral of f over $(-\infty, \infty)$**
>
> Let f be a continuous function over the unbounded interval $(-\infty, \infty)$. Let c be any real number, and suppose that both the improper integrals
>
> $$\int_{-\infty}^{c} f(x)\, dx \quad \text{and} \quad \int_{c}^{\infty} f(x)\, dx$$
>
> are convergent. Then the improper integral of f over $(-\infty, \infty)$ is defined by
>
> $$\int_{-\infty}^{\infty} f(x)\, dx = \int_{-\infty}^{c} f(x)\, dx + \int_{c}^{\infty} f(x)\, dx \qquad (15)$$

In this case, we say that the improper integral on the left in Equation (15) is convergent. If either one of the two improper integrals on the right in (15) is divergent, then the improper integral on the left is not defined.

Note   Usually, we choose $c = 0$.

EXAMPLE 5 Evaluate the improper integral

$$\int_{-\infty}^{\infty} x e^{-x^2}\, dx$$

and give a geometric interpretation of the results.

Solution   Take the number c in Equation (15) to be 0. Let's first evaluate

$$\int_{-\infty}^{0} x e^{-x^2}\, dx = \lim_{a \to -\infty} \int_{a}^{0} x e^{-x^2}\, dx$$

$$= \lim_{a \to -\infty} \left. -\frac{1}{2} e^{-x^2} \right|_{a}^{0}$$

$$= \lim_{a \to -\infty} \left[-\frac{1}{2} + \frac{1}{2} e^{-a^2} \right] = -\frac{1}{2}$$

Next, we evaluate

$$\int_{0}^{\infty} x e^{-x^2}\, dx = \lim_{b \to \infty} \int_{0}^{b} x e^{-x^2}\, dx$$

$$= \lim_{b \to \infty} \left. -\frac{1}{2} e^{-x^2} \right|_{0}^{b}$$

$$= \lim_{b \to \infty} \left(-\frac{1}{2} e^{-b^2} + \frac{1}{2} \right) = \frac{1}{2}$$

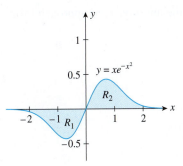

FIGURE 15

$$\int_{-\infty}^{\infty} xe^{-x^2}\, dx = \int_{-\infty}^{0} xe^{-x^2}\, dx + \int_{0}^{\infty} xe^{-x^2}\, dx$$

Therefore,

$$\int_{-\infty}^{\infty} xe^{-x^2}\, dx = \int_{-\infty}^{0} xe^{-x^2}\, dx + \int_{0}^{\infty} xe^{-x^2}\, dx$$

$$= -\frac{1}{2} + \frac{1}{2}$$

$$= 0$$

The graph of $y = xe^{-x^2}$ is sketched in Figure 15. A glance at the figure tells us that the improper integral

$$\int_{-\infty}^{0} xe^{-x^2}\, dx$$

gives the negative of the area of the region R_1, bounded above by the x-axis, below by the curve $y = xe^{-x^2}$, and on the right by the y-axis $(x = 0)$, provided that the integral is convergent.

Also, the improper integral

$$\int_{0}^{\infty} xe^{-x^2}\, dx$$

gives the area of the region R_2 under the curve $y = xe^{-x^2}$ for $x \geq 0$. Since the graph of f is symmetric with respect to the origin, the area of R_1 is equal to the area of R_2. In other words,

$$\int_{-\infty}^{0} xe^{-x^2}\, dx = -\int_{0}^{\infty} xe^{-x^2}\, dx$$

Therefore,

$$\int_{-\infty}^{\infty} xe^{-x^2}\, dx = \int_{-\infty}^{0} xe^{-x^2}\, dx + \int_{0}^{\infty} xe^{-x^2}\, dx$$

$$= -\int_{0}^{\infty} xe^{-x^2}\, dx + \int_{0}^{\infty} xe^{-x^2}\, dx$$

$$= 0$$

as was shown earlier.

Perpetuities

Recall from Section 6.7 that the present value of an annuity is given by

$$PV \approx mP \int_{0}^{T} e^{-rt}\, dt = \frac{mP}{r}\left(1 - e^{-rT}\right) \tag{16}$$

Now, if the payments of an annuity are allowed to continue indefinitely, we have what is called a **perpetuity**. The present value of a perpetuity is given by the improper integral

$$PV = mP \int_{0}^{\infty} e^{-rt}\, dt$$

obtained from Formula (16) by allowing the term of the annuity, T, to approach infinity. Thus,

$$
mP \int_0^\infty e^{-rt}\, dt = \lim_{b \to \infty} mP \int_0^b e^{-rt}\, dt
$$

$$
= mP \lim_{b \to \infty} \int_0^b e^{-rt}\, dt
$$

$$
= mP \lim_{b \to \infty} \left(-\frac{1}{r} e^{-rt} \Big|_0^b \right)
$$

$$
= mP \lim_{b \to \infty} \left(-\frac{1}{r} e^{-rb} + \frac{1}{r} \right) = \frac{mP}{r}
$$

The Present Value of a Perpetuity

The **present value PV of a perpetuity** is given by

$$
PV = \frac{mP}{r} \tag{17}
$$

where m is the number of payments per year, P is the size of each payment, and r is the interest rate per year (compounded continuously).

APPLIED EXAMPLE 6 Endowments The Robinson family wishes to create a scholarship fund at a college. If a scholarship in the amount of $5000 is awarded annually beginning 1 year from now, find the amount of the endowment they are required to make now. Assume that this fund will earn interest at a rate of 4% per year compounded continuously.

Solution   The amount of the endowment, A, is given by the present value of a perpetuity, with $m = 1$, $P = 5000$, and $r = 0.04$. Using Formula (17), we find

$$
A = \frac{(1)(5000)}{0.04}
$$

$$
= 125{,}000
$$

or $125,000.

The improper integral also plays an important role in the study of probability theory, as we will see in Chapter 10.

7.4 Self-Check Exercises

1. Evaluate $\displaystyle\int_{-\infty}^\infty \frac{x^3}{(1 + x^4)^{3/2}}\, dx$.

2. **PRESENT VALUE OF AN INCOME STREAM** Suppose an income stream is expected to continue indefinitely. Then the present value of such a stream can be calculated from the formula for the present value of an income stream by letting T approach infinity. Thus, the required present value is given by

$$
PV = \int_0^\infty P(t)e^{-rt}\, dt
$$

Suppose Marcia has an oil well in her backyard that generates a stream of income given by

$$
P(t) = 20e^{-0.02t}
$$

where $P(t)$ is expressed in thousands of dollars per year and t is the time in years from the present. Assuming that the prevailing interest rate in the foreseeable future is 4%/year compounded continuously, what is the present value of the income stream?

Solutions to Self-Check Exercises 7.4 can be found on page 541.

7.4 **Concept Questions**

1. **a.** Define $\int_a^\infty f(x)\,dx$, where f is continuous on $[a, \infty)$.

   b. Define $\int_{-\infty}^b f(x)\,dx$, where f is continuous on $(-\infty, b]$.

   c. Define $\int_{-\infty}^\infty f(x)\,dx$ where f is continuous on $(-\infty, \infty)$.

2. What is the present value of a perpetuity? Give a formula for computing it.

7.4 **Exercises**

In Exercises 1–4, find the area of the shaded region, if it exists.

1.

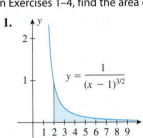

$y = \dfrac{1}{(x-1)^{3/2}}$

2.

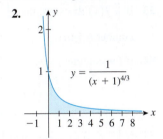

$y = \dfrac{1}{(x+1)^{4/3}}$

3.

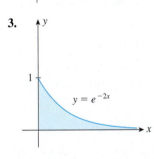

$y = e^{-2x}$

4.

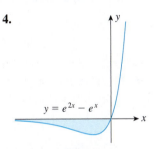

$y = e^{2x} - e^x$

In Exercises 5–14, find the area of the region under the curve $y = f(x)$ over the indicated interval.

5. $f(x) = \dfrac{2}{x^2}; \, x \ge 3$

6. $f(x) = \dfrac{2}{x^3}; \, x \ge 2$

7. $f(x) = \dfrac{1}{(x-2)^2}; \, x \ge 3$

8. $f(x) = \dfrac{2}{(x+1)^3}; \, x \ge 0$

9. $f(x) = \dfrac{1}{x^{3/2}}; \, x \ge 1$

10. $f(x) = \dfrac{3}{x^{5/2}}; \, x \ge 4$

11. $f(x) = \dfrac{1}{(x+1)^{5/2}}; \, x \ge 0$

12. $f(x) = \dfrac{1}{(1-x)^{3/2}}; \, x \le 0$

13. $f(x) = e^{2x}; \, x \le 2$

14. $f(x) = xe^{-x^2}; \, x \ge 0$

15. Find the area of the region bounded by the x-axis and the graph of the function

$$f(x) = \frac{x}{(1 + x^2)^2}$$

16. Find the area of the region bounded by the x-axis and the graph of the function

$$f(x) = \frac{e^x}{(1 + e^x)^2}$$

17. Consider the improper integral

$$\int_0^\infty \sqrt{x}\,dx$$

   a. Evaluate $I(b) = \int_0^b \sqrt{x}\,dx$.

   b. Show that

$$\lim_{b \to \infty} I(b) = \infty$$

   thus proving that the given improper integral is divergent.

18. Consider the improper integral

$$\int_1^\infty x^{-2/3}\,dx$$

   a. Evaluate $I(b) = \int_1^b x^{-2/3}\,dx$.

   b. Show that

$$\lim_{b \to \infty} I(b) = \infty$$

   thus proving that the given improper integral is divergent.

In Exercises 19–46, evaluate each improper integral whenever it is convergent.

19. $\displaystyle\int_1^\infty \frac{3}{x^4}\,dx$

20. $\displaystyle\int_1^\infty \frac{1}{x^3}\,dx$

21. $\displaystyle\int_4^\infty \frac{2}{x^{3/2}}\,dx$

22. $\displaystyle\int_1^\infty \frac{1}{\sqrt{x}}\,dx$

23. $\displaystyle\int_1^\infty \frac{4}{x}\,dx$

24. $\displaystyle\int_2^\infty \frac{3}{x}\,dx$

25. $\displaystyle\int_{-\infty}^0 \frac{1}{(x-2)^3}\,dx$

26. $\displaystyle\int_2^\infty \frac{1}{(x+1)^2}\,dx$

27. $\displaystyle\int_1^\infty \frac{1}{(2x-1)^{3/2}}\,dx$

28. $\displaystyle\int_{-\infty}^0 \frac{1}{(4-x)^{3/2}}\,dx$

29. $\displaystyle\int_0^\infty e^{-x}\,dx$

30. $\displaystyle\int_0^\infty e^{-x/2}\,dx$

31. $\displaystyle\int_{-\infty}^0 e^{2x}\,dx$

32. $\displaystyle\int_{-\infty}^0 e^{3x}\,dx$

33. $\displaystyle\int_1^\infty \frac{e^{\sqrt{x}}}{\sqrt{x}}\,dx$

34. $\displaystyle\int_1^\infty \frac{e^{-\sqrt{x}}}{\sqrt{x}}\,dx$

35. $\displaystyle\int_{-\infty}^{0} xe^x \, dx$

36. $\displaystyle\int_{0}^{\infty} xe^{-2x} \, dx$

37. $\displaystyle\int_{-\infty}^{\infty} x \, dx$

38. $\displaystyle\int_{-\infty}^{\infty} x^3 \, dx$

39. $\displaystyle\int_{-\infty}^{\infty} x^3(1 + x^4)^{-2} \, dx$

40. $\displaystyle\int_{-\infty}^{\infty} x(x^2 + 4)^{-3/2} \, dx$

41. $\displaystyle\int_{-\infty}^{\infty} xe^{1-x^2} \, dx$

42. $\displaystyle\int_{-\infty}^{\infty} \left(x - \frac{1}{2}\right)e^{-x^2+x-1} \, dx$

43. $\displaystyle\int_{-\infty}^{\infty} \frac{e^{-x}}{1 + e^{-x}} \, dx$

44. $\displaystyle\int_{-\infty}^{\infty} \frac{xe^{-x^2}}{1 + e^{-x^2}} \, dx$

45. $\displaystyle\int_{e}^{\infty} \frac{1}{x \ln^3 x} \, dx$

46. $\displaystyle\int_{e^2}^{\infty} \frac{1}{x \ln x} \, dx$

47. THE AMOUNT OF AN ENDOWMENT A university alumni group wishes to provide an annual scholarship in the amount of $1500 beginning next year. If the scholarship fund will earn an interest rate of 5%/year compounded continuously, find the amount of the endowment the alumni are required to make now.

48. THE AMOUNT OF AN ENDOWMENT Mel Thompson wishes to establish a fund to provide a university medical center with an annual research grant of $50,000 beginning next year. If the fund will earn an interest rate of 6%/year compounded continuously, find the amount of the endowment he is required to make now.

49. AMOUNT OF AN INVESTMENT Heidi has an investment that generates an income of $30,000/year in perpetuity. Assuming that her income flows continually and her investment earns interest at the rate of 4%/year compounded continuously, what is the present value of her investment?

50. PERPETUAL NET INCOME STREAMS The present value of a perpetual stream of income that flows continually at the rate of $P(t)$ dollars/year is given by the formula

$$PV = \int_{0}^{\infty} P(t)e^{-rt} \, dt$$

where r is the interest rate compounded continuously. Using this formula, find the present value of a perpetual net income stream that is generated at the rate of

$$P(t) = 10{,}000 + 4000t$$

dollars/year.

Hint: $\displaystyle\lim_{b\to\infty} \frac{b}{e^{rb}} = 0$ if $r > 0$.

51. ESTABLISHING A TRUST FUND Becky Wilkinson wants to establish a trust fund that will provide her children and heirs with a perpetual annuity in the amount of

$$P(t) = 20 + t$$

thousand dollars/year beginning next year. If the trust fund will earn an interest rate of 4%/year compounded continuously, find the amount that she must place in the trust fund now.

Hint: Use the formula given in Exercise 50.

52. PERPETUAL NET INCOME STREAM The perpetual stream of income to be derived from the investment in a proposed new soft drink is assumed to flow continually at the rate of

$$P(t) = 2te^{0.02t}$$

thousand dollars/year. If the prevailing interest rate is assumed to be 5%/year compounded continuously, what is the present value of the investment?

Hint: Use the formula given in Exercise 50.

In Exercises 53–56, determine whether the statement is true or false. If it is true, explain why it is true. If it is false, give an example to show why it is false.

53. If $\displaystyle\int_{a}^{\infty} f(x) \, dx$ exists, then $\displaystyle\int_{b}^{\infty} f(x) \, dx$ exists for every real number $b > a$.

54. If f is continuous on $(-\infty, \infty)$, then

$$\int_{-\infty}^{\infty} f(x) \, dx = \lim_{t\to\infty} \int_{-t}^{t} f(x) \, dx$$

55. If $\displaystyle\int_{-\infty}^{\infty} f(x) \, dx$ exists, then $\displaystyle\int_{0}^{\infty} f(x) \, dx$ exists, and

$$\int_{-\infty}^{\infty} f(x) \, dx = 2\int_{0}^{\infty} f(x) \, dx$$

56. If $\displaystyle\int_{a}^{\infty} f(x) \, dx$ exists, then $\displaystyle\int_{-\infty}^{-a} f(x) \, dx$ exists, and

$$\int_{a}^{\infty} f(x) \, dx = -\int_{-\infty}^{-a} f(x) \, dx$$

57. CAPITAL VALUE The capital value (present sale value) CV of property that can be rented on a perpetual basis for R dollars annually is given by

$$CV \approx \int_{0}^{\infty} Re^{-it} \, dt$$

where i is the prevailing continuous interest rate.
a. Show that $CV \approx R/i$.
b. Find the capital value of property that can be rented at $10,000 annually when the prevailing continuous interest rate is 6%/year.

58. Show that an integral of the form $\displaystyle\int_{a}^{\infty} e^{-px} \, dx$ is convergent if $p > 0$ and divergent if $p \leq 0$.

59. Show that an integral of the form $\displaystyle\int_{-\infty}^{b} e^{px} \, dx$ is convergent if $p > 0$ and divergent if $p \leq 0$.

60. Find the values of p such that $\displaystyle\int_{1}^{\infty} \frac{1}{x^p} \, dx$ is convergent.

7.4 **Solutions to Self-Check Exercises**

1. Write

$$\int_{-\infty}^{\infty} \frac{x^3}{(1+x^4)^{3/2}} \, dx = \int_{-\infty}^{0} \frac{x^3}{(1+x^4)^{3/2}} \, dx + \int_{0}^{\infty} \frac{x^3}{(1+x^4)^{3/2}} \, dx$$

Now,

$$\int_{-\infty}^{0} \frac{x^3}{(1+x^4)^{3/2}} \, dx = \lim_{a \to -\infty} \int_{a}^{0} x^3 (1+x^4)^{-3/2} \, dx$$

$$= \lim_{a \to -\infty} \frac{1}{4} (-2)(1+x^4)^{-1/2} \Big|_{a}^{0} \quad \text{Integrate by substitution.}$$

$$= -\frac{1}{2} \lim_{a \to -\infty} \left[1 - \frac{1}{(1+a^4)^{1/2}} \right]$$

$$= -\frac{1}{2}$$

Similarly, you can show that

$$\int_{0}^{\infty} \frac{x^3}{(1+x^4)^{3/2}} \, dx = \frac{1}{2}$$

Therefore,

$$\int_{-\infty}^{\infty} \frac{x^3}{(1+x^4)^{3/2}} \, dx = -\frac{1}{2} + \frac{1}{2} = 0$$

2. The required present value is given by

$$PV = \int_{0}^{\infty} 20 e^{-0.02t} e^{-0.04t} \, dt$$

$$= 20 \int_{0}^{\infty} e^{-0.06t} \, dt$$

$$= 20 \lim_{b \to \infty} \int_{0}^{b} e^{-0.06t} \, dt$$

$$= -\frac{20}{0.06} \lim_{b \to \infty} e^{-0.06t} \Big|_{0}^{b}$$

$$= -\frac{1000}{3} \lim_{b \to \infty} (e^{-0.06b} - 1)$$

$$= \frac{1000}{3}$$

or approximately \$333,333.

7.5 **Volumes of Solids of Revolution**

Finding the Volume of a Solid of Revolution

In this section we use a Riemann sum to find a formula for computing the volume of a solid that results when a region is revolved about an axis. The solid is called a **solid of revolution**. To find the volume of a solid of revolution, suppose the plane region under the curve defined by a nonnegative continuous function $y = f(x)$ between $x = a$ and $x = b$ is revolved about the x-axis (Figure 16).

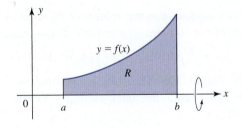

(a) Region R under the curve

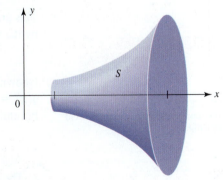

(b) Solid S obtained by revolving R about the x-axis

FIGURE 16

To derive a formula for finding the volume of the solid of revolution, we divide the interval $[a, b]$ into n subintervals, each of equal length, by means of the points $x_0 = a, x_1, x_2, \ldots, x_n = b$, so that the length of each subinterval is given by $\Delta x = (b = a)/n$. Also, let $p_1, p_2, \ldots, p_n$ be points in the subintervals $[x_0, x_1]$, $[x_1, x_2], \ldots, [x_{n-1}, x_n]$, respectively (Figure 17).

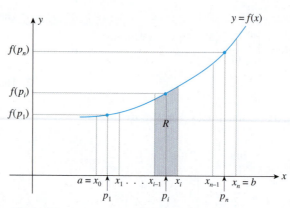

FIGURE 17
The region R is revolved about the x-axis.

Let's concentrate on the part of the solid of revolution that is swept out by revolving the region under the curve between $x = x_{i-1}$ and $x = x_i$ about the x-axis. The volume ΔV_i of this object, shown in Figure 18a, may be approximated by the volume of the disk shown in Figure 18b, of radius $f(p_i)$ and width Δx, obtained by revolving the rectangular region of height $f(p_i)$ and width Δx about the x-axis (Figure 18c).

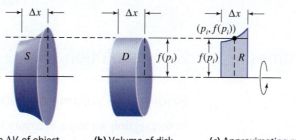

(a) Volume ΔV_i of object   **(b)** Volume of disk   **(c)** Approximating rectangle
FIGURE 18

Since the volume of a disk (cylinder) is equal to $\pi r^2 h$, we see that

$$\Delta V_i \approx \pi [f(p_i)]^2 \Delta x$$

This analysis suggests that the volume of the solid of revolution may be approximated by the sum of the volumes of n suitable disks—namely,

$$\pi [f(p_1)]^2 \Delta x + \pi [f(p_2)]^2 \Delta x + \cdots + \pi [f(p_n)]^2 \Delta x$$

(shown in Figure 19). This last expression is a Riemann sum of the function $g(x) = \pi [f(x)]^2$ over the interval $[a, b]$. Letting n approach infinity, we obtain Formula 18.

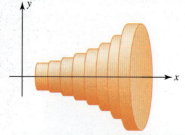

FIGURE 19
The solid of revolution is approximated by n disks.

> **Volume of a Solid of Revolution**
>
> The volume V of the solid of revolution obtained by revolving the region below the graph of a nonnegative function $y = f(x)$ from $x = a$ to $x = b$ about the x-axis is
>
> $$V = \pi \int_a^b [f(x)]^2 \, dx \qquad (18)$$

EXAMPLE 1 Find the volume of the solid of revolution obtained by revolving the region under the curve $y = f(x) = \sqrt{x}$ from $x = 0$ to $x = 2$ about the x-axis.

Solution   The region under the curve and the resulting solid of revolution are shown in Figure 20.

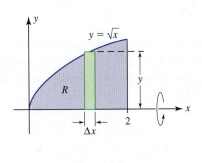

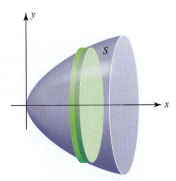

(a) The region under the curve $y = \sqrt{x}$ from $x = 0$ to $x = 2$

(b) The solid of revolution obtained when R is revolved about the x-axis

FIGURE **20**

If R is revolved about the x-axis, we obtain the solid of revolution S.

Using Formula (18), we find that the required volume is given by

$$\pi \int_0^2 (\sqrt{x})^2 \, dx = \pi \int_0^2 x \, dx$$

$$= \frac{\pi}{2} x^2 \Big|_0^2 = \frac{\pi}{2} (4 - 0) = 2\pi$$

EXAMPLE 2 Derive a formula for the volume of a right circular cone of radius r and height h.

Solution   The cone is shown in Figure 21a. It is the solid of revolution obtained by revolving the region R of Figure 21b about the x-axis. The region R is the region under the straight line $y = f(x) = (r/h)x$ between $x = 0$ and $x = h$.

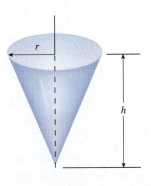

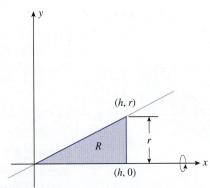

(a) A right circular cone of radius r and height h

(b) R is the region under the straight line $y = \dfrac{r}{h} x$ from $x = 0$ to $x = h$.

FIGURE **21**

Using Formula (18), we see that the required volume is given by

$$V = \pi \int_0^h \left[\left(\frac{r}{h} \right) x \right]^2 dx = \frac{\pi r^2}{h^2} \int_0^h x^2 \, dx$$

$$= \left(\frac{\pi r^2}{3h^2} \right) x^3 \Big|_0^h = \frac{1}{3} \pi r^2 h$$

Next, let's consider the solid of revolution obtained by revolving the region R bounded above by the graph of the nonnegative function $f(x)$ and below by the graph of the nonnegative function $g(x)$, from $x = a$ to $x = b$, about the x-axis (Figure 22).

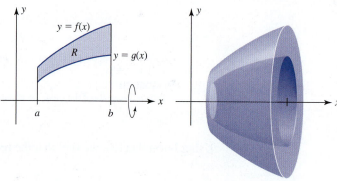

(a) R is the region bounded by the curves $y = f(x)$ and $y = g(x)$ from $x = a$ to $x = b$.

(b) The solid of revolution obtained by revolving R about the x-axis.

FIGURE 22

To derive a formula for computing the volume of this solid of revolution, observe that the required volume is the volume of the solid of revolution obtained by revolving the region under the curve $y = f(x)$ from $x = a$ to $x = b$ about the x-axis *minus* the volume of the solid of revolution obtained by revolving the region under the curve $y = g(x)$ from $x = a$ to $x = b$ about the x-axis. Thus, the required volume is given by

$$V = \pi \int_a^b [f(x)]^2 \, dx - \pi \int_a^b [g(x)]^2 \, dx$$

or

$$V = \pi \int_a^b \left\{ [f(x)]^2 - [g(x)]^2 \right\} dx \tag{19}$$

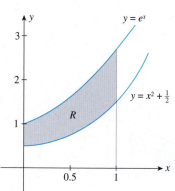

FIGURE 23
R is the region bounded by the curves $y = e^x$ and $y = x^2 + \frac{1}{2}$ from $x = 0$ to $x = 1$.

EXAMPLE 3 Find the volume of the solid of revolution obtained by revolving the region bounded by the curves $y = e^x$ and $y = x^2 + \frac{1}{2}$ from $x = 0$ to $x = 1$ about the x-axis (Figure 23).

Solution Using Formula (19) with $f(x) = e^x$ and $g(x) = x^2 + \frac{1}{2}$, we see that the required volume is given by

$$V = \pi \int_0^1 \left[(e^x)^2 - \left(x^2 + \frac{1}{2} \right)^2 \right] dx$$

$$= \pi \int_0^1 \left(e^{2x} - x^4 - x^2 - \frac{1}{4} \right) dx$$

$$= \pi \left(\frac{1}{2} e^{2x} - \frac{1}{5} x^5 - \frac{1}{3} x^3 - \frac{1}{4} x \right) \Big|_0^1$$

$$= \pi \left[\left(\frac{1}{2} e^2 - \frac{1}{5} - \frac{1}{3} - \frac{1}{4} \right) - \frac{1}{2} \right]$$

$$= \frac{\pi}{60} (30e^2 - 77)$$

or approximately 7.57.

EXAMPLE 4 Find the volume of the solid obtained by revolving the region bounded by the curves $y = f(x) = x^2$ and $y = g(x) = x^3$ about the x-axis.

Solution We first find the point(s) of intersection of the two curves by solving the system of equations

$$\begin{cases} y = x^2 \\ y = x^3 \end{cases}$$

We have

$$x^3 = x^2$$
$$x^3 - x^2 = 0$$
$$x^2(x - 1) = 0$$

so $x = 0$ or $x = 1$, and the points of intersection are $(0, 0)$ and $(1, 1)$ (Figure 24).

Next, observe that $g(x) \le f(x)$ for all values of x between $x = 0$ and $x = 1$, so using Formula (19), we find the required volume to be

$$V = \pi \int_0^1 \left[(x^2)^2 - (x^3)^2 \right] dx$$

$$= \pi \int_0^1 (x^4 - x^6) \, dx$$

$$= \pi \left(\frac{1}{5} x^5 - \frac{1}{7} x^7 \right) \Big|_0^1$$

$$= \pi \left(\frac{1}{5} - \frac{1}{7} \right)$$

$$= \frac{2\pi}{35}$$

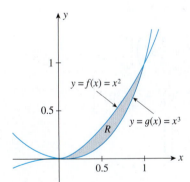

FIGURE 24
R is the region bounded by the curves $y = x^2$ and $y = x^3$.

APPLIED EXAMPLE 5 Fuel Tank of a Space Shuttle The external fuel tank for a space shuttle has a shape that may be obtained by revolving the region under the curve

$$f(x) = \begin{cases} 4\sqrt{10} & \text{if } -120 \le x \le 10 \\ \dfrac{1}{5} \sqrt{x}(30 - x) & \text{if } 10 < x \le 30 \end{cases}$$

from $x = -120$ to $x = 30$ about the x-axis (Figure 25) where all measurements are given in feet. The tank carries liquid hydrogen for fueling the shuttle's three main engines. Estimate the capacity of the tank (231 cubic inches = 1 gallon).

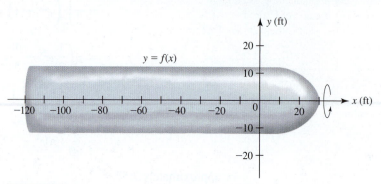

FIGURE 25
The solid of revolution obtained by revolving the region under the curve $y = f(x)$ about the x-axis.

Solution The volume of the tank is given by

$$V = \pi \int_{-120}^{30} [f(x)]^2 \, dx$$

$$= \pi \int_{-120}^{10} (4\sqrt{10})^2 \, dx + \pi \int_{10}^{30} \left[\frac{1}{5} \sqrt{x}(30 - x) \right]^2 \, dx$$

$$= 160\pi \int_{-120}^{10} dx + \frac{\pi}{25} \int_{10}^{30} x(30 - x)^2 \, dx$$

$$= 160\pi x \Big|_{-120}^{10} + \frac{\pi}{25} \int_{10}^{30} (900x - 60x^2 + x^3) \, dx$$

$$= 160\pi(130) + \frac{\pi}{25} \left(450x^2 - 20x^3 + \frac{1}{4}x^4 \right)\Big|_{10}^{30}$$

$$= 20{,}800\pi + \frac{\pi}{25} \left\{ \left[450(30)^2 - 20(30)^3 + \frac{1}{4}(30)^4 \right] \right.$$

$$\left. - \left[450(10)^2 - 20(10)^3 + \frac{1}{4}(10)^4 \right] \right\}$$

$$= 20{,}800\pi + 1600\pi = 22{,}400\pi$$

or approximately 70,372 cubic feet. Therefore, its capacity is approximately $(70{,}372)(12^3)/231$, or approximately 526,419, gallons.

7.5 Self-Check Exercise

Find the volume of the solid of revolution obtained by revolving the region bounded by the graphs of $f(x) = 5 - x^2$ and $g(x) = 1$ about the x-axis.

The solution to Self-Check Exercise 7.5 can be found on page 548.

7.5 Concept Questions

1. If f is a nonnegative function on the interval $[a, b]$ and the region under the graph of f from $x = a$ to $x = b$ is revolved about the x-axis, what is the volume of the resulting solid of revolution?

2. If f and g are nonnegative functions and $f(x) \geq g(x)$ for all x in $[a, b]$, what is the volume of the solid obtained by revolving the region bounded above by the graph of f and below by the graph of g for $a \leq x \leq b$ about the x-axis?

3. Write an expression giving the volume of the solid obtained by revolving the region bounded by the graphs of f and g on $[a, b]$ (shown below) about the x-axis.

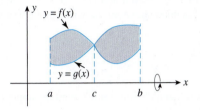

7.5 Exercises

In Exercises 1–4, find the volume of the solid that is obtained by revolving the region about the indicated axis.

1.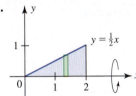

2.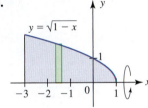

3.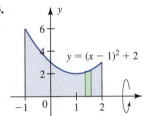

4.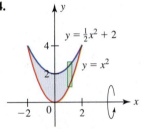

In Exercises 5–14, find the volume of the solid obtained by revolving the region under the curve $y = f(x)$ from $x = a$ to $x = b$ about the x-axis.

5. $y = 3x$; $a = 0, b = 1$

6. $y = x + 1$; $a = 0, b = 2$

7. $y = \sqrt{x}$; $a = 1, b = 4$

8. $y = \sqrt{x + 1}$; $a = 0, b = 3$

9. $y = \sqrt{1 + x^2}$; $a = 0, b = 1$

10. $y = \sqrt{4 - x^2}$; $a = -2, b = 2$

11. $y = 1 - x^2$; $a = -1, b = 1$

12. $y = 2x - x^2$; $a = 0, b = 2$

13. $y = e^x$; $a = 0, b = 1$

14. $y = e^{-3x}$; $a = 0, b = 1$

In Exercises 15–22, find the volume of the solid obtained by revolving the region bounded above by the curve $y = f(x)$ and below by the curve $y = g(x)$ from $x = a$ to $x = b$ about the x-axis.

15. $f(x) = x$ and $g(x) = x^2$; $a = 0, b = 1$

16. $f(x) = 1$ and $g(x) = \sqrt{x}$; $a = 0, b = 1$

17. $f(x) = 4 - x^2$ and $g(x) = 3$; $a = -1, b = 1$

18. $f(x) = 8$ and $g(x) = x^{3/2}$; $a = 0, b = 4$

19. $f(x) = \sqrt{16 - x^2}$ and $g(x) = x$; $a = 0, b = 2\sqrt{2}$

20. $f(x) = e^x$ and $g(x) = \dfrac{1}{x}$; $a = 1, b = 2$

21. $f(x) = e^x$ and $g(x) = e^{-x}$; $a = 0, b = 1$

22. $f(x) = e^x$ and $g(x) = \sqrt{x}$; $a = 1, b = 2$

In Exercises 23–30, find the volume of the solid obtained by revolving the region bounded by the graphs of the functions about the x-axis.

23. $y = x$ and $y = \sqrt{x}$

24. $y = x^{1/3}$ and $y = x^2$

25. $y = \dfrac{1}{2}x + 3$ and $y = x^2$

26. $y = 2x$ and $y = x\sqrt{x + 1}$

27. $y = x^2$ and $y = 4 - x^2$

28. $y = x^2$ and $y = x(4 - x)$

29. $y = \dfrac{1}{x}$, $y = x$, and $y = 2x$

   Hint: You will need to evaluate two integrals.

30. $y = \sqrt{x - 1}$ and $y = (x - 1)^2$

31. By computing the volume of the solid obtained by revolving the region under the semicircle $y = \sqrt{r^2 - x^2}$ from $x = -r$ to $x = r$ about the x-axis, show that the volume of a sphere of radius r is $\frac{4}{3}\pi r^3$ cubic units.

32. VOLUME OF A FOOTBALL Find the volume of the prolate spheroid (a solid of revolution in the shape of a football) obtained by revolving the region under the graph of the function $y = \frac{3}{5}\sqrt{25 - x^2}$ from $x = -5$ to $x = 5$ about the x-axis.

33. CAPACITY OF A MAN-MADE LAKE A man-made lake is approximately circular and has a cross section that is a region bounded above by the x-axis and below by the graph of

$$y = 10\left[\left(\frac{x}{100}\right)^2 - 1\right]$$

(see the accompanying figure). Find the approximate capacity of the lake.

Hint: Find the volume of the solid of revolution obtained by revolving the shaded region about the y-axis. Thus, $V = \pi \int_{-10}^{0} [f(y)]^2\, dy$, where $f(y)$ is obtained by solving the given equation for x in terms of y.

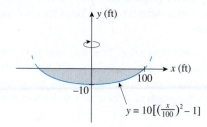

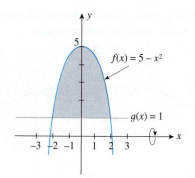

7.5 Solution to Self-Check Exercise

To find the points of intersection of the two curves, we solve the equation $5 - x^2 = 1$, giving $x = \pm 2$ (see the figure). Using Formula (19), we find the required volume:

$$V = \pi \int_{-2}^{2} [(5 - x^2)^2 - 1^2]\, dx$$

$$= 2\pi \int_{0}^{2} (25 - 10x^2 + x^4 - 1)\, dx \quad \text{By symmetry}$$

$$= 2\pi \int_{0}^{2} (x^4 - 10x^2 + 24)\, dx$$

$$= 2\pi \left(\frac{1}{5}x^5 - \frac{10}{3}x^3 + 24x\right)\Big|_{0}^{2}$$

$$= 2\pi \left(\frac{32}{5} - \frac{80}{3} + 48\right)$$

$$= \frac{832\pi}{15}$$

CHAPTER 7 Summary of Principal Formulas and Terms

FORMULAS

1. Integration by parts	$\int u\, dv = uv - \int v\, du$
2. Integration by parts for definite integrals	$\int_{a}^{b} u\, dv = uv\Big|_{a}^{b} - \int_{a}^{b} v\, du$
3. Trapezoidal Rule	$\int_{a}^{b} f(x)\, dx \approx \dfrac{\Delta x}{2}[f(x_0) + 2f(x_1)$ $+ 2f(x_2) + \cdots$ $+ 2f(x_{n-1}) + f(x_n)]$ where $\Delta x = \dfrac{b - a}{n}$

4. Simpson's Rule

$$\int_a^b f(x)\,dx \approx \frac{\Delta x}{3}\left[f(x_0) + 4f(x_1)\right.$$
$$+ 2f(x_2) + 4f(x_3)$$
$$+ 2f(x_4) + \cdots$$
$$\left. + 4f(x_{n-1}) + f(x_n)\right]$$
$$\text{where } \Delta x = \frac{b-a}{n}$$

5. Maximum error for Trapezoidal Rule

$$\frac{M(b-a)^3}{12n^2}, \text{ where } |f''(x)| \le M$$
$$(a \le x \le b)$$

6. Maximum error for Simpson's Rule

$$\frac{M(b-a)^5}{180n^4}, \text{ where } |f^{(4)}(x)| \le M$$
$$(a \le x \le b)$$

7. Improper integral of f over $[a, \infty)$

$$\int_a^\infty f(x)\,dx = \lim_{b \to \infty} \int_a^b f(x)\,dx$$

8. Improper integral of f over $(-\infty, b]$

$$\int_{-\infty}^b f(x)\,dx = \lim_{a \to -\infty} \int_a^b f(x)\,dx$$

9. Improper integral of f over $(-\infty, \infty)$

$$\int_{-\infty}^\infty f(x)\,dx = \int_{-\infty}^c f(x)\,dx + \int_c^\infty f(x)\,dx$$

10. Volume of a solid of revolution

$$V = \pi \int_a^b [f(x)]^2\,dx$$

TERMS

improper integral (532)   divergent integral (534)   solid of revolution (541)

convergent integral (534)   perpetuity (537)

CHAPTER 7   Concept Review Questions

Fill in the blanks.

1. The integration by parts formula is obtained by reversing the _____ Rule. The formula for integration by parts is $\int u\,dv =$ _____. In choosing u and dv, we want du to be simpler than _____ and dv to be _____ _____ _____.

2. To find $I = \int x \ln(x^2 + 1)\,dx$ using the table of integrals, we need to first use the substitution rule with $u =$ _____, so that $du =$ _____ to transform I into the integral $I = \frac{1}{2}\int u \ln u\,du$. We then choose Formula _____ to evaluate this integral.

3. The Trapezoidal Rule states that $\int_a^b f(x)\,dx \approx$ _____ where $\Delta x = \frac{b-a}{n}$. The error E_n in approximating $\int_a^b f(x)\,dx$ by the Trapezoidal Rule satisfies $|E_n| \le$ _____.

4. Simpson's Rule states that $\int_a^b f(x)\,dx \approx$ _____ where $\Delta x = \frac{b-a}{n}$ and n is _____. The error E_n in approximating $\int_a^b f(x)\,dx$ by Simpson's Rule satisfies $|E_n| \le$ _____.

5. The improper integrals $\int_{-\infty}^b f(x)\,dx =$ _____, $\int_a^\infty f(x)\,dx =$ _____, and $\int_{-\infty}^\infty f(x)\,dx =$ _____.

6. The volume V of the solid of revolution obtained by revolving the _____ below the graph of a nonnegative continuous function from _____ to _____ about the x-axis is _____.

In Exercises 1–6, evaluate the integral.

1. $\displaystyle\int 2xe^{-x}\, dx$

2. $\displaystyle\int xe^{4x}\, dx$

3. $\displaystyle\int \ln 5x\, dx$

4. $\displaystyle\int_{1}^{4} \ln 2x\, dx$

5. $\displaystyle\int_{0}^{1} xe^{-2x}\, dx$

6. $\displaystyle\int_{0}^{2} xe^{2x}\, dx$

7. Find the function f given that the slope of the tangent line to the graph of f at any point $(x, f(x))$ is

$$f'(x) = \frac{\ln x}{\sqrt{x}}$$

and that the graph of f passes through the point $(1, -2)$.

8. Find the function f given that the slope of the tangent line to the graph of f at any point $(x, f(x))$ is

$$f'(x) = xe^{-3x}$$

and that the graph of f passes through the point $(0, 0)$.

In Exercises 9–14, use the table of integrals in Section 7.2 to evaluate the integral.

9. $\displaystyle\int \frac{x}{(3 + 2x)^2}\, dx$

10. $\displaystyle\int \frac{2x}{\sqrt{2x + 3}}\, dx$

11. $\displaystyle\int x^2 e^{4x}\, dx$

12. $\displaystyle\int \frac{dx}{(x^2 - 25)^{3/2}}$

13. $\displaystyle\int \frac{dx}{x^2\sqrt{x^2 - 4}}$

14. $\displaystyle\int 8x^3 \ln 2x\, dx$

In Exercises 15–20, evaluate each improper integral whenever it is convergent.

15. $\displaystyle\int_{0}^{\infty} e^{-2x}\, dx$

16. $\displaystyle\int_{-\infty}^{0} e^{3x}\, dx$

17. $\displaystyle\int_{3}^{\infty} \frac{2}{x}\, dx$

18. $\displaystyle\int_{2}^{\infty} \frac{1}{(x + 2)^{3/2}}\, dx$

19. $\displaystyle\int_{2}^{\infty} \frac{dx}{(1 + 2x)^2}$

20. $\displaystyle\int_{1}^{\infty} 3e^{1-x}\, dx$

In Exercises 21–24, use the Trapezoidal Rule and Simpson's Rule to approximate the value of the definite integral.

21. $\displaystyle\int_{1}^{3} \frac{dx}{1 + \sqrt{x}}; n = 4$

22. $\displaystyle\int_{0}^{1} e^{x^2}\, dx; n = 4$

23. $\displaystyle\int_{-1}^{1} \sqrt{1 + x^4}\, dx; n = 4$

24. $\displaystyle\int_{1}^{3} \frac{e^x}{x}\, dx; n = 4$

25. Find a bound on the error in approximating the integral $\displaystyle\int_{0}^{1} \frac{dx}{x + 1}$ with $n = 8$ using (a) the Trapezoidal Rule and (b) Simpson's Rule.

26. PRODUCERS' SURPLUS The supply equation for the GTC Slim-Phone is given by

$$p = 2\sqrt{25 + x^2}$$

where p is the unit price in dollars and x is the quantity demanded per month in units of 10,000. Find the producers' surplus if the market price is $26. Use the table of integrals in Section 7.2 to evaluate the definite integral.

27. COMPUTER GAME SALES The sales of Starr Communication's newest computer game, Laser Beams, are currently

$$te^{-0.05t}$$

units/month (each unit representing 1000 games), where t denotes the number of months since the release of the game. Find an expression that gives the total number of games sold as a function of t. How many games will be sold by the end of the first year?

28. DEMAND FOR COMPUTER SOFTWARE The demand equation for a computer software program is given by

$$p = 2\sqrt{325 - x^2}$$

where p is the unit price in dollars and x is the quantity demanded each month in units of a thousand. Find the consumers' surplus if the market price is $30. Evaluate the definite integral using Simpson's Rule with $n = 10$.

29. OIL SPILLS Using aerial photographs, the Coast Guard was able to determine the dimensions of an oil spill along an embankment on a coastline, as shown in the accompanying figure. Using (a) the Trapezoidal Rule and (b) Simpson's Rule with $n = 10$, estimate the area of the oil spill.

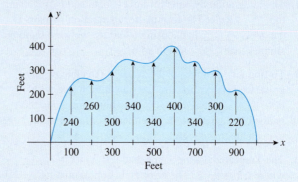

30. SURFACE AREA OF A LAKE A man-made lake located in Lake View Condominiums has the shape depicted in the following figure. The measurements shown were taken at 15-ft intervals. Using Simpson's Rule with $n = 10$, estimate the surface area of the lake.

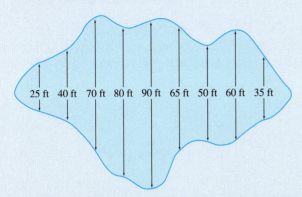

25 ft 40 ft 70 ft 80 ft 90 ft 65 ft 50 ft 60 ft 35 ft

31. PERPETUITIES Lindsey wishes to establish a memorial fund at Newtown Hospital in the amount of $10,000/year beginning next year. If the fund earns interest at a rate of 4%/year compounded continuously, find the amount of endowment that he is required to make now.

32. Find the volume of the solid that is obtained by revolving the region about the indicated axis.

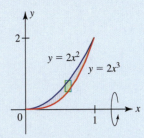

33. Find the volume of the solid of revolution obtained by revolving the region under the curve $f(x) = 1/x$ from $x = 1$ to $x = 3$ about the x-axis.

34. Find the volume of the solid of revolution obtained by revolving the region bounded above by the curve $f(x) = \sqrt{x}$ and below by the curve $g(x) = x^2$ about the x-axis.

CHAPTER 7 Before Moving On . . .

1. Find $\displaystyle\int x^2 \ln x \, dx$.

2. Use the table of integrals to find $\displaystyle\int \frac{dx}{x^2\sqrt{8 + 2x^2}}$.

3. Evaluate $\displaystyle\int_2^4 \sqrt{x^2 + 1}\, dx$ using the Trapezoidal Rule with $n = 5$.

4. Evaluate $\displaystyle\int_1^3 e^{0.2x}\, dx$ using Simpson's Rule with $n = 6$.

5. Evaluate $\displaystyle\int_1^\infty e^{-2x}\, dx$.

6. Find the volume of the solid obtained by revolving the region under the graph of $y = \sqrt{x}$ on $[0, 2]$ about the x-axis.

Calculus of Several Variables

U P TO NOW, we have dealt with functions involving one variable. However, many situations involve functions of two or more variables. For example, the Consumer Price Index (CPI) compiled by the Bureau of Labor Statistics depends on the price of more than 95,000 consumer items. To study such relationships, we need the notion of a function of several variables, the first topic in this chapter. Next, generalizing the concept of the derivative of a function of one variable, we study the *partial derivatives* of a function of two or more variables. Using partial derivatives, we study the rate of change of a function with respect to one variable while holding all other variables constant. We then learn how to find the extremum values of a function of several variables. As an application of optimization theory, we learn how to find an equation of the straight line that "best" fits a set of data points scattered about a straight line. We also extend the concept of the differential of one variable, which was introduced in Section 3.7, to the case of a function of two or more variables. Finally, we generalize the notion of the integral to the case involving a function of two variables.

What should the dimensions of the new swimming pool be? It will be built in an elliptical area located in the rear of the promenade deck. Subject to this constraint, what are the dimensions of the largest pool that can be built? See Example 5, page 610, to see how to solve this problem.

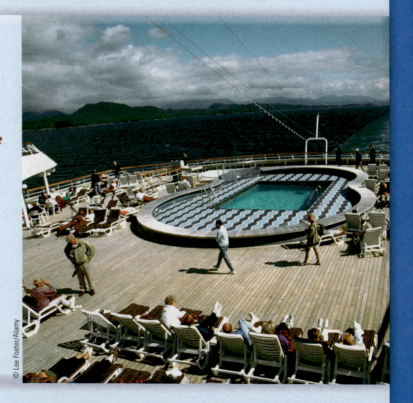

© Lee Foster/Alamy

8.1 Functions of Several Variables

Up to now, our study of calculus has been restricted to functions of one variable. In many practical situations, however, the formulation of a problem results in a mathematical model that involves a function of two or more variables. For example, suppose Ace Novelty determines that the profits are $6, $5, and $4 for three types of souvenirs it produces. Let x, y, and z denote the number of Type A, Type B, and Type C souvenirs to be made; then the company's profit is given by

$$P = 6x + 5y + 4z$$

and P is a function of the three variables, x, y, and z.

Functions of Two Variables

Although this chapter deals with real-valued functions of several variables, most of our definitions and results are stated in terms of a function of two variables. One reason for adopting this approach, as you will soon see, is that there is a geometric interpretation for this special case, which serves as an important visual aid. We can then draw upon the experience gained from studying the two-variable case to help us understand the concepts and results connected with the more general case, which, by and large, is just a simple extension of the lower-dimensional case.

> **A Function of Two Variables**
>
> A real-valued **function of two variables** f consists of
>
> 1. A set A of ordered pairs of real numbers (x, y) called the **domain** of the function.
> 2. A rule that associates with each ordered pair in the domain of f one and only one real number, denoted by $z = f(x, y)$.

The variables x and y are called **independent variables,** and the variable z, which is dependent on the values of x and y, is referred to as a **dependent variable.**

As in the case of a real-valued function of one real variable, the number $z = f(x, y)$ is called the **value of** f at the point (x, y). And, unless specified, the domain of the function f will be taken to be the largest possible set for which the rule defining f is meaningful.

EXAMPLE 1 Let f be the function defined by

$$f(x, y) = x + xy + y^2 + 2$$

Compute $f(0, 0)$, $f(1, 2)$, and $f(2, 1)$.

Solution   We have

$$f(0, 0) = 0 + (0)(0) + 0^2 + 2 = 2$$
$$f(1, 2) = 1 + (1)(2) + 2^2 + 2 = 9$$
$$f(2, 1) = 2 + (2)(1) + 1^2 + 2 = 7$$

The domain of a function of two variables $f(x, y)$ is a set of ordered pairs of real numbers and may therefore be viewed as a subset of the xy-plane.

EXAMPLE 2 Find the domain of each function.

a. $f(x, y) = x^2 + y^2$ **b.** $g(x, y) = \dfrac{2}{x - y}$ **c.** $h(x, y) = \sqrt{1 - x^2 - y^2}$

Solution

a. $f(x, y)$ is defined for all real values of x and y, so the domain of the function f is the set of all points (x, y) in the xy-plane.
b. $g(x, y)$ is defined for all $x \neq y$, so the domain of the function g is the set of all points in the xy-plane except those lying on the line $y = x$ (Figure 1a).
c. We require that $1 - x^2 - y^2 \geq 0$ or $x^2 + y^2 \leq 1$, which is just the set of all points (x, y) lying on and inside the circle of radius 1 with center at the origin (Figure 1b).

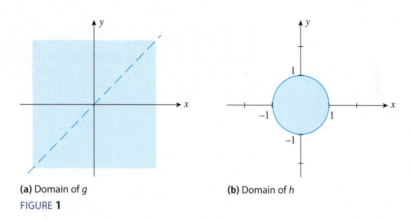

(a) Domain of g (b) Domain of h

FIGURE **1**

💲 APPLIED EXAMPLE 3 Revenue Functions Acrosonic manufactures a book-shelf loudspeaker system that may be bought fully assembled or in a kit. The demand equations that relate the unit prices, p and q, to the quantities demanded weekly, x and y, of the assembled and kit versions of the loudspeaker systems are given by

$$p = 300 - \frac{1}{4}x - \frac{1}{8}y \quad \text{and} \quad q = 240 - \frac{1}{8}x - \frac{3}{8}y$$

a. What is the weekly total revenue function $R(x, y)$?
b. What is the domain of the function R?

Solution

a. The weekly revenue realizable from the sale of x units of the assembled speaker systems at p dollars per unit is given by xp dollars. Similarly, the weekly revenue realizable from the sale of y units of the kits at q dollars per unit is given by yq dollars. Therefore, the weekly total revenue function R is given by

$$R(x, y) = xp + yq$$

$$= x\left(300 - \frac{1}{4}x - \frac{1}{8}y\right) + y\left(240 - \frac{1}{8}x - \frac{3}{8}y\right) \quad (x^2) \text{ See page 8.}$$

$$= -\frac{1}{4}x^2 - \frac{3}{8}y^2 - \frac{1}{4}xy + 300x + 240y$$

b. To find the domain of the function R, let's observe that the quantities x, y, p, and q must be nonnegative. This observation leads to the following system of linear inequalities:

$$300 - \frac{1}{4}x - \frac{1}{8}y \geq 0 \tag{1}$$

$$240 - \frac{1}{8}x - \frac{3}{8}y \geq 0 \tag{2}$$

$$x \geq 0 \tag{3}$$

$$y \geq 0 \tag{4}$$

This system of linear inequalities defines a region D in the xy-plane that is the domain of R. To sketch D, we first draw the line defined by the equation

$$300 - \frac{1}{4}x - \frac{1}{8}y = 0$$

obtained from Inequality (1) by replacing it by an equality (see Figure 2). Observe that this line divides the plane into two half-planes. To find the half-plane determined by Inequality (1), we pick a point lying in one of the half-planes. For simplicity, we pick the origin, $(0, 0)$. This *test point* satisfies Inequality (1), since

$$300 - \frac{1}{4}(0) - \frac{1}{8}(0) \geq 0$$

This shows that the *lower* half-plane (the half-plane containing the test point) is the half-plane described by Inequality (1). Note that the line itself is included in the graph of the inequality, since equality is also allowed in Inequality (1).

Similarly, we can show that Inequality (2) defines the lower half-plane determined by the equation

$$240 - \frac{1}{8}x - \frac{3}{8}y = 0$$

obtained from replacing Inequality (2) by an equation. Once again, the line itself is included in the graph.

Finally, Inequalities (3) and (4) together define the first quadrant with the positive x- and y-axis. Therefore, the region D is obtained by taking the intersection of the two half-planes in the first quadrant.

The domain D of the function R is sketched in Figure 2.

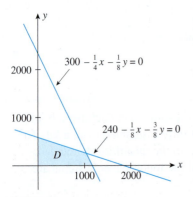

FIGURE 2
The domain of $R(x, y)$

Explore and Discuss

Suppose the total profit of a two-product company is given by $P(x, y)$, where x denotes the number of units of the first product produced and sold and y denotes the number of units of the second product produced and sold. Fix $x = a$, where a is a positive number such that (a, y) is in the domain of P. Describe and give an economic interpretation of the function $f(y) = P(a, y)$. Next, fix $y = b$, where b is a positive number such that (x, b) is in the domain of P. Describe and give an economic interpretation of the function $g(x) = P(x, b)$.

$ APPLIED EXAMPLE 4 Home Mortgage Payments The monthly payment that amortizes a loan of A dollars in t years when the interest rate is r per year compounded monthly is given by

$$P = f(A, r, t) = \frac{Ar}{12\left[1 - \left(1 + \frac{r}{12}\right)^{-12t}\right]}$$

Find the monthly payment for a home mortgage of $270,000 to be amortized over 30 years when the interest rate is 6% per year, compounded monthly.

Solution   Letting $A = 270,000$, $r = 0.06$, and $t = 30$, we find the required monthly payment to be

$$P = f(270,000, 0.06, 30) = \frac{270,000(0.06)}{12[1 - (1 + \frac{0.06}{12})^{-360}]}$$

$$\approx 1618.79$$

or approximately $1618.79.

Graphs of Functions of Two Variables

To graph a function of two variables, we need a three-dimensional coordinate system. This is readily constructed by adding a third axis to the plane Cartesian coordinate system in such a way that the three resulting axes are mutually perpendicular and intersect at O. Observe that, by construction, the zeros of the three number scales coincide at the origin of the **three-dimensional Cartesian coordinate system** (Figure 3).

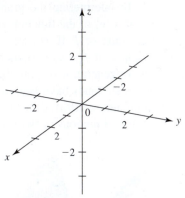

FIGURE **3**
The three-dimensional Cartesian coordinate system

A point in three-dimensional space can now be represented uniquely in this coordinate system by an **ordered triple** of numbers (x, y, z), and, conversely, every ordered triple of real numbers (x, y, z) represents a point in three-dimensional space (Figure 4a). For example, the points $A(2, 3, 4)$, $B(1, -2, -2)$, $C(2, 4, 0)$, and $D(0, 0, 4)$ are shown in Figure 4b.

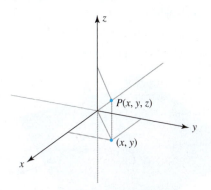

(a) A point in three-dimensional space

(b) Some sample points in three-dimensional space

FIGURE **4**

Now, if $f(x, y)$ is a function of two variables x and y, the domain of f is a subset of the xy-plane. If we denote $f(x, y)$ by z, then the totality of all points (x, y, z), that is, $(x, y, f(x, y))$, makes up the **graph** of the function f and is, except for certain degenerate cases, a surface in three-dimensional space (Figure 5).

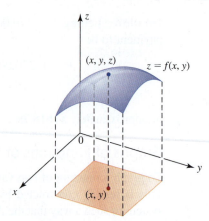

FIGURE 5
The graph of a function in three-dimensional space

In interpreting the graph of a function $f(x, y)$, one often thinks of the value $z = f(x, y)$ of the function at the point (x, y) as the "height" of the point (x, y, z) on the graph of f. If $f(x, y) > 0$, then the point (x, y, z) is $f(x, y)$ units above the xy-plane; if $f(x, y) < 0$, then the point (x, y, z) is $|f(x, y)|$ units below the xy-plane.

In general, it is quite difficult to draw the graph of a function of two variables. But techniques have been developed that enable us to generate such graphs with a minimum of effort, using a computer. Figure 6 shows the computer-generated graphs of some functions of two variables.

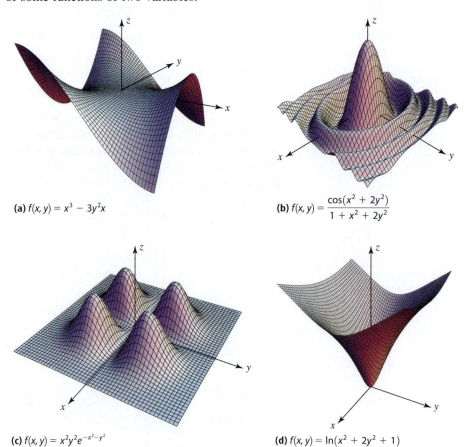

(a) $f(x, y) = x^3 - 3y^2x$

(b) $f(x, y) = \dfrac{\cos(x^2 + 2y^2)}{1 + x^2 + 2y^2}$

(c) $f(x, y) = x^2 y^2 e^{-x^2 - y^2}$

(d) $f(x, y) = \ln(x^2 + 2y^2 + 1)$

FIGURE 6
Four computer-generated graphs of functions of two variables

Level Curves

We can visualize the graph of a function of two variables by using *level curves*. To define the level curve of a function f of two variables, let $z = f(x, y)$ and consider the trace of f in the plane $z = k$ (k, a constant), as shown in Figure 7a. If we project this trace onto the xy-plane, we obtain a curve C with equation $f(x, y) = k$, called a *level curve* of f (Figure 7b).

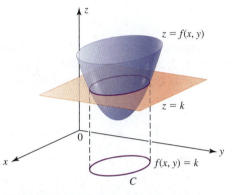

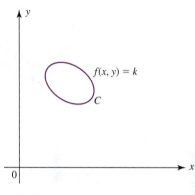

(a) The level curve C with equation $f(x, y) = k$ is the projection of the trace of f in the plane $z = k$ onto the xy-plane.

(b) The level curve C

FIGURE **7**

Level Curves

The **level curves** of a function f of two variables are the curves with equations $f(x, y) = k$, where k is a constant in the range of f.

Notice that the level curve with equation $f(x, y) = k$ is the set of all points in the domain of f corresponding to the points on the surface $z = f(x, y)$ having the same height or depth k. By drawing the level curves corresponding to several admissible values of k, we obtain a *contour map*. The map enables us to visualize the surface represented by the graph of $z = f(x, y)$: We simply lift or depress the level curve to see the "cross sections" of the surface. Figure 8a shows a hill, and Figure 8b shows a contour map associated with that hill.

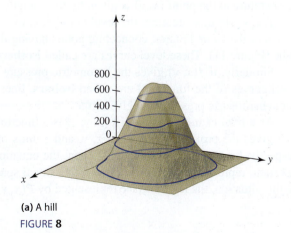

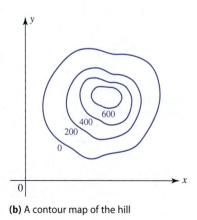

(a) A hill

(b) A contour map of the hill

FIGURE **8**

EXAMPLE 5 Sketch a contour map for the function $f(x, y) = x^2 + y^2$.

Solution   The level curves are the graphs of the equation $x^2 + y^2 = k$ for nonnegative numbers k. Taking $k = 0, 1, 4, 9$, and 16, for example, we obtain

$$k = 0: \quad x^2 + y^2 = 0$$
$$k = 1: \quad x^2 + y^2 = 1$$
$$k = 4: \quad x^2 + y^2 = 4 = 2^2$$
$$k = 9: \quad x^2 + y^2 = 9 = 3^2$$
$$k = 16: \quad x^2 + y^2 = 16 = 4^2$$

The five level curves are concentric circles with center at the origin and radius given by $r = 0, 1, 2, 3$, and 4, respectively (Figure 9a). A sketch of the graph of $f(x, y) = x^2 + y^2$ is included for your reference in Figure 9b.

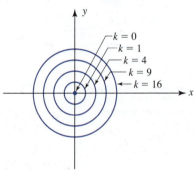

(a) Contour map for $f(x, y) = x^2 + y^2$

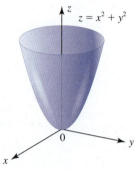

(b) The graph of $f(x, y) = x^2 + y^2$

FIGURE **9**

EXAMPLE 6 Sketch the level curves for the function $f(x, y) = 2x^2 - y$ corresponding to $z = -2, -1, 0, 1$, and 2.

Solution   The level curves are the graphs of the equation $2x^2 - y = k$ or $y = 2x^2 - k$ for $k = -2, -1, 0, 1$, and 2. The required level curves are shown in Figure 10.

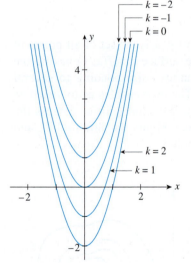

FIGURE **10**
Level curves for $f(x, y) = 2x^2 - y$

Level curves of functions of two variables are found in many practical applications. For example, if $f(x, y)$ denotes the temperature at a location within the continental United States with longitude x and latitude y at a certain point in time, then the temperature at the point (x, y) is given by the "height" of the surface, represented by $z = f(x, y)$. In this situation, the level curve $f(x, y) = k$ is a curve superimposed on a map of the United States, connecting points having the same temperature at a given time (Figure 11). These level curves are called **isotherms.**

Similarly, if $f(x, y)$ gives the barometric pressure at the location (x, y), then the level curves of the function f are called **isobars,** lines connecting points having the same barometric pressure at a given time.

As a final example, suppose $P(x, y, z)$ is a function of three variables x, y, and z that gives the profit realized when x, y, and z units of three products, A, B, and C, respectively, are produced and sold. Then, the equation $P(x, y, z) = k$, where k is a constant, represents a surface in three-dimensional space called a **level surface** of P. In this situation, the level surface represented by $P(x, y, z) = k$ represents the product mix that results in a profit of exactly k dollars. Such a level surface is called an **iso-profit surface.**

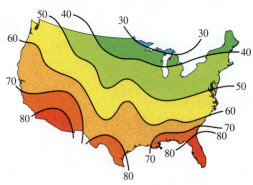

FIGURE **11**
Isotherms: curves connecting points that have the same temperature

---

8.1 Self-Check Exercises

1. Let $f(x, y) = x^2 - 3xy + \sqrt{x + y}$. Compute $f(1, 3)$ and $f(-1, 1)$. Is the point $(-1, 0)$ in the domain of f?

2. Find the domain of $f(x, y) = \dfrac{1}{x} + \dfrac{1}{x - y} - e^{x+y}$.

3. **EFFECT OF ADVERTISING ON REVENUE** Odyssey Travel Agency has a monthly advertising budget of $20,000. Odyssey's management estimates that if they spend x dollars on newspaper advertising and y dollars on television adver-

tising, then the monthly revenue will be

$$f(x, y) = 30x^{1/4}y^{3/4}$$

dollars. What will be the monthly revenue if Odyssey spends $5000/month on newspaper ads and $15,000/month on television ads? If Odyssey spends $4000/month on newspaper ads and $16,000/month on television ads?

Solutions to Self-Check Exercises 8.1 can be found on page 566.

8.1 Concept Questions

1. What is a function of two variables? Give an example of a function of two variables, and state its rule of definition and domain.

2. If f is a function of two variables and (a, b) is in the domain of f with $c = a$ and $d = b$, what can you say about the relationship between $f(a, b)$ and $f(c, d)$?

3. Define (a) the graph of $f(x, y)$ and (b) a level curve of f.

4. Suppose f is a function of two variables, and let $P = f(x, y)$ denote the profit realized when x units of Product A and y units of Product B are produced and sold. Give an interpretation of the level curve of f, defined by the equation $f(x, y) = k$, where k is a positive constant.

8.1 Exercises

1. Let $f(x, y) = 2x + 3y - 4$. Compute $f(0, 0), f(1, 0)$, $f(0, 1), f(1, 2)$, and $f(2, -1)$.

2. Let $g(x, y) = 2x^2 - y^2$. Compute $g(1, 2), g(2, 1)$, $g(1, 1), g(-1, 1)$, and $g(2, -1)$.

3. Let $f(x, y) = x^2 + 2xy - x + 3$. Compute $f(1, 2)$, $f(2, 1), f(-1, 2)$, and $f(2, -1)$.

4. Let $h(x, y) = (x + y)/(x - y)$. Compute $h(0, 1)$, $h(-1, 1), h(2, 1)$, and $h(\pi, -\pi)$.

5. Let $g(s, t) = 3s\sqrt{t} + t\sqrt{s} + 2$. Compute $g(1, 2)$, $g(2, 1)$, $g(0, 4)$, and $g(4, 9)$.

6. Let $f(x, y) = xye^{x^2+y^2}$. Compute $f(0, 0)$, $f(0, 1)$, $f(1, 1)$, and $f(-1, -1)$.

7. Let $h(s, t) = s \ln t - t \ln s$. Compute $h(1, e)$, $h(e, 1)$, and $h(e, e)$.

8. Let $f(u, v) = (u^2 + v^2)e^{uv^2}$. Compute $f(0, 1)$, $f(-1, -1)$, $f(a, b)$, and $f(b, a)$.

9. Let $g(r, s, t) = re^{s/t}$. Compute $g(1, 1, 1)$, $g(1, 0, 1)$, and $g(-1, -1, -1)$.

10. Let $g(u, v, w) = (ue^{vw} + ve^{uw} + we^{uv})/(u^2 + v^2 + w^2)$. Compute $g(1, 2, 3)$ and $g(3, 2, 1)$.

In Exercises 11–18, find the domain of the function.

11. $f(x, y) = 2x + 3y$

12. $g(x, y, z) = x^2 + y^2 + z^2$

13. $h(u, v) = \dfrac{uv}{u - v}$

14. $f(s, t) = \sqrt{s^2 + t^2}$

15. $g(r, s) = \sqrt{rs}$

16. $f(x, y) = e^{-xy}$

17. $h(x, y) = \ln(x + y - 5)$

18. $h(u, v) = \sqrt{4 - u^2 - v^2}$

In Exercises 19–24, sketch the level curves of the function corresponding to each value of z.

19. $f(x, y) = 2x + 3y$; $z = -2, -1, 0, 1, 2$

20. $f(x, y) = -x^2 + y$; $z = -2, -1, 0, 1, 2$

21. $f(x, y) = 2x^2 + y$; $z = -2, -1, 0, 1, 2$

22. $f(x, y) = xy$; $z = -4, -2, 2, 4$

23. $f(x, y) = \sqrt{16 - x^2 - y^2}$; $z = 0, 1, 2, 3, 4$

24. $f(x, y) = e^x - y$; $z = -2, -1, 0, 1, 2$

25. Find an equation of the level curve of $f(x, y) = \sqrt{x^2 + y^2}$ that contains the point $(3, 4)$.

26. Find an equation of the level surface of $f(x, y, z) = 2x^2 + 3y^2 - z$ that contains the point $(-1, 2, -3)$.

In Exercises 27 and 28, match the graph of the surface with one of the contour maps labeled (a) and (b).

(a)

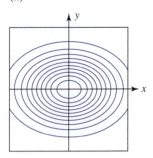

(b)

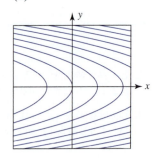

27. $f(x, y) = x + y^2$

28. $f(x, y) = e^{1-2x^2-4y^2}$

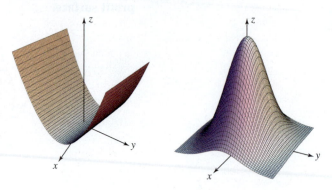

29. Can two level curves of a function f of two variables x and y intersect? Explain.

30. A *level set* of f is the set $S = \{(x, y) \mid f(x, y) = k\}$, where k is in the range of f. Let

$$f(x, y) = \begin{cases} 0 & \text{if } x^2 + y^2 < 1 \\ x^2 + y^2 - 1 & \text{if } x^2 + y^2 \ge 1 \end{cases}$$

Sketch the level set of f for $k = 0$ and 3.

31. The volume of a cylindrical tank of radius r and height h is given by

$$V = f(r, h) = \pi r^2 h$$

Find the volume of a cylindrical tank of radius 1.5 ft and height 4 ft.

32. IQs The IQ (intelligence quotient) of a person whose mental age is m years and whose chronological age is c years is defined as

$$f(m, c) = \frac{100m}{c}$$

What is the IQ of a 9-year-old child who has a mental age of 13.5 years?

33. PRICE-TO-EARNINGS RATIO The current P/E ratio (price-to-earnings ratio) of a stock is defined as

$$R = \frac{P}{E}$$

where P is the current market price per share of the stock and E is the earnings per share for the most recent 12-month period.

a. What is the domain of the function R?

b. The earnings per share of IBM Corporation for 2011 were $13.09, and its price per share on March 19, 2012 was $205.56. What was the P/E ratio of IBM at that time?

Source: IBM Corporate Annual Report.

34. CURRENT DIVIDEND YIELD The current dividend yield for common stock is calculated by using the formula

$$Y = \frac{D}{P}$$

where D is the most recent full-year dividend and P is the current price per share (both measured in dollars).

a. What is the domain of the function Y?

b. The annualized dividend of IBM Corporation for the year 2011 was $3, and its price per share was $205.56 on March 19, 2012. What was the current dividend yield for the common stock of IBM at that time?

Source: IBM Corporation Annual Report.

35. BODY MASS INDEX The body mass index (BMI) is used to identify, evaluate, and treat overweight and obese adults. The BMI value for an adult of weight w (in kilograms) and height h (in meters) is defined to be

$$M = f(w, h) = \frac{w}{h^2}$$

According to federal guidelines, an adult is overweight if he or she has a BMI value greater than 25 but less than 30 and is "obese" if the value is greater than or equal to 30.

a. What is the BMI of an adult who weighs in at 80 kg and stands 1.8 m tall?

b. What is the maximum weight for an adult of height 1.8 m, who is not classified as overweight or obese?

36. POISEUILLE'S LAW Poiseuille's Law states that the resistance R of blood flowing in a blood vessel of length l and radius r is given by

$$R = f(l, r) = \frac{kl}{r^4}$$

where k is a constant that depends on the viscosity of blood. What is the resistance, in terms of k, of blood flowing through an arteriole 4 cm long and of radius 0.1 cm?

37. COST FUNCTION FOR A LOUDSPEAKER SYSTEM Acrosonic manufactures a bookshelf loudspeaker system that may be bought fully assembled or in a kit. The costs in labor and material incurred in manufacturing a fully assembled system and a kit are $200 and $120, respectively. In addition, Acrosonic has fixed costs of $20,000/month.

a. Write the monthly cost function C for Acrosonic in terms of the number of fully assembled systems x and the number of kits y manufactured.

b. What is the domain of C?

c. What is the total cost incurred in manufacturing 1000 fully assembled systems and 200 kits in a month?

38. REVENUE FUNCTION FOR ELECTRIC CARS Bell Motors manufactures two models of electric cars. The demand equations giving the relationship between the unit price, p and q, and the number of cars demanded per year, x and y, of Models S1 and S2 are

$$p = 60{,}000 - 4x - 2y \quad \text{and} \quad q = 50{,}000 - 2x - 4y$$

respectively.

a. What is the total yearly revenue function $R(x, y)$?

b. What is the domain of R? Sketch the domain R.

c. Is the point $(3000, 2000)$ in the domain of R? Interpret your result.

Hint: Show that $x = 3000$ and $y = 2000$ satisfy the system of inequalities obtained in part (b).

d. What is the total revenue realized by Bell Motors if it sells 3000 Model S1s and 2000 Model S2s?

39. REVENUE FUNCTIONS FOR DESKS Country Workshop manufactures both finished and unfinished furniture for the home. The estimated quantities demanded each week of its roll-top desks in the finished and unfinished versions are x and y units when the corresponding unit prices are

$$p = 200 - \frac{1}{5}x - \frac{1}{10}y$$

$$q = 160 - \frac{1}{10}x - \frac{1}{4}y$$

dollars, respectively.

a. What is the weekly total revenue function $R(x, y)$?

b. Find the domain of the function R.

40. REVENUE FUNCTION FOR DESKS For the total revenue function $R(x, y)$ of Exercise 39, compute $R(100, 60)$ and $R(60, 100)$. Interpret your results.

41. REVENUE FUNCTIONS FOR A BOOK PUBLISHER Weston Publishing publishes a deluxe edition and a standard edition of its Spanish–English dictionary. Weston's management estimates that the number of deluxe editions demanded is x copies/day and the number of standard editions demanded is y copies/day when the unit prices are

$$p = 20 - 0.005x - 0.001y$$

$$q = 15 - 0.001x - 0.003y$$

dollars, respectively.

a. Find the daily total revenue function $R(x, y)$.

b. Find the domain of the function R.

42. REVENUE FUNCTION FOR A BOOK PUBLISHER For the total revenue function $R(x, y)$ of Exercise 41, compute $R(300, 200)$ and $R(200, 300)$. Interpret your results.

43. VOLUME OF A GAS The volume of a certain mass of gas is related to its pressure and temperature by the formula

$$V = \frac{30.9T}{P}$$

where the volume V is measured in liters, the temperature T is measured in kelvins (obtained by adding 273 to the Celsius temperature), and the pressure P is measured in millimeters of mercury pressure.

a. Find the domain of the function V.

b. Calculate the volume of the gas at standard temperature and pressure—that is, when $T = 273$ K and $P = 760$ mm of mercury.

44. SURFACE AREA OF A HUMAN BODY An empirical formula by E. F. Dubois relates the surface area S of a human body (in square meters) to its weight W (in kilograms) and its height H (in centimeters). The formula, given by

$$S = 0.007184W^{0.425}H^{0.725}$$

is used by physiologists in metabolism studies.
a. Find the domain of the function S.
b. What is the surface area of a human body that weighs 70 kg and has a height of 178 cm?

45. ESTIMATING THE WEIGHT OF A TROUT A formula for estimating the weight of a trout (from measurements) is

$$W = \frac{LG^2}{800}$$

where L is its length and G is its girth (the distance around the body of the fish at its largest point), both measured in inches. The weight of the fish W is in pounds.
a. What is the domain of the function W?
b. Sue caught a trout and measured its length to be 20 in. and its girth to be 12 in. What is its approximate weight?

46. ESTIMATING THE WEIGHT OF A FISH Refer to Exercise 45. A trout caught by Ashley is 20% longer and has a girth that is 10% shorter than the one caught by Jane.
a. Whose catch is heavier?
b. By how much does the weight of Ashley's catch differ from the weight of Jane's catch?

47. PRODUCTION FUNCTION FOR A COUNTRY Suppose the output of a certain country is given by

$$f(x, y) = 100x^{3/5}y^{2/5}$$

billion dollars if x billion dollars are spent on labor and y billion dollars are spent on capital. Find the output if the country spent $32 billion on labor and $243 billion on capital.

48. PRODUCTION FUNCTION Economists have found that the output of a finished product, $f(x, y)$, is sometimes described by the function

$$f(x, y) = ax^b y^{1-b}$$

where x stands for the amount of money expended on labor, y stands for the amount expended on capital, and a and b are positive constants with $0 < b < 1$.
a. If p is a positive number, show that

$$f(px, py) = pf(x, y)$$

b. Use the result of part (a) to show that if the amount of money expended for labor and capital are both increased by $r\%$, then the output is also increased by $r\%$.

49. ARSON FOR PROFIT A study of arson for profit was conducted by a team of paid civilian experts and police detectives appointed by the mayor of a large city. It was

found that the number of suspicious fires in that city in 2016 was very closely related to the concentration of tenants in the city's public housing and to the level of reinvestment in the area in conventional mortgages by the ten largest banks. In fact, the number of fires was closely approximated by the formula

$$N(x, y) = \frac{100(1000 + 0.03x^2y)^{1/2}}{(5 + 0.2y)^2}$$

$$(0 \le x \le 150; 5 \le y \le 35)$$

where x denotes the number of persons per census tract and y denotes the level of reinvestment in the area in cents per dollar deposited. Using this formula, estimate the total number of suspicious fires in the districts of the city where the concentration of public housing tenants was 100/census tract and the level of reinvestment was 20 cents/dollar deposited.

50. CONTINUOUSLY COMPOUNDED INTEREST If a principal of P dollars is deposited in an account earning interest at the rate of r/year compounded continuously, then the accumulated amount at the end of t years is given by

$$A = f(P, r, t) = Pe^{rt}$$

dollars. Find the accumulated amount at the end of 3 years if a sum of $10,000 is deposited in an account earning interest at the rate of 6%/year.

51. HOME MORTGAGES The monthly payment that amortizes a loan of A dollars in t years when the interest rate is r/year, compounded monthly, is given by

$$P = f(A, r, t) = \frac{Ar}{12[1 - (1 + \frac{r}{12})^{-12t}]}$$

a. What is the monthly payment for a home mortgage of $300,000 that will be amortized over 30 years with an interest rate of 4%/year? An interest rate of 6%/year?
b. Find the monthly payment for a home mortgage of $300,000 that will be amortized over 20 years with an interest rate of 6%/year.

52. HOME MORTGAGES Suppose a home buyer secures a bank loan of A dollars to purchase a house. If the interest rate charged is r/year compounded monthly and the loan is to be amortized in t years, then the principal repayment at the end of i months is given by

$$B = f(A, r, t, i)$$
$$= A\left[\frac{(1 + \frac{r}{12})^i - 1}{(1 + \frac{r}{12})^{12t} - 1}\right] \qquad (0 \le i \le 12t)$$

Suppose the Blakelys borrow a sum of $280,000 from a bank to help finance the purchase of a house and the bank charges interest at a rate of 6%/year. If the Blakelys agree to repay the loan in equal installments over 30 years, how much will they owe the bank after the 60th payment (5 years)? The 240th payment (20 years)?

53. WILSON LOT-SIZE FORMULA The Wilson lot-size formula in economics states that the optimal quantity Q of goods for a store to order is given by

$$Q = f(C, N, h) = \sqrt{\frac{2\,CN}{h}}$$

where C is the cost of placing an order, N is the number of items the store sells per week, and h is the weekly holding cost for each item. Find the most economical quantity of 10-speed bicycles to order if it costs the store $20 to place an order, $5 to hold a bicycle for a week, and the store expects to sell 40 bicycles/week.

54. WIND POWER The power output (in watts) of a certain brand of wind turbine generators is estimated to be

$$P = f(R, V) = 0.772R^2V^3$$

where R is the radius (in meters) of a rotor blade and V is the wind speed (in meters per second). Estimate the power output of a model of these generators if its radius is 30 m and the wind speed is 16 m/s.

55. INTERNATIONAL AMERICA'S CUP CLASS Drafted by an international committee in 1989, the rules for the new International America's Cup Class (IACC) include a formula that governs the basic yacht dimensions. The formula

$$f(L, S, D) \leq 42$$

where

$$f(L, S, D) = \frac{L + 1.25S^{1/2} - 9.80D^{1/3}}{0.388}$$

balances the rated length L (in meters), the rated sail area S (in square meters), and the displacement D (in cubic meters). All changes in the basic dimensions are trade-offs. For example, if you want to pick up speed by increasing the sail area, you must pay for it by decreasing the length or increasing the displacement, both of which slow down the boat. Show that Yacht A of rated length 20.95 m, rated sail area 277.3 m², and displacement 17.56 m³ and the longer and heavier Yacht B with $L = 21.87$, $S = 311.78$, and $D = 22.48$ both satisfy the formula.

Source: americascup.com.

56. FORCE GENERATED BY A CENTRIFUGE A centrifuge is a machine designed for the specific purpose of subjecting materials to a sustained centrifugal force. The actual amount of centrifugal force, F, expressed in dynes (1 gram of force = 980 dynes) is given by

$$F = f(M, S, R) = \frac{\pi^2 S^2 MR}{900}$$

where S is in revolutions per minute (rpm), M is in grams, and R is in centimeters. Show that an object revolving at the rate of 600 rpm in a circle with radius of 10 cm generates a centrifugal force that is approximately 40 times gravity.

57. IDEAL GAS LAW According to the *ideal gas law,* the volume V of an ideal gas is related to its pressure P and temperature T by the formula

$$V = \frac{kT}{P}$$

where k is a positive constant. Describe the level curves of V and give a physical interpretation of your result.

58. ISOQUANTS Let $f(x, y)$ denote the output of a country. If x units of its resources are spent on labor and y units are spent on capital, then the level curve of f with equation $f(x, y) = k$, where k denotes a constant output, is an *isoquant.* Each point (x, y) on the isoquant corresponds to a level of investment in labor, x, and investment in capital, y, that results in the same level of output k. Suppose the output of a certain country is

$$f(x, y) = 100x^{3/4}y^{1/4}$$

billion dollars when x billion dollars are spent on labor and y billion dollars are spent on capital.
   a. What is the output if the country spends $81 billion on labor and $16 billion on capital?
   b. Suppose the output remains constant at the level found in part (a). Find an equation giving the relationship between x and y.
   c. Use the equation found in part (b) to complete the following table.

x	50	60	70	80	90
y					

   Hint: Solve the equation for y in terms of x.
   d. Use the results of part (c) to sketch the isoquant of f corresponding to the constant output found in part (b).

In Exercises 59–64, determine whether the statement is true or false. If it is true, explain why it is true. If it is false, explain why, or give an example to show why it is false.

59. If h is a function of x and y, then there are functions f and g of one variable such that

$$h(x, y) = f(x) + g(y)$$

60. If f is a function of x and y and a is a real number, then

$$f(ax, ay) = af(x, y)$$

61. The domain of $f(x, y) = 1/(x^2 - y^2)$ is $\{(x, y) \,|\, y \neq x\}$.

62. Every point on the level curve $f(x, y) = c$ corresponds to a point on the graph of f that is c units above the xy-plane if $c > 0$ and $|c|$ units below the xy-plane if $c < 0$.

63. f is a function of x and y if and only if for any two points $P_1(x_1, y_1)$ and $P_2(x_2, y_2)$ in the domain of f, $f(x_1, y_1) = f(x_2, y_2)$ implies that $P_1(x_1, y_1) = P_2(x_2, y_2)$.

64. If f is a function of x and y, and k is a real number in the range of f, then there exists at most one point $P(a, b)$ in the domain of f such that $f(a, b) = k$.

8.1 Solutions to Self-Check Exercises

1. $f(1, 3) = 1^2 - 3(1)(3) + \sqrt{1 + 3} = -6$

$f(-1, 1) = (-1)^2 - 3(-1)(1) + \sqrt{-1 + 1} = 4$

The point $(-1, 0)$ is not in the domain of f because the term $\sqrt{x + y}$ is not defined when $x = -1$ and $y = 0$. In fact, the domain of f consists of all real values of x and y that satisfy the inequality $x + y \geq 0$, the shaded half-plane shown in the accompanying figure.

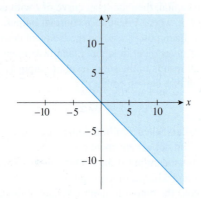

2. Since division by zero is not permitted, we see that $x \neq 0$ and $x - y \neq 0$. Therefore, the domain of f is the set of all points in the xy-plane except for the y-axis ($x = 0$) and the straight line $x = y$.

3. If Odyssey spends \$5000/month on newspaper ads ($x = 5000$) and \$15,000/month on television ads ($y = 15,000$), then its monthly revenue will be given by

$$f(5000, 15,000) = 30(5000)^{1/4}(15,000)^{3/4}$$
$$\approx 341,926.06$$

or approximately \$341,926. If the agency spends \$4000/month on newspaper ads and \$16,000/month on television ads, then its monthly revenue will be given by

$$f(4000, 16,000) = 30(4000)^{1/4}(16,000)^{3/4}$$
$$\approx 339,411.26$$

or approximately \$339,411.

8.2 Partial Derivatives

Partial Derivatives

For a function $f(x)$ of one variable x, there is no ambiguity when we speak about the rate of change of $f(x)$ with respect to x, since x must be constrained to move along the x-axis. The situation becomes more complicated, however, when we study the rate of change of a function of two or more variables. For example, the domain D of a function of two variables $f(x, y)$ is a subset of the plane, so if (a, b) is any point in the domain of f, there are infinitely many directions from which one can approach the point (a, b) (Figure 12). We may therefore ask for the rate of change of f at (a, b) along any of these directions.

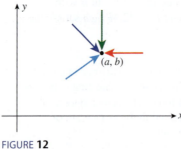

FIGURE **12**
We can approach a point in the plane from infinitely many directions.

However, we will not deal with this general problem. Instead, we will restrict ourselves to studying the rate of change of the function $f(x, y)$ at a point (a, b) in each of two *preferred directions*—namely, the direction parallel to the x-axis and the direction parallel to the y-axis. Let $y = b$, where b is a constant, so that $f(x, b)$ is a function of the one variable x. Since the equation $z = f(x, y)$ is the equation of a surface, the

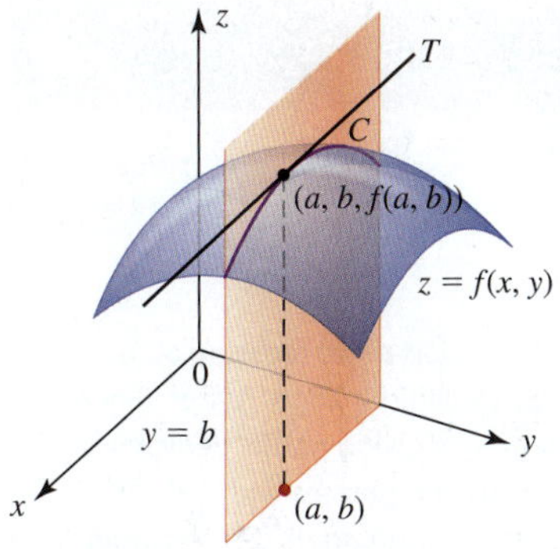

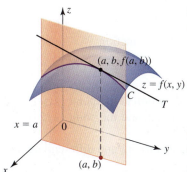

FIGURE 13
The curve C is formed by the intersection of the plane $y = b$ with the surface $z = f(x, y)$.

equation $z = f(x, b)$ is the equation of the curve C on the surface formed by the intersection of the surface and the plane $y = b$ (Figure 13).

Because $f(x, b)$ is a function of one variable x, we may compute the derivative of f with respect to x at $x = a$. This derivative, obtained by keeping the variable y fixed at b and differentiating the resulting function $f(x, b)$ with respect to x, is called the **first partial derivative of f with respect to x at (a, b)**, written

$$\frac{\partial z}{\partial x}(a, b) \quad \text{or} \quad \frac{\partial f}{\partial x}(a, b) \quad \text{or} \quad f_x(a, b)$$

Thus,

$$\frac{\partial z}{\partial x}(a, b) = \frac{\partial f}{\partial x}(a, b) = f_x(a, b) = \lim_{h \to 0} \frac{f(a + h, b) - f(a, b)}{h}$$

provided that the limit exists. The first partial derivative of f with respect to x at (a, b) measures both the slope of the tangent line T to the curve C and the rate of change of the function f in the x-direction when $x = a$ and $y = b$. We also write

$$\left. \frac{\partial f}{\partial x} \right|_{(a, b)} = f_x(a, b)$$

Similarly, we define the **first partial derivative of f with respect to y at (a, b)**, written

$$\frac{\partial z}{\partial y}(a, b) \quad \text{or} \quad \frac{\partial f}{\partial y}(a, b) \quad \text{or} \quad f_y(a, b)$$

as the derivative obtained by keeping the variable x fixed at a and differentiating the resulting function $f(a, y)$ with respect to y. That is,

$$\frac{\partial z}{\partial y}(a, b) = \frac{\partial f}{\partial y}(a, b) = f_y(a, b)$$

$$= \lim_{k \to 0} \frac{f(a, b + k) - f(a, b)}{k}$$

if the limit exists. The first partial derivative of f with respect to y at (a, b) measures both the slope of the tangent line T to the curve C, obtained by holding x constant (Figure 14), and the rate of change of the function f in the y-direction when $x = a$ and $y = b$. We write

$$\left. \frac{\partial f}{\partial y} \right|_{(a, b)} = f_y(a, b)$$

Before looking at some examples, let's summarize these definitions.

FIGURE 14
The first partial derivative of f with respect to y at (a, b) measures the slope of the tangent line T to the curve C with x held constant.

First Partial Derivatives of $f(x, y)$

Suppose $f(x, y)$ is a function of the two variables x and y. Then the **first partial derivative of f** with respect to x at the point (x, y) is

$$\frac{\partial f}{\partial x} = \lim_{h \to 0} \frac{f(x + h, y) - f(x, y)}{h}$$

provided that the limit exists. The first partial derivative of f with respect to y at the point (x, y) is

$$\frac{\partial f}{\partial y} = \lim_{k \to 0} \frac{f(x, y + k) - f(x, y)}{k}$$

provided that the limit exists.

EXAMPLE 1 Find the partial derivatives $\dfrac{\partial f}{\partial x}$ and $\dfrac{\partial f}{\partial y}$ of the function

$$f(x, y) = x^2 - xy^2 + y^3$$

What is the rate of change of the function f in the x-direction at the point $(1, 2)$? What is the rate of change of the function f in the y-direction at the point $(1, 2)$?

Solution   To compute $\dfrac{\partial f}{\partial x}$, think of the variable y as a constant and differentiate the resulting function of x with respect to x. Let's write

$$f(x, y) = x^2 - xy^2 + y^3$$

where the variable y to be treated as a constant is shown in color. Then,

$$\frac{\partial f}{\partial x} = 2x - y^2$$

To compute $\dfrac{\partial f}{\partial y}$, think of the variable x as being fixed—that is, as a constant—and differentiate the resulting function of y with respect to y. In this case,

$$f(x, y) = x^2 - xy^2 + y^3$$

so

$$\frac{\partial f}{\partial y} = -2xy + 3y^2$$

The rate of change of the function f in the x-direction at the point $(1, 2)$ is given by

$$f_x(1, 2) = \left. \frac{\partial f}{\partial x} \right|_{(1, 2)} = 2(1) - 2^2 = -2$$

That is, f decreases 2 units for each unit increase in the x-direction, y being kept constant ($y = 2$). The rate of change of the function f in the y-direction at the point $(1, 2)$ is given by

$$f_y(1, 2) = \left.\frac{\partial f}{\partial y}\right|_{(1, 2)} = -2(1)(2) + 3(2)^2 = 8$$

That is, f increases 8 units for each unit increase in the y-direction, x being kept constant ($x = 1$). ■

Explore and Discuss

Refer to the Explore and Discuss on page 556. Suppose management has decided that the projected sales of the first product is a units. Describe how you might help management decide how many units of the second product the company should produce and sell in order to maximize the company's total profit. Justify your method to management. Suppose, however, that management feels that b units of the second product should be manufactured and sold. How would you help management decide how many units of the first product to manufacture in order to maximize the company's total profit?

EXAMPLE 2 Compute the first partial derivatives of each function.

a. $f(x, y) = \dfrac{xy}{x^2 + y^2}$ **b.** $g(s, t) = (s^2 - st + t^2)^5$

c. $h(u, v) = e^{u^2 - v^2}$ **d.** $f(x, y) = \ln(x^2 + 2y^2)$

Solution

a. To compute $\dfrac{\partial f}{\partial x}$, think of the variable y as a constant. Thus,

$$f(x, y) = \frac{xy}{x^2 + y^2}$$

Then using the Quotient Rule, we have

$$\frac{\partial f}{\partial x} = \frac{(x^2 + y^2)y - xy(2x)}{(x^2 + y^2)^2} = \frac{x^2y + y^3 - 2x^2y}{(x^2 + y^2)^2}$$

$$= \frac{y(y^2 - x^2)}{(x^2 + y^2)^2}$$

upon simplification and factorization. To compute $\dfrac{\partial f}{\partial y}$, think of the variable x as a constant. Thus,

$$f(x, y) = \frac{xy}{x^2 + y^2}$$

Then using the Quotient Rule once again, we obtain

$$\frac{\partial f}{\partial y} = \frac{(x^2 + y^2)x - xy(2y)}{(x^2 + y^2)^2} = \frac{x^3 + xy^2 - 2xy^2}{(x^2 + y^2)^2}$$

$$= \frac{x(x^2 - y^2)}{(x^2 + y^2)^2}$$

b. To compute $\dfrac{\partial g}{\partial s}$, we treat the variable t as if it were a constant. Thus,

$$g(s, t) = (s^2 - st + t^2)^5$$

Using the General Power Rule, we find

$$\frac{\partial g}{\partial s} = 5(s^2 - st + t^2)^4 \cdot (2s - t)$$
$$= 5(2s - t)(s^2 - st + t^2)^4$$

To compute $\frac{\partial g}{\partial t}$, we treat the variable s as if it were a constant. Thus,

$$g(s, t) = (s^2 - st + t^2)^5$$
$$\frac{\partial g}{\partial t} = 5(s^2 - st + t^2)^4 (-s + 2t)$$
$$= 5(2t - s)(s^2 - st + t^2)^4$$

c. To compute $\frac{\partial h}{\partial u}$, think of the variable v as a constant. Thus,

$$h(u, v) = e^{u^2 - v^2}$$

Using the Chain Rule for Exponential Functions, we have

$$\frac{\partial h}{\partial u} = e^{u^2 - v^2} \cdot 2u = 2u e^{u^2 - v^2}$$

Next, we treat the variable u as if it were a constant,

$$h(u, v) = e^{u^2 - v^2}$$

and we obtain

$$\frac{\partial h}{\partial v} = e^{u^2 - v^2} \cdot (-2v) = -2v e^{u^2 - v^2}$$

d. To compute $\frac{\partial f}{\partial x}$, think of the variable y as a constant. Thus,

$$f(x, y) = \ln(x^2 + 2y^2)$$

so the Chain Rule for Logarithmic Functions gives

$$\frac{\partial f}{\partial x} = \frac{2x}{x^2 + 2y^2}$$

Next, treating the variable x as if it were a constant, we find

$$f(x, y) = \ln(x^2 + 2y^2)$$
$$\frac{\partial f}{\partial y} = \frac{4y}{x^2 + 2y^2}$$

To compute the partial derivative of a function of several variables with respect to one variable—say, x—we think of the other variables as if they were constants and differentiate the resulting function with respect to x.

Explore and Discuss

1. Let (a, b) be a point in the domain of $f(x, y)$. Put $g(x) = f(x, b)$, and suppose that g is differentiable at $x = a$. Explain why you can find $f_x(a, b)$ by computing $g'(a)$. How would you go about calculating $f_y(a, b)$ using a similar technique? Give a geometric interpretation of these processes.

2. Let $f(x, y) = x^2 y^3 - 3x^2 y + 2$. Use the method of Problem 1 to find $f_x(1, 2)$ and $f_y(1, 2)$.

EXAMPLE 3 Compute the first partial derivatives of the function

$$w = f(x, y, z) = xyz - xe^{yz} + x \ln y$$

Solution Here we have a function of three variables, x, y, and z, and we are required to compute

$$\frac{\partial f}{\partial x}, \quad \frac{\partial f}{\partial y}, \quad \frac{\partial f}{\partial z}$$

To compute f_x, we think of the other two variables, y and z, as fixed, and we differentiate the resulting function of x with respect to x, thereby obtaining

$$f_x = yz - e^{yz} + \ln y$$

To compute f_y, we think of the other two variables, x and z, as constants, and we differentiate the resulting function of y with respect to y. We then obtain

$$f_y = xz - xze^{yz} + \frac{x}{y}$$

Finally, to compute f_z, we treat the variables x and y as constants and differentiate the function f with respect to z, obtaining

$$f_z = xy - xye^{yz}$$

Exploring with TECHNOLOGY

Refer to the Explore and Discuss on page 570. Let

$$f(x, y) = \frac{e^{\sqrt{xy}}}{(1 + xy^2)^{3/2}}$$

1. Compute $g(x) = f(x, 1)$, and use a graphing utility to plot the graph of g in the viewing window $[0, 2] \times [0, 2]$.
2. Use the differentiation operation of your graphing utility to find $g'(1)$ and hence $f_x(1, 1)$.
3. Compute $h(y) = f(1, y)$, and use a graphing utility to plot the graph of h in the viewing window $[0, 2] \times [0, 2]$.
4. Use the differentiation operation of your graphing utility to find $h'(1)$ and hence $f_y(1, 1)$.

The Cobb–Douglas Production Function

For an economic interpretation of the first partial derivatives of a function of two variables, let's turn our attention to the function

$$f(x, y) = ax^b y^{1-b} \tag{5}$$

where a and b are positive constants with $0 < b < 1$. This function is called the **Cobb–Douglas production function.** Here, x stands for the amount of money expended for labor, y stands for the cost of capital equipment (buildings, machinery, and other tools of production), and the function f measures the output of the finished product (in suitable units) and is called, accordingly, the production function.

The partial derivative f_x is called the **marginal productivity of labor.** It measures the rate of change of production with respect to the amount of money expended for labor, with the level of capital expenditure held constant. Similarly, the partial derivative f_y, called the **marginal productivity of capital,** measures the rate of change of production with respect to the amount expended on capital, with the level of labor expenditure held fixed.

APPLIED EXAMPLE 4 Marginal Productivity A certain country's production in the early years following World War II is described by the function

$$f(x, y) = 30x^{2/3}y^{1/3}$$

units, when x units of labor and y units of capital were used.

a. Compute f_x and f_y.
b. What are the marginal productivity of labor and the marginal productivity of capital when the amounts expended on labor and capital are 125 units and 27 units, respectively?
c. Should the government have encouraged capital investment rather than increasing expenditure on labor to increase the country's productivity?

Solution

a. $f_x = 30 \cdot \dfrac{2}{3} x^{-1/3}y^{1/3} = 20\left(\dfrac{y}{x}\right)^{1/3}$

$f_y = 30x^{2/3} \cdot \dfrac{1}{3} y^{-2/3} = 10\left(\dfrac{x}{y}\right)^{2/3}$

b. The required marginal productivity of labor is given by

$$f_x(125, 27) = 20\left(\frac{27}{125}\right)^{1/3} = 20\left(\frac{3}{5}\right) = 12$$

or 12 units per unit increase in labor expenditure (capital expenditure is held constant at 27 units). The required marginal productivity of capital is given by

$$f_y(125, 27) = 10\left(\frac{125}{27}\right)^{2/3} = 10\left(\frac{25}{9}\right) = 27\tfrac{7}{9}$$

or $27\tfrac{7}{9}$ units per unit increase in capital expenditure (labor outlay is held constant at 125 units).

c. From the results of part (b), we see that a unit increase in capital expenditure resulted in a much faster increase in productivity than a unit increase in labor expenditure would have. Therefore, the government should have encouraged increased spending on capital rather than on labor during the early years of reconstruction. ◼

Substitute and Complementary Commodities

For another application of the first partial derivatives of a function of two variables in the field of economics, let's consider the relative demands of two commodities. We say that the two commodities are **substitute** (competitive) **commodities** if a decrease in the demand for one results in an increase in the demand for the other. Examples of substitute commodities are coffee and tea. Conversely, two commodities are referred to as **complementary commodities** if a decrease in the demand for one results in a decrease in the demand for the other as well. Examples of complementary commodities are automobiles and automobile tires.

We now derive a criterion for determining whether two commodities A and B are substitute or complementary. Suppose the demand equations that relate the quantities demanded, x and y, to the unit prices, p and q, of the two commodities are given by

$$x = f(p, q) \quad \text{and} \quad y = g(p, q)$$

Let's consider the partial derivative $\partial f/\partial p$. Since f is the demand function for Commodity A, we see that for fixed q, f is typically a decreasing function of p—that is, $\partial f/\partial p < 0$. Now, if the two commodities were substitute commodities, then the quantity demanded of Commodity B would increase with respect to p—that is, $\partial g/\partial p > 0$. A similar argument with p fixed shows that if A and B are substitute commodities, then $\partial f/\partial q > 0$. Thus, the two commodities A and B are substitute commodities if

$$\frac{\partial f}{\partial q} > 0 \quad \text{and} \quad \frac{\partial g}{\partial p} > 0$$

Similarly, A and B are complementary commodities if

$$\frac{\partial f}{\partial q} < 0 \quad \text{and} \quad \frac{\partial g}{\partial p} < 0$$

Substitute and Complementary Commodities

Two commodities A and B are substitute commodities if

$$\frac{\partial f}{\partial q} > 0 \quad \text{and} \quad \frac{\partial g}{\partial p} > 0 \tag{6}$$

Two commodities A and B are complementary commodities if

$$\frac{\partial f}{\partial q} < 0 \quad \text{and} \quad \frac{\partial g}{\partial p} < 0 \tag{7}$$

 APPLIED EXAMPLE 5 Substitute and Complementary Commodities
Suppose that the daily demand for butter is given by

$$x = f(p, q) = \frac{3q}{1 + p^2}$$

and the daily demand for margarine is given by

$$y = g(p, q) = \frac{2p}{1 + \sqrt{q}} \qquad (p > 0, q > 0)$$

where p and q denote the prices per pound (in dollars) of butter and margarine, respectively, and x and y are measured in millions of pounds. Determine whether these two commodities are substitute, complementary, or neither.

Solution   We compute

$$\frac{\partial f}{\partial q} = \frac{3}{1 + p^2} \quad \text{and} \quad \frac{\partial g}{\partial p} = \frac{2}{1 + \sqrt{q}}$$

Since

$$\frac{\partial f}{\partial q} > 0 \quad \text{and} \quad \frac{\partial g}{\partial p} > 0$$

for all values of $p > 0$ and $q > 0$, we conclude that butter and margarine are substitute commodities.

Second-Order Partial Derivatives

The first partial derivatives $f_x(x, y)$ and $f_y(x, y)$ of a function $f(x, y)$ of the two variables x and y are also functions of x and y. As such, we may differentiate each of the functions f_x and f_y to obtain the **second-order partial derivatives of f** (Figure 15). Thus, differentiating the function f_x with respect to x leads to the second partial derivative

$$f_{xx} = \frac{\partial^2 f}{\partial x^2} = \frac{\partial}{\partial x}(f_x)$$

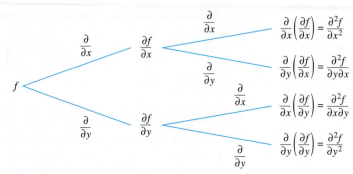

FIGURE 15
A schematic showing the four second-order partial derivatives of f

However, differentiation of f_x with respect to y leads to the second partial derivative

$$f_{xy} = \frac{\partial^2 f}{\partial y \partial x} = \frac{\partial}{\partial y}(f_x)$$

Similarly, differentiation of the function f_y with respect to x and with respect to y leads to

$$f_{yx} = \frac{\partial^2 f}{\partial x \partial y} = \frac{\partial}{\partial x}(f_y)$$

$$f_{yy} = \frac{\partial^2 f}{\partial y^2} = \frac{\partial}{\partial y}(f_y)$$

respectively. Although it is not always true that $f_{xy} = f_{yx}$, they are equal, however, if both f_{xy} and f_{yx} are continuous. We might add that this is the case in most practical applications.

EXAMPLE 6 Find the second-order partial derivatives of the function

$$f(x, y) = x^3 - 3x^2 y + 3xy^2 + y^2$$

Solution   The first partial derivatives of f are

$$f_x = \frac{\partial}{\partial x}(x^3 - 3x^2 y + 3xy^2 + y^2)$$
$$= 3x^2 - 6xy + 3y^2$$

$$f_y = \frac{\partial}{\partial y}(x^3 - 3x^2 y + 3xy^2 + y^2)$$
$$= -3x^2 + 6xy + 2y$$

Therefore,

$$f_{xx} = \frac{\partial}{\partial x}(f_x) = \frac{\partial}{\partial x}(3x^2 - 6xy + 3y^2)$$

$$= 6x - 6y = 6(x - y)$$

$$f_{xy} = \frac{\partial}{\partial y}(f_x) = \frac{\partial}{\partial y}(3x^2 - 6xy + 3y^2)$$

$$= -6x + 6y = 6(y - x)$$

$$f_{yx} = \frac{\partial}{\partial x}(f_y) = \frac{\partial}{\partial x}(-3x^2 + 6xy + 2y)$$

$$= -6x + 6y = 6(y - x)$$

$$f_{yy} = \frac{\partial}{\partial y}(f_y) = \frac{\partial}{\partial y}(-3x^2 + 6xy + 2y)$$

$$= 6x + 2$$

Note that $f_{xy} = f_{yx}$ everywhere.

EXAMPLE 7 Find the second-order partial derivatives of the function

$$f(x, y) = e^{xy^2}$$

Solution   We have

$$f_x = \frac{\partial}{\partial x}(e^{xy^2})$$

$$= y^2 e^{xy^2}$$

$$f_y = \frac{\partial}{\partial y}(e^{xy^2})$$

$$= 2xy e^{xy^2}$$

so the required second-order partial derivatives of f are

$$f_{xx} = \frac{\partial}{\partial x}(f_x) = \frac{\partial}{\partial x}(y^2 e^{xy^2})$$

$$= y^4 e^{xy^2}$$

$$f_{xy} = \frac{\partial}{\partial y}(f_x) = \frac{\partial}{\partial y}(y^2 e^{xy^2})$$

$$= 2y e^{xy^2} + 2xy^3 e^{xy^2} \qquad (x^2) \text{ See page 9.}$$

$$= 2y e^{xy^2}(1 + xy^2)$$

$$f_{yx} = \frac{\partial}{\partial x}(f_y) = \frac{\partial}{\partial x}(2xy e^{xy^2})$$

$$= 2y e^{xy^2} + 2xy^3 e^{xy^2}$$

$$= 2y e^{xy^2}(1 + xy^2)$$

$$f_{yy} = \frac{\partial}{\partial y}(f_y) = \frac{\partial}{\partial y}(2xy e^{xy^2})$$

$$= 2x e^{xy^2} + (2xy)(2xy)e^{xy^2}$$

$$= 2x e^{xy^2}(1 + 2xy^2)$$

Note that $f_{xy} = f_{yx}$ everywhere.

8.2 Self-Check Exercises

1. Compute the first partial derivatives of
   $f(x, y) = x^3 - 2xy^2 + y^2 - 8$.

2. Find the first partial derivatives of
   $f(x, y) = x \ln y + ye^x - x^2$ at $(0, 1)$, and interpret
   your results.

3. Find the second-order partial derivatives of the function
   of Self-Check Exercise 1.

4. **MARGINAL PRODUCTIVITY** A certain country's production is
   described by the function

$$f(x, y) = 60x^{1/3}y^{2/3}$$

   when x units of labor and y units of capital are used.

a. What are the marginal productivity of labor and the
   marginal productivity of capital when the amounts
   expended on labor and capital are 125 units and
   8 units, respectively?

b. Should the government encourage capital investment
   rather than increased expenditure on labor at this time
   to increase the country's productivity?

*Solutions to Self-Check Exercises 8.2 can be found on
page 580.*

8.2 Concept Questions

1. a. What is the partial derivative of $f(x, y)$ with respect to
   x at (a, b)?
   b. Give a geometric interpretation of $f_x(a, b)$ and a practical interpretation of $f_x(a, b)$.

2. a. What are substitute commodities and complementary
   commodities? Give an example of each.
   b. Suppose $x = f(p, q)$ and $y = g(p, q)$ are demand
   functions for two commodities A and B, respectively.

   Give conditions for determining whether A and B are
   substitute or complementary commodities.

3. List all second-order partial derivatives of a function of
   two variables.

4. How many second-order partial derivatives are there for a
   function of three variables f, assuming that all such derivatives exist? List all of them.

8.2 Exercises

1. Let $f(x, y) = x^2 + 2y^2$.
   a. Find $f_x(2, 1)$ and $f_y(2, 1)$.
   b. Interpret the numbers in part (a) as slopes.
   c. Interpret the numbers in part (a) as rates of change.

2. Let $f(x, y) = 9 - x^2 + xy - 2y^2$.
   a. Find $f_x(1, 2)$ and $f_y(1, 2)$.
   b. Interpret the numbers in part (a) as slopes.
   c. Interpret the numbers in part (a) as rates of change.

In Exercises 3–24, find the first partial derivatives of the function.

3. $f(x, y) = 2x + 3y + 5$ 4. $f(x, y) = 2xy$

5. $g(x, y) = 2x^2 + 4y + 1$ 6. $f(x, y) = 1 + x^2 + y^2$

7. $f(x, y) = \dfrac{2y}{x^2}$ 8. $f(x, y) = \dfrac{x}{1 + y}$

9. $g(u, v) = \dfrac{u - v}{u + v}$ 10. $f(x, y) = \dfrac{x^2 - y^2}{x^2 + y^2}$

11. $f(s, t) = (s^2 - st + t^2)^3$ 12. $g(s, t) = s^2t + st^{-3}$

13. $f(x, y) = (x^2 + y^2)^{2/3}$ 14. $f(x, y) = x\sqrt{1 + y^2}$

15. $f(x, y) = e^{xy+1}$ 16. $f(x, y) = (e^x + e^y)^5$

17. $f(x, y) = x \ln y + y \ln x$ 18. $f(x, y) = x^2 e^{y^2}$

19. $g(u, v) = e^u \ln v$ 20. $f(x, y) = \dfrac{e^{xy}}{x + y}$

21. $f(x, y, z) = xyz + xy^2 + yz^2 + zx^2$

22. $g(u, v, w) = \dfrac{2uvw}{u^2 + v^2 + w^2}$

23. $h(r, s, t) = e^{rst}$ 24. $f(x, y, z) = xe^{y/z}$

In Exercises 25–34, evaluate the first partial derivatives of the
function at the given point.

25. $f(x, y) = x^2y + xy^2$; $(1, 2)$

26. $f(x, y) = x^2 + xy + y^2 + 2x - y$; $(-1, 2)$

27. $f(x, y) = x\sqrt{y} + y^2$; $(2, 1)$

28. $g(x, y) = \sqrt{x^2 + y^2}$; $(3, 4)$

29. $f(x, y) = \dfrac{x}{y}$; $(1, 2)$ 30. $f(x, y) = \dfrac{x + y}{x - y}$; $(1, -2)$

31. $f(x, y) = e^{xy}$; $(1, 1)$ 32. $f(x, y) = e^x \ln y$; $(0, e)$

33. $f(x, y, z) = x^2yz^3$; $(1, 0, 2)$

34. $f(x, y, z) = x^2y^2 + z^2$; $(1, 1, 2)$

In Exercises 35–42, find the second-order partial derivatives of the function. In each case, show that the mixed partial derivatives f_{xy} and f_{yx} are equal.

35. $f(x, y) = x^2y + xy^3$

36. $f(x, y) = x^3 + x^2y + x + 4$

37. $f(x, y) = x^2 - 2xy + 2y^2 + x - 2y$

38. $f(x, y) = x^3 + x^2y^2 + y^3 + x + y$

39. $f(x, y) = \sqrt{x^2 + y^2}$

40. $f(x, y) = x\sqrt{y} + y\sqrt{x}$

41. $f(x, y) = e^{-x/y}$

42. $f(x, y) = \ln(1 + x^2y^2)$

43. PRODUCTIVITY OF A COUNTRY The productivity of a South American country is given by the function

$$f(x, y) = 20x^{3/4}y^{1/4}$$

when x units of labor and y units of capital are used.
   a. What are the marginal productivity of labor and the marginal productivity of capital when the amounts expended on labor and capital are 256 units and 16 units, respectively?
   b. Should the government encourage capital investment rather than increased expenditure on labor at this time to increase the country's productivity?

44. PRODUCTIVITY OF A COUNTRY The productivity of a country in Western Europe is given by the function

$$f(x, y) = 40x^{4/5}y^{1/5}$$

when x units of labor and y units of capital are used.
   a. What are the marginal productivity of labor and the marginal productivity of capital when the amounts expended on labor and capital are 32 units and 243 units, respectively?
   b. Should the government encourage capital investment rather than increased expenditure on labor at this time to increase the country's productivity?

45. PRODUCTION FUNCTION FOR CORDLESS LED CANDLES Luminar Corporation manufactures cordless LED window candles. The number of candles it can manufacture per day, P, depends on the amount of labor utilized, x (in work-hours per day), and the expenditure on capital investment, y (in dollars per day), according to the production function

$$P(x, y) = x^2 + 5x + 2xy + 3y^2 + 2y$$

Find the marginal productivities if $x = 400$ and $y = 300$, and interpret your results.

46. PRODUCTIVITY OF A COMPANY The number of souvenir coffee mugs (in hundreds) that Ace Novelty can produce monthly is given by the production function

$$P(x, y) = 0.2x^2 + 2x + 3xy + 0.4y^2 + 3y$$

where x denotes the amount of labor utilized (measured in thousands of work-hours per month) and y denotes the expenditure on capital investment (in thousands of dollars per month). Find the marginal productivities if 10,000 work-hours per month are utilized and a capital investment of \$5000/month is made. Interpret your results.

47. LAND PRICES The rectangular region R shown in the following figure represents a city's financial district. The price of land within the district is approximated by the function

$$p(x, y) = 200 - 10\left(x - \frac{1}{2}\right)^2 - 15(y - 1)^2$$

where $p(x, y)$ is the price of land at the point (x, y) in dollars per square foot and x and y are measured in miles. Compute

$$\frac{\partial p}{\partial x}(0, 1) \quad \text{and} \quad \frac{\partial p}{\partial y}(0, 1)$$

and interpret your results.

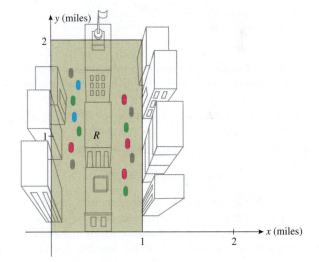

48. REVENUE FUNCTIONS The total weekly revenue (in dollars) of Country Workshop associated with manufacturing and selling their rolltop desks is given by the function

$$R(x, y) = -0.2x^2 - 0.25y^2 - 0.2xy + 200x + 160y$$

where x denotes the number of finished units and y denotes the number of unfinished units manufactured and sold each week. Compute $\partial R/\partial x$ and $\partial R/\partial y$ when $x = 300$ and $y = 250$. Interpret your results.

49. PROFIT FUNCTIONS The monthly profit (in dollars) of Bond and Barker Department Store depends on the level of inventory x (in thousands of dollars) and the floor space y (in thousands of square feet) available for display of the merchandise, as given by the equation

$$P(x, y) = -0.02x^2 - 15y^2 + xy$$
$$+ 39x + 25y - 20{,}000$$

Compute $\partial P/\partial x$ and $\partial P/\partial y$ when $x = 4000$ and $y = 150$. Interpret your results. Repeat with $x = 5000$ and $y = 150$.

50. Hotel Room Demand The number of rooms in demand at the hotels of the Goodwood Hotel Group is given by

$$D(x, y) = 500e^{-0.02(x-120)}(1 - 0.7e^{-0.0001y})$$
$$(120 \le x \le 280, 0 \le y \le 30,000)$$

where x is the daily room rate in dollars and y is the weekly spending on advertisement, also in dollars.
a. Show that if y is held fast, then D is a decreasing function of x and if x is held fast, then D is an increasing function of y. Interpret your results.
b. If the spending on advertisement is held fast at $25,000/week, find the approximate change in the demand for the number of rooms if the daily room rate is cut by $1 from the current $200/night.

51. Arson Study A study of arson for profit conducted for a certain city found that the number of suspicious fires is approximated by

$$N(x, y) = \frac{120\sqrt{1000 + 0.03x^2y}}{(5 + 0.2y)^2}$$
$$(0 \le x \le 150, 5 \le y \le 35)$$

where x denotes the number of persons per census tract and y denotes the level of reinvestment in conventional mortgages by the city's ten largest banks measured in cents per dollars deposited. Compute $\partial N/\partial x$ and $\partial N/\partial y$ when $x = 100$ and $y = 20$, and interpret your results.

52. Surface Area of a Human Body The formula

$$S = 0.007184W^{0.425}H^{0.725}$$

gives the surface area S of a human body (in square meters) in terms of its weight W (in kilograms) and its height H (in centimeters). Compute $\partial S/\partial W$ and $\partial S/\partial H$ when $W = 70$ kg and $H = 180$ cm. Interpret your results.

53. Wind Chill Factor The wind chill temperature is the temperature that you would feel in still air if the presence of wind were taken into consideration. The following table gives the wind chill temperature $T = f(t, s)$ in degrees Fahrenheit in terms of the actual air temperature t in degrees Fahrenheit and the wind speed s in miles per hour (mph).

		\multicolumn{7}{c}{**Wind speed (mph)**}						
	s / t	10	15	20	25	30	35	40
Actual air temperature (°F)	30	21.2	19.0	17.4	16.0	14.9	13.9	13.0
	32	23.7	21.6	20.0	18.7	17.6	16.6	15.8
	34	26.2	24.2	22.6	21.4	20.3	19.4	18.6
	36	28.7	26.7	25.2	24.0	23.0	22.2	21.4
	38	31.2	29.3	27.9	26.7	25.7	24.9	24.2
	40	33.6	31.8	30.5	29.4	28.5	27.7	26.9

a. Estimate the rate of change of the wind chill temperature T with respect to the actual air temperature when the wind speed is constant at 25 mph and the actual air temperature is 34°F.
Hint: Show that it is given by

$$\frac{\partial T}{\partial t}(34, 25) \approx \frac{f(36, 25) - f(34, 25)}{2}$$

b. Estimate the rate of change of the wind chill temperature T with respect to the wind speed when the actual air temperature is constant at 34°F and the wind speed is 25 mph.
Source: National Weather Service.

54. Wind Chill Factor A formula used by meteorologists to calculate the wind chill temperature (the temperature that you would feel in still air if the presence of wind were taken into consideration) is

$$T = f(t, s) = 35.74 + 0.6215t - 35.75s^{0.16} + 0.4275ts^{0.16}$$
$$(s \ge 1)$$

where t is the actual air temperature in degrees Fahrenheit and s is the wind speed in mph.
a. What is the wind chill temperature when the actual air temperature is 32°F and the wind speed is 20 mph?
b. If the temperature is 32°F, by how much approximately will the wind chill temperature change if the wind speed increases from 20 mph to 21 mph?

55. Engine Efficiency The efficiency of an internal combustion engine is given by

$$E = \left(1 - \frac{v}{V}\right)^{0.4}$$

where V and v are the respective maximum and minimum volumes of air in each cylinder.
a. Show that $\partial E/\partial V > 0$, and interpret your result.
b. Show that $\partial E/\partial v < 0$, and interpret your result.

56. Volume of a Gas The volume V (in liters) of a certain mass of gas is related to its pressure P (in millimeters of mercury) and its temperature T (in kelvins) by the law

$$V = \frac{30.9T}{P}$$

Compute $\partial V/\partial T$ and $\partial V/\partial P$ when $T = 300$ and $P = 800$. Interpret your results.

57. Complementary and Substitute Commodities In a survey conducted by a video magazine, it was determined that the demand equation for DVD players is given by

$$x = f(p, q) = 10,000 - 10p + 0.2q^2$$

and the demand equation for Blu-ray players is given by

$$y = g(p, q) = 5000 + 0.8p^2 - 20q$$

where p and q denote the unit prices (in dollars) for the DVD and Blu-ray players, respectively, and x and y denote the number of DVD and Blu-ray players demanded per week. Determine whether these two products are substitute, complementary, or neither.

58. **COMPLEMENTARY AND SUBSTITUTE COMMODITIES** In a survey, it was determined that the demand equation for DVD players is given by

$$x = f(p, q) = 10{,}000 - 10p - e^{0.5q}$$

and the demand equation for blank DVD discs is given by

$$y = g(p, q) = 50{,}000 - 4000q - 10p$$

where p and q denote the unit prices, respectively, and x and y denote the number of DVD players and the number of blank DVD discs demanded each week. Determine whether these two products are substitute, complementary, or neither.

59. **COMPLEMENTARY AND SUBSTITUTE COMMODITIES** Refer to Exercise 39, Exercises 8.1. Show that the finished and unfinished home furniture manufactured by Country Workshop are substitute commodities.

   Hint: Solve the system of equations for x and y in terms of p and q.

For Exercises 60–62, let $x = f(p, q)$ be the demand equation for the commodities A and B, where p and q are the respective unit prices. The *elasticity of demand for A* is

$$E_p = -\frac{p\dfrac{\partial x}{\partial p}}{x}$$

and the *cross elasticity of demand for A with respect to q* is

$$E_q = -\frac{q\dfrac{\partial x}{\partial q}}{x}$$

(see Section 3.4).

60. **ELASTICITY OF DEMAND** The demand equation for Product A is

$$x = 400 - 8p^2 + 0.4q$$

where x is the quantity demanded of Product A, and p and q are the respective unit prices of Product A and a related Product B. Compute E_p and E_q when $p = 6$ and $q = 40$, and interpret your results.

61. **ELASTICITY OF DEMAND** Suppose that the daily demand for butter is given by

$$x = f(p, q) = \frac{3q}{1 + p^2}$$

where p and q denote the prices per pound (in dollars) of butter and margarine, respectively, and x is measured in millions of pounds. Compute E_p and E_q when $p = 5$ and $q = 4$. Interpret your results.

62. **ELASTICITY OF DEMAND** Suppose that the daily demand for margarine is given by

$$x = g(p, q) = \frac{2q}{1 + \sqrt{p}} \qquad (p, q > 0)$$

where p and q denote the prices per pound (in dollars) of margarine and butter, respectively, and x is measured in millions of pounds. Compute E_p and E_q when $p = 4$ and $q = 5$. Interpret your results.

63. According to the *ideal gas law*, the volume V (in liters) of an ideal gas is related to its pressure P (in pascals) and temperature T (in kelvins) by the formula

$$V = \frac{kT}{P}$$

where k is a constant. Show that

$$\frac{\partial V}{\partial T} \cdot \frac{\partial T}{\partial P} \cdot \frac{\partial P}{\partial V} = -1$$

64. **KINETIC ENERGY OF A BODY** The kinetic energy K of a body of mass m and velocity v is given by

$$K = \frac{1}{2}mv^2$$

Show that $\dfrac{\partial K}{\partial m} \cdot \dfrac{\partial^2 K}{\partial v^2} = K$.

65. **COBB–DOUGLAS PRODUCTION FUNCTION** Show that the Cobb–Douglas production function $P = kx^\alpha y^{1-\alpha}$, where $0 < \alpha < 1$, satisfies the equation

$$x\frac{\partial P}{\partial x} + y\frac{\partial P}{\partial y} = P$$

This equation is called Euler's equation.

66. Let $f(x, y, z) = x^2y + xy^2 + yz^3 + xye^{2z}$. Find f_{xz}, f_{yz}, and f_{zz}.

In Exercises 67–72, determine whether the statement is true or false. If it is true, explain why it is true. If it is false, give an example to show why it is false.

67. If $f_x(x, y)$ is defined at (a, b), then $f_y(x, y)$ must also be defined at (a, b).

68. If $f(x, y)$ is continuous at (a, b), then both $f_x(a, b)$ and $f_y(a, b)$ exist.

69. If $f_x(x, y) = 0$ and $f_y(x, y) = 0$ for all x and y, then f must be a constant function.

70. If $f_x(a, b) < 0$, then f is decreasing with respect to x near (a, b).

71. If $f_{xy}(x, y)$ and $f_{yx}(x, y)$ are both continuous for all values of x and y, then $f_{xy} = f_{yx}$ for all values of x and y.

72. If both f_{xy} and f_{yx} are defined at (a, b), then f_{xx} and f_{yy} must be defined at (a, b).

8.2 Solutions to Self-Check Exercises

1. $f_x = \dfrac{\partial f}{\partial x} = 3x^2 - 2y^2$

$f_y = \dfrac{\partial f}{\partial y} = -2x(2y) + 2y$

$\quad = 2y(1 - 2x)$

2. $f_x = \ln y + ye^x - 2x;\, f_y = \dfrac{x}{y} + e^x$

In particular,

$f_x(0, 1) = \ln\,1 + 1e^0 - 2(0) = 1$

$f_y(0, 1) = \dfrac{0}{1} + e^0 = 1$

The results tell us that at the point $(0, 1)$, $f(x, y)$ increases 1 unit for each unit increase in the x-direction, y being kept constant; $f(x, y)$ also increases 1 unit for each unit increase in the y-direction, x being kept constant.

3. From the results of Self-Check Exercise 1,

$f_x = 3x^2 - 2y^2$

Therefore,

$f_{xx} = \dfrac{\partial}{\partial x}(3x^2 - 2y^2) = 6x$

$f_{xy} = \dfrac{\partial}{\partial y}(3x^2 - 2y^2) = -4y$

Also, from the results of Self-Check Exercise 1,

$f_y = 2y(1 - 2x)$

Thus,

$f_{yx} = \dfrac{\partial}{\partial x}[2y(1 - 2x)] = -4y$

$f_{yy} = \dfrac{\partial}{\partial y}[2y(1 - 2x)] = 2(1 - 2x)$

4. a. The marginal productivity of labor when the amounts expended on labor and capital are x and y units, respectively, is given by

$$f_x(x, y) = 60\left(\frac{1}{3}x^{-2/3}\right)y^{2/3} = 20\left(\frac{y}{x}\right)^{2/3}$$

In particular, the required marginal productivity of labor is given by

$$f_x(125, 8) = 20\left(\frac{8}{125}\right)^{2/3} = 20\left(\frac{4}{25}\right)$$

or 3.2 units/unit increase in labor expenditure, capital expenditure being held constant at 8 units. Next, we compute

$$f_y(x, y) = 60x^{1/3}\left(\frac{2}{3}y^{-1/3}\right) = 40\left(\frac{x}{y}\right)^{1/3}$$

and deduce that the required marginal productivity of capital is given by

$$f_y(125, 8) = 40\left(\frac{125}{8}\right)^{1/3} = 40\left(\frac{5}{2}\right)$$

or 100 units/unit increase in capital expenditure, labor expenditure being held constant at 125 units.

b. The results of part (a) tell us that the government should encourage increased spending on capital rather than on labor.

USING TECHNOLOGY   Finding Partial Derivatives at a Given Point

Suppose $f(x, y)$ is a function of two variables and we wish to compute

$$f_x(a, b) = \left.\frac{\partial f}{\partial x}\right|_{(a, b)}$$

Recall that in computing $\partial f/\partial x$, we think of y as being fixed. But in this situation, we are evaluating $\partial f/\partial x$ at (a, b). Therefore, we set y equal to b. Doing this leads to the function g of one variable, x, defined by

$$g(x) = f(x, b)$$

It follows from the definition of the partial derivative that

$$f_x(a, b) = g'(a)$$

Thus, the value of the partial derivative $\partial f/\partial x$ at a given point (a, b) can be found by evaluating the derivative of a function of one variable. In particular, the latter can be found by using the numerical derivative operation of a graphing utility. We find $f_y(a, b)$ in a similar manner.

EXAMPLE 1 Let $f(x, y) = (1 + xy^2)^{3/2}e^{x^2y}$. Find (a) $f_x(1, 2)$ and (b) $f_y(1, 2)$.

Solution

a. Define $g(x) = f(x, 2) = (1 + 4x)^{3/2}e^{2x^2}$. Using the numerical derivative operation to find $g'(1)$, we obtain

$$f_x(1, 2) = g'(1) \approx 429.583225$$

b. Define $h(y) = f(1, y) = (1 + y^2)^{3/2}e^y$. Using the numerical derivative operation to find $h'(2)$, we obtain

$$f_y(1, 2) = h'(2) \approx 181.7468642$$

TECHNOLOGY EXERCISES

For each of the functions in Exercises 1–6, compute

$$\frac{\partial f}{\partial x} \quad \text{and} \quad \frac{\partial f}{\partial y}$$

at the given point:

1. $f(x, y) = \sqrt{x}(2 + xy^2)^{1/3}; \ (1, 2)$

2. $f(x, y) = \sqrt{xy}(1 + 2xy)^{2/3}; \ (1, 4)$

3. $f(x, y) = \dfrac{x + y^2}{1 + x^2y}; \ (1, 2)$

4. $f(x, y) = \dfrac{xy^2}{(\sqrt{x} + \sqrt{y})^2}; \ (4, 1)$

5. $f(x, y) = e^{-xy^2}(x + y)^{1/3}; \ (1, 1)$

6. $f(x, y) = \dfrac{\ln(\sqrt{x} + y^2)}{x^2 + y^2}; \ (4, 1)$

8.3 Maxima and Minima of Functions of Several Variables

Maxima and Minima

In Chapter 4, we saw that the solution of a problem often reduces to finding the extreme values of a function of one variable. In practice, however, situations also arise in which a problem is solved by finding the absolute maximum or absolute minimum value of a function of two or more variables.

For example, suppose Scandi Company manufactures computer desks in both assembled and unassembled versions. Its profit P is therefore a function of the number of assembled units, x, and the number of unassembled units, y, manufactured and sold per week; that is, $P = f(x, y)$. A question of paramount importance to the manufacturer is: How many assembled and unassembled desks should the company manufacture per week to maximize its weekly profit? Mathematically, the problem is solved by finding the values of x and y that will make $f(x, y)$ a maximum.

In this section, we will focus our attention on finding the extrema of a function of two variables. As in the case of a function of one variable, we distinguish between the relative (or local) extrema and the absolute extrema of a function of two variables.

> **Relative Extrema of a Function of Two Variables**
>
> Let f be a function defined on a region R containing the point (a, b). Then f has a **relative maximum** at (a, b) if $f(x, y) \le f(a, b)$ for all points (x, y) that are sufficiently close to (a, b). The number $f(a, b)$ is called a **relative maximum value**. Similarly, f has a **relative minimum** at (a, b), with **relative minimum value** $f(a, b)$, if $f(x, y) \ge f(a, b)$ for all points (x, y) that are sufficiently close to (a, b).

Loosely speaking, f has a relative maximum at (a, b) if the point $(a\,b, f(a, b))$ is the highest point on the graph of f when compared with all nearby points. A similar interpretation holds for a relative minimum.

If the inequalities in this last definition hold for *all* points (x, y) in the domain of f, then f has an **absolute maximum** (or **absolute minimum**) at (a, b) with **absolute maximum value** (or **absolute minimum value**) $f(a, b)$. Figure 16 shows the graph of a function with relative maxima at (a, b) and (e, g) and a relative minimum at (c, d). The absolute maximum of f occurs at (e, g), and the absolute minimum of f occurs at (h, i).

Observe that in the case of a function of one variable, a relative extremum (relative maximum or relative minimum) may or may not be an absolute extremum.

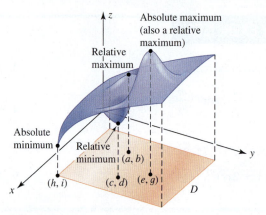

FIGURE **16**
The relative and absolute extrema of the function f over the domain D

Now let's turn our attention to the study of relative extrema of a function. Suppose that a differentiable function $f(x, y)$ of two variables has a relative maximum (relative minimum) at a point (a, b) in the domain of f. From Figure 17, it is clear that at the point (a, b) the slopes of the "tangent lines" to the surface in any direction must be zero. In particular, this implies that both

$$\frac{\partial f}{\partial x}(a, b) \quad \text{and} \quad \frac{\partial f}{\partial y}(a, b)$$

must be zero.

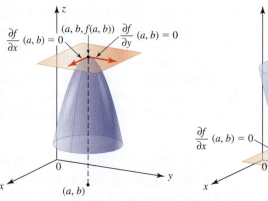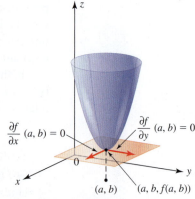

(a) f has a relative maximum at (a, b). 　　**(b)** f has a relative minimum at (a, b).
FIGURE **17**

Lest we be tempted to jump to the conclusion that a differentiable function f satisfying both the conditions

$$\frac{\partial f}{\partial x}(a, b) = 0 \quad \text{and} \quad \frac{\partial f}{\partial y}(a, b) = 0$$

at a point (a, b) must have a relative extremum at the point (a, b), let's examine the graph of the function f depicted in Figure 18. Here, both

$$\frac{\partial f}{\partial x}(a, b) = 0 \quad \text{and} \quad \frac{\partial f}{\partial y}(a, b) = 0$$

but f has neither a relative maximum nor a relative minimum at the point (a, b) because some nearby points are higher and some are lower than the point $(a, b, f(a, b))$. The point $(a, b, f(a, b))$ is called a **saddle point.**

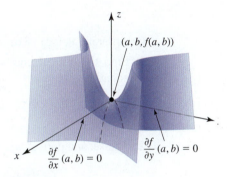

FIGURE 18
The point $(a, b, f(a, b))$ is called a saddle point.

Finally, an examination of the graph of the function f depicted in Figure 19 should convince you that f has a relative maximum at the point (a, b). But both $\partial f / \partial x$ and $\partial f / \partial y$ fail to be defined at (a, b).

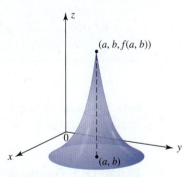

FIGURE 19
f has a relative maximum at (a, b), but neither $\partial f / \partial x$ nor $\partial f / \partial y$ exists at (a, b).

To summarize, a function f of two variables can have a relative extremum only at a point (a, b) in its domain where $\partial f / \partial x$ and $\partial f / \partial y$ both exist and are equal to zero at (a, b) or at least one of the partial derivatives does not exist. As in the case of one variable, we refer to a point in the domain of f that *may* give rise to a relative extremum as a critical point. The precise definition follows.

Critical Point of f

A **critical point** of f is a point (a, b) in the domain of f such that both

$$\frac{\partial f}{\partial x}(a, b) = 0 \quad \text{and} \quad \frac{\partial f}{\partial y}(a, b) = 0$$

or at least one of the partial derivatives does not exist.

To determine the nature of a critical point of a function $f(x, y)$ of two variables, we use the second partial derivatives of f. The resulting test, which helps us to classify these points, is called the **second derivative test** and is incorporated in the following procedure for finding and classifying the relative extrema of f.

Determining Relative Extrema

1. Find the critical points of $f(x, y)$ by solving the system of simultaneous equations

$$f_x(x, y) = 0$$
$$f_y(x, y) = 0$$

2. The second derivative test: Let

$$D(x, y) = f_{xx}(x, y)f_{yy}(x, y) - f^2_{xy}(x, y)$$

Then

   a. $D(a, b) > 0$ and $f_{xx}(a, b) < 0$ implies that $f(x, y)$ has a **relative maximum** at the point (a, b).

   b. $D(a, b) > 0$ and $f_{xx}(a, b) > 0$ implies that $f(x, y)$ has a **relative minimum** at the point (a, b).

   c. $D(a, b) < 0$ implies that $f(x, y)$ has neither a relative maximum nor a relative minimum at the point (a, b). The point $(a, b, f(a, b))$ is called a saddle point.

   d. $D(a, b) = 0$ implies that the test is inconclusive, so some other technique must be used to solve the problem.

Note   We can replace $f_{xx}(a, b)$ by $f_{yy}(a, b)$ in parts 2a and 2b because $D(a, b) > 0$ implies that $f_{xx}(a, b)$ and $f_{yy}(a, b)$ must have the same sign.   ▪

EXAMPLE 1 Find the relative extrema of the function

$$f(x, y) = x^2 + y^2$$

Solution   We have

$$f_x(x, y) = 2x$$
$$f_y(x, y) = 2y$$

To find the critical point(s) of f, we set $f_x(x, y) = 0$ and $f_y(x, y) = 0$ and solve the resulting system of simultaneous equations $2x = 0$ and $2y = 0$. We obtain $x = 0$,

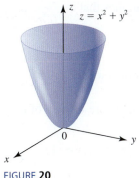

FIGURE 20
The graph of $f(x, y) = x^2 + y^2$

$y = 0$, or $(0, 0)$, as the sole critical point of f. Next, we apply the second derivative test to determine the nature of the critical point $(0, 0)$. We compute

$$f_{xx}(x, y) = 2 \qquad f_{xy}(x, y) = 0 \qquad f_{yy}(x, y) = 2$$

and

$$D(x, y) = f_{xx}(x, y)f_{yy}(x, y) - f_{xy}^2(x, y) = (2)(2) - 0 = 4$$

In particular, $D(0, 0) = 4$. Since $D(0, 0) > 0$ and $f_{xx}(0, 0) = 2 > 0$, we conclude that $f(x, y)$ has a relative minimum at the point $(0, 0)$. The relative minimum value, 0, also happens to be the absolute minimum of f. The graph of the function f, shown in Figure 20, confirms these results.

EXAMPLE 2 Find the relative extrema of the function

$$f(x, y) = 3x^2 - 4xy + 4y^2 - 4x + 8y + 4$$

Solution We have

$$f_x(x, y) = 6x - 4y - 4$$
$$f_y(x, y) = -4x + 8y + 8$$

To find the critical points of f, we set $f_x(x, y) = 0$ and $f_y(x, y) = 0$ and solve the resulting system of simultaneous equations

$$6x - 4y = 4$$
$$-4x + 8y = -8$$

Multiplying the first equation by 2 and the second equation by 3, we obtain the equivalent system

$$12x - 8y = 8$$
$$-12x + 24y = -24$$

Adding the two equations gives $16y = -16$, or $y = -1$. We substitute this value for y into either equation in the system to get $x = 0$. Thus, the only critical point of f is the point $(0, -1)$. Next, we apply the second derivative test to determine whether the point $(0, -1)$ gives rise to a relative extremum of f. We compute

$$f_{xx}(x, y) = 6 \qquad f_{xy}(x, y) = -4 \qquad f_{yy}(x, y) = 8$$

and

$$D(x, y) = f_{xx}(x, y)f_{yy}(x, y) - f_{xy}^2(x, y) = (6)(8) - (-4)^2 = 32$$

Since $D(0, -1) = 32 > 0$ and $f_{xx}(0, -1) = 6 > 0$, we conclude that $f(x, y)$ has a relative minimum at the point $(0, -1)$. The value of $f(x, y)$ at the point $(0, -1)$ is given by

$$f(0, -1) = 3(0)^2 - 4(0)(-1) + 4(-1)^2 - 4(0) + 8(-1) + 4 = 0$$

EXAMPLE 3 Find the relative extrema of the function

$$f(x, y) = 4y^3 + x^2 - 12y^2 - 36y + 2$$

Solution To find the critical points of f, we set $f_x = 0$ and $f_y = 0$ simultaneously, obtaining

$$f_x(x, y) = 2x = 0$$
$$f_y(x, y) = 12y^2 - 24y - 36 = 0$$

Explore and Discuss

Suppose $f(x, y)$ has a relative extremum (relative maximum or relative minimum) at a point (a, b). Let $g(x) = f(x, b)$ and $h(y) = f(a, y)$. Assuming that f and g are differentiable, explain why $g'(a) = 0$ and $h'(b) = 0$. Explain why these results are equivalent to the conditions $f_x(a, b) = 0$ and $f_y(a, b) = 0$.

The first equation implies that $x = 0$. The second equation implies that

$$y^2 - 2y - 3 = 0$$
$$(y + 1)(y - 3) = 0$$

that is, $y = -1$ or 3. Therefore, there are two critical points of the function f: $(0, -1)$ and $(0, 3)$.

Next, we apply the second derivative test to determine the nature of each of the two critical points. We compute

$$f_{xx}(x, y) = 2 \qquad f_{xy}(x, y) = 0 \qquad f_{yy}(x, y) = 24y - 24 = 24(y - 1)$$

Therefore,

$$D(x, y) = f_{xx}(x, y)f_{yy}(x, y) - f^2_{xy}(x, y) = 48(y - 1)$$

For the point $(0, -1)$,

$$D(0, -1) = 48(-1 - 1) = -96 < 0$$

Since $D(0, -1) < 0$, we conclude that the point $(0, -1)$ gives a saddle point of f. For the point $(0, 3)$,

$$D(0, 3) = 48(3 - 1) = 96 > 0$$

Since $D(0, 3) > 0$ and $f_{xx}(0, 3) > 0$, we conclude that the function f has a relative minimum at the point $(0, 3)$. Furthermore, since

$$f(0, 3) = 4(3)^3 + (0)^2 - 12(3)^2 - 36(3) + 2$$
$$= -106$$

we see that the relative minimum value of f is -106.

As in the case of a practical optimization problem involving a function of one variable, the solution to an optimization problem involving a function of several variables calls for finding the *absolute* extremum of the function. Determining the absolute extremum of a function of several variables is more difficult than merely finding the relative extrema of the function. However, in many situations, the absolute extremum of a function actually coincides with the largest relative extremum of the function that occurs in the interior of its domain. We assume that the problems considered here belong to this category. Furthermore, the existence of the absolute extremum (solution) of a practical problem is often deduced from the geometric or practical nature of the problem.

APPLIED EXAMPLE 4 Maximizing Profits The total weekly revenue (in dollars) that Acrosonic realizes in producing and selling its bookshelf loudspeaker systems is given by

$$R(x, y) = -\frac{1}{4}x^2 - \frac{3}{8}y^2 - \frac{1}{4}xy + 300x + 240y$$

where x denotes the number of fully assembled units and y denotes the number of kits produced and sold each week. The total weekly cost attributable to the production of these loudspeakers is

$$C(x, y) = 180x + 140y + 5000$$

dollars, where x and y have the same meaning as before. Determine how many assembled units and how many kits Acrosonic should produce per week to maximize its profit. What is the maximum profit?

Explore and Discuss

1. Refer to the second derivative test. Can the condition $f_{xx}(a, b) < 0$ in part 2a be replaced by the condition $f_{yy}(a, b) < 0$? Explain your answer. How about the condition $f_{xx}(a, b) > 0$ in part 2b?

2. Let $f(x, y) = x^4 + y^4$.
   a. Show that $(0, 0)$ is a critical point of f and that $D(0, 0) = 0$.
   b. Explain why f has a relative (in fact, an absolute) minimum at $(0, 0)$. Does this contradict the second derivative test? Explain your answer.

Solution   The contribution to Acrosonic's weekly profit stemming from the production and sale of the bookshelf loudspeaker systems is given by

$$P(x, y) = R(x, y) - C(x, y)$$
$$= \left(-\frac{1}{4}x^2 - \frac{3}{8}y^2 - \frac{1}{4}xy + 300x + 240y\right) - (180x + 140y + 5000)$$
$$= -\frac{1}{4}x^2 - \frac{3}{8}y^2 - \frac{1}{4}xy + 120x + 100y - 5000$$

To find the relative maximum of the profit function $P(x, y)$, we first locate the critical point(s) of P. Setting $P_x(x, y)$ and $P_y(x, y)$ equal to zero, we obtain

$$P_x = -\frac{1}{2}x - \frac{1}{4}y + 120 = 0$$
$$P_y = -\frac{3}{4}y - \frac{1}{4}x + 100 = 0$$

Solving the first of these equations for y yields

$$y = -2x + 480$$

which, upon substitution into the second equation, yields

$$-\frac{3}{4}(-2x + 480) - \frac{1}{4}x + 100 = 0$$
$$6x - 1440 - x + 400 = 0$$
$$x = 208$$

We substitute this value of x into the equation $y = -2x + 480$ to get

$$y = 64$$

Therefore, the function P has the sole critical point $(208, 64)$. To show that the point $(208, 64)$ is a solution to our problem, we use the second derivative test. We compute

$$P_{xx} = -\frac{1}{2} \qquad P_{xy} = -\frac{1}{4} \qquad P_{yy} = -\frac{3}{4}$$

So

$$D(x, y) = \left(-\frac{1}{2}\right)\left(-\frac{3}{4}\right) - \left(-\frac{1}{4}\right)^2 = \frac{3}{8} - \frac{1}{16} = \frac{5}{16}$$

Since $D(208, 64) > 0$ and $P_{xx}(208, 64) < 0$, the point $(208, 64)$ yields a relative maximum of P. It can be shown that this relative maximum is also the absolute maximum of P. We conclude that Acrosonic can maximize its weekly profit by manufacturing 208 assembled units and 64 kits of their bookshelf loudspeaker systems. The maximum weekly profit realizable from the production and sale of these loudspeaker systems is given by

$$P(208, 64) = -\frac{1}{4}(208)^2 - \frac{3}{8}(64)^2 - \frac{1}{4}(208)(64)$$
$$+ 120(208) + 100(64) - 5000$$
$$= 10,680$$

or $10,680.

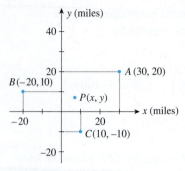

FIGURE 21
Locating a site for a television relay station

APPLIED EXAMPLE 5 Locating a Television Relay Station Site A television relay station will serve Towns A, B, and C, whose relative locations are shown in Figure 21. Determine a site for the location of the station if the sum of the squares of the distances from each town to the site is minimized.

Solution Suppose the required site is located at the point $P(x, y)$. With the aid of the distance formula, we find that the square of the distance from Town A to the site is

$$(x - 30)^2 + (y - 20)^2 \qquad (x^2) \text{ See page 26.}$$

The respective distances from Towns B and C to the site are found in a similar manner, so the sum of the squares of the distances from each town to the site is given by

$$f(x, y) = (x - 30)^2 + (y - 20)^2 + (x + 20)^2$$
$$+ (y - 10)^2 + (x - 10)^2 + (y + 10)^2$$

To find the relative minimum of $f(x, y)$, we first find the critical point(s) of f. Using the Chain Rule to find $f_x(x, y)$ and $f_y(x, y)$ and setting each equal to zero, we obtain

$$f_x = 2(x - 30) + 2(x + 20) + 2(x - 10) = 6x - 40 = 0$$
$$f_y = 2(y - 20) + 2(y - 10) + 2(y + 10) = 6y - 40 = 0$$

from which we deduce that $\left(\frac{20}{3}, \frac{20}{3}\right)$ is the sole critical point of f. Since

$$f_{xx} = 6 \qquad f_{xy} = 0 \qquad f_{yy} = 6$$

we have

$$D(x, y) = f_{xx} f_{yy} - f_{xy}^2 = (6)(6) - 0 = 36$$

Since $D\left(\frac{20}{3}, \frac{20}{3}\right) > 0$ and $f_{xx}\left(\frac{20}{3}, \frac{20}{3}\right) > 0$, we conclude that the point $\left(\frac{20}{3}, \frac{20}{3}\right)$ yields a relative minimum of f. Thus, the required site has coordinates $x = \frac{20}{3}$ and $y = \frac{20}{3}$.

8.3 Self-Check Exercises

1. Let $f(x, y) = 2x^2 + 3y^2 - 4xy + 4x - 2y + 3$.
   a. Find the critical point of f.
   b. Use the second derivative test to classify the nature of the critical point.
   c. Find the relative extremum of f, if it exists.

2. **MAXIMIZING PROFIT** Robertson Controls manufactures two basic models of setback thermostats: a standard mechanical thermostat and a deluxe electronic thermostat. Robertson's monthly revenue (in hundreds of dollars) is

   $$R(x, y) = -\frac{1}{8}x^2 - \frac{1}{2}y^2 - \frac{1}{4}xy + 20x + 60y$$

where x (in units of a hundred) denotes the number of mechanical thermostats manufactured each month and y (in units of a hundred) denotes the number of electronic thermostats manufactured each month. The total monthly cost incurred in producing these thermostats is

$$C(x, y) = 7x + 20y + 280$$

hundred dollars. Find how many thermostats of each model Robertson should manufacture each month to maximize its profits. What is the maximum profit?

Solutions to Self-Check Exercises 8.3 can be found on page 592.

8.3 Concept Questions

1. Explain the terms (a) relative maximum of a function $f(x, y)$ and (b) absolute maximum of a function $f(x, y)$.

2. **a.** What is a critical point of a function $f(x, y)$?
   b. Explain the role of a critical point in determining the relative extrema of a function of two variables.

3. Explain how the second derivative test is used to determine the relative extrema of a function of two variables.

4. In (a)–(d), suppose that (a, b) is a critical point of f. Are the given conditions sufficient for you to determine whether f has a relative maximum, a relative minimum, or a saddle point at (a, b)? Explain.

a. $f_{xx}(a, b) = -2, f_{xy}(a, b) = 3, f_{yy}(a, b) = -5$
b. $f_{xx}(a, b) = 3, f_{xy}(a, b) = 3, f_{yy}(a, b) = 2$
c. $f_{xx}(a, b) = 1, f_{xy}(a, b) = 2, f_{yy}(a, b) = 4$
d. $f_{xx}(a, b) = 2, f_{xy}(a, b) = 2, f_{yy}(a, b) = 4$

8.3 Exercises

In Exercises 1–20, find the critical point(s) of the function. Then use the second derivative test to classify the nature of each point, if possible. Finally, determine the relative extrema of the function.

1. $f(x, y) = 1 - 2x^2 - 3y^2$

2. $f(x, y) = x^2 - xy + y^2 + 1$

3. $f(x, y) = x^2 - y^2 - 2x + 4y + 1$

4. $f(x, y) = 2x^2 + y^2 - 4x + 6y + 3$

5. $f(x, y) = x^2 + 2xy + 2y^2 - 4x + 8y - 1$

6. $f(x, y) = x^2 - 4xy + 2y^2 + 4x + 8y - 1$

7. $f(x, y) = 2x^3 + y^2 - 9x^2 - 4y + 12x - 2$

8. $f(x, y) = 2x^3 + y^2 - 6x^2 - 4y + 12x - 2$

9. $f(x, y) = x^3 + y^2 - 2xy + 7x - 8y + 4$

10. $f(x, y) = 2y^3 - 3y^2 - 12y + 2x^2 - 6x + 2$

11. $f(x, y) = x^3 - 3xy + y^3 - 2$

12. $f(x, y) = x^3 - 2xy + y^2 + 5$

13. $f(x, y) = xy + \dfrac{4}{x} + \dfrac{2}{y}$ 14. $f(x, y) = \dfrac{x}{y^2} + xy$

15. $f(x, y) = x^2 - e^{y^2}$ 16. $f(x, y) = e^{x^2 - y^2}$

17. $f(x, y) = e^{x^2 + y^2}$ 18. $f(x, y) = e^{xy}$

19. $f(x, y) = \ln(1 + x^2 + y^2)$

20. $f(x, y) = xy + \ln x + 2y^2$

21. **Maximizing Profit** The total weekly revenue (in dollars) of the Country Workshop realized in manufacturing and selling its rolltop desks is given by

$$R(x, y) = -0.2x^2 - 0.25y^2 - 0.2xy + 200x + 160y$$

where x denotes the number of finished units and y denotes the number of unfinished units manufactured and sold each week. The total weekly cost attributable to the manufacture of these desks is given by

$$C(x, y) = 100x + 70y + 4000$$

dollars. Determine how many finished units and how many unfinished units the company should manufacture each week to maximize its profit. What is the maximum profit realizable?

22. **Maximizing Profit** The total daily revenue (in dollars) that Weston Publishing realizes in publishing and selling its English-language dictionaries is given by

$$R(x, y) = -0.005x^2 - 0.003y^2 - 0.002xy$$
$$+ 20x + 15y$$

where x denotes the number of deluxe copies and y denotes the number of standard copies published and sold daily. The total daily cost of publishing these dictionaries is given by

$$C(x, y) = 6x + 3y + 200$$

dollars. Determine how many deluxe copies and how many standard copies Weston should publish each day to maximize its profits. What is the maximum profit realizable?

23. **Maximum Price of Land** The rectangular region R shown in the accompanying figure represents the financial district of a city. The price of land within the district is approximated by the function

$$p(x, y) = 200 - 10\left(x - \frac{1}{2}\right)^2 - 15(y - 1)^2$$

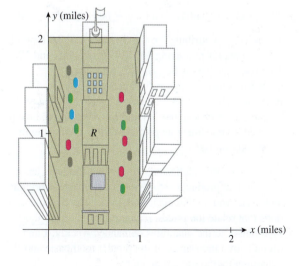

where $p(x, y)$ is the price of land at the point (x, y) in dollars per square foot and x and y are measured in miles. At what point within the financial district is the price of land highest?

24. MAXIMIZING PROFIT C&G Imports imports two brands of white wine, one from Germany and the other from Italy. The German wine costs $4/bottle, and the Italian wine costs $3/bottle. It has been estimated that if the German wine sells for p dollars/bottle and the Italian wine sells for for q dollars/bottle, then

$$2000 - 150p + 100q$$

bottles of the German wine and

$$1000 + 80p - 120q$$

bottles of the Italian wine will be sold each week. Determine the unit price for each brand that will allow C&G to realize the largest possible weekly profit.

25. MAXIMIZING REVENUE The management of Cal Supermarkets has determined that the quantity demanded per week of their 90% lean ground sirloin, x, and the quantity demanded per week of their 80% ground beef, y (both measured in pounds), are related to their unit prices p and q (in dollars), respectively, by the equations

$$x = 6400 - 400p - 200q \quad \text{and} \quad y = 5600 - 200p - 400q$$

a. What is the total revenue function $R(p, q)$?
   Hint: $R(p, q) = xp + yq$
b. What price should Cal Supermarkets charge for each product to maximize its weekly revenue? How many pounds of each product will then be sold? What is the maximum revenue?

26. MAXIMIZING PROFIT Johnson's Household Products has a division that produces two sizes of bar soap. The demand equations that relate the prices p and q (in dollars per hundred bars), to the quantities demanded, x and y (in units of a hundred), of the 3.5-oz size bar soap and the 5-oz bath size bar soap are given by

$$p = 80 - 0.01x - 0.005y \quad \text{and} \quad q = 60 - 0.005x - 0.015y$$

The fixed cost attributed to the division is $10,000/week, and the cost for producing 100 3.5-oz size bars and 100 5-oz bath size bars is $8 and $12, respectively.
a. What is the weekly profit function $P(x, y)$?
b. How many of the 3.5-oz size bars and how many of the 5-oz bath size bars should the division produce per week to maximize its profit? What is the maximum weekly profit?

27. MAXIMIZING PROFIT Johnson's Household Products has a division that produces two types of toothpaste: a regular toothpaste and a whitening tooth paste. The demand equations that relate the prices, p and q (in dollars per thousand units), to the quantities demanded weekly, x and y (in units of a thousand), of the regular toothpaste and the whitening toothpaste are given by

$$p = 3000 - 20x - 10y \quad \text{and} \quad q = 4000 - 10x - 30y$$

respectively. The fixed cost attributed to the division is $20,000/week, and the cost for producing 1000 tubes of regular and 1000 tubes of whitening toothpaste is $400 and $500, respectively.
a. What is the weekly total revenue function $R(x, y)$?
b. What is the weekly total cost function $C(x, y)$?
c. What is the weekly profit function $P(x, y)$?
d. How many tubes of regular and whitening toothpaste should be produced weekly to maximize the division's profit? What is the maximum weekly profit?

28. DETERMINING THE OPTIMAL SITE FOR A POWER STATION An auxiliary electric power station will serve three communities, A, B, and C, whose relative locations are shown in the accompanying figure. Determine where the power station should be located if the sum of the squares of the distances from each community to the site is minimized.

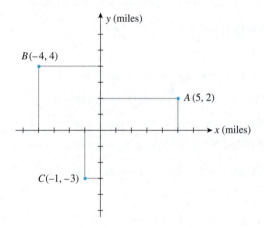

29. LOCATING A RADIO STATION The following figure shows the locations of three neighboring communities. The operators of a newly proposed radio station have decided that the site $P(x, y)$ for the station should be chosen so that the sum of the squares of the distances from the site to each community is minimized. Find the location of the proposed radio station.

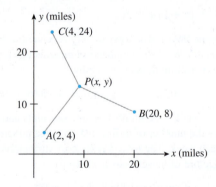

30. PARCEL POST REGULATIONS Postal regulations specify that a parcel sent by parcel post may have a combined length and girth of no more than 130 in. Find the dimensions of a cylindrical package of greatest volume that can be sent through the mail. What is the volume of such a package?

Hint: The length plus the girth is $2\pi r + l$.

31. **PACKAGING** An open rectangular box having a volume of 108 in.^{3} is to be constructed from a tin sheet. Find the dimensions of such a box if the amount of material used in its construction is to be minimal.

    Hint: Let the dimensions of the box be $x \times y \times z$. Then $xyz = 108$, and the amount of material used is given by $S = xy + 2yz + 2xz$. Show that

    $$S = f(x, y) = xy + \frac{216}{x} + \frac{216}{y}$$

    Minimize $f(x, y)$.

32. **PACKAGING** An open rectangular box having a surface area of 300 in.^{2} is to be constructed from a tin sheet. Find the dimensions of the box if the volume of the box is to be as large as possible. What is the maximum volume?

    Hint: Let the dimensions of the box be $x \times y \times z$ (see the figure that follows). Then the surface area is $xy + 2xz + 2yz$, and its volume is xyz.

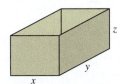

33. **PACKAGING** Postal regulations specify that the combined length and girth of a parcel sent by parcel post may not exceed 130 in. Find the dimensions of the rectangular package that would have the greatest possible volume under these regulations.

    Hint: Let the dimensions of the box be $x \times y \times z$. (see the figure below). Then $2x + 2z + y = 130$, and the volume $V = xyz$. Show that

    $$V = f(x, z) = 130xz - 2x^2z - 2xz^2$$

    Maximize $f(x, z)$.

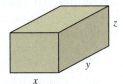

34. **MINIMIZING HEATING AND COOLING COSTS** A building in the shape of a rectangular box is to have a volume of 12,000 ft^3 (see the figure). It is estimated that the annual heating and cooling costs will be \$2/ft^2 for the top, \$4/ft^2 for the front and back, and \$3/ft^2 for the sides. Find the dimensions of the building that will result in a minimal annual heating and cooling cost. What is the minimal annual heating and cooling cost?

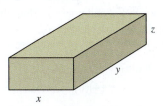

35. **PACKAGING** An open box having a volume of 48 in.^{3} is to be constructed. If the box is to include a partition that is parallel to a side of the box, as shown in the figure, and the amount of material used is to be minimal, what should be the dimensions of the box?

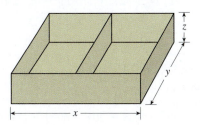

In Exercises 36–42, determine whether the statement is true or false. If it is true, explain why it is true. If it is false, give an example to show why it is false.

36. If $f_x(a, b) = 0$ and $f_y(a, b) = 0$, then f must have a relative extremum at (a, b).

37. If (a, b) is a critical point of f and both the conditions $f_{xx}(a, b) < 0$ and $f_{yy}(a, b) < 0$ hold, then f has a relative maximum at (a, b).

38. If $f(x, y)$ has a relative maximum at (a, b), then $f_x(a, b) = 0$ and $f_y(a, b) = 0$.

39. Let $h(x, y) = f(x) + g(y)$. If $f(x) > 0$ and $g(y) < 0$, then h cannot have a relative maximum or a relative minimum at any point.

40. If $f(x, y)$ satisfies $f_{xx}(a, b) \neq 0, f_{xy}(a, b) = 0, f_{yy}(a, b) \neq 0$, and $f_{xx}(a, b) + f_{yy}(a, b) = 0$ at the critical point (a, b) of f, then f cannot have a relative extremum at (a, b).

41. Suppose $h(x, y) = f(x) + g(y)$, where f and g have continuous second derivatives near a and b, respectively. If a is a critical number of f, b is a critical number of g, and $f''(a)g''(b) > 0$, then h has a relative extremum at (a, b).

42. If f_{xx} and f_{yy} have opposite signs at a critical point (a, b) of f, then f has a saddle point at (a, b).

8.3   Solutions to Self-Check Exercises

1. a. To find the critical point(s) of f, we solve the system of equations

$$f_x = 4x - 4y + 4 = 0$$
$$f_y = -4x + 6y - 2 = 0$$

obtaining $x = -2$ and $y = -1$. Thus, the only critical point of f is the point $(-2, -1)$.

b. We have $f_{xx} = 4$, $f_{xy} = -4$, and $f_{yy} = 6$, so

$$\begin{aligned} D(x, y) &= f_{xx}f_{yy} - f_{xy}^2 \\ &= (4)(6) - (-4)^2 = 8 \end{aligned}$$

Since $D(-2, -1) > 0$ and $f_{xx}(-2, -1) > 0$, we conclude that f has a relative minimum at the point $(-2, -1)$.

c. The relative minimum value of $f(x, y)$ at the point $(-2, -1)$ is

$$\begin{aligned} f(-2, -1) &= 2(-2)^2 + 3(-1)^2 - 4(-2)(-1) \\ &\quad + 4(-2) - 2(-1) + 3 \\ &= 0 \end{aligned}$$

2. Robertson's monthly profit is

$$P(x, y) = R(x, y) - C(x, y)$$
$$= \left(-\frac{1}{8}x^2 - \frac{1}{2}y^2 - \frac{1}{4}xy + 20x + 60y \right) - (7x + 20y + 280)$$
$$= -\frac{1}{8}x^2 - \frac{1}{2}y^2 - \frac{1}{4}xy + 13x + 40y - 280$$

The critical point of P is found by solving the system

$$P_x = -\frac{1}{4}x - \frac{1}{4}y + 13 = 0$$
$$P_y = -\frac{1}{4}x - y + 40 = 0$$

giving $x = 16$ and $y = 36$. Thus, $(16, 36)$ is the critical point of P. Next,

$$P_{xx} = -\frac{1}{4} \qquad P_{xy} = -\frac{1}{4} \qquad P_{yy} = -1$$

and

$$\begin{aligned} D(x, y) &= P_{xx}P_{yy} - P_{xy}^2 \\ &= \left(-\frac{1}{4} \right)(-1) - \left(-\frac{1}{4} \right)^2 = \frac{3}{16} \end{aligned}$$

Since $D(16, 36) > 0$ and $P_{xx}(16, 36) < 0$, the point $(16, 36)$ yields a relative maximum of P. We conclude that the monthly profit is maximized by manufacturing 1600 mechanical and 3600 electronic setback thermostats each month. The maximum monthly profit realizable is

$$\begin{aligned} P(16, 36) &= -\frac{1}{8}(16)^2 - \frac{1}{2}(36)^2 - \frac{1}{4}(16)(36) \\ &\quad + 13(16) + 40(36) - 280 \\ &= 544 \end{aligned}$$

or $54,400.

8.4   The Method of Least Squares

The Method of Least Squares

In Section 1.4, Example 10, we saw how a linear equation can be used to approximate the sales trend for a local sporting goods store. As we saw there, one use of a **trend line** is to predict a store's future sales. Recall that we obtained the line by requiring that it pass through two data points, the rationale being that such a line seems to *fit* the data reasonably well.

In this section, we describe a general method, known as the **method of least squares,** for determining a straight line that, in some sense, *best* fits a set of data points when the points are scattered about a straight line. To illustrate the principle behind the method of least squares, suppose, for simplicity, that we are given five data points,

$$P_1(x_1, y_1), \quad P_2(x_2, y_2), \quad P_3(x_3, y_3), \quad P_4(x_4, y_4), \quad P_5(x_5, y_5)$$

that describe the relationship between the two variables x and y. By plotting these data points, we obtain a graph called a **scatter diagram** (Figure 22).

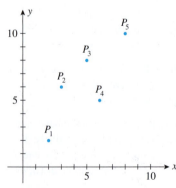

FIGURE 22
A scatter diagram

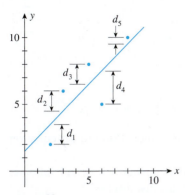

FIGURE 23
The approximating line misses the points by the amounts $d_1, d_2, \ldots, d_5$, respectively.

If we try to fit a straight line to these data points, the line will miss the first, second, third, fourth, and fifth data points by the amounts d_1, d_2, d_3, d_4, and d_5, respectively (Figure 23).

The **principle of least squares** states that the straight line L that fits the data points best is the one chosen by requiring that the sum of the squares of d_1, d_2, d_3, d_4, d_5—that is,

$$d_1^2 + d_2^2 + d_3^2 + d_4^2 + d_5^2$$

be made as small as possible. If we think of the amount d_1 as the error made when the value y_1 is approximated by the corresponding value of y lying on the straight line L, d_2 as the error made when the value y_2 is approximated by the corresponding value of y, and so on, then it can be seen that the least-squares criterion calls for minimizing the sum of the squares of the errors. The line L obtained in this manner is called the **least-squares line**, or **regression line**.

To find a method for computing the regression line L, suppose L has representation $y = f(x) = mx + b$, where m and b are to be determined. Observe that

$$
\begin{aligned}
d_1^2 &+ d_2^2 + d_3^2 + d_4^2 + d_5^2 \\
&= [f(x_1) - y_1]^2 + [f(x_2) - y_2]^2 + [f(x_3) - y_3]^2 \\
&\quad + [f(x_4) - y_4]^2 + [f(x_5) - y_5]^2 \\
&= (mx_1 + b - y_1)^2 + (mx_2 + b - y_2)^2 + (mx_3 + b - y_3)^2 \\
&\quad + (mx_4 + b - y_4)^2 + (mx_5 + b - y_5)^2
\end{aligned}
$$

and may be viewed as a function of the two variables m and b. Thus, the least-squares criterion is equivalent to minimizing the function

$$
\begin{aligned}
f(m, b) &= (mx_1 + b - y_1)^2 + (mx_2 + b - y_2)^2 + (mx_3 + b - y_3)^2 \\
&\quad + (mx_4 + b - y_4)^2 + (mx_5 + b - y_5)^2
\end{aligned}
$$

with respect to m and b. To find the minimum of $f(m, b)$, we use the Chain Rule and compute

$$
\begin{aligned}
\frac{\partial f}{\partial m} &= 2(mx_1 + b - y_1)x_1 + 2(mx_2 + b - y_2)x_2 + 2(mx_3 + b - y_3)x_3 \\
&\quad + 2(mx_4 + b - y_4)x_4 + 2(mx_5 + b - y_5)x_5 \\
&= 2[mx_1^2 + bx_1 - x_1y_1 + mx_2^2 + bx_2 - x_2y_2 + mx_3^2 + bx_3 - x_3y_3 \\
&\quad + mx_4^2 + bx_4 - x_4y_4 + mx_5^2 + bx_5 - x_5y_5] \\
&= 2[(x_1^2 + x_2^2 + x_3^2 + x_4^2 + x_5^2)m + (x_1 + x_2 + x_3 + x_4 + x_5)b \\
&\quad - (x_1y_1 + x_2y_2 + x_3y_3 + x_4y_4 + x_5y_5)]
\end{aligned}
$$

and

$$
\begin{aligned}
\frac{\partial f}{\partial b} &= 2(mx_1 + b - y_1) + 2(mx_2 + b - y_2) + 2(mx_3 + b - y_3) \\
&\quad + 2(mx_4 + b - y_4) + 2(mx_5 + b - y_5) \\
&= 2[(x_1 + x_2 + x_3 + x_4 + x_5)m + 5b - (y_1 + y_2 + y_3 + y_4 + y_5)]
\end{aligned}
$$

Setting

$$\frac{\partial f}{\partial m} = 0 \quad \text{and} \quad \frac{\partial f}{\partial b} = 0$$

gives

$$(x_1^2 + x_2^2 + x_3^2 + x_4^2 + x_5^2)m + (x_1 + x_2 + x_3 + x_4 + x_5)b$$
$$= x_1y_1 + x_2y_2 + x_3y_3 + x_4y_4 + x_5y_5$$

and

$$(x_1 + x_2 + x_3 + x_4 + x_5)m + 5b = y_1 + y_2 + y_3 + y_4 + y_5$$

Solving these two simultaneous equations for m and b then leads to an equation $y = mx + b$ of a straight line.

Before looking at an example, we state a more general result whose derivation is identical to the special case involving the five data points just discussed.

The Method of Least Squares

Suppose we are given n data points:

$$P_1(x_1, y_1), \quad P_2(x_2, y_2), \quad P_3(x_3, y_3), \dots, P_n(x_n, y_n)$$

Then the least-squares (regression) line for the data is given by the linear equation

$$y = f(x) = mx + b$$

where the constants m and b satisfy the equations

$$(x_1^2 + x_2^2 + x_3^2 + \cdots + x_n^2)m + (x_1 + x_2 + x_3 + \cdots + x_n)b$$
$$= x_1y_1 + x_2y_2 + x_3y_3 + \cdots + x_ny_n \tag{8}$$

and

$$(x_1 + x_2 + x_3 + \cdots + x_n)m + nb$$
$$= y_1 + y_2 + y_3 + \cdots + y_n \tag{9}$$

simultaneously. Equations (8) and (9) are called **normal equations**.

EXAMPLE 1 Find an equation of the least-squares line for the data

$$P_1(1, 1), \quad P_2(2, 3), \quad P_3(3, 4), \quad P_4(4, 3), \quad P_5(5, 6)$$

Solution   Here, we have $n = 5$ and

$$x_1 = 1 \qquad x_2 = 2 \qquad x_3 = 3 \qquad x_4 = 4 \qquad x_5 = 5$$
$$y_1 = 1 \qquad y_2 = 3 \qquad y_3 = 4 \qquad y_4 = 3 \qquad y_5 = 6$$

so Equation (8) becomes

$$(1 + 4 + 9 + 16 + 25)m + (1 + 2 + 3 + 4 + 5)b = 1 + 6 + 12 + 12 + 30$$

or

$$55m + 15b = 61 \tag{10}$$

and Equation (9) becomes

$$(1 + 2 + 3 + 4 + 5)m + 5b = 1 + 3 + 4 + 3 + 6$$

or

$$15m + 5b = 17 \tag{11}$$

Solving Equation (11) for b gives

$$b = -3m + \frac{17}{5} \qquad\qquad \textbf{(12)}$$

which, upon substitution into Equation (10), gives

$$55m + 15\left(-3m + \frac{17}{5}\right) = 61$$
$$55m - 45m + 51 = 61$$
$$10m = 10$$
$$m = 1$$

Substituting this value of m into Equation (12) gives

$$b = -3 + \frac{17}{5} = \frac{2}{5} = 0.4$$

Therefore, the required equation of the least-squares line is

$$y = x + 0.4$$

The scatter diagram and the regression line are shown in Figure 24.

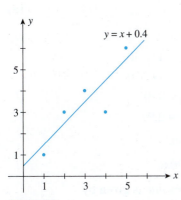

FIGURE 24
The scatter diagram and the least-squares
line $y = x + 0.4$

$ APPLIED EXAMPLE 2 A Firm's Advertising Expense and Profit The proprietor of Leisure Travel Service compiled the following data relating the firm's annual profit to its annual advertising expenditure (both measured in thousands of dollars).

Annual Advertising Expenditure, x	12	14	17	21	26	30
Annual Profit, y	60	70	90	100	100	120

a. Determine an equation of the least-squares line for these data.
b. Draw a scatter diagram and the least-squares line for these data.
c. Use the result obtained in part (a) to predict Leisure Travel's annual profit if the annual advertising budget is $20,000.

Solution

a. The calculations required for obtaining the normal equations may be summarized as follows:

	x	y	x^2	xy
	12	60	144	720
	14	70	196	980
	17	90	289	1,530
	21	100	441	2,100
	26	100	676	2,600
	30	120	900	3,600
Sum	120	540	2,646	11,530

The normal equations are

$$6b + 120m = 540 \tag{13}$$
$$120b + 2646m = 11{,}530 \tag{14}$$

Solving Equation (13) for b gives

$$b = -20m + 90 \tag{15}$$

which, upon substitution into Equation (14), gives

$$120(-20m + 90) + 2646m = 11{,}530$$
$$-2400m + 10{,}800 + 2646m = 11{,}530$$
$$246m = 730$$
$$m \approx 2.97$$

Substituting this value of m into Equation (15) gives

$$b \approx -20(2.97) + 90 = 30.6$$

Therefore, the required equation of the least-squares line is given by

$$y = f(x) \approx 2.97x + 30.6$$

b. The scatter diagram and the least-squares line are shown in Figure 25.
c. Leisure Travel's predicted annual profit corresponding to an annual budget of $20,000 is given by

$$f(20) \approx 2.97(20) + 30.6$$
$$= 90$$

or $90,000.

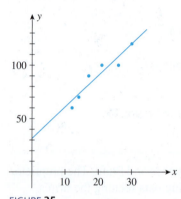

FIGURE 25
The scatter diagram and the least-squares
line $y = 2.97x + 30.6$

APPLIED EXAMPLE 3 Maximizing Profit A market research study conducted for Century Communications provided the following data based on the projected monthly sales x (in thousands) of Century's DVD version of a box-office hit adventure movie with a proposed wholesale unit price of p dollars.

p	38	36	34.5	30	28.5
x	2.2	5.4	7.0	11.5	14.6

a. Find the demand equation if the demand curve is the least-squares line for these data.
b. The total monthly cost function associated with producing and distributing the DVD movies is given by

$$C(x) = 4x + 25$$

where x denotes the number of discs (in thousands) produced and sold and $C(x)$ is in thousands of dollars. Determine the unit wholesale price that will maximize Century's monthly profit.

Solution

a. The calculations required for obtaining the normal equations may be summarized as follows:

	x	p	x^2	xp
	2.2	38	4.84	83.6
	5.4	36	29.16	194.4
	7.0	34.5	49	241.5
	11.5	30	132.25	345
	14.6	28.5	213.16	416.1
Sum	40.7	167	428.41	1280.6

The normal equations are

$$5b + 40.7m = 167$$
$$40.7b + 428.41m = 1280.6$$

Solving this system of linear equations simultaneously, we find that

$$m \approx -0.81 \quad \text{and} \quad b \approx 40.00$$

Therefore, the required equation of the least-squares line is given by

$$p = f(x) \approx -0.81x + 40.00$$

which is the required demand equation, provided $0 \le x \le 49.38$.

b. The approximate total revenue function in this case is given by

$$R(x) = xp = -0.81x^2 + 40.00x$$

Since the total cost function is

$$C(x) = 4x + 25$$

we see that the approximate profit function is

$$P(x) = -0.81x^2 + 40.00x - (4x + 25)$$
$$= -0.81x^2 + 36.00x - 25$$

To find the absolute maximum of $P(x)$ over the closed interval $[0, 49.38]$, we compute

$$P'(x) = -1.62x + 36.00$$

Setting $P'(x) = 0$ gives $x \approx 22.22$ as the only critical point of P. Finally, from the table

x	0	22.22	49.38
$P(x)$	-25	375.00	-222.41

we see that the optimal wholesale price is

$$p = -0.81(22.22) + 40.00 = 22.00$$

or $22 per disc.

8.4 Self-Check Exercises

1. Find an equation of the least-squares line for the data

$P_1(0, 3)$, $P_2(2, 6.5)$, $P_3(4, 10)$, $P_4(6, 16)$, $P_5(7, 16.5)$

2. GLOBAL BOX-OFFICE RECEIPTS Global box-office ticket sales have been growing steadily over the years, reflecting the rapid growth in overseas markets, particularly in China.

The sales (in billions of dollars) from 2007 through 2011 are summarized in the following table.

Year	2007	2008	2009	2010	2011
Sales, y	26.1	27.2	28.9	31.1	32.6

a. Find an equation of the least-squares line for these data. (Let $x = 1$ represent 2007.)

b. Use the result of part (a) to find the global ticket sales for 2014, assuming that the trend continued.

Source: Motion Picture Association of America.

Solutions to Self-Check Exercises 8.4 can be found on page 601.

8.4 Concept Questions

1. Explain the terms (a) *scatter diagram* and (b) *least-squares line*.

2. Explain the *principle of least squares* in your own words.

3. In the method of least squares, we are required to minimize the sum of the *squares* of the errors. Comment on replacing this criterion by (a) minimizing the sum of the errors, and (b) minimizing the sum of the absolute values of the errors.

4. Given the array $(x_1, y_1), (x_2, y_2), \ldots, (x_n, y_n)$, the mean or average of the array is the point $(\bar{x}, \bar{y})$, where

$$\bar{x} = \frac{1}{n}(x_1 + x_2 + \cdots + x_n)$$

and

$$\bar{y} = \frac{1}{n}(y_1 + y_2 + \cdots + y_n)$$

Show that the least-squares line for the array passes through $(\bar{x}, \bar{y})$.

8.4 Exercises

In Exercises 1–6, (a) find an equation of the least-squares line for the data, and (b) draw a scatter diagram for the data and graph the least-squares line.

1.

x	1	2	3	4
y	4	6	8	11

2.

x	1	3	5	7	9
y	9	8	6	3	2

3.

x	1	2	3	4	4	6
y	4.5	5	3	2	3.5	1

4.

x	1	1	2	3	4	4	5
y	2	3	3	3.5	3.5	4	5

5. $P_1(1, 3), P_2(2, 5), P_3(3, 5), P_4(4, 7), P_5(5, 8)$

6. $P_1(1, 8), P_2(2, 6), P_3(5, 6), P_4(7, 4), P_5(10, 1)$

7. COLLEGE ADMISSIONS The following data, compiled by the admissions office at Faber College during the past 5 years, relate the number of college brochures and follow-up letters (x) sent to a preselected list of high school juniors who had taken the PSAT and the number of completed applications (y) received from these students (both measured in units of 1000):

x	4	4.5	5	5.5	6
y	0.5	0.6	0.8	0.9	1.2

a. Determine the equation of the least-squares line for these data.

b. Draw a scatter diagram and the least-squares line for these data.

c. Use the result obtained in part (a) to predict the number of completed applications that might be expected if 6400 brochures and follow-up letters are sent out during the next year.

8. NET SALES The management of Kaldor, a manufacturer of electric motors, submitted the following data in the annual report to its stockholders. The table shows the net sales (in millions of dollars) during the 5 years that have elapsed since the new management team took over. (The first year the firm operated under the new management corresponds to the time period $x = 1$, and the four subsequent years correspond to $x = 2, 3, 4, 5$.)

Year, x	1	2	3	4	5
Net Sales, y	426	437	460	473	477

a. Determine the equation of the least-squares line for these data.

b. Draw a scatter diagram and the least-squares line for these data.

c. Use the result obtained in part (a) to predict the net sales for the upcoming year.

9. **SAT VERBAL SCORES** The following data, compiled by the superintendent of schools in a large metropolitan area, shows the average SAT verbal scores of high school seniors during the 5 years since the district implemented the "back-to-basics" program:

Year, x	1	2	3	4	5
Average Score, y	436	438	428	430	426

a. Determine the equation of the least-squares line for these data.

b. Draw a scatter diagram and the least-squares line for these data.

c. Use the result obtained in part (a) to predict the average SAT verbal score of high school seniors 2 years from now ($x = 7$).

10. **COST OF SUMMER BLOCKBUSTERS** Hollywood is spending more and more to produce its big summer movies each year. The estimated costs of summer big-budget releases (in billions of dollars) for the years 2011 through 2013 are given in the following table:

Year	2011	2012	2013
Spending, y	2.1	2.4	2.7

a. Letting $x = 1$ denote 2011, find an equation of the least-squares line for these data.

b. Use the result of part (a) to estimate the amount of money Hollywood spent in 2015 to produce its big summer movies for that year, assuming the trend continued.

Source: Los Angeles Times.

11. **FACEBOOK USERS** End-of-year data for the number of Facebook users (in millions) from 2008 through 2011 are given in the following table:

Year	2008	2009	2010	2011
Number, y	154.5	381.8	654.5	845.0

a. Letting $x = 0$ denote the end of 2008, find an equation of the least-squares line for these data.

b. Use the result of part (a) to estimate the number of Facebook users at the end of 2015, assuming that the trend continued.

Source: Company reports.

12. **E-BOOK AUDIENCE** The number of adults (in millions) using e-book devices is expected to climb in the years ahead. The projected number of e-book readers in the

United States from 2011 through 2015 is given in the following table:

Year	2011	2012	2013	2014	2015
Number, y	25.3	33.4	39.5	50.0	59.6

a. Letting $x = 0$ denote 2011, find an equation of the least-squares line for these data.

b. Use the result of part (a) to estimate the projected average rate of growth of the number of e-book readers between 2011 and 2015.

Source: Forrester Research, Inc.

13. **MASS-TRANSIT SUBSIDIES** The following table gives the projected state subsidies (in millions of dollars) to the Massachusetts Bay Transit Authority (MBTA) over a 5-year period:

Year, x	1	2	3	4	5
Subsidy, y	20	24	26	28	32

a. Find an equation of the least-squares line for these data.

b. Assuming that the trend continued, estimate the state subsidy to the MBTA for the eighth year ($x = 8$).

Source: Massachusetts Bay Transit Authority.

14. **PERCENTAGE OF THE POPULATION ENROLLED IN SCHOOL** The percentage of the population (aged 3 years or older) who were enrolled in school from 2007 through 2011 is given in the following table:

Year	2007	2008	2009	2010	2011
Percent, y	26.2	26.8	27.5	28.3	28.7

a. Letting $x = 0$ denote 2007, find an equation of the least-squares line for these data.

b. Use the result of part (a) to estimate the percentage of the population (aged 3 or older) who were enrolled in school in 2014, assuming that the trend continued.

Source: U.S. Census Bureau.

15. **REVENUE OF MOODY'S CORPORATION** Moody's Corporation is the holding company for Moody's Investors Service, which has a 40% share in the world credit-rating market. According to company reports, the projected total revenue (in billions of dollars) of the company is as follows:

Year	2004	2005	2006	2007	2008
Revenue, y	1.42	1.73	1.98	2.32	2.65

a. Letting $x = 4$ denote 2004, find an equation of the least-squares line for these data.

b. Use the results of part (a) to estimate the rate of change of the revenue of the company for the period in question.

c. Use the result of part (a) to estimate the total revenue of the company in 2010, assuming that the trend continued.

Source: Company reports.

16. **HOUSEHOLDS WITH SOMEONE UNDER 18** The percentage of households in which someone was under 18 years old from 2007 through 2011 is given in the following table:

Year	2007	2008	2009	2010	2011
Percent, y	34.4	34.1	33.4	33.1	32.7

a. Letting $x = 0$ denote 2007, find an equation of the least-squares line for these data.
b. Use the result of part (a) to estimate the percentage of households in which someone was under 18 years old in 2013, assuming that the trend continued.
Source: U.S. Census Bureau.

17. **GROWTH OF CREDIT UNIONS** Credit union membership is on the rise. The following table gives the number (in millions) of credit union members from 2003 through 2011 in 2-year intervals:

Year	2003	2005	2007	2009	2011
Number, y	82.0	84.7	86.8	89.7	91.8

a. Letting $x = 0$ denote 2003, find an equation of the least-squares line for these data.
b. Assuming that the trend continued, estimate the number of credit union members in 2013 ($x = 5$).
Source: National Credit Union Association.

18. **FIRST-CLASS MAIL VOLUME** As more and more people turn to using the Internet and phones to pay bills and to communicate, replacing letters, the first-class mail volume is expected to decline until 2020. The following table gives the volume (in billions of pieces) of first-class mail from 2007 through 2011:

Year	2007	2008	2009	2010	2011
Value, y	95.9	91.7	83.8	78.2	73.5

a. Letting $x = 1$ denote 2007, find an equation of the least-squares line for these data.
b. Use the results of part (a) to estimate the volume of first-class mail in 2014, assuming that the trend continued through that year.
Source: U.S. Postal Service.

19. **SATELLITE TV SUBSCRIBERS** The number of satellite and telecommunications subscribers has continued to grow over the years. The following table gives the number of subscribers (in millions) from 2006 through 2010:

Year	2006	2007	2008	2009	2010
Number, y	29.4	32.2	34.8	37.7	40.4

a. Letting $x = 0$ denote 2006, find an equation of the least-squares line for these data.
b. Use the result of part (a) to estimate the average rate of growth of the number of subscribers between 2006 and 2010.
Source: SNL Ragan.

20. **ONLINE VIDEO ADVERTISING** Although still a small percentage of all online advertising, online video advertising is growing. The following table gives the projected spending on Web video advertising (in billions of dollars) through 2016:

Year	2011	2012	2013	2014	2015	2016
Spending, y	2.0	3.1	4.5	6.3	7.8	9.3

a. Letting $x = 0$ denote 2011, find an equation of the least-squares line for these data.
b. Use the result of part (a) to estimate the projected rate of growth of video advertising from 2011 through 2016.
Source: eMarketer.

21. **U.S. OUTDOOR ADVERTISING** U.S. outdoor advertising expenditure (in billions of dollars) from 2011 through 2015 is given in the following table ($x = 0$ corresponds to 2011):

Year	2011	2012	2013	2014	2015
Expenditure, y	6.4	6.8	7.1	7.4	7.6

a. Find an equation of the least-squares line for these data.
b. Use the result of part (a) to estimate the rate of change of the advertising expenditures for the period in question.
Source: Outdoor Advertising Association.

22. **ONLINE SALES OF USED AUTOS** The amount (in millions of dollars) of used autos sold online in the United States is expected to grow in accordance with the figures given in the following table ($x = 0$ corresponds to 2011):

Year, x	0	1	2	3	4
Sales, y	12.9	13.9	14.65	15.25	15.85

a. Find an equation of the least-squares line for these data.
b. Use the result of part (a) to estimate the sales of used autos online in 2016, assuming that the predicted trend continued.
Source: comScore Networks, Inc.

23. **BOUNCED-CHECK CHARGES** Overdraft fees have become an important piece of a bank's total fee income. The following table gives the bank revenue from overdraft fees (in billions of dollars) from 2004 through 2009. Here, $x = 4$ corresponds to 2004.

Year, x	4	5	6	7	8	9
Revenue, y	27.5	29	31	34	36	38

a. Find an equation of the least-squares line for these data.
b. Use the result of part (a) to estimate the average rate of increase in overdraft fees over the period under consideration.

c. Assuming that the trend continued, what was the revenue from overdraft fees in 2011?

Source: New York Times.

24. **Male Life Expectancy at 65** The Census Bureau projections of male life expectancy at age 65 in the United States are summarized in the following table ($x = 0$ corresponds to 2000):

Year, x	0	10	20	30	40	50
Years Beyond 65, y	15.9	16.8	17.6	18.5	19.3	20.3

a. Find an equation of the least-squares line for these data.

b. Use the result of part (a) to estimate the life expectancy at 65 of a male in 2040. How does this result compare with the given data for that year?

c. Use the result of part (a) to estimate the life expectancy at 65 of a male in 2030.

Source: U.S. Census Bureau.

25. **Home Health-Care and Equipment Spending** The following table gives the projected spending on home care and durable medical equipment (in billions of dollars) from 2004 through 2016 ($x = 0$ corresponds to 2004):

Year, x	0	2	4	6	8	10	12
Spending, y	60	74	90	106	118	128	150

a. Find an equation of the least-squares line for these data.

b. Use the result of part (a) to approximate the spending on home care and durable medical equipment in 2015.

c. Use the result of part (a) to estimate the rate of change of the spending on home care and durable medical equipment for the period from 2004 through 2016.

Source: National Association of Home Care and Hospice.

26. **Global Defense Spending** The following table gives the projected global defense spending (in trillions of dollars) from the beginning of 2008 ($t = 0$) through 2015 ($t = 7$):

Year, t	0	1	2	3	4	5	6	7
Spending, y	1.38	1.44	1.49	1.56	1.61	1.67	1.74	1.78

a. Find an equation of the least-squares line for these data.

b. Use the result of part (a) to estimate the rate of change in the projected global defense spending from 2008 through 2015.

c. Assuming that the trend continues, what will the global spending on defense be in 2018?

Source: Homeland Security Research.

In Exercises 27–30, determine whether the statement is true or false. If it is true, explain why it is true. If it is false, give an example to show why it is false.

27. The least-squares line must pass through at least one data point.

28. The error incurred in approximating n data points using the least-squares linear function is zero if and only if the n data points lie on a nonvertical straight line.

29. If the data consist of two distinct points, then the least-squares line is just the line that passes through the two points.

30. A data point lies on the least-squares line if and only if the vertical distance between the point and the line is equal to zero.

8.4 Solutions to Self-Check Exercises

1. We first construct the table:

	x	y	x^2	xy
	0	3	0	0
	2	6.5	4	13
	4	10	16	40
	6	16	36	96
	7	16.5	49	115.5
Sum	19	52	105	264.5

The normal equations are

$$5b + 19m = 52$$
$$19b + 105m = 264.5$$

Solving the first equation for b gives

$$b = -3.8m + 10.4$$

which, upon substitution into the second equation, gives

$$19(-3.8m + 10.4) + 105m = 264.5$$
$$-72.2m + 197.6 + 105m = 264.5$$
$$32.8m = 66.9$$
$$m \approx 2.04$$

Substituting this value of m into the expression for b found earlier gives

$$b \approx -3.8(2.04) + 10.4 \approx 2.65$$

Therefore, the required least-squares line has the equation given by

$$y = 2.04x + 2.65$$

2. a. We summarize the calculations as follows:

	x	y	x^2	xy
	1	26.1	1	26.1
	2	27.2	4	54.4
	3	28.9	9	86.7
	4	31.1	16	124.4
	5	32.6	25	163.0
Sum	15	145.9	55	454.6

The normal equations are

$$5b + 15m = 145.9$$
$$15b + 55m = 454.6$$

Solving this system, we find $m = 1.69$ and $b = 24.11$. Therefore, the required equation is

$$y = f(x) = 1.69x + 24.11$$

b. The predicted global sales for 2014 are given by

$$f(8) = 1.69(8) + 24.11$$
$$= 37.63$$

or approximately $37.6 billion.

USING TECHNOLOGY   Finding an Equation of a Least-Squares Line

Graphing Utility

A graphing utility is especially useful in calculating an equation of the least-squares line for a set of data. We simply enter the given data in the form of lists into the calculator and then use the linear regression function to obtain the coefficients of the required equation.

EXAMPLE 1 Find an equation of the least-squares line for the data

x	1.1	2.3	3.2	4.6	5.8	6.7	8.0
y	-5.8	-5.1	-4.8	-4.4	-3.7	-3.2	-2.5

Plot the scatter diagram and the least-squares line for this data.

Solution   First, we enter the data as follows:

$$x_1 = 1.1 \qquad y_1 = -5.8 \qquad x_2 = 2.3 \qquad y_2 = -5.1 \qquad x_3 = 3.2$$
$$y_3 = -4.8 \qquad x_4 = 4.6 \qquad y_4 = -4.4 \qquad x_5 = 5.8 \qquad y_5 = -3.7$$
$$x_6 = 6.7 \qquad y_6 = -3.2 \qquad x_7 = 8.0 \qquad y_7 = -2.5$$

Then, using the linear regression function from the statistics menu, we obtain the output shown in Figure T1a. Therefore, an equation of the least-squares line $(y = ax + b)$ is

$$y = 0.46x - 6.3$$

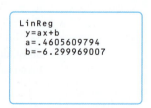

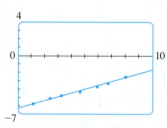

(a) The TI-83/84 linear regression screen

(b) The scatter diagram and least-squares line for the data

FIGURE **T1**

The graph of the least-squares equation and the scatter diagram for the data are shown in Figure T1b.

APPLIED EXAMPLE 2 Erosion of the Middle Class The idea of a large, stable, middle class (defined as those with annual household incomes in 2010 between \$39,000 and \$118,000 for a family of three), is central to America's sense of itself. The following table gives the percentage of middle-income adults (y) in the United States from 1971 through 2011.

Year	1971	1981	1991	2001	2011
Percent, y	61	59	56	54	51

Let t be measured in decades with $t = 0$ corresponding to 1971.
a. Find an equation of the least-squares line for these data.
b. If this trend continues, what will the percentage of middle-income adults be in 2021?
Source: Pew Research Center.

Solution

a. First we enter the data as follows:

$$x_1 = 0 \qquad y_1 = 61 \qquad x_2 = 1 \qquad y_2 = 59 \qquad x_3 = 2$$
$$y_3 = 56 \qquad x_4 = 3 \qquad y_4 = 54 \qquad x_5 = 4 \qquad y_5 = 51$$

Then, using the linear regression function from the statistics menu, we obtain the output shown in Figure T2. Therefore, an equation of this least-squares line is

$$y = -2.5t + 61.2$$

b. The percentage of middle-income adults in 2021 will be

$$y = -(2.5)(5) + 61.2 = 48.7$$

or approximately 48.7%.

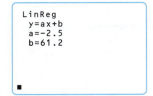

```
LinReg
  y=ax+b
  a=-2.5
  b=61.2

■
```

FIGURE T2
The TI-83/84 linear regression screen

TECHNOLOGY EXERCISES

In Exercises 1–4, find an equation of the least-squares line for the given data.

1.

x	2.1	3.4	4.7	5.6	6.8	7.2
y	8.8	12.1	14.8	16.9	19.8	21.1

2.

x	1.1	2.4	3.2	4.7	5.6	7.2
y	−0.5	1.2	2.4	4.4	5.7	8.1

3.

x	−2.1	−1.1	0.1	1.4	2.5	4.2	5.1
y	6.2	4.7	3.5	1.9	0.4	−1.4	−2.5

4.

x	−1.12	0.1	1.24	2.76	4.21	6.82
y	7.61	4.9	2.74	−0.47	−3.51	−8.94

5. WORLDWIDE CONSULTING SPENDING The following table gives the projected worldwide consulting spending (in billions of dollars) from 2005 through 2009 ($x = 5$ corresponds to 2005):

Year, x	5	6	7	8	9
Spending, y	254	279	300	320	345

a. Find an equation of the least-squares line for these data.
b. Use the results of part (a) to estimate the average rate of increase of worldwide consulting spending over the period under consideration.
c. Use the results of part (a) to estimate the amount of spending in 2010, assuming that the trend continued.
Source: Kennedy Information.

6. U.S. ONLINE BANKING HOUSEHOLDS The following table gives the projected U.S. online banking households as a percentage of all U.S. banking households from 2001 ($x = 1$) through 2007 ($x = 7$):

Year, x	1	2	3	4	5	6	7
Percent, y	21.2	26.7	32.2	37.7	43.2	48.7	54.2

a. Find an equation of the least-squares line for these data.
b. Use the result of part (a) to estimate the percentage of U.S. online banking households in 2010.
Source: Jupiter Research.

7. INFORMATION SECURITY SOFTWARE SPENDING As online attacks persist, spending on information security software continues to rise. The following table gives the forecast for the worldwide sales (in billions of dollars) of information security software through 2007 ($t = 0$ corresponds to 2002):

Year, t	0	1	2	3	4	5
Spending, y	6.8	8.3	9.8	11.3	12.8	14.9

a. Find an equation of the least-squares line for these data.

b. Use the result of part (a) to estimate the spending on information security software in 2008, assuming that the trend continued.

Source: International Data Corporation.

8. ONLINE SPENDING The convenience of shopping on the Web combined with high-speed broadband access services is spurring online spending. The projected online spending per buyer (in dollars) from 2002 ($x = 0$) through 2008 ($x = 6$) is given in the following table:

Year, x	0	1	2	3	4	5	6
Spending, y	501	540	585	631	680	728	779

a. Find an equation of the least-squares line for these data.

b. Use the result of part (a) to estimate the rate of change of spending per buyer between 2002 and 2008.

Source: U.S. Department of Commerce.

9. SALES OF GPS EQUIPMENT The annual sales (in billions of dollars) of global positioning system (GPS) equipment from the year 2000 ($x = 0$) through 2006 are given in the following table:

Year, x	0	1	2	3	4	5	6
Annual Sales, y	7.9	9.6	11.5	13.3	15.2	16.0	18.8

a. Find an equation of the least-squares line for these data.

b. Use the equation found in part (a) to estimate the annual sales of GPS equipment for 2008, assuming that the trend continued.

Source: ABI Research.

8.5 Constrained Maxima and Minima and the Method of Lagrange Multipliers

Constrained Relative Extrema

In Section 8.3, we studied the problem of determining the relative extrema of a function $f(x, y)$ without placing any restrictions on the independent variables x and y—except, of course, that the point (x, y) lie in the domain of f. Such a relative extremum of a function f is referred to as an **unconstrained relative extremum** of f. However, in many practical optimization problems, we must maximize or minimize a function in which the independent variables are subjected to certain further constraints.

In this section, we discuss a powerful method for determining the relative extrema of a function $f(x, y)$ whose independent variables x and y are required to satisfy one or more constraints of the form $g(x, y) = 0$. Such a relative extremum of a function f is called a **constrained relative extremum** of f. We can see the difference between an unconstrained extremum of a function $f(x, y)$ of two variables and a constrained extremum of f, where the independent variables x and y are subjected to a constraint of the form $g(x, y) = 0$, by considering the geometry of the two cases. Figure 26a depicts the graph of a function $f(x, y)$ that has an unconstrained relative minimum at the point $(0, 0)$. However, when the independent variables x and y are subjected to an equality constraint of the form $g(x, y) = 0$, the points (x, y, z) that satisfy both $z = f(x, y)$ and the constraint equation $g(x, y) = 0$ lie on a curve C. Therefore, the constrained relative minimum of f must also lie on C (Figure 26b).

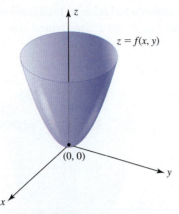

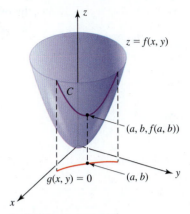

(a) f is not subject to any constraints.   (b) f is subject to a constraint.

FIGURE 26
The function f has an unconstrained minimum value of 0, but it has a constrained minimum value of $f(a, b)$ when subjected to the constraint $g(x, y) = 0$.

Our first example involves an equality constraint $g(x, y) = 0$ in which we solve for the variable y explicitly in terms of x. In this case we may apply the technique used in Chapter 4 to find the relative extrema of a function of one variable.

EXAMPLE 1 Find the relative minimum of the function

$$f(x, y) = 2x^2 + y^2$$

subject to the constraint $g(x, y) = x + y - 1 = 0$.

Solution   Solving the constraint equation for y explicitly in terms of x, we obtain $y = -x + 1$. Substituting this value of y into the function $f(x, y) = 2x^2 + y^2$ results in a function of x,

$$h(x) = 2x^2 + (-x + 1)^2 = 3x^2 - 2x + 1$$

The function h describes the curve C lying on the graph of f on which the constrained relative minimum of f occurs. To find this point, use the technique developed in Chapter 4 to determine the relative extrema of a function of one variable:

$$h'(x) = 6x - 2 = 2(3x - 1)$$

Setting $h'(x) = 0$ gives $x = \frac{1}{3}$ as the sole critical number of the function h. Next, we find

$$h''(x) = 6$$

and, in particular,

$$h''\left(\frac{1}{3}\right) = 6 > 0$$

Therefore, by the second derivative test, the point $x = \frac{1}{3}$ gives rise to a relative minimum of h. Substitute this value of x into the constraint equation $x + y - 1 = 0$ to get $y = \frac{2}{3}$. Thus, the point $\left(\frac{1}{3}, \frac{2}{3}\right)$ gives rise to the required constrained relative minimum of f. Since

$$f\left(\frac{1}{3}, \frac{2}{3}\right) = 2\left(\frac{1}{3}\right)^2 + \left(\frac{2}{3}\right)^2 = \frac{2}{3}$$

the required constrained relative minimum value of f is $\frac{2}{3}$ at the point $\left(\frac{1}{3}, \frac{2}{3}\right)$. It may be shown that $\frac{2}{3}$ is in fact a constrained absolute minimum value of f (Figure 27).

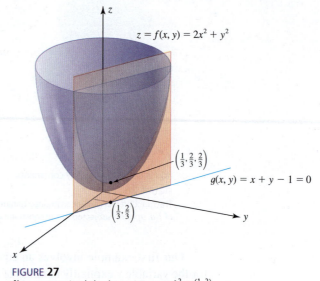

FIGURE 27
f has a constrained absolute minimum of $\frac{2}{3}$ at $\left(\frac{1}{3}, \frac{2}{3}\right)$.

The Method of Lagrange Multipliers

The major drawback of the technique used in Example 1 is that it relies on our ability to solve the constraint equation $g(x, y) = 0$ for y explicitly in terms of x. This is not always easy. Moreover, even when we can solve the constraint equation $g(x, y) = 0$ for y explicitly in terms of x, the resulting function of one variable that is to be optimized may turn out to be unnecessarily complicated. Fortunately, an easier method exists. This method, called the **method of Lagrange multipliers** (Joseph Lagrange, 1736–1813), is as follows:

> **The Method of Lagrange Multipliers**
>
> To find the relative extrema of the function $f(x, y)$ subject to the constraint $g(x, y) = 0$ (assuming that these extreme values exist),
>
> **1.** Form an auxiliary function
>
> $$F(x, y, \lambda) = f(x, y) + \lambda g(x, y)$$
>
> called the Lagrangian function (the variable λ is called the Lagrange multiplier).
>
> **2.** Solve the system that consists of the equations
>
> $$F_x = 0 \qquad F_y = 0 \qquad F_\lambda = 0$$
>
> for all values of x, y, and λ. The solutions of the system are candidates for the extrema of f.
>
> **3.** Evaluate f at each point found in Step 2. The largest value yields the constrained maximum of f, and the smallest value yields the constrained minimum of f.

Let's re-solve Example 1 using the method of Lagrange multipliers.

EXAMPLE 2 Using the method of Lagrange multipliers, find the relative minimum of the function

$$f(x, y) = 2x^2 + y^2$$

subject to the constraint $x + y = 1$.

Solution Write the constraint equation $x + y = 1$ in the form $g(x, y) = x + y - 1 = 0$. Then, form the Lagrangian function

$$F(x, y, \lambda) = f(x, y) + \lambda g(x, y)$$
$$= 2x^2 + y^2 + \lambda(x + y - 1)$$

To find the critical point(s) of the function F, solve the system composed of the equations

$$F_x = 4x + \lambda \quad = 0$$
$$F_y = 2y + \lambda \quad = 0$$
$$F_\lambda = x + y - 1 = 0$$

Solving the first and second equations in this system for x and y in terms of λ, we obtain

$$x = -\frac{1}{4}\lambda \quad \text{and} \quad y = -\frac{1}{2}\lambda$$

which, upon substitution into the third equation, yields

$$-\frac{1}{4}\lambda - \frac{1}{2}\lambda - 1 = 0 \quad \text{or} \quad \lambda = -\frac{4}{3}$$

Therefore, $x = \frac{1}{3}$ and $y = \frac{2}{3}$, and $\left(\frac{1}{3}, \frac{2}{3}\right)$ affords a constrained minimum of the function f, in agreement with the result obtained earlier. ◾

Note A disadvantage of the method of Lagrange multipliers is that there is no test analogous to the second derivative test mentioned in Section 8.3 for determining whether a critical point of a function of two or more variables leads to a relative maximum or relative minimum of the function. Here we have to rely on the geometric or physical nature of the problem to help us draw the necessary conclusions (see Example 2). ◾

The method of Lagrange multipliers may be used to solve a problem involving a function of three or more variables, as illustrated in the next example.

EXAMPLE 3 Use the method of Lagrange multipliers to find the minimum of the function

$$f(x, y, z) = 2xy + 6yz + 8xz$$

subject to the constraint

$$xyz = 12{,}000$$

(*Note:* The existence of the minimum is suggested by the geometry of the problem.)

Solution Write the constraint equation $xyz = 12{,}000$ in the form $g(x, y, x) = xyz - 12{,}000 = 0$. Then the Lagrangian function is

$$F(x, y, z, \lambda) = f(x, y, z) + \lambda g(x, y, z)$$
$$= 2xy + 6yz + 8xz + \lambda(xyz - 12{,}000)$$

To find the critical point(s) of the function F, we solve the system composed of the equations

$$F_x = 2y + 8z + \lambda yz = 0$$
$$F_y = 2x + 6z + \lambda xz = 0$$
$$F_z = 6y + 8x + \lambda xy = 0$$
$$F_\lambda = xyz - 12{,}000 = 0$$

Solving the first three equations of the system for λ in terms of x, y, and z, we have

$$\lambda = -\frac{2y + 8z}{yz}$$

$$\lambda = -\frac{2x + 6z}{xz}$$

$$\lambda = -\frac{6y + 8x}{xy}$$

Equating the first two expressions for λ leads to

$$\frac{2y + 8z}{yz} = \frac{2x + 6z}{xz}$$

$$2xy + 8xz = 2xy + 6yz$$

$$x = \frac{3}{4}y$$

Next, equating the second and third expressions for λ above yields

$$\frac{2x + 6z}{xz} = \frac{6y + 8x}{xy}$$

$$2xy + 6yz = 6yz + 8xz$$

$$z = \frac{1}{4}y$$

Finally, substituting these values of x and z into the equation $xyz - 12{,}000 = 0$, the fourth equation of the first system of equations, we have

$$\left(\frac{3}{4}y\right)(y)\left(\frac{1}{4}y\right) - 12{,}000 = 0$$

$$y^3 = \frac{(12{,}000)(4)(4)}{3} = 64{,}000$$

$$y = 40$$

The corresponding values of x and z are given by $x = \frac{3}{4}(40) = 30$ and $z = \frac{1}{4}(40) = 10$. Therefore, we see that the point $(30, 40, 10)$ gives the constrained minimum of f. The minimum value is

$$f(30, 40, 10) = 2(30)(40) + 6(40)(10) + 8(30)(10) = 7200$$

$ **APPLIED EXAMPLE 4** Maximizing Profit Refer to Example 4, Section 8.3. The total weekly profit (in dollars) that Acrosonic realized in producing and selling its bookshelf loudspeaker systems is given by the profit function

$$P(x, y) = -\frac{1}{4}x^2 - \frac{3}{8}y^2 - \frac{1}{4}xy + 120x + 100y - 5000$$

where x denotes the number of fully assembled units and y denotes the number of kits produced and sold per week. Acrosonic's management decides that production of these loudspeaker systems should be restricted to a total of exactly 230 units each week. Under this condition, how many fully assembled units and how many kits should be produced each week to maximize Acrosonic's weekly profit?

Solution The problem is equivalent to the problem of maximizing the function

$$P(x, y) = -\frac{1}{4}x^2 - \frac{3}{8}y^2 - \frac{1}{4}xy + 120x + 100y - 5000$$

subject to the constraint

$$g(x, y) = x + y - 230 = 0$$

The Lagrangian function is

$$F(x, y, \lambda) = P(x, y) + \lambda g(x, y)$$
$$= -\frac{1}{4}x^2 - \frac{3}{8}y^2 - \frac{1}{4}xy + 120x + 100y$$
$$- 5000 + \lambda(x + y - 230)$$

To find the critical point(s) of F, solve the following system of equations:

$$F_x = -\frac{1}{2}x - \frac{1}{4}y + 120 + \lambda = 0$$

$$F_y = -\frac{3}{4}y - \frac{1}{4}x + 100 + \lambda = 0$$

$$F_\lambda = x + y - 230 = 0$$

Solving the first equation of this system for λ, we obtain

$$\lambda = \frac{1}{2}x + \frac{1}{4}y - 120$$

which, upon substitution into the second equation, yields

$$-\frac{3}{4}y - \frac{1}{4}x + 100 + \frac{1}{2}x + \frac{1}{4}y - 120 = 0$$

$$-\frac{1}{2}y + \frac{1}{4}x - 20 = 0$$

Solving the last equation for y gives

$$y = \frac{1}{2}x - 40$$

Substituting this value of y into the third equation of the system, we have

$$x + \frac{1}{2}x - 40 - 230 = 0$$

$$x = 180$$

The corresponding value of y is $\frac{1}{2}(180) - 40$, or 50. Thus, the required constrained relative maximum of P occurs at the point $(180, 50)$. Again, we can show that the point $(180, 50)$ yields a constrained absolute maximum for P. Thus, Acrosonic's profit is maximized by producing 180 assembled and 50 kit versions of their bookshelf loudspeaker systems. The maximum weekly profit realizable is given by

$$P(180, 50) = -\frac{1}{4}(180)^2 - \frac{3}{8}(50)^2 - \frac{1}{4}(180)(50)$$
$$+ 120(180) + 100(50) - 5000$$
$$= 10,312.5$$

or \$10,312.50.

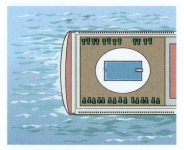

FIGURE 28
A rectangular-shaped pool will be built in the elliptical-shaped poolside area.

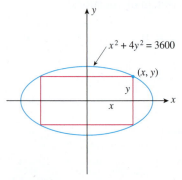

FIGURE 29
We want to find the largest rectangle that can be inscribed in the ellipse described by $x^2 + 4y^2 = 3600$.

 APPLIED EXAMPLE 5 Designing a Cruise-Ship Pool The operators of the *Viking Princess*, a luxury cruise liner, are contemplating the addition of another swimming pool to the ship. The chief engineer has suggested that an area in the form of an ellipse located in the rear of the promenade deck would be suitable for this purpose. This location would provide a poolside area with sufficient space for passenger movement and placement of deck chairs (Figure 28). It has been determined that the shape of the ellipse may be described by the equation $x^2 + 4y^2 = 3600$, where x and y are measured in feet. *Viking*'s operators would like to know the dimensions of the rectangular pool with the largest possible area that would meet these requirements.

Solution To solve this problem, we need to find the rectangle of largest area that can be inscribed in the ellipse with equation $x^2 + 4y^2 = 3600$. Letting the sides of the rectangle be $2x$ and $2y$ feet, we see that the area of the rectangle is $A = 4xy$ (Figure 29). Furthermore, the point (x, y) must be constrained to lie on the ellipse so that it satisfies the equation $x^2 + 4y^2 = 3600$. Thus, the problem is equivalent to the problem of maximizing the function

$$f(x, y) = 4xy$$

subject to the constraint $g(x, y) = x^2 + 4y^2 - 3600 = 0$. The Lagrangian function is

$$F(x, y, \lambda) = f(x, y) + \lambda g(x, y)$$
$$= 4xy + \lambda(x^2 + 4y^2 - 3600)$$

To find the critical point(s) of F, we solve the following system of equations:

$$F_x = 4y + 2\lambda x = 0$$
$$F_y = 4x + 8\lambda y = 0$$
$$F_\lambda = x^2 + 4y^2 - 3600 = 0$$

Solving the first equation of this system for λ, we obtain

$$\lambda = -\frac{2y}{x}$$

which, upon substitution into the second equation, yields

$$4x + 8\left(-\frac{2y}{x}\right)y = 0 \quad \text{or} \quad x^2 - 4y^2 = 0$$

that is, $x = \pm 2y$. Substituting these values of x into the third equation of the system, we have

$$4y^2 + 4y^2 - 3600 = 0$$

or, upon solving $y = \pm\sqrt{450} = \pm 15\sqrt{2}$. The corresponding values of x are $\pm 30\sqrt{2}$. Because both x and y must be nonnegative, we have $x = 30\sqrt{2}$ and $y = 15\sqrt{2}$. Thus, the dimensions of the pool with maximum area are $30\sqrt{2}$ feet $\times$ $60\sqrt{2}$ feet, or approximately 42 feet $\times$ 85 feet.

APPLIED EXAMPLE 6 Cobb–Douglas Production Function Suppose x units of labor and y units of capital are required to produce

$$f(x, y) = 100x^{3/4}y^{1/4}$$

units of a certain product. If each unit of labor costs $200, each unit of capital costs $300, and a total of $60,000 is available for production, determine how many units of labor and how many units of capital should be used to maximize production.

Solution The total cost of x units of labor at $200 per unit and y units of capital at $300 per unit is equal to $200x + 300y$ dollars. But $60,000 is budgeted for production, so $200x + 300y = 60,000$, which we rewrite as

$$g(x, y) = 200x + 300y - 60,000 = 0$$

To maximize $f(x, y) = 100x^{3/4}y^{1/4}$ subject to the constraint $g(x, y) = 0$, we form the Lagrangian function

$$\begin{aligned} F(x, y, \lambda) &= f(x, y) + \lambda g(x, y) \\ &= 100x^{3/4}y^{1/4} + \lambda(200x + 300y - 60,000) \end{aligned}$$

To find the critical point(s) of F, we solve the following system of equations:

$$\begin{aligned} F_x &= 75x^{-1/4}y^{1/4} + 200\lambda = 0 \\ F_y &= 25x^{3/4}y^{-3/4} + 300\lambda = 0 \\ F_\lambda &= 200x + 300y - 60,000 = 0 \end{aligned}$$

Solving the first equation for λ, we have

$$\lambda = -\frac{75x^{-1/4}y^{1/4}}{200} = -\frac{3}{8}\left(\frac{y}{x}\right)^{1/4}$$

which, when substituted into the second equation, yields

$$25\left(\frac{x}{y}\right)^{3/4} + 300\left(-\frac{3}{8}\right)\left(\frac{y}{x}\right)^{1/4} = 0$$

Multiplying the last equation by $\left(\dfrac{x}{y}\right)^{1/4}$ then gives

$$25\left(\frac{x}{y}\right) - \frac{900}{8} = 0$$

$$x = \left(\frac{900}{8}\right)\left(\frac{1}{25}\right)y = \frac{9}{2}y$$

Substituting this value of x into the third equation of the first system of equations, we have

$$200\left(\frac{9}{2}y\right) + 300y - 60,000 = 0$$

from which we deduce that $y = 50$. Hence, $x = 225$. Thus, maximum production is achieved when 225 units of labor and 50 units of capital are used. ∎

When used in the context of Example 6, the negative of the Lagrange multiplier λ is called the **marginal productivity of money**. That is, if one additional dollar is available for production, then approximately $-\lambda$ additional units of a product can be produced. Here,

$$\lambda = -\frac{3}{8}\left(\frac{y}{x}\right)^{1/4} = -\frac{3}{8}\left(\frac{50}{225}\right)^{1/4} \approx -0.257$$

so in this case, the marginal productivity of money is 0.257. For example, if $65,000 is available for production instead of the originally budgeted figure of $60,000, then the maximum production may be boosted from the original

$$f(225, 50) = 100(225)^{3/4}(50)^{1/4}$$

or 15,448 units, to approximately

$$15,448 + 5000(0.257)$$

or 16,733 units.

8.5 Self-Check Exercises

1. Use the method of Lagrange multipliers to find the relative maximum of the function

$$f(x, y) = -2x^2 - y^2$$

subject to the constraint $3x + 4y = 12$.

2. **MAXIMIZING PROFIT** The total monthly profit of Robertson Controls in manufacturing and selling x hundred of its standard mechanical setback thermostats and y hundred of its deluxe electronic setback thermostats each month is given by the total profit function

$$P(x, y) = -\frac{1}{8}x^2 - \frac{1}{2}y^2 - \frac{1}{4}xy + 13x + 40y - 280$$

where P is in hundreds of dollars. If the production of setback thermostats is to be restricted to a total of exactly 4000/month, how many of each model should Robertson manufacture to maximize its monthly profits? What is the maximum monthly profit?

Solutions to Self-Check Exercises 8.5 can be found on page 615.

8.5 Concept Questions

1. What is a constrained relative extremum of a function f?

2. Explain how the method of Lagrange multipliers is used to find the relative extrema of $f(x, y)$ subject to $g(x, y) = 0$.

8.5 Exercises

In Exercises 1–16, use the method of Lagrange multipliers to optimize the function subject to the given constraint.

1. Minimize the function $f(x, y) = x^2 + 3y^2$ subject to the constraint $x + y - 1 = 0$.

2. Minimize the function $f(x, y) = x^2 + y^2 - xy$ subject to the constraint $x + 2y - 14 = 0$.

3. Maximize the function $f(x, y) = 2x + 3y - x^2 - y^2$ subject to the constraint $x + 2y = 9$.

4. Maximize the function $f(x, y) = 16 - x^2 - y^2$ subject to the constraint $x + y - 6 = 0$.

5. Minimize the function $f(x, y) = x^2 + 4y^2$ subject to the constraint $xy = 1$.

6. Minimize the function $f(x, y) = xy$ subject to the constraint $x^2 + 4y^2 = 4$.

7. Maximize the function $f(x, y) = x + 5y - 2xy - x^2 - 2y^2$ subject to the constraint $2x + y = 4$.

8. Maximize the function $f(x, y) = xy$ subject to the constraint $2x + 3y - 6 = 0$.

9. Maximize the function $f(x, y) = xy^2$ subject to the constraint $9x^2 + y^2 = 9$.

10. Minimize the function $f(x, y) = \sqrt{y^2 - x^2}$ subject to the constraint $x + 2y - 5 = 0$.

11. Find the maximum and minimum values of the function $f(x, y) = xy$ subject to the constraint $x^2 + y^2 = 16$.

12. Find the maximum and minimum values of the function $f(x, y) = e^{xy}$ subject to the constraint $x^2 + y^2 = 8$.

13. Find the maximum and minimum values of the function $f(x, y) = xy^2$ subject to the constraint $x^2 + y^2 = 1$.

14. Maximize the function $f(x, y, z) = xyz$ subject to the constraint $2x + 2y + z = 84$.

15. Minimize the function $f(x, y, z) = x^2 + y^2 + z^2$ subject to the constraint $3x + 2y + z = 6$.

16. Find the maximum value of the function $f(x, y, z) = x + 2y - 3z$ subject to the constraint $z = 4x^2 + y^2$.

17. **MAXIMIZING PROFIT** The total weekly profit (in dollars) realized by Country Workshop in manufacturing and selling its rolltop desks is given by the profit function

$$P(x, y) = -0.2x^2 - 0.25y^2 - 0.2xy$$
$$+ 100x + 90y - 4000$$

where x denotes the number of finished units and y denotes the number of unfinished units manufactured and sold each week. The company's management has decided to restrict the manufacture of these desks to a total of exactly 200 units/week. How many finished and how many unfinished units should be manufactured each week to maximize the company's weekly profit?

18. **MAXIMIZING PROFIT** The total daily profit (in dollars) realized by Weston Publishing in publishing and selling its dictionaries is given by the profit function

$$P(x, y) = -0.005x^2 - 0.003y^2 - 0.002xy$$
$$+ 14x + 12y - 200$$

where x stands for the number of deluxe editions and y denotes the number of standard editions sold daily. Weston's management has decided that publication of these dictionaries should be restricted to a total of exactly 400 copies/day. How many deluxe copies and how many standard copies should be published each day to maximize Weston's daily profit?

19. **MINIMIZING CONSTRUCTION COSTS** The management of UNICO Department Store has decided to enclose an 800-ft^2 area outside their building to display potted plants. The enclosed area will be a rectangle, one side of which is provided by the external wall of the store. Two sides of the enclosure will be made of pine board, and the fourth side will be made of galvanized steel fencing material. If the pine board fencing costs $6/running foot and the steel fencing costs $3/running foot, determine the dimensions of the enclosure that will cost the least to erect.

20. **PACKAGING** Find the dimensions of an open rectangular box of maximum volume and having an area of 48 ft^2 that can be constructed from a piece of cardboard. What is the volume of the box?

21. **PACKAGING** Find the dimensions of an open rectangular box of maximum volume and having an area of 12 ft^2 that can be constructed from a piece of cardboard. What is the volume of the box?

22. **MAXIMIZING PROFIT** The Ace Novelty company produces two souvenirs: Type A and Type B. The number of Type A souvenirs, x, and the number of Type B souvenirs, y, that the company can produce weekly are related by the equation $2x^2 + y - 3 = 0$, where x and y are measured in units of a thousand. The profits for a Type A souvenir and a Type B souvenir are $4 and $2, respectively. How many of each type of souvenirs should the company produce to maximize its profit?

23. **POSTAL REGULATIONS** Find the dimensions of a rectangular package having the greatest possible volume and satisfying the postal regulation that specifies that the combined length and girth of an express mail or priority mail package may not exceed 108 in.

24. **PARCEL POST REGULATIONS** Postal regulations specify that a parcel sent by parcel post may have a combined length and girth of no more than 130 in. Find the dimensions of the cylindrical package of greatest volume that may be sent through the mail. What is the volume of such a package?
   Hint: The length plus the girth is $2\pi r + l$, and the volume is $\pi r^2 l$.

25. **MINIMIZING CONTAINER COSTS** The Betty Moore Company requires that its corned beef hash containers have a capacity of 64 in.^{3}, be right circular cylinders, and be made of a tin alloy. Find the radius and height of the least expensive container that can be made if the metal for the side and bottom costs 4¢/in.^{2} and the metal for the pull-off lid costs 2¢/in.^{2}.
   Hint: Let the radius and height of the container be r and h in., respectively. Then, the volume of the container is $\pi r^2 h = 64$, and the cost is given by $C(r, h) = 8\pi rh + 6\pi r^2$.

26. **MINIMIZING CONSTRUCTION COSTS** An open rectangular box is to be constructed from material that costs $3/ft^2 for the bottom and $1/ft^2 for its sides. Find the dimensions of the box of greatest volume that can be constructed for $36.

27. **MINIMIZING CONSTRUCTION COSTS** A closed rectangular box having a volume of 4 ft^3 is to be constructed. If the material for the sides costs $1.00/ft^2 and the material for the top and bottom costs $1.50/ft^2, find the dimensions of the box that can be constructed with minimum cost.

28. **MINIMIZING CONSTRUCTION COSTS** An open rectangular box is to have a volume of 12 ft^3. If the material for its base costs three times as much (per square foot) as the material for its sides, what are the dimensions of the box that can be constructed at the minimum cost?

29. **MINIMIZING CONSTRUCTION COSTS** A rectangular box is to have a volume of 16 ft^3. If the material for its base costs twice as much (per square foot) as the material for its top and sides, find the dimensions of the box that can be constructed at the minimum cost.

30. **MAXIMIZING SALES** Ross–Simons Company has a monthly advertising budget of $60,000. Their marketing department estimates that if they spend x dollars on newspaper advertising and y dollars on television advertising, then the monthly sales will be given by

$$z = f(x, y) = 90x^{1/4}y^{3/4}$$

dollars. Determine how much money Ross–Simons should spend on newspaper ads and on television ads each month to maximize its monthly sales.

31. **MAXIMIZING PRODUCTION** Suppose a company can manufacture $P(x, y)$ units of a certain product by utilizing x units of labor and y units of capital. Furthermore, suppose the unit cost of labor and the unit cost of capital are p and q dollars, respectively ($p > 0, q > 0$). Finally, suppose that the management of the company has allocated C dollars for the manufacture of the product. Use the method of Lagrange multipliers to show that at the maximum level of production,

$$\frac{P_x(x^*, y^*)}{P_y(x^*, y^*)} = \frac{p}{q}$$

where x^* and y^* are the units of labor and capital, respectively, used at that level, and $P_x(x^*, y^*)$ and $P_y(x^*, y^*)$ are not both zero.

32. **MAXIMIZING PRODUCTION** Suppose that the output of the finished product $f(x, y)$ of a company is described by the Cobb–Douglas production function

$$f(x,y) = ax^b y^{1-b}$$

where x is the amount of money expended on labor, y is the amount expended on capital, and a and b are positive constants with $0 < b < 1$. Furthermore, suppose that the unit costs of labor and capital are p and q dollars, respectively. Finally, suppose that the management of the company has allocated a budget of C dollars for the manufacture of the product. Using the result of Exercise 31 or otherwise, find the amount of money to be spent on labor and capital to maximize production.

33. **MAXIMIZING PRODUCTION** John Mills—the proprietor of Mills Engine Company, a manufacturer of model airplane engines—finds that it takes x units of labor and y units of capital to produce

$$f(x, y) = 100x^{3/4} y^{1/4}$$

units of the product. If a unit of labor costs $100, a unit of capital costs $200, and $200,000 is budgeted for production, determine how many units should be expended on labor and how many units should be expended on capital in order to maximize production.

34. **MINIMIZING HEATING AND COOLING COSTS** Use the method of Lagrange multipliers to solve Exercise 34, Exercises 8.3.

35. **MAXIMIZING PRODUCTION** Re-solve Exercise 33 using the result of Exercise 31.

36. **MINIMIZING PRODUCTION COST** Suppose that a company can manufacture $P(x, y)$ units of a certain product by utilizing x units of labor and y units of capital. Furthermore, suppose that the unit costs of labor and capital are p and q dollars, respectively, and the company has decided to manufacture k units of the product, where k is a constant. Use the method of Lagrange multipliers to show that the cost is minimized if

$$\frac{P_x(x^*, y^*)}{P_y(x^*, y^*)} = \frac{p}{q}$$

where x^* and y^* are the units of labor and capital, respectively, used at that level, and $P_x(x^*, y^*)$ and $P_y(x^*, y^*)$ are not both zero.

37. **MINIMIZING PRODUCTION COST** Suppose that the output of the finished product $f(x, y)$ of a company is described by the Cobb–Douglas production function

$$f(x, y) = ax^b y^{1-b}$$

where x is the number of units of labor, y is the number of units expended on capital, and a and b are positive constants with $0 < b < 1$. Furthermore, suppose that the unit costs of labor and capital are p and q dollars, respectively. Finally, suppose that the management of the company has decided to manufacture k units of the product, where k is a constant. Using the results of Exercise 36 or otherwise, find the amount of money to be spent on labor and capital to minimize the cost.

38. **MINIMIZING PRODUCTION COST** John Mills—the proprietor of Mills engine company, a manufacturer of model airplane engines—finds that it take x units of labor and y units of capital to produce

$$f(x, y) = 100x^{3/4} y^{1/4}$$

engines. A unit of labor costs $100 and a unit of capital costs $200. Mr. Mills has decided to manufacture 2000 engines. Using the results of Exercise 37, or otherwise, find the amount of money he should spend on labor and capital to minimize his production cost.

In Exercises 39–42, determine whether the statement is true or false. If it is true, explain why it is true. If it is false, give an example to show why it is false.

39. If (a, b) gives rise to a (constrained) relative extremum of f subject to the constraint $g(x, y) = 0$, then (a, b) also gives rise to the unconstrained relative extremum of f.

40. If (a, b) gives rise to a (constrained) relative extremum of f subject to the constraint $g(x, y) = 0$, then $f_x(a, b) = 0$ and $f_y(a, b) = 0$, simultaneously.

41. Suppose f and g have continuous first partial derivatives in some region D in the plane. If f has an extremum at a point (a, b) subject to the constraint $g(x, y) = c$, then there exists a constant λ such that

$$f_x(a, b) = -\lambda g_x(a, b)$$
$$f_y(a, b) = -\lambda g_y(a, b) \quad \text{and} \quad g(a, b) = 0$$

42. If f is defined everywhere, then the constrained maximum (minimum) of f, if it exists, is always smaller (larger) than the unconstrained maximum (minimum).

8.5 Solutions to Self-Check Exercises

1. Write the constraint equation in the form
$g(x, y) = 3x + 4y - 12 = 0$. Then, the Lagrangian
function is

$$F(x, y, \lambda) = -2x^2 - y^2 + \lambda(3x + 4y - 12)$$

To find the critical point(s) of F, we solve the system

$$F_x = -4x + 3\lambda = 0$$
$$F_y = -2y + 4\lambda = 0$$
$$F_\lambda = 3x + 4y - 12 = 0$$

Solving the first two equations for x and y in terms of λ,
we find $x = \frac{3}{4}\lambda$ and $y = 2\lambda$. Substituting these values of x
and y into the third equation of the system yields

$$3\left(\frac{3}{4}\lambda\right) + 4(2\lambda) - 12 = 0$$

or $\lambda = \frac{48}{41}$. Therefore, $x = \left(\frac{3}{4}\right)\left(\frac{48}{41}\right) = \frac{36}{41}$ and $y = 2\left(\frac{48}{41}\right) = \frac{96}{41}$,
and we see that the point $\left(\frac{36}{41}, \frac{96}{41}\right)$ gives the constrained
maximum of f. The maximum value is

$$f\left(\frac{36}{41}, \frac{96}{41}\right) = -2\left(\frac{36}{41}\right)^2 - \left(\frac{96}{41}\right)^2$$

$$= -\frac{11{,}808}{1681} = -\frac{288}{41}$$

2. We want to maximize

$$P(x, y) = -\frac{1}{8}x^2 - \frac{1}{2}y^2 - \frac{1}{4}xy + 13x + 40y - 280$$

subject to the constraint $g(x, y) = x + y - 40 = 0$.

The Lagrangian function is

$$F(x, y, \lambda) = P(x, y) + \lambda g(x, y)$$

$$= -\frac{1}{8}x^2 - \frac{1}{2}y^2 - \frac{1}{4}xy + 13x$$

$$+ 40y - 280 + \lambda(x + y - 40)$$

To find the critical points of F, solve the following system of equations:

$$F_x = -\frac{1}{4}x - \frac{1}{4}y + 13 + \lambda = 0$$

$$F_y = -\frac{1}{4}x - y + 40 + \lambda = 0$$

$$F_\lambda = x + y - 40 = 0$$

Subtracting the first equation from the second gives

$$-\frac{3}{4}y + 27 = 0 \quad \text{or} \quad y = 36$$

Substituting this value of y into the third equation yields
$x = 4$. Therefore, to maximize its monthly profits, Robertson should manufacture 400 standard and 3600 deluxe
thermostats. The maximum monthly profit is given by

$$P(4, 36) = -\frac{1}{8}(4)^2 - \frac{1}{2}(36)^2 - \frac{1}{4}(4)(36)$$

$$+ 13(4) + 40(36) - 280$$

$$= 526$$

or \$52,600.

8.6 Total Differentials

Increments

Recall that if f is a function of one variable defined by $y = f(x)$, then the *increment*
in y is defined to be

$$\Delta y = f(x + \Delta x) - f(x)$$

where Δx is an increment in x (Figure 30a). The increment of a function of two or more
variables is defined in an analogous manner. For example, if z is a function of two variables defined by $z = f(x, y)$, then the **increment** in z is

$$\Delta z = f(x + \Delta x, y + \Delta y) - f(x, y) \tag{16}$$

where Δx and Δy are the increments in the independent variables x and y, respectively. (See Figure 30b.)

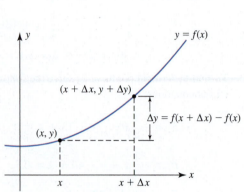

(a) The increment Δy is the change in y as x changes from x to $x + \Delta x$.

(b) The increment Δz is the change in z as x changes from x to $x + \Delta x$ and y changes from y to $y + \Delta y$.

FIGURE 30

EXAMPLE 1 Let $z = f(x, y) = 2x^2 - xy$. Find Δz. Then use your result to find the change in z if (x, y) changes from $(1, 1)$ to $(0.98, 1.03)$.

Solution   Using (16), we obtain

$$
\begin{aligned}
\Delta z &= f(x + \Delta x, y + \Delta y) - f(x, y) \\
&= [2(x + \Delta x)^2 - (x + \Delta x)(y + \Delta y)] - (2x^2 - xy) \\
&= 2x^2 + 4x\Delta x + 2(\Delta x)^2 - xy - x\Delta y - y\Delta x - \Delta x\Delta y - 2x^2 + xy \\
&= (4x - y)\Delta x - x\Delta y + 2(\Delta x)^2 - \Delta x\Delta y
\end{aligned}
$$

Next, to find the increment in z if (x, y) changes from $(1, 1)$ to $(0.98, 1.03)$, we note that $\Delta x = 0.98 - 1 = -0.02$ and $\Delta y = 1.03 - 1 = 0.03$. Therefore, using the result obtained earlier with $x = 1$, $y = 1$, $\Delta x = -0.02$, and $\Delta y = 0.03$, we obtain

$$
\begin{aligned}
\Delta z &= [4(1) - 1](-0.02) - (1)(0.03) + 2(-0.02)^2 - (-0.02)(0.03) \\
&= -0.0886
\end{aligned}
$$

You can verify the correctness of this result by calculating the quantity $f(0.98, 1.03) - f(1, 1)$.

The Total Differential

Recall from Section 3.7 that if f is a function of one variable defined by $y = f(x)$, then the differential of f at x is defined by

$$
dy = f'(x)\, dx
$$

where $dx = \Delta x$ is the differential in x. Furthermore, we saw that

$$
\Delta y \approx dy
$$

if Δx is small (Figure 31a).

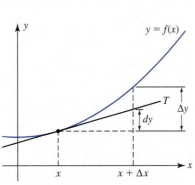

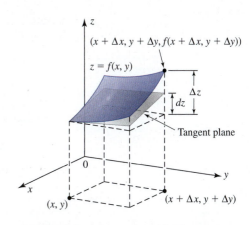

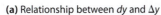

(a) Relationship between dy and Δy

(b) Relationship between dz and Δz. The tangent plane is the analog of tangent line T in the one-variable case.

FIGURE **31**

The concept of the differential extends readily to a function of two or more variables.

Total Differential

Let $z = f(x, y)$ define a differentiable function of x and y.

1. The **differentials** of the independent variables x and y are $dx = \Delta x$ and $dy = \Delta y$.

2. The **differential** of the dependent variable z is

$$dz = \frac{\partial f}{\partial x} dx + \frac{\partial f}{\partial y} dy \qquad (17)$$

Thus, analogous to the one-variable case, the total differential of z is a linear function of dx and dy. Furthermore, it provides us with an approximation of the exact change in z,

$$\Delta z = f(x + \Delta x, y + \Delta y) - f(x, y)$$

corresponding to a net change Δx in x from x to $x + \Delta x$ and a net change Δy in y from y to $y + \Delta y$; that is,

$$\Delta z \approx dz = \frac{\partial f}{\partial x}(x, y)\, dx + \frac{\partial f}{\partial y}(x, y)\, dy \qquad (18)$$

provided that $\Delta x = dx$ and $\Delta y = dy$ are sufficiently small.

EXAMPLE 2 Let $z = 2x^2 y + y^3$.

a. Find the differential dz of z.

b. Find the approximate change in z when x changes from $x = 1$ to $x = 1.01$ and y changes from $y = 2$ to $y = 1.98$.

c. Find the actual change in z when x changes from $x = 1$ to $x = 1.01$ and y changes from $y = 2$ to $y = 1.98$. Compare the result with that obtained in part (b).

Solution

a. Let $f(x, y) = 2x^2y + y^3$. Then the required differential is

$$dz = \frac{\partial f}{\partial x}\,dx + \frac{\partial f}{\partial y}\,dy = 4xy\,dx + (2x^2 + 3y^2)\,dy$$

b. Here $x = 1$, $y = 2$, and $dx = 1.01 - 1 = 0.01$ and $dy = 1.98 - 2 = -0.02$. Therefore,

$$\Delta z \approx dz = 4(1)(2)(0.01) + [2(1) + 3(4)](-0.02) = -0.20$$

c. The actual change in z is given by

$$\begin{aligned}
\Delta z &= f(1.01, 1.98) - f(1, 2) \\
&= [2(1.01)^2(1.98) + (1.98)^3] - [2(1)^2(2) + (2)^3] \\
&= 11.801988 - 12 \\
&\approx -0.1980
\end{aligned}$$

We see that $\Delta z \approx dz$, as expected.

$ APPLIED EXAMPLE 3 Approximating Changes in Revenue The weekly total revenue of Acrosonic Company resulting from the production and sales of x fully assembled bookshelf loudspeaker systems and y kit versions of the same loud-speaker system is

$$R(x, y) = -\frac{1}{4}x^2 - \frac{3}{8}y^2 - \frac{1}{4}xy + 300x + 240y$$

dollars. Determine the approximate change in Acrosonic's weekly total revenue when the level of production is increased from 200 assembled units and 60 kits per week to 206 assembled units and 64 kits per week.

Solution The approximate change in the weekly total revenue is given by the total differential R at $x = 200$ and $y = 60$, $dx = 206 - 200 = 6$ and $dy = 64 - 60 = 4$, that is, by

$$\begin{aligned}
dR &= \frac{\partial R}{\partial x}\,dx + \frac{\partial R}{\partial y}\,dy \bigg|_{\substack{x=200,\,y=60 \\ dx=6,\,dy=4}} \\
&= \left(-\frac{1}{2}x - \frac{1}{4}y + 300\right)\bigg|_{(200,\,60)} \cdot (6) \\
&= + \left(-\frac{3}{4}y - \frac{1}{4}x + 240\right)\bigg|_{(200,\,60)} \cdot (4) \\
&= (-100 - 15 + 300)6 + (-45 - 50 + 240)4 \\
&= 1690
\end{aligned}$$

or $1690.

$ APPLIED EXAMPLE 4 Cobb–Douglas Production Function The production for a certain country in the early years following World War II is described by the function

$$f(x, y) = 30x^{2/3}y^{1/3}$$

units, when x units of labor and y units of capital were utilized. Find the approximate change in output if the amount expended on labor had been decreased from 125 units to 123 units and the amount expended on capital had been increased from 27 to 29 units. Is your result as expected given the result of Example 4c, Section 8.2?

Solution   The approximate change in output is given by the total differential of f at $x = 125$, $y = 27$, $dx = 123 - 125 = -2$, and $dy = 29 - 27 = 2$, that is, by

$$df = \frac{\partial f}{\partial x}\,dx + \frac{\partial f}{\partial y}\,dy\bigg|_{\substack{x=125,\,y=27 \\ dx=-2,\,dy=2}}$$

$$= 20x^{-1/3}y^{1/3}\bigg|_{(125,\,27)} \cdot (-2) + 10x^{2/3}y^{-2/3}\bigg|_{(125,\,27)} \cdot (2)$$

$$= 20\left(\frac{27}{125}\right)^{1/3}(-2) + 10\left(\frac{125}{27}\right)^{2/3}(2)$$

$$= -20\left(\frac{3}{5}\right)(2) + 10\left(\frac{25}{9}\right)(2) = \frac{284}{9}$$

or $31\frac{5}{9}$ units. This result is fully compatible with the result of Example 4c, Section 8.2, in which the recommendation was to encourage increased spending on capital rather than on labor.   ∎

If f is a function of the three variables x, y, and z, then the total differential of $w = f(x, y, z)$ is defined to be

$$dw = \frac{\partial f}{\partial x}\,dx + \frac{\partial f}{\partial y}\,dy + \frac{\partial f}{\partial z}\,dz$$

where $dx = \Delta x$, $dy = \Delta y$, and $dz = \Delta z$ are the actual changes in the independent variables x, y, and z as x changes from $x = a$ to $x = a + \Delta x$, y changes from $y = b$ to $y = b + \Delta y$, and z changes from $z = c$ to $z = c + \Delta z$, respectively.

APPLIED EXAMPLE 5 Error Analysis Find the maximum percentage error in calculating the volume of a rectangular box if an error of at most 1% is made in measuring the length, width, and height of the box.

Solution   Let x, y, and z denote the length, width, and height, respectively, of the rectangular box. Then the volume of the box is given by $V = f(x, y, z) = xyz$ cubic units. Now suppose that the true dimensions of the rectangular box are a, b, and c units, respectively. Since the error committed in measuring the length, width, and height of the box is at most 1%, we have

$$|\Delta x| = |x - a| \le 0.01a$$
$$|\Delta y| = |y - b| \le 0.01b$$
$$|\Delta z| = |z - c| \le 0.01c$$

Therefore, the maximum error in calculating the volume of the box is

$$|\Delta V| \approx |dV| = \left|\frac{\partial f}{\partial x}\,dx + \frac{\partial f}{\partial y}\,dy + \frac{\partial f}{\partial z}\,dz\right|\Bigg|_{x=a,\,y=b,\,z=c}$$

$$= \left| yz\,dx + xz\,dy + xy\,dz \right|\bigg|_{x=a,\,y=b,\,z=c}$$

$$= |bc\,dx + ac\,dy + ab\,dz|$$

$$\le bc|dx| + ac|dy| + ab|dz|$$

$$\le bc(0.01a) + ac(0.01b) + ab(0.01c)$$

$$= (0.03)abc$$

Explore and Discuss

Refer to Example 5, in which we found the maximum percentage error in calculating the volume of the rectangular box to be *approximately* 3%. What is the precise maximum percentage error?

Since the actual volume of the box is abc cubic units, we see that the maximum percentage error in calculating its volume is

$$\frac{|\Delta V|}{V|_{(a,b,c)}} \approx \frac{(0.03)abc}{abc} = 0.03$$

that is, approximately 3%.

8.6 Self-Check Exercise

Let f be a function defined by $z = f(x, y) = 3xy^2 - 4y$. Find the total differential of f at $(-1, 3)$. Then find the approximate change in z when x changes from $x = -1$ to $x = -0.98$ and y changes from $y = 3$ to $y = 3.01$.

The solution to Self-Check Exercise 8.6 can be found on page 623.

8.6 Concept Questions

1. If $z = f(x, y)$, what is the differential of x? The differential of y? What is the total differential of z?

2. Let $z = f(x, y)$. What is the relationship between the actual change, Δz, when x changes from x to $x + \Delta x$ and y changes from y to $y + \Delta y$, and the total differential, dz, of f at (x, y)?

8.6 Exercises

1. Let $z = 2x^2 + 3y^2$, and suppose that (x, y) changes from $(2, -1)$ to $(2.01, -0.98)$.
   a. Compute Δz.   b. Compute dz.
   c. Compare the values of Δz and dz.

2. Let $z = x^2 - 2xy + 3y^2$, and suppose that (x, y) changes from $(2, 1)$ to $(1.97, 1.02)$.
   a. Compute Δz.   b. Compute dz.
   c. Compare the values of Δz and dz.

In Exercises 3–18, find the total differential of the function.

3. $f(x, y) = 2x^2 - 3xy + 4x$

4. $f(x, y) = xy^3 - x^2y^2$

5. $f(x, y) = \sqrt{x^2 + y^2}$

6. $f(x, y) = (x + 3y^2)^{1/3}$

7. $f(x, y) = \dfrac{5y}{x - y}$

8. $f(x, y) = \dfrac{x + y}{x - y}$

9. $f(x, y) = 2x^5 - ye^{-3x}$

10. $f(x, y) = xye^{x+y}$

11. $f(x, y) = x^2e^y + y \ln x$

12. $f(x, y) = \ln(x^2 + y^2)$

13. $f(x, y, z) = xy^2z^3$

14. $f(x, y, z) = x\sqrt{y} + y\sqrt{z}$

15. $f(x, y, z) = \dfrac{x}{y + z}$

16. $f(x, y, z) = \dfrac{x + y}{y + z}$

17. $f(x, y, z) = xyz + xe^{yz}$

18. $f(x, y, z) = \sqrt{e^x + e^y + ze^{xy}}$

In Exercises 19–30, find the approximate change in z when the point (x, y) changes from (x_0, y_0) to (x_1, y_1).

19. $f(x, y) = 4x^2 - xy$; from $(1, 2)$ to $(1.01, 2.02)$

20. $f(x, y) = 2x^2 - 2x^3y^2 - y^3$; from $(-1, 2)$ to $(-0.98, 2.01)$

21. $f(x, y) = x^{2/3}y^{1/2}$; from $(8, 9)$ to $(7.97, 9.03)$

22. $f(x, y) = \sqrt{x^2 + y^2}$; from $(1, 3)$ to $(1.03, 3.03)$

23. $f(x, y) = \dfrac{x}{x - y}$; from $(-3, -2)$ to $(-3.02, -1.98)$

24. $f(x, y) = \dfrac{x - y}{x + y}$; from $(-3, -2)$ to $(-3.02, -1.98)$

25. $f(x, y) = 2xe^{-y}$; from $(4, 0)$ to $(4.03, 0.03)$

26. $f(x, y) = \sqrt{x}e^{y}$; from $(1, 1)$ to $(1.01, 0.98)$

27. $f(x, y) = xe^{xy} - y^2$; from $(-1, 0)$ to $(-0.97, 0.03)$

28. $f(x, y) = xe^{-y} + ye^{-x}$; from $(1, 1)$ to $(1.01, 0.90)$

29. $f(x, y) = x \ln x + y \ln x$; from $(2, 3)$ to $(1.98, 2.89)$

30. $f(x, y) = \ln(xy)^{1/2}$; from $(5, 10)$ to $(5.05, 9.95)$

31. EFFECT OF INVENTORY AND FLOOR SPACE ON PROFIT The monthly profit (in dollars) of Bond and Barker Department Store depends on the level of inventory x (in thousands of dollars) and the floor space y (in thousands of square feet) available for display of the merchandise, as given by

$$P(x, y) = -0.02x^2 - 15y^2 + xy + 39x + 25y - 20{,}000$$

Currently, the level of inventory is $\$4{,}000{,}000$ ($x = 4000$), and the floor space is $150{,}000 \text{ ft}^2$ ($y = 150$). Find the anticipated change in monthly profit if management increases the level of inventory by $\$500{,}000$ and decreases the floor space for display of merchandise by $10{,}000 \text{ ft}^2$.

32. EFFECT OF PRODUCTION ON PROFIT The Country Workshop's total weekly profit (in dollars) realized in manufacturing and selling its rolltop desks is given by

$$P(x, y) = -0.2x^2 - 0.25y^2 - 0.2xy + 100x + 90y - 4000$$

where x stands for the number of finished units and y denotes the number of unfinished units manufactured and sold per week. Currently, the weekly output is 190 finished units and 105 unfinished units. Determine the approximate change in the total weekly profit if the sole proprietor of the Country Workshop decides to increase the number of finished units to 200/week and decrease the number of unfinished units to 100/week.

33. REVENUE OF A TRAVEL AGENCY The Odyssey Travel Agency's monthly revenue (in thousands of dollars) depends on the amount of money x (in thousands of dollars) spent on advertising per month and the number of agents y in its employ in accordance with the rule

$$R(x, y) = -x^2 - 0.5y^2 + xy + 8x + 3y + 20$$

Currently, the amount of money spent on advertising is $\$10{,}000$ per month, and there are 15 agents in the agency's employ. Estimate the change in revenue resulting from an increase of $\$1000$ per month in advertising expenditure and a decrease of 1 agent.

34. EFFECT OF CAPITAL AND LABOR ON PRODUCTIVITY The production of a South American country is given by the function

$$f(x, y) = 20x^{3/4}y^{1/4}$$

when x units of labor and y units of capital are utilized. Find the approximate change in output if the amount expended on labor is decreased from 256 to 254 units and the amount expended on capital is increased from 16 to 18 units.

35. EFFECT OF CAPITAL AND LABOR ON PRODUCTIVITY The productivity of a certain country is given by the function

$$f(x, y) = 30x^{4/5}y^{1/5}$$

when x units of labor and y units of capital are utilized. What is the approximate change in the number of units produced if the amount expended on labor is decreased from 243 to 240 units and the amount expended on capital is increased from 32 to 35 units?

36. MARGINAL PRODUCTIVITY OF MONEY Refer to Applied Example 6, page 610. Suppose x units of labor and y units of capital are required to produce

$$f(x, y) = 100x^{3/4}y^{1/4}$$

units of a certain product. Suppose that each unit of labor costs $\$200$ and each unit of capital costs $\$300$ and a total of $\$60{,}000$ is available for production. Show that at the maximum level of production, the availability of one additional dollar for production can result in the production of $-\lambda$ additional units of the product, where λ is the Lagrange multiplier. (*Note:* The negative of λ is called the **marginal productivity of labor.**)

37. SUSPENSION BRIDGE CABLES The supports of a cable of a suspension bridge are at the same level and at a distance of L ft apart. The supports are a feet higher than the lowest point of the cable (see the figure). If the weight of the cable is negligible and the bridge has a uniform weight of W lb/ft, then the tension (in pounds) in the cable at its lowest point is given by

$$H = \frac{WL^2}{8a}$$

If W, L, and a are measured with possible maximum errors of 1%, 2%, and 2%, respectively, determine the maximum percentage error in calculating H.

38. PRICE–EARNINGS RATIO The price-to-earnings ratio (P/E ratio) of a stock is given by

$$R(x, y) = \frac{x}{y}$$

where x denotes the price per share of the stock and y denotes the earnings per share. Estimate the change in the P/E ratio of a stock if its price increases from $60/share to $62/share while its earnings decrease from $4/share to $3.80/share.

39. LEVERAGED RETURN The return on assets using borrowed money, called *leveraged return,* is given by

$$L = \frac{Y - (1 - D)R}{D}$$

where Y is the return on the asset, R is the cost of borrowed money, and D is the percentage of money the investor must put down to secure the loan. Leanne wants to buy a bond with money that is borrowed from a bank. The bank requires her to pay 20% in cash to secure the loan. Find the approximate change in the leveraged return on the bond if the return on the bond changes from 6%/year to 6.5%/year and the cost of the borrowed money changes from 5%/year to 5.2%/year.

40. ERROR IN CALCULATING THE SURFACE AREA OF A HUMAN The formula

$$S = 0.007184W^{0.425}H^{0.725}$$

gives the surface area S of a human body (in square meters) in terms of its weight W in kilograms and its height H in centimeters. If an error of 1% is made in measuring the weight of a person and an error of 2% is made in measuring the height, what is the percentage error in the measurement of the person's surface area?

41. ERROR IN CALCULATING THE VOLUME OF A STORAGE TANK A storage tank has the shape of a right circular cylinder. Suppose that the radius and height of the tank are measured at 1.5 ft and 5 ft, respectively, with a possible error of 0.05 ft and 0.1 ft, respectively. Use differentials to estimate the error in calculating the capacity of the tank.

42. The dimensions of a closed rectangular box are measured as 30 in., 40 in., and 60 in., with a maximum error of 0.2 in. in each measurement. Use differentials to estimate the maximum error in calculating the volume of the box.

43. Use differentials to estimate the maximum error in calculating the surface area of the box of Exercise 42.

44. EFFECT OF CAPITAL AND LABOR ON PRODUCTIVITY The production of a certain company is given by the function

$$f(x, y) = 50x^{1/3}y^{2/3}$$

when x units of labor and y units of capital are utilized. Find the approximate percentage change in the production of the company if labor is increased by 2% and capital is increased by 1%.

45. The pressure P (in pascals), the volume V (in liters), and the temperature T (in kelvins) of an ideal gas are related by the equation $PV = 8.314T$. Use differentials to find the approximate change in the pressure of the gas if its volume increases from 20 L to 20.2 L and its temperature decreases from 300 K to 295 K.

46. SPECIFIC GRAVITY The specific gravity of an object with density greater than that of water can be determined by using the formula

$$S = \frac{A}{A - W}$$

where A and W are the weights of the object in air and in water, respectively. If the measurements of an object are $A = 2.2$ lb and $W = 1.8$ lb with maximum errors of 0.02 lb and 0.04 lb, respectively, find the approximate maximum error in calculating S.

47. ERROR IN CALCULATING TOTAL RESISTANCE The total resistance R of three resistors with resistance R_1, R_2, and R_3, connected in parallel, is given by the relationship

$$\frac{1}{R} = \frac{1}{R_1} + \frac{1}{R_2} + \frac{1}{R_3}$$

If R_1, R_2, and R_3 are measured at 100, 200, and 300 ohms, respectively, with a maximum error of 1% in each measurement, find the approximate maximum error in the calculated value of R.

48. ERROR IN MEASURING ARTERIAL BLOOD FLOW The flow of blood through an arteriole in cubic centimeters per second is given by

$$V = \frac{\pi p r^4}{8kl}$$

where l is the length (in cm) of the arteriole, r is its radius (in cm), p is the difference in pressure between the two ends of the arteriole (in dyne/cm^2), and k is the viscosity of blood (in dyne-sec/cm^2). Find the approximate percentage change in the flow of blood if an error of 2% is made in measuring the length of the arteriole and an error of 1% is made in measuring its radius. Assume that p and k are constant.

8.6 Solution to Self-Check Exercise

We find

$$\frac{\partial f}{\partial x} = 3y^2 \quad \text{and} \quad \frac{\partial f}{\partial y} = 6xy - 4$$

so that

$$\frac{\partial f}{\partial x}(-1, 3) = 3(3)^2 = 27$$

and

$$\frac{\partial f}{\partial y}(-1, 3) = 6(-1)(3) - 4 = -22$$

Therefore, the total differential is

$$dz = \frac{\partial f}{\partial x}(-1, 3)\, dx + \frac{\partial f}{\partial y}(-1, 3)\, dy$$

$$= 27\, dx - 22\, dy$$

Now $dx = -0.98 - (-1) = 0.02$ and $dy = 3.01 - 3 = 0.01$, so the approximate change in z is

$$dz = 27(0.02) - 22(0.01) = 0.32$$

8.7 Double Integrals

A Geometric Interpretation of the Double Integral

To introduce the notion of the integral of a function of two variables, let's first recall the definition of the definite integral of a continuous function of one variable $y = f(x)$ over the interval $[a, b]$. We first divide the interval $[a, b]$ into n subintervals, each of equal length, by the points $x_0 = a < x_1 < x_2 < \cdots < x_n = b$ and define the **Riemann sum** by

$$S_n = f(p_1)h + f(p_2)h + \cdots + f(p_n)h$$

where $h = (b - a)/n$ and p_i is an arbitrary point in the interval $[x_{i-1}, x_i]$. The definite integral of f over $[a, b]$ is then defined as the limit of the Riemann sum S_n as n tends to infinity, whenever it exists. Furthermore, recall that when f is a nonnegative continuous function on $[a, b]$, then the ith term of the Riemann sum, $f(p_i)h$, is an approximation (by the area of a rectangle) of the area under that part of the graph of $y = f(x)$ between $x = x_{i-1}$ and $x = x_i$, so that the Riemann sum S_n provides us with an approximation of the area under the curve $y = f(x)$ from $x = a$ to $x = b$. The integral

$$\int_a^b f(x)\, dx = \lim_{n \to \infty} S_n$$

gives the *actual* area under the curve from $x = a$ to $x = b$.

Now suppose $f(x, y)$ is a continuous function of two variables defined over a region R. For simplicity, we assume for the moment that R is a rectangular region in the plane (Figure 32). Let's construct a Riemann sum for this function over the rectangle R by

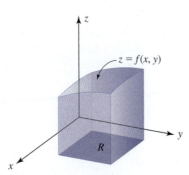

FIGURE 32
$f(x, y)$ is a function defined over a rectangular region R.

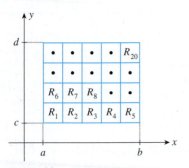

FIGURE 33
Grid with $m = 5$ and $n = 4$

following a procedure that parallels the case for a function of one variable over an interval I. We begin by observing that the analog of a *partition* in the two-dimensional case is a rectangular **grid** composed of mn rectangles, each of length h and width k, as a result of partitioning the side of the rectangle R of length $(b - a)$ into m segments and the side of length $(d - c)$ into n segments. By construction,

$$h = \frac{b - a}{m} \quad \text{and} \quad k = \frac{d - c}{n}$$

A sample grid with $m = 5$ and $n = 4$ is shown in Figure 33.

Let's label the rectangles R_1, R_2, R_3, . . . , R_{mn}. If (x_i, y_i) is *any* point in R_i $(1 \le i \le mn)$, then the **Riemann sum of $f(x, y)$ over the region R** is defined as

$$S(m, n) = f(x_1, y_1)hk + f(x_2, y_2)hk + \cdots + f(x_{mn}, y_{mn})hk$$

If the limit of $S(m, n)$ exists as both m and n tend to infinity, we call this limit the value of the **double integral of $f(x, y)$ over the region R** and denote it by

$$\iint_R f(x, y)\, dA$$

If $f(x, y)$ is a nonnegative function, then it defines a solid S bounded above by the graph of f and below by the rectangular region R. Furthermore, the solid S is the union of the mn solids bounded above by the graph of f and below by the mn rectangular regions corresponding to the partition of R (Figure 34).

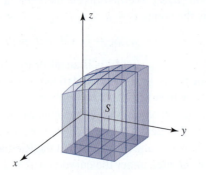
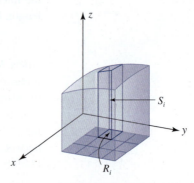

(a) The solid S is the union of mn solids (shown here with $m = 3$ and $n = 4$).

(b) A typical solid S_i is bounded above by the graph of f and lies above R_i.

FIGURE 34

The volume of a typical solid S_i can be approximated by a parallelepiped with base R_i and height $f(x_i, y_i)$ (Figure 35).

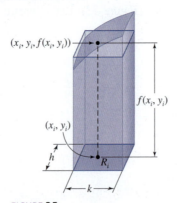

FIGURE 35
The volume of S_i is approximated by the parallelepiped with base R_i and height $f(x_i, y_i)$.

Therefore, the Riemann sum $S(m, n)$ gives us an approximation of the volume of the solid bounded above by the surface $z = f(x, y)$ and below by the plane region R. As both m and n tend to infinity, the Riemann sum $S(m, n)$ approaches the *actual* volume under the solid.

Evaluating a Double Integral over a Rectangular Region

Let's turn our attention to the evaluation of the double integral

$$\int_R \int f(x, y) \, dA$$

where R is the rectangular region shown in Figure 32. As in the case of the definite integral of a function of one variable, it turns out that the double integral can be evaluated without our having to first find an appropriate Riemann sum and then take the limit of that sum. Instead, as we will now see, the technique calls for evaluating two single integrals—the so-called **iterated integrals**—in succession, using a process that might be called "antipartial differentiation." The technique is described in the following result, which we state without proof.

Let R be the rectangle defined by the inequalities $a \le x \le b$ and $c \le y \le d$ (see Figure 33). Then

$$\int_R \int f(x, y) \, dA = \int_c^d \left[\int_a^b f(x, y) \, dx \right] dy \tag{19}$$

where the iterated integrals on the right-hand side are evaluated as follows. We first compute the integral

$$\int_a^b f(x, y) \, dx$$

by treating y as if it were a constant and integrating the resulting function of x with respect to x (dx reminds us that we are integrating with respect to x). In this manner, we obtain a value for the integral that may contain the variable y. Thus,

$$\int_a^b f(x, y) \, dx = g(y)$$

for some function g. Substituting this value into Equation (19) gives

$$\int_c^d g(y) \, dy$$

which may be integrated in the usual manner.

EXAMPLE 1 Evaluate $\int_R \int f(x, y) \, dA$, where $f(x, y) = x + 2y$ and R is the rectangle defined by $1 \le x \le 4$ and $1 \le y \le 2$.

Solution Using Equation (19), we find

$$\int_R \int f(x, y) \, dA = \int_1^2 \left[\int_1^4 (x + 2y) \, dx \right] dy$$

To compute

$$\int_1^4 (x + 2y) \, dx$$

Explore and Discuss

Using a geometric interpretation, evaluate

$$\int_R \int \sqrt{4 - x^2 - y^2} \, dA$$

where $R = \{(x, y) \,|\, x^2 + y^2 \le 4\}$.

we treat y as if it were a constant (remember that dx reminds us that we are integrating with respect to x). We obtain

$$\int_1^4 (x + 2y)\, dx = \frac{1}{2}x^2 + 2xy \Big|_{x=1}^{x=4}$$

$$= \left[\frac{1}{2}(16) + 2(4)y\right] - \left[\frac{1}{2}(1) + 2(1)y\right]$$

$$= \frac{15}{2} + 6y$$

Thus,

$$\int_R\int f(x, y)\, dA = \int_1^2 \left(\frac{15}{2} + 6y\right) dy = \left(\frac{15}{2}y + 3y^2\right)\Big|_1^2$$

$$= (15 + 12) - \left(\frac{15}{2} + 3\right) = 16\tfrac{1}{2}$$

Evaluating a Double Integral over a Plane Region

Up to now, we have assumed that the region over which a double integral is to be evaluated is rectangular. In fact, however, it is possible to compute the double integral of functions over rather arbitrary regions. The next theorem, which we state without proof, expands the number of types of regions over which we may integrate.

THEOREM 1

a. Suppose $g_1(x)$ and $g_2(x)$ are continuous functions on $[a, b]$ and the region R is defined by $R = \{(x, y)\,|\, g_1(x) \le y \le g_2(x);\, a \le x \le b\}$. Then

$$\int_R\int f(x, y)\, dA = \int_a^b \left[\int_{g_1(x)}^{g_2(x)} f(x, y)\, dy\right] dx \qquad (20)$$

(Figure 36a).

b. Suppose $h_1(y)$ and $h_2(y)$ are continuous functions on $[c, d]$ and the region R is defined by $R = \{(x, y)\,|\, h_1(y) \le x \le h_2(y);\, c \le y \le d\}$. Then

$$\int_R\int f(x, y)\, dA = \int_c^d \left[\int_{h_1(y)}^{h_2(y)} f(x, y)\, dx\right] dy \qquad (21)$$

(Figure 36b).

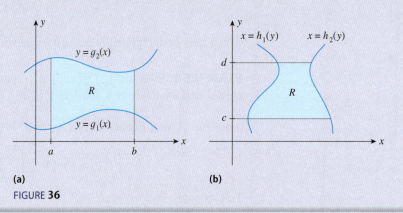

(a)

(b)

FIGURE 36

Notes

1. Observe that the lower and upper limits of integration with respect to y in Equation (20) are given by $y = g_1(x)$ and $y = g_2(x)$. This is to be expected, since for a fixed value of x lying between $x = a$ and $x = b$, y runs between the lower curve defined by $y = g_1(x)$ and the upper curve defined by $y = g_2(x)$ (see Figure 36a). Observe, too, that in the special case when $g_1(x) = c$ and $g_2(x) = d$, the region R is rectangular, and Equation (20) reduces to Equation (19).
2. For a fixed value of y, x runs between $x = h_1(y)$ and $x = h_2(y)$, giving the indicated limits of integration with respect to x in Equation (21) (see Figure 36b).
3. Note that the two curves in Figure 36b are not graphs of functions of x (use the Vertical Line Test), but they are graphs of functions of y. It is this observation that justifies the approach leading to Equation (21).

We now look at several examples.

EXAMPLE 2 Evaluate $\int_R\int f(x, y)\, dA$ given that $f(x, y) = x^2 + y^2$ and R is the region bounded by the graphs of $g_1(x) = x$ and $g_2(x) = 2x$ for $0 \le x \le 2$.

Solution The region under consideration is shown in Figure 37. Using Equation (20), we find

$$\int_R\int f(x, y)\, dA = \int_0^2\left[\int_x^{2x} (x^2 + y^2)\, dy\right] dx$$

$$= \int_0^2\left[\left(x^2 y + \frac{1}{3} y^3\right)\Big|_x^{2x}\right] dx$$

$$= \int_0^2\left[\left(2x^3 + \frac{8}{3} x^3\right) - \left(x^3 + \frac{1}{3} x^3\right)\right] dx$$

$$= \int_0^2 \frac{10}{3} x^3\, dx = \frac{5}{6} x^4\Big|_0^2 = 13\tfrac{1}{3}$$

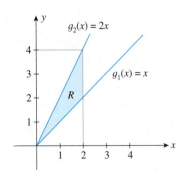

FIGURE **37**
R is the region bounded by $g_1(x) = x$ and $g_2(x) = 2x$ for $0 \le x \le 2$.

EXAMPLE 3 Evaluate $\int_R\int f(x, y)\, dA$, where $f(x, y) = xe^y$ and R is the plane region bounded by the graphs of $y = x^2$ and $y = x$.

Solution The region in question is shown in Figure 38. The points of intersection of the two curves are found by solving the equation $x^2 = x$, giving $x = 0$ and $x = 1$. Using Equation (20), we find

$$\int_R\int f(x, y)\, dA = \int_0^1\left(\int_{x^2}^x xe^y\, dy\right) dx = \int_0^1\left(xe^y\Big|_{x^2}^x\right) dx$$

$$= \int_0^1 (xe^x - xe^{x^2})\, dx = \int_0^1 xe^x\, dx - \int_0^1 xe^{x^2}\, dx$$

Next, integrating the first integral on the right-hand side by parts, we have

$$\int_R\int f(x, y)\, dA = \left[(x - 1)e^x - \frac{1}{2} e^{x^2}\right]\Big|_0^1$$

$$= -\frac{1}{2} e - \left(-1 - \frac{1}{2}\right) = \frac{1}{2}(3 - e)$$

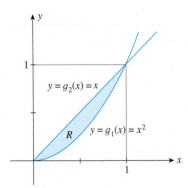

FIGURE **38**
R is the region bounded by $y = x^2$ and $y = x$.

The next example not only illustrates the use of Equation (21) but also shows that it may be the only viable way to evaluate the given double integral.

EXAMPLE 4 Evaluate

$$\iint_R xe^{y^2}\, dA$$

where R is the plane region bounded by the y-axis, $x = 0$, the horizontal line $y = 4$, and the graph of $y = x^2$.

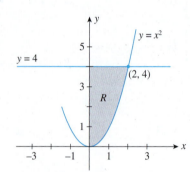

FIGURE 39
R is the region bounded by the y-axis, $x = 0$, $y = 4$, and $y = x^2$.

Solution The region R is shown in Figure 39. The point of intersection of the line $y = 4$ and the graph of $y = x^2$ is found by solving the equation $x^2 = 4$, giving $x = 2$ and the required point $(2, 4)$. Using Equation (20) with $y = g_1(x) = x^2$ and $y = g_2(x) = 4$ we obtain

$$\iint_R xe^{y^2}\, dA = \int_0^2\left[\int_{x^2}^4 xe^{y^2}\, dy\right] dx$$

Evaluation of the integral

$$\int_{x^2}^4 xe^{y^2}\, dy = x\int_{x^2}^4 e^{y^2}\, dy$$

calls for finding the antiderivative of the integrand e^{y^2} in terms of elementary functions, a task that, as was pointed out in Section 7.3, cannot be done. Let's begin afresh and attempt to make use of Equation (21).

Since the equation $y = x^2$ is equivalent to the equation $x = \sqrt{y}$, which expresses x as a function of y, we may write, with $x = h_1(y) = 0$ and $h_2(y) = \sqrt{y}$,

$$\iint_R xe^{y^2}\, dA = \int_0^4\left[\int_0^{\sqrt{y}} xe^{y^2}\, dx\right] dy = \int_0^4\left[\frac{1}{2}x^2e^{y^2}\Big|_0^{\sqrt{y}}\right] dy$$

$$= \int_0^4 \frac{1}{2}ye^{y^2}\, dy = \frac{1}{4}e^{y^2}\Big|_0^4 = \frac{1}{4}(e^{16} - 1) \qquad \blacksquare$$

8.7 Self-Check Exercise

Evaluate $\int_R\int (x + y)\, dA$, where R is the region bounded by the graphs of $g_1(x) = x$ and $g_2(x) = x^{1/3}$ in the first quadrant.

The solution to Self-Check Exercise 8.7 can be found on page 630.

8.7 Concept Questions

1. Give a geometric interpretation of $\int_R\int f(x, y)\,dA$, where f is a nonnegative function on the rectangular region R in the xy-plane.

2. What is an iterated integral? How is $\int_R\int f(x, y)\,dA$ evaluated in terms of iterated integrals, where R is the rectangular region defined by $a \le x \le b$ and $c \le y \le d$?

3. Suppose g_1 and g_2 are continuous and $g_1(x) \le g_2(x)$ on the interval $[a, b]$ and

$R = \{(x, y)\,|\,g_1(x) \le y \le g_2(x), a \le x \le b\}$. What is $\int_R\int f(x, y)\,dA$, where f is a continuous function defined on R?

4. Suppose h_1 and h_2 are continuous and $h_1(y) \le h_2(y)$ on the interval $[c, d]$ and $R = \{(x, y)\,|\,h_1(y) \le x \le h_2(y), c \le y \le d\}$, what is $\int_R\int f(x, y)\,dA$, where f is a continuous function defined on R?

8.7 Exercises

In Exercises 1–25, evaluate the double integral

$$\int_R\int f(x, y)\,dA$$

for the function $f(x, y)$ and the region R.

1. $f(x, y) = y + 2x$; R is the rectangle defined by $1 \le x \le 2$ and $0 \le y \le 1$.

2. $f(x, y) = x + 2y$; R is the rectangle defined by $-1 \le x \le 2$ and $0 \le y \le 2$.

3. $f(x, y) = xy^2$; R is the rectangle defined by $-1 \le x \le 1$ and $0 \le y \le 1$.

4. $f(x, y) = 12xy^2 + 8y^3$; R is the rectangle defined by $0 \le x \le 1$ and $0 \le y \le 2$.

5. $f(x, y) = \dfrac{x}{y}$; R is the rectangle defined by $-1 \le x \le 2$ and $1 \le y \le e^3$.

6. $f(x, y) = \dfrac{xy}{1 + y^2}$; R is the rectangle defined by $-2 \le x \le 2$ and $0 \le y \le 1$.

7. $f(x, y) = 4xe^{2x^2 + y}$; R is the rectangle defined by $0 \le x \le 1$ and $-2 \le y \le 0$.

8. $f(x, y) = \dfrac{y}{x^2}\,e^{y/x}$; R is the rectangle defined by $1 \le x \le 2$ and $0 \le y \le 1$.

9. $f(x, y) = \ln y$; R is the rectangle defined by $0 \le x \le 1$ and $1 \le y \le e$.

10. $f(x, y) = \dfrac{\ln y}{x}$; R is the rectangle defined by $1 \le x \le e^2$ and $1 \le y \le e$.

11. $f(x, y) = x + 2y$; R is bounded by the lines $x = 1$, $y = 0$, and $y = x$.

12. $f(x, y) = xy$; R is bounded by the lines $x = 1$, $y = 0$ and $y = x$.

13. $f(x, y) = 2x + 4y$; R is bounded by $x = 1$, $x = 3$, $y = 0$, and $y = x + 1$.

14. $f(x, y) = 2 - y$; R is bounded by $x = -1$, $x = 1 - y$, $y = 0$, and $y = 2$.

15. $f(x, y) = x + y$; R is bounded by $x = 0$, $x = \sqrt{y}$, and $y = 4$.

16. $f(x, y) = x^2y^2$; R is bounded by $x = 0$, $x = 1$, $y = x^2$, and $y = x^3$.

17. $f(x, y) = y$; R is bounded by $x = 0$, $x = \sqrt{4 - y^2}$, and $y = 0$.

18. $f(x, y) = \dfrac{y}{x^3 + 2}$; R is bounded by the lines $x = 1$, $y = 0$, and $y = x$.

19. $f(x, y) = 2xe^y$; R is bounded by the lines $x = 1$, $y = 0$, and $y = x$.

20. $f(x, y) = 2x$; R is bounded by $x = e^{2y}$, $x = y$, $y = 0$, and $y = 1$.

21. $f(x, y) = ye^x$; R is bounded by $y = \sqrt{x}$ and $y = x$.

22. $f(x, y) = xe^{-y^2}$; R is bounded by $x = 0$, $x = \sqrt{y}$, and $y = 4$.

23. $f(x, y) = e^{y^2}$; R is bounded by $x = 0$, $y = 2x$, and $y = 2$.

24. $f(x, y) = y$; R is bounded by $y = \ln x$, $x = e$, and $y = 0$.

25. $f(x, y) = ye^{x^3}$; R is bounded by $x = \dfrac{y}{2}$, $x = 1$, and $y = 0$.

In Exercises 26 and 27, determine whether the statement is true or false. If it is true, explain why it is true. If it is false, give an example to show why it is false.

26. If $h(x, y) = f(x)g(y)$, where f is continuous on $[a, b]$ and g is continuous on $[c, d]$, then $\int_R\int h(x, y)\,dA = [\int_a^b f(x)\,dx][\int_c^d g(y)\,dy]$, where $R = \{(x, y)\,|\,a \le x \le b; c \le y \le d\}$.

27. If $\int_{R_1}\int f(x, y)\,dA$ exists, where $R_1 = \{(x, y)\,|\,a \le x \le b; c \le y \le d\}$, then $\int_{R_2}\int f(x, y)\,dA$ exists, where $R_2 = \{(x, y)\,|\,c \le x \le d; a \le y \le b\}$.

8.7 Solution to Self-Check Exercise

The region R is shown in the accompanying figure. The points of intersection of the two curves are found by solving the equation $x = x^{1/3}$, giving $x = 0$ and $x = 1$. Using Equation (20), we find

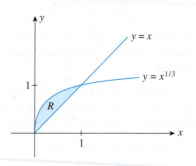

$$\iint_R (x + y) \, dA = \int_0^1 \left[\int_x^{x^{1/3}} (x + y) \, dy \right] dx$$

$$= \int_0^1 \left(xy + \frac{1}{2} y^2 \Big|_x^{x^{1/3}} \right) dx$$

$$= \int_0^1 \left[\left(x^{4/3} + \frac{1}{2} x^{2/3} \right) - \left(x^2 + \frac{1}{2} x^2 \right) \right] dx$$

$$= \int_0^1 \left(x^{4/3} + \frac{1}{2} x^{2/3} - \frac{3}{2} x^2 \right) dx$$

$$= \frac{3}{7} x^{7/3} + \frac{3}{10} x^{5/3} - \frac{1}{2} x^3 \Big|_0^1$$

$$= \frac{3}{7} + \frac{3}{10} - \frac{1}{2} = \frac{8}{35}$$

8.8 Applications of Double Integrals

In this section, we will discuss applications involving the double integral.

Finding the Volume of a Solid by Double Integrals

As we saw earlier, the double integral

$$\iint_R f(x, y) \, dA$$

gives the volume of the solid bounded by the graph of $f(x, y)$ over the region R.

> **The Volume of a Solid Under a Surface**
>
> Let R be a region in the xy-plane and let f be continuous and nonnegative on R. Then, the **volume of a solid under a surface** bounded above by $z = f(x, y)$ and below by R is given by
>
> $$V = \iint_R f(x, y) \, dA$$

EXAMPLE 1 Find the volume of the solid bounded above by the plane $z = f(x, y) = y$ and below by the plane region R defined by $y = \sqrt{1 - x^2}$ $(0 \le x \le 1)$.

Solution The region R is sketched in Figure 40. Observe that $f(x, y) = y \ge 0$ for $(x, y) \in R$. Therefore, the required volume is given by

$$\iint_R y \, dA = \int_0^1 \left(\int_0^{\sqrt{1 - x^2}} y \, dy \right) dx = \int_0^1 \left(\frac{1}{2} y^2 \Big|_0^{\sqrt{1 - x^2}} \right) dx$$

$$= \int_0^1 \frac{1}{2} (1 - x^2) \, dx = \frac{1}{2} \left(x - \frac{1}{3} x^3 \right) \Big|_0^1 = \frac{1}{3}$$

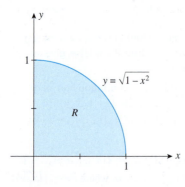

FIGURE 40
The plane region R defined by
$y = \sqrt{1 - x^2}$ $(0 \le x \le 1)$

The solid is shown in Figure 41. Note that it is not necessary to make a sketch of the solid in order to compute its volume.

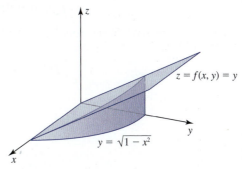

FIGURE 41
The solid bounded above by the plane $z = y$ and below by the plane region defined by $y = \sqrt{1 - x^2}$ $(0 \le x \le 1)$

Population of a City

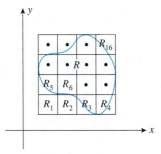

FIGURE 42
The rectangular region R representing a certain district of a city is enclosed by a rectangular grid.

Suppose the plane region R represents a certain district of a city and $f(x, y)$ gives the population density (the number of people per square mile) at any point (x, y) in R. Enclose the set R by a rectangle, and construct a grid for it in the usual manner. In any rectangular region of the grid that has no point in common with R, set $f(x_i, y_i)hk = 0$ (Figure 42). Then, corresponding to any grid covering the set R, the general term of the Riemann sum $f(x_i, y_i)hk$ (population density times area) gives the number of people living in that part of the city corresponding to the rectangular region R_i. Therefore, the Riemann sum gives an approximation of the number of people living in the district represented by R and, in the limit, the double integral

$$\int_R \int f(x, y) \, dA$$

gives the actual number of people living in the district under consideration.

 APPLIED EXAMPLE 2 Population Density of a City The population density (number of people per square mile) of a certain city is described by the function

$$f(x, y) = 10,000e^{-0.2|x|-0.1|y|}$$

where the origin $(0, 0)$ gives the location of the city hall. What is the population inside the rectangular area described by

$$R = \{(x, y) \mid -10 \le x \le 10; \ -5 \le y \le 5\}$$

if x and y are in miles? (See Figure 43.)

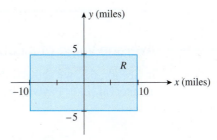

FIGURE 43
The rectangular region R represents a certain district of a city.

Solution By symmetry, it suffices to compute the population in the first quadrant. (Why?) Then, upon observing that in this quadrant

$$f(x, y) = 10{,}000e^{-0.2x-0.1y} = 10{,}000e^{-0.2x}e^{-0.1y}$$

we see that the population in R is given by

$$\iint_R f(x, y)\, dA = 4\int_0^{10}\left(\int_0^5 10{,}000e^{-0.2x}e^{-0.1y}\, dy\right)dx$$

$$= 4\int_0^{10}\left(-100{,}000e^{-0.2x}e^{-0.1y}\Big|_0^5\right)dx$$

$$= 400{,}000(1 - e^{-0.5})\int_0^{10}e^{-0.2x}\, dx$$

$$= 2{,}000{,}000(1 - e^{-0.5})(1 - e^{-2})$$

or approximately 680,438.

Explore and Discuss

1. Consider the improper double integral $\iint_D f(x, y)\, dA$ of the continuous function f of two variables defined over the plane region

   $$D = \{(x, y)\,|\,0 \le x < \infty;\, 0 \le y < \infty\}$$

   Using the definition of improper integrals of functions of one variable (Section 7.4), explain why it makes sense to define

   $$\iint_D f(x, y)\, dA = \lim_{N\to\infty}\int_0^N\left[\lim_{M\to\infty}\int_0^M f(x, y)\, dx\right]dy$$

   $$= \lim_{M\to\infty}\int_0^M\left[\lim_{N\to\infty}\int_0^N f(x, y)\, dy\right]dx$$

   provided that the limits exist.

2. Refer to Example 2. Assuming that the population density of the city is described by

   $$f(x, y) = 10{,}000e^{-0.2|x|-0.1|y|}$$

   for $-\infty < x < \infty$ and $-\infty < y < \infty$, show that the population outside the rectangular region

   $$R = \{(x, y)\,|\,-10 < x < 10;\, -5 < y \le 5\}$$

   of Example 2 is given by

   $$4\iint_D f(x, y)\, dx\, dy - 680{,}438$$

   (Recall that 680,438 is the approximate population inside R.)

3. Use the results of parts 1 and 2 to determine the population of the city outside the rectangular area R. (Assume that there are no other major cities nearby.)

Average Value of a Function

In Section 6.5, we showed that the average value of a continuous function $f(x)$ over an interval $[a, b]$ is given by

$$\frac{1}{b - a}\int_a^b f(x)\, dx$$

That is, the average value of a function over $[a, b]$ is the integral of f over $[a, b]$ divided by the length of the interval. An analogous result holds for a function of two variables $f(x, y)$ over a plane region R. To see this, we enclose R by a rectangle and construct a rectangular grid. Let (x_i, y_i) be any point in the rectangle R_i of area hk. Now, the average value of the mn numbers $f(x_1, y_1), f(x_2, y_2), \ldots, f(x_{mn}, y_{mn})$ is given by

$$\frac{f(x_1, y_1) + f(x_2, y_2) + \cdots + f(x_{mn}, y_{mn})}{mn}$$

which can also be written as

$$\frac{hk}{hk} \left[\frac{f(x_1, y_1) + f(x_2, y_2) + \cdots + f(x_{mn}, y_{mn})}{mn} \right]$$

$$= \frac{1}{(mn)hk} [f(x_1, y_1) + f(x_2, y_2) + \cdots + f(x_{mn}, y_{mn})]hk$$

Now the area of R is approximated by the sum of the mn rectangles (*omitting* those having no points in common with R), each of area hk. Note that this is the denominator of the previous expression. Therefore, taking the limit as m and n both tend to infinity, we obtain the following formula for the *average value of* $f(x, y)$ *over* R.

Average Value of $f(x, y)$ over the Region R

If f is integrable over the plane region R, then its average value over R is given by

$$\frac{\displaystyle\int_R \int f(x, y)\, dA}{\text{Area of } R} \quad \text{or} \quad \frac{\displaystyle\int_R \int f(x, y)\, dA}{\displaystyle\int_R \int dA} \tag{22}$$

Note   If we let $f(x, y) = 1$ for all (x, y) in R, then

$$\int_R \int f(x, y)\, dA = \int_R \int dA = \text{Area of } R$$

EXAMPLE 3 Find the average value of the function $f(x, y) = xy$ over the plane region R bounded by the graph of $y = e^x$ and the lines $x = 0$ and $x = 1$.

Solution The region R is shown in Figure 44. The area of the region R is given by

$$\int_0^1 \left(\int_0^{e^x} dy \right) dx = \int_0^1 \left(y \Big|_0^{e^x} \right) dx$$

$$= \int_0^1 e^x\, dx$$

$$= e^x \Big|_0^1$$

$$= e - 1$$

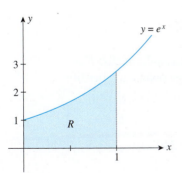

FIGURE 44
The plane region R defined by
$y = e^x \, (0 \le x \le 1)$

We would obtain the same result had we viewed the area of this region as the area of the region under the curve $y = e^x$ from $x = 0$ to $x = 1$. Next, we compute

$$\iint_R f(x, y)\, dA = \int_0^1 \left(\int_0^{e^x} xy\, dy \right) dx$$

$$= \int_0^1 \left(\frac{1}{2} xy^2 \Big|_0^{e^x} \right) dx$$

$$= \int_0^1 \frac{1}{2} xe^{2x}\, dx$$

$$= \frac{1}{4} xe^{2x} - \frac{1}{8} e^{2x} \Big|_0^1 \qquad \text{Integrate by parts.}$$

$$= \left(\frac{1}{4} e^2 - \frac{1}{8} e^2 \right) + \frac{1}{8}$$

$$= \frac{1}{8}\left(e^2 + 1 \right)$$

Therefore, the required average value is given by

$$\frac{\displaystyle\iint_R f(x, y)\, dA}{\displaystyle\iint_R dA} = \frac{\frac{1}{8}(e^2 + 1)}{e - 1} = \frac{e^2 + 1}{8(e - 1)}$$

APPLIED EXAMPLE 4 Population Density (Refer to Example 2.) The population density of a certain city (number of people per square mile) is described by the function

$$f(x, y) = 10{,}000e^{-0.2|x| - 0.1|y|}$$

where the origin gives the location of the city hall. What is the average population density inside the rectangular area described by

$$R = \{(x, y) \,|\, {-}10 \le x \le 10;\ {-}5 \le y \le 5\}$$

where x and y are measured in miles?

Solution From the results of Example 2, we know that

$$\iint_R f(x, y)\, dA \approx 680{,}438$$

From Figure 44, we see that the area of the plane rectangular region R is $(20)(10)$, or 200, square miles. Therefore, the average population density inside R is

$$\frac{\displaystyle\iint_R f(x, y)\, dA}{\displaystyle\iint_R dA} = \frac{680{,}438}{200} = 3402.19$$

or approximately 3402 people per square mile.

8.8 Self-Check Exercise

POPULATION DENSITY OF A COASTAL TOWN The population density (number of people per square mile) of a coastal town located on an island is described by the function

$$f(x, y) = \frac{5000xe^y}{1 + 2x^2} \qquad (0 \le x \le 4; -2 \le y \le 0)$$

where x and y are measured in miles (see the accompanying figure). What is the population inside the rectangular area defined by $R = \{(x, y) \mid 0 \le x \le 4; -2 \le y \le 0\}$? What is the average population density in the area?

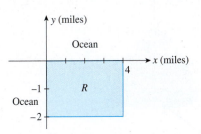

The solution to Self-Check Exercise 8.8 can be found on page 637.

8.8 Concept Questions

1. Give the formula for finding the volume of the solid bounded above by the graph of $z = f(x, y)$ and below by the region R in the xy-plane.

2. What is the average value of $f(x, y)$ over the region R?

3. Suppose a plane region R represents a certain district of a city, $f(x, y)$ gives the population density at any point (x, y) in R, and the set R is enclosed by a rectangle.

a. Explain how a Riemann sum can be used to approximate the number of people living in the district represented by R.

b. What does the general term of the Riemann sum $f(x_i, y_i)hk$ represent?

c. What does $\int_R \int f(x, y) \, dA$ represent?

8.8 Exercises

In Exercises 1–8, use a double integral to find the volume of the solid shown in the figure.

1.

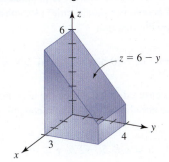

2.

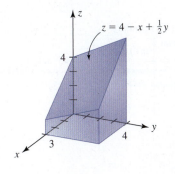

3.

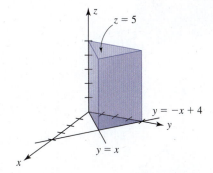

4.

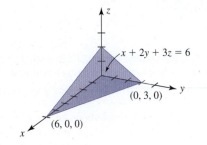

5.

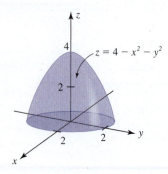

$z = 4 - x^2 - y^2$

6.

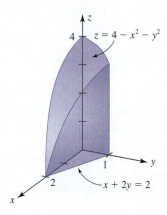

$z = 4 - x^2 - y^2$

$x + 2y = 2$

7.

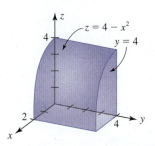

$z = 4 - x^2$

$y = 4$

8.

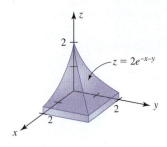

$z = 2e^{-x-y}$

In Exercises 9–16, find the volume of the solid bounded above by the surface $z = f(x, y)$ and below by the plane region R.

9. $f(x, y) = 4 - 2x - y$; $R = \{(x, y)\,|\,0 \le x \le 1; 0 \le y \le 2\}$

10. $f(x, y) = 2x + y$; R is the triangle bounded by $y = 2x$, $y = 0$, and $x = 2$.

11. $f(x, y) = x^2 + y^2$; R is the rectangle with vertices $(0, 0)$, $(1, 0)$, $(1, 2)$, and $(0, 2)$.

12. $f(x, y) = e^{x+2y}$; R is the triangle with vertices $(0, 0)$, $(1, 0)$, and $(0, 1)$.

13. $f(x, y) = 2xe^y$; R is the triangle bounded by $y = x$, $y = 2$, and $x = 0$.

14. $f(x, y) = \dfrac{2y}{1 + x^2}$; R is the region bounded by $y = \sqrt{x}$, $y = 0$, and $x = 4$.

15. $f(x, y) = 2x^2y$; R is the region bounded by the graphs of $y = x$ and $y = x^2$.

16. $f(x, y) = x$; R is the region in the first quadrant bounded by the semicircle $y = \sqrt{16 - x^2}$, the x-axis, and the y-axis.

In Exercises 17–22, find the average value of the function $f(x, y)$ over the plane region R.

17. $f(x, y) = 6x^2y^3$; $R = \{(x, y)\,|\,0 \le x \le 2; 0 \le y \le 3\}$

18. $f(x, y) = x + 2y$; R is the triangle with vertices $(0, 0)$, $(1, 0)$, and $(1, 1)$.

19. $f(x, y) = xy$; R is the triangle bounded by $y = x$, $y = 2 - x$, and $y = 0$.

20. $f(x, y) = e^{-x^2}$; R is the triangle with vertices $(0, 0)$, $(1, 0)$, and $(1, 1)$.

21. $f(x, y) = xe^y$; R is the triangle with vertices $(0, 0)$, $(1, 0)$, and $(1, 1)$.

22. $f(x, y) = \ln x$; R is the region bounded by the graphs of $y = 2x$ and $y = 0$ from $x = 1$ to $x = 3$.
Hint: Use integration by parts.

23. POPULATION DENSITY OF A COASTAL TOWN The population density (number of people per square mile) of a coastal town is described by the function

$$f(x, y) = \frac{10{,}000e^y}{1 + 0.5|x|} \qquad (-10 \le x \le 10; -4 \le y \le 0)$$

where x and y are measured in miles (see the accompanying figure). Find the population inside the rectangular area described by

$$R = \{(x, y)\,|\,-5 \le x \le 5; -2 \le y \le 0\}$$

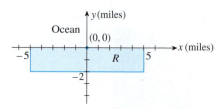

24. AVERAGE POPULATION DENSITY OF A COASTAL TOWN Refer to Exercise 23. Find the average population density inside the rectangular area R.

25. POPULATION DENSITY OF A CITY The population density (number of people per square mile) of a certain city is given by the function

$$f(x, y) = \frac{50,000|xy|}{(x^2 + 20)(y^2 + 36)}$$

where the origin $(0, 0)$ gives the location of the government center. Find the population inside the rectangular area described by

$$R = \{(x, y)\,|\,-15 \le x \le 15;\,-20 \le y \le 20\}$$

26. AVERAGE PROFIT The Country Workshop's total weekly profit (in dollars) realized in manufacturing and selling its rolltop desks is given by the profit function

$$P(x, y) = -0.2x^2 - 0.25y^2 - 0.2xy$$
$$+ 100x + 90y - 4000$$

where x stands for the number of finished units and y stands for the number of unfinished units manufactured and sold each week. Find the average weekly profit if the number of finished units manufactured and sold varies between 180/week and 200/week and the number of unfinished units varies between 100/week and 120/week.

27. AVERAGE PRICE OF LAND The rectangular region R shown in the accompanying figure represents a city's financial district. The price of land in the district is approximated by the function

$$p(x, y) = 200 - 10\left(x - \frac{1}{2}\right)^2 - 15(y - 1)^2$$

where $p(x, y)$ is the price of land at the point (x, y) in dollars/square foot and x and y are measured in miles.

What is the average price of land per square foot in the district?

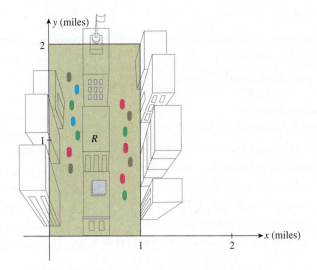

In Exercises 28 and 29, determine whether the statement is true or false. If it is true, explain why it is true. If it is false, give an example to show why it is false.

28. Let R be a region in the xy-plane, and let f and g be continuous functions on R that satisfy the condition $f(x, y) \le g(x, y)$ for all (x, y) in R. Then,

$$\int_R\int [g(x, y) - f(x, y)]\,dA$$

gives the volume of the solid bounded above by the surface $z = g(x, y)$ and below by the surface $z = f(x, y)$.

29. Suppose f is nonnegative and continuous over the plane region R. Then, the average value of f over R can be thought of as the (constant) height of the cylinder with base R and volume that is exactly equal to the volume of the solid under the graph of $z = f(x, y)$. (*Note:* The cylinder referred to here has sides perpendicular to R.)

8.8 Solution to Self-Check Exercise

The population in R is given by

$$\int_R\int f(x, y)\,dA = \int_0^4\left(\int_{-2}^0 \frac{5000xe^y}{1 + 2x^2}\,dy\right)dx$$

$$= \int_0^4\left[\frac{5000xe^y}{1 + 2x^2}\bigg|_{-2}^0\right]dx$$

$$= 5000(1 - e^{-2})\int_0^4 \frac{x}{1 + 2x^2}\,dx$$

$$= 5000(1 - e^{-2})\left[\frac{1}{4}\ln(1 + 2x^2)\bigg|_0^4\right]$$

$$= 5000(1 - e^{-2})\left(\frac{1}{4}\right)\ln 33$$

or approximately 3779 people. The average population density inside R is

$$\frac{\int_R\int f(x, y)\,dA}{\int_R\int dA} \approx \frac{3779}{(2)(4)}$$

or approximately 472 people/mi^2.

Summary of Principal Terms

TERMS

function of two variables (554)	relative maximum (581)	scatter diagram (592)
domain (554)	relative maximum value (581)	least-squares line (593)
three-dimensional Cartesian coordinate system (557)	relative minimum (581)	regression line (593)
level curve (559)	relative minimum value (581)	normal equation (594)
first partial derivatives of f (567)	absolute maximum (582)	constrained relative extremum (604)
Cobb–Douglas production function (571)	absolute minimum (582)	method of Lagrange multipliers (606)
marginal productivity of labor (571)	absolute maximum value (582)	increment (615)
marginal productivity of capital (571)	absolute minimum value (582)	total differential (617)
substitute commodity (572)	saddle point (583)	Riemann sum (623)
complementary commodity (572)	critical point (584)	double integral (624)
second-order partial derivatives of f (574)	second derivative test (584)	volume of a solid under a surface (630)
	method of least squares (592)	

Concept Review Questions

Fill in the blanks.

1. The domain of a function f of two variables is a subset of the _____-plane. The rule of f associates with each _____ _____ in the domain of f one and only one _____ _____, denoted by $z =$ _____.

2. If the function f has rule $z = f(x, y)$, then x and y are called _____ variables, and z is a/an _____ variable. The number z is also called the _____ of f at (x, y).

3. The graph of a function f of two variables is the set of all points (x, y, z), where _____ and (x, y) is the domain of _____. The graph of a function of two variables is a/an _____ in three-dimensional space.

4. The trace of the graph of $f(x, y)$ in the plane $z = k$ is the curve with equation _____ lying in the plane $z = k$. The projection of the trace of f in the plane $z = k$ onto the xy-plane is called the _____ _____ of f. The contour map associated with f is obtained by drawing the _____ _____ of f corresponding to several admissible values of _____.

5. The partial derivative $\partial f/\partial x$ of f at (x, y) can be found by thinking of y as a/an _____ in the expression for f and differentiating this expression with respect to _____ as if it were a function of x alone.

6. The number $f_x(a, b)$ measures the _____ of the tangent line to the curve C obtained by intersecting the graph of f with the plane $y = b$ at the point _____. It also measures the rate of change of f with respect to _____ at the point (a, b) with y held fixed with value _____.

7. A function $f(x, y)$ has a relative maximum at (a, b) if $f(x, y)$ _____ $f(a, b)$ for all points (x, y) that are sufficiently close to _____. The absolute maximum value of

$f(x, y)$ is the number $f(a, b)$ such that $f(x, y)$ _____ $f(a, b)$ for all (x, y) in the _____ of f.

8. A critical point of $f(x, y)$ is a point (a, b) in the _____ of f such that _____ _____ _____ or at least one of the partial derivatives of f does not _____. A critical point of f is a/an _____ for a relative extremum of f.

9. By plotting the points associated with a set of data, we obtain the _____ diagram for the data. The least-squares line is the line obtained by _____ the sum of the squares of the errors made when the y-values of the data points are approximated by the corresponding y-values of the _____ line. The least-squares equation is found by solving the _____ equations.

10. The method of Lagrange multipliers solves the problem of finding the relative extrema of a function $f(x, y)$ subject to the constraint _____. We first form the Lagrangian function $F(x, y, \lambda) =$ _____. Then we solve the system consisting of the three equations _____, _____, and _____ for x, y, and λ. These solutions give the critical points that give rise to the relative _____ of f.

11. For a differentiable function $z = f(x, y)$, the differentials of the _____ variables x and y are _____ and _____, and the differential of the dependent variable z is $dz =$ _____.

12. If $f(x, y)$ is continuous and nonnegative over a region R in the xy-plane and $\int_R \int f(x, y)\, dA$ exists, then the double integral gives the _____ of the _____ bounded by the graph of $f(x, y)$ over the region R.

13. The integral $\int_R \int f(x, y)\, dA$ is evaluated by using _____ integrals. For example, $\int_R \int (2x + y^2)\, dA$, where $R = \{(x, y)\,|\,0 \leq x \leq 1;\, 3 \leq y \leq 5\}$, is equal to $\int_0^1 \int_3^5 (2x + y^2)\, dy\, dx$ or the (iterated) integral _____.

CHAPTER 8   Review Exercises

1. Let $f(x, y) = \dfrac{xy}{x^2 + y^2}$. Compute $f(0, 1)$, $f(1, 0)$, and $f(1, 1)$. Does $f(0, 0)$ exist?

2. Let $f(x, y) = \dfrac{xe^y}{1 + \ln xy}$. Compute $f(1, 1)$, $f(1, 2)$, and $f(2, 1)$. Does $f(1, 0)$ exist?

3. Let $h(x, y, z) = xye^z + \dfrac{x}{y}$. Compute $h(1, 1, 0)$, $h(-1, 1, 1)$, and $h(1, -1, 1)$.

4. Find the domain of the function $f(u, v) = \dfrac{\sqrt{u}}{u - v}$.

5. Find the domain of the function $f(x, y) = \dfrac{x - y}{x + y}$.

6. Find the domain of the function

$$f(x, y) = x\sqrt{y} + y\sqrt{1 - x}$$

7. Find the domain of the function

$$f(x, y, z) = \frac{xy\sqrt{z}}{(1 - x)(1 - y)(1 - z)}$$

In Exercises 8–11, sketch the level curves of the function corresponding to each value of z.

8. $z = f(x, y) = 2x + 3y$; $z = -2, -1, 0, 1, 2$

9. $z = f(x, y) = y - x^2$; $z = -2, -1, 0, 1, 2$

10. $z = f(x, y) = \sqrt{x^2 + y^2}$; $z = 0, 1, 2, 3, 4$

11. $z = f(x, y) = e^{xy}$; $z = 1, 2, 3$

In Exercises 12–21, compute the first partial derivatives of the function.

12. $f(x, y) = x^2y^3 + 3xy^2 + \dfrac{x}{y}$

13. $f(x, y) = x\sqrt{y} + y\sqrt{x}$ **14.** $f(u, v) = \sqrt{uv^2 - 2u}$

15. $f(x, y) = \dfrac{x - y}{y + 2x}$ **16.** $g(x, y) = \dfrac{xy}{x^2 + y^2}$

17. $h(x, y) = (2xy + 3y^2)^5$ **18.** $f(x, y) = (xe^y + 1)^{1/2}$

19. $f(x, y) = (x^2 + y^2)e^{x^2+y^2}$

20. $f(x, y) = \ln(1 + 2x^2 + 4y^4)$

21. $f(x, y) = \ln\left(1 + \dfrac{x^2}{y^2}\right)$

In Exercises 22–27, compute the second-order partial derivatives of the function.

22. $f(x, y) = x^3 - 2x^2y + y^2 + x - 2y$

23. $f(x, y) = x^4 + 2x^2y^2 - y^4$

24. $f(x, y) = (2x^2 + 3y^2)^3$ **25.** $g(x, y) = \dfrac{x}{x + y^2}$

26. $g(x, y) = e^{x^2+y^2}$ **27.** $h(s, t) = \ln\left(\dfrac{s}{t}\right)$

28. Let $f(x, y, z) = x^3y^2z + xy^2z + 3xy - 4z$. Compute $f_x(1, 1, 0)$, $f_y(1, 1, 0)$, and $f_z(1, 1, 0)$, and interpret your results.

In Exercises 29–34, find the critical point(s) of the functions. Then use the second derivative test to classify the nature of each of these points, if possible. Finally, determine the relative extrema of each function.

29. $f(x, y) = 2x^2 + y^2 - 8x - 6y + 4$

30. $f(x, y) = x^2 + 3xy + y^2 - 10x - 20y + 12$

31. $f(x, y) = x^3 - 3xy + y^2$

32. $f(x, y) = x^3 + y^2 - 4xy + 17x - 10y + 8$

33. $f(x, y) = e^{2x^2+y^2}$

34. $f(x, y) = \ln(x^2 + y^2 - 2x - 2y + 4)$

In Exercises 35–38, use the method of Lagrange multipliers to optimize the function subject to the given constraints.

35. Maximize the function $f(x, y) = -3x^2 - y^2 + 2xy$ subject to the constraint $2x + y = 4$.

36. Minimize the function

$$f(x, y) = 2x^2 + 3y^2 - 6xy + 4x - 9y + 10$$

subject to the constraint $x + y = 1$.

37. Find the maximum and minimum values of the function $f(x, y) = 2x - 3y + 1$ subject to the constraint $2x^2 + 3y^2 - 125 = 0$.

38. Find the maximum and minimum values of the function $f(x, y) = e^{x-y}$ subject to the constraint $x^2 + y^2 = 1$.

In Exercises 39 and 40, find the total differential of each function at the given point.

39. $f(x, y) = (x^2 + y^4)^{3/2}$; $(3, 2)$

40. $f(x, y) = xe^{x-y} + x \ln y$; $(1, 1)$

In Exercises 41 and 42, find the total approximate change in z when the point (x, y) changes from (x_0, y_0) to (x_1, y_1).

41. $f(x, y) = 2x^2y^3 + 3y^2x^2 - 2xy$; from $(1, -1)$ to $(1.02, -0.98)$

42. $f(x, y) = 4x^{3/4}y^{1/4}$; from $(16, 81)$ to $(17, 80)$

In Exercises 43–46, evaluate the double integrals.

43. $f(x, y) = 3x - 2y$; R is the rectangle defined by $2 \le x \le 4$ and $-1 \le y \le 2$.

44. $f(x, y) = e^{-x-2y}$; R is the rectangle defined by $0 \le x \le 2$ and $0 \le y \le 1$.

45. $f(x, y) = 2x^2y$; R is bounded by the curves $y = x^2$ and $y = x^3$ in the first quadrant.

46. $f(x, y) = \dfrac{y}{x}$, R is bounded by the lines $x = 2$, $y = 1$, and $y = x$.

In Exercises 47 and 48, find the volume of the solid bounded above by the surface $z = f(x, y)$ and below by the plane region R.

47. $f(x, y) = 4x^2 + y^2$; $R = \{0 \le x \le 2; 0 \le y \le 1\}$

48. $f(x, y) = x + y$; R is the region bounded by $y = x^2$, $y = 4x$, and $y = 4$.

49. Find the average value of the function

$$f(x, y) = xy + 1$$

over the plane region R bounded by $y = x^2$ and $y = 2x$.

50. IQs The IQ (intelligence quotient) of a person whose chronological age is c and whose mental age is m is defined as

$$I(c, m) = \frac{100m}{c}$$

Describe the level curves of I. Sketch the level curves corresponding to $I = 90, 100, 120, 180$. Interpret your results.

51. Revenue Functions A division of Ditton Industries makes a 16-speed and a 10-speed electric blender. The company's management estimates that x units of the 16-speed model and y units of the 10-speed model are demanded daily when the unit prices are

$$p = 80 - 0.02x - 0.1y$$
$$q = 60 - 0.1x - 0.05y$$

dollars, respectively.
a. Find the daily total revenue function $R(x, y)$.
b. Find the domain of the function R.
c. Compute $R(100, 300)$ and interpret your result.

52. Demand for CD Players In a survey conducted by *Home Entertainment* magazine, it was determined that the demand equation for CD players is given by

$$x = f(p, q) = 900 - 9p - e^{0.4q}$$

whereas the demand equation for audio CDs is given by

$$y = g(p, q) = 20{,}000 - 3000q - 4p$$

where p and q denote the unit prices (in dollars) for the CD players and audio CDs, respectively, and x and y denote the number of CD players and audio CDs demanded per week. Determine whether these two products are substitute, complementary, or neither.

53. Estimating Changes in Profit The total daily profit function (in dollars) of Weston Publishing Company realized in publishing and selling its English language dictionaries is given by

$$P(x, y) = -0.0005x^2 - 0.003y^2 - 0.002xy$$
$$+ 14x + 12y - 200$$

where x denotes the number of deluxe copies and y denotes the number of standard copies published and sold daily. Currently, the number of deluxe and standard copies of the dictionaries published and sold daily are 1000 and 1700, respectively. Determine the approximate daily change in the total daily profit if the number of deluxe copies is increased to 1050 and the number of standard copies is decreased to 1650 per day.

54. Average Daily TV-Viewing Time The following data were compiled by the Bureau of Television Advertising in a large metropolitan area, giving the average daily TV-viewing time per household in that area over the years 2006 to 2014:

Year	2006	2008
Daily Viewing Time, y	6 hr 9 min	6 hr 30 min

Year	2010	2012	2014
Daily Viewing Time, y	6 hr 36 min	7 hr	7 hr 16 min

a. Find the least-squares line for these data. (Let $x = 1$ represent 2006 and let y be the viewing time in minutes.)
b. Estimate the average daily TV-viewing time per household in 2016.

55. Female Life Expectancy at 65 The projections of female life expectancy at age 65 years in the United States are summarized in the following table ($x = 0$ corresponds to 2000):

Year, x	0	10	20	30	40	50
Years Beyond 65, y	19.5	20.0	20.6	21.2	21.8	22.4

a. Find an equation of the least-squares line for these data.
b. Use the result of (a) to estimate life expectancy at 65 of a female in 2040. How does this result compare with the given data for that year?
c. Use the result of (a) to estimate the life expectancy at 65 of a female in 2030.

Source: U.S. Census Bureau.

56. MOBILE PHONE USE IN CHINA The number of mobile phone users (in millions) in China from 2007 ($x = 0$) through 2011 is given in the following table:

Year	2007	2008	2009	2010	2011
Number, y	547.2	638.9	750.1	861.2	929.8

a. Find an equation of the least squares line for these data.

b. Use the result of part (a) to estimate the number of mobile phone users in China in 2014, assuming that the trend continued.

Source: HIS iSuppli.

57. MAXIMIZING REVENUE Odyssey Travel Agency's monthly revenue (in dollars) depends on the amount of money x (in thousands of dollars) spent on advertising per month and the number of agents y in its employ in accordance with the rule

$$R(x, y) = -x^2 - 0.5y^2 + xy + 8x + 3y + 20$$

Determine the amount of money the agency should spend per month and the number of agents it should employ to maximize its monthly revenue.

58. MINIMIZING FENCING COSTS The owner of the Rancho Grande wants to enclose a rectangular piece of grazing land along the straight portion of a river and then subdivide it using a fence running parallel to the sides that are perpendicular to the river. No fencing is required along the river. If the material for the sides costs \$3/running yard and the material for the divider costs \$2/running yard, what will be the dimensions of a 303,750-yd^2 pasture if the cost of fencing is kept to a minimum?

59. MAXIMIZING REVENUE The annual profit (in hundreds of dollars) of Apex Travel Agency is given by

$$P(x, y) = -\frac{3}{2}x^2 - 2y^2 + xy + 38x + 178y - 1500$$

where x denotes the number of agents it has on its payroll and y denotes the number of advertisements it places on television. How many agents and how many advertisements should Apex place on television to maximize its annual profit? What is the maximum annual profit?

60. MAXIMIZING THE WEIGHTS OF FISH A pond is stacked with x bass and y trout (in hundreds). The average weights of bass and trout after 1 year are $\left(8 - x - \frac{1}{2}y\right)$ lb and $\left(11 - \frac{1}{2}x - 2y\right)$ lb, respectively. Find the number of each species of fish in the pond that will make the total weight of fish a maximum.

61. COBB–DOUGLAS PRODUCTION FUNCTIONS The production of Q units of a commodity is related to the amount of labor x and the amount of capital y (in suitable units) expended by the equation

$$Q = f(x, y) = x^{3/4}y^{1/4}$$

If an expenditure of 100 units is available for production, how should it be apportioned between labor and capital so that Q is maximized?

Hint: Use the method of Lagrange multipliers to maximize the function Q subject to the constraint $x + y = 100$.

62. COBB–DOUGLAS PRODUCTION FUNCTION Show that the Cobb–Douglas production function $P = kx^ay^{1-a}$, where $0 < a < 1$, satisfies the equation

$$x\frac{\partial P}{\partial x} + y\frac{\partial P}{\partial y} = P$$

CHAPTER 8 Before Moving On ...

1. Find the domain of

$$f(x, y) = \frac{\sqrt{x} + \sqrt{y}}{(1 - x)(2 - y)}$$

2. Find the first and second partial derivatives of

$$f(x, y) = x^2y + e^{xy}$$

3. Find the relative extrema, if any, of

$$f(x, y) = 2x^3 + 2y^3 - 6xy - 5$$

4. Find an equation of the least-squares line for the following data:

x	0	1	2	3	5
y	2.9	5.1	6.8	8.8	13.2

5. Use the method of Lagrange multipliers to find the minimum of $f(x, y) = 3x^2 + 3y^2 + 1$ subject to $x + y = 1$.

6. Let $z = 2x^2 - xy$.

a. Find the differential dz.

b. Compute the value of dz if (x, y) changes from $(1, 1)$ to $(0.98, 1.03)$.

7. Evaluate $\int_R\int (1 - xy)\, dA$, where R is the region bounded by $x = 0$, $x = 1$, $y = x$, and $y = x^2$.

9 Differential Equations

AN EQUATION involving the derivative, or differential, of an unknown function is called a *differential equation*. In this chapter, we show how differential equations are used to solve problems involving the growth of an amount of money earning interest compounded continuously, the growth of a population of bacteria, the decay of radioactive material, and the rate at which a person learns a new subject, to name just a few.

If the amount of fertilizer used to cultivate land on a wheat farm is increased, will the crop yield increase substantially? Researchers at a Midwestern university found that a new experimental fertilizer increased the wheat yield of the land at the university's experimental field station. In Example 3, page 659, you will see how they used a differential equation to estimate the crop yield.

© Fotokostic/Shutterstock.com

9.1 Differential Equations

Models Involving Differential Equations

We first encountered differential equations in Section 6.1. Recall that a **differential equation** is an equation that involves an unknown function and its derivative(s). Here are some examples of differential equations:

$$\frac{dy}{dx} = xe^x \qquad \frac{dy}{dx} + 2y = x^2 \qquad \frac{d^2y}{dt^2} + \left(\frac{dy}{dt}\right)^3 + ty - 8 = 0$$

Differential equations appear in almost every branch of applied mathematics, and the study of these equations remains one of the most active areas of research in mathematics. As you will see in the next few examples, models involving differential equations often arise from the mathematical formulation of practical problems.

Unrestricted Growth Models   We first discussed the unrestricted growth model in Chapter 5. There we saw that the size of a population at any time t, $Q(t)$, increases at a rate that is proportional to $Q(t)$ itself. Thus,

$$\frac{dQ}{dt} = kQ \tag{1}$$

where k is a constant of proportionality. This is a differential equation involving the unknown function Q and its derivative Q'.

Restricted Growth Models   In many applications, the quantity $Q(t)$ does not exhibit unrestricted growth but approaches some definite upper bound. The learning curves and logistic functions we discussed in Chapter 5 are examples of restricted growth models. Let's derive the mathematical models that lead to these functions.

Suppose $Q(t)$ does not exceed some number C, called the *carrying capacity of the environment*. Furthermore, suppose the rate of growth of this quantity is *proportional* to the difference between its upper bound and its current size. The resulting differential equation is

$$\frac{dQ}{dt} = k(C - Q) \tag{2}$$

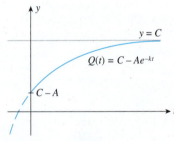

FIGURE 1
$Q(t)$ describes a learning curve.

where k is a constant of proportionality. Observe that if the initial population is small relative to C, then the rate of growth of Q is relatively large. But as $Q(t)$ approaches C, the difference $C - Q(t)$ approaches zero, as does the rate of growth of Q. In Section 9.3, you will see that the solution of the differential Equation (2) is a function that describes a learning curve (Figure 1).

Next, let's consider a restricted growth model in which the rate of growth of a quantity $Q(t)$ is *jointly proportional* to its current size and the difference between its upper bound and its current size; that is,

$$\frac{dQ}{dt} = kQ(C - Q) \tag{3}$$

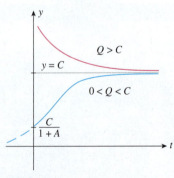

FIGURE 2
Two solutions of the logistic equation

$$Q(t) = \frac{C}{1 + Ae^{-Ckt}}$$

where k is a constant of proportionality. Observe that when $Q(t)$ is small relative to C, the rate of growth of Q is approximately proportional to Q. But as $Q(t)$ approaches C, the growth rate slows down to zero. If $Q > C$, then $dQ/dt < 0$ and the quantity is decreasing with time, with the decay rate slowing down as Q approaches C. We will show later that the solution of the differential Equation (3) is just the logistic function we discussed in Chapter 5. Its graph is shown in Figure 2.

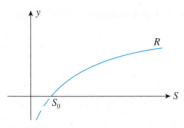

FIGURE 3
The solution to the differential equation (4) describes the response to a stimulus.

FIGURE 4
The rate of change of the amount of salt at time t = (Rate of salt flowing in) − (Rate of salt flowing out)

Stimulus Response   In the quantitative theory of psychology, one model that describes the relationship between a stimulus S and the resulting response R is the Weber–Fechner Law. This law asserts that the rate of change of a reaction R is inversely proportional to the stimulus S. Mathematically, this law may be expressed as

$$\frac{dR}{dS} = \frac{k}{S} \tag{4}$$

where k is a constant of proportionality. Furthermore, suppose that the threshold level, the lowest level of stimulation at which sensation is detected, is S_0. Then we have the condition $R = 0$ when $S = S_0$; that is, $R(S_0) = 0$. The graph of R is shown in Figure 3.

Mixture Problems   Our next example is a typical mixture problem. Suppose a tank initially contains 10 gallons of pure water. Brine containing 3 pounds of salt per gallon flows into the tank at a rate of 2 gallons per minute, and the well-stirred mixture flows out of the tank at the same rate. How much salt is in the tank at any given time?

Let's formulate this problem mathematically. Suppose $A(t)$ denotes the amount of salt in the tank at any time t. Then the derivative dA/dt, the rate of change of the amount of salt at any time t, must satisfy the condition

$$\frac{dA}{dt} = (\text{Rate of salt flowing in}) - (\text{Rate of salt flowing out})$$

(Figure 4). But the rate at which salt flows into the tank is given by

$$(2 \text{ gal/min})(3 \text{ lb/gal}) \qquad \text{\color{red}(Rate of flow)} \times \text{\color{red}(Concentration)}$$

or 6 pounds per minute. Since the rate at which the solution leaves the tank is the same as the rate at which the brine is poured into it, the tank contains 10 gallons of the mixture at any time t. Since the salt content at any time t is A pounds, the concentration of the salt in the mixture is $(A/10)$ pounds per gallon. Therefore, the rate at which salt flows out of the tank is given by

$$(2 \text{ gal/min})\left(\frac{A}{10} \text{ lb/gal}\right)$$

or $A/5$ pounds per minute. Therefore, we are led to the differential equation

$$\frac{dA}{dt} = 6 - \frac{A}{5} \tag{5}$$

An additional condition arises from the fact that initially there is no salt in the solution. This condition may be expressed mathematically as $A = 0$ when $t = 0$ or $A(0) = 0$.

In Section 9.3 we will solve each of the differential equations we have introduced here.

Solutions of Differential Equations

Suppose we are given a differential equation involving the derivative(s) of a function y. Recall that a **solution** to a differential equation is any function $f(x)$ that satisfies the differential equation. Thus, $y = f(x)$ is a solution of the differential equation provided that the replacement of y and its derivative(s) by the function $f(x)$ and its corresponding derivatives reduces the given differential equation to an identity for all values of x.

EXAMPLE 1 Show that the function $f(x) = e^{-x} + x - 1$ is a solution of the differential equation

$$y' + y = x$$

Solution Let

$$y = f(x) = e^{-x} + x - 1$$

Then

$$y' = f'(x) = -e^{-x} + 1$$

Substituting these equations into the left side of the given differential equation yields

$$\overbrace{(-e^{-x} + 1)}^{y'} + \overbrace{(e^{-x} + x - 1)}^{y} = -e^{-x} + 1 + e^{-x} + x - 1 = x$$

which is equal to the right side of the given equation for all values of x. Therefore, $f(x) = e^{-x} + x - 1$ is a solution of the given differential equation.   ▮

In Example 1, we verified that $y = e^{-x} + x - 1$ is a solution of the differential equation $y' + y = x$. This is by no means the only solution of this differential equation, as the next example shows.

EXAMPLE 2 Show that any function of the form $f(x) = ce^{-x} + x - 1$, where c is a constant, is a solution of the differential equation

$$y' + y = x$$

Solution Let

$$y = f(x) = ce^{-x} + x - 1$$

Then

$$y' = f'(x) = -ce^{-x} + 1$$

Substituting these equations into the left side of the given differential equation yields

$$\overbrace{-ce^{-x} + 1}^{y'} + \overbrace{ce^{-x} + x - 1}^{y} = x$$

and we have verified the assertion.   ▮

It can be shown that *every* solution of the differential equation $y' + y = x$ must have the form $y = ce^{-x} + x - 1$, where c is a constant; therefore, this is a **general solution** of the differential equation $y' + y = x$. Figure 5 shows a family of solutions of this differential equation for selected values of c.

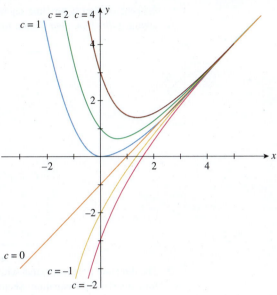

FIGURE 5
Some solutions of $y' + y = x$

Recall that the solution obtained by assigning a specific value to the constant c is called a **particular solution** of the differential equation. For example, the particular solution $y = e^{-x} + x - 1$ of Example 1 is obtained from the general solution by taking $c = 1$. In practice, a particular solution of a differential equation is obtained from the general solution of the differential equation by requiring that the solution and/or its derivative(s) satisfy certain conditions at one or more values of x.

EXAMPLE 3 Use the results of Example 2 to find the particular solution of the equation $y' + y = x$ that satisfies the condition $y(0) = 0$; that is, $f(0) = 0$, where f denotes the solution.

Solution From the results of Example 2, we see that

$$y = f(x) = ce^{-x} + x - 1$$

is a solution of the given differential equation for all constants c. Using the given condition, we see that

$$f(0) = ce^0 + 0 - 1 = c - 1 = 0 \quad \text{or} \quad c = 1$$

Therefore, the required particular solution is $y = e^{-x} + x - 1$.　　　　■

Explore and Discuss

Suppose that $y = f(x)$ is a solution of the differential equation $dy/dx = F(x, y)$.

1. If (a, b) is a point in the domain of F, explain why $F(a, b)$ gives the slope of f at $x = a$.
2. For the differential equation $dy/dx = x/y$, compute $F(x, y)$ for selected integral values of x and y. (For example, try $x = 0, \pm 1, \pm 2$ and $y = \pm 1, \pm 2, \pm 3$.) Verify that if you draw

(continued)

a lineal element (a tiny line segment) having slope $F(x, y)$ through each point (x, y), you obtain a *direction field* similar to the one shown in the figure:

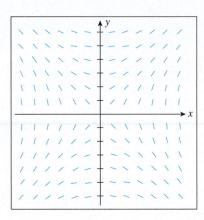

3. The direction field associated with the differential equation hints at the solution curves for the differential equation. Sketch a few solution curves for the differential equation. [You will be asked to verify your answer to part (3) in the next section.]

9.1 Self-Check Exercises

1. Consider the differential equation

$$xy' + 2y = 4x^2$$

a. Show that $y = x^2 + (c/x^2)$ is a solution of the differential equation.

b. Find the particular solution of the differential equation that satisfies $y(1) = 4$.

2. **POPULATION OF A SPECIES** The population of a certain species grows at a rate directly proportional to the square root of its size. If the initial population is N_0, find the population at any time t. Formulate but do not solve the problem.

Solutions to Self-Check Exercises 9.1 can be found on page 650.

9.1 Concept Questions

1. Define the following terms in your own words.
   a. A differential equation.
   b. The general solution of a differential equation.
   c. A particular solution of a differential equation.

2. Give a differential equation that describes the situation:
   a. The size of a population at any time t, $Q(t)$, increases at a rate that is proportional to $Q(t)$ itself.
   b. The size of a population at any time t, $Q(t)$, does not exceed some number C, and the rate of growth of $Q(t)$ is proportional to the difference between its upper bound and its current size.

3. a. Does the equation of Question 2(a) describe restricted or unrestricted growth?
   b. Does the equation of Question 2(b) describe restricted or unrestricted growth?

4. Given a typical mixture problem in which $A(t)$ denotes the amount of salt in the tank at any time t, what condition must the derivative dA/dt, the rate of change of the amount of salt at any time t, satisfy?

9.1 Exercises

In Exercises 1–12, verify that y is a solution of the differential equation.

1. $y = x^2$; $xy' + y = 3x^2$

2. $y = e^x$; $y' - y = 0$

3. $y = \dfrac{1}{2} + ce^{-x^2}$, c any constant; $y' + 2xy = x$

4. $y = Ce^{kx}$, C any constant; $\dfrac{dy}{dx} = ky$

5. $y = e^{-2x}$; $y'' + y' - 2y = 0$

6. $y = C_1e^x + C_2e^{2x}$; $y'' - 3y' + 2y = 0$

7. $y = C_1e^{-2x} + C_2xe^{-2x}$; $y'' + 4y' + 4y = 0$

8. $y = C_1 + C_2x^{1/3}$; $3xy'' + 2y' = 0$

9. $y = \dfrac{C_1}{x} + C_2\dfrac{\ln x}{x}$; $x^2y'' + 3xy' + y = 0$

10. $y = C_1e^x + C_2xe^x + C_3x^2e^x$; $y''' - 3y'' + 3y' - y = 0$

11. $y = C - Ae^{-kt}$, A and C constants; $\dfrac{dy}{dt} = k(C - y)$

12. $y = \dfrac{C}{1 + Ae^{-Ckt}}$, A and C constants; $\dfrac{dy}{dt} = ky(C - y)$

In Exercises 13–18, verify that y is a general solution of the differential equation. Then find a particular solution of the differential equation that satisfies the side condition.

13. $y = Cx^2 - 2x$; $y' - 2\left(\dfrac{y}{x}\right) = 2$; $y(1) = 10$

14. $y = Ce^{-x^2}$; $y' = -2xy$; $y(0) = y_0$

15. $y = \dfrac{C}{x}$; $y' + \left(\dfrac{1}{x}\right)y = 0$; $y(1) = 1$

16. $y = Ce^{2x} - 2x - 1$; $y' - 2y - 4x = 0$; $y(0) = 3$

17. $y = \dfrac{Ce^x}{x} + \dfrac{1}{2}xe^x$; $y' + \left(\dfrac{1 - x}{x}\right)y = e^x$; $y(1) = -\dfrac{1}{2}e$

18. $y = C_1x^3 + C_2x^2$; $x^2y'' - 4xy' + 6y = 0$; $y(2) = 0$ and $y'(2) = 4$

19. RADIOACTIVE DECAY A radioactive substance decays at a rate directly proportional to the amount present. If the substance is present in the amount of Q_0 g initially ($t = 0$), find the amount present at any time t. Formulate but do not solve the problem in terms of a differential equation with a side condition.

20. SUPPLY AND DEMAND Let $S(t)$ denote the supply of a certain commodity as a function of time t. Suppose the rate of change of the supply is proportional to the difference between the demand $D(t)$ and the supply. Find a differential equation that describes this situation.

21. NET INVESTMENT The management of a company has decided that the level of its investment should not exceed C dollars. Furthermore, management has decided that the rate of net investment (the rate of change of the total capital invested) should be proportional to the difference between C and the total capital invested. Formulate but do not solve the problem in terms of a differential equation.

22. LAMBERT'S LAW OF ABSORPTION Lambert's Law of Absorption states that the percentage of incident light, L, absorbed in passing through a thin layer of material, x, is proportional to the thickness of the material. If, for a certain material, x_0 in. of the material reduces the light to half its intensity, how much additional material is needed to reduce the intensity to a quarter of its initial value? Formulate but do not solve the problem in terms of a differential equation with a side condition.

23. CONCENTRATION OF A DRUG IN THE BLOODSTREAM The rate at which the concentration of a drug in the bloodstream decreases is proportional to the concentration at any time t. Initially, the concentration of the drug in the bloodstream is C_0 g/mL. What is the concentration of the drug in the bloodstream at any time t? Formulate but do not solve the problem in terms of a differential equation with a side condition.

24. AMOUNT OF GLUCOSE IN THE BLOODSTREAM Suppose glucose is infused into the bloodstream at a constant rate of C g/min and, at the same time, the glucose is converted and removed from the bloodstream at a rate proportional to the amount of glucose present. Show that the amount of glucose $A(t)$ present in the bloodstream at any time t is governed by the differential equation

$$A' = C - kA$$

where k is a constant.

25. NEWTON'S LAW OF COOLING Newton's Law of Cooling states that the temperature of a body drops at a rate that is proportional to the difference between the temperature y of the body and the constant temperature C of the surrounding medium. (Assume that the temperature of the body is initially greater than C.) Show that Newton's Law of Cooling may be expressed as the differential equation

$$\dfrac{dy}{dt} = -k(y - C) \qquad y(0) = y_0$$

where y_0 denotes the temperature of the body before immersion in the medium.

26. **FICK'S LAW** Suppose a cell of volume V cc is surrounded by a homogeneous chemical solution of concentration C g/cc. Let y denote the concentration of the solute inside the cell at any time t, and suppose that, initially, the concentration is y_0. Fick's Law, named after the German physiologist Adolf Fick (1829–1901), states that the rate of change of the concentration of solute inside the cell at any time t is proportional to the difference between the concentration of the solute outside the cell and the concentration inside the cell and inversely proportional to the volume of the cell. Show that Fick's Law may be expressed as the differential equation

$$\frac{dy}{dt} = \frac{k}{V}(C - y) \qquad y(0) = y_0$$

where k is a constant. (*Note:* The constant of proportionality k depends on the area and permeability of the cell membrane.)

27. **ALLOMETRIC LAWS** Allometry is the study of the relative growth of a part of an organism in relation to the growth of an entire organism. Suppose $x(t)$ denotes the weight of an animal's organ at time t and $g(t)$ denotes the weight of another organ in the same animal at the same time t. An allometric law states that the relative growth rate of one organ, $(dx/dt)/x$, is proportional to the relative growth rate of the other, $(dy/dt)/y$. Show that this allometric law may be stated in terms of the differential equation

$$\frac{1}{x}\frac{dx}{dt} = k\frac{1}{y}\frac{dy}{dt}$$

where k is a constant.

28. **GOMPERTZ GROWTH CURVE** Suppose a quantity $Q(t)$ does not exceed some number C; that is, $Q(t) \leq C$ for all t. Suppose further that the rate of growth of $Q(t)$ is jointly proportional to its current size and the difference between C and the natural logarithm of its current size. What is the size of the quantity $Q(t)$ at any time t? Show that the mathematical formulation of this problem leads to the differential equation

$$\frac{dQ}{dt} = kQ(C - \ln Q) \qquad Q(0) = Q_0$$

where Q_0 denotes the size of the quantity present initially. The graph of $Q(t)$ is called the *Gompertz growth curve*. This model, like the ones leading to the learning curve and the logistic curve, describes restricted growth.

In Exercises 29–32, determine whether the statement is true or false. If it is true, explain why it is true. If it is false, explain why, or give an example to show why it is false.

29. The function $f(x) = x^2 + 2x + \dfrac{1}{x}$ is a solution of the differential equation $xy' + y = 3x^2 + 4x$.

30. The function $f(x) = \dfrac{1}{4}e^{3x} + ce^{-x}$ is a solution of the differential equation $y' + y = e^{3x}$.

31. The function $f(x) = 2 + ce^{-x^3}$ is a solution of the differential equation $y' + 3x^2y = x^2$.

32. The function $f(x) = 1 + cx^{-2}$ is a solution of the differential equation $xy' + 2y = 3$.

<h2>9.1   Solutions to Self-Check Exercises</h2>

1. **a.** We compute

$$y' = 2x - \frac{2c}{x^3}$$

Substituting this value of y' into the left side of the given differential equation gives

$$x\left(2x - \frac{2c}{x^3}\right) + 2\left(x^2 + \frac{c}{x^2}\right)$$

$$= 2x^2 - \frac{2c}{x^2} + 2x^2 + \frac{2c}{x^2} = 4x^2$$

which equals the expression on the right side of the differential equation, and this verifies the assertion.

b. Using the given condition, we have

$$4 = 1^2 + \frac{c}{1^2} \quad \text{or} \quad c = 3$$

and the required particular solution is

$$y = x^2 + \frac{3}{x^2}$$

2. Let N denote the size of the population at any time t. Then the required differential equation is

$$\frac{dN}{dt} = kN^{1/2}$$

and the initial condition is $N(0) = N_0$.

9.2 Separation of Variables

The Method of Separation of Variables

Differential equations are classified according to their basic form. A compelling reason for this categorization is that different methods are used to solve different types of equations.

The **order** of a differential equation is the order of the highest derivative of the unknown function appearing in the equation. A differential equation may be classified by its order. For example, the differential equations

$$y' = xe^x \quad \text{and} \quad y' + 2y = x^2$$

are **first-order equations**, whereas the differential equation

$$\frac{d^2y}{dt^2} + \left(\frac{dy}{dt}\right)^3 + ty - 8 = 0$$

is a second-order equation. For the remainder of this chapter, we restrict our study to first-order differential equations.

In this section, we describe a method for solving an important class of first-order differential equations: those that can be written in the form

$$\frac{dy}{dx} = f(x)g(y)$$

where $f(x)$ is a function of x only and $g(y)$ is a function of y only. We refer to these equations as **separable differential equations** because the variables can be separated. Equations (1) through (5) are first-order separable differential equations. For example, Equation (3)

$$\frac{dQ}{dt} = kQ(C - Q)$$

has the form $dQ/dt = f(t)g(Q)$, where $f(t) = k$ and $g(Q) = Q(C - Q)$, so it is separable. On the other hand, the differential equation

$$\frac{dy}{dx} = xy^2 + 2$$

is *not* separable.

Separable first-order equations can be solved by using the **method of separation of variables.**

Method of Separation of Variables

Suppose we are given a first-order separable differential equation in the form

$$\frac{dy}{dx} = f(x)g(y) \tag{6}$$

Step 1 *Separate the variables* by writing Equation (6) in the form

$$\frac{dy}{g(y)} = f(x)\, dx \tag{7}$$

Step 2 Integrate each side of Equation (7) with respect to the appropriate variable.

We will justify this method at the end of this section.

Solving Separable Differential Equations

EXAMPLE 1 Find the general solution of the differential equation $\dfrac{dy}{dx} = \dfrac{y}{x}$.

Solution

Step 1   Since the differential equation has the form

$$\frac{dy}{dx} = \frac{1}{x} \cdot y = f(x)g(y)$$

where $f(x) = 1/x$ and $g(y) = y$, it is separable. Separating the variables, we obtain

$$\frac{dy}{y} = \frac{dx}{x}$$

Step 2   Integrating each side of the last equation with respect to the appropriate variable, we have

$$\int \frac{dy}{y} = \int \frac{dx}{x}$$

or

$$\ln|y| + C_1 = \ln|x| + C_2$$

where C_1 and C_2 are arbitrary constants. If we choose C such that $C_2 - C_1 = \ln|C|$, then

$$\ln|y| = \ln|x| + \ln|C|$$
$$= \ln(|C||x|) = \ln|Cx| \qquad \text{\small ln } A + \text{\small ln } B = \text{\small ln } AB$$

so the general solution is $y = Cx$.

EXAMPLE 2 Find the general solution of the differential equation

$$y' = \frac{xy}{x^2 + 1}$$

Solution

Step 1   Observe that the differential equation has the form

$$\frac{dy}{dx} = \left(\frac{x}{x^2 + 1}\right)y = f(x)g(y)$$

where $f(x) = x/(x^2 + 1)$ and $g(y) = y$ and is therefore separable. Separating the variables, we obtain

$$\frac{dy}{y} = \left(\frac{x}{x^2 + 1}\right)dx$$

Step 2   Integrating each side of the last equation with respect to the appropriate variable, we have

$$\int \frac{dy}{y} = \int \frac{x}{x^2 + 1}dx$$

or

$$\ln|y| + C_1 = \frac{1}{2}\ln(x^2 + 1) + C_2$$

$$\ln|y| = \frac{1}{2}\ln(x^2 + 1) + C_2 - C_1$$

where C_1 and C_2 are arbitrary constants of integration. If we choose C such that $C_2 - C_1 = \ln|C|$, then we have

$$\ln|y| = \frac{1}{2}\ln(x^2 + 1) + \ln|C|$$

$$= \ln\sqrt{x^2 + 1} + \ln|C|$$

$$= \ln|C\sqrt{x^2 + 1}| \qquad \text{\textcolor{red}{$\ln A + \ln B = \ln AB$}}$$

so the general solution is

$$y = C\sqrt{x^2 + 1}$$

Exploring with TECHNOLOGY

Refer to Example 2, in which it was shown that the general solution of the given differential equation is $y = C\sqrt{x^2 + 1}$. Use a graphing utility to plot the graphs of the members of this family of solutions corresponding to $C = -3, -2, -1, 0, 1, 2,$ and 3. Use the standard viewing window.

EXAMPLE 3 Find the particular solution of the differential equation

$$ye^x + (y^2 - 1)y' = 0$$

that satisfies the condition $y(0) = 1$.

Solution

Step 1   Writing the given differential equation in the form

$$ye^x + (y^2 - 1)\frac{dy}{dx} = 0 \quad \text{or} \quad (y^2 - 1)\frac{dy}{dx} = -ye^x$$

and separating the variables, we obtain

$$\frac{y^2 - 1}{y}dy = -e^x\,dx$$

Step 2   Integrating each side of this equation with respect to the appropriate variable, we have

$$\int \frac{y^2 - 1}{y}dy = -\int e^x\,dx$$

$$\int \left(y - \frac{1}{y}\right)dy = -\int e^x\,dx$$

$$\frac{1}{2}y^2 - \ln|y| = -e^x + C_1$$

$$y^2 - \ln y^2 = -2e^x + C \qquad \text{\textcolor{red}{$C = 2C_1$}}$$

Using the condition $y(0) = 1$, we have

$$1 - \ln 1 = -2 + C \quad \text{or} \quad C = 3$$

Therefore, the required solution is

$$y^2 - \ln y^2 = -2e^x + 3$$

Example 3 is an initial-value problem. In general, an **initial-value problem** consists of a differential equation with one or more side conditions specified at a point. Also observe that the solution of Example 3 appeared as an implicit equation involving x and y. This often happens when we solve separable differential equations.

EXAMPLE 4 Find an equation satisfying the following conditions: (1) the slope of the tangent line to the graph of the equation at any point $P(x, y)$ is given by $-x/(2y)$ and (2) the graph of the equation passes through the point $P(1, 2)$.

Solution The slope of the tangent line at any point $P(x, y)$ on the graph of the equation is given by

$$y' = \frac{dy}{dx} = -\frac{x}{2y}$$

which is a separable differential equation. Separating the variables, we obtain

$$2y \, dy = -x \, dx$$

which, upon integration, yields

$$y^2 = -\frac{1}{2}x^2 + C_1$$

or

$$x^2 + 2y^2 = C \qquad C = 2C_1$$

where C is an arbitrary constant.

To evaluate C, we use the second condition, which implies that when $x = 1$, $y = 2$. This gives

$$1^2 + 2(2^2) = C \quad \text{or} \quad C = 9$$

Hence the required equation is

$$x^2 + 2y^2 = 9$$

The graph of this equation appears in Figure 6.

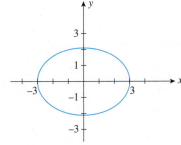

FIGURE **6**
The graph of $x^2 + 2y^2 = 9$

Justification of the Method of Separation of Variables

Let's consider the separable Equation (6) in its general form:

$$\frac{dy}{dx} = f(x)g(y)$$

If $g(y) \neq 0$, we may rewrite the equation in the form

$$\frac{1}{g(y)}\frac{dy}{dx} - f(x) = 0$$

Now suppose that G is an antiderivative of $1/g$ and F is an antiderivative of f. Using the Chain Rule, we see that

$$\frac{d}{dx}[G(y) - F(x)] = G'(y)\frac{dy}{dx} - F'(x) = \frac{1}{g(y)}\frac{dy}{dx} - f(x)$$

Therefore,

$$\frac{d}{dx}[G(y) - F(x)] = 0$$

so

$$G(y) - F(x) = C \qquad C, \text{ a constant}$$

But the last equation is equivalent to

$$G(y) = F(x) + C \quad \text{or} \quad \int\frac{dy}{g(y)} = \int f(x)\,dx$$

which is precisely the result of Step 2 in the method of separation of variables.

9.2 Self-Check Exercise

Find the solution of the differential equation $y' = 2x^2y + 2x^2$ that satisfies the condition $y(0) = 0$.

The solution to Self-Check Exercise 9.2 can be found on page 656.

9.2 Concept Questions

1. a. What is the order of a differential equation? Give an example.
  b. What is a separable equation? Give an example of a differential equation that is separable and one that is *not* separable.

2. Explain how you would solve a separable equation.

3. What is an initial-value problem?

9.2 Exercises

In Exercises 1–16, solve the first-order differential equation by separating variables.

1. $y' = \dfrac{x + 1}{y^2}$

2. $y' = \dfrac{x^2}{y}$

3. $y' = \dfrac{e^x}{y^2}$

4. $y' = -\dfrac{x}{y}$

5. $y' = 2y$

6. $y' = 2(y - 1)$

7. $y' = xy^2$

8. $y' = \dfrac{2y}{x + 1}$

9. $y' = -2(3y + 4)$

10. $y' = \dfrac{2y + 3}{x^2}$

11. $y' = \dfrac{x^2 + 1}{3y^2}$

12. $y' = \dfrac{xe^x}{2y}$

13. $y' = \sqrt{\dfrac{y}{x}}$

14. $y' = \dfrac{xy^2}{\sqrt{1 + x^2}}$

15. $y' = \dfrac{y\ln x}{x}$

16. $y' = \dfrac{(x - 4)y^4}{x^3(y^2 - 3)}$

In Exercises 17–28, find the solution of the initial-value problem.

17. $y' = \dfrac{2x}{y};\ y(1) = -2$

18. $y' = xe^{-y};\ y(0) = 1$

19. $y' = 2 - y;\ y(0) = 3$

20. $y' = \dfrac{y}{x};\ y(1) = 1$

21. $y' = 3xy - 2x;\ y(0) = 1$

22. $y' = xe^{x^2}y;\ y(0) = 1$

23. $y' = \dfrac{xy}{x^2 + 1};\ y(0) = 1$

24. $y' = x^2y^{-1/2};\ y(1) = 1$

25. $y' = xye^x;\ y(1) = 1$

26. $y' = 2xe^{-y};\ y(0) = 1$

27. $y' = 3x^2 e^{-y}$; $y(0) = 1$ **28.** $y' = \dfrac{y^2}{x-2}$; $y(3) = 1$

29. Find a function f given that (1) the slope of the tangent line to the graph of f at any point $P(x, y)$ is given by $dy/dx = (3x^2)/(2y)$ and (2) the graph of f passes through the point $(1, 3)$.

30. Find a function f given that (1) the slope of the tangent line to the graph of f at any point $P(x, y)$ is given by $dy/dx = 3xy$ and (2) the graph of f passes through the point $(0, 2)$.

31. Exponential Decay Use separation of variables to solve the differential equation

$$\frac{dQ}{dt} = -kQ \qquad Q(0) = Q_0$$

where k and Q_0 are positive constants, describing exponential decay.

32. Fick's Law Refer to Exercise 26, Section 9.1. Use separation of variables to solve the differential equation

$$\frac{dy}{dt} = \frac{k}{V}(C - y) \qquad y(0) = y_0$$

where k, V, C, and y_0 are constants with $C - y > 0$. Find $\lim_{t \to \infty} y$, and interpret your result.

33. Concentration of Glucose in the Bloodstream Refer to Exercise 24, Section 9.1. Use separation of variables to solve the differential equation $A' = C - kA$, where C and k are positive constants.
Hint: Rewrite the given differential equation in the form

$$\frac{dA}{dt} = k\left(\frac{C}{k} - A\right)$$

34. Allometric Laws Refer to Exercise 27, Section 9.1. Use separation of variables to solve the differential equation

$$\frac{1}{x}\frac{dx}{dt} = k\frac{1}{y}\frac{dy}{dt}$$

where k is a constant.

In Exercises 35–42, determine whether the statement is true or false. If it is true, explain why it is true. If it is false, explain why, or give an example to show why it is false.

35. If $y = f(x)$ is a solution of a first-order differential equation, then $y = Cf(x)$ is also a solution.

36. If $y = f(x)$ is a solution of a first-order differential equation, then $y = f(x) + C$ is also a solution.

37. The differential equation $y' = xy + 2x - y - 2$ is separable.

38. The differential equation $y' = x^2 - y^2$ is separable.

39. If the differentiable equation $M(x, y)\, dx + N(x, y)\, dy = 0$ can be written so that $M(x, y) = f(x)g(y)$ and $N(x, y) = F(x)G(y)$ for functions f, g, F, and G, then it is separable.

40. The differential equation

$$(x^2 + 2)\, dx + (2x - 4xy)\, dy = 0$$

is separable.

41. The differential equation $y\, dx - (y - xy^2)\, dy = 0$ is separable.

42. The differential equation $\dfrac{dy}{dx} = \dfrac{f(x)g(y)}{F(x) + G(y)}$ is separable.

9.2   Solution to Self-Check Exercise

Writing the differential equation in the form

$$\frac{dy}{dx} = 2x^2(y + 1)$$

and separating variables, we obtain

$$\frac{dy}{y + 1} = 2x^2\, dx$$

Integrating each side of the last equation with respect to the appropriate variable, we have

$$\int \frac{dy}{y + 1} = \int 2x^2\, dx \quad \text{or} \quad \ln|y + 1| = \frac{2}{3}x^3 + C_1$$

$$|y + 1| = e^{(2/3)x^3 + C_1} = C_2 e^{(2/3)x^3} \qquad C_2 = e^{C_1}$$

$$y + 1 = C e^{(2/3)x^3} \qquad\qquad C = \pm C_2$$

$$y = -1 + C e^{(2/3)x^3}$$

Using the initial condition $y(0) = 0$, we have

$$0 = -1 + C \quad \text{or} \quad C = 1$$

So

$$y = e^{(2/3)x^3} - 1$$

9.3 Applications of Separable Differential Equations

In this section, we look at some applications of first-order separable differential equations. We begin by revisiting some of the applications discussed in Section 9.1.

Unrestricted Growth Models

The differential equation describing an unrestricted growth model is given by

$$\frac{dQ}{dt} = kQ$$

where $Q(t)$ represents the size of a certain population at time t and k is a positive constant. Separating the variables in this differential equation and integrating, we have

$$\int \frac{dQ}{Q} = \int k\,dt$$

$$\ln|Q| = kt + C_1$$

$$|Q| = e^{kt+C_1} = e^{kt}e^{C_1} = C_2 e^{kt} \qquad C_2 = e^{C_1}$$

$$Q = Ce^{kt} \qquad\qquad C = \pm C_2$$

Thus, we may write the solution as

$$Q(t) = Ce^{kt}$$

Observe that if the quantity present initially is denoted by Q_0, then $Q(0) = Q_0$. Applying this condition yields the equation

$$Ce^0 = Q_0 \quad\text{or}\quad C = Q_0$$

Therefore, the model for unrestricted exponential growth with initial population Q_0 is given by

$$Q(t) = Q_0 e^{kt} \tag{8}$$

(Figure 7).

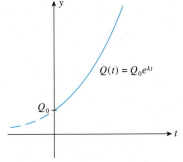

$Q(t) = Q_0 e^{kt}$

FIGURE 7
An unrestricted growth model

APPLIED EXAMPLE 1 Growth of Bacteria Under ideal laboratory conditions, the rate of growth of bacteria in a culture is proportional to the size of the culture at any time t. Suppose that 10,000 bacteria are present initially in a culture and 60,000 are present 2 hours later. How many bacteria will there be in the culture at the end of 4 hours?

Solution Let $Q(t)$ denote the number of bacteria present in the culture at time t, where t is measured in hours. Then

$$\frac{dQ}{dt} = kQ$$

where k is a constant of proportionality. Solving this separable first-order differential equation, we obtain

$$Q(t) = Q_0 e^{kt} \qquad \text{Equation (8)}$$

where Q_0 denotes the initial bacteria population. Since $Q_0 = 10,000$, we have

$$Q(t) = 10,000 e^{kt}$$

Next, the condition that 60,000 bacteria are present 2 hours later translates into $Q(2) = 60,000$, or

$$60,000 = 10,000e^{2k}$$
$$e^{2k} = 6$$
$$e^{k} = 6^{1/2}$$

Thus, the number of bacteria present at any time t is given by

$$Q(t) = 10,000e^{kt} = 10,000(e^{k})^{t}$$
$$= (10,000)6^{t/2}$$

In particular, the number of bacteria present in the culture at the end of 4 hours is given by

$$Q(4) = 10,000(6^{4/2})$$
$$= 360,000$$

APPLIED EXAMPLE 2 Weber–Fechner Law Derive the Weber–Fechner Law describing the relationship between a stimulus S and the resulting response R by solving the differential equation $dR/dS = k/S$ subject to the condition $R = 0$ when $S = S_0$, where S_0 is the threshold level.

Solution   The differential equation

$$\frac{dR}{dS} = \frac{k}{S}$$

is separable. Separating the variables, we have

$$dR = k\frac{dS}{S}$$

Integrating both sides of the equation, we have

$$\int dR = k\int \frac{dS}{S}$$
$$R = k\ln S + C$$

where C is an arbitrary constant. Using the condition $R = 0$ when $S = S_0$ gives

$$0 = k\ln S_0 + C$$
$$C = -k\ln S_0$$

Substituting this value of C in the expression for R leads to

$$R = k\ln S - k\ln S_0$$
$$= k\ln \frac{S}{S_0}$$

the required relationship between R and S. The graph of R is shown in Figure 8.

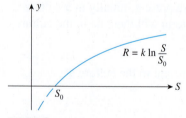

FIGURE 8
The Weber–Fechner Law

$R = k\ln \dfrac{S}{S_0}$

Restricted Growth Models

From Section 9.1, we see that a differential equation describing a restricted growth model is given by

$$\frac{dQ}{dt} = k(C - Q) \qquad (9)$$

where both k and C are positive constants. To solve this separable first-order differential equation, we first separate the variables and then integrate, obtaining

$$\int \frac{dQ}{C - Q} = \int k \, dt$$

$$-\ln|C - Q| = kt + C_1 \qquad \text{\textcolor{red}{C_1, an arbitrary constant}}$$

$$\ln|C - Q| = -kt - C_1$$

$$|C - Q| = e^{-kt - C_1} = e^{-kt}e^{-C_1} = C_2 e^{-kt} \qquad \text{\textcolor{red}{$C_2 = e^{-C_1}$}}$$

$$C - Q = Ae^{-kt} \qquad \text{\textcolor{red}{$A = \pm C_2$}}$$

$$Q = C - Ae^{-kt} \qquad\qquad\qquad\qquad\qquad \textbf{(10)}$$

This is the equation of the learning curve (Figure 9) studied in Chapter 5.

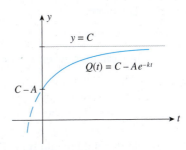

FIGURE 9
A restricted exponential growth model

APPLIED EXAMPLE 3 Yield of a Wheat Field In an experiment conducted by researchers in the Agriculture Department of a Midwestern university, it was found that the maximum yield of wheat in the university's experimental field station was 150 bushels per acre. Furthermore, the researchers discovered that the rate at which the yield of wheat increased was governed by the differential equation

$$\frac{dQ}{dx} = k(150 - Q)$$

where $Q(x)$ denotes the yield in bushels per acre and x is the amount in pounds of an experimental fertilizer used per acre of land. Data obtained in the experiment indicated that 10 pounds of fertilizer per acre of land would result in a yield of 80 bushels of wheat per acre, whereas 20 pounds of fertilizer per acre of land would result in a yield of 120 bushels of wheat per acre. Determine the yield if 30 pounds of fertilizer were used per acre.

Solution The given differential equation has the same form as Equation (10) with $C = 150$. Solving it directly or using Equation (10), we see that the yield per acre is given by

$$Q(x) = 150 - Ae^{-kx}$$

The first condition implies that $Q(10) = 80$, that is,

$$150 - Ae^{-10k} = 80$$

or $A = 70e^{10k}$. Therefore,

$$Q(x) = 150 - 70e^{10k}e^{-kx}$$
$$= 150 - 70e^{-k(x-10)}$$

The second condition implies that $Q(20) = 120$, or

$$150 - 70e^{-k(20-10)} = 120$$
$$70e^{-10k} = 30$$
$$e^{-10k} = \frac{3}{7}$$

Taking the logarithm of each side of the equation, we find

$$\ln e^{-10k} = \ln\left(\frac{3}{7}\right)$$
$$-10k = \ln 3 - \ln 7 \approx -0.8473$$
$$k \approx 0.085$$

Therefore,

$$Q(x) = 150 - 70e^{-0.085(x-10)}$$

In particular, when $x = 30$, we have

$$Q(30) = 150 - 70e^{-0.085(20)}$$
$$= 150 - 70e^{-1.7}$$
$$\approx 137$$

So the yield would be 137 bushels per acre if 30 pounds of fertilizer were used per acre. The graph of Q is shown in Figure 10.

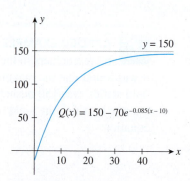

FIGURE 10
Q is a function relating crop yield to the amount of fertilizer used.

Next, let's consider a differential equation describing another type of restricted growth:

$$\frac{dQ}{dt} = kQ(C - Q)$$

where k and C are positive constants. Separating variables and integrating each side of the resulting equation with respect to the appropriate variable, we have

$$\int \frac{1}{Q(C - Q)} \, dQ = \int k \, dt$$

As it stands, the integrand on the left side of this equation is not in a form that can be easily integrated. However, observe that

$$\frac{1}{Q(C - Q)} = \frac{1}{C}\left[\frac{1}{Q} + \frac{1}{C - Q}\right]$$

as you may verify by adding the terms between the brackets on the right-hand side. Making use of this identity, we have

$$\int \frac{1}{C}\left[\frac{1}{Q} + \frac{1}{C - Q}\right] dQ = \int k \, dt$$

$$\int \frac{dQ}{Q} + \int \frac{dQ}{C - Q} = Ck \int dt$$

$$\ln|Q| - \ln|C - Q| = Ckt + b \qquad \text{\color{red}{b, an arbitrary constant}}$$

$$\ln\left|\frac{Q}{C - Q}\right| = Ckt + b$$

$$\left|\frac{Q}{C - Q}\right| = e^{Ckt+b} = e^b e^{Ckt}$$

$$\frac{Q}{C - Q} = Be^{Ckt} \qquad \text{\color{red}{$B = \pm e^b$}}$$

$$Q = CBe^{Ckt} - QBe^{Ckt}$$

$$(1 + Be^{Ckt})Q = CBe^{Ckt}$$

and

$$Q = \frac{CBe^{Ckt}}{1 + Be^{Ckt}}$$

or

$$Q(t) = \frac{C}{1 + Ae^{-Ckt}} \qquad \text{\color{red}{$A = \dfrac{1}{B}$}} \tag{11}$$

(see Figure 2, page 644). In its final form, this function is equivalent to the logistic function encountered in Chapter 5.

APPLIED EXAMPLE 4 Spread of a Flu Epidemic During a flu epidemic, 5% of the 5000 army personnel stationed at Fort MacArthur had contracted influenza at time $t = 0$. Furthermore, the rate at which they were contracting influenza was jointly proportional to the number of personnel who had already contracted the disease and the noninfected population. If 20% of the personnel had contracted the flu by the 10th day, find the number of personnel who had contracted the flu by the 13th day.

Solution Let $Q(t)$ denote the number of army personnel who had contracted the flu after t days. Then

$$\frac{dQ}{dt} = kQ(5000 - Q)$$

We may solve this separable differential equation directly, or we may use the result for the more general problem obtained earlier. Opting for the latter, we use Equation (11) with $C = 5000$ to obtain

$$Q(t) = \frac{5000}{1 + Ae^{-5000kt}}$$

The condition that 5% of the population had contracted influenza at time $t = 0$ implies that

$$Q(0) = \frac{5000}{1 + A} = 250$$

from which we see that $A = 19$. Therefore,

$$Q(t) = \frac{5000}{1 + 19e^{-5000kt}}$$

Next, the condition that 20% of the population had contracted influenza by the 10th day implies that

$$Q(10) = \frac{5000}{1 + 19e^{-50,000k}} = 1000$$

or

$$1 + 19e^{-50,000k} = 5$$

$$e^{-50,000k} = \frac{4}{19}$$

$$-50,000k = \ln 4 - \ln 19$$

and

$$k = -\frac{1}{50,000}(\ln 4 - \ln 19)$$

$$\approx 0.0000312$$

Therefore,

$$Q(t) = \frac{5000}{1 + 19e^{-0.156t}}$$

In particular, the number of army personnel who had contracted the flu by the 13th day is given by

$$Q(13) = \frac{5000}{1 + 19e^{-0.156(13)}} = \frac{5000}{1 + 19e^{-2.028}} \approx 1428$$

or approximately 29%. The graph of Q is shown in Figure 11.

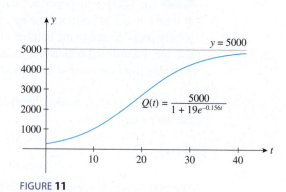

FIGURE 11
An epidemic model

Explore and Discuss

Consider the model for restricted growth described by the differential equation $dQ/dt = kQ(C - Q)$ with solution given by $C/(1 + Ae^{-Ckt})$.

1. Show that the rate of growth of Q is greatest at $t = (\ln A)/(kC)$.
   Hint: Use the differential equation and Equation (11).

2. Refer to Example 4. At what time is the number of influenza cases increasing at the greatest rate?

APPLIED EXAMPLE 5 A Mixture Problem A tank initially contains 10 gallons of pure water. Brine containing 3 pounds of salt per gallon flows into the tank at a rate of 2 gallons per minute, and the well-stirred mixture flows out of the tank at the same rate. How much salt is present at the end of 10 minutes? How much salt is present in the long run?

Solution The problem was formulated mathematically on page 645, and we were led to the differential equation

$$\frac{dA}{dt} = 6 - \frac{A}{5} = \frac{30 - A}{5}$$

subject to the condition $A(0) = 0$. Separating variables, we have

$$\frac{dA}{30 - A} = \frac{1}{5} dt$$

Next, we integrate both sides of the last equation to obtain

$$\int \frac{dA}{30 - A} = \int \frac{1}{5} dt$$

$$-\ln|30 - A| = \frac{1}{5} t + b \qquad \text{\textit{b}, a constant}$$

$$\ln|30 - A| = -\frac{1}{5} t - b$$

$$30 - A = e^{-b} e^{-t/5} \qquad (0 \le A < 30)$$

$$A = 30 - Ce^{-t/5} \qquad C = e^{-b}$$

The condition $A(0) = 0$ implies that

$$0 = 30 - C$$

giving $C = 30$, so

$$A(t) = 30(1 - e^{-t/5})$$

The amount of salt present after 10 minutes is

$$A(10) = 30(1 - e^{-2}) \approx 25.94$$

or 25.94 pounds. The amount of salt present in the long run is

$$\lim_{t \to \infty} A(t) = \lim_{t \to \infty} 30(1 - e^{-t/5}) = 30$$

or 30 pounds (Figure 12).

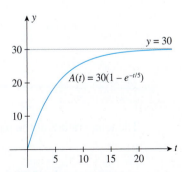

FIGURE 12
The solution of the differential
equation $\dfrac{dA}{dt} = 6 - \dfrac{A}{5}$

APPLIED EXAMPLE 6 Newton's Law of Cooling Newton's Law of Cooling states that the temperature of an object drops at a rate that is proportional to the difference between the temperature of the object and that of the surrounding medium. Suppose that an apple pie is taken out of the oven at a temperature of 200°F and placed on the counter in a room where the temperature is 70°F. If the temperature of the apple pie is 150°F after 5 minutes, find its temperature $y(t)$ as a function of time t. What will the temperature of the pie be 10 minutes after it is taken out of the oven?

Solution Let y denote the temperature of the apple pie. Since the temperature of the pie drops at a rate that is proportional to the difference between the temperature of the pie and the temperature of the room, we see that y satisfies the differential equation

$$\frac{dy}{dt} = -k(y - 70)$$

where the positive number k is the constant of proportionality. To solve the differential equation, we separate variables and integrate, obtaining

$$\int \frac{dy}{y - 70} = \int -k \, dt$$

$$\ln|y - 70| = -kt + d \qquad \text{\textit{d}, an arbitrary constant}$$

$$y - 70 = e^{-kt+d} = Ae^{-kt} \qquad A = e^d, y > 70$$

$$y = 70 + Ae^{-kt}$$

The condition that the pie is 200°F when it is taken out of the oven translates into the initial condition $y(0) = 200$. Using this condition, we have

$$200 = 70 + Ae^0 = 70 + A$$

or $A = 130$. Therefore,

$$y = 70 + 130e^{-kt}$$

To determine the value of k, we use the fact that $y(5) = 150$, obtaining

$$150 = 70 + 130e^{-5k}$$

$$130e^{-5k} = 80$$

$$e^{-5k} = \frac{80}{130}$$

$$-5k = \ln \frac{80}{130}$$

$$k = -\frac{1}{5} \ln \frac{80}{130}$$

$$\approx 0.097$$

Therefore,

$$y(t) = 70 + 130e^{-0.097t}$$

The temperature of the pie 10 minutes after it has been taken out of the oven is

$$y(10) = 70 + 130e^{-0.097(10)} \approx 119.28$$

or approximately 119.3°F.

An Application to Economics

APPLIED EXAMPLE 7 Elasticity of Demand Recall from Section 3.4 that if $x = f(p)$ is a demand equation, then the elasticity of demand at price p is given by

$$E(p) = -\frac{pf'(p)}{f(p)}$$

Find all demand functions that have unit elasticity.

Solution A demand function has unit elasticity if and only if $E(p) = 1$. So we need to solve the equation

$$-\frac{pf'(p)}{f(p)} = 1$$

Rewriting $x = f(p)$, we obtain

$$\frac{-p\dfrac{dx}{dp}}{x} = 1 \quad \text{or} \quad \frac{dx}{dp} = -\frac{x}{p}$$

This equation is separable. Separating the variables leads to

$$\frac{dx}{x} = -\frac{dp}{p}$$

Integrating both sides of the equation, we have

$$\int \frac{dx}{x} = -\int \frac{dp}{p}$$

$$\ln|x| = -\ln|p| + C_1$$

where C_1 is an arbitrary constant. Since x and p are positive, we can write the last equation as

$$\ln x + \ln p = \ln xp = C_1 \qquad \text{\small ln } A + \ln B = \ln AB$$

or

$$xp = e^{C_1} = c \qquad \text{\small } c = e^{C_1}$$

So the required demand functions have the form

$$x = \frac{c}{p}$$

where c is a constant.

9.3 Self-Check Exercise

RATE OF GROWTH OF MONEY Suppose that money deposited into a bank grows at a rate that is proportional to the amount accumulated. If the amount on deposit initially is P dollars, find an expression for the accumulated amount A after t years. Reconcile your result with the continuous compound interest formula, $A = Pe^{rt}$.

The solution to Self-Check Exercise 9.3 can be found on page 669.

9.3   Concept Questions

1. Consider the differential equation $\dfrac{dQ}{dt} = kQ$ $(k > 0)$ describing unrestricted growth. Assume that $Q(0) = Q_0 > 0$.
   a. What does the equation say about the rate of growth of Q with respect to t? What happens to the rate of growth as t approaches infinity?
   b. Reconcile your answer to part (a) with the solution $Q(t) = Q_0 e^{kt}$ of the differential equation.

2. Consider the differential equation $\dfrac{dR}{dS} = \dfrac{k}{S}$ $(k > 0)$ describing the relationship between a stimulus S and the resulting response R (the Weber–Fechner Law).

   a. What can you say about the rate of change of R with respect to S? What happens to the rate of change as S approaches infinity?
   b. Reconcile your answers to part (a) with the solution $R = k \ln \dfrac{S}{S_0}$ of the differential equation. Assume that $S(0) = S_0 > 0$.

9.3   Exercises

1. **CHEMICAL DECOMPOSITION** The rate of decomposition of a certain chemical substance is directly proportional to the amount present at any time t. If y_0 g of the chemical are present at time $t = 0$, find an expression for the amount present at any time t.

2. **GROWTH OF BACTERIA** Under ideal laboratory conditions, the rate of growth of bacteria in a culture is proportional to the size of the culture at any time t. Suppose that 2000 bacteria are present initially in the culture and 5000 are present 1 hr later. How many bacteria will be in the culture at the end of 2 hr?

3. **WORLD POPULATION GROWTH** The world population at the beginning of 1980 was 4.5 billion. Assuming that the population grew at the rate of approximately 1.3%/year, find a function Q that expresses the world population (in billions) as a function of time t (in years). What was the world population at the beginning of 2015?

4. **POPULATION GROWTH** The population of a certain community is increasing at a rate directly proportional to the population at any time t. In the last 3 years, the population has doubled. How long will it take for the population to triple?

5. **LAMBERT'S LAW OF ABSORPTION** According to Lambert's Law of Absorption, the percentage of incident light, L, absorbed in passing through a thin layer of material, x, is proportional to the thickness of the material. If $\frac{1}{2}$ in. of a certain material reduces the light to half of its intensity, how much additional material is needed to reduce the intensity to one fourth of its initial value?

6. **SAVINGS ACCOUNTS** An amount of money deposited in a savings account grows at a rate proportional to the amount present. (It can be shown that an amount of money grows in this manner if it earns interest

compounded continuously.) Suppose $10,000 is deposited in an account earning interest at the rate of 4%/year compounded continuously.
   a. What is the accumulated amount after 5 years?
   b. How long does it take for the original deposit to double in value?

7. **CHEMICAL REACTIONS** In a certain chemical reaction, a substance is converted into another substance at a rate proportional to the square of the amount of the first substance present at any time t. Initially $(t = 0)$, 50 g of the first substance was present; 1 hr later, only 10 g of it remained. Find an expression that gives the amount of the first substance present at any time t. What is the amount present after 2 hr?

8. **NEWTON'S LAW OF COOLING** Newton's Law of Cooling states that the rate at which the temperature of an object changes is directly proportional to the difference between the temperature of the object and that of the surrounding medium. A horseshoe heated to a temperature of 100°C is immersed in a large tank of water at a (constant) temperature of 30°C at time $t = 0$. Three minutes later, the temperature of the horseshoe is reduced to 70°C. Derive an expression that gives the temperature of the horseshoe at any time t. What is the temperature of the horseshoe 5 min after it has been immersed in the water?

9. **NEWTON'S LAW OF COOLING** Newton's Law of Cooling states that the rate at which the temperature of an object changes is directly proportional to the difference between the temperature of the object and that of the surrounding medium. A cup of coffee is prepared with boiling water (212°F) and left to cool on the counter in a room where the temperature is 72°F. If the temperature of the coffee is 140°F after 2 min, determine when the coffee will be cool enough to drink (say, 110°F).

10. **EXPONENTAL DECAY** A radioactive isotope with an initial mass of 100 mg decays at a rate that is proportional to its mass. Five years later, its mass is 60 mg. Find an expression giving the amount of the isotope remaining at any time t. What is the amount remaining after 10 years?

11. **RADIOACTIVE DECAY** If 4 g of a radioactive substance is present at time $t = 1$ (years) and 1 g is present at $t = 6$, how much was present initially? What is the half-life of the substance?
    Hint: See Example 2, page 393.

12. **LEARNING CURVES** The American Court Reporter Institute finds that the average student taking Elementary Machine Shorthand will progress at a rate given by

    $$\frac{dQ}{dt} = k(80 - Q)$$

    in a 20-week course, where $Q(t)$ measures the number of words of dictation a student can take per minute after t weeks in the course. If the average student can take 50 words of dictation per minute after 10 weeks in the course, how many words per minute can the average student take after completing the course?

13. **TRAINING PERSONNEL** The personnel manager of Gibraltar Insurance Company estimates that the number of insurance claims an experienced clerk can process in a day is 40. Furthermore, the rate at which a clerk can process insurance claims per day during the tth week of training is proportional to the difference between the maximum number possible (40) and the number he or she can process per day in the tth week. If the number of claims the average trainee can process after 2 weeks on the job is 10/day, determine how many claims the average trainee can process after 6 weeks on the job.

14. **EFFECT OF IMMIGRATION ON POPULATION GROWTH** Suppose a country's population at any time t grows in accordance with the rule

    $$\frac{dP}{dt} = kP + I$$

    where P denotes the population at any time t, k is a positive constant reflecting the natural growth rate of the population, and I is a constant giving the (constant) rate of immigration into the country. If the total population of the country at time $t = 0$ is P_0, find an expression for the population at any time t.

15. **EFFECT OF IMMIGRATION ON POPULATION GROWTH** Refer to Exercise 14. The population of the United States in the year 1980 ($t = 0$) was 226.5 million. Suppose the natural growth rate is 0.8% annually ($k = 0.008$), net immigration is allowed at the rate of 0.5 million people/year ($I = 0.5$) and $P_0 = 226.5$. What was the U.S. population in 2015?

16. **SUPPLY AND DEMAND** Assume that the rate of change of the supply of a commodity is proportional to the difference between the demand and the supply, so that

    $$\frac{dS}{dt} = k(D - S)$$

    where k is a constant of proportionality. Suppose that D is constant and $S(0) = S_0$. Find a formula for $S(t)$.

17. **SUPPLY AND DEMAND** Assume that the rate of change of the unit price of a commodity is proportional to the difference between the demand and the supply, so that

    $$\frac{dp}{dt} = k(D - S)$$

    where k is a constant of proportionality. Suppose that $D = 50 - 2p$, $S = 5 + 3p$, and $p(0) = 4$. Find a formula for $p(t)$.

18. **SINKING FUNDS** The proprietor of Carson Hardware Store has decided to set up a sinking fund for the purpose of purchasing a computer server 2 years from now. It is expected that the server will cost $30,000. The fund grows at the rate of

    $$\frac{dA}{dt} = rA + P$$

    where A denotes the size of the fund at any time t, r is the annual interest rate earned by the fund compounded continuously, and P is the amount (in dollars) paid into the fund by the proprietor per year (assume that this is done on a frequent basis in small deposits over the year so that it is essentially continuous). If the fund earns 5% interest per year compounded continuously, determine the size of the yearly investment the proprietor should pay into the fund.

19. **SPREAD OF A RUMOR** The rate at which a rumor spreads through an Alpine village of 400 residents is jointly proportional to the number of residents who have heard it and the number who have not. Initially, 10 residents heard the rumor, but 2 days later, this number had increased to 80. Find the number of people who will have heard the rumor after 1 week.

20. **GROWTH OF A FRUIT FLY COLONY** A biologist has determined that the maximum number of fruit flies that can be sustained in a carefully controlled environment (with a limited supply of space and food) is 400. Suppose that the rate at which the population of the colony increases obeys the rule

    $$\frac{dQ}{dt} = kQ(C - Q)$$

    where C is the carrying capacity (400) and Q denotes the number of fruit flies in the colony at time t. If the initial population of fruit flies in the experiment is 10 and it grows to 45 after 10 days, determine the population of the colony of fruit flies at the end of the 20th day.

21. **NET INVESTMENT** Refer to Exercise 21, Section 9.1. The management of a company has decided that the level of its investment should not exceed C dollars. Furthermore, management has decided that the rate of net investment (the rate of change of the total capital invested) should be proportional to the difference between C and the total capital invested. Formulate and solve the differential equation.

22. **DOMAR GROWTH MODEL** According to the Domar growth model, the value of an investment, I, at time t is given by

$$\frac{dI}{dt} = csI$$

where c is a positive constant called the *marginal propensity to consume* and s is a constant called the *output capital ratio*. Solve the differential equation given that the initial level of investment is $I(0) = I_0$.

23. **UTILITY FUNCTIONS** A person's utility function associated with a product, $U(x)$, is the "perceived value" that he or she derives from owning x units of the product. Suppose a person's utility function (in appropriate units) associated with money is related to the amount of money in her possession, x, by the differential equation

$$\frac{dU}{dx} = \frac{k}{x + 1}$$

where k is a positive constant.
a. Solve the differential equation given that $U(0) = 0$, and make a rough sketch of the graph of U.
b. What happens to the person's perceived value of money as he or she possesses more of it?

24. **DISCHARGING WATER FROM A TANK** A container that has a constant cross section A is filled with water to height H. (See the accompanying figure.) The water is discharged through an opening of cross section B at the base of the container. By using Torricelli's Law, it can be shown that the height h of the water at time t satisfies the initial-value problem

$$\frac{dh}{dt} = -\frac{B}{A}\sqrt{2gh} \qquad h(0) = H$$

a. Find an expression for h.
b. Find the time T it takes for the tank to empty.
c. Find T if $A = 4$ (ft^2), $B = 1$ (in.^{2}), $H = 16$ (ft), and $t = 32$ (ft/sec^2).

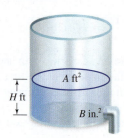

25. **GOMPERTZ GROWTH CURVES** Refer to Exercise 28, Section 9.1. Consider the differential equation

$$\frac{dQ}{dt} = kQ(C - \ln Q)$$

with the initial condition $Q(0) = Q_0$. The solution $Q(t)$ describes restricted growth and has a graph known as the Gompertz curve. Using separation of variables, solve this differential equation.

26. **GOMPERTZ GROWTH CURVES** Consider the Gompertz differential equation

$$\frac{dP}{dt} = cP \ln\left(\frac{L}{P}\right)$$

where c is a positive constant and L is the carrying capacity of the environment.
a. Solve the differential equation with the initial condition $P(0) = P_0$.
b. Find $\lim\limits_{t \to \infty} P(t)$.
c. Show that $P(t)$ is increasing most rapidly when $P = L/e$.
d. Show that $P(t)$ is increasing most rapidly when

$$t = \frac{\ln \ln\left(\dfrac{L}{P_0}\right)}{c}$$

27. **CONCENTRATION OF A DRUG IN THE BLOODSTREAM** Suppose that the rate at which the concentration of a drug in the bloodstream decreases is proportional to the concentration at time t. Initially, there is no drug in the bloodstream. At time $t = 0$ a drug having a concentration of C_0 g/mL is introduced into the bloodstream.
a. What is the concentration of the drug in the bloodstream at the end of T hr?
b. If at time T another dosage having the concentration of C_0 g/mL is infused into the bloodstream, what is the concentration of the drug at the end of $2T$ hr?
c. If the process were continued, what would the concentration of the drug be at the end of NT hr?
d. Find the concentration of the drug in the bloodstream in the long run.
Hint: Evaluate $\lim\limits_{N \to \infty} y(NT)$, where $y(NT)$ denotes the concentration of the drug at the end of NT hr.

28. **SPREAD OF DISEASE** A simple mathematical model in epidemiology for the spread of a disease assumes that the rate at which the disease spreads is jointly proportional to the number of infected people and the number of uninfected people. Suppose that there are a total of N people in the population, of whom N_0 are infected initially. Show that the number of infected people after t weeks, $x(t)$, is given by

$$x(t) = \frac{N}{1 + \left(\dfrac{N - N_0}{N_0}\right)e^{-kNt}}$$

where k is a positive constant.

29. Von Bertalanffy Growth Model The von Bertalanffy growth model is used to predict the length of commercial fish. The model is described by the differential equation

$$\frac{dx}{dt} = k(L - x)$$

where $x(t)$ is the length of the fish at time t, k is a positive constant called the von Bertalanffy growth rate, and L is the maximum length of the fish.
   a. Find $x(t)$ given that the length of the fish at $t = 0$ is x_0.
   b. At the time the larvae hatch, the North Sea haddock are about 0.4 cm long, and the average haddock grows to a length of 10 cm after 1 year. Find an expression for the length of the North Sea haddock at time t.
   c. Plot the graph of x. Take $L = 100$ (cm).
   d. On average, the haddock that are caught today are between 40 cm and 60 cm long. What are the ages of the haddock that are caught?

30. Chemical Reaction Rates Two chemical solutions, one containing N molecules of chemical A and another containing M molecules of chemical B, are mixed together at time $t = 0$. The molecules from the two chemicals combine to form another chemical solution containing y molecules of chemical AB. The rate at which the AB molecules are formed, dy/dt, is called the *reaction rate*

and is jointly proportional to $(N - y)$ and $(M - y)$. Thus,

$$\frac{dy}{dt} = k(N - y)(M - y)$$

where k is a constant. (We assume that the temperature of the chemical mixture remains constant during the interaction.) Solve this differential equation with the initial condition $y(0) = 0$ assuming that $N - y > 0$ and $M - y > 0$.
Hint: Use the identity

$$\frac{1}{(N - y)(M - y)} = \frac{1}{M - N}\left(\frac{1}{N - y} - \frac{1}{M - y}\right)$$

31. Mixture Problems A tank initially contains 20 gal of pure water. Brine containing 2 lb of salt per gallon flows into the tank at a rate of 3 gal/min, and the well-stirred mixture flows out of the tank at the same rate. How much salt is present in the tank at any time t? How much salt is present at the end of 20 min? How much salt is present in the long run?

32. Mixture Problems A tank initially contains 50 gal of brine, in which 10 lb of salt is dissolved. Brine containing 2 lb of dissolved salt per gallon flows into the tank at the rate of 2 gal/min, and the well-stirred mixture flows out of the tank at the same rate. How much salt is present in the tank at the end of 10 min?

9.3 Solution to Self-Check Exercise

Since the rate of growth of the money is proportional to the current amount, we have the initial-value problem

$$\begin{cases} \dfrac{dA}{dt} = kA \\ A(0) = P \end{cases}$$

Using Equation (8), we have

$$A(t) = A(0)e^{kt} = Pe^{kt}$$

Therefore, the accumulated amount after t years is given by

$$A(t) = Pe^{kt}$$

dollars.

If we compare this result with the formula $A = Pe^{rt}$, we see that the formulas are identical when the growth constant k is taken to be equal to r, the nominal interest rate. This shows that money deposited into a bank with interest compounded continuously grows according to the law of natural growth.

9.4 Approximate Solutions of Differential Equations

Euler's Method

As in the case of definite integrals, there are many differential equations whose exact solutions cannot be found by using any of the available methods. Here again, we resort to approximate solutions.

Many numerical methods have been developed for the efficient computation of approximate solutions to differential equations. In this section, we will look at a method for solving the problem

$$\frac{dy}{dx} = F(x, y) \qquad y(x_0) = y_0 \tag{12}$$

Euler's method, named after Leonhard Euler (1707–1783), describes a way of finding an approximate solution of Equation (12). Basically, the technique calls for approximating the actual solution $y = f(x)$ at certain selected values of x. The values of f between two adjacent values of x are then found by linear interpolation. This situation is depicted geometrically in Figure 13. Thus, in Euler's method, the actual solution curve of the differential equation is approximated by a suitable polygonal curve.

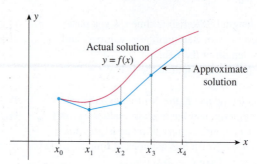

FIGURE 13
Using Euler's method, the actual solution curve of the differential equation is approximated by a polygonal curve.

To describe the method, let h be a small positive number and let $x_n = x_0 + nh$, where $n = 1, 2, 3, \ldots$; that is,

$$x_1 = x_0 + h \qquad x_2 = x_0 + 2h \qquad x_3 = x_0 + 3h \qquad \ldots$$

Thus, the points $x_0, x_1, x_2, x_3, \ldots$ are spaced evenly apart, and the distance between any two adjacent points is h units.

We begin by finding an approximation y_1 to the value of the actual solution, $f(x_1)$ at $x = x_1$. Observe that the *initial* condition $y(x_0) = y_0$ of (12) tells us that the point (x_0, y_0) lies on the solution curve. Euler's method calls for approximating the part of the graph of f on the interval $[x_0, x_1]$ by the straight-line segment that is tangent to the graph of f at (x_0, y_0). To find an equation of this straight-line segment, observe that the slope of this line segment is equal to $F(x_0, y_0)$. So using the point-slope form of an equation of a line, we see that the required equation is

$$y - y_0 = F(x_0, y_0)(x - x_0) \qquad \text{(x^2) See page 37.}$$

or

$$y = y_0 + F(x_0, y_0)(x - x_0)$$

Therefore, the approximation y_1 to $f(x_1)$ is obtained by replacing x by x_1. Thus,

$$y_1 = y_0 + F(x_0, y_0)(x_1 - x_0)$$
$$= y_0 + F(x_0, y_0)h \qquad \text{Since } x_1 - x_0 = h$$

This situation is depicted in Figure 14.

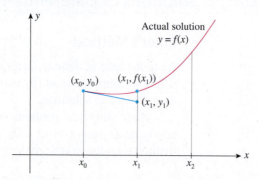

FIGURE 14
$y_1 = y_0 + hF(x_0, y_0)$ is the number used to approximate $f(x_1)$.

Next, to find an approximation y_2 to the value of the actual solution, $f(x_2)$, at $x = x_2$, we repeat the preceding procedure but this time we take the slope of the straight-line segment on $[x_1, x_2]$ to be $F(x_1, y_1)$. We obtain

$$y_2 = y_1 + F(x_1, y_1)h$$

(See Figure 15.)

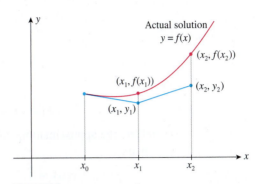

FIGURE 15
$y_2 = y_1 + hF(x_1, y_1)$ is the number used to approximate $f(x_2)$.

Continuing in this manner, we see that $y_1, y_2, \ldots, y_n$ can be found by the general formula

$$y_n = y_{n-1} + hF(x_{n-1}, y_{n-1}) \qquad (n = 1, 2, \ldots)$$

We now summarize this procedure.

Euler's Method

Suppose we are given the differential equation

$$\frac{dy}{dx} = F(x, y)$$

subject to the initial condition $y(x_0) = y_0$ and we wish to find an approximation of $y(b)$, where b is a number greater than x_0 and n is a positive integer. Compute

$$h = \frac{b - x_0}{n}$$

$$x_1 = x_0 + h \qquad x_2 = x_0 + 2h \qquad x_3 = x_0 + 3h \qquad \ldots \qquad x_n = x_0 + nh = b$$

and

$$y_0 = y(x_0)$$
$$y_1 = y_0 + hF(x_0, y_0)$$
$$y_2 = y_1 + hF(x_1, y_1)$$
$$\vdots$$
$$y_n = y_{n-1} + hF(x_{n-1}, y_{n-1})$$

Then y_n gives an approximation of the true value $y(b)$ of the solution to the initial-value problem at $x = b$.

Solving Differential Equations With Euler's Method

EXAMPLE 1 Use Euler's method with $n = 8$ to obtain an approximation of the solution of the initial-value problem

$$y' = x - y \qquad y(0) = 1$$

when $x = 2$.

Solution Here, $x_0 = 0$ and $b = 2$, so taking $n = 8$, we find

$$h = \frac{2 - 0}{8} = \frac{1}{4}$$

and

$$x_0 = 0 \qquad x_1 = \frac{1}{4} \qquad x_2 = \frac{1}{2} \qquad x_3 = \frac{3}{4} \qquad x_4 = 1$$

$$x_5 = \frac{5}{4} \qquad x_6 = \frac{3}{2} \qquad x_7 = \frac{7}{4} \qquad x_8 = b = 2$$

Also,

$$F(x, y) = x - y \quad \text{and} \quad y_0 = y(0) = 1$$

Therefore, the approximations of the actual solution at the points $x_0, x_1, x_2, \ldots,$ $x_n = b$ are

$$y_0 = y(0) = 1$$

$$y_1 = y_0 + hF(x_0, y_0) = 1 + \frac{1}{4}(0 - 1) = \frac{3}{4}$$

$$y_2 = y_1 + hF(x_1, y_1) = \frac{3}{4} + \frac{1}{4}\left(\frac{1}{4} - \frac{3}{4}\right) = \frac{5}{8}$$

$$y_3 = y_2 + hF(x_2, y_2) = \frac{5}{8} + \frac{1}{4}\left(\frac{1}{2} - \frac{5}{8}\right) = \frac{19}{32}$$

$$y_4 = y_3 + hF(x_3, y_3) = \frac{19}{32} + \frac{1}{4}\left(\frac{3}{4} - \frac{19}{32}\right) = \frac{81}{128}$$

$$y_5 = y_4 + hF(x_4, y_4) = \frac{81}{128} + \frac{1}{4}\left(1 - \frac{81}{128}\right) = \frac{371}{512}$$

$$y_6 = y_5 + hF(x_5, y_5) = \frac{371}{512} + \frac{1}{4}\left(\frac{5}{4} - \frac{371}{512}\right) = \frac{1753}{2048}$$

$$y_7 = y_6 + hF(x_6, y_6) = \frac{1753}{2048} + \frac{1}{4}\left(\frac{3}{2} - \frac{1753}{2048}\right) = \frac{8331}{8192}$$

$$y_8 = y_7 + hF(x_7, y_7) = \frac{8331}{8192} + \frac{1}{4}\left(\frac{7}{4} - \frac{8331}{8192}\right) = \frac{39{,}329}{32{,}768}$$

Thus, the approximate value of $y(2)$ is

$$\frac{39{,}329}{32{,}768} \approx 1.2002$$

EXAMPLE 2 Use Euler's method with (a) $n = 5$ and (b) $n = 10$ to approximate the solution of the initial-value problem

$$y' = -2xy^2 \qquad y(0) = 1$$

on the interval $[0, 0.5]$. Find the actual solution of the initial-value problem. Finally, sketch the graphs of the approximate solutions and the actual solution for $0 \le x \le 0.5$ on the same set of axes.

Solution

a. Here, $x_0 = 0$ and $b = 0.5$. Taking $n = 5$, we find

$$h = \frac{0.5 - 0}{5} = 0.1$$

and $x_0 = 0$, $x_1 = 0.1$, $x_2 = 0.2$, $x_3 = 0.3$, $x_4 = 0.4$, and $x_5 = b = 0.5$. Also,

$$F(x, y) = -2xy^2 \quad \text{and} \quad y_0 = y(0) = 1$$

Therefore,

$$y_0 = y(0) = 1$$
$$y_1 = y_0 + hF(x_0, y_0) = 1 + 0.1(-2)(0)(1)^2 = 1$$
$$y_2 = y_1 + hF(x_1, y_1) = 1 + 0.1(-2)(0.1)(1)^2 = 0.98$$
$$y_3 = y_2 + hF(x_2, y_2) = 0.98 + 0.1(-2)(0.2)(0.98)^2 \approx 0.9416$$
$$y_4 = y_3 + hF(x_3, y_3) = 0.9416 + 0.1(-2)(0.3)(0.9416)^2 \approx 0.8884$$
$$y_5 = y_4 + hF(x_4, y_4) = 0.8884 + 0.1(-2)(0.4)(0.8884)^2 \approx 0.8253$$

b. Here, $x_0 = 0$ and $b = 0.5$. Taking $n = 10$, we find

$$h = \frac{0.5 - 0}{10}$$

and $x_0 = 0$, $x_1 = 0.05$, $x_2 = 0.10$, $\ldots$, $x_9 = 0.45$, and $x_{10} = 0.5 = b$. Proceeding as in part (a), we obtain the approximate solutions listed in the following table:

x	0.00	0.05	0.10	0.15	0.20	0.25
y_n	1.0000	1.0000	0.9950	0.9851	0.9705	0.9517

x	0.30	0.35	0.40	0.45	0.50
y_n	0.9291	0.9032	0.8746	0.8440	0.8119

To obtain the actual solution of the differential equation, we separate variables, obtaining

$$\frac{dy}{y^2} = -2x \, dx$$

Integrating each side of the last equation with respect to the appropriate variable, we have

$$\int \frac{dy}{y^2} = -\int 2x \, dx$$

or

$$-\frac{1}{y} + C_1 = -x^2 + C_2$$

$$\frac{1}{y} = x^2 + C \qquad C = C_1 - C_2$$

$$y = \frac{1}{x^2 + C}$$

Using the condition $y(0) = 1$, we have

$$1 = \frac{1}{0 + C} \quad \text{or} \quad C = 1$$

Therefore, the required solution is given by

$$y = \frac{1}{x^2 + 1}$$

The graphs of the approximate solutions and the actual solution are sketched in Figure 16.

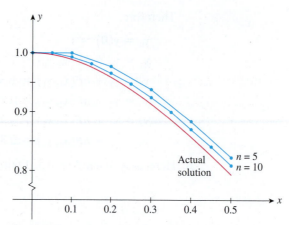

FIGURE 16
The approximate solutions and the actual solution to an initial-value problem

---

9.4 Self-Check Exercise

Use Euler's method with $n = 5$ to obtain approximations to the solution of the initial-value problem

$$y' = 2x + y \qquad y(0) = 1$$

when $x = 1$.

The solution to Self-Check Exercise 9.4 can be found on page 675.

9.4 Concept Questions

1. Describe the ideas behind Euler's method for finding the approximations to the solution of the initial-value problem

$$\frac{dy}{dx} = F(x, y) \qquad y(x_0) = y_0$$

in your own words.

2. Give the procedure for Euler's method.

9.4 Exercises

A calculator is recommended for this exercise set. In Exercises 1–10, use Euler's method with (a) $n = 4$ and (b) $n = 6$ to obtain approximations to the solution of the initial-value problem when $x = b$.

1. $y' = x + y, y(0) = 1; b = 1$

2. $y' = x - 2y, y(0) = 1; b = 2$

3. $y' = 2x - y + 1, y(0) = 2; b = 2$

4. $y' = 2xy, y(0) = 1; b = 0.5$

5. $y' = -2xy^2, y(0) = 1; b = 0.5$

6. $y' = x^2 + y^2, y(0) = 1; b = 1.5$

7. $y' = \sqrt{x + y}, y(1) = 1; b = 1.5$

8. $y' = (1 + x^2)^{-1}, y(0) = 0; b = 1$

9. $y' = \dfrac{x}{y}, y(0) = 1; b = 1$

10. $y' = xy^{1/3}, y(0) = 1; b = 1$

In Exercises 11–15, use Euler's method with $n = 5$ to obtain approximations to the solution to the initial-value problem over the indicated interval.

11. $y' = \dfrac{1}{2}xy$, $y(0) = 1$; $0 \le x \le 1$

12. $y' = x^2 y$, $y(0) = 2$; $0 \le x \le 0.6$

13. $y' = 2x - y + 1$, $y(0) = 2$; $0 \le x \le 1$

14. $y' = x + y^2$, $y(0) = 0$; $0 \le x \le 0.5$

15. $y' = x^2 + y$, $y(0) = 1$; $0 \le x \le 0.5$

16. GROWTH OF SERVICE INDUSTRIES It has been estimated that service industries, which currently make up 30% of the nonfarm workforce in a certain country, will continue to grow at the rate of

$$R(t) = 5e^{1/(t+1)}$$

percent per decade t decades from now. Estimate the percentage of the nonfarm workforce in the service industries one decade from now.

Hint: Show that the desired answer is $P(1)$, where P is the solution of the initial-value problem

$$P' = 5e^{1/(t+1)} \qquad P(0) = 30$$

Then use Euler's method with $n = 10$ to approximate the solution.

9.4 Solution to Self-Check Exercise

Here, $x_0 = 0$ and $b = 1$ so taking $n = 5$, we find

$$h = \frac{1-0}{5} = \frac{1}{5}$$

and

$$x_0 = 0 \qquad x_1 = \frac{1}{5} \qquad x_2 = \frac{2}{5}$$
$$x_3 = \frac{3}{5} \qquad x_4 = \frac{4}{5} \qquad x_5 = b = 1$$

Also,

$$F(x, y) = 2x + y \quad \text{and} \quad y_0 = y(0) = 1$$

Therefore, the approximations to the actual solution at the points $x_0, x_1, x_2, \ldots, x_5 = 1$ are

$$y_0 = y(0) = 1$$

$$y_1 = y_0 + hF(x_0, y_0) = 1 + \frac{1}{5}(0 + 1) = \frac{6}{5}$$

$$y_2 = y_1 + hF(x_1, y_1) = \frac{6}{5} + \frac{1}{5}\left(\frac{2}{5} + \frac{6}{5}\right) = \frac{38}{25}$$

$$y_3 = y_2 + hF(x_2, y_2) = \frac{38}{25} + \frac{1}{5}\left(\frac{4}{5} + \frac{38}{25}\right) = \frac{248}{125}$$

$$y_4 = y_3 + hF(x_3, y_3) = \frac{248}{125} + \frac{1}{5}\left(\frac{6}{5} + \frac{248}{125}\right) = \frac{1638}{625}$$

$$y_5 = y_4 + hF(x_4, y_4) = \frac{1638}{625} + \frac{1}{5}\left(\frac{8}{5} + \frac{1638}{625}\right) = \frac{10{,}828}{3125}$$

Thus, the approximate value of $y(1)$ is

$$\frac{10{,}828}{3125} \approx 3.4650$$

CHAPTER 9 **Summary of Principal Terms**

TERMS

differential equation (644)

general solution of a differential equation (646)

particular solution of a differential equation (647)

first-order differential equation (651)

separable differential equation (651)

method of separation of variables (651)

initial-value problem (654)

Euler's method (670)

CHAPTER 9 Concept Review Questions

Fill in the blanks.

1. **a.** An equation that involves an unknown function and its derivatives is called a/an _____ _____.
   b. A solution of a differential equation is any function that _____ the differential equation.

2. **a.** The solution of a differential equation that involves a constant c is called a _____ solution of the differential equation.
   b. The solution obtained by assigning a specific value to c is called a/an _____ solution.

3. **a.** The order of a differential equation is the order of the _____ derivative of the _____ function in the equation.
   b. A separable differential equation has the form
   $$\frac{dy}{dx} = \underline{\quad}; \text{ the equation } y' = \frac{xy}{x^2 + y^2} \text{ is } \underline{\quad}\ \underline{\quad};$$
   the equation $\dfrac{dy}{dx} = \dfrac{e^y}{1 + x^2}$ is _____.

   c. To solve a separable equation, we first _____ the variables and then integrate each term with respect to the appropriate _____.

4. **a.** The differential equation describing an unrestricted growth model is given by $\dfrac{dQ}{dt} = \underline{\quad}.$
   b. The model for unrestricted exponential growth with initial population Q_0 is given by $Q(t) = \underline{\quad}.$
   c. The differential equations $\dfrac{dQ}{dt} = k(C - Q)$ and $\dfrac{dQ}{dt} = kQ(C - Q)$ describe _____ exponential growth.

5. Euler's method is used to find an _____ solution of an initial-value problem. The _____ solution curve of a differential equation is approximated by a/an _____ curve.

CHAPTER 9 Review Exercises

In Exercises 1–3, verify that y is a solution of the differential equation.

1. $y = C_1 e^{2x} + C_2 e^{-3x};\ y'' + y' - 6y = 0$

2. $y = 2e^{2x} + 3x - 2;\ y'' - y' - 2y = -6x + 1$

3. $y = Cx^{-4/3};\ 4xy^3\,dx + 3x^2y^2\,dy = 0$

In Exercises 4 and 5, verify that y is a general solution of the differential equation, and find a particular solution of the differential equation satisfying the initial condition.

4. $y = \dfrac{1}{x^2 - C};\ \dfrac{dy}{dx} = -2xy^2;\ y(0) = 1$

5. $y = (9x + C)^{-1/3};\ \dfrac{dy}{dx} = -3y^4;\ y(0) = \dfrac{1}{2}$

In Exercises 6–11, solve the differential equation.

6. $y' = \dfrac{x^3 + 1}{y^2}$

7. $\dfrac{dy}{dt} = 2(4 - y)$

8. $y' = \dfrac{y \ln x}{x}$

9. $y' = 3x^2y^2 + y^2;\ y(0) = -2$

10. $y' = x^2(1 - y);\ y(0) = -2$

11. $\dfrac{dy}{dx} = -\dfrac{3}{2}x^2y;\ y(0) = 3$

12. Find a function f given that (1) the slope of the tangent line to the graph of f at any point $P(x, y)$ is given by
$$y' = -\frac{4xy}{x^2 + 1}$$
and (2) the graph of f passes through the point $(1, 1)$.

In Exercises 13–16, use Euler's method with (a) $n = 4$ and (b) $n = 6$ to obtain an approximation of the initial-value problem when $x = b$.

13. $y' = x + y^2,\ y(0) = 0;\ b = 1$

14. $y' = x^2 + 2y^2,\ y(0) = 0;\ b = 1$

15. $y' = 1 + 2xy^2,\ y(0) = 0;\ b = 1$

16. $y' = e^x + y^2,\ y(0) = 0;\ b = 1$

In Exercises 17 and 18, use Euler's method with $n = 5$ to obtain an approximate solution to the initial-value problem over the indicated interval.

17. $y' = 2xy,\ y(0) = 1;\ 0 \le x \le 1$

18. $y' = x^2 + y^2,\ y(0) = 1;\ 0 \le x \le 1$

19. **RESALE VALUE OF A MACHINE** The resale value of a certain machine decreases at a rate proportional to the machine's current value. The machine was purchased at \$50,000 and 2 years later was worth \$32,000.
    a. Find an expression for the resale value of the machine at any time t.
    b. Find the value of the machine after 5 years.

20. BRENTANO–STEVENS LAW The Brentano–Stevens Law, which describes the rate of change of a response R to a stimulus S, is given by

$$\frac{dR}{dS} = k \cdot \frac{R}{S}$$

where k is a positive constant. Solve this differential equation.

21. CONTINUOUS COMPOUND INTEREST HAL Corporation invests P dollars/year (assume that this is done on a frequent basis in small deposits over the year so that it is essentially continuous) into a fund earning interest at the rate of r per year compounded continuously. Then the size of the fund A grows at a rate given by

$$\frac{dA}{dt} = rA + P$$

Suppose $A = 0$ when $t = 0$. Determine the size of the fund after t years. What is the size of the fund after 5 years if $P = 50{,}000$ and $r = 0.06$?

22. FUTURE VALUE OF AN ANNUITY The future value S of an annuity (a stream of payments made continuously) satisfies the equation

$$\frac{dS}{dt} = rS + d$$

where r denotes the interest rate compounded continuously and d is a positive constant giving the rate at which payments are made into the account.
   a. If the future value of an annuity at time $t = 0$ is $\$S_0$, find an expression for the future value of the annuity at any time t.
   b. If the future value of an annuity at $t = 0$ is $\$10{,}000$, the interest rate is 6% compounded continuously, and a constant stream of payments of $\$2000$/year are made into the account, what is the future value of the annuity after 5 years?

23. DEMAND FOR A COMMODITY Suppose the demand D for a certain commodity is constant and the rate of change in the supply S over time t is proportional to the difference between the demand and the supply, so that

$$\frac{dS}{dt} = k(D - S)$$

Find an expression for the supply at any time t if the supply at $t = 0$ is S_0.

24. NEWTON'S LAW OF COOLING Barbara placed an 11-lb rib roast, which had been standing at room temperature (68°F), into a 350°F oven at 4 P.M. At 6 P.M. the temperature of the roast was 118°F. At what time would the temperature of the roast have been 150°F (medium rare)?
Hint: Use Newton's Law of Cooling (or heating).

25. ATMOSPHERIC PRESSURE The atmospheric pressure P with respect to h decreases at a rate that is proportional to P, provided that the temperature is constant.
   a. Find an expression for atmospheric pressure as a function of altitude.
   b. If the atmospheric pressure is 15 psi at ground level and 10 psi at an altitude of 10,000 ft, what is the atmospheric pressure at 20,000 ft?

26. A LEARNING CURVE The American Court Reporting Institute finds that the average student taking the Advanced Stenotype course will progress at a rate given by

$$\frac{dQ}{dt} = k(120 - Q)$$

in a 20-week course, where $Q(t)$ measures the number of words of dictation the student can take per minute after t weeks in the course. [Assume that $Q(0) = 60$.] If the average student can take 90 words of dictation per minute after 10 weeks in the course, how many words per minute can the average student take after completing the course?

27. SPREAD OF A RUMOR A rumor to the effect that a rent increase was imminent was first heard by four residents of the Chatham West Condominium Complex. The rumor spread through the complex of 200 single-family dwellings at a rate jointly proportional to the number of families who had heard it and the number who had not. Two days later, the number of families who had heard the rumor had increased to 40. Find how many families had heard the rumor after 5 days.

28. A MIXTURE PROBLEM A tank initially contains 40 gal of pure water. Brine containing 3 lb of salt per gallon flows into the tank at a rate of 4 gal/min, and the well-stirred mixture flows out of the tank at the same rate. How much salt is in the tank at any time t? How much salt is in the tank in the long run?

CHAPTER 9 Before Moving On . . .

1. a. Show that $y = 2x^2 + cx$ is a general solution of the differential equation $xy' - y = 2x^2$.

   b. Find the particular solution of the differential equation in part (a) that satisfies $y(1) = 2$.

2. Find the solution of the differential equation

$$(1 + x)y\, dx + x\, dy = 0$$

that satisfies the condition $y(1) = 1$.

3. POPULATION GROWTH The population of a town grows at a rate that is proportional to the current population. Suppose that the initial population was 5000 and that it had doubled in size after 5 years. Find an expression for the population P at year t. How long would it take for the population to reach 12,000?

4. Use Euler's method with $n = 5$ to approximate the solution of the initial-value problem $y' = y^2 - x^2$, $y(0) = 1$, at $x = 0.5$.

10 Probability and Calculus

THE SYSTEMATIC STUDY of probability began in the seventeenth century when certain aristocrats wanted to discover superior strategies to use in the gaming rooms of Europe. Some of the best mathematicians of the period were engaged in this pursuit. Since then, applications of probability have evolved in virtually every sphere of human endeavor that contains an element of uncertainty.

In this chapter, we take a look at the role of calculus in the study of *probability* involving a *continuous random variable*. We see how probability can be used to find the average life span of a certain brand of light bulbs, the average waiting time for patients in a health clinic, and the percentage of a current Mediterranean population who have serum cholesterol levels between 160 and 180 mg/dL—to name but a few applications.

If you arrive at a random time between 7 P.M. and 7:15 P.M. in front of the Bellagio Hotel in Las Vegas and find the fountains not performing, what is the probability that you will have to wait at least 5 minutes before the next "Fountains of Bellagio" show begins? In Example 4, page 685, we use a probability density function to help us answer this question.

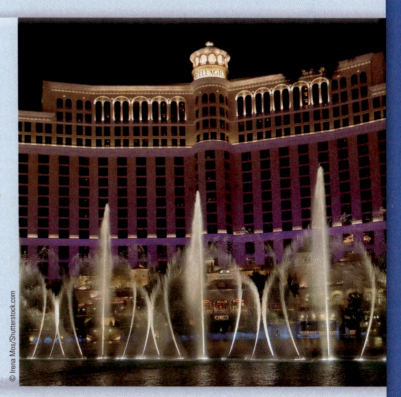

© Irena Mos/Shutterstock.com

10.1 Probability Distributions of Random Variables

Probability

We begin by discussing some terminology and notation that are important in the study of probability.

> **Experiment, Outcome, Sample Point, Sample Space, and Event**
>
> An **experiment** is an activity with observable results called **outcomes**, or **sample points.** The totality of all outcomes is the **sample space** of the experiment. A subset of the sample space is called an **event** of the experiment.

Now, given an event associated with an experiment, our primary objective is to determine the likelihood that this event will occur. This likelihood, or **probability of the event,** is a number between 0 and 1 and may be viewed as the proportionate number of times that the event will occur if the experiment associated with the event is repeated indefinitely under independent and similar conditions. Consider the simple experiment of tossing an unbiased coin and observing whether it lands "heads" (H) or "tails" (T). Since the coin is *unbiased*, we see that the probability of each outcome is $\frac{1}{2}$, abbreviated

$$P(\text{H}) = \frac{1}{2} \quad \text{and} \quad P(\text{T}) = \frac{1}{2}$$

Discrete Random Variables

We often assign numerical values to the outcomes of an experiment. For example, suppose an experiment consists of throwing a die and observing the face that lands up. If we let X denote the outcome of the experiment, then X assumes one of the values 1, 2, 3, 4, 5, or 6. Because the values assumed by X depend on the outcomes of a chance experiment, the outcome X is referred to as a **random variable.** In this case, the random variable X is also said to be **finite discrete,** since it can assume only a finite number of integer values.

The function P that associates with each value of a random variable its probability of occurrence is called a *probability function.* It is *discrete* because its domain consists of a finite set. In general, a discrete probability function is defined as follows.

> **Discrete Probability Function**
>
> A **discrete probability function** P with domain $\{x_1, x_2, \ldots, x_n\}$ satisfies these conditions:
>
> **1.** $0 \le P(x_i) \le 1, \quad \text{for } 1 \le i \le n$
> **2.** $P(x_1) + P(x_2) + \cdots + P(x_n) = 1$

Note These conditions state that the probability assigned to an outcome must be non-negative and less than or equal to 1 and that the sum of all probabilities must be 1. ■

Histograms

We can exhibit a discrete probability function or probability distribution graphically by constructing a **histogram.** First locate the values of the random variable on a horizontal line. Then, above each number, erect a rectangle whose height is equal to the probability associated with that value of the random variable.

TABLE 1

Number of Cars	Frequency of Occurrence
0	2
1	9
2	16
3	12
4	8
5	6
6	4
7	2
8	1

For example, consider the data in Table 1, the number of cars observed waiting in line at 2-minute intervals between 3 P.M. and 5 P.M. on a certain Friday at the drive-in teller of the Westwood Savings Bank and the corresponding frequency of occurrence. If we divide each number on the right of the table by 60 (the sum of these numbers), then we obtain the respective probabilities associated with the random variable X, when X assumes the values 0, 1, 2, . . . , 8. For example,

$$P(X = 0) = \frac{2}{60} \approx .03$$

$$P(X = 1) = \frac{9}{60} = .15$$
$$\vdots$$

The resulting probability distribution is shown in Table 2.

The histogram associated with this probability distribution is shown in Figure 1. Observe that the area of a rectangle in a histogram is associated with a value of a random variable X and that this area gives precisely the probability associated with that value of X. This follows because each rectangle, by construction, has width 1 and height corresponding to the probability associated with the value of the random variable.

Another consequence arising from the method of construction of a histogram is that the probability associated with more than one value of the random variable X is given by the sum of the areas of the rectangles associated with those values of X. For example, the probability that three or four cars are in line is given by

$$P(X = 3) + P(X = 4)$$

which may be obtained from the histogram by adding the areas of the rectangles associated with the values 3 and 4 of the random variable X. Thus, the required probability is

$$P(X = 3) + P(X = 4) = .20 + .13 = .33$$

TABLE 2

Probability Distribution for the Random Variable X

x	$P(X = x)$
0	.03
1	.15
2	.27
3	.20
4	.13
5	.10
6	.07
7	.03
8	.02

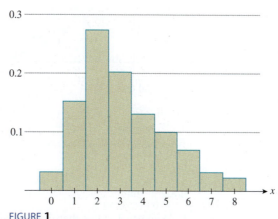

FIGURE 1
Probability distribution of the number of cars waiting in line

Continuous Random Variables

A random variable X that can assume any value in an interval is called a **continuous random variable**. Examples of continuous random variables are the life span of a light bulb, the length of a telephone call, the length of an infant at birth, the daily amount of rainfall in Boston, and the life span of a certain plant species. For the remainder of this section, we are interested primarily in continuous random variables.

Consider an experiment in which the associated random variable X has the interval $[a, b]$ as its sample space. Then an event of the experiment is any subset of $[a, b]$. For example, if X denotes the life span of a light bulb, then the sample space associated with the experiment is $[0, \infty)$, and the event that a light bulb selected at random has a life span between 500 and 600 hours, inclusive, is described by the interval $[500, 600]$ or, equivalently, by the inequality $500 \le X \le 600$. The probability that the light bulb will have a life span of between 500 and 600 hours is denoted by $P(500 \le X \le 600)$.

In general, we will be interested in computing $P(a \le X \le b)$, the probability that a random variable X assumes a value in the interval $a \le X \le b$. This computation is based on the notion of a probability density function, which we now introduce.

Probability Density Function

A **probability density function** of a random variable X in an interval I, where I may be bounded or unbounded, is a nonnegative function f having the following properties.

1. The total area of the region under the graph of f is equal to 1 (Figure 2a).
2. The probability that an observed value of the random variable X lies in the interval $[a, b]$ is given by

$$P(a \le X \le b) = \int_a^b f(x)\, dx$$

(Figure 2b).

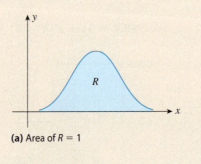

(a) Area of $R = 1$

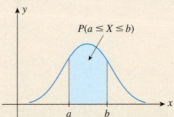

(b) $P(a \le X \le b)$ is the probability that an outcome of an experiment will lie between a and b.

FIGURE 2

Notes

1. Property 1 states that the probability that a continuous random variable takes on a value lying in its range is 1, a certainty, which is expected.
2. Property 2 states that the probability that the random variable X assumes a value in an interval $a \le X \le b$ is given by the area of the region between the graph of f and the x-axis from $X = a$ to $X = b$. Because the area under one point on the graph of f is equal to zero, we see that $P(a \le X \le b) = P(a < X \le b) = P(a \le X < b) = P(a < X < b)$.
3. A probability density function of a random variable X may be constructed by using methods that range from theoretical considerations of the problem on one extreme to an interpretation of data associated with the experiment on the other.

EXAMPLE 1 Show that each of the following functions satisfies the nonnegativity condition and Property 1 of probability density functions.

a. $f(x) = \dfrac{2}{27} x(x - 1)$ $(1 \le x \le 4)$

b. $f(x) = \dfrac{1}{3} e^{(-1/3)x}$ $(0 \le x < \infty)$

Solution

a. Since the factors x and $(x - 1)$ are both nonnegative on $[1, 4]$, we see that $f(x) \ge 0$ on $[1, 4]$. Next, we compute

$$\int_1^4 \frac{2}{27} x(x - 1)\, dx = \frac{2}{27} \int_1^4 (x^2 - x)\, dx$$

$$= \frac{2}{27} \left(\frac{1}{3} x^3 - \frac{1}{2} x^2 \right) \Big|_1^4$$

$$= \frac{2}{27} \left[\left(\frac{64}{3} - 8 \right) - \left(\frac{1}{3} - \frac{1}{2} \right) \right]$$

$$= \frac{2}{27} \left(\frac{27}{2} \right)$$

$$= 1$$

showing that Property 1 of probability density functions holds as well.

b. First, $f(x) = \frac{1}{3} e^{(-1/3)x} \ge 0$ for all values of x in $[0, \infty)$. Next,

$$\int_0^\infty \frac{1}{3} e^{(-1/3)x}\, dx = \lim_{b \to \infty} \int_0^b \frac{1}{3} e^{(-1/3)x}\, dx$$

$$= \lim_{b \to \infty} -e^{(-1/3)x} \Big|_0^b$$

$$= \lim_{b \to \infty} \left(-e^{(-1/3)b} + 1 \right)$$

$$= 1$$

so the area under the graph of $f(x) = \frac{1}{3} e^{(-1/3)x}$ on $[0, \infty)$ is equal to 1, as we set out to show. ∎

EXAMPLE 2

a. Determine the value of the constant k such that the function $f(x) = kx^2$ is a probability density function on the interval $[0, 5]$.

b. If X is a continuous random variable with the probability density function given in part (a), compute the probability that X will assume a value between 1 and 2.

c. Find the probability that X will assume a value at 3.

Solution

a. We compute

$$\int_0^5 kx^2\, dx = k \int_0^5 x^2\, dx$$

$$= \frac{k}{3} x^3 \Big|_0^5$$

$$= \frac{125}{3} k$$

Since this value must be equal to 1, we find that $k = \frac{3}{125}$.

b. The required probability is given by

$$P(1 \le X \le 2) = \int_1^2 f(x) \, dx = \int_1^2 \frac{3}{125} x^2 \, dx$$

$$= \frac{1}{125} x^3 \bigg|_1^2 = \frac{1}{125} (8 - 1)$$

$$= \frac{7}{125}$$

The graph of the probability density function f and the area corresponding to the probability $P(1 \le X \le 2)$ are shown in Figure 3.

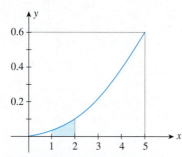

FIGURE 3
$P(1 \le X \le 2)$ for the probability
density function $y = \dfrac{3}{125} x^2$

c. The required probability is given by

$$P(X = 3) = \int_3^3 f(x) \, dx$$

$$= \int_3^3 \frac{3}{125} x^2 \, dx = 0$$

Note   Observe that the probability that a continuous random variable X will assume a value at a number a is zero, since $P(X = a) = \int_a^a f(x) \, dx = 0$. (See page 682.)

EXAMPLE 3 Let f be a constant function defined on an interval $[a, b]$. What properties must f satisfy in order for f to be a probability density function on $[a, b]$? Find an expression for f.

Solution   Let $f(x) = k$ for x in the interval $[a, b]$, where k is a constant. Clearly, k must be positive on $[a, b]$. Next, we compute

$$\int_a^b f(x) \, dx = \int_a^b k \, dx = kx \bigg|_a^b = k(b - a)$$

Since the area of f under $[a, b]$ must be equal to 1, we have

$$k(b - a) = 1$$

or

$$k = \frac{1}{b - a}$$

So the required function is defined by

$$f(x) = \frac{1}{b-a} \qquad (a \le x \le b)$$

Let's summarize the result of Example 3.

Uniform Density Function

Let f be the constant probability density function

$$f(x) = \frac{1}{b-a} \qquad (a \le x \le b)$$

defined on an interval $[a, b]$, where $a < b$. Then f is called a **uniform density function** on $[a, b]$.

A random variable that has a uniform density function is said to be **uniformly distributed.** An application involving a uniform distribution follows.

APPLIED EXAMPLE 4 Musical Fountains "The Fountains of Bellagio" is a choreographed water display set to light and music that takes place in front of the Bellagio Hotel in Las Vegas. In the evening, the shows take place every 15 minutes from 7 P.M. to midnight. The duration of each show is 7 minutes. If Joan arrives at a random time between 7 P.M. and 7:15 P.M. for an evening show and finds the fountains not performing, find the probability that Joan will have to wait at least 5 minutes before the next show starts.

Solution Let X denote the time (in minutes) that Joan has to wait for the next show. Then X is uniformly distributed over the interval $[0, 8]$. (The fountains are not performing for a duration of 8 minutes each time the show is not on.) So the uniform density function associated with this problem is

$$f(x) = \frac{1}{8} \qquad (0 \le x \le 8)$$

and the desired probability is

$$P(5 \le X \le 8) = \int_5^8 \frac{1}{8}\,dx = \frac{1}{8}x\Big|_5^8 = \frac{1}{8}(8-5) = \frac{3}{8}$$

APPLIED EXAMPLE 5 Life Span of Light Bulbs TKK Products manufactures a 200-watt electric light bulb. Laboratory tests show that the life spans of these light bulbs have a distribution described by the probability density function

$$f(x) = 0.001e^{-0.001x} \qquad (0 \le x < \infty)$$

Determine the probability that a light bulb will have a life span of (a) 500 hours or less, (b) more than 500 hours, and (c) more than 1000 hours but less than 1500 hours.

Solution   Let X denote the life span of a light bulb.

a. The probability that a light bulb will have a life span of 500 hours or less is given by

$$P(0 \le X \le 500) = \int_0^{500} 0.001e^{-0.001x}\, dx$$

$$= -e^{-0.001x}\Big|_0^{500} = -e^{-0.5} + 1$$

$$\approx 0.3935$$

b. The probability that a light bulb will have a life span of more than 500 hours is given by

$$P(X > 500) = \int_{500}^{\infty} 0.001e^{-0.001x}\, dx$$

$$= \lim_{b \to \infty} \int_{500}^{b} 0.001e^{-0.001x}\, dx$$

$$= \lim_{b \to \infty} -e^{-0.001x}\Big|_{500}^{b}$$

$$= \lim_{b \to \infty} \left(-e^{-0.001b} + e^{-0.5} \right)$$

$$= e^{-0.5} \approx 0.6065$$

This result may also be obtained by observing that

$$P(X > 500) = 1 - P(X \le 500)$$

$$= 1 - 0.3935 \qquad \text{\color{red}Use the result from part (a).}$$

$$\approx 0.6065$$

c. The probability that a light bulb will have a life span of more than 1000 hours but less than 1500 hours is given by

$$P(1000 < X < 1500) = \int_{1000}^{1500} 0.001e^{-0.001x}\, dx$$

$$= -e^{-0.001x}\Big|_{1000}^{1500}$$

$$= -e^{-1.5} + e^{-1}$$

$$\approx -0.2231 + 0.3679$$

$$= 0.1448$$

The probability density function of Example 5 is called an **exponential density function.** More generally, we have the following.

Exponential Density Function (with Parameter k)

Let f be the function defined by

$$f(x) = ke^{-kx} \qquad (0 \le x < \infty)$$

where k is a positive constant. Then f is called an **exponential density function** with parameter k on $[0, \infty)$.

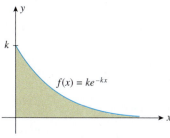

You are asked to verify that f is a probability density function in Exercise 61. The graph of f is shown in Figure 4. The random variable that is associated with f is said to be **exponentially distributed.** Exponential random variables are used to represent the life span of an electronic component, the duration of a telephone call, the waiting time in a doctor's office, and the time between successive flight arrivals and departures in an airport, to mention but a few applications.

Joint Probability Density Functions

Sometimes the outcomes of an experiment are associated with more than one random variable. For example, we might be interested in the relationship between the weight and height of newborn infants or between the price and tread life of automobile tires. To study such problems, we need to extend the concept of a probability density function of a random variable to functions of more than one variable. In the case involving two random variables, we have the following definition.

> ### Joint Probability Density Function
>
> A **joint probability density function** of the random variables X and Y on a region D is a nonnegative function $f(x, y)$ having the property
>
> $$\iint_D f(x, y)\, dA = 1$$
>
> Thus, the volume of the solid under the graph of f is equal to 1.
>
> The probability that the observed values of the random variables X and Y lie in a region $R \subset D$ is given by
>
> $$P[(X, Y) \text{ in } R] = \iint_R f(x, y)\, dA$$

EXAMPLE 6 Show that the function $f(x, y) = xy$ is a joint probability density function on $D = \{(x, y) \mid 0 \le x \le 1; 0 \le y \le 2\}$.

Solution First, observe that $f(x, y) = xy$ is nonnegative on D. Next, we compute

$$
\begin{aligned}
\iint_D f(x, y)\, dA &= \int_0^2 \int_0^1 xy\, dx\, dy \\
&= \int_0^2 \left(\frac{1}{2} x^2 y \Big|_0^1 \right) dy \\
&= \int_0^2 \frac{1}{2} y\, dy = \frac{1}{4} y^2 \Big|_0^2 = 1
\end{aligned}
$$

Therefore, f is a joint probability density function on D.

EXAMPLE 7 Let $f(x, y) = xy$ be the joint probability density function for the random variables X and Y on $D = \{(x, y) \mid 0 \le x \le 1; 0 \le y \le 2\}$. Find
(a) $P(0 \le X \le \frac{1}{2}; 1 \le Y \le 2)$ and (b) $P(X + Y \le 1)$.

Solution

a. The required probability is given by

$$P\left(0 \le X \le \frac{1}{2}; 1 \le Y \le 2\right) = \int_1^2 \int_0^{1/2} xy \, dx \, dy$$

$$= \int_1^2 \left(\frac{1}{2} x^2 y \Big|_0^{1/2}\right) dy$$

$$= \int_1^2 \frac{1}{8} y \, dy = \frac{1}{16} y^2 \Big|_1^2 = \frac{3}{16}$$

b. The region $R = \{(x, y) \mid x + y \le 1\}$ is shown in Figure 5. The required probability is

$$P(X + Y \le 1) = \int_0^1 \int_0^{1-x} xy \, dy \, dx$$

$$= \int_0^1 \left(\frac{1}{2} xy^2 \Big|_0^{1-x}\right) dx$$

$$= \int_0^1 \frac{1}{2} x(1 - x)^2 \, dx$$

$$= \frac{1}{2} \int_0^1 (x - 2x^2 + x^3) \, dx$$

$$= \frac{1}{2} \left(\frac{1}{2} x^2 - \frac{2}{3} x^3 + \frac{1}{4} x^4\right) \Big|_0^1$$

$$= \frac{1}{2} \left(\frac{1}{2} - \frac{2}{3} + \frac{1}{4}\right) = \frac{1}{24}$$

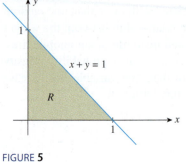

FIGURE 5

EXAMPLE 8 Let $f(x, y) = 2e^{-x-2y}$ be the joint probability density function for the random variables X and Y on $D = \{(x, y) \mid x \ge 0 \text{ and } y \ge 0\}$.

a. Find the probability that X will assume a value between 0 and 1 and that Y will assume a value between 1 and 2.

b. Find the probability that X will assume a value greater than 1 and that Y will assume a value less than 2.

Solution

a. The required probability is given by

$$P(0 \le X \le 1; 1 \le Y \le 2) = \int_1^2 \int_0^1 2e^{-x-2y} \, dx \, dy$$

$$= \int_1^2 \left(-2e^{-x-2y} \Big|_0^1\right) dy$$

$$= \int_1^2 (-2e^{-1-2y} + 2e^{-2y}) \, dy$$

$$= \int_1^2 2(1 - e^{-1})e^{-2y} \, dy \qquad \text{Simplify the integrand.}$$

$$= 2(1 - e^{-1})\left(-\frac{1}{2} e^{-2y}\right) \Big|_1^2$$

$$= -(1 - e^{-1})e^{-2y} \Big|_1^2$$

$$= -(1 - e^{-1})(e^{-4} - e^{-2}) \approx .0740$$

b. The required probability is given by

$$P(X > 1; 0 \le Y < 2) = \int_0^2 \int_1^\infty 2e^{-x-2y} \, dx \, dy$$

$$= \int_0^2 \left(\lim_{b \to \infty} \int_1^b 2e^{-x-2y} \, dx \right) dy$$

$$= \int_0^2 \left(\lim_{b \to \infty} -2e^{-x-2y} \Big|_1^b \right) dy$$

$$= \int_0^2 \left[\lim_{b \to \infty} \left(-2e^{-b-2y} + 2e^{-1-2y} \right) \right] dy$$

$$= \int_0^2 2e^{-1-2y} \, dy$$

$$= -e^{-1-2y} \Big|_0^2 = -e^{-5} + e^{-1} \approx .3611 \quad \blacksquare$$

10.1 Self-Check Exercises

1. Determine the value of the constant k such that the function $f(x) = k(4x - x^2)$ is a probability density function on the interval $[0, 4]$.

2. Suppose that X is a continuous random variable with the probability density function of Self-Check Exercise 1. Find the probability that X will assume a value between 1 and 3.

Solutions to Self-Check Exercises 10.1 can be found on page 692.

10.1 Concept Questions

1. Define the following terms in your own words:
   a. Experiment
   b. Sample point (outcome)
   c. Event
   d. Probability of an event

2. What is the difference between a finite discrete random variable and a continuous random variable?

3. **a.** What is a probability density function?
   b. What is a joint probability density function?

10.1 Exercises

In Exercises 1–10, show that the function is a probability density function on the specified interval.

1. $f(x) = \dfrac{1}{16} x; (2 \le x \le 6)$

2. $f(x) = \dfrac{2}{9} (3x - x^2); (0 \le x \le 3)$

3. $f(x) = \dfrac{3}{8} x^2; (0 \le x \le 2)$

4. $f(x) = \dfrac{3}{32} (x - 1)(5 - x); (1 \le x \le 5)$

5. $f(x) = 20(x^3 - x^4); (0 \le x \le 1)$

6. $f(x) = \dfrac{8}{7x^2}; (1 \le x \le 8)$

7. $f(x) = \dfrac{3}{14} \sqrt{x}; (1 \le x \le 4)$

8. $f(x) = \dfrac{12 - x}{72}; (0 \le x \le 12)$

9. $f(x) = \dfrac{x}{(x^2 + 1)^{3/2}}; (0 \le x < \infty)$

10. $f(x) = 4xe^{-2x^2}; (0 \le x < \infty)$

In Exercises 11–18, find the value of the constant k such that the function is a probability density function on the indicated interval.

11. $f(x) = k; [1, 4]$ 12. $f(x) = kx; [0, 4]$

13. $f(x) = k(4 - x); [0, 4]$ 14. $f(x) = kx^3; [0, 1]$

15. $f(x) = k\sqrt{x}; [0, 4]$ 16. $f(x) = \dfrac{k}{x}; [1, 5]$

17. $f(x) = \dfrac{k}{x^3}; [1, \infty)$ **18.** $f(x) = ke^{-x/2}; [0, \infty)$

In Exercises 19–28, f is the probability density function for the random variable X defined on the given interval. Find the indicated probabilities.

19. $f(x) = \dfrac{1}{12}x; [1, 5]$

   a. $P(2 \leq X \leq 4)$ **b.** $P(1 \leq X \leq 4)$
   c. $P(X \geq 2)$ **d.** $P(X = 2)$

20. $f(x) = \dfrac{1}{9}x^2; [0, 3]$

   a. $P(1 \leq X \leq 2)$ **b.** $P(1 < X \leq 3)$
   c. $P(X \leq 2)$ **d.** $P(X = 1)$

21. $f(x) = \dfrac{3}{32}(4 - x^2); [-2, 2]$

   a. $P(-1 \leq X \leq 1)$ **b.** $P(X \leq 0)$
   c. $P(X > -1)$ **d.** $P(X = 0)$

22. $f(x) = \dfrac{3}{16}\sqrt{x}; [0, 4]$

   a. $P(1 < X < 3)$ **b.** $P(X \leq 2)$
   c. $P(X = 2)$ **d.** $P(X \geq 1)$

23. $f(x) = \dfrac{1}{4\sqrt{x}}; [1, 9]$

   a. $P(X \geq 4)$ **b.** $P(1 \leq X < 8)$
   c. $P(X = 3)$ **d.** $P(X \leq 4)$

24. $f(x) = \dfrac{1}{2}e^{-x/2}; [0, \infty)$

   a. $P(X \leq 4)$ **b.** $P(1 < X < 2)$
   c. $P(X = 50)$ **d.** $P(X \geq 2)$

25. $f(x) = 4xe^{-2x^2}; [0, \infty)$
   a. $P(0 \leq X \leq 4)$ **b.** $P(X \geq 1)$

26. $f(x) = \dfrac{1}{9}xe^{-x/3}; [0, \infty)$

   a. $P(0 \leq X \leq 3)$ **b.** $P(X \geq 1)$

27. $f(x) = \begin{cases} x & \text{if } 0 \leq x \leq 1 \\ 2 - x & \text{if } 1 \leq x \leq 2 \end{cases}; [0, 2]$

   a. $P(\frac{1}{2} \leq X \leq 1)$ **b.** $P(\frac{1}{2} \leq X \leq \frac{3}{2})$
   c. $P(X \geq 1)$ **d.** $P(X \leq \frac{3}{2})$

28. $f(x) = \begin{cases} \dfrac{3}{40}\sqrt{x} & \text{if } 0 \leq x \leq 4 \\ \dfrac{12}{5x^2} & \text{if } 4 < x < \infty \end{cases}; [0, \infty)$

   a. $P(1 \leq X \leq 4)$ **b.** $P(0 \leq X \leq 5)$

In Exercises 29–32, show that the function is a joint probability density function on D.

29. $f(x, y) = \dfrac{1}{4}; D = \{0 \leq x \leq 2; 1 \leq y \leq 3\}$

30. $f(x, y) = \dfrac{1}{4}(x + 2y); D = \{0 \leq x \leq 2; 0 \leq y \leq 1\}$

31. $f(x, y) = \dfrac{1}{3}xy; D = \{0 \leq x \leq 2; 1 \leq y \leq 2\}$

32. $f(x, y) = 4(1 - x)(2 - y); D = \{0 \leq x \leq 1; 1 \leq y \leq 2\}$

In Exercises 33–36, find the value of the constant k so that the function is a joint probability density function on D.

33. $f(x, y) = kx^2y; D = \{0 \leq x \leq 1; 1 \leq y \leq 2\}$

34. $f(x, y) = k\sqrt{x}(2 - y); D = \{1 \leq x \leq 2; 0 \leq y \leq 2\}$

35. $f(x, y) = k(x - x^2)e^{-2y}; D = \{0 \leq x \leq 1; 1 \leq y < \infty\}$

36. $f(x, y) = kxye^{-(x^2+y^2)}; D = \{0 < x < \infty; 0 < y < \infty\}$

In Exercises 37–40, f is a joint probability density function for the random variables X and Y on D. Find the indicated probabilities.

37. $f(x, y) = xy; D = \{(x, y) \mid 0 \leq x \leq 1; 0 \leq y \leq 2\}$
   a. $P(0 \leq X \leq 1; 0 \leq Y \leq 1)$
   b. $P(X + 2Y \leq 1)$

38. $f(x, y) = \dfrac{1}{12}(x + y); D = \{(x, y) \mid 0 \leq x \leq 2; 1 \leq y \leq 3\}$
   a. $P(0 \leq X \leq 1; 1 \leq Y \leq 2)$
   b. $P(Y \geq 1; X + Y \leq 3)$

39. $f(x, y) = \dfrac{9}{224}\sqrt{xy}; D = \{(x, y) \mid 1 \leq x \leq 4; 0 \leq y \leq 4\}$
   a. $P(1 \leq X \leq 2; 0 \leq Y \leq 1)$
   b. $P(1 \leq X \leq 4; 0 \leq Y \leq \sqrt{X})$

40. $f(x, y) = 3\sqrt{x}e^{-2y}; D = \{(x, y) \mid 0 \leq x \leq 1; 0 \leq y < \infty\}$
   a. $P(0 \leq X \leq 1; 1 \leq Y < \infty)$
   b. $P(\frac{1}{2} \leq X \leq 1; 0 \leq Y < \infty)$

41. WAITING TIME AT A HEALTH CLINIC The average waiting time for patients arriving at the Newtown Health Clinic between 1 P.M. and 4 P.M. on a weekday is an exponentially distributed random variable X with probability density function $f(x) = \frac{1}{15}e^{-x/15}$.
   a. What is the probability that a patient arriving at the clinic between 1 P.M. and 4 P.M. will have to wait between 10 and 12 min?
   b. What is the probability that a patient arriving at the clinic between 1 P.M. and 4 P.M. will have to wait more than 15 min?

42. LIFE SPAN OF A PLANT SPECIES The life span of a certain plant species (in days) is described by the probability density function

$$f(x) = \dfrac{1}{100}e^{-x/100} \qquad (0 \leq x < \infty)$$

   a. Find the probability that a plant of this species will live for 100 days or less.
   b. Find the probability that a plant of this species will live more than 120 days.
   c. Find the probability that a plant of this species will live more than 60 days but less than 140 days.

43. **WAITING TIME AT A BAKERY** The proprietress of Ann's Bakery finds that the waiting time (in minutes) for a customer to be served during the hours between 12 noon and 1 P.M. is an exponentially distributed random variable X with associated probability function $f(t) = \frac{1}{2} e^{-t/2}$.
    a. Find the probability that a customer arriving between noon and 1 P.M. will have to wait at most 3 min.
    b. Find the probability that a customer arriving between noon and 1 P.M. will have to wait between 2 and 3 min.
    c. Find the probability that a customer arriving between noon and 1 P.M. will have to wait at least 3 min.

44. **FREQUENCY OF ROAD REPAIRS** The fraction of streets in the downtown section of a certain city that need repairs in a given year is a random variable with a distribution described by the probability density function

$$f(x) = 12x^2(1 - x) \qquad (0 \le x \le 1)$$

Find the probability that at most half of the streets will need repairs in any given year.

45. **RELIABILITY OF ROBOTS** National Welding uses industrial robots in some of its assembly-line operations. Management has determined that the lengths of time (in hours) between breakdowns are exponentially distributed with probability density function $f(t) = 0.001e^{-0.001t}$.
    a. What is the probability that a robot selected at random will break down between 600 and 800 hr of use?
    b. What is the probability that a robot will break down after 1200 hr of use?

46. **WAITING TIME AT AN EXPRESSWAY TOLLBOOTH** Suppose the time intervals (in seconds) between arrivals of successive cars at an expressway tollbooth during rush hour are exponentially distributed with probability density function $f(t) = \frac{1}{8} e^{-t/8}$. Find the probability that the time interval between arrivals of successive cars is more than 8 sec.

47. **TIME INTERVALS BETWEEN PHONE CALLS** A study conducted by UniMart, a mail-order department store, reveals that the time intervals (in minutes) between incoming telephone calls on its toll-free 800 line between 10 A.M. and 2 P.M. are exponentially distributed with probability density function $f(t) = \frac{1}{30} e^{-t/30}$. What is the probability that the time interval between successive calls is more than 2 min?

48. **RELIABILITY OF MICROPROCESSORS** The microprocessors manufactured by United Motor Works, which are used in automobiles to regulate fuel consumption, are guaranteed against defects for 20,000 mi of use. Tests conducted in the laboratory under simulated driving conditions reveal that the distances driven (in miles) before the microprocessors break down are exponentially distributed with probability density function $f(t) = 0.00001e^{-0.00001t}$. What is the probability that a microprocessor selected at random will fail during the warranty period?

49. **COMMUTING** Bill takes the commuter train to work every day. During the morning commute, a train arrives every 15 min. If Bill arrives at the station at a random time for the morning commute, what is the probability that he will have to wait:
    a. At least 5 min for a train?
    b. Between 5 and 8 min for a train?

50. **MUSICAL FOUNTAINS** Refer to Example 4. "The Fountains of Bellagio" is a choreographed water display set to light and music that takes place in front of the Bellagio Hotel in Las Vegas. In the evening, the shows take place every 15 min from 7 P.M. to midnight. The duration of each show is 7 min. If Joan arrives at a random time between 7 P.M. and 7:15 P.M. for an evening show and finds the fountains not performing, find the probability that Joan will have to wait at most 5 min before the next show starts.

51. **THE FOUNTAINS OF BELLAGIO** Refer to Example 4. "The Fountains of Bellagio" is a choreographed water display set to light and music that takes place in front of the Bellagio Hotel in Las Vegas. In the evening, the shows take place every 15 min from 7 P.M. to midnight. The duration of each show is 7 min. If Joan arrives at a random time between 8 P.M. and 8:30 P.M. for an evening show, find the probability that Joan will have to wait:
    a. Less than 5 min before the next show begins.
    b. More than 10 min before the next show begins.

52. **PROBABILITY OF SNOWFALL** The amount of snowfall (in feet) in a remote region of Alaska in the month of January is a continuous random variable with probability density function $f(x) = \frac{2}{9}x(3 - x), 0 \le x \le 3$. Find the probability that the amount of snowfall will be between 1 and 2 ft; more than 1 ft.

53. **PROBABILITY OF RAINFALL** The amount of rainfall (in inches) on a tropical island in the month of August is a continuous random variable with probability density function

$$f(x) = \frac{1}{16}x(6 - x) \qquad (0 \le x \le 6)$$

Find the probability that the amount of rainfall in August will be less than 2 in.

54. **LIFE EXPECTANCY OF PLASMA TVS** The life expectancy (in years) of a certain brand of plasma TV is a continuous random variable with probability density function

$$f(t) = 9(9 + t^2)^{-3/2} \qquad (0 \le t < \infty)$$

Find the probability that a randomly chosen plasma TV will last more than 4 years.
Hint: Integrate using a table of integrals.

55. PRODUCT RELIABILITY The tread life (in thousands of miles) of a certain make of tire is a continuous random variable with probability density function

$$f(x) = 0.02e^{-0.02x} \qquad (0 \le x < \infty)$$

a. Find the probability that a randomly selected tire of this make will have a tread life of at most 30,000 mi.
b. Find the probability that a randomly selected tire of this make will have a tread life between 40,000 and 60,000 mi.
c. Find the probability that a randomly selected tire of this make will have a tread life of at least 70,000 mi.

56. PRODUCT RELIABILITY The *reliability function* of a product, $R(x)$, is the probability that the product will not fail in the time interval $[0, x]$.

a. Suppose that the time to failure of a product is a random variable X with probability density function $f(x)$. Find a formula for $R(x)$ in terms of $f(x)$.
b. Suppose that f is exponentially distributed with parameter k, that is,

$$f(x) = ke^{-kx} \qquad (x \ge 0)$$

Find an expression for R.
c. Suppose that a component has a failure rate of 0.05 per thousand hours ($k = 0.05$). What is the probability that the component will survive at least 10,000 hr of operation?

57. CABLE TV SUBSCRIBERS The management of Telstar Cable estimates that the number of new subscribers (in thousands) to their service next year in Foxboro, X, and the number of new subscribers (in thousands) next year in Sharon, Y, have a distribution given by the joint probability density function

$$f(x, y) = \frac{9}{4000} xy \sqrt{25 - x^2}(4 - y);$$

$$D = \{(x, y) \mid 0 \le x \le 5; 0 \le y \le 4\}$$

Find the probability that the number of new subscribers next year in Foxboro will be between 2000 and 2500 and the number of new subscribers in Sharon will be between 1000 and 2000.

58. CABLE TV SUBSCRIBERS Refer to Exercise 57. What is the probability that the total number of new subscribers next year in Foxboro will be less than 3000 and the number of new subscribers next year in Sharon will be less than 2000?

59. PRODUCT RELIABILITY Hal has a tablet PC and a smartphone. Suppose that the random variables X and Y give the time to failure (in years) of the tablet PC and the smartphone, respectively. If the joint probability density function for X and Y is

$$f(x, y) = 0.06e^{-0.2x}e^{-0.3y} \qquad (x \ge 0 \text{ and } y \ge 0)$$

what is the probability that both Hal's tablet PC and smartphone will last 4 years or longer?

60. Let $f(x) = \frac{3}{8}x^2$ be a probability density function on $[0, 2]$. Find c such that $P(X \le c) = .1$.

61. Show that the exponential density function

$$f(x) = ke^{-kx}$$

where k is a positive constant, satisfies the conditions for being a probability density function on the interval $[0, \infty)$.

62. Suppose that f is the probability density function of the random variable X on the interval $[A, B]$. Show that if $A \le b \le c \le B$, then $P(X \le b) \le P(X \le c)$.

In Exercises 63–66, determine whether the statement is true or false. If it is true, explain why it is true. If it is false, give an example to show why it is false.

63. If $\int_a^b f(x)\, dx = 1$, then f is a probability density function on $[a, b]$.

64. If f is a probability density function on an interval $[a, b]$, then f is a probability density function on $[c, d]$ for any real numbers c and d satisfying $a < c < d < b$.

65. If f is a probability density function on $[a, b]$, then $0 \le f(x) \le 1$ on $[a, b]$.

66. The function $f(x) = \frac{3}{2}x^2 - 3x$ is a probability density function on $[1, 3]$.

10.1   Solutions to Self-Check Exercises

1. We compute

$$\int_0^4 k(4x - x^2)\, dx = k\left(2x^2 - \frac{1}{3}x^3\right)\Big|_0^4$$

$$= k\left(32 - \frac{64}{3}\right)$$

$$= \frac{32}{3}k$$

Since this value must be equal to 1, we have $k = \frac{3}{32}$.

2. The required probability is given by

$$P(1 < X < 3) = \int_1^3 f(x)\, dx$$

$$= \int_1^3 \frac{3}{32}(4x - x^2)\, dx$$

$$= \frac{3}{32}\left(2x^2 - \frac{1}{3}x^3\right)\Big|_1^3$$

$$= \frac{3}{32}\left[(18 - 9) - \left(2 - \frac{1}{3}\right)\right] = \frac{11}{16}$$

USING **TECHNOLOGY**

Graphing a Histogram

Graphing Utility

A graphing utility can be used to plot the histogram for a given set of data, as illustrated in the following example.

APPLIED EXAMPLE 1 A survey of 90,000 households conducted in a certain year revealed the following percentage of women who wear a shoe size within the given ranges.

Shoe Size	<5	$5–5\frac{1}{2}$	$6–6\frac{1}{2}$	$7–7\frac{1}{2}$	$8–8\frac{1}{2}$	$9–9\frac{1}{2}$	$10–10\frac{1}{2}$	$>10\frac{1}{2}$
Women (%)	1	5	15	27	29	14	7	2

Source: Footwear Market Insights survey.

Let X denote the random variable taking on the values 1 through 8, where 1 corresponds to a shoe size less than 5, 2 corresponds to a shoe size of $5–5\frac{1}{2}$, and so on.

a. Plot a histogram for the given data.
b. What percentage of women in the survey wear a shoe size within the ranges $7–7\frac{1}{2}$ or $8–8\frac{1}{2}$?

Solution

a. Enter the values of X as $x_1 = 1$, $x_2 = 2$, . . . , $x_8 = 8$ and the corresponding values of Y as $y_1 = 1$, $y_2 = 5$, . . . , $y_8 = 2$. Then using the **DRAW** function from the Statistics menu, we draw the histogram shown in Figure T1.

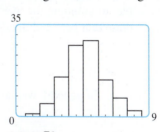

FIGURE T1
The histogram for the given data, using the viewing window [0, 9] × [0, 35]

b. The probability that a woman participating in the survey wears a shoe size within the ranges $7–7\frac{1}{2}$ or $8–8\frac{1}{2}$ is given by

$$P(X = 4) + P(X = 5) = .27 + .29 = .56$$

This tells us that 56% of the women wear a shoe size within the ranges $7–7\frac{1}{2}$ or $8–8\frac{1}{2}$.

TECHNOLOGY EXERCISES

1. Graph the histogram associated with the data given in Table 2, page 681. Compare your graph with that given in Figure 1, page 681.

Family Size	2	3	4	5	6	7	8
Frequency of Occurrence	350	200	245	125	66	10	4

2. **DISTRIBUTION OF FAMILIES BY SIZE** A survey was conducted by the Public Housing Authority in a certain community among 1000 families to determine the distribution of families by size. The results follow:

a. Find the probability distribution of the random variable X, where X denotes the number of persons in a randomly chosen family.
b. Graph the histogram corresponding to the probability distribution found in part (a).

3. **WAITING LINES** The data below were obtained in a study conducted by the manager of Sav-More Supermarket. In this study, the number of customers waiting in line at the express checkout at the beginning of each 3-min interval between 9 A.M. and 12 noon on Saturday was observed.

Customers	0	1	2	3	4	5
Frequency of Occurrence	1	4	2	7	14	8

Customers	6	7	8	9	10
Frequency of Occurrence	10	6	3	4	1

a. Find the probability distribution of the random variable X, where X denotes the number of customers observed waiting in line.

b. Graph the histogram representing this probability distribution.

4. **TELEVISION PILOTS** After the private screening of a new television pilot, audience members were asked to rate the new show on a scale of 1 to 10 (10 being the highest rating). From a group of 140 people, these responses were obtained:

Rating	1	2	3	4	5
Frequency of Occurrence	1	4	3	11	23

Rating	6	7	8	9	10
Frequency of Occurrence	21	28	29	16	4

Let the random variable X denote the rating given to the show by a randomly chosen audience member. Find the probability distribution, and graph the histogram associated with these data.

10.2 Expected Value and Standard Deviation

Expected Value

The average value of a set of numbers is a familiar notion to most people. For example, to compute the average of the four numbers 12, 16, 23, and 37, we simply add these numbers and divide the resulting sum by 4, giving the average as

$$\frac{12 + 16 + 23 + 37}{4} = \frac{88}{4}$$

or 22. In general, we have the following definition.

> **Average, or Mean**
>
> The **average**, or **mean**, of the n numbers $x_1, x_2, \ldots, x_n$ is $\bar{x}$ (read "x bar"), where
>
> $$\bar{x} = \frac{x_1 + x_2 + \cdots + x_n}{n}$$

APPLIED EXAMPLE 1 Waiting Times The number of cars observed waiting in line at the beginning of each 2-minute interval between 3 P.M. and 5 P.M. on a certain Friday at the drive-in teller of Westwood Savings Bank and the corresponding frequency of occurrence are shown in Table 3. Find the average number of cars observed waiting in line at the beginning of each 2-minute interval during the 2-hour period.

TABLE 3									
Cars	0	1	2	3	4	5	6	7	8
Frequency of Occurrence	2	9	16	12	8	6	4	2	1

Solution Observe from Table 3 that the number 0 (of cars) occurs twice, the number 1 occurs nine times, and so on. Altogether, there are

$$2 + 9 + 16 + 12 + 8 + 6 + 4 + 2 + 1 = 60$$

numbers to be averaged. Therefore, the required average is given by

$$\frac{0 \cdot 2 + 1 \cdot 9 + 2 \cdot 16 + 3 \cdot 12 + 4 \cdot 8 + 5 \cdot 6 + 6 \cdot 4 + 7 \cdot 2 + 8 \cdot 1}{60} \approx 3.1 \qquad \textbf{(1)}$$

or approximately 3.1 cars.

Let's reconsider the expression in Equation (1) that gives the average of the frequency distribution shown in Table 3. Dividing each term by the denominator, we can rewrite the expression in the form

$$0 \cdot \left(\frac{2}{60}\right) + 1 \cdot \left(\frac{9}{60}\right) + 2 \cdot \left(\frac{16}{60}\right) + 3 \cdot \left(\frac{12}{60}\right) + 4 \cdot \left(\frac{8}{60}\right) + 5 \cdot \left(\frac{6}{60}\right)$$
$$+ 6 \cdot \left(\frac{4}{60}\right) + 7 \cdot \left(\frac{2}{60}\right) + 8 \cdot \left(\frac{1}{60}\right)$$

Observe that each term in the sum is a product of two factors. The first factor is the value assumed by the random variable X, where X denotes the number of cars observed waiting in line and the second factor is just the probability associated with that value of the random variable. This observation suggests the following general method for calculating the expected value (that is, the average, or mean) of a random variable X that assumes a finite number of values from the knowledge of its probability distribution.

> **Expected Value of a Discrete Random Variable X**
>
> Let X denote a random variable that assumes the values $x_1, x_2, \ldots, x_n$ with associated probabilities $p_1, p_2, \ldots, p_n$, respectively. Then the **expected value** of X, $E(X)$, is given by
>
> $$E(X) = x_1 p_1 + x_2 p_2 + \cdots + x_n p_n \qquad \textbf{(2)}$$

Note The numbers $x_1, x_2, \ldots, x_n$ may be positive, zero, or negative. For example, such a number will be positive if it represents a profit and negative if it represents a loss.

TABLE 4

Probability Distribution for the Random Variable X

x	$P(X = x)$
0	.03
1	.15
2	.27
3	.20
4	.13
5	.10
6	.07
7	.03
8	.02

 APPLIED EXAMPLE 2 Waiting Times Refer to Example 1. Let the random variable X denote the number of cars observed waiting in line. Use the probability distribution for the random variable X (Table 4) to solve Example 1 again.

Solution Let X denote the number of cars observed waiting in line. Then the average number of cars observed waiting in line is given by the expected value of X—that is, by

$$E(X) = 0 \cdot (.03) + 1 \cdot (.15) + 2 \cdot (.27) + 3 \cdot (.20) + 4 \cdot (.13)$$
$$+ 5 \cdot (.10) + 6 \cdot (.07) + 7 \cdot (.03) + 8 \cdot (.02)$$
$$= 3.1$$

or 3.1 cars, which agrees with the earlier results.

The expected value of a random variable X is a measure of the central tendency of the probability distribution associated with X. In repeated trials of an experiment with a random variable X, the average of the observed values of X gets closer and closer to the expected value of X as the number of trials gets larger and larger. Geometrically, the expected value of a random variable X has the following simple interpretation: If a laminate is made of the histogram of a probability distribution associated with a discrete random variable X, then the expected value of X corresponds to the point on the base of the laminate at which the laminate will balance perfectly when the point is directly over a fulcrum (Figure 6).

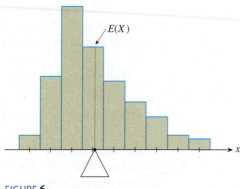

FIGURE 6
Expected value of a random variable X

Now, suppose X is a continuous random variable and f is the probability density function associated with it. For simplicity, let's first assume that $a \leq X \leq b$. Divide the interval $[a, b]$ into n subintervals of equal length $\Delta x = (b - a)/n$ by means of the $(n + 1)$ points $x_0 = a, x_1, x_2, \ldots, x_n = b$ (Figure 7). To find an approximation of the average value, or expected value, of X on the interval $[a, b]$, let's treat X as if it were a discrete random variable that takes on the values $x_1, x_2, \ldots, x_n$ with probabilities $p_1, p_2, \ldots, p_n$. Then,

$$E(X) \approx x_1 p_1 + x_2 p_2 + \cdots + x_n p_n$$

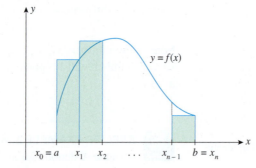

FIGURE 7
Approximating the expected value of a random variable X
on [a, b] by a Riemann sum

But p_1 is the probability that X is in the interval $[x_0, x_1]$, and this is just the area under the graph of f from $x = x_0$ to $x = x_1$, which may be approximated by $f(x_1) \Delta x$. The probabilities $p_2, \ldots, p_n$ may be approximated in a similar manner. Thus,

$$E(X) \approx x_1 f(x_1) \Delta x + x_2 f(x_2) \Delta x + \cdots + x_n f(x_n) \Delta x$$

which is seen to be the Riemann sum of the function $g(x) = xf(x)$ over the interval $[a, b]$. Letting n approach infinity, we obtain the following formula:

Expected Value of a Continuous Random Variable
Suppose the function f defined on the interval $[a, b]$ is the probability density function associated with a continuous random variable X. Then the **expected value** of X is

$$E(X) = \int_a^b xf(x)\, dx \tag{3}$$

If either $a = -\infty$ or $b = \infty$, then the integral in Equation (3) becomes an improper integral.

The expected value of a random variable plays an important role in many practical applications. For example, if X represents the life span of a certain brand of electronic components, then the expected value of X gives the average life span of these components. If X measures the waiting time in a doctor's office, then $E(X)$ gives the average waiting time, and so on.

APPLIED EXAMPLE 3 Life Span of Light Bulbs Show that if a continuous random variable X is exponentially distributed with the probability density function

$$f(x) = ke^{-kx} \qquad (0 \le x < \infty)$$

then the expected value $E(X)$ is equal to $1/k$. Using this result, determine the average life span of a 200-watt light bulb manufactured by TKK Products of Example 5, page 685.

Solution   We compute

$$E(X) = \int_0^\infty xf(x)\, dx$$

$$= \int_0^\infty kxe^{-kx}\, dx$$

$$= k \lim_{b \to \infty} \int_0^b xe^{-kx}\, dx$$

Integrating by parts with

$$u = x \quad \text{and} \quad dv = e^{-kx}\, dx$$

so that

$$du = dx \quad \text{and} \quad v = -\frac{1}{k}e^{-kx}$$

we have

$$E(X) = k \lim_{b \to \infty} \left(-\frac{1}{k}xe^{-kx} \bigg|_0^b + \frac{1}{k}\int_0^b e^{-kx}\, dx \right)$$

$$= k \lim_{b \to \infty} \left[-\left(\frac{1}{k}\right)be^{-kb} - \frac{1}{k^2}e^{-kx} \bigg|_0^b \right]$$

$$= k \lim_{b \to \infty} \left[-\left(\frac{1}{k}\right)be^{-kb} - \frac{1}{k^2}e^{-kb} + \frac{1}{k^2} \right]$$

$$= -\lim_{b \to \infty} \frac{b}{e^{kb}} - \frac{1}{k}\lim_{b \to \infty}\frac{1}{e^{kb}} + \frac{1}{k}\lim_{b \to \infty} 1$$

Now, by taking a sequence of values of b that approaches infinity—for example, $b = 10, 100, 1000, 10{,}000, \ldots$—we see that, for a fixed k,

$$\lim_{b \to \infty} \frac{b}{e^{kb}} = 0$$

(This limit can be proved by using l'Hôpital's Rule. See Appendix B.) Therefore,

$$E(X) = \frac{1}{k}$$

as we set out to show. From Example 5 of Section 10.1, we have $k = 0.001$. So we see that the average life span of the TKK light bulbs is $1/(0.001) = 1000$ hours.

Before considering another example, let's summarize the important result obtained in Example 3.

The Expected Value of an Exponential Random Variable

If a continuous random variable X is exponentially distributed with probability density function

$$f(x) = ke^{-kx} \qquad (0 \le x < \infty)$$

then the **expected (average) value** of X is given by

$$E(X) = \frac{1}{k}$$

Exploring with TECHNOLOGY

Refer to Example 3. Plot the graphs of $f(x) = x/e^{kx}$ for different positive values of k, using the appropriate viewing windows to demonstrate that $\lim\limits_{b \to \infty} 1/e^{kb} = 0$ for a fixed $k > 0$. Repeat for the function $f(x) = 1/e^{kx}$ to demonstrate that $\lim\limits_{b \to \infty} 1/e^{kb} = 0$.

APPLIED EXAMPLE 4 Airport Traffic On a typical Monday morning, the time between successive arrivals of planes at Jackson International Airport is an exponentially distributed random variable X with expected value of 10 (minutes).

a. Find the probability density function associated with X.
b. What is the probability that between 6 and 8 minutes will elapse between successive arrivals of planes?
c. What is the probability that the time between successive arrivals of planes will be more than 15 minutes?

Solution

a. Since X is exponentially distributed, the associated probability density function has the form $f(x) = ke^{-kx}$. Next, since the expected value of X is 10, we see that

$$E(X) = \frac{1}{k} = 10$$

$$k = \frac{1}{10}$$

$$= 0.1$$

so the required probability density function is

$$f(x) = 0.1e^{-0.1x}$$

b. The probability that between 6 and 8 minutes will elapse between successive arrivals is given by

$$P(6 \le X \le 8) = \int_6^8 0.1e^{-0.1x}\, dx = -e^{-0.1x}\Big|_6^8$$

$$= -e^{-0.8} + e^{-0.6}$$

$$\approx 0.10$$

c. The probability that the time between successive arrivals will be more than 15 minutes is given by

$$P(X > 15) = \int_{15}^{\infty} 0.1e^{-0.1x}\,dx$$

$$= \lim_{b\to\infty} \int_{15}^{b} 0.1e^{-0.1x}\,dx$$

$$= \lim_{b\to\infty} \left(-e^{-0.1x}\Big|_{15}^{b} \right)$$

$$= \lim_{b\to\infty} \left(-e^{-0.1b} + e^{-1.5} \right) = e^{-1.5}$$

$$\approx 0.22$$

Variance

The mean, or expected value, of a random variable enables us to express an important property of the probability distribution associated with the random variable in terms of a single number. But the knowledge of the location, or central tendency, of a probability distribution alone is usually not enough to give a reasonably accurate picture of the probability distribution. Consider, for example, the two probability distributions whose histograms appear in Figure 8. Both distributions have the same expected value, or mean, of $\mu = 4$ (the Greek letter μ is read "mu"). Note that the probability distribution with the histogram shown in Figure 8a is closely concentrated about its mean μ, whereas the one with the histogram shown in Figure 8b is widely dispersed or spread about its mean.

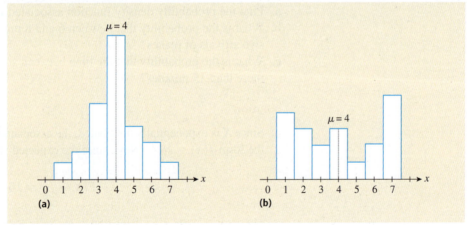

FIGURE 8
The histograms of two probability distributions

As another example, suppose that Olivia has ten packages of Brand *A* potato chips and ten packages of Brand *B* potato chips. After carefully measuring the weights of each package, she obtains the following results:

Weight (in ounces)									
Brand *A* 16.1	16.0	15.8	16.0	15.9	16.1	15.9	16.0	16.0	16.2
Brand *B* 16.3	15.7	15.8	16.2	15.9	16.1	15.7	16.2	16.0	16.1

In Example 7, we verify that the mean weights for each of the two brands is 16 ounces. However, a cursory examination of the data now shows that the weights of the Brand *B* packages exhibit much greater dispersion about the mean than do those of Brand *A*.

One measure of the degree of dispersion, or spread, of a probability distribution about its mean is given by the *variance* of the random variable associated with the probability distribution. A probability distribution with a small spread about its mean will have a small variance, whereas one with a larger spread will have a larger variance. Thus, the variance of the random variable associated with the probability distribution whose histogram appears in Figure 8a is smaller than the variance of the random variable associated with the probability distribution whose histogram is shown in Figure 8b (see Example 1). Also, as we will see in Example 7, the variance of the random variable associated with the weights of the Brand *A* potato chips is smaller than that of the random variable associated with the weights of the Brand *B* potato chips.

We now define the variance of a random variable.

Variance of a Random Variable *X*

Suppose a random variable has the probability distribution

x	x_1	x_2	x_3	$\cdots$	x_n
$P(X = x)$	p_1	p_2	p_3	$\cdots$	p_n

and expected value

$$E(X) = \mu$$

Then the **variance** of the random variable *X* is

$$\text{Var}(X) = p_1(x_1 - \mu)^2 + p_2(x_2 - \mu)^2 + \cdots + p_n(x_n - \mu)^2 \qquad (4)$$

Let's look a little closer at Equation (4). First, note that the numbers

$$x_1 - \mu, x_2 - \mu, \ldots, x_n - \mu \qquad (5)$$

measure the **deviations** of $x_1, x_2, \ldots, x_n$ from μ, respectively. Thus, the numbers

$$(x_1 - \mu)^2, (x_2 - \mu)^2, \ldots, (x_n - \mu)^2 \qquad (6)$$

measure the squares of the deviations of $x_1, x_2, \ldots, x_n$ from μ, respectively. Next, by multiplying each of the numbers in (6) by the probability associated with each value of the random variable X, the numbers are weighted accordingly so that their sum is a measure of the variance of X about its mean. An attempt to define the variance of a random variable about its mean in a similar manner using the deviations in (5), rather than their squares, would not be fruitful, since some of the deviations may be positive whereas others may be negative and hence (because of cancellations) the sum will not give a satisfactory measure of the variance of the random variable.

EXAMPLE 5 Find the variance of the random variable X and of the random variable Y whose probability distributions are shown in the following table. These are the probability distributions associated with the histograms shown in Figure 8a–b.

x	$P(X = x)$	y	$P(Y = y)$
1	.05	1	.2
2	.075	2	.15
3	.2	3	.1
4	.375	4	.15
5	.15	5	.05
6	.1	6	.1
7	.05	7	.25

Solution The mean of the random variable X is given by

$$\mu_X = (1)(.05) + (2)(.075) + (3)(.2) + (4)(.375) + (5)(.15)$$
$$+ (6)(.1) + (7)(.05)$$
$$= 4$$

Therefore, using Equation (4) and the data from the probability distribution of X, we find that the variance of X is given by

$$\text{Var}(X) = (.05)(1 - 4)^2 + (.075)(2 - 4)^2 + (.2)(3 - 4)^2$$
$$+ (.375)(4 - 4)^2 + (.15)(5 - 4)^2$$
$$+ (.1)(6 - 4)^2 + (.05)(7 - 4)^2$$
$$= 1.95$$

Next, we find that the mean of the random variable Y is given by

$$\mu_Y = (1)(.2) + (2)(.15) + (3)(.1) + (4)(.15) + (5)(.05)$$
$$+ (6)(.1) + (7)(.25)$$
$$= 4$$

so the variance of Y is given by

$$\text{Var}(Y) = (.2)(1 - 4)^2 + (.15)(2 - 4)^2 + (.1)(3 - 4)^2$$
$$+ (.15)(4 - 4)^2 + (.05)(5 - 4)^2$$
$$+ (.1)(6 - 4)^2 + (.25)(7 - 4)^2$$
$$= 5.2$$

Note that $\text{Var}(X)$ is smaller than $\text{Var}(Y)$, which confirms our earlier observations about the spread (or dispersion) of the probability distribution of X and Y, respectively.

Standard Deviation

Because Equation (4), which gives the variance of the random variable X, involves the squares of the deviations, the unit of measurement of $\text{Var}(X)$ is the square of the unit of measurement of the values of X. For example, if the values assumed by the random variable X are measured in units of a gram, then $\text{Var}(X)$ will be measured in units involving the *square* of a gram. To remedy this situation, one normally works with the square root of $\text{Var}(X)$ rather than $\text{Var}(X)$ itself. The former is called the standard deviation of X.

Standard Deviation of a Random Variable X

The **standard deviation of a random variable** X denoted σ (pronounced "sigma"), is defined by

$$\sigma = \sqrt{\text{Var}(X)}$$
$$= \sqrt{p_1(x_1 - \mu)^2 + p_2(x_2 - \mu)^2 + \cdots + p_n(x_n - \mu)^2} \qquad (7)$$

where $x_1, x_2, \ldots, x_n$ denote the values assumed by the random variable X and $p_1 = P(X = x_1), p_2 = P(X = x_2), \ldots, p_n = P(X = x_n)$.

EXAMPLE 6 Find the standard deviations of the random variables X and Y of Example 5.

Solution From the results of Example 5, we have $\text{Var}(X) = 1.95$ and $\text{Var}(Y) = 5.2$. Taking their respective square roots, we have

$$\sigma_X = \sqrt{1.95}$$
$$\approx 1.40$$
$$\sigma_Y = \sqrt{5.2}$$
$$\approx 2.28$$

$ **APPLIED EXAMPLE 7** Packaging Let X and Y denote the random variables whose values are the weights of the Brand A and Brand B potato chips, respectively (see page 700). Compute the means and standard deviations of X and Y and interpret your results.

Solution The probability distributions of X and Y may be computed from the given data as follows:

Brand A			Brand B		
x	Relative Frequency of Occurrence	$P(X = x)$	y	Relative Frequency of Occurrence	$P(Y = y)$
15.8	1	.1	15.7	2	.2
15.9	2	.2	15.8	1	.1
16.0	4	.4	15.9	1	.1
16.1	2	.2	16.0	1	.1
16.2	1	.1	16.1	2	.2
			16.2	2	.2
			16.3	1	.1

The means of X and Y are given by

$$\mu_X = (.1)(15.8) + (.2)(15.9) + (.4)(16.0) + (.2)(16.1)$$
$$+ (.1)(16.2)$$
$$= 16$$
$$\mu_Y = (.2)(15.7) + (.1)(15.8) + (.1)(15.9) + (.1)(16.0)$$
$$+ (.2)(16.1) + (.2)(16.2) + (.1)(16.3)$$
$$= 16$$

Therefore,

$$\text{Var}(X) = (.1)(15.8 - 16)^2 + (.2)(15.9 - 16)^2 + (.4)(16 - 16)^2$$
$$+ (.2)(16.1 - 16)^2 + (.1)(16.2 - 16)^2$$
$$= 0.012$$
$$\text{Var}(Y) = (.2)(15.7 - 16)^2 + (.1)(15.8 - 16)^2 + (.1)(15.9 - 16)^2$$
$$+ (.1)(16.0 - 16)^2 + (.2)(16.1 - 16)^2 + (.2)(16.2 - 16)^2$$
$$+ (.1)(16.3 - 16)^2$$
$$= 0.042$$

Explore and Discuss

A useful alternative formula for the variance is

$$\sigma^2 = E(X^2) - \mu^2$$

where $E(X^2)$ is the expected value of X^2.

1. Establish the validity of the formula.

2. Use the formula to verify the calculations in Example 7.

so the standard deviations are

$$\sigma_X = \sqrt{\text{Var}(X)}$$
$$= \sqrt{0.012}$$
$$\approx 0.11$$
$$\sigma_Y = \sqrt{\text{Var}(Y)}$$
$$= \sqrt{0.042}$$
$$\approx 0.20$$

The mean of X and that of Y are equal to 16. Therefore, the average weight of a package of potato chips of either brand is 16 ounces. However, the standard deviation of Y is greater than that of X. This tells us that the weights of the packages of Brand B potato chips are more widely dispersed about the common mean of 16 than are those of Brand A.

Explore and Discuss

Suppose the mean weight of m packages of Brand A potato chips is μ_1 and the standard deviation from the mean of their weight distribution is σ_1. Also suppose the mean weight of n packages of Brand B potato chips is μ_2 and the standard deviation from the mean of their weight distribution is σ_2.

1. Show that the mean of the combined weights of packages of Brand A and Brand B is

$$\mu = \frac{m\mu_1 + n\mu_2}{m + n}$$

2. If $\mu_1 = \mu_2$, show that the standard deviation from the mean of the combined-weight distribution is

$$\sigma = \left(\frac{m\sigma_1^2 + n\sigma_2^2}{m + n} \right)^{1/2}$$

3. Refer to Example 5, page 701. Using the results of parts 1 and 2, find the mean and the standard deviation of the combined-weight distribution.

The following example illustrates the technique for finding the mean and standard deviation for grouped data.

APPLIED EXAMPLE 8 Married Males The following table gives the number of married males in the United States aged 15 years and over but less than 65 years in 2011:

Age (in years)	15–19	20–34	35–44	45–54	55–64
Number (in thousands)	11,220	32,206	20,308	21,990	18,346

Source: U.S. Census Bureau.

Find the mean and the standard deviation for these data.

Solution

Let X denote the random variable that measures the number of married males. Taking X to be the midpoint of a group interval, we obtain the following probability distribution:

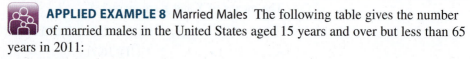

x	17	27	39.5	49.5	59.5
$P(X = x)$	$\left(\dfrac{11,220}{104,070} \right)$	$\left(\dfrac{32,206}{104,070} \right)$	$\left(\dfrac{20,308}{104,070} \right)$	$\left(\dfrac{21,990}{104,070} \right)$	$\left(\dfrac{18,346}{104,070} \right)$

The mean of X is

$$\mu = \left(\frac{11{,}220}{104{,}070}\right)(17) + \left(\frac{32{,}206}{104{,}070}\right)(27) + \left(\frac{20{,}308}{104{,}070}\right)(39.5)$$
$$+ \left(\frac{21{,}990}{104{,}070}\right)(49.5) + \left(\frac{18{,}346}{104{,}070}\right)(59.5)$$
$$\approx 38.8446$$

Next, we see that

$$\text{Var}(X) = \left(\frac{11{,}220}{104{,}070}\right)(17 - 38.8446)^2 + \left(\frac{32{,}206}{104{,}070}\right)(27 - 38.8446)^2$$
$$+ \left(\frac{20{,}308}{104{,}070}\right)(39.5 - 38.8446)^2 + \left(\frac{21{,}990}{104{,}070}\right)(49.5 - 38.8446)^2$$
$$+ \left(\frac{18{,}346}{104{,}070}\right)(59.5 - 38.8446)^2$$
$$\approx 194.1483$$

So the standard variation for these data is

$$\sigma = \sqrt{194.1483} \approx 13.93$$

Now suppose that X is a *continuous* random variable. Using an argument similar to that used earlier when we extended the result for the expected value from the discrete to the continuous case, we have the following definitions.

Variance and Standard Deviation of a Continuous Random Variable

Let X be a continuous random variable with probability density function $f(x)$ on $[a, b]$. Then the **variance** of X is

$$\text{Var}(X) = \int_a^b (x - \mu)^2 f(x)\, dx \tag{8}$$

and the **standard deviation** of X is

$$\sigma = \sqrt{\text{Var}(X)} \tag{9}$$

EXAMPLE 9 Find the expected value, variance, and standard deviation of the random variable X associated with the probability density function

$$f(x) = \frac{32}{15x^3}$$

on $[1, 4]$.

Solution Using Formula (3), we find the mean of X:

$$\mu = \int_a^b xf(x)\, dx = \int_1^4 x \cdot \frac{32}{15x^3}\, dx$$
$$= \frac{32}{15}\int_1^4 x^{-2}\, dx = \frac{32}{15}\left(-\frac{1}{x}\right)\Big|_1^4$$
$$= \frac{32}{15}\left(-\frac{1}{4} + 1\right) = \frac{8}{5}$$

Next, using Formula (8), we find

$$\text{Var}(X) = \int_a^b (x - \mu)^2 f(x) \, dx = \int_1^4 \left(x - \frac{8}{5} \right)^2 \cdot \frac{32}{15x^3} \, dx$$

$$= \frac{32}{15} \int_1^4 \left(x^2 - \frac{16}{5} x + \frac{64}{25} \right) \frac{1}{x^3} \, dx$$

$$= \frac{32}{15} \int_1^4 \left(\frac{1}{x} - \frac{16}{5} x^{-2} + \frac{64}{25} x^{-3} \right) dx$$

$$= \frac{32}{15} \left[\ln x + \frac{16}{5x} - \frac{32}{25x^2} \right]_1^4$$

$$= \frac{32}{15} \left[\left(\ln 4 + \frac{16}{5 \cdot 4} - \frac{32}{25 \cdot 16} \right) - \left(\ln 1 + \frac{16}{5} - \frac{32}{25} \right) \right]$$

$$= \frac{32}{15} \left(\ln 4 + \frac{4}{5} - \frac{2}{25} - \frac{16}{5} + \frac{32}{25} \right)$$

$$= \frac{32}{15} \left(\ln 4 - \frac{6}{5} \right) \approx 0.40$$

Finally, using Formula (9), we find the required standard deviation to be

$$\sigma = \sqrt{\text{Var}(X)} \approx \sqrt{0.40} \approx 0.63$$

Alternative Formula for Variance   Using Formula (8) to calculate the variance of a continuous random variable can be rather tedious. The following formula often makes this task easier:

$$\text{Var}(X) = \int_a^b x^2 f(x) \, dx - \mu^2 \tag{10}$$

This equation follows from these computations:

$$\text{Var}(X) = \int_a^b (x - \mu)^2 f(x) \, dx$$

$$= \int_a^b (x^2 - 2\mu x + \mu^2) f(x) \, dx$$

$$= \int_a^b x^2 f(x) \, dx - 2\mu \int_a^b x f(x) \, dx + \mu^2 \int_a^b f(x) \, dx$$

$$= \int_a^b x^2 f(x) \, dx - 2\mu \cdot \mu + \mu^2 \quad \text{Since } \int_a^b x f(x) \, dx = \mu \quad \text{and} \quad \int_a^b f(x) \, dx = 1$$

$$= \int_a^b x^2 f(x) \, dx - \mu^2$$

EXAMPLE 10   Use Formula (10) to calculate the variance of the random variable of Example 9.

Solution   Using Formula (10), we have

$$\text{Var}(X) = \int_a^b x^2 f(x) \, dx - \mu^2 = \int_1^4 x^2 \cdot \frac{32}{15x^3} \, dx - \left(\frac{8}{5} \right)^2$$

$$= \frac{32}{15} \int_1^4 \frac{1}{x} \, dx - \frac{64}{25} = \frac{32}{15} \ln x \Big|_1^4 - \frac{64}{25}$$

$$= \frac{32}{15} \ln 4 - \frac{64}{25} \approx 0.40$$

as obtained earlier.

Self-Check Exercise

Find the expected value, variance, and standard deviation of the random variable X associated with the probability density function $f(x) = 6(x - x^2)$ on $[0, 1]$.

The solution to Self-Check Exercise 10.2 can be found on page 709.

10.2 Concept Questions

1. a. What is the average or mean of a set of n numbers?
   b. What is the expected value of a discrete random variable?
   c. What is the expected value of a continuous random variable?

2. a. Define the variance $\text{Var}(X)$ of a discrete random variable X.
   b. What is the standard deviation σ of a discrete random variable X?

3. What are the variance and standard deviation of a continuous random variable?

10.2 Exercises

In Exercises 1–14, find the mean, variance, and standard deviation of the random variable x associated with the probability density function over the indicated interval.

1. $f(x) = \dfrac{1}{3}$; $[3, 6]$

2. $f(x) = \dfrac{1}{4}$; $[2, 6]$

3. $f(x) = \dfrac{3}{125} x^2$; $[0, 5]$

4. $f(x) = \dfrac{3}{8} x^2$; $[0, 2]$

5. $f(x) = \dfrac{3}{32} (x - 1)(5 - x)$; $[1, 5]$

6. $f(x) = 20(x^3 - x^4)$; $[0, 1]$

7. $f(x) = \dfrac{8}{7x^2}$; $[1, 8]$

8. $f(x) = \dfrac{4}{3x^2}$; $[1, 4]$

9. $f(x) = \dfrac{3}{14} \sqrt{x}$; $[1, 4]$

10. $f(x) = \dfrac{5}{2} x^{3/2}$; $[0, 1]$

11. $f(x) = \dfrac{3}{x^4}$; $[1, \infty)$

12. $f(x) = 3.5x^{-4.5}$; $[1, \infty)$

13. $f(x) = \dfrac{1}{4} e^{-x/4}$; $[0, \infty)$

Hint: $\lim\limits_{x \to \infty} x^n e^{kx} = 0$, if $k < 0$

14. $f(x) = \dfrac{1}{9} x e^{-x/3}$; $[0, \infty)$

Hint: $\lim\limits_{x \to \infty} x^n e^{kx} = 0$, if $k < 0$

15. SHOPPING HABITS The amount of time (in minutes) a shopper spends browsing in the magazine section of a supermarket is a continuous random variable with probability density function

$$f(t) = \frac{2}{25} t \qquad (0 \le t \le 5)$$

How much time is a shopper chosen at random expected to spend in the magazine section?

16. PROBABILITY OF RAINFALL The amount of rainfall (in inches) on a tropical island in the month of August is a continuous random variable with probability density function

$$f(x) = \frac{1}{36} x(6 - x) \qquad (0 \le x \le 6)$$

What is the expected amount of rainfall on any day in August on that island? Find the variance and standard deviation.

17. REACTION TIME OF A MOTORIST The amount of time (in seconds) it takes a motorist to react to a road emergency is a continuous random variable with probability density function

$$f(t) = \frac{9}{4t^3} \qquad (1 \le t \le 3)$$

What is the expected reaction time for a motorist chosen at random? Find the variance and the standard deviation.

18. DEMAND FOR BUTTER The quantity demanded (in thousands of pounds) of a certain brand of butter each week is a continuous random variable with probability density function

$$f(x) = \frac{6}{125} x(5 - x) \qquad (0 \le x \le 5)$$

What is the expected demand for this brand of butter each week? Find the variance and the standard deviation.

19. EXPECTED SNOWFALL The amount of snowfall in feet in a remote region of Alaska in the month of January is a continuous random variable with probability density function

$$f(x) = \frac{2}{9} x(3 - x) \qquad (0 \le x \le 3)$$

Find the amount of snowfall one can expect in any given month of January in Alaska. Find the variance and the standard deviation.

20. GAS SALES The amount of gas (in thousands of gallons) Al's Gas Station sells on a typical Monday is a continuous random variable with probability density function

$$f(x) = 4(x - 2)^3 \qquad (2 \le x \le 3)$$

How much gas can the gas station expect to sell each Monday? Find the variance and the standard deviation.

21. LIFE SPAN OF PLASMA TVS The life span (in years) of a certain brand of plasma TV is a continuous random variable with probability density function

$$f(t) = 9(9 + t^2)^{-3/2} \qquad (0 \le t < \infty)$$

How long is one of these plasma TVs expected to last?

22. AIRPORT TRAFFIC On a typical Wednesday morning, the time x (in minutes) between successive arrivals of planes at the Carson Municipal Airport is described by the probability density function

$$f(x) = \frac{1}{10} e^{-x/10} \qquad (0 \le x < \infty)$$

Suppose you saw a plane arriving while you were at the Carson Municipal Airport on a Wednesday morning. How long would it be before you could expect to see another plane arrive?

23. Recall that a random variable X is said to be *uniformly distributed* over the interval $[a, b]$ if it has probability density function

$$f(x) = \frac{1}{b - a} \qquad (a \le x \le b)$$

Find $E(X)$ and $V(X)$, and interpret your result.

24. COMMUTING Refer to Exercise 23. Bill takes the commuter train to work every day. During the morning commute, a train arrives every 15 min. If Bill arrives at the station at a random time for the morning commute, what will his average waiting time be?

25. EXPECTED DELIVERY TIME Refer to Exercise 23. A restaurant receives a delivery of pastries from a supplier each morning at a time that varies uniformly between 6 A.M. and 7 A.M.

  a. What is the probability that the delivery on a given morning will arrive between 6:30 A.M. and 6.45 A.M.?
  b. What is the expected time of delivery?

26. The probability density function f associated with a continuous random variable X has the form $f(x) = ax + b/x \, (1 \le x \le e)$. If $E(X) = 2$, find the values of a and b.

27. The probability function f associated with a continuous random variable X has the form $f(x) = ax^2 + bx \, (0 \le x \le 1)$. If $E(X) = 0.6$, find the values of a and b.

28. Find conditions on a, b, and c such that the function f defined by

$$f(x) = e^{-ax} (bx + c)$$

is a probability function on the interval $[0, \infty)$.
  Hint: Use integration by parts.

In Exercises 29–34, find the median of the random variable X with the probability density function defined on the indicated interval I. The median of X is defined to be the number m such that $P(X \le m) = \frac{1}{2}$. Observe that half of the X-values lie below m and the other half lie above m.

29. $f(x) = \frac{1}{6}$; $[2, 8]$ **30.** $f(x) = \frac{2}{15} x$; $[1, 4]$

31. $f(x) = \frac{3}{16} \sqrt{x}$; $[0, 4]$ **32.** $f(x) = \frac{1}{6\sqrt{x}}$; $[1, 16]$

33. $f(x) = \frac{1}{x^2}$; $[1, \infty)$ **34.** $f(x) = \frac{1}{2} e^{-x/2}$; $[0, \infty)$

In Exercises 35 and 36, determine whether the statement is true or false. If it is true, explain why it is true. If it is false, give an example to show why it is false.

35. If f is a probability density function of a continuous random variable X in the interval $[a, b]$, then the expected value of X is given by $\int_a^b x^2 f(x) \, dx$.

36. If f is a probability density function of a continuous random variable X in the interval $[a, b]$, then

$$\text{Var}(X) = \int_a^b x^2 f(x) \, dx - \left[\int_a^b x f(x) \, dx \right]^2$$

10.2 Solution to Self-Check Exercise

Using Formula (3), we find

$$\mu = \int_0^1 xf(x) = 6\int_0^1 (x^2 - x^3)\, dx$$

$$= 6\left[\frac{1}{3}x^3 - \frac{1}{4}x^4\right]\Big|_0^1$$

$$= 6\left(\frac{1}{3} - \frac{1}{4}\right) = \frac{1}{2}$$

Using Formula (8), we find

$$\text{Var}(X) = \int_0^1 (x - \mu)^2 f(x)\, dx$$

$$= \int_0^1 \left(x - \frac{1}{2}\right)^2 6(x - x^2)\, dx$$

$$= 6\int_0^1 \left(x^2 - x + \frac{1}{4}\right)(x - x^2)\, dx$$

$$= 6\int_0^1 \left(-x^4 + 2x^3 - \frac{5}{4}x^2 + \frac{1}{4}x\right) dx$$

$$= 6\left(-\frac{1}{5}x^5 + \frac{1}{2}x^4 - \frac{5}{12}x^3 + \frac{1}{8}x^2\right)\Big|_0^1$$

$$= 6\left(-\frac{1}{5} + \frac{1}{2} - \frac{5}{12} + \frac{1}{8}\right) = 0.05$$

Using Formula (9), we see that

$$\sigma = \sqrt{\text{Var}(X)} = \sqrt{0.05} \approx 0.22$$

USING TECHNOLOGY

Finding the Mean and Standard Deviation

The calculation of the mean and standard deviation of a random variable is facilitated by the use of a graphing utility.

 APPLIED EXAMPLE 1 Age Distribution of Company Directors A survey conducted in a certain year of the Fortune 1000 companies revealed the following age distribution of the company directors:

Age (in years)	20–24	25–29	30–34	35–39	40–44	45–49	50–54
Directors	1	6	28	104	277	607	1142

Age (in years)	55–59	60–64	65–69	70–74	75–79	80–84	85–89
Directors	1413	1424	494	159	62	31	5

Source: Directorship.

Let X denote the random variable taking on the values 1 through 14, where 1 corresponds to the age bracket 20–24, 2 corresponds to the age bracket 25–29, and so on.

a. Plot a histogram for the given data.
b. Find the mean and the standard deviation of these data. Interpret your results.

Solution

a. Enter the values of X as $x_1 = 1$, $x_2 = 2, \ldots, x_{14} = 14$ and the corresponding values of Y as $y_1 = 1$, $y_2 = 6, \ldots, y_{14} = 5$. Then using the **DRAW** function from the Statistics menu of a graphing utility, we obtain the histogram shown in Figure T1.

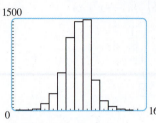

1500

0   16

FIGURE T1
The histogram for the given data,
using the viewing window
$[0, 16] \times [0, 1500]$

b. Using the appropriate function from the Statistics menu, we find that $\bar{x} \approx 7.9193$ and $\sigma x \approx 1.6378$; that is, the mean of X is $\mu \approx 7.9$, and the standard deviation is $\sigma \approx 1.6$. Interpreting our results, we see that the average age of the directors is in the 55- to 60-year-old bracket.

TECHNOLOGY EXERCISES

1. a. Graph the histogram associated with the random variable X in Example 5, page 701.
   b. Find the mean and the standard deviation for the random variable X.

2. a. Graph the histogram associated with the random variable Y in Example 5, page 701.
   b. Find the mean and the standard deviation for the random variable Y.

3. DRIVING AGE REQUIREMENTS The minimum age requirement for a regular driver's license differs from state to state. The frequency distribution for this age requirement in the 50 states is given in the following table:

Minimum Age	15	16	17	18	19	21
Frequency of Occurrence	1	15	4	28	1	1

   a. Describe a random variable X that is associated with these data.
   b. Find the probability distribution for the random variable X.
   c. Graph the histogram associated with these data.
   d. Find the mean and the standard deviation for these data.

4. The distribution of the number of chocolate chips in a cookie is shown in the following table:

Number of Chocolate Chips, x	0	1	2	3	4	5
$P(X = x)$	.01	.03	.05	.11	.13	.24

Number of Chocolate Chips, x	6	7	8
$P(X = x)$	.22	.16	.05

   a. Find the probability distribution for the random variable X.
   b. Graph the histogram associated with these data.
   c. Find the mean and the standard deviation for these data.

5. A sugar refiner uses a machine to pack sugar in 5-lb cartons. To check the machine's accuracy, cartons are selected at random and weighed. The results follow:

4.98	5.02	4.96	4.97	5.03
4.96	4.98	5.01	5.02	5.06
4.97	5.04	5.04	5.01	4.99
4.98	5.04	5.01	5.03	5.05
4.96	4.97	5.02	5.04	4.97
5.03	5.01	5.00	5.01	4.98

   a. Describe a random variable X that is associated with these data.
   b. Find the probability distribution for the random variable X.
   c. Find the mean and the standard deviation for these data.

6. The scores of 25 students in a mathematics examination follow:

90 85 74 92 68

94 66 87 85 70

72 68 73 72 69

66 58 70 74 88

90 98 71 75 68

a. Describe a random variable X that is associated with these data.

b. Find the probability distribution for the random variable X.

c. Find the mean and the standard deviation for these data.

10.3 Normal Distributions

Normal Distributions

In Section 10.2 we saw the useful role played by exponential density functions in many applications. In this section, we look at yet another class of continuous probability distributions, known as **normal distributions.** The normal distribution is without doubt the most important of all the probability distributions. Many phenomena, such as the heights of people in a given population, the weights of newborn infants, the IQs of college students, and the actual weights of 16-ounce packages of cereals, have probability distributions that are approximately normal. The normal distribution also provides us with an accurate approximation to the distributions of many random variables associated with random sampling problems.

The general **normal probability density function** with mean μ and standard deviation σ is defined to be

$$f(x) = \frac{e^{-(1/2)[(x-\mu)/\sigma]^2}}{\sigma\sqrt{2\pi}} \qquad (-\infty < x < \infty) \qquad \textbf{(11)}$$

The graph of a normal distribution, which is bell shaped, is called a **normal curve** (Figure 9).

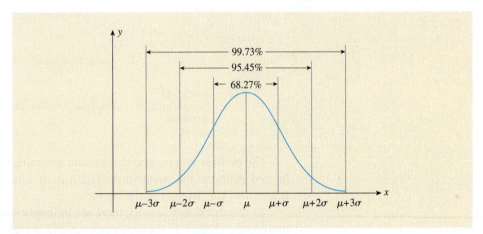

FIGURE 9
A normal curve

The normal curve (and therefore the corresponding normal distribution) is completely determined by its mean μ and standard deviation σ. In fact, the normal curve has the following characteristics, which are described in terms of these two parameters.

1. The curve has a peak at $x = \mu$.

2. The curve is symmetric with respect to the vertical line $x = \mu$.

3. The curve always lies above the x-axis but approaches the x-axis as x extends indefinitely in either direction.

4. The area under the curve is 1.

5. For any normal curve, 68.27% of the area under the curve lies within 1 standard deviation of the mean (that is, between $\mu - \sigma$ and $\mu + \sigma$), 95.45% of the area lies within 2 standard deviations of the mean, and 99.73% of the area lies within 3 standard deviations of the mean.

Figure 10 shows two normal curves with different means μ_1 and μ_2 but the same standard deviation. Figure 11 shows two normal curves with the same mean but different standard deviations σ_1 and σ_2. (Which number is smaller?)

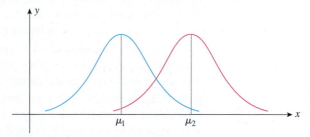

FIGURE 10
Two normal curves that have the same standard deviation but different means

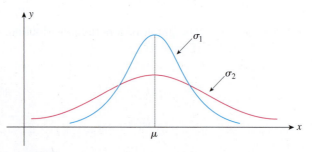

FIGURE 11
Two normal curves that have the same mean but different standard deviations

The mean μ of a normal distribution determines where the center of the curve is located, whereas the standard deviation σ of a normal distribution determines the peakedness (or flatness) of the curve.

As this discussion reveals, there are infinitely many normal curves corresponding to different choices of the parameters μ and σ that characterize such curves. Fortunately, any normal curve may be transformed into any other normal curve (as we will see later), so in the study of normal curves, it suffices to single out one such particular curve for special attention. The normal curve with mean $\mu = 0$ and standard deviation $\sigma = 1$ is called the **standard normal curve.** The corresponding distribution is called the **standard normal distribution.** The random variable itself is called the **standard normal random variable** and is commonly denoted by Z.

Consider the probability density function

$$f(x) = \frac{1}{\sqrt{2\pi}} e^{-x^2/2}$$

which is Equation (11) with $\mu = 0$ and $\sigma = 1$.

1. Use a graphing utility to plot the graph of f, using the viewing window $[-4, 4] \times [0, 0.5]$.
2. Use the numerical integration function of a graphing utility to find the area of the region under the graph of f on the intervals $[-1, 1]$, $[-2, 2]$, and $[-3, 3]$, thereby verifying property 5 of normal distributions for the special case in which $\mu = 0$ and $\sigma = 1$.

Computations of Probabilities Associated with Normal Distributions

Areas under the standard normal curve have been extensively computed and tabulated. Appendix C gives the areas of the regions under the standard normal curve to the left of the number z; these areas correspond to probabilities of the form $P(Z < z)$ or $P(Z \leq z)$. The next several examples illustrate the use of this table in computations involving the probabilities associated with the standard normal variable.

EXAMPLE 1 Let Z be the standard normal variable. Make a sketch of the appropriate region under the standard normal curve, and then find the values of

a. $P(Z < 1.24)$ **b.** $P(Z > 0.5)$
c. $P(0.24 < Z < 1.48)$ **d.** $P(-1.65 < Z < 2.02)$

Solution

a. The region under the standard normal curve associated with the probability $P(Z < 1.24)$ is shown in Figure 12. To find the area of the required region using Appendix C, we first locate the number 1.2 in the column and the number 0.04 in the row, both headed by z, and read off the number 0.8925 appearing in the body of the table. Thus,

$$P(Z < 1.24) = .8925$$

FIGURE 12
$P(Z < 1.24)$

b. The region under the standard normal curve associated with the probability $P(Z > 0.5)$ is shown in Figure 13a. Observe, however, that the required area is, by virtue of the symmetry of the standard normal curve, equal to the shaded area shown in Figure 13b. Thus,

$$P(Z > 0.5) = P(Z < -0.5)$$
$$= .3085$$

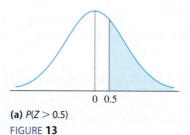

(a) $P(Z > 0.5)$

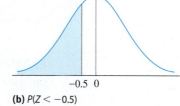

(b) $P(Z < -0.5)$

FIGURE 13

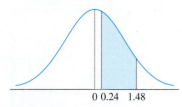

FIGURE 14
$P(0.24 < Z < 1.48)$

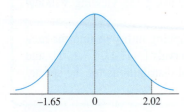

FIGURE 15
$P(-1.65 < Z < 2.02)$

c. The probability $P(0.24 < Z < 1.48)$ is equal to the shaded area shown in Figure 14. This area is obtained by subtracting the area under the curve to the left of $z = 0.24$ from the area under the curve to the left of $z = 1.48$; that is,

$$P(0.24 < Z < 1.48) = P(Z < 1.48) - P(Z < 0.24)$$
$$= .9306 - .5948$$
$$= .3358$$

d. The probability $P(-1.65 < Z < 2.02)$ is given by the shaded area shown in Figure 15. We have

$$P(-1.65 < Z < 2.02) = P(Z < 2.02) - P(Z < -1.65)$$
$$= .9783 - .0495$$
$$= .9288$$

Exploring with TECHNOLOGY

We can calculate the areas under the standard normal curve using the function normalcdf(. This will give a more accurate value than one obtained from the table. To call the function, press **2nd** **VARS** on the TI-83/84, then select 2:normal cdf(. For example, to compute $P(0.24 < Z < 1.48)$, enter

$$\texttt{normalcdf(.24,1.48)}$$

The TI-83/84 screen is shown in Figure 16. The answer (to three decimal places) agrees with the result obtained in Example 1c.

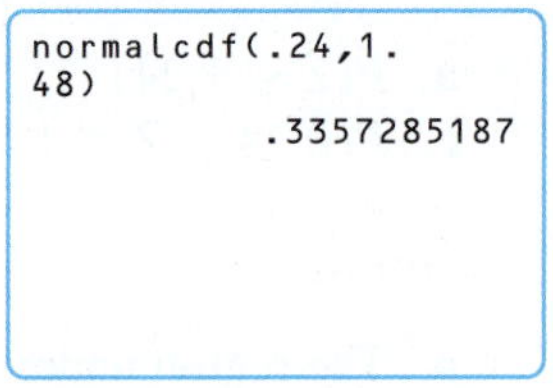

```
normalcdf(.24,1.
48)
            .3357285187
```

FIGURE 16

To find $P(Z < 1.24)$, we write $P(Z < 1.24) = .5 + P(0 < Z < 1.24)$ and enter

$$\texttt{.5+normalcdf(0,1.24)}$$

The TI-83/84 screen is shown in Figure 17. The answer agrees with the result obtained in Example 1a.

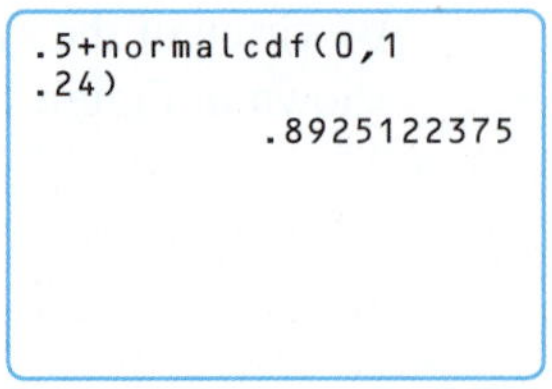

```
.5+normalcdf(0,1
.24)
            .8925122375
```

FIGURE 17

EXAMPLE 2 Let Z be the standard normal random variable. Find the value of z if z satisfies

a. $P(Z < z) = .9474$ **b.** $P(Z > z) = .9115$ **c.** $P(-z < Z < z) = .7888$

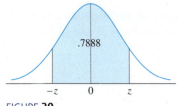

FIGURE **18**
$P(Z < z) = .9474$

Solution

a. Refer to Figure 18. We want the value of Z such that the area of the region under the standard normal curve and to the left of $Z = z$ is .9474. Locating the number .9474 in Appendix C, and reading back, we find that $z = 1.62$.

b. Since $P(Z > z)$, or equivalently, the area of the region to the right of z is greater than 0.5, it follows that z must be negative (Figure 19); hence, $-z$ is positive. Furthermore, the area of the region to the right of z is the same as the area of the region to the left of $-z$. Therefore,

$$P(Z > z) = P(Z < -z)$$
$$= .9115$$

Looking up the table, we find $-z = 1.35$, so $z = -1.35$.

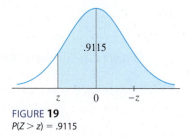

FIGURE **19**
$P(Z > z) = .9115$

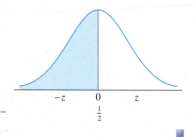

FIGURE **20**
$P(-z < Z < z) = .7888$

c. The region associated with $P(-z < Z < z)$ is shown in Figure 20. Observe that by symmetry, the area of this region is just double that of the area of the region between $Z = 0$ and $Z = z$; that is,

$$P(-z < Z < z) = 2P(0 < Z < z)$$

Furthermore,

$$P(0 < Z < z) = P(Z < z) - \frac{1}{2}$$

(Figure 21). Therefore,

$$\frac{1}{2}P(-z < Z < z) = P(Z < z) - \frac{1}{2}$$

or, solving for $P(Z < z)$, we have

$$P(Z < z) = \frac{1}{2} + \frac{1}{2}P(-z < Z < z)$$
$$= \frac{1}{2}(1 + .7888)$$
$$= .8944$$

Consulting the table, we find $z = 1.25$.

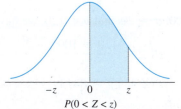

$P(0 < Z < z)$

$=$

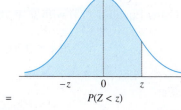

$P(Z < z)$

$-$

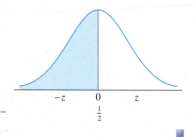

$\frac{1}{2}$

FIGURE **21**

We now turn our attention to the computation of probabilities associated with normal distributions whose means and standard deviations are not necessarily equal to 0 and 1, respectively. As was mentioned earlier, any normal curve may be transformed into the standard normal curve. In particular, it may be shown that if X is a normal random variable with mean μ and standard deviation σ, then it can be transformed into the standard normal random variable Z by means of the substitution

$$Z = \frac{X - \mu}{\sigma}$$

The area of the region under the normal curve (with random variable X) between $x = a$ and $x = b$ is *equal* to the area of the region under the standard normal curve between $z = (a - \mu)/\sigma$ and $z = (b - \mu)/\sigma$. In terms of probabilities associated with these distributions, we have

$$P(a < X < b) = P\left(\frac{a - \mu}{\sigma} < Z < \frac{b - \mu}{\sigma}\right) \tag{12}$$

(Figure 22). Similarly, we have

$$P(X < b) = P\left(Z < \frac{b - \mu}{\sigma}\right) \tag{13}$$

$$P(X > a) = P\left(Z > \frac{a - \mu}{\sigma}\right) \tag{14}$$

Thus, with the help of Equations (12)–(14), computations of probabilities associated with any normal distribution may be reduced to the computations of areas of regions under the standard normal curve.

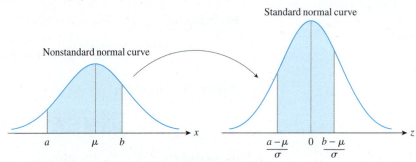

Area under the curve between a and b = Area under the curve between $\dfrac{a - \mu}{\sigma}$ and $\dfrac{b - \mu}{\sigma}$

FIGURE **22**

EXAMPLE 3 Suppose X is a normal random variable with $\mu = 100$ and $\sigma = 20$. Find the values of

a. $P(X < 120)$ **b.** $P(X > 70)$ **c.** $P(75 < X < 110)$

Solution

a. Using Equation (13) with $\mu = 100$, $\sigma = 20$, and $b = 120$, we have

$$P(X < 120) = P\left(Z < \frac{120 - 100}{20}\right)$$

$$= P(Z < 1) = .8413 \qquad \text{Use the table of values of Z in Appendix C.}$$

b. Using Equation (14) with $\mu = 100$, $\sigma = 20$, and $a = 70$, we have

$$P(X > 70) = P\left(Z > \frac{70 - 100}{20}\right)$$
$$= P(Z > -1.5) = P(Z < 1.5) = .9332$$

c. Using Equation (12) with $\mu = 100$, $\sigma = 20$, $a = 75$, and $b = 110$, we have

$$P(75 < X < 110) = P\left(\frac{75 - 100}{20} < Z < \frac{110 - 100}{20}\right)$$
$$= P(-1.25 < Z < 0.5)$$
$$= P(Z < 0.5) - P(Z < -1.25) \qquad \text{See Figure 23.}$$
$$= .6915 - .1056 = .5859$$

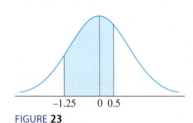

−1.25   0 0.5

FIGURE 23

Applications Involving Normal Random Variables

Next, we look at some applications involving the normal distribution.

APPLIED EXAMPLE 4 Birth Weights of Infants The medical records of infants delivered at the Kaiser Memorial Hospital show that the infants' birth weights in pounds are normally distributed with a mean of 7.4 and a standard deviation of 1.2. Find the probability that an infant selected at random from among those delivered at the hospital weighed more than 9.2 pounds at birth.

Solution Let X be the normal random variable denoting the birth weights of infants delivered at the hospital. Then the probability that an infant selected at random has a birth weight of more than 9.2 pounds is given by $P(X > 9.2)$. To compute $P(X > 9.2)$, we use Equation (14) with $\mu = 7.4$, $\sigma = 1.2$, and $a = 9.2$. We find that

$$P(X > 9.2) = P\left(Z > \frac{9.2 - 7.4}{1.2}\right) \qquad P(X > a) = P\left(Z > \frac{a - \mu}{\sigma}\right)$$
$$= P(Z > 1.5)$$
$$= P(Z < -1.5)$$
$$= .0668$$

Thus, the probability that an infant delivered at the hospital weighs more than 9.2 pounds is .0668.

APPLIED EXAMPLE 5 Packaging Idaho Natural Produce Corporation ships potatoes to its distributors in bags whose weights are normally distributed with a mean weight of 50 pounds and standard deviation of 0.5 pound. If a bag of potatoes is selected at random from a shipment, what is the probability that it weighs

a. more than 51 pounds?
b. less than 49 pounds?
c. between 49 and 51 pounds?

Solution Let X denote the weight of a bag of potatoes packed by the company. Then the mean and standard deviation of X are $\mu = 50$ and $\sigma = 0.5$, respectively.

a. The probability that a bag selected at random weighs more than 51 pounds is given by

$$P(X > 51) = P\left(Z > \frac{51 - 50}{0.5}\right) \qquad P(X > a) = P\left(Z > \frac{a - \mu}{\sigma}\right)$$
$$= P(Z > 2)$$
$$= P(Z < -2)$$
$$= .0228$$

b. The probability that a bag selected at random weighs less than 49 pounds is given by

$$P(X < 49) = P\left(Z < \frac{49 - 50}{0.5}\right) \qquad P(X < b) = P\left(Z < \frac{b - \mu}{\sigma}\right)$$
$$= P(Z < -2)$$
$$= .0228$$

c. The probability that a bag selected at random weighs between 49 and 51 pounds is given by

$$P(49 < X < 51)$$

$$= P\left(\frac{49 - 50}{0.5} < Z < \frac{51 - 50}{0.5}\right) \qquad \begin{aligned} & P(a < X < b) \\ & = P\left(\frac{a - \mu}{\sigma} < Z < \frac{b - \mu}{\sigma}\right) \end{aligned}$$

$$= P(-2 < Z < 2)$$
$$= P(Z < 2) - P(Z < -2)$$
$$= .9772 - .0228$$
$$= .9544$$

APPLIED EXAMPLE 6 College Admissions Eligibility The grade point average (GPA) of the senior class of Jefferson High School is normally distributed with a mean of 2.7 and a standard deviation of 0.4. If a senior in the top 10% of his or her class is eligible for admission to any of the nine campuses of the state university system, what is the minimum GPA that a senior should have to ensure eligibility for university admission?

Solution Let X denote the GPA of a randomly selected senior at Jefferson High School, and let x denote the minimum GPA that will ensure his or her eligibility for admission to the university. Since only the top 10% are eligible for admission, x must satisfy the equation

$$P(X \geq x) = .1$$

Using Equation (14) with $\mu = 2.7$ and $\sigma = 0.4$, we find that

$$P(X \geq x) = P\left(Z \geq \frac{x - 2.7}{0.4}\right) = .1 \qquad P(X > a) = P\left(Z > \frac{a - \mu}{\sigma}\right)$$

This is equivalent to the equation

$$P\left(Z < \frac{x - 2.7}{0.4}\right) = .9 \qquad \text{Why?}$$

Consulting Appendix C, we find that

$$\frac{x - 2.7}{0.4} \approx 1.28$$

Upon solving for x, we obtain

$$x = (1.28)(0.4) + 2.7$$
$$\approx 3.2$$

Thus, to ensure eligibility for admission to one of the nine campuses of the state university system, a senior at Jefferson High School should have a minimum of 3.2 GPA.

10.3 Self-Check Exercises

1. Let Z be a standard normal variable.
   a. Find the value of $P(-1.2 < Z < 2.1)$ by first making a sketch of the appropriate region under the standard normal curve.
   b. Find the value of z if z satisfies $P(-z < Z < z) = .8764$.

2. Let X be a normal random variable with $\mu = 80$ and $\sigma = 10$. Find the values of:
   a. $P(X < 100)$ b. $P(X > 60)$ c. $P(70 < X < 90)$

3. CHOLESTEROL LEVELS The serum cholesterol levels in milligrams per deciliter (mg/dL) in a current Mediterranean population are found to be normally distributed with a mean of 160 and a standard deviation of 50. Scientists at the National Heart, Lung, and Blood Institute consider this pattern ideal for a minimal risk of heart attacks. Find the percentage of the population having blood cholesterol levels between 160 and 180 mg/dL.

Solutions to Self-Check Exercises 10.3 can be found on page 721.

10.3 Concept Questions

1. Consider the following normal curve with mean μ and standard deviation σ:

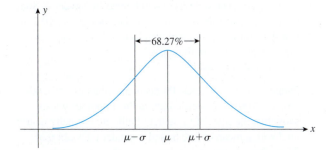

   a. What is the x-coordinate of the peak of the curve?
   b. What can you say about the symmetry of the curve?
   c. Does the curve always lie above the x-axis? What happens to the curve as x extends indefinitely to the left or right?
   d. What is the value of the area under the curve?
   e. Between what values does 68.27% of the area under the curve lie?

2. a. What is the difference between a normal curve and a standard normal curve?
   b. If X is a normal random variable with mean μ and standard deviation σ, write $P(a < X < b)$ in terms of the probabilities associated with the standard normal random variable Z.

10.3 Exercises

In Exercises 1–6, find the value of the probability of the standard normal variable Z corresponding to the shaded area under the standard normal curve.

1. $P(Z < 1.45)$

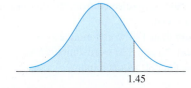

2. $P(Z > 1.11)$

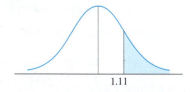

3. $P(Z < -1.75)$

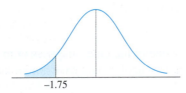

4. $P(0.30 < Z < 1.83)$

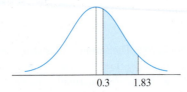

5. $P(-1.32 < Z < 1.74)$

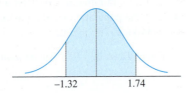

6. $P(-2.35 < Z < -0.51)$

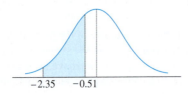

In Exercises 7–14, (a) make a sketch of the area under the standard normal curve corresponding to the probability and (b) find the value of the probability of the standard normal random variable Z corresponding to this area.

7. $P(Z < 1.38)$ **8.** $P(Z > 2.27)$

9. $P(Z < -0.64)$ **10.** $P(0.45 < Z < 1.75)$

11. $P(Z > -1.26)$ **12.** $P(-1.48 < Z < 1.54)$

13. $P(0.68 < Z < 2.02)$ **14.** $P(-1.41 < Z < -0.24)$

15. Let Z be the standard normal variable. Find the values of z if z satisfies:
  a. $P(Z < z) = .8907$ **b.** $P(Z < z) = .2090$

16. Let Z be the standard normal variable. Find the values of z if z satisfies:
  a. $P(Z > z) = .9678$ **b.** $P(-z < Z < z) = .8354$

17. Let Z be the standard normal variable. Find the values of z if z satisfies:
  a. $P(Z > -z) = .9713$ **b.** $P(Z < -z) = .9713$

18. Suppose X is a normal random variable with $\mu = 380$ and $\sigma = 20$. Find the value of:
  a. $P(X < 405)$ **b.** $P(400 < X < 430)$ **c.** $P(X > 400)$

19. Suppose X is a normal random variable with $\mu = 50$ and $\sigma = 5$. Find the value of:
  a. $P(X < 60)$ **b.** $P(X > 43)$ **c.** $P(46 < X < 58)$

20. Suppose X is a normal random variable with $\mu = 500$ and $\sigma = 75$. Find the value of:
  a. $P(X < 750)$ **b.** $P(X > 350)$ **c.** $P(400 < X < 600)$

21. MEDICAL RECORDS The medical records of infants delivered at Kaiser Memorial Hospital show that the infants' lengths at birth (in inches) are normally distributed with a mean of 20 and a standard deviation of 2.6. Find the probability that an infant selected at random from among those delivered at the hospital measures:
  a. More than 22 in.   **b.** Less than 18 in.
  c. Between 19 and 21 in.

22. FACTORY WORKERS' WAGES According to the data released by the Chamber of Commerce of a certain city, the weekly wages of factory workers are normally distributed with a mean of $720 and a standard deviation of $60. What is the probability that a factory worker selected at random from the city makes a weekly wage:
  a. Of less than $720?   **b.** Of more than $912?
  c. Between $660 and $780?

23. BOTTLING JAM The weights of jam bottled by Snyder & Sons are normally distributed with a mean of 16 oz and a standard deviation of 0.2 oz. If a bottle of this jam is selected at random from a shipment of this jam, what is the probability that it weighs:
  a. More than 16.1 oz?   **b.** Less than 15.8 oz?

24. BOTTLING JAM The weights of jam bottled by Snyder & Sons are normally distributed with a standard deviation of 0.2 oz. If just 1% of the jars of jam weighs less than 16 oz, what is the average weight of these jars?
  Hint: You need to find μ.

25. PRODUCT RELIABILITY TKK Products manufactures 50-, 60-, 75-, and 100-watt electric light bulbs. Laboratory tests show that the lives of these light bulbs are normally distributed with a mean of 750 hr and a standard deviation of 75 hr. What is the probability that a TKK light bulb selected at random will burn:
  a. For more than 900 hr?
  b. For less than 600 hr?
  c. Between 750 and 900 hr?
  d. Between 600 and 800 hr?

26. EDUCATION On average, a student takes dictation at a speed of 100 words/minute midway through an advanced court reporting course at the American Institute of Court Reporting. Assuming that the dictation speeds of the students are normally distributed and that the standard deviation is 20 words/minute, what is the probability that a student randomly selected from the course can take dictation at a speed:
  a. Of more than 120 words/minute?
  b. Between 80 and 120 words/minute?
  c. Of less than 80 words/minute?

27. IQs The IQs of students at Wilson Elementary School were measured recently and found to be normally distributed with a mean of 100 and a standard deviation of 15. What is the probability that a student selected at random will have an IQ:

a. Of 140 or higher?   **b.** Of 120 or higher?
c. Between 100 and 120?   **d.** Of 90 or less?

28. PRODUCT RELIABILITY The tread lives of the Super Titan radial tires under normal driving conditions are normally distributed with a mean of 40,000 mi and a standard deviation of 2000 mi. What is the probability that a tire selected at random will have a tread life of more than 35,000 mi? What is the probability that a tire selected at random will have a tread life between 35,000 and 45,000 mi?

29. FEMALE FACTORY WORKERS' WAGES According to data released by the Chamber of Commerce of a certain city, the weekly wages (in dollars) of female factory workers are normally distributed with a mean of 675 and a standard deviation of 50. Find the probability that a female factory worker selected at random from the city makes a weekly wage of $650 to $750.

30. CIVIL SERVICE EXAMS To be eligible for further consideration, applicants for certain civil service positions must first pass a written qualifying examination on which a score of 70 or more must be obtained. In a recent examination, it was found that the scores were normally distributed with a mean of 60 points and a standard deviation of 10 points. Determine the percentage of applicants who passed the written qualifying examination.

31. WARRANTIES The general manager of the service department of MCA Television has estimated that the time that elapses between the dates of purchase and the dates on which the 50-in. plasma TVs manufactured by the company first require service is normally distributed with a mean of 22 months and a standard deviation of 4 months. If the company gives a 1-year warranty on parts and labor for these TVs, determine the percentage of these TVs manufactured and sold by the company that will require service before the warranty period runs out.

32. GRADE DISTRIBUTIONS The scores on an economics examination are normally distributed with a mean of 72 and a standard deviation of 16. If the instructor assigns a grade of A to 10% of the class, what is the lowest score a student may have and still obtain an A?

33. GRADE DISTRIBUTIONS The scores on a sociology examination are normally distributed with a mean of 70 and a standard deviation of 10. If the instructor assigns As to 15%, Bs to 25%, Cs to 40%, Ds to 15%, and Fs to 5% of the class, find the cutoff points for grades A–D.

34. HIGHWAY SPEEDS The speeds (in miles per hour) of motor vehicles on a certain stretch of Route 3A as clocked at a certain place along the highway are normally distributed with a mean of 64.2 mph and a standard deviation of 8.44 mph. What is the probability that a motor vehicle selected at random is traveling at:

a. More than 65 mph?
b. Less than 60 mph?
c. Between 65 and 70 mph?

10.3 Solutions to Self-Check Exercises

1. a. The probability $P(-1.2 < Z < 2.1)$ is given by the shaded area in the accompanying figure:

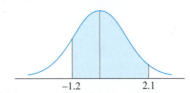

We have

$$P(-1.2 < Z < 2.1) = P(Z < 2.1) - P(Z < -1.2)$$
$$= .9821 - .1151$$
$$= .867$$

b. The region associated with $P(-z < Z < z)$ is shown in the accompanying figure:

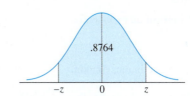

Observe that we have the following relationship:

$$P(Z < z) = \frac{1}{2}[1 + P(-z < Z < z)]$$

(see Example 2c). With $P(-z < Z < z) = .8764$, we find that

$$P(Z < z) = \frac{1}{2}(1 + .8764)$$
$$= .9382$$

Consulting Appendix C, we find that $z = 1.54$.

2. Using the transformations (12), (13), and (14) and the table of values of Z, we have

a. $P(X < 100) = P\left(Z < \dfrac{100 - 80}{10}\right)$

$= P(Z < 2)$

$= .9772$

b. $P(X > 60) = P\left(Z > \dfrac{60 - 80}{10}\right)$

$= P(Z > -2)$

$= P(Z < 2)$

$= .9772$

c. $P(70 < X < 90) = P\left(\dfrac{70 - 80}{10} < Z < \dfrac{90 - 80}{10}\right)$

$= P(-1 < Z < 1)$

$= P(Z < 1) - P(Z < -1)$

$= .8413 - .1587$

$= .6826$

3. Let X be the normal random variable denoting the serum cholesterol levels in milligrams per deciliter in the current Mediterranean population under consideration. Then the percentage of the population having blood cholesterol levels between 160 and 180 mg/dL is given by $P(160 < X < 180)$. To compute $P(160 < X < 180)$, we use Equation (12) with $\mu = 160$, $\sigma = 50$, $a = 160$, and $b = 180$. We find

$P(160 < X < 180) = P\left(\dfrac{160 - 160}{50} < Z < \dfrac{180 - 160}{50}\right)$

$= P(0 < Z < 0.4)$

$= P(Z < 0.4) - P(Z < 0)$

$= .6554 - .5000$

$= .1554$

Thus, approximately 15.5% of the population has blood cholesterol levels between 160 and 180 mg/dL.

CHAPTER 10	**Summary of Principal Formulas and Terms**

FORMULAS

1. Uniform density function	$f(x) = \dfrac{1}{b - a} \quad (a \le x \le b)$
2. Exponential density function	$f(x) = ke^{-kx}$
3. Expected value of a finite discrete random variable	$E(X) = x_1 p_1 + x_2 p_2 + \cdots + x_n p_n$
4. Expected value of a continuous random variable	$E(X) = \displaystyle\int_a^b xf(x)\,dx$
5. Expected value of an exponential random variable	$E(X) = \dfrac{1}{k}$
6. Variance of a continuous random variable	$\text{Var}(X) = \displaystyle\int_a^b (x - \mu)^2 f(x)\,dx$
7. Standard deviation of a continuous random variable	$\sigma = \sqrt{\text{Var}(X)}$

TERMS

experiment (680)

outcomes (sample points) (680)

sample space (680)

event (680)

probability of an event (680)

random variable (680)

discrete probability function (680)

histogram (680)

continuous random variable (681)

probability density function (682)

uniform density function (685)

exponential density function (686)

joint probability density function (687)

average (mean) (694)

expected value (695, 697, 699)

variance (701)

standard deviation of a random variable (702)

normal distribution (711)

normal curve (711)

standard normal random variable (712)

CHAPTER 10   Concept Review Questions

Fill in the blanks.

1. **a.** An activity with observable results is called a/an

    _____.

    b. The observable results are called _____ or _____

    _____.

    c. The totality of all outcomes is called a/an _____

    _____.

    d. The probability of an event is a number between

    _____ and _____.

2. **a.** A numerical value assigned to the outcome of a
    chance experiment is called a/an _____ variable.

    b. A random variable that assumes only a finite number
    of integer values is said to be _____ _____.

    c. A discrete probability function P with domain
    $\{x_1, x_2, \ldots, x_n\}$ satisfies the two conditions
    _____ $\leq P(x_i) \leq$ _____ for $1 \leq i \leq n$ and
    $P(x_1) + P(x_2) + \cdots + P(x_n) =$ _____.

3. **a.** A random variable that can assume any value in an
    interval is called a/an _____ random variable.

    b. A probability density function of a continuous random
    variable is a/an _____ function such that the total area
    of the region under the graph of f is equal to _____.
    The probability that an observed value of the random
    variable X lies in the interval $[a, b]$ is given by
    $P(a \leq X \leq b) =$ _____.

4. A joint probability density function of X and Y on D is a/an
    _____ _____ function $f(x, y)$ such that $\iint\limits_D f(x, y)\, dA =$ _____.

    The probability that the observed values of the random
    variables X and Y lie in a region $R \subset D$ is given by
    $P[(X, Y) \in R] =$ _____.

5. **a.** The average or mean of n numbers $x_1, x_2, \ldots, x_n$ is
    $\bar{x} =$ _____.

    b. The expected value of a random variable X is a mea-
    sure of the _____ _____ of the probability distribution
    associated with _____.

   c. The expected value of a discrete random variable X is
    $E(X) =$ _____, where $x_1, x_2, \ldots, x_n$ are assumed by
    _____ and $p_1, p_2, \ldots, p_n$ are the associated _____.

    d. The expected value of a continuous random variable X
    is $E(X) =$ _____, where f is the _____ _____ function
    defined on _____ that is associated with X.

6. **a.** The variance of a random variable is a measure of the
    degree of _____ of a probability distribution about its
    mean.

    b. If X denotes a random variable that assumes the
    values $x_1, x_2, \ldots, x_n$ with associated probabilities
    $p_1, p_2, \ldots, p_n$ and $\mu = E(X)$, then the variance of the
    random variable X is $\text{Var}(X) =$ _____.

    c. The standard deviation of a random variable X is
    $\sigma =$ _____.

7. If x is a continuous random variable with probability
    density function $f(x)$ on $[a, b]$, then the variance of X is
    $\text{Var}(X) =$ _____, and the standard deviation of X is
    $\sigma =$ _____.

8. A normal curve has the following properties.
    a. The curve has a peak at _____.
    b. The curve is symmetric with respect to the line _____.
    c. The curve always lies _____ the x-axis and approaches
    _____ _____ as x extends indefinitely in either
    direction.
    d. The area under the curve is _____.
    e. About 68.27% of the area under the curve lies within
    1 _____ _____ of the mean, 99.45% of the area lies
    within _____ standard deviations of the mean, and
    about 99.73% of the area lies within _____ standard
    deviations of the mean.

CHAPTER 10   Review Exercises

In Exercises 1–4, show that the function is a probability density
function on the given interval.

1. $f(x) = \dfrac{1}{28}(2x + 3)$; $[0, 4]$ 2. $f(x) = \dfrac{3}{16}\sqrt{x}$; $[0, 4]$

3. $f(x) = \dfrac{1}{4}$; $[7, 11]$ 4. $f(x) = \dfrac{4}{x^5}$; $[1, \infty)$

In Exercises 5–8, find the value of the constant k such that the
function is a probability density function on the given interval.

5. $f(x) = kx^2$; $[0, 9]$ 6. $f(x) = \dfrac{k}{\sqrt{x}}$; $[1, 16]$

7. $f(x) = \dfrac{k}{x^2}$; $[1, 3]$ 8. $f(x) = \dfrac{k}{x^{2.5}}$; $[1, \infty)$

In Exercises 9–12, f is a probability density function defined on
the given interval. Find the indicated probabilities.

9. $f(x) = \dfrac{1}{4}$; $[1, 5]$

    a. $P(2 \leq X \leq 4)$ **b.** $P(X \leq 3)$ **c.** $P(X \geq 2)$

10. $f(x) = \dfrac{2}{21}x$; $[2, 5]$

    a. $P(X \leq 4)$ **b.** $P(X = 4)$ **c.** $P(3 \leq X \leq 4)$

11. $f(x) = \dfrac{3}{16}\sqrt{x}; [0, 4]$

   a. $P(1 \le X \le 3)$ **b.** $P(X \le 3)$ **c.** $P(X = 2)$

12. $f(x) = \dfrac{1}{x^2}; [1, \infty)$

   a. $P(X \le 10)$ **b.** $P(2 \le X \le 4)$ **c.** $P(X \ge 2)$

In Exercises 13–16, find the mean, variance, and standard deviation of the random variable X associated with the probability density function f over the given interval.

13. $f(x) = \dfrac{1}{5}; [2, 7]$ **14.** $f(x) = \dfrac{1}{28}(2x + 3); [0, 4]$

15. $f(x) = \dfrac{1}{4}(3x^2 + 1); [-1, 1]$

16. $f(x) = \dfrac{4}{x^5}; [1, \infty)$

In Exercises 17–20, Z is the standard normal random variable. Find the given probability.

17. $P(Z < 2.24)$ **18.** $P(Z > -1.24)$

19. $P(0.24 \le Z \le 1.28)$ **20.** $P(-1.37 \le Z \le 1.37)$

21. Show that the function

$$f(x, y) = \frac{1}{64}x^{1/2}y^{1/3}; D = \{0 \le x \le 4; 0 \le y \le 8\}$$

   is a joint probability density function on D.

22. Given that

$$f(x, y) = x(2 - y); D = \{(x, y)\,|\,0 \le x \le 1; 0 \le y \le 2\}$$

   is a joint probability density function on D, find the indicated probabilities.

   a. $P(0 \le X \le \frac{1}{2}; 0 \le Y \le 1)$
   b. $P(Y \le 2X)$

23. Suppose X is a normal random variable with $\mu = 80$ and $\sigma = 8$. Find these values:

   a. $P(X \le 84)$ **b.** $P(X \ge 70)$ **c.** $P(75 \le X \le 85)$

24. Suppose X is a normal random variable with $\mu = 45$ and $\sigma = 3$. Find these values:

   a. $P(X \le 50)$ **b.** $P(X \ge 40)$ **c.** $P(40 \le X \le 50)$

25. PACKAGING COFFEE The weights of coffee in 1-lb packages sold at Star Supermarket are normally distributed with a mean of 16 oz and a standard deviation of 0.1 oz. If a package of the coffee is selected at random from the shelf, what is the probability that it weighs:

   a. More than 16.3 oz?
   b. Less than 15.7 oz?
   c. Between 15.7 and 16.3 oz?

26. LENGTH OF A HOSPITAL STAY Records at Centerville Hospital indicate that the length of time in days that a maternity patient stays in the hospital has a probability density function given by

$$f(t) = \frac{1}{4}e^{-t/4}$$

   a. What is the probability that a woman entering the maternity wing will be there more than 6 days?
   b. What is the probability that a woman entering the maternity wing will be there less than 2 days?
   c. What is the average length of time that a woman entering the maternity wing stays in the hospital?

27. LIFE SPAN OF A PLANT The life span of a certain plant species (in days) is described by the probability density function

$$f(x) = \frac{1}{100}e^{-x/100}$$

   If a plant of this species is selected at random, how long can the plant be expected to live?

28. EXPECTED DELIVERY TIME A restaurant receives a delivery of pastries from a supplier each morning at a time that varies uniformly between 6 A.M. and 7:30 A.M.

   a. What is the probability that the delivery on a given morning will arrive between 6:30 A.M. and 7 A.M.?
   b. What is the expected time of delivery?

29. LIFE SPAN OF A CAR BATTERY The life span (in years) of a certain make of car battery is an exponentially distributed random variable with an expected value of 5. Find the probability that the life span of a battery is (a) less than 4 years, (b) more than 6 years, and (c) between 2 and 4 years.

30. The probability density function associated with a continuous random variable X has the form $f(x) = a + bx^2$ $(0 \le x \le 1)$. If $E(X) = 0.6$, find the values of a and b.

CHAPTER 10 Before Moving On . . .

1. Find the value of the constant k such that the function $f(x) = k(x + 2x^2)$ is a probability density function on the interval $[0, 1]$.

2. If $f(x) = 0.1e^{-0.1x}$ is a probability density function on the interval $[0, \infty)$, find $P(5 < X < 10)$.

3. Find the mean, variance, and standard deviation of the random variable X associated with the probability density function $f(x) = \dfrac{1}{4\sqrt{x}}$ over the interval $[1, 9]$.

4. Suppose X is a normal random variable with $\mu = 40$ and $\sigma = 4$. Find the following values:

   a. $P(X < 50)$ **b.** $P(X > 35)$ **c.** $P(30 < X < 50)$

11 Taylor Polynomials and Infinite Series

I N THIS CHAPTER, we see how certain functions can be represented by a *power series*. A power series involves infinitely many terms, but when truncated, it is just a polynomial. By approximating a function with a *Taylor polynomial,* we are often able to obtain approximate solutions to problems that we cannot otherwise solve.

We also look at *Newton's method* for finding the zeros of a function. For example, Newton's method can be used to find the critical numbers of a function, which, as you may recall, are candidates for the solution of the optimization problems considered in Chapter 4.

What effect will a proposed $20 billion tax cut have on the economy over the long run? In Example 6, page 751, we will see how the sum of an infinite series can be used to answer this question.

Sean Locke/iStockphoto.com

11.1   Taylor Polynomials

As we saw earlier, it is not always possible to obtain an exact solution to a problem; in such cases, we have to settle for an approximate solution. In this section, we see how a function can be approximated near a given point by a polynomial. Polynomials, as we have seen time and again, are easy to work with; for example, they are easy to evaluate, differentiate, and integrate. So by using a polynomial rather than the function itself, we can often obtain an approximate solution to a problem that we might otherwise not be able to solve.

Taylor Polynomials

Suppose we are given a differentiable function f and a number a in the domain of f. Then the polynomial of degree 0 that best approximates f near $x = a$ is the constant polynomial

$$P_0(x) = f(a)$$

which coincides with f at a (Figure 1a).

Now unless f is a constant function, it is possible in many cases to obtain a better approximation of f near $x = a$ by using a polynomial function of degree 1. Recall that the linear function

$$L(x) = f(a) + f'(a)(x - a) \qquad (x^2) \text{ See page 140.}$$

is just an equation of the tangent line to the graph of the function f at the point $(a, f(a))$ (Figure 1b). As such, the value of the function L coincides with the value of f at a, and its slope coincides with the slope of f at a; that is,

$$L(a) = f(a) \quad \text{and} \quad L'(a) = f'(a)$$

Let's write

$$L(x) = P_1(x)$$

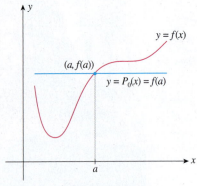

(a) $P_0(x) = f(a)$ is a zero-degree polynomial that approximates f near $x = a$.

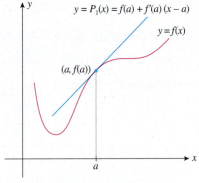

(b) $P_1(x) = f(a) + f'(a)(x - a)$ is a first-degree polynomial that approximates f near $x = a$.

FIGURE 1

Then the polynomial approximation of degree 1 of f near $x = a$ is

$$P_1(x) = f(a) + f'(a)(x - a)$$

This discussion suggests that yet a better approximation to f at a may be found by using a polynomial of degree 2, $P_2(x)$, and requiring that its value, slope, and concavity coincide with those of f at a. In other words, $P_2(x)$ should satisfy the three conditions

$$P_2(a) = f(a), \quad P_2'(a) = f'(a), \quad P_2''(a) = f''(a)$$

The third condition ensures that the graph of the polynomial bends in the right way, at least near $x = a$. Pursuing this line of reasoning, we are led to the search for a polynomial of degree n in $x - a$,

$$P_n(x) = a_0 + a_1(x - a) + a_2(x - a)^2 \\ + a_3(x - a)^3 + \cdots + a_n(x - a)^n$$

(where $a_0, a_1, \ldots, a_n$ are constants), that satisfies the conditions

$$P_n(a) = f(a), \quad P_n'(a) = f'(a), \quad P_n''(a) = f''(a), \quad \ldots, \quad P_n^{(n)}(a) = f^{(n)}(a) \quad \textbf{(1)}$$

To find the required polynomial, we compute

$$P_n'(x) = a_1 + 2a_2(x - a) + 3a_3(x - a)^2 + \cdots + na_n(x - a)^{n-1}$$
$$P_n''(x) = 2a_2 + 3 \cdot 2a_3(x - a) + \cdots + n(n - 1)a_n(x - a)^{n-2}$$
$$P_n'''(x) = 3 \cdot 2a_3 + 4 \cdot 3 \cdot 2a_4(x - a) + \cdots + n(n - 1)(n - 2)a_n(x - a)^{n-3}$$
$$\vdots$$
$$P_n^{(n)}(x) = n(n - 1)(n - 2) \cdots (1)a_n$$

Setting $x = a$ in each of the expressions for $P_n(x), P_n'(x), P_n''(x), \ldots, P_n^{(n)}(x)$ in succession and using the conditions in (1), we find

$$P_n(a) = a_0 = f(a)$$
$$P_n'(a) = a_1 = f'(a)$$
$$P_n''(a) = 2a_2 = f''(a)$$
$$P_n'''(a) = 3 \cdot 2a_3 = f'''(a)$$
$$\vdots$$
$$P_n^{(n)}(a) = n(n - 1)(n - 2) \cdots (1)a_n = f^{(n)}(a)$$

from which we deduce that

$$a_0 = f(a), \quad a_1 = f'(a), \quad a_2 = \frac{1}{2}f''(a), \quad a_3 = \frac{1}{3 \cdot 2}f'''(a), \quad \ldots,$$

$$a_n = \frac{1}{n(n - 1)(n - 2) \cdots (1)}f^{(n)}(a)$$

Let's introduce the expression $n!$ (read "n factorial"), defined by

$$n! = n(n - 1)(n - 2)(n - 3) \cdots 3 \cdot 2 \cdot 1 \quad \text{(for } n \geq 1)$$
$$0! = 1$$

Thus,

$$1! = 1$$
$$2! = 2 \cdot 1 = 2$$
$$3! = 3 \cdot 2 \cdot 1 = 6$$
$$4! = 4 \cdot 3 \cdot 2 \cdot 1 = 24$$
$$5! = 5 \cdot 4 \cdot 3 \cdot 2 \cdot 1 = 120$$

and so on. Using this notation, we may write the coefficients of $P_n(x)$ as

$$a_0 = f(a), \quad a_1 = f'(a), \quad a_2 = \frac{1}{2!}f''(a), \quad a_3 = \frac{1}{3!}f'''(a), \quad \ldots, \quad a_n = \frac{1}{n!}f^{(n)}(a)$$

so the required polynomial is

$$P_n(x) = f(a) + f'(a)(x-a) + \frac{f''(a)}{2!}(x-a)^2 + \cdots + \frac{f^{(n)}(a)}{n!}(x-a)^n$$

> **The *n*th Taylor Polynomial**
>
> Suppose that the function f and its first n derivatives are defined at a. Then the nth Taylor polynomial of f at a is the polynomial
>
> $$P_n(x) = f(a) + f'(a)(x-a)$$
> $$+ \frac{f''(a)}{2!}(x-a)^2 + \cdots + \frac{f^{(n)}(a)}{n!}(x-a)^n \qquad (2)$$
>
> which coincides with $f(x), f'(x), \ldots, f^{(n)}(x)$ at a; that is,
>
> $$P_n(a) = f(a), \quad P_n'(a) = f'(a), \quad \ldots, \quad P_n^{(n)}(a) = f^{(n)}(a)$$

Using a Taylor Polynomial to Approximate a Function

In many instances, the Taylor polynomial $P_n(x)$ provides us with a good approximation of $f(x)$ near $x = a$.

EXAMPLE 1 Find the Taylor polynomials P_1, P_2, P_3, and P_4 of $f(x) = e^x$ at $x = 0$, and sketch the graph of each polynomial superimposed upon the graph of $f(x) = e^x$.

Solution Here, $a = 0$, and since

$$f(x) = f'(x) = f''(x) = f'''(x) = f^{(4)}(x) = e^x$$

we find

$$f(0) = f'(0) = f''(0) = f'''(0) = f^{(4)}(0) = 1$$

Using Equation (2) with $n = 1, 2, 3$, and 4 in succession, we find that the required Taylor polynomials are

$$P_1(x) = f(0) + f'(0)(x - 0) = 1 + x$$

$$P_2(x) = f(0) + f'(0)(x - 0) + \frac{f''(0)}{2!}(x - 0)^2 = 1 + x + \frac{1}{2}x^2$$

$$P_3(x) = f(0) + f'(0)(x - 0) + \frac{f''(0)}{2!}(x - 0)^2 + \frac{f'''(0)}{3!}(x - 0)^3$$

$$= 1 + x + \frac{1}{2}x^2 + \frac{1}{6}x^3$$

$$P_4(x) = f(0) + f'(0)(x - 0) + \frac{f''(0)}{2!}(x - 0)^2 + \frac{f'''(0)}{3!}(x - 0)^3$$

$$+ \frac{f^{(4)}(0)}{4!}(x - 0)^4 = 1 + x + \frac{1}{2}x^2 + \frac{1}{6}x^3 + \frac{1}{24}x^4$$

The graphs of these polynomials are shown in Figure 2. Observe that the approximation of $f(x)$ near $x = 0$ improves as the degree of the approximating Taylor polynomial increases.

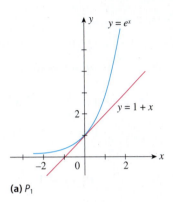

(a) P_1

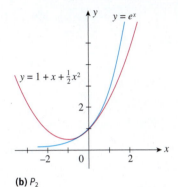

(b) P_2

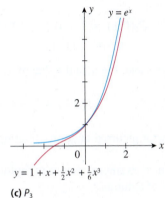

(c) P_3

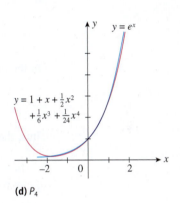

(d) P_4

FIGURE **2**
The graphs of the first four Taylor polynomials of $f(x) = e^x$ at $x = 0$ superimposed upon the graph of $f(x) = e^x$.

Exploring with TECHNOLOGY

Let $f(x) = xe^{-x}$.

1. Find the first four Taylor polynomials of f at $x = 0$.
2. Use a graphing utility to plot the graphs of f, P_1, P_2, P_3, and P_4 on the same set of axes in the viewing window $[-0.5, 1] \times [-0.5, 0.5]$.
3. Comment on the approximation of f by the polynomial P_n near $x = 0$ for $n = 1, 2, 3,$ and 4. What happens to the approximation if x is "far" from the origin?

EXAMPLE 2 Find the nth Taylor polynomial of $f(x) = 1/(1 - x)$ at $x = 0$. Compute $P_5(0.1)$, and compare this result with $f(0.1)$.

Solution   Here, $a = 0$. We find

$$f(x) = \frac{1}{1 - x} = (1 - x)^{-1}$$

$$f'(x) = -(1 - x)^{-2}(-1) = (1 - x)^{-2}$$

$$f''(x) = -2(1 - x)^{-3}(-1) = 2(1 - x)^{-3}$$

$$f'''(x) = 3 \cdot 2(1 - x)^{-4}$$

$$f^{(4)}(x) = 4 \cdot 3 \cdot 2(1 - x)^{-5}$$

$$\vdots$$

$$f^{(n)}(x) = n(n - 1)(n - 2) \cdots (1)(1 - x)^{-n-1}$$

so

$$f(0) = 1, \ f'(0) = 1, \ f''(0) = 2, \ f'''(0) = 3!, \quad f^{(4)}(0) = 4!, \ \ldots, \ f^{(n)}(0) = n!$$

Therefore, the nth Taylor polynomial is

$$P_n(x) = 1 + \frac{1}{1}x + \frac{2}{2!}x^2 + \frac{3!}{3!}x^3 + \frac{4!}{4!}x^4 + \cdots + \frac{n!}{n!}x^n$$

$$= 1 + x + x^2 + x^3 + x^4 + \cdots + x^n$$

In particular,

$$P_5(x) = 1 + x + x^2 + x^3 + x^4 + x^5$$

so

$$P_5(0.1) = 1 + 0.1 + (0.1)^2 + (0.1)^3 + (0.1)^4 + (0.1)^5$$

$$= 1.11111$$

On the other hand, the actual value of $f(x)$ at $x = 0.1$ is

$$f(0.1) = \frac{1}{1 - 0.1} = \frac{1}{0.9} = 1.111\ldots$$

Thus, the error incurred in approximating $f(0.1)$ by $P_5(0.1)$ is $0.00000111\ldots$. ◼

In the next example, we use a Taylor polynomial to approximate the value of the square root of a number.

EXAMPLE 3 Obtain the second Taylor polynomial of $f(x) = \sqrt{x}$ at $x = 25$, and use it to approximate the value of $\sqrt{26.5}$.

Solution Here, $a = 25$ and $n = 2$. Since

$$f(x) = x^{1/2}, \ \ f'(x) = \frac{1}{2}x^{-1/2} = \frac{1}{2\sqrt{x}}, \ \ f''(x) = -\frac{1}{4}x^{-3/2} = -\frac{1}{4x^{3/2}}$$

we find that

$$f(25) = 5, \ \ f'(25) = \frac{1}{10}, \ \ f''(25) = -\frac{1}{500}$$

Using Formula (2), we obtain the required polynomial

$$P_2(x) = f(25) + f'(25)(x - 25) + \frac{f''(25)}{2!}(x - 25)^2$$

$$= 5 + \frac{1}{10}(x - 25) - \frac{1}{1000}(x - 25)^2$$

Next, using $P_2(x)$ as an approximation of $f(x)$ near $x = 25$, we find

$$\sqrt{26.5} = f(26.5) \approx P_2(26.5)$$

$$= 5 + \frac{1}{10}(26.5 - 25) - \frac{1}{1000}(26.5 - 25)^2$$

$$= 5.14775$$

The exact value of $\sqrt{26.5}$, rounded off to five decimal places, is 5.14782. Thus, the error incurred in the approximation is 0.00007. ◼

The next example shows how a Taylor polynomial can be used to approximate an integral that involves an integrand whose antiderivative cannot be expressed as an elementary function.

Explore and Discuss

Suppose you want to find an approximation to the value of $\sqrt[3]{26.5}$ using a second Taylor polynomial of a function f expanded about a.

1. How would you choose f and a?

2. Approximate $\sqrt[3]{26.5}$, and compare your result with that obtained by using a calculator.

APPLIED EXAMPLE 4 Growth of Service Industries It has been estimated that service industries, which currently make up 30% of the nonfarm workforce in a certain country, will continue to grow at the rate of

$$R(t) = 5e^{1/(t+1)}$$

percent per decade, t decades from now. Estimate the percentage of the nonfarm workforce in service industries one decade from now.

Solution The percentage of the nonfarm workforce in service industries t decades from now will be given by

$$P(t) = \int 5e^{1/(t+1)}\, dt \qquad [P(0) = 30]$$

This integral cannot be expressed in terms of an elementary function. To obtain an approximate solution to the problem, let's first make the substitution

$$u = \frac{1}{t+1}$$

so

$$t + 1 = \frac{1}{u} \qquad t = \frac{1}{u} - 1 \qquad \text{and} \qquad dt = -\frac{1}{u^2}\, du$$

The integral becomes

$$F(u) = 5\int e^u\left(-\frac{du}{u^2}\right) = -5\int \frac{e^u}{u^2}\, du$$

Next, let's approximate e^u at $u = 0$ by a fourth-degree Taylor polynomial. From Example 1, we have

$$e^u \approx 1 + u + \frac{u^2}{2!} + \frac{u^3}{3!} + \frac{u^4}{4!}$$

Thus,

$$F(u) \approx -5\int \frac{1}{u^2}\left(1 + u + \frac{u^2}{2} + \frac{u^3}{6} + \frac{u^4}{24}\right) du$$

$$= -5\int\left(\frac{1}{u^2} + \frac{1}{u} + \frac{1}{2} + \frac{u}{6} + \frac{u^2}{24}\right) du$$

$$= -5\left(-\frac{1}{u} + \ln u + \frac{1}{2}u + \frac{u^2}{12} + \frac{u^3}{72}\right) + C$$

Therefore,

$$P(t) \approx -5\left[-(t+1) + \ln\left(\frac{1}{t+1}\right) + \frac{1}{2(t+1)} + \frac{1}{12(t+1)^2} + \frac{1}{72(t+1)^3}\right] + C$$

Using the condition $P(0) = 30$, we find

$$P(0) \approx -5\left(-1 + \ln 1 + \frac{1}{2} + \frac{1}{12} + \frac{1}{72}\right) + C = 30$$

or $C \approx 27.99$. So

$$P(t) \approx -5\left[-(t+1) + \ln\left(\frac{1}{t+1}\right) + \frac{1}{2(t+1)} + \frac{1}{12(t+1)^2} + \frac{1}{72(t+1)^3}\right]$$
$$+ 27.99$$

In particular, the percentage of the nonfarm workforce in service industries one decade from now will be given by

$$P(1) \approx -5\left[-2 + \ln\left(\frac{1}{2}\right) + \frac{1}{4} + \frac{1}{48} + \frac{1}{576}\right] + 27.99$$

$$\approx 40.09$$

or approximately 40.1%.

Errors in Taylor Polynomial Approximations

The *error* incurred in approximating $f(x)$ by its Taylor polynomial $P_N(x)$ at a is

$$R_N(x) = f(x) - P_N(x)$$

A formula for $R_N(x)$, called the *remainder* associated with $P_N(x)$, is now stated without proof.

THEOREM 1

The Remainder Theorem

Suppose a function f has derivatives up to order $(N + 1)$ on an interval I containing the number a. Then for each x in I, there exists a number c lying between a and x such that

$$R_N(x) = \frac{f^{(N+1)}(c)}{(N + 1)!}(x - a)^{N+1}$$

Usually, the exact value of c is unknown. But from a practical point of view, this is not a serious drawback because what we need is just a bound on the error incurred when a Taylor polynomial of f is used to approximate $f(x)$. For this purpose, we have the following result, which is a consequence of the Remainder Theorem.

Error Bound for Taylor Polynomial Approximations

Suppose a function f has derivatives up to order $(N + 1)$ on an interval I containing the number a. Then for each x in I,

$$|R_N(x)| \le \frac{M}{(N + 1)!}|x - a|^{N+1} \tag{3}$$

where M is any number such that $|f^{(N+1)}(t)| \le M$ for all t lying between a and x.

EXAMPLE 5 Refer to Example 2. Use Formula (3) to obtain an upper bound for the error incurred in approximating $f(0.1)$ by $P_5(0.1)$, where $f(x) = 1/(1 - x)$ and $P_5(x)$ is the fifth Taylor polynomial of f at $x = 0$. Compare this result with the actual error incurred in the approximation.

Solution   First, we need to find a number M such that

$$|f^{(6)}(t)| \le M$$

for all t lying in the interval $[0, 0.1]$. From the solution of Example 2, we find that

$$f^{(6)}(t) = (6)(5)(4)(3)(2)(1)(1 - t)^{-6-1}$$

$$= \frac{720}{(1 - t)^7}$$

Observe that $f^{(6)}(t)$ is increasing on the interval $[0, 0.1]$. [Just look at $f^{(7)}(t)$.] Consequently, its maximum is attained at the right endpoint $t = 0.1$. Thus, we may take

$$M = \frac{720}{(1 - 0.1)^7} \approx 1505.3411$$

Next, with $a = 0$, $x = 0.1$, and $M = 1505.3411$, Formula (3) gives

$$|R_5(0.1)| \leq \frac{1505.3411}{(5 + 1)!} (0.1)^{5+1}$$

$$\approx 0.0000021$$

Finally, we refer once again to Example 2 to see that the actual error incurred in approximating $f(0.1)$ by $P_5(0.1)$ was $0.00000111. . .$, and this is less than the error bound 0.0000021 just obtained. ∎

Explore and Discuss

Refer to the Explore and Discuss question on page 730. Obtain an upper bound for the error incurred in approximating $\sqrt[3]{26.5}$, using the Taylor polynomial you have chosen.

11.1 Self-Check Exercises

1. Let $f(x) = \dfrac{1}{1 + x}$.

   a. Find the nth Taylor polynomial of f at $x = 0$.

   b. Use the third Taylor polynomial of f at $x = 0$ to approximate $f(0.1)$. Compare this result with $f(0.1)$.

2. Refer to Exercise 1. Find a bound on the error in the approximation of $f(0.1)$.

Solutions to Self-Check Exercises 11.1 can be found on page 735.

11.1 Concept Questions

1. **a.** What is a Taylor polynomial?

   b. Suppose

   $$P_n(x) = a_0 + a_1(x - a) + a_2(x - a)^2 + \cdots + a_n(x - a)^n$$

   is a Taylor polynomial. What is $a_i (0 \leq i \leq n)$ in terms of f and/or its derivatives?

2. **a.** What is the error incurred if $f(x)$ is approximated by its Nth-order Taylor polynomial $P_N(x)$? Write an expression for the error (remainder) $R_N(x)$.

   b. Write an expression giving the bound for the error when $f(x)$ is approximated by $P_N(x)$.

11.1 Exercises

In Exercises 1–10, find the first three Taylor polynomials of the function at the indicated number.

1. $f(x) = e^{-x}$; $x = 0$

2. $f(x) = e^{2x}$; $x = 0$

3. $f(x) = \dfrac{1}{x + 2}$; $x = 0$

4. $f(x) = \dfrac{1}{1 - x}$; $x = 2$

5. $f(x) = \dfrac{1}{x}$; $x = 1$

6. $f(x) = \dfrac{1}{x + 2}$; $x = 1$

7. $f(x) = \sqrt{1 - x}$; $x = 0$

8. $f(x) = \sqrt{x}$; $x = 4$

9. $f(x) = \ln(1 - x)$; $x = 0$

10. $f(x) = xe^x$; $x = 0$

In Exercises 11–20, find the nth Taylor polynomial of the function at the indicated number.

11. $f(x) = x^4$ at $x = 2$, $n = 2$

12. $f(x) = x^5$ at $x = -1$, $n = 3$

13. $f(x) = \ln x$ at $x = 1$, $n = 4$

14. $f(x) = \dfrac{1}{x}$ at $x = 2$, $n = 4$

15. $f(x) = e^x$ at $x = 1$, $n = 4$

16. $f(x) = e^{2x}$ at $x = 1$, $n = 4$

17. $f(x) = (1 - x)^{1/3}$ at $x = 0, n = 3$

18. $f(x) = \sqrt{x + 1}$ at $x = 0, n = 3$

19. $f(x) = \dfrac{1}{2x + 3}$ at $x = 0, n = 3$

20. $f(x) = x^2 e^x$ at $x = 0, n = 2$

21. Find the nth Taylor polynomial of $f(x) = \dfrac{1}{1 + 2x}$ at $x = 0$.
Compute $P_4(0.1)$, and compare this result with $f(0.1)$.

22. Find the third Taylor polynomial of $f(x) = \sqrt{3x + 1}$ at $x = 0$. Compute $P_3(0.2)$, and compare this result with $f(0.2)$.

23. Find the fourth Taylor polynomial of $f(x) = e^{-x/2}$ at $x = 0$, and use it to estimate $e^{-0.1}$.

24. Find the third Taylor polynomial of $f(x) = \ln x$ at $x = 1$, and use it to estimate $\ln 1.1$.

25. Find the second Taylor polynomial of $f(x) = \sqrt{x}$ at $x = 16$, and use it to estimate the value of $\sqrt{15.6}$.

26. Find the second Taylor polynomial of $f(x) = x^{1/3}$ at $x = 8$, and use it to estimate the value of $\sqrt[3]{8.1}$.

27. Find the third Taylor polynomial of $f(x) = \ln(x + 1)$ at $x = 0$, and use it to estimate the value of

$$\int_0^{1/2} \ln(x + 1)\, dx$$

Compare your result with the exact value found by using integration by parts.

28. Use the fourth Taylor polynomial of $f(x) = \ln(1 + x)$ to approximate

$$\int_{0.1}^{0.2} \frac{\ln(x + 1)}{x}\, dx$$

29. Use the second Taylor polynomial of $f(x) = \sqrt{x}$ at $x = 4$ to approximate $\sqrt{4.06}$, and find a bound for the error in the approximation.

30. Use the third Taylor polynomial of $f(x) = e^{-x}$ at $x = 0$ to approximate $e^{-0.2}$, and find a bound for the error in the approximation.

31. Use the third Taylor polynomial of $f(x) = \dfrac{1}{1 - x}$ at $x = 0$ to approximate $f(0.2)$. Find a bound for the error in the approximation, and compare your results to the exact value of $f(0.2)$.

32. Use the second Taylor polynomial of $f(x) = \sqrt{x + 1}$ at $x = 0$ to approximate $f(0.1)$, and find a bound for the error in the approximation.

33. Use the second Taylor polynomial of $f(x) = \ln x$ at $x = 1$ to approximate $\ln 1.1$, and find a bound for the error in the approximation.

34. Use the second Taylor polynomial of $f(x) = \ln x$ at $x = 1$ to approximate $\ln 0.9$, and find a bound for the error in the approximation.

35. Let $f(x) = e^{-x}$.
  a. Find a bound in the error incurred in approximating $f(x)$ by the third Taylor polynomial, $P_3(x)$, of f at $x = 0$ in the interval $[0, 1]$.
  b. Estimate $\int_0^1 e^{-x}\, dx$ by computing $\int_0^1 P_3(x)\, dx$.
  c. What is the bound on the error in the approximation in (b)?
  Hint: Compute $\int_0^1 B\, dx$, where B is a bound on $|R_3(x)|$ for x in $[0, 1]$.
  d. What is the actual error?

36. **a.** Find a bound on the error incurred in approximating $f(x) = \ln x$ by the fourth Taylor polynomial, $P_4(x)$, of f at $x = 1$ in the interval $[1, 1.5]$.
  b. Approximate the area under the graph of f from $x = 1$ to $x = 1.5$ by computing

$$\int_1^{1.5} P_4(x)\, dx$$

  c. What is the bound on the error in the approximation in part (b)?
  Hint: Compute $\int_1^{1.5} B\, dx$, where B is a bound on $|R_4(x)|$ for x in $[1, 1.5]$.
  d. What is the actual error?
  Hint: Use integration by parts to evaluate $\int_1^{1.5} \ln x\, dx$.

37. Use the fourth Taylor polynomial at $x = 0$ to approximate the area under the graph of $f(x) = e^{-x^2/2}$ from $x = 0$ to $x = 0.5$.
Hint: $e^u \approx 1 + u + \frac{1}{2}u^2$

38. Use the fourth Taylor polynomial at $x = 0$ to approximate the area under the graph of $f(x) = \ln(1 + x^2)$ from $x = -\frac{1}{2}$ to $x = \frac{1}{2}$.
Hint: First find the Taylor polynomial of $f(u) = \ln(1 + u)$; then use the substitution $u = x^2$.

39. Growth of Service Industries It has been estimated that service industries, which currently make up 30% of the nonfarm workforce in a certain country, will continue to grow at the rate of

$$R(t) = 6e^{1/(2t + 1)}$$

percent per decade t decades from now. Estimate the percentage of the nonfarm workforce in service industries two decades from now.

40. Vacation Trends The Travel Data Center of a European country has estimated that the percentage of vacations traditionally taken during the months of July, August, and September will continue to decline at the rate of

$$R(t) = -2e^{1/(t + 1)}$$

percent per decade ($t = 0$ corresponds to the year 2005) as more and more vacationers switch to fall and winter vacations. There were 37% summer vacationers in that country in 2005. Estimate the percentage of vacations that were taken in the summer months in 2015.

41. **CONTINUING EDUCATION ENROLLMENT** The registrar of Kellogg University estimates that the total student enrollment in the Continuing Education division will be given by

$$N(t) = -\frac{20,000}{\sqrt{1 + 0.2t}} + 21,000$$

where $N(t)$ denotes the number of students enrolled in the division t years from now. Use the second Taylor polynomial of N at $t = 0$ to approximate the average enrollment at Kellogg University between $t = 0$ and $t = 2$.

42. **CONCENTRATION OF CARBON MONOXIDE IN THE AIR** According to a joint study conducted by Oxnard's Environmental Management Department and a state government agency, the concentration of carbon monoxide (CO) in the air due to automobile exhaust t years from now is given by

$$C(t) = 0.01(0.2t^2 + 4t + 64)^{2/3}$$

parts per million. Use the second Taylor polynomial of C at $t = 0$ to approximate the average level of concentration of CO in the air between $t = 0$ and $t = 2$.

In Exercises 43–46, determine whether the statement is true or false. If it is true, explain why it is true. If it is false, explain why, or give an example to show why it is false.

43. If f is a polynomial of degree 4 and $P_4(x)$ is the fourth Taylor polynomial of f at a, then $f(x) \neq P_4(x)$ for at least one value of x.

44. The function $f(x) = 1/(x - 2)$ has a Taylor polynomial at a for every value of a except $a = 2$.

45. Suppose f is a polynomial of degree N and R_N is the remainder associated with P_N, the Nth Taylor polynomial of f at a. Then $R_N(x) = 0$ for every value of x.

46. Let P_N denote the Nth Taylor polynomial of the function $f(x) = e^{-x}$ at a. Then $R_N(x) = f(x) - P_N(x) \neq 0$ for all $x \neq 0$.

11.1 Solutions to Self-Check Exercises

1. a. We rewrite $f(x)$ as

$$f(x) = (1 + x)^{-1}$$

and compute

$$f'(x) = -(1 + x)^{-2}$$
$$f''(x) = (-1)(-2)(1 + x)^{-3} = 2(1 + x)^{-3}$$
$$f'''(x) = 2(-3)(1 + x)^{-4} = -3 \cdot 2(1 + x)^{-4}$$
$$f^{(4)}(x) = -3 \cdot 2(-4)(1 + x)^{-5} = 4 \cdot 3 \cdot 2(1 + x)^{-5}$$
$$\vdots$$
$$f^{(n)}(x) = (-1)^n n!(1 + x)^{-(n+1)}$$

Evaluating $f(x)$ and its derivatives at $x = 0$ gives

$$f(0) = 1, \quad f'(0) = -1, \quad f''(0) = 2, \quad f'''(0) = -3!,$$
$$f^{(4)}(0) = 4!, \quad \ldots, \quad f^{(n)}(0) = (-1)^n n!$$

Therefore, the nth Taylor polynomial is

$$P_n(x) = 1 + \frac{(-1)}{1!}x + \frac{2}{2!}x^2 + \frac{-3!}{3!}x^3$$
$$+ \frac{4!}{4!}x^4 + \cdots + \frac{(-1)^n n!}{n!}x^n$$
$$= 1 - x + x^2 - x^3 + x^4 - \cdots + (-1)^n x^n$$

b. The third Taylor polynomial of f at $x = 0$ is

$$P_3(x) = 1 - x + x^2 - x^3$$

So

$$P_3(0.1) = 1 - 0.1 + (0.1)^2 - (0.1)^3$$
$$= 0.909$$

The actual value of $f(x)$ at $x = 0.1$ is

$$f(0.1) = \frac{1}{1 + 0.1} = \frac{1}{1.1} = 0.909090\ldots$$

Thus, the error incurred in approximating $f(0.1)$ by $P_3(0.1)$ is $0.00009090\ldots$.

2. First, we need to find a number M such that

$$|f^{(4)}(t)| \leq M$$

for all t lying in the interval $[0, 0.1]$. From the solution to Self-Check Exercise 1, we see that

$$f^{(5)}(x) = (-1)^5 5! (1 + x)^{-6} = -5! (1 + x)^{-6}$$

Since $f^{(5)}(x)$ is negative for x in $[0, 0.1]$, $f^{(4)}(x)$ is decreasing on $[0, 0.1]$. Therefore, the maximum of the function

$$f^{(4)}(x) = 4! (1 + x)^{-5}$$

is attained at $x = 0$, and its value is

$$f^{(4)}(0) = 4! = 24$$

So we can take $M = 24$. Next, with $a = 0$, $x = 0.1$, and $M = 24$, Formula (3) gives

$$|R_3(0.1)| \leq \frac{24}{(3 + 1)!} |0.1 - 0|^{3+1}$$
$$= \frac{24}{24} (0.1)^4 = 0.0001$$

the required error bound.

11.2 Infinite Sequences

An idealized superball is dropped from a height of 1 meter onto a flat surface. Suppose that each time the ball hits the surface, it rebounds to two thirds of its previous height. If we let a_1 denote the initial height of the ball, a_2 denote the maximum height attained on the first rebound, a_3 denote the maximum height attained on the second rebound, and so on, then we have

$$a_1 = 1, \quad a_2 = \frac{2}{3}, \quad a_3 = \frac{4}{9}, \quad a_4 = \frac{8}{27}, \quad \ldots$$

(see Figure 3).

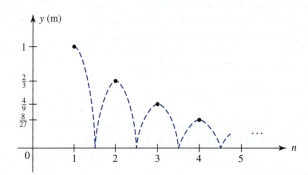

FIGURE 3
The ball rebounds to two thirds of its previous height upon hitting the surface.

This list of numbers, $a_1, a_2, a_3, \ldots$, is an example of an *infinite sequence*, or simply a *sequence*. If we define the function f by $f(x) = (2/3)^{x-1}$ and allow x to take on the positive integral values $x = 1, 2, 3, \ldots, n, \ldots$, then we see that the sequence $a_1, a_2, a_3, \ldots$ may be viewed as the functional values of f at these numbers. Thus,

$$f(1) = 1, \quad f(2) = \frac{2}{3}, \quad f(3) = \frac{4}{9}, \quad \ldots, \quad f(n) = \left(\frac{2}{3}\right)^{n-1}, \quad \ldots$$
$$\quad\;\downarrow \qquad\qquad \downarrow \qquad\qquad \downarrow \qquad\qquad\qquad \downarrow$$
$$\quad\; a_1 \qquad\qquad a_2 \qquad\qquad a_3 \qquad\qquad\qquad a_n \;\; \ldots$$

This discussion motivates the following definition.

> **Infinite Sequence**
>
> An **infinite sequence,** denoted by $\{a_n\}$, is a function whose domain is the set of positive integers. The functional values $a_1, a_2, a_3, \ldots, a_n, \ldots$ are the **terms** of the sequence, and the term a_n is called the **nth term** of the sequence.

Notes

1. The sequence $\{a_n\}$ is also denoted by $\{a_n\}_{n=1}^{\infty}$.
2. Sometimes it is convenient to begin a sequence with a_k. In this case, the sequence is $\{a_n\}_{n=k}^{\infty}$, and its terms are $a_k, a_{k+1}, a_{k+2}, \ldots, a_n, \ldots$.

EXAMPLE 1 List the terms of the sequence.

a. $\left\{\dfrac{n}{n+1}\right\}$ **b.** $\left\{\dfrac{\sqrt{n}}{2^{n-1}}\right\}$ **c.** $\{(-1)^n\sqrt{n-2}\}_{n=2}^{\infty}$

Solution

a. Here, $a_n = f(n) = \dfrac{n}{n+1}$. Thus,

$$a_1 = f(1) = \frac{1}{1+1} = \frac{1}{2}, \quad a_2 = f(2) = \frac{2}{2+1} = \frac{2}{3}, \quad a_3 = f(3) = \frac{3}{3+1} = \frac{3}{4}$$

and so forth. Then the given sequence can be written as

$$\left\{\frac{n}{n+1}\right\} = \left\{\frac{1}{2}, \frac{2}{3}, \frac{3}{4}, \frac{4}{5}, \ldots, \frac{n}{n+1}, \ldots\right\}$$

b. $\left\{\dfrac{\sqrt{n}}{2^{n-1}}\right\} = \left\{\dfrac{\sqrt{1}}{2^0}, \dfrac{\sqrt{2}}{2^1}, \dfrac{\sqrt{3}}{2^2}, \dfrac{\sqrt{4}}{2^3}, \ldots, \dfrac{\sqrt{n}}{2^{n-1}}, \ldots\right\}$

c. $\{(-1)^n\sqrt{n-2}\}_{n=2}^{\infty}$

$$= \{(-1)^2\sqrt{0}, (-1)^3\sqrt{1}, (-1)^4\sqrt{2}, (-1)^5\sqrt{3}, \ldots, (-1)^n\sqrt{n-2}, \ldots\}$$
$$= \{0, -\sqrt{1}, \sqrt{2}, -\sqrt{3}, \ldots, (-1)^n\sqrt{n-2}, \ldots\}$$

Notice that n starts from 2 in this example. (See Note 2, page 736.)

We can often guess at the nth term of a sequence by studying the first few terms of the sequence and recognizing the pattern that emerges.

EXAMPLE 2 Find an expression for the nth term of each sequence.

a. $\left\{2, \dfrac{3}{\sqrt{2}}, \dfrac{4}{\sqrt{3}}, \dfrac{5}{\sqrt{4}}, \ldots\right\}$ **b.** $\left\{1, \dfrac{1}{8}, \dfrac{1}{27}, \dfrac{1}{64}, \ldots\right\}$ **c.** $\left\{1, -\dfrac{1}{2}, \dfrac{1}{3}, -\dfrac{1}{4}, \ldots\right\}$

Solution

a. The terms of the sequence may be written in the form

$$a_1 = \frac{1+1}{\sqrt{1}}, \quad a_2 = \frac{2+1}{\sqrt{2}}, \quad a_3 = \frac{3+1}{\sqrt{3}}, \quad a_4 = \frac{4+1}{\sqrt{4}}, \quad \ldots$$

from which we see that $a_n = \dfrac{n+1}{\sqrt{n}}$.

b. Here,

$$a_1 = \frac{1}{1^3}, \quad a_2 = \frac{1}{2^3}, \quad a_3 = \frac{1}{3^3}, \quad a_4 = \frac{1}{4^3}, \quad \ldots$$

so $a_n = \dfrac{1}{n^3}$.

c. Note that $(-1)^r$ is equal to 1 if r is an even integer and is equal to -1 if r is an odd integer. Using this result, we obtain

$$a_1 = \frac{(-1)^0}{1}, \quad a_2 = \frac{(-1)^1}{2}, \quad a_3 = \frac{(-1)^2}{3}, \quad a_4 = \frac{(-1)^3}{4}, \quad \ldots$$

We conclude that the nth term is $a_n = (-1)^{n-1}/n$.

Graphs of Infinite Sequences

Since an infinite sequence is a function, we can sketch its graph. Because the domain of the function is the set of positive integers, the graph of a sequence consists of an infinite collection of points in the xy-plane.

EXAMPLE 3 Sketch the graph of the sequence.

a. $\{n + 1\}$ **b.** $\left\{\dfrac{n + 1}{n}\right\}$ **c.** $\{(-1)^n\}$

Solution

a. The terms of the sequence $\{n + 1\}$ are $2, 3, 4, 5, \ldots$. Recalling that these numbers are precisely the functional values of $f(n) = n + 1$ for $n = 1, 2, 3, \ldots$, we obtain the following table of values of f:

n	1	2	3	4	5	$\cdots$
$f(n)$	2	3	4	5	6	$\cdots$

from which we construct the graph of $\{n + 1\}$ shown in Figure 4a.

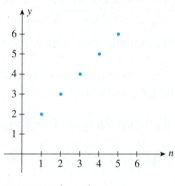

(a) Graph of $\{n + 1\}$

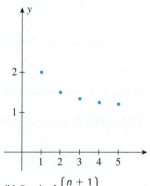

(b) Graph of $\left\{\dfrac{n + 1}{n}\right\}$

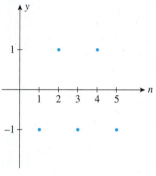

(c) Graph of $\{(-1)^n\}$

FIGURE 4

b. From the following table of values of f,

n	1	2	3	4	5	$\cdots$
$f(n)$	2	$\frac{3}{2}$	$\frac{4}{3}$	$\frac{5}{4}$	$\frac{6}{5}$	$\cdots$

we sketch the graph shown in Figure 4b.

c. We use the following table of values for the sequence

n	1	2	3	4	5	$\cdots$
$f(n)$	-1	1	-1	1	-1	$\cdots$

to sketch the graph of $\{(-1)^n\}$ shown in Figure 4c.

Note Notice that the function $f(n) = n + 1$ defining the sequence $\{n + 1\}$ may be viewed as the function $f(x) = x + 1$ ($-\infty < x < \infty$) with x *restricted* to the set of positive integers. Similarly, the function defining the sequence $\{(n + 1)/n\}$ is just the function $f(x) = (x + 1)/x$ with a similar restriction on its domain. Finally, the function defining the sequence $\{(-1)^n\}$ can be viewed as the function $f(x) = \cos \pi x$ with x restricted to the set of positive integers.* These observations suggest that it is possible to study the properties of a sequence by analyzing the properties of a corresponding function defined for *all* values of x in some suitable domain of f.

---

*The trigonometric function $f(x) = \cos \pi x$ will be discussed in Chapter 12.

The Limit of a Sequence

Given a sequence $\{a_n\}$, we may ask whether the terms a_n of the sequence approach some specific number L as n gets larger and larger. If they do, we say that the sequence a_n *converges* to L.

> **Limit of a Sequence**
>
> Let $\{a_n\}$ be a sequence. We say that the sequence $\{a_n\}$ **converges** and has the limit L, written
>
> $$\lim_{n \to \infty} a_n = L$$
>
> if the terms of the sequence, a_n, can be made as close to L as we please by taking n sufficiently large. If a sequence is not convergent, it is said to be **divergent.**

EXAMPLE 4 Determine whether the infinite sequence converges or diverges.

a. $\left\{\dfrac{1}{n}\right\}$ **b.** $\{\sqrt{n}\}$ **c.** $\{(-1)^n\}$

Solution

a. The terms of the sequence,

$$1, \quad \frac{1}{2}, \quad \frac{1}{3}, \quad \frac{1}{4}, \quad \frac{1}{5}, \quad \cdots, \quad \frac{1}{n}, \quad \cdots$$

approach the number $L = 0$ as n gets larger and larger. We conclude that

$$\lim_{n \to \infty} a_n = \lim_{n \to \infty} \frac{1}{n} = 0$$

Compare this with

$$\lim_{x \to \infty} \frac{1}{x} = 0$$

b. The terms of the sequence,

$$\sqrt{1}, \quad \sqrt{2}, \quad \sqrt{3}, \quad \sqrt{4}, \quad \sqrt{5}, \quad \ldots, \quad \sqrt{n}, \quad \ldots$$

get larger and larger without bound as n gets larger and larger. Consequently, they do not approach any finite number L as n tends to infinity. Therefore, the sequence is divergent. Compare this with

$$\lim_{x \to \infty} \sqrt{x} = \infty$$

c. The terms of the sequence are

$$a_1 = -1, \quad a_2 = 1, \quad a_3 = -1, \quad a_4 = 1, \quad a_5 = -1, \quad \ldots, \quad a_n = (-1)^n, \quad \ldots$$

Thus, no matter how large n is, there are terms that are equal to -1 (those with odd-numbered subscripts) and also terms that are equal to 1 (those with even-numbered subscripts). This implies that there cannot be a *unique* real number L such that a_n is arbitrarily close to L no matter how large n is. Therefore, the sequence is divergent.

Refer to Example 4 and the Note on page 738.

1. Plot the graph of $f(x) = 1/x$, using the viewing window $[0, 10] \times [0, 3]$, and thus verify graphically that $\lim_{n \to \infty} (1/n) = 0$.

2. Plot the graph of $f(x) = \sqrt{x}$, using an appropriate viewing window, and thus verify graphically that $\lim_{n \to \infty} \sqrt{n}$ does not exist.

3. Plot the graph of $f(x) = \cos \pi x$, using the viewing window $[0, 10] \times [-2, 2]$, and thus verify graphically that $\lim_{n \to \infty} (-1)^n$ does not exist. (*Note:* Trigonometric functions will be studied in Chapter 12.)

The following properties of sequences, which parallel those of the limit of $f(x)$ at infinity, are helpful in computing limits of sequences.

Limit Properties of Sequences

Suppose

$$\lim_{n \to \infty} a_n = A \quad \text{and} \quad \lim_{n \to \infty} b_n = B$$

Then

1. $\displaystyle\lim_{n \to \infty} ca_n = c \lim_{n \to \infty} a_n = cA$ *c*, a constant

2. $\displaystyle\lim_{n \to \infty} (a_n \pm b_n) = \lim_{n \to \infty} a_n \pm \lim_{n \to \infty} b_n = A \pm B$ Sum Rule

3. $\displaystyle\lim_{n \to \infty} a_n b_n = \left(\lim_{n \to \infty} a_n\right)\left(\lim_{n \to \infty} b_n\right) = AB$ Product Rule

4. $\displaystyle\lim_{n \to \infty} \frac{a_n}{b_n} = \frac{\displaystyle\lim_{n \to \infty} a_n}{\displaystyle\lim_{n \to \infty} b_n} = \frac{A}{B}$ provided that $B \neq 0$ and $b_n \neq 0$ Quotient Rule

EXAMPLE 5 Evaluate:

a. $\displaystyle\lim_{n \to \infty} \left(\frac{1}{2}\right)^n$ **b.** $\displaystyle\lim_{n \to \infty} \frac{2n^2 + 1}{3n^2 + n + 2}$

Solution

a. $\displaystyle\lim_{n \to \infty} \left(\frac{1}{2}\right)^n = \lim_{n \to \infty} \frac{1}{2^n}$

$\displaystyle = \frac{1}{\lim_{n \to \infty} 2^n} = 0$

b. Dividing the numerator and denominator by n^2, we obtain the following:

$$\lim_{n \to \infty} \frac{2n^2 + 1}{3n^2 + n + 2} = \lim_{n \to \infty} \frac{2 + \dfrac{1}{n^2}}{3 + \dfrac{1}{n} + \dfrac{2}{n^2}}$$

$$= \frac{\displaystyle\lim_{n \to \infty} \left(2 + \frac{1}{n^2} \right)}{\displaystyle\lim_{n \to \infty} \left(3 + \frac{1}{n} + \frac{2}{n^2} \right)}$$

$$= \frac{2}{3}$$

Explore and Discuss

Consider the sequences $\{a_n\}$ and $\{b_n\}$ defined by $a_n = (-1)^{n+1}$ and $b_n = (-1)^n$.

1. Show that $\displaystyle\lim_{n \to \infty} a_n$ and $\displaystyle\lim_{n \to \infty} b_n$ do not exist.
2. Show that $\displaystyle\lim_{n \to \infty} (a_n + b_n) = 0$.
3. Do the results of parts 1 and 2 contradict the limit properties of sequences listed on page 740? Explain your answer.

APPLIED EXAMPLE 6 Quality Control Of the spark plugs manufactured by the Parts Division of United Motors Corporation, 2% are defective. It can be shown that the probability of getting at least one defective plug in a random sample of n spark plugs is $f(n) = 1 - (0.98)^n$. Consider the sequence $\{a_n\}$ defined by $a_n = f(n)$.

a. Write down the terms a_5, a_{10}, a_{25}, a_{100}, and a_{200} of the sequence $\{a_n\}$.
b. Evaluate $\displaystyle\lim_{n \to \infty} a_n$, and interpret your results.

Solution

a. The required terms of the sequence are

$$0.10, \quad 0.18, \quad 0.40, \quad 0.87, \quad 0.98$$

For example, the probability of getting at least one defective plug in a random sample of 25 is .40—that is, a 40% chance.

b. $\displaystyle\lim_{n \to \infty} a_n = \lim_{n \to \infty} [1 - (0.98)^n] = \lim_{n \to \infty} 1 - \lim_{n \to \infty} (0.98)^n$

$$= 1 - 0 = 1$$

The result tells us that if the sample is large enough, we will almost certainly pick at least one defective plug!

11.2 Self-Check Exercises

1. Consider the sequence $\left\{ \dfrac{n}{n^2 + 1} \right\}$.

   a. Sketch the graph of the sequence.
   b. Show that the sequence is decreasing, that is,
      $a_1 > a_2 > a_3 > \cdots$.
      Hint: Consider $f(x) = x/(x^2 + 1)$, and show that f is decreasing by computing f'.

2. Determine whether the sequence $\left\{ \dfrac{3n^2 + 2n + 1}{n^2 + 4} \right\}$ converges or diverges. If it converges, find its limit.

Solutions to Self-Check Exercises 11.2 can be found on page 743.

11.2 Concept Question

Explain each of the following terms in your own words and give an example of each.

a. Sequence   **b.** Convergent sequence
c. Divergent sequence   **d.** Limit of a sequence

11.2 Exercises

In Exercises 1–9, write down the first five terms of the sequence.

1. $\{a_n\} = \{2^{n-1}\}$

2. $\{a_n\} = \left\{\dfrac{2n}{1+n^2}\right\}$

3. $\{a_n\} = \left\{\dfrac{n-1}{n+1}\right\}$

4. $\{a_n\} = \left\{\left(-\dfrac{1}{3}\right)^n\right\}$

5. $\{a_n\} = \left\{\dfrac{2^{n-1}}{n!}\right\}$

6. $\{a_n\} = \left\{\dfrac{(-1)^n}{(2n)!}\right\}$

7. $\{a_n\} = \left\{\dfrac{e^n}{n^3}\right\}$

8. $\{a_n\} = \left\{\dfrac{\sqrt{n}}{\sqrt{n}+1}\right\}$

9. $\{a_n\} = \left\{\dfrac{3n^2 - n + 1}{2n^2 + 1}\right\}$

In Exercises 10–21, find the nth term of the sequence. (Assume that the "obvious" pattern continues.)

10. $\dfrac{1}{2}, \dfrac{1}{4}, \dfrac{1}{6}, \dfrac{1}{8}, \ldots$

11. $1, 4, 7, 10, \ldots$

12. $\dfrac{1}{3}, \dfrac{1}{7}, \dfrac{1}{11}, \dfrac{1}{15}, \ldots$

13. $1, \dfrac{1}{8}, \dfrac{1}{27}, \dfrac{1}{64}, \ldots$

14. $1, \dfrac{2}{3}, \dfrac{4}{9}, \dfrac{8}{27}, \ldots$

15. $2, \dfrac{8}{5}, \dfrac{32}{25}, \dfrac{128}{125}, \ldots$

16. $1, \dfrac{5}{4}, \dfrac{7}{5}, \dfrac{9}{6}, \ldots$

17. $1, -\dfrac{1}{2}, \dfrac{1}{4}, -\dfrac{1}{8}, \ldots$

18. $1 + \dfrac{1}{2}, 1 + \dfrac{1}{3}, 1 + \dfrac{1}{4}, 1 + \dfrac{1}{5}, \ldots$

19. $\dfrac{1}{2 \cdot 3}, \dfrac{2}{3 \cdot 4}, \dfrac{3}{4 \cdot 5}, \dfrac{4}{5 \cdot 6}, \ldots$

20. $1, \dfrac{2}{1 \cdot 3}, \dfrac{4}{1 \cdot 3 \cdot 5}, \dfrac{8}{1 \cdot 3 \cdot 5 \cdot 7}, \ldots$

21. $1, e, \dfrac{e^2}{2}, \dfrac{e^3}{6}, \dfrac{e^4}{24}, \dfrac{e^5}{120}, \ldots$

In Exercises 22–29, sketch the graph of the sequence.

22. $\{n^2\}$

23. $\left\{\dfrac{2n}{n+1}\right\}$

24. $\left\{\dfrac{(-1)^n}{n}\right\}$

25. $\{\sqrt{n}\}$

26. $\{\ln n\}$

27. $\{e^n\}$

28. $\{ne^{-n}\}$

29. $\{n - \sqrt{n}\}$

In Exercises 30–44, determine the convergence or divergence of the sequence $\{a_n\}$. If the sequence converges, find its limit.

30. $a_n = \dfrac{n}{n^2 + 1}$

31. $a_n = \dfrac{n+1}{2n}$

32. $a_n = \sqrt[3]{n}$

33. $a_n = \dfrac{(-1)^n}{\sqrt{n}}$

34. $a_n = \dfrac{1}{n+1} - \dfrac{1}{n+2}$

35. $a_n = \dfrac{\sqrt{n} - 1}{\sqrt{n} + 1}$

36. $a_n = \dfrac{3n^2 + n - 1}{6n^2 + n + 1}$

37. $a_n = \dfrac{2n^4 - 1}{n^3 + 2n + 1}$

38. $a_n = \dfrac{1 + (-1)^n}{3^n}$

39. $a_n = 2 - \dfrac{1}{2^n}$

40. $a_n = \dfrac{2n}{n!}$

41. $a_n = \dfrac{2^n}{3^n}$

42. $a_n = \dfrac{2^n - 1}{2^n}$

43. $a_n = \dfrac{n}{\sqrt{2n^2 + 3}}$

44. $a_n = \dfrac{2^n}{n!}$

45. AUTOMOBILE MICROPROCESSORS Of the microprocessors manufactured by a microelectronics firm for use in regulating fuel consumption in automobiles, $1\frac{1}{2}\%$ are defective. It can be shown that the probability of getting at least one defective microprocessor in a random sample of n microprocessors is $f(n) = 1 - (0.985)^n$. Consider the sequence $\{a_n\}$ defined by $a_n = f(n)$.
a. Write down the terms $a_1, a_{10}, a_{100},$ and a_{1000} of the sequence $\{a_n\}$.
b. Evaluate $\lim\limits_{n \to \infty} a_n$, and interpret your results.

46. SAVINGS ACCOUNTS An amount of $1000 is deposited in a bank that pays 4% interest/year compounded daily (take the number of days in a year to be 365). Let a_n denote the total amount on deposit after n days, assuming that no deposits or withdrawals are made during the period in question.
a. Find the formula for a_n.
   Hint: See Section 5.3.
b. Compute $a_1, a_{10}, a_{50},$ and a_{100}.
c. What amount is on deposit after 1 year?

47. ACCUMULATED AMOUNT Suppose that $100 is deposited into an account earning interest at 6%/year compounded monthly. Let a_n denote the amount on deposit (called the accumulated amount or the future value) at the end of the nth month.

$$\lim_{n\to\infty} \frac{2n^2 + 1}{3n^2 + n + 2} = \lim_{n\to\infty} \frac{2 + \dfrac{1}{n^2}}{3 + \dfrac{1}{n} + \dfrac{2}{n^2}}$$

$$= \frac{\lim\limits_{n\to\infty}\left(2 + \dfrac{1}{n^2}\right)}{\lim\limits_{n\to\infty}\left(3 + \dfrac{1}{n} + \dfrac{2}{n^2}\right)}$$

$$= \frac{2}{3}$$

Explore and Discuss

Consider the sequences $\{a_n\}$ and $\{b_n\}$ defined by $a_n = (-1)^{n+1}$ and $b_n = (-1)^n$.

1. Show that $\lim\limits_{n\to\infty} a_n$ and $\lim\limits_{n\to\infty} b_n$ do not exist.

2. Show that $\lim\limits_{n\to\infty} (a_n + b_n) = 0$.

3. Do the results of parts 1 and 2 contradict the limit properties of sequences listed on page 740? Explain your answer.

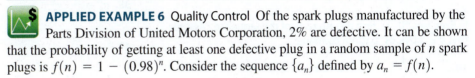 **APPLIED EXAMPLE 6** Quality Control Of the spark plugs manufactured by the Parts Division of United Motors Corporation, 2% are defective. It can be shown that the probability of getting at least one defective plug in a random sample of n spark plugs is $f(n) = 1 - (0.98)^n$. Consider the sequence $\{a_n\}$ defined by $a_n = f(n)$.

a. Write down the terms a_5, a_{10}, a_{25}, a_{100}, and a_{200} of the sequence $\{a_n\}$.
b. Evaluate $\lim\limits_{n\to\infty} a_n$, and interpret your results.

Solution

a. The required terms of the sequence are

$$0.10, \quad 0.18, \quad 0.40, \quad 0.87, \quad 0.98$$

For example, the probability of getting at least one defective plug in a random sample of 25 is .40—that is, a 40% chance.

b. $\lim\limits_{n\to\infty} a_n = \lim\limits_{n\to\infty} \left[1 - (0.98)^n\right] = \lim\limits_{n\to\infty} 1 - \lim\limits_{n\to\infty} (0.98)^n$

$$= 1 - 0 = 1$$

The result tells us that if the sample is large enough, we will almost certainly pick at least one defective plug!

11.2 Self-Check Exercises

1. Consider the sequence $\left\{\dfrac{n}{n^2 + 1}\right\}$.

   a. Sketch the graph of the sequence.
   b. Show that the sequence is decreasing, that is,
     $a_1 > a_2 > a_3 > \cdots$.
     Hint: Consider $f(x) = x/(x^2 + 1)$, and show that f is decreasing by computing f'.

2. Determine whether the sequence $\left\{\dfrac{3n^2 + 2n + 1}{n^2 + 4}\right\}$ converges or diverges. If it converges, find its limit.

Solutions to Self-Check Exercises 11.2 can be found on page 743.

11.2   Concept Question

Explain each of the following terms in your own words and give an example of each.

a. Sequence   **b.** Convergent sequence
c. Divergent sequence   **d.** Limit of a sequence

11.2   Exercises

In Exercises 1–9, write down the first five terms of the sequence.

1. $\{a_n\} = \{2^{n-1}\}$

2. $\{a_n\} = \left\{ \dfrac{2n}{1 + n^2} \right\}$

3. $\{a_n\} = \left\{ \dfrac{n - 1}{n + 1} \right\}$

4. $\{a_n\} = \left\{ \left(-\dfrac{1}{3} \right)^n \right\}$

5. $\{a_n\} = \left\{ \dfrac{2^{n-1}}{n!} \right\}$

6. $\{a_n\} = \left\{ \dfrac{(-1)^n}{(2n)!} \right\}$

7. $\{a_n\} = \left\{ \dfrac{e^n}{n^3} \right\}$

8. $\{a_n\} = \left\{ \dfrac{\sqrt{n}}{\sqrt{n} + 1} \right\}$

9. $\{a_n\} = \left\{ \dfrac{3n^2 - n + 1}{2n^2 + 1} \right\}$

In Exercises 10–21, find the nth term of the sequence. (Assume that the "obvious" pattern continues.)

10. $\dfrac{1}{2}, \dfrac{1}{4}, \dfrac{1}{6}, \dfrac{1}{8}, \ldots$

11. $1, 4, 7, 10, \ldots$

12. $\dfrac{1}{3}, \dfrac{1}{7}, \dfrac{1}{11}, \dfrac{1}{15}, \ldots$

13. $1, \dfrac{1}{8}, \dfrac{1}{27}, \dfrac{1}{64}, \ldots$

14. $1, \dfrac{2}{3}, \dfrac{4}{9}, \dfrac{8}{27}, \ldots$

15. $2, \dfrac{8}{5}, \dfrac{32}{25}, \dfrac{128}{125}, \ldots$

16. $1, \dfrac{5}{4}, \dfrac{7}{5}, \dfrac{9}{6}, \ldots$

17. $1, -\dfrac{1}{2}, \dfrac{1}{4}, -\dfrac{1}{8}, \ldots$

18. $1 + \dfrac{1}{2}, 1 + \dfrac{1}{3}, 1 + \dfrac{1}{4}, 1 + \dfrac{1}{5}, \ldots$

19. $\dfrac{1}{2 \cdot 3}, \dfrac{2}{3 \cdot 4}, \dfrac{3}{4 \cdot 5}, \dfrac{4}{5 \cdot 6}, \ldots$

20. $1, \dfrac{2}{1 \cdot 3}, \dfrac{4}{1 \cdot 3 \cdot 5}, \dfrac{8}{1 \cdot 3 \cdot 5 \cdot 7}, \ldots$

21. $1, e, \dfrac{e^2}{2}, \dfrac{e^3}{6}, \dfrac{e^4}{24}, \dfrac{e^5}{120}, \ldots$

In Exercises 22–29, sketch the graph of the sequence.

22. $\{n^2\}$

23. $\left\{ \dfrac{2n}{n + 1} \right\}$

24. $\left\{ \dfrac{(-1)^n}{n} \right\}$

25. $\{\sqrt{n}\}$

26. $\{\ln n\}$

27. $\{e^n\}$

28. $\{ne^{-n}\}$

29. $\{n - \sqrt{n}\}$

In Exercises 30–44, determine the convergence or divergence of the sequence $\{a_n\}$. If the sequence converges, find its limit.

30. $a_n = \dfrac{n}{n^2 + 1}$

31. $a_n = \dfrac{n + 1}{2n}$

32. $a_n = \sqrt[3]{n}$

33. $a_n = \dfrac{(-1)^n}{\sqrt{n}}$

34. $a_n = \dfrac{1}{n + 1} - \dfrac{1}{n + 2}$

35. $a_n = \dfrac{\sqrt{n} - 1}{\sqrt{n} + 1}$

36. $a_n = \dfrac{3n^2 + n - 1}{6n^2 + n + 1}$

37. $a_n = \dfrac{2n^4 - 1}{n^3 + 2n + 1}$

38. $a_n = \dfrac{1 + (-1)^n}{3^n}$

39. $a_n = 2 - \dfrac{1}{2^n}$

40. $a_n = \dfrac{2n}{n!}$

41. $a_n = \dfrac{2^n}{3^n}$

42. $a_n = \dfrac{2^n - 1}{2^n}$

43. $a_n = \dfrac{n}{\sqrt{2n^2 + 3}}$

44. $a_n = \dfrac{2^n}{n!}$

45. **AUTOMOBILE MICROPROCESSORS** Of the microprocessors manufactured by a microelectronics firm for use in regulating fuel consumption in automobiles, $1\frac{1}{2}\%$ are defective. It can be shown that the probability of getting at least one defective microprocessor in a random sample of n microprocessors is $f(n) = 1 - (0.985)^n$. Consider the sequence $\{a_n\}$ defined by $a_n = f(n)$.
   a. Write down the terms a_1, a_{10}, a_{100}, and a_{1000} of the sequence $\{a_n\}$.
   b. Evaluate $\lim\limits_{n \to \infty} a_n$, and interpret your results.

46. **SAVINGS ACCOUNTS** An amount of $1000 is deposited in a bank that pays 4% interest/year compounded daily (take the number of days in a year to be 365). Let a_n denote the total amount on deposit after n days, assuming that no deposits or withdrawals are made during the period in question.
   a. Find the formula for a_n.
      Hint: See Section 5.3.
   b. Compute a_1, a_{10}, a_{50}, and a_{100}.
   c. What amount is on deposit after 1 year?

47. **ACCUMULATED AMOUNT** Suppose that $100 is deposited into an account earning interest at 6%/year compounded monthly. Let a_n denote the amount on deposit (called the accumulated amount or the future value) at the end of the nth month.

a. Show that $a_1 = 100(1.005)$, $a_2 = 100(1.005)^2$, and $a_3 = 100(1.005)^3$.
b. Find the accumulated amount a_n.
c. Find the 24th term of the sequence $\{a_n\}$, and interpret your result.

48. **TRANSMISSION OF DISEASE** In the early stages of an epidemic, the number of persons who have contracted the disease on the $(n + 1)$st day, a_{n+1}, is related to the number of persons who have the disease on the nth day, a_n, by the equation

$$a_{n+1} = (1 + aN - b)a_n$$

where N denotes the total population and a and b are positive constants that depend on the nature of the disease. Put $r = 1 + aN - b$.
a. Write down the first n terms of the sequence $\{a_n\}$.
b. Evaluate $\lim_{n \to \infty} a_n$ for each of the three cases $r < 1$, $r = 1$, and $r > 1$.
c. Interpret the results of part (b).

In Exercises 49–52, determine whether the statement is true or false. If it is true, explain why it is true. If it is false, give an example to show why it is false.

49. If $\lim_{n \to \infty} a_n = L$ and $\lim_{n \to \infty} b_n = 0$, then $\lim_{n \to \infty} a_n b_n = 0$.

50. If $\{a_n\}$ and $\{b_n\}$ are sequences such that $\lim_{n \to \infty} (a_n + b_n)$ exists, then both $\lim_{n \to \infty} a_n$ and $\lim_{n \to \infty} b_n$ must exist.

51. If $\{a_n\}$ is bounded (that is $|a_n| \leq M$ for some positive real number M, for $n = 1, 2, 3, \ldots$) and $\{b_n\}$ converges, then $\lim_{n \to \infty} a_n b_n$ exists.

52. If $\lim_{n \to \infty} a_n b_n$ exists, then both $\lim_{n \to \infty} a_n$ and $\lim_{n \to \infty} b_n$ must exist.

11.2 Solutions to Self-Check Exercises

1. a. From the table,

n	1	2	3	4	5	$\cdots$
$f(n)$	$\frac{1}{2}$	$\frac{2}{5}$	$\frac{3}{10}$	$\frac{4}{17}$	$\frac{5}{26}$	$\cdots$

we obtain the graph shown in the figure.

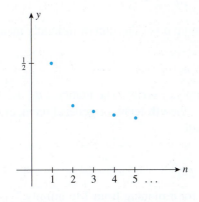

b. We compute

$$f'(x) = \frac{(x^2 + 1)(1) - x(2x)}{(x^2 + 1)^2} = \frac{1 - x^2}{(x^2 + 1)^2} < 0$$

for $x > 1$, so f is decreasing on $(1, \infty)$. We conclude that $\{a_n\}$ is decreasing.

2. We compute

$$\lim_{n \to \infty} \frac{3n^2 + 2n + 1}{n^2 + 4} = \lim_{n \to \infty} \frac{3 + \dfrac{2}{n} + \dfrac{1}{n^2}}{1 + \dfrac{4}{n^2}} \quad \text{Divide numerator and denominator by } n^2.$$

$$= \frac{3}{1} = 3$$

11.3 Infinite Series

The Sum of an Infinite Series

Consider again the example involving the bouncing ball. Earlier, we found a sequence describing the maximum height traveled by the ball on each rebound after hitting a surface. The question that follows naturally is: How do we find the total distance traveled by the ball? To answer this question, recall that the initial height and the heights attained on each subsequent rebound are

$$1, \quad \frac{2}{3}, \quad \left(\frac{2}{3}\right)^2, \quad \left(\frac{2}{3}\right)^3, \quad \cdots$$

meters, respectively (Figure 5).

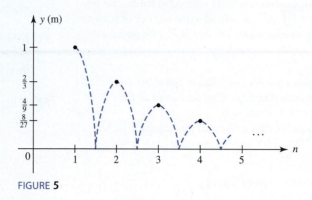

FIGURE 5

Observe that the distance traveled by the ball when it first hits the surface is 1 meter. When the ball hits the surface the second time, it will have traveled a total distance of

$$1 + 2\left(\frac{2}{3}\right) \quad \text{or} \quad 1 + \frac{4}{3}$$

meters. When it hits the surface the third time, it will have traveled a distance of

$$1 + 2\left(\frac{2}{3}\right) + 2\left(\frac{2}{3}\right)^2 \quad \text{or} \quad 1 + \frac{4}{3} + \frac{8}{9}$$

meters. Continuing in this fashion, we see that the total distance traveled by the ball is

$$1 + 2\left(\frac{2}{3}\right) + 2\left(\frac{2}{3}\right)^2 + 2\left(\frac{2}{3}\right)^3 + \cdots \tag{4}$$

meters. Observe that this last expression involves the sum of infinitely many terms. In general, an expression of the form

$$a_1 + a_2 + a_3 + \cdots + a_n + \cdots$$

is called an **infinite series** or, more simply, a **series**. The numbers $a_1, a_2, a_3, \ldots$ are called the **terms** of the series; a_n is called the **nth term,** or **general term,** of the series. The series itself is denoted by the symbol

$$\sum_{n=1}^{\infty} a_n$$

and is read "the sum of the numbers a_n for n running from 1 to infinity."

How do we define the "sum" of an infinite series, if it exists? To answer this question, we use the same technique we have employed several times before: using quantities we can compute to help us define new ones. For example, in defining the slope of the tangent line to the graph of a function, we take the limit of the slopes of secant lines (quantities we can compute); and in defining the area under the graph of a function, we take the limit of the sum of the areas of rectangles (again, quantities we can compute). Here, we define the sum of an infinite series as the limit of a sequence of *finite* sums (quantities that we can compute).

We can get an inkling of how this may be done by examining the series in (4), which gives the total distance traveled by the ball. Define the sequence $\{S_N\}$ by

$$S_1 = 1$$

$$S_2 = 1 + 2\left(\frac{2}{3}\right)$$

$$S_3 = 1 + 2\left(\frac{2}{3}\right) + 2\left(\frac{2}{3}\right)^2$$

$$\vdots$$

$$S_N = 1 + 2\left(\frac{2}{3}\right) + 2\left(\frac{2}{3}\right)^2 + \cdots + 2\left(\frac{2}{3}\right)^{N-1}$$

$$\vdots$$

giving the total vertical distance traveled by the ball when it hits the surface the first time, the second time, the third time, ..., and the Nth time, ..., respectively. If the series in (4) has a sum S (the total distance traveled by the ball), then the terms of the sequence $\{S_N\}$ form a sequence of increasingly accurate approximations to S. This suggests that we define $S = \lim_{N \to \infty} S_N$. (We will complete the solution to this problem in Example 3.)

Motivated by this discussion, we define the Nth partial sum of an infinite series.

Nth Partial Sum of an Infinite Series

Given an infinite series $\displaystyle\sum_{n=1}^{\infty} a_n = a_1 + a_2 + a_3 + \cdots + a_n + \cdots$, the **Nth partial sum** of the series is $\displaystyle S_N = \sum_{n=1}^{N} a_n = a_1 + a_2 + a_3 + \cdots + a_N$. If the sequence of partial sums $\{S_N\}$ **converges** to the number S—that is, if $\lim_{N \to \infty} S_N = S$—then the series $\sum a_n$ **converges** and has **sum** S, written

$$\sum_{n=1}^{\infty} a_n = a_1 + a_2 + a_3 + \cdots + a_n + \cdots = S$$

If $\{S_N\}$ diverges, then the series $\sum a_n$ **diverges**.

 Be sure to note the difference between a sequence and a series. A sequence is a *succession* of terms, whereas a series is a *sum* of terms.

EXAMPLE 1

a. Show that the following infinite series diverges:

$$\sum_{n=1}^{\infty} (-1)^{n+1} = 1 - 1 + 1 - 1 + 1 - 1 + \cdots$$

b. Show that the following infinite series converges:

$$\sum_{n=1}^{\infty} \frac{1}{n(n+1)} = \frac{1}{1 \cdot 2} + \frac{1}{2 \cdot 3} + \frac{1}{3 \cdot 4} + \cdots$$

Solution

a. The partial sums of the given infinite series are

$$S_1 = 1, \quad S_2 = 1 - 1 = 0, \quad S_3 = 1 - 1 + 1 = 1, \quad S_4 = 1 - 1 + 1 - 1 = 0, \quad \ldots$$

The sequence of partial sums $\{S_N\}$ evidently diverges, so the given infinite series diverges.

b. Let's write

$$\frac{1}{n(n+1)} = \frac{1}{n} - \frac{1}{n+1}$$

(Verify this equality.) The Nth partial sum of the given series is

$$
\begin{aligned}
S_N &= \sum_{n=1}^{N} \frac{1}{n(n+1)} = \sum_{n=1}^{N} \left(\frac{1}{n} - \frac{1}{n+1} \right) \\
&= \left(1 - \frac{1}{2} \right) + \left(\frac{1}{2} - \frac{1}{3} \right) + \left(\frac{1}{3} - \frac{1}{4} \right) + \cdots + \left(\frac{1}{N} - \frac{1}{N+1} \right) \\
&= 1 + \left(-\frac{1}{2} + \frac{1}{2} \right) + \left(-\frac{1}{3} + \frac{1}{3} \right) + \cdots + \left(-\frac{1}{N} + \frac{1}{N} \right) - \frac{1}{N+1} \\
&= 1 - \frac{1}{N+1}
\end{aligned}
$$

Since

$$\lim_{N \to \infty} S_N = \lim_{N \to \infty} \left(1 - \frac{1}{N+1} \right) = 1$$

we conclude that the given series converges and has a sum equal to 1; that is,

$$\sum_{n=1}^{\infty} \frac{1}{n(n+1)} = 1$$

The series in Example 1(b) is called a **telescoping series** because all the terms between the first and the last in the third expression for S_N "collapse."

Exploring with TECHNOLOGY

Refer to Example 1b.

1. Verify graphically the results of Example 1b by plotting the points (n, S_n) for $n = 1, 2, \ldots, 6$, where

$$S_1 = \frac{1}{1 \cdot 2}, \quad S_2 = \frac{1}{1 \cdot 2} + \frac{1}{2 \cdot 3}, \quad \ldots, \quad S_6 = \frac{1}{1 \cdot 2} + \frac{1}{2 \cdot 3} + \cdots + \frac{1}{6 \cdot 7}$$

and $S = 1$ in the viewing window $[0, 6] \times [0, 1]$.

2. Refer to the abbreviated form for S_N, namely, $S_N = \left(1 - \dfrac{1}{N+1} \right)$. By

plotting the graphs of $y_1 = \left(1 - \dfrac{1}{x+1} \right)$ and $y_2 = 1$ in the viewing

window $[0, 50] \times [0, 1.1]$, verify graphically that $\lim_{N \to \infty} S_N = 1$ and thus the

result $\displaystyle\sum_{n=1}^{\infty} \frac{1}{n(n+1)} = 1$, as obtained in Example 1b.

Explore and Discuss

1. An example of a divergent infinite series is

$$\sum_{n=0}^{\infty} 1 = 1 + 1 + 1 + \cdots$$

Show that the series is divergent by establishing the following:
a. The Nth partial sum of the infinite series is $S_N = N$.
b. $\lim_{N \to \infty} S_N = \infty$, so $\{S_N\}$ is divergent, and the desired result follows.

2. Since the *terms* of the infinite series in part 1 do not decrease, it is evident that the partial sums of the series must grow without bound. Therefore, it is not difficult to see that the infinite series cannot converge. Now consider the infinite series

$$\sum_{n=1}^{\infty} \frac{1}{n} = 1 + \frac{1}{2} + \frac{1}{3} + \cdots$$

Even though the terms of this series (called the *harmonic series*) approach zero as n goes to infinity, the harmonic series is divergent. Show that this result is true by establishing the following:

a. Observe that $S_2 = 1 + \dfrac{1}{2} > \dfrac{1}{2} + \dfrac{1}{2} = 1$

and

$$S_4 = 1 + \frac{1}{2} + \frac{1}{3} + \frac{1}{4}$$

$$= S_2 + \frac{1}{3} + \frac{1}{4} > 1 + \left(\frac{1}{4} + \frac{1}{4}\right) = \frac{3}{2}$$

Using a similar argument, show that $S_8 > \frac{4}{2}$ and $S_{16} > \frac{5}{2}$. Conclude that, in general,

$$S_{2^n} > \frac{n+1}{2}$$

b. Explain why $S_1 < S_2 < S_3 < \cdots < S_n < \cdots$.
c. Use the results of parts (a) and (b) to explain why the harmonic series is divergent.

Geometric Series

In general, it is not easy to determine whether a given infinite series is convergent or divergent. It is even more difficult to determine the sum of an infinite series that is known to be convergent. But there is an important and useful series whose sum, if it exists, is easy to find.

> **Geometric Series**
>
> A **geometric series** with common ratio r is a series of the form
>
> $$\sum_{n=0}^{\infty} ar^n = a + ar + ar^2 + ar^3 + \cdots + ar^n + \cdots \quad (a \neq 0) \quad \textbf{(5)}$$

The ratio here refers to the ratio of two consecutive terms.

Note that we begin the summation for this series with $n = 0$ instead of $n = 1$. To determine the conditions under which the geometric series (5) converges and find its sum, let's consider the Nth partial sum of the infinite series

$$S_N = a + ar + ar^2 + \cdots + ar^N$$

Multiplying both sides of the equation by r gives

$$rS_N = ar + ar^2 + ar^3 + \cdots + ar^{N+1}$$

Subtracting the second equation from the first yields

$$S_N - rS_N = a - ar^{N+1}$$
$$(1 - r)S_N = a(1 - r^{N+1})$$
$$S_N = \frac{a(1 - r^{N+1})}{1 - r}$$

provided that $r \neq 1$. You are asked to show that if $|r| \geq 1$, then the series (5) diverges (Exercise 44). On the other hand, observe that if $|r| < 1$ (that is, if $-1 < r < 1$), then

$$\lim_{n \to \infty} r^{n+1} = 0$$

For example, if $r = \frac{1}{2}$, then

$$\lim_{n \to \infty} r^{n+1} = \lim_{n \to \infty}\left(\frac{1}{2}\right)^{n+1} = \lim_{n \to \infty} \frac{1}{2^{n+1}} = 0$$

Using this fact, together with the properties of limits stated earlier, we see that

$$\lim_{N \to \infty} S_N = \lim_{N \to \infty} \frac{a(1 - r^{N+1})}{1 - r}$$

$$= \frac{a}{1 - r} \lim_{N \to \infty} (1 - r^{N+1})$$

$$= \frac{a}{1 - r}$$

These results are summarized in Theorem 2.

THEOREM 2

If $|r| < 1$, then the geometric series

$$\sum_{n=0}^{\infty} ar^n = a + ar + ar^2 + \cdots$$

converges, and its sum is $\dfrac{a}{1 - r}$; that is,

$$\sum_{n=0}^{\infty} ar^n = a + ar + ar^2 + \cdots = \frac{a}{1 - r} \qquad (6)$$

The series diverges if $|r| \geq 1$.

EXAMPLE 2 Show that the infinite series is a geometric series, and find its sum if it is convergent.

a. $\displaystyle\sum_{n=0}^{\infty} \frac{1}{2^n}$ **b.** $\displaystyle\sum_{n=0}^{\infty} 3\left(\frac{5}{2}\right)^n$ **c.** $\displaystyle\sum_{n=1}^{\infty} 5\left(-\frac{3}{4}\right)^n$

Solution

a. Observe that

$$\frac{1}{2^n} = \left(\frac{1}{2}\right)^n$$

so

$$\sum_{n=0}^{\infty} \frac{1}{2^n} = \sum_{n=0}^{\infty} \left(\frac{1}{2}\right)^n$$

is a geometric series with $a = 1$ and $r = \frac{1}{2}$. Since $|r| = \frac{1}{2} < 1$, we conclude that the series is convergent. Finally, using (6), we have

$$\sum_{n=0}^{\infty} \left(\frac{1}{2}\right)^n = 1 + \frac{1}{2} + \frac{1}{4} + \cdots = \frac{1}{1 - \frac{1}{2}} = 2$$

b. This is a geometric series with $a = 3$ and $r = \frac{5}{2}$. Since

$$|r| = \left|\frac{5}{2}\right| = \frac{5}{2} > 1$$

we deduce that the series is divergent.

c. The summation here begins with $n = 1$. However, we may rewrite the series as

$$\sum_{n=1}^{\infty} 5\left(-\frac{3}{4}\right)^n = 5\left(-\frac{3}{4}\right) + 5\left(-\frac{3}{4}\right)^2 + 5\left(-\frac{3}{4}\right)^3 + \cdots$$

$$= 5\left(-\frac{3}{4}\right)\left[1 + \left(-\frac{3}{4}\right) + \left(-\frac{3}{4}\right)^2 + \cdots\right]$$

$$= \sum_{n=0}^{\infty} \left(\frac{-15}{4}\right)\left(-\frac{3}{4}\right)^n$$

which is just a geometric series with $a = -\frac{15}{4}$ and $r = -\frac{3}{4}$. Since $|r| = \left|-\frac{3}{4}\right| = \frac{3}{4} < 1$, the series is convergent. Using Formula (6), we find

$$\sum_{n=1}^{\infty} 5\left(-\frac{3}{4}\right)^n = \sum_{n=0}^{\infty} \left(-\frac{15}{4}\right)\left(-\frac{3}{4}\right)^n = \frac{-\frac{15}{4}}{1 - \left(-\frac{3}{4}\right)} = -\frac{\frac{15}{4}}{\frac{7}{4}} = -\frac{15}{7}$$

APPLIED EXAMPLE 3 A Bouncing Ball Complete the solution of the bouncing ball problem introduced at the beginning of this section. Recall that the total vertical distance traveled by the ball is given by

$$1 + 2\left(\frac{2}{3}\right) + 2\left(\frac{2}{3}\right)^2 + 2\left(\frac{2}{3}\right)^3 + \cdots$$

meters.

Solution If we denote the total vertical distance traveled by the ball by d, then

$$d = 1 + \sum_{n=0}^{\infty} \left(\frac{4}{3}\right)\left(\frac{2}{3}\right)^n$$

The expression after the first term is a geometric series with $a = \frac{4}{3}$ and $r = \frac{2}{3}$. Using Theorem 2, we obtain

$$d = 1 + \frac{\frac{4}{3}}{1 - \frac{2}{3}} = 1 + 4 = 5$$

and conclude that the total distance traveled by the ball is 5 meters.

Properties of Infinite Series

The following properties of infinite series enable us to perform algebraic operations on convergent series.

> **Properties of Infinite Series**
>
> If $\displaystyle\sum_{n=1}^{\infty} a_n$ and $\displaystyle\sum_{n=1}^{\infty} b_n$ are convergent infinite series and c is a constant, then
>
> **1.** $\displaystyle\sum_{n=1}^{\infty} ca_n = c\sum_{n=1}^{\infty} a_n$ **2.** $\displaystyle\sum_{n=1}^{\infty} (a_n \pm b_n) = \sum_{n=1}^{\infty} a_n \pm \sum_{n=1}^{\infty} b_n$

Thus, we may multiply each term of a convergent series by a constant c, which results in a convergent series whose sum is c times the sum of the original series. We may also add (subtract) the corresponding terms of two convergent series, which gives a convergent series whose sum is the sum (difference) of the sums of the original series.

EXAMPLE 4 Find the sum of the following series if it exists:

$$\sum_{n=0}^{\infty} \frac{2 \cdot 3^n - 2^n}{5^n} = 1 + \frac{4}{5} + \frac{14}{25} + \cdots$$

Solution   Note that this series starts with $n = 0$. We can write

$$\sum_{n=0}^{\infty} \frac{2 \cdot 3^n - 2^n}{5^n} = \sum_{n=0}^{\infty} \left(\frac{2 \cdot 3^n}{5^n} - \frac{2^n}{5^n}\right)$$

Now observe that

$$\sum_{n=0}^{\infty} \frac{2 \cdot 3^n}{5^n} \quad \text{and} \quad \sum_{n=0}^{\infty} \frac{2^n}{5^n}$$

are convergent geometric series with ratios $|r| = \frac{3}{5} < 1$ and $|r| = \frac{2}{5} < 1$, respectively. Therefore, using Property 2 of infinite series, we have

$$\sum_{n=0}^{\infty} \frac{2 \cdot 3^n - 2^n}{5^n} = \sum_{n=0}^{\infty} \frac{2 \cdot 3^n}{5^n} - \sum_{n=0}^{\infty} \frac{2^n}{5^n}$$

$$= 2\sum_{n=0}^{\infty} \frac{3^n}{5^n} - \sum_{n=0}^{\infty} \frac{2^n}{5^n}$$

$$= 2\sum_{n=0}^{\infty} \left(\frac{3}{5}\right)^n - \sum_{n=0}^{\infty} \left(\frac{2}{5}\right)^n$$

$$= 2\left(\frac{1}{1 - \frac{3}{5}}\right) - \left(\frac{1}{1 - \frac{2}{5}}\right)$$

$$= 2\left(\frac{5}{2}\right) - \frac{5}{3} = \frac{10}{3}$$

We now consider some additional applications of geometric series.

EXAMPLE 5 Find the rational number that has the repeated decimal representation $0.222\ldots$.

Solution   By definition, the decimal representation

$$0.222\ldots = \frac{2}{10} + \frac{2}{100} + \frac{2}{1000} + \cdots$$

$$= \frac{2}{10}\left(1 + \frac{1}{10} + \frac{1}{100} + \cdots\right)$$

$$= \sum_{n=0}^{\infty} \left(\frac{2}{10}\right)\left(\frac{1}{10}\right)^n$$

which is a geometric series with $a = \frac{2}{10}$ and $r = \frac{1}{10}$. Since $|r| = r = \frac{1}{10} < 1$, the series converges. In fact, using Formula (6), we have

$$0.222\ldots = \frac{\left(\frac{2}{10}\right)}{1 - \frac{1}{10}} = \frac{\frac{2}{10}}{\frac{9}{10}} = \frac{2}{9}$$

The following example illustrates a phenomenon in economics known as the **multiplier effect**.

APPLIED EXAMPLE 6 The Multiplier Effect Suppose the average wage earner saves 10% of her take-home pay and spends the other 90%. Estimate the impact that a proposed $20 billion tax cut will have on the economy over the long run in terms of the additional spending generated.

Solution   Of the $20 billion received by the original beneficiaries of the proposed tax cut, $(0.9)(20)$ billion dollars will be spent. Of the $(0.9)(20)$ billion dollars reinjected into the economy, 90% of it, or $(0.9)(0.9)(20)$ billion dollars, will find its way into the economy again. This process will go on ad infinitum, so this one-time proposed tax cut will result in additional spending over the years in the amount of

$$(0.9)(20) + (0.9)^2(20) + (0.9)^3(20) + \cdots = (0.9)(20)\left[1 + 0.9 + 0.9^2 + 0.9^3 + \cdots\right]$$
$$= 18\left[\frac{1}{1 - 0.9}\right]$$
$$= 180$$

or $180 billion.

A **perpetuity** is a sequence of payments made at regular time intervals and continuing on forever. The **capital value of a perpetuity** is the sum of the present values of all future payments. The following example illustrates these concepts.

APPLIED EXAMPLE 7 Establishing a Scholarship Fund The Robinson family wishes to establish a scholarship fund at a college. If a scholarship in the amount of $5000 is to be awarded on an annual basis beginning next year, find the amount of the endowment they are required to make now. Assume that this fund will earn interest at a rate of 10% per year compounded continuously.

Solution   The amount of the endowment, A, is given by the sum of the present values of the amounts awarded annually in perpetuity. Now, the present value of the amount of the first award is equal to

$$5000e^{-0.1(1)}$$

(see Section 6.7). The present value of the amount of the second award is

$$5000e^{-0.1(2)}$$

and so on. Continuing, we see that the present value of the amount of the nth award is

$$5000e^{-0.1(n)}$$

Therefore, the amount of the endowment is

$$A = 5000e^{-0.1(1)} + 5000e^{-0.1(2)} + \cdots + 5000e^{-0.1(n)} + \cdots$$

To find the sum of the infinite series on the right-hand side, let

$$r = e^{-0.1}$$

Then

$$A = 5000r^1 + 5000r^2 + \cdots + 5000r^n + \cdots$$
$$= 5000r(1 + r + r^2 + \cdots + r^n + \cdots)$$

The series within the parentheses is a geometric series with $|r| = e^{-0.1} \approx 0.905 < 1$, so using (6), we find

$$A = 5000r\left(\frac{1}{1-r}\right) = \frac{5000e^{-0.1}}{1 - e^{-0.1}} \approx 47{,}541.66$$

Thus, the amount of the endowment is $47,541.66.

Our final example is an application of a geometric series in the field of medicine.

APPLIED EXAMPLE 8 Residual Drug in the Bloodstream A patient will be given 5 units of a certain drug daily for an indefinite period of time. For this particular drug, it is known that the fraction of a dose that remains in the patient's body after t days is given by $e^{-0.3t}$. Determine the residual amount of the drug that may be expected to be in the patient's body after extended treatment.

Solution The amount of the drug in the patient's body 1 day after the first dose is administered and prior to administration of the second dose is $5e^{-0.3}$ units. The amount of the drug in the patient's body 2 days later and prior to administration of the third dose consists of the residuals from the first two doses. Of the first dose, $5e^{-(0.3)2}$ units of the drug are left in the patient's body, and of the second dose, $5e^{-0.3}$ units of the drug are left. Thus, the amount of the drug 2 days later is given by

$$5e^{-0.3} + 5e^{-0.3(2)}$$

units. Continuing, we see that the amount of the drug left in the patient's body in the long run and just prior to administration of a fresh dose is given by

$$R = 5e^{-0.3} + 5e^{-0.3(2)} + 5e^{-0.3(3)} + \cdots$$

To find the sum of the infinite series, we let

$$r = e^{-0.3}$$

Then

$$A = 5r + 5r^2 + 5r^3 + \cdots = 5r(1 + r + r^2 + \cdots)$$
$$= \frac{5r}{1-r} = \frac{5e^{-0.3}}{1 - e^{-0.3}} \approx 14.29$$

Therefore, after extended treatment, the residual amount of drug in the patient's body is approximately 14.29 units.

11.3 Self-Check Exercises

1. Determine whether the geometric series

$$\sum_{n=0}^{\infty} 5\left(-\frac{1}{3}\right)^n = 5 - 5\left(\frac{1}{3}\right) + 5\left(\frac{1}{9}\right) - \cdots$$

is convergent or divergent. If it is convergent, find its sum.

2. **THE MULTIPLIER EFFECT** Suppose the average wage earner in a certain country saves 12% of his take-home pay and spends the other 88%. Estimate the impact that a proposed $10 billion tax cut will have on the economy over the long run due to the additional spending generated.

Solutions to Self-Check Exercises 11.3 can be found on page 754.

11.3 Concept Questions

1. Explain the difference between the following:
   a. A sequence and a series
   b. A convergent sequence and a convergent series
   c. A divergent sequence and a divergent series
   d. The limit of a sequence and the sum of a series

2. Suppose $\sum_{n=1}^{\infty} a_n = 6$.

   a. Evaluate $\lim_{N \to \infty} S_N$, where S_N is the Nth partial sum of $\sum_{n=1}^{\infty} a_n$.

   b. Find $\sum_{n=2}^{\infty} a_n$ if it is known that $a_1 = \frac{1}{2}$.

11.3 Exercises

In Exercises 1–4, find the Nth partial sum of the infinite series and evaluate its limit to determine whether the series converges or diverges. If the series is convergent, find its sum.

1. $\sum_{n=1}^{\infty} (-2)^n$

2. $\sum_{n=1}^{\infty} \left(\frac{1}{n+3} - \frac{1}{n+4} \right)$

3. $\sum_{n=1}^{\infty} \frac{1}{n^2 + 3n + 2}$

4. $\sum_{n=2}^{\infty} \left(\frac{1}{\ln n} - \frac{1}{\ln(n+1)} \right)$

Hint: $\dfrac{1}{n^2 + 3n + 2} = \dfrac{1}{n+1} - \dfrac{1}{n+2}$

In Exercises 5–16, determine whether the geometric series converges or diverges. If it converges, find its sum.

5. $\sum_{n=0}^{\infty} \left(\frac{1}{3} \right)^n$

6. $\sum_{n=0}^{\infty} 4 \left(-\frac{2}{3} \right)^n$

7. $\sum_{n=0}^{\infty} 2(1.01)^n$

8. $\sum_{n=0}^{\infty} 3(0.9)^n$

9. $\sum_{n=0}^{\infty} \frac{(-2)^n}{3^n}$

10. $\sum_{n=0}^{\infty} \frac{3}{2^n}$

11. $\sum_{n=0}^{\infty} \frac{2^n}{3^{n+2}}$

12. $\sum_{n=0}^{\infty} \frac{3^{n+1}}{4^{n-1}}$

13. $\sum_{n=0}^{\infty} e^{-0.2n}$

14. $\sum_{n=0}^{\infty} 2e^{-0.1n}$

15. $\sum_{n=0}^{\infty} \left(-\frac{3}{\pi} \right)^n$

16. $\sum_{n=1}^{\infty} \frac{e^n}{3^{n+1}}$

In Exercises 17–26, determine whether the series converges or diverges. If it converges, find its sum.

17. $8 + 4 + \dfrac{1}{2} + \dfrac{1}{4} + \dfrac{1}{8} + \cdots$ (Assume that the "obvious" pattern continues.)

18. $1 + 0.2 + 0.04 + 0.008 + \cdots$ (Assume that the "obvious" pattern continues.)

19. $3 - \dfrac{1}{3} + \dfrac{1}{9} - \dfrac{1}{27} + \cdots$ (Assume that the "obvious" pattern continues.)

20. $5 - 1.01 + (1.01)^2 - (1.01)^3 + \cdots$ (Assume that the "obvious" pattern continues.)

21. $\sum_{n=0}^{\infty} \frac{3 + 2^n}{3^n}$

22. $\sum_{n=0}^{\infty} \frac{2^n - 3^n}{4^n}$

23. $\sum_{n=0}^{\infty} \frac{3 \cdot 2^n + 4^n}{3^n}$

24. $\sum_{n=0}^{\infty} \frac{(2 \cdot 3^n) - (3 \cdot 5^n)}{7^n}$

25. $\sum_{n=1}^{\infty} \left[\left(\frac{e}{\pi} \right)^n + \left(\frac{\pi}{e^2} \right)^n \right]$

26. $\sum_{n=1}^{\infty} \left[\frac{1}{2^n} - \frac{1}{n(n+1)} \right]$

Hint: $\dfrac{1}{n(n+1)} = \dfrac{1}{n} - \dfrac{1}{n+1}$

In Exercises 27–30, express the decimal as a rational number.

27. $0.3333\ldots$

28. $0.121212\ldots$

29. $1.213213213\ldots$

30. $6.23143143143\ldots$

In Exercises 31–34, find the values of x for which the series converges, and find the sum of the series.

Hint: First show that the series is a geometric series.

31. $\sum_{n=0}^{\infty} (-x)^n$

32. $\sum_{n=0}^{\infty} (x - 2)^n$

33. $\sum_{n=1}^{\infty} 2^n (x - 1)^n$

34. $\sum_{n=0}^{\infty} \frac{x^{2n}}{3^n}$

35. **EFFECT OF A TAX CUT ON SPENDING** Suppose that the average wage earner saves 9% of her take-home pay and spends the other 91%. Estimate the impact that a proposed $30 billion tax cut will have on the economy over the long run due to the additional spending generated.

36. **A BOUNCING BALL** A ball is dropped from a height of 10 m. After hitting the ground, it rebounds to a height of 5 m and then continues to rebound at one half of its former height thereafter. Find the total vertical distance traveled by the ball before it comes to rest.

37. **WINNING A TOSS** Peter and Paul take turns tossing a pair of dice. The first to throw a 7 wins. If Peter starts the game, then it can be shown that his chances of winning are given by

$$p = \frac{1}{6} + \left(\frac{1}{6} \right) \left(\frac{5}{6} \right)^2 + \left(\frac{1}{6} \right) \left(\frac{5}{6} \right)^4 + \cdots$$

Find p.

38. ENDOWMENTS Hal Corporation wants to establish a fund to provide the art center of a large metropolitan area with an annual grant of $250,000 beginning next year. If the fund will earn interest at a rate of 10%/year compounded continuously, find the amount of endowment the corporation must make at this time.

39. TRANSMISSION OF DISEASE Refer to Exercise 48, page 743. It can be shown that the total number of individuals, S_n, who have contracted the disease some time between the first and nth day is approximated by

$$S_n = a_1 + aNa_1 + aNa_2 + \cdots + aNa_{n-1}$$

Show that if $r < 1$, where $r = 1 + aN - b$, then the total number of persons who will have contracted the disease at some stage of the epidemic is no larger than $a_1 b/(b - aN)$.

40. CAPITAL VALUE OF A PERPETUITY Find a formula for the capital value of a perpetuity involving payments of P dollars each, paid at the end of each of m periods/year into a fund that earns interest at the nominal rate of r%/year compounded m times/year by verifying the following:

a. The present value of the nth payment is

$$P\left(1 + \frac{r}{m}\right)^{-n}$$

b. The capital value is

$$A = P\left(1 + \frac{r}{m}\right)^{-1} + P\left(1 + \frac{r}{m}\right)^{-2} + P\left(1 + \frac{r}{m}\right)^{-3} + \cdots$$

c. Using the fact that the series in part (b) is a geometric series, its sum is

$$A = \frac{mP}{r}$$

41. CAPITAL VALUE OF A PERPETUITY Find a formula for the capital value of a perpetuity involving payments of P dollars paid at the end of each investment period into a fund that earns interest at the rate of r%/year compounded continuously.
Hint: Refer to Exercise 40.

42. RESIDUAL DRUG IN THE BLOODSTREAM Ten units of a certain drug are administered to a patient on a daily basis. The fraction of this drug that remains in the patient's bloodstream after t days is given by

$$f(t) = e^{-t/4}$$

After extended treatment with the drug, what is the residual amount of the drug in the patient's bloodstream just before a new dose is administered?

43. RESIDUAL DRUG IN THE BLOODSTREAM Suppose that a dose of C units of a certain drug is administered to a patient and the fraction of the dose remaining in the patient's bloodstream t hr after the dose is administered is given by Ce^{-kt}, where k is a positive constant.

a. Show that the residual concentration of the drug in the bloodstream just before a new dose is administered after extended treatment with the drug is given by

$$R = \frac{Ce^{-kt}}{1 - e^{-kt}}$$

b. If the highest concentration of this particular drug that is considered safe is S units, find the minimal time that must exist between doses.
Hint: $C + R \le S$

44. Let

$$\sum_{n=0}^{\infty} ar^n = a + ar + ar^2 + ar^3 + \cdots + ar^n + \cdots$$

be a geometric series with common ratio r. Show that if $|r| \ge 1$, then the series diverges.

In Exercises 45–48, determine whether the statement is true or false. If it is true, explain why it is true. If it is false, give an example to show why it is false.

45. If $\displaystyle\sum_{n=0}^{\infty} (a_n + b_n)$ converges, then both $\displaystyle\sum_{n=0}^{\infty} a_n$ and $\displaystyle\sum_{n=0}^{\infty} b_n$ must converge.

46. If $\displaystyle\sum_{n=0}^{\infty} a_n$ converges and $\displaystyle\sum_{n=0}^{\infty} b_n$ converges, then $\displaystyle\sum_{n=0}^{\infty} (ca_n + db_n)$ also converges, where c and d are constants.

47. If $|r| < 1$, then $\displaystyle\sum_{n=0}^{\infty} |r^n| = \frac{1}{1 - |r|}$.

48. If $|r| > 1$, then $\displaystyle\sum_{n=1}^{\infty} \frac{1}{r^n} = \frac{1}{r - 1}$.

11.3 Solutions to Self-Check Exercises

1. This is a geometric series with $r = -\frac{1}{3}$ and $a = 5$. Since $\left|-\frac{1}{3}\right| < 1$, we see that the series is convergent. Its sum is

$$\frac{5}{1 - \left(-\frac{1}{3}\right)} = \frac{5}{\frac{4}{3}} = \frac{15}{4} = 3\frac{3}{4}$$

2. Of the $10 billion received by the original beneficiaries of the proposed cut, $(0.88)(10)$ billion dollars will be spent. Of the $(0.88)(10)$ billion dollars reinjected into the economy, 88% of it, or $(0.88)(0.88)(10)$ billion dollars, will find its way into the economy again. This process will go

on ad infinitum, so this one-time proposed tax cut will result in additional spending over the years in the amount of

$$(0.88)(10) + (0.88)^2(10) + (0.88)^3(10) + \cdots$$
$$= (0.88)(10)[1 + 0.88 + (0.88)^2 + \cdots]$$
$$= 8.8\left[\frac{1}{1 - 0.88}\right]$$
$$\approx 73.3$$

or $73.3 billion.

11.4 Series with Positive Terms

The convergence or divergence of a telescoping series or a geometric series is relatively easy to determine because we can find a simple formula for the Nth partial sum S_N of these series. Very often, obtaining a simple formula for the Nth partial sum of an infinite series is difficult or impossible, and we are forced to look for alternative ways to investigate the convergence or divergence of the series.

In this section, we look at several tests for determining the convergence or divergence of an infinite series by examining its nth term a_n. These tests will confirm the convergence of a series without yielding a value for its sum. Often, this is all that is needed. Once it has been ascertained that a series is convergent, we can approximate its sum to any degree of accuracy desired by adding up the terms of its Nth partial sum S_N, provided that N is chosen large enough.

Our first test tells us how to identify a divergent series.

The Test for Divergence

The following theorem tells us that the terms of a convergent series must ultimately approach zero.

THEOREM 3

If $\displaystyle\sum_{n=1}^{\infty} a_n$ converges, then $\displaystyle\lim_{n\to\infty} a_n = 0$.

To prove this result, let

$$S_n = a_1 + a_2 + \cdots + a_{n-1} + a_n = S_{n-1} + a_n$$

Then

$$a_n = S_n - S_{n-1}$$

Since $\displaystyle\sum_{n=1}^{\infty} a_n$ is convergent, the sequence $\{S_n\}$ is convergent. Let $\displaystyle\lim_{n\to\infty} S_n = S$. Then

$$\lim_{n\to\infty} a_n = \lim_{n\to\infty} (S_n - S_{n-1}) = \lim_{n\to\infty} S_n - \lim_{n\to\infty} S_{n-1} = S - S = 0$$

An important consequence of Theorem 3 is the following useful test for *divergence*.

THEOREM 4

The Test for Divergence

If $\displaystyle\lim_{n\to\infty} a_n$ does not exist or $\displaystyle\lim_{n\to\infty} a_n \neq 0$, then $\displaystyle\sum_{n=1}^{\infty} a_n$ diverges.

Observe that the test for divergence does *not* say that if $\displaystyle\lim_{n\to\infty} a_n = 0$, then $\displaystyle\sum_{n=1}^{\infty} a_n$ must converge. In other words, the converse of Theorem 4 is not true in general. For example, $\displaystyle\lim_{n\to\infty} \frac{1}{n} = 0$, yet, as we will see later, the **harmonic series** $\displaystyle\sum_{n=1}^{\infty} \frac{1}{n}$ is divergent. (See Example 3.) In short, the test for divergence rules out convergence for a series whose nth term does not approach zero but yields no information if the nth term of a series does approach zero; that is, the series may or may not converge.

EXAMPLE 1 Show that the following series are divergent:

a. $\displaystyle\sum_{n=1}^{\infty}(-1)^{n-1}$ **b.** $\displaystyle\sum_{n=1}^{\infty}\frac{2n^2+1}{3n^2-1}$

Solution

a. Here, $a_n = (-1)^{n-1}$. Since

$$\lim_{n\to\infty} a_n = \lim_{n\to\infty} (-1)^{n-1}$$

does not exist, we conclude by the test for divergence that the series diverges.

b. Here,

$$\lim_{n\to\infty} a_n = \lim_{n\to\infty}\frac{2n^2+1}{3n^2-1} = \lim_{n\to\infty}\frac{2+\dfrac{1}{n^2}}{3-\dfrac{1}{n^2}} = \frac{2}{3} \neq 0$$

so by the test for divergence, the series diverges.

We now look at several tests that tell us whether a series is convergent. These tests apply only to series with positive terms.

The Integral Test

The integral test ties the convergence or divergence of an infinite series $\displaystyle\sum_{n=1}^{\infty} a_n$ to the convergence or divergence of the improper integral $\int_1^{\infty} f(x)\,dx$, where $f(n) = a_n$.

> **THEOREM 5**
>
> **The Integral Test**
>
> Suppose that f is a continuous, positive, and decreasing function on $[1, \infty)$. If $f(n) = a_n$ for all $n \geq 1$, then
>
> $$\sum_{n=1}^{\infty} a_n \quad \text{and} \quad \int_1^{\infty} f(x)\,dx$$
>
> either both converge or both diverge.

Let's give an intuitive justification for this theorem. Figure 6a shows the graph of f and the inscribed rectangles $a_2, a_3, \ldots, a_N$. If you examine the figure, you will see that the height of the first rectangle is $a_2 = f(2)$.

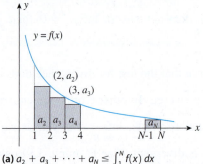   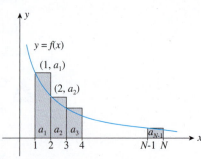

(a) $a_2 + a_3 + \cdots + a_N \leq \int_1^N f(x)\,dx$ **(b)** $\int_1^N f(x)\,dx \leq a_1 + a_2 + \cdots + a_{N-1}$

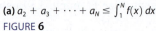

FIGURE **6**

Since this rectangle has width 1, the area of the rectangle is also $a_2 = f(2)$. Similarly, the area of the second rectangle is a_3, and so on. Comparing the sum of the areas of the first $(N-1)$ inscribed rectangles with the area under the graph of f over the interval $[1, N]$, we see that

$$a_2 + a_3 + \cdots + a_N \le \int_1^N f(x)\, dx$$

which implies that

$$S_N = a_1 + a_2 + a_3 + \cdots + a_N \le a_1 + \int_1^N f(x)\, dx$$

If $\int_1^\infty f(x)\, dx$ is convergent and has value L, then

$$S_N \le a_1 + \int_1^N f(x)\, dx \le a_1 + L$$

This shows that $\{S_N\}$ is bounded above. Also,

$$S_{N+1} = S_N + a_{N+1} \ge S_N \qquad \text{Because } a_{N+1} = f(N+1) \ge 0$$

shows that $\{S_N\}$ is increasing as well.

Intuitively, such a sequence must converge to a number no greater than its upper bound. (Although this result can be demonstrated with mathematical rigor, we will not do so here.) In other words $\sum_{n=1}^\infty a_n$ is convergent.

Next, by examining Figure 6b, you will see that

$$\int_1^N f(x)\, dx \le a_1 + a_2 + \cdots + a_{N-1} = S_{N-1}$$

So if $\int_1^\infty f(x)\, dx$ diverges to infinity, then $\lim_{N \to \infty} S_N = \infty$, so $\sum_{n=1}^\infty a_n$ is divergent.

Notes

1. The integral test simply tells us whether a series converges or diverges. If it indicates that a series converges, we may not conclude that the (finite) value of the improper integral used in conjunction with the test is the *sum* of the convergent series.

2. Since the convergence of an infinite series is not affected by the omission or addition of a finite number of terms to the series, we sometimes study the series $\sum_{n=N}^\infty a_n = a_N + a_{N+1} + \cdots$ rather than the series $\sum_{n=1}^\infty a_n$. In this case, the series is compared to the improper integral $\int_N^\infty f(x)\, dx$, as we will see in Example 4. ■

EXAMPLE 2 Use the integral test to determine whether $\sum_{n=1}^\infty \dfrac{1}{n^2}$ converges or diverges.

Solution   Here,

$$a_n = f(n) = \frac{1}{n^2}$$

so we consider the function $f(x) = 1/x^2$. Since f is continuous, positive, and decreasing on $[1, \infty)$, we may use the integral test.

Now

$$\int_1^\infty \frac{1}{x^2}\, dx = \lim_{b\to\infty} \int_1^b x^{-2}\, dx = \lim_{b\to\infty}\left[-\frac{1}{x}\bigg|_1^b\right]$$

$$= \lim_{b\to\infty}\left(-\frac{1}{b} + 1\right) = 1$$

Since $\int_1^\infty \frac{1}{x^2}\, dx$ converges, we conclude that $\sum_{n=1}^\infty \frac{1}{n^2}$ converges as well.

EXAMPLE 3 Use the integral test to determine whether the harmonic series $\sum_{n=1}^\infty \frac{1}{n}$ converges or diverges.

Solution Here, $a_n = f(n) = 1/n$, so we consider the function $f(x) = 1/x$. Since f is continuous, positive, and decreasing on $[1, \infty)$, we may use the integral test. But as you may verify,

$$\int_1^\infty \frac{1}{x}\, dx = \infty$$

We conclude that $\sum_{n=1}^\infty \frac{1}{n}$ diverges.

EXAMPLE 4 Use the integral test to determine whether $\sum_{n=2}^\infty \frac{\ln n}{n}$ converges or diverges.

Solution Here, $a_n = (\ln n)/n$, so we consider the function $f(x) = (\ln x)/x$. Observe that f is continuous and positive on $[2, \infty)$. Next, we compute

$$f'(x) = \frac{x\left(\dfrac{1}{x}\right) - \ln x}{x^2} = \frac{1 - \ln x}{x^2}$$

Note that $f'(x) < 0$ if $\ln x > 1$, that is, if $x > e$. This shows that f is decreasing on $[3, \infty)$. (See the Note on page 757.) Therefore, we may use the integral test. Now

$$\int_3^\infty \frac{\ln x}{x}\, dx = \lim_{b\to\infty} \int_3^b \frac{\ln x}{x}\, dx = \lim_{b\to\infty}\left[\frac{1}{2}(\ln x)^2 \bigg|_3^b\right]$$

$$= \lim_{b\to\infty} \frac{1}{2}\left[(\ln b)^2 - (\ln 3)^2\right] = \infty$$

and we conclude that $\sum_{n=2}^\infty \frac{\ln n}{n}$ diverges.

The *p*-Series

The following series will play an important role in our work later on.

***p*-Series**

A *p-series* is a series of the form

$$\sum_{n=1}^\infty \frac{1}{n^p} = 1 + \frac{1}{2^p} + \frac{1}{3^p} + \cdots + \frac{1}{n^p} + \cdots$$

where p is a constant.

Observe that for $p = 1$, the p-series is just the harmonic series $\sum\limits_{n=1}^{\infty} \dfrac{1}{n}$. The conditions for the convergence or divergence of the p-series can be found by applying the integral test to the series.

THEOREM 6

Convergence of p-Series

The p-series $\sum\limits_{n=1}^{\infty} \dfrac{1}{n^p}$ converges if $p > 1$ and diverges if $p \leq 1$.

Proof   If $p < 0$, then

$$\lim_{n \to \infty} \frac{1}{n^p} = \infty$$

If $p = 0$, then

$$\lim_{n \to \infty} \frac{1}{n^p} = 1$$

In either case, $\lim\limits_{n \to \infty} \dfrac{1}{n^p} \neq 0$, so the p-series diverges by the test for divergence. If $p > 0$, then the function $f(x) = 1/x^p$ is continuous, positive, and decreasing on $[1, \infty)$. It can be shown that $\displaystyle\int_1^{\infty} \dfrac{1}{x^p}\, dx$ converges if $p > 1$ and diverges if $p \leq 1$ (see Exercise 53). Using this result and the integral test, we conclude that $\sum\limits_{n=1}^{\infty} \dfrac{1}{n^p}$ converges if $p > 1$ and diverges if $0 < p \leq 1$. Therefore, $\sum\limits_{n=1}^{\infty} \dfrac{1}{n^p}$ converges if $p > 1$ and diverges if $p \leq 1$.

EXAMPLE 5 Determine whether each series converges or diverges.

a. $\sum\limits_{n=1}^{\infty} \dfrac{1}{n^2}$ **b.** $\sum\limits_{n=1}^{\infty} \dfrac{1}{\sqrt{n}}$ **c.** $\sum\limits_{n=1}^{\infty} n^{-1.001}$

Solution

a. This is a p-series with $p = 2 > 1$, so by Theorem 6, the series converges.

b. Rewriting the series in the form $\sum\limits_{n=1}^{\infty} \dfrac{1}{n^{1/2}}$, we see that the series is a p-series with $p = \frac{1}{2} < 1$, so by Theorem 6, it diverges.

c. We rewrite the series in the form $\sum\limits_{n=1}^{\infty} \dfrac{1}{n^{1.001}}$, which we recognize to be a p-series with $p = 1.001 > 1$, and we conclude accordingly that the series converges. ◼

The Comparison Test

The convergence or divergence of a given series $\sum a_n$ can sometimes be determined by comparing its terms with the terms of a *test series* that is known to be convergent or divergent. This is the basis for the comparison test for series that follows. In the rest of this section, we assume that all series under consideration have positive terms.

Suppose that the terms of a series $\sum a_n$ are smaller than the corresponding terms of a series $\sum b_n$. This situation is illustrated in Figure 7, where the respective terms are represented by rectangles, each of width 1 and appropriate height.

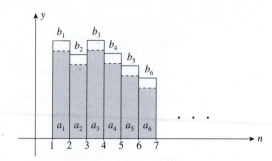

FIGURE 7
Each rectangle representing a_n is contained in the rectangle representing b_n.

If $\sum b_n$ is convergent, the total area of the rectangles representing this series is finite. Since each rectangle representing the terms of the series $\sum a_n$ is contained in a corresponding rectangle representing the terms of $\sum b_n$, the total area of the rectangles representing $\sum a_n$ must also be finite; that is, the series $\sum a_n$ must be convergent. A similar argument would seem to suggest that if all the terms of a series $\sum a_n$ are larger than the corresponding terms of a series $\sum b_n$ that is known to be divergent, then $\sum a_n$ must itself be divergent. These observations lead to the following theorem.

THEOREM 7

The Comparison Test

Suppose that $\sum a_n$ and $\sum b_n$ are series with positive terms.

a. If $\sum b_n$ is convergent and $a_n \le b_n$ for all n, then $\sum a_n$ is also convergent.

b. If $\sum b_n$ is divergent and $a_n \ge b_n$ for all n, then $\sum a_n$ is also divergent.

Here is an intuitive justification for Theorem 7. Let

$$S_N = \sum_{n=1}^{N} a_n \quad \text{and} \quad T_N = \sum_{n=1}^{N} b_n$$

be the Nth terms of the sequence of partial sums of $\sum a_n$ and $\sum b_n$, respectively. Since both series have positive terms, $\{S_N\}$ and $\{T_N\}$ are increasing.

1. If $\displaystyle\sum_{n=1}^{\infty} b_n$ is convergent, then there exists a number L such that $\lim_{N \to \infty} T_N = L$ and $T_N \le L$ for all N. Since $a_n \le b_n$ for all n, we have $S_N \le T_N$, and this implies that $S_N \le L$ for all N. We have shown that $\{S_N\}$ is increasing and bounded above, so, as before, we can argue intuitively that S_N and therefore $\sum a_n$ converges.

2. If $\sum b_n$ is divergent, then $\lim_{N \to \infty} T_N = \infty$, since $\{T_N\}$ is increasing. But $a_n \ge b_n$ for all n, and this implies that $S_N \ge T_N$, which in turn implies that $\lim_{N \to \infty} S_N = \infty$. Therefore, $\sum a_n$ diverges.

Note   Since the convergence or divergence of a series is not affected by the omission of a finite number of terms of the series, the condition $a_n \le b_n$ (or $a_n \ge b_n$) for all n can be replaced by the condition that these inequalities hold for all $n \ge k$ for some integer k.

To use Theorem 7, we need a catalog of test series whose convergence and divergence are known. In what follows, we will use the geometric series and the *p*-series as test series.

EXAMPLE 6 Determine whether the series

$$\sum_{n=1}^{\infty} \frac{1}{n^2 + 2}$$

converges or diverges.

Solution Let

$$a_n = \frac{1}{n^2 + 2}$$

Observe that when *n* is large, $n^2 + 2$ behaves like n^2, so a_n behaves like

$$b_n = \frac{1}{n^2}$$

This observation suggests that we compare $\sum a_n$ with the test series $\sum b_n$, which is a convergent *p*-series with $p = 2$. Now

$$0 < \frac{1}{n^2 + 2} < \frac{1}{n^2} \qquad (n \geq 1)$$

and the given series is indeed "smaller" than the test series $\sum_{n=1}^{\infty} \frac{1}{n^2}$. Since the test series converges, we conclude by the comparison test that

$$\sum_{n=1}^{\infty} \frac{1}{n^2 + 2}$$

also converges.

EXAMPLE 7 Determine whether the series

$$\sum_{n=1}^{\infty} \frac{1}{3 + 2^n}$$

converges or diverges.

Solution Let

$$a_n = \frac{1}{3 + 2^n}$$

Observe that if *n* is large, $3 + 2^n$ behaves like 2^n, so a_n behaves like $b_n = 1/2^n$. This observation suggests that we compare $\sum a_n$ with $\sum b_n$. Now the series $\sum 1/2^n = \sum (1/2)^n$ is a geometric series with $|r| = 1/2 < 1$, so it is convergent. Since

$$a_n = \frac{1}{3 + 2^n} < \frac{1}{2^n} = b_n \qquad (n \geq 1)$$

the comparison test tells us that the given series is convergent.

EXAMPLE 8 Determine whether the series

$$\sum_{n=2}^{\infty} \frac{1}{\sqrt{n} - 1}$$

is convergent or divergent.

Solution   Let

$$a_n = \frac{1}{\sqrt{n} - 1}$$

If n is large, $\sqrt{n} - 1$ behaves like $\sqrt{n}$, so a_n behaves like $b_n = 1/\sqrt{n}$. Now the series

$$\sum_{n=2}^{\infty} b_n = \sum_{n=2}^{\infty} \frac{1}{\sqrt{n}} = \sum_{n=2}^{\infty} \frac{1}{n^{1/2}}$$

is a p-series with $p = 1/2 < 1$, so it is divergent. Since

$$a_n = \frac{1}{\sqrt{n} - 1} > \frac{1}{\sqrt{n}} = b_n \qquad (\text{for } n \geq 2) \qquad \text{See the Note following Theorem 7.}$$

the comparison test implies that the given series is divergent.

11.4   Self-Check Exercises

1. Show that the series $\sum_{n=1}^{\infty} \frac{\sqrt{n^2 + 1}}{2n + 3}$ is divergent.

2. Use the integral test to determine whether the series

   $$\sum_{n=2}^{\infty} \frac{1}{n \ln n} \text{ converges or diverges.}$$

Solutions to Self-Check Exercises 11.4 can be found on page 764.

11.4   Concept Questions

1. Consider the series $\sum_{n=1}^{\infty} \frac{1}{n} = \frac{1}{1} + \frac{1}{2} + \frac{1}{3} + \frac{1}{4} + \frac{1}{5} + \cdots$,
   and $f(x) = \frac{1}{x}$. The graph of f and the rectangles that lie
   below the curve are shown in the accompanying figure.

   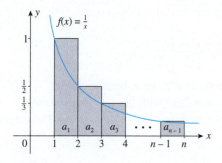

   a. Compute $a_1 = f(1)$, $a_2 = f(2)$, $a_3 = f(3)$, ...,
      $a_n = f(n)$.

   b. Explain why $S_{N-1} = \frac{1}{1} + \frac{1}{2} + \frac{1}{3} + \cdots + \frac{1}{N-1} \geq$
      $\int_1^N \frac{1}{x} \, dx.$

   c. Show that $\int_1^{\infty} \frac{1}{x} \, dx$ is divergent, and conclude that
      $$\sum_{n=1}^{\infty} \frac{1}{n} \text{ diverges.}$$
      Note: This is the harmonic series that was shown to be divergent in Example 3.

2. Let $\sum a_n$ and $\sum b_n$ be series with positive terms.
   a. If $\sum b_n$ is convergent and $a_n \geq b_n$ for all n, what can
      you say about the convergence or divergence of $\sum a_n$?
      Give examples.
   b. If $\sum b_n$ is divergent and $a_n \leq b_n$ for all n, what can
      you say about the convergence or divergence of $\sum a_n$?
      Give examples.

3. Let $\sum a_n$, $\sum b_n$, and $\sum c_n$ be series with positive terms.
   a. If $\sum a_n$ is convergent and $b_n + c_n \leq a_n$ for all n, what
      can you say about the convergence or divergence of
      $\sum b_n$ and $\sum c_n$?
   b. If $\sum a_n$ is divergent and $b_n + c_n \geq a_n$ for all n, what
      can you say about the convergence or divergence of
      $\sum b_n$ and $\sum c_n$?

11.4 Exercises

In Exercises 1–10, show that the series is divergent.

1. $\dfrac{1}{2} + \dfrac{2}{3} + \dfrac{3}{4} + \cdots$

2. $1 - \dfrac{3}{2} + \dfrac{9}{4} - \dfrac{27}{8} + \cdots$

3. $\displaystyle\sum_{n=1}^{\infty} \dfrac{2n}{3n+1}$

4. $\displaystyle\sum_{n=1}^{\infty} \dfrac{n^2}{2n^2+1}$

5. $\displaystyle\sum_{n=1}^{\infty} 2(1.5)^n$

6. $\displaystyle\sum_{n=0}^{\infty} \dfrac{(-1)^n 3^n}{2^{n-1}}$

7. $\displaystyle\sum_{n=1}^{\infty} \dfrac{1}{2+3^{-n}}$

8. $\displaystyle\sum_{n=1}^{\infty} \dfrac{n}{\sqrt{2n^2+1}}$

9. $\displaystyle\sum_{n=0}^{\infty} \left(-\dfrac{\pi}{3}\right)^n$

10. $\displaystyle\sum_{n=1}^{\infty} \dfrac{3^{n+1}}{e^n}$

In Exercises 11–20, use the integral test to determine whether the series is convergent or divergent.

11. $\displaystyle\sum_{n=1}^{\infty} \dfrac{1}{n+1}$

12. $\displaystyle\sum_{n=1}^{\infty} \dfrac{3}{2n-1}$

13. $\displaystyle\sum_{n=1}^{\infty} \dfrac{n}{2n^2+1}$

14. $\displaystyle\sum_{n=1}^{\infty} ne^{-n^2}$

15. $\displaystyle\sum_{n=1}^{\infty} ne^{-n}$

16. $\displaystyle\sum_{n=1}^{\infty} \dfrac{1}{n(2n-1)}$

17. $\displaystyle\sum_{n=1}^{\infty} \dfrac{n}{(n^2+1)^{3/2}}$

18. $\displaystyle\sum_{n=2}^{\infty} \dfrac{1}{n\sqrt{\ln n}}$

19. $\displaystyle\sum_{n=9}^{\infty} \dfrac{1}{n(\ln n)^3}$

20. $\displaystyle\sum_{n=0}^{\infty} \dfrac{1}{e^n+1}$

In Exercises 21–26, determine whether the p-series is convergent or divergent.

21. $\displaystyle\sum_{n=1}^{\infty} \dfrac{1}{n^3}$

22. $\displaystyle\sum_{n=1}^{\infty} \dfrac{1}{n^{2/3}}$

23. $\displaystyle\sum_{n=1}^{\infty} \dfrac{1}{n^{1.01}}$

24. $\displaystyle\sum_{n=1}^{\infty} \dfrac{1}{n^e}$

25. $\displaystyle\sum_{n=1}^{\infty} n^{-\pi}$

26. $\displaystyle\sum_{n=1}^{\infty} n^{-0.98}$

In Exercises 27–36, use the comparison test to determine whether the series is convergent or divergent.

27. $\displaystyle\sum_{n=1}^{\infty} \dfrac{1}{2n^2+1}$

28. $\displaystyle\sum_{n=1}^{\infty} \dfrac{1}{n^2+2n}$

29. $\displaystyle\sum_{n=3}^{\infty} \dfrac{1}{n-2}$

30. $\displaystyle\sum_{n=2}^{\infty} \dfrac{1}{n^{2/3}-1}$

31. $\displaystyle\sum_{n=2}^{\infty} \dfrac{1}{\sqrt{n^2-1}}$

32. $\displaystyle\sum_{n=0}^{\infty} \dfrac{1}{\sqrt{n^3+1}}$

33. $\displaystyle\sum_{n=0}^{\infty} \dfrac{2^n}{3^n+1}$

34. $\displaystyle\sum_{n=3}^{\infty} \dfrac{3^n}{2^n-4}$

35. $\displaystyle\sum_{n=2}^{\infty} \dfrac{\ln n}{n}$

36. $\displaystyle\sum_{n=1}^{\infty} \dfrac{1}{n^n}$

In Exercises 37–48, determine whether the series is convergent or divergent.

37. $\displaystyle\sum_{n=0}^{\infty} \dfrac{1}{\sqrt{n+1}}$

38. $\displaystyle\sum_{n=1}^{\infty} \dfrac{n}{\sqrt{2n^2+1}}$

39. $\displaystyle\sum_{n=2}^{\infty} \dfrac{1}{n\sqrt{n^2+1}}$

40. $\displaystyle\sum_{n=2}^{\infty} \dfrac{\sqrt{n^2+1}}{n^2}$

41. $\displaystyle\sum_{n=1}^{\infty} \left(\dfrac{1}{n\sqrt{n}} + \dfrac{2}{n^2}\right)$

42. $\displaystyle\sum_{n=1}^{\infty} \left[\left(\dfrac{2}{3}\right)^n + \dfrac{1}{n^{3/2}}\right]$

43. $\displaystyle\sum_{n=2}^{\infty} \dfrac{\ln n}{\sqrt{n}}$

44. $\displaystyle\sum_{n=2}^{\infty} \dfrac{\ln n}{n^{2.1}}$

45. $\displaystyle\sum_{n=2}^{\infty} \dfrac{1}{n(\ln n)^2}$

46. $\displaystyle\sum_{n=1}^{\infty} \dfrac{e^{1/n}}{n^2}$

47. $\displaystyle\sum_{n=1}^{\infty} \dfrac{1}{\sqrt{n}+4}$

Hint: $\dfrac{1}{\sqrt{n}+4} > \dfrac{1}{3\sqrt{n}}$ if $n > 4$

48. $\displaystyle\sum_{n=1}^{\infty} \dfrac{1}{4n^2-1}$

Hint: $\dfrac{1}{4n^2-1} < \dfrac{1}{2n^2}$ if $n \geq 1$

In Exercises 49 and 50, find all values of p for which the series is convergent.

49. $\displaystyle\sum_{n=2}^{\infty} \dfrac{1}{n(\ln n)^p}$

50. $\displaystyle\sum_{n=1}^{\infty} \dfrac{\ln n}{n^p}$

51. Find the value(s) of a for which the series

$$\sum_{n=1}^{\infty} \left(\dfrac{a}{n+1} - \dfrac{1}{n+2}\right)$$

converges. Justify your answer.

52. Consider the series $\displaystyle\sum_{n=0}^{\infty} e^{-n}$

a. Evaluate $\int_0^{\infty} e^{-x}\, dx$, and use the integral test to show that the given series is convergent.

b. Show that the given series is a geometric series, and find its sum.

c. Conclude that although the convergence of $\int_0^{\infty} e^{-x}\, dx$ implies convergence of the infinite series, its value does not give the sum of the infinite series.

53. Show that $\displaystyle\int_1^{\infty} \dfrac{1}{x^p}\, dx$ converges if $p > 1$ and diverges if $p \leq 1$.

54. Suppose that $\sum_{n=1}^{\infty} a_n$ is a convergent series with positive terms. Let $f(n) = a_n$, where f is a continuous and decreasing function for $x \geq N$, where N is some positive integer. Show that the error incurred in approximating the sum of the given series by the Nth partial sum of the series

$$S_N = \sum_{n=1}^{N} a_n$$

is less than $\int_N^{\infty} f(x)\, dx$.

In Exercises 55–60, determine whether the statement is true or false. If it is true, explain why it is true. If it is false, give an example to show why it is false.

55. The series $\sum_{n=1}^{\infty} \dfrac{x}{n}$ converges only for $x = 0$.

56. If $\lim_{n \to \infty} a_n = 0$, then $\sum_{n=0}^{\infty} a_n$ converges.

57. If $\sum_{n=0}^{\infty} a_n$ diverges, then $\lim_{n \to \infty} a_n \neq 0$.

58. $\int_1^{\infty} \dfrac{2}{(x^2 + 1)^{1.1}}\, dx$ converges.

59. Suppose that $\sum a_n$ and $\sum b_n$ are series with positive terms. If $\sum a_n$ is convergent and $b_n \geq a_n$ for all n, then $\sum b_n$ is divergent.

60. Suppose that $\sum a_n$ and $\sum b_n$ are series with positive terms. If $\sum b_n$ is divergent and $a_n \leq b_n$, for all n, then $\sum a_n$ may or may not converge.

11.4 Solutions to Self-Check Exercises

1. Here,

$$\lim_{n \to \infty} a_n = \lim_{n \to \infty} \frac{\sqrt{n^2 + 1}}{2n + 3}$$

$$= \lim_{n \to \infty} \frac{\sqrt{1 + \dfrac{1}{n^2}}}{2 + \dfrac{3}{n}} = \frac{1}{2} \neq 0$$

so by the divergence test, the series diverges.

2. Here, $a_n = 1/(n \ln n)$, so we consider the function $f(x) = 1/(x \ln x)$. Observe that f is continuous and positive on $[2, \infty)$. Next, we compute

$$f'(x) = \frac{(x \ln x)(0) - (1)(\ln x + 1)}{[x(\ln x)]^2}$$

$$= -\frac{\ln x + 1}{[x(\ln x)]^2}$$

Note that $f(x) < 0$ if $\ln x > -1$, that is, if $x > 1/e$. This shows that f is decreasing on $[2, \infty)$. Therefore, we may use the integral test.

$$\int_2^{\infty} \frac{1}{x \ln x}\, dx = \lim_{b \to \infty} \int_2^{b} \frac{1}{x \ln x}\, dx$$

$$= \lim_{b \to \infty} \left[\ln(\ln x)\right]_2^{b}$$

$$= \lim_{b \to \infty} \left[\ln(\ln b) - \ln(\ln 2)\right]$$

$$= \infty$$

and we conclude that $\sum_{n=2}^{\infty} \dfrac{1}{n \ln n}$ diverges.

11.5 Power Series and Taylor Series

Power Series

Until now, we have dealt with series with constant terms. In this section we will study infinite series of the form

$$\sum_{n=0}^{\infty} a_n x^n = a_0 + a_1 x + a_2 x^2 + a_3 x^3 + \cdots + a_n x^n + \cdots$$

where x is a variable. More generally, we will consider series of the form

$$\sum_{n=0}^{\infty} a_n (x - a)^n = a_0 + a_1(x - a) + a_2(x - a)^2 + a_3(x - a)^3 + \cdots + a_n(x - a)^n + \cdots$$

from which $\displaystyle\sum_{n=0}^{\infty} a_n x^n$ may be obtained as a special case by putting $a = 0$. We may view such series as generalizations of the notion of a polynomial to an infinite series.

Examples of power series are

$$\sum_{n=0}^{\infty} x^n = 1 + x + x^2 + x^3 + \cdots$$

$$\sum_{n=0}^{\infty} \frac{(-1)^n x^n}{n!} = 1 - x + \frac{x^2}{2!} - \frac{x^3}{3!} + \cdots$$

and

$$\sum_{n=0}^{\infty} \frac{(-1)^n \left(x - \frac{1}{4}\right)^{2n+1}}{(2n+1)!} = \left(x - \frac{1}{4}\right) - \frac{\left(x - \frac{1}{4}\right)^3}{3!} + \frac{\left(x - \frac{1}{4}\right)^5}{5!} - \cdots$$

Observe that if we truncate each of these series, we obtain a polynomial.

Power Series

Let x be a variable. A **power series in x** is a series of the form

$$\sum_{n=0}^{\infty} a_n x^n = a_0 + a_1 x + a_2 x^2 + a_3 x^3 + \cdots + a_n x^n + \cdots \qquad (7)$$

where the a_n's are constants and are called the **coefficients** of the series. More generally, a **power series in $(x - a)$**, where a is a constant, is a series of the form

$$\sum_{n=0}^{\infty} a_n (x - a)^n = a_0 + a_1 (x - a) + a_2 (x - a)^2 \qquad (8)$$
$$+ a_3 (x - a)^3 + \cdots + a_n (x - a)^n + \cdots$$

Notes

1. A power series in $(x - a)$ is also called a **power series centered at a.** Thus, a power series in x is just a series centered at the origin.
2. To simplify the notation used for a power series, we have adopted the convention that $(x - a)^0 = 1$, even when $x = a$. ◼

We can view a power series as a function f defined by the rule

$$f(x) = \sum_{n=0}^{\infty} a_n (x - a)^n$$

Interval of Convergence

Let's examine some important properties of the power series given in (8). Observe that if x is assigned a value, then this power series becomes an infinite series with constant terms. Accordingly, we can determine, at least theoretically, whether this infinite series converges or diverges. It can be shown that the set of all values of x for which this power series *converges* is an interval centered at a, called the **interval of convergence** of the power series. This observation suggests that we may view the power series (8) as a function f whose domain coincides with the interval of convergence of the series and whose functional values are the sums of the infinite series obtained by allowing x to take on all values in the interval of convergence. In this case, we also say that the function f is **represented by the power series** (8).

How do we determine the interval of convergence of a power series? Theorem 8 provides the answer to this question. (We omit the proof.)

THEOREM 8

Suppose that we are given the power series.

$$\sum_{n=0}^{\infty} a_n(x - a)^n$$

Let

$$R = \lim_{n \to \infty} \left| \frac{a_n}{a_{n+1}} \right|$$

a. If $R = 0$, the series converges only for $x = a$.

b. If $0 < R < \infty$, the series converges for x in the interval $(a - R, a + R)$ and diverges for x outside the interval $[a - R, a + R]$ (Figure 8).

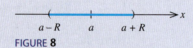

FIGURE 8

c. If $R = \infty$, the series converges for all x.

Notes

1. The domain of convergence of a power series is an interval. This **interval of convergence** is determined by R, the **radius of convergence.** Depending on whether $R = 0$, $0 < R < \infty$, or $R = \infty$, the interval of convergence may just be the degenerate interval consisting of the point a, a bona fide interval $(a - R, a + R)$, or the entire real line.

2. Inside its interval of convergence, a power series represents a function f. The domain of f coincides with its interval of convergence; that is,

$$f(x) = \sum_{n=0}^{\infty} a_n(x - a)^n \qquad x \in (a - R, a + R)$$

3. In general, it is difficult to determine whether a power series is convergent at an endpoint. For this reason, we restrict our attention to points inside the interval of convergence of a power series.

EXAMPLE 1 Find the radius of convergence and the interval of convergence of the power series

$$\sum_{n=0}^{\infty} \frac{(x - 1)^n}{2^n}$$

Show that $f(2)$ exists, and find its value, where

$$f(x) = \sum_{n=0}^{\infty} \frac{(x - 1)^n}{2^n}$$

Solution   Since $a_n = 1/2^n$, we have

$$R = \lim_{n \to \infty} \left| \frac{a_n}{a_{n+1}} \right| = \lim_{n \to \infty} \left| \frac{\dfrac{1}{2^n}}{\dfrac{1}{2^{n+1}}} \right|$$

$$= \lim_{n \to \infty} \frac{2^{n+1}}{2^n} = \lim_{n \to \infty} 2 = 2$$

Since $a = 1$, we see that the interval of convergence of the given series is $(-1, 3)$. Thus, the given power series defines a function

$$f(x) = \sum_{n=0}^{\infty} \frac{(x - 1)^n}{2^n} \qquad x \in (-1, 3)$$

in the interval $(-1, 3)$. Since $2 \in (-1, 3)$, $f(2)$ exists. In fact,

$$f(2) = \sum_{n=0}^{\infty} \frac{(2 - 1)^n}{2^n} = \sum_{n=0}^{\infty} \frac{1}{2^n}$$

This is a geometric series with $a = 1$ and $r = \frac{1}{2}$, so

$$f(2) = \frac{1}{1 - \frac{1}{2}} = 2$$

EXAMPLE 2 Find the interval of convergence of each of the following power series:

a. $\displaystyle\sum_{n=0}^{\infty} n^3 (x + 2)^n$ **b.** $\displaystyle\sum_{n=0}^{\infty} n! \, (x - 1)^n$ **c.** $\displaystyle\sum_{n=0}^{\infty} \frac{x^n}{n!}$

Solution

a. Here, $a_n = n^3$, so

$$R = \lim_{n \to \infty} \left| \frac{a_n}{a_{n+1}} \right| = \lim_{n \to \infty} \frac{n^3}{(n + 1)^3}$$

$$= \lim_{n \to \infty} \frac{1}{\left(1 + \dfrac{1}{n}\right)^3} = 1 \qquad \textcolor{red}{\text{Dividing numerator and denominator by } n^3}$$

Since $a = -2$, we find that the interval of convergence of the series is $(-3, -1)$.

b. Here, $a_n = n!$, so

$$R = \lim_{n \to \infty} \left| \frac{a_n}{a_{n+1}} \right| = \lim_{n \to \infty} \left| \frac{n!}{(n + 1)!} \right|$$

$$= \lim_{n \to \infty} \frac{1}{n + 1} = 0$$

Therefore, the series converges only at $a = 1$.

c. Here, $a_n = 1/n!$ and

$$R = \lim_{n \to \infty} \left| \frac{a_n}{a_{n+1}} \right| = \lim_{n \to \infty} \left| \frac{\dfrac{1}{n!}}{\dfrac{1}{(n + 1)!}} \right| = \lim_{n \to \infty} (n + 1) = \infty$$

Therefore, the interval of convergence of the series is $(-\infty, \infty)$; that is, it converges for any value of x.

Explore and Discuss

Suppose the power series $\displaystyle\sum_{n=0}^{\infty} a_n (x - a)^n$ has a radius of convergence R. What can you deduce about the radius of convergence of the series $\displaystyle\sum_{n=0}^{\infty} a_n (x - a)^{2n}$?

Finding a Taylor Series

Now that we have defined a power series, we are in a position to see what happens when the number of terms of a Taylor polynomial is allowed to increase without bound. In other words, we wish to study the expression

$$f(a) + f'(a)(x - a) + \frac{f''(a)}{2!}(x - a)^2 + \cdots + \frac{f^{(n)}(a)}{n!}(x - a)^n + \cdots \quad (9)$$

called the **Taylor series of** f at a. If the series (9) is truncated after $(n + 1)$ terms, the result is the Taylor polynomial of degree n of f at a (see Section 11.1). Note that when $a = 0$, (9) reduces to

$$f(0) + f'(0)x + \frac{f''(0)}{2!}x^2 + \cdots + \frac{f^{(n)}(0)}{n!}x^n + \cdots \quad (10)$$

which is referred to as the **Maclaurin series of** f. Thus, the Maclaurin series is a special case of the Taylor series with $a = 0$.

Observe that the Taylor series (9) has the form of the infinite series

$$\sum_{n=0}^{\infty} a_n(x - a)^n = a_0 + a_1(x - a) + a_2(x - a)^2 + \cdots + a_n(x - a)^n + \cdots \quad (11)$$

In fact, comparing the Taylor series (9) with the infinite series (11), we see that

$$a_0 = f(a), \quad a_1 = f'(a), \quad a_2 = \frac{f''(a)}{2!}, \quad \ldots, \quad a_n = \frac{f^{(n)}(a)}{n!}, \quad \ldots$$

Thus, the Taylor series is just a power series with coefficients that involve the values of some function f and its derivatives at $x = a$.

Theorem 8 guarantees that a power series represents a function whose domain is the interval of convergence of the series, $(a - R, a + R)$, where $R > 0$. Next we show that the power series of a function defined in this manner must be a Taylor series. But first, we need the following theorem. (We omit the proof.)

THEOREM 9

Suppose the function f is defined by

$$f(x) = \sum_{n=0}^{\infty} a_n(x - a)^n = a_0 + a_1(x - a) + a_2(x - a)^2 + \cdots$$

with radius of convergence $R > 0$. Then

$$f'(x) = \sum_{n=1}^{\infty} na_n(x - a)^{n-1} = a_1 + 2a_2(x - a) + 3a_3(x - a)^2 + \cdots$$

on the interval $(a - R, a + R)$.

Thus, f' may be found by differentiating the power series term by term.

We now prove the statement asserted earlier. Suppose that f is represented by a power series centered about $x = a$; that is,

$$f(x) = a_0 + a_1(x - a) + a_2(x - a)^2 + a_3(x - a)^3 + \cdots + a_n(x - a)^n + \cdots$$

Then, applying Theorem 9 repeatedly, we find

$$f'(x) = a_1 + 2a_2(x - a) + 3a_3(x - a)^2 + \cdots + na_n(x - a)^{n-1} + \cdots$$

$$f''(x) = 2a_2 + 3 \cdot 2a_3(x - a) + 4 \cdot 3a_4(x - a)^2 + \cdots$$
$$+ n(n - 1)a_n(x - a)^{n-2} + \cdots$$

$$f'''(x) = 3 \cdot 2a_3 + 4 \cdot 3 \cdot 2a_4(x - a) + \cdots$$
$$+ n(n - 1)(n - 2)a_n(x - a)^{n-3} + \cdots$$
$$= 3! \, a_3 + 4! \, a_4(x - a) + \cdots + n(n - 1)(n - 2)a_n(x - a)^{n-3} + \cdots$$

$$\vdots$$

$$f^{(n)}(x) = n! \, a_n + (n + 1)! \, a_{n+1}(x - a) + \cdots$$

Evaluating $f(x)$ and each of these derivatives at $x = a$ yields

$$f(a) = a_0$$
$$f'(a) = a_1$$
$$f''(a) = 2a_2 = 2! \, a_2$$
$$f'''(a) = 3 \cdot 2a_3 = 3! \, a_3$$
$$\vdots$$
$$f^{(n)}(a) = n! \, a_n$$

Thus,

$$a_0 = f(a), \quad a_1 = f'(a), \quad a_2 = \frac{f''(a)}{2!}, \quad a_3 = \frac{f'''(a)}{3!}, \quad \ldots, \quad a_n = \frac{f^{(n)}(a)}{n!}, \quad \ldots$$

as we set out to show.

Next, consider the converse problem. More precisely, suppose that we are given a function f that has derivatives of *all* orders in an open interval I. Can we find a power series representation of f in I? Theorem 10 answers this question.

THEOREM 10

Taylor Series Representation of a Function

If a function f has derivatives of all orders in an open interval $I = (a - R, a + R)$ $(R > 0)$ centered at a, then

$$f(x) = \sum_{n=0}^{\infty} \frac{f^{(n)}(a)}{n!}(x - a)^n$$

if and only if

$$\lim_{N \to \infty} R_N(x) = 0$$

for all x in I, where $R_N(x) = f(x) - P_N(x)$.

Thus, f has a power series representation in the form of a Taylor series provided that the error term associated with the Taylor polynomial tends to zero as the number of terms of the Taylor polynomial increases without bound. (A proof of this theorem is sketched in Exercise 33.)

In what follows, *we assume that the functions under consideration have Taylor series representations.*

EXAMPLE 3 Find the Taylor series of the function $f(x) = \dfrac{1}{x - 1}$ at $x = 2$.

Solution Here, $a = 2$ and

$$f(x) = \frac{1}{x - 1} \qquad\qquad f(2) = 1$$

$$f'(x) = -\frac{1}{(x - 1)^2} \qquad\qquad f'(2) = -1$$

$$f''(x) = \frac{2}{(x - 1)^3} \qquad\qquad f''(2) = 2$$

$$f'''(x) = -\frac{3 \cdot 2}{(x - 1)^4} \qquad\qquad f'''(2) = -6 = -3!$$

$$\vdots \qquad\qquad\qquad \vdots$$

$$f^{(n)}(x) = (-1)^n \frac{n!}{(x - 1)^{n+1}} \qquad f^{(n)}(2) = (-1)^n n!$$

$$\vdots \qquad\qquad\qquad \vdots$$

Therefore, the required Taylor series is

$$\frac{1}{x - 1} = f(2) + f'(2)(x - 2) + \frac{f''(2)}{2!}(x - 2)^2 + \frac{f'''(2)}{3!}(x - 2)^3 + \cdots$$

$$+ \frac{f^{(n)}(2)}{n!}(x - 2)^n + \cdots$$

$$= 1 - (x - 2) + \frac{2!}{2!}(x - 2)^2 - \frac{3!}{3!}(x - 2)^3 + \cdots$$

$$+ (-1)^n \frac{n!}{n!}(x - 2)^n + \cdots$$

$$= 1 - (x - 2) + (x - 2)^2 - (x - 2)^3 + \cdots + (-1)^n(x - 2)^n + \cdots$$

$$= \sum_{n=0}^{\infty} (-1)^n(x - 2)^n$$

To find the radius of convergence of the series, we compute

$$R = \lim_{n \to \infty} \left| \frac{a_n}{a_{n+1}} \right| = \lim_{n \to \infty} \left| \frac{(-1)^n}{(-1)^{n+1}} \right|$$

$$= \lim_{n \to \infty} 1 = 1$$

Thus, the series converges in the interval $(1, 3)$. We see that the function $f(x) = 1/(x - 1)$ is represented by the Taylor series in the interval $(1, 3)$.

⚠ The representation of the function $f(x) = 1/(x - 1)$ by its Taylor series at $x = 2$ is valid only in the interval $(1, 3)$, despite the fact that f itself has a domain that is the set of all real numbers except $x = 1$. This serves to remind us of the local nature of this representation.

Exploring with TECHNOLOGY

Refer to Example 3, where it was shown that the Taylor series at $x = 2$ representing $f(x) = 1/(x - 1)$ is

$$P(x) = \sum_{n=0}^{\infty} (-1)^n(x - 2)^n = 1 - (x - 2) + (x - 2)^2 - (x - 2)^3 + \cdots$$

for $1 < x < 3$. This means that if c is any number satisfying $1 < c < 3$, then $f(c) = 1/(c - 1) = P(c)$. In particular, this means that $\lim_{N \to \infty} |P_N(c) - f(c)| = 0$, where $\{P_N(x)\}$ is the sequence of partial sums of $P(x)$.

1. Plot the graphs of f, P_0, P_1, P_2, P_3, P_4, P_5, and P_6 on the same set of axes, using the viewing window $[2.5, 3.1] \times [-0.1, 1.1]$.

2. Do the results of part 1 give a visual confirmation of the statement $\lim_{N \to \infty} P_N(c) = f(c)$ for the special case in which $c = 2.8$? Explain what happens when $c = 3$.

EXAMPLE 4 Find the Taylor series of the function $f(x) = \ln x$ at $x = 1$.

Solution Here, $a = 1$ and

$$f(x) = \ln x \qquad\qquad f(1) = 0$$

$$f'(x) = \frac{1}{x} \qquad\qquad f'(1) = 1$$

$$f''(x) = -\frac{1}{x^2} \qquad\qquad f''(1) = -1$$

$$f'''(x) = \frac{2}{x^3} = \frac{2!}{x^3} \qquad\qquad f'''(1) = 2!$$

$$f^{(4)}(x) = -\frac{3 \cdot 2}{x^4} = -\frac{3!}{x^4} \qquad\qquad f^{(4)}(1) = -3!$$

$$\vdots \qquad\qquad\qquad \vdots$$

$$f^{(n)}(x) = (-1)^{n+1} \frac{(n-1)!}{x^n} \qquad f^{(n)}(1) = (-1)^{n+1}(n-1)!$$

$$\vdots \qquad\qquad\qquad \vdots$$

Therefore, the required Taylor series is

$$f(x) = \ln x = f(1) + f'(1)(x - 1) + \frac{f''(1)}{2!}(x - 1)^2 + \frac{f'''(1)}{3!}(x - 1)^3$$

$$+ \cdots + \frac{f^{(n)}(1)}{n!}(x - 1)^n + \cdots$$

$$= (x - 1) - \frac{1}{2!}(x - 1)^2 + \frac{2!}{3!}(x - 1)^3 - \frac{3!}{4!}(x - 1)^4 + \cdots$$

$$+ (-1)^{n+1} \frac{(n-1)!}{n!}(x - 1)^n + \cdots$$

$$= (x - 1) - \frac{1}{2}(x - 1)^2 + \frac{1}{3}(x - 1)^3 - \frac{1}{4}(x - 1)^4 + \cdots$$

$$+ \frac{(-1)^{n+1}}{n}(x - 1)^n + \cdots$$

$$= \sum_{n=1}^{\infty} \frac{(-1)^{n+1}}{n}(x - 1)^n$$

Since

$$R = \lim_{n\to\infty}\left|\frac{a_n}{a_{n+1}}\right| = \lim_{n\to\infty}\left|\frac{\dfrac{(-1)^{n+1}}{n}}{\dfrac{(-1)^{n+2}}{n+1}}\right|$$

$$= \lim_{n\to\infty}\frac{n+1}{n} = \lim_{n\to\infty}\left(1 + \frac{1}{n}\right) = 1$$

we see that the power series representation of $f(x) = \ln x$ is valid in the interval $(0, 2)$. It can be shown that the representation is valid at $x = 2$ as well.

EXAMPLE 5 Find the Taylor series of the function $f(x) = e^x$ at $x = 0$.

Solution   Here, $a = 0$, and since

$$f(x) = f'(x) = f''(x) = \cdots = e^x$$

we see that

$$f(0) = f'(0) = f''(0) = \cdots = e^0 = 1$$

Therefore, the required Taylor series is

$$f(x) = e^x = f(0) + f'(0)(x - 0) + \frac{f''(0)}{2!}(x - 0)^2 + \cdots + \frac{f^{(n)}(0)}{n!}(x - 0)^n + \cdots$$

$$= 1 + x + \frac{x^2}{2!} + \frac{x^3}{3!} + \cdots + \frac{x^n}{n!} + \cdots$$

$$= \sum_{n=0}^{\infty}\frac{x^n}{n!}$$

Since

$$R = \lim_{n\to\infty}\left|\frac{a_n}{a_{n+1}}\right| = \lim_{n\to\infty}\left|\frac{\dfrac{1}{n!}}{\dfrac{1}{(n+1)!}}\right|$$

$$= \lim_{n\to\infty}(n + 1) = \infty$$

we see that the series representation is valid for all x.

11.5   Self-Check Exercises

1. Find the interval of convergence of the power series $\sum_{n=0}^{\infty}\dfrac{x^n}{n^2 + 1}$. (Disregard the endpoints.)

2. Find the Taylor series of the function $f(x) = e^{-x}$ at $x = 1$, and determine its interval of convergence.

Solutions to Self-Check Exercises 11.5 can be found on page 774.

11.5 Concept Questions

1. a. What is a power series in x?
  b. What is a power series in $(x - a)$?

2. a. What is the radius of convergence of a power series?
  b. What is the interval of convergence of a power series?
  c. How do you find the radius and the interval of convergence of a power series?

3. Suppose that $\displaystyle\sum_{n=0}^{\infty} a_n x^n$ has radius of convergence 2. What can you say about the convergence or divergence of

$$\sum_{n=0}^{\infty} a_n \left(\frac{3}{2}\right)^n ?$$

4. Suppose that $\displaystyle\sum_{n=0}^{\infty} a_n(x - 2)^n$ diverges for $x = 0$. What can you say about the convergence or divergence of $\displaystyle\sum_{n=0}^{\infty} a_n 5^n$? What about $\displaystyle\sum_{n=0}^{\infty} a_n \left(\frac{1}{2}\right)^n$?

5. a. What is a Taylor series?
  b. Suppose that $\displaystyle f(x) = \sum_{n=0}^{\infty} a_n(x - c)^n$ for x in $(-R, R)$, where $R > 0$. What is $f^{(n)}(c)$?
  c. The Taylor series of f at $x = 1$ is $\displaystyle\sum_{n=0}^{\infty} (-1)^n \frac{(x-1)^{n+1}}{n+1}$. What is $f^{(5)}(1)$?

11.5 Exercises

In Exercises 1–20, find the radius of convergence and the interval of convergence of the power series.

1. $\displaystyle\sum_{n=0}^{\infty} (x - 1)^n$

2. $\displaystyle\sum_{n=0}^{\infty} \left(\frac{x}{2}\right)^n$

3. $\displaystyle\sum_{n=1}^{\infty} n^2 x^n$

4. $\displaystyle\sum_{n=0}^{\infty} \frac{(n+1)(x+2)^n}{2^n}$

5. $\displaystyle\sum_{n=0}^{\infty} \frac{(-1)^n x^n}{4^n}$

6. $\displaystyle\sum_{n=0}^{\infty} \frac{(2x)^n}{3^n}$

7. $\displaystyle\sum_{n=0}^{\infty} \frac{(x-1)^n}{n!\, 2^n}$

8. $\displaystyle\sum_{n=0}^{\infty} (2n)!\, x^n$

9. $\displaystyle\sum_{n=0}^{\infty} \frac{(-1)^n n!(x+2)^n}{2^n}$

10. $\displaystyle\sum_{n=2}^{\infty} \frac{x^n}{n(n+1)}$

11. $\displaystyle\sum_{n=2}^{\infty} \frac{(x+3)^n}{(n+1)^2}$

12. $\displaystyle\sum_{n=0}^{\infty} \frac{n!\,(x+1)^n}{(3n)!}$

13. $\displaystyle\sum_{n=1}^{\infty} \frac{2n(x-3)^n}{(n+1)!}$

14. $\displaystyle\sum_{n=0}^{\infty} \frac{(-2x)^{2n}}{4^n(n+1)}$

15. $\displaystyle\sum_{n=1}^{\infty} \frac{n(-2x)^n}{n+1}$

16. $\displaystyle\sum_{n=0}^{\infty} \frac{x^{2n+1}}{(2n+1)!}$

17. $\displaystyle\sum_{n=0}^{\infty} \frac{n!\,(x+1)^n}{3^n}$

18. $\displaystyle\sum_{n=0}^{\infty} (n+1)(x+2)^n$

19. $\displaystyle\sum_{n=0}^{\infty} \frac{n^3(x-3)^n}{3^n}$

20. $\displaystyle\sum_{n=1}^{\infty} \frac{(-1)^{n+1}(x-2)^n}{n2^n}$

In Exercises 21–32, find the Taylor series of the function at the indicated number, and give its radius and interval of convergence. (Disregard the endpoints.)

21. $f(x) = \dfrac{1}{x}; x = 1$

22. $f(x) = \dfrac{1}{x+1}; x = 0$

23. $f(x) = \dfrac{1}{x+1}; x = 2$

24. $f(x) = \dfrac{1}{1-x}; x = 0$

25. $f(x) = \dfrac{1}{1-x}; x = 2$

26. $f(x) = \ln(x+1); x = 0$

27. $f(x) = \sqrt{x}; x = 1$

28. $f(x) = \sqrt{1-x}; x = 0$

29. $f(x) = e^{2x}; x = 0$

30. $f(x) = e^{2x}; x = 1$

31. $f(x) = \dfrac{1}{\sqrt{x+1}}; x = 0$

32. $f(x) = \sqrt{x+1}; x = 1$

33. Prove Theorem 10.
  Hint: The Nth partial sum of the Taylor series is $S_N(x) = P_N(x)$, where $P_N(x)$ is the Nth Taylor polynomial. Show that $\displaystyle\lim_{N\to\infty} S_N(x) = f(x)$ if and only if $\displaystyle\lim_{N\to\infty} R_N(x) = 0$.

In Exercises 34–36, determine whether the statement is true or false. If it is true, explain why it is true. If it is false, give an example to show why it is false.

34. If $\displaystyle\sum_{n=0}^{\infty} a_n(x-2)^n$ converges for $x = 4$, then it converges for $x = 1$.

35. If $\displaystyle\sum_{n=0}^{\infty} a_n(x-a)^n$ has radius of convergence R, then $\displaystyle\sum_{n=0}^{\infty} na_n(x-a)^n$ has radius of convergence R.

36. If $\displaystyle\sum_{n=0}^{\infty} a_n(x-a)^n$ has radius of convergence R, then $\displaystyle\sum_{n=0}^{\infty} a_n^2(x-a)^n$ has radius of convergence $\sqrt{R}$.

11.5 Solutions to Self-Check Exercises

1. We first find the radius of convergence of the power series. Since $a_n = \dfrac{1}{n^2 + 1}$, we have

$$R = \lim_{n \to \infty} \left| \frac{a_n}{a_{n+1}} \right| = \lim_{n \to \infty} \frac{\dfrac{1}{n^2 + 1}}{\dfrac{1}{(n + 1)^2 + 1}}$$

$$= \lim_{n \to \infty} \frac{(n + 1)^2 + 1}{n^2 + 1}$$

$$= \lim_{n \to \infty} \frac{\left(1 + \dfrac{1}{n}\right)^2 + \dfrac{1}{n^2}}{1 + \dfrac{1}{n^2}} \quad \text{\textcolor{red}{Dividing numerator and denominator by } } n^2$$

$$= 1$$

Therefore, the interval of convergence of the series is $(-1, 1)$.

2. Here, $a = 1$ and

$$
\begin{array}{ll}
f(x) = e^{-x} & f(1) = e^{-1} \\
f'(x) = -e^{-x} & f'(1) = -e^{-1} \\
f''(x) = e^{-x} & f''(1) = e^{-1} \\
f'''(x) = -e^{-x} & f'''(1) = -e^{-1} \\
\quad \vdots & \quad \vdots \\
f^{(n)}(x) = (-1)^n e^{-x} & f^{(n)}(1) = (-1)^n e^{-1} \\
\quad \vdots & \quad \vdots
\end{array}
$$

Therefore, the required Taylor series is

$$e^{-x} = f(1) + f'(1)(x - 1) + \frac{f''(1)}{2!}(x - 1)^2$$

$$+ \frac{f'''(1)}{3!}(x - 1)^3 + \cdots + \frac{f^{(n)}(1)}{n!}(x - 1)^n + \cdots$$

$$= e^{-1} - \frac{e^{-1}}{1!}(x - 1) + \frac{e^{-1}}{2!}(x - 1)^2$$

$$- \frac{e^{-1}}{3!}(x - 1)^3 + \cdots + \frac{(-1)^n e^{-1}}{n!}(x - 1)^n + \cdots$$

$$= \frac{1}{e} - \frac{1}{e}(x - 1) + \frac{1}{2! \, e}(x - 1)^2 - \frac{1}{3! \, e}(x - 1)^3 + \cdots$$

$$+ \frac{(-1)^n}{n! \, e}(x - 1)^n + \cdots$$

$$= \sum_{n=0}^{\infty} \frac{(-1)^n}{n! \, e}(x - 1)^n$$

The interval of convergence is $(-\infty, \infty)$. (See Example 5.)

11.6 More on Taylor Series

A Useful Technique for Finding Taylor Series

In Section 11.5, we found the power series representation of certain functions. This representation turned out to be the Taylor series of f at a. The method that we used to compute the series relies solely on our ability to find the higher-order derivatives of the function f. This method, however, is rather tedious.

In this section, we show how the Taylor series of a function can often be found by manipulating some well-known power series. For this purpose, we first catalog the power series of some of the most commonly used functions (Table 1). Each of these representations was derived in Sections 11.1 and 11.5.

TABLE 1

Power Series Representations for Some Common Functions

1. $\dfrac{1}{1 - x} = 1 + x + x^2 + \cdots + x^n + \cdots \qquad (-1 < x < 1)$

2. $e^x = 1 + x + \dfrac{1}{2!}x^2 + \dfrac{1}{3!}x^3 + \cdots + \dfrac{1}{n!}x^n + \cdots \qquad (-\infty < x < \infty)$

3. $\ln x = (x - 1) - \dfrac{1}{2}(x - 1)^2 + \dfrac{1}{3}(x - 1)^3 - \cdots + \dfrac{(-1)^{n+1}}{n}(x - 1)^n + \cdots \qquad (0 < x \leq 2)$

EXAMPLE 1 Find the Taylor series of each function at the indicated number.

a. $f(x) = \dfrac{1}{1 + x}$; $x = 0$ **b.** $f(x) = \dfrac{1}{1 + x}$; $x = 2$

c. $f(x) = \dfrac{1}{1 - 3x}$; $x = 0$ **d.** $f(x) = \dfrac{x}{1 + x^2}$; $x = 0$

Solution

a. We write

$$f(x) = \frac{1}{1 - (-x)}$$

and use the power series representation of $1/(1 - x)$ with x replaced by $-x$ in Formula (1) in Table 1 to obtain

$$f(x) = \frac{1}{1 + x}$$

$$= 1 + (-x) + (-x)^2 + \cdots + (-x)^n + \cdots$$

$$= 1 - x + x^2 - x^3 + \cdots + (-1)^n x^n + \cdots \qquad (-1 < x < 1)$$

b. We write

$$f(x) = \frac{1}{1 + x} = \frac{1}{3 + (x - 2)} = \frac{1}{3\left[1 + \left(\dfrac{x - 2}{3}\right)\right]} = \frac{1}{3}\left[\frac{1}{1 + \left(\dfrac{x - 2}{3}\right)}\right]$$

From the result of part (a), we have

$$\frac{1}{1 + u} = 1 - u + u^2 - u^3 + \cdots + (-1)^n u^n + \cdots \qquad (-1 < u < 1)$$

Thus, with $u = \dfrac{x - 2}{3}$, we find

$$f(x) = \frac{1}{3}\left[\frac{1}{1 + \left(\dfrac{x - 2}{3}\right)}\right]$$

$$= \frac{1}{3}\left[1 - \left(\frac{x - 2}{3}\right) + \left(\frac{x - 2}{3}\right)^2 - \left(\frac{x - 2}{3}\right)^3 + \cdots + (-1)^n\left(\frac{x - 2}{3}\right)^n + \cdots\right]$$

$$= \frac{1}{3} - \frac{1}{3^2}(x - 2) + \frac{1}{3^3}(x - 2)^2 - \frac{1}{3^4}(x - 2)^3 + \cdots + \frac{(-1)^n}{3^{n+1}}(x - 2)^n + \cdots$$

which converges when

$$-1 < \frac{x - 2}{3} < 1 \quad \text{or} \quad -1 < x < 5$$

c. Here, we simply replace x in the power series representation of $1/(1 - x)$ by $3x$ to get

$$f(x) = \frac{1}{1 - 3x} = 1 + 3x + (3x)^2 + (3x)^3 + \cdots + (3x)^n + \cdots$$

$$= 1 + 3x + 3^2 x^2 + 3^3 x^3 + \cdots + 3^n x^n + \cdots$$

The series converges for $-1 < 3x < 1$, or $-\frac{1}{3} < x < \frac{1}{3}$.

d. We write

$$f(x) = \frac{x}{1 + x^2} = x\left[\frac{1}{1 - (-x^2)}\right] \qquad \text{Use Formula (1) in Table 1.}$$

$$= x[1 + (-x^2) + (-x^2)^2 + (-x^2)^3 + \cdots + (-x^2)^n + \cdots]$$

$$= x - x^3 + x^5 - x^7 + \cdots + (-1)^n x^{2n+1} + \cdots$$

The series converges for $-1 < x < 1$.

EXAMPLE 2 Find the Taylor series of the function at the indicated number.

a. $f(x) = xe^{-x}; x = 0$ **b.** $f(x) = \ln(1 + x); x = 0$

Solution

a. First, we replace x with $-x$ in the expression

$$e^x = 1 + x + \frac{x^2}{2!} + \frac{x^3}{3!} + \cdots + \frac{x^n}{n!} + \cdots \qquad \text{Use Formula (2) in Table 1.}$$

to obtain

$$e^{-x} = 1 + (-x) + \frac{(-x)^2}{2!} + \frac{(-x)^3}{3!} + \cdots + \frac{(-x)^n}{n!} + \cdots$$

$$= 1 - x + \frac{x^2}{2!} - \frac{x^3}{3!} + \cdots + \frac{(-1)^n x^n}{n!} + \cdots$$

Then multiplying both sides of this expression by x gives the required expression

$$f(x) = xe^{-x} = x - x^2 + \frac{x^3}{2!} - \frac{x^4}{3!} + \cdots + \frac{(-1)^n x^{n+1}}{n!} + \cdots$$

b. From Table 1, we obtain the following:

$$\ln x = (x - 1) - \frac{1}{2}(x - 1)^2 + \frac{1}{3}(x - 1)^3 - \cdots$$

$$+ \frac{(-1)^{n+1}}{n}(x - 1)^n + \cdots \qquad (0 < x \le 2)$$

Replacing x with $1 + x$ in this expression, we obtain the required expression

$$f(x) = \ln(1 + x) = x - \frac{1}{2}x^2 + \frac{1}{3}x^3 - \cdots + \frac{(-1)^{n+1}}{n}x^n + \cdots$$

valid for $-1 < x \le 1$. (Why?)

Explore and Discuss

The formula

$$\ln x = (x - 1) - \frac{1}{2}(x - 1)^2 + \frac{1}{3}(x - 1)^3 - \cdots \qquad (0 < x \le 2) \qquad \textbf{(A)}$$

from Table 1 can be used to compute the value of $\ln x$ for $0 < x \le 2$. However, the restriction on x and the slow convergence of the series limit its effectiveness from the computational point of view. A more effective formula, first obtained by the Scottish mathematician James Gregory (1638–1675), follows:

1. Using Formula (A), derive the formulas

$$\ln(1 + x) = x - \frac{x^2}{2} + \frac{x^3}{3} - \frac{x^4}{4} - \cdots \qquad (-1 < x \le 1)$$

and

$$\ln(1 - x) = -x - \frac{x^2}{2} - \frac{x^3}{3} - \frac{x^4}{4} - \cdots \qquad (-1 \leq x < 1)$$

2. Use part 1 to show that

$$\ln\left(\frac{1 + x}{1 - x}\right) = 2\left(x + \frac{x^3}{3} + \frac{x^5}{5} + \frac{x^7}{7} + \cdots\right) \qquad (-1 < x < 1)$$

3. To compute the natural logarithm of a positive number p, let

$$p = \frac{1 + x}{1 - x}$$

and show that

$$x = \frac{p - 1}{p + 1} \qquad (-1 < x < 1)$$

4. Use parts 2 and 3 to show that

$$\ln 2 = 2\left[\left(\frac{1}{3}\right) + \frac{\left(\frac{1}{3}\right)^3}{3} + \frac{\left(\frac{1}{3}\right)^5}{5} + \frac{\left(\frac{1}{3}\right)^7}{7} + \cdots\right]$$

and this yields $\ln 2 \approx 0.6931$ when we add the first four terms of the series. This approximation of $\ln 2$ is accurate to four decimal places.

5. Compare this method of computing $\ln 2$ with that of using Formula (A) directly.

In Theorem 10 in Section 11.5, we saw that a power series may be differentiated term by term to yield another power series whose interval of convergence coincides with that of the original series. The following theorem tells us that we may integrate a power series term by term.

THEOREM 11

Suppose

$$f(x) = \sum_{n=0}^{\infty} a_n(x - a)^n$$
$$= a_0 + a_1(x - a) + a_2(x - a)^2 + a_3(x - a)^3 + \cdots$$
$$+ a_n(x - a)^n + \cdots \qquad x \in (a - R, a + R)$$

Then

$$\int f(x)\, dx = \sum_{n=0}^{\infty} \frac{a_n}{n + 1}(x - a)^{n+1} + C$$
$$= a_0(x - a) + \frac{a_1}{2}(x - a)^2 + \frac{a_2}{3}(x - a)^3 + \cdots$$
$$+ \frac{a_n}{n + 1}(x - a)^{n+1} + \cdots + C \qquad x \in (a - R, a + R)$$

Theorems 10 and 11 can also be used to help us find the power series representation of a function starting from the series representation of some appropriate function, as the next two examples show.

EXAMPLE 3 Differentiate the power series for the function $1/(1 - x)$ at $x = 0$ to obtain a Taylor series representation of the function $f(x) = 1/(1 - x)^2$ at $x = 0$. (See Table 1.)

Solution   From Table 1, we have

$$\frac{1}{1 - x} = 1 + x + x^2 + x^3 + \cdots + x^n + \cdots \qquad (-1 < x < 1)$$

Differentiating both sides of the equation with respect to x and using Theorem 10, we obtain

$$\frac{d}{dx}\left(\frac{1}{1 - x}\right) = \frac{d}{dx}(1 + x + x^2 + x^3 + \cdots + x^n + \cdots)$$

or

$$f(x) = \frac{1}{(1 - x)^2}$$

$$= 1 + 2x + 3x^2 + \cdots + nx^{n-1} + \cdots \qquad (-1 < x < 1)$$

EXAMPLE 4 Integrate the Taylor series for the function $1/(1 + x)$ at $x = 0$ to obtain a power series representation for the function $f(x) = \ln(1 + x)$ centered at $x = 0$. (See Table 1.) Compare your result with that of Example 2b.

Solution   From Table 1, we see that

$$\frac{1}{1 - x} = 1 + x + x^2 + x^3 + \cdots + x^n + \cdots \qquad (-1 < x < 1)$$

Replacing x by $-x$ gives

$$\frac{1}{1 - (-x)} = 1 + (-x) + (-x)^2 + (-x)^3 + \cdots + (-x)^n + \cdots$$

or

$$\frac{1}{1 + x} = 1 - x + x^2 - x^3 + \cdots + (-1)^n x^n + \cdots \qquad (-1 < x < 1)$$

Finally, integrating both sides of this equation with respect to x and using Theorem 11, we obtain

$$\int \frac{1}{1 + x}\,dx = \int \left[1 - x + x^2 - x^3 + \cdots + (-1)^n x^n + \cdots\right]dx$$

or

$$f(x) = \ln(1 + x) = x - \frac{1}{2}x^2 + \frac{1}{3}x^3 - \cdots + \frac{(-1)^n}{n + 1}x^{n+1} + \cdots + C$$

But $f(0) = \ln 1 = 0$ implies that $C = 0$. So

$$f(x) = x - \frac{1}{2}x^2 + \frac{1}{3}x^3 - \cdots + \frac{(-1)^{n+1}}{n}x^n + \cdots \qquad (-1 < x \le 1)$$

This result is the same as that of Example 2b, as expected.

In many practical applications, all that is required is an approximation of the actual solution to a problem. Thus, rather than working with the Taylor series of a function at a point that is *equal* to $f(x)$ inside its interval of convergence, one often works with the truncated Taylor series. The truncated Taylor series is a Taylor polynomial that gives an acceptable approximation to the values of $f(x)$ in the neighborhood of the point about which the function is expanded, provided that the degree of the polynomial or, equivalently, the number of terms of the Taylor series retained is large enough.

Before proceeding, we want to point out how much easier it is to obtain the Taylor polynomial approximation of a function using the method of this section rather than obtaining it directly as was done in Section 11.1.

APPLIED EXAMPLE 5 Serum Cholesterol Levels The serum cholesterol levels (in mg/dL) in a current Mediterranean population are found to be normally distributed with a probability density function given by

$$f(x) = \frac{1}{50\sqrt{2\pi}}\, e^{-\frac{1}{2}[(x-160)/50]^2}$$

Scientists at the National Heart, Lung, and Blood Institute consider this pattern ideal for a minimal risk of heart attacks. Find the percentage of the population who have serum cholesterol levels between 160 and 180 mg/dL.

Solution The required probability is given by

$$P(160 \le x \le 180) = \frac{1}{50\sqrt{2\pi}} \int_{160}^{180} e^{-\frac{1}{2}[(x-160)/50]^2}\, dx$$

(see Section 10.3). Let's approximate the integrand by a sixth-degree Taylor polynomial about $x = 160$. From Table 1, we have

$$e^x = 1 + x + \frac{1}{2!}x^2 + \frac{1}{3!}x^3 + \cdots$$

Replacing x with $-\frac{1}{2}[(x-160)/50]^2$, we obtain

$$e^{-\frac{1}{2}[(x-160)/50]^2} \approx 1 - \frac{1}{2}\left(\frac{x-160}{50}\right)^2 + \frac{1}{2!}\left[-\frac{1}{2}\left(\frac{x-160}{50}\right)^2\right]^2$$
$$+ \frac{1}{3!}\left[-\frac{1}{2}\left(\frac{x-160}{50}\right)^2\right]^3$$
$$= 1 - \frac{(x-160)^2}{5000} + \frac{(x-160)^4}{5\cdot 10^7} - \frac{(x-160)^6}{7.5\cdot 10^{11}}$$

Therefore,

$$P(160 \le x \le 180)$$
$$\approx \frac{1}{50\sqrt{2\pi}} \int_{160}^{180}\left[1 - \frac{(x-160)^2}{5000} + \frac{(x-160)^4}{5\cdot 10^7} - \frac{(x-160)^6}{7.5\cdot 10^{11}}\right] dx$$
$$\approx \frac{1}{50\sqrt{2\pi}}\left[x - \frac{(x-160)^3}{15{,}000} + \frac{(x-160)^5}{2.5\cdot 10^8} - \frac{(x-160)^7}{5.25\cdot 10^{12}}\right]\Bigg|_{160}^{180}$$
$$\approx \frac{1}{50\sqrt{2\pi}}\left[(180 - 0.53333 + 0.0128 - 0.00024) - 160\right]$$
$$\approx 0.1554$$

so approximately 15.5% of the population has blood cholesterol levels between 160 and 180 mg/dL.

11.6 Self-Check Exercises

1. Find the Taylor series of the function $f(x) = xe^{-x^2}$ at $x = 0$.

2. Use the result from Self-Check Exercise 1 to write the seventh Taylor polynomial of f about $x = 0$, and use this polynomial to approximate

$$\int_0^{0.5} xe^{-x^2}\, dx$$

Compare this result with the exact value of the integral.

Solutions to Self-Check Exercises 11.6 can be found on page 781.

11.6 Concept Questions

In Exercises 1–3, rewrite *f(x)* in a form that will enable you to use the power series representations of functions in Table 1 to find the Taylor series of *f* at the indicated number. Do not find the series.

1. $f(x) = \dfrac{x}{1 + 2x}; x = 0$ 2. $f(x) = \dfrac{1}{2x + 3}; x = 1$

3. $f(x) = \dfrac{2x}{4 + 3x^2}; x = 0$

11.6 Exercises

In Exercises 1–20, find the Taylor series of each function at the indicated number. Give the interval of convergence for each series.

1. $f(x) = \dfrac{1}{1 - x}; x = 2$ 2. $f(x) = \dfrac{1}{1 + x}; x = 1$

3. $f(x) = \dfrac{1}{1 + 3x}; x = 0$ 4. $f(x) = \dfrac{x}{1 - 2x}; x = 0$

5. $f(x) = \dfrac{1}{4 - 3x}; x = 0$ 6. $f(x) = \dfrac{1}{4 - 3x}; x = 1$

7. $f(x) = \dfrac{1}{1 - x^2}; x = 0$ 8. $f(x) = \dfrac{x^2}{1 + x^3}; x = 0$

9. $f(x) = e^{-x}; x = 0$ 10. $f(x) = e^x; x = 1$

11. $f(x) = x^2e^{-x^2}; x = 0$ 12. $f(x) = xe^{x/2}; x = 0$

13. $f(x) = \dfrac{1}{2}(e^x + e^{-x}); x = 0$

14. $f(x) = \dfrac{1}{2}(e^x - e^{-x}); x = 0$

15. $f(x) = \ln(1 + 2x); x = 0$

16. $f(x) = \ln\left(1 + \dfrac{x}{2}\right); x = 0$

17. $f(x) = \ln(1 + x^2); x = 0$

18. $f(x) = \ln(1 + 2x); x = 2$

19. $f(x) = (x - 2)\ln x; x = 2$

20. $f(x) = x^2 \ln\left(1 + \dfrac{x}{2}\right); x = 0$

21. Differentiate the power series for $\ln(1 + x)$ at $x = 0$ to obtain a series representation for the function $f(x) = 1/(1 + x)$.

22. Differentiate the power series for $1/(1 + x)$ at $x = 0$ to obtain a series representation for the function $f(x) = 1/(1 + x)^2$.

23. Integrate the power series for $1/(1 + x)$ to obtain a power series representation for the function $f(x) = \ln(1 + x)$.

24. Integrate the power series for $2x/(1 + x^2)$ to obtain a power series representation for the function $f(x) = \ln(1 + x^2)$.

25. Use a sixth-degree Taylor polynomial to approximate

$$\int_0^{0.5} \dfrac{1}{\sqrt{1 + x^2}}\, dx$$

Hint: $P_6(x) = 1 - \dfrac{1}{2}x^2 + \dfrac{3}{8}x^4 - \dfrac{5}{16}x^6$

26. Use a sixth-degree Taylor polynomial to approximate

$$\int_0^{0.4} \ln(1 + x^2)\, dx$$

27. Use an eighth-degree Taylor polynomial to approximate

$$\int_0^1 e^{-x^2}\, dx$$

28. Use a sixth-degree Taylor polynomial to approximate

$$\int_0^{0.5} \dfrac{\ln(1 + x)}{x}\, dx$$

29. Use an eighth-degree Taylor polynomial of
$f(x) = 1/(1 + x^2)$ at $x = 0$ and the relationship

$$\pi = 4 \int_0^1 \frac{dx}{1 + x^2}$$

to obtain an approximation of π.

30. Factory Worker Wages According to data released by a city's Chamber of Commerce, the weekly wages of factory workers are normally distributed according to the probability density function

$$f(x) = \frac{1}{50\sqrt{2\pi}} e^{-\frac{1}{2}[(x - 700)/50]^2}$$

Find the probability that a worker selected at random from the city has a weekly wage of $650–$750.

31. MP3 Player Reliability The General Manager of the Service Department of MCA Media Company has estimated that the time that elapses between the dates of purchase and the dates on which the MP3 players manufactured by the company first require service is normally distributed according to the probability density function

$$f(x) = \frac{1}{10\sqrt{2\pi}} e^{-\frac{1}{2}[(x - 30)/10]^2}$$

where x is measured in months. Determine the percentage of players manufactured and sold by MCA that may require service 28–32 months after purchase.

32. Demand for Sports Watches The demand equation for the Tempus sports watch is given by

$$p = 50e^{-0.1(x + 1)^2}$$

where x (in units of a thousand) is the quantity demanded per week and p is the unit wholesale price in dollars. National Importers, the supplier of the watches, will make x thousand units available in the market if the unit wholesale price is

$$p = 10 + 5x^2$$

dollars. Use a Taylor polynomial approximation of

$$p = f(x) = 50e^{-0.1(x + 1)^2}$$

to find the consumers' surplus.
Hint: The equilibrium quantity is 1689 units.

11.6 Solutions to Self-Check Exercises

1. First, we replace x in the expression

$$e^x = 1 + x + \frac{x^2}{2!} + \frac{x^3}{3!} + \cdots + \frac{x^n}{n!} + \cdots$$

(see Table 1) with $-x^2$ to obtain

$$e^{-x^2} = 1 + (-x^2) + \frac{(-x^2)^2}{2!}$$

$$+ \frac{(-x^2)^3}{3!} + \cdots + \frac{(-x^2)^n}{n!} + \cdots$$

$$= 1 - x^2 + \frac{x^4}{2!} - \frac{x^6}{3!} + \cdots + \frac{(-1)^n x^{2n}}{n!} + \cdots$$

Multiplying both sides of this expression by x gives

$$f(x) = xe^{-x^2} = x - x^3 + \frac{x^5}{2!}$$

$$- \frac{x^7}{3!} + \cdots + \frac{(-1)^n x^{2n+1}}{n!} + \cdots$$

$$= \sum_{n=0}^{\infty} \frac{(-1)^n x^{2n+1}}{n!}$$

2. Using the result from Self-Check Exercise 1, we see that the seventh Taylor polynomial of f about $x = 0$ is

$$P_7(x) = x - x^3 + \frac{x^5}{2!} - \frac{x^7}{3!}$$

$$= x - x^3 + \frac{1}{2}x^5 - \frac{1}{6}x^7$$

Therefore,

$$\int_0^{0.5} xe^{-x^2} \, dx \approx \int_0^{0.5} \left(x - x^3 + \frac{1}{2}x^5 - \frac{1}{6}x^7 \right) dx$$

$$= \frac{1}{2}x^2 - \frac{1}{4}x^4 + \frac{1}{12}x^6 - \frac{1}{48}x^8 \Big|_0^{0.5}$$

$$\approx 0.1105957$$

The exact value of the integral is

$$\int_0^{0.5} xe^{-x^2} \, dx = -\frac{1}{2}e^{-x^2} \Big|_0^{0.5} \qquad \text{Use the substitution } u = x^2.$$

$$= -\frac{1}{2}(e^{-0.25} - 1)$$

which is approximately 0.1105996. Thus, the error is 0.0000039.

Newton's Method

Previously, we have had occasion to find the zeros of a function f or, equivalently, the **roots** of the equation $f(x) = 0$. For example, the x-intercepts of a function f are precisely the values of x that satisfy $f(x) = 0$; the critical numbers of f include the roots of the equation $f'(x) = 0$; and the candidates for the inflection points of f include the roots of the equation $f''(x) = 0$.

For linear and quadratic functions or for polynomial functions that are easily factored, the zeros of f are readily found. In practice, however, we often encounter functions with zeros that cannot be found as readily. For example, the function

$$f(t) = -t^3 + 96t^2 + 5$$

that gives the altitude (in feet) of a rocket t seconds into flight (Example 8, page 167), is not easily factored, so its zeros cannot be found by elementary algebraic methods. (We will find the zeros of this function in Example 2.) Another example of a function with zeros that are not easily found is $g(x) = e^{2x} - 3x - 2$. In this section, we develop an algorithm, based on the approximation of a function f by a first-degree Taylor polynomial, to approximate the value of a zero of f to any desired degree of accuracy.

Suppose f has a zero at $x = c$ (Figure 9). Let x_0 be an initial estimate of the actual zero c of f. Now the first-degree Taylor polynomial of f at $x = x_0$ is

$$P_1(x) = f(x_0) + f'(x_0)(x - x_0)$$

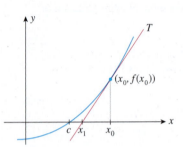

FIGURE 9
x_0 is our first guess in estimating the value of c, the zero of f.

which is just the equation of the tangent line to the graph of $y = f(x)$ at the point $(x_0, f(x_0))$. Since the tangent line T is an approximation of the graph of $y = f(x)$ near $x = x_0$ (and x_0 is assumed to be close to c!), we expect the zero of $P_1(x)$ to be close to the zero c of f. But the zero of the linear function $P_1(x)$ is found by setting $P_1(x) = 0$. Thus,

$$f(x_0) + f'(x_0)(x - x_0) = 0$$
$$f'(x_0)(x - x_0) = -f(x_0)$$
$$x - x_0 = -\frac{f(x_0)}{f'(x_0)}$$
$$x = x_0 - \frac{f(x_0)}{f'(x_0)}$$

This provides us with another estimate of the zero of f:

$$x_1 = x_0 - \frac{f(x_0)}{f'(x_0)}$$

In general, the estimate x_1 is better than the initial estimate x_0.

This process may be repeated by replacing the initial estimate x_0 with the recent estimate x_1. This leads to yet another estimate,

$$x_2 = x_1 - \frac{f(x_1)}{f'(x_1)}$$

which is usually better than x_1. In this manner, we generate a sequence of approximations $x_0, x_1, x_2, \ldots, x_n, x_{n+1}, \ldots$, with

$$x_{n+1} = x_n - \frac{f(x_n)}{f'(x_n)}$$

which, in most instances, approaches the zero c of f.

Our discussion leads us to the following algorithm for finding an approximation to the root c of $f(x) = 0$.

> **Newton's Method (also called the *Newton–Raphson method*)**
>
> 1. Pick an initial estimate x_0 of the root c.
> 2. Find a new estimate using the iterative formula
>
> $$x_{n+1} = x_n - \frac{f(x_n)}{f'(x_n)} \qquad (n = 0, 1, 2, \ldots) \tag{12}$$
>
> 3. Compute $|x_n - x_{n+1}|$. If this number is less than a prescribed positive number, stop. The required approximation to the root $x = c$ is $x = x_{n+1}$.

We refer to the repetitive use of Equation (12) as an **iteration.**

Solving Equations Using Newton's Method

EXAMPLE 1 Use Newton's method to approximate the zero of $f(x) = x^2 - 2$. Start the iteration with initial guess $x_0 = 1$, and terminate the process when two successive approximations differ by less than 0.00001.

Solution We have

$$f(x) = x^2 - 2 \qquad f'(x) = 2x$$

so by Equation (12), the required iterative formula is

$$x_{n+1} = x_n - \frac{x_n^2 - 2}{2x_n} = \frac{x_n^2 + 2}{2x_n}$$

With $x_0 = 1$, we find

$$x_1 = \frac{1^2 + 2}{2(1)} = 1.5$$

$$x_2 = \frac{(1.5)^2 + 2}{2(1.5)} \approx 1.416667$$

$$x_3 = \frac{(1.416667)^2 + 2}{2(1.416667)} \approx 1.414216$$

$$x_4 = \frac{(1.414216)^2 + 2}{2(1.414216)} \approx 1.414214$$

Since $x_3 - x_4 = 0.000002 < 0.00001$, we terminate the process. The sequence generated converges to $\sqrt{2}$, which is one of the two roots of the equation $x^2 - 2 = 0$. Note that, to six places, $\sqrt{2} = 1.414214$!

 APPLIED EXAMPLE 2 Flight of a Rocket Refer to Example 8, page 167. The altitude in feet of a rocket t seconds into flight is given by

$$s = f(t) = -t^3 + 96t^2 + 5 \qquad (t \geq 0)$$

Use Newton's method to find the time T when the rocket hits the earth, accurate to two decimal places.

Solution   The rocket hits the earth when the altitude is equal to zero. So we are required to solve the equation

$$s = f(t) = -t^3 + 96t^2 + 5 = 0$$

Let's use Newton's method with initial guess $t_0 = 100$ (Figure 10). Here,

$$f'(t) = -3t^2 + 192t$$

so the required iterative formula takes the form

$$
\begin{aligned}
t_{n+1} &= t_n - \frac{-t_n^3 + 96t_n^2 + 5}{-3t_n^2 + 192t_n} \\
&= t_n - \frac{t_n^3 - 96t_n^2 - 5}{3t_n^2 - 192t_n} \\
&= \frac{3t_n^3 - 192t_n^2 - t_n^3 + 96t_n^2 + 5}{3t_n^2 - 192t_n} \\
&= \frac{2t_n^3 - 96t_n^2 + 5}{3t_n^2 - 192t_n}
\end{aligned}
$$

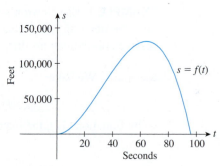

FIGURE **10**
The rocket's altitude t seconds into flight is given by $f(t)$.

We have

$$
\begin{aligned}
t_1 &= \frac{2(100)^3 - 96(100)^2 + 5}{3(100)^2 - 192(100)} \approx 96.297 \\
t_2 &= \frac{2(96.297)^3 - 96(96.297)^2 + 5}{3(96.297)^2 - 192(96.297)} \approx 96.002 \\
t_3 &= \frac{2(96.002)^3 - 96(96.002)^2 + 5}{3(96.002)^2 - 192(96.002)} \approx 96.001
\end{aligned}
$$

Thus, $T \approx 96.00$. So the rocket hits the earth approximately 96 seconds after liftoff.

The next example shows how we can use Newton's method to find the equilibrium quantity and price of a certain commodity.

 APPLIED EXAMPLE 3 Market Equilibrium The demand equation for the Tempus sports watch is given by

$$p = 50e^{-0.1(x+1)^2}$$

where x thousand units are demanded per week and p is the unit wholesale price in dollars. National Importers, the supplier of the watches, will make x thousand units available in the market if the unit wholesale price is

$$p = 10 + 5x^2$$

dollars. Use Newton's method to find the equilibrium quantity and price. Terminate the process when two successive approximations differ by less than 0.0001.

Solution   We determine the equilibrium point by finding the point of intersection of the demand curve and the supply curve (Figure 11). To solve the system of equations

$$p = 50e^{-0.1(x+1)^2}$$
$$p = 10 + 5x^2$$

we substitute the second equation into the first, obtaining

$$10 + 5x^2 = 50e^{-0.1(x+1)^2}$$
$$10 + 5x^2 - 50e^{-0.1(x+1)^2} = 0$$

To solve the last equation, we use Newton's method with

$$\begin{aligned} f(x) &= 10 + 5x^2 - 50e^{-0.1(x+1)^2} \\ &= 5[2 + x^2 - 10e^{-0.1(x+1)^2}] \\ f'(x) &= 10x - 50e^{-0.1(x+1)^2} \cdot [-0.2(x+1)] \\ &= 10[x + (x+1)e^{-0.1(x+1)^2}] \end{aligned}$$

leading to the iterative formula

$$x_{n+1} = x_n - \frac{2 + x_n^2 - 10e^{-0.1(x_n+1)^2}}{2[x_n + (x_n+1)e^{-0.1(x_n+1)^2}]}$$

Referring to Figure 11, we see that a reasonable initial estimate is $x_0 = 2$. We have

$$x_1 = 2 - \frac{2 + 2^2 - 10e^{-0.1(3)^2}}{2[2 + 3e^{-0.1(3)^2}]} \approx 1.69962$$

$$x_2 = 1.69962 - \frac{2 + (1.69962)^2 - 10e^{-0.1(2.69962)^2}}{2[1.69962 + 2.69962e^{-0.1(2.69962)^2}]} \approx 1.68899$$

$$x_3 = 1.68899 - \frac{2 + (1.68899)^2 - 10e^{-0.1(2.68899)^2}}{2[1.68899 + 2.68899e^{-0.1(2.68899)^2}]} \approx 1.68898$$

Since $x_3 - x_4 < 0.0001$, the equilibrium quantity is approximately 1689 watches. The equilibrium wholesale price is

$$p = 10 + 5(1.68898)^2 \approx 24.263$$

or approximately $24.26 per watch.

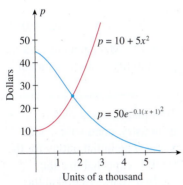

$p = 10 + 5x^2$

$p = 50e^{-0.1(x+1)^2}$

Units of a thousand

FIGURE 11
The equilibrium point is the point of intersection of the demand curve and the supply curve.

The Internal Rate of Return on an Investment

Yet another use of Newton's method is to find the internal rate of return on an investment. Suppose a company has an initial outlay of C dollars in an investment that yields returns of $R_1, R_2, \ldots, R_n$ dollars at the end of the first, second, $\ldots$, nth periods, respectively. Then the *net present value* of the investment is

$$\frac{R_1}{1+r} + \frac{R_2}{(1+r)^2} + \frac{R_3}{(1+r)^3} + \cdots + \frac{R_n}{(1+r)^n} - C$$

where r denotes the interest rate per period earned on the investment. Now the **internal rate of return** on the investment is defined as the rate of return for which the net present value of the investment is equal to zero; that is, it is the value of r that satisfies the equation

$$\frac{R_1}{1+r} + \frac{R_2}{(1+r)^2} + \frac{R_3}{(1+r)^3} + \cdots + \frac{R_n}{(1+r)^n} - C = 0$$

or, equivalently, the equation

$$C(1+r)^n - R_1(1+r)^{n-1} - R_2(1+r)^{n-2} - R_3(1+r)^{n-3} - \cdots - R_n = 0$$

obtained by multiplying both sides of the former by $(1+r)^n$. Management uses the internal rate of return to determine the worthiness or profitability of an investment.

APPLIED EXAMPLE 4 Internal Rate of Return The management of A-1 Rental—a tool and equipment rental service for industry, contractors, and homeowners—is contemplating purchasing new equipment. The initial outlay for the equipment, which has a useful life of 4 years, is $45,000. It is expected that the investment will yield returns of $15,000 at the end of the first year, $18,000 at the end of the second year, $14,000 at the end of the third year, and $10,000 at the end of the fourth year. Use Newton's method to find the internal rate of return on this investment. Terminate the process when two successive approximations differ by less than 0.0001.

Solution Here, $n = 4$, $C = 45,000$, $R_1 = 15,000$, $R_2 = 18,000$, $R_3 = 14,000$, and $R_4 = 10,000$. So we are required to solve the equation

$$45,000(1+r)^4 - 15,000(1+r)^3 - 18,000(1+r)^2 - 14,000(1+r) - 10,000 = 0$$

for r. Letting $x = 1 + r$, we simplify this equation and write

$$f(x) = 45,000x^4 - 15,000x^3 - 18,000x^2 - 14,000x - 10,000 = 0$$

To solve the equation $f(x) = 0$ using Newton's method, we first compute

$$f'(x) = 180,000x^3 - 45,000x^2 - 36,000x - 14,000$$

The required iterative formula is

$$x_{n+1} = x_n - \frac{45,000x_n^4 - 15,000x_n^3 - 18,000x_n^2 - 14,000x_n - 10,000}{180,000x_n^3 - 45,000x_n^2 - 36,000x_n - 14,000}$$

Starting with the initial estimate $x_0 = 1.1$, we find

$$x_1 = 1.1 - \frac{45,000(1.1)^4 - 15,000(1.1)^3 - 18,000(1.1)^2 - 14,000(1.1) - 10,000}{180,000(1.1)^3 - 45,000(1.1)^2 - 36,000(1.1) - 14,000}$$

$$\approx 1.10958$$

$x_2 \approx 1.10941$ and $x_3 \approx 1.10941$

Therefore, we may take $x \approx 1.1094$. Since $r \approx 0.1094$, the rate of return on the investment is approximately 10.94% per year.

Having seen how effective Newton's method can be for finding the zeros of a function, we want to point out that there are situations in which the method fails and that care must be exercised in applying it. Figure 12a illustrates a situation in which $f'(x_n) = 0$ for some n (in this case, $n = 2$). Since the iterative Formula (12) involves division by $f'(x_n)$, it should be clear why the method fails to work in this case.

However, if you choose a different initial estimate x_0, the situation may yet be salvaged (Figure 12b).

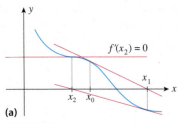 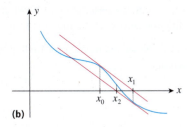

(a)   (b)

FIGURE 12
In (a), Newton's method fails to work because $f'(x_2) = 0$, but this situation is remedied in (b) by selecting a different initial estimate x_0.

Explore and Discuss

For a concrete example of a situation similar to that depicted in Figure 12, consider the function $f(x) = x^3 - 1.5x^2 - 6x + 2$.

1. Show that Newton's method fails to work if we choose $x_0 = -1$ or $x_0 = 2$ for an initial estimate.

2. Using the initial estimates $x_0 = -2.5$, $x_0 = 1$, and $x_0 = 2.5$, show that the three roots of $f(x) = 0$ are -2, 0.313859, and 3.186141, respectively.

3. Using a graphing utility, plot the graph of f in the viewing window $[-3, 4] \times [-10, 7]$. Verify the results of part 2, using **TRACE** and **ZOOM** or the root-finding function of your calculator.

The next situation, shown in Figure 13, is more serious, and the method will not work for any choice of the initial estimate x_0 other than the actual zero of f because the sequence, $x_1, x_2, \ldots x_n$ diverges.

FIGURE 13
Newton's method fails here because the sequence of estimates diverges.

Explore and Discuss

For a concrete example of a situation similar to that depicted in Figure 13, consider the function $f(x) = x^{1/3}$.

1. Show that Newton's iteration for solving the equation $f(x) = 0$ is $x_{n+1} = -2x_n$, with x_0 being an initial guess.

2. Show that the sequence $x_0, x_1, x_2, \ldots$ diverges for any choice of x_0 other than zero and therefore Newton's method does not lead to the unique solution $x = 0$ for the problem under consideration.

3. Illustrate this situation geometrically.

11.7 Self-Check Exercises

1. Use three iterations of Newton's method on an appropriate function f and initial guess x_0 to obtain an estimate of $\sqrt[3]{10}$, accurate to five decimal places.

2. **POPULATION GROWTH** A study prepared for a certain Sunbelt town's Chamber of Commerce projected that the town's population in the next 3 years will grow according to the model

$$P(t) = 50{,}000 + 30t^{3/2} + 20t$$

where $P(t)$ denotes the population t months from now. Use Newton's method to estimate the time when the population will reach 55,000, accurate to two decimal places.

Solutions to Self-Check Exercises 11.7 can be found on page 790.

11.7 Concept Questions

1. Give a geometric description of Newton's method for finding the zero(s) of the function f. Illustrate graphically.

2. Describe Newton's method for finding the zero(s) of a function f.

3. Does Newton's method always work for any choice of the initial estimate x_0 of the root of $f(x) = 0$? Explain graphically.

11.7 Exercises

A calculator is recommended for this exercise set. In Exercises 1–6, estimate the value of each radical to five decimal places by using three iterations of Newton's method with the indicated initial guess for each function.

1. $\sqrt{3}$; $f(x) = x^2 - 3$; $x_0 = 1.5$

2. $\sqrt{5}$; $f(x) = x^2 - 5$; $x_0 = 2$

3. $\sqrt{7}$; $f(x) = x^2 - 7$; $x_0 = 2.5$

4. $\sqrt[3]{6}$; $f(x) = x^3 - 6$; $x_0 = 2$

5. $\sqrt[3]{14}$; $f(x) = x^3 - 14$; $x_0 = 2.5$

6. $\sqrt[4]{50}$; $f(x) = x^4 - 50$; $x_0 = 2.5$

In Exercises 7–10, use Newton's method to find the zero(s) of f to four decimal places by solving the equation $f(x) = 0$. Use the initial estimate(s) x_0.

7. $f(x) = -x^3 - 2x + 2$; $x_0 = 1$

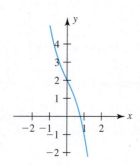

8. $f(x) = 2x^3 - 15x^2 + 36x - 20$; $x_0 = 1$

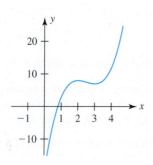

9. $f(x) = \dfrac{3}{2}x^4 - 2x^3 - 6x^2 + 8$; $x_0 = 1$ and $x_0 = 3$

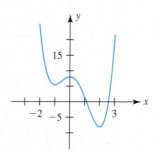

10. $f(x) = x - \sqrt{1 - x^2}$; $x_0 = 0.5$

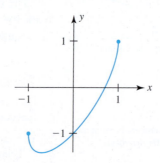

In Exercises 11–18, use Newton's method to approximate the indicated zero of each function. Continue with the iteration until two successive approximations differ by less than 0.0001.

11. The zero of $f(x) = x^2 - x - 3$ between $x = 2$ and $x = 3$

12. The zero of $f(x) = x^3 + x - 3$ between $x = 1$ and $x = 2$

13. The zero of $f(x) = x^3 + 2x^2 + x - 5$ between $x = 1$ and $x = 2$

14. The zero of $f(x) = x^5 + x - 1$ between $x = 0$ and $x = 1$

15. The zero of $f(x) = \sqrt{x + 1} - x$ between $x = 1$ and $x = 2$

16. The zero of $f(x) = e^{-x} - x$ between $x = 0$ and $x = 1$

17. The zero of $f(x) = e^x - 1/x$ between $x = 0$ and $x = 1$

18. The zero of $f(x) = \ln x^2 - 0.7x + 1$ between $x = 6$ and $x = 7$

19. Let $f(x) = 2x^3 - 9x^2 + 12x - 2$.

    a. Show that $f(x) = 0$ has a root between $x = 0$ and $x = 1$.

       Hint: Compute $f(0)$ and $f(1)$, and use the fact that f is continuous.

    b. Use Newton's method to find the zero of f in the interval $(0, 1)$, accurate to four decimal places.

20. Let $f(x) = x^3 - x - 1$.

    a. Show that $f(x) = 0$ has a root between $x = 1$ and $x = 2$.

       Hint: See Exercise 19.

    b. Use Newton's method to find the zero of f in the interval $(1, 2)$, accurate to four decimal places.

21. Let $f(x) = x^3 - 3x - 1$.

    a. Show that f has a zero between $x = 1$ and $x = 2$.

       Hint: See Exercise 19.

    b. Use Newton's method to find the zero of f, accurate to four decimal places.

22. Let $f(x) = x^4 - 4x^3 + 10$.

    a. Show that $f(x) = 0$ has a root between $x = 1$ and $x = 2$.

       Hint: See Exercise 19.

    b. Use Newton's method to find the zero of f, accurate to four decimal places.

In Exercises 23–28, make a rough sketch of the graphs of each of the given pairs of functions. Use your sketch to approximate the point(s) of intersection of the two graphs, and then apply Newton's method to refine the approximation of the x-coordinate of the point of intersection. Express your answers accurate to four decimal places.

23. $f(x) = 2\sqrt{x + 3}; g(x) = 2x - 1$

24. $f(x) = e^{-x^2}; g(x) = x^2$

25. $f(x) = e^{-x}; g(x) = x - 1$

26. $f(x) = \ln x; g(x) = 2 - x$

27. $f(x) = \sqrt{x}; g(x) = e^{-x}$

28. $f(x) = 2 - x^2; g(x) = \ln x$

29. MINIMIZING AVERAGE COST A division of Ditton Industries manufactures the Futura model microwave oven. Given that the daily cost of producing these microwave ovens (in dollars) is given by

$$C(x) = 0.0002x^3 - 0.06x^2 + 120x + 5000$$

where x stands for the number of units produced, find the level of production that minimizes the daily average cost per unit.

30. ALTITUDE OF A ROCKET The altitude (in feet) of a rocket t sec into flight is given by

$$s = f(t) = -2t^3 + 114t^2 + 480t + 1 \qquad (t \geq 0)$$

Find the time T, accurate to three decimal places, when the rocket hits the earth.

31. TEMPERATURE The temperature at 6 A.M. on a certain December day was measured at 15.6°F. In the next t hr, the temperature was given by

$$T = -0.05t^3 + 0.4t^2 + 3.8t + 15.6 \qquad (0 \leq t \leq 15)$$

where T is measured in degrees Fahrenheit. At what time was the temperature 0°F?

32. INTERNAL RATE OF RETURN ON AN INVESTMENT The proprietor of Qwik Digital Photo Lab recently purchased $12,000 of new photo-printing equipment. She expects that this investment, which has a useful life of 4 years, will yield returns of $4000 at the end of the first year, $5000 at the end of the second year, $4000 at the end of the third year, and $3000 at the end of the fourth year. Find the internal rate of return on the investment, accurate to four decimal places.

33. INTERNAL RATE OF RETURN ON AN INVESTMENT Executive Limousine Service recently acquired limousines worth $120,000. The projected returns over the next 3 years, the time period the limousines will be in service, are $80,000 at the end of the first year, $60,000 at the end of the second year, and $40,000 at the end of the third year. Find the internal rate of return on the investment, accurate to four decimal places.

34. INTERNAL RATE OF RETURN ON AN INVESTMENT Suppose an initial outlay of $C in an investment yields returns of $R at the end of each period over N periods.

    a. Show that the internal rate of return on the investment, r, may be obtained by solving the equation

$$Cr + R[(1 + r)^{-N} - 1] = 0$$

       Hint: $1 + x + x^2 + \cdots + x^{N-1} = \dfrac{1 - x^{N+1}}{1 - x}$

    b. Show that r can be found by performing the iteration

$$r_{n+1} = r_n - \frac{Cr_n + R[(1 + r_n)^{-N} - 1]}{C - NR(1 + r_n)^{-N-1}}$$

$$[r_0 \text{ (positive), an initial guess}]$$

       Hint: Apply Newton's method to the function $f(r) = Cr + R[(1 + r)^{-N} - 1]$.

35. HOME MORTGAGES Refer to Exercise 34. The Flemings secured a loan of $400,000 from a bank to finance the purchase of a house. They have agreed to repay the loan in equal monthly installments of $2577 over 25 years. The bank charges interest at the rate of $12r$/year on the unpaid balance, and interest computations are made at the end of each month. Find r and the annual interest rate, accurate to four decimal places.

36. HOME MORTGAGES Refer to Exercise 34. The Blakelys borrowed a sum of $350,000 from a bank to help finance the purchase of a house. The bank charges interest at the rate of $12r$/year on the unpaid balance, with interest being computed at the end of each month. The Blakelys have agreed to repay the loan in equal monthly installments of $1965.36 over 30 years. What is the annual rate of interest charged by the bank?

37. CAR LOANS Refer to Exercise 34. The price of Jane's new car is $32,000. Suppose that she makes a down payment of 25% toward the purchase of the car and secures financing for the balance over 4 years. If the monthly payment is $608.72, what is the rate of interest charged by the finance company?

38. REAL ESTATE INVESTMENT GROUPS Refer to Exercise 34. A group of private investors purchased a condominium complex for $2 million. They made an initial down payment of 10% and have obtained financing for the balance. If the loan is amortized over 15 years with quarterly repayments of $65,039, determine the interest rate charged by the bank. Assume that interest is calculated at the end of each quarter and is based on the unpaid balance.

39. DEMAND FOR WRISTWATCHES The quantity of Sicard wristwatches demanded per month is related to the unit price by the equation

$$p = d(x) = \frac{50}{0.01x^2 + 1} \qquad (1 \le x \le 20)$$

where p is measured in dollars and x is measured in thousands. The supplier is willing to make x thousand wristwatches available per month when the price per watch is given by $p = s(x) = 0.1x + 20$ dollars. Find the equilibrium quantity and price.

40. DEMAND FOR 4G TABLET COMPUTERS The weekly demand for the Pulsar 4G tablet computer is given by the demand equation

$$p = -0.05x + 600 \qquad (0 \le x \le 12,000)$$

where p denotes the wholesale unit price in dollars and x denotes the quantity demanded. The weekly total cost function associated with the manufacture of these computers is given by

$$C = 0.000002x^3 - 0.03x^2 + 400x + 80,000$$

where $C(x)$ denotes the total cost incurred in the production of x computers. Find the break-even level(s) of operation for the company.
Hint: Solve the equation $P(x) = 0$, where P is the total profit function.

41. a. Show that using Newton's method for finding the nth root of a real number a leads to the iterative formula

$$x_{i+1} = \left(\frac{n-1}{n}\right)x_i + \frac{a}{nx_i^{n-1}}$$

b. Use the result of part (a) to find $\sqrt[4]{42}$, accurate to three decimal places.

11.7 Solutions to Self-Check Exercises

1. Since $\sqrt[3]{10}$ is the only (real) cube root of the equation $x^3 - 10 = 0$, let's take $f(x) = x^3 - 10$. We have

$$f'(x) = 3x^2$$

so the required iteration formula is

$$x_{n+1} = x_n - \frac{f(x_n)}{f'(x_n)}$$

$$= x_n - \frac{x_n^3 - 10}{3x_n^2}$$

$$= \frac{2x_n^3 + 10}{3x_n^2}$$

Taking $x_0 = 2$, we find

$$x_1 = \frac{2(2^3) + 10}{3(2^2)} \approx 2.166667$$

$$x_2 = \frac{2(2.166667)^3 + 10}{3(2.166667)^2} \approx 2.154504$$

$$x_3 = \frac{2(2.154504)^3 + 10}{3(2.154504)^2} \approx 2.154435$$

Therefore, $\sqrt[3]{10} \approx 2.15444$.

2. We need to solve the equation $P(t) = 55,000$, or

$$50,000 + 30t^{3/2} + 20t = 55,000$$

Rewriting, we have

$$30t^{3/2} + 20t - 5000 = 0$$

$$3t^{3/2} + 2t - 500 = 0$$

Thus, the problem reduces to that of finding the zero of the function

$$f(t) = 3t^{3/2} + 2t - 500$$

To solve the equation $f(t) = 0$ using Newton's method, we first find

$$f'(t) = \frac{9}{2}t^{1/2} + 2 = \frac{9t^{1/2} + 4}{2}$$

So the required iteration formula is

$$t_{n+1} = t_n - \frac{3t_n^{3/2} + 2t_n - 500}{\dfrac{9t_n^{1/2} + 4}{2}}$$

$$= t_n - \frac{6t_n^{3/2} + 4t_n - 1000}{9t_n^{1/2} + 4}$$

$$= \frac{9t_n^{3/2} + 4t_n - (6t_n^{3/2} + 4t_n - 1000)}{9t_n^{1/2} + 4}$$

$$= \frac{3t_n^{3/2} + 1000}{9t_n^{1/2} + 4}$$

Using the initial guess $t_0 = 24$, we find

$$t_1 = \frac{3(24)^{3/2} + 1000}{9(24)^{1/2} + 4} \approx 28.129$$

$$t_2 = \frac{3(28.129)^{3/2} + 1000}{9(28.129)^{1/2} + 4} \approx 27.981$$

$$t_3 = \frac{3(27.981)^{3/2} + 1000}{9(27.981)^{1/2} + 4} \approx 27.981$$

So $t \approx 27.98$. Thus, the population will reach 55,000 approximately 28 months from now.

USING TECHNOLOGY

Newton's Method

The value of successive approximations in Newton's method can be generated by using a TI-83/84 as shown in the following example.

EXAMPLE 1 Use Newton's method to approximate the zero of $f(x) = x^3 - e^x + 1$ between $x = 1$ and $x = 2$. Use $x_0 = 1.2$ as the initial estimate. Continue with the iteration until two successive approximations differ by less than 0.00001.

Solution We have

$$f(x) = x^3 - e^x + 1 \qquad f'(x) = 3x^2 - e^x$$

Begin by entering the following functions into the calculator: Press $\boxed{Y=}$. Then enter

$$Y_1 = X^3 - e^{\wedge}(X) + 1$$

Press $\boxed{ENTER}$, then enter

$$Y_2 = 3X^2 - e^{\wedge}(X)$$

Next, return to the home screen (press $\boxed{2nd}$ $\boxed{QUIT}$). To enter the initial estimate $x_0 = 1.2$, we assign the value 1.2 to the variable X by pressing

$$\boxed{1.2} \quad \boxed{STO \blacktriangleright} \quad \boxed{X, T, \theta, n} \quad \text{followed by} \quad \boxed{ENTER}$$

To obtain the first iteration and assign its value to X, we press

$$X - Y_1/Y_2 \quad \boxed{STO \blacktriangleright} \quad X \qquad (\text{Use } \boxed{VARS} \text{ to select Y1 and Y2.})$$

Pressing $\boxed{ENTER}$ in succession gives the required iteration (see Figure T1). The required zero is approximately 1.545007279.

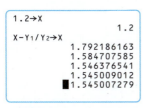

FIGURE T1
The TI 83/84 screen showing the iteration obtained by using Newton's method

TECHNOLOGY EXERCISES

In Exercises 1–6, use Newton's method to approximate the indicated zero of each function. Continue with the iteration until two successive approximations differ by less than 0.0001.

1. The zero of $f(x) = x^3 + x - 4$ between $x = 0$ and $x = 2$. Take $x_0 = 1$.

2. The zero of $f(x) = x^3 + 2x^2 + x - 6$ between $x = 1$ and $x = 2$. Take $x_0 = 1.5$.

3. The zero of $f(x) = x^3 + 2x + 2$ between $x = -1$ and $x = 0$. Take $x_0 = -0.5$.

4. The zero of $f(x) = x^4 + x - 4$ between $x = 0$ and $x = 2$. Take $x_0 = 1$.

5. The zero of $f(x) = x^5 + x - 1$ between $x = 0$ and $x = 1$. Take $x_0 = 0.5$.

6. The zero of $f(x) = x^5 + 2x^4 + 2x - 4$ between $x = 0$ and $x = 1$. Take $x_0 = 0.5$.

In Exercises 7 and 8, approximate the zero of the function in the indicated interval to five decimal places.

7. $f(x) = x^3 + 3x^2 - 3$ in $[-2, 0]$

8. $f(x) = x^3 - x - 1$ in $[1, 2]$

CHAPTER 11   Summary of Principal Formulas and Terms

FORMULAS

1. nth Taylor polynomial of f at a	$P_n(x) = f(a) + f'(a)(x - a)$ $\qquad + \dfrac{f''(a)}{2!}(x - a)^2 + \cdots$ $\qquad + \dfrac{f^n(a)}{n!}(x - a)^n$
2. Infinite series	$\displaystyle\sum_{n=1}^{\infty} a_n = a_1 + a_2 + a_3 + \cdots$
3. Geometric series	$\displaystyle\sum_{n=0}^{\infty} ar^n = a + ar + ar^2 + \cdots$ $\qquad = \dfrac{a}{1 - r} \qquad (|r| < 1)$
4. Power series in x	$\displaystyle\sum_{n=0}^{\infty} a_n x^n = a_0 + a_1 x + a_2 x^2 + a_3 x^3$ $\qquad + \cdots + a_n x^n + \cdots$
5. Power series in $(x - a)$	$\displaystyle\sum_{n=0}^{\infty} a_n(x - a)^n = a_0 + a_1(x - a)$ $\qquad + a_2(x - a)^2 + a_3(x - a)^3$ $\qquad + \cdots + a_n(x - a)^n + \cdots$
6. Taylor series of f at a	$f(x) = f(a) + f'(a)(x - a)$ $\qquad + \dfrac{f''(a)}{2!}(x - a)^2$ $\qquad + \dfrac{f'''(a)}{3!}(x - a)^3 + \cdots$
7. Iterative formula for Newton's method	$x_{n+1} = x_n - \dfrac{f(x_n)}{f'(x_n)} \qquad (n = 0, 1, 2, \ldots)$

TERMS

error bound for Taylor polynomials (732)

infinite sequence (736)

limit of a sequence (739)

convergent sequence (739)

divergent sequence (739)

infinite series (744)

partial sum of an infinite series (745)

convergent series (745)

divergent series (745)

geometric series (747)

multiplier effect (751)

perpetuity (751)

capital value of a perpetuity (751)

p-series (758)

power series in x (765)

power series in $(x - a)$ (765)

interval of convergence (765)

Taylor series of f (768)

Maclaurin series of f (768)

iteration (783)

internal rate of return (786)

CHAPTER 11   Concept Review Questions

Fill in the blanks.

1. a. The nth Taylor polynomial of f at a is the polynomial $P_n(x) = $ _____.

b. In terms of f and its derivatives, we have $P_n(a) = $ _____, $P_n'(a) = $ _____, $\ldots$, $P_n^{(n)}(a) = $ _____.

2. If f has derivatives up through order $N + 1$ in an interval I containing the number a, then for each x in _____, there exists a number c lying between _____ and _____ such that $R_N(x) = $ _____, where $R_N(x) = f(x) - P_N(x)$.

3. a. An infinite sequence is a/an _____ whose domain is the set of positive _____. The term a_n = _____ is called the _____ _____ of the sequence.

b. If a_n can be made as close to the number L as we please by taking n sufficiently _____, then $\{a_n\}$ is said to _____ to L.

4. If $\lim\limits_{n\to\infty} a_n = A$, $\lim\limits_{n\to\infty} b_n = B$, and c is any real number, then $\lim\limits_{n\to\infty} ca_n =$ _____, $\lim\limits_{n\to\infty}(a_n + b_n) =$ _____, $\lim\limits_{n\to\infty} a_n b_n =$ _____, and $\lim\limits_{n\to\infty} \dfrac{a_n}{b_n} =$ _____, provided that _____.

5. a. A series $\sum\limits_{n=1}^{\infty} a_n$ converges and has sum S if its sequence of _____ _____ converges to S.

b. The series $\sum\limits_{n=0}^{\infty} ar^n$ is called a/an _____ series. It converges if $|r| <$ _____, and it diverges if $|r| \geq$ _____.

6. a. If $\sum\limits_{n=1}^{\infty} a_n$ converges, then $\lim\limits_{n\to\infty} a_n =$ _____. If $\lim\limits_{n\to\infty} a_n$ does not exist or if $\lim\limits_{n\to\infty} a_n \neq 0$, then $\sum\limits_{n=1}^{\infty} a_n$ _____.

b. If $\sum\limits_{n=1}^{\infty} a_n = A$ and $\sum\limits_{n=1}^{\infty} b_n = B$ and c is any real number, then $\sum\limits_{n=1}^{\infty}(ca_n + b_n) =$ _____.

7. a. If f is a continuous, positive, and decreasing function on $[1, \infty)$ and $f(n) = a_n$ for all n, then $\sum\limits_{n=1}^{\infty} a_n$ and $\int_1^{\infty} f(x)\, dx$ are either both _____ or both _____.

b. The p-series is the series _____, and it converges if _____ and diverges if _____.

8. Let $\sum a_n$ and $\sum b_n$ be series with positive terms.
a. If $a_n \leq b_n$ for all n and $\sum b_n$ converges, then $\sum a_n$ _____.

b. If $\sum b_n$ diverges and _____ for all n, then $\sum a_n$ also diverges.

9. a. A power series in $(x - a)$ is a series of the form _____.
b. For a power series in $(x - a)$, exactly one of the following is true: It converges only at _____; it converges for all _____; or it converges for x in the interval _____, where R is some positive number. In the last case, the series diverges for x _____ this interval.

10. a. The Taylor series of a function f at a is $\sum\limits_{n=0}^{\infty} a_n(x - a)^n$, where $a_n =$ _____.
b. If f has derivatives of all orders in I and $R_N(x) = f(x) - P_N(x)$, then $f(x) = \sum\limits_{n=0}^{\infty} \dfrac{f^{(n)}}{n!}(x - a)^n$ if and only if _____. In this case, f is said to be represented by _____ _____ _____ of f at $x = a$.

CHAPTER 11 Review Exercises

In Exercises 1–4, find the fourth Taylor polynomial of the function f at the indicated number.

1. $f(x) = \dfrac{1}{x + 2}$ at $x = -1$ **2.** $f(x) = e^{-x}$ at $x = 1$

3. $f(x) = \ln(1 + x^2)$ at $x = 0$ **4.** $f(x) = \dfrac{1}{(1 + x)^2}$ at $x = 0$

5. Find the second Taylor polynomial of $f(x) = \sqrt[3]{x}$ at $x = 8$, and use it to estimate the value of $\sqrt[3]{7.8}$.

6. Find the sixth Taylor polynomial of $f(x) = e^{-x^2}$ at $x = 0$, and use it to estimate the value of
$$\int_0^1 e^{-x^2}\, dx$$
accurate to four decimal places.

7. Use the second Taylor polynomial of $f(x) = \sqrt[3]{x}$ at $x = 27$ to approximate $\sqrt[3]{26.98}$, and find a bound for the error in the approximation.

8. Use the third Taylor polynomial of $f(x) = 1/(1 + x)$ to approximate $f(0.1)$. Find a bound for the error in the approximation, and compare your results to the exact value of $f(0.1)$.

9. Use a fifth-degree Taylor polynomial to estimate e^{-1}.
Hint: Use Table 1.

10. Use a fourth-degree Taylor polynomial to estimate $\int_0^{0.2} e^{-x^2/2}\, dx$.
Hint: Use Table 1.

In Exercises 11–16, the nth term of a sequence is given. Determine whether the sequence converges or diverges. If the sequence converges, find its limit.

11. $a_n = \dfrac{n}{3n - 2}$ **12.** $a_n = \dfrac{1}{n + 1} - \dfrac{1}{n + 3}$

13. $a_n = \dfrac{2n^2 + 1}{3n^2 - 1}$ **14.** $a_n = \dfrac{(-1)^{n-1}n}{n + 1}$

15. $a_n = 1 - \dfrac{1}{2^n}$ **16.** $a_n = \dfrac{1 + \sqrt{n}}{1 - \sqrt{n}}$

In Exercises 17–20, find the sum of the geometric series if it converges.

17. $\sum\limits_{n=1}^{\infty} \dfrac{2^n}{3^n}$ **18.** $\sum\limits_{n=1}^{\infty} 2^{-n}3^{-n+1}$

19. $\sum\limits_{n=1}^{\infty} (-1)^{n-1}\left(\dfrac{1}{\sqrt{2}}\right)^n$ **20.** $\sum\limits_{n=1}^{\infty} \left(\dfrac{1}{e}\right)^n$

In Exercises 21 and 22, find the sum of the series.

21. $\sum\limits_{n=0}^{\infty} \left(\dfrac{1}{3^n} - \dfrac{1}{4^{n+1}}\right)$ **22.** $\sum\limits_{n=1}^{\infty} \left[\left(\dfrac{3}{5}\right)^n - \dfrac{1}{n(n + 1)}\right]$

23. Express the repeating decimal 1.424242. . . as a rational number.

24. Express the repeating decimal 3.142142142. . . as a rational number.

In Exercises 25–32, determine whether the series is convergent or divergent.

25. $\displaystyle\sum_{n=1}^{\infty} \frac{n^2 + 1}{2n^2 - 1}$

26. $\displaystyle\sum_{n=1}^{\infty} \frac{n + 1}{2n^2 + 4n}$

27. $\displaystyle\sum_{n=1}^{\infty} \frac{n}{2n^3 + 1}$

28. $\displaystyle\sum_{n=1}^{\infty} \frac{1}{\sqrt{n^3 + n}}$

29. $\displaystyle\sum_{n=1}^{\infty} \left(\frac{1}{n}\right)^{1.1}$

30. $\displaystyle\sum_{n=1}^{\infty} \frac{n^3}{n^5 + 2}$

31. $\displaystyle\sum_{n=2}^{\infty} \frac{1}{n(\ln n)^{3/2}}$

32. $\displaystyle\sum_{n=0}^{\infty} \frac{e^{-n}}{n + 1}$

In Exercises 33–36, find the radius of convergence and the interval of convergence of each power series. (Disregard endpoints.)

33. $\displaystyle\sum_{n=0}^{\infty} \frac{x^n}{n^2 + 2}$

34. $\displaystyle\sum_{n=1}^{\infty} \frac{(-1)^{n-1}}{\sqrt{n}} x^n$

35. $\displaystyle\sum_{n=1}^{\infty} \frac{(x - 1)^n}{n(n + 1)}$

36. $\displaystyle\sum_{n=2}^{\infty} \frac{e^n}{n^2} (x - 2)^n$

In Exercises 37–40, find the Taylor series of the function at the indicated number. Give the interval of convergence for each series.

37. $f(x) = \dfrac{1}{2x - 1}; x = 0$ 38. $f(x) = e^{-x}; x = 1$

39. $f(x) = \ln(1 + 2x); x = 0$

40. $f(x) = x^2 e^{-2x}; x = 0$

41. Find the radius of convergence of the series

$$\sum_{n=1}^{\infty} \frac{n^n}{(2n)!} (x - 1)^n.$$

42. Find a power series representation of $\displaystyle\int \frac{e^{-x}}{x} dx$.

43. Find a power series representation of $\displaystyle\int \frac{e^x - 1}{x} dx$.

44. Apply Newton's method to an appropriate function $f(x)$ to obtain an estimate of $\sqrt[3]{12}$, accurate to four decimal places.

45. Use Newton's method to find the root of the equation $x^3 + x^2 - 1 = 0$ between $x = 0$ and $x = 1$, accurate to four decimal places.

46. Use Newton's method to find the point of intersection of the graphs of the functions $y = f(x) = 2x$ and $y = g(x) = e^{-x}$, accurate to four decimal places.

47. A suitcase released from rest at the top of a plane metal slide moves a distance of

$$x = f(t) = 27(t + 3e^{-t/3} - 3)$$

feet in t sec. If the metal slide is 24 ft long, how long does it take the suitcase to reach the bottom?
Hint: Use Newton's method.

48. **ESTABLISHING A TRUST FUND** Sam Simpson wishes to establish a trust fund that will provide each of his two children with an annual income of $10,000 beginning next year and continuing throughout their lifetimes. Find the amount of money he is required to place in trust now if the fund will earn interest at the rate of 6%/year compounded continuously.

49. **EFFECT OF A TAX CUT** Suppose the average wage earner in a certain country saves 8% of her take-home pay and spends the other 92%.
   a. Estimate the impact that a proposed $10 billion tax cut will have on the economy over the long run in terms of the additional spending generated.
   b. Estimate the impact that a $10 billion tax cut will have on the economy over the long run if, at the same time, legislation is enacted to boost the rate of savings of the average taxpayer from 8% to 10%.

50. **HEIGHTS OF WOMEN** The heights of 4000 women who participated in a recent survey were found to be normally distributed according to the probability density function

$$f(x) = \frac{1}{2.5\sqrt{2\pi}} e^{-\frac{1}{2}[(x - 64.5)/2.5]^2}$$

Use the third Taylor polynomial of f to find the percentage of these women who have heights between 63.5 in. and 65.5 in.

CHAPTER 11   Before Moving On . . .

1. Find the first three Taylor polynomials of $f(x) = xe^{-x}$ at $x = 0$.

2. Determine whether the sequence converges or diverges. If the sequence converges, find its limit.
   a. $\left\{\dfrac{2n^2}{3n^2 + 2n + 1}\right\}$ b. $\left\{\dfrac{2n}{\sqrt{n^2 + 1}}\right\}$

3. Determine whether the series converges or diverges. If it converges, find its sum.
   a. $\displaystyle\sum_{n=0}^{\infty} \frac{3 - 2^n}{5^n}$ b. $\displaystyle\sum_{n=1}^{\infty} \left(\frac{1}{n + 2} - \frac{1}{n + 3}\right)$

4. Determine whether the series is convergent or divergent.
   a. $\displaystyle\sum_{n=1}^{\infty} \frac{1}{2n^2 + n + 3}$ b. $\displaystyle\sum_{n=2}^{\infty} \frac{(\ln n)^2}{n}$

5. Find the Taylor series of $f(x) = \dfrac{1}{1 + x}$ at $x = -2$, and give its radius of convergence.

12 Trigonometric Functions

DURING THE GOLDEN PERIOD of Greek civilization, Apollonius (262–200 B.C.E.) developed the trigonometric techniques necessary for calculating the radii of various circles. This was done in an effort to describe planetary motion, a physical process that exhibits a cyclical, or periodic, mode of behavior. Many other real-life phenomena—such as business cycles, earthquake vibrations, respiratory cycles, sales trends, and sound waves—exhibit cyclical behavior patterns.

In this chapter, we extend our study of calculus to an important class of functions called the trigonometric functions. These functions are periodic and hence lend themselves readily to describing many natural phenomena.

The revenue of McMenamy's Fish Shanty follows a cyclical pattern. When is the revenue of the restaurant increasing most rapidly? In Example 7, page 819, we answer this question.

© Nick Hanna/Alamy

Measurement of Angles

Angles

An **angle** in the plane is generated by rotating a ray about its endpoint. The starting position of the ray is called the **initial side** of the angle, the final position of the ray is called the **terminal side,** and the point of intersection of the two sides is called the **vertex** of the angle (see Figure 1a).

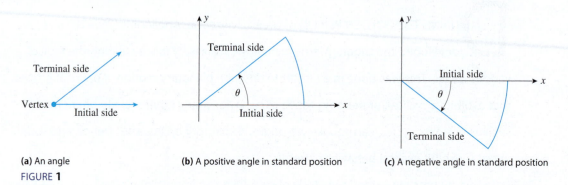

(a) An angle
        (b) A positive angle in standard position
        (c) A negative angle in standard position

FIGURE 1

In a rectangular coordinate system an angle θ (the Greek *theta*) is in **standard position** if its vertex is centered at the origin and its initial side coincides with the positive x-axis. An angle is **positive** if it is generated by a counterclockwise rotation and **negative** if it is generated by a clockwise rotation (Figure 1b–c).

When we say that an angle lies in a certain quadrant, we are referring to the quadrant in which the terminal side lies. The angle shown in Figure 2a lies in Quadrant II, and the angles shown in Figures 2b and 2c lie in Quadrant IV.

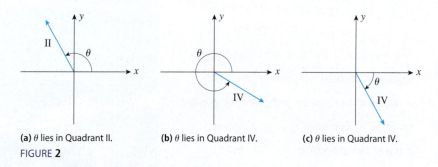

(a) θ lies in Quadrant II.
    (b) θ lies in Quadrant IV.
    (c) θ lies in Quadrant IV.

FIGURE 2

Degree and Radian Measure

An angle may be measured in either degrees or radians. A **degree** is the measure of the angle formed by $\frac{1}{360}$ of one complete revolution. If we rotate an initial ray in standard position through one complete revolution, we obtain an angle of $360°$.

A **radian** is the measure of the central angle subtended by an arc equal in length to the radius of the circle. In Figure 3, if s is the length of the arc subtended by a central angle θ in a circle of radius r, then

$$\theta = \frac{s}{r} \text{ radians} \tag{1}$$

For convenience we often work with the **unit circle,** that is, the circle of radius 1 centered at the origin. On the unit circle, an angle of 1 radian is subtended by an arc of length 1 (see Figure 4). To specify the units of measure for the angle θ in Figure 4,

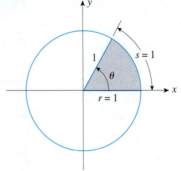

FIGURE **3**

$\theta = \dfrac{s}{r}$ radians

FIGURE **4**

$\theta = \dfrac{s}{r} = 1$ radian

we write $\theta = 1$ radian or $\theta = 1$. By convention, if the unit of measure is not specifically stated, we assume that it is radians.

Since the circumference of the unit circle is 2π and the central angle subtended by one complete revolution is $360°$, we see that

$$2\pi \text{ radians (rad)} = 360°$$

or

$$1 \text{ rad} = \left(\frac{180}{\pi}\right)^{\circ}$$

and

$$1° = \frac{\pi}{180} \text{ rad}$$

These relationships suggest the following useful conversion rules.

Converting Degrees and Radians

To convert degrees to radians, multiply by $\dfrac{\pi}{180}$.   **(2)**

To convert radians to degrees, multiply by $\dfrac{180}{\pi}$.   **(3)**

EXAMPLE 1 Convert each angle to radian measure:

a. $30°$ **b.** $45°$ **c.** $300°$ **d.** $450°$ **e.** $-240°$

Solution   Using Formula (2), we have

a. $30 \cdot \dfrac{\pi}{180} = \dfrac{\pi}{6}$, or $\dfrac{\pi}{6}$ rad

b. $45 \cdot \dfrac{\pi}{180} = \dfrac{\pi}{4}$, or $\dfrac{\pi}{4}$ rad

c. $300 \cdot \dfrac{\pi}{180} = \dfrac{5\pi}{3}$, or $\dfrac{5\pi}{3}$ rad

d. $450 \cdot \dfrac{\pi}{180} = \dfrac{5\pi}{2}$, or $\dfrac{5\pi}{2}$ rad

e. $-240 \cdot \dfrac{\pi}{180} = -\dfrac{4\pi}{3}$, or $-\dfrac{4\pi}{3}$ rad

EXAMPLE 2 Convert each angle to degree measure:

a. $\dfrac{\pi}{2}$ radians   **b.** $\dfrac{5\pi}{4}$ radians   **c.** $-\dfrac{3\pi}{4}$ radians   **d.** $\dfrac{7\pi}{2}$ radians

Solution Using Formula (3), we have

a. $\dfrac{\pi}{2} \cdot \dfrac{180}{\pi} = 90$, or 90 degrees

b. $\dfrac{5\pi}{4} \cdot \dfrac{180}{\pi} = 225$, or 225 degrees

c. $-\dfrac{3\pi}{4} \cdot \dfrac{180}{\pi} = -135$, or -135 degrees

d. $\dfrac{7\pi}{2} \cdot \dfrac{180}{\pi} = 630$, or 630 degrees

More than one angle may have the same initial and terminal sides. We call such angles **coterminal.** For example the angle $\frac{4\pi}{3}$ has the same initial and terminal sides as the angle $\theta = -\frac{2\pi}{3}$ (see Figure 5).

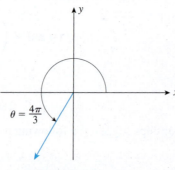 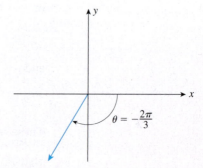

(a) $\theta = \dfrac{4\pi}{3}$ **(b)** $\theta = -\dfrac{2\pi}{3}$

FIGURE 5
Coterminal angles

An angle may be greater than 2π rad. For example, an angle of 3π rad is generated by rotating a ray in a counterclockwise direction through one and a half revolutions (Figure 6a). Similarly, an angle of $-\frac{5\pi}{2}$ rad is generated by rotating a ray in a clockwise direction through one and a quarter revolutions (Figure 6b).

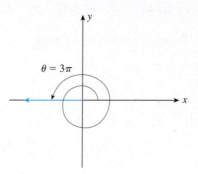 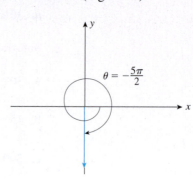

(a) $\theta = 3\pi$ **(b)** $\theta = -\dfrac{5\pi}{2}$

FIGURE 6
Angles generated by more than one revolution

The radian and degree measures of several common angles are given in Table 1. Be sure that you familiarize yourself with these values.

TABLE 1

Degrees	0°	30°	45°	60°	90°	120°	135°	150°	180°	270°	360°
Radians	0	$\dfrac{\pi}{6}$	$\dfrac{\pi}{4}$	$\dfrac{\pi}{3}$	$\dfrac{\pi}{2}$	$\dfrac{2\pi}{3}$	$\dfrac{3\pi}{4}$	$\dfrac{5\pi}{6}$	π	$\dfrac{3\pi}{2}$	2π

12.1 Self-Check Exercises

1. a. Convert 315° to radian measure.
  b. Convert $-\frac{5\pi}{4}$ radians to degree measure.

2. Make a sketch of the angle $-\frac{2\pi}{3}$ radians.

Solutions to Self-Check Exercises 12.1 can be found on page 801.

12.1 Concept Questions

1. Define the following terms in your own words: (a) degree and (b) radian.

2. When is an angle in standard position?

3. How do you convert degrees to radians and radians to degrees? Illustrate with examples.

12.1 Exercises

In Exercises 1–4, express the angle shown in the figure in radian measure.

1.

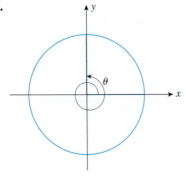

2.

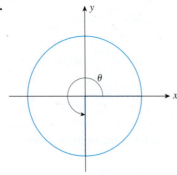

3.

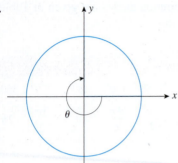

4.

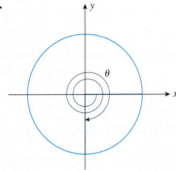

5. Identify the quadrant in which the angle lies.
  a. 220°
  b. −110°
  c. 460°
  d. −310°

6. Identify the quadrant in which the angle lies.
  a. $\dfrac{13\pi}{6}$ radians
  b. $-\dfrac{11\pi}{4}$ radians
  c. $\dfrac{17\pi}{3}$ radians
  d. $-\dfrac{25\pi}{12}$ radians

In Exercises 7–12, convert the angle to radian measure.

7. 75°
8. 330°

9. 160°
10. −210°

11. 630°
12. −420°

In Exercises 13–18, convert the angle to degree measure.

13. $\dfrac{2\pi}{3}$ radians
14. $\dfrac{7\pi}{6}$ radians

15. $-\dfrac{3\pi}{2}$ radians
16. $-\dfrac{13\pi}{12}$ radians

17. $\dfrac{22\pi}{18}$ radians
18. $-\dfrac{21\pi}{6}$ radians

In Exercises 19–22, make a sketch of the angle on a unit circle.

19. 225°
20. −120°

21. $\dfrac{7\pi}{3}$ radians
22. $-\dfrac{13\pi}{6}$ radians

In Exercises 23–26, determine an angle that is a coterminal angle of the angle θ. Use degree measure. (Answers may vary.)

23.

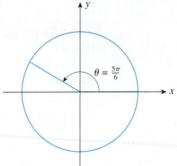

24.

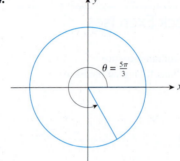

25.

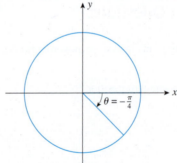

26.

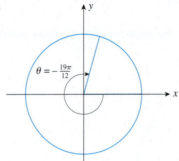

In Exercises 27–30, determine whether the statement is true or false. If it is true, explain why it is true. If it is false, explain why, or give an example to show why it is false.

27. The angle $3630°$ lies in the first quadrant.

28. The angle $\frac{103\pi}{6}$ radians lies in the second quadrant.

29. If θ is any angle, then $\theta + n(360)$, where n is a nonzero integer, is coterminal with θ (all angles measured in degrees).

30. If x, y, and z are measured in degrees, then $(x + y + z)$ degrees is equal to $\frac{\pi}{180}(x + y + z)$ radians.

12.1 Solutions to Self-Check Exercises

1. a. Using Formula (2), we find that the required radian measure is

$$315 \cdot \frac{\pi}{180} = \frac{7\pi}{4} \quad \text{or} \quad \frac{7\pi}{4} \text{ radians}$$

b. Using Formula (3), we find that the required degree measure is

$$\frac{-5\pi}{4} \cdot \frac{180}{\pi} = -225 \quad \text{or} \quad -225°$$

2.

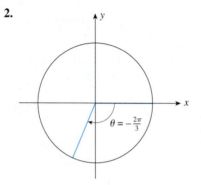

12.2 The Trigonometric Functions and Their Graphs

The Trigonometric Functions

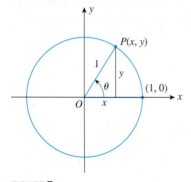

FIGURE 7
P is a point on the unit circle with coordinates $x = \cos\theta$ and $y = \sin\theta$.

Let $P(x, y)$ be a point on the unit circle such that the radius $\overline{OP}$ forms an angle of θ radians with respect to the positive x-axis (see Figure 7).

We define the **sine** of the angle θ, written $\sin\theta$, to be the y-coordinate of P. Similarly, the **cosine** of the angle θ, written $\cos\theta$, is defined to be the x-coordinate of P. The other trigonometric functions, tangent, cosecant, secant, and cotangent of θ—written $\tan\theta$, $\csc\theta$, $\sec\theta$, and $\cot\theta$, respectively—are defined in terms of the sine and cosine functions.

> **Trigonometric Functions**
>
> Let $P(x, y)$ be a point on the unit circle such that the radius $\overline{OP}$ forms an angle of θ radians with respect to the positive x-axis. Then
>
> $$\cos\theta = x \qquad\qquad \sin\theta = y \qquad\qquad \tan\theta = \frac{y}{x} = \frac{\sin\theta}{\cos\theta}$$
>
> $$\sec\theta = \frac{1}{x} = \frac{1}{\cos\theta} \qquad \csc\theta = \frac{1}{y} = \frac{1}{\sin\theta} \qquad \cot\theta = \frac{x}{y} = \frac{\cos\theta}{\sin\theta}$$

As you work with trigonometric functions, it will be helpful to remember the values of the sine, cosine, and tangent of some important angles, such as $\theta = 0, \frac{\pi}{6}, \frac{\pi}{4}, \frac{\pi}{3}, \frac{\pi}{2}$, and so on. These values may be found by using elementary geometry and algebra.

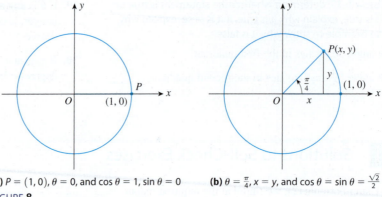

(a) $P = (1, 0)$, $\theta = 0$, and $\cos \theta = 1$, $\sin \theta = 0$ (b) $\theta = \frac{\pi}{4}$, $x = y$, and $\cos \theta = \sin \theta = \frac{\sqrt{2}}{2}$

FIGURE **8**

For example, if $\theta = 0$, then the point P has coordinates $(1, 0)$ (see Figure 8a), and we see that

$$\sin 0 = y = 0 \qquad \cos 0 = x = 1 \qquad \tan 0 = \frac{y}{x} = 0$$

As another example, if $\theta = \frac{\pi}{4}$, $x = y$. (See Figure 8b.) By the Pythagorean Theorem, we have

$$x^2 + y^2 = 2x^2 = 1$$

and $x = y = \frac{\sqrt{2}}{2}$. Therefore,

$$\sin \frac{\pi}{4} = y = \frac{\sqrt{2}}{2} \qquad \cos \frac{\pi}{4} = x = \frac{\sqrt{2}}{2} \qquad \tan \frac{\pi}{4} = \frac{y}{x} = 1$$

The sign of a trigonometric function of an angle θ is determined by the quadrant in which the terminal side of θ lies. Figure 9 shows the functions that are positive in each quadrant. The trigonometric functions that are positive in each quadrant can be remembered with the mnemonic device ASTC: **A**ll **S**tudents **T**ake **C**alculus. The signs of the other functions are easy to remember, since they are all negative.

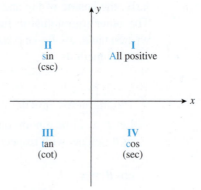

FIGURE **9**
The trigonometric functions that are positive
in each quadrant

To evaluate the trigonometric functions in quadrants other than the first quadrant, we use a reference angle. A **reference angle** for an angle θ is the acute angle formed by the x-axis and the terminal side of θ. Reference angles for each quadrant are depicted in Figure 10.

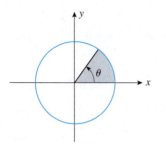

(a) Reference angle is θ.

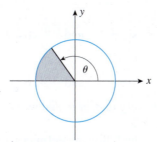

(b) Reference angle is $\pi - \theta$.

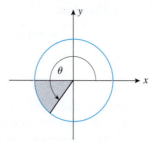

(c) Reference angle is $\theta - \pi$.

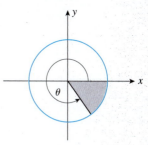

(d) Reference angle is $2\pi - \theta$.

FIGURE 10

The next example illustrates how we find the trigonometric functions of an angle.

EXAMPLE 1   Find the sine, cosine, and tangent of $\frac{5\pi}{4}$.

Solution   We first determine the reference angle for the given angle. As is indicated in Figure 11, the reference angle is $\frac{5\pi}{4} - \pi = \frac{\pi}{4}$, or 45°. Since $\sin 45° = \frac{\sqrt{2}}{2}$ and the sine is negative in Quadrant III, we conclude that $\sin \frac{5\pi}{4} = \frac{-\sqrt{2}}{2}$. Similarly, since $\cos 45° = \frac{\sqrt{2}}{2}$ and the cosine is negative in Quadrant III, we conclude that $\cos \frac{5\pi}{4} = \frac{-\sqrt{2}}{2}$. Finally, since $\tan 45° = 1$ and the tangent is positive in Quadrant III, we conclude that $\tan \frac{5\pi}{4} = 1$. ◼

The values of the trigonometric functions that we found in Example 1 are *exact*. The *approximate* value of any trigonometric function can be found by using a calculator. If you use a calculator, be sure to set the mode correctly. For example, to find $\sin \frac{5\pi}{4}$, first set the calculator in radian mode and then enter $\sin \frac{5\pi}{4}$. The result will be

$$\sin \frac{5\pi}{4} \approx -0.7071068$$

The number of digits in your answer will depend on the calculator that you use. As we saw in Example 1, the *exact* value of $\sin \frac{5\pi}{4}$ is $\frac{-\sqrt{2}}{2}$. Notice that we do not need to use reference angles when we use a calculator.

Referring once again to the unit circle, which is reproduced in Figure 12, we see that an angle of 2π rad corresponds to one complete revolution on the unit circle. Since $P(x, y) = (\cos \theta, \sin \theta)$ is the point where the terminal side of θ intersects the unit circle, we see that the values of $\sin \theta$ and $\cos \theta$ repeat themselves in subsequent revolutions.

FIGURE 11
The reference angle for $\theta = \frac{5\pi}{4}$ is $\frac{\pi}{4}$, or 45°.

Reference angle: $\frac{\pi}{4}$

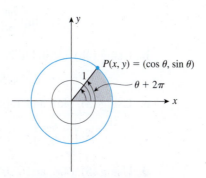

FIGURE 12
The x and y coordinates of the point P are the same for θ and $\theta + 2\pi$.

Therefore,

$$\sin(\theta + 2\pi) = \sin\theta \quad \text{and} \quad \cos(\theta + 2\pi) = \cos\theta \qquad \textbf{(4a)}$$

and

$$\sin(\theta + 2n\pi) = \sin\theta \quad \text{and} \quad \cos(\theta + 2n\pi) = \cos\theta \qquad \textbf{(4b)}$$

for every real number θ and every integer n, and we say that the sine and cosine functions are periodic with period 2π.

More generally, we have the following definition of a periodic function.

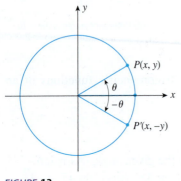

FIGURE 13
$\sin(-\theta) = -\sin\theta$ and
$\cos(-\theta) = \cos\theta.$

> **Periodic Function**
>
> A function f is **periodic** if there is a number $p > 0$ such that
>
> $$f(x + p) = f(x)$$
>
> for all x in the domain of f. The smallest such number p is called the **period** of f.

Also, from Figure 13, we see that

$$\sin(-\theta) = -\sin\theta \quad \text{and} \quad \cos(-\theta) = \cos\theta \qquad \textbf{(5)}$$

The values of the sine and cosine of some common angles are given in Table 2.

TABLE 2

θ in radians	0	$\dfrac{\pi}{6}$	$\dfrac{\pi}{4}$	$\dfrac{\pi}{3}$	$\dfrac{\pi}{2}$	$\dfrac{2\pi}{3}$	$\dfrac{3\pi}{4}$	$\dfrac{5\pi}{6}$	π	$\dfrac{7\pi}{6}$	$\dfrac{5\pi}{4}$	$\dfrac{4\pi}{3}$	$\dfrac{3\pi}{2}$	$\dfrac{5\pi}{3}$	$\dfrac{7\pi}{4}$	$\dfrac{11\pi}{6}$	2π
$\sin\theta$	0	$\dfrac{1}{2}$	$\dfrac{\sqrt{2}}{2}$	$\dfrac{\sqrt{3}}{2}$	1	$\dfrac{\sqrt{3}}{2}$	$\dfrac{\sqrt{2}}{2}$	$\dfrac{1}{2}$	0	$-\dfrac{1}{2}$	$-\dfrac{\sqrt{2}}{2}$	$-\dfrac{\sqrt{3}}{2}$	-1	$-\dfrac{\sqrt{3}}{2}$	$-\dfrac{\sqrt{2}}{2}$	$-\dfrac{1}{2}$	0
$\cos\theta$	1	$\dfrac{\sqrt{3}}{2}$	$\dfrac{\sqrt{2}}{2}$	$\dfrac{1}{2}$	0	$-\dfrac{1}{2}$	$-\dfrac{\sqrt{2}}{2}$	$-\dfrac{\sqrt{3}}{2}$	-1	$-\dfrac{\sqrt{3}}{2}$	$-\dfrac{\sqrt{2}}{2}$	$-\dfrac{1}{2}$	0	$\dfrac{1}{2}$	$\dfrac{\sqrt{2}}{2}$	$\dfrac{\sqrt{3}}{2}$	1

EXAMPLE 2 Evaluate:

a. $\sin\dfrac{7\pi}{2}$ **b.** $\cos 5\pi$ **c.** $\sin\left(-\dfrac{5\pi}{2}\right)$ **d.** $\cos\left(-\dfrac{11\pi}{4}\right)$

Solution

a. Using (4a) and Table 2, we have

$$\sin\left(\frac{7\pi}{2}\right) = \sin\left(2\pi + \frac{3\pi}{2}\right) = \sin\frac{3\pi}{2} = -1$$

b. Using (4b) and Table 2, we have

$$\cos 5\pi = \cos(4\pi + \pi) = \cos\pi = -1$$

c. Using (5), (4a), and Table 2, we have

$$\sin\left(-\frac{5\pi}{2}\right) = -\sin\left(\frac{5\pi}{2}\right) = -\sin\left(2\pi + \frac{\pi}{2}\right) = -\sin\frac{\pi}{2} = -1$$

d. Using (5), (4a), and Table 2, we have

$$\cos\left(-\frac{11\pi}{4}\right) = \cos\frac{11\pi}{4} = \cos\left(2\pi + \frac{3\pi}{4}\right) = \cos\frac{3\pi}{4} = -\frac{\sqrt{2}}{2}$$

Graphs of the Trigonometric Functions

To draw the graph of the function $y = f(x) = \sin x$, we first note that $\sin x$ is defined for every real number x, so the domain of the sine function is $(-\infty, \infty)$. Next, since the sine function is periodic with period 2π, it suffices to concentrate on sketching that part of the graph of $y = \sin x$ on the interval $[0, 2\pi]$ and repeating it as necessary. With the help of Table 2, we sketch the graph of $y = \sin x$ (Figure 14).

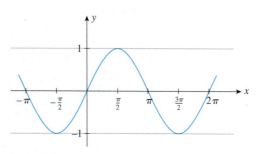

FIGURE 14
The graph of $y = \sin x$

In a similar manner, we can sketch the graph of $y = \cos x$ (Figure 15).

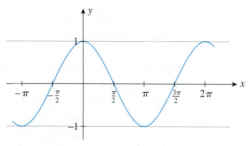

FIGURE 15
The graph of $y = \cos x$

Exploring with TECHNOLOGY

Plot the graphs of $f(x) = \sin x$ and $g(x) = \cos x$ in the viewing window $[-10, 10] \times [-1.1, 1.1]$.

1. What do the graphs suggest about the relationship between $f(x)$ and $g(x)$?
2. Confirm your observation by plotting the graphs of $h(x) = \sin(x + \frac{\pi}{2})$ and $g(x) = \cos x$ in the viewing window $[-10, 10] \times [-1.1, 1.1]$.

To sketch the graph of $y = \tan x$, we first note that $\tan x = (\sin x)/(\cos x)$. So $\tan x$ is not defined when $\cos x = 0$, that is, when $x = \frac{\pi}{2} \pm n\pi$ $(n = 0, 1, 2, 3, \ldots)$. The function is defined at all other points, so the domain of the tangent function is the set of all real numbers with the exception of the points just noted. Next, we can show that the vertical lines with equation $x = \frac{\pi}{2} \pm n\pi$ $(n = 0, 1, 2, 3, \ldots)$ are vertical asymptotes of the graph of $f(x) = \tan x$. For example, since we can readily verify with the help of a calculator that

$$\lim_{x \to (\pi/2)^-} \tan x = \infty \quad \text{and} \quad \lim_{x \to (\pi/2)^+} \tan x = -\infty$$

we conclude that $x = \frac{\pi}{2}$ is a vertical asymptote of the graph of $y = \tan x$. Finally, using Table 2, we sketch the graph of $y = \tan x$ (Figure 16). Observe that the tangent function is periodic with period π. It follows that

$$\tan(x + n\pi) = \tan x \tag{6}$$

whenever n is an integer.

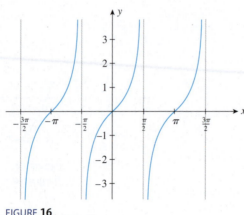

FIGURE **16**
The graph of $y = \tan x$

The graphs of $y = \sec x$, $y = \csc x$, and $y = \cot x$ may be sketched in a similar manner (see Exercises 23–25).

Let's look more closely at the graphs shown in Figures 14 and 15. Notice that the graphs of $y = \sin x$ and $y = \cos x$ oscillate between $y = -1$ and $y = 1$. In general, the graphs of the functions $y = A \sin x$ and $y = A \cos x$ oscillate between $y = -A$ and $y = A$, and we say that their **amplitude** is $|A|$. The graphs of $y = 4 \sin x$ and $y = \frac{1}{4} \sin x$ are shown in Figure 17a–b. Observe that the factor 4 in $y = 4 \sin x$ has the effect of "stretching" the graph of $y = \sin x$ between the values of -4 and 4, whereas the factor $\frac{1}{4}$ in $y = \frac{1}{4} \sin x$ has the effect of "compressing" the graph between $-\frac{1}{4}$ and $\frac{1}{4}$.

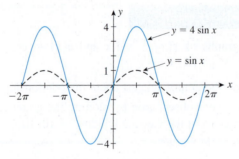

(a) The graph of $y = 4 \sin x$ superimposed upon the graph of $y = \sin x$

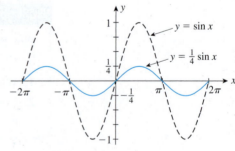

(b) The graph of $y = \frac{1}{4} \sin x$ superimposed upon the graph of $y = \sin x$

FIGURE **17**

Next, let's compare the graphs of $y = \cos 2x$ and $y = \cos(x/2)$ with the graph of $y = \cos x$ (see Figure 18a–b). Notice here that the factor of 2 has the effect of "speeding up" the graph of the cosine: The period is decreased from 2π to π. In contrast, the factor of $\frac{1}{2}$ has the effect of "slowing down" the graph of the cosine: The period is increased from 2π to 4π. In general, the period of both $y = \sin Bx$ and $y = \cos Bx$ is $2\pi/|B|$ if $B \neq 0$.

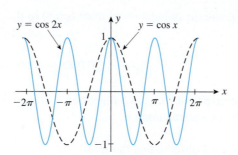

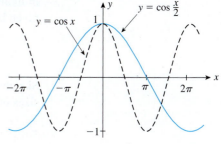

(a) The graph of $y = \cos 2x$ superimposed upon the graph of $y = \cos x$

(b) The graph of $y = \cos \frac{x}{2}$ superimposed upon the graph of $y = \cos x$

FIGURE **18**

We now summarize these definitions.

Period and Amplitude of $A \sin Bx$ and $A \cos Bx$

The graphs of

$$f(x) = A \sin Bx \qquad \text{and} \qquad f(x) = A \cos Bx$$

where $A \neq 0$ and $B \neq 0$, have period $2\pi/|B|$ and amplitude $|A|$.

EXAMPLE 3 Sketch the graph of $y = 3 \sin \frac{1}{2} x$.

Solution The function $y = 3 \sin \frac{1}{2} x$ has the form $y = A \sin Bx$, where $A = 3$ and $B = \frac{1}{2}$. This tells us that the amplitude of the graph is 3 and the period is $2\pi/|\frac{1}{2}| = 4\pi$. Using the graph of the sine curve, we sketch the graph of $y = 3 \sin \frac{1}{2} x$ over one period $[0, 4\pi]$. (See Figure 19.) Next, the periodic properties of the sine function allow us to extend the graph in either direction by completing another cycle as shown.

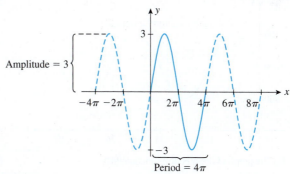

FIGURE **19**
The graph of $y = 3 \sin \frac{1}{2} x$ has amplitude 3 and period 4π.

The Predator–Prey Population Model

We will now look at a mathematical model of a phenomenon that exhibits cyclical behavior: the so-called **predator–prey population model.**

 APPLIED EXAMPLE 4 Predator–Prey Population The population of owls (predators) in a certain region over a 2-year period is estimated to be

$$P_1(t) = 1000 + 100 \sin\left(\frac{\pi t}{12}\right)$$

in month t, and the population of mice (prey) in the same area at time t is given by

$$P_2(t) = 20{,}000 + 4000 \cos\left(\frac{\pi t}{12}\right)$$

Sketch the graphs of these two functions, and explain the relationship between the sizes of the two populations.

Solution   First observe that the term

$$100 \sin\left(\frac{\pi t}{12}\right) \qquad A \sin Bx$$

has period

$$\frac{2\pi}{\frac{\pi}{12}} = 24 \qquad \frac{2\pi}{|B|}$$

and amplitude $|A| = 100$. Similarly, the term

$$4000 \cos\left(\frac{\pi t}{12}\right)$$

has period 24 and amplitude 4000. Next, recall that the sine and cosine functions oscillate between -1 and $+1$, so $P_1(t)$ is seen to oscillate between $[1000 + 100(-1)]$, or 900, and $[1000 + 100(1)]$, or 1100, while $P_2(t)$ oscillates between $[20{,}000 + 4000(-1)]$, or 16,000, and $[20{,}000 + 4000(1)]$, or 24,000. Finally, plotting a few points on each graph for, say, $t = 0, 2, 3$, and so on, we obtain the graphs of the functions P_1 and P_2 as shown in Figure 20.

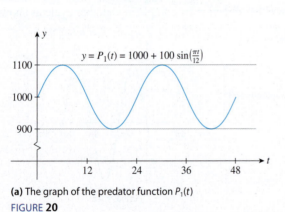

(a) The graph of the predator function $P_1(t)$

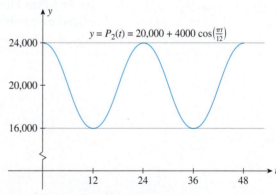

(b) The graph of the prey function $P_2(t)$

FIGURE 20

From the graphs, we see that at time $t = 0$, the predator population stands at 1000 owls. As it increases, the prey population decreases from 24,000 mice at that instant of time. Eventually, this decrease in the food supply causes the predator population to decrease, which in turn causes an increase in the prey population. But as the prey population increases, resulting in an increase in food supply, the predator population once again increases. The cycle is complete and starts all over again. ▪

Trigonometric Identities

Equations that express the relationships between trigonometric functions, such as

$$\sin(-\theta) = -\sin\theta \quad \text{and} \quad \cos(-\theta) = \cos\theta$$

are called **trigonometric identities.** Some other important trigonometric identities are listed in Table 3. Each identity holds true for every value of θ in the common domain of the specified functions. The proofs of these identities may be found in any elementary trigonometry book.

TABLE 3

Trigonometric Identities

Pythagorean Identities	Half-Angle Formulas	Sum and Difference Formulas
$\sin^2 \theta + \cos^2 \theta = 1$	$\cos^2 \theta = \dfrac{1}{2}(1 + \cos 2\theta)$	$\sin(A \pm B) = \sin A \cos B \pm \cos A \sin B$
$\tan^2 \theta + 1 = \sec^2 \theta$	$\sin^2 \theta = \dfrac{1}{2}(1 - \cos 2\theta)$	$\cos(A \pm B) = \cos A \cos B \mp \sin A \sin B$
$\cot^2 \theta + 1 = \csc^2 \theta$		

Double-Angle Formulas	Cofunctions of Complementary Angles
$\sin 2A = 2 \sin A \cos A$	$\sin \theta = \cos\left(\dfrac{\pi}{2} - \theta\right)$
$\cos 2A = \cos^2 A - \sin^2 A$	$\cos \theta = \sin\left(\dfrac{\pi}{2} - \theta\right)$

As we will see later, these identities are useful in simplifying trigonometric expressions and equations and in deriving other trigonometric relationships. They are also used to verify other trigonometric identities, as illustrated in the next example.

EXAMPLE 5 Verify the identity

$$\sin \theta \, (\csc \theta - \sin \theta) = \cos^2 \theta$$

Solution We verify this identity by showing that the expression on the left side of the equation can be transformed into the expression on the right side. Thus,

$$\sin \theta \, (\csc \theta - \sin \theta) = \sin \theta \csc \theta - \sin^2 \theta$$
$$= \sin \theta \, \frac{1}{\sin \theta} - \sin^2 \theta$$
$$= 1 - \sin^2 \theta$$
$$= \cos^2 \theta$$

Exploring with TECHNOLOGY

1. You can confirm some of the trigonometric identities graphically in many ways. For example, explain why an identity given in the form $f(\theta) = g(\theta)$ over some domain D is equivalent to the following statements:
   a. The graphs of f and g coincide when viewed in any appropriate viewing window.
   b. The graphs of f and $-g$ are reflections of each other with respect to the θ-axis when viewed in any appropriate viewing window.
   c. The graph of $h = f - g$ is the graph of the zero function $h(\theta) = 0$ in the common domain of f and g.

2. Use the observations made in part 1 to verify graphically the identity $\sin^2 \theta + \cos^2 \theta = 1$ by taking $f(x) = \sin^2 x + \cos^2 x$ and $g(x) = 1$.
   Hint: Enter $f(x)$ as $y_1 = (\sin x)^2 + (\cos x)^2$, and use the viewing window $[-10, 10] \times [-1.1, 1.1]$.

3. Verify graphically the half-angle formula $\cos^2 \theta = \frac{1}{2}(1 + \cos 2\theta)$.

4. Verify graphically the formula $\sin \theta = \cos\left(\frac{\pi}{2} - \theta\right)$.

12.2 Self-Check Exercises

1. Evaluate $\cos\left(-\frac{13\pi}{3}\right)$.

2. Solve the equation $\cos\theta = -\frac{\sqrt{2}}{2}$ for $0 \le \theta \le 2\pi$.

3. Sketch the graph of $y = 2\cos x$.

Solutions to Self-Check Exercises 12.2 can be found on page 812.

12.2 Concept Questions

1. Define $\sin\theta$, $\cos\theta$, $\tan\theta$, $\csc\theta$, $\sec\theta$, and $\cot\theta$ in terms of x and y, where (x, y) is a point on the unit circle.

2. Write the values of $\sin\theta$, $\cos\theta$, and $\tan\theta$ for $\theta = 0, \frac{\pi}{6}, \frac{\pi}{4}, \frac{\pi}{3}, \frac{\pi}{2}, \frac{2\pi}{3}, \frac{5\pi}{6}$, and π.

3. Complete the equation:
   a. $\sin^2\theta + \cos^2\theta = $ _____
   b. $\sec^2\theta = $ _____
   c. $\cot^2\theta + 1 = $ _____

4. Complete the equation:
   a. $\dfrac{1 + \cos 2\theta}{2} = $ _____
   b. $\dfrac{1 - \cos 2\theta}{2} = $ _____
   c. $\sin(A \pm B) = $ _____

5. Complete the equation:
   a. $\cos(A \pm B) = $ _____
   b. $\sin 2A = $ _____
   c. $\cos 2A = $ _____

12.2 Exercises

In Exercises 1–10, evaluate the trigonometric function.

1. $\sin 3\pi$

2. $\cos\left(-\dfrac{3\pi}{2}\right)$

3. $\sin\dfrac{9\pi}{2}$

4. $\cos\dfrac{13\pi}{6}$

5. $\sin\left(-\dfrac{4\pi}{3}\right)$

6. $\cos\left(-\dfrac{5\pi}{4}\right)$

7. $\tan\dfrac{\pi}{6}$

8. $\cot\left(-\dfrac{\pi}{3}\right)$

9. $\sec\left(-\dfrac{5\pi}{8}\right)$

10. $\csc\dfrac{9\pi}{4}$

In Exercises 11–14, find the six trigonometric functions of the angle.

11. $\dfrac{\pi}{2}$

12. $-\dfrac{\pi}{6}$

13. $\dfrac{5\pi}{3}$

14. $-\dfrac{3\pi}{4}$

In Exercises 15–22, find all values of θ that satisfy the equation over the interval $[0, 2\pi]$.

15. $\sin\theta = -\dfrac{1}{2}$

16. $\tan\theta = 1$

17. $\cot\theta = -\sqrt{3}$

18. $\csc\theta = \sqrt{2}$

19. $\sec\theta = -1$

20. $\cos\theta = \sin\theta$

21. $\sin\theta = \sin\left(-\dfrac{4\pi}{3}\right)$

22. $\cos\theta = \cos\left(-\dfrac{\pi}{6}\right)$

In Exercises 23–26, sketch the graph of the function over the interval $[0, 2\pi]$.

23. $y = \csc x$

24. $y = \sec x$

25. $y = \cot x$

26. $y = \tan 2x$

In Exercises 27–30, sketch the graph of the function over the interval $[0, 2\pi]$ by comparing it to the graph of $y = \sin x$.

27. $y = \sin 2x$

28. $y = 2\sin x$

29. $y = -\sin x$

30. $y = -2\sin 2x$

In Exercises 31–38, determine the amplitude and the period for the function. Sketch the graph of the function over one period.

31. $y = \sin(x - \pi)$

32. $y = \cos(x + \pi)$

33. $y = \sin\left(x + \dfrac{\pi}{2}\right)$

34. $y = \cos\left(x - \dfrac{\pi}{4}\right)$

35. $y = \cos x + 2$

36. $y = 2 - \sin x$

37. $y = -2\cos 3x$

38. $y = -3\sin(-4x)$

In Exercises 39–46, verify each identity.

39. $\cos^2\theta - \sin^2\theta = 2\cos^2\theta - 1$

40. $1 - 2\sin^2\theta = 2\cos^2\theta - 1$

41. $(\sec\theta + \tan\theta)(1 - \sin\theta) = \cos\theta$

42. $\dfrac{\sin^2\theta}{1 + \cos^2\theta} = \dfrac{1 - \cos^2\theta}{2 - \sin^2\theta}$

43. $(1 + \cot^2 \theta) \tan^2 \theta = \sec^2 \theta$

44. $\dfrac{\sec \theta - \cos \theta}{\tan \theta} = \sin \theta$ **45.** $\dfrac{\csc \theta}{\tan \theta + \cot \theta} = \cos \theta$

46. $\tan(A + B) = \dfrac{\tan A + \tan B}{1 - \tan A \tan B}$

47. The accompanying figure shows a right triangle ABC superimposed over a unit circle in the xy-coordinate system. By considering similar triangles, show that

$$\sin \theta = \frac{BC}{AC} = \frac{\text{Opposite side}}{\text{Hypotenuse}}$$

$$\cos \theta = \frac{AB}{AC} = \frac{\text{Adjacent side}}{\text{Hypotenuse}}$$

$$\tan \theta = \frac{BC}{AB} = \frac{\text{Opposite side}}{\text{Adjacent side}}$$

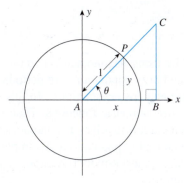

48. Refer to Exercise 47. Find the values of the six trigonometric functions of θ, where θ is the angle shown in the right triangle in the accompanying figure.

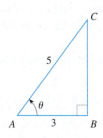

49. Refer to Exercise 47. Find the values of the six trigonometric functions of θ, where θ is the angle shown in the right triangle in the accompanying figure.

50. PREDATOR–PREY POPULATION The population of foxes (predators) in a certain region over a 2-year period is estimated to be

$$P_1(t) = 400 + 50 \sin\left(\frac{\pi t}{12}\right)$$

in month t, and the population of rabbits (prey) in the same region at time t is given by

$$P_2(t) = 3000 + 500 \cos\left(\frac{\pi t}{12}\right)$$

Sketch the graphs of each of these two functions, and explain the relationship between the sizes of the two populations.

51. BLOOD PRESSURE The arterial blood pressure of an individual in a state of relaxation is given by

$$P(t) = 100 + 20 \sin 6t$$

where $P(t)$ is measured in millimeters of mercury (Hg) and t is the time in seconds.
  a. Show that the individual's systolic pressure (maximum blood pressure) is 120 and his diastolic pressure (minimum blood pressure) is 80.
  b. Find the values of t when the individual's blood pressure is highest and lowest.

52. FLIGHT TAKEOFF After takeoff, an airplane climbs at an angle of 20° at a speed of 200 ft/sec. How long does it take for the airplane to reach an altitude of 10,000 ft?

53. SPEED OF A JOGGER A man located at a point A on one bank of a river that is 1000 ft wide observed a woman jogging on the opposite bank. When the jogger was first spotted, the angle between the river bank and the man's line of sight was 30°. One minute later, the angle was 40°. How fast was the woman running if she maintained a constant speed?

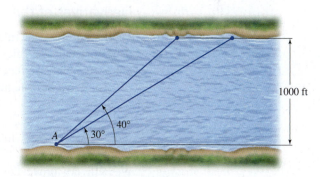

In Exercises 54–58, determine whether the statement is true or false. If it is true, explain why it is true. If it is false, give an example to show why it is false.

54. $\sin(a - b) = -\sin(b - a)$

55. If $\sin \theta = -\dfrac{\sqrt{3}}{2}$ and $0 \le \theta \le 2\pi$, then $\theta = \dfrac{5\pi}{3}$.

56. If $\tan \theta = \dfrac{1}{3}$, then $\sin \theta = \dfrac{\sqrt{10}}{10}$.

57. $\cos 2\theta = 2 \cos^2 \theta - 1$

58. $\sin 3\theta = 3 \sin \theta + 4 \sin^3 \theta$

12.2 Solutions to Self-Check Exercises

1. $\cos\left(-\dfrac{13\pi}{3}\right) = \cos\dfrac{13\pi}{3}$

$$= \cos\left(4\pi + \dfrac{\pi}{3}\right)$$

$$= \cos\dfrac{\pi}{3} = \dfrac{1}{2}$$

2. $\cos\theta$ is negative for θ in Quadrants II and III. From Table 2, we see that the required values of θ are $\frac{3\pi}{4}$ and $\frac{5\pi}{4}$.

3. Here, $A = 2$ and $B = 1$, so the amplitude of the graph is 2 and the period is 2π. Transforming the graph of $y = \cos x$, we obtain the graph of $y = 2\cos x$.

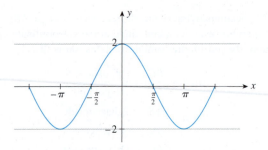

12.3 Differentiation of Trigonometric Functions

Many real-world problems are modeled by using trigonometric functions. We have already seen that the graphs of predator and prey populations are periodic and hence can be described by a combination of sine and cosine functions. In this section, we will develop rules for differentiating the trigonometric functions. Then we will use these rules to solve problems whose formulation involves trigonometric functions, such as finding the rate of change of predator–prey populations (Example 5), finding the time when a revenue function is increasing most rapidly (Example 7), and finding the maximum capacity of a trough (Example 8).

Derivatives of the Sine and Cosine Functions

First, we develop the rule for differentiating the sine and cosine functions. Then, using these rules and the rules of differentiation, we derive the rules for differentiating the other trigonometric functions.

To derive the rule for differentiating the sine function, we need the following two results:

$$\lim_{h\to 0}\frac{\sin h}{h} = 1 \quad\text{and}\quad \lim_{h\to 0}\frac{\cos h - 1}{h} = 0 \tag{7}$$

We will not prove the first limit, but its plausibility can be seen by examining Table 4, which was constructed with the aid of a calculator.

TABLE 4

Values of $\dfrac{\sin h}{h}$ for Selected Values of h (in Radians) Approaching Zero

h	± 0.5	± 0.1	± 0.01	± 0.001
$\dfrac{\sin h}{h}$	0.9588511	0.9983342	0.9999833	0.9999998

The second limit can be derived from the first by using the appropriate trigonometric identity and the properties of limits (see Exercise 76).

You can obtain a visual confirmation of the limits

$$\lim_{h \to 0} \frac{\sin h}{h} = 1 \quad \text{and} \quad \lim_{h \to 0} \frac{\cos h - 1}{h} = 0$$

in Equation (7) as follows:

1. Plot the graph of $f(x) = (\sin x)/x$, using the viewing window $[-1, 1] \times [0, 2]$. Then use **ZOOM** and **TRACE** to find the values of $f(x)$ for

values of x close to $x = 0$. What happens when you try to evaluate $f(0)$? Explain your answer.

2. Plot the graph of $g(x) = (\cos x - 1)/x$, using the viewing window $[-1, 1] \times [-0.5, 0.5]$. Then use **ZOOM** and **TRACE** to show that $g(x)$ is close to zero if x is close to zero.

Caution: Convincing as these arguments are, they do not prove the stated results.

We are now in the position to derive the formula for the derivative of $\sin x$. Suppose that $f(x) = \sin x$. Then,

$$f'(x) = \lim_{h \to 0} \frac{f(x + h) - f(x)}{h} \qquad \text{\textcolor{red}{Definition of the derivative}}$$

$$= \lim_{h \to 0} \frac{\sin(x + h) - \sin x}{h}$$

$$= \lim_{h \to 0} \frac{\sin x \cos h + \cos x \sin h - \sin x}{h} \qquad \text{\textcolor{red}{Use the Sine Rule for sums.}}$$

$$= \lim_{h \to 0} \frac{\sin x \, (\cos h - 1) + \cos x \sin h}{h}$$

$$= \lim_{h \to 0} \sin x \left(\frac{\cos h - 1}{h} \right) + \lim_{h \to 0} \cos x \left(\frac{\sin h}{h} \right)$$

$$= \sin x \lim_{h \to 0} \left(\frac{\cos h - 1}{h} \right) + \cos x \lim_{h \to 0} \left(\frac{\sin h}{h} \right)$$

$$= (\sin x)(0) + (\cos x)(1) \qquad \text{\textcolor{red}{Use Equation (7).}}$$

$$= \cos x$$

With the help of the Chain Rule, this result can be generalized as follows: Suppose that

$$y = h(x) = \sin f(x)$$

where $f(x)$ is a differentiable function of x. Then letting $u = f(x)$ so that $y = \sin u$, we apply the Chain Rule and obtain

$$h'(x) = \frac{dy}{du} \cdot \frac{du}{dx} = (\cos u) \frac{du}{dx} = [\cos f(x)] f'(x)$$

Derivatives of the Sine and Generalized Sine Functions

Rule 1a: $\dfrac{d}{dx} (\sin x) = \cos x$

Rule 1b: $\dfrac{d}{dx} [\sin f(x)] = [\cos f(x)] f'(x)$

Observe that Rule 1b reduces to Rule 1a when $f(x) = x$, as expected.

The relationship between the function $f(x) = \sin x$ and its derivative $f'(x) = \cos x$ can be seen by sketching the graphs of both functions (see Figure 21). Here, we interpret $f'(x)$ as the slope of the tangent line to the graph of f at the point $(x, f(x))$.

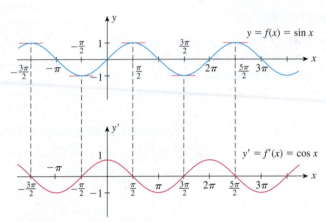

FIGURE 21
The graphs of $f(x) = \sin x$ and its derivative $f'(x) = \cos x$

EXAMPLE 1 Differentiate each of the following functions:

a. $f(x) = x^2 \sin x$

b. $g(x) = \sin(2x + 1)$

c. $h(x) = (x + \sin x^2)^{10}$

Solution

a. Using the Product Rule followed by Rule 1a, we find

$$f'(x) = 2x \sin x + x^2 \frac{d}{dx}(\sin x)$$
$$= 2x \sin x + x^2 \cos x = x(2 \sin x + x \cos x)$$

b. Using Rule 1b, we find

$$g'(x) = [\cos(2x + 1)] \frac{d}{dx}(2x + 1) = 2 \cos(2x + 1)$$

c. We first use the General Power Rule, followed by Rule 1b. We obtain

$$h'(x) = 10(x + \sin x^2)^9 \frac{d}{dx}(x + \sin x^2)$$
$$= 10(x + \sin x^2)^9 \left[1 + \cos x^2 \frac{d}{dx}(x^2) \right]$$
$$= 10(x + \sin x^2)^9 (1 + 2x \cos x^2)$$

To derive the rule for differentiating the cosine function, we make use of the following relationships between the cosine function and the sine function:

$$\cos x = \sin\left(\frac{\pi}{2} - x\right) \quad \text{and} \quad \sin x = \cos\left(\frac{\pi}{2} - x\right)$$

(see Table 3, page 809). Now, with the help of Rule 1b, we see that if $f(x) = \cos x$, then

$$f'(x) = \frac{d}{dx} \cos x$$

$$= \frac{d}{dx} \sin\left(\frac{\pi}{2} - x\right)$$

$$= \cos\left(\frac{\pi}{2} - x\right) \frac{d}{dx}\left(\frac{\pi}{2} - x\right)$$

$$= (\sin x)(-1)$$

$$= -\sin x$$

This result may be generalized immediately by using the Chain Rule. In fact, if

$$y = h(x) = \cos f(x)$$

where $f(x)$ is a differentiable function of x, then with $u = f(x)$, we have $y = \cos u$, so

$$\frac{dy}{dx} = \frac{dy}{du} \cdot \frac{du}{dx} = (-\sin u)\frac{du}{dx} = -[\sin f(x)]f'(x)$$

Derivatives of the Cosine and Generalized Cosine Functions

Rule 2a: $\dfrac{d}{dx}(\cos x) = -\sin x$

Rule 2b: $\dfrac{d}{dx}[\cos f(x)] = -[\sin f(x)]f'(x)$

Observe that Rule 2b reduces to Rule 2a when $f(x) = x$, as expected.

EXAMPLE 2 Find the derivative of each function:

a. $f(x) = \cos(2x^2 - 1)$ **b.** $g(x) = \sqrt{\cos 2x}$ **c.** $h(x) = e^{\sin 2x + \cos 3x}$

Solution

a. Using Rule 2b, we find

$$f'(x) = -\sin(2x^2 - 1)\frac{d}{dx}(2x^2 - 1)$$

$$= -[\sin(2x^2 - 1)]4x$$

$$= -4x \sin(2x^2 - 1)$$

b. We first rewrite $g(x)$ as $g(x) = (\cos 2x)^{1/2}$. Using the Power Rule followed by the Chain Rule (Rule 2b), we find

$$g'(x) = \frac{1}{2}(\cos 2x)^{-1/2}\frac{d}{dx}(\cos 2x)$$

$$= \frac{1}{2}(\cos 2x)^{-1/2}(-\sin 2x)\frac{d}{dx}(2x)$$

$$= \frac{1}{2}(\cos 2x)^{-1/2}(-\sin 2x)(2)$$

$$= -\frac{\sin 2x}{\sqrt{\cos 2x}}$$

Explore and Discuss

Derive Rule 2a,

$$\frac{d}{dx}(\cos x) = -\sin x$$

directly from the definition of the derivative and the relationships given in Equation (7).

c. Using the corollary to the Chain Rule for Exponential Functions, we find

$$h'(x) = e^{\sin 2x + \cos 3x} \cdot \frac{d}{dx}(\sin 2x + \cos 3x)$$

$$= e^{\sin 2x + \cos 3x}\left[(\cos 2x)\frac{d}{dx}(2x) - (\sin 3x)\frac{d}{dx}(3x)\right]$$

$$= (2\cos 2x - 3\sin 3x)e^{\sin 2x + \cos 3x}$$

Derivatives of the Other Trigonometric Functions

We are now in a position to find the rule for differentiating the tangent function. Using the Quotient Rule, we see that if $f(x) = \tan x$, then

$$f'(x) = \frac{d}{dx}\tan x = \frac{d}{dx}\left(\frac{\sin x}{\cos x}\right)$$

$$= \frac{(\cos x)\dfrac{d}{dx}\sin x - (\sin x)\dfrac{d}{dx}\cos x}{\cos^2 x}$$

$$= \frac{(\cos x)(\cos x) - (\sin x)(-\sin x)}{\cos^2 x}$$

$$= \frac{\cos^2 x + \sin^2 x}{\cos^2 x} = \frac{1}{\cos^2 x} = \sec^2 x$$

Once again, this result may be generalized with the help of the Chain Rule, and we are led to the next two rules.

Derivatives of the Tangent and the Generalized Tangent Functions

Rule 3a: $\dfrac{d}{dx}(\tan x) = \sec^2 x$

Rule 3b: $\dfrac{d}{dx}[\tan f(x)] = [\sec^2 f(x)]f'(x)$

The rules for differentiating the remaining three trigonometric functions are derived in a similar manner. We now state the generalized versions of these rules. Here, $f(x)$ is a differentiable function of x.

Derivatives of the Cosecant, Secant, and Cotangent Functions

Rule 4: $\dfrac{d}{dx}[\csc f(x)] = -[\csc f(x)][\cot f(x)]f'(x)$

Rule 5: $\dfrac{d}{dx}[\sec f(x)] = [\sec f(x)][\tan f(x)]f'(x)$

Rule 6: $\dfrac{d}{dx}[(\cot f(x)] = -[\csc^2 f(x)]f'(x)$

Note   As an aid to remembering the signs of the derivatives of the trigonometric functions, observe that functions beginning with a "c" [$\cos x$, $\csc x$, $\cot x$, $\cos f(x)$, $\csc f(x)$, and $\cot f(x)$] have a minus sign attached to their derivatives.

EXAMPLE 3 Differentiate $y = (\sec x)(x + \tan x)$.

Solution   Using the Product Rule and Rules 3 and 5, we have

$$\frac{dy}{dx} = \frac{d}{dx}[(\sec x)(x + \tan x)]$$

$$= (\sec x)\frac{d}{dx}(x + \tan x) + (x + \tan x)\frac{d}{dx}(\sec x)$$

$$= (\sec x)(1 + \sec^2 x) + (x + \tan x)(\sec x \tan x)$$

$$= (\sec x)(1 + \sec^2 x + x \tan x + \tan^2 x)$$

$$= (\sec x)(2 + x \tan x + 2\tan^2 x) \qquad \sec^2 x = 1 + \tan^2 x$$

EXAMPLE 4 Find an equation of the tangent line to the graph of the function $f(x) = \tan 2x$ at the point $\left(\frac{\pi}{8}, 1\right)$.

Solution   The slope of the tangent line at any point on the graph of f is given by

$$f'(x) = 2\sec^2 2x$$

In particular, the slope of the tangent line at the point $\left(\frac{\pi}{8}, 1\right)$ is given by

$$f'\left(\frac{\pi}{8}\right) = 2\sec^2\left(\frac{\pi}{4}\right)$$

$$= 2(\sqrt{2})^2 = 4$$

Therefore, a required equation is given by

$$y - 1 = 4\left(x - \frac{\pi}{8}\right)$$

$$y = 4x + \left(1 - \frac{\pi}{2}\right)$$

The techniques developed in Chapter 4 may be used to study the properties of functions involving trigonometric functions, as the following examples show.

APPLIED EXAMPLE 5 Predator–Prey Populations The owl population in a certain area is estimated to be

$$P_1(t) = 1000 + 100\sin\left(\frac{\pi t}{12}\right)$$

in month t, and the mouse population in the same area at time t is given by

$$P_2(t) = 20{,}000 + 4000\cos\left(\frac{\pi t}{12}\right)$$

Find the rate of change of each population when $t = 2$.

Solution   The rate of change of the owl population at any time t is given by

$$P_1'(t) = 100\left[\cos\left(\frac{\pi t}{12}\right)\right]\left(\frac{\pi}{12}\right)$$

$$= \frac{25\pi}{3}\cos\left(\frac{\pi t}{12}\right)$$

and the rate of change of the mouse population at any time t is given by

$$P_2'(t) = 4000\left[-\sin\left(\frac{\pi t}{12}\right)\right]\left(\frac{\pi}{12}\right)$$

$$= -\frac{1000\pi}{3}\sin\left(\frac{\pi t}{12}\right)$$

In particular, when $t = 2$, the rate of change of the owl population is

$$P_1'(2) = \frac{25\pi}{3}\cos\frac{\pi}{6}$$

$$= \frac{25\pi}{3}\left(\frac{\sqrt{3}}{2}\right) \approx 22.7$$

That is, the predator population is increasing at the rate of approximately 22.7 owls per month. Similarly, the rate of change of the mouse population is

$$P_2'(2) = -\frac{1000\pi}{3}\sin\frac{\pi}{6}$$

$$= -\frac{1000\pi}{3}\left(\frac{1}{2}\right) \approx -523.6$$

That is, the prey population is decreasing at the rate of approximately 523.6 mice per month. ▪

EXAMPLE 6 Sketch the graph of the function $y = f(x) = \sin^2 x$ on the interval $[0, \pi]$ by first obtaining the following information:

a. The intervals where f is increasing and where it is decreasing
b. The absolute extrema of f
c. The concavity of f
d. The inflection points of f

Solution

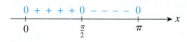

FIGURE 22
The sign diagram for f'

a. $f'(x) = 2\sin x \cos x = \sin 2x$. Setting $f'(x) = 0$ gives $x = 0, \frac{\pi}{2}$, or π. The sign diagram for f' (Figure 22) shows that f is increasing on $\left(0, \frac{\pi}{2}\right)$ and decreasing on $\left(\frac{\pi}{2}, \pi\right)$.

b. From the result of part (a) and the following table,

x	0	$\dfrac{\pi}{2}$	π
$f(x)$	0	1	0

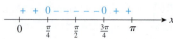

FIGURE 23
The sign diagram for f''

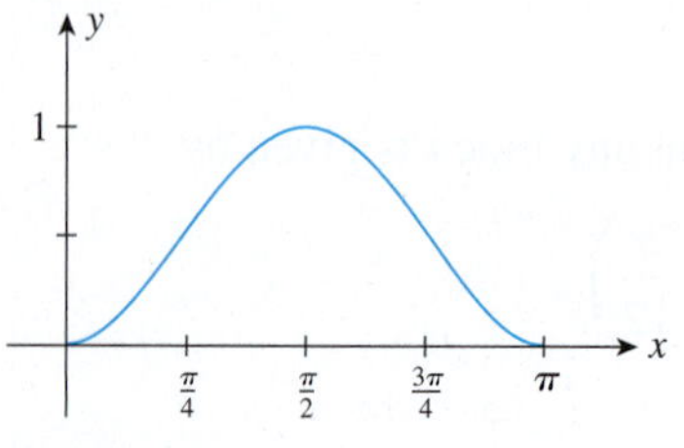

we see that the endpoints $x = 0$ and $x = \pi$ yield the absolute minimum value of f, namely, 0, while the absolute maximum value of f is 1, which occurs at $x = \frac{\pi}{2}$.

c. We compute $f''(x) = 2\cos 2x$. Setting $f''(x) = 0$ gives $2x = \frac{\pi}{2}$ or $\frac{3\pi}{2}$; that is, $x = \frac{\pi}{4}$ or $\frac{3\pi}{4}$. From the sign diagram for f'', shown in Figure 23, we conclude that f is concave upward on the intervals $\left(0, \frac{\pi}{4}\right)$ and $\left(\frac{3\pi}{4}, \pi\right)$ and concave downward on the interval $\left(\frac{\pi}{4}, \frac{3\pi}{4}\right)$.

d. From the results of part (c), we see that $\left(\frac{\pi}{4}, \frac{1}{2}\right)$ and $\left(\frac{3\pi}{4}, \frac{1}{2}\right)$ are inflection points of f.

FIGURE 24
The graph of $y = \sin^2 x$

Finally, the graph of f is sketched in Figure 24. ▪

APPLIED EXAMPLE 7 Restaurant Revenue The revenue of McMenamy's Fish Shanty, located at a popular summer resort, is approximately

$$R(t) = 2\left[5 - 4\cos\left(\frac{\pi t}{6}\right)\right] \qquad (0 \le t \le 12)$$

thousand dollars during the tth week ($t = 1$ corresponds to the first week of June). When is the weekly revenue increasing most rapidly?

Solution The revenue function R is increasing at the rate of

$$R'(t) = -8\left[-\sin\left(\frac{\pi t}{6}\right)\right]\left(\frac{\pi}{6}\right)$$
$$= \frac{4\pi}{3}\sin\left(\frac{\pi t}{6}\right)$$

thousand dollars per week. We want to maximize R'. So we compute

$$R''(t) = \frac{4\pi}{3}\left[\cos\left(\frac{\pi t}{6}\right)\right]\left(\frac{\pi}{6}\right)$$
$$= \frac{2\pi^2}{9}\cos\left(\frac{\pi t}{6}\right)$$

Setting $R''(t) = 0$ gives

$$\cos\left(\frac{\pi t}{6}\right) = 0$$

from which we see that $t = 3$ and $t = 9$ are critical numbers of $R'(t)$. Since $R'(t)$ is a continuous function on the closed interval $[0, 12]$, the absolute maximum must occur at a critical number or at an endpoint of the interval. From the following table, we see that the weekly revenue is increasing most rapidly when $t = 3$, that is, in the third week of June (Figure 25).

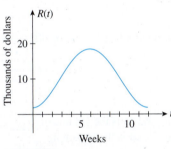

FIGURE 25
The graph of $R(t) = 2\left[5 - 4\cos\left(\frac{\pi t}{6}\right)\right]$

t	0	3	9	12
$R'(t)$	0	$\dfrac{4\pi}{3}$	$-\dfrac{4\pi}{3}$	0

 APPLIED EXAMPLE 8 Maximizing the Capacity of a Trough A trough with a trapezoidal cross section is to be constructed with a 1-foot base and sides that are 6 feet long and 1 foot wide (Figure 26). Find the angle of inclination θ that maximizes the capacity of the trough.

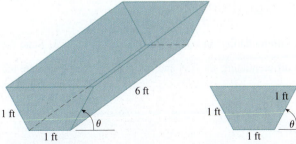

(a) We want to maximize the capacity of the trough with the given dimensions.

(b) A cross section of the trough has the shape of a trapezoid.

FIGURE 26

Solution   The volume of the trough is the product of the area of its cross section and its length. Since the length is constant, it suffices to maximize the area of its cross section. But the cross section is a trapezoid with area given by one half the sum of the parallel sides times its height, or

$$A = \frac{1}{2}[1 + (1 + 2\cos\theta)]\sin\theta$$

$$= (1 + \cos\theta)\sin\theta$$

To find the absolute maximum of the continuous function A over the closed interval $[0, \frac{\pi}{2}]$, we first compute the derivative of A, obtaining

$$A' = -\sin^2\theta + (1 + \cos\theta)\cos\theta$$

$$= -\sin^2\theta + \cos\theta + \cos^2\theta$$

$$= (\cos^2\theta - 1) + \cos\theta + \cos^2\theta \qquad \text{Use the identity } \sin^2\theta + \cos^2\theta = 1.$$

$$= 2\cos^2\theta + \cos\theta - 1$$

$$= (2\cos\theta - 1)(\cos\theta + 1)$$

Setting $A' = 0$ gives $\cos\theta = \frac{1}{2}$, or $\cos\theta = -1$; that is, $\theta = \frac{\pi}{3}$ (the other values of θ lie outside the interval $[0, \frac{\pi}{2}]$). Evaluating the function A at the critical number $\theta = \frac{\pi}{3}$ of A and at the endpoints of the interval, we have

θ	0	$\frac{\pi}{3}$	$\frac{\pi}{2}$
A	0	$\frac{3\sqrt{3}}{4}$	1

from which we see that the volume of the trough is maximized when $\theta = \frac{\pi}{3}$, or 60°.   ◼

12.3   Self-Check Exercises

1. Find the derivative of $f(x) = x\cos 2x$.

2. If $f(x) = x + \tan x$, find $f'(\frac{\pi}{4})$.

3. **VOLUME OF AIR INHALED DURING RESPIRATION** Suppose the volume of air inhaled by a person during respiration is given by

$$V(t) = \frac{6}{5\pi}\left[1 - \sin\left(\frac{\pi t}{2}\right)\right] \qquad (t \geq 0)$$

liters at time t (in seconds). When is the volume of inhaled air at a maximum? What is the maximum volume?

Solutions to Self-Check Exercises 12.3 can be found on page 824.

12.3   Concept Questions

1. State the rules for differentiating $\sin x$ and $\cos x$.

2. State the rules for differentiating $\sin f(x)$ and $\cos f(x)$.

3. State the rules for differentiating $\tan x$ and $\tan f(x)$.

4. State the rules for differentiating $\csc f(x)$, $\sec f(x)$, and $\cot f(x)$.

12.3   Exercises

In Exercises 1–30, find the derivative of the function.

1. $f(x) = \cos 3x$

2. $f(x) = \sin 5x$

3. $f(x) = 2\cos \pi x$

4. $f(x) = \pi \sin 2x$

5. $f(x) = \sin(x^2 + 1)$

6. $f(x) = \cos \pi x^2$

7. $f(x) = \tan 2x^2$

8. $f(x) = \cot \sqrt{x}$

9. $f(x) = x \sin x$

10. $f(x) = x^2 \cos x$

11. $f(x) = 2 \sin 3x + 3 \cos 2x$

12. $f(x) = 2 \cot 2x + \sec 3x$

13. $f(x) = x^2 \cos 2x$

14. $f(x) = \sqrt{x} \sin \pi x$

15. $f(x) = \sin \sqrt{x^2 - 1}$

16. $f(x) = \csc(x^2 + 1)$

17. $f(x) = e^x \sec x$

18. $f(x) = e^{-x} \csc x$

19. $f(x) = x \cos \dfrac{1}{x}$

20. $f(x) = x^2 \sin \dfrac{1}{x}$

21. $f(x) = \dfrac{x - \sin x}{1 + \cos x}$

22. $f(x) = \dfrac{\sin 2x}{1 + \cos 3x}$

23. $f(x) = \sqrt{\tan x}$

24. $f(x) = \sqrt{\cos x + \sin x}$

25. $f(x) = \dfrac{\sin x}{x}$

26. $f(x) = \dfrac{\cos x}{x^2 + 1}$

27. $f(x) = \tan^2 x$

28. $f(x) = \cot \sqrt{x}$

29. $f(x) = e^{\cot x}$

30. $f(x) = e^{\tan x + \sec x}$

In Exercises 31–34, find the second derivative of the function.

31. $f(x) = \sin x$

32. $y = 3 \cos x - x \sin x$

33. $h(t) = (t^2 + 1) \sin t$

34. $g(x) = \sec x$

35. Find an equation of the tangent line to the graph of the function $f(x) = \cot 2x$ at the point $\left(\frac{\pi}{4}, 0\right)$.

36. Find an equation of the tangent line to the graph of the function $f(x) = e^{\sec x}$ at the point $\left(\frac{\pi}{4}, e^{\sqrt{2}}\right)$.

In Exercises 37–40, find the rate of change of y with respect to x at the indicated value of x.

37. $y = x^2 \sec x$; $x = \dfrac{\pi}{4}$

38. $y = \csc x - 2 \cos x$; $x = \dfrac{\pi}{6}$

39. $y = \dfrac{\sin x}{1 - \cos x}$; $x = \dfrac{\pi}{2}$

40. $y = \dfrac{x \tan x}{\sec x}$; $x = 0$

In Exercises 41–44, find the x-coordinate(s) of the point(s) on the graph of the function at which the tangent line has the indicated slope.

41. $f(x) = \sin x$; $m_{\tan} = 1$

42. $g(x) = x + \sin x$; $m_{\tan} = 1$

43. $h(x) = \csc x$; $m_{\tan} = 0$

44. $f(x) = \cot x$; $m_{\tan} = -2$

45. Determine the intervals where the function

$$f(x) = e^x \cos x \qquad (0 < x < 2\pi)$$

is increasing and where it is decreasing.

46. Determine the point(s) of inflection of the function

$$f(x) = x + \sin x \qquad (-2\pi < x < 2\pi)$$

In Exercises 47–50, sketch the graph of the function over the specified interval by obtaining the following information:

a. The intervals where f is increasing and where it is decreasing
b. The relative extrema of f
c. The concavity of f
d. The inflection points of f

47. $f(x) = \sin x + \cos x$ $(0 \le x \le 2\pi)$

48. $f(x) = x - \sin x$ $(0 \le x \le 2\pi)$

49. $f(x) = 2 \sin x + \sin 2x$ $(0 \le x \le 2\pi)$

50. $f(x) = e^x \cos x$ $(0 \le x \le 2\pi)$

51. a. Find the Taylor series about $x = 0$ for the function $f(x) = \sin x$.
b. Use the result of part (a) to show that

$$\lim_{x \to 0} \frac{\sin x}{x} = 1$$

52. a. Find the Taylor series about $x = 0$ for the function $f(x) = \cos x$.
b. Use the result of part (a) to show that

$$\lim_{x \to 0} \frac{\cos x - 1}{x} = 0$$

53. PREDATOR–PREY POPULATIONS The wolf population in a certain northern region is estimated to be

$$P_1(t) = 8000 + 1000 \sin\left(\frac{\pi t}{24}\right)$$

in month t, and the caribou population in the same region is given by

$$P_2(t) = 40{,}000 + 12{,}000 \cos\left(\frac{\pi t}{24}\right)$$

Find the rate of change of each population when $t = 12$.

54. STOCK PRICES The closing price (in dollars) per share of stock of Tempco Electronics on the tth day it was traded is approximated by

$$P(t) = 20 + 12 \sin\left(\frac{\pi t}{30}\right) - 6 \sin\left(\frac{\pi t}{15}\right) + 4 \sin\left(\frac{\pi t}{10}\right)$$
$$-3 \sin\left(\frac{2\pi t}{15}\right) \qquad (0 \le t \le 24)$$

where $t = 0$ corresponds to the time the stock was first listed on a major stock exchange. What was the rate of change of the stock's price at the close of the 15th day of trading? What was the closing price on that day?

55. NUMBER OF DAYLIGHT HOURS IN BOSTON The number of hours of daylight on a certain day in Boston is approximated by

$$f(t) = 3 \sin\left[\frac{2\pi(t - 79)}{365}\right] + 12$$

where $t = 0$ corresponds to January 1. Compute $f'(79)$, and interpret your result.

56. WATER LEVEL IN A HARBOR The water level (in feet) in Boston Harbor during a certain 24-hr period is approximated by the formula

$$H = 4.8 \sin\left[\frac{\pi}{6}(t - 10)\right] + 7.6 \qquad (0 \le t \le 24)$$

where $t = 0$ corresponds to 12 midnight. When is the water level rising and when is it falling? Find the relative extrema of H, and interpret your results.

57. AVERAGE DAILY TEMPERATURE The average daily temperature (in degrees Fahrenheit) at a tourist resort in Cameron Highlands is approximated by

$$T = 62 - 18 \cos\left[\frac{2\pi(t - 23)}{365}\right]$$

on the tth day ($t = 1$ corresponds to January 1). Which day was the warmest day of the year? The coldest day?

58. VOLUME OF AIR INHALED DURING RESPIRATION Suppose the volume of air inhaled by a person during respiration is given by

$$V(t) = \frac{6}{5\pi}\left[1 - \cos\left(\frac{\pi t}{2}\right)\right]$$

liters at time t (in seconds). When is the volume of inhaled air at a maximum? What is the maximum volume?

59. RESTAURANT REVENUE The revenue of McMenamy's Fish Shanty located at a popular summer resort is approximately

$$R(t) = 2\left[5 - 4\cos\left(\frac{\pi t}{6}\right)\right] \qquad (0 \le t \le 12)$$

thousand dollars during the tth week ($t = 1$ corresponds to the first week of June). When does the restaurant realize the greatest revenue?

60. TELEVISING A ROCKET LAUNCH A major network is televising the launching of a rocket. A camera tracking the liftoff of the rocket is located at point A, as shown in the accompanying figure, where ϕ (phi) denotes the angle of elevation of the camera at A. How fast is ϕ changing at the instant when the rocket is at a distance of 13,000 ft from the camera and this distance is increasing at the rate of 480 ft/sec?

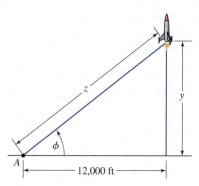

61. MAXIMUM AIR FLOW DURING RESPIRATION Refer to Exercise 58. Find the rate of flow of air into and out of the lungs. At what time is the rate of flow of air at a maximum? A minimum?

62. A SLIDING LADDER A 20-ft ladder leaning against a wall begins to slide. How fast is the angle between the ladder and the wall changing at the instant of time when the bottom of the ladder is 12 ft from the wall and sliding away from the wall at the rate of 5 ft/sec?

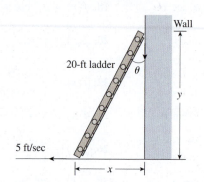

63. TRACKING A SUSPECT A police cruiser hunting for a suspect pulls over and stops at a point 20 ft from a straight wall. The flasher on top of the cruiser revolves at a constant rate of 90°/sec, and the light beam casts a spot of light as it strikes the wall. How fast is the spot of light moving along the wall at a point 30 ft from the point on the wall closest to the cruiser?

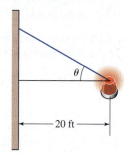

64. PERCENTAGE ERROR IN MEASURING HEIGHT From a point on level ground 150 ft from the base of a derrick, José measures the angle of elevation to the top of the derrick as 60°. If José's measurement is subject to an error of ±1%, find the percentage error in the measured height of the derrick.

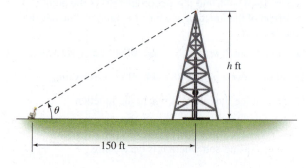

65. MAXIMIZING DRAINAGE CAPACITY The cross section of a drain is a trapezoid, as shown in the figure. The sides and the bottom of the trapezoid are each 5 ft long. Determine the angle θ such that the drain will have a maximal cross-sectional area.

66. SURFACE AREA OF A HONEYCOMB The accompanying figure depicts a prism-shaped single cell in a honeycomb. The front end of the prism is a regular hexagon, and the back is formed by the sides of the cell coming together at a point. It can be shown that the surface area of a cell is given by

$$S(\theta) = 6ab + \frac{3}{2}b^2\left(\frac{\sqrt{3} - \cos \theta}{\sin \theta}\right) \qquad (0 < \theta < \tfrac{\pi}{2})$$

where θ is the angle between one of the (three) upper surfaces and the altitude. Show that the surface area is minimized if $\cos \theta = \frac{1}{\sqrt{3}}$, or $\theta \approx 54.7°$. Measurements of actual honeycombs have confirmed that this is, in fact, the angle found in beehives.

67. FINDING THE POSITION OF A PLANET As shown in the accompanying figure, the position of a planet that revolves about the sun with an elliptical orbit can be located by calculating the central angle θ. Suppose that the central angle sustained by a planet on a certain day satisfies the equation

$$\theta - 0.5 \sin \theta = 1$$

Use Newton's method to find θ.

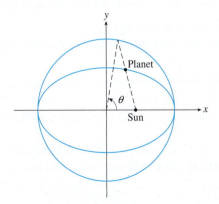

68. FLOW OF BLOOD Suppose that some of the fluid flowing along a pipe of radius R is diverted to a pipe of smaller radius r attached to the former at an angle θ (see the figure). Such is the case when blood flowing along an artery is pumped into an arteriole. What should be the angle θ so that the energy loss due to friction in moving the fluid is minimal? (Assume that the minimum exists.) Solve the problem using the following steps:

a. Poiseuille's Law states that the loss of energy due to friction in nonturbulent flow is proportional to the length of the path and inversely proportional to the fourth power of the radius. Use this law to show that the energy loss in moving the fluid from P to S via Q is

$$L = \frac{kd_1}{R^4} + \frac{kd_2}{r^4}$$

where k is a constant.

b. Suppose that a and b are fixed. Find d_1 and d_2 in terms of a and b. Then use this result together with the result from part (a) to show that

$$E = k\left(\frac{a - b \cot \theta}{R^4} + \frac{b \csc \theta}{r^4}\right)$$

c. Using the technique of this section, show that E is minimized when the following equation is satisfied:

$$\cos \theta = \frac{r^4}{R^4}$$

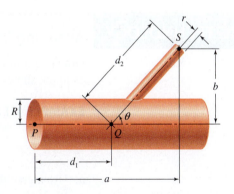

In Exercises 69–72, determine whether the statement is true or false. If it is true, explain why it is true. If it is false, explain why, or give an example to show why it is false.

69. $\displaystyle\lim_{x \to 0} \frac{\tan x}{x} = 1$

Hint: Use Equation (7).

70. The graph of $f(x) = x - \sin x$ is nondecreasing on $(-\infty, \infty)$.

71. The function $f(x) = \cos x$ has a relative minimum at a point $x = a$ where $g(x) = \sin x$ has a relative maximum.

72. The graph of $f(x) = \sin x + \cos x$ is concave downward on the interval $(0, \frac{\pi}{2})$.

73. Prove Rule 4:

If $\quad h(x) = \csc f(x)$

then $\quad h'(x) = -[\csc f(x)][\cot f(x)]f'(x)$

Hint: $\csc x = \dfrac{1}{\sin x}$

74. Prove Rule 5:

If $\quad h(x) = \sec f(x)$

then $\quad h'(x) = [\sec f(x)][\tan f(x)]f'(x)$

Hint: $\sec x = \dfrac{1}{\cos x}$

75. Prove Rule 6:

If $\quad h(x) = \cot f(x)$

then $\quad h'(x) = -[\csc^2 f(x)]f'(x)$

Hint: $\cot x = \dfrac{1}{\tan x} = \dfrac{\cos x}{\sin x}$

76. Prove that $\lim\limits_{h \to 0} \dfrac{\cos h - 1}{h} = 0$.

Hint: Multiply by $\dfrac{\cos h + 1}{\cos h + 1}$.

12.3 Solutions to Self-Check Exercises

1. $f'(x) = x \dfrac{d}{dx} \cos 2x + (\cos 2x) \dfrac{d}{dx} x$

$\qquad = x(-\sin 2x)(2) + (\cos 2x)(1)$

$\qquad = \cos 2x - 2x \sin 2x$

2. $f'(x) = 1 + \sec^2 x$. Therefore,

$$f'\left(\frac{\pi}{4}\right) = 1 + \sec^2\left(\frac{\pi}{4}\right) = 1 + (\sqrt{2})^2 = 3$$

3. We compute

$$V'(t) = \frac{6}{5\pi}\left[-\cos\left(\frac{\pi t}{2}\right)\right]\left(\frac{\pi}{2}\right) = -\frac{3}{5}\cos\left(\frac{\pi t}{2}\right)$$

Setting $V'(t) = 0$ and solving the resulting equation give $t = 1 + 2n$ $(n = 0, 1, 2, \ldots)$ as critical numbers of V. The numbers $t = 1, 3, 5, 7, \ldots$ give rise to the relative extrema of V, which coincide with the absolute extrema of V with values 0 and $\frac{12}{5\pi}$. Thus, the volume of inhaled air is at a maximum when $t = 3, 7, 11, \ldots$, and its value is $\frac{12}{5\pi}$ liters.

USING TECHNOLOGY   Analyzing Trigonometric Functions

Graphing utilities can be used to analyze complicated trigonometric functions, as this example shows.

EXAMPLE 1 Let

$$f(x) = \frac{\sqrt{x} \cos 2x}{\sqrt{x^2 + \sin 3x}} \qquad (0 < x \le 2)$$

a. Use a graphing utility to plot the graph of f in the viewing window $[0, 2] \times [-2, 1]$.

b. Use the numerical derivative operation of a graphing utility to find the rate of change of $f(x)$ at $x = 1$.

c. Find the inflection points of f.

d. Evaluate $\lim\limits_{x \to 0^+} f(x)$. Explain why $f(0)$ is not defined. What happens if you use the evaluation function of the graphing utility to find $f(0)$?

e. Find the absolute maximum and absolute minimum values of f on the interval $[0.5, 2]$.

Solution

a. The graph of f in the viewing window $[0, 2] \times [-2, 1]$ is shown in Figure T1.

b. We find $f'(1) \approx -2.0628$.

c. The inflection points are $(1.0868, -0.5734)$ and $(1.9244, -0.5881)$. (*Note:* Move the cursor near $x = 1$ to locate the first inflection point and near $x = 1.5$ to locate the other.)

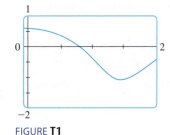

FIGURE **T1**

d. Using **ZOOM-IN**, we find $\lim\limits_{x \to 0^+} f(x) \approx 0.5774$. $f(0)$ is not defined because both the numerator and the denominator of $f(x)$ approach zero as x approaches zero. If we use the evaluation function of the graphing utility to find $f(0)$, the calculator will not give a value for y.

e. Using the operation to find the absolute minimum and the absolute maximum in the viewing window $[0.5, 2] \times [-2, 1]$, we find the absolute minimum value to be -1.0823 and the absolute maximum value to be 0.3421. ■

TECHNOLOGY EXERCISES

In Exercises 1–6, use the numerical derivative operation of a graphing utility to find the rate of change of $f(x)$ at the value of x. Give your answer accurate to four decimal places.

1. $f(x) = x^2\sqrt{1 + \sin^2 x}; \; x = 0.5$

2. $f(x) = \dfrac{1 + \cos x}{\sqrt{1 + \sin 3x}}; \; x = \dfrac{\pi}{4}$

3. $f(x) = \dfrac{x \tan x}{\sqrt{1 + x^2}}; \; x = 0.4$

4. $f(x) = e^{\cot x}(1 + x^2)^{3/2}; \; x = 1$

5. $f(x) = \cos[\csc(\sqrt{x} + 1)]; \; x = \dfrac{\pi}{3}$

6. $f(x) = \dfrac{\ln(\cos x)}{(1 + e^{-x})^{3/2}}; \; x = 0.8$

In Exercises 7–10, use a graphing utility to find the absolute maximum and the absolute minimum values of f in the given interval. Express your answers accurate to four decimal places.

7. $f(x) = \dfrac{\sin x}{x}; \; [1, 6]$

8. $f(x) = \dfrac{\cos x^2}{\sqrt{x}}; \; [1, 3]$

9. $f(x) = \dfrac{x \sec x}{1 + \cot x}; \; [0.5, 1]$

10. $f(x) = \dfrac{x + \cos x}{1 + 0.5 \sin x}; \; [0, 2]$

11. STOCK PRICES Refer to Exercise 54, page 821.
  a. Plot the graph of P, using the viewing window $[0, 20] \times [15, 35]$.
  b. What was the rate of change of the stock's price at the close of the 23rd day of trading?
  c. What was the closing price on that day?

12. AVERAGE DAILY TEMPERATURE Refer to Exercise 57, page 822.
  a. Plot the graph of T, using the viewing window $[0, 365] \times [20, 80]$.
  b. Using **TRACE** and **ZOOM** or using the graphing utility's operation for finding the absolute extrema of a function, verify the solution to the problem.

13. VOLUME OF AIR INHALED DURING RESPIRATION Refer to Exercise 58, page 822.
  a. Plot the graph of V, using the viewing window $[0, 12] \times [0, 2]$.
  b. Verify the solution to the problem using a graphing utility.

14. NUMBER OF DAYLIGHT HOURS IN BOSTON The number of hours of daylight on a particular day of the year in Boston is approximated by the function

$$f(t) = 3 \sin\left[\frac{2\pi}{365}(t - 79)\right] + 12$$

where $t = 0$ corresponds to January 1.
  a. Sketch the graph of f on the interval $[0, 365]$.
  b. When does the longest day occur? When does the shortest day occur?

15. EFFECT OF AN EARTHQUAKE ON A STRUCTURE To study the effect an earthquake has on a structure, engineers look at the way a beam bends when subjected to an earth tremor. The equation

$$D = a - a \cos\left(\frac{\pi h}{2L}\right) \qquad (0 \le h \le L)$$

where L is the length of a beam and a is the maximum deflection from the vertical, has been used by engineers to calculate the deflection D at a point on the beam h feet from the ground (see the figure). Suppose a 10-ft vertical beam has a maximum deflection of $\frac{1}{2}$ ft when subjected to an external force. Using differentials, estimate the difference in the deflection between the point midway on the beam and the point $\frac{1}{10}$ ft above it.

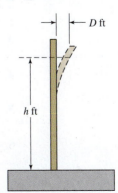

16. A Swinging Door The figure shows the top of a swinging door equipped with a spring that acts to close the door and a hydraulic mechanism that acts as a damper opposing the movement of the door. The door is released from rest at an angle of $\frac{\pi}{3}$ radians from the equilibrium position, and the angle of the door t sec after release is described by the equation

$$\theta(t) = \frac{\pi}{3} e^{-2t}(\cos \sqrt{2}t + \sqrt{2} \sin \sqrt{2}t)$$

a. How fast is θ changing half a second after the door is released?

b. Evaluate $\lim_{t \to \infty} \theta(t)$, and interpret your result.

c. Plot the graph of θ, and interpret your result.

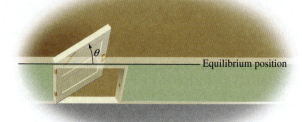

Equilibrium position

12.4 Integration of Trigonometric Functions

Integrating Trigonometric Functions

Each equation pertaining to a rule of differentiation given in Section 12.3 may be integrated to yield a corresponding integration formula. For example, integrating the equation $f'(x) = \cos x$ of Rule 1 with respect to x yields

$$\int \cos x \, dx = \int f'(x) \, dx = f(x) + C = \sin x + C$$

where C is a constant of integration. The six rules of integration that are obtainable in this manner follow.

Trigonometric Integration Formulas

1. $\displaystyle\int \sin x \, dx = -\cos x + C$

2. $\displaystyle\int \cos x \, dx = \sin x + C$

3. $\displaystyle\int \sec^2 x \, dx = \tan x + C$

4. $\displaystyle\int \csc^2 x \, dx = -\cot x + C$

5. $\displaystyle\int \sec x \tan x \, dx = \sec x + C$

6. $\displaystyle\int \csc x \cot x \, dx = -\csc x + C$

EXAMPLE 1 Find $\int \cos 3x \, dx$.

Solution Let's put $u = 3x$ so that $du = 3 \, dx$ and $dx = \frac{1}{3} \, du$. Then

$$\int \cos 3x \, dx = \frac{1}{3} \int \cos u \, du$$

$$= \frac{1}{3} \sin u + C \qquad \text{Use Rule 2.}$$

$$= \frac{1}{3} \sin 3x + C$$

EXAMPLE 2 Find $\int \sec(2x + 1) \tan(2x + 1) \, dx$.

Solution Put $u = 2x + 1$ so that $du = 2 \, dx$, or $dx = \frac{1}{2} \, du$. Then

$$\int \sec(2x + 1) \tan(2x + 1) \, dx = \frac{1}{2} \int \sec u \tan u \, du$$

$$= \frac{1}{2} \sec u + C$$

$$= \frac{1}{2} \sec(2x + 1) + C$$

EXAMPLE 3 Find $\int \dfrac{\sin x}{1 + \cos x} \, dx$.

Solution Put $u = 1 + \cos x$ so that $du = -\sin x \, dx$, or $\sin x \, dx = -du$. Then

$$\int \frac{\sin x}{1 + \cos x} \, dx = -\int \frac{du}{u}$$

$$= -\ln|u| + C$$

$$= -\ln|1 + \cos x| + C$$

EXAMPLE 4 Evaluate $\displaystyle\int_0^{\pi/2} x \cos 2x \, dx$.

Solution We first integrate the indefinite integral

$$\int x \cos 2x \, dx$$

by parts with

$$u = x \quad \text{and} \quad dv = \cos 2x \, dx$$

so that

$$du = dx \quad \text{and} \quad v = \frac{1}{2} \sin 2x$$

Therefore,

$$\int x \cos 2x \, dx = uv - \int v \, du$$

$$= \frac{1}{2} x \sin 2x - \frac{1}{2} \int \sin 2x \, dx$$

$$= \frac{1}{2} x \sin 2x + \frac{1}{4} \cos 2x + C$$

We have used the method of substitution to evaluate the integral on the right. Therefore,

$$\int_0^{\pi/2} x \cos 2x \, dx = \frac{1}{2} x \sin 2x + \frac{1}{4} \cos 2x \Big|_0^{\pi/2}$$

$$= \left[\frac{1}{2}\left(\frac{\pi}{2}\right) \sin \pi + \frac{1}{4} \cos \pi \right] - \left(0 + \frac{1}{4} \cos 0 \right)$$

$$= -\frac{1}{4} - \frac{1}{4} = -\frac{1}{2}$$

Explore and Discuss

Refer to the Explore and Discuss question on page 472 (Chapter 6).

1. Show that if f is even and g is odd, then the function fg is odd.

2. Refer to Example 4, and use the result of part (1) to show that

$$\int_{-\pi/2}^{\pi/2} x \cos 2x \, dx = 0$$

3. What is the area of the region bounded by the graph of $f(x) = x \cos 2x$, the x-axis, and the lines $x = -\frac{\pi}{2}$ and $x = \frac{\pi}{2}$?

EXAMPLE 5 Find the area under the curve of $y = \sin 2t$ from $t = 0$ to $t = \frac{\pi}{4}$.

Solution The required area, shown in Figure 27, is given by

$$\int_0^{\pi/4} \sin 2t \, dt = -\frac{1}{2} \cos 2t \Big|_0^{\pi/4} = -\frac{1}{2} \cos \frac{\pi}{2} + \frac{1}{2} \cos 0 = \frac{1}{2}$$

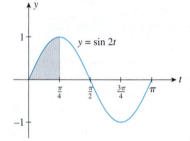

FIGURE 27
The area under the curve $y = \sin 2t$ from $t = 0$ to $t = \frac{\pi}{4}$

APPLIED EXAMPLE 6 Stock Prices The weekly closing price of HAL Corporation stock in week t is approximated by

$$f(t) = 30 + t \sin\left(\frac{\pi t}{6}\right) \qquad (0 \le t \le 15)$$

where $f(t)$ is the price (in dollars) per share. Find the average weekly closing price of the stock over the 15-week period.

Solution The average weekly closing price of the stock over the 15-week period in question is given by

$$A = \frac{1}{15 - 0} \int_0^{15} \left[30 + t \sin\left(\frac{\pi t}{6}\right) \right] dt$$

$$= \frac{1}{15} \int_0^{15} 30 \, dt + \frac{1}{15} \int_0^{15} t \sin\left(\frac{\pi t}{6}\right) dt$$

$$= 30 + \frac{1}{15} \int_0^{15} t \sin\left(\frac{\pi t}{6}\right) dt$$

Integrating by parts with

$$u = t \quad \text{and} \quad dv = \sin\left(\frac{\pi t}{6}\right) dt$$

so that

$$du = dt \quad \text{and} \quad v = -\frac{6}{\pi}\cos\left(\frac{\pi t}{6}\right)$$

we have

$$A = 30 + \frac{1}{15}\left[-\left(\frac{6t}{\pi}\right)\cos\left(\frac{\pi t}{6}\right)\Big|_0^{15} + \frac{6}{\pi}\int_0^{15}\cos\left(\frac{\pi t}{6}\right)dt\right]$$

$$= 30 + \frac{1}{15}\left\{-\left(\frac{6}{\pi}\right)(15)\cos\frac{15\pi}{6} + \frac{6}{\pi}(0)\cos 0 + \left[\left(\frac{6}{\pi}\right)^2\sin\left(\frac{\pi t}{6}\right)\Big|_0^{15}\right]\right\}$$

$$= 30 + \frac{1}{15}\left[-\left(\frac{6}{\pi}\right)(15)(0) + \left(\frac{6}{\pi}\right)^2\sin\frac{15\pi}{6} - \left(\frac{6}{\pi}\right)^2\sin 0\right]$$

$$= 30 + \frac{1}{15}\left(\frac{6}{\pi}\right)^2(1) \approx 30.24$$

or approximately \$30.24 per share.

> ### Explore and Discuss
>
> Without actually evaluating the integral, explain the following, using the properties of the trigonometric functions:
>
> **1.** $\displaystyle\int_0^{\pi/2} \sin x \, dx = \int_{10\pi}^{21\pi/2} \sin x \, dx$ **2.** $\displaystyle\int_{-3}^3 (2 + \cos 3x - \sin 5x) \, dx \geq 0$

Additional Trigonometric Integration Formulas

The six integration formulas listed at the beginning of this section were immediate consequences of the six corresponding rules of differentiation of Section 12.3. We now complete the list of integration formulas for the basic trigonometric functions by giving the integration formulas for the tangent, secant, cosecant, and cotangent functions.

> ### Additional Trigonometric Integration Formulas
>
> **7.** $\displaystyle\int \tan x \, dx = -\ln|\cos x| + C$
>
> **8.** $\displaystyle\int \sec x \, dx = \ln|\sec x + \tan x| + C$
>
> **9.** $\displaystyle\int \csc x \, dx = \ln|\csc x - \cot x| + C$
>
> **10.** $\displaystyle\int \cot x \, dx = \ln|\sin x| + C$

To prove Formula 7, we write

$$\int \tan x \, dx = \int \frac{\sin x}{\cos x} \, dx$$

Putting $u = \cos x$ so that $du = -\sin x \, dx$, we find

$$\int \tan x \, dx = -\int \frac{du}{u} = -\ln|u| + C$$

$$= -\ln|\cos x| + C$$

as we set out to show.

12.4   Self-Check Exercises

1. Evaluate $\displaystyle\int_0^{\pi/2} \frac{\cos x}{1 + \sin x}\, dx$.

2. Fox Population The rate of change of the population of foxes in a certain region in month t is

$$\frac{25\pi}{6} \cos\left(\frac{\pi t}{12}\right)$$

foxes per month. If the population at $t = 0$ is 400, find an expression giving the number of foxes in month t.

Solutions to Self-Check Exercises 12.4 can be found on page 832.

12.4   Concept Questions

1. State the rules for integrating $\sin x$, $\cos x$, and $\sec^2 x$.

2. State the rules for integrating $\csc^2 x$, $\sec x \tan x$, and $\csc x \cot x$.

12.4   Exercises

In Exercises 1–32, find or evaluate each integral.

1. $\displaystyle\int \sin 3x\, dx$

2. $\displaystyle\int \cos(x + \pi)\, dx$

3. $\displaystyle\int (3 \sin x + 4 \cos x)\, dx$

4. $\displaystyle\int (x^2 - \cos 2x)\, dx$

5. $\displaystyle\int \sec^2 2x\, dx$

6. $\displaystyle\int \csc^2 3x\, dx$

7. $\displaystyle\int x \cos(x^2)\, dx$

8. $\displaystyle\int x \sec(x^2) \tan(x^2)\, dx$

9. $\displaystyle\int \csc \pi x \cot \pi x\, dx$

10. $\displaystyle\int \sec 2x \tan 2x\, dx$

11. $\displaystyle\int_{-\pi/2}^{\pi/2} (\sin x + \cos x)\, dx$

12. $\displaystyle\int_0^{\pi} (2 \cos x + 3 \sin x)\, dx$

13. $\displaystyle\int_0^{\pi/6} \tan 2x\, dx$

14. $\displaystyle\int_{\pi/8}^{\pi/4} \cot 2x\, dx$

15. $\displaystyle\int \sin^3 x \cos x\, dx$

16. $\displaystyle\int \cos^2 x \sin x\, dx$

17. $\displaystyle\int \sec \pi x\, dx$

18. $\displaystyle\int \csc(1 - x)\, dx$

19. $\displaystyle\int_0^{\pi/12} \sec 3x\, dx$

20. $\displaystyle\int_0^{\pi/8} \tan 2x\, dx$

21. $\displaystyle\int \sqrt{\cos x} \sin x\, dx$

22. $\displaystyle\int (\sin x)^{1/3} \cos x\, dx$

23. $\displaystyle\int \cos 3x \sqrt{1 - 2 \sin 3x}\, dx$

24. $\displaystyle\int \frac{\cos x}{\sqrt{1 + \sin x}}\, dx$

25. $\displaystyle\int \tan^3 x \sec^2 x\, dx$

26. $\displaystyle\int \sec^5 x \tan x\, dx$

27. $\displaystyle\int \csc^2 x\, (\cot x - 1)^3\, dx$

28. $\displaystyle\int \sec^2 x \sqrt{1 + \tan x}\, dx$

29. $\displaystyle\int_1^{e^\pi} \frac{\sin(\ln x)}{x}\, dx$

30. $\displaystyle\int_0^{\pi/2} \frac{\cos x}{1 + \sin x}\, dx$

31. $\displaystyle\int \sin(\ln x)\, dx$

Hint: Integrate by parts.

32. $\displaystyle\int_1^{e^{\pi/2}} \cos(\ln x)\, dx$

Hint: Integrate by parts.

In Exercises 33–36, find the area of the region under the graph of the function f from $x = a$ to $x = b$.

33. $f(x) = \cos \dfrac{x}{4}$; $a = 0$, $b = \pi$

34. $f(x) = x + \sin x$; $a = 0$, $b = \dfrac{\pi}{4}$

35. $f(x) = \tan x$; $a = 0$, $b = \dfrac{\pi}{4}$

36. $f(x) = \cot x + \csc x$; $a = \dfrac{\pi}{4}$, $b = \dfrac{\pi}{2}$

37. Find the area of the region bounded above by the graph of $y = x$ and below by the graph of $y = \sin x$ from $x = 0$ to $x = \pi$.

38. Find the area of the region bounded above by the graph of $y = \cos 2x$ and below by the graph of $y = x$ from $x = 0$ to $x = \frac{\pi}{8}$. Make a sketch of the region.

In Exercises 39–42, solve the differential equation or initial-value problem.

39. $\dfrac{dy}{dx} = \dfrac{\sin x}{\cos y}$; $y(0) = 0$ **40.** $\dfrac{dx}{dt} = -e^x \cos t$

41. $\csc y \, dx + \sec x \, dy = 0$

42. $\cot y \, dx - x \, dy = 0$; $y(1) = 0$

43. STOCK PRICES The weekly closing price of TMA Corporation stock in week t is approximated by

$$f(t) = 80 + 3t \cos\left(\frac{\pi t}{6}\right) \qquad (0 \le t \le 15)$$

where $f(t)$ is the price (in dollars) per share. Find the average weekly closing price of the stock over the 15-week period.

44. AVERAGE DAILY TEMPERATURE The average daily temperature at a tourist resort in the Cameron Highlands is approximately

$$T = 62 - 18 \cos\left[\frac{2\pi(t - 23)}{365}\right]$$

degrees Fahrenheit on the tth day ($t = 1$ corresponds to January 1). What is the average temperature at the resort in the month of January?
Hint: Find the average value of T over the interval $[0, 31]$.

45. RESTAURANT REVENUE The revenue of McMenamy's Fish Shanty, located at a popular summer resort, is approximately

$$R(t) = 2\left[5 - 4 \cos\left(\frac{\pi t}{6}\right)\right] \qquad (0 \le t \le 12)$$

thousand dollars during the tth week ($t = 0$ corresponds to the first week of June). What is the total revenue realized by the restaurant over the 12-week period starting June 1?

46. STOCK PRICES Refer to Exercise 54, Section 12.3. What was the average price per share of Tempco Electronics stock over the first 20 days that it was traded on the stock exchange?

47. VOLUME OF AIR INHALED DURING RESPIRATION Suppose that the rate of air flow into and out of a person's lungs during respiration is

$$R(t) = 0.6 \sin\left(\frac{\pi t}{2}\right)$$

liters per second, where t is the time in seconds. Find an expression for the volume of air V in the person's lungs at any time t. Assume that $V(0) = 0$.

48. AVERAGE VOLUME OF INHALED AIR DURING RESPIRATION Refer to Exercise 47. What is the average volume of inspired air in the person's lungs during one respiratory cycle?
Hint: Find the average value of the function giving the volume of inhaled air (obtained in the solution to Exercise 47) over the interval $[0, 4]$.

49. PREDATOR–PREY POPULATIONS The wolf and caribou populations in a certain northern region are given by

$$P_1(t) = 8000 + 1000 \sin\left(\frac{\pi t}{24}\right)$$

and

$$P_2(t) = 40{,}000 + 12{,}000 \cos\left(\frac{\pi t}{24}\right)$$

respectively, at time t, where t is measured in months. What are the average wolf and caribou populations over the time interval $[0, 6]$?

50. GROWTH OF A FRUIT FLY COLONY A biologist has determined that by varying the food supply available to a population of *Drosophila* (fruit flies) in a controlled experiment, the rate at which the population grows is given by

$$\frac{dQ}{dt} = 0.0001(4 + 5 \cos 2t)Q(400 - Q)$$

where Q denotes the population of *Drosophila* at any time t (in days). If ten fruit flies are present initially, determine the number of fruit flies that will be in the colony after 20 days.
Hint: Solve the separable differential equation using the following identity:

$$\frac{1}{Q(C - Q)} = \frac{1}{C}\left[\frac{1}{Q} + \frac{1}{C - Q}\right]$$

51. INSECT POPULATION The population P of an insect at time t (in days) is described by the differential equation

$$\frac{dP}{dt} = kP \cos at$$

where k and a are positive constants. Find $P(t)$ given that the insect population at $t = 0$ is P_0.
Hint: See Section 9.2.

52. a. Find the value of k such that the function $f(x) = k \sin\left(\frac{\pi x}{5}\right)$ is a probability density function on the interval $[0, 5]$.
  b. Find the mean and variance of the random variable X associated with the probability density function f of part (a).
  Hint: Integrate by parts.

53. Use the sixth Taylor polynomial approximation of $\dfrac{\sin x}{x}$,

$$\frac{\sin x}{x} \approx 1 - \frac{x^2}{6} + \frac{x^4}{120} - \frac{x^6}{5040}$$

to obtain an approximation of the integral

$$\int_0^{\pi/2} \frac{\sin x}{x} \, dx$$

54. Use the 12th Taylor polynomial approximation of $\cos x^2$,

$$\cos x^2 \approx 1 - \frac{x^4}{2} + \frac{x^8}{24} - \frac{x^{12}}{720}$$

to obtain an approximation of the Fresnel cosine integral

$$\int_0^x \cos t^2 \, dt$$

(It is important in the study of the theory of diffraction.) In particular, estimate

$$\int_0^1 \cos t^2 \, dt$$

In Exercises 55–58, determine whether the statement is true or false. If it is true, explain why it is true. If it is false, give an example to show why it is false.

55. $\displaystyle\int_a^b \sin x \, dx = \int_{a+2\pi}^{b+2\pi} \sin x \, dx$

56. $\displaystyle\int_a^b \cos x \, dx = \int_a^{b+2\pi} \cos x \, dx$

57. $\displaystyle\int_{-\pi/2}^{\pi/2} \sqrt[3]{x} \sin 2x \, dx > 0$

58. $\displaystyle\int_{-\pi/2}^{\pi/2} |\sin x| \, dx = \int_{-\pi/2}^{\pi/2} |\cos x| \, dx$

12.4 Solutions to Self-Check Exercises

1. Let $u = 1 + \sin x$ so that $du = \cos x \, dx$. Then

$$\int \frac{\cos x}{1 + \sin x} \, dx = \int \frac{du}{u} = \ln|u| + C$$
$$= \ln|1 + \sin x| + C$$

Therefore,

$$\int_0^{\pi/2} \frac{\cos x}{1 + \sin x} \, dx = \ln|1 + \sin x| \, \Big|_0^{\pi/2} = \ln 2$$

2. The population of foxes in month t is

$$P = \int \frac{25\pi}{6} \cos\left(\frac{\pi t}{12}\right) dt$$
$$= \frac{25\pi}{6} \int \cos\left(\frac{\pi t}{12}\right) dt$$

Let $u = \frac{\pi t}{12}$ so that $du = \frac{\pi}{12} \, dt$. Therefore,

$$P = \left(\frac{25\pi}{6}\right)\left(\frac{12}{\pi}\right) \int \cos u \, du = 50 \sin u + C$$
$$= 50 \sin\left(\frac{\pi t}{12}\right) + C$$

To evaluate C, note that $P = 400$ when $t = 0$. That is,

$$400 = 50(0) + C \quad \text{or} \quad C = 400$$

Therefore,

$$P(t) = 400 + 50 \sin\left(\frac{\pi t}{12}\right)$$

USING TECHNOLOGY   Evaluating Integrals of Trigonometric Functions

The numerical integration operation of a graphing utility can be used to evaluate definite integrals involving trigonometric functions.

EXAMPLE 1 Use a graphing utility to find the area of the region R that is completely enclosed by the graphs of the functions

$$f(x) = \cos x + \sin 2x \quad \text{and} \quad g(x) = 1 - x^2$$

Solution   The graphs of f and g in the viewing window $[-2, 2] \times [-2, 2]$ are shown in Figure T1.

Using the intersection operation of a graphing utility, we find that the x-coordinates of the points of intersection of the two graphs are ≈ -1.1554 and 0.

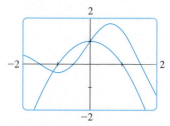

FIGURE **T1**

Since the graph of g lies above that of f on the interval $[-1.1554, 0]$, we see that the area of R is given by

$$A \approx \int_{-1.1554}^{0} [(1 - x^2) - (\cos x + \sin 2x)] \, dx$$

$$= \int_{-1.1554}^{0} (1 - x^2 - \cos x - \sin 2x) \, dx$$

$$\approx 0.5635$$

TECHNOLOGY EXERCISES

In Exercises 1–4, use a graphing utility to find the area of the region under the graph of f on the interval $[a, b]$. Express your answer accurate to four decimal places.

1. $f(x) = \sqrt{x} \sin 2x; \; [0, 1]$

2. $f(x) = \dfrac{x}{\sin x}; \; [1, 2]$

3. $f(x) = \dfrac{\cos x}{\sqrt{1 + x^2}}; \; [0, 1]$

4. $f(x) = (\cos x) \ln\left(\dfrac{1 + x}{1 - x}\right); \; [0, \frac{1}{2}]$

In Exercises 5–12, use a graphing calculator to evaluate the definite integral. Express your answer accurate to four decimal places.

5. $\displaystyle\int_{0}^{0.3} (0.2 \cos x - 0.3 \sin 2x + 1.2 \tan 3x) \, dx$

6. $\displaystyle\int_{0}^{1.1} \sqrt{1 + x^2} \sec x \, dx$

7. $\displaystyle\int_{0}^{2\pi} \dfrac{dx}{10 + 3 \cos x}$

8. $\displaystyle\int_{0}^{\pi/2} \sqrt{1 + \dfrac{1}{2} \sin^2 x} \, dx$

9. $\displaystyle\int_{0.5}^{1} \ln(\sin x) \, dx$

10. $\displaystyle\int_{1}^{2} e^{\cot x} \, dx$

11. $\displaystyle\int_{0}^{1} \cos x^2 \, dx$

12. $\displaystyle\int_{0}^{3} \dfrac{\cos x}{(1 + \sin^2 x)^{3/2}} \, dx$

In Exercises 13–16, use a graphing utility (a) to plot the graphs of the functions f and g and (b) to find the area of the region enclosed by these graphs and the vertical lines $x = a$ and $x = b$. Express your answer accurate to four decimal places.

13. $f(x) = x \sin 2x; \; g(x) = \dfrac{1}{2}x^2 - 2; \; a = 0, b = 1$

14. $f(x) = \tan^2 x; \; g(x) = 5 - \sqrt{1 + \sqrt{x}}; \; a = 0, b = 1$

15. $f(x) = e^{\cos x}; \; g(x) = \ln(\tan x); \; a = 0.2, b = 0.8$

16. $f(x) = \cot x; \; g(x) = \sqrt{1 + 2 \sin x}; \; a = 1, b = 2$

In Exercises 17–20, use a graphing utility (a) to plot the graphs of the functions f and g and (b) to find the area of the region completely enclosed by the graphs of these functions.

17. $f(x) = 2 \cos 3x$ and $g(x) = x^2$

18. $f(x) = \cos^2 x$ and $g(x) = e^{x^2} - 1$

19. $f(x) = \sin 2x$ and $g(x) = x - \sqrt{x}$

20. $f(x) = e^{\sin x}$ and $g(x) = 0.5x^2$

21. BOSTON HARBOR WATER LEVEL The water level (in feet) in Boston Harbor during a certain 24-hr period is approximated by the formula

$$H = 4.8 \sin\left[\dfrac{\pi}{6} (t - 10)\right] + 7.6 \qquad (0 \le t \le 24)$$

where $t = 0$ corresponds to 12 midnight. What is the average water level in Boston Harbor over the 24-hr period on that day?

22. NUMBER OF DAYLIGHT HOURS IN CHICAGO The number of hours of daylight at any time t in Chicago is approximated by

$$L(t) = 2.8 \sin\left[\dfrac{2\pi}{365} (t - 79)\right] + 12$$

where t is measured in days and $t = 0$ corresponds to January 1. What is the daily average number of hours of daylight in Chicago over the year? Over the summer months from June 21 ($t = 171$) through September 20 ($t = 262$)?

CHAPTER 12   Summary of Principal Formulas and Terms

FORMULAS

1. Degree-radian conversion	360 degrees $= 2\pi$ radians
2. Derivatives of the trigonometric functions:	
Sine	$\dfrac{d}{dx}(\sin u) = \cos u \dfrac{du}{dx}$
Cosine	$\dfrac{d}{dx}(\cos u) = -\sin u \dfrac{du}{dx}$
Tangent	$\dfrac{d}{dx}(\tan u) = \sec^2 u \dfrac{du}{dx}$
Cosecant	$\dfrac{d}{dx}(\csc u) = -\csc u \cot u \dfrac{du}{dx}$
Secant	$\dfrac{d}{dx}(\sec u) = \sec u \tan u \dfrac{du}{dx}$
Cotangent	$\dfrac{d}{dx}(\cot u) = -\csc^2 u \dfrac{du}{dx}$

3. Integrals of the trigonometric functions:

$\displaystyle\int \sin u \, du = -\cos u + C$	$\displaystyle\int \csc u \cot u \, du = -\csc u + C$
$\displaystyle\int \cos u \, du = \sin u + C$	$\displaystyle\int \tan u \, du = -\ln|\cos u| + C$
$\displaystyle\int \sec^2 u \, du = \tan u + C$	$\displaystyle\int \sec u \, du = \ln|\sec u + \tan u| + C$
$\displaystyle\int \csc^2 u \, du = -\cot u + C$	$\displaystyle\int \csc u \, du = \ln|\csc u - \cot u| + C$
$\displaystyle\int \sec u \tan u \, du = \sec u + C$	$\displaystyle\int \cot u \, du = \ln|\sin u| + C$

TERMS

angle (796)	negative angle (796)	periodic function (804)
initial side (796)	degree (796)	period of a function (804)
terminal side (796)	radian (796)	amplitude (806)
vertex (796)	unit circle (796)	predator–prey population model (807)
standard position (796)	coterminal angle (798)	trigonometric identity (809)
positive angle (796)	reference angle (802)	

CHAPTER 12   Concept Review Questions

Fill in the blanks.

1. a. The measure of the central angle subtended by an arc equal in length to the radius of the circle is called a/an _____.

b. The angle subtended by one complete revolution is _____ radians or _____ degrees.

2. a. If x is the number of degrees, then _____ gives the number of radians.
  b. If x is the number of radians, then _____ gives the number of degrees.

3. If $P(x, y)$ is a point on the unit circle such that the radius of $\overline{OP}$ forms an angle of θ radians with respect to the positive x-axis, then $\cos \theta =$ _____, $\sin \theta =$ _____, $\tan \theta =$ _____, $\csc \theta =$ _____, $\sec \theta =$ _____, and $\cot \theta =$ _____.

4. a. $\sin^2 \theta +$ _____ $= 1$ **b.** $\tan^2 \theta +$ _____ $= \sec^2 \theta$
  c. $\cot^2 \theta +$ _____ $= \csc^2 \theta$ **d.** $\cos^2 \theta = \frac{1}{2}(1 +$ _____$)$

5. a. $\sin^2 \theta = \frac{1}{2}(1 -$ _____$)$
  b. $\sin(A + B) = \sin A \cos B$ _____ $\cos A \sin B$
  c. $\sin 2A = 2$ _____
  d. $\cos 2A = \cos^2 A -$ _____

6. a. $\frac{d}{dx}[\sin f(x)] =$ _____ **b.** $\frac{d}{dx}[\cos f(x)] =$ _____
  c. $\frac{d}{dx}[\tan f(x)] =$ _____ **d.** $\frac{d}{dx}[\csc f(x)] =$ _____
  e. $\frac{d}{dx}[\sec f(x)] =$ _____ **f.** $\frac{d}{dx}[\cot f(x)] =$ _____

7. a. $\int \sin x \, dx =$ _____ **b.** $\int \cos x \, dx =$ _____
  c. $\int \sec^2 x \, dx =$ _____ **d.** $\int \csc^2 x \, dx =$ _____
  e. $\int \sec x \tan x \, dx =$ _____ **f.** $\int \csc x \cot x \, dx =$ _____

CHAPTER 12 Review Exercises

In Exercises 1–3, convert the angle to radian measure.

1. $120°$ **2.** $450°$ **3.** $-225°$

In Exercises 4–6, convert the angle to degree measure.

4. $\frac{11\pi}{6}$ radians **5.** $-\frac{5\pi}{2}$ radians **6.** $-\frac{7\pi}{4}$ radians

In Exercises 7 and 8, find all values of θ that satisfy the equation over the interval $[0, 2\pi]$.

7. $\cos \theta = \frac{1}{2}$ **8.** $\cot \theta = -\sqrt{3}$

In Exercises 9–20, find the derivative of the function f.

9. $f(x) = \sin 3x$ **10.** $f(x) = 2 \cos \frac{x}{2}$

11. $f(x) = 2 \sin x - 3 \cos 2x$

12. $f(x) = \sec^2 \sqrt{x}$

13. $f(x) = e^{-x} \tan 3x$ **14.** $f(x) = (1 - \csc 2x)^2$

15. $f(x) = 4 \sin x \cos x$ **16.** $f(x) = \frac{\cos x}{1 - \cos x}$

17. $f(x) = \frac{1 - \tan x}{1 - \cot x}$ **18.** $f(x) = \ln(\cos^2 x)$

19. $f(x) = \sin(\sin x)$ **20.** $f(x) = e^{\sin x} \cos x$

In Exercises 21 and 22, find the amplitude and period for the graph of each function. Sketch the graph of the function over one period.

21. $y = 4 \sin(x - \pi)$ **22.** $y = 3 \cos 2x$

23. Find an equation of the tangent line to the graph of the function $f(x) = \tan^2 x$ at the point $\left(\frac{\pi}{4}, 1\right)$.

24. Find the function f given that its derivative is $f'(x) = \sqrt{x} + \sin x$ and its graph passes through the point $(0, 2)$.

25. Sketch the graph of the function $y = f(x) = \cos^2 x$ on the interval $[0, \pi]$ by obtaining the following information:
  a. The intervals where f is increasing and where it is decreasing
  b. The relative extrema of f
  c. The concavity of f
  d. The inflection points of f

In Exercises 26–35, find or evaluate the integral.

26. $\int \cos\left(\frac{2x}{3}\right) dx$ **27.** $\int \sin^2 2x \, dx$

28. $\int x \csc(x^2) \cot(x^2) \, dx$

29. $\int x \cos x \, dx$

30. $\int \sin^2 x \cos x \, dx$ **31.** $\int e^x \sin e^x \, dx$

32. $\int \frac{\cos x}{\sin^2 x} dx$ **33.** $\int_0^{\pi/4} \tan x \, dx$

34. $\int_{\pi/6}^{\pi/2} \frac{\cos x}{1 - \cos^2 x} dx$

35. $\int_0^{\pi} \sin x \, (1 + \cos^3 x) \, dx$

36. Find the area of the region bounded by the curves $f(x) = \sin x$ and $g(x) = \cos x$ from $x = \frac{\pi}{4}$ to $x = \frac{5\pi}{4}$.

37. PREDATOR–PREY POPULATIONS The population of foxes in a certain region over a 2-year period is estimated to be

$$P_1(t) = 400 + 50 \sin\left(\frac{\pi t}{12}\right)$$

in month t, and the population of rabbits in the same region in month t is given by

$$P_2(t) = 3000 + 500 \cos\left(\frac{\pi t}{12}\right)$$

Find the rate of change of the populations when $t = 4$.

38. HOTEL OCCUPANCY RATE The occupancy rate (in percent) of a large hotel in Maui in month t is described by the function

$$R(t) = 60 + 37 \sin^2\left(\frac{\pi t}{12}\right) \qquad (0 \le t \le 12)$$

where $t = 0$ corresponds to the beginning of June. When is the occupancy rate highest? When is the occupancy rate increasing most rapidly?

39. HOTEL OCCUPANCY RATE Refer to Exercise 38. Determine the average occupancy rate of the hotel over a 1-year period from June 1 through May 31 of the following year.
Hint: $\sin^2 x = \dfrac{1 - \cos 2x}{2}$

40. RESPIRATORY CYCLES The volume of air inhaled by a person during respiration is given by

$$V(t) = \frac{6}{5\pi}\left[1 - \cos\left(\frac{\pi t}{2}\right)\right]$$

liters at time t (in seconds). What is the average volume of air inhaled by a person over one cycle from $t = 0$ to $t = 4$?

CHAPTER 12 Before Moving On . . .

1. Sketch the graph of $f(x) = 3 \cos 2x - 1$ on $[0, \pi]$.

2. Verify the identity $\cos^2 x = \dfrac{1 + \cos 2x}{2}$.

3. Find the derivative of $f(x) = e^{-2x} \sin 3x$.

4. Evaluate $\displaystyle\int_0^{\pi/4} \cos^2 x \, dx$.

Hint: Use the identity in Exercise 2.

5. Find the area of the region under the graph of $f(x) = \sin^2 x \cos x$ from $x = 0$ to $x = \frac{\pi}{4}$.

APPENDIX

A.1 The Inverse of a Function

Consider the position function

$$s = f(t) = 4t^2 \qquad (0 \le t \le 30) \tag{1}$$

giving the position of a maglev at any time t in its domain $[0, 30]$. The graph of f is shown in Figure 1. Equation (1) enables us to compute algebraically the position of the maglev at any time t. Geometrically, we can find the position of the maglev at any given time t by following the path indicated in Figure 1.

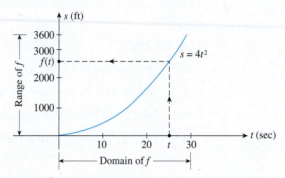

FIGURE 1
Each t in the domain of f is associated with the (unique) position $s = f(t)$ of the maglev.

Now consider the reverse problem: Knowing the position function of the maglev, can we find some way of obtaining the time it takes for the maglev to reach a given position? Geometrically, this problem is easily solved: Locate the point on the s-axis corresponding to the given position. Follow the path considered earlier but traced in the *opposite* direction. This path associates the given position s with the desired time t.

Algebraically, we can obtain a formula for the time t it takes for the maglev to get to the position s by solving Equation (1) for t in terms of s. Thus,

$$t = \frac{1}{2}\sqrt{s}$$

(We reject the negative root because t lies in $[0, 30]$.) Observe that the function g defined by

$$t = g(s) = \frac{1}{2}\sqrt{s}$$

has domain $[0, 3600]$ (the range of f) and range $[0, 30]$ (the domain of f) (Figure 2).

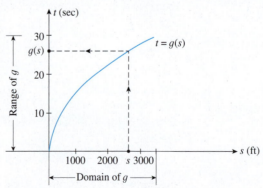

FIGURE 2
Each s in the domain of g is associated with the (unique) time $t = g(s)$.

The functions f and g have the following properties:

1. The domain of g is the range of f and vice versa.

2. $(g \circ f)(t) = g[f(t)] = \dfrac{1}{2}\sqrt{f(t)} = \dfrac{1}{2}\sqrt{4t^2} = t$

and

$$(f \circ g)(s) = f[g(s)] = 4[g(s)]^2 = 4\left(\dfrac{1}{2}\sqrt{s}\right)^2 = s$$

In other words, one undoes what the other does. This is to be expected because f maps t onto $s = f(t)$ and g maps $s = f(t)$ back onto t.

The functions f and g are said to be *inverses* of each other. More generally, we have the following definition.

Inverse Functions

A function g is the inverse of the function f if

$$f[g(x)] = x \text{ for every } x \text{ in the domain of } g$$
$$g[f(x)] = x \text{ for every } x \text{ in the domain of } f$$

Equivalently, g is the inverse of f if the following condition is satisfied:

$$y = f(x) \quad \text{if and only if} \quad x = g(y)$$

for every x in the domain of f and for every y in the domain of g.

Note The inverse of f is normally denoted by f^{-1} (read "f inverse"). ■

 Do not confuse $f^{-1}(x)$ with $[f(x)]^{-1} = \dfrac{1}{f(x)}$!

EXAMPLE 1 Show that the functions $f(x) = x^{1/3}$ and $g(x) = x^3$ are inverses of each other.

Solution First, observe that the domain and range of both f and g are $(-\infty, \infty)$. Therefore, both composite functions $f \circ g$ and $g \circ f$ are defined. Next, we compute

$$(f \circ g)(x) = f[g(x)] = [g(x)]^{1/3} = (x^3)^{1/3} = x$$

and

$$(g \circ f)(x) = g[f(x)] = [f(x)]^3 = (x^{1/3})^3 = x$$

Since $f[g(x)] = g[f(x)] = x$, we conclude that f and g are inverses of each other. In short, $f^{-1}(x) = x^3$. ■

Interpreting Our Results We can view f as a cube root–extracting machine and g as a "cubing" machine. In this light, it is easy to see that one function does undo what the other does. So f and g are indeed inverses of each other.

A.2   The Graphs of Inverse Functions

The graphs of $f(x) = x^{1/3}$ and $f^{-1}(x) = x^3$ are shown in Figure 3.

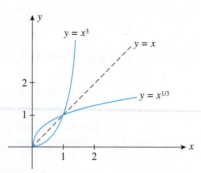

FIGURE 3
The functions $f(x) = x^{1/3}$ and $f^{-1}(x) = x^3$ are inverses of each other. Their graphs are symmetric with respect to the line $y = x$.

They seem to suggest that the graphs of inverse functions are mirror images of each other with respect to the line $y = x$. This is true in general, but we will not prove it here.

The Graphs of Inverse Functions

The graph of f^{-1} is the reflection of the graph of f with respect to the line $y = x$ and vice versa.

A.3   Functions That Have Inverses

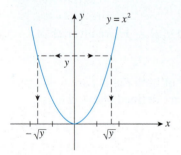

FIGURE 4
Each value of y is associated with two values of x.

Not every function has an inverse. Consider, for example, the function f defined by $y = x^2$ with domain $(-\infty, \infty)$ and range $[0, \infty)$. From the graph of f shown in Figure 4, we can see that each value of y in the range $[0, \infty)$ of f is associated with exactly *two* numbers $x = \pm\sqrt{y}$ (except for $y = 0$) in the domain $(-\infty, \infty)$ of f. This implies that f does not have an inverse because the uniqueness requirement of a function cannot be satisfied in this case. Observe that any horizontal line $y = c$ ($c > 0$) intersects the graph of f at more than one point.

Next, consider the function g defined by the same rule as that of f, namely $y = x^2$, but with domain restricted to $[0, \infty)$. From the graph of g shown in Figure 5, you can see that each value of y in the range $[0, \infty)$ of g is mapped onto exactly *one* number $x = \sqrt{y}$ in the domain $[0, \infty)$ of g.

Thus, in this case, we can define the inverse function of g, from the range $[0, \infty)$ of g onto the domain $[0, \infty)$ of g. To find the rule for g^{-1}, we solve the equation $y = x^2$ for x in terms of y. Thus, $x = \sqrt{y}$ (because $x \geq 0$), and so $g^{-1}(y) = \sqrt{y}$, or, since y is a dummy variable, we can write $g^{-1}(x) = \sqrt{x}$. Also, observe that every horizontal line intersects the graph of g at no more than one point.

Why does g have an inverse but f does not? Observe that f takes on the same value twice; that is, there are two values of x that are mapped onto each value of y (except $y = 0$). On the other hand, g never takes on the same value more than once. The function g is said to be *one-to-one*.

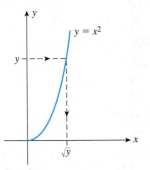

FIGURE 5
Each value of y is associated with exactly one value of x.

One-to-One Function

A function f is **one-to-one** if it never takes on the same value more than once. In other words, if x_1 and x_2 are any two numbers in the domain of f with $x_1 \neq x_2$, then $f(x_1) \neq f(x_2)$.

Geometrically, a function is one-to-one if every horizontal line intersects its graph at no more than one point. This is called the **Horizontal Line Test.**

The next theorem tells us when an inverse function exists.

THEOREM 1

The Existence of an Inverse Function

A function has an inverse if and only if it is one-to-one.

A.4 Finding the Inverse of a Function

Here is a summary of the steps for finding the inverse of a function (if it exists).

Guidelines for Finding the Inverse of a Function

1. Write $y = f(x)$.
2. Solve for x in terms of y (if possible).
3. Interchange x and y to obtain $y = f^{-1}(x)$.

EXAMPLE 1 Find the inverse of the function defined by $f(x) = \dfrac{1}{\sqrt{2x - 3}}$.

Solution To find the rule for this inverse, write

$$y = \frac{1}{\sqrt{2x - 3}} \qquad (y > 0)$$

and then solve the equation for x:

$$y^2 = \frac{1}{2x - 3} \qquad \text{Square both sides.}$$

$$2x - 3 = \frac{1}{y^2} \qquad \text{Take reciprocals.}$$

$$2x = \frac{1}{y^2} + 3 = \frac{3y^2 + 1}{y^2}$$

$$x = \frac{3y^2 + 1}{2y^2}$$

Finally, interchanging x and y, we obtain

$$y = \frac{3x^2 + 1}{2x^2}$$

giving the rule for f^{-1} as

$$f^{-1}(x) = \frac{3x^2 + 1}{2x^2}$$

The graphs of both f and f^{-1} are shown in Figure 6.

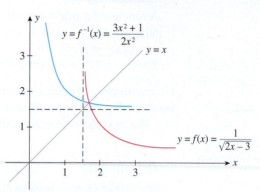

FIGURE 6
The graphs of f and f^{-1}. Notice that they are reflections of each other about the line $y = x$.

Exercises

In Exercises 1–6, show that f and g are inverses of each other by showing that $f[g(x)] = x$ and $g[f(x)] = x$.

1. $f(x) = \frac{1}{3}x^3; g(x) = \sqrt[3]{3x}$ **2.** $f(x) = \frac{1}{x}; g(x) = \frac{1}{x}$

3. $f(x) = 2x + 3; g(x) = \frac{x - 3}{2}$

4. $f(x) = x^2 + 1, (x \leq 0); g(x) = -\sqrt{x - 1}$

5. $f(x) = 4(x + 1)^{2/3}, (x \geq -1); g(x) = \frac{1}{8}(x^{3/2} - 8),$ $(x \geq 0)$

6. $f(x) = \frac{1 + x}{1 - x}; g(x) = \frac{x - 1}{x + 1}$

In Exercises 7–12, you are given the graph of a function f. Determine whether f is one-to-one.

7.

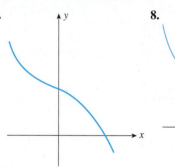

8.

9.

10.

11.

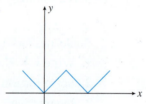

12.

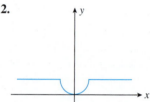

In Exercises 13–18, find the inverse of f. Then sketch the graphs of f and f^{-1} on the same set of axes.

13. $f(x) = 3x - 2$ **14.** $f(x) = x^2, x \leq 0$

15. $f(x) = x^3 + 1$ **16.** $f(x) = 2\sqrt{x} + 3$

17. $f(x) = \sqrt{9 - x^2}, (x \geq 0)$ **18.** $f(x) = x^{3/5} + 1$

19. A HOT-AIR BALLOON A hot-air balloon rises vertically from the ground so that its height after t sec is $h = \frac{1}{2}t^2 + \frac{1}{2}t$ ft $(0 \leq t \leq 60)$.

  a. Find the inverse of the function $f(t) = \frac{1}{2}t^2 + \frac{1}{2}t$, and explain what it represents.

  b. Use the result of part (a) to find the time when the altitude of the balloon is between 120 and 210 ft.

20. AGING POPULATION The population of Americans age 55 years and older as a percentage of the total population is approximated by the function

$$f(t) = 10.72(0.9t + 10)^{0.3} \quad (0 \leq t \leq 20)$$

where t is measured in years and $t = 0$ corresponds to the year 2000.

  a. Find the rule for f^{-1}.

  b. Evaluate $f^{-1}(25)$ and interpret your result.

  Source: U.S. Census Bureau.

B.1   Indeterminate Forms

In Section 2.4, we encountered the limit

$$\lim_{t \to 2} \frac{4(t^2 - 4)}{t - 2} \tag{1}$$

when we attempted to find the velocity of the maglev at time $t = 2$. Observe that both the numerator and the denominator of this expression approach zero as t approaches two.

More generally, if $\lim_{x \to a} f(x) = 0$ and $\lim_{x \to a} g(x) = 0$, then the limit

$$\lim_{x \to a} \frac{f(x)}{g(x)}$$

is called an **indeterminate form of the type 0/0.** As the name implies, the undefined expression 0/0 does not provide us with a definitive answer concerning the existence of the limit or its value, if the limit exists.

Recall that we evaluated the limit in (1) through algebraic sleight of hand. Thus,

$$\lim_{t \to 2} \frac{4(t^2 - 4)}{t - 2} = \lim_{t \to 2} \frac{4(t + 2)(t - 2)}{t - 2} = \lim_{t \to 2} 4(t + 2) = 16$$

This example raises the following question: Given an indeterminate form of the type 0/0, is there a more general and efficient method for resolving whether the limit

$$\lim_{x \to a} \frac{f(x)}{g(x)}$$

exists? If so, what is the limit?

B.2   The Indeterminate Forms 0/0 and ∞/∞ and l'Hôpital's Rule

To gain insight into the nature of an indeterminate form of the type 0/0, let's consider the following limits:

a. $\lim_{x \to 0^+} \dfrac{x^2}{x}$ **b.** $\lim_{x \to 0^+} \dfrac{2x}{3x}$ **c.** $\lim_{x \to 0^+} \dfrac{x}{x^2}$

Each of these limits is an indeterminate form of the type 0/0. We can evaluate each limit as follows:

a. $\lim_{x \to 0^+} \dfrac{x^2}{x} = \lim_{x \to 0^+} x = 0$ **b.** $\lim_{x \to 0^+} \dfrac{2x}{3x} = \lim_{x \to 0^+} \dfrac{2}{3} = \dfrac{2}{3}$ **c.** $\lim_{x \to 0^+} \dfrac{x}{x^2} = \lim_{x \to 0^+} \dfrac{1}{x} = \infty$

Let's examine each limit in greater detail. In part (a), the numerator $f_1(x) = x^2$ goes to zero faster than the denominator $g_1(x) = x$, when x is close to zero. So it is plausible that the ratio $f_1(x)/g_1(x)$ should approach zero as x approaches zero. In part (b), the numerator $f_2(x) = 2x$ goes to zero at $(2x)/(3x) = 2/3$, or two thirds of the rate that $g_2(x) = 3x$ goes to zero, so the answer seems reasonable. Finally, in part (c), the denominator $g_3(x) = x^2$ goes to zero faster than the numerator $f_3(x) = x$, and consequently, we expect the ratio to "blow up."

These three examples suggest that the existence or nonexistence of the limit, as well as the value of the limit, depends on how fast the numerator $f(x)$ and the denominator $g(x)$ go to zero. This observation suggests the following technique for evaluating these indeterminate forms: Because both $f(x)$ and $g(x)$ go to zero as x approaches zero, we cannot determine the limit of the quotient by using the Quotient Rule for limits. So we might consider the limit of the ratio of their *derivatives*, $f'(x)$ and $g'(x)$, since the derivatives measure how fast $f(x)$ and $g(x)$ change. In other words, it might be plausible that if both $f(x) \to 0$ and $g(x) \to 0$ as $x \to 0$, then

$$\lim_{x \to 0} \frac{f(x)}{g(x)} = \lim_{x \to 0} \frac{f'(x)}{g'(x)}$$

Let's try this on the limit in Expression (1). For this limit, we have

$$\lim_{t \to 2} \frac{4(t^2 - 4)}{t - 2} = \lim_{t \to 2} \frac{\frac{d}{dt}[4(t^2 - 4)]}{\frac{d}{dt}(t - 2)} = \lim_{t \to 2} \frac{8t}{1} = 16$$

which is the value we obtained before!

This method, which we have arrived at intuitively, is given validity by the theorem known as l'Hôpital's Rule. The theorem is named after the French mathematician Guillaume Francois Antoine de l'Hôpital (1661–1704), who published the first calculus text in 1696. But before stating l'Hôpital's Rule, we need to define another type of indeterminate form.

If $\lim_{x \to a} f(x) = \pm\infty$ and $\lim_{x \to a} g(x) = \pm\infty$, then the limit

$$\lim_{x \to a} \frac{f(x)}{g(x)}$$

is also an indeterminate form of the type ∞/∞, $-\infty/\infty$, $\infty/-\infty$, or $-\infty/-\infty$. To see why this limit is an indeterminate form, we simply write

$$\lim_{x \to a} \frac{f(x)}{g(x)} = \lim_{x \to a} \frac{\dfrac{1}{g(x)}}{\dfrac{1}{f(x)}}$$

which has the form $0/0$ and is therefore indeterminate. We refer to each of these limits as an **indeterminate form of the type ∞/∞**, since the sign provides little useful information.

THEOREM 1

l'Hôpital's Rule

Suppose f and g are differentiable on an open interval I that contains a, with the possible exception of a itself and $g'(x) \neq 0$ for all x in I, with the possible exception of a. If $\lim_{x \to a} \dfrac{f(x)}{g(x)}$ has an indeterminate form of the type $0/0$ or ∞/∞, then

$$\lim_{x \to a} \frac{f(x)}{g(x)} = \lim_{x \to a} \frac{f'(x)}{g'(x)}$$

provided that the limit on the right exists or is infinite.

⚠ The expression $f'(x)/g'(x)$ is the *ratio* of the derivatives of $f(x)$ and $g(x)$—it is *not* obtained from f/g by using the Quotient Rule for differentiating.

Notes

1. l'Hôpital's Rule is valid for one-sided limits as well as limits at infinity or negative infinity; that is, we can replace $x \to a$ by any of the symbols $x \to a^+$, $x \to a^-$, $x \to -\infty$, and $x \to \infty$.

2. Before applying l'Hôpital's Rule, check to see that the limit has one of the indeterminate forms. For example,

$$\lim_{x \to 0^+} \frac{1}{x} = \infty$$

If we had applied l'Hôpital's Rule to evaluate the limit without first ascertaining that it had an indeterminate form, we would have obtained the erroneous result

$$\lim_{x \to 0^+} \frac{1}{x} = \lim_{x \to 0^+} \frac{0}{1} = 0$$

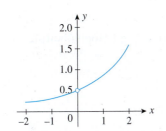

FIGURE 1
The graph of $y = (e^x - 1)/2x$ gives a visual confirmation of the result of Example 1.

EXAMPLE 1 Evaluate $\displaystyle\lim_{x \to 0} \frac{e^x - 1}{2x}$.

Solution   We have an indeterminate form of the type $0/0$. Applying l'Hôpital's Rule, we obtain

$$\lim_{x \to 0} \frac{e^x - 1}{2x} = \lim_{x \to 0} \frac{\frac{d}{dx}(e^x - 1)}{\frac{d}{dx}(2x)} = \lim_{x \to 0} \frac{e^x}{2} = \frac{1}{2}$$

(See Figure 1.)

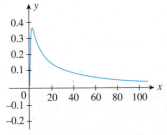

FIGURE 2
The graph of $y = (\ln x)/x$ shows that $y \to 0$ as $x \to \infty$.

EXAMPLE 2 Evaluate $\displaystyle\lim_{x \to \infty} \frac{\ln x}{x}$.

Solution   We have an indeterminate form of the type ∞/∞. Applying l'Hôpital's Rule, we obtain

$$\lim_{x \to \infty} \frac{\ln x}{x} = \lim_{x \to \infty} \frac{\frac{d}{dx}(\ln x)}{\frac{d}{dx}(x)} = \lim_{x \to \infty} \frac{1}{x} = 0$$

(See Figure 2.)

EXAMPLE 3 Evaluate $\displaystyle\lim_{x \to 1^+} \frac{\sin \pi x}{\sqrt{x - 1}}$.

Solution   We have an indeterminate form of the type $0/0$. Applying l'Hôpital's Rule, we obtain

$$\lim_{x \to 1^+} \frac{\sin \pi x}{(x - 1)^{1/2}} = \lim_{x \to 1^+} \frac{\pi \cos \pi x}{\frac{1}{2}(x - 1)^{-1/2}}$$

$$= \lim_{x \to 1^+} 2\pi (\cos \pi x) \sqrt{x - 1}$$

$$= 0$$

(See Figure 3.)

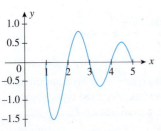

FIGURE 3
The graph of $y = (\sin \pi x)/(x - 1)^{1/2}$ shows that $\displaystyle\lim_{x \to 1^+} (\sin \pi x)/(\sqrt{x - 1}) = 0$.

To resolve a limit involving an indeterminate form, we sometimes need to apply l'Hôpital's Rule more than once. This is illustrated in the next two examples.

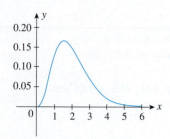

FIGURE 4
The graph of $y = x^3/e^{2x}$ shows that $y \to 0$ as $x \to \infty$.

EXAMPLE 4 Evaluate $\lim\limits_{x\to\infty} \dfrac{x^3}{e^{2x}}$.

Solution   Applying l'Hôpital's Rule (three times), we obtain

$$\lim_{x\to\infty} \frac{x^3}{e^{2x}} = \lim_{x\to\infty} \frac{3x^2}{2e^{2x}} \qquad \text{Type: } \infty/\infty$$

$$= \lim_{x\to\infty} \frac{6x}{4e^{2x}} \qquad \text{Type: } \infty/\infty$$

$$= \lim_{x\to\infty} \frac{6}{8e^{2x}} = 0$$

(See Figure 4.)

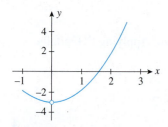

FIGURE 5
The graph of $y = x^3/(x - \tan x)$ shows that $y \to -3$ as $x \to 0$. Note that y is not defined at $x = 0$.

EXAMPLE 5 Evaluate $\lim\limits_{x\to 0} \dfrac{x^3}{x - \tan x}$.

Solution   We have an indeterminate form of the type 0/0. Using l'Hôpital's Rule repeatedly, we obtain

$$\lim_{x\to 0} \frac{x^3}{x - \tan x} = \lim_{x\to 0} \frac{3x^2}{1 - \sec^2 x} \qquad \text{Type: } 0/0$$

$$= \lim_{x\to 0} \frac{6x}{-2 \sec^2 x \tan x} \qquad \text{Type: } 0/0$$

$$= \lim_{x\to 0} \frac{6}{-4 \sec^2 x \tan^2 x - 2 \sec^4 x} = \frac{6}{-2} = -3$$

(See Figure 5.)

Exercises

In Exercises 1–26, evaluate the limit using l'Hôpital's Rule if appropriate.

1. $\lim\limits_{x\to 1} \dfrac{x - 1}{x^2 - 1}$

2. $\lim\limits_{x\to -1} \dfrac{x^2 - 2x - 3}{x + 1}$

3. $\lim\limits_{x\to 2} \dfrac{x^3 - 8}{x - 2}$

4. $\lim\limits_{x\to 1} \dfrac{x^7 - 1}{x^4 - 1}$

5. $\lim\limits_{x\to 0} \dfrac{e^x - 1}{x^2 + x}$

6. $\lim\limits_{x\to 1} \dfrac{\ln x}{x - 1}$

7. $\lim\limits_{t\to\pi} \dfrac{\sin t}{\pi - t}$

8. $\lim\limits_{x\to 0} \dfrac{e^x - 1}{x + \sin x}$

9. $\lim\limits_{\theta\to 0} \dfrac{\tan 2\theta}{\theta}$

10. $\lim\limits_{x\to 0} \dfrac{\sin 2x}{x}$

11. $\lim\limits_{x\to\infty} \dfrac{x + \cos x}{2x + 1}$

12. $\lim\limits_{\theta\to 0} \dfrac{\theta + \sin \theta}{\tan \theta}$

13. $\lim\limits_{x\to 0} \dfrac{\sin x - x \cos x}{\tan^3 x}$

14. $\lim\limits_{u\to\pi} \dfrac{2 \sin^2 u}{1 + \cos u}$

15. $\lim\limits_{x\to\infty} \dfrac{\sqrt{x}}{\ln x}$

16. $\lim\limits_{x\to\infty} \dfrac{e^x}{x^4}$

17. $\lim\limits_{x\to\infty} \dfrac{(\ln x)^3}{x^2}$

18. $\lim\limits_{x\to 1} \dfrac{x^{1/2} - x^{1/3}}{x - 1}$

19. $\lim\limits_{x\to\infty} \dfrac{\ln(1 + e^x)}{x^2}$

20. $\lim\limits_{x\to 1} \dfrac{a^{\ln x} - x}{\ln x}$

21. $\lim\limits_{x\to -1} \dfrac{\sqrt{x + 2} + x}{\sqrt[3]{2x + 1} + 1}$

22. $\lim\limits_{x\to 0} \dfrac{\ln(x^2 + 1)}{\cos x - 1}$

23. $\lim\limits_{x\to 0^+} \dfrac{e^{x^2} + x - 1}{1 - \sqrt{1 - x^2}}$

24. $\lim\limits_{x\to 0} \dfrac{\ln(1 + x) - \tan x}{x^2}$

25. $\lim\limits_{x\to 0} \dfrac{\sin x - x}{e^x - e^{-x} - 2x}$

26. $\lim\limits_{x\to 0} \dfrac{e^{x^2} - 1}{1 - \cos x}$

If $\lim\limits_{x\to a} f(x) = \infty$ and $\lim\limits_{x\to a} g(x) = \infty$, then $\lim\limits_{x\to a} [f(x) - g(x)]$ is said to be an *indeterminate form of the type* $\infty - \infty$. An indeterminate form of this type can be expressed as one of the type 0/0 or ∞/∞ by algebraic manipulation. In Exercises 27–30, use this observation to evaluate the limits.

27. $\lim\limits_{x\to 0^+} \left(\dfrac{1}{x} - \dfrac{1}{e^x - 1} \right)$

28. $\lim\limits_{x\to 0^+} \left(\dfrac{1}{x} - \dfrac{1}{1 - \cos x} \right)$

29. $\lim\limits_{t\to(\pi/2)^-} (\tan t - \sec t)$

30. $\lim\limits_{x\to 1^+} \left(\dfrac{1}{\ln x} - \dfrac{1}{x - 1} \right)$

In Exercises 31–34, determine whether the statement is true or false. If it is true, explain why it is true. If it is false, explain why, or give an example to show why it is false.

31. $\lim\limits_{x \to 0} \dfrac{x}{e^x} = \lim\limits_{x \to 0} \dfrac{1}{e^x} = 1$

32. Suppose f and g satisfy the hypothesis for l'Hôpital's Rule. If $\lim\limits_{x \to a} f(x) = \lim\limits_{x \to a} g(x) = 0$, then

$$\lim_{x \to a} \frac{f(x)}{g(x)} = \lim_{x \to a} \frac{d}{dx}\left[\frac{f(x)}{g(x)}\right].$$

33. Suppose f and g satisfy the hypothesis for l'Hôpital's Rule. If $\lim\limits_{x \to a} f(x) = \lim\limits_{x \to a} g(x) = 0$, then $\lim\limits_{x \to a} \dfrac{f(x)}{g(x)}$ does not exist.

34. Suppose f and g satisfy the hypothesis for l'Hôpital's Rule. If $\lim\limits_{x \to a} f(x) = \lim\limits_{x \to a} g(x) = 0$, $\lim\limits_{x \to a} f'(x) = L$, and

$$\lim_{x \to a} g'(x) = M \neq 0, \text{ then } \lim_{x \to a} \frac{f(x)}{g(x)} = \frac{L}{M}.$$

C.1   The Standard Normal Distribution

TABLE C.1

The Standard Normal Distribution

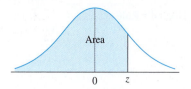

Area

0   z

$$F_z(z) = P(Z \le z)$$

z	0.00	0.01	0.02	0.03	0.04	0.05	0.06	0.07	0.08	0.09
−3.4	0.0003	0.0003	0.0003	0.0003	0.0003	0.0003	0.0003	0.0003	0.0003	0.0002
−3.3	0.0005	0.0005	0.0005	0.0004	0.0004	0.0004	0.0004	0.0004	0.0004	0.0003
−3.2	0.0007	0.0007	0.0006	0.0006	0.0006	0.0006	0.0006	0.0005	0.0005	0.0005
−3.1	0.0010	0.0009	0.0009	0.0009	0.0008	0.0008	0.0008	0.0008	0.0007	0.0007
−3.0	0.0013	0.0013	0.0013	0.0012	0.0012	0.0011	0.0011	0.0011	0.0010	0.0010
−2.9	0.0019	0.0018	0.0017	0.0017	0.0016	0.0016	0.0015	0.0015	0.0014	0.0014
−2.8	0.0026	0.0025	0.0024	0.0023	0.0023	0.0022	0.0021	0.0021	0.0020	0.0019
−2.7	0.0035	0.0034	0.0033	0.0032	0.0031	0.0030	0.0029	0.0028	0.0027	0.0026
−2.6	0.0047	0.0045	0.0044	0.0043	0.0041	0.0040	0.0039	0.0038	0.0037	0.0036
−2.5	0.0062	0.0060	0.0059	0.0057	0.0055	0.0054	0.0052	0.0051	0.0049	0.0048
−2.4	0.0082	0.0080	0.0078	0.0075	0.0073	0.0071	0.0069	0.0068	0.0066	0.0064
−2.3	0.0107	0.0104	0.0102	0.0099	0.0096	0.0094	0.0091	0.0089	0.0087	0.0084
−2.2	0.0139	0.0136	0.0132	0.0129	0.0125	0.0122	0.0119	0.0116	0.0113	0.0110
−2.1	0.0179	0.0174	0.0170	0.0166	0.0162	0.0158	0.0154	0.0150	0.0146	0.0143
−2.0	0.0228	0.0222	0.0217	0.0212	0.0207	0.0202	0.0197	0.0192	0.0188	0.0183
−1.9	0.0287	0.0281	0.0274	0.0268	0.0262	0.0256	0.0250	0.0244	0.0239	0.0233
−1.8	0.0359	0.0352	0.0344	0.0336	0.0329	0.0322	0.0314	0.0307	0.0301	0.0294
−1.7	0.0446	0.0436	0.0427	0.0418	0.0409	0.0401	0.0392	0.0384	0.0375	0.0367
−1.6	0.0548	0.0537	0.0526	0.0516	0.0505	0.0495	0.0485	0.0475	0.0465	0.0455
−1.5	0.0668	0.0655	0.0643	0.0630	0.0618	0.0606	0.0594	0.0582	0.0571	0.0559
−1.4	0.0808	0.0793	0.0778	0.0764	0.0749	0.0735	0.0722	0.0708	0.0694	0.0681
−1.3	0.0968	0.0951	0.0934	0.0918	0.0901	0.0885	0.0869	0.0853	0.0838	0.0823
−1.2	0.1151	0.1131	0.1112	0.1093	0.1075	0.1056	0.1038	0.1020	0.1003	0.0985
−1.1	0.1357	0.1335	0.1314	0.1292	0.1271	0.1251	0.1230	0.1210	0.1190	0.1170
−1.0	0.1587	0.1562	0.1539	0.1515	0.1492	0.1469	0.1446	0.1423	0.1401	0.1379
−0.9	0.1841	0.1814	0.1788	0.1762	0.1736	0.1711	0.1685	0.1660	0.1635	0.1611
−0.8	0.2119	0.2090	0.2061	0.2033	0.2005	0.1977	0.1949	0.1922	0.1894	0.1867
−0.7	0.2420	0.2389	0.2358	0.2327	0.2296	0.2266	0.2236	0.2206	0.2177	0.2148
−0.6	0.2743	0.2709	0.2676	0.2643	0.2611	0.2578	0.2546	0.2514	0.2483	0.2451
−0.5	0.3085	0.3050	0.3015	0.2981	0.2946	0.2912	0.2877	0.2843	0.2810	0.2776
−0.4	0.3446	0.3409	0.3372	0.3336	0.3300	0.3264	0.3228	0.3192	0.3156	0.3121
−0.3	0.3821	0.3783	0.3745	0.3707	0.3669	0.3632	0.3594	0.3557	0.3520	0.3483
−0.2	0.4207	0.4168	0.4129	0.4090	0.4052	0.4013	0.3974	0.3936	0.3897	0.3859
−0.1	0.4602	0.4562	0.4522	0.4483	0.4443	0.4404	0.4364	0.4325	0.4286	0.4247
−0.0	0.5000	0.4960	0.4920	0.4880	0.4840	0.4801	0.4761	0.4721	0.4681	0.4641

TABLE C.1 (*continued*)

The Standard Normal Distribution

$$F_z(z) = P(Z \leq z)$$

z	0.00	0.01	0.02	0.03	0.04	0.05	0.06	0.07	0.08	0.09
0.0	0.5000	0.5040	0.5080	0.5120	0.5160	0.5199	0.5239	0.5279	0.5319	0.5359
0.1	0.5398	0.5438	0.5478	0.5517	0.5557	0.5596	0.5636	0.5675	0.5714	0.5753
0.2	0.5793	0.5832	0.5871	0.5910	0.5948	0.5987	0.6026	0.6064	0.6103	0.6141
0.3	0.6179	0.6217	0.6255	0.6293	0.6331	0.6368	0.6406	0.6443	0.6480	0.6517
0.4	0.6554	0.6591	0.6628	0.6664	0.6700	0.6736	0.6772	0.6808	0.6844	0.6879
0.5	0.6915	0.6950	0.6985	0.7019	0.7054	0.7088	0.7123	0.7157	0.7190	0.7224
0.6	0.7257	0.7291	0.7324	0.7357	0.7389	0.7422	0.7454	0.7486	0.7517	0.7549
0.7	0.7580	0.7611	0.7642	0.7673	0.7704	0.7734	0.7764	0.7794	0.7823	0.7852
0.8	0.7881	0.7910	0.7939	0.7967	0.7995	0.8023	0.8051	0.8078	0.8106	0.8133
0.9	0.8159	0.8186	0.8212	0.8238	0.8264	0.8289	0.8315	0.8340	0.8365	0.8389
1.0	0.8413	0.8438	0.8461	0.8485	0.8508	0.8531	0.8554	0.8577	0.8599	0.8621
1.1	0.8643	0.8665	0.8686	0.8708	0.8729	0.8749	0.8770	0.8790	0.8810	0.8830
1.2	0.8849	0.8869	0.8888	0.8907	0.8925	0.8944	0.8962	0.8980	0.8997	0.9015
1.3	0.9032	0.9049	0.9066	0.9082	0.9099	0.9115	0.9131	0.9147	0.9162	0.9177
1.4	0.9192	0.9207	0.9222	0.9236	0.9251	0.9265	0.9278	0.9292	0.9306	0.9319
1.5	0.9332	0.9345	0.9357	0.9370	0.9382	0.9394	0.9406	0.9418	0.9429	0.9441
1.6	0.9452	0.9463	0.9474	0.9484	0.9495	0.9505	0.9515	0.9525	0.9535	0.9545
1.7	0.9554	0.9564	0.9573	0.9582	0.9591	0.9599	0.9608	0.9616	0.9625	0.9633
1.8	0.9641	0.9649	0.9656	0.9664	0.9671	0.9678	0.9686	0.9693	0.9699	0.9706
1.9	0.9713	0.9719	0.9726	0.9732	0.9738	0.9744	0.9750	0.9756	0.9761	0.9767
2.0	0.9772	0.9778	0.9783	0.9788	0.9793	0.9798	0.9803	0.9808	0.9812	0.9817
2.1	0.9821	0.9826	0.9830	0.9834	0.9838	0.9842	0.9846	0.9850	0.9854	0.9857
2.2	0.9861	0.9864	0.9868	0.9871	0.9875	0.9878	0.9881	0.9884	0.9887	0.9890
2.3	0.9893	0.9896	0.9898	0.9901	0.9904	0.9906	0.9909	0.9911	0.9913	0.9916
2.4	0.9918	0.9920	0.9922	0.9925	0.9927	0.9929	0.9931	0.9932	0.9934	0.9936
2.5	0.9938	0.9940	0.9951	0.9943	0.9945	0.9946	0.9948	0.9949	0.9951	0.9952
2.6	0.9953	0.9955	0.9956	0.9957	0.9959	0.9960	0.9961	0.9962	0.9963	0.9964
2.7	0.9965	0.9966	0.9967	0.9968	0.9969	0.9970	0.9971	0.9972	0.9973	0.9974
2.8	0.9974	0.9975	0.9976	0.9977	0.9977	0.9978	0.9979	0.9979	0.9980	0.9981
2.9	0.9981	0.9982	0.9982	0.9983	0.9984	0.9984	0.9985	0.9985	0.9986	0.9986
3.0	0.9987	0.9987	0.9987	0.9988	0.9988	0.9989	0.9989	0.9989	0.9990	0.9990
3.1	0.9990	0.9991	0.9991	0.9991	0.9992	0.9992	0.9992	0.9992	0.9993	0.9993
3.2	0.9993	0.9993	0.9994	0.9994	0.9994	0.9994	0.9994	0.9995	0.9995	0.9995
3.3	0.9995	0.9995	0.9995	0.9996	0.9996	0.9996	0.9996	0.9996	0.9996	0.9997
3.4	0.9997	0.9997	0.9997	0.9997	0.9997	0.9997	0.9997	0.9997	0.9997	0.9998

Answers

Exercises 1.1, page 13

1.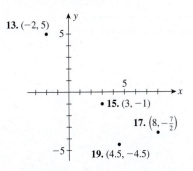

(number line: 0, 3, 6 with segment between 3 and 6)

3. (number line: −1, 0, 4)

5. (number line: 0)

7. 9   **9.** 1   **11.** 4   **13.** 7   **15.** $\frac{1}{5}$ **17.** 2

19. 2   **21.** 1   **23.** True   **25.** False   **27.** False

29. False   **31.** False   **33.** $\dfrac{1}{(xy)^2}$ **35.** $\dfrac{1}{x^{5/6}}$ **37.** $\dfrac{1}{(s+t)^3}$

39. $x^{13/3}$ **41.** $\dfrac{1}{x^3}$ **43.** x **45.** $\dfrac{9}{x^2 y^4}$ **47.** $\dfrac{y^8}{x^{10}}$

49. $2x^{11/6}$ **51.** $-2xy^2$ **53.** $2x^{4/3}y^{1/2}$ **55.** 2.828

57. 5.196   **59.** 31.62   **61.** 316.2   **63.** $\dfrac{3\sqrt{x}}{2x}$

65. $\dfrac{2\sqrt{3y}}{3}$ **67.** $\dfrac{\sqrt[3]{x^2}}{x}$ **69.** $\dfrac{2x}{3\sqrt{x}}$ **71.** $\dfrac{2y}{\sqrt{2xy}}$

73. $\dfrac{xz}{y\sqrt[3]{xz^2}}$ **75.** $9x^2 + 3x + 1$ **77.** $4y^2 + y + 8$

79. $-x - 1$ **81.** $\frac{2}{3} + e - e^{-1}$ **83.** $6\sqrt{2} + 8 + \frac{1}{2}\sqrt{x} - \frac{11}{4}\sqrt{y}$

85. $x^2 + 6x - 16$ **87.** $a^2 + 10a + 25$ **89.** $x^2 + 4xy + 4y^2$

91. $4x^2 - y^2$ **93.** $x(6x^3 + 4x^2 - 3x + 12)$ **95.** $\dfrac{6(4x^2 + 3)}{\sqrt{2x^2 + 3}}$

97. $-1000e^{-0.1t}(10 + t)$ **99.** $2x^3(2x^2 - 6x - 3)$

101. $7a^2(a^2 + 7ab - 6b^2)$ **103.** $e^{-x}(1 - x)$ **105.** $\frac{1}{2}x^{-5/2}(4 - 3x)$

107. $(2a + b)(3c - 2d)$ **109.** $(2a + b)(2a - b)$

111. $-2(3x + 5)(2x - 1)$ **113.** $3(x - 4)(x + 2)$

115. $2(3x - 5)(2x + 3)$ **117.** $(3x - 4y)(3x + 4y)$

119. $(x^2 + 5)(x^4 - 5x^2 + 25)$ **121.** $x^3 - xy^2$

123. $4(x - 1)(3x - 1)(2x + 2)^3 = 32(x - 1)(3x - 1)(x + 1)^3$

125. $4(x - 1)(3x - 1)(2x + 2)^3 = 32(x - 1)(3x - 1)(x + 1)^3$

127. $2x(x^2 + 2)^2(5x^4 + 20x^2 + 17)$

129. -4 and 3 **131.** -1 and $\frac{1}{2}$ **133.** 2 and 2

135. -2 and $\frac{3}{4}$ **137.** $\frac{1}{2} + \frac{1}{4}\sqrt{10}$ and $\frac{1}{2} - \frac{1}{4}\sqrt{10}$

139. $-1 + \frac{1}{2}\sqrt{10}$ and $-1 - \frac{1}{2}\sqrt{10}$ **141.** $0.7t^2 + 350t$

143. a. 107,772   **b.** 38,103   **c.** 9639

145. $100x(80 - x)$ **147.** True

Exercises 1.2, page 23

1. $\dfrac{x - 1}{x - 2}$ **3.** $\dfrac{3(2t + 1)}{2t - 1}$ **5.** $-\dfrac{7}{(4x - 1)^2}$ **7.** -8

9. $\dfrac{3x - 1}{2}$ **11.** $\dfrac{t + 20}{3t + 2}$ **13.** $-\dfrac{x(2x - 13)}{(2x - 1)(2x + 5)}$

15. $-\dfrac{x + 27}{(x - 3)^2(x + 3)}$ **17.** $\dfrac{x + 1}{x - 1}$ **19.** $\dfrac{4x^2 + 7}{\sqrt{2x^2 + 7}}$

21. $\dfrac{5(1 - t^2)}{(t^2 + 1)^2}$ **23.** $\dfrac{2(3x^2 - 1)}{(x^2 + 1)^3}$ **25.** $\dfrac{3(2x + 1)^2}{(3x + 2)^4}$

27. $100\left[\dfrac{10t^2 - 1000}{(t^2 + 20t + 100)^2}\right] = 1000\left[\dfrac{t - 10}{(t + 10)^3}\right]$ **29.** $\dfrac{\sqrt{3} + 1}{2}$

31. $\dfrac{\sqrt{x} + \sqrt{y}}{x - y}$ **33.** $\dfrac{(\sqrt{a} + \sqrt{b})^2}{a - b}$ **35.** $\dfrac{x}{3\sqrt{x}}$

37. $-\dfrac{2}{3(1 + \sqrt{3})}$ **39.** $-\dfrac{x + 1}{\sqrt{x + 2}(1 - \sqrt{x + 2})}$

41. False   **43.** False   **45.** $(-\infty, 2)$ **47.** $(-\infty, -5]$

49. $(-4, 6)$ **51.** $(-\infty, -3) \cup (3, \infty)$ **53.** $(-2, 3)$

55. $[-3, 5]$ **57.** $(-\infty, 1] \cup [\frac{3}{2}, \infty)$ **59.** $(-\infty, -3] \cup (2, \infty)$

61. $(-\infty, 0] \cup (1, \infty)$ **63.** 4   **65.** 2   **67.** $5\sqrt{3}$ **69.** $\pi + 1$

71. 2   **73.** False   **75.** False   **77.** True   **79.** False

81. True   **83.** False   **85.** $[362, 488.7]$ **87.** $12,300

89. $52,000   **91.** $|x - 0.5| \le 0.01$

93. Between 1000 and 4000 units

95. Between 98.04% and 98.36% of the toxic pollutants

97. Between 10:18 A.M. and 12:42 P.M.   **99.** False   **101.** True

Exercises 1.3, page 30

1. $(3, 3)$; Quadrant I   **3.** $(2, -2)$; Quadrant IV

5. $(-4, -6)$; Quadrant III   **7.** A **9.** $E, F,$ and G

11. F **13–19.** See the accompanying figure.

13. $(-2, 5)$

15. $(3, -1)$

17. $\left(8, -\frac{7}{2}\right)$

19. $(4.5, -4.5)$

21. 5   **23.** $\sqrt{61}$ **25.** $(-8, -6)$ and $(8, -6)$

29. $(x - 2)^2 + (y + 3)^2 = 25$ **31.** $x^2 + y^2 = 25$

33. $(x - 2)^2 + (y + 3)^2 = 34$

35. a. $(4, 4)$ **b.** 10 mi   **c.** 5.66 mi   **37.** No

39. Freight train; $4400   **41.** Model C

43. a. $\sqrt{400t^2 + 625\left(t + \frac{1}{2}\right)^2}$ mi   **b.** 58.31 mi   **45. b.** $\left(\frac{1}{2}, -\frac{3}{2}\right)$

47. True   **49.** False

Exercises 1.4, page 42

1. e   **3.** a   **5.** f   **7.** $\frac{1}{2}$ **9.** Not defined   **11.** 5

13. $\frac{5}{6}$ **15.** $\frac{d - b}{c - a}$ $(a \neq c)$ **17. a.** 4   **b.** -8

19. Parallel   **21.** Perpendicular   **23.** -5 **25.** $y = -3$

27. $y = 2x - 10$ **29.** $y = 2$ **31.** $y = 3x - 2$

33. $y = x + 1$ **35.** $y = 3x + 4$ **37.** $y = 5$

39. $y = \frac{1}{2}x$; $m = \frac{1}{2}$; $b = 0$ **41.** $y = \frac{2}{3}x - 3$; $m = \frac{2}{3}$; $b = -3$

43. $y = -\frac{1}{2}x + \frac{7}{2}$; $m = -\frac{1}{2}$; $b = \frac{7}{2}$ **45.** $y = \frac{1}{2}x + 3$

47. $y = -2x + 2$ **49.** $y = -6$ **51.** $y = b$

53. $y = \frac{2}{3}x - \frac{2}{3}$ **55.** $k = 8$

57.

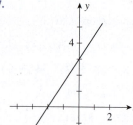

59.

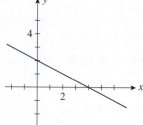

61.

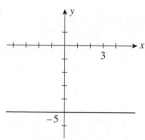

65. $y = -2x - 4$ **67.** $y = \frac{1}{8}x - \frac{1}{2}$ **69.** Yes

71. The points do not lie on a straight line.

73. a.

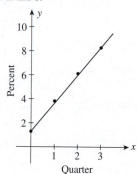

b. 1.9467; 70.082

c. The capacity utilization has been increasing by 1.9467% each year since 1990, when it stood at 70.082%.

d. Shortly after April, 2005

75. a. $y = 0.55x$ **b.** 2000   **77.** 89.6% of men's wages

79. a. and b.

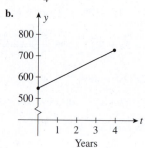

c. $y = 2.3x + 1.3$ **d.** 10.5%

81. a. $y = \dfrac{181}{4}x + 545$

b.

c. 907

83. True   **85.** False   **87.** True

Chapter 1 Concept Review, page 48

1. ordered; abscissa (x-coordinate); ordinate (y-coordinate)

2. a. x; y **b.** third   **3.** $\sqrt{(c - a)^2 + (d - b)^2}$

4. $(x - a)^2 + (y - b)^2 = r^2$

5. a. $\dfrac{y_2 - y_1}{x_2 - x_1}$ **b.** undefined   **c.** 0   **d.** positive

6. $m_1 = m_2; m_1 = -\dfrac{1}{m_2}$

7. a. $y - y_1 = m(x - x_1)$; point-slope form
b. $y = mx + b$; slope-intercept

8. a. $Ax + By + C = 0$ (A, B not both zero)   **b.** $-\dfrac{a}{b}$

Chapter 1 Review Exercises, page 48

1. $[-2, \infty)$ **2.** $[-1, 2]$ **3.** $(-\infty, -4) \cup (5, \infty)$

4. $(-\infty, -5) \cup (5, \infty)$ **5.** 4   **6.** 1   **7.** $\pi - 6$

8. $8 - 3\sqrt{3}$ **9.** $\frac{27}{8}$ **10.** 25   **11.** $\frac{1}{144}$ **12.** -32

13. $\frac{1}{4}$ **14.** $3\sqrt[3]{3}$ **15.** $4(x^2 + y)^2$ **16.** $\dfrac{a^{15}}{b^{11}}$ **17.** $\dfrac{2x}{3z}$

18. $-x^{1/2}$ **19.** $6xy^7$ **20.** $9x^2y^4$

21. $-2\pi r^2(\pi r - 50)$ or $2\pi r^2(50 - \pi r)$

22. $2vw(v^2 + w^2 + u^2)$ **23.** $(4 - x)(4 + x)$

24. $6t(2t - 3)(t + 1)$ **25.** $-\frac{3}{4}$ and $\frac{1}{2}$ **26.** -2 and $\frac{1}{3}$

27. $0, -3, 1$ **28.** $\dfrac{\sqrt{2}}{2}$ and $-\dfrac{\sqrt{2}}{2}$ **29.** $[-2, \frac{1}{2}]$ **30.** $(-2, -\frac{3}{2})$

31. $(-1, 4)$ **32.** $\frac{3}{2}; \frac{2}{3}$ **33.** $1 + \sqrt{6}, 1 - \sqrt{6}$

34. $-2 + \dfrac{\sqrt{2}}{2}; -2 - \dfrac{\sqrt{2}}{2}$ **35.** $\dfrac{180}{(t + 6)^2}$

36. $\dfrac{15x^2 + 24x + 2}{4(x + 2)(3x^2 + 2)}$ **37.** $\dfrac{78x^2 - 8x - 27}{3(2x^2 - 1)(3x - 1)}$

38. $\dfrac{2(x + 2)}{\sqrt{x + 1}}$ **39.** $\dfrac{1}{\sqrt{x} + 1}$ **40.** $\dfrac{x - \sqrt{x}}{2x}$ **41.** 5

42. 5   **43.** 2   **44.** $x = -2$ **45.** $y = 4$ **46.** $y = -\frac{1}{10}x + \frac{19}{5}$

47. $y = -\frac{4}{5}x + \frac{12}{5}$ **48.** $y = \frac{5}{2}x + 9$ **49.** $y = \frac{3}{4}x + \frac{11}{2}$

50. $y = -\frac{3}{4}x + \frac{9}{2}$ **51.** $y = -\frac{3}{5}x + \frac{12}{5}$ **52.** $y = 3x + 7$

53. $y = -\frac{3}{2}x - 7$ **54.** No   **55.** $k = 3$

56.

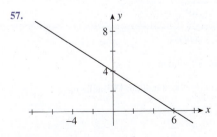

57.

58. \$100   **59.** \$400   **60.** Between 1 sec and 3 sec

61. a. and **b.**

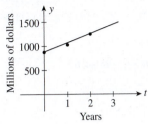

c. $y = 182t + 887$ **d.** \$2161 million

62. a. and **b.**

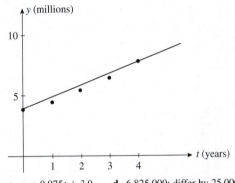

c. $y = 0.975t + 3.9$ **d.** 6,825,000; differ by 25,000

Chapter 1 Before Moving On, page 50

1. a. $\sqrt{3} + \sqrt{2} - \pi$ **b.** -3 **2. a.** $12x^5y$ **b.** $\dfrac{b^5}{a^3}$

3. a. $\dfrac{2x\sqrt{y}}{3y}$ **b.** $\dfrac{x(\sqrt{x} + 4)}{x - 16}$

4. a. $\dfrac{1 - 3x^2}{2\sqrt{x}(x^2 + 1)^2}$ **b.** $\dfrac{6\sqrt{x} + 2}{x + 2}$ or $\dfrac{6}{\sqrt{x + 2}}$

5. $\dfrac{x - y}{(\sqrt{x} - \sqrt{y})^2}$ **6. a.** $2x(3x + 2)(2x - 3)$ **b.** $(2b + 3c)(x - y)$

7. a. $-\frac{1}{4}$ and 1   **b.** $\dfrac{5 \pm \sqrt{13}}{6}$ **8.** $4\sqrt{5}$

9. $y = \frac{7}{5}x - \frac{3}{5}$ **10.** $y = -\frac{1}{3}x + \frac{4}{3}$

CHAPTER 2

Exercises 2.1, page 59

1. $21, -9, 5a + 6, -5a + 6, 5a + 21$

3. $-3, 6, 3a^2 - 6a - 3, 3a^2 + 6a - 3, 3x^2 - 6$

5. $2a + 2h + 5, -2a + 5, 2a^2 + 5, 2a - 4h + 5, 4a - 2h + 5$

7. $\dfrac{8}{15}, 0, \dfrac{2a}{a^2 - 1}, \dfrac{2(2 + a)}{a^2 + 4a + 3}, \dfrac{2(t + 1)}{t(t + 2)}$

9. $8, \dfrac{2a^2}{\sqrt{a - 1}}, \dfrac{2(x + 1)^2}{\sqrt{x}}, \dfrac{2(x - 1)^2}{\sqrt{x - 2}}$

11. $5, 1, 1$ **13.** $\frac{5}{2}, 3, 3, 9$

15. a. -2 **b.** (i) $x = 2$; (ii) $x = 1$ **c.** $[0, 6]$ **d.** $[-2, 6]$

17. Yes   **19.** Yes   **21.** 7   **23.** $(-\infty, \infty)$ **25.** $(-\infty, 0) \cup (0, \infty)$

27. $(-\infty, \infty)$ **29.** $(-\infty, 5]$ **31.** $(-\infty, -1) \cup (-1, 1) \cup (1, \infty)$

33. $[-3, \infty)$ **35.** $(-\infty, -2) \cup (-2, 1]$

37. a. $(-\infty, \infty)$ **b.** $6, 0, -4, -6, -\frac{25}{4}, -6, -4, 0$
c.

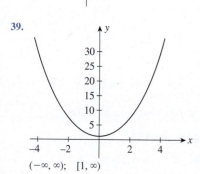

39.

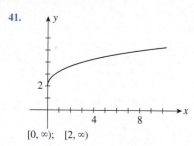

$(-\infty, \infty)$; $[1, \infty)$

41.

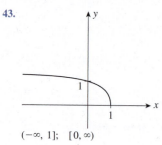

$[0, \infty)$; $[2, \infty)$

43.

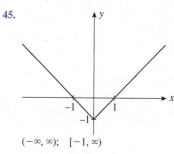

$(-\infty, 1]$; $[0, \infty)$

45.

$(-\infty, \infty)$; $[-1, \infty)$

47.

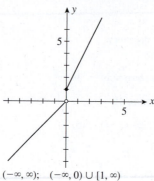

$(-\infty, \infty)$; $(-\infty, 0) \cup [1, \infty)$

49.

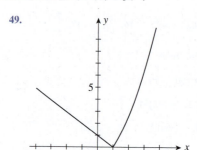

$(-\infty, \infty)$; $[0, \infty)$

51. Yes   **53.** No   **55.** Yes   **57.** Yes

59. 10π in.   **61.** $6 billion; $43.5 billion; $81 billion

63. a. From the beginning of 2001 until the end of 2005
b. After the beginning of 2006 until the end of 2010
c. 2006; both were approximately $900

65. a. $f(t) = \begin{cases} 0.0185t + 0.58 & \text{if } 0 \le t \le 20 \\ 0.015t + 0.65 & \text{if } 20 < t \le 30 \end{cases}$

b. 0.0185/year from 1960 through 1980; 0.015/year from 1980 through 1990
c. 1983

67. a. $I(x) = 1.053x$ **b.** $1600.56

69. $S(r) = 4\pi r^2$

71. a. 0.3 year/year   **b.** 41.2 years   **c.** 42.4 years

73. a. $V = -12{,}000n + 120{,}000$
b.

c. $48,000   **d.** $12,000/year

75. a. 0.46 million   **b.** 3.1 million

77. 16.8 years; 10.18 years   **79.** 8 million; 13.3 million

81. 0.77

83. $(0, \infty)$

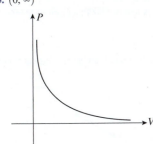

85. a. $0.6 trillion; $0.6 trillion
   b. $0.96 trillion; $1.2 trillion

87. $f(x) = \begin{cases} 2.32 & \text{if } 0 < x < 4 \\ 2.50 & \text{if } 4 \le x < 5 \\ 2.68 & \text{if } 5 \le x < 6 \\ 2.86 & \text{if } 6 \le x < 7 \\ 3.04 & \text{if } 7 \le x < 8 \\ 3.22 & \text{if } 8 \le x < 9 \\ 3.40 & \text{if } 9 \le x < 10 \\ 3.58 & \text{if } 10 \le x < 11 \\ 3.76 & \text{if } 11 \le x < 12 \\ 3.94 & \text{if } 12 \le x < 13 \\ 4.12 & \text{if } x = 13 \end{cases}$

   a. $(0, 13]$
   b.

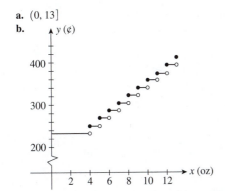

89. a. $D(t) = \begin{cases} 14t & \text{if } 0 \le t \le 2 \\ 2\sqrt{74t^2 - 100t + 100} & \text{if } 2 < t \end{cases}$
   b. 76.16 mi

91. False   **93.** False   **95.** False   **97.** False

Using Technology Exercises 2.1, page 68

1. a.

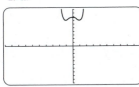

b.

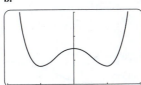

3. a.

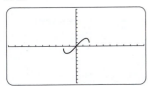

b.

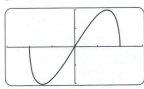

5.

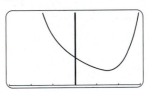

7.

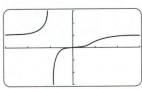

9. 18.5505   **11.** 4.1616

13. a.

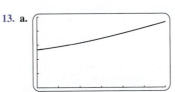

   b. $9.4 billion, $13.9 billion

15. a.

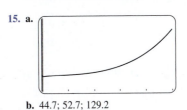

   b. 44.7; 52.7; 129.2

Exercises 2.2, page 74

1. $f(x) + g(x) = x^3 + x^2 + 3$

3. $f(x)g(x) = x^5 - 2x^3 + 5x^2 - 10$

5. $\dfrac{f(x)}{g(x)} = \dfrac{x^3 + 5}{x^2 - 2}$

7. $\dfrac{f(x)g(x)}{h(x)} = \dfrac{x^5 - 2x^3 + 5x^2 - 10}{2x + 4}$

9. $f(x) + g(x) = x - 1 + \sqrt{x + 1}$

11. $f(x)g(x) = (x - 1)\sqrt{x + 1}$

13. $\dfrac{g(x)}{h(x)} = \dfrac{\sqrt{x + 1}}{2x^3 - 1}$

15. $\dfrac{f(x)g(x)}{h(x)} = \dfrac{(x - 1)\sqrt{x + 1}}{2x^3 - 1}$

17. $\dfrac{f(x) - h(x)}{g(x)} = \dfrac{x - 2x^3}{\sqrt{x + 1}}$

19. $f(x) + g(x) = x^2 + \sqrt{x} + 3$;
   $f(x) - g(x) = x^2 - \sqrt{x} + 7$;
   $f(x)g(x) = (x^2 + 5)(\sqrt{x} - 2); \dfrac{f(x)}{g(x)} = \dfrac{x^2 + 5}{\sqrt{x} - 2}$

21. $f(x) + g(x) = \dfrac{(x - 1)\sqrt{x + 3} + 1}{x - 1}$;
   $f(x) - g(x) = \dfrac{(x - 1)\sqrt{x + 3} - 1}{x - 1}$;
   $f(x)g(x) = \dfrac{\sqrt{x + 3}}{x - 1}; \dfrac{f(x)}{g(x)} = (x - 1)\sqrt{x + 3}$

23. $f(x) + g(x) = \dfrac{2(x^2 - 2)}{(x - 1)(x - 2)};$

$f(x) - g(x) = \dfrac{-2x}{(x - 1)(x - 2)};$

$f(x)g(x) = \dfrac{(x + 1)(x + 2)}{(x - 1)(x - 2)}; \dfrac{f(x)}{g(x)} = \dfrac{(x + 1)(x - 2)}{(x - 1)(x + 2)}$

25. $f(g(x)) = x^4 + x^2 + 1; g(f(x)) = (x^2 + x + 1)^2$

27. $f(g(x)) = \sqrt{x^2 - 1} + 1; g(f(x)) = x + 2\sqrt{x}$

29. $f(g(x)) = \dfrac{x}{x^2 + 1}; g(f(x)) = \dfrac{x^2 + 1}{x}$ **31.** 49

33. $\dfrac{\sqrt{5}}{5}$ **35.** $f(x) = 2x^3 + x^2 + 1$ and $g(x) = x^5$

37. $f(x) = x^2 - 1$ and $g(x) = \sqrt{x}$

39. $f(x) = x^2 - 1$ and $g(x) = \dfrac{1}{x}$

41. $f(x) = 3x^2 + 2$ and $g(x) = \dfrac{1}{x^{3/2}}$ **43.** $3h$ **45.** $-h(2a + h)$

47. $2a + h$ **49.** $3a^2 + 3ah + h^2 - 1$ **51.** $-\dfrac{1}{a(a + h)}$

53. The total revenue in dollars from both restaurants at time t

55. The value in dollars of Nancy's shares of IBM at time t

57. The carbon monoxide pollution from cars in parts per million at time t

59. $C(x) = 0.6x + 12,100$

61. $0.0075t^2 + 0.13t + 0.17; D$ gives the difference in year t between the deficit without the rescue package and the deficit with the rescue package.

63. a. 23; In 2002, 23% of reported serious crimes ended in the arrests or in the identification of the suspects.
b. 18; In 2007, 18% of reported serious crimes ended in the arrests or in the identification of the suspects.

65. a. $C(x) = 0.000001x^3 - 0.01x^2 + 50x + 20,000$
b. $P(x) = -0.000001x^3 - 0.01x^2 + 100x - 20,000$
c. $132,000

67. a. $17.6111t^3 - 180.357t^2 + 1132.39t + 9871$
b. $3862.98; $10,113.49; $13,976.46   **c.** $13,976.46

69. a. 55%; 98.2%   **b.** $444,700; $1,167,600

71. a. $s(x) = f(x) + g(x) + h(x)$

73. True   **75.** False   **77.** True

Exercises 2.3, page 88

1. Yes; $y = -\frac{2}{3}x + 2$ **3.** Yes; $y = \frac{1}{2}x + 2$

5. Yes; $y = \frac{1}{2}x + \frac{9}{4}$ **7.** No   **9.** Polynomial function; degree 6

11. Polynomial function; degree 6

13. Some other function   **15.** $m = -1; b = 2$

17. a. $C(x) = 8x + 40,000$ **b.** $R(x) = 12x$
c. $P(x) = 4x - 40,000$ **d.** A loss of $8000; a profit of $8000

19. $43,200   **21.** 104 mg

23. a. $f(t) = -13.2t + 400$ **b.** 373.6 million metric tons equivalent

25. a.

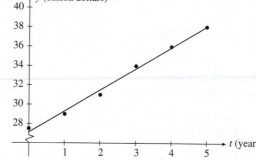

b. $40.26 billion   **c.** $2.19 billion/year

27. $128,000   **29. a.** 21.76 min   **b.** 111.4 min

31. a. $789.45   **b.** $2740.80

33. 123,780,000 kWh; 175,820,000 kWh   **35.** 11%; 36%

37. a. 0.06 million terabytes   **b.** 3.66 million terabytes

39. a.

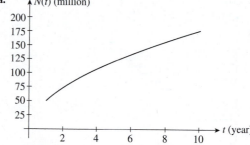

b. 177,000,000

41. 582,650; 2,532,700   **43. a.** 344; 4896   **b.** $6616

45. b. 2003

47. a. $T = \frac{1}{4}N + 40$ **b.** 248 chirps/min

49. a. 43.25; 41; 1547   **b.** 3350   **51.** $72,000

53. $5 billion; $152 billion

55. a.

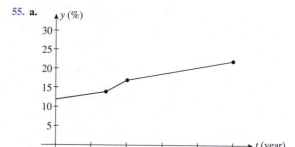

b. 13%; 22%

57. a.

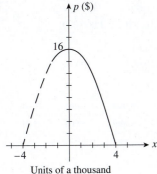

Units of a thousand

b. 3000 units

59. a.

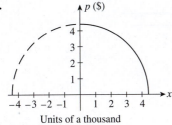

Units of a thousand

b. 3000

61. a.

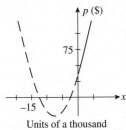

Units of a thousand

b. $76

63. a.

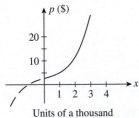

Units of a thousand

b. $15

65. a.

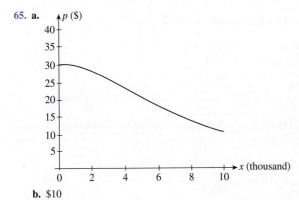

b. $10

67. a.

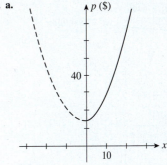

Units of a thousand

b. $20

69. a. $\dfrac{b - d}{c - a}; \dfrac{bc - ad}{c - a}$

b. If the unit price is increased, then the equilibrium quantity decreases while the equilibrium price increases.

c. If the upper bound for the unit price of a commodity is lowered, then both the equilibrium quantity and the equilibrium price drop.

71. 2500; $67.50   **73.** 11,000; $3   **75.** 8000; $80

77. $f(x) = 2x + \dfrac{500}{x}; x > 0$ **79.** $f(x) = 0.5x^2 + \dfrac{8}{x}; x > 0$

81. $f(x) = (22 + x)(36 - 2x)$ bushels/acre

83. a. $P(x) = (10,000 + x)(5 - 0.0002x)$ **b.** $60,800

85. False   **87.** False

Using Technology Exercises 2.3, page 98

1. $(-3.0414, 0.1503); (3.0414, 7.4497)$

3. $(-2.3371, 2.4117); (6.0514, -2.5015)$

5. $(-1.0219, -6.3461); (1.2414, -1.5931); (5.7805, 7.9391)$

7. a.

b. 438 wall clocks; $40.92

9. a. $f(t) = 1.85t + 16.9$

b.

c.

t	1	2	3	4	5	6
y	18.8	20.6	22.5	24.3	26.2	28.0

d. 31.7 gal

11. a. $f(t) = -0.221t^2 + 4.14t + 64.8$
b.

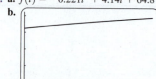

c. 77.8 million

13. a. $f(t) = 2.4t^2 + 15t + 31.4$
b.

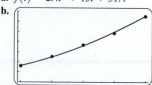

15. a. $f(t) = -0.00081t^3 + 0.0206t^2 + 0.125t + 1.69$
b.

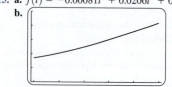

c. $1.8 trillion; $2.7 trillion; $4.2 trillion

17. a. $-0.0056t^3 + 0.112t^2 + 0.51t + 8$
b.

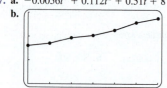

c. $8 billion, $10.4 billion, $13.9 billion

19. a. $f(t) = 0.00125t^4 - 0.0051t^3 - 0.0243t^2 + 0.129t + 1.71$
b.

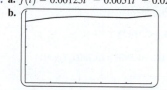

c. 1.71 mg; 1.81 mg; 1.85 mg; 1.84 mg; 1.83 mg; 1.89 mg
d. 2.13 mg/cigarette

Exercises 2.4, page 115

1. $\lim\limits_{x \to -2} f(x) = 3$ **3.** $\lim\limits_{x \to 3} f(x) = 3$ **5.** $\lim\limits_{x \to -2} f(x) = 3$

7. The limit does not exist.

9.

x	1.9	1.99	1.999
$f(x)$	4.61	4.9601	4.9960

x	2.001	2.01	2.1
$f(x)$	5.004	5.0401	5.41

$\lim\limits_{x \to 2} (x^2 + 1) = 5$

11.

x	-0.1	-0.01	-0.001
$f(x)$	-1	-1	-1

x	0.001	0.01	0.1
$f(x)$	1	1	1

The limit does not exist.

13.

x	0.9	0.99	0.999
$f(x)$	100	10,000	1,000,000

x	1.001	1.01	1.1
$f(x)$	1,000,000	10,000	100

The limit does not exist.

15.

x	0.9	0.99	0.999	1.001	1.01	1.1
$f(x)$	2.9	2.99	2.999	3.001	3.01	3.1

$\lim\limits_{x \to 1} \dfrac{x^2 + x - 2}{x - 1} = 3$

17.

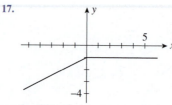

$\lim\limits_{x \to 0} f(x) = -1$

19.

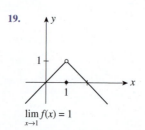

$\lim\limits_{x \to 1} f(x) = 1$

21.

$\lim\limits_{x \to 0} f(x) = 0$

23. 3   **25.** 3   **27.** -1   **29.** 2

31. -4   **33.** $\frac{5}{4}$ **35.** 2   **37.** $\sqrt{171} = 3\sqrt{19}$

39. $\frac{3}{2}$ **41.** -1   **43.** -6   **45.** 2

47. $\frac{1}{6}$ **49.** 2   **51.** -1   **53.** -10

55. The limit does not exist.   **57.** $\frac{5}{3}$ **59.** $\frac{1}{2}$ **61.** $\frac{1}{3}$

63. $\lim\limits_{x \to \infty} f(x) = \infty$; $\lim\limits_{x \to -\infty} f(x) = \infty$ **65.** 0; 0

67. $\lim\limits_{x \to \infty} f(x) = -\infty$; $\lim\limits_{x \to -\infty} f(x) = -\infty$

69.

x	1	10	100	1000
$f(x)$	0.5	0.009901	0.0001	0.000001

x	-1	-10	-100	-1000
$f(x)$	0.5	0.009901	0.0001	0.000001

$\lim\limits_{x \to \infty} f(x) = 0$ and $\lim\limits_{x \to -\infty} f(x) = 0$

71.

x	1	5	10	100
$f(x)$	12	360	2910	2.99×10^6

x	1000	-1	-5
$f(x)$	2.999×10^9	6	-390

x	-10	-100	-1000
$f(x)$	-3090	-3.01×10^6	-3.0×10^9

$\lim_{x \to \infty} f(x) = \infty$ and $\lim_{x \to -\infty} f(x) = -\infty$

73. 3   **75.** 3   **77.** $\lim_{x \to -\infty} f(x) = -\infty$ **79.** 0

81. a. \$0.5 million; \$0.75 million; \$1.17 million; \$2 million;
\$4.5 million; \$9.5 million
b. The limit does not exist; as the percent of pollutant to be removed
approaches 100, the cost becomes astronomical.

83. \$2.20; the average cost of producing x DVDs will approach
\$2.20/disc in the long run.

85. a. \$24 million; \$60 million; \$83.1 million   **b.** \$120 million

87. a. 137.3¢/mi; 59.8¢/mi; 45.1¢/mi; 39.8¢/mi; 37.3¢/mi
b.

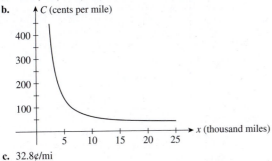

c. 32.8¢/mi

89. False   **91.** True   **93.** True

95. a moles/liter/second   **97.** No

Using Technology Exercises 2.4, page 121

1. 5   **3.** 3   **5.** $\frac{2}{3}$ **7.** $e^2 \approx 7.38906$

11. a.

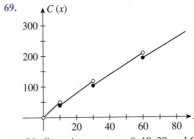

b. 25,000

Exercises 2.5, page 130

1. 3; 2; the limit does not exist.

3. The limit does not exist; 2; the limit does not exist.

5. 0; 2; the limit does not exist.

7. -2; 2; the limit does not exist.   **9.** True   **11.** True

13. False   **15.** True   **17.** False   **19.** True   **21.** 6

23. $-\frac{1}{4}$ **25.** The limit does not exist.   **27.** -1 **29.** 0

31. -4 **33.** The limit does not exist.   **35.** 4   **37.** 0; 0

39. $x = 0$; conditions 2 and 3   **41.** Continuous everywhere

43. $x = 0$; condition 3   **45.** $(-\infty, \infty)$ **47.** $(-\infty, \infty)$

49. $\left(-\infty, \frac{1}{2}\right) \cup \left(\frac{1}{2}, \infty\right)$ **51.** $(-\infty, -2) \cup (-2, 1) \cup (1, \infty)$

53. $(-\infty, \infty)$ **55.** $(-\infty, \infty)$ **57.** -1 and 1   **59.** 1 and 2

61. f is discontinuous at $x = 4, 5, \ldots, 13$.

63. Michael makes progress toward solving the problem until $x = x_1$.
Between $x = x_1$ and $x = x_2$, he makes no further progress. But at
$x = x_2$ he suddenly achieves a breakthrough, and at $x = x_3$ he
proceeds to complete the problem.

65. Conditions 2 and 3 are not satisfied at each of these points.

67.

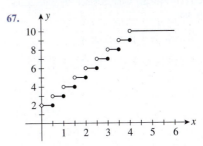

f is discontinuous at $x = \frac{1}{2}, 1, 1\frac{1}{2}, \ldots, 4$.

69.

C is discontinuous at $x = 0, 10, 30,$ and 60.

71. a. ∞; as the time taken to excite the tissue is made shorter and
shorter, the strength of the electric current gets stronger and
stronger.
b. b; as the time taken to excite the tissue is made longer and longer,
the strength of the electric current gets weaker and weaker and
approaches b.

73. 3

75. a. f is a polynomial of degree 2.   **b.** $f(1) = 3$ and $f(3) = -1$

77. a. f is a polynomial of degree 3.   **b.** $f(-1) = -4$ and $f(1) = 4$

79. 0.59   **81.** 1.34

83. c. $\frac{1}{2}; \frac{7}{2}$; Joan sees the ball on its way up $\frac{1}{2}$ sec after it was thrown and
again 3 sec later.

85. False   **87.** False   **89.** False   **91.** False   **93.** False

95. False   **97.** False

Using Technology Exercises 2.5, page 136

1. $x = 0, 1$ **3.** $x = 0, \frac{1}{2}$ **5.** $x = -\frac{1}{2}, 2$ **7.** $x = -2, 1$

9. 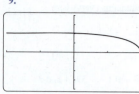 **11.**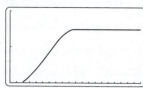

Exercises 2.6, page 149

1. 1.5 lb/month; 0.58 lb/month; 1.25 lb/month

3. 3.1%/hr; −21.2%/hr

5. a. Car A **b.** They are traveling at the same speed.
c. Car B **d.** Both cars covered the same distance; they are again side-by-side.

7. a. P_2 **b.** P_1 **c.** Bactericide B; Bactericide A

9. 0   **11.** 2   **13.** $6x$ **15.** $-2x + 3$ **17.** $2; y = 2x + 7$

19. $6; y = 6x - 3$ **21.** $\frac{1}{9}; y = \frac{1}{9}x - \frac{2}{3}$

23. a. $4x$ **b.** $y = 4x - 1$
c.

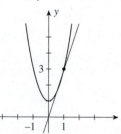

25. a. $2x - 2$ **b.** $(1, 0)$
c.

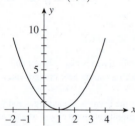

d. 0

27. a. 6; 5.5; 5.1   **b.** 5
c. The computations in part (a) illustrate that as h approaches zero, the average rate of change approaches the instantaneous rate of change.

29. a. 130 ft/sec; 128.2 ft/sec; 128.02 ft/sec   **b.** 128 ft/sec
c. The computations in part (a) illustrate that as the time intervals over which the average velocity are computed become smaller and smaller, the average velocity approaches the instantaneous velocity of the car at $t = 20$.

31. a. 5 sec   **b.** 80 ft/sec   **c.** 160 ft/sec

33. a. $-\frac{1}{6}$ L/atm   **b.** $-\frac{1}{4}$ L/atm

35. a. $-\frac{2}{3}x + 7$
b. $333 per $1000 spent on advertising; −$13,000 per $1000 spent on advertising

37. $6 billion/year; $10 billion/year

39. a. $f'(h)$ gives the instantaneous rate of change of the temperature with respect to height at a given height h, in °F per foot.
b. Negative   **c.** ≈ -0.05°F

41. Average rate of change of the seal population over $[a, a + h]$; instantaneous rate of change of the seal population at $x = a$

43. Average rate of change of the country's industrial production over $[a, a + h]$; instantaneous rate of change of the country's industrial production at $x = a$

45. Average rate of change of atmospheric pressure with respect to altitude over $[a, a + h]$; instantaneous rate of change of atmospheric pressure with respect to altitude at $x = a$

47. a. Yes   **b.** No   **c.** No   **49. a.** Yes   **b.** Yes   **c.** No

51. a. No   **b.** No   **c.** No

53. 32.1, 30.939, 30.814, 30.8014, 30.8001, 30.8000; 30.8 ft/sec

55. False

57.

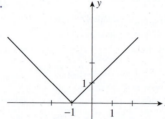

59. $a = 2, b = -1$

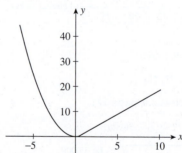

Using Technology Exercises 2.6, page 155

1. a. 9
b.

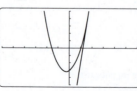

c. $y = 9x - 11$

3. a. $\frac{1}{12}$
b.

c. $y = \frac{1}{12}x + \frac{4}{3}$

5. a. 4
b.

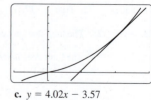

c. $y = 4x - 1$

7. a. 4.02
b.

c. $y = 4.02x - 3.57$

9. a.

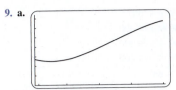

  b. 2.88 million/decade

Chapter 2 Concept Review, page 156

1. domain; range; B

2. domain; $f(x)$; vertical, point

3. $f(x) \pm g(x)$; $f(x)g(x)$; $\dfrac{f(x)}{g(x)}$; $A \cap B$; $A \cap B$; 0

4. $g[f(x)]$; f; $f(x)$; g

5. a. $P(x) = a_n x^n + a_{n-1}x^{n-1} + \cdots + a_1 x + a_0$
    ($a_n \neq 0$, n, a positive integer)
  b. linear; quadratic; cubic
  c. quotient; polynomials
  d. x^r (r a real number)

6. $f(x)$; L; a

7. a. L^r **b.** $L \pm M$
  c. LM **d.** $\dfrac{L}{M}$; $M \neq 0$

8. a. L; x **b.** M; negative; absolute

9. a. right   **b.** left   **c.** L; L

10. a. continuous   **b.** discontinuous   **c.** every

11. a. a; a; $g(a)$ **b.** everywhere   **c.** $Q(x)$

12. a. $[a, b]$; $f(c) = M$ **b.** $f(x) = 0$; (a, b)

13. a. $f'(a)$ **b.** $y = f(a) + m(x - a)$

14. a. $\dfrac{f(a + h) - f(a)}{h}$ **b.** $\displaystyle\lim_{h \to 0} \dfrac{f(a + h) - f(a)}{h}$

Chapter 2 Review Exercises, page 157

1. a. $(-\infty, 9]$ **b.** $(-\infty, -1) \cup \left(-1, \tfrac{3}{2}\right) \cup \left(\tfrac{3}{2}, \infty\right)$

2. a. $(-\infty, -3) \cup (-3, 2]$ **b.** $(-\infty, \infty)$

3. a. 0   **b.** $3a^2 + 17a + 20$ **c.** $12a^2 + 10a - 2$
  d. $3a^2 + 6ah + 3h^2 + 5a + 5h - 2$

4. a. $4x^2 - 2x + 6$ **b.** $2x^2 + 8xh + 8h^2 - x - 2h + 1$

5. a.

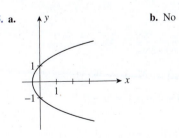

  b. No   **c.** Yes

6.

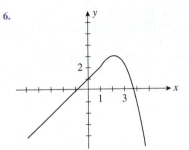

7. a. $\dfrac{2x + 3}{x}$ **b.** $\dfrac{1}{x(2x + 3)}$ **c.** $\dfrac{1}{2x + 3}$ **d.** $\dfrac{2}{x} + 3$

8. a. $2x^2 + 7$; $4x^2 - 4x + 5$ **b.** $\dfrac{3(x + 1)}{3x + 4}$; $\dfrac{1}{7 - 3x}$

  c. $\dfrac{1}{\sqrt{x + 1}} - 3$; $\dfrac{1}{\sqrt{x - 2}}$

9. a. $f(x) = 2x^2 + x + 1$; $g(x) = \dfrac{1}{x^3}$
  b. $f(x) = x^2 + x + 4$; $g(x) = \sqrt{x}$

10. $-\tfrac{3}{8}$ **11.** -3 **12.** 2   **13.** -21 **14.** 0   **15.** -1

16. The limit does not exist.   **17.** 7   **18.** $\tfrac{9}{2}$ **19.** 1   **20.** $\tfrac{1}{2}$

21. 1   **22.** 1   **23.** $\tfrac{3}{2}$ **24.** The limit does not exist.

25.

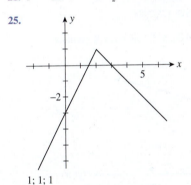

1; 1; 1

26.

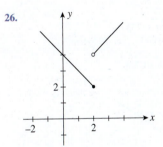

4; 2; the limit does not exist.

27. $x = 2$ **28.** $x = -\tfrac{1}{2}, 1$ **29.** $x = -1$ **30.** $x = 0$

31. a. 3; 2.5; 2.1   **b.** 2   **32.** 4   **33.** $\tfrac{3}{2}$; $y = \tfrac{3}{2}x + 5$

34. -4; $y = -4x + 4$ **35.** $\dfrac{1}{x^2}$

36. a. Yes   **b.** No   **37.** 54,000

38. a. $S(t) = t + 2.4$ **b.** \$5.4 million

39. a. $C(x) = 6x + 30,000$ **b.** $R(x) = 10x$
  c. $P(x) = 4x - 30,000$ **d.** (6000); $2000; $18,000

40. $\left(6, \dfrac{21}{2}\right)$ **41.** $P(x) = 8x - 20,000$ **42.** $6000; $22

43. 117 mg   **44.** $400,000   **45.** $45,000

46. 400; 800   **47.** 990; 2240

48.

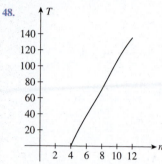

As the length of the list increases, the time taken to learn the list increases by a very large amount.

49. After $5\frac{1}{2}$ years   **50.** 5000; $20

51. 20.1 years; 13 years

52. a. $250 billion   **b.** $1.366 trillion

53. 648,000; 902,000; 1,345,200; 1,762,800

54. a. $16.4 billion; $17.6 billion; $18.3 billion; $18.8 billion; $19.3 billion

b.

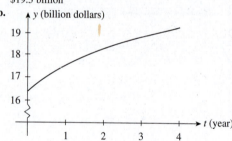

55. a. $f(t) = 267$; $g(t) = 2t^2 + 46t + 733$
  b. $f(t) + g(t) = 2t^2 + 46t + 1000$ **c.** 1936 tons

56. a. 59.8%; 58.9%; 59.2%; 60.7%, 61.7%
b.

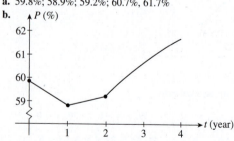

c. 60.7%

57. a. $f(r) = \pi r^2$ **b.** $g(t) = 2t$ **c.** $h(t) = 4\pi t^2$
  d. 3600π ft$^2 \approx 11,300$ ft^2

58. $x(20 - 2x)^2$ in.^{3} **59.** $100x^2 + \dfrac{1350}{x}$

60. $C(x) = \begin{cases} 5 & \text{if } 1 \le x \le 100 \\ 9 & \text{if } 100 < x \le 200 \\ 12.50 & \text{if } 200 < x \le 300 \\ 15.00 & \text{if } 300 < x \le 400 \\ 7 + 0.02x & \text{if } x > 400 \end{cases}$

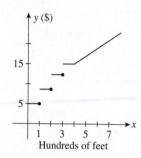

The function is discontinuous at $x = 100, 200,$ and 300.

61. 20

62. a. $C'(x)$ gives the instantaneous rate of change of the total manufacturing cost C in dollars with respect to the quantity produced when x units of the product are produced.
  b. Positive
  c. $20

63. True   **64.** False

Chapter 2 Before Moving On, page 160

1. a. 3   **b.** 2   **c.** $\frac{17}{4}$

2. a. $\dfrac{1}{x + 1} + x^2 + 1$ **b.** $\dfrac{x^2 + 1}{x + 1}$ **c.** $\dfrac{1}{x^2 + 2}$
  d. $\dfrac{1}{(x + 1)^2} + 1$

3. $108x^2 - 4x^3$ **4.** 2   **5. a.** 0   **b.** 1; no

6. $-1; y = -x$

CHAPTER 3

Exercises 3.1, page 169

1. 0   **3.** $5x^4$ **5.** $3.1x^{2.1}$ **7.** $6x$

9. $2\pi r$ **11.** $\dfrac{3}{x^{2/3}}$ **13.** $\dfrac{3}{2\sqrt{x}}$ **15.** $-84x^{-13}$

17. $10x - 3$ **19.** $-3x^2 + 4x$ **21.** $0.06x - 0.4$

23. $4x - 4 - \dfrac{3}{x^2}$ **25.** $16x^3 - \frac{15}{2}x^{3/2}$ **27.** $-\dfrac{5}{x^2} - \dfrac{8}{x^3}$

29. $-\dfrac{16}{t^5} + \dfrac{9}{t^4} - \dfrac{2}{t^2}$ **31.** $3 - \dfrac{5}{2\sqrt{x}}$ **33.** $-\dfrac{4}{x^3} + \dfrac{1}{x^{4/3}}$

35. a. 20   **b.** -4 **c.** 20   **37.** 3   **39.** 11

41. $m = 5; y = 5x - 4$ **43.** $m = 3; y = 3x - 7$

45. a. $(0, 0)$

b.

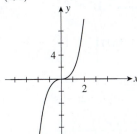

47. a. $(-2, -7), (2, 9)$

b. $y = 12x + 17$ and $y = 12x - 15$

c.

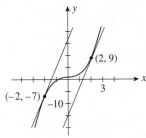

49. a. $(0, 0); \left(1, -\frac{13}{12}\right)$ **b.** $(0, 0); \left(2, -\frac{8}{3}\right); \left(-1, -\frac{5}{12}\right)$
c. $(0, 0); \left(4, \frac{80}{3}\right); \left(-3, \frac{81}{4}\right)$

51. a. $\dfrac{16\pi}{9}$ cm³/cm **b.** $\dfrac{25\pi}{4}$ cm³/cm

53. a. 491.5 million **b.** 476.8 million/year

55. a. 49.6%; 36.9%; 32.1% **b.** -3.3%/decade; -2.3%/decade

57. a. 44.4% **b.** 1%/decade

59. a. $120 - 30t$ **b.** 120 ft/sec **c.** 240 ft

61. a. \$18,115.41 **b.** \$1290.12/year

63. a. 15 points/year; 12.6 points/year; 0 points/year
b. 10 points/year

65. 63,000 people/year; 60,000 people/year; 55,000 people/year; 48,000 people/year; yes

67. a. $-6t^2 + 24t$ **b.** 0 ft/sec; 24 ft/sec; 0 ft/sec; -72 ft/sec
c. 69 ft

69. 155 people/month; 200 people/month

71. $-0.006x^2 + 1.2x + 1$; (a)

73. a. 12%; 31.2% **b.** 0.8%/year; 1.4%/year

75. a. $G(t) = \begin{cases} -0.0002t^2 + 0.032t + 0.1 & \text{if } 0 \le t < 5 \\ 0.0002t^2 - 0.006t + 0.28 & \text{if } 5 \le t < 10 \\ -0.0012t^2 + 0.082t - 0.46 & \text{if } 10 \le t < 15 \end{cases}$
b. 2800 jobs/year; 53,200 jobs/year

77. False

Using Technology Exercises 3.1, page 175

1. 1 **3.** 0.4226 **5.** 0.1613

7. a.

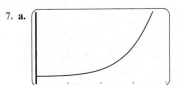

b. 3.4295 ppm/40 years; 164.239 ppm/40 years

Exercises 3.2, page 181

1. $2x(2x) + (x^2 + 1)(2)$, or $6x^2 + 2$

3. $(t - 1)(2) + (2t + 1)(1)$, or $4t - 1$

5. $(3x + 1)(2x) + (x^2 - 2)(3)$, or $9x^2 + 2x - 6$

7. $(x^3 - 1)(1) + (x + 1)(3x^2)$, or $4x^3 + 3x^2 - 1$

9. $(w^3 - w^2 + w - 1)(2w) + (w^2 + 2)(3w^2 - 2w + 1)$, or
$5w^4 - 4w^3 + 9w^2 - 6w + 2$

11. $(5x^2 + 1)(x^{-1/2}) + (2x^{1/2} - 1)(10x)$, or $\dfrac{25x^2 - 10x\sqrt{x} + 1}{\sqrt{x}}$

13. $\dfrac{(x^2 - 5x + 2)(x^2 + 2)}{x^2} + \dfrac{(x^2 - 2)(2x - 5)}{x}$, or $\dfrac{3x^4 - 10x^3 + 4}{x^2}$

15. $\dfrac{-1}{(x - 2)^2}$ **17.** $\dfrac{2(2x + 1) - (2x - 1)(2)}{(2x + 1)^2}$, or $\dfrac{4}{(2x + 1)^2}$

19. $-\dfrac{2x + 1}{(x^2 + x + 2)^2}$ **21.** $\dfrac{s^2 + 2s + 4}{(s + 1)^2}$

23. $\dfrac{\left(\frac{1}{2} x^{-1/2}\right)[x^2 + 1 - 4x^{3/2}\left(x^{1/2} + 1\right)]}{(x^2 + 1)^2}$, or $\dfrac{-3x^2 - 4x^{3/2} + 1}{2\sqrt{x}(x^2 + 1)^2}$

25. $\dfrac{2x^3 + 2x^2 + 2x - 2x^3 - x^2 - 4x - 2}{(x^2 + x + 1)^2}$, or $\dfrac{x^2 - 2x - 2}{(x^2 + x + 1)^2}$

27. $\dfrac{(x - 2)(3x^2 + 2x + 1) - (x^3 + x^2 + x + 1)}{(x - 2)^2}$, or
$\dfrac{2x^3 - 5x^2 - 4x - 3}{(x - 2)^2}$

29. $\dfrac{(x^2 - 4)(x^2 + 4)(2x + 8) - (x^2 + 8x - 4)(4x^3)}{(x^2 - 4)^2(x^2 + 4)^2}$, or
$\dfrac{-2x^5 - 24x^4 + 16x^3 - 32x - 128}{(x^2 - 4)^2(x^2 + 4)^2}$

31. 8 **33.** -9 **35.** $2(3x^2 - x + 3)$; 10

37. $\dfrac{-3x^4 + 2x^2 - 1}{(x^4 - 2x^2 - 1)^2}$; $-\dfrac{1}{2}$ **39.** 60; $y = 60x - 102$

41. $-\dfrac{1}{2}$; $y = -\dfrac{1}{2} x + \dfrac{3}{2}$ **43.** 8 **45.** $y = 7x - 5$

47. $\left(\frac{1}{3}, \frac{50}{27}\right); (1, 2)$ **49.** $\left(\frac{4}{3}, -\frac{770}{27}\right); (2, -30)$

51. $y = -\dfrac{1}{2} x + 1$; $y = 2x - \dfrac{3}{2}$

53. 0.125, 0.5, 2, 50; the cost of removing (essentially) all of the pollutant is prohibitively high.

55. -5000/min; -1600/min; 7000; 4000

57. a. $\dfrac{50x}{0.01x^2 + 1}$ **b.** $\dfrac{50(1 - 0.01x^2)}{(0.01x^2 + 1)^2}$

c. 6.69, 0, −3.70; the revenue is increasing at the rate of approximately $6700/thousand watches/week when the level of sales is 8000 watches/week; the rate of change of the revenue is $0/thousand watches/week when the level of sales is 10,000 watches/week, and the revenue is decreasing at the rate of approximately $3700/thousand watches/week when the sales are 12,000 watches/week.

59. a. 4.7%/year **b.** Dropping at the rate of 0.67%/year/year

61. a. $\dfrac{180}{(t + 6)^2}$ **b.** 3.7; 2.2; 1.8; 1.1

c.   Yes

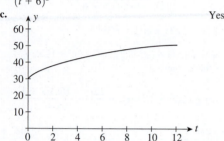

d. 50 words/min

63. Dropping at the rate of 0.0375 ppm/year; dropping at the rate of 0.006 ppm/year

65. a. $xD'(x) + D(x)$ **b.** $a - 2bx$

67. a. 0 **b.** c **c.** $\dfrac{kc}{(k + s)^2}$

69. False **71.** False

Using Technology Exercises 3.2, page 186

1. 0.8750 **3.** 0.0774

5. −0.5000 **7.** 31,312/year

Exercises 3.3, page 194

1. $6(2x - 1)^2$ **3.** $10x(x^2 + 2)^4$

5. $3(2x - x^2)^2(2 - 2x)$, or $6x^2(1 - x)(2 - x)^2$ **7.** $\dfrac{-4}{(2x + 1)^3}$

9. $5x(x^2 - 4)^{3/2}$ **11.** $\dfrac{3}{2\sqrt{3x - 2}}$ **13.** $\dfrac{-2x}{3(1 - x^2)^{2/3}}$

15. $-\dfrac{6}{(2x + 3)^4}$ **17.** $\dfrac{-1}{(2t - 4)^{3/2}}$ **19.** $\dfrac{3(16x^3 + 1)}{2(4x^4 + x)^{5/2}}$

21. $-2(3x^2 + 2x + 1)^{-3}(6x + 2)$ or $-4(3x + 1)(3x^2 + 2x + 1)^{-3}$

23. $3(x^2 + 1)^2(2x) - 2(x^3 + 1)(3x^2)$, or $6x(2x^2 - x + 1)$

25. $3(t^{-1} - t^{-2})^2(-t^{-2} + 2t^{-3})$

27. $\dfrac{1}{2\sqrt{x - 1}} + \dfrac{1}{2\sqrt{x + 1}}$

29. $2x^2(4)(3 - 4x)^3(-4) + (3 - 4x)^4(4x)$, or $(-12x)(4x - 1)(3 - 4x)^3$

31. $8(x - 1)^2(2x + 1)^3 + 2(x - 1)(2x + 1)^4$, or $6(x - 1)(2x - 1)(2x + 1)^3$

33. $3\left(\dfrac{x + 3}{x - 2}\right)^2\left[\dfrac{(x - 2)(1) - (x + 3)(1)}{(x - 2)^2}\right]$, or $-\dfrac{15(x + 3)^2}{(x - 2)^4}$

35. $\dfrac{3}{2}\left(\dfrac{t}{2t + 1}\right)^{1/2}\left[\dfrac{(2t + 1)(1) - t(2)}{(2t + 1)^2}\right]$, or $\dfrac{3t^{1/2}}{2(2t + 1)^{5/2}}$

37. $\dfrac{1}{2}\left(\dfrac{u + 1}{3u + 2}\right)^{-1/2}\left[\dfrac{(3u + 2)(1) - (u + 1)(3)}{(3u + 2)^2}\right]$, or $-\dfrac{1}{2\sqrt{u + 1}\,(3u + 2)^{3/2}}$

39. $\dfrac{(x^2 - 1)^4(2x) - x^2(4)(x^2 - 1)^3(2x)}{(x^2 - 1)^8}$, or $\dfrac{(-2x)(3x^2 + 1)}{(x^2 - 1)^5}$

41. $\dfrac{2x(x^2 - 1)^3(3x^2 + 1)^2[9(x^2 - 1) - 4(3x^2 + 1)]}{(x^2 - 1)^8}$, or $-\dfrac{2x(3x^2 + 13)(3x^2 + 1)^2}{(x^2 - 1)^5}$

43. $\dfrac{(2x + 1)^{-1/2}[(x^2 - 1) - (2x + 1)(2x)]}{(x^2 - 1)^2}$, or $-\dfrac{3x^2 + 2x + 1}{\sqrt{2x + 1}\,(x^2 - 1)^2}$

45. $\dfrac{(t^2 + 1)^{1/2}(\frac{1}{2})(t + 1)^{-1/2}(1) - (t + 1)^{1/2}(\frac{1}{2})(t^2 + 1)^{-1/2}(2t)}{t^2 + 1}$, or $-\dfrac{t^2 + 2t - 1}{2\sqrt{t + 1}\,(t^2 + 1)^{3/2}}$

47. $4(3x + 1)^3(3)(x^2 - x + 1)^3 + (3x + 1)^4(3)(x^2 - x + 1)^2(2x - 1)$, or $3(3x + 1)^3(x^2 - x + 1)^2(10x^2 - 5x + 3)$

49. $\frac{4}{3}u^{1/3}$; $6x$; $8x(3x^2 - 1)^{1/3}$

51. $-\dfrac{2}{3u^{5/3}}$; $6x^2 - 1$; $-\dfrac{2(6x^2 - 1)}{3(2x^3 - x + 1)^{5/3}}$

53. $\frac{1}{2}u^{-1/2} - \frac{1}{2}u^{-3/2}$; $3x^2 - 1$; $\dfrac{(3x^2 - 1)(x^3 - x - 1)}{2(x^3 - x)^{3/2}}$

55. $2f'(2x + 1)$ **57.** −12 **59.** 6 **61.** No

63. $y = -33x + 57$ **65.** $y = \frac{43}{5}x - \frac{54}{5}$

67. 0.333 million/week; 0.305 million/week; 16 million; 22.7 million

69. 102,000/year

71. a. 30%; 10.3%; the probability of survival 1 year after diagnosis is approximately 30%, and after 2 years is approximately 10.3%.
b. −33.8%/year; −10.4%/year; after 1 year, the probability of survival is dropping at the rate of approximately 34%/year, and after 2 years is dropping at the rate of approximately 10.4%/year.

73. a. $0.0267(0.2t^2 + 4t + 64)^{-1/3}(0.1t + 1)$ **b.** 0.0090 ppm/year

75. a. $0.03[3t^2(t - 7)^4 + t^3(4)(t - 7)^3]$, or $0.21t^2(t - 3)(t - 7)^3$
b. 90.72; 0; −90.72; at 8 A.M. the level of nitrogen dioxide is increasing; at 10 A.M. the level stops increasing; at 11 A.M. the level is decreasing.

77.

$$300\left[\frac{(t+25)\frac{1}{2}(\frac{1}{2}t^2+2t+25)^{-1/2}(t+2)-(\frac{1}{2}t^2+2t+25)^{1/2}(1)}{(t+25)^2}\right],$$

or $\dfrac{3450t}{(t+25)^2\sqrt{\frac{1}{2}t^2+2t+25}}$; 2.9 beats/min/sec, 0.7 beats/min/sec, 0.2 beats/min/sec, 179 beats/min

79. 160π ft^2/sec

81.

$$(1.42)\left[\frac{(3t^2+80t+550)(14t+140)-(7t^2+140t+700)(6t+80)}{(3t^2+80t+550)^2}\right],$$

or $\dfrac{1.42(140t^2+3500t+21{,}000)}{(3t^2+80t+550)^2}$; 31,312 jobs/year/month

83. $\dfrac{2500p}{9\sqrt{t}\sqrt{810{,}000-p^2}(1+\frac{1}{8}\sqrt{t})^2}$; 19 computers/month

85. True   **87.** True

Using Technology Exercises 3.3, page 198

1. 0.5774   **3.** 0.9390   **5.** −4.9498

7. 5,414,500 people/year; 2,513,600 people/year

Exercises 3.4, page 209

1. a. $C(x)$ is always increasing because as the number of units x produced increases, the amount of money that must be spent on production also increases.
 b. 4000

3. a. $1.80; $1.60   **b.** $1.80; $1.60

5. a. $100+\dfrac{200{,}000}{x}$ **b.** $-\dfrac{200{,}000}{x^2}$
 c. $\overline{C}(x)$ approaches $100 if the production level is very high.

7. $\dfrac{2000}{x}+2-0.0001x$; $-\dfrac{2000}{x^2}-0.0001$

9. a. $8000-200x$ **b.** 200, 0, −200   **c.** $40

11. a. $-0.04x^2+600x-300{,}000$
 b. $-0.08x+600$ **c.** 200; −40
 d. The profit increases as production increases, peaking at 7500 units; beyond this level, profit falls.

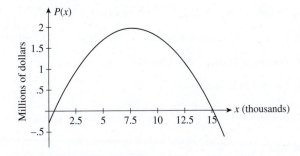

13. a. $600x-0.05x^2$; $-0.000002x^3-0.02x^2+200x-80{,}000$
 b. $0.000006x^2-0.06x+400$; $600-0.1x$;
  $-0.000006x^2-0.04x+200$
 c. 304; 400; 96

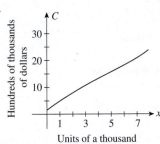

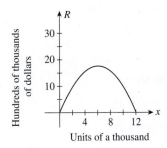

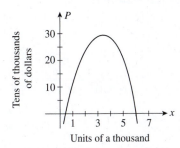

15. a. $0.000002x^2-0.03x+400+\dfrac{80{,}000}{x}$
 b. $0.000004x-0.03-\dfrac{80{,}000}{x^2}$
 c. −0.0132; 0.0092; the marginal average cost is negative (average cost is decreasing) when 5000 units are produced and positive (average cost is increasing) when 10,000 units are produced.
 d.

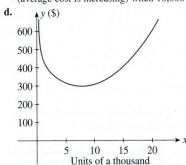

17. a. $\dfrac{50x}{0.01x^2+1}$ **b.** $\dfrac{50-0.5x^2}{(0.01x^2+1)^2}$
 c. $44,380; when the level of production is 2000 units, the revenue increases at the rate of $44,380 per additional 1000 units produced.

19. 1.21   **21.** 0.288

23. 81.82%/unit change in x **25.** −84.85%/unit change in x

27. $\frac{5}{3}$; elastic   **29.** 1; unitary   **31.** 0.104; inelastic

33. a. $100\dfrac{P(t)I'(t) - I(t)P'(t)}{P(t)I(t)}$

b. $\dfrac{50 - 2t}{25 + t}$ **c.** 1.7%/year

35. a. Inelastic; elastic   **b.** When $p = 8.66$
c. Increase   **d.** Increase

37. a. Inelastic   **b.** Increase

39. $\dfrac{2p^2}{9 - p^2}$; for $p < \sqrt{3}$, demand is inelastic; for $p = \sqrt{3}$, demand is unitary; and for $p > \sqrt{3}$, demand is elastic.

41. True

Exercises 3.5, page 218

1. $8x - 2$; 8   **3.** $6x^2 - 6x$; $6(2x - 1)$

5. $4t^3 - 6t^2 + 12t - 3$; $12(t^2 - t + 1)$

7. $10x(x^2 + 2)^4$; $10(x^2 + 2)^3(9x^2 + 2)$

9. $6t(2t^2 - 1)(6t^2 - 1)$; $6(60t^4 - 24t^2 + 1)$

11. $14x(2x^2 + 2)^{5/2}$; $28(2x^2 + 2)^{3/2}(6x^2 + 1)$

13. $(x^2 + 1)(5x^2 + 1)$; $4x(5x^2 + 3)$

15. $\dfrac{1}{(2x + 1)^2}$; $-\dfrac{4}{(2x + 1)^3}$ **17.** $\dfrac{2}{(s + 1)^2}$; $-\dfrac{4}{(s + 1)^3}$

19. $-\dfrac{3}{2(4 - 3u)^{1/2}}$; $-\dfrac{9}{4(4 - 3u)^{3/2}}$ **21.** $72x - 24$ **23.** $-\dfrac{6}{x^4}$

25. $\frac{81}{8}(3s - 2)^{-5/2}$ **27.** $192(2x - 3)$ **29.** 128 ft/sec; 32 ft/sec²

31. a. 19%   **b.** 5.1%/year
c. 0.76%/year/year; the percentage of vehicles equipped with transmissions that have seven speeds or more is increasing at the rate of 0.76%/year/year.

33. a. and b.

t	0	1	2	3	4	5	6	7	8
$N'(t)$	0	2.7	4.8	6.3	7.2	7.5	7.2	6.3	4.8
$N''(t)$				0.6	0	-0.6	-1.2	-1.8	

35. 2736; 756 records/month; 148 records/month/month

37. The proportion of the U.S. population that was obese was increasing at an increasing rate from 1991 through 2004.

39. -3.09; 0.35; 10 min after the start of the test, the smoke remaining is decreasing at a rate of 3.09%/min, but the rate at which the rate of smoke is decreasing is decreasing at the rate of 0.35%/min/min.

41. -0.01%/year²; the rate of change of the percentage of Americans aged 55 and over decreases at the rate of 0.01%/year².

43. True   **45.** True   **49.** $f(x) = x^{n+1/2}$

Using Technology Exercises 3.5, page 221

1. -18 **3.** 15.2762   **5.** -0.6255 **7.** 0.1973

Exercises 3.6, page 231

1. a. $-\frac{1}{2}$ **b.** $-\frac{1}{2}$ **3. a.** $-\dfrac{1}{x^2}$ **b.** $-\dfrac{y}{x}$

5. a. $2x - 1 + \dfrac{4}{x^2}$ **b.** $3x - 2 - \dfrac{y}{x}$

7. a. $\dfrac{1 - x^2}{(1 + x^2)^2}$ **b.** $-2y^2 + \dfrac{y}{x}$ **9.** $-\dfrac{x}{y}$ **11.** $\dfrac{x}{2y}$

13. $1 - \dfrac{y}{x}$ **15.** $-\dfrac{y}{x}$ **17.** $-\dfrac{\sqrt{y}}{\sqrt{x}}$ **19.** $2\sqrt{x + y} - 1$

21. $-\dfrac{y^3}{x^3}$ **23.** $\dfrac{2\sqrt{xy} - y}{x - 2\sqrt{xy}}$ **25.** $\dfrac{6x - 3y - 1}{3x + 1}$

27. $\dfrac{2(2x - y^{3/2})}{3x\sqrt{y} - 4y}$ **29.** $-\dfrac{2x^2 + 2xy + y^2}{x^2 + 2xy + 2y^2}$ **31.** $y = 2$

33. $y = -\frac{3}{2}x + \frac{5}{2}$ **35.** $\dfrac{2y}{x^2}$ **37.** $\dfrac{2y(y - x)}{(2y - x)^3}$

39. a. $\dfrac{dV}{dt} = \pi r\left(r\dfrac{dh}{dt} + 2h\dfrac{dr}{dt}\right)$ **b.** 3.6π cu in./sec

41. Dropping at the rate of 111 tires/week

43. Increasing at the rate of 44 headphones/week

45. Dropping at the rate of 3.7¢/carton/week   **47.** 0.37; inelastic

49. a. $1440 billion
b. decrease by approximately $11.4 billion; $11.4 billion/$1 billion

51. 60π ft²/sec   **53.** 3.14 ft/sec   **55.** 13.4 ft/sec

57. 7.69 ft/sec   **61.** 196.8 ft/sec   **63.** 9 ft/sec

65. 19.2 ft/sec   **67.** $\frac{1}{7}$ L/sec   **69.** False   **71.** True   **73.** True

Exercises 3.7, page 240

1. $4x\, dx$ **3.** $(3x^2 - 1)\, dx$ **5.** $\dfrac{dx}{2\sqrt{x + 1}}$ **7.** $\dfrac{6x + 1}{2\sqrt{x}}\, dx$

9. $\dfrac{x^2 - 2}{x^2}\, dx$ **11.** $\dfrac{-x^2 + 2x + 1}{(x^2 + 1)^2}\, dx$ **13.** $\dfrac{6x - 1}{2\sqrt{3x^2 - x}}\, dx$

15. a. $2x\, dx$ **b.** 0.04   **c.** 0.0404

17. a. $-\dfrac{dx}{x^2}$ **b.** -0.05 **c.** -0.05263

19. 3.167   **21.** 7.0357   **23.** 1.983   **25.** 0.298

27. 2.50375   **29.** ± 8.64 cm³   **31.** 18.85 ft³

33. It will drop by approximately 40%.   **35.** 274 sec   **37.** 283,937

39. Decrease of $1.33   **41.** $\pm$\$64,800   **43.** 167 tires/week

45. Decrease of 11 crimes/year

47. a. $100,000\left(1 + \dfrac{r}{12}\right)^{119} dr$ **b.** $134.60; $269.20; $403.80

49. True

Using Technology Exercises 3.7, page 244

1. 7.5787   **3.** 0.03122   **5.** -0.01988

7. $44/month; $58.67/month; $73.34/month   **9.** 625

Chapter 3 Concept Review, page 246

1. a. 0   **b.** nx^{n-1} **c.** $cf'(x)$ **d.** $f'(x) \pm g'(x)$

2. a. $f(x)g'(x) + g(x)f'(x)$ **b.** $\dfrac{g(x)f'(x) - f(x)g'(x)}{[g(x)]^2}$

3. a. $g'[f(x)]f'(x)$ **b.** $n[f(x)]^{n-1}f'(x)$

4. marginal cost; marginal revenue; marginal profit; marginal average cost

5. a. $-p\dfrac{f'(p)}{f(p)}$ **b.** elastic; unitary; inelastic

6. both sides; $\dfrac{dy}{dx}$ **7.** y; $\dfrac{dy}{dt}$; a **8.** $\dfrac{-f(t)f'(t)}{g(t)}$; $\dfrac{-f(t)g'(t)}{g(t)}$

9. a. $x_2 - x_1$ **b.** $f(x + \Delta x) - f(x)$ **10.** Δx; Δx; x; $f'(x)\,dx$

Chapter 3 Review Exercises, page 247

1. $15x^4 - 8x^3 + 6x - 2$ **2.** $24x^5 + 8x^3 + 6x$

3. $\dfrac{6}{x^4} - \dfrac{3}{x^2}$ **4.** $4t - 9t^2 + \frac{1}{2}t^{-3/2}$ **5.** $-\dfrac{1}{t^{3/2}} - \dfrac{6}{t^{5/2}}$

6. $2x - \dfrac{2}{x^2}$ **7.** $1 - \dfrac{2}{t^2} - \dfrac{6}{t^3}$ **8.** $4s + \dfrac{4}{s^2} - \dfrac{1}{s^{3/2}}$

9. $2x + \dfrac{3}{x^{5/2}}$ **10.** $\dfrac{(2x-1)(1) - (x+1)(2)}{(2x-1)^2}$, or $-\dfrac{3}{(2x-1)^2}$

11. $\dfrac{(2t^2+1)(2t) - t^2(4t)}{(2t^2+1)^2}$, or $\dfrac{2t}{(2t^2+1)^2}$

12. $\dfrac{(t^{1/2}+1)\frac{1}{2}t^{-1/2} - t^{1/2}(\frac{1}{2}t^{-1/2})}{(t^{1/2}+1)^2}$, or $\dfrac{1}{2\sqrt{t}\,(\sqrt{t}+1)^2}$

13. $\dfrac{(x^{1/2}+1)(\frac{1}{2}x^{-1/2}) - (x^{1/2}-1)(\frac{1}{2}x^{-1/2})}{(x^{1/2}+1)^2}$ or $\dfrac{1}{\sqrt{x}(\sqrt{x}+1)^2}$

14. $\dfrac{(2t^2+1)(1) - t(4t)}{(2t^2+1)^2}$, or $\dfrac{1-2t^2}{(2t^2+1)^2}$

15. $\dfrac{(x^2-1)(4x^3+2x) - (x^4+x^2)(2x)}{(x^2-1)^2}$, or $\dfrac{2x(x^4-2x^2-1)}{(x^2-1)^2}$

16. $3(4x+1)(2x^2+x)^2$ **17.** $8(3x^3-2)^7(9x^2)$, or $72x^2(3x^3-2)^7$

18. $5(x^{1/2}+2)^4 \cdot \frac{1}{2}x^{-1/2}$, or $\dfrac{5(\sqrt{x}+2)^4}{2\sqrt{x}}$

19. $\frac{1}{2}(2t^2+1)^{-1/2}(4t)$, or $\dfrac{2t}{\sqrt{2t^2+1}}$

20. $\frac{1}{3}(1-2t^3)^{-2/3}(-6t^2)$, or $-2t^2(1-2t^3)^{-2/3}$

21. $-4(3t^2-2t+5)^{-3}(3t-1)$, or $-\dfrac{4(3t-1)}{(3t^2-2t+5)^3}$

22. $-\frac{3}{2}(2x^3-3x^2+1)^{-5/2}(6x^2-6x)$, or $-9x(x-1)(2x^3-3x^2+1)^{-5/2}$

23. $2\left(x+\dfrac{1}{x}\right)\left(1-\dfrac{1}{x^2}\right)$, or $\dfrac{2(x^2+1)(x^2-1)}{x^3}$

24. $\dfrac{(2x^2+1)^2(1) - (1+x)2(2x^2+1)(4x)}{(2x^2+1)^4}$, or $-\dfrac{6x^2+8x-1}{(2x^2+1)^3}$

25. $(t^2+t)^4(4t) + 2t^2 \cdot 4(t^2+t)^3(2t+1)$, or $4t^2(5t+3)(t^2+t)^3$

26. $(2x+1)^3 \cdot 2(2x^2+x)(2x+1) + (x^2+x)^2\,3(2x+1)^2(2)$, or $2(2x+1)^2(x^2+x)(7x^2+7x+1)$

27. $x^{1/2} \cdot 3(x^2-1)^2(2x) + (x^2-1)^3 \cdot \frac{1}{2}x^{-1/2}$, or $\dfrac{(13x^2-1)(x^2-1)^2}{2\sqrt{x}}$

28. $\dfrac{(x^3+2)^{1/2}(1) - x \cdot \frac{1}{2}(x^3+2)^{-1/2} \cdot 3x^2}{x^3+2}$, or $\dfrac{4-x^3}{2(x^3+2)^{3/2}}$

29. $\dfrac{(4x-3)\frac{1}{2}(3x+2)^{-1/2}(3) - (3x+2)^{1/2}(4)}{(4x-3)^2}$, or $-\dfrac{12x+25}{2\sqrt{3x+2}(4x-3)^2}$

30. $\dfrac{(t+1)^3\frac{1}{2}(2t+1)^{-1/2}(2) - (2t+1)^{1/2} \cdot 3(t+1)^2(1)}{(t+1)^6}$, or $-\dfrac{5t+2}{\sqrt{2t+1}(t+1)^4}$

31. $2(12x^2 - 9x + 2)$ **32.** $-\dfrac{1}{4x^{3/2}} + \dfrac{3}{4x^{5/2}}$

33. $\dfrac{(t^2+4)^2(-2t) - (4-t^2)2(t^2+4)(2t)}{(t^2+4)^4}$, or $\dfrac{2t(t^2-12)}{(t^2+4)^3}$

34. $2(15x^4 + 12x^2 + 6x + 1)$

35. $2(2x^2+1)^{-1/2} + 2x\left(-\dfrac{1}{2}\right)(2x^2+1)^{-3/2}(4x)$, or $\dfrac{2}{(2x^2+1)^{3/2}}$

36. $(t^2+1)^2(14t) + (7t^2+1)(2)(t^2+1)(2t)$, or $6t(t^2+1)(7t^2+3)$

37. $\dfrac{2x}{y}$ **38.** $\dfrac{2x^2-y}{x}$ **39.** $-\dfrac{2x}{y^2-1}$ **40.** $-\dfrac{x(1+2y^2)}{y(2x^2+1)}$

41. $\dfrac{x-2y}{2x+y}$ **42.** $\dfrac{4y-6xy-1}{3x^2-4x-2}$ **43.** $\dfrac{2(x^4-1)}{x^3}\,dx$

44. $-\dfrac{3x^2}{2(x^3+1)^{3/2}}\,dx$

45. a. $\dfrac{2x}{\sqrt{2(x^2+2)}}\,dx$ **b.** 0.1333   **c.** 0.1335; differ by 0.0002

46. 2.9926

47. a. $(2, -25)$ and $(-1, 14)$ **b.** $y = -4x - 17$; $y = -4x + 10$

48. a. $\left(-2, \frac{25}{3}\right)$ and $\left(1, -\frac{13}{6}\right)$ **b.** $y = -2x + \frac{13}{3}$; $y = -2x - \frac{1}{6}$

49. $y = -\dfrac{\sqrt{3}}{3}x + \dfrac{4}{3}\sqrt{3}$ **50.** $y = 112x - 80$

51. $-\dfrac{48}{(2x-1)^4}$; $\left(-\infty; \frac{1}{2}\right) \cup \left(\frac{1}{2}, \infty\right)$

52. a. \$3.1 billion; \$10 billion   **b.** \$0.68 billion/year; \$2.08 billion/year

53. a. 75.3 million   **b.** 1.3 million viewers/year

54. a. 15%; 32%   **b.** 0.51%/year; 1.04%/year

55. a. 20,430   **b.** 225 cameras/year

56. a. \$33.2 billion   **b.** \$7.7 billion/year

57. a. 4445   **b.** 494 people/year

58. a. 1,024,000   **b.** 133,000 copies/week

59. 200 subscribers/week

60. $-14.346(1 + t)^{-1.45}$; $-2.92¢$/min/year; $19.45¢$/min

61. ≈ 75 years; 0.07 year/year

62. a. \$2.20; \$2.20   **b.** $\dfrac{2500}{x} + 2.2$; $-\dfrac{2500}{x^2}$

   c. $\lim\limits_{x \to \infty} \left(\dfrac{2500}{x} + 2.2 \right) = 2.2$

63. \$0.9487/1000 radios

64. A decrease of \$15/1000 units

65. a. \$39.07   **b.** \$39

66. a. $-0.02x^2 + 600x$ **b.** $-0.04x + 600$
   c. 200; the sale of the 10,001st phone will bring a revenue of \$200.

67. a. $2000x - 0.04x^2$; $-0.000002x^3 - 0.02x^2 + 1000x - 120,000$;
   $0.000002x^2 - 0.02x + 1000 + \dfrac{120,000}{x}$
   b. $0.000006x^2 - 0.04x + 1000$; $2000 - 0.08x$;
   $-0.000006x^2 - 0.04x + 1000$; $0.000004x - 0.02 - \dfrac{120,000}{x^2}$
   c. 934; 1760; 826
   d. -0.0048; 0.010125; at a production level of 5000, the average
   cost is decreasing by 0.48¢/unit; at a production level of 8000, the
   average cost is increasing by 1.0125¢/unit.

68. a. $80 + \dfrac{150,000}{x}$ **b.** $-\dfrac{150,000}{x^2}$
   c. If the production level is very high, then the unit cost approaches
   \$80/desk.

69. a. $\frac{1}{3}$; inelastic   **b.** 1; unitary   **c.** 3; elastic

70. $\dfrac{25}{2(25 - \sqrt{p})}$; for $p > 156.25$, demand is elastic; for $p = 156.25$,
   demand is unitary; and for $p < 156.25$, demand is inelastic.

71. a. Inelastic   **b.** Increase

72. a. Elastic   **b.** Decrease

73. 1.2; -1.2; the GDP is increasing at the rate of \$1.2 billion/year;
   the rate of change of the GDP is decreasing at a rate of
   \$1.2 billion/year/year.

74. $\frac{17}{3}$ (ft/sec); $\frac{76}{27}$ (ft/sec²)

Chapter 3 Before Moving On, page 250

1. $6x^2 - \dfrac{1}{x^{2/3}} - \dfrac{10}{3x^{5/3}}$ **2.** $\dfrac{4x^2 - 1}{\sqrt{2x^2 - 1}}$ **3.** $-\dfrac{2x^2 + 2x - 1}{(x^2 + x + 1)^2}$

4. $-\dfrac{1}{2(x+1)^{3/2}}$; $\dfrac{3}{4(x+1)^{5/2}}$; $-\dfrac{15}{8(x+1)^{7/2}}$ **5.** $\dfrac{-y^2 + 2xy - 3x^2}{2xy - x^2}$

6. a. $\dfrac{2x^2 + 5}{\sqrt{x^2 + 5}}\, dx$ **b.** 0.0433

CHAPTER 4

Exercises 4.1, page 264

1. Decreasing on $(-\infty, 0)$ and increasing on $(0, \infty)$

3. Increasing on $(-\infty, -1)$ and $(1, \infty)$ and decreasing on $(-1, 1)$

5. Decreasing on $(-\infty, 0)$ and $(2, \infty)$ and increasing on $(0, 2)$

7. Decreasing on $(-\infty, -1)$ and $(1, \infty)$ and increasing on $(-1, 1)$

9. Increasing on $(20.2, 20.6)$ and $(21.7, 21.8)$, constant on $(19.6, 20.2)$
   and $(20.6, 21.1)$, and decreasing on $(21.1, 21.7)$ and $(21.8, 22.7)$

11. a. f is decreasing on $(0, 4)$ **b.** f is constant on $(4, 12)$
   c. f is increasing on $(12, 24)$

13. a. 3, 5, and 7
   b.

$$+ \; + \; + \; 0 \; - \; 0 \; + \; 0 \; + \;\Big\downarrow\; - \; - \; - \; \text{sign of } f'$$
$$\underset{0 \qquad\quad 3 \qquad 5 \qquad 7 \quad 9}{\xrightarrow{\hspace{4cm}} x}$$

   f' not defined
   c. Relative maxima at $(3, 3)$ and $(9, 6)$; relative minimum at $(5, 1)$

15. Increasing on $(-\infty, \infty)$

17. Decreasing on $\left(-\infty, -\frac{1}{4}\right)$ and increasing on $\left(-\frac{1}{4}, \infty\right)$

19. Decreasing on $\left(-\infty, -\dfrac{\sqrt{3}}{3}\right)$ and $\left(\dfrac{\sqrt{3}}{3}, \infty\right)$ and increasing on
   $\left(-\dfrac{\sqrt{3}}{3}, \dfrac{\sqrt{3}}{3}\right)$

21. Increasing on $(-\infty, -2)$ and $(0, \infty)$ and decreasing on $(-2, 0)$

23. Increasing on $(-\infty, -1)$ and $(3, \infty)$ and decreasing on $(-1, 3)$

25. Decreasing on $(-\infty, 3)$ and increasing on $(3, \infty)$

27. Decreasing on $\left(-\infty, -\frac{3}{2}\right)$ and $\left(-\frac{3}{2}, \infty\right)$

29. Increasing on $(-1, 1)$ and decreasing on $(-\infty, -1)$ and $(1, \infty)$

31. Decreasing on $(-\infty, 0)$ and increasing on $(0, \infty)$

33. Decreasing on $(-\infty, 5)$ and increasing on $(5, \infty)$

35. Decreasing on $\left(-1, -\frac{2}{3}\right)$ and increasing on $\left(-\frac{2}{3}, \infty\right)$

37. Increasing on $(-\infty, 0)$ and $(2, \infty)$; decreasing on $(0, 1)$ and $(1, 2)$

39. Relative maximum: $f(0) = 1$; relative minima: $f(-1) = 0$ and
   $f(1) = 0$

41. Relative maximum: $f(-1) = 2$; relative minimum: $f(1) = -2$

43. Relative maximum: $f(1) = 3$; relative minimum: $f(2) = 2$

45. Relative maximum: $f(-3) = -\frac{9}{2}$; relative minimum: $f(3) = \frac{9}{2}$

47. Decreasing on (a, c), where $f'(x) < 0$; increasing on (c, d), where
   $f'(x) > 0$; constant on (d, e), where $f'(x) = 0$; decreasing on (e, b),
   where $f'(x) < 0$

49. Increasing on $(0, c)$; neither increasing nor decreasing at c; decreasing on (c, b)

51. (a)   **53.** (d)   **55.** Relative minimum: $g\left(-\frac{3}{2}\right) = \frac{23}{4}$

57. Relative maximum: $h(3) = 15$

59. Relative minimum: $f(0) = 2$

61. Relative maximum: $f(-1) = 8$; relative minimum: $f(1) = 4$

63. Relative maximum: $f(0) = 0$; relative minima: $f(-1) = -\frac{1}{2}$ and $f(1) = -\frac{1}{2}$

65. Relative minimum: $g(3) = -15$

67. Relative minimum: $F(-2) = 84$; relative maximum: $F(2) = -44$

69. None   **71.** Relative minimum: $g(10) = 610$

73. None   **75.** None

79. a. f is increasing on $(0, 10)$.
 b. Managed services grew from 1999 through 2009.

81. Rising on $(0, 33)$ and descending on $(33, T)$ for some positive number T.

83. f is decreasing on $(0, 1)$ and increasing on $(1, 4)$. The average speed decreases from 6 A.M. to 7 A.M. and then picks up from 7 A.M. to 10 A.M.

85. The U.S. public debt outstanding was increasing throughout the period under consideration.

87. a. Increasing on $(0, 10)$ **b.** Sales will be increasing.

91. A is increasing on $(0, 4)$ and decreasing on $(4, 5)$. The cash in the trust funds will be increasing from 2005 to 2045 and decreasing from 2045 to 2055.

93. a. 30%; 41.4%
 b. The percentage of small and lower-midsize vehicles was increasing from 2005 through 2015.

95. a. $R(x) = ax - bx^2$
 b. Increasing on $\left(0, \dfrac{a}{2b}\right)$; stationary at $x = \dfrac{a}{2b}$; and decreasing on $\left(\dfrac{a}{2b}, \dfrac{a}{b}\right)$
 c. $\dfrac{a}{2b}$; $\dfrac{a^2}{4b}$

97. False   **99.** True   **101.** False   **103.** True

107. $a = -4$; $b = 24$

109. a. $2x$ if $x \le 0$; $-\dfrac{2}{x^3}$ if $x > 0$ **b.** No

Using Technology Exercises 4.1, page 271

1. a. f is decreasing on $(-\infty, -0.2934)$ and increasing on $(-0.2934, \infty)$.
 b. Relative minimum: $f(-0.2934) = -2.5435$

3. a. f is increasing on $(-\infty, -1.6144)$ and $(0.2390, \infty)$ and decreasing on $(-1.6144, 0.2390)$.
 b. Relative maximum: $f(-1.6144) = 26.7991$; relative minimum: $f(0.2390) = 1.6733$

5. a. f is decreasing on $(-\infty, -1)$ and $(0.33, \infty)$ and increasing on $(-1, 0.33)$.
 b. Relative maximum: $f(0.33) = 1.11$; relative minimum: $f(-1) = -0.63$

7. a. f is decreasing on $(-1, -0.71)$ and increasing on $(-0.71, 1)$.
 b. Relative minimum: $f(-0.71) = -1.41$

9. a.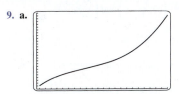

11. Increasing on $(0, 4.5)$ and decreasing on $(4.5, 11)$; 11:30 A.M.; 164 PSI

Exercises 4.2, page 282

1. Concave downward on $(-\infty, 0)$ and concave upward on $(0, \infty)$; inflection point: $(0, 0)$

3. Concave downward on $(-\infty, 0)$ and $(0, \infty)$

5. Concave upward on $(-\infty, 0)$ and $(1, \infty)$ and concave downward on $(0, 1)$; inflection points: $(0, 0)$ and $(1, -1)$

7. Concave downward on $(-\infty, -2)$, $(-2, 2)$, and $(2, \infty)$

9. a. Concave upward on $(0, 2)$, $(4, 6)$, $(7, 9)$, and $(9, 12)$ and concave downward on $(2, 4)$ and $(6, 7)$
 b. $\left(2, \frac{5}{2}\right)$, $(4, 2)$, $(6, 2)$, and $(7, 3)$

11. (a)   **13.** (b)

15. a. $D_1'(t) > 0$, $D_2'(t) > 0$, $D_1''(t) > 0$, and $D_2''(t) < 0$ on $(0, 12)$
 b. With or without the proposed promotional campaign, the deposits will increase; with the promotion, the deposits will increase at an increasing rate; without the promotion, the deposits will increase at a decreasing rate.

17. (c)   **19.** (d)

21. a. Between 8 A.M. and 10 A.M., the rate of change of the rate of change of the number of smartphones assembled is increasing; between 10 A.M. and 12 noon, it is decreasing.
 b. At 10 A.M.

23. At the time t_0, corresponding to its t-coordinate, the restoration process is working at its peak.

29. Concave upward on $(-\infty, \infty)$

31. Concave upward on $(-\infty, 0)$; concave downward on $(0, \infty)$

33. Concave upward on $(-\infty, 0)$ and $(3, \infty)$; concave downward on $(0, 3)$

35. Concave downward on $(-\infty, 0)$ and $(0, \infty)$

37. Concave downward on $(-\infty, 4)$

39. Concave downward on $(-\infty, 2)$; concave upward on $(2, \infty)$

41. Concave upward on $\left(-\infty, -\dfrac{\sqrt{6}}{3}\right)$ and $\left(\dfrac{\sqrt{6}}{3}, \infty\right)$; concave downward on $\left(-\dfrac{\sqrt{6}}{3}, \dfrac{\sqrt{6}}{3}\right)$

43. Concave downward on $(-\infty, 1)$; concave upward on $(1, \infty)$

45. Concave upward on $(-\infty, 0)$ and $(0, \infty)$

47. Concave upward on $(-\infty, 2)$; concave downward on $(2, \infty)$

49. $(0, -2)$ **51.** $(1, -20)$ **53.** $(0, 1)$ and $\left(\frac{2}{3}, \frac{11}{27}\right)$

55. $(0, 0)$ **57.** $(1, 2)$ **59.** $\left(-\frac{\sqrt{3}}{3}, \frac{3}{2}\right)$ and $\left(\frac{\sqrt{3}}{3}, \frac{3}{2}\right)$

61. Relative maximum: $f(1) = 5$ **63.** None

65. Relative maximum: $f(-1) = -\frac{7}{3}$, relative minimum: $f(5) = -\frac{115}{3}$

67. Relative maximum: $g(-3) = -6$; relative minimum: $g(3) = 6$

69. None   **71.** Relative minimum: $f(-2) = 12$

73. Relative maximum: $g(1) = \frac{1}{2}$; relative minimum: $g(-1) = -\frac{1}{2}$

75. Relative maximum: $f(0) = 0$; relative minimum: $f\left(\frac{4}{3}\right) = \frac{256}{27}$

77.

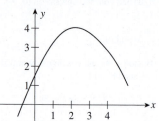

79.

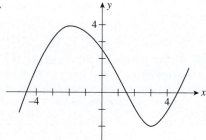

81.

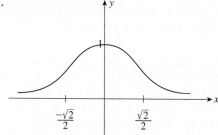

83. a. N is increasing on $(0, 12)$.
   b. $N''(t) < 0$ on $(0, 6)$ and $N''(t) > 0$ on $(6, 12)$
   c. The rate of growth of the number of help-wanted advertisements was decreasing over the first 6 months of the year and increasing over the last 6 months.

85. $f(t)$ increases at an increasing rate until the water level reaches the middle of the vase at which time (corresponding to the inflection point) $f(t)$ is increasing at the fastest rate. The water rises faster when the vase is narrower. After that, $f(t)$ increases at a decreasing rate until the vase is filled.

87. b. The rate of increase of the average state cigarette tax was decreasing from 2001 to 2007.

89. a. The average amount of atmospheric carbon dioxide was increasing from 1958 through 2013.
   b. The rate of increase of the amount of atmospheric carbon dioxide was increasing from 1958 through 2013.

91. a. Concave downward on $(0, 7.27)$ and concave upward on $(7.27, 21)$
   b. $(7.27, 5.16)$; the U.S. public debt was increasing at a decreasing rate from 1990 through the end of the first quarter in 1997 and then continued to increase but at an accelerated pace from that point on.

93. a. 20.1; 25.1; 25.63   **b.** Early 1984

101. The level of ozone peaked at approximately 4 P.M. on that day.

103. b. 44   **105.** False   **107.** False

Exercises 4.3, page 298

1. Horizontal asymptote: $y = 0$

3. Horizontal asymptote: $y = 0$; vertical asymptote: $x = 0$

5. Horizontal asymptote: $y = 0$; vertical asymptotes: $x = -1$ and $x = 1$

7. Horizontal asymptote: $y = 3$; vertical asymptote: $x = 0$

9. Horizontal asymptotes: $y = 1$ and $y = -1$

11. Horizontal asymptote: $y = 0$; vertical asymptote: $x = 0$

13. Horizontal asymptote: $y = 0$; vertical asymptote: $x = 0$

15. Horizontal asymptote: $y = 1$; vertical asymptote: $x = -2$

17. None

19. Horizontal asymptote: $y = 1$; vertical asymptotes: $t = -4$ and $t = 4$

21. Horizontal asymptote: $y = 0$; vertical asymptotes: $x = -2$ and $x = 3$

23. Horizontal asymptote: $y = 2$; vertical asymptote: $t = 2$

25. Horizontal asymptote: $y = 1$; vertical asymptotes: $x = -2$ and $x = 2$

27. None   **29.** f is the derivative function of the function g.

31.

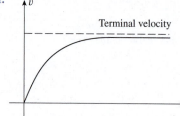

33.

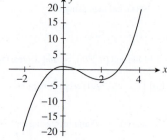

35.

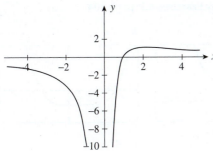

37.

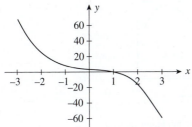

39.

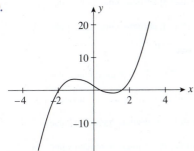

41.

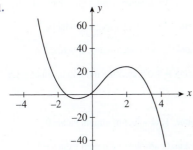

43.

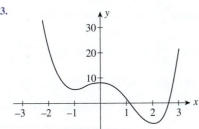

45.

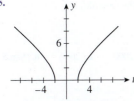

47.

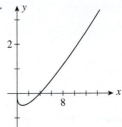

49.

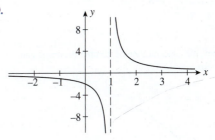

51.

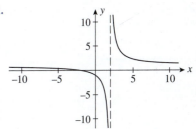

53.

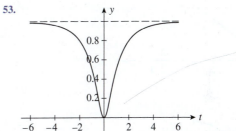

55.

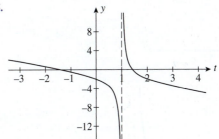

57.

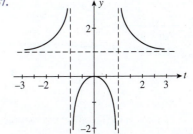

59.

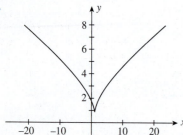

61. a. $x = 100$ **b.** No

63. a. $y = 0$
b. As time passes, the concentration of the drug decreases and approaches zero.

65.

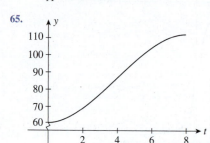

67.

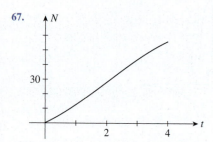

69.

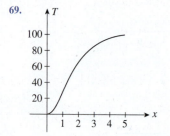

71.

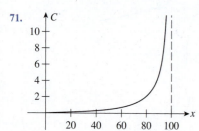

73. False

Using Technology Exercises 4.3, page 304

1.

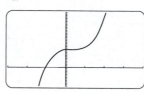

3.

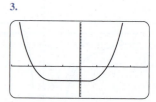

5. -0.9733; 2.3165, 4.6569 **7.** 1.5142

9.

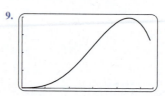

Exercises 4.4, page 313

1. None **3.** Absolute minimum value: 0

5. Absolute maximum value: 3; absolute minimum value: -2

7. Absolute maximum value: 3; absolute minimum value: $-\frac{27}{16}$

9. Absolute minimum value: $-\frac{41}{8}$

11. No absolute extrema **13.** Absolute maximum value: 1

15. Absolute maximum value: 5; absolute minimum value: -4

17. Absolute maximum value: 10; absolute minimum value: 1

19. Absolute maximum value: 19; absolute minimum value: -1

21. Absolute maximum value: 16; absolute minimum value: -1

23. Absolute maximum value: 3; absolute minimum value: $\frac{5}{3}$

25. Absolute maximum value: $\frac{65}{4}$; absolute minimum value: 5

27. Absolute maximum value ≈ 1.04; absolute minimum value: -1.5

29. No absolute extrema

31. Absolute maximum value: 1; absolute minimum value: 0

33. Absolute maximum value: 0; absolute minimum value: -3

35. Absolute maximum value: $\dfrac{\sqrt{2}}{4} \approx 0.35$; absolute minimum value: $-\frac{1}{3}$

37. Absolute maximum value: $\dfrac{\sqrt{2}}{2}$; absolute minimum value: $-\dfrac{\sqrt{2}}{2}$

39. 144 ft **41. a.** 18.1%; 2009 **b.** 20.81%; 2013

43. 7 A.M.; 30 mph **45.** 268.7 ft **47.** 5000 copies **49.** 3333

51. 110

53. a. $0.000002x^2 + 5 + \dfrac{400}{x}$ **b.** 464 cases/day
c. 464 cases/day **d.** same

55. a. $0.0025x + 80 + \dfrac{10,000}{x}$ **b.** 2000 **c.** 2000 **d.** Same

57. 10,000 watches

59. a. 2 days after the organic waste was dumped into the pond
  b. 3.5 days after the organic waste was dumped into the pond

65. 749,833 new inmates

69. b. 9.075 billion   **71.** $\frac{3}{2}u$ ft/sec

75. $R = r; \dfrac{E^2}{4r}$ watts   **77.** 7.4 lb   **79.** False

81. True   **83.** True

87. c.

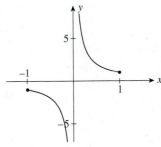

Using Technology Exercises 4.4, page 320

1. Absolute maximum value: 145.9; absolute minimum value: −4.3834

3. Absolute maximum value: 16; absolute minimum value: −0.1257

5. Absolute maximum value: 2.8889; absolute minimum value: 0

7. a.

9. b. approximately 1145

Exercises 4.5, page 327

1. 25 ft × 25 ft

3. 750 yd × 1500 yd; 1,125,000 yd²

5. $10\sqrt{2}$ ft × $40\sqrt{2}$ ft

7. $\frac{16}{3}$ in. × $\frac{16}{3}$ in. × $\frac{4}{3}$ in.

9. 3 in. × 3 in. × 4 in.   **11.** 2 ft × 2 ft × 5 ft   **13.** 5 in. × 10 in.

15. $r = 1.5$ in., $h = 4$ in.

17. $\frac{2}{3}\sqrt[3]{9}$ ft × $\sqrt[3]{9}$ ft × $\frac{2}{3}\sqrt[3]{9}$ ft

19. 250; $62,500; $250

21. 85; $28,900; $340   **23.** 60 mph

25. $w \approx 13.86$ in.; $h \approx 19.60$ in.

27. $x = 2250$ ft   **29.** $x \approx 2.68$

31. 440 ft; 140 ft; 184,874 ft²   **33.** 45; 44,445

Chapter 4 Concept Review, page 332

1. a. $f(x_1) < f(x_2)$ **b.** $f(x_1) > f(x_2)$

2. a. increasing   **b.** $f'(x) < 0$ **c.** constant

3. a. $f(x) \le f(c)$ **b.** $f(x) \ge f(c)$

4. a. domain; = 0; exist   **b.** critical number
  c. relative extremum

5. a. $f'(x)$ **b.** > 0 **c.** concavity
  d. relative maximum; relative extremum

6. $\pm\infty; \pm\infty$ **7.** 0; 0   **8.** b; b

9. a. $f(x) \le f(c)$; absolute maximum value
  b. $f(x) \ge f(c)$; open interval

10. continuous; absolute; absolute

Chapter 4 Review Exercises, page 333

1. a. f is increasing on $(-\infty, \infty)$ **b.** No relative extrema
  c. Concave down on $(-\infty, 1)$; concave up on $(1, \infty)$
  d. $\left(1, -\frac{17}{3}\right)$

2. a. f is increasing on $(-\infty, \infty)$ **b.** No relative extrema
  c. Concave down on $(-\infty, 2)$; concave up on $(2, \infty)$ **d.** $(2, 0)$

3. a. f is increasing on $(-1, 0)$ and $(1, \infty)$ and decreasing on
    $(-\infty, -1)$ and $(0, 1)$
  b. Relative maximum value: 0; relative minimum value: −1
  c. Concave up on $\left(-\infty, -\frac{\sqrt{3}}{3}\right)$ and $\left(\frac{\sqrt{3}}{3}, \infty\right)$; concave down
    on $\left(-\frac{\sqrt{3}}{3}, \frac{\sqrt{3}}{3}\right)$
  d. $\left(-\frac{\sqrt{3}}{3}, -\frac{5}{9}\right); \left(\frac{\sqrt{3}}{3}, -\frac{5}{9}\right)$

4. a. f is increasing on $(-\infty, -2)$ and $(2, \infty)$ and decreasing on
    $(-2, 0)$ and $(0, 2)$
  b. Relative maximum value: −4; relative minimum value: 4
  c. Concave down on $(-\infty, 0)$; concave up on $(0, \infty)$ **d.** None

5. a. f is increasing on $(-\infty, 0)$ and $(2, \infty)$; decreasing on $(0, 1)$
    and $(1, 2)$
  b. Relative maximum value: 0; relative minimum value: 4
  c. Concave up on $(1, \infty)$; concave down on $(-\infty, 1)$ **d.** None

6. a. f is increasing on $(1, \infty)$ **b.** No relative extrema
  c. Concave down on $(1, \infty)$ **d.** None

7. a. f is decreasing on $(-\infty, \infty)$ **b.** No relative extrema
  c. Concave down on $(-\infty, 1)$; concave up on $(1, \infty)$ **d.** $(1, 0)$

8. a. f is increasing on $(1, \infty)$ **b.** No relative extrema
  c. Concave down on $\left(1, \frac{4}{3}\right)$; concave up on $\left(\frac{4}{3}, \infty\right)$
  d. $\left(\frac{4}{3}, \frac{4\sqrt{3}}{9}\right)$

9. a. f is increasing on $(-\infty, -1)$ and $(-1, \infty)$
  b. No relative extrema
  c. Concave down on $(-1, \infty)$; concave up on $(-\infty, -1)$
  d. None

10. a. f is decreasing on $(-\infty, 0)$ and increasing on $(0, \infty)$

b. Relative minimum value: -1

c. Concave down on $\left(-\infty, -\dfrac{\sqrt{3}}{3}\right)$ and $\left(\dfrac{\sqrt{3}}{3}, \infty\right)$; concave up on $\left(-\dfrac{\sqrt{3}}{3}, \dfrac{\sqrt{3}}{3}\right)$

d. $\left(-\dfrac{\sqrt{3}}{3}, -\dfrac{3}{4}\right); \left(\dfrac{\sqrt{3}}{3}, -\dfrac{3}{4}\right)$

11.

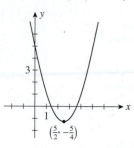

$\left(\frac{5}{2}, -\frac{5}{4}\right)$

12.

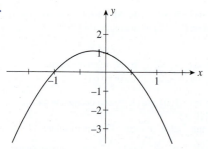

13.

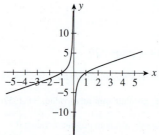

14.

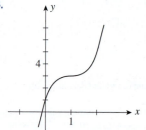

15.

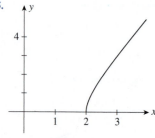

16.

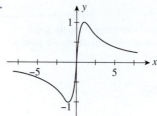

17.

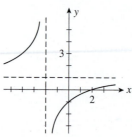

18.

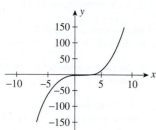

19. Horizontal asymptote: $y = 0$; vertical asymptote: $x = -\dfrac{3}{2}$

20. Horizontal asymptote: $y = 2$; vertical asymptote: $x = -1$

21. Horizontal asymptote: $y = 0$; vertical asymptotes: $x = -2$, $x = 4$

22. Horizontal asymptote: $y = 1$; vertical asymptote: $x = 1$

23. Absolute minimum value: $-\dfrac{25}{8}$

24. Absolute minimum value: 0

25. Absolute maximum value: 5; absolute minimum value: 0

26. Absolute maximum value: $\dfrac{5}{3}$; absolute minimum value: 1

27. Absolute maximum value: -16; absolute minimum value: -32

28. Absolute maximum value: $\dfrac{1}{2}$; absolute minimum value: 0

29. Absolute maximum value: $\dfrac{8}{3}$; absolute minimum value: 0

30. Absolute maximum value: $\dfrac{215}{9}$; absolute minimum value: 7

31. Absolute maximum value: $\dfrac{1}{2}$; absolute minimum value: $-\dfrac{1}{2}$

32. No absolute extrema

33. a. The sign of $R_1'(t)$ is negative; the sign of $R_2'(t)$ is positive. The sign of $R_1''(t)$ is negative and the sign of $R_2''(t)$ is positive.

b. The revenue of the neighborhood bookstore is decreasing at an increasing rate while the revenue of the new branch of the national bookstore is increasing at an increasing rate.

34. The rumor spread initially with increasing speed. The rate at which the rumor is spread reaches a maximum at the time corresponding to the t-coordinate of the point P on the curve. Thereafter, the speed at which the rumor is spread decreases.

35. $4000

36. b. The sales continued to accelerate through the years under consideration.

37. a. The amount of AMT paid was increasing over those years.
b. The amount of AMT paid accelerated over those years.

38. a. 506,000; 125,480
b. The number of measles deaths was dropping from 1999 through 2005.
c. April 2002; approximately -41 thousand deaths/year/year

39. $(100, 4600)$; sales increase at an increasing rate until $100,000 is spent on advertising; after that, any additional expenditure results in increased sales but at a slower rate of increase.

40. a. Decreasing on $(0, 21.4)$; increasing on $(21.4, 30)$
b. The percentage of men 65 years and older in the workforce was decreasing from 1970 until mid-1991 and increasing from mid-1991 through 2000.

41. $(267, 11{,}874)$; the rate of increase is lowest when 267 calculators are produced.

43. a. 13.0%, 22.2%

44. a. $I'(t) = -\dfrac{200t}{(t^2 + 10)^2}$

b. $I''(t) = \dfrac{-200(10 - 3t^2)}{(t^2 + 10)^3}$; concave up on $\left(\dfrac{\sqrt{10}}{3}, \infty\right)$; concave down on $\left(0, \dfrac{\sqrt{10}}{3}\right)$

c.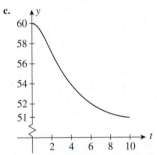

d. The rate of decline in the environmental quality of the wildlife was increasing the first 1.8 years. After that time, the rate of decline decreased.

45. 168 discs   **46.** 3000 cameras

47. a. $0.001x + 100 + \dfrac{4000}{x}$ **b.** 2000 guitars   **48.** 10 A.M.

49. a. Decreasing on $(0, 12.7)$; increasing on $(12.7, 30)$
b. 7.9
c. The percent of women 65 years and older in the workforce was decreasing from 1970 to Sept. 1982 and increasing from Sept. 1982 to 2000. It reached a minimum value of 7.9% in Sept. 1982.

50. As the age of the driver increases from 16 to 27 years, the predicted number of crash fatalities drops.

52. The number of applicants was decreasing from 1997–1998 to 2001–2002, increasing from 2001–2002 to 2006–2007, and constant from 2006–2007 to 2008–2009.

53. a. $f'(x) = \begin{cases} -2x & \text{if } x \neq 0 \\ 0 & \text{if } x = 0 \end{cases}$ **b.** No

54. 74.07 in.³   **55.** Radius: 2 ft; height: 8 ft

56. 1 ft × 2 ft × 2 ft   **57.** 20,000 cases

58. If $a > 0$, f is decreasing on $\left(-\infty, -\dfrac{b}{2a}\right)$ and increasing on $\left(-\dfrac{b}{2a}, \infty\right)$; if $a < 0$, f is increasing on $\left(-\infty, -\dfrac{b}{2a}\right)$ and decreasing on $\left(-\dfrac{b}{2a}, \infty\right)$.

59. $a = -4$; $b = 11$ **60.** $c \geq \frac{3}{2}$

62. a. $f'(x) = 3x^2$ if $x \neq 0$ **b.** No

Chapter 4 Before Moving On, page 336

1. Decreasing on $(-\infty, 0)$ and $(2, \infty)$; increasing on $(0, 1)$ and $(1, 2)$

2. Relative minimum: $(1, -10)$

3. Concave downward on $\left(-\infty, \frac{1}{4}\right)$; concave upward on $\left(\frac{1}{4}, \infty\right)$; $\left(\frac{1}{4}, \frac{83}{96}\right)$

4.

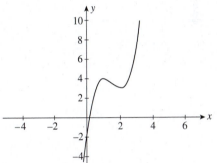

5. Absolute minimum value: -5; absolute maximum value: 80

6. $r = h = \dfrac{1}{\sqrt[3]{\pi}}$ (ft)

CHAPTER 5

Exercises 5.1, page 342

1. a. 16   **b.** 27   **3. a.** 3   **b.** $\sqrt{5}$

5. a. -3 **b.** 8   **7. a.** 25   **b.** $4^{1.8}$

9. a. $4x^3$ **b.** $5xy^2\sqrt{x}$ **11. a.** $\dfrac{2}{a}$ **b.** $\frac{1}{3}b^2$

13. a. $8x^9y^6$ **b.** $16x^4y^4z^6$ **15. a.** $\dfrac{64x^6}{y^4}$ **b.** $(x + y)(x - y)$

17. 3   **19.** 3   **21.** 3   **23.** $\frac{5}{4}$ **25.** 1 or 2

27.

29.

31.

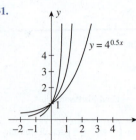

$y = 4^{0.5x}$

33.

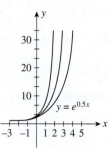

$y = e^{0.5x}$

35.

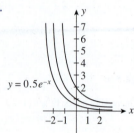

$y = 0.5e^{-x}$

37. $f(x) = 100\left(\frac{6}{5}\right)^x$ **39.** 54.6

41. a.

t	0	1	2	3
$f(t)$	64.0	77.2	93.2	112.5

b.

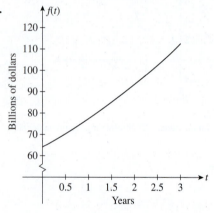

43. a. 115,423,000; 140,977,00; 313,751,000

b.

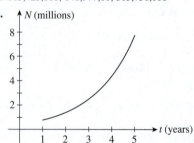

45. 34,210,000

47. a. 0 g/cm³   **b.** 0.2367 g/cm³   **c.** 0.7598 g/cm³   **d.** 0 g/cm³

49. False   **51.** True   **53.** True

Using Technology Exercises 5.1, page 345

1.

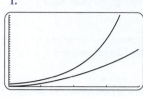

3.

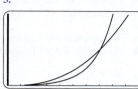

5.

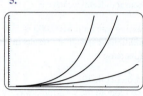

7.

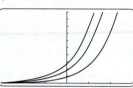

9.

11. a.

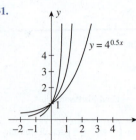

b. 0.08 g/cm³   **c.** 0.12 g/cm³   **d.** 0.2 g/cm³

13. a.

b. 20 sec   **c.** 35.1 sec

Exercises 5.2, page 351

1. $\log_2 64 = 6$ **3.** $\log_4 \frac{1}{16} = -2$ **5.** $\log_{1/3} \frac{1}{3} = 1$

7. $\log_{32} 16 = \frac{4}{5}$ **9.** $\log_{10} 0.001 = -3$ **11.** 1.0792

13. 1.2042   **15.** 1.6813   **17.** $\ln a^2 b^3$ **19.** $\ln \dfrac{3\sqrt{xy}}{\sqrt[3]{z}}$

21. $\log x + 4\log(x + 1)$ **23.** $\frac{1}{2}\log(x + 1) - \log(x^2 + 1)$

25. $\ln x - x^2$ **27.** $-\frac{3}{2}\ln x - \frac{1}{2}\ln(1 + x^2)$

29.

$y = \log_3 x$

31.

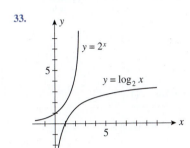

33.

35. 5.1986   **37.** −0.0912   **39.** −8.0472   **41.** −4.9041

43. $-2\ln\left(\dfrac{A}{B}\right)$ **45.** $f(x) = 2 + 2.8854\ln x$ **47.** 106 mm

49. a. $10^3 I_0$ **b.** 100,000 times greater
c. 10,000,000 times greater

51. a. 170°F   **b.** 15.54 min   **53.** 2022

55. a. 9.1 sec   **b.** 20.3 sec   **57.** 1:30 P.M.

59. False   **61.** True   **63.** True   **65. a.** $\ln 2$

Exercises 5.3, page 365

1. $3714.87   **3.** $190,573.37

5. a. 6.09%/year   **b.** 5.095%/year

7. a. $32,829.86   **b.** $32,789.85

9. $6107.01   **11.** $9231.20   **13.** $16,705.40   **15.** $44,206.85

17. 13.59%/year   **19.** 19.21%/year   **21.** 22.17%/year

23. 4.4 years   **25.** 27.5 years   **27.** 13.86%/year   **29.** 13.9 years

31. 2.16   **33.** Both are worth the same: $13,200

35. No   **37.** $20,471.64   **39.** $14,505.43

41. $731,250   **43.** 92¢   **45.** $87,740.82   **47.** $23,227.22

49. $80,000e^{(\sqrt{t/2})-0.09t}$; $151,718   **51.** Investment A **53.** 12.75%

57. Bank B **61.** 2.105%/year   **63.** True

Using Technology Exercises 5.3, page 370

1. $5872.78   **3.** 8.95%/year   **5.** $29,743.30

Exercises 5.4, page 376

1. $3e^{3x}$ **3.** $-e^{-t}$ **5.** $e^x + 2x$ **7.** $x^2 e^x(x+3)$

9. $\dfrac{e^x(x-1)}{x^2}$ **11.** $3(e^x - e^{-x})$ **13.** $-\dfrac{2}{e^w}$ **15.** $6e^{3x-1}$

17. $-2xe^{-x^2}$ **19.** $\dfrac{3e^{-1/x}}{x^2}$ **21.** $25e^x(e^x+1)^{24}$ **23.** $\dfrac{e^{\sqrt{x}}}{2\sqrt{x}}$

25. $e^{3x+2}(3x-2)$ **27.** $\dfrac{2e^x}{(e^x+1)^2}$ **29.** $16e^{-4x} + 9e^{3x}$

31. $6e^{3x}(3x+2)$ **33.** $y = 2x - 2$

35. f is increasing on $(-\infty, 0)$ and decreasing on $(0, \infty)$.

37. Concave downward on $(-\infty, 0)$; concave upward on $(0, \infty)$

39. $(1, e^{-2})$ **41.** $y = e^{-1/2}(-\sqrt{2}x + 2)$; $y = e^{-1/2}(\sqrt{2}x + 2)$

43. Absolute maximum value: 1; absolute minimum value: e^{-1}

45. Absolute minimum value: −1; absolute maximum value: $2e^{-3/2}$

47.

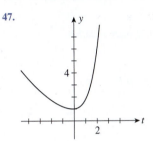

49.

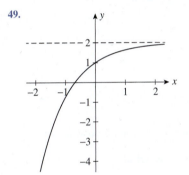

51. $\dfrac{2x}{4e^{2y} - 3y^2}$ **53.** $-\dfrac{e^y\left(\dfrac{dy}{dx}\right)^2}{1 + e^y}$ or $-\dfrac{e^y}{(1 + e^y)^3}$ **55.** $-\dfrac{1}{e}$

57. a. 188.7 million   **b.** 12.6 million viewers/year

59. At the beginning of 2008, the value of stolen drugs was increasing at the rate of $66.64 million/year.

61. a. −$6065/day/day; −$3679/day/day; −$2231/day/day; −$1353/day/day
b. 2 days

63. b. 4505/year/year; 273 cases/year/year

65. a. 1.43%; 23.51%
b. The frequency of Alzheimer's disease increases with age in the age range under consideration.
c. The frequency of Alzheimer's disease increases at an increasing rate with age in the age range under consideration.

67. a. $100xe^{-0.0001x}$ **b.** $100(1 - 0.0001x)e^{-0.0001x}$ **c.** $0/pair

69. a. −1.34 cents/bottle   **b.** $217.03/bottle

71. a. 30   **b.** $\dfrac{297{,}000e^{-x}}{(1 + 99e^{-x})^2}$

c.

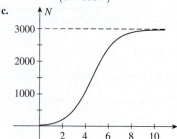

73. a. 0.02p

b. Inelastic if $0 < p < 50$, unitary if $p = 50$, and elastic if $p > 50$

c. Decrease   **d.** Decrease

75. a. 45.6 kg   **b.** 4.2 kg   **77.** 7.72 years; $160,208

79. $2\frac{1}{2}$ hours after drinking; 0.21%

81. a. 70°F   **b.** −14.7°F/min   **c.** 30°F

83. a. 2.15 g/min   **b.** 60 g

85. b.

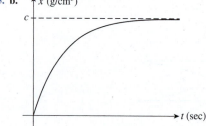

87. a. Decreasing at the rate of 70 mg/day; decreasing at the rate of 43 mg/day

b. 1 day after the first dose   **c.** 125 mg

89. False   **91.** False

Using Technology Exercises 5.4, page 381

1. 5.4366   **3.** 12.3929   **5.** 0.1861

7. a. 50   **c.**

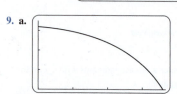

9. a.

b. $160,000; $87.07/month; $126,928.78; $334.18/month

Exercises 5.5, page 387

1. $\dfrac{5}{x}$ **3.** $\dfrac{1}{x+1}$ **5.** $\dfrac{8}{x}$ **7.** $\dfrac{1}{2x}$ **9.** $\dfrac{-2}{x}$

11. $\dfrac{8x - 5}{4x^2 - 5x + 3}$ **13.** $\dfrac{1}{x(x+1)}$ **15.** $x(1 + 2\ln x)$

17. $\dfrac{2(1 - \ln x)}{x^2}$ **19.** $\dfrac{3}{u - 2}$ **21.** $\dfrac{1}{2x\sqrt{\ln x}}$

23. $\dfrac{2\ln x}{x}$ **25.** $\dfrac{3x^2}{x^3 + 1}$ **27.** $\dfrac{(x\ln x + 1)e^x}{x}$

29. $\dfrac{e^{2t}[2(t + 1)\ln(t + 1) + 1]}{t + 1}$ **31.** $\dfrac{1 - 2\ln x}{x^3}$ **33.** $\dfrac{1}{x\ln x}$

35. $-\dfrac{1}{x^2}$ **37.** $\dfrac{2(2 - x^2)}{(x^2 + 2)^2}$ **39.** $3 + 2\ln x$

41. $(x + 1)(5x + 7)(x + 2)^2$

43. $(x - 1)(x + 1)^2(x + 3)^3(9x^2 + 14x - 7)$

45. $\dfrac{(2x^2 - 1)^4(38x^2 + 40x + 1)}{2(x + 1)^{3/2}}$ **47.** $3^x \ln 3$

49. $(x^2 + 1)^{x-1}[2x^2 + (x^2 + 1)\ln(x^2 + 1)]$ **51.** $(\ln x + 1)y$

53. $y = x - 1$

55. f is decreasing on $(-\infty, 0)$ and increasing on $(0, \infty)$.

57. Concave up: $(-\infty, -1)$ and $(1, \infty)$; concave down: $(-1, 0)$ and $(0, 1)$

59. $(-1, \ln 2)$ and $(1, \ln 2)$ **61.** $y = 4x - 3$

63. Absolute minimum value: 1; absolute maximum value: $3 - \ln 3$

65. $\dfrac{y(x - 1)}{x(1 - y)}$ **67.** $\dfrac{3 + 2xy}{x^3}$ **69.** $-\frac{1}{2}$

71. 0.0580%/kg; 0.0133%/kg

73. b. W is concave downward on $(1, 6)$.

75. a. $C\left(1 - \dfrac{2}{N}\right)^n \ln\left(1 - \dfrac{2}{N}\right)$ **b.** $\ln\left(1 - \dfrac{2}{N}\right)$

77. 2.04%/month   **79.** 150 consultants; $3.22 million

81. b. 70%/year

85. a. 6   **b.** 276,310 times the standard reference intensity

87. $\dfrac{(A - Bx)yx'}{(Dy - C)x}$

89.

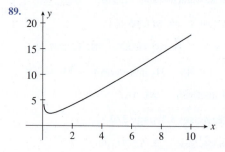

91. b. $\ln 3(3^x)$ **93.** $x^2(x\ln 2 + 3)2^x$ **95.** $x\left(\dfrac{1}{\ln 10} + 2\log_{10} x\right)$

97. False

Exercises 5.6, page 399

1. a. 0.02 **b.** 300

c.

t	0	10	20	100	1000
Q	300	366	448	2217	1.46×10^{11}

3. a. $Q(t) = 100e^{0.035t}$ **b.** 266 min **c.** $Q(t) = 1000e^{0.035t}$

5. a. 54.93 years **b.** 14.25 billion

7. 8.7 lb/in.^{2}; -0.0004 lb/in.^{2}/ft

9. $Q(t) = 100e^{-0.049t}$; 70.7 g; 3.45 g/day

11. 13,412 years ago

13. a. 50/year **b.** 30

15. a. $S(t) = 100e^{0.081t}$ **b.** \$127.5 million

17.

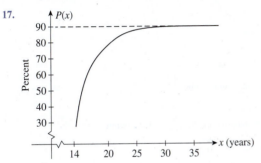

a. 27.8% **b.** 75.8% **c.** 90.0%

19.

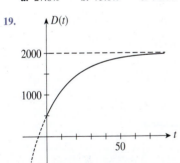

a. 573 computers/month; 1177 computers/month;
1548 computers/month; 1925 computers/month
b. 2000 computers/month **c.** 46 computers/month

21. a. 122 cm **b.** 14 cm/year **c.** 200 cm

23. a. 10 **b.** 400 **c.** 154 **d.** 15/day

25. 10 million barrels/day **27.** 1080; 280/hr **29.** 15 1b

31. a. $\dfrac{\ln \frac{b}{a}}{b - a}$ min **b.** 0 g/cm^3 **33. b.** 5599 years

35. b. Q increases most rapidly at $t = \dfrac{\ln B}{k}$.

37. 0.14

Using Technology Exercises 5.6, page 403

1. a.

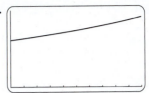

b. 666 million, 926.8 million **c.** 38.3 million/year/year

3. a.

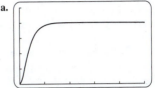

b. 400; Starr eventually will sell 400,000 copies of Laser Beams.

5. a.

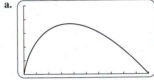

b. 3.68 cm

7. a.

b. 0 **c.** 0.237 g/cm^3
d. 0.760 g/cm^3 **e.** 0

Chapter 5 Concept Review, page 405

1. power; 0; 1; exponential

2. a. $(-\infty, \infty)$; $(0, \infty)$
b. $(0, 1)$; $(-\infty, \infty)$

3. a. $(0, \infty)$; $(-\infty, \infty)$; $(1, 0)$
b. <1; >1

4. a. x
b. x

5. accumulated amount; principal; nominal interest rate; number of conversion periods; term

6. $\left(1 + \frac{r}{m}\right)^m - 1$ **7.** Pe^{rt}

8. a. $e^{f(x)}f'(x)$ **b.** $\dfrac{f'(x)}{f(x)}$

9. a. initially; growth
b. decay
c. time; one half

10. a. horizontal asymptote; C
b. horizontal asymptote; A, carrying capacity

Chapter 5 Review Exercises, page 406

1. a. and **b.**

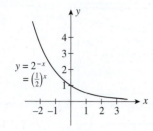

$$y = 2^{-x} = \left(\tfrac{1}{2}\right)^x$$

2. $\log_{2/3} \tfrac{27}{8} = -3$ **3.** $\log_{16} 0.125 = -\tfrac{3}{4}$ **4.** $x = \tfrac{15}{2}$ **5.** $x = 2$

6. $x + y + z$ **7.** $x + 2y - z$ **8.** $y + 2z$

9.

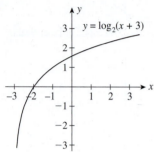

$y = \log_2(x + 3)$

10.

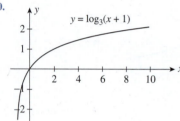

$y = \log_3(x + 1)$

11. a. \$11,274.86   **b.** \$11,274.97   **12.** 6.12%/year

13. 6.8 years   **14.** 7.77%/year

15. $(2x + 1)e^{2x}$ **16.** $\dfrac{e^t}{2\sqrt{t}} + \sqrt{t}e^t + 1$ **17.** $\dfrac{1 - 4t}{2\sqrt{t}e^{2t}}$

18. $\dfrac{e^x(x^2 + x + 1)}{\sqrt{1 + x^2}}$ **19.** $\dfrac{2(e^{2x} + 2)}{(1 + e^{-2x})^2}$ **20.** $4xe^{2x^2 - 1}$

21. $(1 - 2x^2)e^{-x^2}$ **22.** $3e^{2x}(1 + e^{2x})^{1/2}$ **23.** $(x + 1)^2 e^x$

24. $\ln t + 1$ **25.** $\dfrac{2xe^{x^2}}{e^{x^2} + 1}$ **26.** $\dfrac{\ln x - 1}{(\ln x)^2}$

27. $\dfrac{x - x\ln x + 1}{x(x + 1)^2}$ **28.** $(x + 2)e^x$ **29.** $\dfrac{4e^{4x}}{e^{4x} + 3}$

30. $\dfrac{(r^3 - r^2 + r + 1)e^r}{(1 + r^2)^2}$ **31.** $\dfrac{1 + e^x(1 - x\ln x)}{x(1 + e^x)^2}$

32. $\dfrac{(2x^2 + 2x^2 \cdot \ln x - 1)e^{x^2}}{x(1 + \ln x)^2}$ **33.** $-\dfrac{9}{(3x + 1)^2}$ **34.** $\dfrac{1}{x}$

35. 0   **36.** -2 **37.** $6x(x^2 + 2)^2(3x^3 + 2x + 1)$

38. $\dfrac{(4x^3 - 5x^2 + 2)(x^2 - 2)}{(x - 1)^2}$ **39.** $y = \dfrac{1}{e^2}(-2x + 3)$

40. $y = \dfrac{1}{e}$

41.

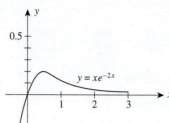

$y = xe^{-2x}$

42.

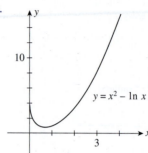

$y = x^2 - \ln x$

43. Absolute maximum value: $\dfrac{1}{e}$

44. Absolute maximum value: $\dfrac{\ln 2}{2}$; absolute minimum value: 0

45. 12%   **46.** \$23,376.16

47. \$93,880.89

48. 9.58 years

49. a. $Q(t) = 2000e^{0.01831t}$ **b.** 162,000

50. 0.0004332

51.

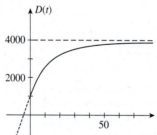

a. 1175, 2540, 3289   **b.** 4000

52. 200 g; -21.2 g/year

53. a. 1986 kWh/year

54. 1.8; -0.11; -0.23; -0.13; the rate of change of the amount of oil used was 1.8 barrels per \$1000 of output per decade in 1965; the amount of oil used was decreasing at the rate of 0.11 barrels per \$1000 of output per decade in 1966, and so on.

55. a. \$9/unit   **b.** \$8/unit/week   **c.** \$18/unit

56. 970

57. a. 3 billion   **b.** 9 billion   **c.** 698.567 million/decade

58. a. 12.5/1000 live births; 9.3/1000 live births; 6.9/1000 live births

b.

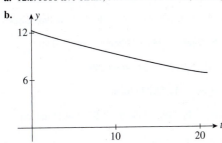

59. 5000; $36,788

60. a. 0 g/cm³　　**b.** 0.0361 g/cm³　　**c.** 0.08 g/cm³

d.

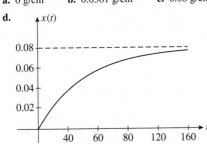

Chapter 5 Before Moving On, page 408

1. −0.9589　　**2.** $4130.37　　**3.** $\dfrac{e^{\sqrt{x}}}{2\sqrt{x}}$　　**4.** 1 + ln 2

5. $e^{2x}\left(\dfrac{4x^2 \ln 3x + 4x - 1}{x^2}\right)$　　**6.** After 8.7 min

CHAPTER 6

Exercises 6.1, page 418

5. b. $y = 2x + C$
c.

7. b. $y = \frac{1}{3}x^3 + C$
c.

9. $6x + C$　　**11.** $\frac{1}{4}x^4 + C$　　**13.** $-\dfrac{1}{3x^3} + C$

15. $\frac{3}{5}x^{5/3} + C$　　**17.** $-\dfrac{4}{x^{1/4}} + C$　　**19.** $-\dfrac{1}{x^2} + C$

21. $\frac{2}{3}\pi t^{3/2} + C$　　**23.** $3x - 2x^2 + C$

25. $\dfrac{1}{3}x^3 + \dfrac{1}{2}x^2 - \dfrac{1}{2x^2} + C$　　**27.** $5e^x + C$

29. $x + \frac{1}{2}x^2 + e^x + C$　　**31.** $x^4 + \dfrac{2}{x} - x + C$

33. $\frac{2}{7}x^{7/2} + \frac{4}{5}x^{5/2} - \frac{1}{2}x^2 + C$　　**35.** $\frac{2}{3}x^{3/2} + 4\sqrt{x} + C$

37. $\frac{1}{9}u^3 + \frac{1}{3}u^2 - \frac{1}{3}u + C$　　**39.** $\frac{2}{3}t^3 - \frac{3}{2}t^2 - 2t + C$

41. $\dfrac{1}{3}x^3 - 2x - \dfrac{1}{x} + C$　　**43.** $\frac{1}{3}s^3 + s^2 + s + C$

45. $e^t + \dfrac{t^{e+1}}{e+1} + C$　　**47.** $\dfrac{1}{2}x^2 + x - \ln|x| - \dfrac{1}{x} + C$

49. $\ln|x| + \dfrac{4}{\sqrt{x}} - \dfrac{1}{x} + C$　　**51.** $\frac{3}{2}x^2 + x + \frac{1}{2}$

53. $x^3 + 2x^2 - x - 5$　　**55.** $x - \dfrac{1}{x} + 3$　　**57.** $x + \ln|x|$

59. $\sqrt{x}$　　**61.** $e^x + \frac{1}{2}x^2 + 2$　　**63.** Branch A

65. $14.3t + 90.1$; 147.3 million　　**67.** $0.1t^2 + 3t$

69. 5000 units; $34,000

71. a. $2.509t^2 - 3.204t + 1.8$　　**b.** 102.313 terawatt-hours
c. 136.744 terawatt-hours

73. 2816.1 acres

75. a. $-16t^2 + 4t + 400$　　**b.** $t = 5.13$ sec
c. 160.16 ft/sec downward

77. 21,960　　**79.** $-t^3 + 96t^2$; 59,400 ft

81. $3370　　**83. a.** $\dfrac{16\sqrt{2}}{3}t^{3/2} - 8t^4$　　**b.** 2.2 in.

85. a. $7.031t^{0.842} - 0.031$　　**b.** 27.23%　　**87.** 1.9424 m²

89. a. $0.2098t^5 - 1.056375t^4 + 4.255t^3 - 12.85595t^2$
　　$+ 32.28t + 52.9710$
b. 75.80 cm

91. $9\frac{7}{9}$ ft/sec²; 396 ft　　**93.** 0.924 ft/sec²

95. a. At least 36 ft/sec²　　**97.** False　　**99.** False

Exercises 6.2, page 430

1. $\frac{1}{5}(4x + 3)^5 + C$　　**3.** $\frac{1}{3}(x^3 - 2x)^3 + C$

5. $-\dfrac{1}{2(2x^2 + 3)^2} + C$　　**7.** $\frac{2}{3}(t^3 + 2)^{3/2} + C$

9. $\frac{1}{10}(x^2 - 1)^{10} + C$　　**11.** $-\frac{1}{5}\ln|1 - x^5| + C$

13. $2\ln|x - 2| + C$　　**15.** $\frac{1}{2}\ln(0.3x^2 - 0.4x + 2) + C$

17. $\frac{1}{3}\ln|3x^2 - 1| + C$　　**19.** $-\frac{1}{2}e^{-2x} + C$　　**21.** $-e^{2-x} + C$

23. $-\frac{1}{2}e^{-x^2} + C$　　**25.** $e^x + e^{-x} + C$　　**27.** $2\ln(1 + e^x) + C$

29. $2e^{\sqrt{x}} + C$　　**31.** $-\dfrac{1}{6(e^{3x} + x^3)^2} + C$　　**33.** $\frac{1}{8}(e^{2x} + 1)^4 + C$

35. $\frac{1}{2}(\ln 5x)^2 + C$ **37.** $3 \ln|\ln x| + C$ **39.** $\frac{2}{3}(\ln x)^{3/2} + C$

41. $\frac{1}{2}e^{x^2} - \frac{1}{2}\ln(x^2 + 2) + C$

43. $\frac{2}{3}(\sqrt{x} - 1)^3 + 3(\sqrt{x} - 1)^2 + 8(\sqrt{x} - 1) + 4\ln|\sqrt{x} - 1| + C$

45. $\dfrac{(6x + 1)(x - 1)^6}{42} + C$

47. $4\sqrt{x} - x - 4\ln(1 + \sqrt{x}) + C$

49. $-\frac{1}{252}(1 - v)^7(28v^2 + 7v + 1) + C$

51. $\frac{1}{2}[(2x - 1)^5 + 5]$ **53.** $e^{-x^2+1} - 1$

55. $21,000 - \dfrac{20,000}{\sqrt{1 + 0.2t}}$; 6858 **57.** 180.7 million GBP

59. 62,286 **61.** $50.02(1 + 1.09t)^{0.1}$; 80

63. a. $135e^{0.067t}$ **b.** 188.722 million

65. a. $19.1301e^{0.163(t+1)} + 3.5699$ **b.** \$40.3 billion

67. $\frac{r}{a}(1 - e^{-at})$ **69.** True **71.** True

Exercises 6.3, page 442

1. 4.27

3. a. 6

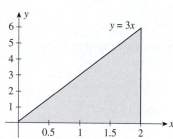

b. 4.5 **c.** 5.25 **d.** Yes

5. a. 4

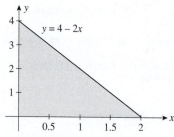

b. 4.8 **c.** 4.4 **d.** Yes

7. a. 18.5 **b.** 18.64 **c.** 18.66 **d.** $\approx 18\frac{2}{3}$

9. a. 25 **b.** 21.12 **c.** 19.88 **d.** ≈ 19.9

11. a. 0.0625 **b.** 0.16 **c.** 0.2025 **d.** ≈ 0.2

13. 4.64 **15.** 0.95 **17.** 9400 ft² **19.** False

Exercises 6.4, page 453

1. 6 **3.** 8 **5.** 12 **7.** 9 **9.** ln 2 **11.** $17\frac{1}{3}$ **13.** $18\frac{1}{4}$

15. $e^2 - 1$ **17.** 6 **19.** 24 **21.** $\frac{56}{3}$ **23.** $\frac{4}{3}$ **25.** $\frac{45}{2}$

27. $\frac{7}{12}$ **29.** ln 4 **31.** 56 **33.** $\frac{256}{15}$ **35.** $\frac{2}{3}$

37. $\frac{8}{3}$ **39.** $\frac{39}{2}$

41. a. A decrease of 0.319 million cases **b.** 1.219 million cases

43. a. \$4100 **b.** \$900 **45. a.** \$2800 **b.** \$219.20

47. $10,133\frac{1}{3}$ ft **49.** \$7182 **51.** 46%; 24%

53. a. $0.86t^{0.96} + 0.04$ **b.** \$6.37 billion

55. a. \$8.86 billion **b.** \$19.57 billion **57.** 149.14 million

59. $\frac{23}{15}$ **61.** False **63.** False

Using Technology Exercises 6.4, page 456

1. 6.1787 **3.** 0.7873 **5.** -0.5888 **7.** 2.7044

9. 3.9973 **11.** 46%; 24% **13.** 60,156

Exercises 6.5, page 463

1. 10 **3.** $\frac{19}{15}$ **5.** $\frac{484}{15}$ **7.** $\sqrt{3} - 1$ **9.** $\frac{1562}{5}$

11. $\frac{32}{15}$ **13.** $\frac{272}{15}$ **15.** $e^4 - 1$ **17.** $\frac{1}{2}e^2 + \frac{5}{6}$ **19.** 0

21. ln 4 **23.** $\frac{1}{3}(\ln 19 - \ln 3)$ **25.** $2e^4 - 2e^2 - \ln 2$

27. $\frac{1}{2}(e^{-4} - e^{-8} - 1)$ **29.** 6 **31.** $\frac{1}{2}$ **33.** $2(\sqrt{e} - \frac{1}{e})$

35. 5 **37.** $\frac{17}{3}$ **39.** -1 **41.** $\frac{13}{6}$ **43.** $\frac{1}{4}(e^4 - 1)$

45. 64 ft **47.** $\approx$\$2.24 million

49. 120.3 billion metric tons **51.** \$3.24 billion/year

53. a. 160.7 billion gal/year **b.** 150.1 billion gal/year/year

55. 12.53 N-sec; 626.5 N **57.** $\frac{2}{3}kR^2$ cm/sec **59.** 0.071 mg/cm³

61. $\$6\frac{1}{3}$ million **63.** 262.9 million cards/year

65. a. $146.711e^{0.0456t} - 0.72$ **b.** 364.5 million
  c. 11.4 million/year

75. Property 5 **77.** 0 **79. a.** -1 **b.** 5 **c.** -13

81. True **83.** False **85.** True

Using Technology Exercises 6.5, page 468

1. 7.71667 **3.** 17.56487 **5.** 159/bean stem

7. 0.48 g/cm³/day

Exercises 6.6, page 475

1. 108 **3.** $\frac{2}{3}$ **5.** $2\frac{2}{3}$ **7.** $1\frac{1}{2}$ **9.** 3 **11.** $3\frac{1}{3}$

13. 27 **15.** $2(e^2 - e^{-1})$ **17.** $12\frac{2}{3}$ **19.** $3\frac{1}{3}$ **21.** $4\frac{3}{4}$

23. $12 - \ln 4$ **25.** $e^2 - e - \ln 2$ **27.** $2\frac{1}{2}$ **29.** $7\frac{1}{3}$ **31.** $\frac{3}{2}$

33. $e^3 - 4 + \dfrac{1}{e}$ **35.** $\frac{125}{6}$ **37.** $\frac{1}{12}$ **39.** $\frac{71}{6}$ **41.** 18

43. S is the additional revenue that Odyssey Travel could realize by switching to the new agency; $S = \int_0^b [g(x) - f(x)] \, dx$

45. Shortfall $= \int_{2010}^{2050} [f(t) - g(t)] \, dt$

47. a. $A_2 - A_1$ **b.** The distance car 2 is ahead of car 1 after T sec

49. $840 - \int_0^{12} f(t)\, dt$ **51.** 42.8 billion metric tons   **53.** 57,179

55. True   **57.** False

Using Technology Exercises 6.6, page 480

1. a.

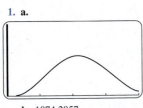

b. 1074.2857

3. a.

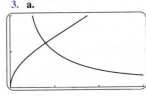

b. 0.9961

5. a.

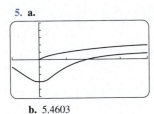

b. 5.4603

7. a.

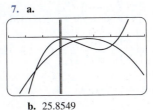

b. 25.8549

9. a.

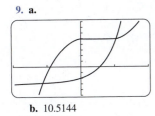

b. 10.5144

11. a.

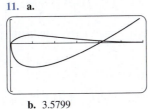

b. 3.5799

13. 207.43

Exercises 6.7, page 490

1. $11,667   **3.** $6667   **5.** $11,667

7. a. 1257/month   **b.** $48,583   **9.** $199,548

11. Consumers' surplus: $13,333; producers' surplus: $11,667

13. $515,224.45   **15.** $824,200   **17.** $91,916

19. a. $131,996   **b.** $113,610   **21.** $40,212

23. $47,916   **25.** $142,423   **27.** $28,330   **29.** 0.360

31. a.

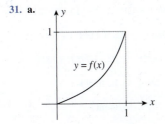

b. 0.175; 0.816

33. a. 0.31; 0.49   **b.** College teachers

Using Technology Exercises 6.7, page 493

1. Consumers' surplus: $18,000,000; producers' surplus: $11,700,000

3. Consumers' surplus: $33,120; producers' surplus: $2880

5. Investment A

Chapter 6 Concept Review, page 496

1. a. $F'(x) = f(x)$ **b.** $F(x) + C$

2. a. $c\int f(x)\, dx$ **b.** $\int f(x)\, dx \pm \int g(x)\, dx$

3. a. unknown   **b.** function   **4.** $g'(x)\, dx;\ \int f(u)\, du$

5. a. $\int_a^b f(x)\, dx$ **b.** minus

6. a. $F(b) - F(a)$; antiderivative   **b.** $\int_a^b f'(x)\, dx$

7. a. $\dfrac{1}{b-a}\int_a^b f(x)\, dx$ **b.** area; area   **8.** $\int_a^b [f(x) - g(x)]\, dx$

9. a. $\int_0^{\bar{x}} D(x)\, dx - \bar{p}\,\bar{x}$ **b.** $\bar{p}\,\bar{x} - \int_0^{\bar{x}} S(x)\, dx$

10. a. $e^{rT}\int_0^T R(t)e^{-rt}\, dt$ **b.** $\int_0^T R(t)e^{-rt}\, dt$

11. $\dfrac{mP}{r}(e^{rT} - 1)$ **12.** $2\int_0^1 [x - f(x)]\, dx$

Chapter 6 Review Exercises, page 496

1. $\frac{1}{4}x^4 + \frac{2}{3}x^3 - \frac{1}{2}x^2 + C$ **2.** $\frac{1}{12}x^4 - \frac{2}{3}x^3 + 8x + C$

3. $\frac{1}{5}x^5 - \frac{1}{2}x^4 - \frac{1}{x} + C$ **4.** $\frac{3}{4}x^{4/3} - \frac{2}{3}x^{3/2} + 4x + C$

5. $\frac{1}{2}x^4 + \frac{2}{5}x^{5/2} + C$ **6.** $\frac{2}{7}x^{7/2} - \frac{1}{3}x^3 + \frac{2}{3}x^{3/2} - x + C$

7. $\frac{1}{3}x^3 - \frac{1}{2}x^2 + 2\ln|x| + 5x + C$ **8.** $\frac{1}{3}(2x + 1)^{3/2} + C$

9. $\frac{3}{8}(3x^2 - 2x + 1)^{4/3} + C$ **10.** $\dfrac{(x^3 + 2)^{11}}{33} + C$

11. $\frac{1}{2}\ln(x^2 - 2x + 5) + C$ **12.** $-e^{-2x} + C$

13. $\frac{1}{2}e^{x^2 + x + 1} + C$ **14.** $\dfrac{1}{e^{-x} + x} + C$ **15.** $\frac{1}{6}(\ln x)^6 + C$

16. $(\ln x)^2 + C$ **17.** $\dfrac{(11x^2 - 1)(x^2 + 1)^{11}}{264} + C$

18. $\frac{2}{15}(3x - 2)(x + 1)^{3/2} + C$ **19.** $\frac{2}{3}(x + 4)\sqrt{x - 2} + C$

20. $2(x - 2)\sqrt{x + 1} + C$ **21.** $\frac{1}{2}$ **22.** -6 **23.** $\frac{17}{3}$

24. 242   **25.** -80 **26.** $\frac{132}{5}$ **27.** $\frac{1}{2}\ln 5$ **28.** $\frac{1}{15}$

29. 4   **30.** $1 - \dfrac{1}{e^2}$ **31.** $\dfrac{e - 1}{2(1 + e)}$ **32.** $\frac{1}{2}$

33. $f(x) = x^3 - 2x^2 + x + 1$ **34.** $f(x) = \sqrt{x^2 + 1}$

35. $f(x) = x + e^{-x} + 1$ **36.** $f(x) = \frac{1}{2}(\ln x)^2 - 2$

37. a. It gives the distance Car A is ahead of Car B.
    b. $t = 10$, $\int_0^{10} [f(t) - g(t)]\, dt$

38. a. It gives the amount by which the revenue of Branch A exceeds that of Branch B.
    b. $t = 10$, $\int_0^{10} [f(t) - g(t)]\, dt$

39. -4.28 **40.** 60.1%

41. $V(t) = 1900(t - 10)^2 + 10,000$; $40,400   **42.** $6740

43. a. $-0.015x^2 + 60x$ **b.** $p = -0.015x + 60$

44. a. $0.05t^3 - 1.8t^2 + 14.4t + 24$ **b.** 56°F

45. a. $-0.01t^3 + 0.109t^2 - 0.032t + 0.1$ **b.** 1.076 billion

46. 3.375 ppm **47.** $3100 **48.** 37.7 million

49. $N(t) = 15,000\sqrt{1 + 0.4t} + 85,000$; 112,659

50. $p = \dfrac{240}{5 - x} - 30$

51. a. $S(t) = 205.89 - 89.89e^{-0.176t}$ **b.** $161.43 billion

52. $3000t - 50,000(1 - e^{-0.04t})$; 16,939

53. 26,027 **54.** 15 **55.** $\frac{1}{2}(e^4 - 1)$

56. $\frac{2}{3}$ **57.** $\frac{9}{2}$ **58.** $e^2 - 3$ **59.** $\frac{3}{10}$ **60.** $\frac{1}{2}$

61. 234,500 barrels **62.** $\frac{1}{3}$ **63.** 26°F

64. 49.7 ft/sec **65.** 67,600/year **66.** $270,000

67. Consumers' surplus: $2083; producers' surplus: $3333

68. $197,652 **69.** $174,420 **70.** $505,696

71. a.

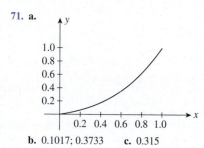

b. 0.1017; 0.3733 **c.** 0.315

72. 90,888

Chapter 6 Before Moving On, page 500

1. $\frac{1}{2}x^4 + \frac{2}{3}x^{3/2} + 2\ln|x| - 4\sqrt{x} + C$ **2.** $e^x + \frac{1}{2}x^2 + 1$

3. $\sqrt{x^2 + 1} + C$ **4.** $\frac{1}{3}(2\sqrt{2} - 1)$ **5.** $\frac{9}{2}$

CHAPTER 7

Exercises 7.1, page 507

1. $\frac{1}{4}e^{2x}(2x - 1) + C$ **3.** $2(x - 4)e^{x/4} + C$

5. $\frac{1}{2}e^{2x} - 2(x - 1)e^x + \frac{1}{3}x^3 + C$ **7.** $xe^x + C$

9. $\dfrac{2(x + 2)}{\sqrt{x + 1}} + C$ **11.** $\frac{2}{3}x(x - 5)^{3/2} - \frac{4}{15}(x - 5)^{5/2} + C$

13. $\dfrac{x^2}{4}(2\ln 2x - 1) + C$ **15.** $\dfrac{x^4}{16}(4\ln x - 1) + C$

17. $\frac{2}{9}x^{3/2}(3\ln\sqrt{x} - 1) + C$ **19.** $-\dfrac{1}{x}(\ln x + 1) + C$

21. $(x^2 + 1)e^x + C$ **23.** $x(\ln x - 1) + C$

25. $-(x^2 + 2x + 2)e^{-x} + C$

27. $\frac{1}{4}x^2[2(\ln x)^2 - 2\ln x + 1] + C$ **29.** $3\ln 3 - 2$

31. $3(4\ln 4 - 3)$ **33.** $\frac{1}{4}(3e^4 + 1)$ **35.** $-\frac{1}{2}xe^{-2x} - \frac{1}{4}e^{-2x} + \frac{13}{4}$

37. $5\ln 5 - 4$ **39.** 1485 ft **41.** $-20e^{-0.1t}(t + 10) + 200$

43. 696 **45.** 101,606 **47.** 235.74 million GBP

49. $840,254 **51.** $141,199 **53.** $143,477 **55.** 0.432

57. a. 0 lb **b.** 176.9 lb **c.** 120.6 lb

59. True **61.** True

Exercises 7.2, page 515

1. $\frac{2}{9}(2 + 3x - 2\ln|2 + 3x|) + C$

3. $\frac{3}{32}[(1 + 2x)^2 - 4(1 + 2x) + 2\ln|1 + 2x|] + C$

5. $2\left[\dfrac{x}{8}\left(\dfrac{9}{4} + 2x^2\right)\sqrt{\dfrac{9}{4} + x^2} - \dfrac{81}{128}\ln\left|x + \sqrt{\dfrac{9}{4} + x^2}\right|\right] + C$

7. $\ln\left|\dfrac{\sqrt{1 + 4x} - 1}{\sqrt{1 + 4x} + 1}\right| + C$ **9.** $\frac{1}{2}\ln 3$ **11.** $\dfrac{x}{9\sqrt{9 - x^2}} + C$

13. $\dfrac{x}{8}(2x^2 - 4)\sqrt{x^2 - 4} - 2\ln|x + \sqrt{x^2 - 4}| + C$

15. $\sqrt{4 - x^2} - 2\ln\left|\dfrac{2 + \sqrt{4 - x^2}}{x}\right| + C$

17. $\frac{1}{4}(2x - 1)e^{2x} + C$ **19.** $\ln|\ln(1 + x)| + C$

21. $\dfrac{1}{3}\left[\dfrac{1}{1 + 3e^x} + \ln(1 + 3e^x)\right] + C$

23. $6[e^{x/2} - \ln(1 + e^{x/2})] + C$

25. $\frac{4}{9}(2 + 3\ln x - 2\ln|2 + 3\ln x|) + C$

27. $e - 2$ **29.** $\dfrac{x^3}{9}(3\ln x - 1) + C$

31. $x[(\ln x)^3 - 3(\ln x)^2 + 6\ln x - 6] + C$ **33.** 27,136

35. 44; 49 **37.** 69.6 years **39.** $10,288 **41.** $112,533

43. $1,901,507 **45.** 0.1722 **47.** 0.472; 0.368; Country B

Exercises 7.3, page 528

1. 2.7037; 2.6667; $\frac{8}{3}$ **3.** 0.2656; 0.2500; $\frac{1}{4}$

5. 0.6970; 0.6933; $\ln 2 \approx 0.6931$ **7.** 0.5090; 0.5004; $\frac{1}{2}$

9. 5.2650; 5.3046; $\frac{16}{3}$ **11.** 0.6336; 0.6321; $-e^{-1} + 1 \approx 0.6321$

13. 0.3837; 0.3863; $2\ln 2 - 1 \approx 0.3863$ **15.** 1.1170; 1.1114

17. 1.3973; 1.4052 **19.** 0.8806; 0.8818 **21.** 3.7757; 3.7625

23. a. 3.6 **b.** 0.0324 **25. a.** 0.013 **b.** 0.00043

27. a. 0.0078125 **b.** 0.000285 **29.** 52.84 mi

31. a. 64.9°F **b.** 64.7°F **33.** 21.65 mpg

35. $25.9 billion **37.** 664.27 million barrels **39.** $22,134

41. a. $51,558 **b.** $51,708 **43.** 0.48

45. $407,667 **47.** 2698.9 ft³/sec **49.** 30%

51. False **53.** True

Exercises 7.4, page 539

1. 2   **3.** $\frac{1}{2}$ **5.** $\frac{2}{3}$ **7.** 1   **9.** 2   **11.** $\frac{2}{3}$ **13.** $\frac{1}{2}e^4$

15. 1   **17. a.** $\frac{2}{3}b^{3/2}$ **19.** 1   **21.** 2   **23.** Divergent

25. $-\frac{1}{8}$ **27.** 1   **29.** 1   **31.** $\frac{1}{2}$ **33.** Divergent

35. -1 **37.** Divergent   **39.** 0   **41.** 0

43. Divergent   **45.** $\frac{1}{2}$ **47.** \$30,000   **49.** \$750,000

51. \$1,125,000   **53.** True   **55.** False

57. b. \$166,667

Exercises 7.5, page 547

1. $\dfrac{2\pi}{3}$ **3.** $\dfrac{153\pi}{5}$ **5.** 3π **7.** $\dfrac{15\pi}{2}$

9. $\dfrac{4\pi}{3}$ **11.** $\dfrac{16\pi}{15}$ **13.** $\dfrac{\pi}{2}(e^2 - 1)$

15. $\dfrac{2\pi}{15}$ **17.** $\dfrac{136\pi}{15}$ **19.** $\dfrac{64\sqrt{2}\pi}{3}$

21. $\dfrac{\pi}{2}(e^2 - 2 + e^{-2})$ **23.** $\dfrac{\pi}{6}$

25. $\dfrac{6517\pi}{240}$ **27.** $\dfrac{64\sqrt{2}\pi}{3}$

29. $\dfrac{8(\sqrt{2}-1)\pi}{3}$ **31.** $\dfrac{4\pi}{3}r^3$

33. $50{,}000\pi$ ft^3

Chapter 7 Concept Review, page 549

1. Product; $uv - \int v\,du$; u; easy to integrate   **2.** $x^2 + 1$; $2x\,dx$; (27)

3. $\dfrac{\Delta x}{2}[f(x_0) + 2f(x_1) + \cdots + 2f(x_{n-1}) + f(x_n)]$; $\dfrac{M(b-a)^3}{12n^2}$

4. $\dfrac{\Delta x}{3}[f(x_0) + 4f(x_1) + 2f(x_2) + 4f(x_3) + 2f(x_4)$
$+ \cdots + 4f(x_{n-1}) + f(x_n)]$; even; $\dfrac{M(b-a)^5}{180n^4}$

5. $\lim\limits_{a\to-\infty} \int_a^b f(x)\,dx$; $\lim\limits_{b\to\infty} \int_a^b f(x)\,dx$; $\int_{-\infty}^c f(x)\,dx + \int_c^\infty f(x)\,dx$

6. region; $x = a$; $x = b$; $V = \pi\int_a^b [f(x)]^2\,dx$

Chapter 7 Review Exercises, page 550

1. $-2(1 + x)e^{-x} + C$ **2.** $\frac{1}{16}(4x - 1)e^{4x} + C$

3. $x(\ln 5x - 1) + C$ **4.** $11\ln 2 - 3$ **5.** $\frac{1}{4}(1 - 3e^{-2})$

6. $\frac{1}{4}(1 + 3e^4)$ **7.** $2\sqrt{x}(\ln x - 2) + 2$

8. $-\frac{1}{3}xe^{-3x} - \frac{1}{9}e^{-3x} + \frac{1}{9}$

9. $\dfrac{1}{4}\left(\dfrac{3}{3 + 2x} + \ln|3 + 2x|\right) + C$

10. $\frac{2}{3}(x - 3)\sqrt{2x + 3} + C$ **11.** $\frac{1}{32}e^{4x}(8x^2 - 4x + 1) + C$

12. $-\dfrac{x}{25\sqrt{x^2 - 25}} + C$ **13.** $\dfrac{1}{4}\dfrac{\sqrt{x^2 - 4}}{x} + C$

14. $\frac{1}{2}x^4(4\ln 2x - 1) + C$ **15.** $\frac{1}{2}$ **16.** $\frac{1}{3}$

17. Divergent   **18.** 1   **19.** $\frac{1}{10}$ **20.** 3

21. 0.8421; 0.8404   **22.** 1.4907; 1.4637   **23.** 2.2379; 2.1791

24. 8.1310; 8.0409   **25. a.** 0.002604   **b.** 0.000033

26. \$1,157,641   **27.** $-20e^{-0.05t}(t + 20) + 400$; 48,761

28. \$41,100   **29.** 274,000 ft^2; 278,667 ft^2 **30.** 7850 ft^2

31. \$250,000   **32.** $\frac{8\pi}{35}$ **33.** $\frac{2\pi}{3}$ **34.** $\frac{3\pi}{10}$

Chapter 7 Before Moving On, page 551

1. $\frac{1}{3}x^3\ln x - \frac{1}{9}x^3 + C$ **2.** $-\dfrac{\sqrt{8 + 2x^2}}{8x} + C$ **3.** 6.3367

4. 3.0036   **5.** $\dfrac{1}{2e^2}$ **6.** 2π

CHAPTER 8

Exercises 8.1, page 561

1. $f(0, 0) = -4$; $f(1, 0) = -2$; $f(0, 1) = -1$; $f(1, 2) = 4$;
$f(2, -1) = -3$

3. $f(1, 2) = 7$; $f(2, 1) = 9$; $f(-1, 2) = 1$; $f(2, -1) = 1$

5. $g(1, 2) = 4 + 3\sqrt{2}$; $g(2, 1) = 8 + \sqrt{2}$; $g(0, 4) = 2$; $g(4, 9) = 56$

7. $h(1, e) = 1$; $h(e, 1) = -1$; $h(e, e) = 0$

9. $g(1, 1, 1) = e$; $g(1, 0, 1) = 1$; $g(-1, -1, -1) = -e$

11. All real values of x and y

13. All real values of u and v except those satisfying the equation $u = v$

15. All real values of r and s satisfying $rs \geq 0$

17. All real values of x and y satisfying $x + y > 5$

19.
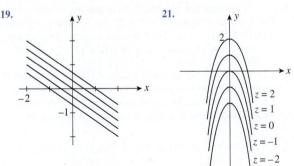

21.

23.
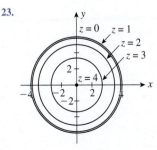

25. $\sqrt{x^2 + y^2} = 5$ **27.** (b)   **29.** No   **31.** 9π ft^3

33. a. P and E are real numbers, with $E \neq 0$.   **b.** 15.7

35. a. 24.69   **b.** 81 kg

37. a. $200x + 120y + 20,000$
   b. The set of all x and y such that $x \geq 0$ and $y \geq 0$
   c. $244,000

39. a. $-\frac{1}{5}x^2 - \frac{1}{4}y^2 - \frac{1}{5}xy + 200x + 160y$
   b. The set of all points (x, y) satisfying $200 - \frac{1}{5}x - \frac{1}{10}y \geq 0$,
   $160 - \frac{1}{10}x - \frac{1}{4}y \geq 0, x \geq 0, y \geq 0$

41. a. $-0.005x^2 - 0.003y^2 - 0.002xy + 20x + 15y$
   b. The set of all ordered pairs (x, y) for which
   $20 - 0.005x - 0.001y \geq 0$
   $15 - 0.001x - 0.003y \geq 0, x \geq 0, y \geq 0$

43. a. The set of all ordered pairs (P, T), where P and T are positive
   numbers
   b. 11.10 L

45. a. The domain of W is the set of all ordered pairs (L, G), where
   $L > 0$ and $G > 0$.
   b. 3.6 lb

47. $7200 billion   **49.** 103

51. a. $1432.25; $1798.65
   b. $2149.29

53. 18

57. The level curves of V have equation $\dfrac{kT}{P} = C$ (C, a positive constant).
   The level curves are a family of straight lines $T = \left(\dfrac{C}{k}\right)P$ lying in
   the first quadrant, since k, T, and P are positive. Every point on the
   level curve $V = C$ gives the same volume C.

59. False   **61.** False   **63.** False

Exercises 8.2, page 576

1. a. 4; 4
   b. $f_x(2, 1) = 4$ says that the slope of the tangent line to the curve of
   intersection of the surface $z = x^2 + 2y^2$ and the plane $y = 1$ at the
   point $(2, 1, 6)$ is 4. $f_y(2, 1) = 4$ says that the slope of the tangent
   line to the curve of intersection of the surface $z = x^2 + 2y^2$ and
   the plane $x = 2$ at the point $(2, 1, 6)$ is 4.
   c. $f_x(2, 1) = 4$ says that the rate of change of $f(x, y)$ with respect to
   x with y held fixed with a value of 1 is 4 units/unit change in x.
   $f_y(2, 1) = 4$ says that the rate of change of $f(x, y)$ with respect to
   y with x held fixed with a value of 2 is 4 units/unit change in y.

3. $f_x = 2; f_y = 3$ **5.** $g_x = 4x; g_y = 4$ **7.** $f_x = -\dfrac{4y}{x^3}; f_y = \dfrac{2}{x^2}$

9. $g_u = \dfrac{2v}{(u + v)^2}; g_v = -\dfrac{2u}{(u + v)^2}$

11. $f_s = 3(2s - t)(s^2 - st + t^2)^2; f_t = 3(2t - s)(s^2 - st + t^2)^2$

13. $f_x = \dfrac{4x}{3(x^2 + y^2)^{1/3}}; f_y = \dfrac{4y}{3(x^2 + y^2)^{1/3}}$

15. $f_x = ye^{xy+1}; f_y = xe^{xy+1}$

17. $f_x = \ln y + \dfrac{y}{x}; f_y = \dfrac{x}{y} + \ln x$ **19.** $g_u = e^u \ln v; g_v = \dfrac{e^u}{v}$

21. $f_x = yz + y^2 + 2xz; f_y = xz + 2xy + z^2; f_z = xy + 2yz + x^2$

23. $h_r = ste^{rst}; h_s = rte^{rst}; h_t = rse^{rst}$ **25.** $f_x(1, 2) = 8; f_y(1, 2) = 5$

27. $f_x(2, 1) = 1; f_y(2, 1) = 3$ **29.** $f_x(1, 2) = \frac{1}{2}; f_y(1, 2) = -\frac{1}{4}$

31. $f_x(1, 1) = e; f_y(1, 1) = e$

33. $f_x(1, 0, 2) = 0; f_y(1, 0, 2) = 8; f_z(1, 0, 2) = 0$

35. $f_{xx} = 2y; f_{xy} = 2x + 3y^2 = f_{yx}; f_{yy} = 6xy$

37. $f_{xx} = 2; f_{xy} = f_{yx} = -2; f_{yy} = 4$

39. $f_{xx} = \dfrac{y^2}{(x^2 + y^2)^{3/2}}; f_{xy} = f_{yx} = -\dfrac{xy}{(x^2 + y^2)^{3/2}};$
   $f_{yy} = \dfrac{x^2}{(x^2 + y^2)^{3/2}}$

41. $f_{xx} = \dfrac{1}{y^2}e^{-x/y}; f_{xy} = \dfrac{y - x}{y^3}e^{-x/y} = f_{yx};$
   $f_{yy} = \dfrac{x}{y^3}\left(\dfrac{x}{y} - 2\right)e^{-x/y}$

43. a. 7.5; 40   **b.** Yes

45. 1405; 2602; this says that an increase of 1 work-hour/day with capital
   expenditures held fixed at $300/day will result in an increase of
   approximately 1405 candles/day. An increase of $1/day in capital
   expenditures with work-hours/day held fixed at 400 will result in an
   increase of approximately 2602 candles/day.

47. $p_x = 10$—at $(0, 1)$, the price of land is changing at the rate of
   $10/ft^2/mile change to the east; $p_y = 0$—at $(0, 1)$, the price of land is
   constant/mile change to the north.

49. 29; -475; this says that when the floor space is held fixed at 150,000
   ft^2, the monthly profit increases at the rate of $29 per thousand-dollar
   increase in the inventory when the inventory is $4,000,000, and when
   the inventory is held fixed at $4,000,000, the monthly profit de-
   creases at the rate of $475 per thousand-square-foot increase in floor
   space when the floor space is 150,000 ft^2; -11; 525

51. 1.06; -2.85; If the level of reinvestment is held constant at 20 cents
   per dollar deposited, the number of fires will grow at the rate of
   approximately 1 fire per increase of 1 person per census tract when
   the number of people per census tract is 100. If the number of people
   per census tract is held constant at 100 per tract, the number of fires
   will decrease at a rate of approximately 2.9 per increase of 1 cent per
   dollar deposited for reinvestment when the level of reinvestment is
   20 cents per dollar deposited.

53. a. 1.3°F/°F   **b.** -0.22°F/mi/hr

57. Substitute commodities

61. 1.923, -1; an increase of 1% in the price of butter will result in a
   1.9% drop in the demand for butter with the price of margarine held
   fixed at $4/lb; an increase of 1% in the price of margarine will result
   in a 1% increase in the demand for butter with the price of butter held
   fixed at $5/lb.

67. False   **69.** True   **71.** True

Using Technology Exercises 8.2, page 581

1. 1.3124; 0.4038

3. −1.8889; 0.7778

5. −0.3863; −0.8497

Exercises 8.3, page 589

1. $(0, 0)$; relative maximum value: $f(0, 0) = 1$

3. $(1, 2)$; saddle point: $f(1, 2) = 4$

5. $(8, -6)$; relative minimum value: $f(8, -6) = -41$

7. $(1, 2)$ and $(2, 2)$; saddle point: $f(1, 2) = -1$; relative minimum value: $f(2, 2) = -2$

9. $\left(-\frac{1}{3}, \frac{11}{3}\right)$ and $(1, 5)$; saddle point: $f\left(-\frac{1}{3}, \frac{11}{3}\right) = -\frac{319}{27}$; relative minimum value: $f(1, 5) = -13$

11. $(0, 0)$ and $(1, 1)$; saddle point: $f(0, 0) = -2$; relative minimum value: $f(1, 1) = -3$

13. $(2, 1)$; relative minimum value: $f(2, 1) = 6$

15. $(0, 0)$; saddle point: $f(0, 0) = -1$

17. $(0, 0)$; relative minimum value: $f(0, 0) = 1$

19. $(0, 0)$; relative minimum value: $f(0, 0) = 0$

21. 200 finished units and 100 unfinished units; $10,500

23. Price of land ($200/ft^2) is highest at $\left(\frac{1}{2}, 1\right)$.

25. a. $400(-p^2 - q^2 - pq + 16p + 14q)$
b. $6/lb for ground sirloin, $4/lb for ground beef; 3200 lb ground sirloin, 2800 lb of ground beef; $30,400

27. a. $-20x^2 - 30y^2 - 20xy + 3000x + 4000y$
b. $400x + 500y + 20,000$
c. $-20x^2 - 30y^2 - 20xy + 2600x + 3500y - 20,000$
d. 43,000 regular tubes; 44,000 whitening tubes; $112,900

29. At $\left(\frac{26}{3}, 12\right)$

31. 6 in. × 6 in. × 3 in.

33. $21\frac{2}{3}$ in. × $43\frac{1}{3}$ in. × $21\frac{2}{3}$ in.

35. 6 in. × 4 in. × 2 in.

37. False   **39.** False

41. True

Exercises 8.4, page 598

1. a. $y = 2.3x + 1.5$
b.

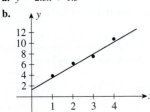

3. a. $y = -0.77x + 5.74$
b.

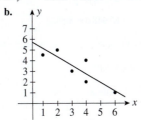

5. a. $y = 1.2x + 2$
b.

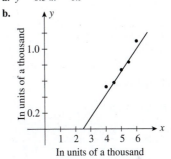

7. a. $y = 0.34x - 0.9$
b.

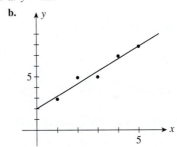

c. 1276 applications

9. a. $y = -2.8x + 440$
b.

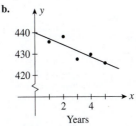

c. 420

11. a. $y = 234.4x + 157.3$ **b.** 1798.1 million

13. a. $y = 2.8x + 17.6$ **b.** $40 million

15. a. $y = 0.305x + 0.19$ **b.** $0.305 billion/year   **c.** $3.24 billion

17. a. $y = 2.46x + 82.1$ **b.** 94.4 million

19. a. $y = 2.75x + 29.4$ **b.** 2.75 million/year

21. a. $y = 0.3x + 6.46$ **b.** $0.3 billion/year

23. a. $y = 2.19x + 18.38$ **b.** $2.19 billion/year   **c.** $42.5 billion

25. a. $y = 7.25x + 60.21$ **b.** $139.96 billion   **c.** $7.25 billion/year

27. False   **29.** True

Using Technology Exercises 8.4, page 603

1. $y = 2.3596x + 3.8639$ **3.** $y = -1.1948x + 3.5525$

5. a. $y = 22.3x + 143.5$ **b.** \$22.3 billion/year   **c.** \$366.5 billion

7. a. $y = 1.5857t + 6.6857$ **b.** \$16.2 billion

9. a. $y = 1.7571x + 7.9143$ **b.** \$22 billion

Exercises 8.5, page 612

1. Min. of $\frac{3}{4}$ at $\left(\frac{3}{4}, \frac{1}{4}\right)$ **3.** Max. of $-\frac{7}{4}$ at $\left(2, \frac{7}{2}\right)$

5. Min. of 4 at $\left(\sqrt{2}, \frac{\sqrt{2}}{2}\right)$ and $\left(-\sqrt{2}, -\frac{\sqrt{2}}{2}\right)$

7. Max. of $-\frac{3}{4}$ at $\left(\frac{3}{2}, 1\right)$

9. Max. of $2\sqrt{3}$ at $\left(\frac{\sqrt{3}}{3}, -\sqrt{6}\right)$ and $\left(\frac{\sqrt{3}}{3}, \sqrt{6}\right)$

11. Max. of 8 at $(2\sqrt{2}, 2\sqrt{2})$ and $(-2\sqrt{2}, -2\sqrt{2})$; min. of -8 at $(2\sqrt{2}, -2\sqrt{2})$ and $(-2\sqrt{2}, 2\sqrt{2})$

13. Max. of $\frac{2\sqrt{3}}{9}$ at $\left(\frac{\sqrt{3}}{3}, \pm\frac{\sqrt{6}}{3}\right)$; min.: $-\frac{2\sqrt{3}}{9}$ at $\left(-\frac{\sqrt{3}}{3}, \pm\frac{\sqrt{6}}{3}\right)$

15. Min. of $\frac{18}{7}$ at $\left(\frac{9}{7}, \frac{6}{7}, \frac{3}{7}\right)$

17. 140 finished units, 60 unfinished units

19. $10\sqrt{2}$ ft $\times 40\sqrt{2}$ ft   **21.** 2 ft $\times$ 2 ft $\times$ 1 ft

23. 18 in. $\times$ 18 in. $\times$ 36 in.   **25.** $r = \frac{4}{3}\sqrt[3]{\frac{18}{\pi}}$ in.; $h = 2\sqrt[3]{\frac{18}{\pi}}$ in.

27. $\frac{2}{3}\sqrt[3]{9}$ ft $\times \frac{2}{3}\sqrt[3]{9}$ ft $\times \sqrt[3]{9}$ ft   **29.** $\frac{2}{3}\sqrt[3]{36}$ ft $\times \frac{2}{3}\sqrt[3]{36}$ ft $\times \sqrt[3]{36}$ ft

33. 1500 units on labor and 250 units on capital

35. 1500 units on labor, 250 units on capital

37. $\frac{kp}{a}\left[\frac{bq}{(1-b)p}\right]^{1-b}; \frac{kq}{a}\left[\frac{(1-b)p}{bq}\right]^{b}$

39. False   **41.** True

Exercises 8.6, page 620

1. a. -0.0386 **b.** -0.04 **3.** $(4x - 3y + 4)\,dx - 3x\,dy$

5. $\frac{x}{\sqrt{x^2 + y^2}}\,dx + \frac{y}{\sqrt{x^2 + y^2}}\,dy$

7. $-\frac{5y}{(x - y)^2}\,dx + \frac{5x}{(x - y)^2}\,dy$

9. $(10x^4 + 3ye^{-3x})\,dx - e^{-3x}\,dy$

11. $\left(2xe^y + \frac{y}{x}\right)dx + (x^2e^y + \ln x)\,dy$

13. $y^2z^3\,dx + 2xyz^3\,dy + 3xy^2z^2\,dz$

15. $\frac{1}{y + z}\,dx - \frac{x}{(y + z)^2}\,dy - \frac{x}{(y + z)^2}\,dz$

17. $(yz + e^{yz})\,dx + xz(1 + e^{yz})\,dy + xy(1 + e^{yz})\,dz$

19. 0.04   **21.** -0.01 **23.** -0.10 **25.** -0.18

27. 0.06   **29.** -0.1401 **31.** An increase of \$19,250/month

33. An increase of \$5000/month   **35.** 43   **37.** $\leq 7\%$

39. 1.7%   **41.** 3.1 ft³   **43.** 104 in.²

45. 3.3256 Pa   **47.** 0.55 ohm

Exercises 8.7, page 629

1. $\frac{7}{2}$ **3.** 0   **5.** $\frac{9}{2}$ **7.** $(e^2 - 1)(1 - e^{-2})$ **9.** 1

11. $\frac{2}{3}$ **13.** $\frac{188}{3}$ **15.** $\frac{84}{5}$

17. $\frac{8}{3}$ **19.** 1   **21.** $\frac{1}{2}(3 - e)$

23. $\frac{1}{4}(e^4 - 1)$ **25.** $\frac{2}{3}(e - 1)$ **27.** False

Exercises 8.8, page 635

1. 48   **3.** 20   **5.** 25.13   **7.** $\frac{64}{3}$ **9.** 4   **11.** $\frac{10}{3}$

13. $2(e^2 - 1)$ **15.** $\frac{2}{35}$ **17.** 54   **19.** $\frac{1}{3}$ **21.** 1

23. 43,329   **25.** 312,455   **27.** \$194/ft²   **29.** True

Chapter 8 Concept Review, page 638

1. xy; ordered pair; real number; $f(x, y)$

2. independent; dependent; value

3. $z = f(x, y)$; f; surface

4. $f(x, y) = k$; level curve; level curves; k

5. constant; x **6.** slope; $(a, b, f(a, b))$; x; b

7. $\leq$; (a, b); $\leq$; domain

8. domain; $f_x(a, b) = 0$ and $f_y(a, b) = 0$; exist; candidate

9. scatter; minimizing; least-squares; normal

10. $g(x, y) = 0$; $f(x, y) + \lambda g(x, y)$; $F_x = 0$; $F_y = 0$; $F_\lambda = 0$; extrema

11. independent; $dx = \Delta x$; $dy = \Delta y$; $\frac{\partial f}{\partial x}\,dx + \frac{\partial f}{\partial y}\,dy$

12. volume; solid   **13.** iterated; $\int_3^5 \int_0^1 (2x + y^2)\,dx\,dy$

Chapter 8 Review Exercises, page 639

1. $0, 0, \frac{1}{2}$; no   **2.** $e, \frac{e^2}{1 + \ln 2}, \frac{2e}{1 + \ln 2}$; no

3. $2, -(e + 1), -(e + 1)$

4. The set of all ordered pairs (u, v) such that $u \neq v$ and $u \geq 0$

5. The set of all ordered pairs (x, y) such that $y \neq -x$

6. The set of all ordered pairs (x, y) such that $x \leq 1$ and $y \geq 0$

7. The set of all ordered triplets (x, y, z) such that $z \geq 0$ and $x \neq 1$, $y \neq 1$, and $z \neq 1$

8. $2x + 3y = z$

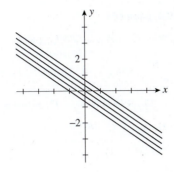

9. $z = y - x^2$

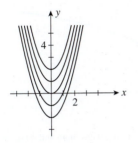

10. $z = \sqrt{x^2 + y^2}$

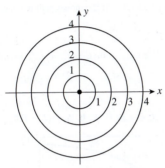

11. $z = e^{xy}$

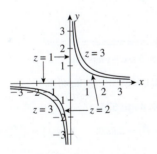

12. $f_x = 2xy^3 + 3y^2 + \dfrac{1}{y}; \; f_y = 3x^2y^2 + 6xy - \dfrac{x}{y^2}$

13. $f_x = \sqrt{y} + \dfrac{y}{2\sqrt{x}}; \; f_y = \dfrac{x}{2\sqrt{y}} + \sqrt{x}$

14. $f_u = \dfrac{v^2 - 2}{2\sqrt{uv^2 - 2u}}; \; f_v = \dfrac{uv}{\sqrt{uv^2 - 2u}}$

15. $f_x = \dfrac{3y}{(y + 2x)^2}; \; f_y = -\dfrac{3x}{(y + 2x)^2}$

16. $g_x = \dfrac{y(y^2 - x^2)}{(x^2 + y^2)^2}; \; g_y = \dfrac{x(x^2 - y^2)}{(x^2 + y^2)^2}$

17. $h_x = 10y(2xy + 3y^2)^4; \; h_y = 10(x + 3y)(2xy + 3y^2)^4$

18. $f_x = \dfrac{e^y}{2(xe^y + 1)^{1/2}}; \; f_y = \dfrac{xe^y}{2(xe^y + 1)^{1/2}}$

19. $f_x = 2x(1 + x^2 + y^2)e^{x^2 + y^2}; \; f_y = 2y(1 + x^2 + y^2)e^{x^2 + y^2}$

20. $f_x = \dfrac{4x}{1 + 2x^2 + 4y^4}; \; f_y = \dfrac{16y^3}{1 + 2x^2 + 4y^4}$

21. $f_x = \dfrac{2x}{x^2 + y^2}; \; f_y = -\dfrac{2x^2}{y(x^2 + y^2)}$

22. $f_{xx} = 6x - 4y; \; f_{xy} = -4x = f_{yx}; \; f_{yy} = 2$

23. $f_{xx} = 12x^2 + 4y^2; \; f_{xy} = 8xy = f_{yx}; \; f_{yy} = 4x^2 - 12y^2$

24. $f_{xx} = 12(2x^2 + 3y^2)(10x^2 + 3y^2);$
$f_{xy} = 144xy(2x^2 + 3y^2) = f_{yx};$
$f_{yy} = 18(2x^2 + 3y^2)(2x^2 + 15y^2)$

25. $g_{xx} = \dfrac{-2y^2}{(x + y^2)^3}; \; g_{xy} = \dfrac{2y(x - y^2)}{(x + y^2)^3} = g_{yx};$
$g_{yy} = \dfrac{2x(3y^2 - x)}{(x + y^2)^3}$

26. $g_{xx} = 2(1 + 2x^2)e^{x^2 + y^2}; \; g_{xy} = 4xye^{x^2 + y^2} = g_{yx};$
$g_{yy} = 2(1 + 2y^2)e^{x^2 + y^2}$

27. $h_{ss} = -\dfrac{1}{s^2}; \; h_{st} = h_{ts} = 0; \; h_{tt} = \dfrac{1}{t^2}$

28. $3; \; 3; \; -2$

29. $(2, 3);$ relative minimum value: $f(2, 3) = -13$

30. $(8, -2);$ saddle point at $f(8, -2) = -8$

31. $(0, 0)$ and $\left(\tfrac{3}{2}, \tfrac{9}{4}\right);$ saddle point at $f(0, 0) = 0;$ relative minimum value:
$f\left(\tfrac{3}{2}, \tfrac{9}{4}\right) = -\tfrac{27}{16}$

32. $\left(-\tfrac{1}{3}, \tfrac{13}{3}\right), (3, 11);$ saddle point at $f\left(-\tfrac{1}{3}, \tfrac{13}{3}\right) = -\tfrac{445}{27};$ relative minimum
value: $f(3, 11) = -35$

33. $(0, 0);$ relative minimum value: $f(0, 0) = 1$

34. $(1, 1);$ relative minimum value: $f(1, 1) = \ln 2$

35. $f\left(\tfrac{12}{11}, \tfrac{20}{11}\right) = -\tfrac{32}{11}$ **36.** $f\left(-\tfrac{1}{22}, \tfrac{23}{22}\right) = \tfrac{175}{44}$

37. Relative maximum value: $f(5, -5) = 26;$ relative minimum value:
$f(-5, 5) = -24$

38. Relative maximum value: $f\left(\dfrac{\sqrt{2}}{2}, -\dfrac{\sqrt{2}}{2}\right) = e^{\sqrt{2}};$ relative

minimum value: $f\left(-\dfrac{\sqrt{2}}{2}, \dfrac{\sqrt{2}}{2}\right) = e^{-\sqrt{2}}$

39. $45 \, dx + 240 \, dy$ **40.** $2 \, dx$ **41.** 0.04 **42.** $\dfrac{227}{54}$

43. 48 **44.** $\tfrac{1}{2}(e^{-2} - 1)^2$ **45.** $\tfrac{2}{63}$ **46.** $\tfrac{1}{4}(3 - 2\ln 2)$

47. $\tfrac{34}{3}$ **48.** $\tfrac{54}{5}$ **49.** 3

50. $k = \dfrac{100 \, m}{c}$

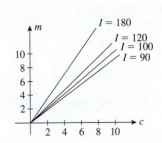

51. a. $R(x, y) = -0.02x^2 - 0.2xy - 0.05y^2 + 80x + 60y$
b. The set of all points satisfying $0.02x + 0.1y \leq 80$,
$0.1x + 0.05y \leq 60$, $x \geq 0$, $y \geq 0$
c. 15,300; the revenue realized from the sale of 100 16-speed and 300 10-speed electric blenders is $15,300.

52. Complementary   **53.** $490

54. a. $y = 8.2x + 361.2$
b. 7 hr 31 min

55. a. $y = 0.059x + 19.45$
b. 21.9 years; almost the same   **c.** 21.2 years

56. a. $y = 98.75x + 547.94$ **b.** 1239.2 million

57. The company should spend $11,000 on advertising and employ 14 agents to maximize its revenue.

58. 337.5 yd $\times$ 900 yd

59. 30 agents, 52 advertisements; $369,800/year

60. 300 bass, 200 trout

61. 75 units on labor; 25 units on capital

Chapter 8 Before Moving On, page 641

1. All real values of x and y satisfying $x \geq 0$, $x \neq 1$, $y \geq 0$, $y \neq 2$

2. $f_x = 2xy + ye^{xy}; f_{xx} = 2y + y^2 e^{xy}; f_{xy} = 2x + (xy + 1)e^{xy} = f_{yx};$
$f_y = x^2 + xe^{xy}; f_{yy} = x^2 e^{xy}$

3. Relative minimum value: $f(1, 1) = -7$ **4.** $y = 2.04x + 2.88$

5. $f\left(\frac{1}{2}, \frac{1}{2}\right) = \frac{5}{2}$ **6. a.** $(4x - y)\,dx - x\,dy$ **b.** -0.09 **7.** $\frac{1}{8}$

CHAPTER 9

Exercises 9.1, page 649

13. $y = 12x^2 - 2x$ **15.** $y = \dfrac{1}{x}$ **17.** $y = -\dfrac{e^x}{x} + \dfrac{1}{2}xe^x$

19. $\dfrac{dQ}{dt} = -kQ; Q(0) = Q_0$ **21.** $\dfrac{dA}{dt} = k(C - A)$

23. $\dfrac{dC}{dt} = -kC; C(0) = C_0$ **29.** True   **31.** False

Exercises 9.2, page 655

1. $y^3 = \frac{3}{2}x^2 + 3x + C$ **3.** $y^3 = 3e^x + C$ **5.** $y = Ce^{2x}$

7. $y = -\dfrac{2}{x^2 + 2C}$ **9.** $y = -\frac{4}{3} + Ce^{-6x}$

11. $y^3 = \frac{1}{3}x^3 + x + C$ **13.** $y^{1/2} - x^{1/2} = C$

15. $y = Ce^{(\ln x)^2/2}$ **17.** $y = -\sqrt{2x^2 + 2}$

19. $y = 2 + e^{-x}$ **21.** $y = \frac{2}{3} + \frac{1}{3}e^{3x^2/2}$ **23.** $y = \sqrt{x^2 + 1}$

25. $y = e^{(x-1)e^x}$ **27.** $y = \ln(x^3 + e)$

29. $y = \sqrt{x^3 + 8}$ **31.** $Q(t) = Q_0 e^{-kt}$ **33.** $A = \dfrac{C}{k} - De^{-kt}$

35. False   **37.** True   **39.** True   **41.** False

Exercises 9.3, page 666

1. $y = y_0 e^{-kt}$ **3.** $Q(t) = 4.5e^{0.013t}$; 7.1 billion   **5.** $\frac{1}{2}$ in.

7. $Q(t) = \dfrac{50}{4t + 1}$; 5.56 g   **9.** 3.6 min   **11.** 5.28 g; 2.5 years

13. 23/day   **15.** 319.9 million   **17.** $p(t) = 9 - 5e^{-5kt}$

19. 395   **21.** $A(t) = C - Be^{-kt}$

23. a. $U(x) = k\ln(x + 1)$

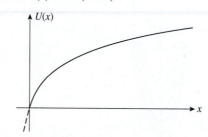

b. As a person accumulates more and more money, the person's perceived value of the money will increase but at a slower rate.

25. $Q(t) = e^C e^{-(C - \ln Q_0)e^{-kt}}$

27. a. $C_0 e^{-kT}$ g/mL   **b.** $C_0(e^{-kT} + e^{-2kT})$ g/mL
c. $C_0(e^{-kT} + e^{-2kT} + \cdots + e^{-NkT})$ g/mL   **d.** $\dfrac{C_0 e^{-kT}}{1 - e^{-kT}}$ g/mL

29. a. $L - (L - x_0)e^{-kt}$ **b.** $L - (L - 0.4)\left(\dfrac{L - 10}{L - 0.4}\right)^t$

c.
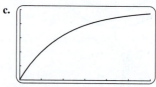

d. Between 5 and 9 years old

31. $x(t) = 40(1 - e^{-3t/20})$; 38 lb; 40 lb

Exercises 9.4, page 674

1. a. $y(1) = \frac{369}{128} \approx 2.8828$ **b.** $y(1) = \frac{70,993}{23,328} \approx 3.043$

3. a. $y(2) = \frac{51}{16} \approx 3.1875$ **b.** $y(2) = \frac{793}{243} \approx 3.2634$

5. a. $y(0.5) \approx 0.8324$ **b.** $y(0.5) \approx 0.8207$

7. a. $y(1.5) \approx 1.7831$ **b.** $y(1.5) \approx 1.7920$

9. a. $y(1) \approx 1.3390$ **b.** $y(1) \approx 1.3654$

11.

x	0.0	0.2	0.4	0.6	0.8	1
y_n	1	1	1.02	1.0608	1.1244	1.2144

13.

x	0.0	0.2	0.4	0.6	0.8	1
y_n	2	1.8	1.72	1.736	1.8288	1.9830

15.

x	0.0	0.1	0.2	0.3	0.4	0.5
y_n	1	1.1	1.211	1.3361	1.4787	1.6426

Chapter 9 Concept Review, page 676

1. a. differential equation   **b.** satisfies

2. a. general   **b.** particular

3. a. highest; unknown   **b.** $f(x)g(y)$; not separable; separable
  c. separate; variable

4. a. kQ **b.** Q_0e^{kt} **c.** restricted

5. approximate; actual; polygonal

Chapter 9 Review Exercises, page 676

4. $y = \dfrac{1}{x^2 + 1}$ **5.** $y = (9x + 8)^{-1/3}$ **6.** $y^3 = \frac{3}{4}x^4 + 3x + C$

7. $y = 4 - Ce^{-2t}$ **8.** $y = Ce^{(\ln x)^2/2}$ **9.** $y = -\dfrac{2}{2x^3 + 2x + 1}$

10. $y = 1 - 3e^{-(1/3)x^3}$ **11.** $y = 3e^{-x^3/2}$ **12.** $y = \dfrac{4}{(x^2 + 1)^2}$

13. a. $y(1) \approx 0.3849$ **b.** $y(1) \approx 0.4361$

14. a. $y(1) \approx 0.2219$ **b.** $y(1) = 0.2628$

15. a. $y(1) \approx 1.3258$ **b.** $y(1) \approx 1.4570$

16. a. $y(1) \approx 1.9083$ **b.** $y(1) = 2.1722$

17.

x	0.0	0.2	0.4	0.6	0.8	1
y_n	1	1	1.08	1.2528	1.5535	2.0506

18.

x	0	0.2	0.4	0.6	0.8	1
y_n	1	1.2	1.496	1.9756	2.8282	4.5560

19. a. $S = 50{,}000(0.8)^t$ **b.** $\$16{,}384$ **20.** $R = CS^k$

21. $A = \dfrac{P}{r}(e^{rt} - 1);\ \$291{,}549.01$

22. a. $S(t) = \dfrac{1}{r}[(rS_0 + d)e^{rt} - d]$ **b.** $\$25{,}160.55$

23. $S = D - (D - S_0)e^{-kt}$ **24.** 7:30 P.M.

25. a. $P = Ce^{-kh}, P > 0$ **b.** 6.7 psi   **26.** 105 words/min

27. a. 183   **28.** $x = 120(1 - e^{-0.1t})$; 120 lb

Chapter 9 Before Moving On, page 678

1. b. $y = 2x^2$ **2.** $y = \dfrac{1}{xe^{x-1}}$ **3.** $P(t) = 5000e^{0.1386t}$; 6.3 years

4. 1.7639

CHAPTER 10

Exercises 10.1, page 689

11. $k = \frac{1}{3}$ **13.** $k = \frac{1}{8}$ **15.** $k = \frac{3}{16}$ **17.** $k = 2$

19. a. $\frac{1}{2}$ **b.** $\frac{5}{8}$ **c.** $\frac{7}{8}$ **d.** 0

21. a. $\frac{11}{16}$ **b.** $\frac{1}{2}$ **c.** $\frac{27}{32}$ **d.** 0

23. a. $\frac{1}{2}$ **b.** $\frac{1}{2}(2\sqrt{2} - 1)$ **c.** 0   **d.** $\frac{1}{2}$

25. a. 1   **b.** .135

27. a. .375   **b.** .75   **c.** .5   **d.** .875

33. $k = 2$ **35.** $k = 12e^2$ **37. a.** $\frac{1}{4}$ **b.** $\frac{1}{96}$

39. a. $\frac{1}{56}(2\sqrt{2} - 1)$ **b.** .2575

41. a. .06   **b.** .37

43. a. .777   **b.** .145   **c.** .223   **45. a.** .099   **b.** .30

47. .02   **49. a.** $\frac{2}{3}$ **b.** $\frac{1}{5}$ **51. a.** $\frac{1}{3}$ **b.** $\frac{1}{3}$

53. .5833   **55. a.** .4512   **b.** .1481   **c.** .2466

57. .0414   **59.** .6152   **63.** False   **65.** False

Using Technology Exercises 10.1, page 693

1.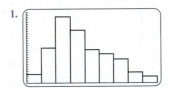

3. a.

x	0	1	2	3	4
$P(X = x)$	0.02	0.07	0.03	.12	.23

x	5	6	7	8	9	10
$P(X = x)$	.13	.17	.10	.05	.07	.02

b.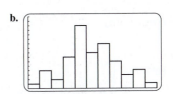

Exercises 10.2, page 707

1. $\mu = \frac{9}{2}$; $\text{Var}(X) = \frac{3}{4}$; $\sigma \approx 0.8660$

3. $\mu = \frac{15}{4}$; $\text{Var}(X) = \frac{15}{16}$; $\sigma \approx 0.9682$

5. $\mu = 3$; $\text{Var}(X) = 0.8$; $\sigma \approx 0.8944$

7. $\mu \approx 2.3765$; $\text{Var}(X) \approx 2.3522$; $\sigma \approx 1.5337$

9. $\mu = \frac{93}{35}$; $\text{Var}(X) \approx 0.7151$; $\sigma \approx 0.846$

11. $\mu = \frac{3}{2}$; $\text{Var}(X) = \frac{3}{4}$; $\sigma = \frac{1}{2}\sqrt{3}$

13. $\mu = 4$; $\text{Var}(X) = 16$; $\sigma = 4$

15. $3\frac{1}{3}$ min   **17.** 1.5 sec; $\frac{9}{4}(\ln 3 - 1)$; $\frac{3}{2}\sqrt{\ln 3 - 1}$

19. 1.5 ft; $\frac{9}{20}$; 0.671   **21.** 3 years   **23.** $\dfrac{b + a}{2}$; $\dfrac{(b - a)^2}{12}$

25. a. $\frac{1}{4}$ **b.** 6:30 A.M.   **27.** $a = -2.4$; $b = 3.6$

29. $m = 5$ **31.** $m \approx 2.52$ **33.** $m = \frac{1}{2}$ **35.** False

Using Technology Exercises 10.2, page 710

1. a.
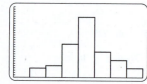

b. $\mu = 4$ and $\sigma \approx 1.40$

3. a. X gives the minimum age requirement for a regular driver's license.

b.

x	15	16	17	18	19	21
$P(X = x)$	.02	.30	.08	.56	.02	.02

c.
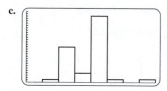

d. $\mu = 17.34$; $\sigma = 1.11$

5. a. Let X denote the random variable that gives the weight of a carton of sugar.

b.

x	4.96	4.97	4.98	4.99	5.00	5.01
$P(X = x)$	$\frac{3}{30}$	$\frac{4}{30}$	$\frac{4}{30}$	$\frac{1}{30}$	$\frac{1}{30}$	$\frac{5}{30}$

x	5.02	5.03	5.04	5.05	5.06
$P(X = x)$	$\frac{3}{30}$	$\frac{3}{30}$	$\frac{4}{30}$	$\frac{1}{30}$	$\frac{1}{30}$

c. $\mu \approx 5.00$; $\text{Var}(X) \approx 0.0009$; $\sigma \approx 0.03$

Exercises 10.3, page 719

1. .9265   **3.** .0401   **5.** .8657

7. a.
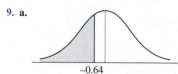
b. .9162

9. a.
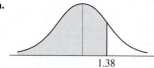
b. .2611

11. a.
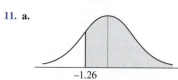
b. .8962

13. a.

b. .2266

15. a. 1.23   **b.** −0.81   **17. a.** 1.9   **b.** −1.9

19. a. .9772   **b.** .9192   **c.** .7333

21. a. .2206   **b.** .2206   **c.** .2960   **23. a.** .3085   **b.** .1587

25. a. .0228   **b.** .0228   **c.** .4772   **d.** .7258

27. a. .0038   **b.** .0918   **c.** .4082   **d.** .2514

29. .6247   **31.** 0.62%   **33.** A: 80; B: 73; C: 62; D: 54

Chapter 10 Concept Review, page 723

1. a. experiment   **b.** outcomes, sample points
c. sample space   **d.** 0, 1

2. a. random   **b.** finite discrete   **c.** 0, 1, 1

3. a. continuous   **b.** nonnegative, 1, $\int_a^b f(x)\, dx$

4. nonnegative, 1, $\iint\limits_R f(x, y)\, dA$

5. a. $(x_1 + x_2 + \cdots + x_n)/n$ **b.** central tendency, X
c. $x_1 p_1 + x_2 p_2 + \cdots + x_n p_n$, X, probabilities
d. $\int_a^b x f(x)\, dx$, probability density, $[a, b]$

6. a. dispersion
b. $p_1(x_1 - \mu)^2 + p_2(x_2 - \mu)^2 + \cdots + p_n(x_n - \mu)^2$
c. $\sqrt{\text{Var}(X)}$

7. $\int_a^b (x - \mu)^2 f(x)\, dx$, $\sqrt{\text{Var}(X)}$

8. a. $x = \mu$ **b.** $x = \mu$ **c.** above, the x-axis   **d.** 1
e. standard deviation, 2, 3

Chapter 10 Review Exercises, page 723

5. $\frac{1}{243}$ **6.** $\frac{1}{6}$ **7.** $\frac{3}{2}$ **8.** 1.5

9. a. $\frac{1}{2}$ **b.** $\frac{1}{2}$ **c.** $\frac{3}{4}$ **10. a.** $\frac{4}{7}$ **b.** 0   **c.** $\frac{1}{3}$

11. a. .52   **b.** .65   **c.** 0   **12. a.** $\frac{9}{10}$ **b.** $\frac{1}{4}$ **c.** $\frac{1}{2}$

13. $\mu = \frac{9}{2}$; $\text{Var}(X) \approx 2.083$; $\sigma \approx 1.44$

14. $\mu = \frac{50}{21}$; $\text{Var}(X) \approx 1.1882$; $\sigma \approx 1.0900$

15. $\mu = 0$; $\text{Var}(X) = \frac{7}{15}$; $\sigma \approx 0.6831$

16. $\mu = \frac{4}{3}$; $\text{Var}(X) = \frac{2}{9}$; $\sigma \approx 0.4714$ **17.** .9875

18. .8925   **19.** .3049   **20.** .8294   **22. a.** $\frac{3}{16}$ **b.** $\frac{5}{6}$

23. a. .6915   **b.** .8944   **c.** .4681

24. a. .9525   **b.** .9525   **c.** .9050

25. a. .0013   **b.** .0013   **c.** .9974

26. a. .2231   **b.** .3935   **c.** 4 days

27. 100 days   **28. a.** $\frac{1}{3}$ **b.** 6:45 A.M.

29. a. .5507   **b.** .3012   **c.** .2210

30. $a = \frac{3}{5}, b = \frac{6}{5}$

Chapter 10 Before Moving On, page 724

1. $\frac{6}{7}$ **2.** .2387   **3.** $\frac{13}{3}$, 5.42, 2.33

4. a. .9938   **b.** .8944   **c.** .9876

CHAPTER 11

Exercises 11.1, page 733

1. $P_1(x) = 1 - x; P_2(x) = 1 - x + \frac{1}{2}x^2; P_3(x) = 1 - x + \frac{1}{2}x^2 - \frac{1}{6}x^3$

3. $P_1(x) = \frac{1}{2} - \frac{1}{4}x; P_2(x) = \frac{1}{2} - \frac{1}{4}x + \frac{1}{8}x^2;$
$P_3(x) = \frac{1}{2} - \frac{1}{4}x + \frac{1}{8}x^2 - \frac{1}{16}x^3$

5. $P_1(x) = 1 - (x - 1); P_2(x) = 1 - (x - 1) + (x - 1)^2;$
$P_3(x) = 1 - (x - 1) + (x - 1)^2 - (x - 1)^3$

7. $P_1(x) = 1 - \frac{1}{2}x; P_2(x) = 1 - \frac{1}{2}x - \frac{1}{8}x^2;$
$P_3(x) = 1 - \frac{1}{2}x - \frac{1}{8}x^2 - \frac{1}{16}x^3$

9. $P_1(x) = -x; P_2(x) = -x - \frac{1}{2}x^2; P_3(x) = -x - \frac{1}{2}x^2 - \frac{1}{3}x^3$

11. $P_2(x) = 16 + 32(x - 2) + 24(x - 2)^2$

13. $P_4(x) = (x - 1) - \frac{1}{2}(x - 1)^2 + \frac{1}{3}(x - 1)^3 - \frac{1}{4}(x - 1)^4$

15. $P_4(x) = e + e(x - 1) + \frac{1}{2}e(x - 1)^2 + \frac{1}{6}e(x - 1)^3 + \frac{1}{24}e(x - 1)^4$

17. $P_3(x) = 1 - \frac{1}{3}x - \frac{1}{9}x^2 - \frac{5}{81}x^3$

19. $P_3(x) = \frac{1}{3} - \frac{2}{9}x + \frac{4}{27}x^2 - \frac{8}{81}x^3$

21. $P_n(x) = 1 - 2x + 4x^2 - 8x^3 + \cdots + (-2)^n x^n; 0.8336; 0.8333\ldots$

23. $P_4(x) = 1 - \frac{1}{2}x + \frac{1}{8}x^2 - \frac{1}{48}x^3 + \frac{1}{384}x^4; 0.90484$

25. $P_2(x) = 4 + \frac{1}{8}(x - 16) - \frac{1}{512}(x - 16)^2; 3.94969$

27. $P_3(x) = x - \frac{1}{2}x^2 + \frac{1}{3}x^3; 0.109; 0.108$

29. 2.01494375; 0.00000042

31. 1.248; 0.0048828; 1.25   **33.** 0.095; 0.00033

35. a. 0.04167   **b.** 0.625   **c.** 0.04167   **d.** 0.007121

37. 0.47995   **39.** 48%   **41.** 2600

43. False   **45.** True

Exercises 11.2, page 742

1. 1, 2, 4, 8, 16   **3.** $0, \frac{1}{3}, \frac{2}{4}, \frac{3}{5}, \frac{4}{6}$ **5.** $1, 1, \frac{4}{6}, \frac{8}{24}, \frac{16}{120}$

7. $e, \frac{e^2}{8}, \frac{e^3}{27}, \frac{e^4}{64}, \frac{e^5}{125}$ **9.** $1, \frac{11}{9}, \frac{25}{19}, \frac{15}{11}, \frac{71}{51}$ **11.** $a_n = 3n - 2$

13. $a_n = \frac{1}{n^3}$ **15.** $a_n = \frac{2^{2n-1}}{5^{n-1}}$

17. $a_n = \frac{(-1)^{n-1}}{2^{n-1}} = \left(-\frac{1}{2}\right)^{n-1}$

19. $a_n = \frac{n}{(n+1)(n+2)}$ **21.** $a_n = \frac{e^{n-1}}{(n-1)!}$

23.
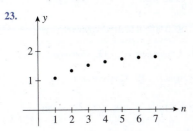

25.
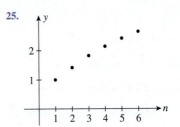

27.
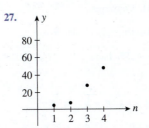

29.
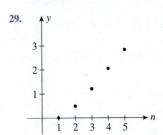

31. Converges; $\frac{1}{2}$ **33.** Converges; 0   **35.** Converges; 1

37. Diverges   **39.** Converges; 2   **41.** Converges; 0

43. Converges; $\dfrac{\sqrt{2}}{2}$

45. a. $a_1 = 0.015, a_{10} \approx 0.14027, a_{100} \approx 0.77939, a_{1000} \approx 0.999999727$
b. 1

47. b. $a_n = 100(1.005)^n$
c. $a_{24} \approx 112.72$. The accumulated amount at the end of 2 years is $112.72.

49. True   **51.** False

Exercises 11.3, page 753

1. $\lim\limits_{N \to \infty} S_N$ does not exist; divergent

3. $S_N = \frac{1}{2} - \frac{1}{N+2}; \frac{1}{2}$ **5.** Converges; $\frac{3}{2}$ **7.** Diverges

9. Converges; $\frac{3}{5}$ **11.** Converges; $\frac{1}{3}$ **13.** Converges; 5.52

15. Converges; $\dfrac{\pi}{\pi + 3}$ **17.** Converges; 13   **19.** Converges; $2\frac{3}{4}$

21. Converges; $7\frac{1}{2}$ **23.** Diverges   **25.** $\dfrac{e^3 - 2\pi e + \pi^2}{(\pi - e)(e^2 - \pi)}$

27. $\frac{1}{3}$ **29.** $\frac{404}{333}$ **31.** $-1 < x < 1; \dfrac{1}{1+x}$

33. $\frac{1}{2} < x < \frac{3}{2}; \dfrac{2(x-1)}{3-2x}$ **35.** $303 billion   **37.** $\frac{6}{11}$

41. $\dfrac{P}{e^r - 1}$ **43. b.** $\dfrac{1}{k}\ln\dfrac{S}{S-C}$ hr   **45.** False   **47.** True

Exercises 11.4, page 763

11. Diverges **13.** Diverges **15.** Converges

17. Converges **19.** Converges **21.** Converges

23. Converges **25.** Converges **27.** Converges

29. Diverges **31.** Diverges **33.** Converges **35.** Diverges

37. Diverges **39.** Converges **41.** Converges

43. Diverges **45.** Converges **47.** Diverges **49.** $p > 1$

51. $a = 1$ **55.** True **57.** False **59.** False

Exercises 11.5, page 773

1. $R = 1; (0, 2)$ **3.** $R = 1; (-1, 1)$ **5.** $R = 4; (-4, 4)$

7. $R = \infty; (-\infty, \infty)$ **9.** $R = 0; x = -2$ **11.** $R = 1; (-4, -2)$

13. $R = \infty; (-\infty, \infty)$ **15.** $R = \frac{1}{2}; \left(-\frac{1}{2}, \frac{1}{2}\right)$ **17.** $R = 0; x = -1$

19. $R = 3; (0, 6)$ **21.** $\sum_{n=0}^{\infty} (-1)^n (x - 1)^n; R = 1; (0, 2)$

23. $\sum_{n=0}^{\infty} (-1)^n \frac{(x - 2)^n}{3^{n+1}}; R = 3; (-1, 5)$

25. $\sum_{n=0}^{\infty} (-1)^{n+1} (x - 2)^n; R = 1; (1, 3)$

27. $1 + \frac{1}{2}(x - 1) + \sum_{n=2}^{\infty} (-1)^{n+1} \frac{1 \cdot 3 \cdot 5 \cdots (2n - 3)}{n! \, 2^n}(x - 1)^n;$
$R = 1; (0, 2)$

29. $\sum_{n=0}^{\infty} \frac{2^n}{n!} x^n; R = \infty; (-\infty, \infty)$

31. $\sum_{n=0}^{\infty} (-1)^n \frac{1 \cdot 3 \cdot 5 \cdots (2n - 1)}{n! \, 2^n} x^n; R = 1; (-1, 1)$ **35.** True

Exercises 11.6, page 780

1. $\sum_{n=0}^{\infty} (-1)^{n+1} (x - 2)^n; (1, 3)$ **3.** $\sum_{n=0}^{\infty} (-1)^n 3^n x^n; \left(-\frac{1}{3}, \frac{1}{3}\right)$

5. $\sum_{n=0}^{\infty} \frac{3^n}{4^{n+1}} x^n; \left(-\frac{4}{3}, \frac{4}{3}\right)$ **7.** $\sum_{n=0}^{\infty} x^{2n}; (-1, 1)$

9. $\sum_{n=0}^{\infty} (-1)^n \frac{x^n}{n!}; (-\infty, \infty)$ **11.** $\sum_{n=0}^{\infty} (-1)^n \frac{x^{2n+2}}{n!}; (-\infty, \infty)$

13. $\sum_{n=0}^{\infty} \frac{x^{2n}}{(2n)!}; (-\infty, \infty)$

15. $\sum_{n=1}^{\infty} (-1)^{n-1} \frac{2^n x^n}{n}; \left(-\frac{1}{2}, \frac{1}{2}\right]$ **17.** $\sum_{n=1}^{\infty} (-1)^{n+1} \frac{x^{2n}}{n}; (-1, 1]$

19. $(\ln 2)(x - 2) + \sum_{n=1}^{\infty} (-1)^{n-1} \left(\frac{1}{n2^n}\right)(x - 2)^{n+1}; (0, 4]$

21. $1 - x + x^2 - \cdots + (-1)^n x^n + \cdots$

23. $x - \frac{1}{2} x^2 + \frac{1}{3} x^3 - \cdots + \frac{(-1)^{n+1}}{n} x^n + \cdots$

25. 0.4812 **27.** 0.7475 **29.** 3.34 **31.** 15.85%

Exercises 11.7, page 788

1. 1.73205 **3.** 2.64575 **5.** 2.41014 **7.** 0.7709

9. $1.1219; 2.5745$ **11.** 2.30278 **13.** 1.11634

15. 1.61803 **17.** 0.56714 **19. b.** 0.1936 **21. b.** 1.8794

23. 2.9365 **25.** 1.2785 **27.** 0.4263 **29.** 294 units/day

31. $8:39$ P.M. **33.** 26.82%/year **35.** $0.0050; 6\%$/year

37. 10%/year **39.** $11{,}671$ units; $\$21.17$/unit **41. b.** 2.546

Using Technology Exercises 11.7, page 791

1. 1.37880 **3.** -0.77092 **5.** 0.75488 **7.** -1.34730

Chapter 11 Concept Review, page 792

1. a. $f(a) + f'(a)(x - a) + \cdots + \frac{f^{(n)}(a)}{n!}(x - a)^n$

b. $f(a), f'(a), \frac{f^{(n)}(a)}{n!}$

2. $I, a, x, \frac{f^{(n+1)}(c)}{(n + 1)!}(x - a)^{n+1}$

3. a. function, integers, $f(n)$, nth term **b.** large, converge

4. $cA, A + B, AB, \frac{A}{B}, B \neq 0$

5. a. partial sums **b.** geometric, $1, 1$

6. a. 0, diverges **b.** $cA + B$

7. a. convergent, divergent **b.** $\sum_{n=1}^{\infty} \frac{1}{n^p}, p > 1, p \leq 1$

8. a. converges **b.** $a_n \geq b_n$

9. a. $\sum_{n=0}^{\infty} a_n(x - a)^n$ **b.** $x = a, x, (a - R, a + R)$, outside

10. a. $\frac{f^{(n)}(a)}{n!}$ **b.** $\lim_{N \to \infty} R_N(x) = 0$, the Taylor series

Chapter 11 Review Exercises, page 793

1. $P_4(x) = 1 - (x + 1) + (x + 1)^2 - (x + 1)^3 + (x + 1)^4$

2. $P_4(x) = e^{-1}\left[1 - (x - 1) + \frac{(x - 1)^2}{2!} - \frac{(x - 1)^3}{3!} + \frac{(x - 1)^4}{4!}\right]$

3. $P_4(x) = x^2 - \frac{1}{2} x^4$ **4.** $P_4(x) = 1 - 2x + 3x^2 - 4x^3 + 5x^4$

5. $P_2(x) = 2 + \frac{1}{12}(x - 8) - \frac{1}{288}(x - 8)^2; 1.983$

6. $P_6(x) = 1 - x^2 + \frac{x^4}{2} - \frac{x^6}{6}; 0.7429$

7. $2.9992591; 8 \times 10^{-11}$ **8.** $0.909; 0.0001; 0.9090909\ldots$

9. 0.37 **10.** 0.20 **11.** Converges; $\frac{1}{3}$ **12.** Converges; 0

13. Converges; $\frac{2}{3}$ **14.** Diverges **15.** Converges; 1

16. Converges; -1 **17.** 2 **18.** $\frac{3}{5}$ **19.** $\sqrt{2} - 1$ **20.** $\frac{1}{e - 1}$

21. $\frac{7}{6}$ **22.** $\frac{1}{2}$ **23.** $\frac{141}{99}$ **24.** $\frac{3139}{999}$ **25.** Diverges

26. Diverges   **27.** Converges   **28.** Converges

29. Converges   **30.** Converges   **31.** Converges

32. Converges   **33.** $R = 1; (-1, 1)$ **34.** $R = 1; (-1, 1)$

35. $R = 1; (0, 2)$

36. $R = \frac{1}{e}; \left(2 - \frac{1}{e}, 2 + \frac{1}{e} \right)$

37. $-1 - 2x - 4x^2 - 8x^3 - \cdots - 2^n x^n - \cdots; \left(-\frac{1}{2}, \frac{1}{2} \right)$

38. $\frac{1}{e} - \frac{1}{e}(x - 1) + \frac{1}{2!e}(x - 1)^2 - \cdots + \frac{(-1)^n}{n!e}(x - 1)^n + \cdots;$
$(-\infty, \infty)$

39. $2x - 2x^2 + \frac{8}{3}x^3 - \cdots + \frac{(-1)^{n+1}2^n}{n}x^n + \cdots; \left(-\frac{1}{2}, \frac{1}{2} \right]$

40. $x^2 - 2x^3 + 2x^4 - \cdots + \frac{(-1)^n 2^n x^{n+2}}{n!} + \cdots; (-\infty, \infty)$

41. $(-\infty, \infty)$ **42.** $\ln|x| + \sum_{n=1}^{\infty}(-1)^n \frac{x^n}{n! \, n} + C$ **43.** $\sum_{n=1}^{\infty} \frac{x^n}{n! \, n} + C$

44. 2.2894   **45.** 0.7549   **46.** (0.3517, 0.7035)

47. 2.65 sec   **48.** $323,433.33

49. a. $115 billion   **b.** $90 billion   **50.** 31.08%

Chapter 11 Before Moving On, page 794

1. $P_1(x) = x, P_2(x) = x - x^2,$ and $P_3(x) = x - x^2 + \frac{x^3}{2}$

2. a. $\frac{2}{3}$ **b.** 2   **3. a.** $\frac{25}{12}$ **b.** $\frac{1}{3}$

4. a. Convergent   **b.** Divergent   **5.** $-\sum_{n=0}^{\infty}(x + 2)^n; 1$

CHAPTER 12

Exercises 12.1, page 799

1. $\frac{5\pi}{2}$ radians   **3.** $-\frac{3\pi}{2}$ radians

5. a. III   **b.** III   **c.** II   **d.** I

7. $\frac{5\pi}{12}$ radians   **9.** $\frac{8\pi}{9}$ radians   **11.** $\frac{7\pi}{2}$ radians

13. 120°   **15.** −270°   **17.** 220°

19.
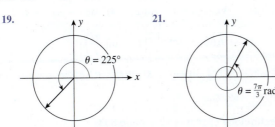

21.

23. −210°   **25.** 315°   **27.** True   **29.** True

Exercises 12.2, page 810

1. 0   **3.** 1   **5.** $\frac{\sqrt{3}}{2}$ **7.** $\frac{\sqrt{3}}{3}$ **9.** −2.6131

11. $\sin\frac{\pi}{2} = 1, \cos\frac{\pi}{2} = 0, \tan\frac{\pi}{2}$ is undefined, $\csc\frac{\pi}{2} = 1$,

$\sec\frac{\pi}{2}$ is undefined, $\cot\frac{\pi}{2} = 0$

13. $\sin\frac{5\pi}{3} = -\frac{\sqrt{3}}{2}, \cos\frac{5\pi}{3} = \frac{1}{2}, \tan\frac{5\pi}{3} = -\sqrt{3}, \csc\frac{5\pi}{3} = -\frac{2\sqrt{3}}{3}$,

$\sec\frac{5\pi}{3} = 2, \cot\frac{5\pi}{3} = -\frac{\sqrt{3}}{3}$

15. $\frac{7\pi}{6}, \frac{11\pi}{6}$ **17.** $\frac{5\pi}{6}, \frac{11\pi}{6}$ **19.** π **21.** $\frac{\pi}{3}, \frac{2\pi}{3}$

23.
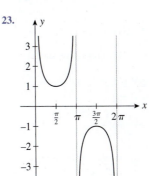

25.
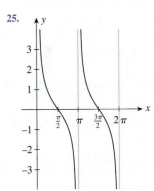

27.
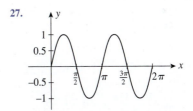

29.
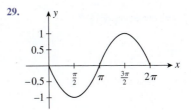

31. $A = 1, P = 2\pi$ **33.** $A = 1, P = 2\pi$

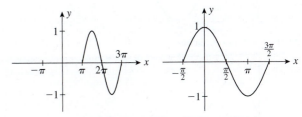

35. $A = 1, P = 2\pi$ **37.** $A = 2, P = \dfrac{2\pi}{3}$

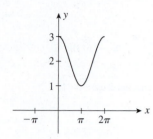
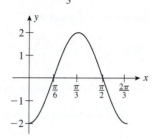

49. $\sin\theta = \frac{5}{13}, \cos\theta = \frac{12}{13}, \tan\theta = \frac{5}{12}, \csc\theta = \frac{13}{5}, \sec\theta = \frac{13}{12}, \cot\theta = \frac{12}{5}$

51. b. $\dfrac{\pi(4n+1)}{12}$ $(n = 0, 1, 2, \ldots)$; $\dfrac{\pi(4n+3)}{12}$ $(n = 0, 1, 2, \ldots)$

53. 9 ft/sec   **55.** False   **57.** True

Exercises 12.3, page 820

1. $-3\sin 3x$ **3.** $-2\pi\sin\pi x$ **5.** $2x\cos(x^2+1)$

7. $4x\sec^2(2x^2)$ **9.** $x\cos x + \sin x$ **11.** $6(\cos 3x - \sin 2x)$

13. $2x(\cos 2x - x\sin 2x)$ **15.** $\dfrac{x\cos\sqrt{x^2-1}}{\sqrt{x^2-1}}$

17. $e^x\sec x\,(1 + \tan x)$ **19.** $\cos\dfrac{1}{x} + \dfrac{1}{x}\sin\dfrac{1}{x}$

21. $\dfrac{x\sin x}{(1+\cos x)^2}$ **23.** $\dfrac{\sec^2 x}{2\sqrt{\tan x}}$ **25.** $\dfrac{x\cos x - \sin x}{x^2}$

27. $2\tan x\sec^2 x$ **29.** $-\csc^2 x\cdot e^{\cot x}$ **31.** $-\sin x$

33. $4t\cos t - t^2\sin t + \sin t$ **35.** $y = -2x + \dfrac{\pi}{2}$

37. $\dfrac{\sqrt{2}\pi(8+\pi)}{16}$ **39.** -1

41. $2k\pi, k = 0, +1, +2, \ldots$ **43.** $\dfrac{(2k+1)\pi}{2}, k = 0, \pm 1, \pm 2, \ldots$

45. Increasing on $\left(0, \frac{\pi}{4}\right)$ and $\left(\frac{5\pi}{4}, 2\pi\right)$, decreasing on $\left(\frac{\pi}{4}, \frac{5\pi}{4}\right)$

47. $f(x) = \sin x + \cos x$

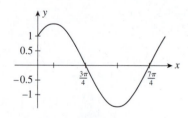

49. $f(x) = 2\sin x + \sin 2x$

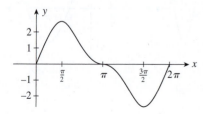

51. a. $f(x) = \sin x = x - \dfrac{x^3}{3!} + \dfrac{x^5}{5!} - \dfrac{x^7}{7!} + \cdots$

$+ (-1)^n\dfrac{x^{2n+1}}{(2n+1)!} + \cdots$

53. 0 wolves/month, -1571 caribou/month   **55.** 0.05

57. Warmest day is July 25; coldest day is January 23

59. Sixth week

61. Maximum when $t = 1, 5, 9, 13, \ldots$,
minimum when $t = 3, 7, 11, 15, \ldots$

63. 70.7 ft/sec   **65.** 60°   **67.** 1.4987 radians

69. True   **71.** False

Using Technology Exercises 12.3, page 825

1. 1.2038   **3.** 0.7762   **5.** -0.2368 **7.** 0.8415; -0.2172

9. 1.1271; 0.2013

11. a.

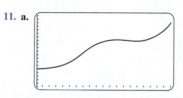

b. $0.90   **c.** $37.86

13. a.

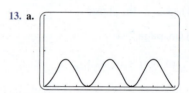

15. 0.006 ft

Exercises 12.4, page 830

1. $-\frac{1}{3}\cos 3x + C$ **3.** $-3\cos x + 4\sin x + C$

5. $\frac{1}{2}\tan 2x + C$ **7.** $\frac{1}{2}\sin(x^2) + C$ **9.** $-\frac{1}{\pi}\csc\pi x + C$

11. 2   **13.** $-\frac{1}{2}\ln\frac{1}{2}$ **15.** $\frac{1}{4}\sin^4 x + C$

17. $\frac{1}{\pi}\ln|\sec\pi x + \tan\pi x| + C$ **19.** $\frac{1}{3}\ln(1 + \sqrt{2})$

21. $-\frac{2}{3}(\cos x)^{3/2} + C$ **23.** $-\frac{1}{9}(1 - 2\sin 3x)^{3/2} + C$

25. $\frac{1}{4}\tan^4 x + C$ **27.** $-\frac{1}{4}(\cot x - 1)^4 + C$ **29.** 2

31. $\frac{1}{2}x[\sin(\ln x) - \cos(\ln x)] + C$ **33.** $2\sqrt{2}$ **35.** $\frac{1}{2}\ln 2$

37. $\frac{1}{2}(\pi^2 - 4)$ **39.** $\sin y = 1 - \cos x$ **41.** $\sin x - \cos y = C$

43. $85   **45.** $120,000   **47.** $\dfrac{1.2}{\pi}\left[1 - \cos\left(\dfrac{\pi t}{2}\right)\right]$

49. 8373 wolves, 50,804 caribou   **51.** $P(t) = P_0 e^{(k/a)\sin at}$

53. 1.37074   **55.** True   **57.** False

Using Technology Exercises 12.4, page 833

1. 0.5419   **3.** 0.7544   **5.** 0.2231   **7.** 0.6587

9. -0.2032 **11.** 0.9045

13. a.

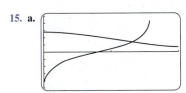

b. 2.2687

15. a.

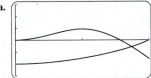

b. 1.8239

17. a.

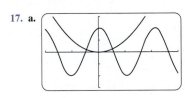

b. 1.2484

19. a.

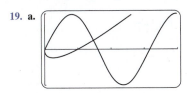

b. 1.0983

21. 7.6 ft

Chapter 12 Concept Review, page 834

1. a. radian   **b.** 2π, 360   **2. a.** $\dfrac{\pi}{180}x$ **b.** $\dfrac{180}{\pi}x$

3. $x, y, \dfrac{y}{x}, \dfrac{1}{y}, \dfrac{1}{x}, \dfrac{x}{y}$ **4. a.** $\cos^2\theta$ **b.** 1   **c.** 1   **d.** $\cos 2\theta$

5. a. $\cos 2\theta$ **b.** $+$ **c.** $\sin A \cos A$ **d.** $\sin^2 A$

6. a. $[\cos f(x)]f'(x)$ **b.** $-[\sin f(x)]f'(x)$
   c. $[\sec^2 f(x)]f'(x)$ **d.** $-[\csc f(x)][\cot f(x)]f'(x)$
   e. $[\sec f(x)][\tan f(x)]f'(x)$ **f.** $-[\csc^2 f(x)]f'(x)$

7. a. $-\cos x + C$ **b.** $\sin x + C$ **c.** $\tan x + C$
   d. $-\cot x + C$ **e.** $\sec x + C$ **f.** $-\csc x + C$

Chapter 12 Review Exercises, page 835

1. $\dfrac{2\pi}{3}$ radians   **2.** $\dfrac{5\pi}{2}$ radians   **3.** $-\dfrac{5\pi}{4}$ radians   **4.** $330°$

5. $-450°$ **6.** $-315°$ **7.** $\dfrac{\pi}{3}$ or $\dfrac{5\pi}{3}$ **8.** $150°$ or $330°$

9. $3\cos 3x$ **10.** $-\sin\frac{x}{2}$ **11.** $2(\cos x + 3\sin 2x)$

12. $\dfrac{\sec^2\sqrt{x}\tan\sqrt{x}}{\sqrt{x}}$ **13.** $e^{-x}(3\sec^2 3x - \tan 3x)$

14. $4(1 - \csc 2x)(\csc 2x \cot 2x)$ **15.** $4\cos 2x$ or $4(\cos^2 x - \sin^2 x)$

16. $-\dfrac{\sin x}{(1 - \cos x)^2}$

17. $\dfrac{(\cot x - 1)\sec^2 x - (1 - \tan x)\csc^2 x}{(1 - \cot x)^2} = -\sec^2 x$

18. $-2\tan x$ **19.** $\cos(\sin x)\cos x$ **20.** $e^{\sin x}(-\sin x + \cos^2 x)$

21. $A = 4, P = 2\pi$ **22.** $A = 3, P = \pi$

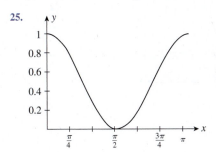

23. $y = 4x + 1 - \pi$ **24.** $\frac{2}{3}x^{3/2} - \cos x + 3$

25.

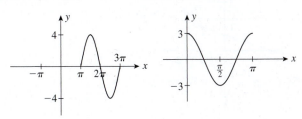

   a. Increasing on $\left(\frac{\pi}{2}, \pi\right)$ and decreasing on $\left(0, \frac{\pi}{2}\right)$
   b. Relative minimum: $f\left(\frac{\pi}{2}\right) = 0$
   c. Concave up on $\left(\frac{\pi}{4}, \frac{3\pi}{4}\right)$ and concave down on
     $\left(0, \frac{\pi}{4}\right)$ and $\left(\frac{3\pi}{4}, \pi\right)$
   d. $\left(\frac{\pi}{4}, \frac{1}{2}\right)$ and $\left(\frac{3\pi}{4}, \frac{1}{2}\right)$

26. $\frac{3}{2}\sin\left(\frac{2x}{3}\right) + C$ **27.** $\frac{1}{2}x - \frac{1}{8}\sin 4x + C$

28. $-\frac{1}{2}\csc(x^2) + C$ **29.** $x\sin x + \cos x + C$

30. $\frac{1}{3}\sin^3 x + C$ **31.** $-\cos e^x + C$ **32.** $-\csc x + C$

33. $\frac{1}{2}\ln 2$ **34.** 1   **35.** 2   **36.** $2\sqrt{2}$

37. Fox population is increasing at the rate of ≈ 6.5 foxes/month; rabbit population is decreasing at the rate of ≈ 113.4 rabbits/month

38. December 1; September 1   **39.** 78.5%   **40.** $\dfrac{6}{5\pi}$ L/sec

Chapter 12 Before Moving On, page 836

1.

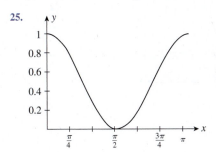

3. $(3\cos 3x - 2\sin 3x)e^{-2x}$ **4.** $\dfrac{\pi + 2}{8}$ **5.** $\dfrac{\sqrt{2}}{12}$

APPENDIX A

Exercises, page 842

7. Yes **9.** No **11.** No

13. $\frac{1}{3}(x + 2)$

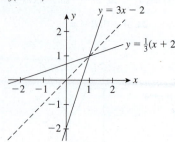

15. $\sqrt[3]{x - 1}$

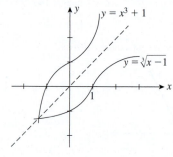

17. $\sqrt{9 - x^2}$ $(x \geq 0)$

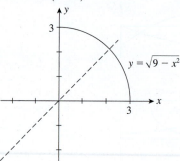

19. a. $-\frac{1}{2} + \frac{1}{2}\sqrt{1 + 8h}$; f^{-1} represents the time at which the balloon is at height h.

b. Between 15 and 20 sec

APPENDIX B

Exercises, page 846

1. $\frac{1}{2}$ **3.** 12 **5.** 1 **7.** 1 **9.** 2 **11.** $\frac{1}{2}$ **13.** $\frac{1}{3}$

15. ∞ **17.** 0 **19.** 0 **21.** $\frac{9}{4}$ **23.** ∞ **25.** $-\frac{1}{2}$

27. $\frac{1}{2}$ **29.** 0 **31.** False **33.** False

INDEX

How-To Technology Index

Basic Rules of Differentiation

1. $\dfrac{d}{dx}(c) = 0,\quad c,\text{ a constant}$

2. $\dfrac{d}{dx}(u^n) = nu^{n-1}\dfrac{du}{dx}$

3. $\dfrac{d}{dx}(u \pm v) = \dfrac{du}{dx} \pm \dfrac{dv}{dx}$

4. $\dfrac{d}{dx}(cu) = c\dfrac{du}{dx},\quad c,\text{ a constant}$

5. $\dfrac{d}{dx}(uv) = u\dfrac{dv}{dx} + v\dfrac{du}{dx}$

6. $\dfrac{d}{dx}\left(\dfrac{u}{v}\right) = \dfrac{v\dfrac{du}{dx} - u\dfrac{dv}{dx}}{v^2}$

7. $\dfrac{d}{dx}(e^u) = e^u\dfrac{du}{dx}$

8. $\dfrac{d}{dx}(\ln u) = \dfrac{1}{u}\cdot\dfrac{du}{dx}$

9. $\dfrac{d}{dx}(\sin u) = \cos u\dfrac{du}{dx}$

10. $\dfrac{d}{dx}(\cos u) = -\sin u\dfrac{du}{dx}$

11. $\dfrac{d}{dx}(\tan u) = \sec^2 u\dfrac{du}{dx}$

12. $\dfrac{d}{dx}(\sec u) = \sec u \tan u\dfrac{du}{dx}$

13. $\dfrac{d}{dx}(\csc u) = -\csc u \cot u\dfrac{du}{dx}$

14. $\dfrac{d}{dx}(\cot u) = -\csc^2 u\dfrac{du}{dx}$

Basic Rules of Integration

1. $\displaystyle\int du = u + C$

2. $\displaystyle\int kf(u)\, du = k\int f(u)\, du,\quad k,\text{ a constant}$

3. $\displaystyle\int [f(u) \pm g(u)]\, du = \int f(u)\, du \pm \int g(u)\, du$

4. $\displaystyle\int u^n\, du = \dfrac{u^{n+1}}{n+1} + C,\quad n \neq -1$

5. $\displaystyle\int e^u\, du = e^u + C$

6. $\displaystyle\int \dfrac{du}{u} = \ln|u| + C$

7. $\displaystyle\int \sin u\, du = -\cos u + C$

8. $\displaystyle\int \cos u\, du = \sin u + C$

9. $\displaystyle\int \sec^2 u\, du = \tan u + C$

10. $\displaystyle\int \csc^2 u\, du = -\cot u + C$

11. $\displaystyle\int \sec u \tan u\, du = \sec u + C$

12. $\displaystyle\int \csc u \cot u\, du = -\csc u + C$

13. $\displaystyle\int \tan u\, du = -\ln|\cos u| + C$

14. $\displaystyle\int \cot u\, du = \ln|\sin u| + C$

15. $\displaystyle\int \sec u\, du = \ln|\sec u + \tan u| + C$

16. $\displaystyle\int \csc u\, du = \ln|\csc u - \cot u| + C$

17. $\displaystyle\int u\, dv = uv - \int v\, du$

List of Applications

$ BUSINESS AND ECONOMICS

(continued)

 **SOCIAL SCIENCES**

LIFE SCIENCES

GENERAL INTEREST